Organic Reactions

Organic Reactions

VOLUME 35

JOHN WILEY & SONS, INC.

New York · Chichester  *Brisbane · Toronto · Singapore*

Published by John Wiley & Sons, Inc.

Library of Congress Catalog Card Number 42–20265

ISBN 0-471-83253-7

Printed in the United States of America

10 9 8 7 6 5 4 3 2 1

PREFACE TO THE SERIES

In the course of nearly every program of research in organic chemistry the investigator finds it necessary to use several of the better-known synthetic reactions. To discover the optimum conditions for the application of even the most familiar one to a compound not previously subjected to the reaction often requires an extensive search of the literature; even then a series of experiments may be necessary. When the results of the investigation are published, the synthesis, which may have required months of work, is usually described without comment. The background of knowledge and experience gained in the literature search and experimentation is thus lost to those who subsequently have occasion to apply the general method. The student of preparative organic chemistry faces similar difficulties. The textbooks and laboratory manuals furnish numerous examples of the application of various syntheses, but only rarely do they convey an accurate conception of the scope and usefulness of the processes.

For many years American organic chemists have discussed these problems. The plan of compiling critical discussions of the more important reactions thus was evolved. The volumes of *Organic Reactions* are collections of chapters each devoted to a single reaction, or a definite phase of a reaction, of wide applicability. The authors have had experience with the processes surveyed. The subjects are presented from the preparative viewpoint, and particular attention is given to limitations, interfering influences, effects of structure, and the selection of experimental techniques. Each chapter includes several detailed procedures illustrating the significant modifications of the method. Most of these procedures have been found satisfactory by the author or one of the editors, but unlike those in *Organic Syntheses* they have not been subjected to careful testing in two or more laboratories.

Each chapter contains tables that include all the examples of the reaction under consideration that the author has been able to find. It is inevitable, however, that in the search of the literature some examples will be missed, especially when the reaction is used as one step in an extended synthesis. Nevertheless, the investigator will be able to use the tables and their accompanying bibliographies in place of most or all of the literature search so often required.

Because of the systematic arrangement of the material in the chapters and the entries in the table, users of the books will be able to find information desired by reference to the table of contents of the appropriate chapter. In the interest of economy the entries in the indices have been kept to a minimum, and, in particular, the compounds listed in the tables are not repeated in the indices.

The success of this publication, which will appear periodically, depends upon the cooperation of organic chemists and their willingness to devote time and effort to the preparation of the chapters. They have manifested their interest already by the almost unanimous acceptance of invitations to contribute to the work. The editors will welcome their continued interest and their suggestions for improvements in *Organic Reactions.*

Chemists who are considering the preparation of a manuscript for submission to *Organic Reactions* are urged to write either secretary before they begin work.

A. HAROLD BLATT

January 9, 1903–March 19, 1986

A. Harold Blatt, long-time Secretary to the Editorial Board of *Organic Reactions* and Treasurer of *Organic Reactions*, Inc., died on March 19, 1986 in Melbourne, Florida at the age of 83 as a result of a stroke.

Harold Blatt was born in Cincinnati, Ohio and received his academic training at Harvard University; he was awarded the B.S. degree in 1923, the M.A. degree in 1925, and the Ph.D. degree in 1926. After stays at the Collège de France in Paris, Harvard University, and the University of Buffalo as a research associate, Professor Blatt joined the faculty at Howard University as Associate Professor in 1932. In 1939 he went to the newly formed Queens College (now part of The City University of New York), where he left a lasting imprint with respect to the appreciation of quality in education and scholarly activity. His teaching and research efforts were interrupted during World War II, when he worked in the Office of Scientific Research and Development. His work on projects related to the war effort was recognized by the awarding of the Presidential Certificate of Merit and the Naval Ordinance Award. Harold Blatt's book with James Bryant Conant, *The Chemistry of Organic Compounds,* was widely used as an introductory text in the 1940s and 1950s and was reprinted in Spanish and French. Professor Blatt was elected Chairman of the Chemistry Department at Queens College for three consecutive terms beginning in 1961. He retired in 1971 after 32 years of distinguished service, and he received the Distinguished Teacher of the Year Award in 1972 from the Alumni Association. His students and colleagues remember him fondly as a scientist whose enthusiasm for continued learning in chemistry was inspiring. Many of his scientific contributions are still being cited; notable examples are his work in the field of oxime chemistry and his review of the Fries reaction in Volume 1 of *Organic Reactions.*

After he retired from Queens College, Harold Blatt moved to Melbourne, Florida, where he helped to organize the new Chemistry Department at the Florida Institute of Technology.

Harold Blatt's work with *Organic Syntheses* and *Organic Reactions* began in 1937 and 1944, respectively. His colleagues on the Editorial and Advisory Boards of *Organic Reactions* and *Organic Syntheses* remember him for his highly dedicated work, sharp editorial eye, and the magnificent wit with which he induced authors of errant phrases to mend their ways. Many of the younger editors relied on Harold Blatt's advice and even acquired part of his clear, concise writing style from their work with him. Since Harold Blatt was primarily responsible for establishing how the detailed tabular surveys of *Organic Reactions* chapters should be set up, he has left us with a heritage to follow in terms of editorial style and format.

Professor Blatt is survived by his son, Dr. Joel Blatt, and a sister.

ROBERT BITTMAN

FRANK C. McGREW

January 28, 1914–November 20, 1986

Frank C. McGrew, member of the Advisory Board of *Organic Reactions*, died on November 20, 1986 in San Francisco, California at the age of 72 while undergoing surgery for an aneurysm.

Frank was born in Seward, Nebraska. He attended the University of Nebraska, receiving a B.S. degree in 1933 and an M.S. degree in 1934 under Professor Cliff S. Hamilton. In 1937 he was awarded a Ph.D. degree in organic chemistry at the University of Illinois under Professor Roger Adams. He joined the Chemical Department (later the Central Research Department) of Du Pont at the Experimental Station outside Wilmington, Delaware, where his research interests turned to polymer chemistry. In 1950 he was transferred to Du Pont's Polychemicals Department, which was responsible for a significant share of Du Pont's synthetic polymer business, as assistant director of research, and became its director of research in 1958. He moved to Geneva, Switzerland, in 1964 as director of technical and business analysis for Du Pont's International Department. He retired in 1972 and entered the consulting business. He remained in Geneva until moving to Kennett Square, Pennsylvania, in 1986.

Frank served on the Science Advisory Panel for the Army from 1958 to 1964, and was a civilian associate with the Office for Scientific Research and Development. Frank had what Kipling called "satiable curiosity"; he dug deeply into any subject that intrigued him, always wanting to satisfy himself that he understood why things were so. His deep interest in the relationships between chemical structure and properties of polymers led him to establish in the Polychemicals Department an outstanding program of research in fundamental polymer science. He had a keen eye for technical talent, and some of the people he put into that program became world-recognized experts in their branches of polymer science. Frank also had a strong practical streak. His department made the nylon intermediates adipic acid and hexamethylenediamine, the latter from the former. Frank figured that hexamethylenediamine could be made at lower cost by anti-Markownikov addition of hydrogen cyanide to butadiene, followed by hydrogenation of the 1,4-dicyano-2-butene. He initiated a project that eventually succeeded in accomplishing this unusual addition and that became the basis for a low-cost commercial process.

Frank served on the Editorial Board of *Organic Reactions* from 1950 to 1959. After moving to the Advisory Board, he maintained an active interest in *Organic Reactions*, even after moving to Geneva.

Frank is survived by his wife Helga, of Geneva; two sons, Michael A. of Columbus, Ohio, and Patrick F. of Geneva; two daughters, Carol P. Getty of Kansas City, Missouri and Mary M. Lee of Rehoboth Beach, Delaware; three sisters, Betty Shroeder of Edwards, Missouri, Nelda Chrisman of Lodi, California, and Carol Jones of Seward, Nebraska; and eight grandchildren.

ROBERT M. JOYCE

CUMULATIVE CHAPTER TITLES BY VOLUME

Volume 1 (1942)

1. **The Reformatsky Reaction,** Ralph L. Shriner
2. **The Arndt-Eistert Reaction,** W. E. Bachmann and W. S. Struve
3. **Chloromethylation of Aromatic Compounds,** R. C. Fuson and C. H. McKeever
4. **The Amination of Heterocyclic Bases by Alkali Amides,** Martin T. Leffler
5. **The Bucherer Reaction,** Nathan L. Drake
6. **The Elbs Reaction,** Louis F. Fieser
7. **The Clemmensen Reduction,** Elmore L. Martin
8. **The Perkin Reaction and Related Reactions,** John R. Johnson
9. **The Acetoacetic Ester Condensation and Certain Related Reactions,** Charles R. Hauser and Boyd E. Hudson, Jr.
10. **The Mannich Reaction,** F. F. Blicke
11. **The Fries Reaction,** A. H. Blatt
12. **The Jacobsen Reaction,** Lee Irvin Smith

Volume 2 (1944)

1. **The Claisen Rearrangement,** D. Stanley Tarbell
2. **The Preparation of Aliphatic Fluorine Compounds,** Albert L. Henne
3. **The Cannizzaro Reaction,** T. A. Geissman
4. **The Formation of Cyclic Ketones by Intramolecular Acylation,** William S. Johnson
5. **Reduction with Aluminum Alkoxides (The Meerwein-Ponndorf-Verley Reduction),** A. L. Wilds

6. **The Preparation of Unsymmetrical Biaryls by the Diazo Reaction and the Nitrosoacetylamine Reaction,** Werner E. Bachmann and Roger A. Hoffman

7. **Replacement of the Aromatic Primary Amino Group by Hydrogen,** Nathan Kornblum

8. **Periodic Acid Oxidation,** Ernest L. Jackson

9. **The Resolution of Alcohols,** A. W. Ingersoll

10. **The Preparation of Aromatic Arsonic and Arsinic Acids by the Bart, Béchamp, and Rosenmund Reactions,** Cliff S. Hamilton and Jack F. Morgan

Volume 3 (1946)

1. **The Alkylation of Aromatic Compounds by the Friedel-Crafts Method,** Charles C. Price

2. **The Willgerodt Reaction,** Marvin Carmack and M. A. Spielman

3. **Preparation of Ketenes and Ketene Dimers,** W. E. Hanford and John C. Sauer

4. **Direct Sulfonation of Aromatic Hydrocarbons and Their Halogen Derivatives,** C. M. Suter and Arthur W. Weston

5. **Azlactones,** H. E. Carter

6. **Substitution and Addition Reactions of Thiocyanogen,** John L. Wood

7. **The Hofmann Reaction,** Everett S. Wallis and John F. Lane

8. **The Schmidt Reaction,** Hans Solff

9. **The Curtius Reaction,** Peter A. S. Smith

Volume 4 (1948)

1. **The Diels-Alder Reaction with Maleic Anhydride,** Milton C. Kloetzel

2. **The Diels-Alder Reaction: Ethylenic and Acetylenic Dienophiles,** H. L. Holmes

3. **The Preparation of Amines by Reductive Alkylation,** William S. Emerson

4. **The Acyloins,** S. M. McElvain

5. **The Synthesis of Benzoins,** Walter S. Ide and Johannes S. Buck

6. **Synthesis of Benzoquinones by Oxidation,** James Cason

7. **The Rosenmund Reduction of Acid Chlorides to Aldehydes,** Erich Mosettig and Ralph Mozingo

8. **The Wolff-Kishner Reduction,** David Todd

Volume 5 (1949)

1. **The Synthesis of Acetylenes,** Thomas L. Jacobs
2. **Cyanoethylation,** H. A. Bruson
3. **The Diels-Alder Reaction: Quinones and Other Cyclenones,** Lewis W. Butz and Anton W. Rytina
4. **Preparation of Aromatic Fluorine Compounds from Diazonium Fluoborates: The Schiemann Reaction,** Arthur Roe
5. **The Friedel and Crafts Reaction with Aliphatic Dibasic Acid Anhydrides,** Ernst Berliner
6. **The Gattermann-Koch Reaction,** Nathan N. Crouse
7. **The Leuckart Reaction,** Maurice L. Moore
8. **Selenium Dioxide Oxidation,** Norman Rabjohn
9. **The Hoesch Synthesis,** Paul E. Spoerri and Adrien S. DuBois
10. **The Darzens Glycidic Ester Condensation,** Melvin S. Newman and Barney J. Magerlein

Volume 6 (1951)

1. **The Stobbe Condensation,** William S. Johnson and Guido H. Daub
2. **The Preparation of 3,4-Dihydroisoquinolines and Related Compounds by the Bischler-Napieralski Reaction,** Wilson M. Whaley and Tuticorin R. Govindachari
3. **The Pictet-Spengler Synthesis of Tetrahydroisoquinolines and Related Compounds,** Wilson M. Whaley and Tuticorin R. Govindachari
4. **The Synthesis of Isoquinolines by the Pomeranz-Fritsch Reaction,** Walter J. Gensler
5. **The Oppenauer Oxidation,** Carl Djerassi
6. **The Synthesis of Phosphonic and Phosphinic Acids,** Gennady M. Kosolapoff
7. **The Halogen-Metal Interconversion Reaction with Organolithium Compounds,** Reuben G. Jones and Henry Gilman
8. **The Preparation of Thiazoles,** Richard H. Wiley, D. C. England, and Lyell C. Behr
9. **The Preparation of Thiophenes and Tetrahydrothiophenes,** Donald E. Wolf and Karl Folkers
10. **Reductions by Lithium Aluminum Hydride,** Weldon G. Brown

Volume 7 (1953)

1. **The Pechmann Reaction,** Suresh Sethna and Ragini Phadke
2. **The Skraup Synthesis of Quinolines,** R. H. F. Manske and Marshall Kalka
3. **Carbon-Carbon Alkylations with Amines and Ammonium Salts,** James H. Brewster and Ernest L. Eliel
4. **The von Braun Cyanogen Bromide Reaction,** Howard A. Hageman
5. **Hydrogenolysis of Benzyl Groups Attached to Oxygen, Nitrogen, or Sulfur,** Walter H. Hartung and Robert Simonoff
6. **The Nitrosation of Aliphatic Carbon Atoms,** Oscar Touster
7. **Epoxidation and Hydroxylation of Ethylenic Compounds with Organic Peracids,** Daniel Swern

Volume 8 (1954)

1. **Catalytic Hydrogenation of Esters to Alcohols,** Homer Adkins
2. **The Synthesis of Ketones from Acid Halides and Organometallic Compounds of Magnesium, Zinc, and Cadmium,** David A. Shirley
3. **The Acylation of Ketones to Form β-Diketones or β-Keto Aldehydes,** C. R. Hauser, Frederic W. Swamer, and Joe T. Adams
4. **The Sommelet Reaction,** S. J. Angyal
5. **The Synthesis of Aldehydes from Carboxylic Acids,** Erich Mosettig
6. **The Metalation Reaction with Organolithium Compounds,** Henry Gilman
7. **β-Lactones,** Harold E. Zaugg
8. **The Reaction of Diazomethane and Its Derivatives with Aldehydes and Ketones,** C. David Gutsche

Volume 9 (1957)

1. **The Cleavage of Non-enolizable Ketones with Sodium Amide,** K. E. Hamlin and Arthur W. Weston
2. **The Gattermann Synthesis of Aldehydes,** William E. Truce
3. **The Baeyer-Villiger Oxidation of Aldehydes and Ketones,** C. H. Hassall
4. **The Alkylation of Esters and Nitriles,** A. C. Cope, H. L. Holmes, and H. O. House
5. **The Reaction of Halogens with Silver Salts of Carboxylic Acids,** C. V. Wilson

Volume 19 (1972)

1. **Conjugate Addition Reactions of Organocopper Reagents,** Gary H. Posner
2. **Formation of Carbon-Carbon Bonds via π-Allylnickel Compounds,** Martin F. Semmelhack
3. **The Thiele-Winter Acetoxylation of Quinones,** J. F. W. McOmie and J. M. Blatchly
4. **Oxidative Decarboxylation of Acids by Lead Tetraacetate,** Roger A. Sheldon and Jay K. Kochi

Volume 20 (1973)

1. **Cyclopropanes from Unsaturated Compounds, Methylene Iodide, and Zinc-Copper Couple,** H. E. Simmons, T. L. Cairns, Susan A. Vladuchick, and Connie M. Hoiness
2. **Sensitized Photooxygenation of Olefins,** R. W. Denny and A. Nickon
3. **The Synthesis of 5-Hydroxyindoles by the Nenitzescu Reaction,** George R. Allen, Jr.
4. **The Zinin Reduction of Nitroarenes,** H. K. Porter

Volume 21 (1974)

1. **Fluorination with Sulfur Tetrafluoride,** G. A. Boswell, Jr., W. C. Ripka, R. M. Scribner, and C. W. Tullock
2. **Modern Methods to Prepare Monofluoroaliphatic Compounds,** William A. Sheppard

Volume 22 (1975)

1. **The Claisen and Cope Rearrangements,** Sara Jane Rhoads and N. Rebecca Raulins
2. **Substitution Reactions Using Organocopper Reagents,** Gary H. Posner
3. **Clemmensen Reduction of Ketones in Anhydrous Organic Solvents,** E. Vedejs
4. **The Reformatsky Reaction,** .Michael W. Rathke

Volume 23 (1976)

1. **Reduction and Related Reactions of α,β-Unsaturated Compounds with Metals in Liquid Ammonia,** Drury Caine
2. **The Acyloin Condensation,** Jordan J. Bloomfield, Dennis C. Owsley, and Janice M. Nelke
3. **Alkenes from Tosylhydrazones,** Robert H. Shapiro

2. **Reductive Dehalogenation of Polyhalo Ketones with Low-Valent Metals and Related Reducing Agents,** Ryoji Noyori and Yoshihiro Hayakawa

3. **Base-Promoted Isomerizations of Epoxides,** Jack K. Crandall and Marcel Apparu

Volume 30 (1984)

1. **Photocyclization of Stilbenes and Related Molecules,** Frank B. Mallory and Clelia W. Mallory

2. **Olefin Synthesis via Deoxygenation of Vicinal Diols,** Eric Block

Volume 31 (1984)

1. **Addition and Substitution Reactions of Nitrile-Stabilized Carbanions,** Siméon Arseniyadis, Keith S. Kyler, and David S. Watt

Volume 32 (1984)

1. **The Intramolecular Diels-Alder Reaction,** Engelbert Ciganek

2. **Synthesis Using Alkyne-Derived Alkenyl- and Alkynylaluminum Compounds,** George Zweifel and Joseph A. Miller

Volume 33 (1985)

1. **Formation of Carbon-Carbon and Carbon-Heteroatom Bonds via Organoboranes and Organoborates,** Ei-Ichi Negishi and Michael J. Idacavage

2. **The Vinylcyclopropane-Cyclopentene Rearrangement,** Tomáš Hudlický, Toni M. Kutchan, and Saiyid M. Naqvi

Volume 34 (1985)

1. **Reductions by Metal Alkoxyaluminum Hydrides,** Jaroslav Málek

2. **Fluorination by Sulfur Tetrafluoride,** Chia-Lin J. Wang

CONTENTS

CHAPTER 1

THE BECKMANN REACTIONS: REARRANGEMENTS, ELIMINATION–ADDITIONS, FRAGMENTATIONS, AND REARRANGEMENT–CYCLIZATIONS

ROBERT E. GAWLEY

Department of Chemistry, University of Miami, Coral Gables, Florida

CONTENTS

ACKNOWLEDGMENTS

The author is grateful to Professor A. I. Meyers and his colleagues in the Chemistry Department of Colorado State University for their hospitality during a sabbatical leave (1984), where much of the work on this chapter was accomplished. Thanks also go to Sanjay Chemburkar, Georgina Hart-Iwatiw, and Greg Smith for their help in compiling the experimental procedures and tables, and to Jeannine Peeples for the meticulous care with which she prepared the final manuscript. Financial support from the National Science Foundation is also acknowledged.

INTRODUCTION

The Beckmann rearrangement, the acid-mediated isomerization of oximes to amides (Eq. 1), was discovered by Beckmann in 1886.[1] As one of the oldest and most familiar transformations in organic chemistry, it has been reviewed several times.[2–10] What has become known as the *Beckmann fragmentation* (Eq. 2) was in fact first observed by Wallach in 1889[11] but was not developed extensively until the 1960s. It has been referred to by several names over the years and has also been reviewed.[5,12] The rearrangement cyclization (Eq. 3) is the intramolecular cyclization of a nitrilium ion generated by Beckmann rearrangement from an oxime. Within the context of an aromatic terminator, the process was first reported by Goldschmidt in 1895.[13] Wallach reported the first case of an aliphatic terminator in 1901,[14] although the reported structure was incorrect, and Perkin was apparently the first to observe a heteroatom terminator, but he also reported an incorrect structure.[15] The process has been exploited only quite recently. This chapter is an update of these reactions, last reviewed in this series in 1960.[5] Not covered are transformations that have been labeled Beckmann-type reactions but are mechanistically unrelated. The most prominent of these is the so-called photochemical Beckmann rearrangement, first observed by De Mayo in 1963[16] and discussed in a review several years ago.[17] Worthy of mention, however, is the fact that many oximes that suffer fragmentation under acidic conditions undergo rearrangement when photolyzed.

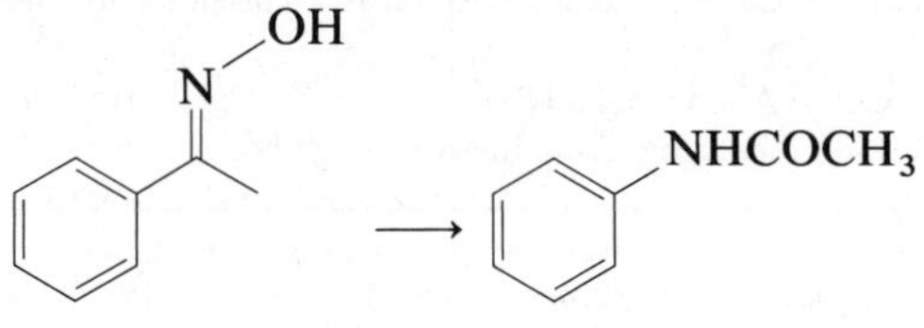

OH N NHCOCH$_3$ (Eq. 1)

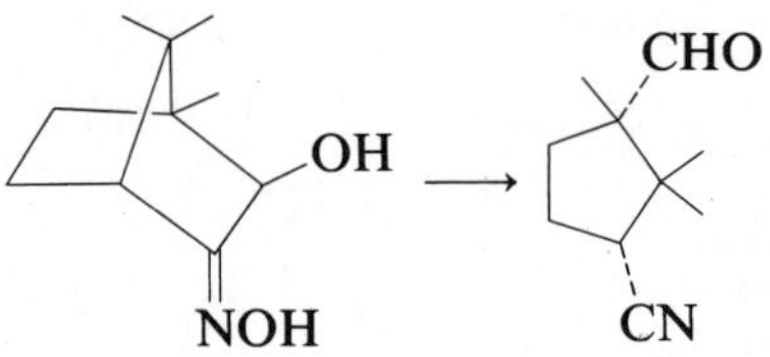

CHO OH NOH CN (Eq. 2)

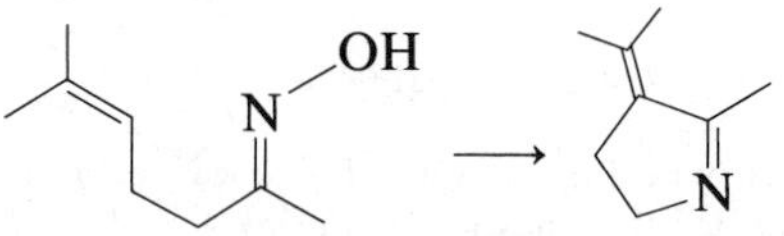

OH N N (Eq. 3)

Throughout this review, the (*E*) and (*Z*) nomenclature is used to describe oxime geometry.[18]

MECHANISM

Scheme 1 illustrates a mechanistic sequence by which the three reactions proceed under most sets of conditions. Key to the stereospecificity of the sequence is the site of attack of the acid, A^+. Attack at nitrogen leads to **1**, which may then isomerize to **2** or rearrange to **3**. The interconversion of **1** and **2** is responsible for instances of nonstereospecificity in the Beckmann rearrangement. Some reagents esterify the oxime and thus form species **3** directly, thereby avoiding the possibility of nonstereospecificity. Nitrogen–oxygen bond cleavage may occur with simultaneous migration or cleavage of the group *anti* to the oxygen to afford either rearrangement or fragmentation products. Moreover, these products may interconvert in certain instances. The initial rearrangement product has been shown to be an imidate of the type **4** in some instances and a free nitrilium ion in others. The imidate may undergo a Chapman rearrangement[19] to an *N*-substituted amide that hydrolyzes on workup, or the imidate itself may be hydrolyzed. The fragmentation process may proceed in a stepwise process, as shown, or there may be a stereospecific elimination from **3** to an alkene and a nitrile. When there is a nucleophile in one of

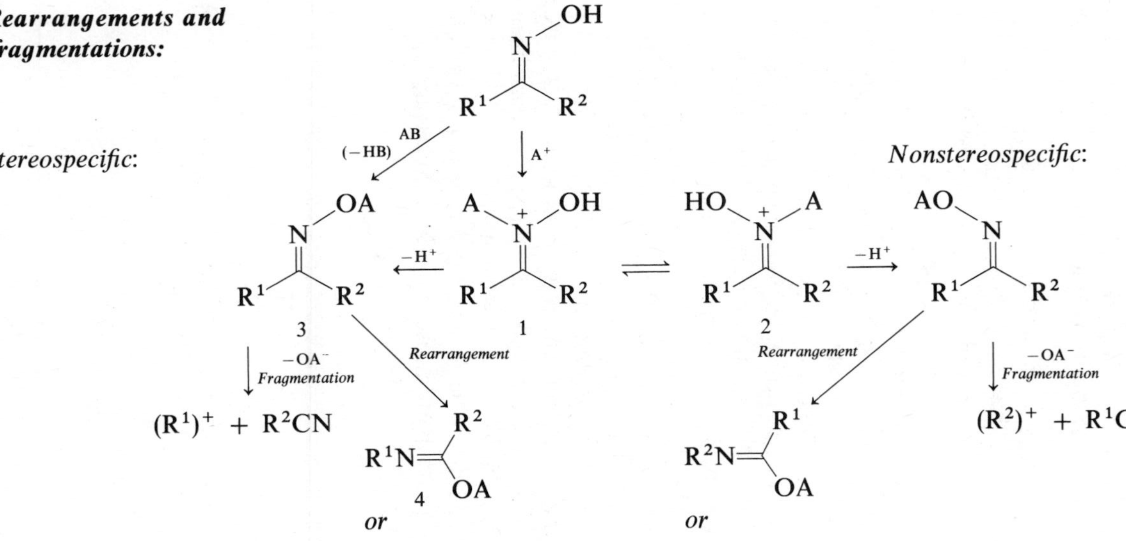

Rearrangements and fragmentations:
Stereospecific:
Nonstereospecific:
AB
(−HB)
A⁺
−H⁺
−H⁺
Rearrangement
Rearrangement
−OA⁻
Fragmentation
−OA⁻
Fragmentation
(R¹)⁺ + R²CN
(R²)⁺ + R¹CN
or
or
R¹—N≡C R² + OA⁻
R²—N≡C—R¹ + OA⁻
Cyclizations:
endo
exo

Scheme 1

the substituents of the oxime, the possibility of ring closure arises. Two modes are possible: when the nucleophile is *anti* to the oxime hydroxyl (i.e., in R^1), the cyclization is *endo*, and a heterocycle is formed; when the nucleophile is *syn* to the oxime hydroxyl (i.e., in R^2), the cyclization is *exo*, and an *N*-alkyl cycloalkanimine is formed, which usually hydrolyzes to the corresponding ketone on workup. The path taken by a specific molecule depends on the structure and the conditions employed.

Rates

Data from two different studies of the sulfuric acid mediated rearrangement of a series of *meta*- and *para*- substituted acetophenone oximes have shown that the rates provide a better correlation with σ than with σ^+, indicating the absence of a resonance-stabilized positive charge in the transition state.[20–22] Moreover, the reaction constants are small: -1.5[20] and -1.9,[21,22] indicating minor substituent effects. In another study, the rate of rearrangement of 2,4,6-trimethylacetophenone oxime was shown to be 3000 times faster than that of *p*-methylacetophenone oxime.[23] The reactive species is the oxime *O*-sulfonic acid, which is formed in a pre-equilibrium from the oxime protonated on nitrogen;[23,24] both species, as well as a postulated *N*-aryl nitrilium ion, can be observed by NMR spectroscopy. In contrast, the reactive species in the perchloric acid mediated rearrangement is thought to be the protonated oxime.[23]

A rationale for these observations is that the reaction is accelerated when the C=N bond is orthogonal to the benzene ring. Thus when the C=N and aromatic groups are coplanar, a concerted 1,2-sigmatropic rearrangement occurs, and resonance stabilization of the developing positive charge is not possible. When the C=N and aromatic groups are forced out of coplanarity by the *ortho* methyl groups, participation assists the N—O cleavage, increases the rate, and affords a discrete intermediate.

Coplanar:

X NOSO$_3$H →

[X N OSO$_3$H]‡ → X $\overset{+}{N}$≡C—

Orthogonal:

NOSO$_3$H → N + → $\overset{+}{N}$≡C—

In aliphatic ketoximes, the structure and solvent may play a role in the mechanism. For example, the relative rates of rearrangement of a series of cycloalkanone oximes in acetic acid and chloroform are consistent with variable transition state geometries (i.e., position of the transition state along the reaction coordinate) and varying degrees of neighboring group participation between different ring sizes.[25] The same authors note the effect of catalytic amounts of acid on the rearrangement of cyclopentanone oxime tosylate. The addition of 11 mol % of perchloric acid to the substrate in acetic acid *retards* the rate by about 20%. In contrast, the addition of trifluoroacetic acid to the same tosylate in chloroform affords a threefold rate *acceleration*. Thus the site of protonation is solvent dependent: perchloric acid in acetic acid protonates the oxime tosylate on nitrogen, whereas trifluoroacetic acid in chloroform protonates on oxygen.

Intermediates

Whether an imidate (e.g., **4**, Scheme 1) is involved also appears to depend on the system and solvent. In studies of the Ritter reaction,[26] it was suggested that free nitrilium ions are favored over imidates in sulfuric acid concentrations above 93%.[27] When ^{18}O-enriched sulfuric acid is used in the rearrangement of acetophenone oxime, the acetanilide produced contains the same isotopic enrichment as the solvent, whereas neither the acetophenone oxime nor the acetanilide incorporates labeled oxygen under the reaction conditions, thus corroborating the intermediacy of a free nitrilium ion.[20] Free nitrilium ions have even been isolated: treatment of the *N*-chloroimines of either benzophenone or pivalophenone with antimony pentachloride in carbon tetrachloride affords the antimony hexachloride salts of *N*-phenylbenzonitrile and *N*-*tert*-butylbenzonitrile in 90 and 80% yields, respectively.[28]

The sulfur trioxide mediated rearrangement of cyclohexanone oxime proceeds through the oxime sulfonic acid, which has been isolated and characterized.[29] When this compound is heated in petroleum to 65°, it rearranges to caprolactam-*N*-sulfonic acid, and the authors postulate the intermediacy of an imidate, which undergoes a Chapman rearrangement.[19]

$NOSO_3H$ $\xrightarrow{65°}$ [OSO_3H, N] $\longrightarrow$ O, N, SO_3H

When 2-arylcyclohexanone oxime tosylates are solvolyzed in pyridine, a species thought to be pyridinium tosylate **5** is observed by NMR. The authors suggest that rearrangement of the tosylate to its imidate isomer is the rate-determining step.[30]

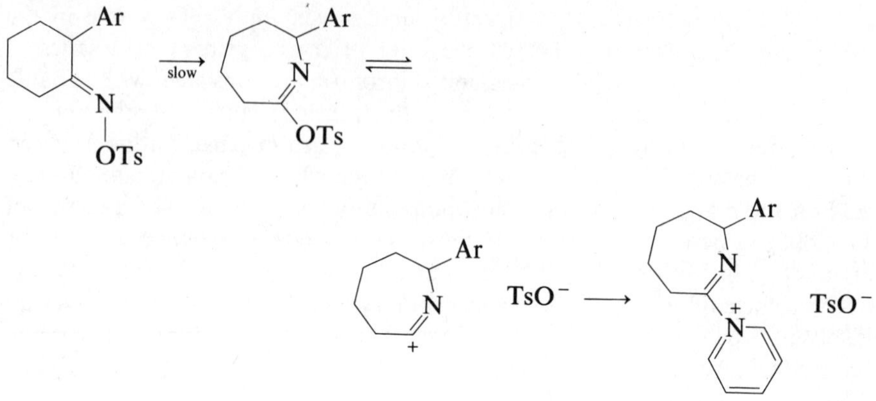

When (*E*,*E*)-benzylidenecyclohexanone oxime is treated with *p*-toluenesulfonyl chloride in pyridine, pyridinium tosylate **6** can be isolated as a crystalline solid.[31] When cyclohexanone oxime tosylate is treated with any of several mild Lewis acids (e.g., silica gel or alumina) in carbon tetrachloride, *O*-tosylcaprolactim (**7**) is observed by IR spectroscopy.[32]

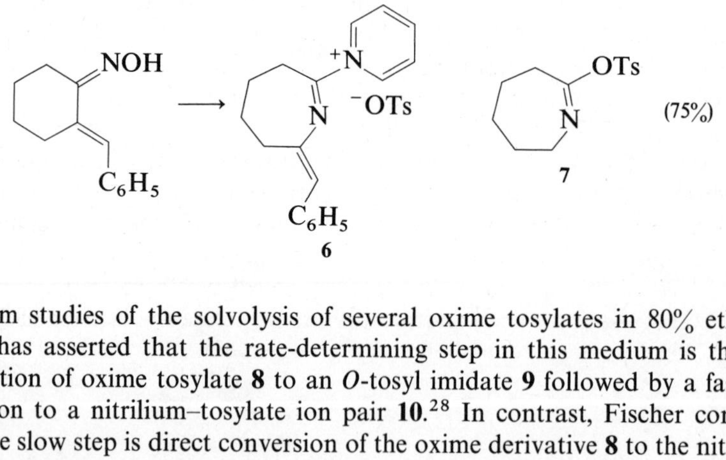

From studies of the solvolysis of several oxime tosylates in 80% ethanol, Grob has asserted that the rate-determining step in this medium is the isomerization of oxime tosylate **8** to an *O*-tosyl imidate **9** followed by a fast dissociation to a nitrilium–tosylate ion pair **10**.[28] In contrast, Fischer contends that the slow step is direct conversion of the oxime derivative **8** to the nitrilium ion **10**, which may then combine to an imidate, fragment, and so on.[33] In a series of rearrangements of oxime picrates, various components of the system were enriched with ^{18}O, but the data obtained can be explained by either sequence of steps.[34]

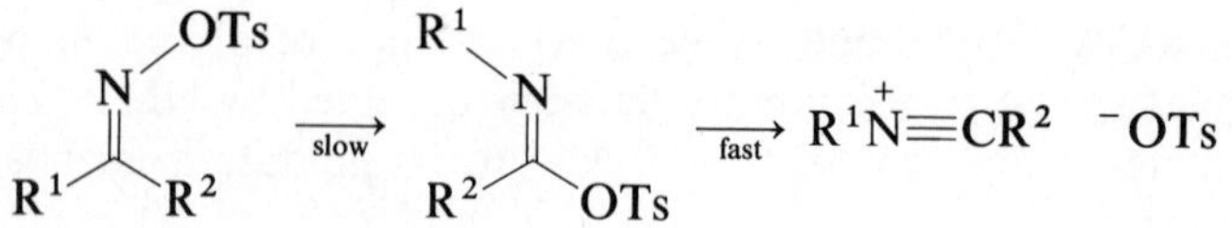

STEREOCHEMISTRY

Rearrangement

In general, the Beckmann rearrangement of ketoximes is stereospecific, involving migration of the group *anti* to the leaving group on nitrogen. This route is operative for aryl–alkyl ketoximes in the gas phase under conditions of chemical ionization mass spectrometry.[35,36] A similar tendency for *anti* migration occurs for both (*E*)- and (*Z*)-aldoximes in the gas phase;[36] but in solution, rearrangement of aldoximes almost always gives primary amides as opposed to *N*-alkylformamides, independent of the oxime geometry. Specific exceptions to these generalizations are discussed later.

The reaction of a ketone with a derivative of hydroxylamine, such as hydroxylamine *O*-sulfonic acid, may proceed via rearrangement of the initially formed carbinolamine derivative without dehydration to the oxime. In this case, migratory aptitude determines the product structure.[37,38]

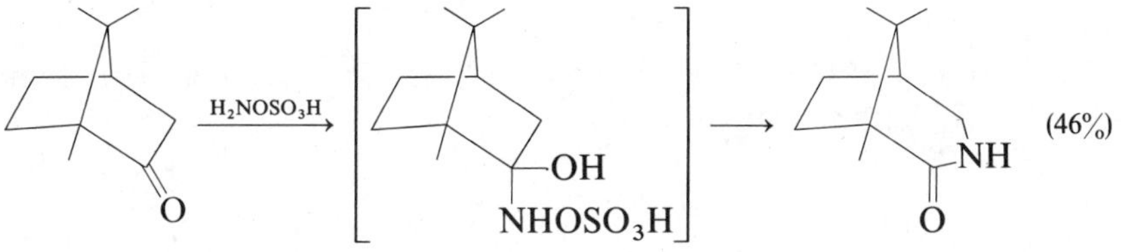

The generalized "nonstereospecific" pathway illustrated in Scheme 1 can be documented very well by the rearrangement of 2-pentanone oxime tosylates in a variety of media.[39] Thus, in the presence of a protic acid, both the thermodynamic mixture of geometric isomers and the pure *Z* isomer rearrange via the *E* isomer only. In contrast, the *Z* isomer rearranges stereospecifically on treatment with alumina.

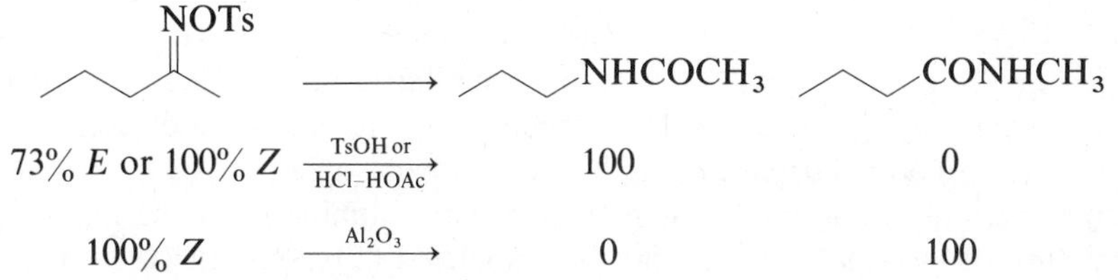

Oxime derivatives of cyclohexenones are often difficult to rearrange stereospecifically. This trend has been particularly well documented in the steroid series. For example, the only product isolated from the rearrangement of oximes **11–15**, in the geometric configuration indicated, results from migration of the methylene carbon. Particularly convincing examples are **14**, which is 50% *E*, and **15**, which is 100% *E*, but that rearrange only from their *Z* configurations.

R¹ON → HN, O

	R^1	R^2	R^3		
11 (25% Z)	Ts	C_8H_{17}	H	HCl, CH_3OH	(87%)[40]
12 (100% Z)	H	C_8H_{17}	H	$SOCl_2$, ether	(20%)[41]
12 "				p-$CH_3CONHC_6H_4SO_2Cl$	(11%)[42]
13 (33% Z)	Ts	OH	H	HCl, CH_3OH	(85%)[40]
14 (50% Z)	H	OH	CH_3	$SOCl_2$, dioxane	(97%)[43]
15 (100% E)	H	O_2CCH_3	CH_3	", "	(59%)[43]

The *Z* isomer of **16** rearranges readily to **17** in 73% yield, but the *E* isomer is unreactive under similar conditions.[44]

C_8H_{17} HON **16** —$SOCl_2$, 10°, 20 min→ O= HN **17** (73%)

Nonsteroidal cyclohexenone oximes exhibit similar properties. Polyphosphoric acid (PPA) rearranges the (*Z*)-oxime of isophorone to lactam **18** in low yield, whereas the (*E*)-oxime is converted to isomeric lactam **19** in good yield under the same conditions.[45] A number of similarly substituted cyclohexenone oximes rearrange (presumably stereospecifically) to mixtures of lactams on treatment with polyphosphoric acid.[46] A mixture of isophorone oxime *O*-mesityl sulfonate geometric isomers, when treated with alumina in methanol, affords a mixture of lactam **18** (50%) and the unreacted (*E*)-oxime mesityl sulfonate isomer (28%), verifying the reluctance of the vinyl group to migrate under routine conditions.[47]

19 (72%) ←PPA, E isomer— NOH —PPA, Z isomer→ **18** (21%)

As mentioned above, (*E*, *E*)-benzylidenecyclohexanone oxime rearranges when treated with *p*-toluenesulfonyl chloride in pyridine to the pyridinium salt **6**, dilute hydrolysis of which affords lactam **20**.[31]

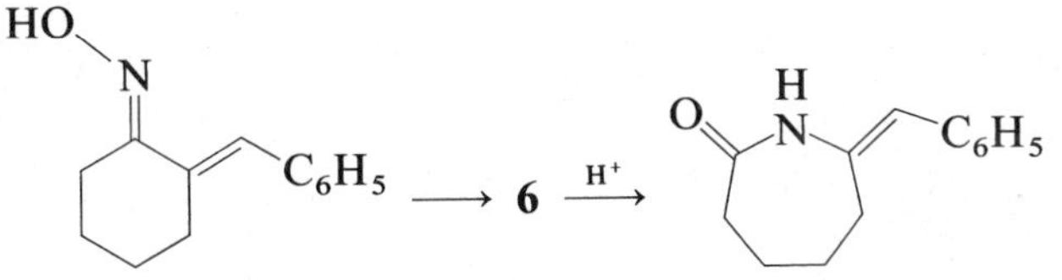

20 (75%)

The conclusion, then, is that there is considerable strain in the transition state involving vinyl group migration within the constraints of certain six-membered ring systems. It has been suggested that the rearrangement of acyclic enone oximes proceeds via an azirene intermediate, which would impart considerable strain in a five- or six-membered ring.[48,49]

Examples of stereospecific rearrangements of aldoximes are summarized in the "Scope and Limitations" section of this chapter.

Fragmentations

Whether a given ketoxime fragments or rearranges depends quite heavily on the stability of the carbonium ion formed as a result of the fragmentation. Fragmentation is to be expected as the major course of reaction when a particularly stable ion is so formed. Thus an oxime whose hydroxyl is *anti* to a quaternary carbon, for example, can be expected to fragment unless carefully controlled conditions are utilized to encourage rearrangement. For example, triterpene oxime **21** fragments to nitrile **23** in 84–90% yield when heated with *p*-toluenesulfonyl chloride in pyridine. When **21** is treated with either the same reagent or phosphoryl chloride at room temperature for 2 days, lactam **22** is obtained in 57–75% yield, accompanied by 20–30% of nitrile **23**.[50]

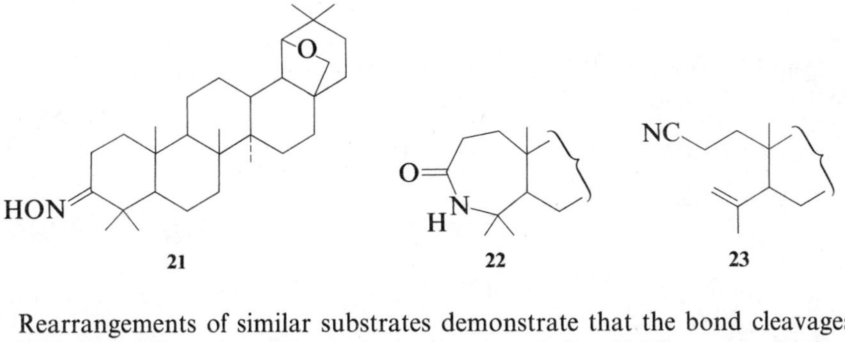

21 **22** **23**

Rearrangements of similar substrates demonstrate that the bond cleavages operate under the rules of stereoelectronic control. Specifically, oxime **24**, isotopically labeled at one of the 4-methyl positions, rearranges to *O*-tosyllactim **25** and then fragments stereospecifically to give alkenyl nitrile **26** in which the label is found on the olefinic carbon.[51,52] When the 5,6-dehydro derivative,

27, of the same oxime is fragmented, the alkenyl nitrile contains the label in the methyl group.[53] The rationale for these observations is that in both compounds, a proton from the pseudoequatorial methyl group is lost from the *O*-tosylcaprolactim intermediate.

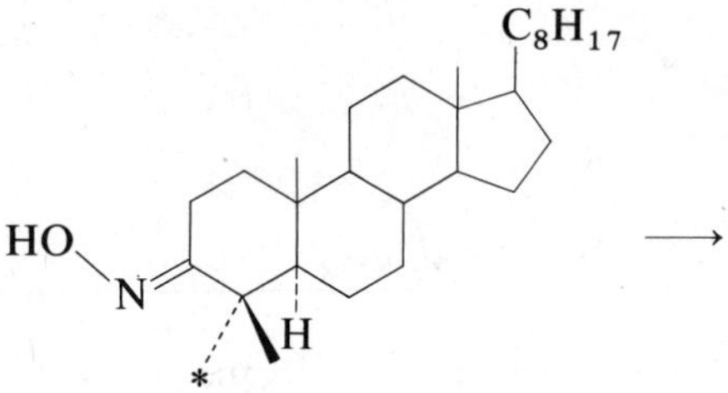

C_8H_{17}

HO N H

* = CD_3 or $^{14}CH_3$

24

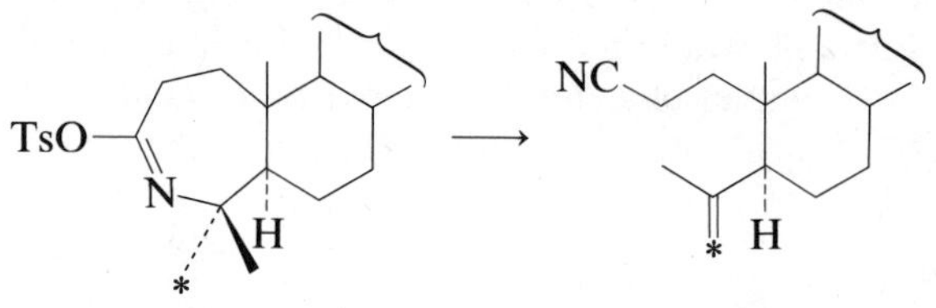

TsO N H *

NC * H

25 **26**

C_8H_{17}

HON CD_3 CH_3

NC CD_3

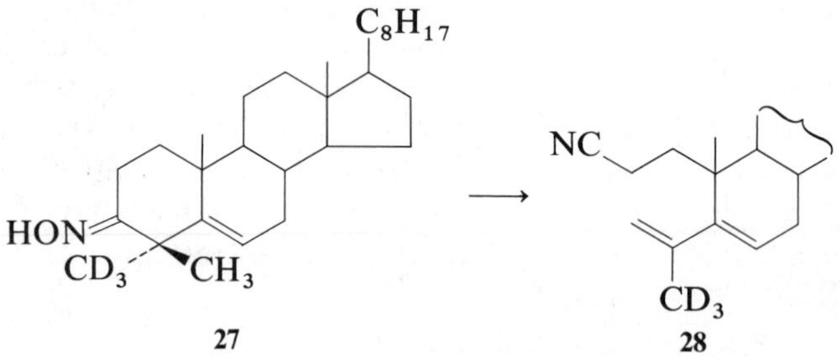

27 **28**

Another example is provided by oxime **29**, in which the C-5 proton is equatorial and, therefore, antiperiplanar to the fragmenting C—C bond.[54]

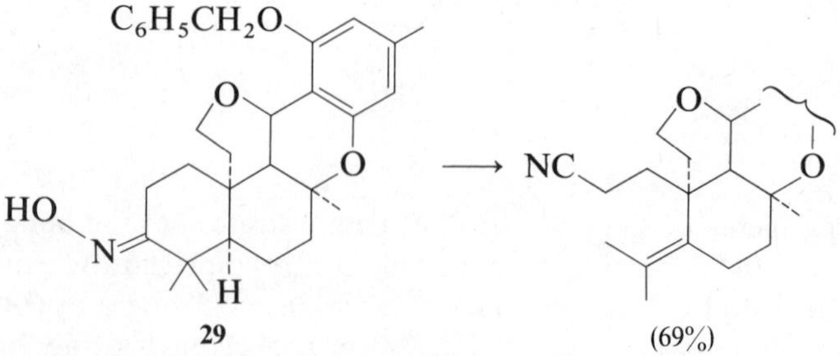

$C_6H_5CH_2O$ O O HO N H

NC O O

29 (69%)

From studies of a series of rigid caged ketone oximes, Grob has concluded that degree of deviation from optimum bond angles is a useful measure of fragmentation aptitude.[55]

As should be expected, the presence of cation-stabilizing heteroatoms influences the degree of fragmentation. But again, oxime stereochemistry is important. For example, the rate constant for fragmentation of the *Z* isomer of oxime benzoate **30** in 80% ethanol at 70° is 1800 times larger than the rate constant for the *E* isomer.[56] Rate studies on a series of related compounds indicate that the energy of activation is roughly 5 kcal/mol greater when the amino substituent is *syn* to the leaving group (*Z* isomer).

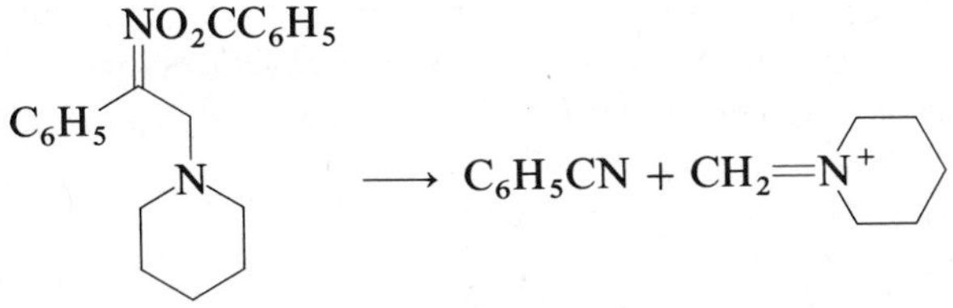

$NO_2CC_6H_5$; C_6H_5 ; N ⟶ $C_6H_5CN + CH_2{=}N^+$

30

Monooximes of 1,2-diketones usually fragment via nucleophilic attack on the carbonyl carbon to give a tetrahedral intermediate such as **31**.[57] When the carbonyl carbon is sterically hindered as in **32**, fragmentation occurs via acylium ion **33**.[58]

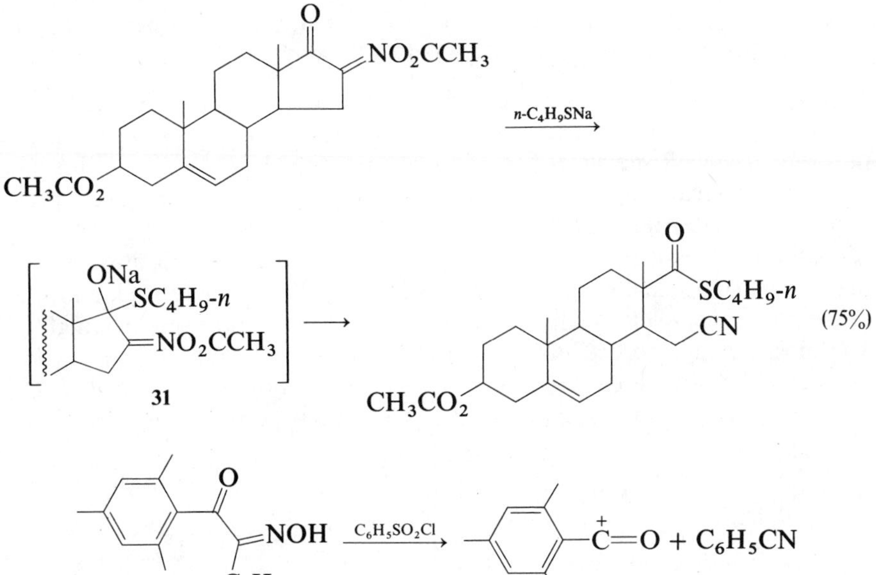

O ; NO_2CCH_3 ; CH_3CO_2 $\xrightarrow{n\text{-}C_4H_9SNa}$

[ONa ; SC_4H_9-*n* ; $=NO_2CCH_3$] **31** ⟶ O ; SC_4H_9-*n* ; CN ; CH_3CO_2 (75%)

O ; NOH ; C_6H_5 **32** $\xrightarrow{C_6H_5SO_2Cl}$ $\overset{+}{C}{=}O + C_6H_5CN$ **33**

Rearrangement–Cyclization

The rearrangement–cyclization has been shown to be stereospecific. Because it involves the intramolecular capture of the nitrilium ion or imidate species, whether the cyclization proceeds via the *endo* or *exo* modes (see Scheme 1) is entirely dependent on oxime geometry.

SCOPE AND LIMITATIONS

Rearrangements

This section deals with Beckmann rearrangements that are *rearrangements* in the strictest terms: the functional group formula is not changed. Thus —C(=NOH)— or a derivative is converted into —C(=O)NH—. The following section ("Elimination–Additions") deals with reactions in which the course of the conversion is intercepted along its path and that, therefore, are not rearrangements in the same sense.

Ketoximes. Table I lists a large number of examples of the Beckmann rearrangement, most of which are relatively routine conversions. Most conditions utilized in these reactions are compatible with functional groups such as carboxylic acids, esters, alcohols, ethers, and alkyl halides but may not be compatible with some of the more sensitive protecting groups. Because there are few general methods available for the controlled manipulation of oxime geometry, Beckmann rearrangements have not seen extensive use in complex total syntheses even though the rearrangement is stereospecific. This section contains a smattering of examples, chosen to afford the reader some insight into the types of reagents that have been used to effect the Beckmann rearrangement and some of the molecules that have been subjected to these rearrangements.

Reagents. Phosphorus pentachloride continues to be a popular reagent for the Beckmann rearrangement, but since numerous examples are listed in the previous *Organic Reactions* review,[5] it is not discussed here. A popular medium for Beckmann rearrangements of compounds that are not acid-sensitive is polyphosphoric acid. After studying the rates of rearrangement of a series of substituted acetophenone oximes, Pearson noted that there is no substituent effect, that the rearrangements are 12–35 times faster than in sulfuric acid, and that the rearrangement will usually occur at or near room temperature overnight.[59] The most common procedure is to heat the oxime at about 130° for a few minutes, although a recent procedure suggests the use of xylene as a cosolvent.[60] A typical example is the rearrangement of adamantylacetone oxime, **34**.[61]

NOH $\xrightarrow{\text{PPA}}$ NHCOCH$_3$ (85%)

34

Mesityl oxide oxime, when treated with polyphosphoric acid, affords a 16% yield of *N*-isobutylacetamide (a reduction product), indicating that polyphosphoric acid may not be the best medium for rearrangement of unsaturated oximes.[62] Several polyphosphoric acid mediated rearrangements have been shown (or postulated) to proceed via a fragmentation–recombination pathway.[63–67] For example, whereas oximes **35** and **37** rearrange to amides **36** and **38**, a mixture of oximes **35** and **37** affords not only **36** and **38** but also crossover products **39** and **40** via the corresponding fragmentation products: benzonitrile, acetonitrile, cumyl and *t*-butyl cations.[64]

$$\underset{\mathbf{35}}{C_6H_5C(CH_3)_2C({=}NOH)C_6H_5} \xrightarrow{\text{PPA}} \underset{\mathbf{36}}{C_6H_5C(CH_3)_2NHCOC_6H_5}$$

$$\underset{\mathbf{37}}{t\text{-}C_4H_9C({=}NOH)CH_3} \xrightarrow{\text{PPA}} \underset{\mathbf{38}}{t\text{-}C_4H_9NHCOCH_3}$$

$$\mathbf{35} + \mathbf{37} \xrightarrow{\text{PPA}} \mathbf{36} + \mathbf{38} + \underset{\mathbf{39}}{t\text{-}C_4H_9NHCOC_6H_5} + \underset{\mathbf{40}}{C_6H_5C(CH_3)_2NHCOCH_3}$$

Another example is the polyphosphoric acid mediated rearrangement of oxime **41**, which occurs at 25° and involves inversion of configuration in the process.[66,67] The observation that decalin alcohol **42** affords the same product when subjected to the Ritter reaction indicates a fragmentation–recombination mechanism is operative.[26] An additional complication is the fact that, at 125°, the major product from **41** is *cis*-decalin via a disproportionation reaction.[66,67] Rearrangement of **41** with *p*-toluenesulfonyl chloride in pyridine occurs with retention of configuration at the migrating carbon, indicating that the rearrangement is not occurring by the fragmentation–recombination mechanism under these conditions.[66,67]

HON · H · **41** — PPA, 25° → (61%) NHCOCH$_3$ · H ← H_2SO_4, CH_3CN — OH · H · **42**

Some of the early literature on the Beckmann rearrangement invokes the intermediacy of a free nitrenium ion. This mechanism was sometimes used to explain *syn* migration when it was observed. Lansbury synthesized a series of substituted tetralone and indanone oximes and demonstrated fairly convincingly that a nitrenium ion can be generated only under very unusual circumstances.[68–71] Dimethylindanone oxime **43**, for example, affords a 1:2 ratio of lactams **44** and **45** in 48% yield, probably via a free nitrenium ion.[70,71]

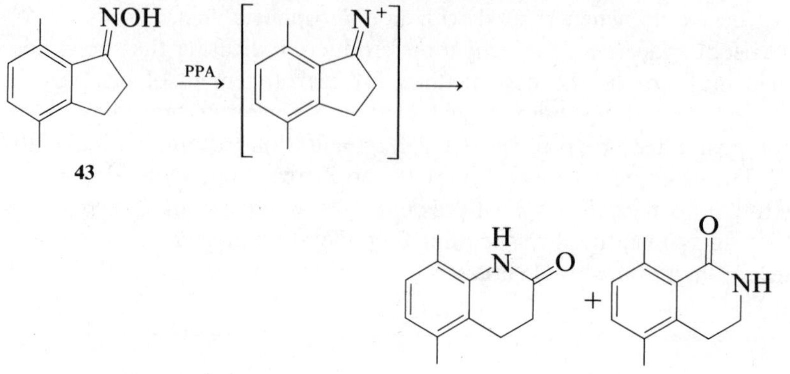

NOH PPA N+ 43

H N O + O NH (48%)

44 (1:2) 45

Trimethylsilyl polyphosphate (PPSE) is an effective medium for the Beckmann rearrangement at room temperature. An example is the rearrangement of oxime **46** in methylene chloride, occurring in 63% yield.[72]

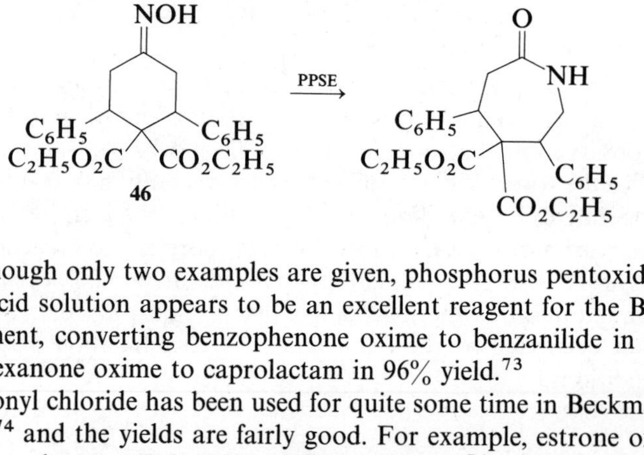

NOH PPSE O NH (63%)

C_6H_5 C_6H_5 $C_2H_5O_2C$ $CO_2C_2H_5$ **46** C_6H_5 $C_2H_5O_2C$ C_6H_5 $CO_2C_2H_5$

Although only two examples are given, phosphorus pentoxide–methanesulfonic acid solution appears to be an excellent reagent for the Beckmann rearrangement, converting benzophenone oxime to benzanilide in 95% yield and cyclohexanone oxime to caprolactam in 96% yield.[73]

Thionyl chloride has been used for quite some time in Beckmann rearrangements,[74] and the yields are fairly good. For example, estrone oxime (**47**) rearranges to lactam **48** in 83% yield in dioxane,[74] while *p*-nitroacetophenone oxime rearranges to *p*-nitroacetanilide in 76% yield in carbon tetrachloride.[75]

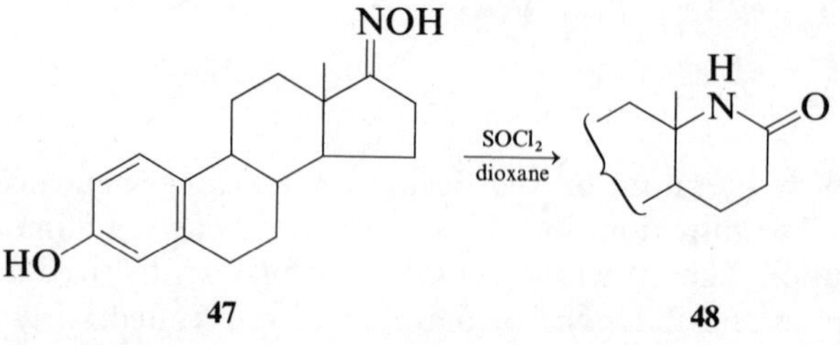

NOH $SOCl_2$ dioxane H N O (83%)

HO **47** **48**

More commonly used than thionyl chloride are benzenesulfonyl chloride and *p*-toluenesulfonyl chloride. The rearrangement can be effected either with or without isolation of the intermediate sulfonate esters. Two papers are often cited as sources of the procedures for the rearrangement.[76,77] Oximes that are prone to fragmentation can often be rearranged by use of these reagents. For example, spiro oxime **49** rearranges to lactam **50** in 73% yield when treated with benzenesulfonyl chloride. In contrast, treatment with phosphorus pentachloride affords alkenyl nitrile **51** in 92% yield, while polyphosphoric acid affords amide **52** (note reduction) in 46% yield.[78]

OH $\xrightarrow{C_6H_5SO_2Cl}$ NH, O (73%)

49 **50**

$(CH_2)_5CN$ $\xleftarrow{PCl_5}$ **49** $\xrightarrow{PPA}$ $(CH_2)_5CONH_2$

51 (92%) (46%) **52**

The sensitivity of fragmentation-prone systems is illustrated by the reaction of oxime tosylate **53**, which gives either lactam **54** when treated with triethylamine at room temperature or nitrile acid **55** when treated with dilute sodium hydroxide at 0°, both in 80% ethanol.[79,80]

$\xleftarrow[80\%\ C_2H_5OH]{(C_2H_5)_3N}$ $\xrightarrow[80\%\ C_2H_5OH]{NaOH}$

O, N, H, O (77%) **54**

O, N, TsO **53**

CO_2H, NC (91%) **55**

Once formed, the oxime tosylates can be readily solvolyzed in 80% ethanol containing one equivalent of triethylamine.[28] A typical example is the rearrangement of oxime tosylate **56**, which affords amide **57** in 99% yield.

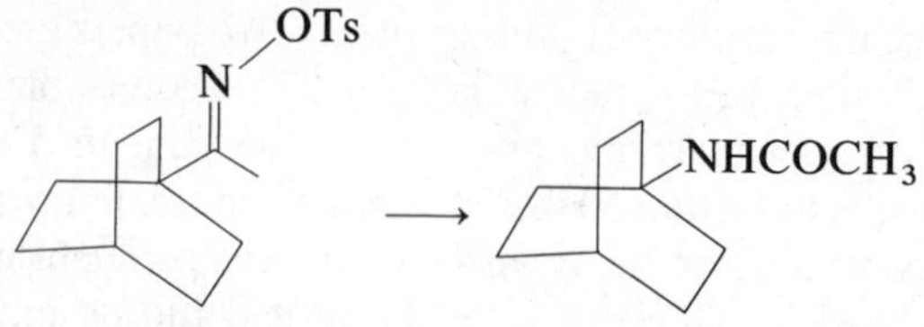

OTs N NHCOCH$_3$ (99%)

56 57

Hydroxylaminesulfonic acid[81] or *O*-mesitylenesulfonylhydroxylamine[82] can be used to convert ketones to lactams directly. It was mentioned previously, however, that these types of rearrangements may not proceed via the oxime sulfonate. For example, the former reagent converts cyclododecanone into lactam **58** in 76% yield,[81] while the latter reagent converts ketone **59** into a 4:1 mixture of lactams **60** and **61**.[82]

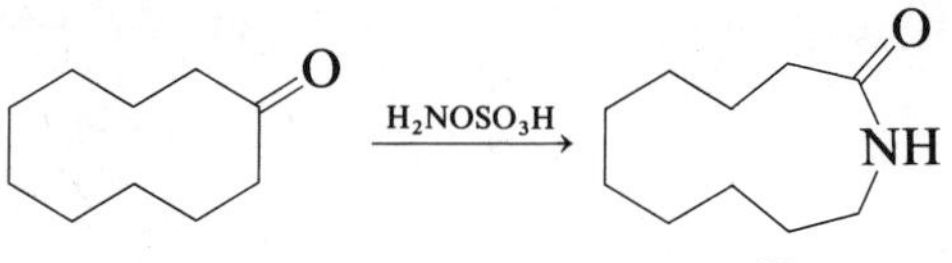

H_2NOSO_3H NH (76%)

58

C_5H_{11}-*n* OSO_2NH_2 HN

CH_3CO_2 CH_3CO_2 CH_3CO_2 CH_3CO_2

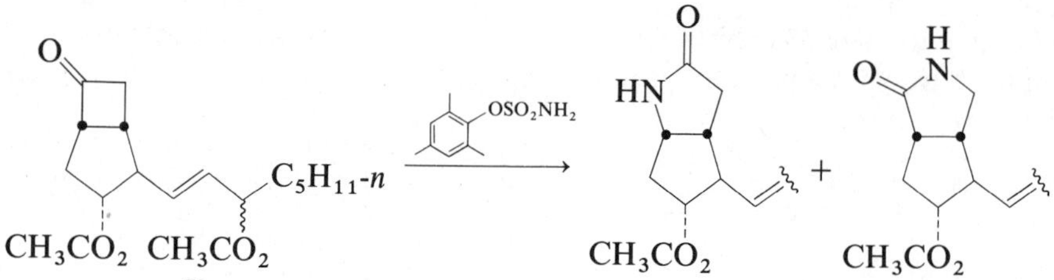

59 **60** (4:1) **61**

Anhydrous hydrogen fluoride converts cyclohexanone and benzophenone oxime benzoates into caprolactam and benzanilide in 72 and 90% yields, respectively.[83]

Triphenylphosphine–carbon tetrachloride can effect the Beckmann rearrangement. For example, 2-octanone oxime is converted to *N*-hexylacetamide in 80% yield by refluxing in either tetrahydrofuran or carbon tetrachloride.[84]

Trimethylsilyl iodide[85] induces the Beckmann rearrangement, probably *via* the corresponding imidoyl iodide.[86] For example, acetophenone oxime affords acetanilide in 55% yield and benzophenone oxime affords benzanilide in 80% yield, but oximes of cyclohexanone, acetone, 2-octanone, and pivalophenone are unreactive.[85]

Lactone **62** in carbon tetrachloride at room temperature converts cyclohexanone oxime to *N*-acylcaprolactam **63** in quantitative yield.[87]

NOH, 62, CCl_4 → 63 (100%)

N-Cyanato imines rearrange, as illustrated by the conversion of benzophenone derivative **64** to imidoyl isocyanate **65** or imidoyl urethane **66** by heating briefly in carbon tetrachloride or ethanol, respectively.[88]

65 ← (CCl_4, heat) 64 → (C_2H_5OH, heat) 66

Substrates. Cyclohexane-1,3-[89] and -1,4-dione[90] dioximes (**67** and **68**) are rearranged to diamine–diacid lactam **69** and amino acid "dimer" lactam **70** by the action of fuming sulfuric acid.

67 → (oleum, H_2SO_4) 69 (20%)

68 → (oleum, H_2SO_4) 70 (29%)

Cyclooctane-1,5-dione dioxime ditosylate (**71**) rearranges to diacid–diamine lactam **72**,[91] while cyclodecane-1,6-dione dioxime ditosylate (**73**) rearranges to a mixture of lactams **74** and **75**.[90]

71 → ($KHCO_3$, dioxane) 72 (47%)

NOTs / NOTs (73) $\xrightarrow[\text{CH}_3\text{CO}_2\text{H}]{\text{NaO}_2\text{CCH}_3}$ 74 (CONH / NHCO) + 75 (CONH / CONH) (61%)

These are probably stereospecific reactions that reflect the stereochemistry of the dioxime substrates, which is partly governed by packing forces in the crystal lattices.

There are a number of examples of rearrangement of oxime derivatives to imidates, which then undergo further rearrangement to *N*-substituted amides, reminiscent of the Chapman rearrangement.[19] This type of transformation is not a Chapman rearrangement in the strictest sense, but is broadly referred to as such in the literature. For example, as mentioned earlier, cyclohexanone oxime sulfonic acid rearranges to caprolactam-*N*-sulfonic acid, probably *via* an imidate which then rearranges again.[29] When benzophenone oxime is treated with a substituted benzoic acid in the presence of triphenylphosphine and diethyl azodiformate, *N*-benzoylbenzanilides are formed.[92] The process probably proceeds as follows: esterification of benzophenone oxime, rearrangement to a benzoyl imidate, and then further rearrangement to the substituted benzanilide.

$(C_6H_5)_2C{=}NO_2CC_6H_4NO_2\text{-}p \longrightarrow [C_6H_5N{=}C(O_2CC_6H_4NO_2\text{-}p)C_6H_5] \longrightarrow C_6H_5CON(C_6H_5)COC_6H_4NO_2\text{-}p$ (87%)

Similarly, oxime picrates (C=NOPic) produce *N*-2,4,6-trinitrophenyl amides or lactams [CON(Pic)R].[93–97] For example, benzocyclooctane picrate **76** affords lactam **77** in 74% yield when refluxed in methylene chloride.[96]

PicO–N= (76) $\xrightarrow[\text{heat}]{\text{CH}_2\text{Cl}_2}$ O, N–Pic (77) (74%)

The Beckmann rearrangement can be carried out on some oxime derivatives of transition-metal compounds. For example, benzoylferrocene oxime is rearranged to amide **78** in 23% yield by treatment with *p*-toluenesulfonyl chloride in pyridine,[98] in 18% yield with alkaline benzenesulfonyl chloride,[99] and in 60–70% yield by refluxing in trichloroacetonitrile.[100]

$$C_5H_5FeC_5H_4C(=NOH)C_6H_5 \longrightarrow \underset{\mathbf{78}}{C_5H_5FeC_5H_4CONHC_6H_5} \quad (18\text{–}70\%)$$

(*Z*)-Benzoylcyclopentadienylmanganese tricarbonyl oxime (**79**) yields amide **80** on treatment with phosphorus pentachloride and pyridine. The *E* isomer under the same conditions affords only $(CO)_3Mn(C_5H_5N)_2Cl^-$.[101–102]

$$\underset{\mathbf{79}}{(CO)_3MnC_5H_4C(=NOH)C_6H_5} \xrightarrow[C_5H_5N]{PCl_5} \underset{\mathbf{80}}{(CO)_3MnC_5H_4CONHC_6H_5}$$

Organotin substrates also rearrange.[103,104] For example, oxime **81** rearranges to amide **82** on treatment with *p*-toluenesulfonyl chloride.[104]

$$\underset{\mathbf{81}}{(C_6H_5)_3Sn(CH_2)_3C(=NOH)CH_3} \xrightarrow{TsCl} \underset{\mathbf{82}}{(C_6H_5)_3Sn(CH_2)_3NHCOCH_3}$$

The oximes of benzoyl[105] and acetyl[106] *closo*-carboranes rearrange to acetamido carboranes on treatment with phosphorus pentachloride.

In certain instances, oximes with other functional groups react further after they rearrange. For instance, oximes of certain keto acids may undergo cyclization after rearrangement. Oxime acid **83** undergoes normal rearrangement to the corresponding amide when treated with polyphosphoric acid, but if water is added and heating is continued, amide hydrolysis and cyclization occur to give lactam **84** in 51% yield.[107]

$$\underset{\mathbf{83}}{CH_3C(=NOH)CH_2CH_2CH(CH(CH_3)_2)CH_2CO_2H} \xrightarrow[2.\ H_2O]{1.\ PPA} \underset{\mathbf{84}}{\text{4-isopropyl-2-piperidinone}} \quad (51\%)$$

A number of oxime acids of general structure **85**, when treated with polyphosphoric acid alone, undergo rearrangement and cyclization to *N*-acyl lactams **86** in 55–70% yield.[108]

$$\mathrm{Ar^1C(=NOH)CH_2CH(Ar^2)CH_2CO_2H}\ (\mathbf{85}) \xrightarrow{\text{PPA}} \mathbf{86}\ (N\text{-}\mathrm{COAr^1}\text{, 4-}\mathrm{Ar^2}\text{-pyrrolidin-2-one})\quad (55\text{–}70\%)$$

Alkenyl oximes may rearrange to either unsaturated amides or *N*-vinyl amides (enamides). For example, chalcone oxime **87** is converted to *N*-phenylcinnamamide (**88**) in 60–70% yield by refluxing in trichloroacetonitrile,[100] in 57% yield by stirring with trimethylsilyl polyphosphate in methylene chloride,[72] or by treating its tosylate ester with silica gel in chloroform.[109]

$$\underset{\mathbf{87}}{\mathrm{C_6H_5C({=}NOH)CH{=}CHC_6H_5}} \longrightarrow \underset{\mathbf{88}}{\mathrm{C_6H_5NHCOCH{=}CHC_6H_5}}\quad (57\text{–}70\%)$$

Alkenyl oxime **89** is converted to *N*-vinylacetamide **90** in 94% yield with phosphorus pentachloride.[110–112]

$$\underset{\mathbf{89}}{\mathrm{C_6H_5CH{=}C(CH_3)C({=}NOH)CH_3}} \xrightarrow{\mathrm{PCl_5}} \underset{\mathbf{90}}{\mathrm{C_6H_5CH{=}C(CH_3)NHCOCH_3}}\quad (94\%)$$

The yields are often quite low, however, as is seen in the phosphorus pentachloride mediated rearrangement of oxime **91**, which occurs in only 20% yield.[113]

$$\underset{\mathbf{91}}{\mathrm{(2\text{-thienyl})CH{=}C(C_6H_5)C({=}NOH)CH_3}} \xrightarrow{\mathrm{PCl_5}} \mathrm{(2\text{-thienyl})CH{=}C(C_6H_5)NHCOCH_3}\quad (20\%)$$

Unsubstituted vinyl oximes may undergo Michael-type reactions following rearrangement, as illustrated by the rearrangement of oxime **92**, which is accompanied by 1,4 addition of chloride ion to the amide product, providing an 81% yield of chloroamide **93**.[114]

$$\underset{\mathbf{92}}{\mathrm{C_6H_5C({=}NOH)CH{=}CH_2}} \xrightarrow{\mathrm{PCl_5}} \underset{\mathbf{93}}{\mathrm{C_6H_5NHCO(CH_2)_2Cl}}\quad (81\%)$$

The rearrangement to an enamide, which is labile to hydrolysis, can be used to degrade steroid side chains. A typical example is the conversion of oxime **94** to ketone **95** in 49% yield on treatment with *p*-acetamidophenylsulfonyl chloride in pyridine.[115]

$$\mathbf{94} \xrightarrow[C_5H_5N]{p\text{-}(CH_3CONH)C_6H_4SO_2Cl} \mathbf{95} \quad (49\%)$$

Oximes of propargyl ketones also undergo Michael addition of chloride ion following Beckmann rearrangement.[116] For example, propargyl ketoxime **96** yields only chlorocrotonamide **97** when treated with phosphorus pentachloride in ether.

$$\underset{\mathbf{96}}{CH_3C{\equiv}C\text{-}C({=}NOH)\text{-}C_3H_7\text{-}i} \xrightarrow{PCl_5} \underset{\mathbf{97}}{CH_3CCl{=}CHCONHC_3H_7\text{-}i}$$

The oxime tosylate of benzoyl cyanide (**98**) rearranges to urethane **99** in 60–62% yield when treated with either potassium hydroxide or sodium ethoxide in ethanol. The diethyl acetal of phenyl isocyanate is an intermediate, and the mechanism probably involves initial attack of ethoxide ion on the oxime carbon.[117]

$$\underset{\mathbf{98}}{C_6H_5C({=}NOTs)CN} \xrightarrow[C_2H_5OH]{NaOC_2H_5} \underset{\mathbf{99}}{C_6H_5NHCO_2C_2H_5} \quad (60\text{–}62\%)$$

N-(*p*-Toluenesulfonyl)benzamide oxime (**100**) rearranges to *N*-(*p*-toluenesulfonyl)-*N*-phenylurea (**101**) in 50% yield on treatment with benzenesulfonyl chloride.[118]

$TsNHC(=NOH)C_6H_5$ (100) $\xrightarrow{C_6H_5SO_2Cl}$ $TsNHC(=O)NHC_6H_5$ (101) (50%)

A unique example of a retrograde Beckmann rearrangement involves treatment of lactam **102** with 48% hydrobromic acid in acetone to afford a mixture of three products: oxime **103**, ketone **104** (the major product), and hydrogen bromide addition product **105**.[119]

C_8H_{17}

102 $\xrightarrow{HBr}$ **103**, X = NOH; **104**, X = O + **105** (Br)

Haloimines. As mentioned above, chloroimines can be used as precursors of nitrilium salts.[28] The reaction of chloroimines is stereospecific: treatment of (*E*)-*p*-chlorobenzophenone-*N*-chloroimine (**106**) with silver tetrafluoroborate affords only *N*-(*p*-chlorophenyl)benzamide (**107**), whereas the *Z* isomer of the same chloroimine affords only *N*-phenyl-*p*-chlorobenzamide.[120]

$p\text{-}ClC_6H_4C(=NCl)C_6H_5$ (106) $\xrightarrow{AgBF_4}$ $p\text{-}ClC_6H_4NHCOC_6H_5$ (107)

Fluoroimines behave similarly. Rearrangement of fluoroimine **108** in sulfuric acid provides amide **109** in low yield.[121]

$(F_2N)(F_2N)C(CH_3)C(=NF)C_6H_5$ (108) $\xrightarrow{H_2SO_4}$ $(F_2N)(F_2N)C(CH_3)NHCOC_6H_5$ (109) (<30%)

Nitrones. The rearrangement of nitrones to amides is quite an old reaction: an early example, involving an aldehyde nitrone, was reported by Beckmann in 1890.[122] The conversion of aldehyde nitrones to amides was reviewed recently.[123] Barton renewed interest in the process in 1971, when he showed that nitrones of unsaturated ketones in the steroid series produced exclusively enamides, in contrast to the unsaturated amides produced in the Beckmann rearrangement.[124,125] For example, nitrone **110**, when treated with *p*-toluenesulfonyl chloride in pyridine, affords enamide **111** in 70% yield.

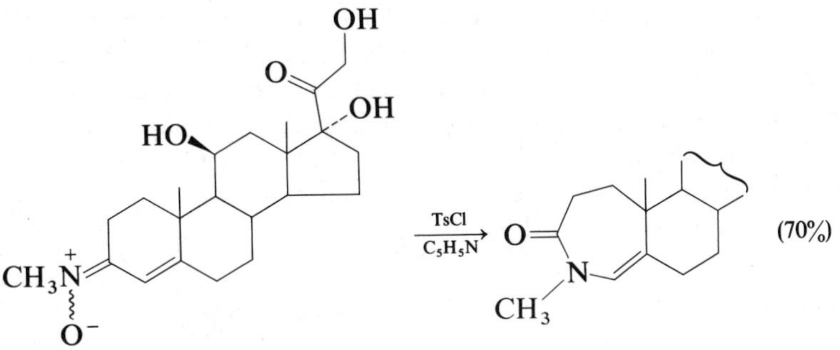

110 $\xrightarrow[C_5H_5N]{TsCl}$ **111** (70%)

The rearrangement is not stereospecific, affording a mixture of lactams from the rearrangement of a saturated 3-keto steroid nitrone.[125] However, the migratory aptitude of the nitrone substituent can often be used to advantage. As part of a synthesis of mesembrine, oxime **112** treated with phosphorus pentoxide in methanesulfonic acid gives a 93% yield of lactam **113**, while treatment of the oxime tosylate or of the *O*-mesityl oxime with alumina gives, at best, mixtures of the desired lactam **114** with **113**.[126] By contrast, nitrone **115** affords only lactam **116** in 32% yield when treated with *p*-toluenesulfonyl chloride in pyridine.[126,127]

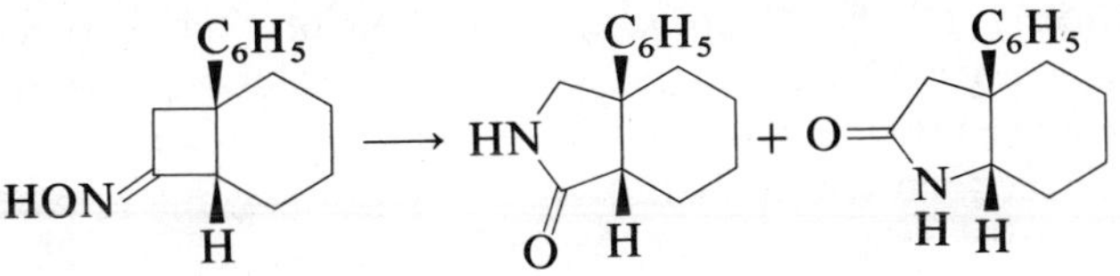

112 ⟶ **113** + **114**

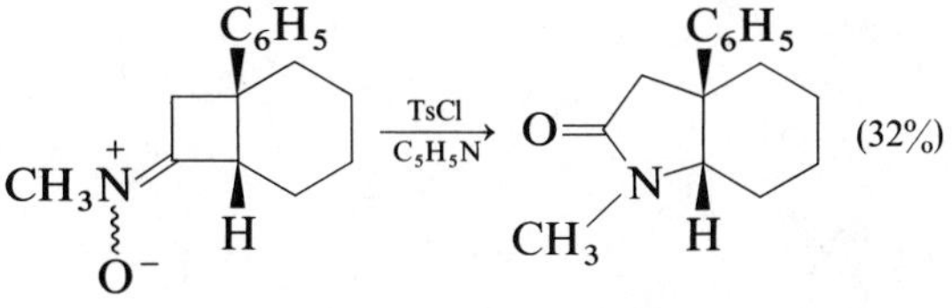

115 $\xrightarrow[C_5H_5N]{TsCl}$ **116** (32%)

The mechanism proposed for the rearrangement involves nucleophilic attack on the nitrone carbon followed by loss of tosylate and hydrolysis.[124] Thus relative migratory aptitudes govern the structure of the lactam or amide obtained, as is the case for the oxime sulfonic acids mentioned earlier.

Aldoximes. Several reagents convert aldoximes to primary amides in a nonstereospecific manner. For example, acetaldoxime, benzaldoxime, and cinnamaldoxime rearrange to acetamide, benzamide, and cinnamamide in yields of 89, 92, and 79%, respectively, when refluxed with silica gel in xylene.[128] Several aliphatic aldoximes are converted to the corresponding amides when treated with boron trifluoride in either acetic acid or ether. For example, hexanaldoxime affords hexanamide in 71% yield when treated with boron trifluoride in acetic acid and in 80% yield when heated with boron trifluoride etherate.[129] Heptanaldoxime is converted to heptanamide in 90% yield by the action of phosphorus pentoxide in methanesulfonic acid.[73] Copper converts benzaldoxime to benzamide in 86% yield.[130] Cupric acetate in dioxane converts oxime **117** to amide **118** in 47% yield.[131]

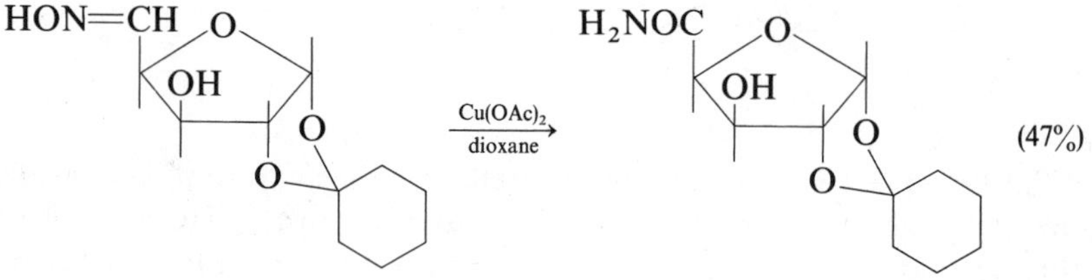

HON=CH, OH, O, O → ($Cu(OAc)_2$, dioxane) H_2NOC, OH, O, O (47%)

117 **118**

Catalytic amounts of nickel(II) acetylacetonate, nickel(II) acetate, or palladium(II) acetate convert acetaldoxime to acetamide (90% yield), (*Z*)-benzaldoxime to benzamide (70% yield), and (*E*)-benzaldoxime to benzamide (45% yield).[132,133]

As mentioned in the "Stereochemistry" section, the rearrangement of aldoximes in the gas phase is stereospecific.[36] In solution, rearrangements of aldoximes are usually not stereospecific and may, in fact, proceed by fragmentation to a nitrile followed by hydrolysis. There are only a few exceptions. For example, porphyrin oxime acetate **119** rearranges to formamide **120** in 28% yield when heated with acetic anhydride.[134]

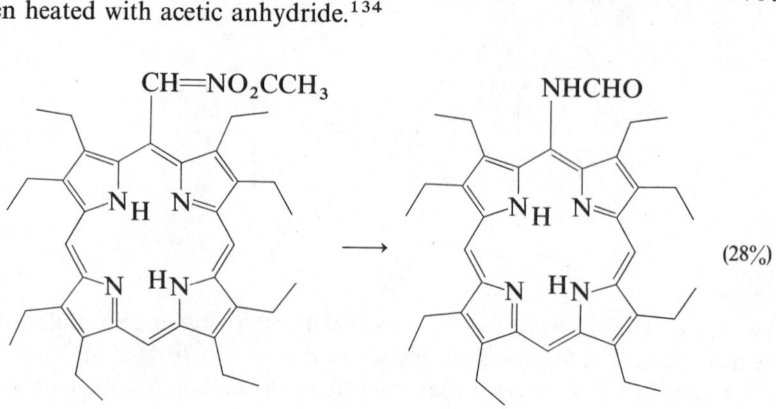

$CH{=}NO_2CCH_3$, NH, N, N, HN → NHCHO, NH, N, N, HN (28%)

119 **120**

3,5-Di-*tert*-butyl-4-hydroxybenzaldoxime rearranges to the corresponding formanilide when treated with sulfuric acid.[135] (*E*)-Benzaldoxime, when treated with 10 mol % of sulfonamide **121**, affords a 75% yield of formanilide.[136]

$$C_6H_5CH{=}NOH \xrightarrow[\textbf{121}]{CH_3O_2C\bar{N}SO_2\overset{+}{N}(C_2H_5)_3} C_6H_5NHCHO \quad (75\%)$$

Mixed orthoesters such as **122**, when refluxed with boron trifluoride and mercury(II) oxide in ether, afford reasonable yields of isonitriles.[137]

$$\underset{\textbf{122}}{C_6H_5CH{=}N{-}O{-}C(OC_2H_5)_2C_2H_5} \xrightarrow[HgO]{BF_3} C_6H_5NC \quad (50\%)$$

Elimination–Additions

Perhaps the most interesting development in Beckmann rearrangement chemistry stems from the realization and exploitation of the fact that the intermediate nitrilium ions, imidates, imidoyl halides, and similar compounds can be trapped with nucleophiles other than water. The products resulting from such trapping are imine derivatives, which may be further manipulated into a variety of amine derivatives (Eq. 4). The oxime derivative rearranges to either imidate **123** or nitrilium ion **124**, which is attacked by a nucleophile, Y^-, to give imine **125**, which is either isolated or reacted with another nucleophile Z^-.

$$R^1R^2C{=}NOA \longrightarrow \underset{\textbf{123}}{R^2C(OA){=}NR^1} \text{ or } \underset{\textbf{124}}{R^1\overset{+}{N}{\equiv}CR^2} \xrightarrow{Y^-} \underset{\textbf{125}}{R^2C(Y){=}NR^1} \xrightarrow[H^+]{Z^-} R^2C(NHR^1)(Y)(Z)$$

(Eq. 4)

There are a few examples where deprotonation can be induced to give a dehydration product. For example, oxime mesylate **126** affords ketenimine **127** in 86% yield when heated to 100° in toluene.[138]

$$\underset{\textbf{126}}{i\text{-}C_3H_7C(C_6H_4CH_3\text{-}p){=}NOMs} \xrightarrow[100^\circ]{C_6H_5CH_3} \underset{\textbf{127}}{(CH_3)_2C{=}C{=}NC_6H_4CH_3\text{-}p} \quad (86\%)$$

A number of examples exist where the nucleophile Y^- is either unreacted substrate or solvent. Oxime **128** "traps itself" when treated with *p*-toluenesulfonyl chloride in pyridine.[139]

C_8H_{17} HON O **128** $\xrightarrow[C_5H_5N]{TsCl}$ O NO N O C_8H_{17} C_8H_{17} (63%)

An example of the isolation of a pyridinium tosylate is mentioned at the beginning of this chapter.[31] Cycloheptanone oxime tosylate affords lactim ether **129** in 62% yield when stirred overnight with methanol.[140]

NOTs $\xrightarrow{CH_3OH}$ N OCH_3 **129** (62%)

Benzophenone *N*-chloroimine produces ethyl imidate **130** when treated with silver tetrafluoroborate and sodium ethoxide in dimethoxyethane.[120]

NCl C_6H_5 C_6H_5 $\xrightarrow[NaOC_2H_5]{AgBF_4}$ OC_2H_5 C_6H_5 C_6H_5 N **130**

Isolation of imidoyl chlorides from the reaction of ketoximes with phosphorus pentachloride is reported.[141–143] For example, 4,4′-dibromobenzophenone oxime affords an 80% yield of imidoyl chloride **131**.[141] Similarly, imidoyl chlorides can be obtained through the use of triphenylphosphine and carbon tetrachloride.[144,145]

$$p\text{-BrC}_6\text{H}_4\text{C(=NOH)C}_6\text{H}_4\text{Br-}p \xrightarrow[\text{CCl}_4]{(\text{C}_6\text{H}_5)_3\text{P}} p\text{-BrC}_6\text{H}_4\text{N=CClC}_6\text{H}_4\text{Br-}p \quad (80\%)$$

131

Similar conversions are reported for the oximes of *N*-(*p*-toluenesulfonyl)-benzamides.[118] For example, when treated with phosphorus pentachloride in ether, amide oxime **132** provides chloroformamidine **133** in 84% yield.

$$\text{TsNHC(=NOH)C}_6\text{H}_5 \xrightarrow{\text{PCl}_5} \text{TsN=CClNHC}_6\text{H}_5 \quad (84\%)$$

132 **133**

Although imidoyl iodides are not sufficiently stable to be isolated, they can be observed spectroscopically in the reaction of oxime derivatives with trimethylsilyl iodide or, more reliably, diethylaluminum iodide.[86] Cyclododecanone oxime mesylate is converted to imidoyl iodide **134**, and 3-pentanone oxime carbonate **135** affords imidoyl iodide **136** by the action of trimethylsilyl iodide, as observed by NMR.

NOMs $\xrightarrow{(\text{CH}_3)_3\text{SiI}}$ I, N (96%)

134

$$(\text{C}_2\text{H}_5)_2\text{C=NOCO}_2\text{C}_2\text{H}_5 \xrightarrow{(\text{CH}_3)_3\text{SiI}} \text{C}_2\text{H}_5\text{N=CIC}_2\text{H}_5 \quad (97\%)$$

135 **136**

The imidoyl iodides can be reacted with such external nucleophiles as thiolates or Grignard reagents, and the imines so produced can be isolated or further reacted with another nucleophile. For example, acetophenone oxime carbonate (**137**) affords ethylthioimidate **138** when treated sequentially with trimethylsilyl iodide and lithium ethanethiolate.[86] 3-Pentanone oxime mesylate affords amine **139** when treated sequentially with diethylaluminum iodide, phenylmagnesium bromide, and diisobutylaluminum hydride (DIBAL).[86]

$$\text{C}_6\text{H}_5\text{C(CH}_3\text{)=NOCO}_2\text{C}_2\text{H}_5 \xrightarrow[\text{2. C}_2\text{H}_5\text{SLi}]{\text{1. (CH}_3)_3\text{SiI}} \text{C}_6\text{H}_5\text{N=C(CH}_3\text{)SC}_2\text{H}_5 \quad (67\%)$$

137 **138**

NOMs $\xrightarrow[\text{2. C}_6\text{H}_5\text{MgBr}]{\text{1. (C}_2\text{H}_5)_2\text{AlI}}$ N, C_6H_5 $\xrightarrow{\text{DIBAL}}$ N, H, C_6H_5 (63%)

139

Carbonyl diimidazole converts benzophenone oxime and acetophenone oxime into the corresponding imidoyl imidazoles (e.g., **140**).[146]

$C_6H_5C(=NOH)C_6H_5$ → (carbonyldiimidazole) → $C_6H_5N=C(C_6H_5)(N\text{-imidazolyl})$ (95%)

140

Yamamoto has demonstrated a more efficient method for the preparation of thioimidates in the reaction of oxime mesylates with alkylthioalanes or arylthioalanes. For example, cyclohexanone oxime mesylate affords caprolactim thioether (**141**) in 62% yield when treated with diisobutylaluminum ethanethiolate, and acetophenone oxime mesylate affords thioimidate **142** in 88% yield when treated with dimethylalane phenylthiolate.[147]

NOMs → $(i\text{-}C_4H_9)_2AlSC_2H_5$ → SC_2H_5 (62%)

141

$C_6H_5C(=NOMs)CH_3$ → $(CH_3)_2AlSC_6H_5$ → $C_6H_5N=C(CH_3)SC_6H_5$ (88%)

142

5-Hexen-2-one oxime mesylate is unreactive under the conditions prescribed for the preceding conversions[147] but does rearrange when treated with a similar reagent that is a stronger Lewis acid: ethylchloroaluminum phenylthiolate (prepared from diethylchloroalane and thiophenol).[148]

NOMs → $C_2H_5Al(Cl)SC_6H_5$ → SC_6H_5

143

This methodology is also useful for the preparation of selenoimidates.[147] For example, cyclododecanone oxime mesylate is converted to selenoimidate **144** in 57% yield by the action of diisobutylaluminum selenophenylate.

NOMs → $(i\text{-}C_4H_9)_2AlSeC_6H_5$ → N, SeC_6H_5 (56%)

144

Perhaps the simplest nucleophile that might be added to the nitrilium ion or imidate is hydride. What appears to be an early example of this type of transformation is the conversion of oxime **145** to amine **146** with lithium aluminum hydride (LAH).[149]

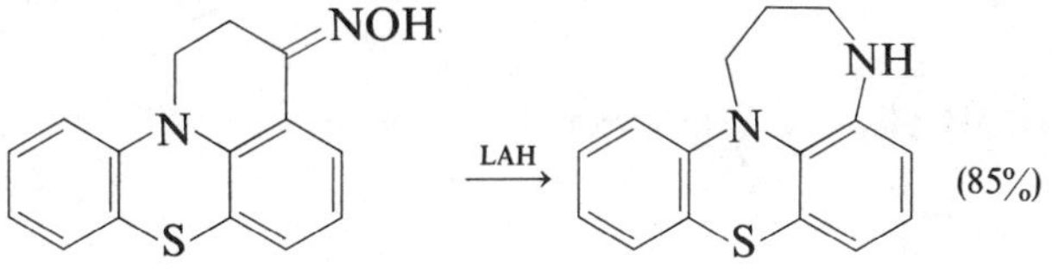

$$\textbf{145}\ (\text{NOH, N, S}) \xrightarrow{\text{LAH}} \textbf{146}\ (\text{NH, N, S}) \quad (85\%)$$

It was later shown that the reaction, which produces primary amines as byproducts, is not stereospecific and proceeds through a hydroxylamine[150] and possibly a nitrene.[151] The mixed reagent, lithium aluminum hydride–aluminum chloride, increases the proportion of secondary amine but proceeds through the hydroxylamine as well.[150] Diisobutylaluminum hydride in ethereal solvents also gives mixtures[152,153] but in methylene chloride gives only rearranged products.[147,153,154] A typical example is the conversion of indanone oxime mesylate **147** to 1,2,3,4-tetrahydroquinoline.[147,153]

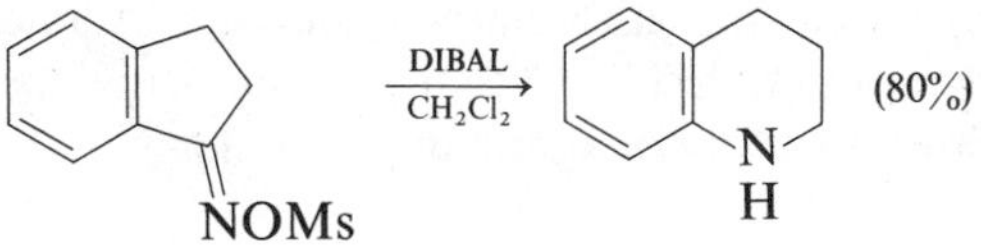

$$\underset{\textbf{147}}{\text{NOMs}} \xrightarrow[\text{CH}_2\text{Cl}_2]{\text{DIBAL}} \text{NH} \quad (80\%)$$

Imidoyl cyanides may be formed when a rearrangement is conducted in the presence of cyanide ion. For example, the oxime of Michler's ketone (**148**) affords an 87% yield of imidoyl cyanide **149** when treated with benzenesulfonyl chloride and four molar equivalents of potassium cyanide.[155]

$$\underset{\textbf{148}}{[p\text{-}(CH_3)_2NC_6H_4]_2C{=}NOH} \xrightarrow[\text{KCN}]{C_6H_5SO_2Cl}$$

$$\underset{\textbf{149}}{{}_6H_4N{=}C(CN)C_6H_4N(CH_3)_2} \quad (87\%)$$

A more general procedure requires treatment of an oxime mesylate, such as **150**, with trimethylsilyl cyanide and diethylaluminum chloride. Imidoyl cyanide **151** is produced in 91% yield.[147]

$$\underset{\textbf{150}}{(n\text{-}C_5H_{11})_2C{=}NOMs} \xrightarrow[(C_2H_5)_2AlCl]{(CH_3)_3SiCN} \underset{\textbf{151}}{n\text{-}C_5H_{11}N{=}C(CN)C_5H_{11}\text{-}n} \quad (91\%)$$

A more unusual example of cyanide trapping occurs when cyclobutanedione dioxime **152** is treated with benzenesulfonyl chloride.[156] The fragmentation–cyclization sequence shown produces oxime **153**, which then rearranges and traps cyanide in a straightforward fashion.

HON= =NOH ⟶ HON= —CN —OH ⟶ O=N— —OH ⟶

152

HON= O ⟶ NC O N

153

Grignard reagents may effect Beckmann rearrangement of an oxime sulfonate and add to the resultant nitrilium ion *in situ*. The product is an imine, which may be reduced or reacted with a second Grignard reagent. For example, cyclohexanone oxime mesylate can be converted to either amine **154** in 63% yield by sequential treatment with *n*-butylmagnesium bromide and diisobutylaluminum hydride or amine **155** in 72% yield by the sequence of methylmagnesium bromide followed by allyl magnesium bromide.[157]

N H (63%) **154** ⟵ 1. $n\text{-}C_4H_9MgBr$ 2. DIBAL — NOMs — 1. CH_3MgBr 2. $CH_2{=}CHCH_2MgBr$ ⟶ N H **155** (72%)

Organoaluminum reagents are also effective for these types of transformations. For example, undecylcyclopentanone oxime mesylate (**156**) is converted to imine **157** in 54% yield when treated with trimethylalane. Reduction with lithium aluminum hydride–trimethylalane affords solenopsin A in 97% yield.[147,158]

$n\text{-}C_{11}H_{23}$ NOMs **156** —$(CH_3)_3Al$⟶ $n\text{-}C_{11}H_{23}$ N **157** (57%) —LAH, $(CH_3)_3Al$⟶ $n\text{-}C_{11}H_{23}$ N H H solenopsin A (97%)

A direct comparison can be made: cyclohexanone oxime mesylate can be converted to amine **155** in 60% yield by the sequence of trimethylalane followed by allylmagnesium bromide.[147]

Other examples of this methodology include a two-step, 58% yield synthesis of coniine from cyclopentanone oxime tosylate,[147] and the synthesis of pumiliotoxin-C from oxime tosylate **158** in 60% yield,[147,153] by the action of tripropylalane, followed by diisobutylaluminum hydride.

NOTs

1. $(n\text{-}C_3H_7)_3Al$
2. DIBAL

N
H

(58%)

coniine

H H NOTs

158

1. $(n\text{-}C_3H_7)_3Al$
2. DIBAL

H N H H H

(60%)

pumiliotoxin-C

Trapping by enol silyl ethers is also possible.[159] For example, cyclohexanone oxime mesylate and silyl enol ether **159** afford ketoimine **160** in 42% yield when treated with ethyldichloroalane. Most ketoimines tautomerize, as demonstrated by the condensation of cycloheptanone oxime mesylate and cyclopentanone enol silyl ether, which affords vinylogous amide **161** in 74% yield. A limitation of the process is that the oximes of cyclopentanone and aryl ketones do not work.

NOMs + $OSi(CH_3)_3$ $\xrightarrow{C_2H_5AlCl_2}$ N O (42%)

159 **160**

NOMs + $OSi(CH_3)_3$ $\xrightarrow{(C_2H_5)_2AlCl}$ NH O (74%)

161

Fragmentations

Ketoximes. It is worth restating that most Beckmann fragmentations occur from the nitrilium or imidate species, as discussed earlier.[28,33] Methods of generating the imidate species other than by fragmentation of an oxime underscore this point. For example, when amide **162** is treated with *p*-toluenesulfonyl chloride in pyridine, nitrile **163** is obtained in 61% yield.[160] This process probably proceeds by way of an intermediate imidoyl tosylate.

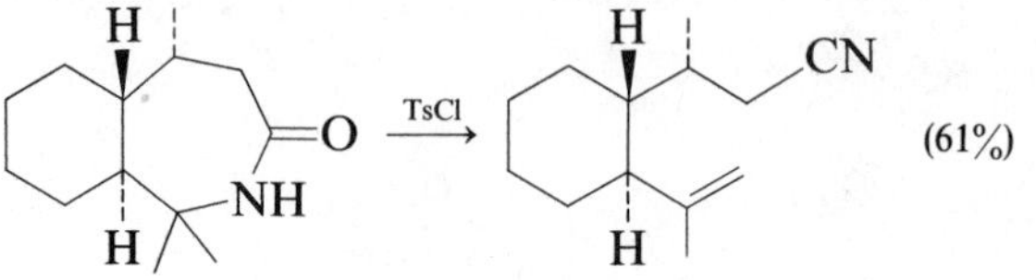

TsCl (61%)

162 **163**

The largest class of structural types that undergo fragmentation, either by design or as a side reaction, are oximes that have quaternary centers adjacent to the oxime carbon. The steric bulk of the quaternary center usually dictates the oxime geometry and, therefore, the structure of the nitrilium ion or imidate. In the absence of some functional group that directs the fate of the cation, a number of products may be isolated.

When the oxime is acyclic, the fragmentation produces two or more products: a nitrile (or an amide if the nitrile is hydrolyzed under the reaction conditions) and the product or products derived from the cation. The reaction is preparatively useful only when the nitrile is the desired product. For example, tetrahydropyran oxime **164** affords a 69% yield of nitrile **165** when treated with phosphorus pentachloride in ether,[161] whereas oxime **166** fragments to give a 98% yield of two isomeric olefins when treated with polyphosphoric acid.[162]

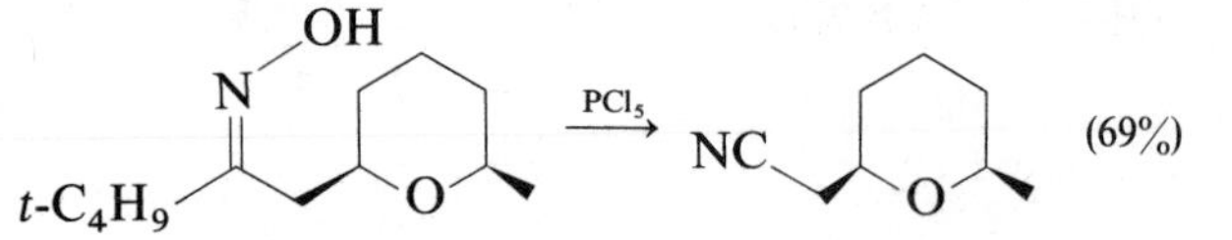

PCl_5 (69%)

164 **165**

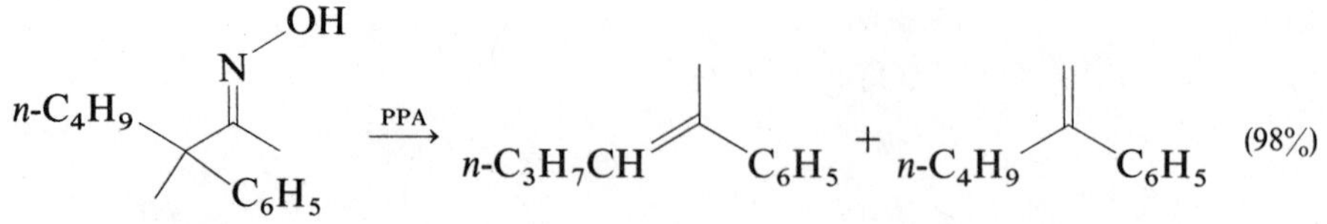

PPA n-C_3H_7CH ... C_6H_5 + n-C_4H_9 ... C_6H_5 (98%)

166

The conditions of the reaction affect the propensity of the system to fragment. For example, oxime **167** affords nitrile **168** in 90% yield when refluxed with *p*-toluenesulfonyl chloride in pyridine.[163] In a similar system, treatment with thionyl chloride in ether for 5 minutes at −20° affords 35% fragmentation product and 32% rearrangement product.[164]

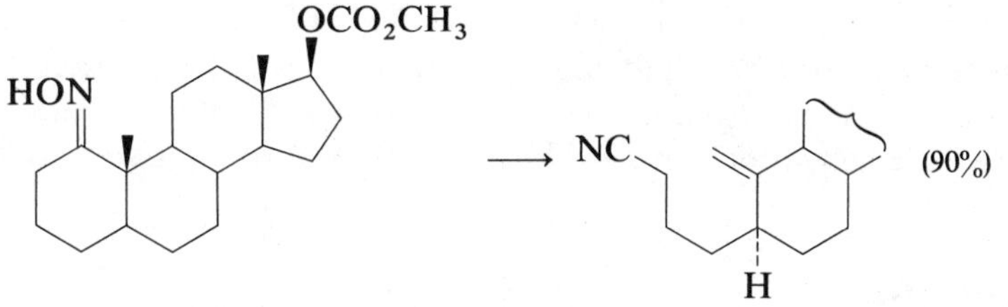

OCO_2CH_3

HON

→ NC (90%)

H

167 168

2,2-Disubstituted cyclic ketoximes fragment readily, although nonspecific deprotonation of the cation may lead to a mixture of products. For example, spiro oxime **169** fragments to nitrile **170** in 92% yield when treated with phosphorus pentachloride in benzene. The same oxime, when treated with polyphosphoric acid, affords *saturated* amide **171**.[78]

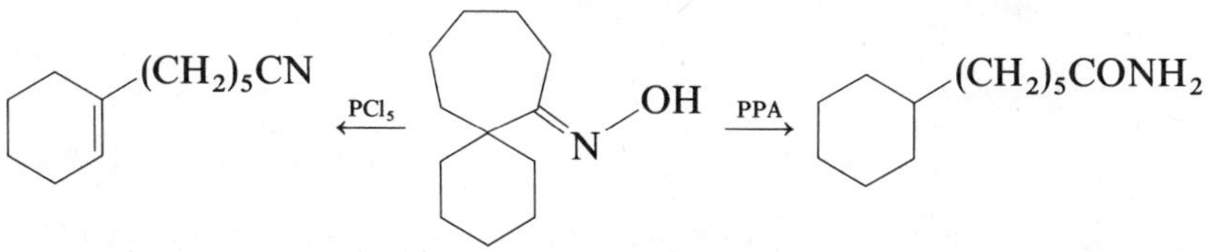

$(CH_2)_5CN$ $\xleftarrow{PCl_5}$ N OH $\xrightarrow{PPA}$ $(CH_2)_5CONH_2$

170 (92%) **169** **171** (46%)

2,2-Diphenylcycloheptanone oxime (**172**) fragments to nitrile **173** in 98% yield when treated with thionyl chloride in benzene.[165] Oxime **174** fragments when treated with polyphosphoric acid at 125°, but the likely product (nitrile **175**) cyclizes and hydrolyzes, producing amide **176** in 70% yield.[166]

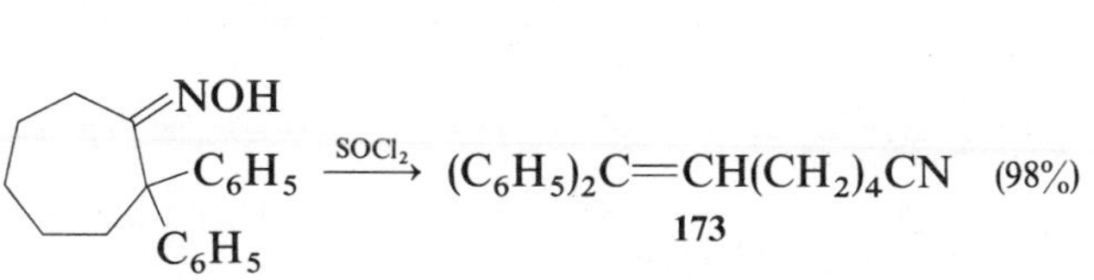

NOH

C_6H_5 $\xrightarrow{SOCl_2}$ $(C_6H_5)_2C{=}CH(CH_2)_4CN$ (98%)

C_6H_5 **173**

172

HON CN $CONH_2$

$\xrightarrow[125°]{PPA}$ → (70%)

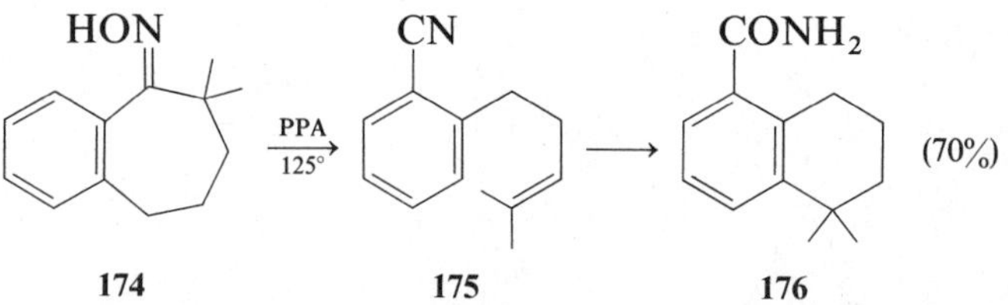

174 **175** **176**

Oximes of camphor and its derivatives are readily fragmented,[167–173] but the deprotonation of the fragmented cation is seldom highly stereoselective. Camphor derivatives can be used as chiral starting materials in enantiospecific syntheses,[169,172] but a proton source is required to equilibrate the olefinic products. For example, when oxime **177** is treated with *p*-toluenesulfonyl chloride in pyridine, a mixture of olefinic nitriles **178** and **179** is obtained in 70% yield.[171] When **177** is treated with a mixture of trifluoroacetic acid and trifluoroacetic anhydride, nitrile **179** is the only product obtained in 80% yield.[172]

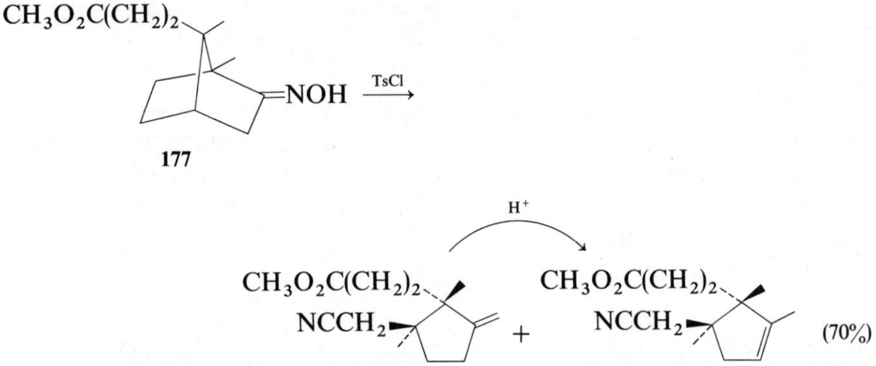

$CH_3O_2C(CH_2)_2$ =NOH $\xrightarrow{TsCl}$

177

H^+

$CH_3O_2C(CH_2)_2$ $NCCH_2$ + $CH_3O_2C(CH_2)_2$ $NCCH_2$ (70%)

178 **179**

Occasionally, ring strain provides the driving force for fragmentation, although rearrangement is sometimes still a major pathway. Oxime benzenesulfonate **180**, for example, when heated in aqueous acetone, fragments to hydroxy nitrile **181** in 31% yield, accompanied by 19% rearrangement.[174] Four-membered ring compounds appear to be more likely to fragment.[175,176] For example, oxime **182** fragments to nitrile **183** in 61% yield when treated with *m*-nitrobenzoyl chloride in methylene chloride.[176]

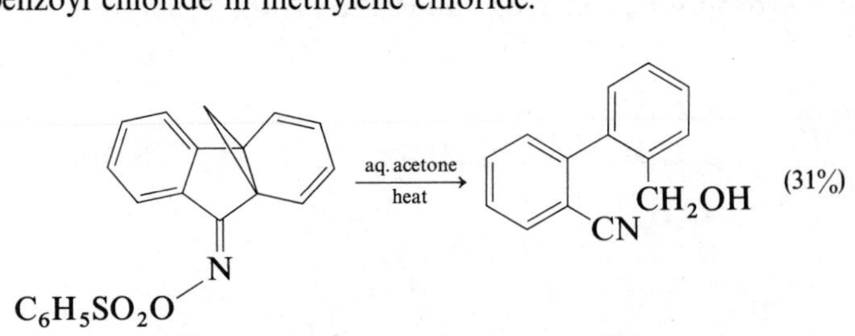

N $C_6H_5SO_2O$ $\xrightarrow[\text{heat}]{\text{aq. acetone}}$ CH_2OH CN (31%)

180 **181**

NOH C_2H_5O $\xrightarrow[CH_2Cl_2]{m\text{-}O_2NC_6H_4COCl}$ CN OC_2H_5 (61%)

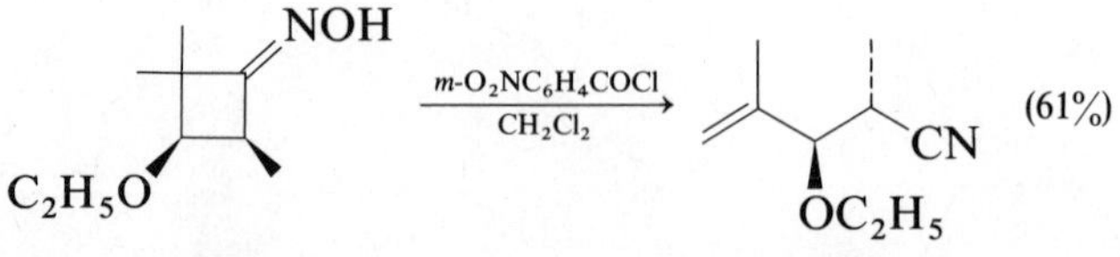

182 **183**

The presence of an appropriately placed oxygen atom facilitates fragmentation by stabilizing the cation formed. When hydroxycholestanone oxime (**184**) is treated with thionyl chloride in ether at $-20°$, ketonitrile **185** is obtained in 94% yield.[177] The same transformation is effected in 86% yield by treatment with dichlorocarbene.[178]

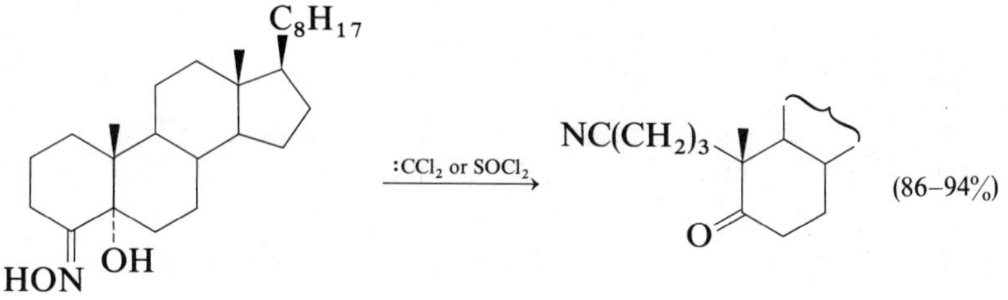

C_8H_{17}

HON OH

184

$\xrightarrow{:CCl_2 \text{ or } SOCl_2}$

$NC(CH_2)_3$

O

185

(86–94%)

Dichlorocarbene also fragments benzoin oxime (**186**) into benzaldehyde and benzonitrile in 80 and 76% yields, respectively.[178] The same fragmentation is effected with polyphosphoric acid, producing the same two products in 33 and 26% yields, respectively.[179]

$$C_6H_5CHOHC(=NOH)C_6H_5 \xrightarrow{:CCl_2} C_6H_5CHO + C_6H_5CN$$

186 (80%) (76%)

Fragmentation of hydroxy oxime **187** with dichlorocarbene affords nitrile aldehyde **188** in 85% yield.[178] Heating for 2 minutes with dilute sulfuric acid gives the same product in 95% yield.[180]

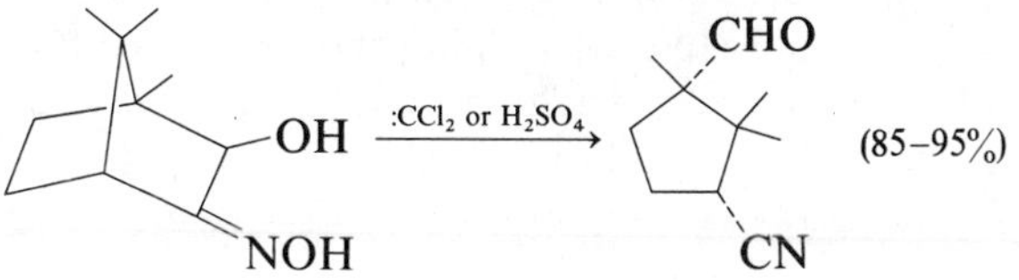

OH NOH

187

$\xrightarrow{:CCl_2 \text{ or } H_2SO_4}$

CHO CN

188

(85–95%)

Fragmentation of hydroxy oxime **189** with *p*-toluenesulfonyl chloride at room temperature gives a 98% yield of nitrile aldehyde **190**.[181] Similar conditions afford only a 23% yield of nitrile aldehydes **192** from oxime **191**, in which the hydroxyl group is more remote, accompanied by 25% lactam.[182,183]

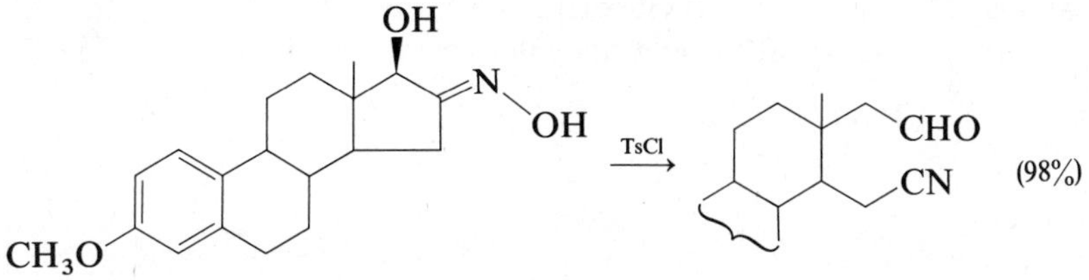

OH N OH

CH_3O

189

$\xrightarrow{TsCl}$

CHO CN

190

(98%)

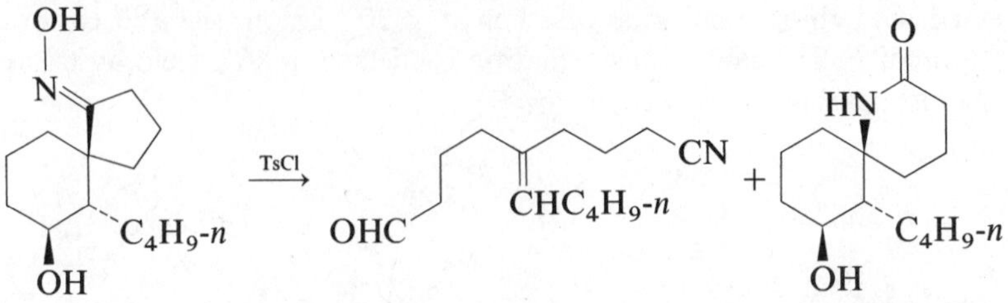

$\xrightarrow{\text{TsCl}}$

191 → 192 (23%) + (25%)

It is of interest to note that hydroxy oximes usually rearrange instead of fragmenting when photolyzed.[184] Exceptions to this statement have been noted, however.[185]

Alkoxy oximes fragment under mild conditions as well. For example, oxime **193** affords nitrile **194** in 93% yield when stirred with acetic anhydride in pyridine.[186]

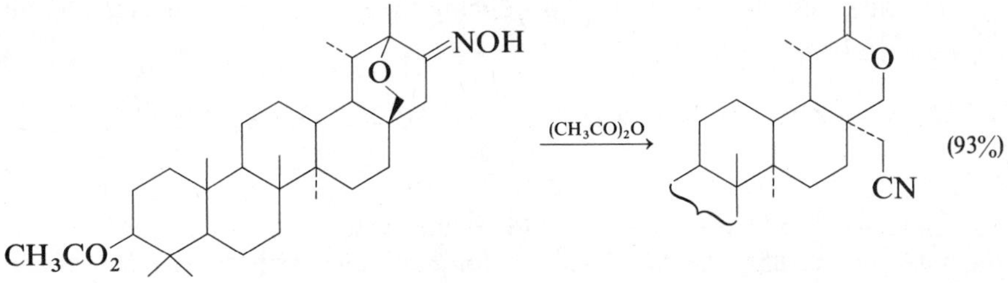

$\xrightarrow{(CH_3CO)_2O}$ (93%)

193 → 194

Oximes of 3-ketotetrahydrofurans and 3-ketotetrahydropyrans fragment readily when treated with thionyl chloride or phosphorus pentachloride.[187] For example, oxime **195** affords a 95% yield of benzophenone and valerolactone when treated with phosphorus pentachloride in chloroform.[187]

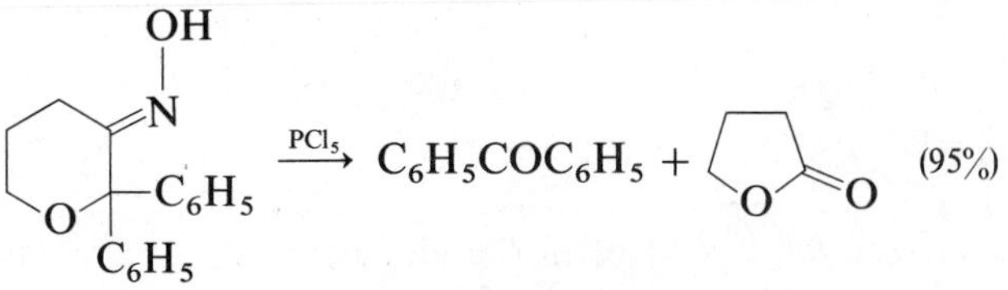

195 $\xrightarrow{PCl_5}$ $C_6H_5COC_6H_5$ + (95%)

Acetals of the monooximes of 1,2-diones fragment readily. For example, acetal oxime **196** fragments (via its orthoester) when treated with methyl orthoformate and methanesulfonic acid, affording ester nitrile **197** in 97% yield.[188,189]

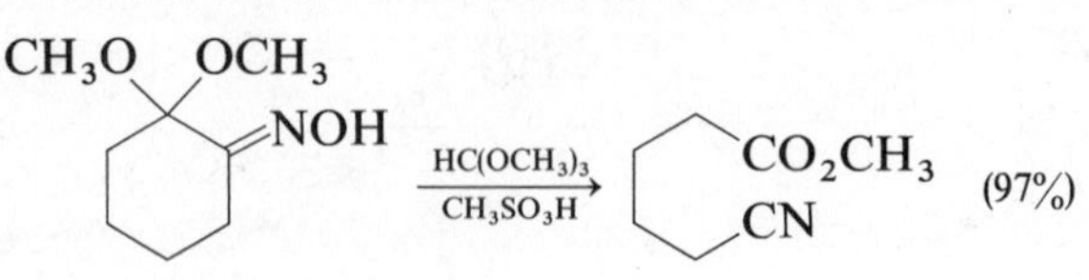

196 $\xrightarrow[CH_3SO_3H]{HC(OCH_3)_3}$ **197** (97%)

Oximes of 3-ketopyrrolidines and 3-ketopiperidines fragment like their oxygen counterparts.[190] For example, piperidine oxime **198** affords *N*-tosylaminonitrile **199** *in 46% yield when treated with p*-toluenesulfonyl chloride and aqueous base.

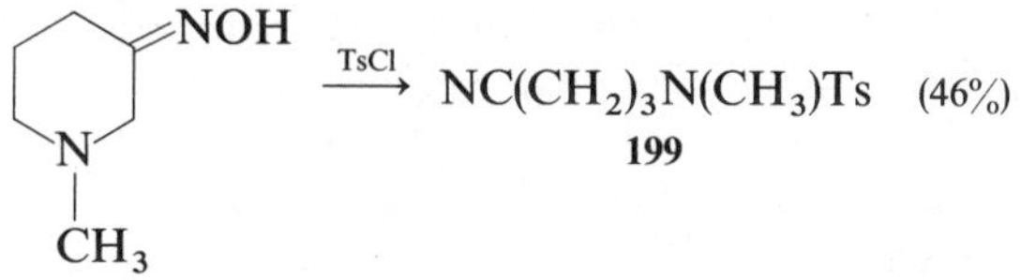

$$\xrightarrow{\text{TsCl}} NC(CH_2)_3N(CH_3)Ts \quad (46\%)$$

198 **199**

Acyclic amino ketoximes fragment as well; the following examples demonstrate the importance of stereoelectronic effects on the yield and the conditions required for the fragmentation. The *E* isomer of oxime benzoate **200** affords a 99% yield of *p*-nitrobenzonitrile when heated in ethanol at 80° for 1 hour. By contrast, the *Z* isomer gives only a 40% yield after 200 hours at the same temperature.[95]

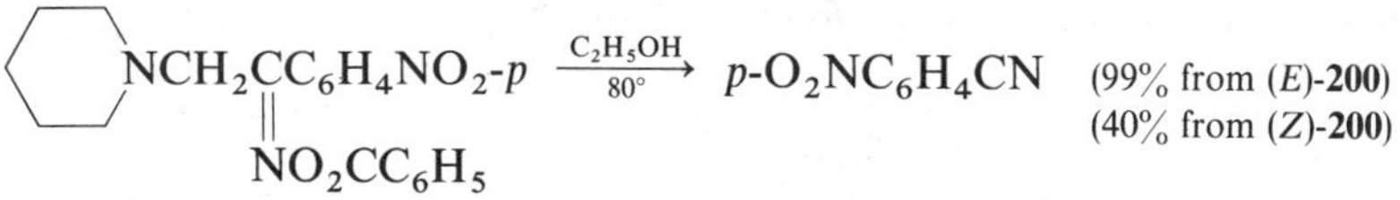

$$NCH_2C(=NO_2CC_6H_5)C_6H_4NO_2\text{-}p \xrightarrow[80^\circ]{C_2H_5OH} p\text{-}O_2NC_6H_4CN$$

(99% from (*E*)-**200**)
(40% from (*Z*)-**200**)

200

An unusual double fragmentation followed by a Friedel–Crafts alkylation is initiated by a Beckmann fragmentation of aspidospermidine oxime **201**. The fragmentation product **202** further fragments to iminium ion **203** and then cyclizes to the eburnamenine nitrile **204** in 55% overall yield.[191]

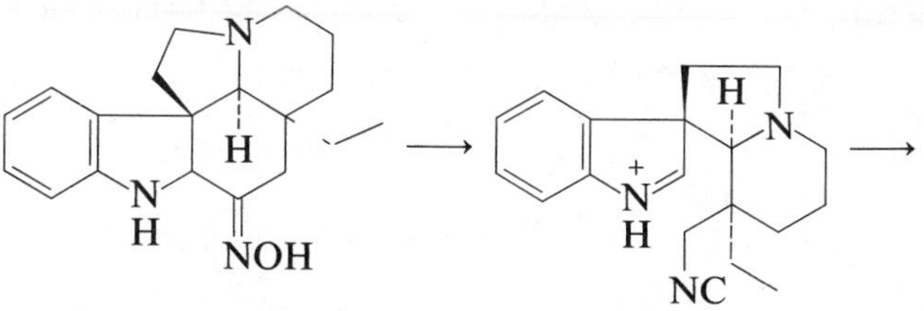

201 → **202** 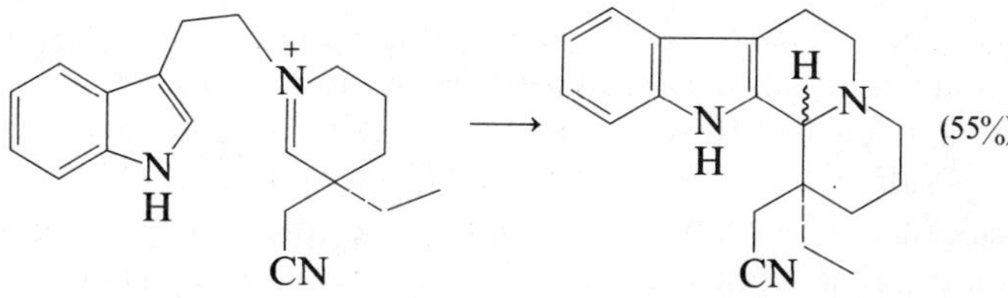 → **203** → **204** (55%)

The oximes of alkylthio ketones are prone to fragment when the geometry of the oxime hydroxyl is *anti* to the sulfur. Although the mechanism of sulfur participation has been disputed,[192,193] it appears that donation of electrons by sulfur to the migrating (or fragmenting) carbon is the most likely explanation of the effect. Oxime **205** provides a 72% yield of nitrile **206** when treated with thionyl chloride in pyridine[193] and a 92% yield of nitrile **207** when treated sequentially with *p*-toluenesulfonyl chloride and ethanol.[194]

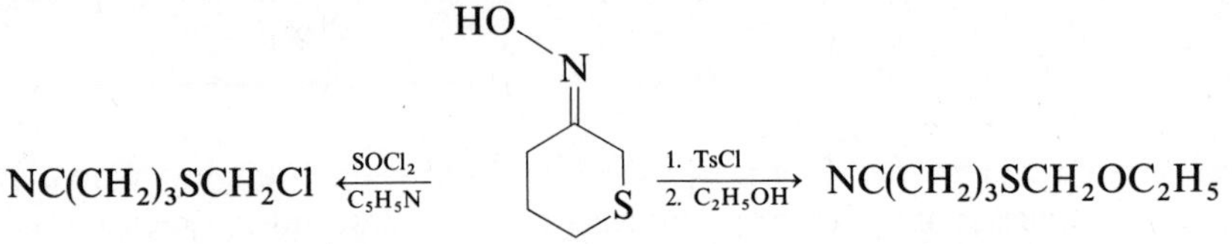

$$NC(CH_2)_3SCH_2Cl \xleftarrow[C_5H_5N]{SOCl_2} \textbf{205} \xrightarrow[2.\ C_2H_5OH]{1.\ TsCl} NC(CH_2)_3SCH_2OC_2H_5$$

206 (72%) **205** **207** (92%)

Sulfur participation of quite a different sort is described for oxime **208**. Specifically, the *E* isomer fragments while the *Z* isomer rearranges.[195,196] The mechanism proposed for the fragmentation is as shown.

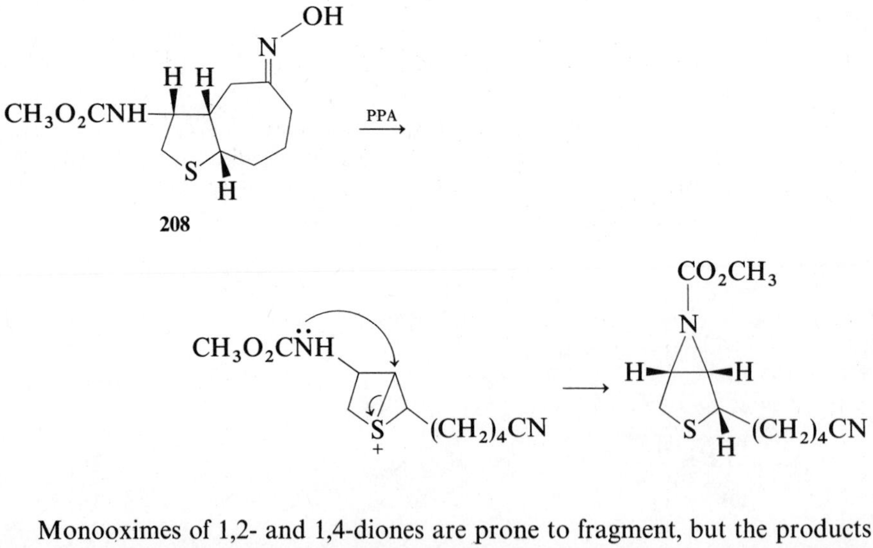

$$\textbf{208} \xrightarrow{PPA}$$

$CH_3O_2C\ddot{N}H$ … $(CH_2)_4CN$ → CO_2CH_3, N, H, H, S, H, $(CH_2)_4CN$

Monooximes of 1,2- and 1,4-diones are prone to fragment, but the products obtained depend on the conditions employed. For example, heptanedione monooxime **209** may produce three products depending on the reagent.[197] Specifically, nitrile **210** is obtained in 60–80% yield when trifluoroacetic acid, benzenesulfonyl chloride, benzoyl chloride, phosphorus pentachloride, or acetyl chloride is used. Sulfuric acid may yield either amide **211** or acid **212**, depending on the vigor of the conditions. Polyphosphoric acid affords a mixture of amide **211** (24%) and acid **212** (28%).

Oximes of 3-ketopyrrolidines and 3-ketopiperidines fragment like their oxygen counterparts.[190] For example, piperidine oxime **198** affords *N*-tosylaminonitrile **199** *in* 46% *yield when treated with p*-toluenesulfonyl chloride and aqueous base.

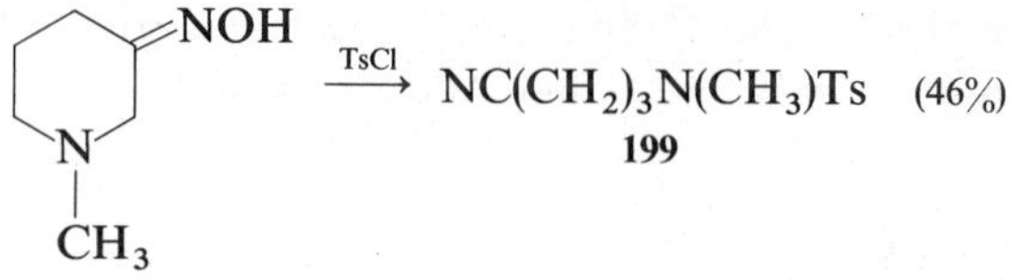

$$\xrightarrow{\text{TsCl}} \mathrm{NC(CH_2)_3N(CH_3)Ts} \quad (46\%)$$

NOH, N, CH_3

198 **199**

Acyclic amino ketoximes fragment as well; the following examples demonstrate the importance of stereoelectronic effects on the yield and the conditions required for the fragmentation. The *E* isomer of oxime benzoate **200** affords a 99% yield of *p*-nitrobenzonitrile when heated in ethanol at 80° for 1 hour. By contrast, the *Z* isomer gives only a 40% yield after 200 hours at the same temperature.[95]

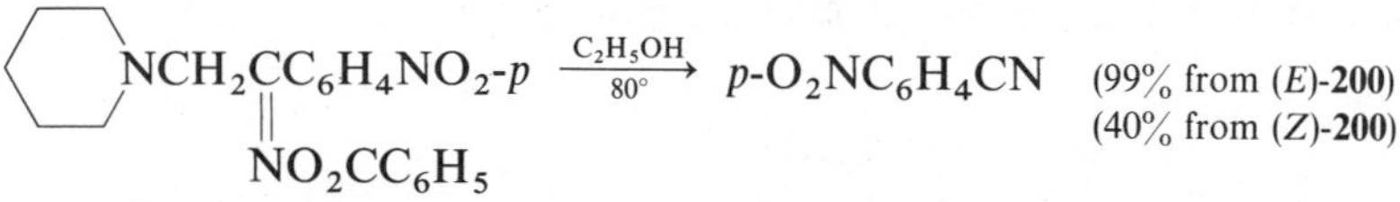

$$\mathrm{NCH_2CC_6H_4NO_2\text{-}}p \xrightarrow[80^\circ]{\mathrm{C_2H_5OH}} p\text{-}\mathrm{O_2NC_6H_4CN} \quad \begin{array}{l}(99\%\text{ from }(E)\text{-}\mathbf{200})\\(40\%\text{ from }(Z)\text{-}\mathbf{200})\end{array}$$

$NO_2CC_6H_5$

200

An unusual double fragmentation followed by a Friedel–Crafts alkylation is initiated by a Beckmann fragmentation of aspidospermidine oxime **201**. The fragmentation product **202** further fragments to iminium ion **203** and then cyclizes to the eburnamenine nitrile **204** in 55% overall yield.[191]

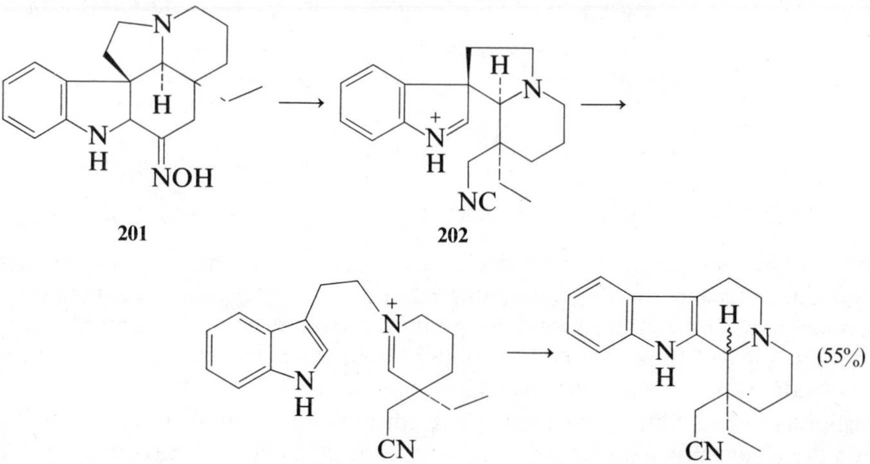

N, H, N, H, NOH → H, N, $\overset{+}{N}$, H, NC →

201 **202**

N, H, $\overset{+}{N}$, CN → H, N, N, H, CN (55%)

203 **204**

The oximes of alkylthio ketones are prone to fragment when the geometry of the oxime hydroxyl is *anti* to the sulfur. Although the mechanism of sulfur participation has been disputed,[192,193] it appears that donation of electrons by sulfur to the migrating (or fragmenting) carbon is the most likely explanation of the effect. Oxime **205** provides a 72% yield of nitrile **206** when treated with thionyl chloride in pyridine[193] and a 92% yield of nitrile **207** when treated sequentially with *p*-toluenesulfonyl chloride and ethanol.[194]

$NC(CH_2)_3SCH_2Cl$ ← ($SOCl_2$, C_5H_5N) — **205** — (1. TsCl, 2. C_2H_5OH) → $NC(CH_2)_3SCH_2OC_2H_5$

206 (72%) **205** **207** (92%)

Sulfur participation of quite a different sort is described for oxime **208**. Specifically, the *E* isomer fragments while the *Z* isomer rearranges.[195,196] The mechanism proposed for the fragmentation is as shown.

CH_3O_2CNH– (**208**) —PPA→

208

$CH_3O_2C\ddot{N}H$ … $(CH_2)_4CN$ → CO_2CH_3, N, H, H, S, H, $(CH_2)_4CN$

Monooximes of 1,2- and 1,4-diones are prone to fragment, but the products obtained depend on the conditions employed. For example, heptanedione monooxime **209** may produce three products depending on the reagent.[197] Specifically, nitrile **210** is obtained in 60–80% yield when trifluoroacetic acid, benzenesulfonyl chloride, benzoyl chloride, phosphorus pentachloride, or acetyl chloride is used. Sulfuric acid may yield either amide **211** or acid **212**, depending on the vigor of the conditions. Polyphosphoric acid affords a mixture of amide **211** (24%) and acid **212** (28%).

$n\text{-}C_4H_9C(=NOH)COCH_3 \longrightarrow n\text{-}C_4H_9CN + n\text{-}C_4H_9CONH_2 + n\text{-}C_4H_9CO_2H$

209 → **210** + **211** + **212**

Cleavage may be effected by nucleophilic attack: potassium *tert*-butoxide produces ester nitrile **214** from oxime acetate **213**.[57]

213 $\xrightarrow{KOC_4H_9\text{-}t}$ **214**

Oximes of keto acids and keto amides fragment although not quite as readily as those of diones. For example, oxime acid **215** affords nitrile **216** when refluxed in dilute sulfuric acid for 3 hours.[198] Oxime amide **217** produces aminonitrile **218** on heating with *p*-toluenesulfonyl chloride for 30 minutes in dioxane.[199]

215 $\xrightarrow{\text{dil. } H_2SO_4}$ **216** (95%)

217 $\xrightarrow{TsCl}$ **218** (68%)

Monooximes of 1,4-diones may fragment by either of two mechanisms. Oxime tosylate **219**, when treated with potassium *tert*-butoxide in tetrahydrofuran, affords an 85% yield of ketonitrile **220** by an elimination mechanism. In oxime tosylate **221**, this mechanism is impossible. Nevertheless, sodium hydroxide in anhydrous ethanol effects a double fragmentation by nucleophilic attack on the carbonyl, affording nitrile acid **222** in quantitative yield.[79,80,200]

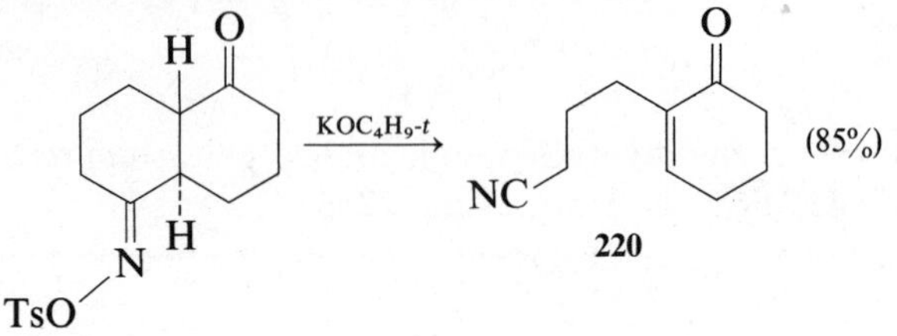

H O
$\xrightarrow{KOC_4H_9\text{-}t}$
O
NC
(85%)
H
N
TsO
219
220

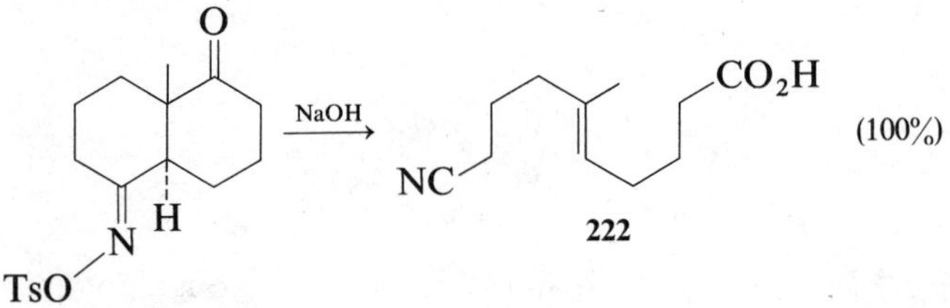

O
$\xrightarrow{NaOH}$
CO_2H
(100%)
NC
H
N
TsO
221
222

Oxime imines of 1,2-diones behave similarly. Oxime **223** fragments to nitrile amide **224** in 97% yield when stirred with *p*-toluenesulfonyl chloride in pyridine.[201]

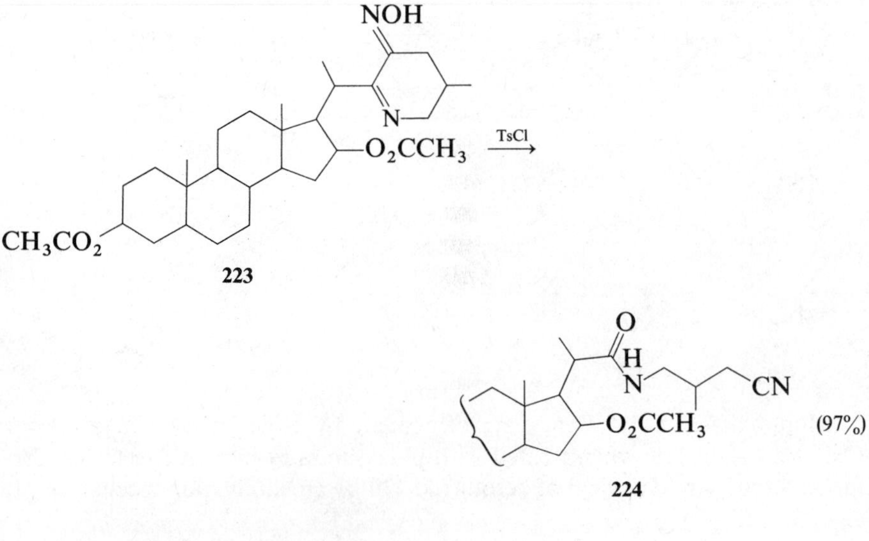

NOH
N
$-O_2CCH_3$ $\xrightarrow{TsCl}$
CH_3CO_2
223

O
H
N
CN
$-O_2CCH_3$
(97%)
224

Nitroso phenols react similarly, perhaps because they are tautomers of the monooximes of diones. For example, nitrosonaphthol **225** fragments to nitrile acid **226** in 54% yield when treated with thionyl chloride in sulfur dioxide[202] and in 49% yield when treated with *p*-toluenesulfonyl chloride in pyridine.[203]

NO OH $\xrightarrow{SOCl_2 \text{ or } TsCl}$ CN CO_2H (49–54%)

225 **226**

Silicon can direct the stereochemical course of a Beckmann fragmentation in a stereospecific manner. Oxime acetate **227** fragments to the *cis*-olefin nitrile **228** in 90% yield when treated with trimethylsilyl trifluoromethanesulfonate; its epimer, **229**, affords *trans*-olefin nitrile **230** in 94% yield under the same conditions.[204]

CH_3CO_2 N $Si(CH_3)_3$ $\xrightarrow{(CH_3)_3SiOSO_2CF_3}$ CN (90%)

227 **228**

CH_3CO_2 N $Si(CH_3)_3$ $\xrightarrow{(CH_3)_3SiOSO_2CF_3}$ CN (94%)

230

229

Aldoximes. In the 25 years covered by this review, over 40 different reagents have been reported to effect the conversion of aldoximes to nitriles. Indeed, this transformation is the one to be expected when an aldoxime is treated with a Lewis acid or a dehydrating agent. Most reagents work equally well, or nearly so, with either oxime diastereomer. The reader can best ascertain the scope of this reaction and the variety of reagents used by scanning Table IV ("Aldoxime Fragmentations").

Rearrangement–Cyclizations

The intramolecular capture of the nitrilium ion or imidate produced in a Beckmann rearrangement or fragmentation was first observed in the nineteenth century. However, synthetically useful procedures have emerged only recently.

Heteroatom Terminators. One of the first examples of a rearrangement cyclization was observed by Perkin in 1890, although the structure postulated, aminal **231**, was corrected to oxazoline **232** in 1960.[15,205] Thus treatment of alkenyl oxime **233** with a mixture of hydrogen chloride, acetic acid, and acetic anhydride rearranges the oxime, hydrates the double bond, and cyclizes by oxygen capture of the intermediate nitrilium ion or imidate.

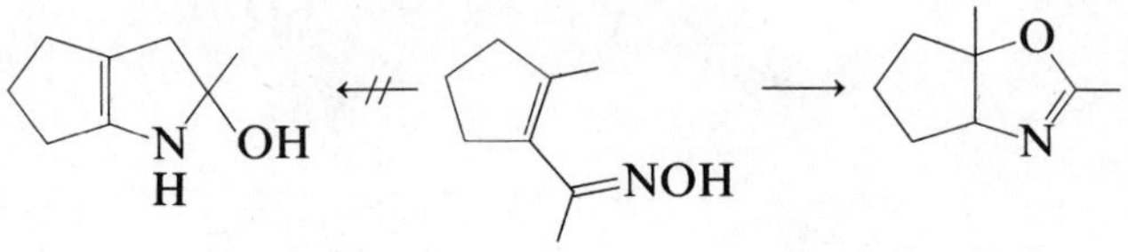

N H OH — NOH — O N

231 **233** **232**

Mesityl oxide oxime (**234**) affords oxazoline **235** in 80% yield by a similar route when treated with sulfuric acid.[206] When a phenolic oxygen is present, oxazole formation appears to be quite facile. For example, oxime **236** affords benzoxazole **237** in 83% yield when treated with phosphoryl chloride.[207]

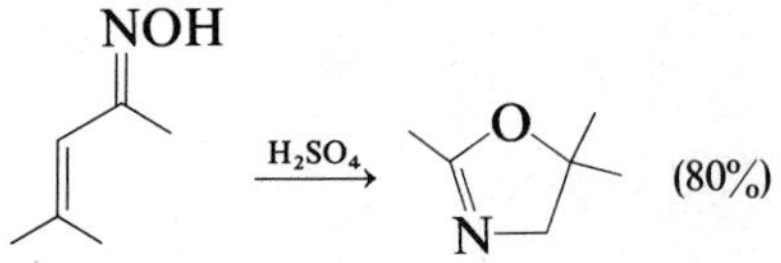

NOH $\xrightarrow{H_2SO_4}$ O N (80%)

234 **235**

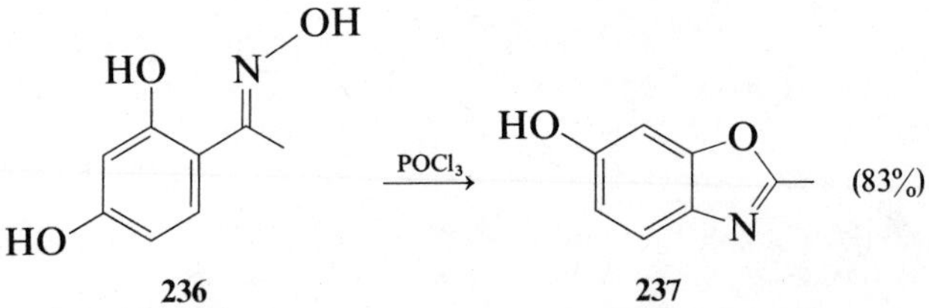

HO N OH, HO $\xrightarrow{POCl_3}$ HO O N (83%)

236 **237**

(*E*,*E*)-Benzil dioxime affords a 99% yield of oxadiazine **238** when heated for 12 minutes with polyphosphoric acid.[179]

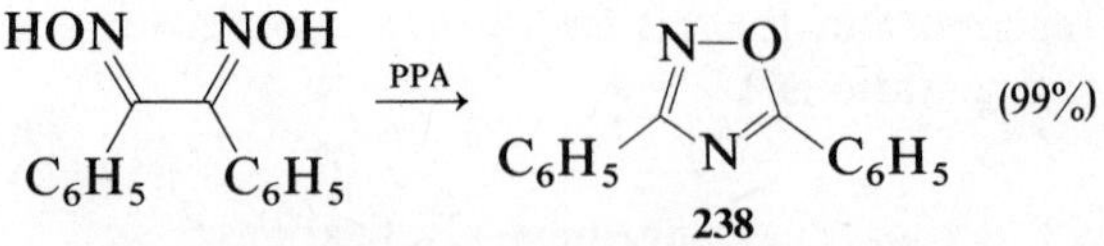

HON NOH C_6H_5 C_6H_5 $\xrightarrow{PPA}$ N—O C_6H_5 N C_6H_5 (99%)

238

Thiophenols also serve as terminators: oxime **239** cyclizes to benzothiazole **240** in 70% yield when treated with polyphosphoric acid.[208]

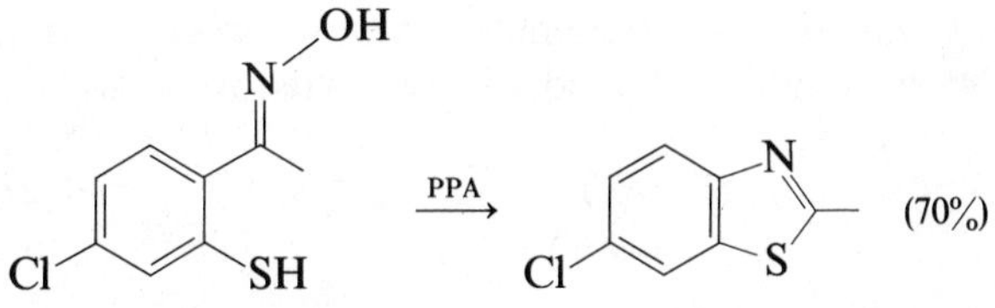

239 → (PPA) 240 (70%)

Nitrogen is reported as a terminator in only a few cases. In one of these, diphenylhydrazone **241** affords imidazole **242** in 80% yield on treatment with polyphosphoric acid.[209]

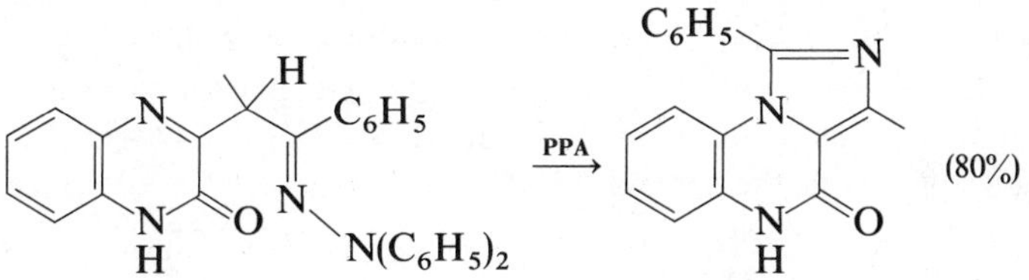

241 → (PPA) 242 (80%)

The formation of valerolactam from cyclodecanedione dioxime **243** can be rationalized by a rearrangement cyclization followed by hydrolysis.[210]

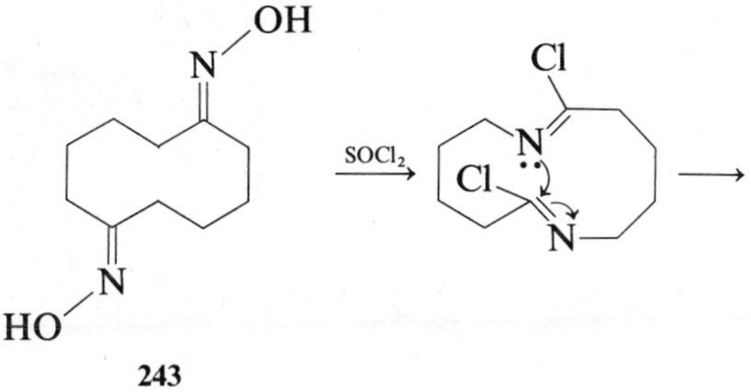

243 → ($SOCl_2$) → → (H_2O) (33%)

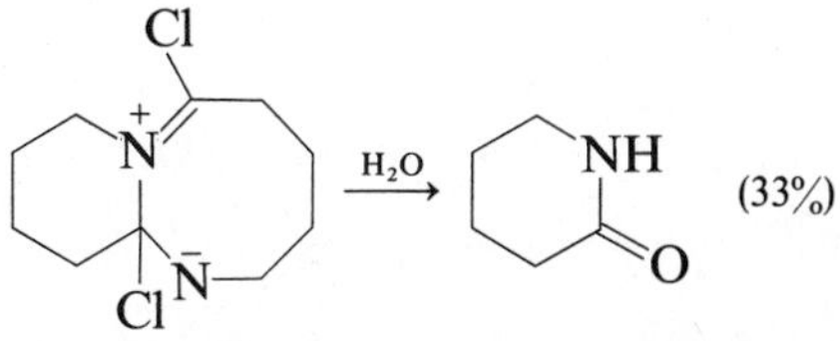

Aromatic Terminators. The first rearrangement-cyclization involving an aromatic terminator, the conversion of benzylideneacetone to 1-methylisoquinoline, was reported by Goldschmidt in 1895.[13] The process is fairly general, comparable to the Bischler–Napieralski reaction.[211]

One paper reports several examples of indole aldoximes undergoing a rearrangement cyclization.[212] This process is surprising in view of the reluctance of aldoximes to rearrange to formamides. Nevertheless, aldoxime **244** affords isoquinoline **245** in undisclosed yield when refluxed in ethanolic sulfuric acid.

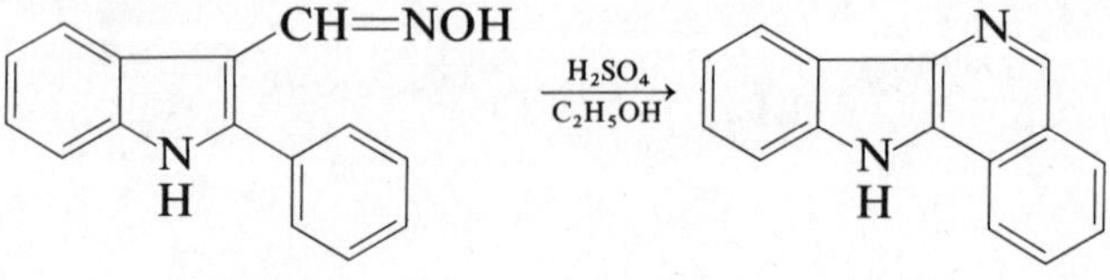

CH=NOH $\xrightarrow[C_2H_5OH]{H_2SO_4}$

244 **245**

Isoquinolines and dihydroisoquinolines can be routinely prepared by rearrangement-cyclization, although the yields are modest, usually in the 30–50% range. For example, oxime **246** affords isoquinoline **247** in 32% yield (via double-bond isomerization) on treatment with phosphorus pentachloride followed by phosphorus pentoxide.[213] The imidoyl chloride is formed in the first step[214] and cyclizes in the second.

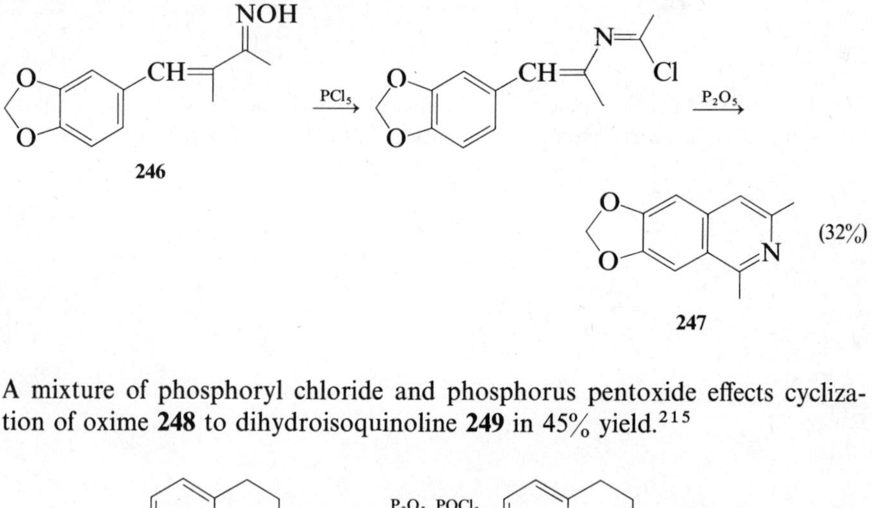

$\xrightarrow{PCl_5}$ $\xrightarrow{P_2O_5}$

246

(32%)

247

A mixture of phosphoryl chloride and phosphorus pentoxide effects cyclization of oxime **248** to dihydroisoquinoline **249** in 45% yield.[215]

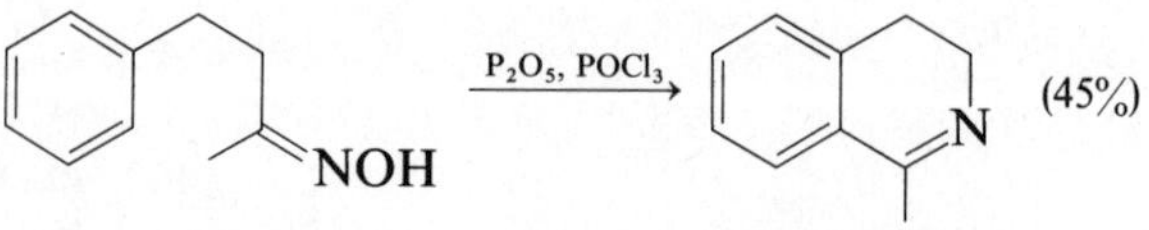

$\xrightarrow{P_2O_5,\ POCl_3}$ (45%)

248 **249**

Because the rearrangement step is stereospecific, the yield of cyclization product is limited by the amount of the appropriate geometric isomer in the starting material. An illustrative example is the reaction of oxime **250**, in which the *E* isomer affords phenylphenanthridine (**251**) via an *endo* cyclization pathway (see Scheme 1), while the *Z* isomer affords fluorenone (**252**) via an *exo* pathway.[215]

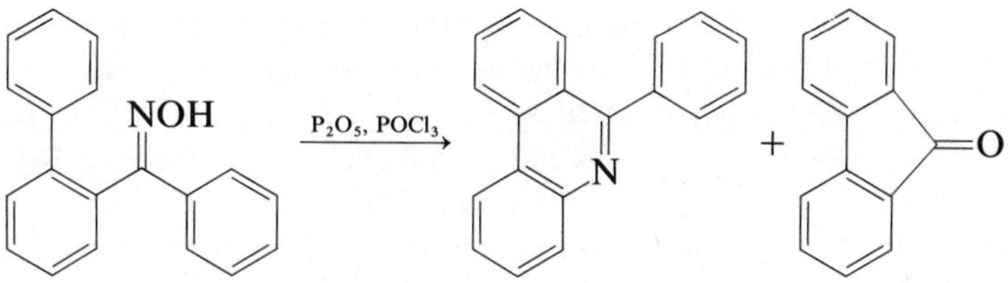

NOH $\xrightarrow{P_2O_5,\ POCl_3}$ N + =O

250 251 252

Olefinic Terminators. Wallach reported in 1901[14] that oxime **253** undergoes rearrangement cyclization to dihydropyridine **254** on treatment with phosphorus pentoxide, but the structure of the reaction product was later corrected to pyrroline **255**.[216] More recently, trimethylsilyl polyphosphate has been used to effect the transformation.[217] The mesylate of **253** can be converted to **255** by using either stannic chloride or diethylaluminum chloride.[218]

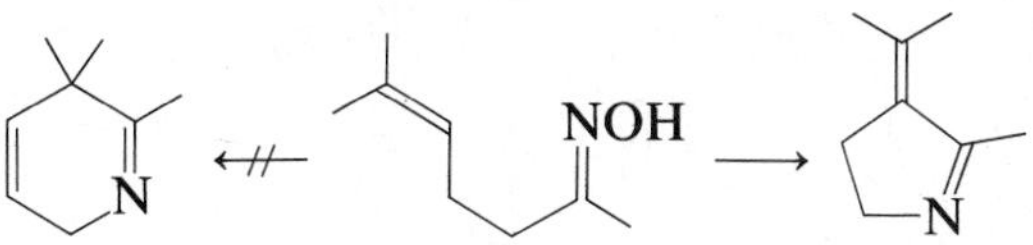

N NOH N

254 253 255

The styryl group serves as a terminator for the rearrangement cyclization, but the chlorovinyl group does not except under forcing conditions.[217] Thus oxime **256** affords pyrroline **257** on treatment with either phosphorus pentoxide or trimethylsilyl polyphosphate, but chlorovinyl oxime **258** cyclizes only on treatment with phosphorus pentoxide. Treatment with trimethylsilyl polyphosphate stops at the intermediate imidate stage, affording only amide **259** on workup.

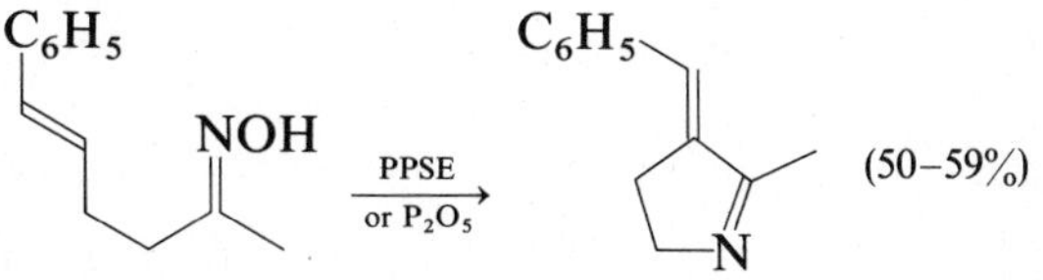

C_6H_5 NOH $\xrightarrow[\text{or } P_2O_5]{\text{PPSE}}$ C_6H_5 N (50–59%)

256 257

Cl (50%) N $\xleftarrow{P_2O_5}$ Cl NOH $\xrightarrow{\text{PPSE}}$ Cl $NHCOCH_3$

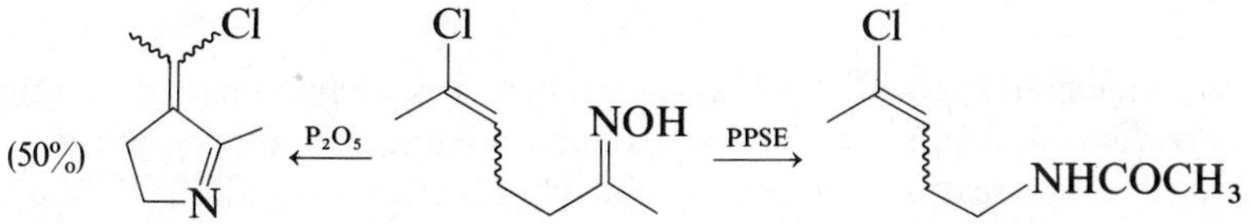

258 259

Alkylation of the oxime dianion prior to rearrangement cyclization is a good way to introduce substituents into the heterocycle regiospecifically.[217–219] For example, oxime **260** is constructed in this way. Alkylation of the dianion of (*E*)-**253** with methyl iodide affords **260**, which cyclizes to pyrroline **261** in good yield.[217,219] (*Z*)-2-Butanone oxime dianion is alkylated with cinnamyl chloride to (*Z*)-**262**; after equilibration of the geometric isomers, the oxime is cyclized to pyrroline **263** in 82% yield.[217,219]

(*E*)-**253** → 1. n-C_4H_9Li, 2. CH_3I → **260** → P_2O_5 → **261** (72%)

262 → P_2O_5 → **263** (82%); reagent: $C_6H_5CH{=}CHCH_2Cl$

Oxime **264** is also constructed in this way. After equilibration of the geometric isomers, mesylation of the oxime, rearrangement, and reduction gives tetrahydropyridine **265**. Reduction completes a short synthesis of solenopsin B.[218]

264 → 1. $(C_2H_5)_2AlCl$, 2. DIBAL → **265** (53%) → H_2 → solenopsin B

Rearrangement cyclization of large-ring ketoximes can be used to synthesize *ansa* pyridines and piperidines. For example, oxime **266** affords pyridine **267** in 25% yield on treatment with phosphoryl chloride in pyridine.[220] Similarly, oxime mesylate **268** is converted to muscopyridine in 80% yield by sequential treatment with trimethylsilyl triflate and manganese dioxide.[218]

HON

$POCl_3$

(25%)

266 **267**

OMs

N

1. $(CH_3)_3SiOTf$
2. MnO_2

(80%)

268 muscopyridine

Utilization of trimethylalane as the Lewis acid mediates trapping of the cyclic carbocation by a methyl group; the tetrahydropyridine so formed may then be reduced to a piperidine. The process is illustrated by the conversion of oxime mesylate **269** to *ansa* piperidine **270**, which occurs in 59% overall yield.[218]

OMs

N

1. $(CH_3)_3Al$
2. DIBAL

N
H

(59%)

269 **270**

Cyclization in the *exo* mode (see Scheme 1) affords carbocycles instead of heterocycles. For example, oxime mesylate **271** affords cyclohexenylamine **272** in 65% yield when treated sequentially with stannic chloride and diisobutylaluminum hydride. In line with reactivities described above, sequential treatment of **271** with trimethylalane and diisobutylaluminum hydride affords cyclohexylamine **273** via trapping of the cyclic cation.[218]

MsO

N

1. $SnCl_4$
2. DIBAL

C_6H_5

1. $(CH_3)_3Al$
2. DIBAL

NHC_6H_5 NHC_6H_5

272 (65%) **271** **273** (63%)

Exo cyclization of oxime mesylate **274** produces a cation that cyclizes in a Friedel-Crafts fashion on the phenyl nucleus that migrated to nitrogen. Diisobutylaluminum hydride reduction of the resultant imine affords **275** in 51% yield.[218]

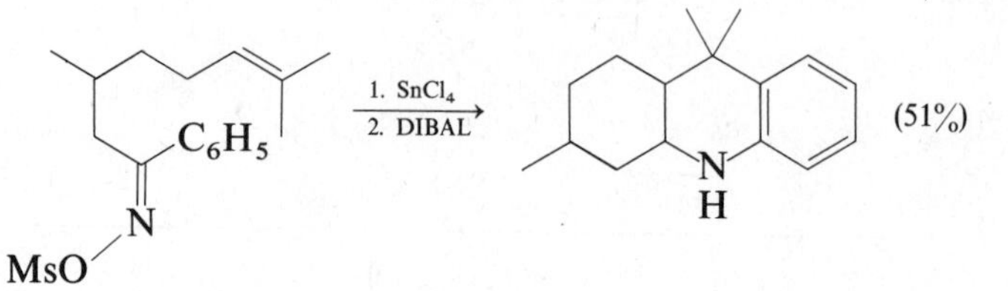

1. $SnCl_4$ 2. DIBAL (51%)

C_6H_5 N MsO N H

274 **275**

A number of cyclizations of nitrilium ions generated from oximes occur via a fragmentation–Ritter reaction pathway. These are processes that are conducted in a protic acid and thus are subject to many of the limitations of the Ritter reaction itself. In particular, the olefin produced in the fragmentation must be loath to rearrange. An example is the conversion of spiro oxime **276** to octalone **277** in 94% yield upon treatment with polyphosphoric acid at 125° for 10 minutes.[221]

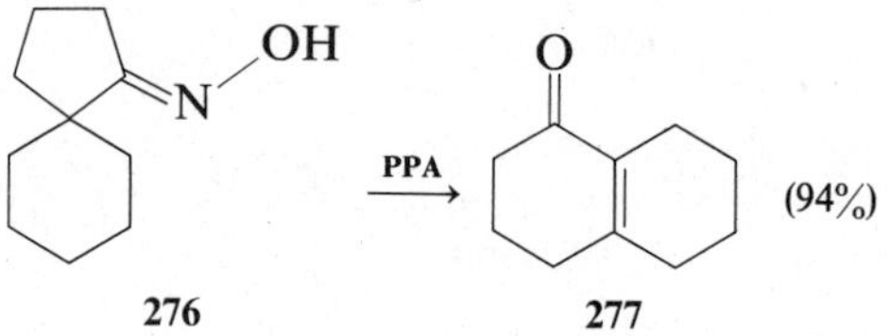

OH N O PPA (94%)

276 **277**

COMPARISON WITH OTHER METHODS

The Beckmann rearrangement consists of the conversion of the ketone to an amide, a conversion that may be accomplished in a number of ways: the Schmidt reaction, the photochemical Beckmann, or the nitrone Beckmann. Of these, only the Beckmann rearrangement is stereospecific. The Schmidt reaction[6,8,222,223] is the conversion of a ketone into an amide by reaction with hydrazoic acid. The structure of the amide obtained from an unsymmetrical ketone depends on the relative migratory aptitudes of the two groups. This can be seen by comparing the products formed from the Schmidt and Beckmann reactions of ketone **278**. Schmidt reaction affords an equimolar mixture of both possible products,[224] whereas solvolysis of the oxime tosylate **279**, in which the tosylate is *anti* to the isopropyl group for steric reasons, affords only the acetanilide stereospecifically.[28]

$$\underset{\textbf{278}}{C_6H_5C(O)C_3H_7\text{-}i} \xrightarrow{HN_3} C_6H_5NHCOC_3H_7\text{-}i + C_6H_5CONHC_3H_7\text{-}i \quad (57\%)$$

1:1

$$\underset{\textbf{279}}{C_6H_5C(=NOTs)C_3H_7\text{-}i} \xrightarrow[\text{reflux}]{80\%\ \text{ethanol}} C_6H_5NHCOC_3H_7\text{-}i \quad (98\%)$$

The nitrone Beckmann,[123–125] which affords *N*-methyl amides from *N*-methylnitrones, sometimes gives lactams structurally different from those obtained otherwise. For example, oximes of unsaturated steroidal ketoximes produce unsaturated amides on rearrangement (**11–15**),[40–43] while the corresponding nitrones afford enamides exclusively (**110**).[124,125] Certain cyclobutanone oximes also fail to rearrange stereospecifically, while the corresponding nitrones afford only a single product (**112** and **115**).[126,127]

The photochemical Beckmann rearrangement[17] converts an oxime to an oxaziridine, which then rearranges to an amide under conditions of stereoelectronic control, governed by orbital alignments in the oxaziridine intermediate.[225] Usually, this means that photochemical rearrangement of an unsymmetrical ketoxime gives mixture of lactams, independent of oxime geometry. For example, lactam **280** is the only rearrangement product obtained on treatment of oxime **281** with thionyl chloride,[226] while photolysis in methanol affords an equimolar mixture of **280** and **282**.[227]

280 $\xleftarrow{SOCl_2}$ **281** $\xrightarrow{h\nu}$ **280** + **282** (36%)

(1:1)

The photochemical process has the advantage that oximes that would normally fragment under even the most mild Beckmann conditions rearrange when photolyzed. For example, oxime **283** affords only fragmentation products under (unstated) Beckmann conditions but provides a 60% yield of lactam **284** when photolyzed in methanol.[227]

283 $\xrightarrow{h\nu}$ **284** (60%)

Those reactions classified as elimination–additions, ketoxime fragmentations, and rearrangement-cyclizations are not easily accomplished in any other way. Aldehydes can be converted to nitriles by the Schmidt reaction.[8]

The stereospecificity of the Beckmann reactions is both a blessing and a curse. It is a blessing if one has in hand the correct diastereomer for the desired transformation and a curse if one does not. These reactions suffer from the fact that the stereochemistry of the oxime cannot be controlled easily. Occasionally, it is possible to separate the two geometric isomers. More often, however, the undesired stereoisomer is wasted. At best, regiochemical alkylations of oxime dianions[228–230] afford stereochemically pure *Z* oximes, but this is of little consequence unless the starting oxime is symmetrical.

EXPERIMENTAL CONDITIONS AND PROCEDURES

Several of the procedures described below require oxime sulfonates as starting materials. The following general procedures can be used to prepare tosylates or mesylates from the corresponding oxime.

General Method for Preparation of Oxime Tosylates (Reaction of a Ketoxime with *p*-Toluenesulfonyl Chloride).[147] To a solution of the oxime (10 mmol) in 10 mL of pyridine at −20° was added 2.3 g (12 mmol) of *p*-toluenesulfonyl chloride portionwise over a period of 5–10 minutes. The resulting mixture was stirred at approximately −20–0° for several hours. The reaction progress was monitored by thin-layer chromatography (TLC). When the mixture was poured with stirring into ice and water, most oxime tosylates crystallized immediately. Filtration, washing several times with cold water, and recrystallization of the crude product afforded the pure oxime tosylates, which were stored in a freezer.

General Method for Preparation of Oxime Mesylates (Reaction of a Ketoxime with Methanesulfonyl Chloride).[147] To a solution of the oxime (10 mmol) and 2.1 mL (15 mmol) of triethylamine in 50 mL of methylene chloride at −20° was added 0.85 mL (11 mmol) of methanesulfonyl chloride over a period of 5–10 minutes. After stirring for an additional 30 minutes, the reaction mixture was transferred to a separatory funnel with the aid of more methylene chloride and washed sequentially with cold 1 *M* hydrochloric acid, saturated sodium bicarbonate, and brine. The organic layer was then dried with sodium sulfate and concentrated. The crude mesylates, which are thermally labile, are usually pure enough for further reaction but may be recrystallized (*carefully*!) from ether–hexane or methylene chloride-hexane at low temperature.

Rearrangements

The previous *Organic Reactions* review of the Beckmann rearrangement[5] contains general procedures using phosphorus pentachloride and sulfuric acid and specific procedures employing polyphosphoric acid, benzenesulfonyl chloride–sodium hydroxide, nitromethane, trifluoroacetic acid, hydrochloric acid in acetic acid, and Raney nickel. Readers interested in procedures using these reagents may consult the prior review or select a reference from Table I. Paralleling the text, the following procedures are categorized by the type of transformation.

1-Isopropyl-5-methyl-2-azabicyclo[4.1.0]heptan-3-one (Rearrangement of a Ketoxime with *p*-Toluenesulfonyl Chloride in Pyridine).[231] To a solution of 25 g (0.13 mol) of *p*-toluenesulfonyl chloride in 30 mL of pyridine at 0° was added a solution of 14.4 g (0.086 mol) of dihydroumbellulone oxime in 30 mL of pyridine. After standing for 1 hour, the reaction mixture was heated on a steam bath for 30 minutes, cooled, and allowed to stand for 2 hours. The resulting mixture was then poured into a slurry of 25 mL of H_2SO_4 and 120 g of ice. After 2 hours, the crystalline product was filtered and washed with water. Ether extraction of the filtrate afforded additional product. Total yield was 13 g (90%), mp 102.2–102.7°.

3-Aza-A-homocholestan-4-one (Rearrangement of a Ketoxime Tosylate with Alumina).[77] A solution of 2.0 g of cholestanone oxime tosylate was taken up in a minimum amount of benzene and applied to the top of an alumina column (activity I or II, alkali- or acid-washed, 25 g of alumina per gram of ester) packed with hexane. Excess *p*-toluenesulfonyl chloride was eluted with hexane; elution with benzene containing increasing amounts of chloroform effected elution of the lactam in 83% yield, mp 271–273°.

***p*-Methylacetanilide (Rearrangement of a Ketoxime with Thionyl Chloride in Carbon Tetrachloride).**[75] To a solution of 0.745 g (5 mmol) of *p*-methylacetophenone oxime in 10 mL of carbon tetrachloride at 0° was added a solution of 0.38 mL (5 mmol) of thionyl chloride in 10 mL of carbon tetrachloride. The reaction was stirred for 2 hours at 0°, washed with water, concentrated to half volume, and cooled. The crystalline product was isolated by filtration: yield 0.52 g (70%).

4,4-Dicarboethoxy-3,5-diphenylcaprolactam [Rearrangement of a Ketoxime with Trimethylsilyl Polyphosphate (PPSE)].[72]

Preparation of Trimethylsilyl Polyphosphate. A mixture of 10 g of phosphorus pentoxide and 21 mL of hexamethyldisiloxane was refluxed in 40 mL of an organic solvent such as methylene chloride, chloroform, benzene, or carbon tetrachloride until dissolution occurred (usually <1 hour).

Rearrangement Procedure. A solution of 0.609 g (1.49 mmol) of 4,4-dicarboethoxy-3,5-diphenylcyclohexanone oxime (**46**) in 4.5 mL of a methylene chloride solution of trimethylsilyl polyphosphate was stirred for 12 hours at room temperature, and then quenched with 7 mL of water. The aqueous layer was extracted with methylene chloride; the combined organic layers were dried with sodium sulfate and condensed. The crude product was purified by silica gel chromatography, yielding 0.384 g (63%) of lactam.

***N*-Phenylbenzamide (Rearrangement of a Ketoxime with Polyphosphoric Acid in Xylene).**[60] A solution of 1.97 g (10 mmol) of benzophenone oxime in 20 mL of xylene was added to a stirred suspension of 6 g of polyphosphoric acid in xylene at 100°. After heating at 100° for 2 hours, the reaction mixture was cooled and quenched with 40 mL of cold (0°) water. The organic layer was separated, dried with sodium sulfate, and concentrated to afford a quantitative yield of the product, mp 162°.

***N*-*n*-Hexylacetamide (Rearrangement of a Ketoxime with Triphenylphosphine–Carbon Tetrachloride).**[84] A solution of 1.4 g (10 mmol) of 2-octanone oxime, 5.25 g (20 mmol) triphenylphosphine, and 50 mL of carbon tetrachloride was refluxed for 2 hours. Concentration of the reaction mixture and distillation of the semisolid residue afforded 0.7–0.85 g (50–61%) of product, bp 84–86°/0.1 mm.

***N*-Methylferrocenecarboxamide (Rearrangement of an Organometallic Ketoxime with Trichloroacetonitrile).**[232] A solution of 2.43 g (10 mmol) of acetylferrocene oxime and 7.2 g (50 mmol) of trichloroacetonitrile in 50 mL of ether was refluxed for 1 hour, cooled, and filtered. The filtrate was diluted with 150 mL of hexane and left overnight at 0°. The crude product was separated by filtration and recrystallized from ethanol to give 1.8 g (74%) of amide, mp 124–125° (decomp.).

2-Azacyclononanone (Rearrangement of a Ketone with Hydroxylaminesulfonic Acid).[81] To a solution of 1.26 g (10 mmol) of cyclooctanone in 10 mL of 95–97% formic acid was added dropwise a solution of 1.7 g (15 mmol) of hydroxylaminesulfonic acid in 5 mL of the same solvent. The reaction mixture was refluxed for 5 hours, cooled, quenched with ice water, neutralized with 5% sodium hydroxide, and extracted with chloroform. The combined organic layers were dried with sodium sulfate and condensed. The residue was distilled to give the lactam in 61% yield, bp 138°/4 mm.

Cinnamamide (Rearrangement of an Aldoxime to a Primary Amide with Silica Gel).[128] A mixture of 2.94 g (20 mmol) of cinnamaldehyde oxime and TLC-grade silica gel (0.24 g, activated at 130–140°, pH of the silica gel suspension = 6.5–7.0) in 25 mL of anhydrous xylene was refluxed for 66 hours. The solution was filtered while hot and partially concentrated. The crude precipitate was recrystallized from benzene–ethanol, affording the product in a yield of 79%, mp 146–147°.

Elimination–Additions

***O*-Methylenantholactim (Methanolysis of a Ketoxime Tosylate).**[140] A solution of 10 g (36 mmol) of cycloheptanone oxime tosylate in 500 mL of 99.98% methanol was allowed to stand for 16 hours at 25°. The solution was brought to pH 8.0 with sodium methylate in methanol and condensed under vacuum.

The residue was taken up in ether, filtered, condensed, and distilled to afford the lactim ether in 62% yield, bp 78°/20 mm.

4,4′-Dinitrobenzimidoyl Chloride (Conversion of an Oxime to an Imidoyl Chloride).[141] 4,4′-Dinitrobenzophenone oxime (7.0 g, 24 mmol) was dissolved in a minimum amount of benzene (225 mL) at 75°. Phosphorus pentachloride (6.4 g, 31 mmol) was added at such a rate as to produce a vigorous reaction that did not become violent. After the addition was complete (10–12 minutes), an aspirator vacuum was applied to remove the solvent and phosphoryl chloride. The solid yellow residue was recrystallized from benzene, affording 6.1 g (82%) of the product, mp 130.5–131°.

2-Oxoheptanonitrile *n*-Pentylimine (Conversion of an Oxime Mesylate to an Imidoyl Cyanide).[147] A solution of 6-undecanone oxime mesylate (263 mg, 1 mmol) in 10 mL of methylene chloride was cooled to −78° and treated successively with trimethylsilyl cyanide (0.15 mL, 1.1 mmol) and diethylaluminum chloride (1.1 mL of a 1 *M* hexane solution, 1.1 mmol). The reaction was warmed to −20°, stirred for 1 hour, and poured into ice-cold 10% sodium hydroxide. Extraction with methylene chloride and purification of the crude product by chromatography on silanized silica gel (ether:hexane, 1:5) afforded 177 mg (91%) of the product as a colorless oil, IR: 2220, 1639 cm^{-1}.

1-Thiomethoxy-*N*-phenylacetaldehyde Imine (Conversion of an Oxime Mesylate to a Thioimidate).[147]

Preparation of Diisobutylaluminum Methanethiolate. Dimethyl sulfide (0.081 mL, 1.1 mmol) was added dropwise at 0° to a solution of diisobutylaluminum hydride (1.1 mL of a 1.0 *M* solution, 1.1 mmol) in hexane. The resulting solution was stirred at 0° for 30 minutes and used immediately.

Rearrangement Procedure. To a stirred solution of 231 mg (1 mmol) of acetophenone oxime mesylate in methylene chloride at −78° was added 1.1 mmol of the aluminum thiolate solution. After 5 minutes, the reaction was warmed to 0° and stirred for 1 hour. The reaction was quenched by successive treatment with sodium fluoride (185 mg, 4.4 mmol) and water (0.06 mL, 3.3 mmol). Vigorous stirring of the resulting suspension was continued at 0° for 20 minutes and the mixture was filtered. Concentration of the filtrate and silica gel chromatography (eluant ethyl acetate) yielded 148 mg of the imino thioether (90%) as a colorless oil, IR: 1622; ^{1}H NMR (CCl_4): δ 2.33 (s, 3*H*, SCH_3), 1.95 (s, 3*H*, CCH_3).

2-Methyl-6-undecyl-3,4,5,6-tetrahydropyridine [Rearrangement of an Oxime Mesylate and Trapping by a Carbon Nucleophile (Organoaluminum Reagent)].[147] To a solution of 1.51 g (4.54 mmol) of 2-undecylcyclopentanone oxime mesylate in methylene chloride at −78° was added 6.8 mL of a 2 *M* solution of trimethylaluminum in toluene (13.6 mmol). After 5 minutes, the reaction was warmed to 25° and stirred for 1 hour. Workup by the sodium fluoride method described in the previous experiment and silica gel chromatography (isopropylamine:ether 1:200) afforded 0.65 g (57%) of the imine as a light yellow oil. IR: 1660 cm^{-1}. This material can be reduced with lithium aluminum hydride-trimethylaluminum to give solenopsin A.

2-*n*-Butylazacycloheptane [Rearrangement of an Oxime Mesylate, Trapping by a Carbon Nucleophile (Grignard Reagent), and Reduction with Diisobutylaluminum Hydride].[157] To a solution of 185 mg (1 mmol) of cyclohexanone oxime mesylate in 5 mL of dry toluene at −78° was added a solution of *n*-butylmagnesium bromide (1.5 mmol, 0.5 mL of a 3 *M* ether solution). The reaction was stirred at −78° for 5 minutes and at 0° for 1 hour. Diisobutylaluminum hydride (2 mL of a 1 *M* hexane solution, 2 mmol) was added and the mixture was stirred at 0° for 1 hour. The reaction was poured into 40 mL of 5% sodium hydroxide, shaken well, and centrifuged to remove the white gel. Extractive workup with methylene chloride followed by column chromatography on silica gel (isopropylamine–ether–hexane) furnished 98 mg (63%) of the product as a colorless oil.

Pumiliotoxin-C [Rearrangement of an Oxime Tosylate, Trapping by a Carbon Nucleophile (Organoaluminum Reagent), and Reduction with Diisobutylaluminum Hydride].[147] To a solution of 384 mg (1.2 mmol) of *cis*-4β-methylhexahydroindanone oxime tosylate in methylene chloride at 25° was added 3.6 mmol of tri-*n*-propylaluminum (1.8 mL of a 2 *M* toluene solution). The resulting mixture was stirred for 30 minutes at 25,° treated with 4.8 mmol of diisobutylaluminum hydride (4.8 mL of a 1 *M* hexane solution), and stirred at 25° for 2 hours. The reaction was quenched by diluting with methylene chloride, adding sodium fluoride (0.6 g, 14.4 mmol) and water (0.2 mL, 10.8 mmol), stirring the resulting suspension vigorously at 0° for 20 minutes, filtering, and concentrating the filtrate. Silica gel chromatography of the residue (isopropylamine:ether: methylene chloride 1:30:30) produced 135 mg (60%) of the alkaloid as a colorless oil.

2-Allyl-2-methylazacycloheptane [Rearrangement of an Oxime Mesylate, Trapping by a Carbon Nucleophile (Organoaluminum Reagent), and Addition of a Grignard Reagent].[147] To a solution of 191 mg (1 mmol) of cyclohexanone oxime mesylate in 5 mL of methylene chloride at −78° was added 2 mmol of trimethylaluminum (1 mL of a 2 *M* solution in toluene). After 5 minutes, the reaction mixture was warmed to 0° and stirred for 30 minutes. The solution was cooled to −78° and treated with 2 mmol of allylmagnesium bromide (1.67 mL of a 1.2 *M* ether solution) and stirred at 0° for 1 hour. The reaction was quenched by pouring into 30 mL of 10% sodium hydroxide solution, shaking and centrifuging to remove the white gel. Extractive workup with methylene chloride and silica gel chromatography of the residue (isopropylamine:ether 1:50) furnished 92 mg (60%) of the product as a colorless oil.

2-(2-Oxo-*n*-octylidene)-6-methylazacycloheptane [Rearrangement of an Oxime Mesylate and Trapping with a Carbon Nucleophile (Enol Silyl Ether)].[159] To a solution of 205 mg (1 mmol) of (*E*)-2-methylcyclohexanone oxime mesylate and 220 mg (1.1 mmol) of 2-(trimethylsilyloxy)-1-octene in methylene chloride at −78° was added 3 mmol of diethylaluminum chloride (3 mL of a 1 *M* hexane solution). After 30 minutes at −78°, the reaction was stirred for 1 hour at 20° and quenched with 10% sodium hydroxide. Methylene chloride extraction and silica gel chromatography (ether:hexane 1:2) afforded 213 mg (90%) of the vinylogous amide as a colorless liquid.

Ketoxime Fragmentations

7-Cyanoheptanal (Fragmentation of an Alkoxy Oxime with Phosphorus Pentachloride)].[233] The fragmentation of 2-methoxycyclooctanone oxime with phosphorus pentachloride, affording the title compound in 85% yield, is described in *Organic Syntheses.*[233]

Senecionitrile (Fragmentation of an Alkoxy Oxime with Thionyl Chloride)[187] To a solution of 4.3 g (27 mmol) of 2,2,5,5-tetramethyltetrahydrofuran-3-one oxime in 40 mL of anhydrous ether at 0° was added dropwise 10 mL (0.137 mol) of thionyl chloride. The resulting solution was kept overnight at room temperature, quenched with methanol, and fractionally distilled. The product boiled at 140–146°. The yield was 1.37 g (62%).

Methyl 3-[3-(2,3β-dimethyl-4β-cyanomethylcyclopentenyl)]propionate (179) (Fragmentation of a Derivative of Camphor Oxime with Trifluoroacetic Anhydride).[172] To a solution of 10 g (41.8 mmol) of oxime ester **177** in 10 mL of methylene chloride at 0° was added dropwise 10 mL (70.3 mmol) of trifluoroacetic anhyride. The reaction mixture was slowly warmed to room temperature; after a total of 4 hours, the olefin isomers were equilibrated by the addition of 10 mL (41.67 mmol) of trifluoroacetic acid. The reaction mixture was stirred for 24 hours and concentrated *in vacuo*. The residue was taken up in ether and washed with brine and potassium bicarbonate. The aqueous layers were back-extracted with ether; the combined organic layers were dried with sodium sulfate and concentrated. Vacuum distillation afforded 3.44 g of **179** (80%) as a colorless oil, bp 88–90°/ 0.003 mm.

***cis*-5-Heptenonitrile (Fragmentation of a Trimethylsilyl ketoxime Acetate with Trimethylsilyl Triflate).**[204] To a solution of 0.12 g (0.5 mmol) of *cis*-2-methyl-3-trimethylsilylcyclohexanone oxime acetate (**227**) in anhydrous methylene chloride at 0° was added dropwise 0.11 g (0.05 mmol) of trimethylsilyl triflate. The mixture was stirred for 4 hours, quenched with 0.1 mL of triethylamine and aqueous sodium bicarbonate and extracted with ether (10 mL). The concentrated ether solution was purified by silica gel chromatography to afford 53 mg (90%) of *cis*-5-heptenonitrile.

Aldoxime Fragmentations

p-Methylbenzonitrile (Fragmentation of an Aldoxime with Thionyl Chloride).[75] To a solution of 0.37 mL (5 mmol) of thionyl chloride in 10 mL of carbon tetrachloride was added a solution of 0.675 g (5 mmol) of *p*-methylbenzaldoxime in 10 mL of carbon tetrachloride. The reaction mixture was stirred at room temperature for 12 hours, washed with 30 mL of water, concentrated, and distilled to afford 509 mg (87%) of *p*-methylbenzonitrile, IR: 2210 cm^{-1}.

Isobutyronitrile (Fragmentation of an Aldoxime with Titanium Tetrachloride).[234] To 200 mL of absolute dioxane at 0–10° was added 11 mL (0.1 mol) of titanium tetrachloride in 25 mL carbon tetrachloride to afford a yellow precipitate. To this suspension was added 16 mL (0.2 mol) of dry pyridine in 35 mL of dry dioxane followed by 4.35 g (0.05 mol) of isobutyraldehyde oxime in 20 mL of dioxane. The reaction was stirred for 43 hours, quenched with 50 mL of water and diluted with ether. The aqueous layer was separated and extracted with ether. The combined organic layers were washed with brine, dried with magnesium sulfate, and distilled to give an 81% yield of the nitrile, bp 107–108°.

Cyclohexylcarbonitrile (Fragmentation of an Aldoxime with Selenium Dioxide).[235] A mixture of 2.22 g (0.02 mol) of selenium dioxide and 2.54 g (0.02 mol) of cyclohexane carboxaldoxime in 40 mL of chloroform was refluxed for 3 hours, cooled, and treated with anhydrous calcium chloride. The reaction mixture was filtered through diatomaceous earth and concentrated *in vacuo*. Distillation afforded the nitrile in 82% yield, bp 66°/9 mm.

Benzonitrile (*In situ* Condensation–Fragmentation of an Aldehyde with Hydroxylamine in Formic Acid).[236] A solution of 1.07 g (0.01 mol) of benzaldehyde and 0.9 g (0.01 mol) of hydroxylamine hydrochloride in 10 g of 95–98% formic acid was refluxed for 30 minutes. After cooling, the mixture was diluted with ice water (100 mL), neutralized with 5% sodium hydroxide, and extracted with ether. The ethereal extracts were dried with magnesium sulfate and concentrated to give 1.02 g (99%) of the nitrile, bp 192°/760 mm.

Rearrangement–Cyclizations

5-Chloro-2-methylbenzothiazole (Rearrangement–Cyclization of a Ketoxime with a Thiol Terminator Using Polyphosphoric Acid).[208] To 50 g of hot (120–130°) polyphosphoric acid was added 10 g (50 mmol) of 5-chloro-2-mercaptoacetophenone oxime in several portions. The mixture was stirred at this temperature for 1.5 hours, cooled, and poured into water. Ether extraction afforded 6.4 g (70%) of the crude benzothiazole, which was recrystallized from petroleum ether, mp 63–65°.

6-Hydroxy-2-methylbenzoxazole (Rearrangement–Cyclization of a Ketoxime with a Hydroxy Terminator Using Phosphoryl Chloride).[207] To a stirred solution of 4.2 g (25 mmol) of 2,4-dihydroxyacetophenone oxime in 5 mL of dimethylacetamide and 15 mL of acetonitrile was added 2.4 mL (26 mmol) of phosphoryl chloride at such a rate as to keep the temperature below 30°. After stirring the reaction for 30 minutes, the solution was poured into 200 mL of ice water containing 6 g of sodium acetate. The crude benzoxazole was collected by filtration (3.08 g, 83%) and recrystallized from acetonitrile, mp 194–196°.

Indolo[3,2-*b*]isoquinoline (Rearrangement–Cyclization of an Aldoxime with an Aromatic Ring Terminator Using Sulfuric Acid).[212] A solution of 2.12 g (9 mmol) of 2-phenylindole-3-carboxaldoxime and 1.6 mL of sulfuric acid in 90 mL of ethanol was refluxed for 1 hour, cooled, and poured into 50 mL of ice water. The resulting solid was filtered and recrystallized from benzene to give the product, mp 208–209°.

1-Methyl-3,4-dihydroisoquinoline (Rearrangement–Cyclization of a Ketoxime with an Aromatic Ring Terminator Using Phosphorus Pentoxide and Phosphoryl Chloride).[215] To a cold solution (0°) of 23.8 g (0.17 mol) of phosphorus pentoxide and 51.5 g (0.34 mol) of phosphoryl chloride in 50 mL of sulfur dioxide in a 300 mL glass pressure apparatus equipped with metal joints and stopcocks was added an ice-cold solution of 5 g (0.034 mol) of benzylacetone oxime in 20 mL of sulfur dioxide. The vessel was closed and the reaction mixture heated at 70° for 12 hours, cooled, and opened. The solvent was allowed to evaporate and the residue was poured into ice water. The aqueous phase was washed with ether to remove the Beckmann rearrangement product, rendered alkaline with sodium hydroxide, and steam distilled. The distillate was again rendered alkaline and extracted with ether. After drying with magnesium sulfate, the concentrated ether layer was vacuum distilled (bp 95–105°/5–6 mm) to afford 2.0 g (45%) of product; picrate mp 188–190°.

3-Benzylidene-2-methyl-Δ^1-pyrroline (Rearrangement–Cyclization of a Ketoxime with a Styryl Terminator Using Trimethylsilyl Polyphosphate (PPSE).[217]

Preparation of Trimethylsilyl Polyphosphate. A mixture of 1.5 g (11 mmol) of phosphorus pentoxide, 3 mL (5 mmol) of hexamethyldisiloxane, and 7 mL of carbon tetrachloride was refluxed for 1.5 hours and cooled.

Rearrangement–Cyclization Procedure. To the colorless trimethylsilyl polyphosphate solution was added 0.183 g (1 mmol) of 6-phenylhex-5-en-2-one oxime. The reaction was refluxed for 7 hours and cooled. The solution was decanted and the gummy precipitate taken up in two successive 5 mL portions of water. The carbon tetrachloride layer was washed with 5 mL of 10% hydrochloric acid, which was added to the other water layers. The combined aqueous layers were cooled to 0°, brought to pH 9 with 50% sodium hydroxide (1.5 mL) and extracted with chloroform. The dried and condensed organic phase yielded an oil that was purified by silica gel chromatography (10% acetone in hexane), giving 105 mg (64%) of product, IR: 1645 cm^1. ^1H NMR ($CDCl_3$): δ 2.19 (t, $^5J = 2.0$ Hz, 3*H*) 2.6–2.9(br m, 2*H*), 3.7–4.1(br m, 2*H*), 6.67 (t, $^4J = 2.7$ Hz, 1*H*), 7.1–7.5(m, 5*H*).

***N*-(3-Methyl-2-cyclohexenyl)aniline (Rearrangement-Cyclization of a Ketoxime Mesylate with an Olefin Terminator Using Stannic Chloride, and Reduction of the Product with Diisobutylaluminum Hydride).**[218] To a solution of 281 mg (1 mmol) of 2-methyl-6-phenylhex-1-en-2-one oxime mesylate in 10 mL of methylene chloride at $-20°$ was added 0.13 mL (1.1 mmol) of stannic chloride. After 5 minutes the reaction was warmed to 0° for 1 hour and quenched by pouring it into 30 mL of 10% sodium hydroxide. The mixture was extracted with methylene chloride and dried with sodium sulfate. The filtered organic layers were concentrated to a volume of 10 mL, cooled to 0°, and treated with 4 mL of a 1 *M* solution (4 mmol) of diisobutylaluminum hydride in hexane. After stirring for 1 hour, the solution was diluted with 20 mL of methylene chloride and quenched with 672 mg (16 mmol) of sodium fluoride and 0.22 mL (12 mmol) of water. The resulting suspension was stirred vigorously for 30 minutes at 25°, filtered, and concentrated. Purification of the resulting oil by silica gel chromatography (eluant ether:hexane 1:20) afforded 121 mg (65%) of the product as a colorless oil.

$\Delta^{9,10}$-1-Octalone (Fragmentation–Cyclization of a Spiro Ketoxime with an Olefin Terminator Using Polyphosphoric Acid).[221] A mixture of 1.04 g (6.2 mmol) of spiro[4,5]decan-1-one oxime and 30 g of polyphosphoric acid was heated at 125–130° for 10 minutes, poured onto ice, and rendered alkaline with sodium hydroxide. Chloroform workup afforded 0.88 g (94%) of a yellow oil that was purified by alumina chromatography (ether eluant); 2,4-dinitrophenylhydrazone, mp 264.5–265.5°.

TABULAR SURVEY

The tables are arranged to correspond to the major sections in the "Scope and Limitations" section, with the exception of "Fragmentations," which are subdivided into ketoxime and aldoxime fragmentations. Within each table, entries are categorized by increasing number of carbons and hydrogens in the ketone precursor to the oxime. Thus an oxime and its acetate, tosylate, and so on are all found together. When the stereochemistry of the C=N bond is indicated in the literature, it is shown in the tables. Substrates that give both fragmentation and rearrangement products, for example, are entered in both of the appropriate tables. The literature coverage is from 1958 to mid-1984. As with any work of this kind, considerable effort has been made to be thorough. My apologies are offered to those whose work has been inadvertently omitted.

Abbreviations used in the tables are as follows:

Acac	acetylacetone
C_8H_{17}	(on a steroid D ring) $CH(CH_3)(CH_2)_3C_3H_7$-*i*
DCC	dicyclohexylcarbodiimide
DIBAL	diisobutylaluminum hydride
DMAC	*N,N*-dimethylacetamide

DMF	*N,N*-dimethylformamide
DMSO	dimethyl sulfoxide
HMPA	hexamethylphosphoric triamide
LAH	lithium aluminum hydride
Ms	methanesulfonyl
Pet	petroleum
Pic	picryl (2,4,6-trinitrophenyl)
PPA	polyphosphoric acid
PPE	ethyl polyphosphate
PPSE	trimethylsilyl polyphosphate
THF	tetrahydrofuran
Ts	*p*-toluenesulfonyl

TABLE I. Rearrangements

No. of Carbon Atoms	Substrate	Reagent and Conditions	Product(s) and Yield(s) (%)	Refs.
C_2	$CH_3CH{=}NOH$	Silica gel, xylene, reflux, 57 h	CH_3CONH_2 (89)	128
		Pd(Acac)$_2$(cat.), CH_3CN, C_6H_6, anisole, 80°, 3.5 h	" (90)	132, 133
C_3	$C_2H_5CH{=}NOH$	BF_3, CH_3CO_2H	$C_2H_5CONH_2$ (87)	129
	$(CH_3)_2C{=}NOPic$	1,4-Dichlorobutane, 80°	$CH_3CON(CH_3)Pic$ (—)	94
C_4	$CH_3C(NF_2)_2C({=}NF)CH_3$	H_2SO_4	$CH_3C(NF_2)_2NHCOCH_3$ (<30)	121
	n-$C_3H_7CH{=}NOH$	BF_3, CH_3CO_2H	n-$C_3H_7CONH_2$ (63)	129
	OTs N	$(C_2H_5)_3N$, 80% ethanol, reflux	$CH_3CONHC_2H_5$ (91) " (55)	28
	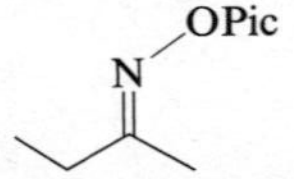 OPic N	1,4-Dichlorobutane, 80°	$CH_3CON(C_2H_5)Pic$ (—)	94
	m-$HCB_{10}H_9$(9-$C(=NOH)CH_3$)CH	PCl_5, ether, 20°, 3 h	m-$HCB_{10}H_9$(9-$NHCOCH_3$)CH (47)	106
C_5	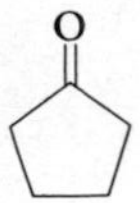 O	H_2NOSO_3H, HCO_2H, reflux	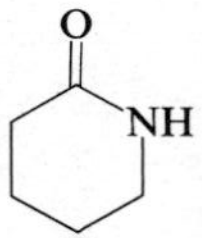 O NH (65)	81

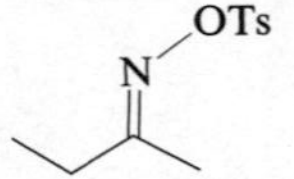

Reactant	Conditions	Product(s) (Yield, %)	Refs.
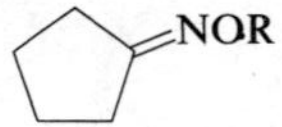 Cyclopentanone oxime derivative (=NOR), R=H	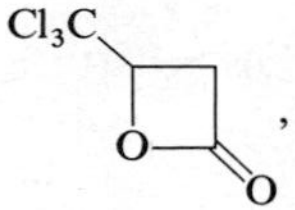 Cl_3C-β-lactone, CCl_4, 14 h	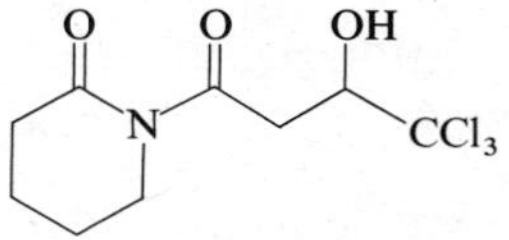 O=C(piperidin-2-one-N)-$CH_2CH(OH)CCl_3$ (100)	87
	SO_3, SO_2, −10°	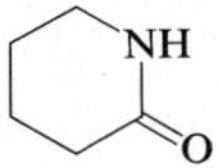 2-Piperidone (NH) (91–93)	237
R=Ts	(−)	" (80)	238
	Y-type zeolite, 120°	" (—)	239, 240
	SiO_2, $CHCl_3$, 5°	" (45)	109
	CH_3OH, reflux, 3 h	" (30) 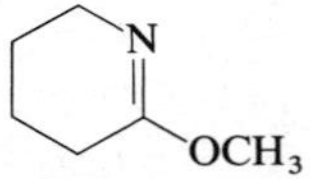 + 2-methoxy-3,4,5,6-tetrahydropyridine (OCH_3) (40)	140
	CH_3CO_2H, 35°, 2 h	$HO_2C(CH_2)_4NHTs$ (8)	140
	$(C_2H_5)_3N$, 80% ethanol, 23° 3.5 h	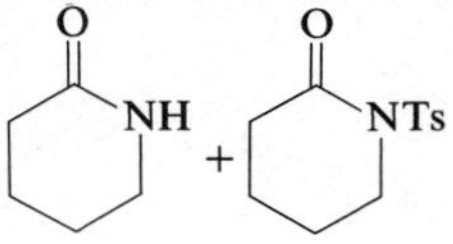 2-Piperidone (NH) I + N-Ts-2-piperidone (NTs) II (—)	28
	80% ethanol, 23°, 3.5 h	I (—)	28
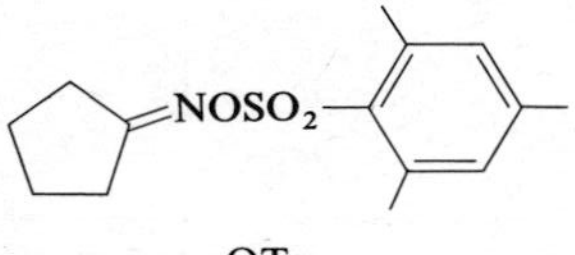 Cyclopentanone =$NOSO_2$-mesityl	Alumina, CH_3OH	I (81)	47
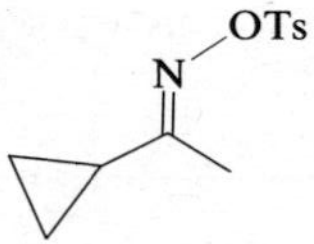 Cyclopropyl methyl ketone oxime (N–OTs)	$(C_2H_5)_3N$, 80% ethanol, reflux	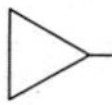 Cyclopropyl–$NHCOCH_3$ (93)	28

TABLE I. REARRANGEMENTS (*Continued*)

No. of Carbon Atoms	Substrate	Reagent and Conditions	Product(s) and Yield(s) (%)	Refs.
	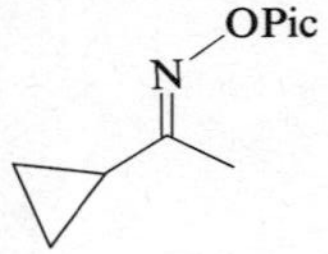 [cyclopropyl methyl ketone O-picryl oxime: cyclopropyl–C(CH₃)=N–OPic]	1,4-Dichlorobutane, 80°	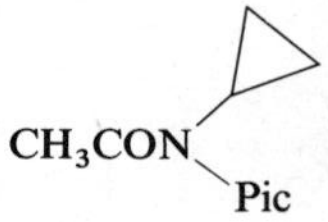 CH_3CON(cyclopropyl)(Pic) (—)	93
	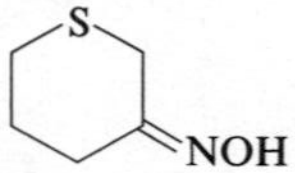 [thian-3-one oxime, =NOH] (Z)	1. TsCl, pyridine, 0° 2. $(C_2H_5)_3N$, dioxane, 23°	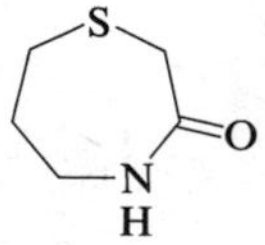 [1,4-thiazepan-3-one, S, N–H, C=O] (81)	194
	(*E*)	"	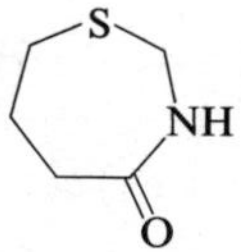 [1,4-thiazepan-5-one, S, NH, C=O] (27)	194
	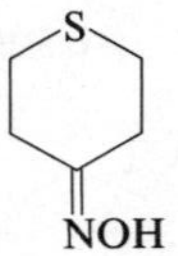 [thian-4-one oxime, =NOH]	"	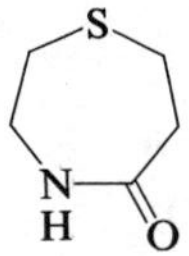 [1,4-thiazepan-5-one, S, N–H, C=O] (98)	194
		(—)	" (46)	241
	n-$C_4H_9CH{=}NOH$	BF_3, CH_3CO_2H	n-$C_4H_9CONH_2$ (59)	129
	i-$C_3H_7C(CH_3){=}NOR$ R = H	$(C_6H_5)_3P$, CCl_4, THF, reflux, 6 h	i-$C_3H_7NHCOCH_3$ (60–70) I	84
	R = Ts	$(C_2H_5)_3N$, 80% ethanol, reflux	I (94)	28
	R = Ts		I + i-$C_3H_7CONHCH_3$ II	

		Alumina	I/II = 88/12 (65–80)	39
		TsOH	I/II = 93/7 (45–60)	39
		HCl, CH_3CO_2H	I/II = 93/7 (45–60)	39
		SiO_2, $CHCl_3$, 5°	I (60)	109
	R = Pic	1,4-Dichlorobutane, 80°	$CH_3CON(Pic)C_3H_7$-*i* (—)	94
	n-$C_3H_7C(CH_3)$=NOR R=H	$(C_6H_5)_3P$, CCl_4, THF, reflux, 6 h	*n*-$C_3H_7NHCOCH_3$ (40) I	84
	R = Ts		I + *n*-$C_3H_7CONHCH_3$ II	
	(*E*)/(*Z*) = 73/27	Alumina	I/II = 3/1 (65–80)	39
	(*E*)/(*Z*) = 73/27	TsOH	I (45–60)	39
	(*E*)/(*Z*) = 73/27	HCl, CH_3CO_2H	I (45–60)	39
	(*Z*)	Alumina	II (65–80)	39
	(*Z*)	TsOH	I (65–80)	39
	R + Pic	1,4-Dichlorobutane, 80°	$CH_3CON(Pic)C_3H_7$-*n* (—)	93
	$(C_2H_5)_2C$=NOH	SO_3, SO_2, −10°	$C_2H_5CONHC_2H_5$ (77–88)	237
C_6	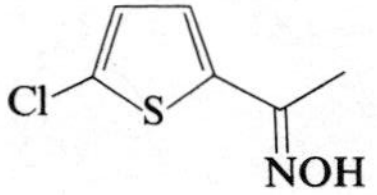 Cl, S, NOH (*E*)	PCl_5, ether	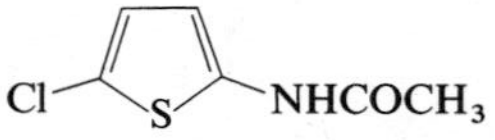 Cl, S, $NHCOCH_3$ (32)	242
	(*Z*)	"	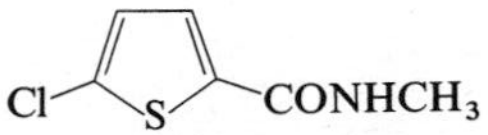 Cl, S, $CONHCH_3$ (27)	242
	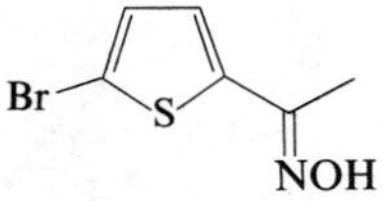 Br, S, NOH	" 0°, 20 h	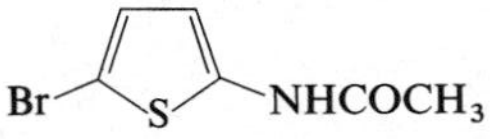 Br, S, $NHCOCH_3$ (74)	243
	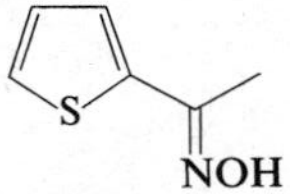 S, NOH (*E*)	1. $C_6H_5SO_2Cl$ 2. Alumina	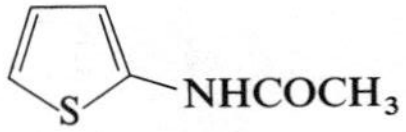 S, $NHCOCH_3$ (90)	77
	(*E*)	PCl_5, ether, 0°, 1 h	" (82)	243

TABLE I. REARRANGEMENTS (*Continued*)

No. of Carbon Atoms	Substrate	Reagent and Conditions	Product(s) and Yield(s) (%)	Refs.
	(*Z*) [structure: 3-acetylthiophene oxime, NOH, S]	1. $C_6H_5SO_2Cl$ 2. Alumina	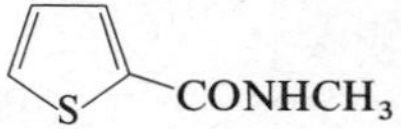 [structure: S] $CONHCH_3$ (—)	77
	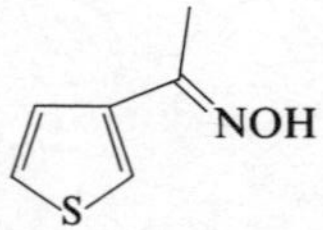	PCl_5, ether, 0°, 8 h	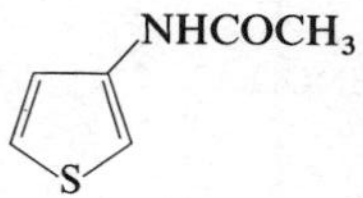 [structure: S] $NHCOCH_3$ (72)	243
	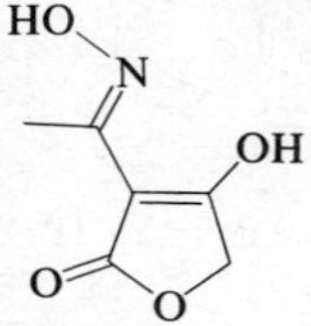 [structure: HO, N, OH, O, O]	$C_6H_5SO_2Cl$, NaOH, 50°	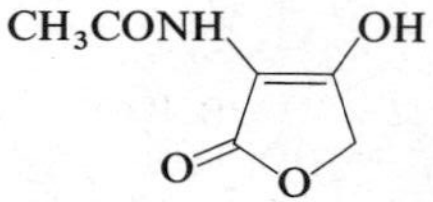 [structure: CH_3CONH, OH, O, O] (43)	244
	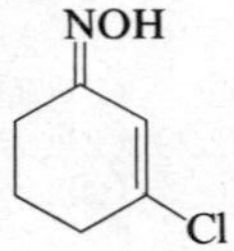 [structure: NOH, Cl] (*E*) or (*E*/*Z*)	PPA, 120°, 1 h	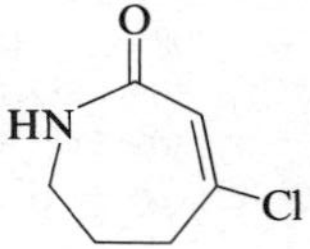 [structure: HN, O, Cl] (26–28)	245
	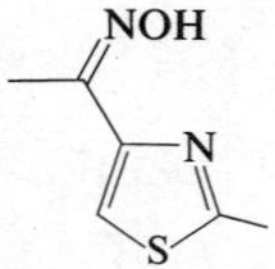 [structure: NOH, N, S]	PCl_5, ether	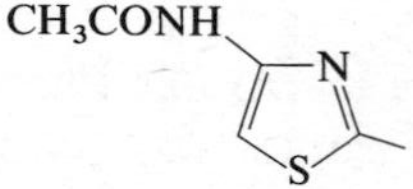 [structure: CH_3CONH, N, S] (31)	246
	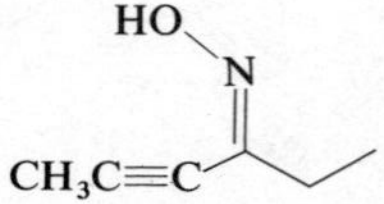 [structure: HO, N] $CH_3C{\equiv}C$	"	$CH_3C(Cl){=}CHCONHC_2H_5$ (—)	116

Reactant	Conditions	Product(s) (Yield, %)	Refs.
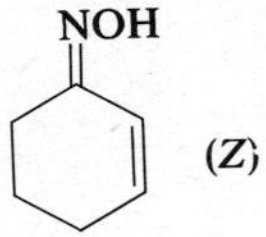 NOH (*Z*)	Oleum, $CHCl_3$	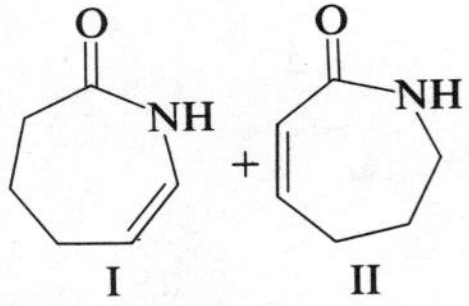 NH I + NH II (—)	247
(*Z*)	Oleum, 140°	I (—)	247
(*E*)	”	I (—)	247
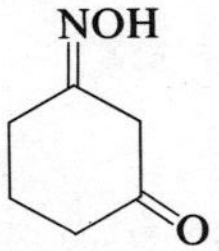 NOH, O	PPA, 120°, 1h	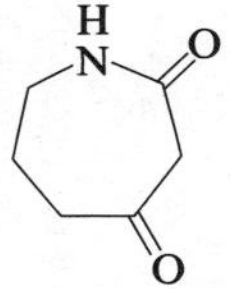 H N, O, O (30)	245
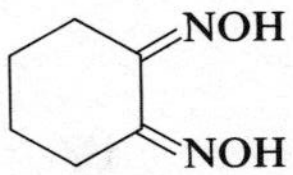 NOH, NOH	Br_2, SO_2, 25°, 1 h	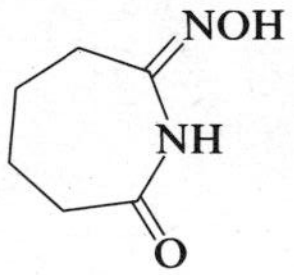 NOH, NH, O (32)	248
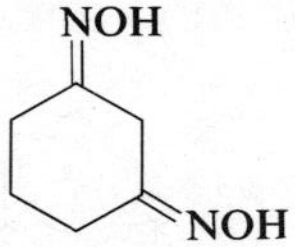 NOH, NOH	1. H_2SO_4, CH_3CO_2H 2. H_2SO_4, oleum, 110°, 10 min	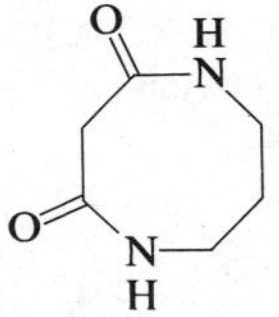 O, H N, O, N H (20)	89
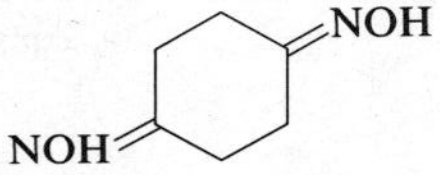 NOH, NOH	H_2SO_4, 18% oleum, 110°, 5 min	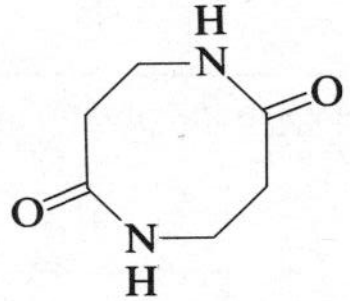 H N, O, O, N H (29)	90

TABLE I. Rearrangements (*Continued*)

No. of Carbon Atoms	Substrate	Reagent and Conditions	Product(s) and Yield(s) (%)	Refs.
	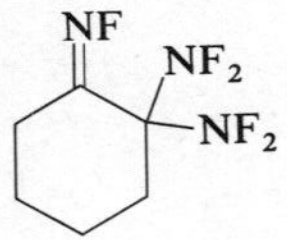 NF, NF_2, NF_2	H_2SO_4	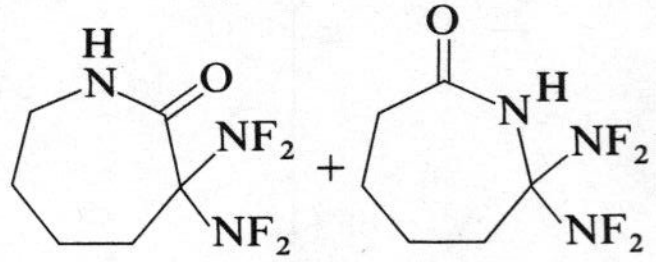 H, N, O, NF_2, NF_2 (major) + O, N, H, NF_2, NF_2 (minor) (25–40)	121
	$ClCH{=}CH(CH_2)_2C(CH_3){=}NOH$	PPSE, CCl_4, reflux	$ClCH{=}CH(CH_2)_2NHCOCH_3$ (—)	217
	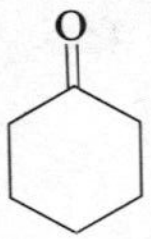 O	$NH_2OH \cdot HCl$, CF_3SO_3H, HCO_2H, reflux, 6 h	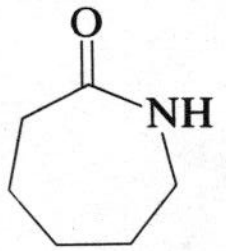 O, NH (60)	249
		H_2NOSO_3H, HCO_2H, reflux, 3 h	" (82)	81
	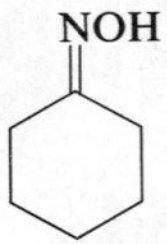 NOH	PPA, xylene, 100°, 4 h	" (90)	60
		$(C_6H_5)_3P$, Cl_2	" (86)	250
		$(C_6H_5)_3P$, Br_2	" (74)	250
		$(C_6H_5)_3P$, I_2	" (39)	250
		PPSE, C_6H_6, 25°, 15 h	" (69)	72
		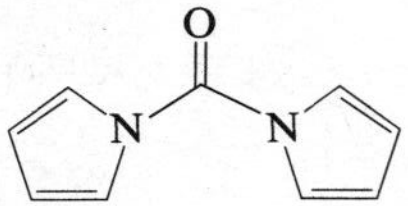 O, N, N	" (55)	146

HCO_2H, reflux, 3 h	" (42)	251
1. $([(CH_3)_2N]_3P)_2O \cdot (BF_4)_2$ CH_3CN, reflux, 20 h 2. H_2O	" (20)	252
IF_5, CH_2Cl_2	" (29)	253
Y-type zeolite	" (—)	239
Cl_2, HBr, SO_2	" (68)	254
Cl_2, NaBr, SO_2	" (28)	254
Cl_2, KBr, SO_2	" (76)	254
Cl_2, NBS, SO_2	" (75)	254
Cl_2, KI, SO_2	" (16)	254
Cl_2, KCl, SO_2	" (13)	254
Cl_2, PCl_5, SO_2	" (22)	254
Cl_2, S_2Cl_2, SO_2	" (32)	254
Cl_2, $SOCl_2$, SO_2	" (12)	254
Cl_2, $AlCl_3$, SO_2	" (20)	254
Cl_2, $FeCl_3$, SO_2	" (12)	254
Cl_2, $SnCl_4$, SO_2	" (7)	254
Cl_2, red P, SO_2	" (45)	254
Cl_2, S, SO_2	" (32)	254
Cl_2, Fe, SO_2	" (18)	254
Cl_2, KNO_3, SO_2	" (12)	254
Cl_2, KNO_2, SO_2	" (15)	254
Cl_2, KCN, SO_2	" (12)	254
SO_3, SO_2, $-10°$	" (80–96)	237
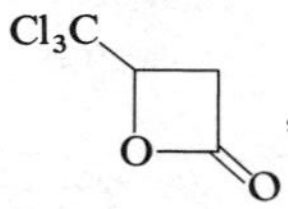 Cl_3C (β-lactone), CCl_4, 25°, 14 h	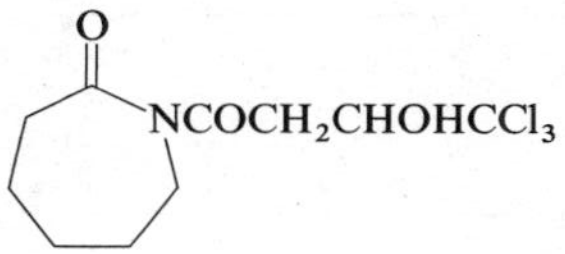 O, $NCOCH_2CHOHCCl_3$ (100)	87

TABLE I. REARRANGEMENTS (*Continued*)

No. of Carbon Atoms	Substrate	Reagent and Conditions	Product(s) and Yield(s) (%)	Refs.
		$SO_3 \cdot$dioxane, 0–65°	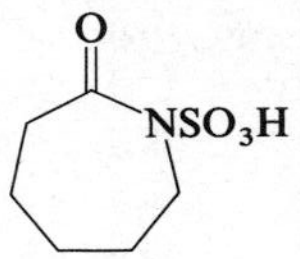 O, NSO_3H (—)	29
		$SO_3 \cdot$dioxane, CH_3CHCl_2	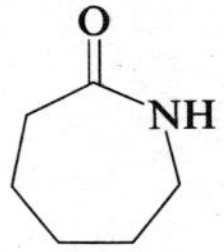 O, NH (80)	255
		$SO_3 \cdot (C_2H_5O)_3P{=}O, CH_3CHCl_2$	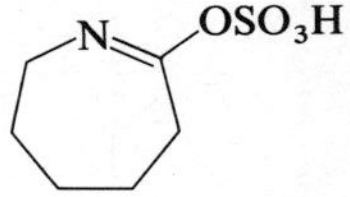 N, OSO_3H (80)	255
		B_2O_3–$Ca_{10}(PO_4)_6(OH)_2$, 300°	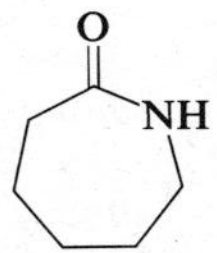 O, NH (67)	256
		$Al(NO_3)_3 \cdot 9H_2O$, $AlCl_3 \cdot 6H_2O$, H_3PO_4, propylene oxide, 320°	" (73)	257
	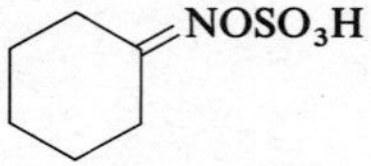 $NOSO_3H$	H_2O, 80 h	" (53)	258

Reactant	Conditions	Product(s) and Yield(s) (%)	Refs.
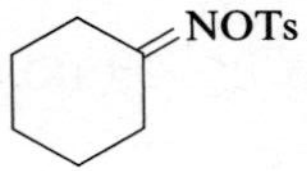 NOTs (cyclohexanone oxime tosylate)	SO_3, $ZnCl_2$, CH_3CHCl_2	" (87)	259
	SO_3, $SnCl_4$, CH_3CHCl_2, $-4°$	" (87)	259
	SO_3, CH_3CHCl_2, $-6°$	" (80)	259
	SO_2	" (—)	260
	CH_3CO_2H, 35°, 1 h	" (100)	140
	CH_3OH, 20°, 15 h	" (48) 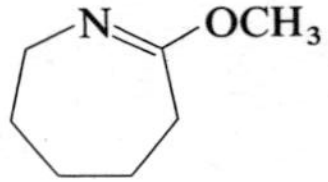 + N, OCH_3 (22)	140
	SiO_2, $CHCl_3$, 5°	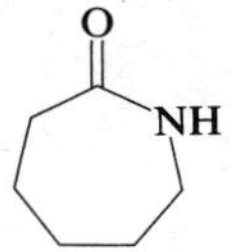 O, NH (70)	109
	$CHCl_3$, 25°, 1 year	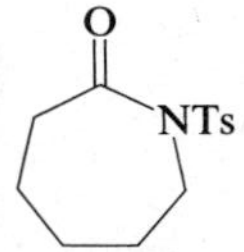 O, NTs (—)	140
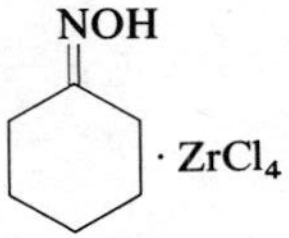 NOH $\cdot ZrCl_4$	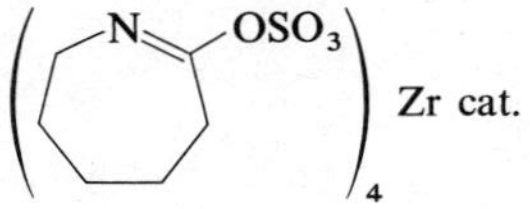 $\left(\text{N}, OSO_3\right)_4$ Zr cat.	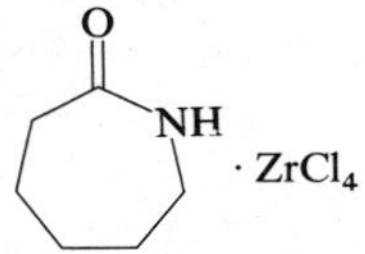 O, NH $\cdot ZrCl_4$ (—)	261
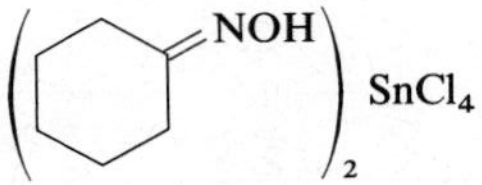 $\left(\text{NOH}\right)_2 SnCl_4$	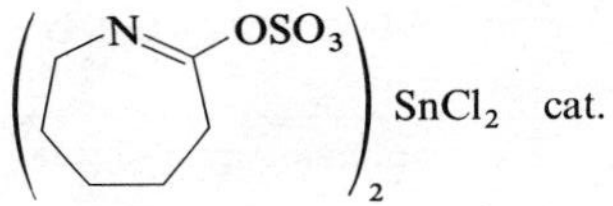 $\left(\text{N}, OSO_3\right)_2$ $SnCl_2$ cat.	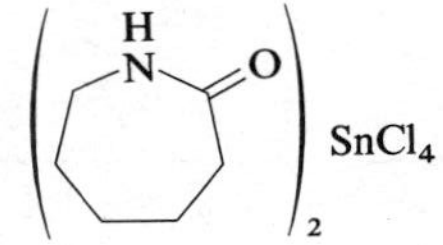 $\left(\text{H, N, O}\right)_2 SnCl_4$ (100)	261

TABLE I. REARRANGEMENTS (*Continued*)

No. of Carbon Atoms	Substrate	Reagent and Conditions	Product(s) and Yield(s) (%)	Refs.
	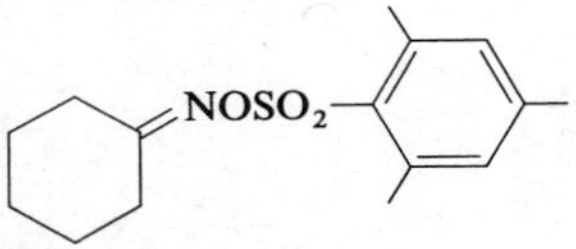 Cyclohexanone oxime O-mesitylenesulfonate ($=NOSO_2$-mesityl)	Alumina, CH_3OH	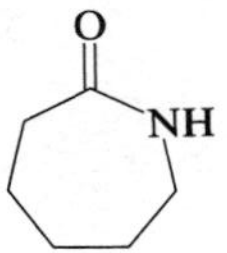 Caprolactam (NH, C=O) (77)	47
	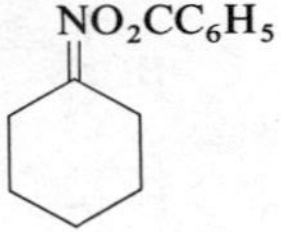 Cyclohexanone oxime benzoate ($=NO_2CC_6H_5$)	HF, 25°, 24 h	" (72)	83
	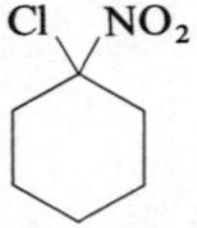 1-Chloro-1-nitrocyclohexane (Cl, NO_2)	$(C_6H_5)_3P$, C_6H_6	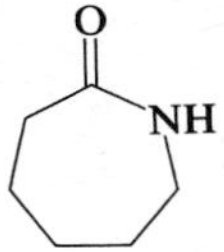 Caprolactam (NH, C=O) (77)	250
	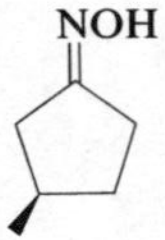 3-Methylcyclopentanone oxime (NOH)	PPA, 150°, 15 min	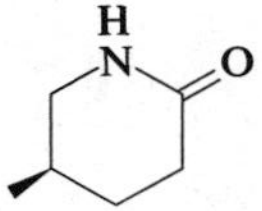 5-Methyl-2-piperidone (H-N, C=O) (70)	107
	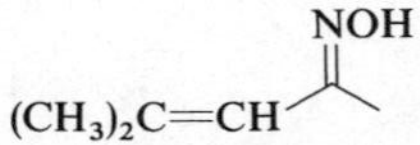 $(CH_3)_2C{=}CH$–C(=NOH)CH_3	PPA	$(CH_3)_2C{=}CHCONHCH_3$ (8) + $(CH_3)_2CHCH_2NHCOCH_3$ (16)	62, 262
	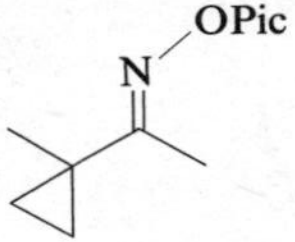 1-Methylcyclopropyl methyl ketone O-picryl oxime (N–OPic)	1,4-Dichlorobutane, 80°	 CH_3CON(Pic)(1-methylcyclopropyl) (—)	93

	Reactant	Conditions	Product(s) (% yield)	Refs.
	1-cyclobutylethanone O-Pic oxime (N–OPic)	"	$CH_3CON(Pic)$-cyclobutyl (—)	93
	1-methyl-4-(hydroxyimino)phosphorinane 1-oxide (NOH; P(=O)CH_3)	PPA, 100°, 20 min	1-methyl-1,4-azaphosphepan-5-one 1-oxide (NH, C=O; P(=O)CH_3) (18)	263
	n-$C_5H_{11}CH{=}NOH$	BF_3, CH_3CO_2H	n-$C_5H_{11}CONH_2$ (71)	129
		$BF_3\cdot$ether, reflux, 1h	" (80)	129
	t-$C_4H_9C(CH_3){=}NOH$	PPA, 120°	t-$C_4H_9NHCOCH_3$ (96)	66
		PPA, 130°, 10 min	" (—)	64
		PCl_5, C_6H_6	" (86)	66
		TsCl, pyridine	" (46)	66
	t-$C_4H_9C(CH_3){=}N$–OR, R = Ts	$(C_2H_5)_3N$, 80% ethanol, reflux	t-$C_4H_9NHCOCH_3$ (80)	28
		SiO_2, $CHCl_3$, 5°	" (20)	109
	R = Pic	1,4-Dichlorobutane, 80°	$CH_3CON(Pic)C_4H_9$-t (—)	93, 94
	i-$C_4H_9C(CH_3){=}N$–OPic	"	$CH_3CON(Pic)C_4H_9$-i (—)	93
C_7	p-$ClC_6H_4CH{=}NOH$	SiO_2, xylene, reflux, 59 h	p-$ClC_6H_4CONH_2$ (91)	128

TABLE I. REARRANGEMENTS (*Continued*)

No. of Carbon Atoms	Substrate	Reagent and Conditions	Product(s) and Yield(s) (%)	Refs.
	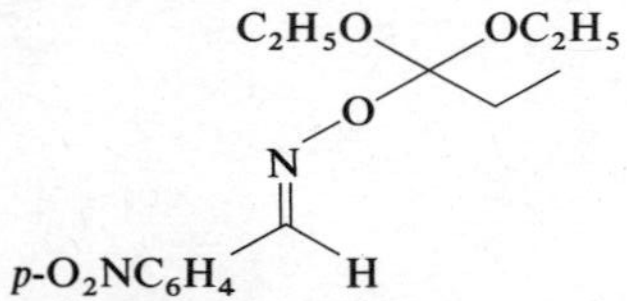 C_2H_5O, OC_2H_5 (diethoxy ketal of p-$O_2NC_6H_4CH{=}N{-}O{-}$, H)	BF_3, HgO, ether, reflux, 3 h	p-$O_2NC_6H_4NC$ (70)	137
	$C_6H_5CH{=}NOH$	Cu, heat, 10–45 min	$C_6H_5CONH_2$ (86)	130
		SiO_2, xylene, reflux, 69 h	" (92)	128
		Alumina, xylene, reflux, 80 h	" (40)	128
	(*E*)	$Ni(Acac)_2$ (cat.), 80°, 4.5 h	" (45)	132, 133
	(*Z*)	" 80°, 30 min	" (70)	132, 133
	(*E*)	$CH_3O_2C\bar{N}SO_2\overset{+}{N}(C_2H_5)_3$, 90°, 30 min	C_6H_5NHCHO (75)	136
	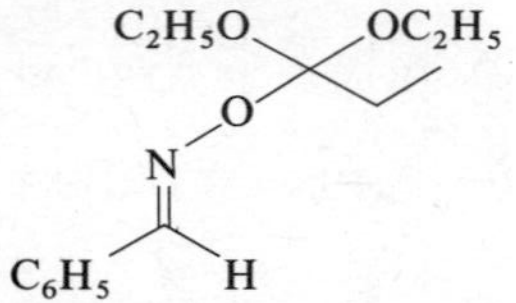 C_2H_5O, OC_2H_5 (diethoxy ketal of $C_6H_5CH{=}N{-}O{-}$, H)	BF_3, HgO, ether, reflux, 3 h	C_6H_5NC (50)	137
	$C_6H_5CO_2H$	Benzophenone, $C_2H_5O_2CN{=}NCO_2C_2H_5$, THF, 25°, 18 h	$C_6H_5CON(C_6H_5)COC_6H_5$ (88)	92
	p-$HOC_6H_4CH{=}NOH$	SiO_2, xylene, reflux, 61 h	p-$HOC_6H_4CONH_2$ (84)	128
	o-$HOC_6H_4CH{=}NOH$	Cu, anisole	o-$HOC_6H_4CONH_2$ (25)	130
		SiO_2, xylene, reflux, 73 h	" (83)	128
	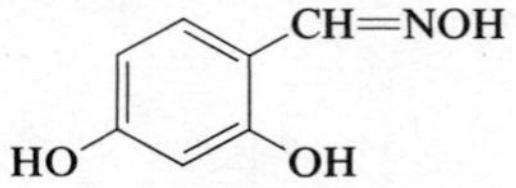 2,4-$(HO)_2C_6H_3CH{=}NOH$ (CH=NOH, OH, HO on ring)	SiO_2, xylene, reflux, 68 h	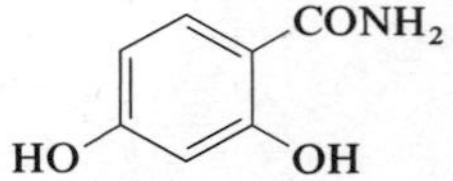 2,4-$(HO)_2C_6H_3CONH_2$ (CONH₂, OH, HO on ring) (61)	128

Reactant	Conditions	Product(s) and Yield(s) (%)	Refs.
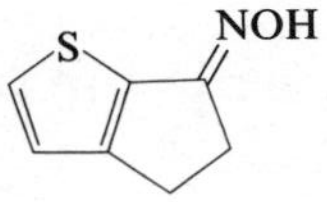 S, NOH	PPA, 135°, 15 min	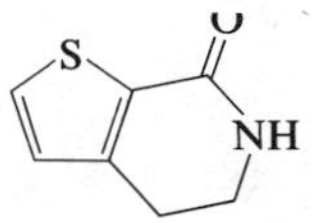 S, O, NH (20)	264
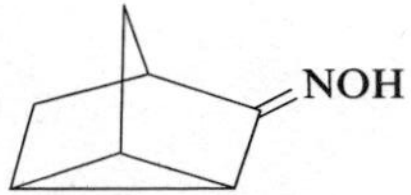 NOH	$C_6H_5SO_2Cl$, NaOH	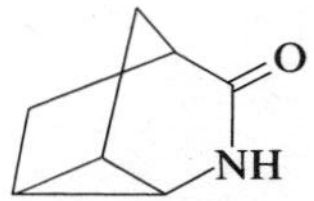 O, NH (38)	265
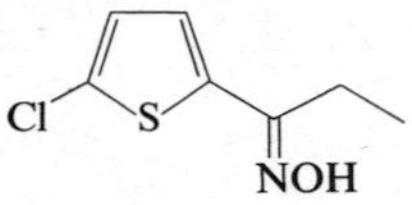 Cl, S, NOH	PPA, 135°, 15 min	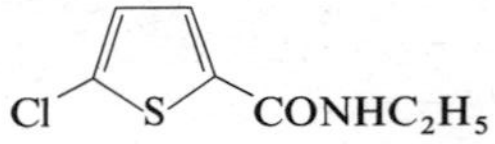 Cl, S, $CONHC_2H_5$ (50)	264
	PCl_5, ether, 25°, 30 min	" (20)	264
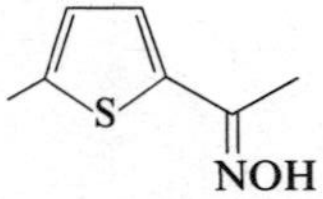 S, NOH	PCl_5, ether, 0°, 6 h	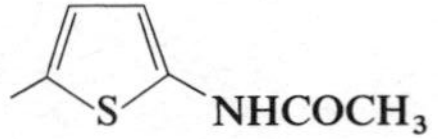 S, $NHCOCH_3$ (87)	243
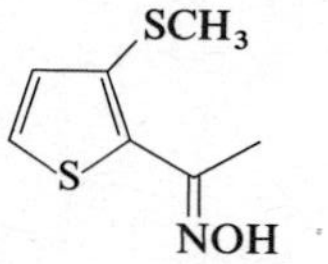 SCH_3, S, NOH	PCl_5, ether, 25°, 1 h	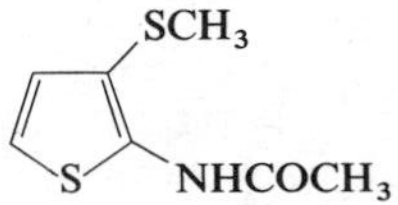 SCH_3, S, $NHCOCH_3$ (84)	266
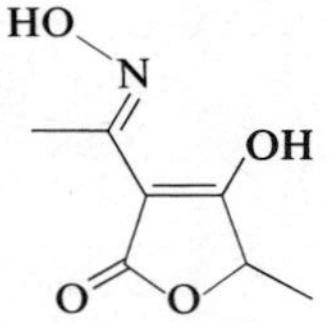 HO, N, OH, O, O	$C_6H_5SO_2Cl$, NaOH, 50°	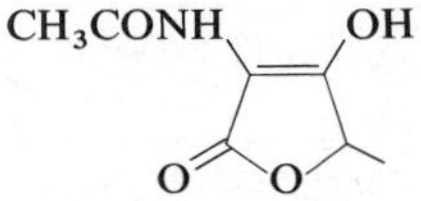 CH_3CONH, OH, O, O (50)	244
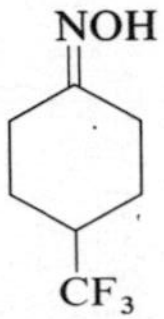 NOH, CF_3	H_2SO_4, 120°, 30 min	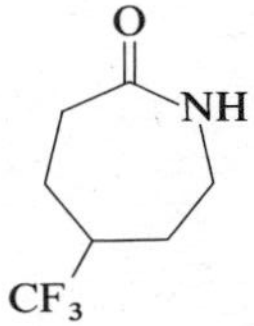 O, NH, CF_3 (45)	267

TABLE I. REARRANGEMENTS (*Continued*)

No. of Carbon Atoms	Substrate	Reagent and Conditions	Product(s) and Yield(s) (%)	Refs.
	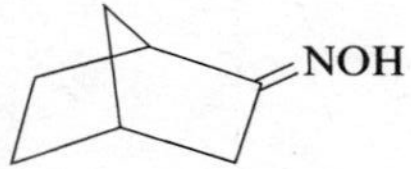 NOH	$C_6H_5SO_2Cl$, NaOH	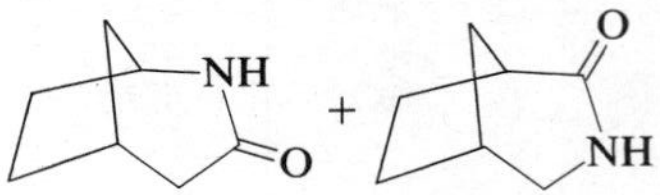 NH, O (I) + O, NH (II) (38)	265
		"	"Lactams" (70) (presumably I and II)	268
		H_2SO_4, 110°	I + II (17)	269
	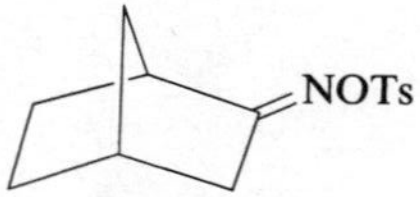 NOTs	Ethanol, reflux	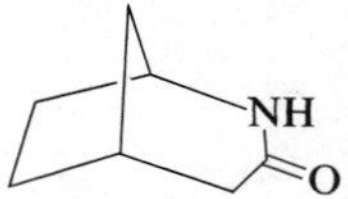 NH, O (—)	270
	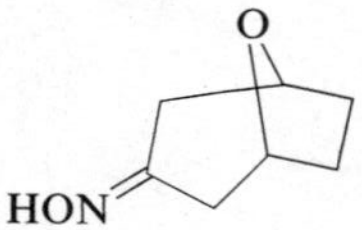 O, HON	P_2O_5, CH_3SO_3H, 80°, 2 h	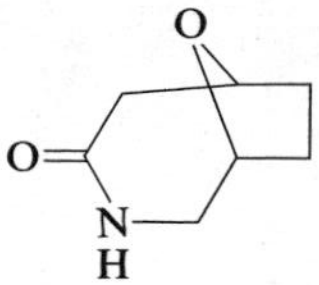 O, O=, N H (60)	271
	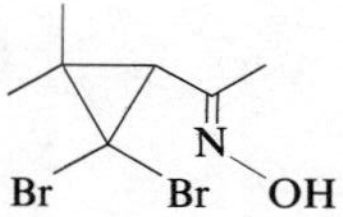 Br, Br, N, OH	PCl_5	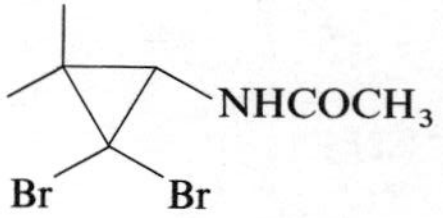 Br, Br, $NHCOCH_3$ (—)	272
	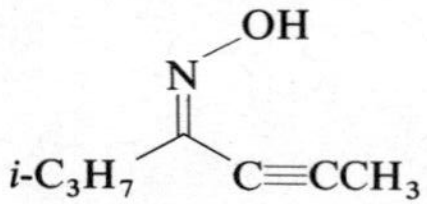 OH, N, i-C_3H_7, $C{\equiv}CCH_3$	PCl_5, ether	$CH_3CCl{=}CHCONHC_3H_7$-i (—)	116

Reactant	Conditions	Product(s) and Yield(s) (%)	Refs.
N–OTs (3-methylcyclohex-2-enone oxime tosylate)	CH_3OH, reflux	HN, O ·TsOH (80)	31
N–OTs, OCH_3	CH_3OH, alumina, 25°	H N, O, OCH_3 (87)	245
NOH, OCH_3 (*E*/*Z*)	PPA, 120°, 1 h	H N, O, O (66)	245
(*Z*)	PPA, 120°, 1 h	" (77)	245
NOH, NOH	H_2SO_4, oleum, 110°, 10 min	HN, O, NH, O (18)	89
N, =NOH	PPA, 130°, 10 min	N, NH, O (85)	273, 274
	5% Oleum, heat, 2 min	" (32)	273, 274

TABLE I. REARRANGEMENTS (*Continued*)

No. of Carbon Atoms	Substrate	Reagent and Conditions	Product(s) and Yield(s) (%)	Refs.
	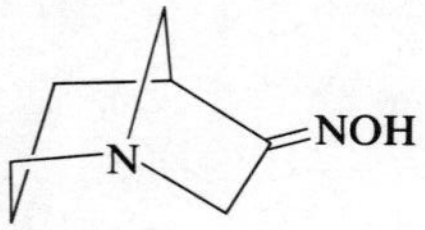 [structure] =NOH	PPA, 130°, 10 min	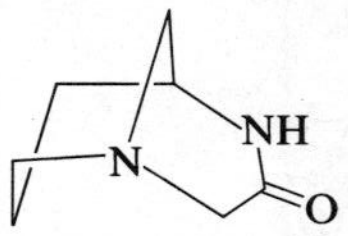 [structure] NH, O (78) 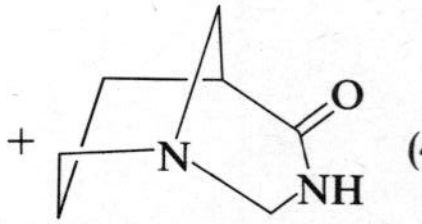 + [structure] O, NH (4)	273
	$CH_3CCl{=}CH(CH_2)_2C(CH_3){=}NOH$	PPSE, CCl_4, reflux	$CH_3CCl{=}CH(CH_2)_2NHCOCH_3$ (—)	217
	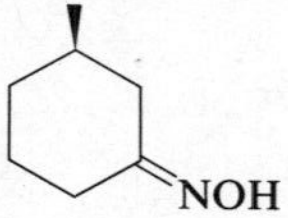 [structure] NOH	$C_6H_5SO_2Cl$, NaOH, H_2O, acetone	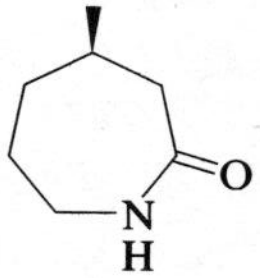 [structure] O, N H (80)	275
	(*E*)	"	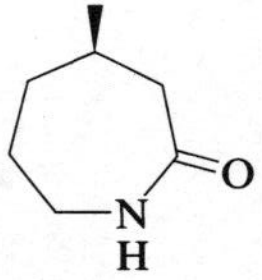 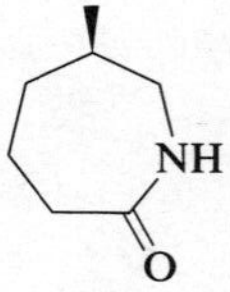 [structure] I + [structure] II (87) (I/II = 1/2.44)	275
	(*Z*)	"	I (80)	275

Reactant	Conditions	Product(s) and Yield(s) (%)	Refs.
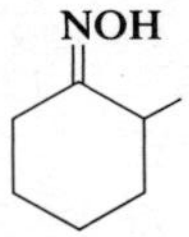 NOH (Z)	85% H_2SO_4	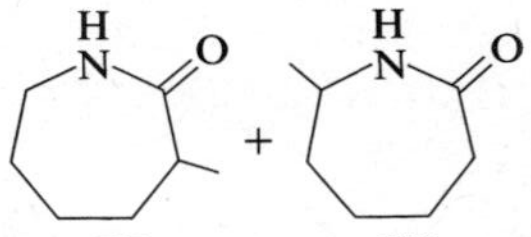 III + IV (III/IV = 9/1) (24)	276
(E)	"	IV (69)	276
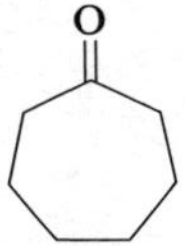 O	H_2NOSO_3H, HCO_2H, reflux, 6 h	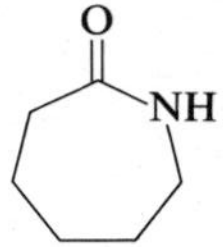 O, NH (75)	81
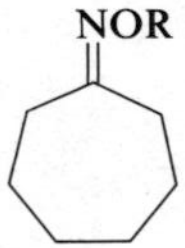 NOR, R = H	(—)	" (88)	238
R=Ts	CH_3CO_2H, 25°, 2 h	" (—)	140
R = H	PCl_5, xylene, 80°, 2 h	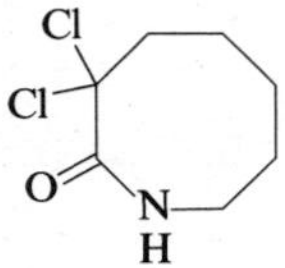 Cl, Cl, O, N, H (55)	277
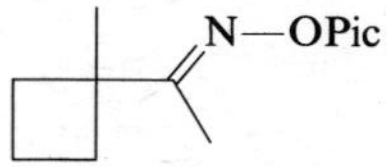 N—OPic	1,4-Dichlorobutane, 80°	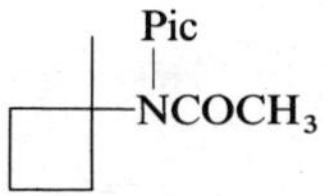 Pic, $NCOCH_3$ (—)	93
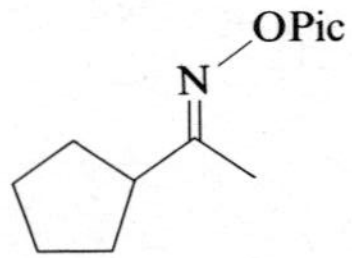 OPic, N	"	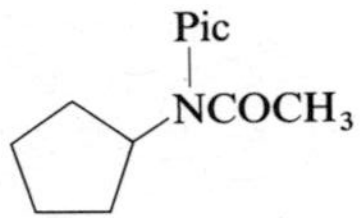 Pic, $NCOCH_3$ (—)	93

TABLE I. REARRANGEMENTS (*Continued*)

No. of Carbon Atoms	Substrate	Reagent and Conditions	Product(s) and Yield(s) (%)	Refs.
	n-$C_3H_7CH{=}CHC({=}NOH)CH_3$	H_2SO_4, 135°	n-$C_3H_7CH{=}CHNHCOCH_3$ (—)	278
	$CH_3C({=}NOH)CH(NH_2)CH_2CO_2C_2H_5$	1. H_2SO_4, 110°, 2 h 2. H_2O, reflux, 6 h	$HO_2CCH(NH_2)CH_2CO_2H$ (45)	279
	n-$C_6H_{13}CH{=}NOH$	BF_3, CH_3CO_2H	n-$C_6H_{13}CONH_2$ (80)	129
		BF_3·ether, reflux, 1 h	" (85)	129
		P_2O_5, CH_3SO_3H, 100°, 1 h	" (90)	73
	n-$C_4H_9C(C_2H_5){=}NOH$	$(C_6H_5)_3P$, CCl_4, reflux, 1.5 h	n-$C_4H_9NHCOC_2H_5$ (80)	84
	n-$C_5H_{11}C(CH_3){=}NOPic$	1,4-Dichlorobutane, 80°	$CH_3CON(Pic)C_5H_{11}$-n (—)	93
	t-$C_4H_9CH_2C(CH_3){=}NOPic$	"	$CH_3CON(Pic)CH_2C_4H_9$-t (—)	93
	$(C_2H_5)_2CHC(CH_3){=}NOPic$	"	$CH_3CON(Pic)CH(C_2H_5)_2$ (—)	93
	$C_2H_5C(CH_3)_2C(CH_3){=}NOPic$	"	$CH_3CON(Pic)C(CH_3)_2C_2H_5$ (—)	93

	Reactant	Conditions	Product(s) (Yield, %)	Refs.
C_8	4-Cl-C_6H_4-C(CN)=N-OTs	KOH, ethanol, 60°	p-$ClC_6H_4NHCO_2C_2H_5$ (64)	117
		$NaOC_2H_5$, ethanol, reflux	" (63)	117
	4-O_2N-C_6H_4-C(CN)=N-OTs	KOH, ethanol	p-$O_2NC_6H_4NHCO_2C_2H_5$ (25)	117
	C_6H_5-C(CN)=N-OTs	KOH, ethanol, 60°	$C_6H_5NHCO_2H_5$ (62)	117
		$NaOC_2H_5$, ethanol, reflux	" (60)	117
	3,4-methylenedioxyphenyl-CH=NOH	SiO_2, xylene, reflux, 64 h	3,4-methylenedioxyphenyl-$CONH_2$ (93)	128
	2-Br-5-NO_2-C_6H_3-C(CH_3)=N-OH	PPA, 100°, 3 min	2-Br-5-NO_2-C_6H_3-$NHCOCH_3$ (76)	280

TABLE I. REARRANGEMENTS (*Continued*)

No. of Carbon Atoms	Substrate	Reagent and Conditions	Product(s) and Yield(s) (%)	Refs.
	2-Br-5-$NO_2C_6H_3C(CH_3)$=NOH (structure)	PPA, 100°, 1 h	2-Br-5-$NO_2C_6H_3CONHCH_3$ (structure) (—)	280
	m-$FC_6H_4C(CH_3)$=NOH	85% H_2SO_4, 80°	*m*-$FC_6H_4NHCOCH_3$ (—)	281
	p-$ClC_6H_4C(CH_3)$=NOH	$SOCl_2$, CCl_4, 44°, 9h	*p*-$ClC_6H_4NHCOCH_3$ (70)	75
		PPSE, CH_2Cl_2, 3.5 h	" (92)	72
		HMPA, 225-235°, 10 min	" (75)	282
	3-Cl-2-$HOC_6H_3C(CH_3)$=NOH (structure)	$POCl_3$	3-Cl-2-$HOC_6H_3NHCOCH_3$ (structure) (—)	283
	5-Cl-2-$HOC_6H_3C(CH_3)$=NOH (structure)	$POCl_3$	5-Cl-2-$HOC_6H_3NHCOCH_3$ (structure) (–)	284
	p-$BrC_6H_4C(CH_3)$=NOH (structure)	$SOCl_2$, CCl_4	*p*-$BrC_6H_4NHCOCH_3$ (72)	75

Reactant	Conditions	Product(s) and Yield(s) (%)	Refs.
	HMPA, 225–235°, 10 min	" (89)	282
	$SOCl_2$, CCl_4, 44°, 9 h	" (76)	75
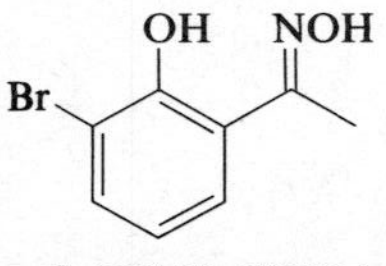 Br OH NOH	$POCl_3$	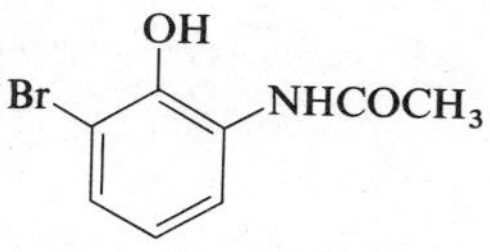 Br OH $NHCOCH_3$ (50)	285
p-$O_2NC_6H_4C(CH_3){=}NOH$	$SOCl_2$, CCl_4	p-$O_2NC_6H_4NHCOCH_3$ (76)	75
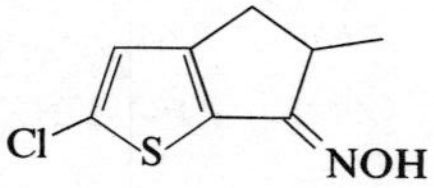 Cl S NOH	PPA, 135°, 15 min	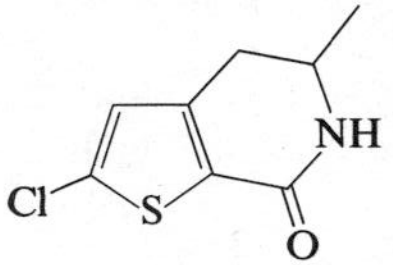 Cl S NH O (52)	264
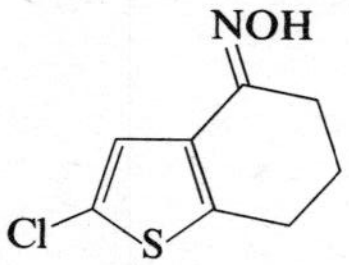 Cl S NOH	PPA	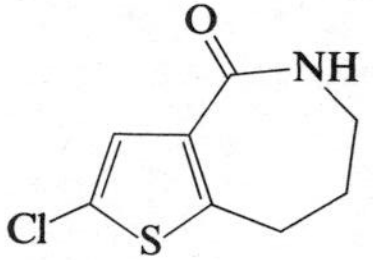 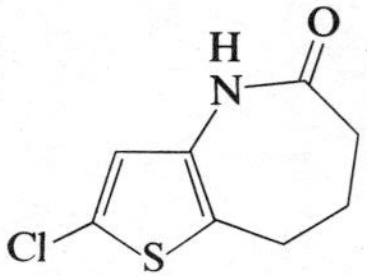 Cl S O NH + Cl S H N O (—)	286
$C_6H_5COCH_3$	$NH_2OH{\cdot}HCl$, CF_3SO_3H, HCO_2H, reflux, 2.5 h	$C_6H_5NHCOCH_3$ (80)	249
$C_6H_5C(CH_3){=}NOH$	PPA, xylene, 100°, 3 h	" (95)	60
	PPE, $CHCl_3$, reflux	" (78)	287
	PPSE, C_6H_6, 25°, 6 h	" (96)	72
	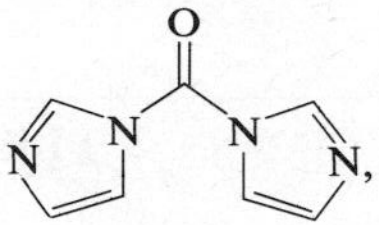 N N O N N, octane, reflux	" (100)	288

TABLE I. REARRANGEMENTS (*Continued*)

No. of Carbon Atoms	Substrate	Reagent and Conditions	Product(s) and Yield(s) (%)	Refs.
		HCO_2H, reflux, 6 h	$C_6H_5NHCOCH_3$ (90)	251
		$(CH_3)_2N^+{=}CCl_2Cl^-$, $CHCl_3$	" (86)	289
		$([(CH_3)_2N]_2P)_2O\cdot(BF_4)_2$, CH_3CN, reflux, 4.5 h	" (75)	252
		$SOCl_2$, CCl_4	" (70)	75
		Cl_3CCN, reflux, 5–6 h	" (60–70)	100
		Cl_3CCN, ether, reflux	" (—)	232
		$(CH_3)_3SiI$, $CHCl_3$, 56°, 4 h	" (55)	85
		HMPA, 225–235°, 10 min	" (52)	282
		IF_5, CH_2Cl_2, 3–15°, 2 h	" (22)	253
		$SO_3\cdot(C_2H_5O)_3PO$, CH_3CHCl_2	" (—)	255
		86.7% H_2SO_4, 80°	"	281
		99% H_2SO_4, 60°	p-$HO_3SC_6H_4NHCOCH_3$ (—)	290
		HCl, pyridine	$C_6H_5NHCOCH_3$ + $C_6H_5CONHCH_3$ (52)	291
	$C_6H_5C(CH_3){=}NOTs$	$(C_2H_5)_3N$, 80% ethanol, reflux	" (99)	28
		SiO_2, $CHCl_3$, 5°, 30 min	" (70)	109
	$C_6H_5C(CH_3){=}NOSO_3X$ X = H	H_2O, 80 h	" (10)	258
	X = NH_4	H_2O, reflux	" (80)	258
	$C_6H_5C(CH_3){=}N{-}OPic$	80% ethanol, 80°, 6 h	" (100)	95
	p-$CH_3C_6H_4CH{=}N{-}O{-}C(OC_2H_5)_2CH_2CH_3$	BF_3, HgO, ether, reflux, 3 h	p-$CH_3C_6H_4NC$ (73)	137

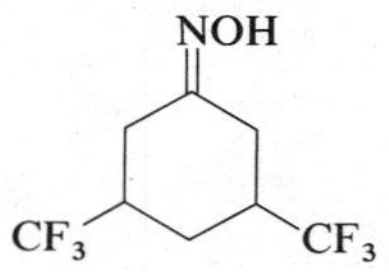	H_2SO_4, 120°, 30 min	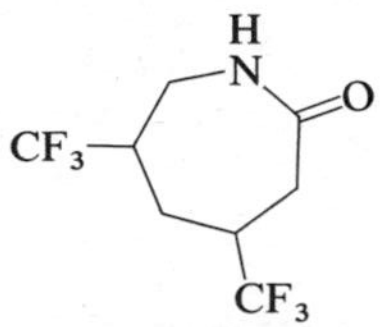 (38)	267
p-$CH_3OC_6H_4CH{=}NOH$	SiO_2, xylene, reflux, 52 h	p-$CH_3OC_6H_4CONH_2$ (81)	128
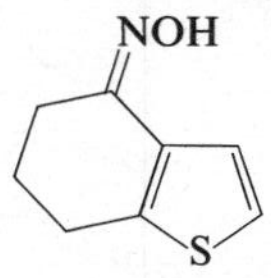	PPA, 135°, 15 min	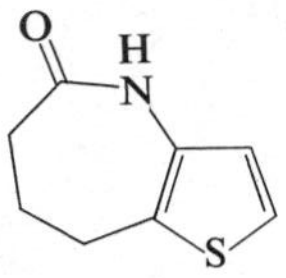 (63)	264
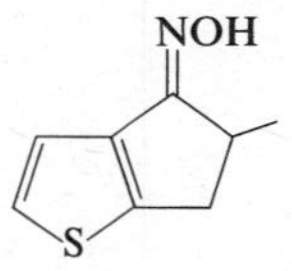	”	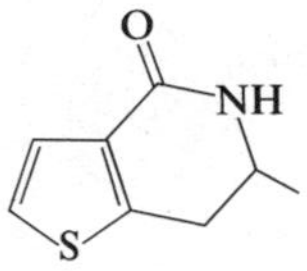 (60)	264
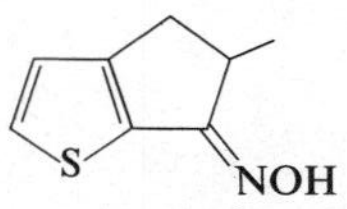	”	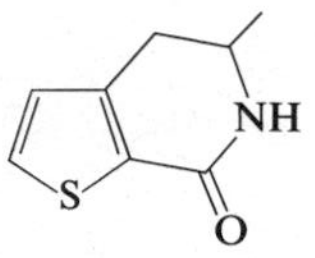 (47)	264
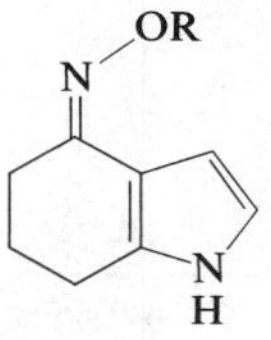 R = H	CH_3CO_2H	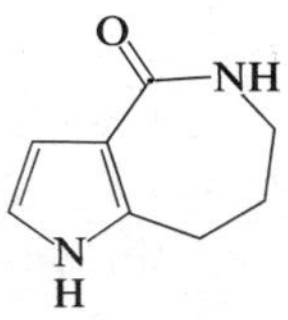 (90)	292
R = Ts	”	” (90)	292

TABLE I. REARRANGEMENTS (*Continued*)

No. of Carbon Atoms	Substrate	Reagent and Conditions	Product(s) and Yield(s) (%)	Refs.
	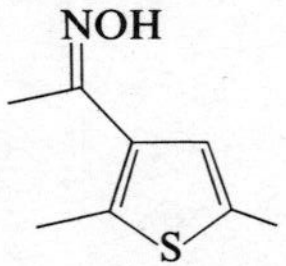 NOH, S	PCl_5, ether, 0°, 4 h	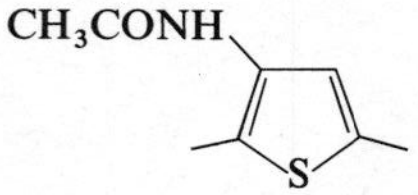 CH_3CONH, S (63)	243
	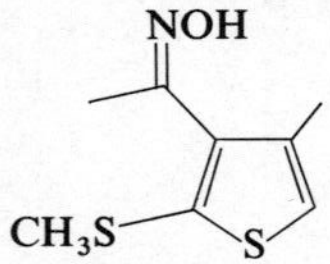 NOH, CH_3S, S	PCl_5, ether, 1 h	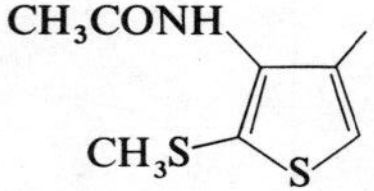 CH_3CONH, CH_3S, S (83)	266
	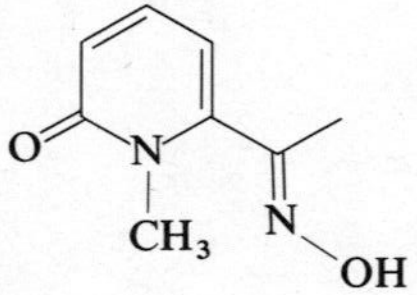 O, N, CH_3, N, OH	PCl_5, ether, 0°, 5 h	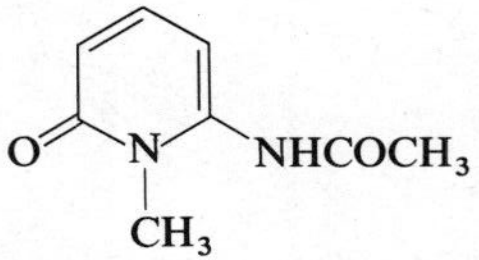 O, N, CH_3, $NHCOCH_3$ (44)	293
	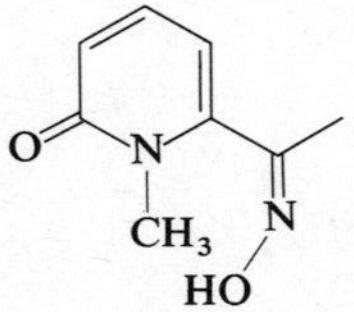 O, N, CH_3, N, HO	"	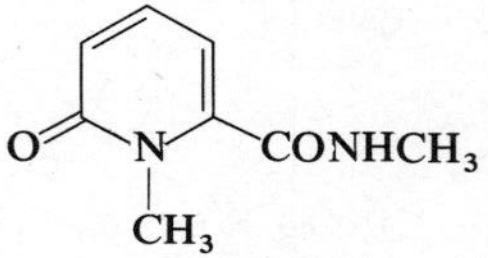 O, N, CH_3, $CONHCH_3$ (40)	293
	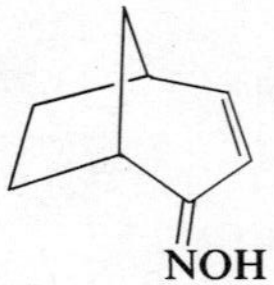 NOH	1. TsCl, reflux, 24 h 2. HCl, CH_3CO_2H, 100°, 1.5 min	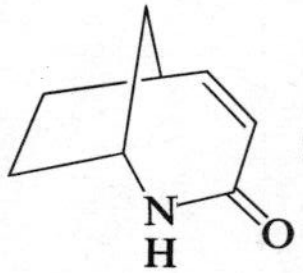 N, H, O (93)	294

Reactant	Conditions	Product(s) and Yield(s) (%)	Refs.
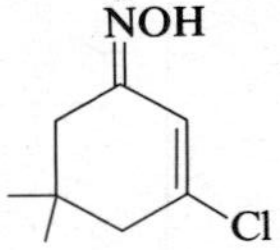 NOH, Cl	PPA, 120°, 1 h	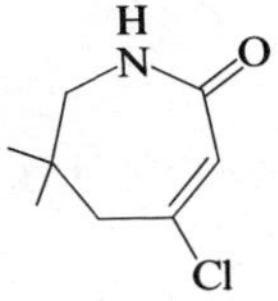 H, N, O, Cl (34)	245
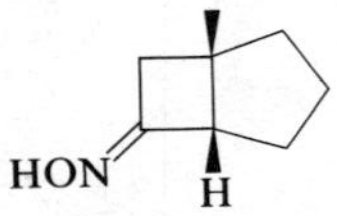 HON, H	PPA, 47°, 15 h	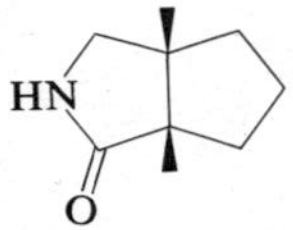 HN, O (50)	126
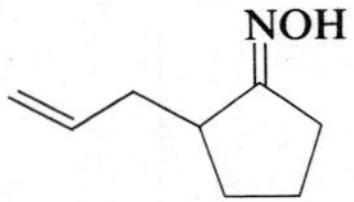 NOH	TsCl, pyridine	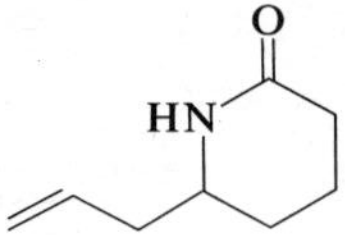 HN, O (70)	295
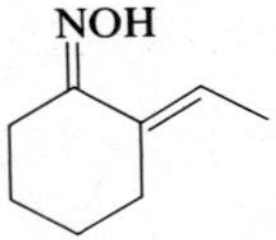 NOH	"	$H_2NCO(CH_2)_4COC_2H_5$ 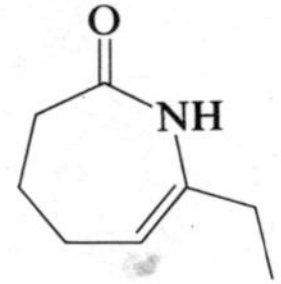 + NH, O (—)	31
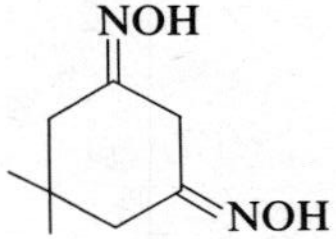 NOH, NOH	H_2SO_4, oleum, 110°, 10 min	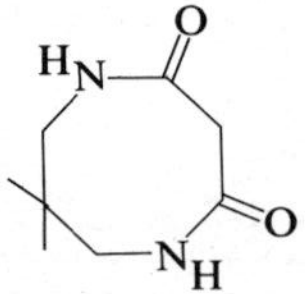 HN, O, NH, O (15)	89
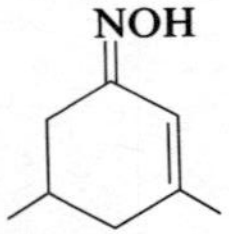 NOH	PPA, 100°, 45 min	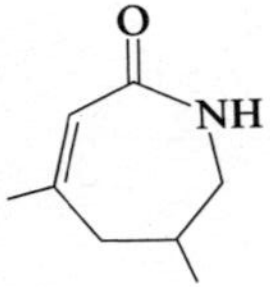 NH, O (38) 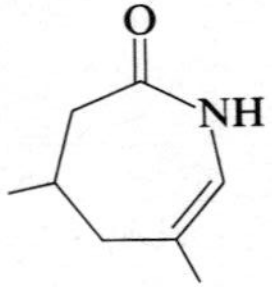 + NH, O (18)	46

TABLE I. REARRANGEMENTS (*Continued*)

No. of Carbon Atoms	Substrate	Reagent and Conditions	Product(s) and Yield(s) (%)	Refs.
	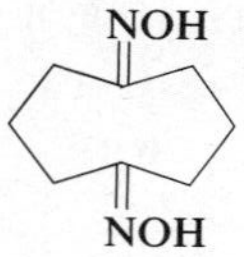 [cyclooctane-1,5-dione dioxime: NOH, NOH]	1. TsCl, pyridine 2. NaO_2CCH_3, dioxane, H_2O, 70°, 2 h	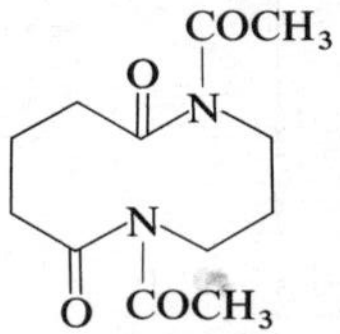 [N,N′-diacetyl diazacyclodecanedione: $COCH_3$, O, N; N, O, $COCH_3$] (88)	91
		1. TsCl, pyridine 2. $KHCO_3$, dioxane, H_2O, 70°, 2 h	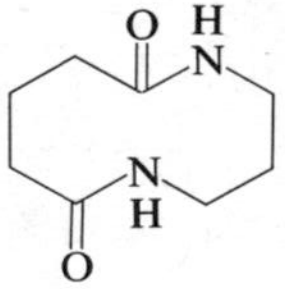 [diazacyclodecanedione: O, H, N; N, H, O] (47)	91
	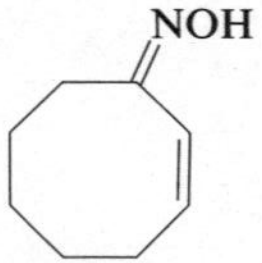 [cyclooct-2-enone oxime: NOH]	TsCl, pyridine	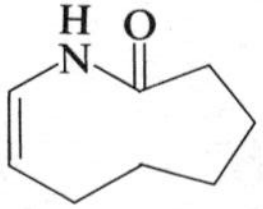 [unsaturated lactam: H, N, O] (—)	31
	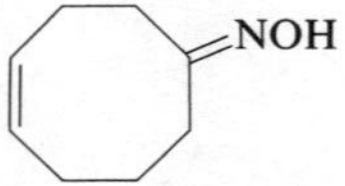 [cyclooct-4-enone oxime: NOH]	1. TsCl 2. K_2CO_3,THF, H_2O	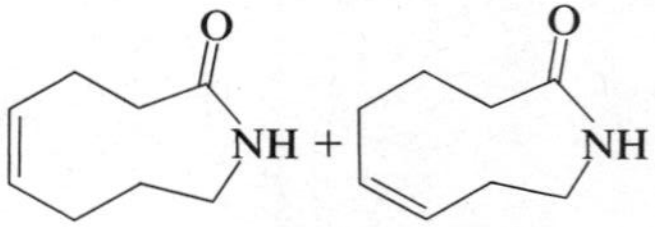 [unsaturated lactam: O, NH] + [unsaturated lactam: O, NH] (68)	296
	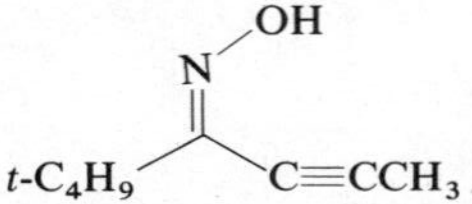 $t\text{-}C_4H_9C(=NOH)C{\equiv}CCH_3$	PCl_5, ether	$CH_3CCl{=}CHCONHC_4H_9\text{-}t$ (—)	116

Reactant	Reagents and Conditions	Product(s) (Yield, %)	Refs.
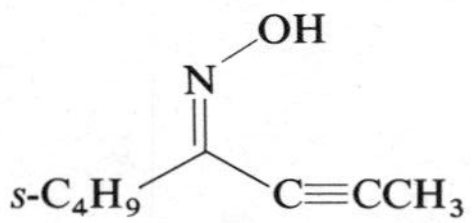 s-$C_4H_9C(=NOH)C{\equiv}CCH_3$	"	$CH_3CCl{=}CHCONHC_4H_9$-s (—)	116
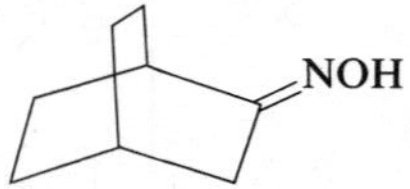	$C_6H_5SO_2Cl$, NaOH	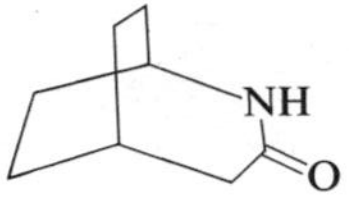 NH, O (84)	297
NOH	"	" (25)	265
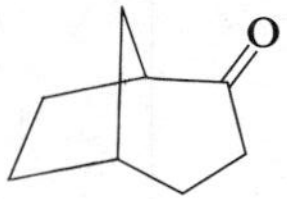 O	H_2NOSO_3H, HCO_2H, reflux, 3 h	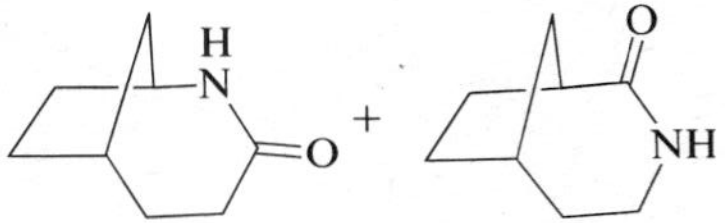 H N, O + O, NH; I, II; I/II = 95/5 (97)	38
	$H_2NOH \cdot HCl$, H_2SO_4, 116°, 30 min	I/II = 31/69 (53)	38
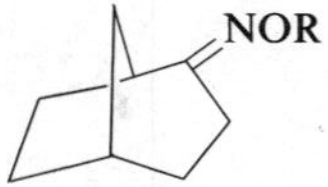 NOR		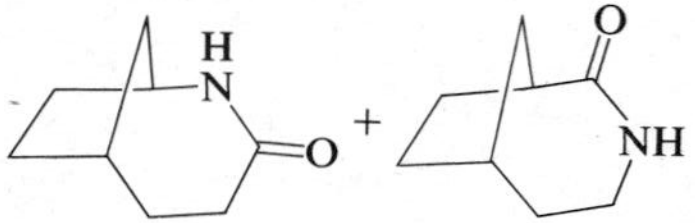 H N, O + O, NH; I, II	
R = H	$C_6H_5SO_2Cl$, NaOH	I (33)	265
R = H	$C_6H_5SO_2Cl$, NaOH, 30°, 90 min	I/II = 57/43 (48)	38
R = H	BF_3, $Cl_2CHCHCl_2$, 110°, 12 h	I/II = 73/27 (44)	38
R = H	PPSE, 25°, 21 h	I/II = 54/46 (46)	38
R = H	H_2SO_4, HCO_2H, reflux 3 h	I/II = 86/13 (39)	38
R = H	H_2SO_4, 116°, 30 min	I/II = 50/50 (77)	38
R = Ts	HCl, CH_3CO_2H, 95°, 20 min	I/II = 59/41 (60)	38

TABLE I. REARRANGEMENTS (*Continued*)

No. of Carbon Atoms	Substrate	Reagent and Conditions	Product(s) and Yield(s) (%) (52)	Refs.
	 NOH	TsCl, pyridine	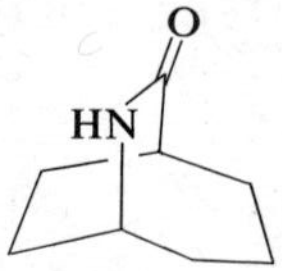 HN, O (64)	298
	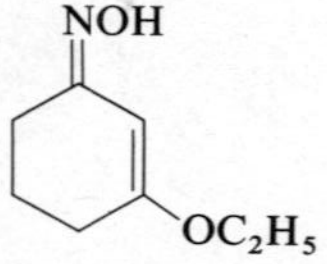 NOH, OC_2H_5	PPA, 120°, 1 h	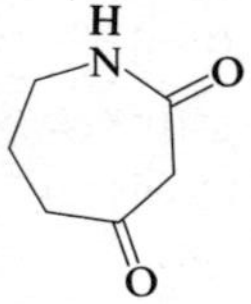 H N, O, O (88)	245
	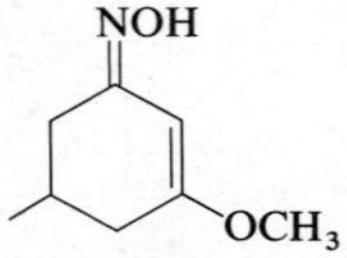 NOH, OCH_3	"	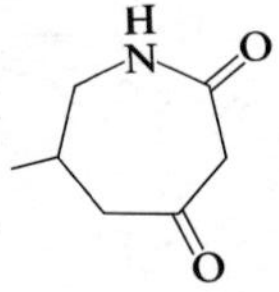 H N, O, O (78)	245
	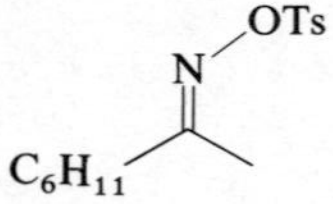 OTs, N, C_6H_{11}	$(C_2H_5)_3N$, 80% ethanol, reflux	$C_6H_{11}NHCOCH_3$ (97)	28
	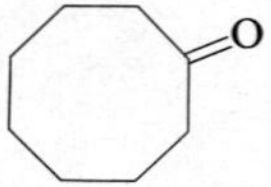 O	H_2NOSO_3H, HCO_2H, reflux, 5 h	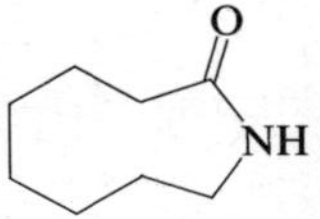 O, NH (61)	81

Reactant	Conditions	Product(s) and Yield(s) (%)	Refs.
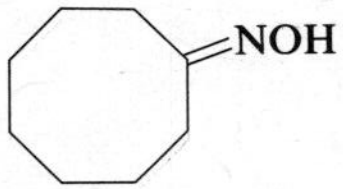 NOH	H_2SO_4	" (89)	299
	PPSE, C_6H_6, 15 h	"	72
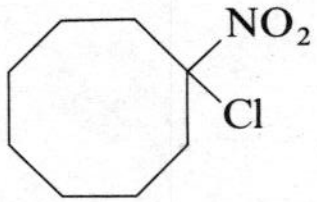 NO_2 Cl	$(C_6H_5)_3P$, C_6H_6	" (42)	250
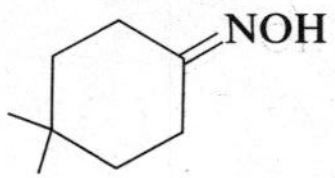 NOH	H_2SO_4	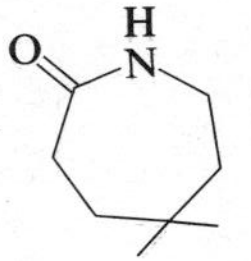 O H N (80)	300
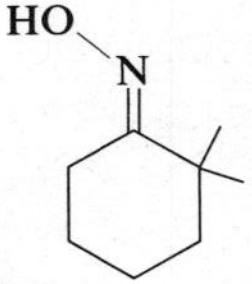 HO N	$SOCl_2$, C_6H_6, 24 h	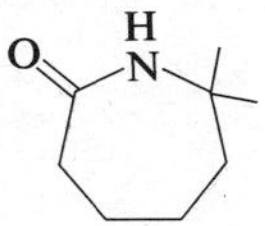 O H N (—)	226
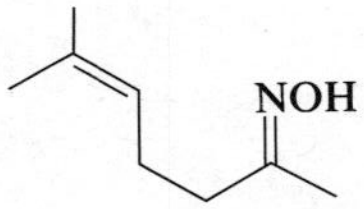 NOH	$(C_6H_5)_3P–CCl_4$, CCl_4, reflux, 4 h	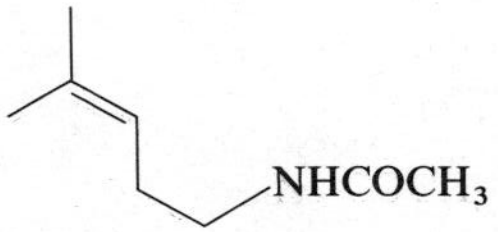 $NHCOCH_3$ (50)	84
	$(C_6H_5)_3P–CCl_4$, THF, reflux, 2–8 h	" (60–70)	84
	$(C_6H_5)_3P–CCl_4$, THF, 94 h	" (70)	84
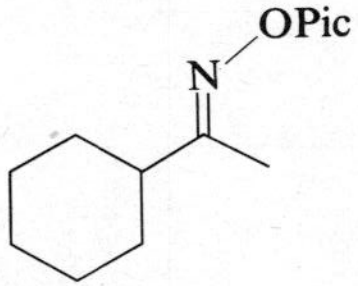 OPic N	1,4-Dichlorobutane, 80°	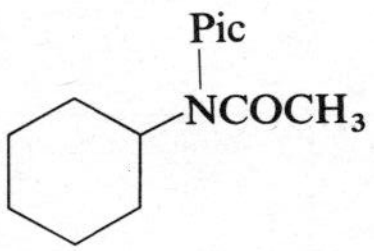 Pic $NCOCH_3$ (—)	93

TABLE I. REARRANGEMENTS (*Continued*)

No. of Carbon Atoms	Substrate	Reagent and Conditions	Product(s) and Yield(s) (%)	Refs.
	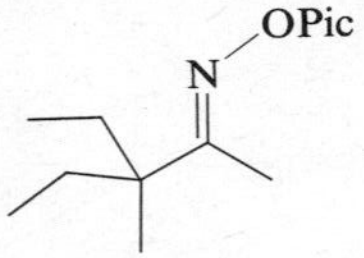 [structure: N–OPic oxime]	1,4-Dichlorobutane, 80°	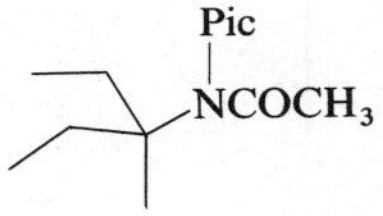 [structure: N(Pic)COCH$_3$] (—)	93
	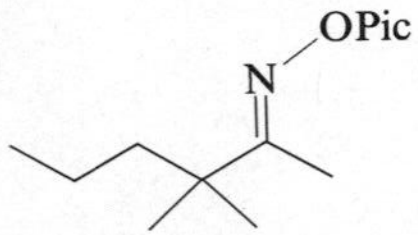 [structure: N–OPic oxime]	"	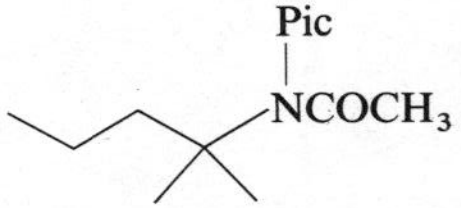 [structure: N(Pic)COCH$_3$] (—)	93
	n-$C_6H_{13}C(CH_3)$=NOH	$(C_6H_5)_3P$—CCl_4, reflux	n-$C_6H_{13}NHCOCH_3$ (80)	84
		$(C_6H_5)_3P$—CCl_4, THF, reflux	" (80)	84
C_9	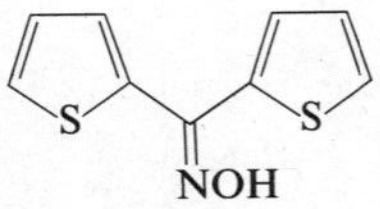 [structure: di-2-thienyl ketoxime, NOH]	PPA	 [structure: 2-thienyl-CONH-2-thienyl] (45)	301
	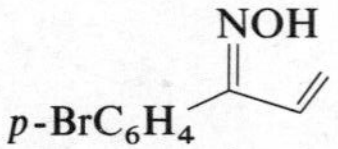 [structure: p-BrC_6H_4C(=NOH)CH=CH$_2$]	PCl_5, ether, C_6H_6, 0°, 40 min	p-$BrC_6H_4NHCO(CH_2)_2Cl$ (90)	114
	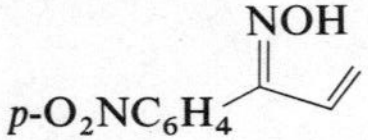 [structure: p-$O_2NC_6H_4$C(=NOH)CH=CH$_2$]	"	p-$O_2NC_6H_4NHCO(CH_2)_2Cl$ (60)	114
	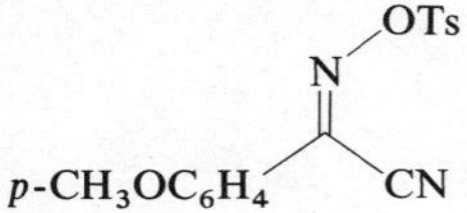 [structure: p-$CH_3OC_6H_4$C(=N–OTs)CN]	KOH, ethanol, 60°	p-$CH_3OC_6H_4NHCO_2C_2H_5$ (77)	117

	$NaOC_2H_5$, ethanol	" (64)	117
$C_6H_5CH{=}CHCH{=}NOH$	SiO_2, xylene, reflux, 66 h	$C_6H_5CH{=}CHCONH_2$ (79)	128
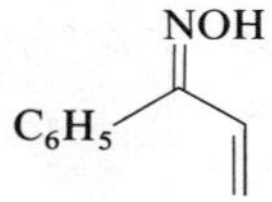 NOH, C_6H_5	PCl_5, ether, C_6H_6, 0°, 40 min	$C_6H_5NHCO(CH_2)_2Cl$ (81)	114
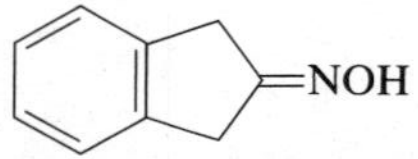 =NOH	PCl_5, ether, 15 h	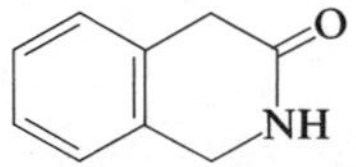 O, NH (70)	302
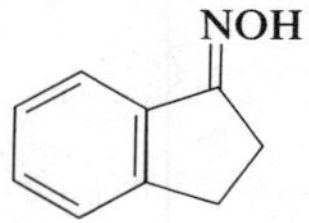 NOH	PPA, 110–120°, 5-10 min	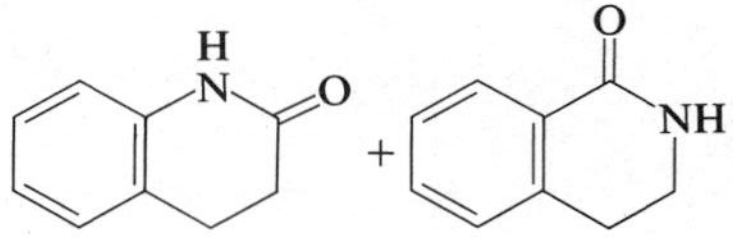 H, N, O + O, NH 9:1 (20)	70
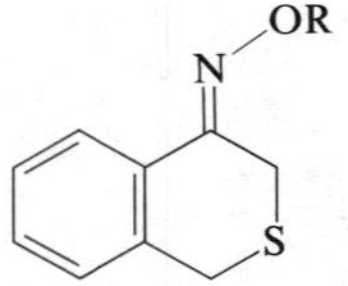 OR, N, S R = H	TsCl, pyridine, reflux, 24 h	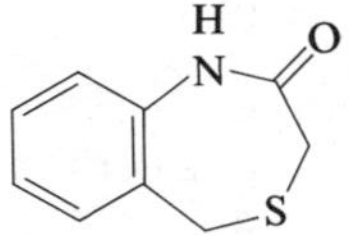 H, N, O, S (81)	193
R = H	PPA, 140°, 20 min	" (65)	193
R = Ts	KO_2CCH_3, ethanol, H_2O, reflux, 30 h	" (65)	193
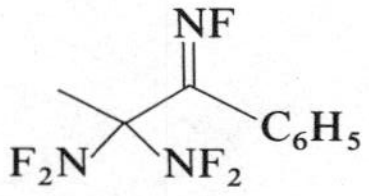 NF, C_6H_5, F_2N, NF_2	H_2SO_4	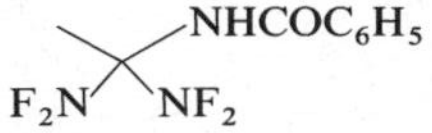 $NHCOC_6H_5$, F_2N, NF_2 (30)	121
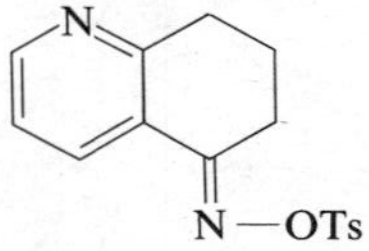 N, N—OTs	KO_2CCH_3, ethanol, reflux, 20 h	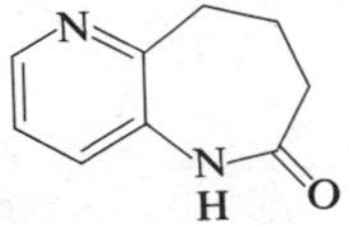 N, N, H, O (84)	303

TABLE I. REARRANGEMENTS (*Continued*)

No. of Carbon Atoms	Substrate	Reagent and Conditions	Product(s) and Yield(s) (%)	Refs.
	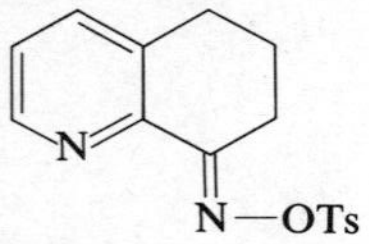 N, N—OTs	KO_2CCH_3, ethanol, reflux, 20 h	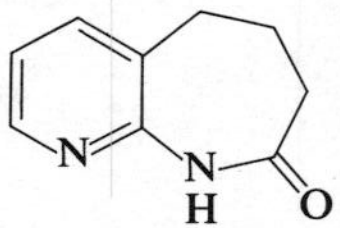 N, N–H, O (76)	303
	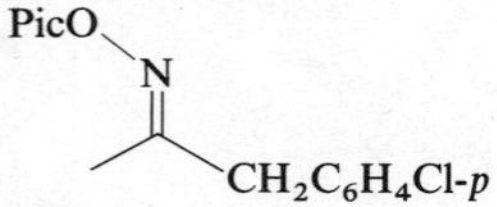 PicO, N, $CH_2C_6H_4Cl$-*p*	1,4-Dichlorobutane	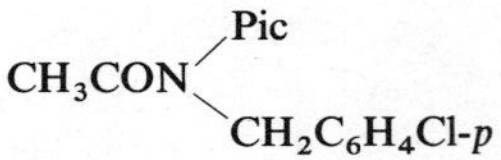 CH_3CON(Pic)$CH_2C_6H_4Cl$-*p* (—)	94
	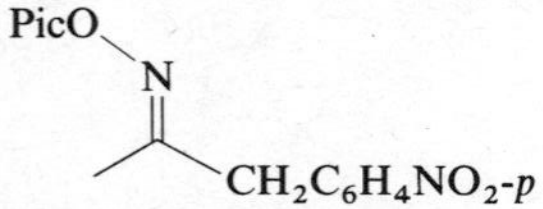 PicO, N, $CH_2C_6H_4NO_2$-*p*	"	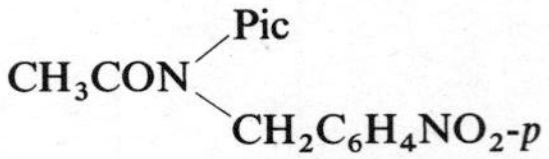 CH_3CON(Pic)$CH_2C_6H_4NO_2$-*p* (—)	94
	p-$CH_3C_6H_4COCH_3$	$NH_2OH \cdot HCl$, CF_3SO_3H, HCO_2H, reflux, 7 h	*p*-$CH_3C_6H_4NHCOCH_3$ (80)	249
	p-$CH_3C_6H_4C(CH_3){=}NOH$	$SOCl_2$, CCl_4, 0.5°, 2 h	*p*-$CH_3C_6H_4NHCOCH_3$ (70)	75
		HMPA, 225–235°, 10 min	" (85)	282
	p-$CH_3OC_6H_4C(CH_3){=}NOH$	$SOCl_2$, CCl_4, 0.5°, 2 h	*p*-$CH_3OC_6H_4NHCOCH_3$ (64)	75
		HMPA, 225–235°, 10 min	" (71)	282
		HCO_2H, reflux, 6 h	" (85)	251
		86% H_2SO_4, 80°	" (—)	281
	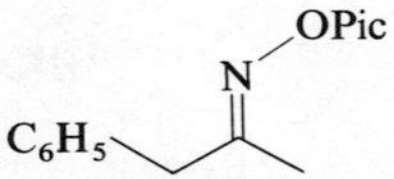 C_6H_5, N, OPic	1,4-Dichlorobutane	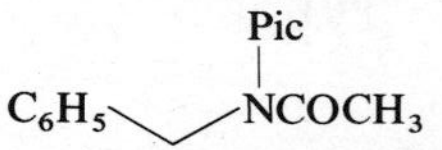 C_6H_5, Pic, $NCOCH_3$ (—)	94
	$C_6H_5CH_2C(CH_3){=}NOH$ (*E*)	P_2O_5, CH_3SO_3H, 100°, 30 min	$C_6H_5CH_2NHCOCH_3$ (—)	304
	(*Z*)	"	$C_6H_5CH_2CONHCH_3$ (—)	304

	99% H_2SO_4	p-$HO_3SC_6H_4CH_2NHCOCH_3$ (—)	290
$C_6H_5COC_2H_5$	$NH_2OH \cdot HCl$, CF_3SO_3H, HCO_2H	$C_6H_5NHCOC_2H_5$ (96)	249
$C_6H_5C(C_2H_5){=}NOH$	Cl_3CCN, reflux, 5–6 h	" (60–70)	100
	HCO_2H, reflux, 6 h	" (80)	251
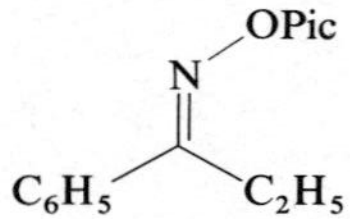	$C_2H_4Cl_2$, reflux	$C_6H_5N(Pic)COC_2H_5$ (96)	97
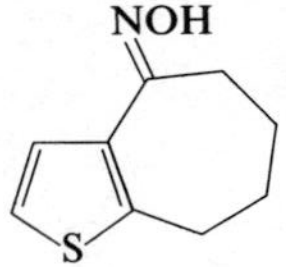	PPA	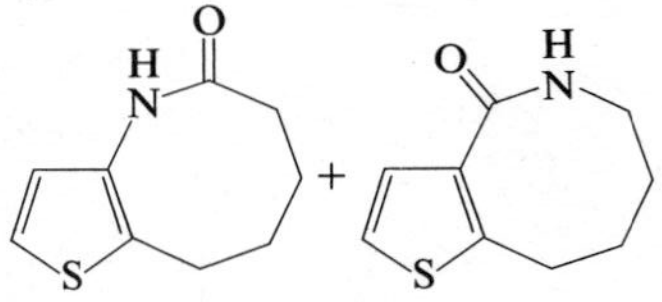 (98:2) (8)	305
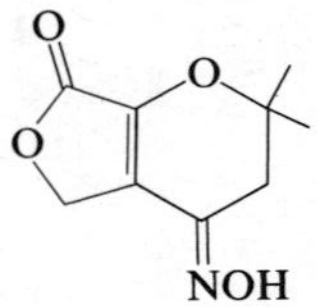	PPA, 110–130°, 20 min	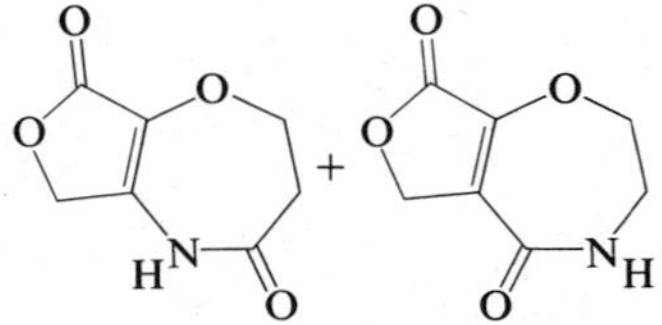 (35)	306
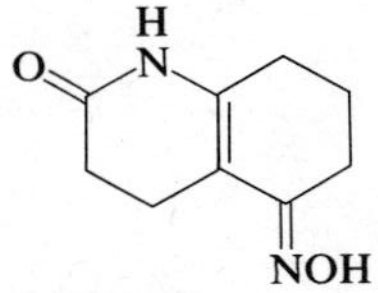	PPA, 120°, 2 h	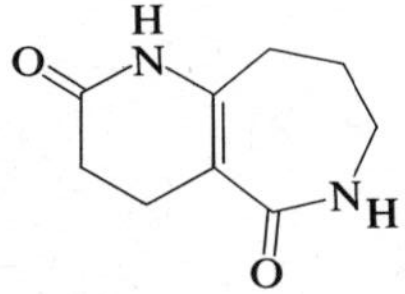 (61)	307
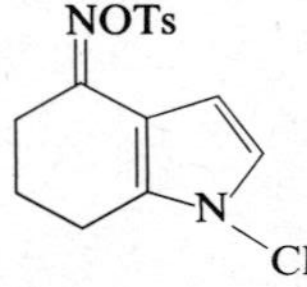 (*E*)	DMF, piperidine	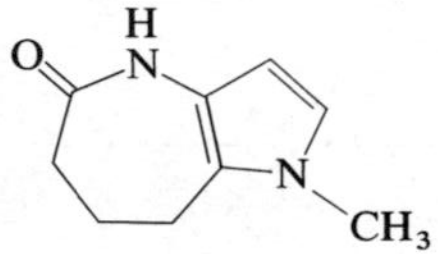 (—)	292

TABLE I. REARRANGEMENTS (*Continued*)

No. of Carbon Atoms	Substrate	Reagent and Conditions	Product(s) and Yield(s) (%)	Refs.
	(*Z*)	DMF, piperidine	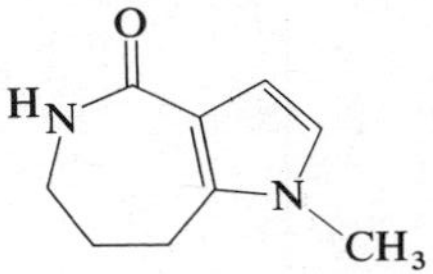 [structure: HN, O, N–CH_3] (80)	292
	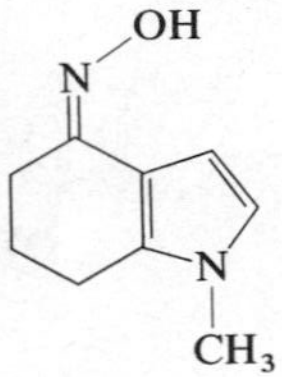 [structure: N–OH, N–CH_3]	PPA, 130°, 30 min	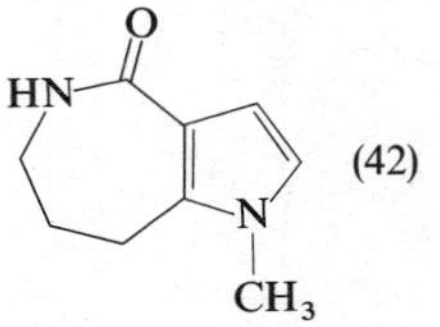 [structure: HN, O, N–CH_3] (42)	308
	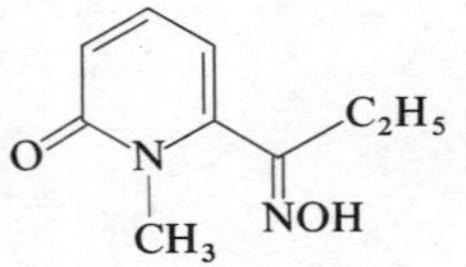 [structure: O, N–CH_3, C_2H_5, NOH] (*E*)	PCl_5, ether, 0°, 5 h	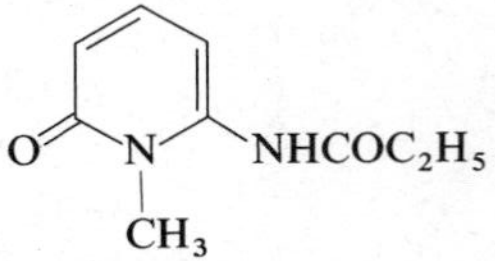 [structure: O, N–CH_3, $NHCOC_2H_5$] (92)	293
	(*Z*)	"	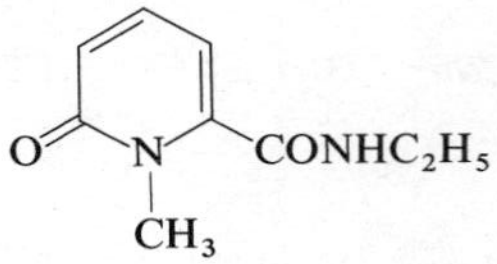 [structure: O, N–CH_3, $CONHC_2H_5$] (84)	293
	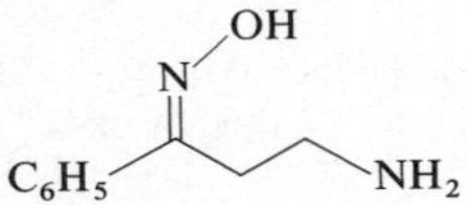 [structure: C_6H_5, N–OH, NH_2]	PCl_5, ether, 30 min	$C_6H_5NHCO(CH_2)_2NH_2$ (85)	309

Oxime	Conditions	Product(s) (Yield, %)	Refs.
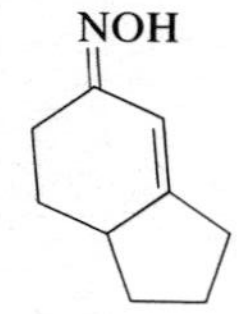 NOH	PPA, 135°	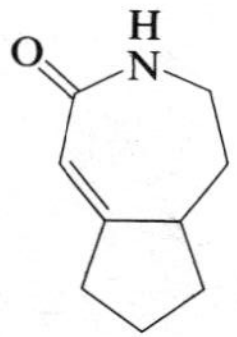 O, H N (52)	310
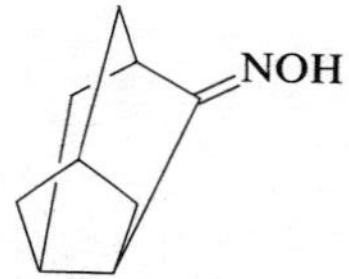 NOH (*E*)	PPE, $CHCl_3$	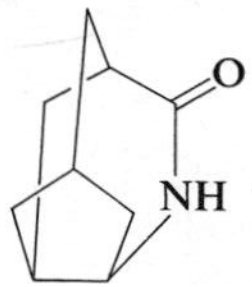 O, NH (—)	55
(*Z*)	”	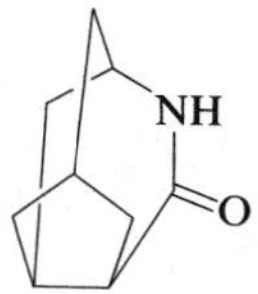 NH, O (—)	55
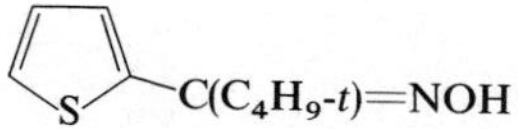 S C(C_4H_9-*t*)=NOH	1. $C_6H_5SO_2Cl$ 2. Alumina	 S $CONHC_4H_9$-*t* (40)	77
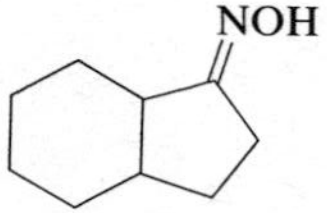 NOH	TsCl, NaOH, dioxane	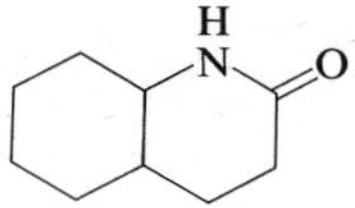 H N O (100)	190
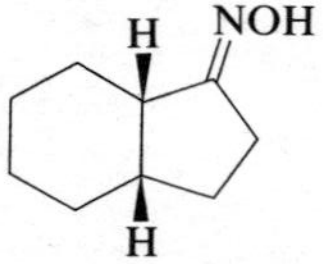 H NOH H	$SOCl_2$, dioxane	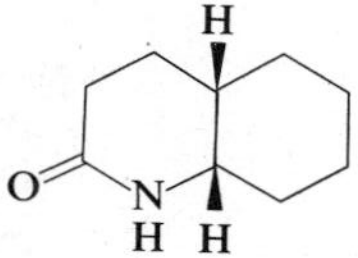 H O N H H (66)	311

TABLE I. REARRANGEMENTS (*Continued*)

No. of Carbon Atoms	Substrate	Reagent and Conditions	Product(s) and Yield(s) (%)	Refs.
	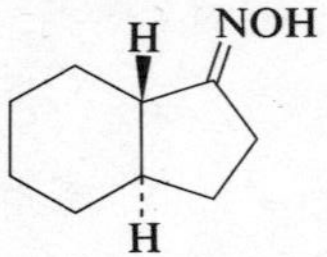 NOH	$SOCl_2$, dioxane	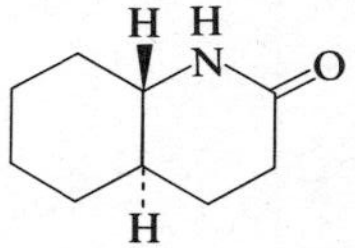 (66)	311
	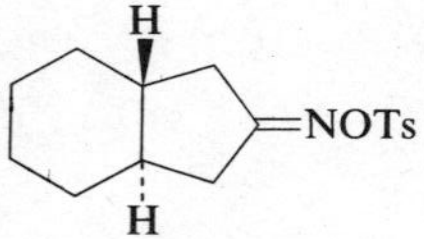 =NOTs	Alumina, C_6H_6	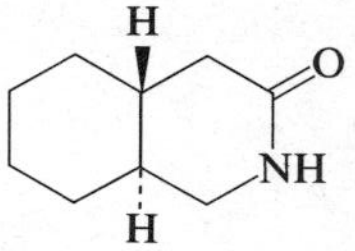 (88)	312
	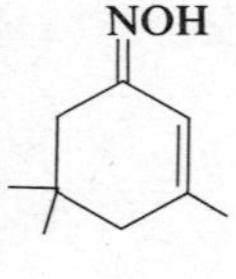 NOH	PPA, 120°	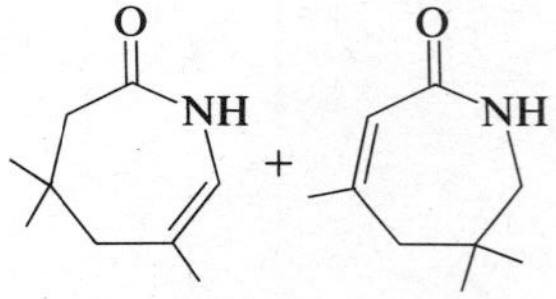 I + II (—)	46
	(*E*)	PPA, 131–135°, 10 min	I (72)	45
	(*Z*)	"	II (21)	45
	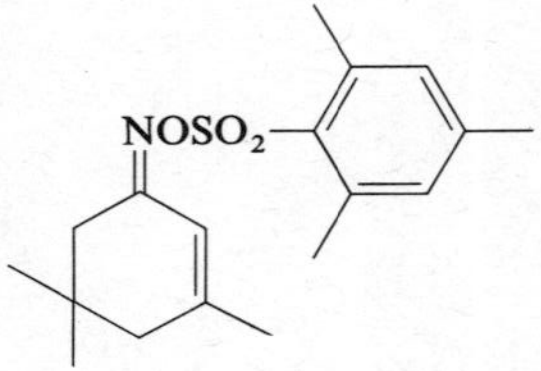 $NOSO_2$	Alumina, CH_3OH	II (50) +	47

Reactant	Conditions	Product (Yield, %)	Refs.
		SO_2O, N (28)	
NOH, OC_2H_5	PPA, 120°, 1 h	HN, O, O (91)	245
NOH, OCH_3	PPA, 120°, 1 h	HN, O, O (79)	245
NOH	1. TsCl, pyridine, CH_2Cl_2 2. K_2CO_3, THF, H_2O, 15 h	O, NH (59)	296
N, OH	TsCl, pyridine, 18 h	O, NH (48)	313

TABLE I. REARRANGEMENTS (*Continued*)

No. of Carbon Atoms	Substrate	Reagent and Conditions	Product(s) and Yield(s) (%)	Refs.
	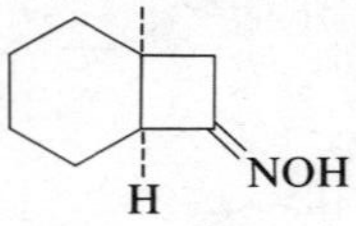 H NOH	PPA, 180–185°, 2 min	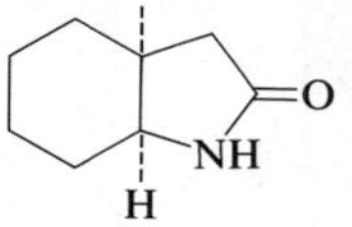 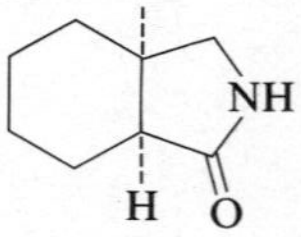 =O + NH NH (59) H I H O II	126
	(*Z*)	(—)	II (—)	127
	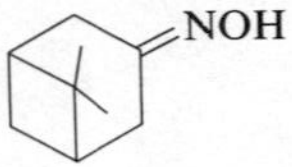 NOH	TsCl, NaOH, acetone	 NH O (—)	314
	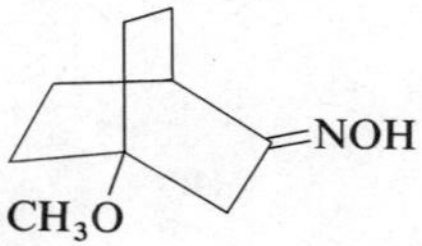 CH_3O NOH	$C_6H_5SO_2Cl$, NaOH	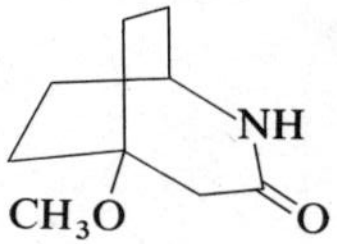 CH_3O NH O (—)	297
		TsCl, pyridine	" (92)	297
	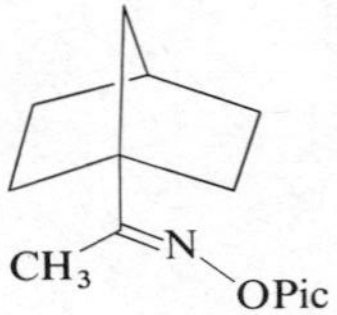 CH_3 N OPic	1,4-Dichlorobutane, 80°	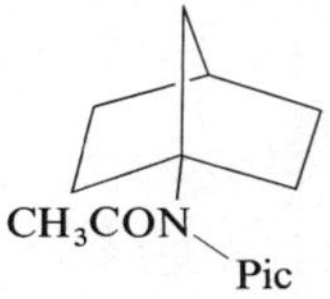 CH_3CON Pic (—)	93

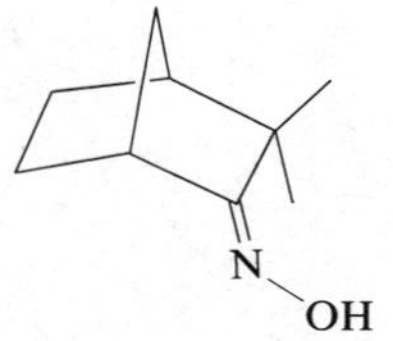	TsCl, NaOH, acetone	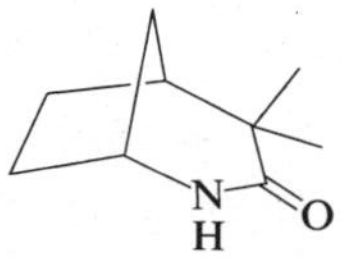	(—)	315
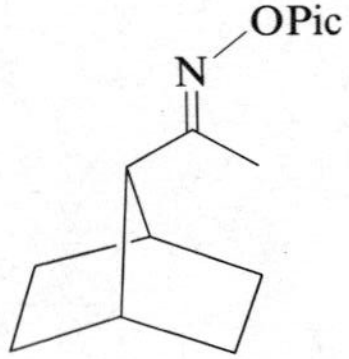	1,4-Dichlorobutane, 80°	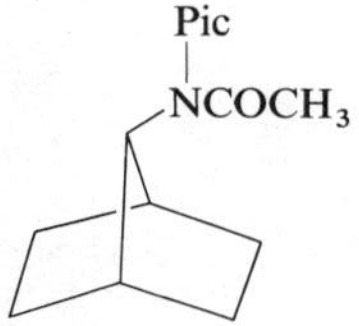	(—)	93
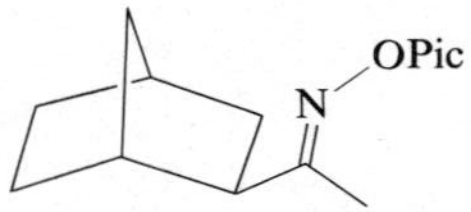	"		(—)	94
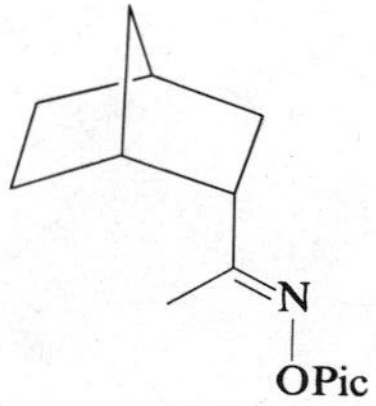	"	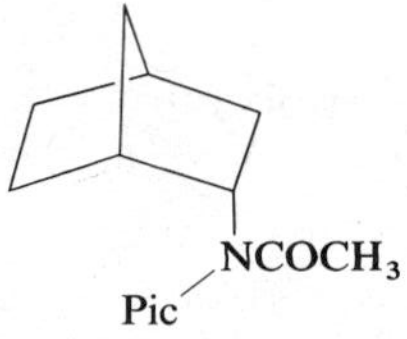	(—)	94
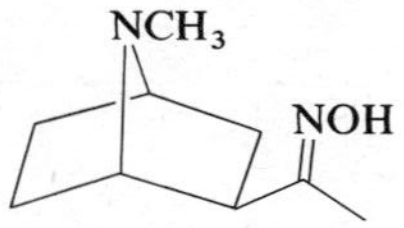	TsCl, dioxane	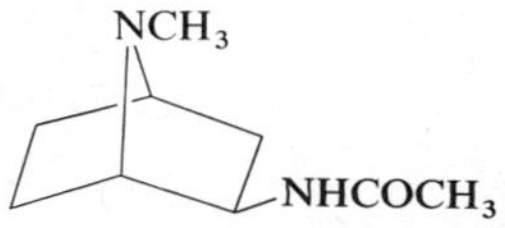	(100)	316

TABLE I. REARRANGEMENTS (*Continued*)

No. of Carbon Atoms	Substrate	Reagent and Conditions	Product(s) and Yield(s) (%)	Refs.
	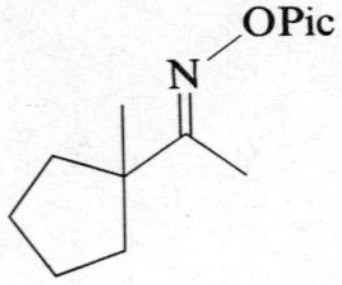 [structure: OPic, N]	1,4-Dichlorobutane, 80°	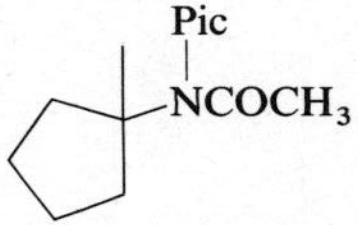 [structure: Pic, $NCOCH_3$] (—)	93
	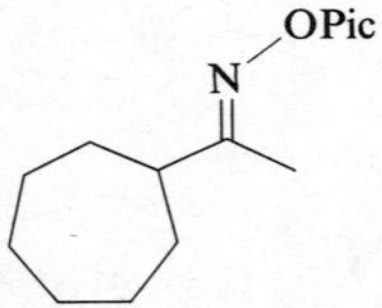 [structure: OPic, N]	"	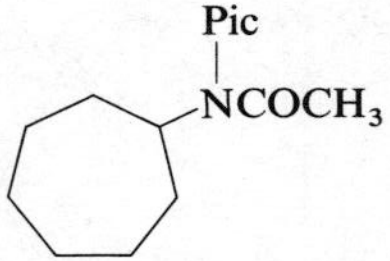 [structure: Pic, $NCOCH_3$] (—)	93
	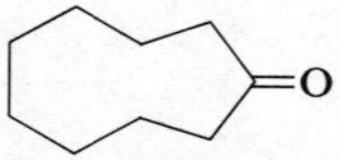 [structure: O]	H_2NOSO_3H, HCO_2H, reflux, 7 h	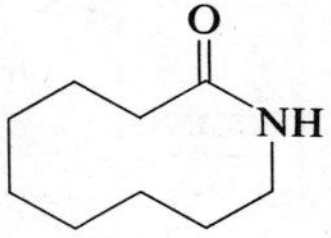 [structure: O, NH] (83)	81
	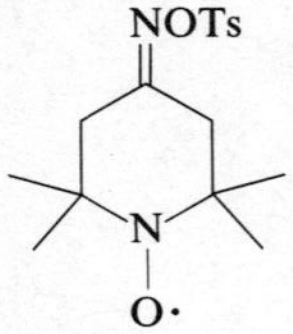 [structure: NOTs, N, O·]	Alumina, C_6H_6, 20°, 3 h	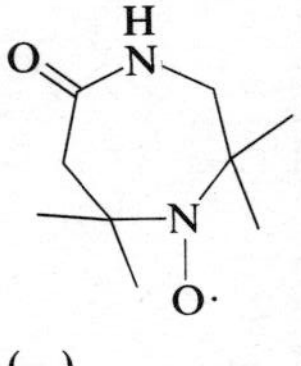 [structure: O, H, N, N, O·] (67)	317
		SiO_2, 2% CH_3OH–$CHCl_3$	(—)	317

	Reactant	Conditions	Product(s) (Yield %)	Refs.
	$t\text{-}C_4H_9C(CH_3)_2C(CH_3)=NOPic$	1,4-Dichlorobutane, 80°	$t\text{-}C_4H_9C(CH_3)_2N(Pic)COCH_3$ (—)	93
	$(C_2H_5)_3CC(CH_3)=NOPic$	"	$(C_2H_5)_3CN(Pic)COCH_3$ (—)	93
C_{10}	$CH_3C(=NOH)C{\equiv}CC_6H_5$	PCl_5, ether	$C_6H_5CCl{=}CHCONHCH_3$ (—)	116
	1-C_6H_5-pyrazol-4-yl-$CH{=}NOH$ (*Z*)	H_2SO_4, 100°, 1 h	1-C_6H_5-pyrazole-4-$CONH_2$ (1)	318
	(*Z*)	PPA, 100°, 1 h	" (85)	318
	(*E*)	H_2SO_4, 100°, 1 h	" (4)	318
	(*E*)	PPA, 100°, 1 h	" (84)	318
	5,6-Methylenedioxyindan-1-one oxime (=NOH)	PPA, 100°, 75 min	6,7-Methylenedioxy-3,4-dihydro-2(1*H*)-quinolinone (N–H, C=O) (25)	319
	3-$COCH_3$-indole (NH)	$H_2NOH\cdot HCl$, ethanol, reflux, 3 d	3-$NHCOCH_3$-indole (NH) (10)	320

TABLE I. REARRANGEMENTS (*Continued*)

No. of Carbon Atoms	Substrate	Reagent and Conditions	Product(s) and Yield(s) (%)	Refs.
	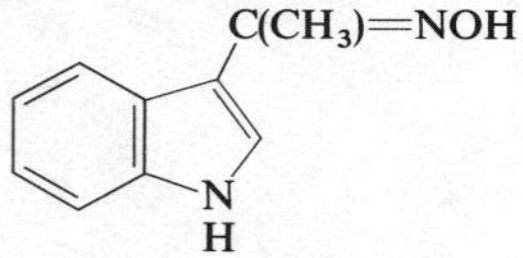 3-indolyl-$C(CH_3)=NOH$ (*E*)	CH_3CO_2H, reflux, 2 h	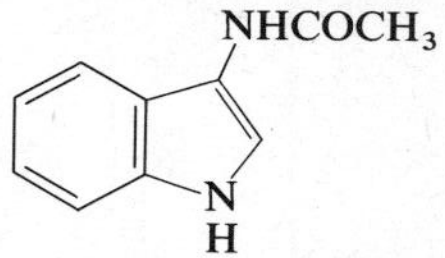 3-indolyl-$NHCOCH_3$ (48)	320
	(*Z*)	PCl_5, ether	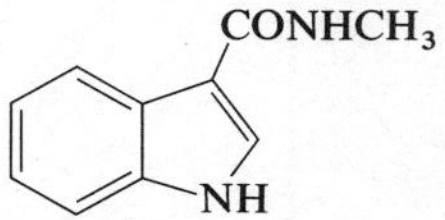 3-indolyl-$CONHCH_3$ (—)	320
	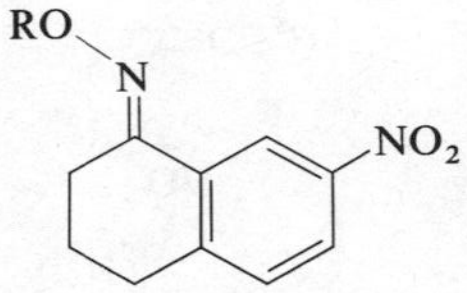 7-nitro-1-tetralone oxime (RO–N=), R=H	PPA, 130°, 30 min	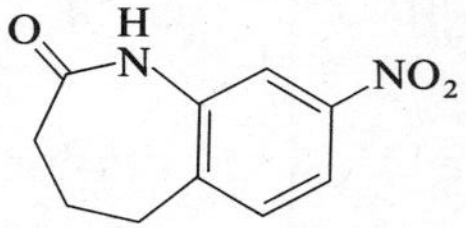 8-nitro-1,3,4,5-tetrahydro-2H-1-benzazepin-2-one (68)	308
	R=Ts	NaO_2CCH_3, ethanol, H_2O, 30 h	" (51)	308
	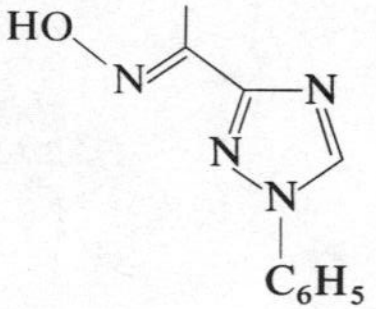 HO–N=C(CH_3)-(1-C_6H_5-1,2,4-triazol-3-yl)	PCl_5, $CHCl_3$, 0°, 2 d	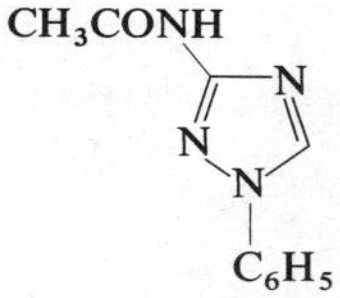 CH_3CONH-(1-C_6H_5-1,2,4-triazol-3-yl) (20)	321

Reactant	Conditions	Product(s) and Yield(s) (%)	Refs.
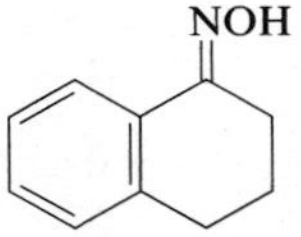 NOH	PPA, 110–120°, 5–10 min	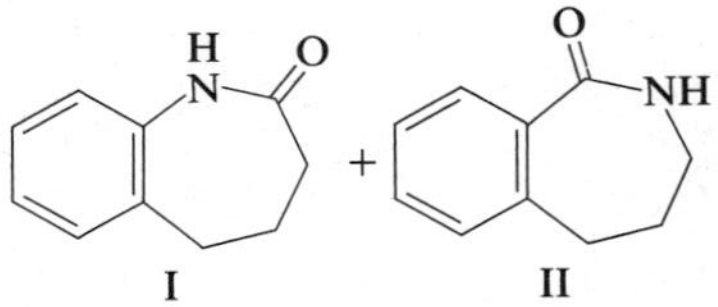 H N O + O NH I II 99:1 (65)	70
	PPA	I + II (—)	322
$C_6H_5CH{=}CHC({=}NOH)CH_3$	PCl_5, ether, 0°, 15 min	$C_6H_5CH{=}CHNHCOCH_3$ (—)	111
p-$CH_3OC_6H_4C({=}NOH)CH{=}CH_2$	PCl_5, ether, C_6H_6, 0°, 40 min	p-$CH_3OC_6H_4NHCO(CH_2)_2Cl$ (79)	114
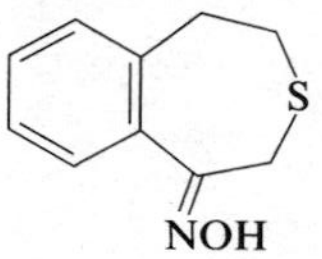 S NOH	PPA, 100°, 4 h	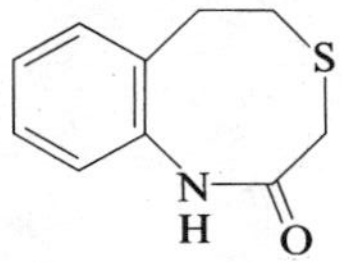 S N H O (40)	323
(E)	PCl_5, C_6H_6, 15 h	" (25)	324
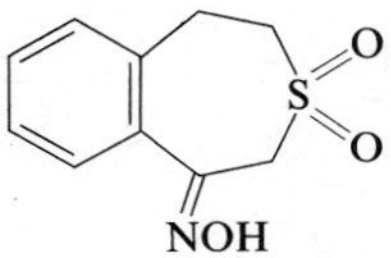 S O O NOH	PPA, 100°, 4 h	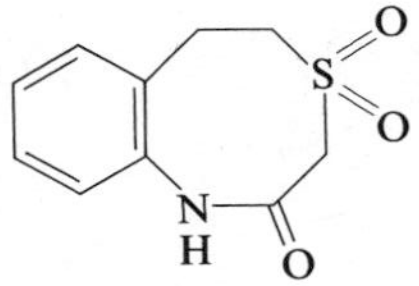 O S O N H O (67)	323
(E)	PCl_5, ether, THF, 0°–25 h	" (31)	324
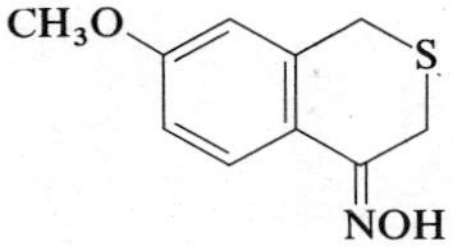 CH_3O S NOH	PPA, 100°, 30 min	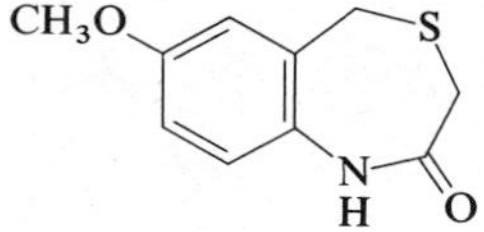 CH_3O S N H O (75)	325
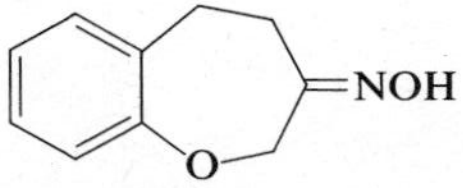 =NOH O	PPA, 90°, 10 min	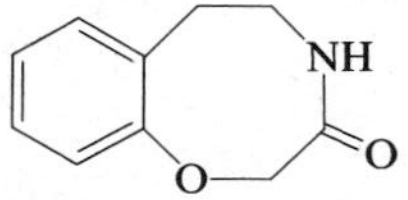 NH O O (15)	326

TABLE I. REARRANGEMENTS (*Continued*)

No. of Carbon Atoms	Substrate	Reagent and Conditions	Product(s) and Yield(s) (%)	Refs.
	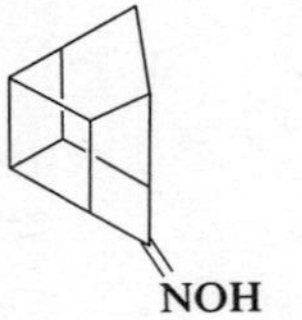 NOH	PPE, $CHCl_3$, reflux, 2 h	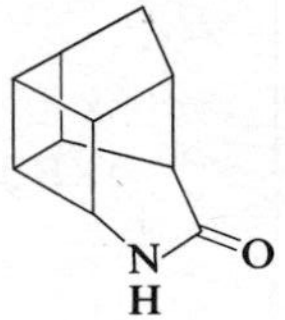 N H O (11) 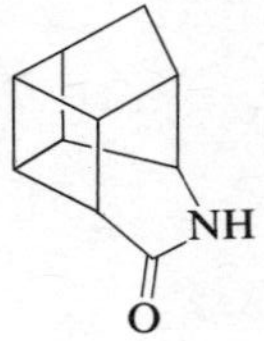 + NH O (48)	327
	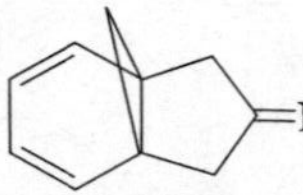 =NOR R = H	1. TsCl, pyridine 2. 2,6-Dimethylpyridine, 40°, 3 d	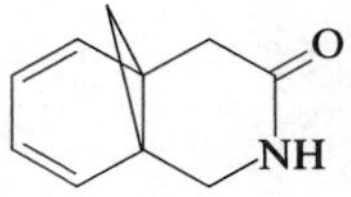 O NH (30)	328
	R = $SO_2C_6H_5$	THF, H_2O	" (60)	329
	OPic N C_6H_5 C_3H_7-*n*	$C_2H_4Cl_2$, warm	Pic $C_6H_5NCOC_3H_7$-*n* (88)	97
	TsO N C_6H_5 C_3H_7-*i*	$(C_2H_5)_3N$, 80% ethanol, reflux	$C_6H_5CONHC_3H_7$-*i* (98)	28
	NOPic C *i*-C_3H_7 C_6H_5 (*Z*)	$C_2H_4Cl_2$, reflux, 20 h	Pic *i*-$C_3H_7NCOC_6H_5$ (84)	97
	(*E*)	$C_2H_4Cl_2$, warm	Pic $C_6H_5NCOC_3H_7$-*i* (81)	97

Reactant	Conditions	Product(s) and Yield(s) (%)	Refs.
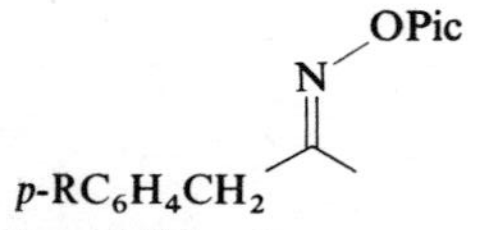 p-$RC_6H_4CH_2C(=NOPic)CH_3$, R = CH_3	1,4-Dichlorobutane, 80°	p-$RC_6H_4CH_2N(Pic)COCH_3$ (—)	94
R = OCH_3	"	" (—)	94
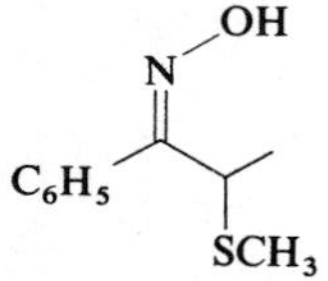 $C_6H_5C(=NOH)CH(CH_3)SCH_3$	1. TsCl, pyridine 2. $(C_2H_5)_3N$, dioxane, 14 h	$C_6H_5NHCOCH(CH_3)SCH_3$ (99)	330
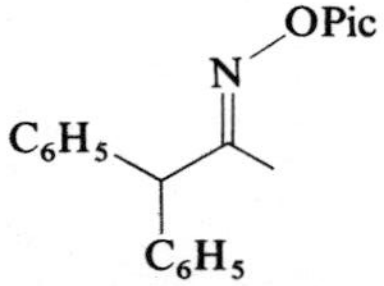 $C_6H_5CH(C_6H_5)C(=NOPic)CH_3$	1,4-Dichlorobutane, 80°	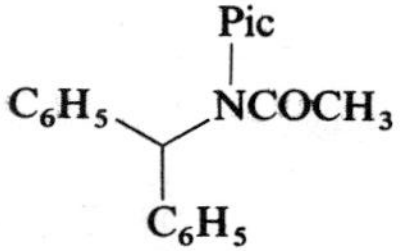 $C_6H_5CH(C_6H_5)N(Pic)COCH_3$ (—)	93
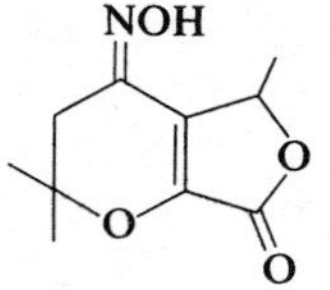 NOH (oxime of bicyclic furanone/dimethylpyranone)	PPA, 110–130°, 20 min	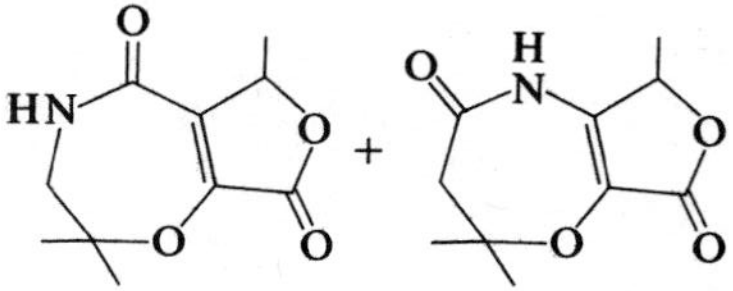 HN lactam + H–N lactam (isomeric ring-expanded lactams) (20)	306
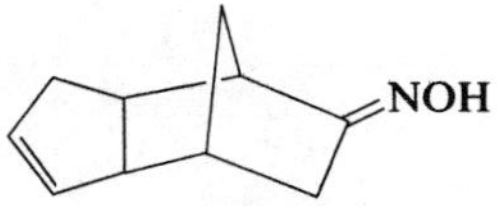 NOH (tricyclic ketoxime)	$C_6H_5SO_2Cl$, NaOH	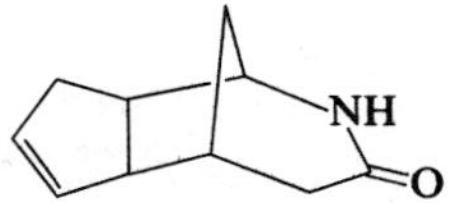 NH lactam (31)	265
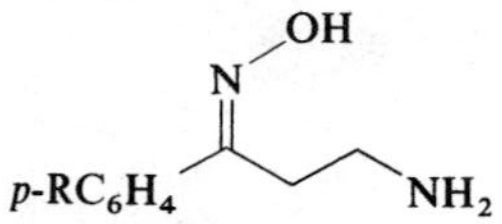 p-$RC_6H_4C(=NOH)CH_2CH_2NH_2$, R = CH_3	PCl_5, ether, 30 min	p-$RC_6H_4NHCO(CH_2)_2NH_2$ (84)	309
R = OCH_3	"	" (72)	309

TABLE I. REARRANGEMENTS (*Continued*)

No. of Carbon Atoms	Substrate	Reagent and Conditions	Product(s) and Yield(s) (%)	Refs.
	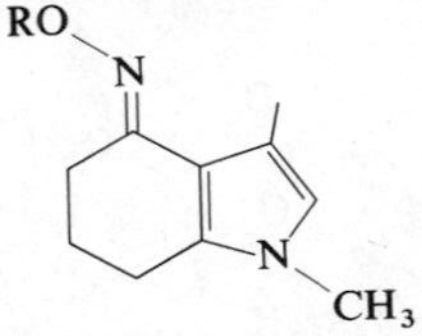 RO–N= (structure) R = H	PPA, 130°, 30 min	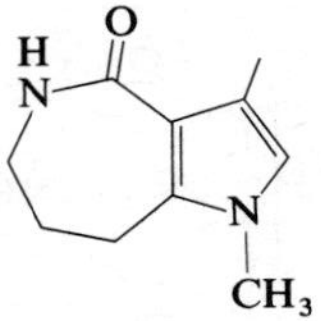 (structure) (31)	308
	R = Ts	NaO_2CCH_3, ethanol, H_2O, 30 h	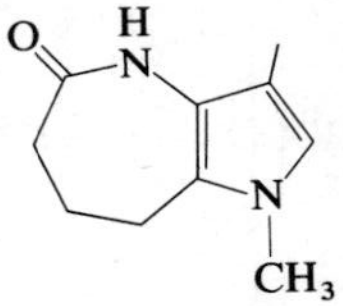 (structure) (62)	308
	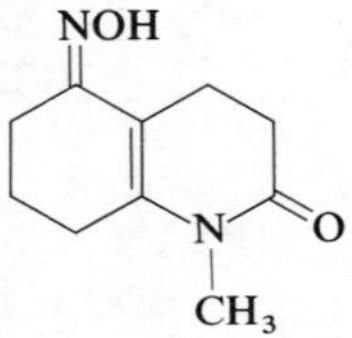 NOH, N–CH_3, O (structure)	PPA, 120°, 2 h	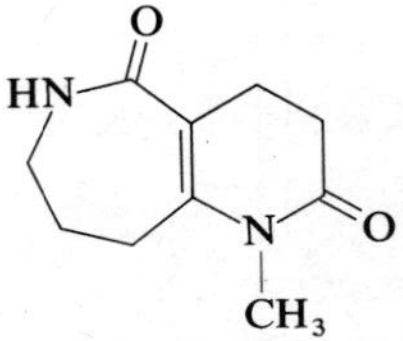 HN, O, N–CH_3 (structure) (46)	307
	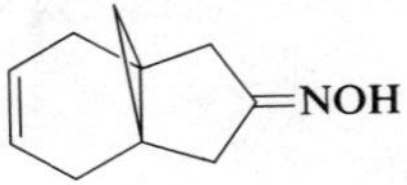 =NOH (structure)	PCl_5, ether, 0°, 5 h	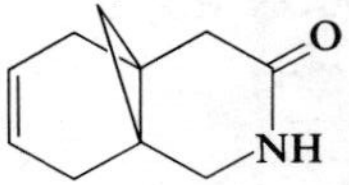 O, NH (structure) (48)	331

Reactant	Conditions	Product (Yield, %)	Refs.
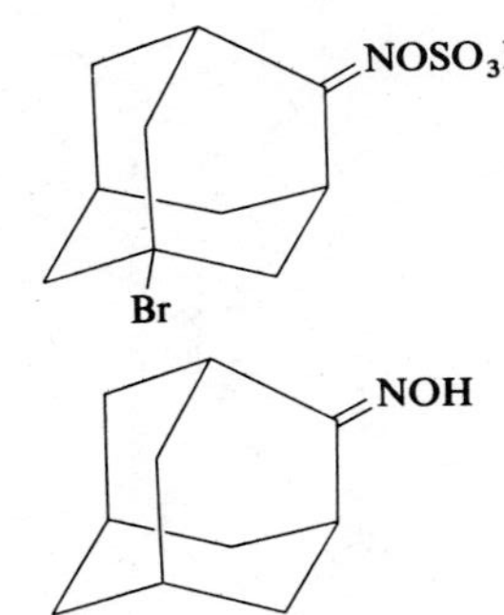	H_2O, 50 h	O, NH, Br (75)	258
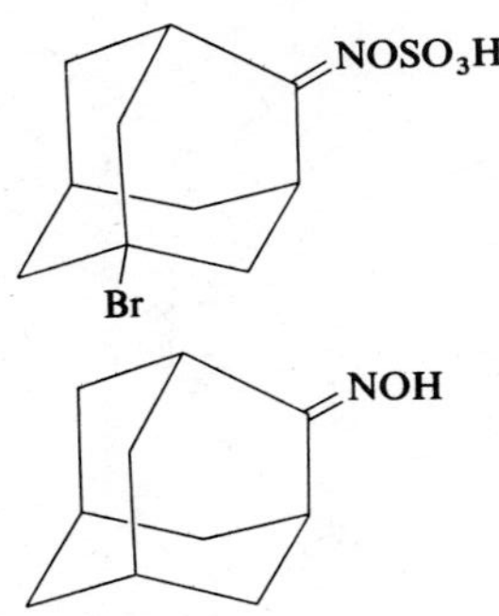	$SOCl_2$, ether, 0°, 1.5 h	NH, O (85)	332
	PPE, $CHCl_3$, reflux, 7 min	" (65)	333
	PPA	" (57)	334
	HCl, CH_3CN	" (40)	334
	PCl_5, ether	" (82)	334
	H_2SO_4, 20°, 4 h	" (32)	334
	24% HBr	" (—)	335
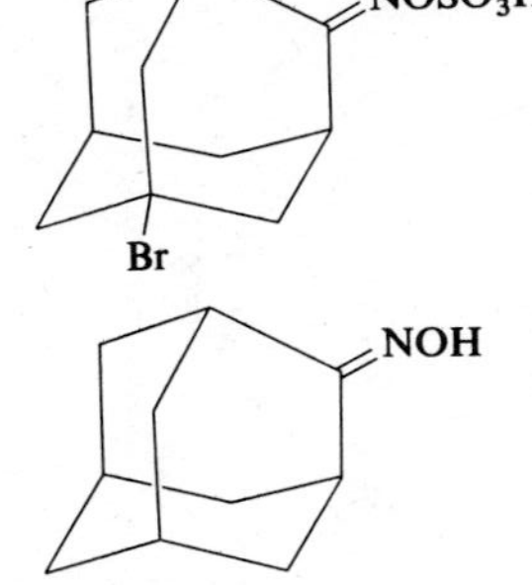 R = H	H_2O, 15 min	" (51)	258
R = NH_4	150–170°, 7 min	" (58)	258
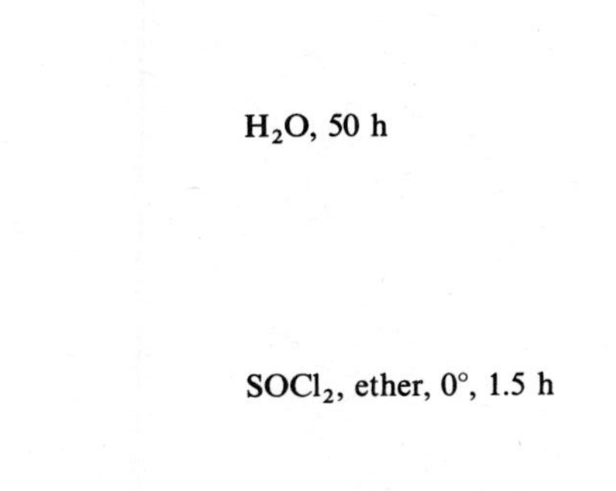	H_2O, 40 h	O, NH (70)	258

TABLE I. REARRANGEMENTS (*Continued*)

No. of Carbon Atoms	Substrate	Reagent and Conditions	Product(s) and Yield(s) (%)	Refs.
	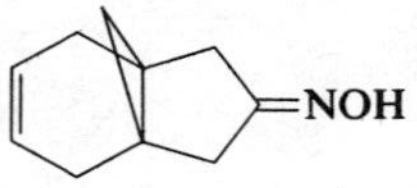 =NOH	PCl_5, ether, 0°, 5 h	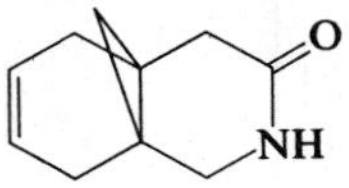 O, NH (48)	331
	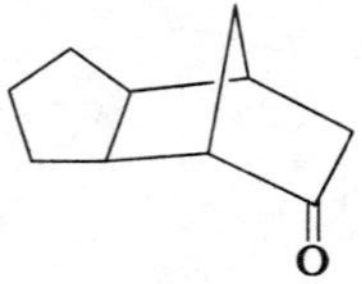 O	$(H_2NOH)_2SO_4$, H_2SO_4, 116°, 10 min	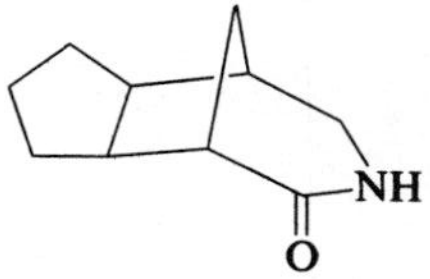 NH, O (25)	336
	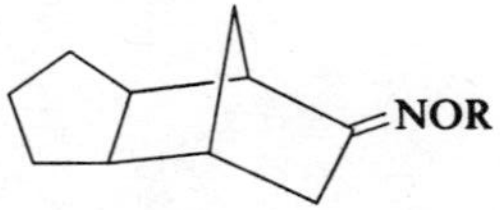 =NOR R = H	$C_6H_5SO_2Cl$, NaOH	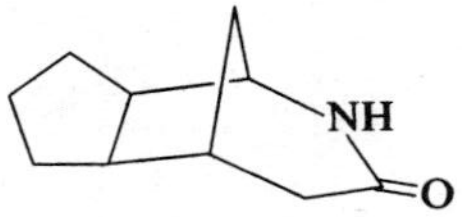 NH, O (44)	265
	R = $COCH_3$	HCl, CH_3CO_2H	" (—)	270
	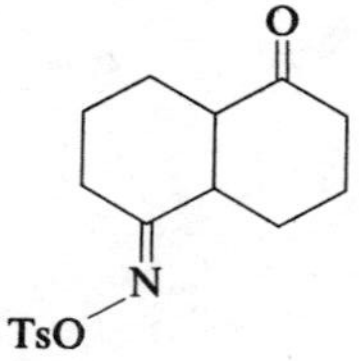 O, TsO–N	80% ethanol	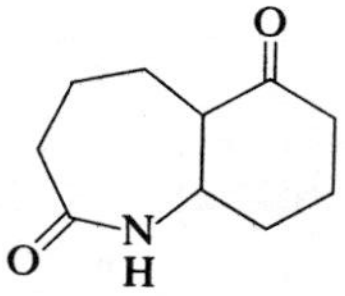 O, O, N H (—)	337
	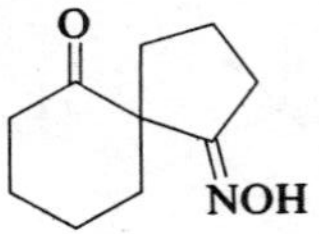 O, NOH	PPA	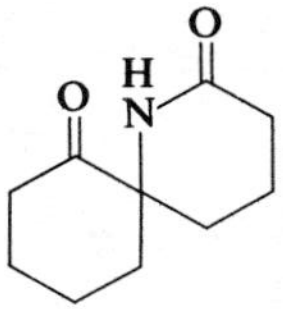 O, H N, O (90)	338

Reactant	Conditions	Product(s) and Yield(s) (%)	Refs.
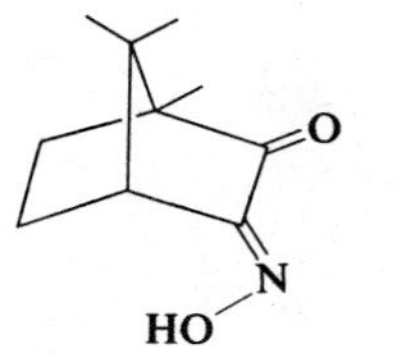 HO–N=	$(CH_3CO)_2O$, CH_3CO_2H, reflux, 17 h	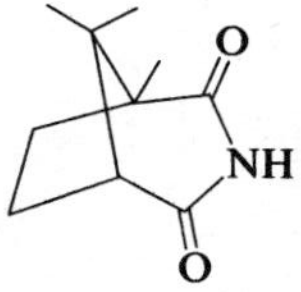 NH (47)	65
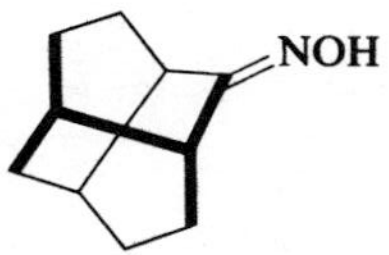 =NOH	$C_6H_5SO_2Cl$, NaOH, 15 h	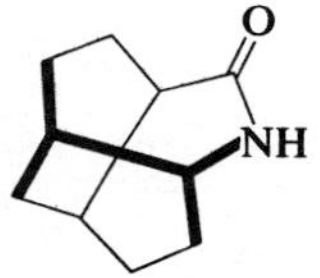 NH (29)	339
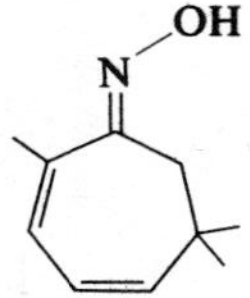 N–OH	TsCl, NaOH, acetone	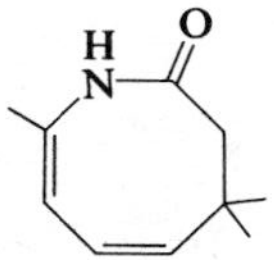 (—)	340
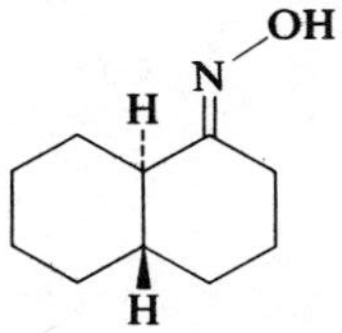 N–OH, H, H	TsCl, 35–40°	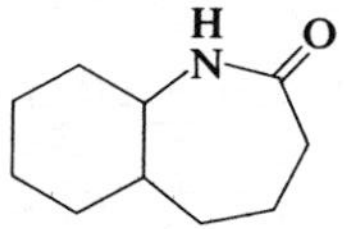 (91)	200
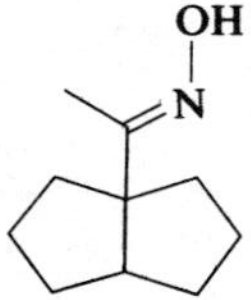 N–OH	PPA, 10 h	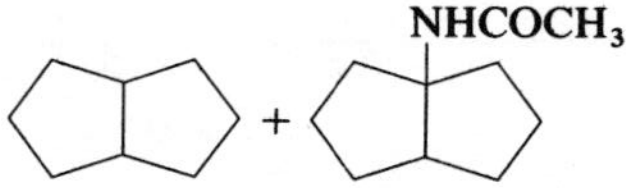 I (8) + II (75), II = $NHCOCH_3$	341

TABLE I. REARRANGEMENTS (*Continued*)

No. of Carbon Atoms	Substrate	Reagent and Conditions	Product(s) and Yield(s) (%)	Refs.
		HCO_2H, reflux, 6 h	I (12) + II (70)	341
		H_2SO_4, 15 h	I (10) + II (67)	341
		PCl_5, $CHCl_3$	I (15) + II (45)	341
	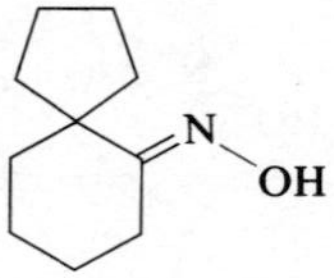 [spiro[4.5]decan-6-one oxime: =N–OH]	$SOCl_2$	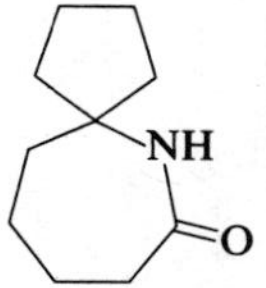 [spiro lactam: NH, =O] (7)	221
		$C_6H_5SO_2Cl$, NaOH	" (30)	221
		P_2O_5, C_6H_6, 70°	" (84)	342
		PCl_5, C_6H_6	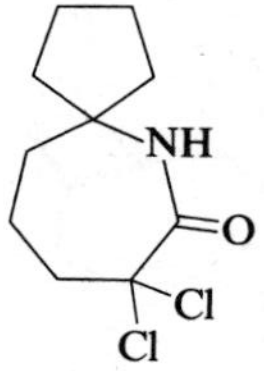 [spiro lactam: NH, =O, Cl, Cl] (10)	342
	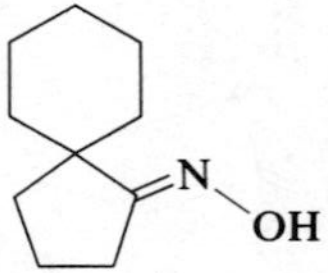 [spiro[4.5]decan-1-one oxime: =N–OH]	$SOCl_2$	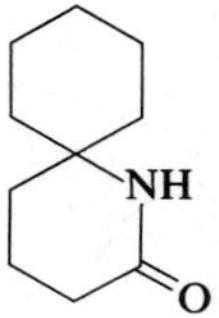 [spiro lactam: NH, =O] (32)	221

Reactant	Conditions	Product(s) and Yield(s) (%)	Refs.
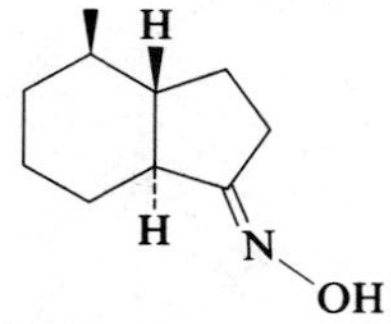	TsCl, pyridine	" (34)	221
	TsCl, NaOH, THF, 15 h	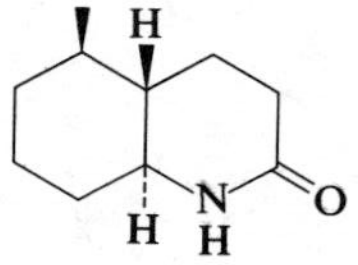 (69)	147
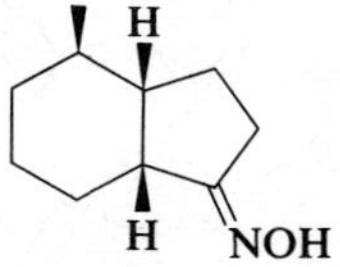	"	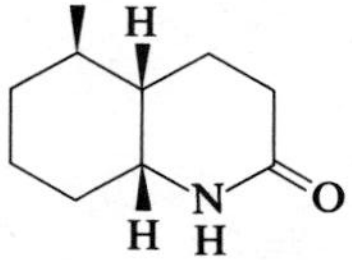 (66)	147
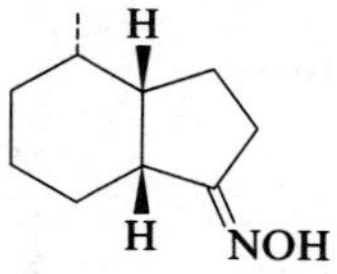	"	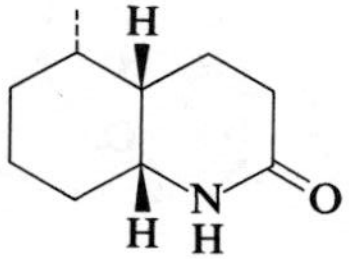 (64)	147
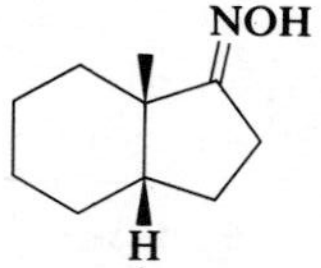	$SOCl_2$, dioxane	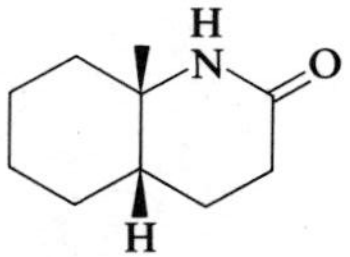 (40)	311
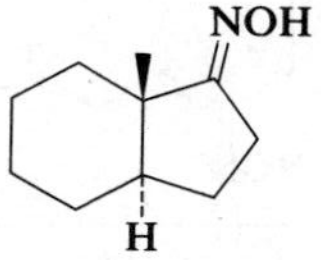	PCl_5, ether, 0°	" (—)	343
	$SOCl_2$, dioxane	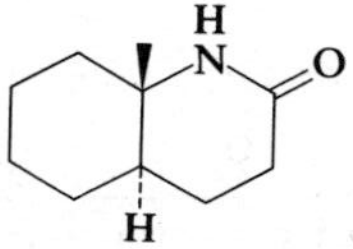 (37)	311

TABLE I. REARRANGEMENTS (*Continued*)

No. of Carbon Atoms	Substrate	Reagent and Conditions	Product(s) and Yield(s) (%)	Refs.
	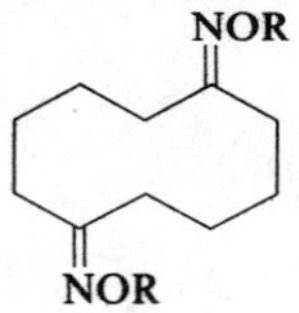 NOR, NOR; R=H	SO_3, SO_2	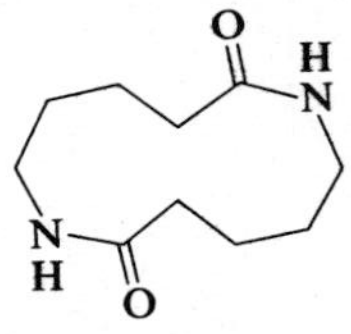 O, H N, N H, O; I (75)	210
	R = Ts	NaO_2CCH_3, CH_3CO_2H	I 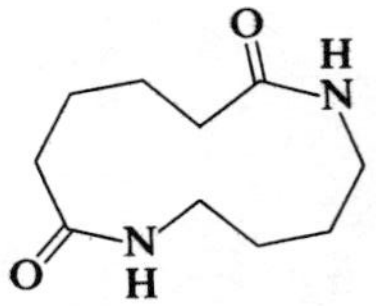 + O, H N, N H, O (61)	90
	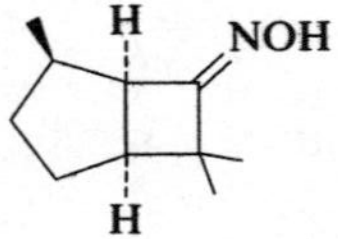 H, NOH, H	TsCl, pyridine, 50°, 30 h	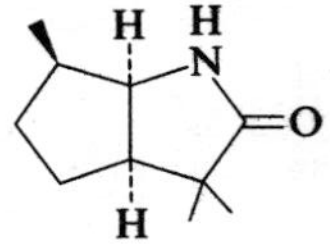 H H N, O, H (18)	344
	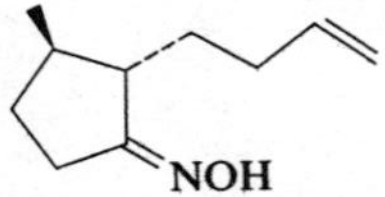 NOH	PCl_5, ether	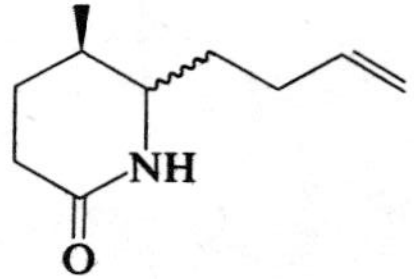 NH, O (60)	295
	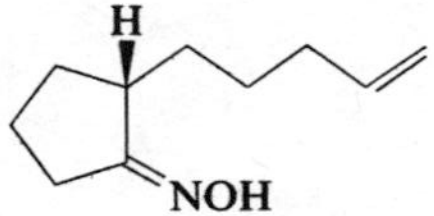 H, NOH	"	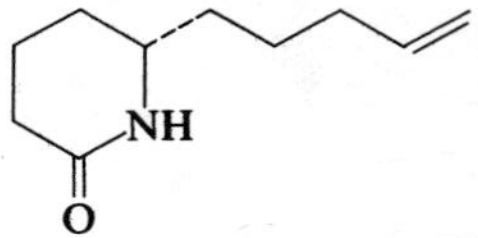 NH, O (45)	295

HO_2CCH_2 … NOH	H_2SO_4, 115°	HO_2CCH_2 … $NHCOCH_3$ (70)	345
H, H, N–OH	TsCl, NaOH, acetone	H, NH, H, O (–)	346
OH, N, H, H	"	O, H, HN, H (—)	346
N–OH	"	NH, O (—)	314
N–OH	"	NH, O (—)	314
NOH	TsCl, pyridine, reflux, 30 min	N, H, O (72)	231

TABLE I. REARRANGEMENTS (*Continued*)

No. of Carbon Atoms	Substrate	Reagent and Conditions	Product(s) and Yield(s) (%)	Refs.
	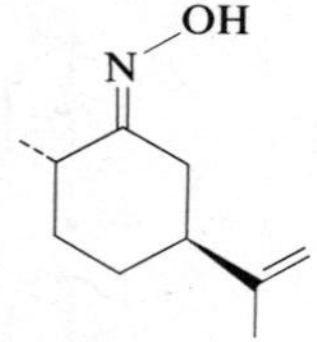 N—OH	TsCl, NaOH, acetone	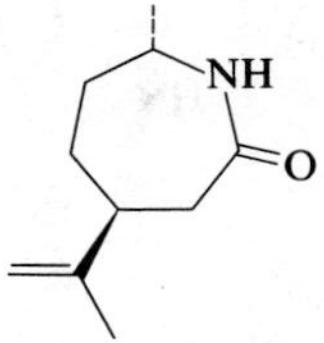 NH, O (—)	347
	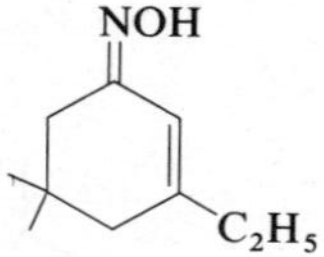 NOH, C_2H_5	PPA, 120°	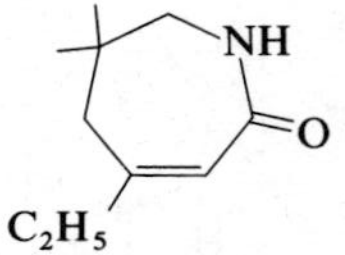 NH, O, C_2H_5 (40) 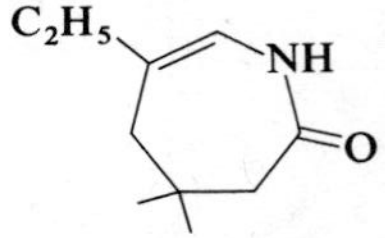 + C_2H_5, NH, O (17)	46
	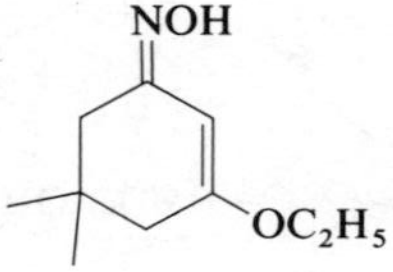 NOH, OC_2H_5	PPA, 120°, 1 h	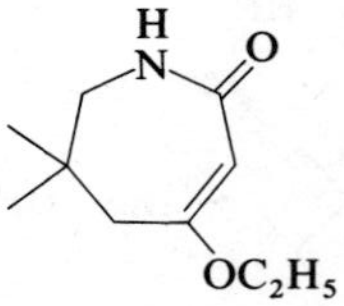 H N, O, OC_2H_5 (81)	245

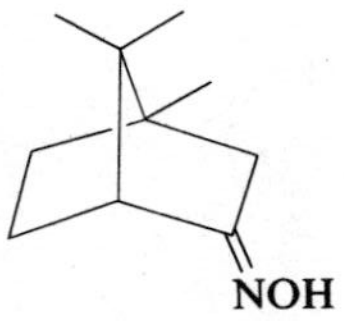

Reactant	Conditions	Product (Yield, %)	Refs.
NOH	TsCl, pyridine, 50°, 3 h	NH, O (43)	344
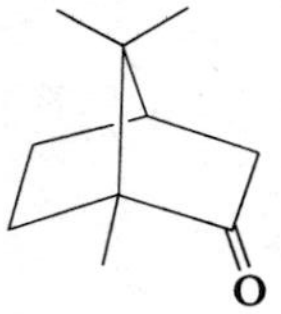 O	H_2NOSO_3H, HCO_2H, reflux	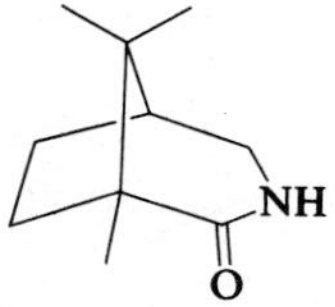 NH, O (46)	37
	H_2NOSO_3H, CH_3CO_2H, reflux	" (48)	37
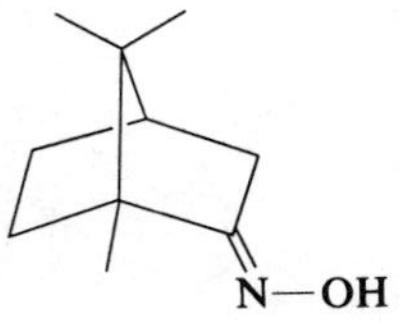 N—OH	HCl, 110°, 10 h	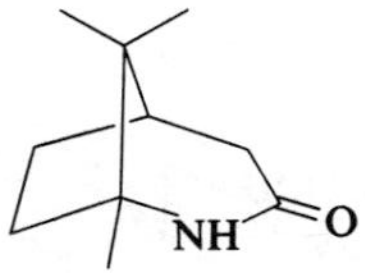 NH, O (—)	168
	TsCl, NaOH, acetone	" (—)	315
	TsCl, pyridine, 50°, 3 h	" (43)	344
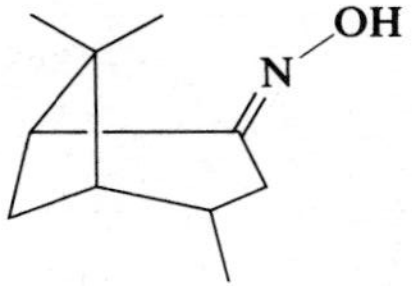 N, OH	TsCl, pyridine, 50°, 5 h	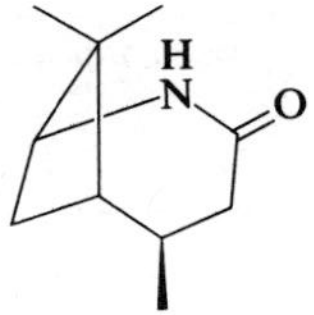 H, N, O (27)	344
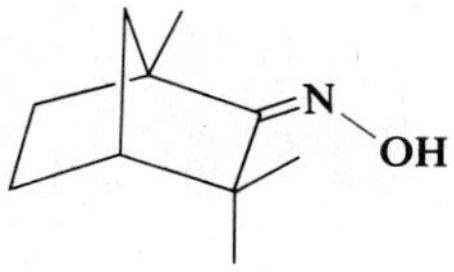 N, OH	TsCl, pyridine	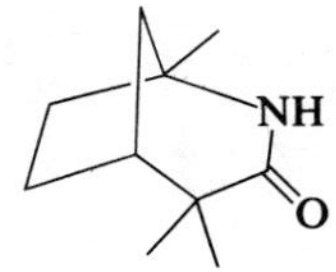 NH, O (33)	348

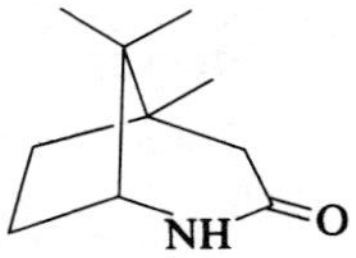

TABLE I. REARRANGEMENTS (*Continued*)

No. of Carbon Atoms	Substrate	Reagent and Conditions	Product(s) and Yield(s) (%)	Refs.
		PCl_5, petroleum ether	" (8)	348
		H_2SO_4	" (7)	348
	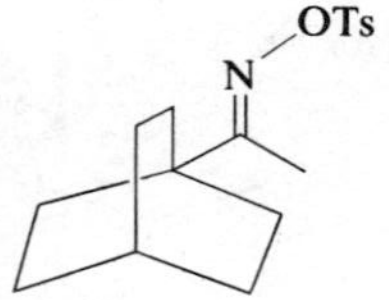 OTs, N	$(C_2H_5)_3N$, 80% ethanol, reflux	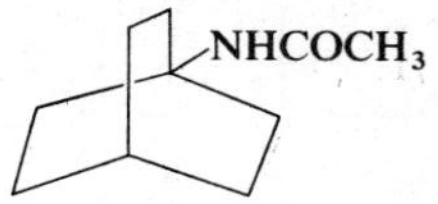 $NHCOCH_3$ (99)	28
		80% ethanol, reflux, 15 h	" (99)	349
	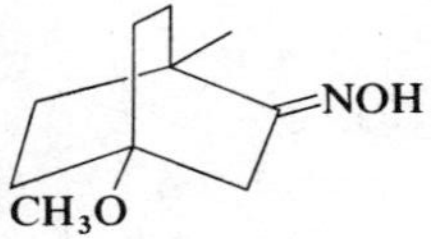 NOH, CH_3O	$C_6H_5SO_2Cl$, NaOH	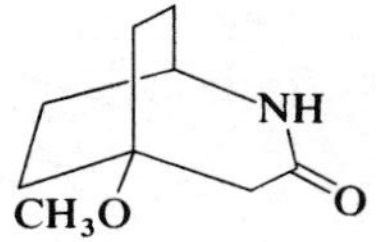 NH, O, CH_3O (23)	297
		TsCl, pyridine	" (27)	297
	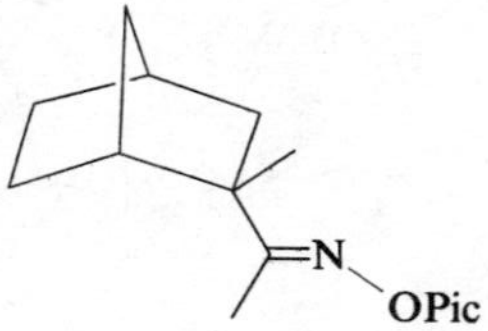 N, OPic	1,4-Dichlorobutane, 80°	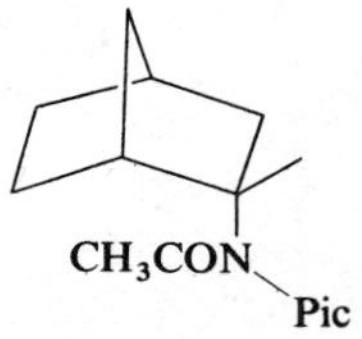 CH_3CON, Pic (—)	94

Reactant	Conditions	Product(s) and Yield(s) (%)	Refs.
(bicyclic oxime, =N–OPic)	1,4-Dichlorobutane, 80°	$CH_3CONPic$ (—)	93
(bicyclic oxime, N–OPic)	"	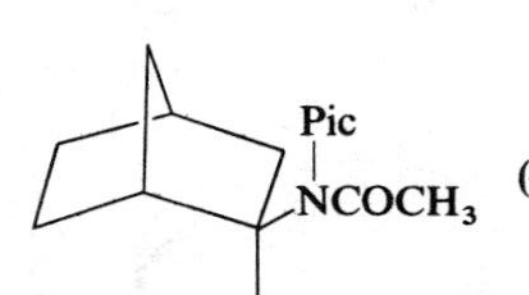 Pic, $NCOCH_3$ (—)	94
(bicyclic oxime, =N–OPic)	"	$CH_3CONPic$ (—)	94
(bicyclic oxime, =N–OPic)	"	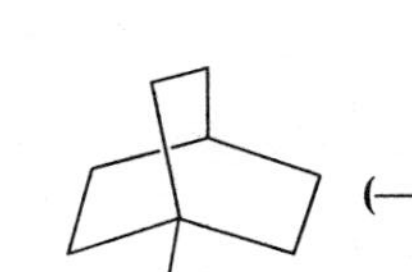 $CH_3CONPic$ (—)	93
NCH_3, CCH_3, NOH	TsCl, dioxane	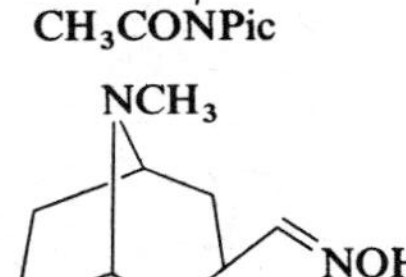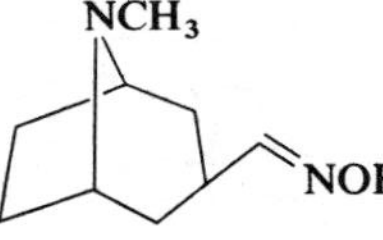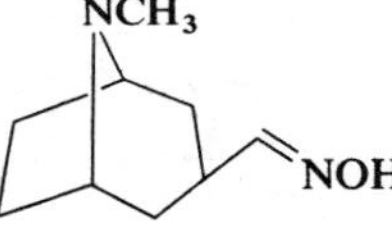 NCH_3, NOH (96)	316
CH_3—CH—CH (O) $B_{10}H_9C(C_6H_5)$=NOH	PCl_5, C_6H_6	CH_3—CH—CH (O) $B_{10}H_9NHCOC_6H_5$ (—)	105

TABLE I. REARRANGEMENTS (*Continued*)

No. of Carbon Atoms	Substrate	Reagent and Conditions	Product(s) and Yield(s) (%)	Refs.
	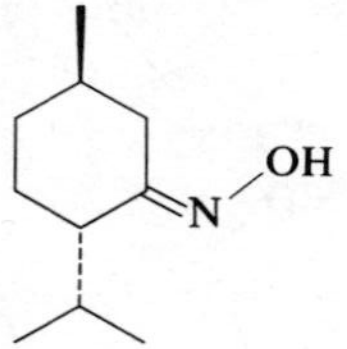 N–OH	$C_6H_5SO_2Cl$, NaOH, H_2O, acetone	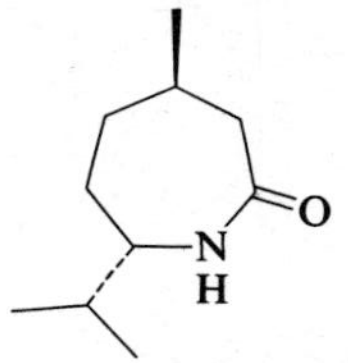 NH, O (—)	275
	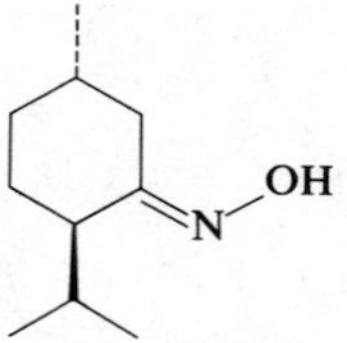 N–OH	TsCl, NaOH, acetone	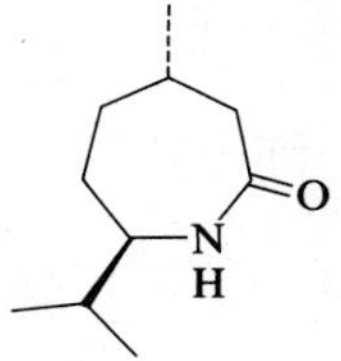 NH, O (—)	347
	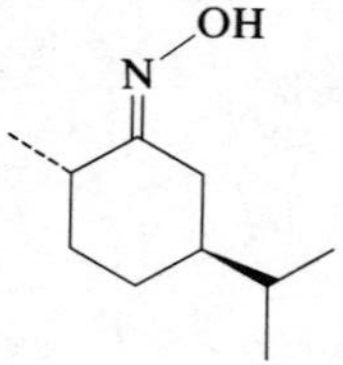 N–OH	"	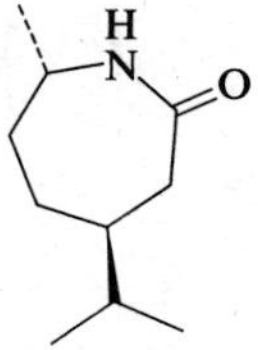 NH, O (—)	347
	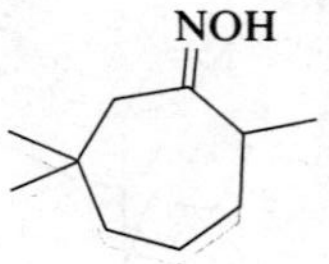 NOH	"	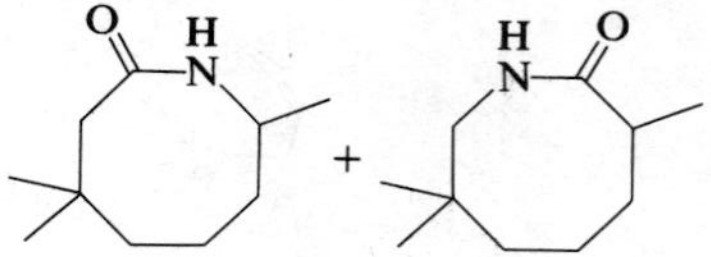 O, NH + NH, O (—)	350

Reactant	Conditions	Product(s) and Yield(s) (%)	Refs.
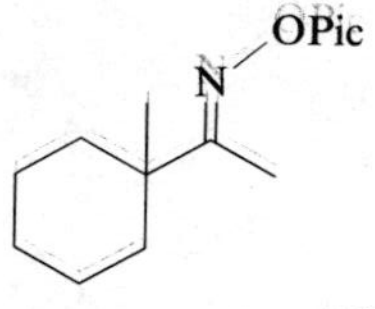 N–OPic	1,4-Dichlorobutane, 80°	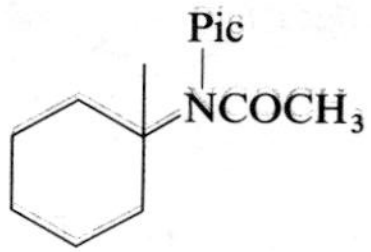 Pic, $NCOCH_3$ (—)	93
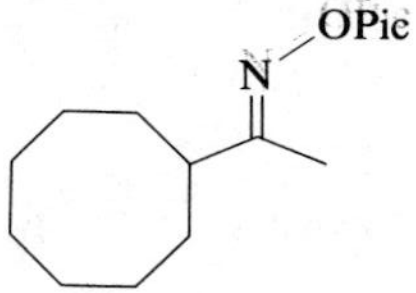 N–OPic	"	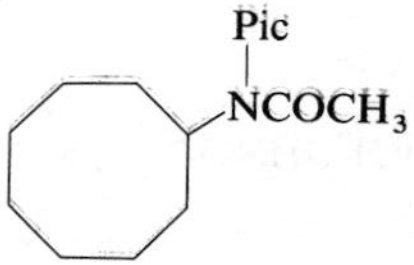 Pic, $NCOCH_3$ (—)	93
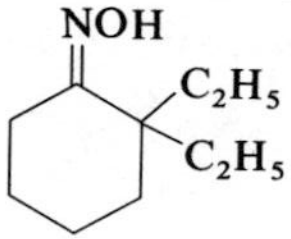 NOH, C_2H_5, C_2H_5	PPA, 115°, 10 min	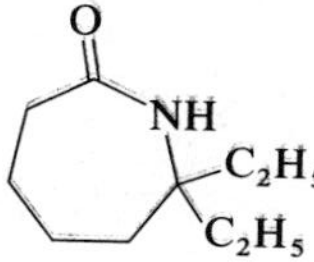 O, NH, C_2H_5, C_2H_5 (29)	351
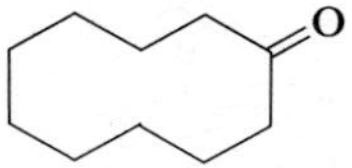 O	H_2NOSO_3H, HCO_2H, reflux, 3 h	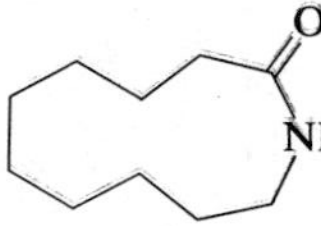 O, NH (76)	81
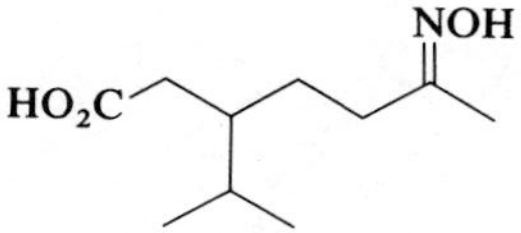 HO_2C, NOH	PPA, 140°, 20 min	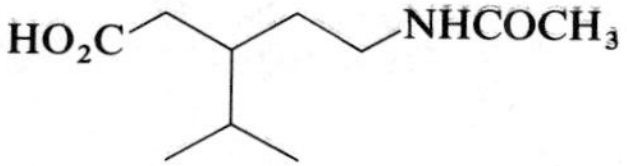 HO_2C, $NHCOCH_3$ (75)	107
	1. PPA, 140°, 20 min 2. H_2O, 100°, 3 h	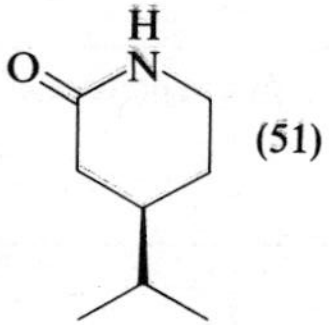 O, H, N (51)	107

TABLE I. REARRANGEMENTS (*Continued*)

No. of Carbon Atoms	Substrate	Reagent and Conditions	Product(s) and Yield(s) (%)	Refs.
	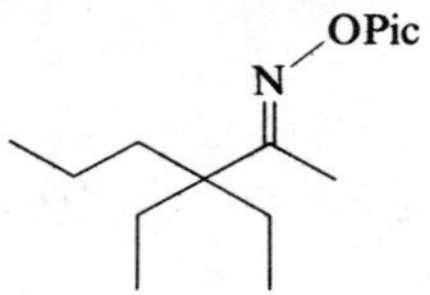 OPic, N	1,4-Dichlorobutane, 80°	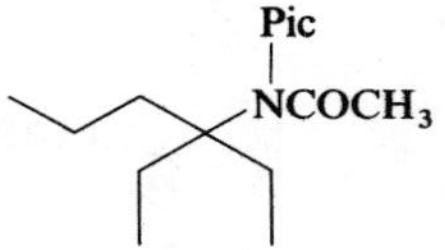 Pic, $NCOCH_3$ (—)	93
	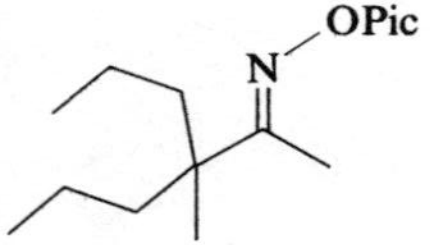 OPic, N	"	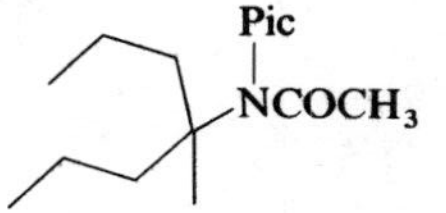 Pic, $NCOCH_3$ (—)	93
C_{11}	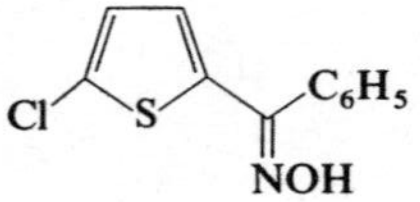 Cl, S, C_6H_5, NOH	PCl_5, ether	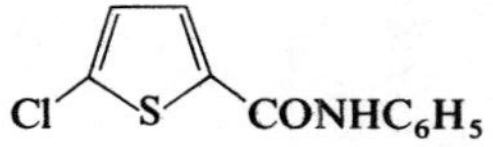 Cl, S, $CONHC_6H_5$ (75)	301
	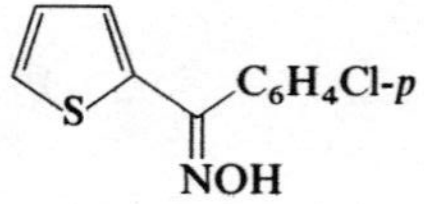 S, C_6H_4Cl-*p*, NOH	"	 S, $CONHC_6H_4Cl$-*p* (24)	352
	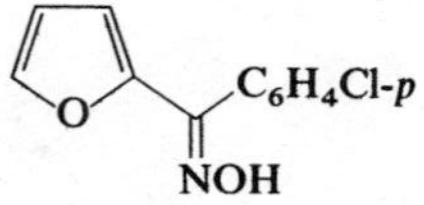 O, C_6H_4Cl-*p*, NOH	"	 O, $CONHC_6H_4Cl$-*p* (20)	352
	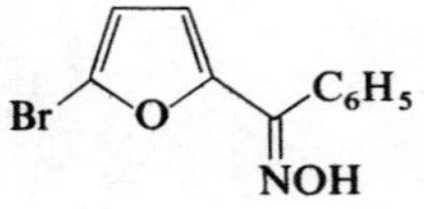 Br, O, C_6H_5, NOH	PPA	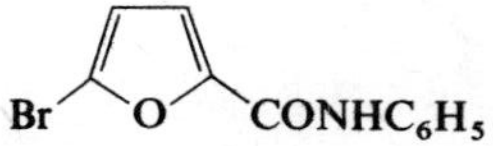 Br, O, $CONHC_6H_5$ (20)	301

Reactant	Conditions	Product(s) (Yield, %)		Refs.
2-thienyl–C(=NOH)–C_6H_5	PCl_5, ether	2-thienyl–$CONHC_6H_5$	(91)	352
	$C_6H_5SO_2Cl$, alumina	"	(90)	77
2-furyl–C(=NOH)–C_6H_5	PCl_5, ether	2-furyl–$CONHC_6H_5$	(60)	352
	PPA, 90-115, 15 min	"	(40)	352
	TsCl, pyridine	"	(—)	353
O_2N–imidazolyl(N–CH_3)–C(=NOH)–C_6H_4Cl-*p*	PPA, 100°, 3h	O_2N–imidazolyl(N–CH_3)–$CONHC_6H_4Cl$-*p*	(75)	354
2-[C(=NOH)CH_3]-3-OH-inden-1-one	PPA, 100°, 10 min	2-$NHCOCH_3$-indane-1,3-dione	(60-64)	355
	CH_3CO_2H	"	(—)	355
	$(CH_3CO)_2O$, cat. pyridine, reflux, 1h	2-$COCH_3$-2-$NHCOCH_3$-indane-1,3-dione	(—)	355
	$(C_2H_5CO)_2O$, cat. pyridine	2-COC_2H_5-2-$NHCOCH_3$-indane-1,3-dione	(—)	355

TABLE I. REARRANGEMENTS (*Continued*)

No. of Carbon Atoms	Substrate	Reagent and Conditions	Product(s) and Yield(s) (%)	Refs.
	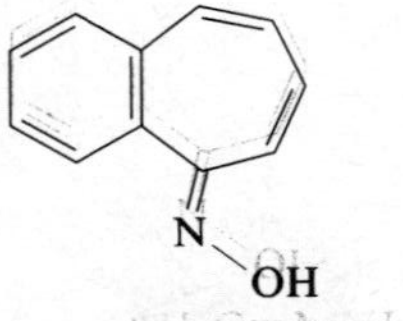 N—OH (oxime)	$C_6H_5SO_2Cl$, pyridine, 0°	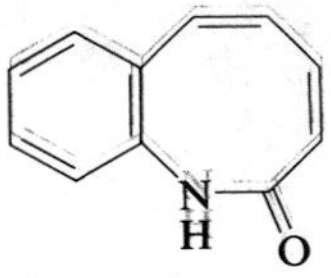 N H, O (68)	356
	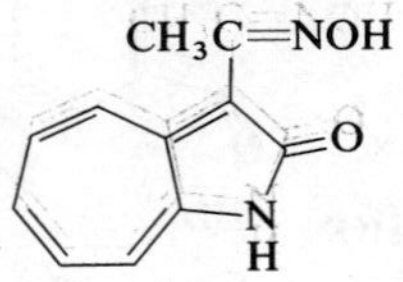 $CH_3C{=}NOH$, O, N H	$(CH_3CO)_2O$, reflux, 4 h	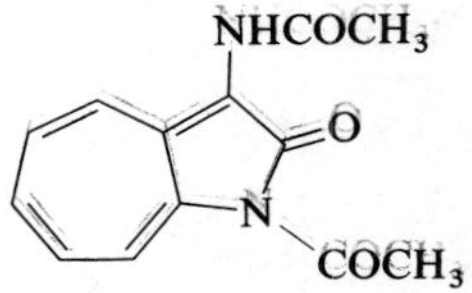 $NHCOCH_3$, O, N, $COCH_3$ (100)	357
	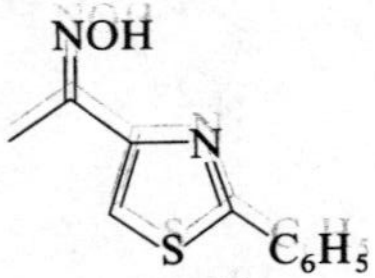 NOH, N, S, C_6H_5	PCl_5, ether	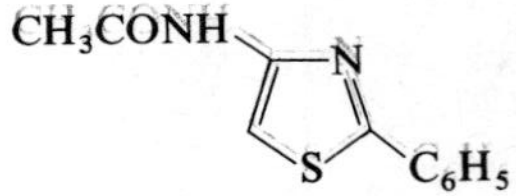 CH_3CONH, N, S, C_6H_5 (34)	246
	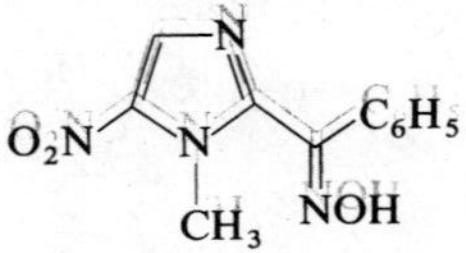 N, O_2N, N, CH_3, C_6H_5, NOH	PPA, 100°, 3 h	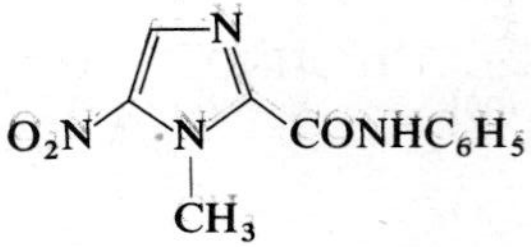 N, O_2N, N, CH_3, $CONHC_6H_5$ (92)	354
	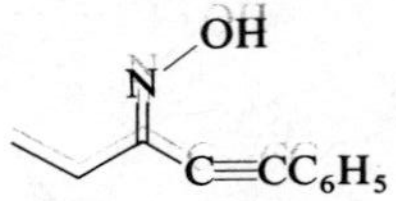 OH, N, $C{\equiv}CC_6H_5$	PCl_5, ether	$C_6H_5CCl{=}CHCONHC_2H_5$ (—)	116

Oxime	Conditions	Product (Yield, %)	Ref.
NOH; N, N, C_6H_5	PCl_5, $CHCl_3$, 4 d	NHCOCH$_3$; N, N, C_6H_5 — I + CONHCH$_3$; N, N, C_6H_5 — II I (72-80) II (17-21)	318
	TsCl, pyridine, reflux, 6 h	I (55-56) + II (32-35)	318
	PPA, 130°, 30 min	I (65-71) + II (27)	318
p-$CH_3OC_6H_4$; N, N, N; N, OH	PCl_5, $CHCl_3$, 2 d	p-$CH_3OC_6H_4$; N, N, N; NH_2 (20)	321
CH_3C; NOH; Cl, Cl; N=; H, N, N, H	PCl_5, CS_2	CH_3CONH—; Cl, Cl; N=; H, N, N, H (69)	358
$COCH_3$; N, CH_3	$H_2NOH \cdot HCl$, ethanol reflux, 3 d	$NHCOCH_3$; N, CH_3 (20)	320
=NOH; N, CH_3 (E)	CH_3CO_2H, reflux, 2 h	" (40)	320

TABLE I. REARRANGEMENTS (*Continued*)

No. of Carbon Atoms	Substrate	Reagent and Conditions	Product(s) and Yield(s) (%)	Refs.
	(*Z*)	PCl_5, ether, 24 h	1-methyl-3-($CONHCH_3$)indole (100)	320
	m-$XC_6H_4CH=C(CH_3)C(=NOH)CH_3$ X = F	PCl_5, ether, dioxane, 0°, 6 h	m-$XC_6H_4CH=C(CH_3)NHCOCH_3$ (60)	359
	X = Cl	"	" (—)	111, 359
	p-$ClC_6H_4CH=C(CH_3)C(=NOH)CH_3$	PCl_5, ether, 0°, 15 min	p-$ClC_6H_4CH=C(CH_3)NHCOCH_3$ (—)	111,112
	o-$ClC_6H_4CH=C(CH_3)C(=NOH)CH_3$	PCl_5, ether, 0°, 24 h	o-$ClC_6H_4CH=C(CH_3)NHCOCH_3$ (—)	112
	p-$O_2NC_6H_4CH=C(CH_3)C(=NOH)CH_3$	PCl_5, ether, dioxane, 0°, 6 h	p-$O_2NC_6H_4CH=C(CH_3)NHCOCH_3$ (55)	112, 359
	m-$O_2NC_6H_4CH=C(CH_3)C(=NOH)CH_3$	PCl_5, ether, 0°, 24 h	m-$O_2NC_6H_4CH=C(CH_3)NHCOCH_3$ (—)	112

Oxime	Conditions	Product(s) and Yield(s) (%)	Refs.
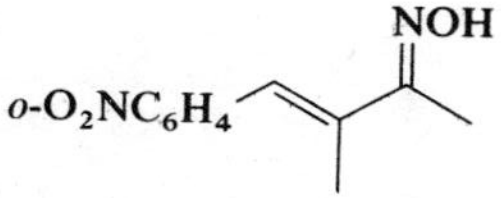 $o\text{-}O_2NC_6H_4$, NOH (structure)	PCl_5, ether, 0°, 24 h	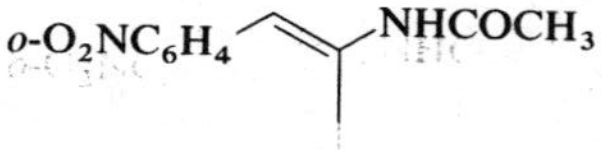 $o\text{-}O_2NC_6H_4$, $NHCOCH_3$ (structure) (—)	112
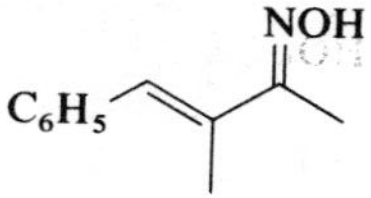 C_6H_5, NOH (structure)	”	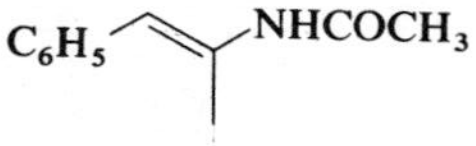 C_6H_5, $NHCOCH_3$ (structure) (93)	111, 112
	PCl_5, dioxane, 5°, 15 min	(94)	110
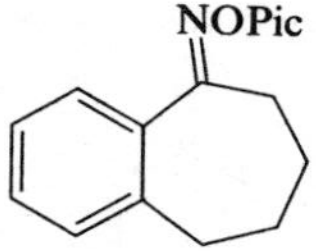 NOPic (structure) (*E*)	$C_2H_4Cl_2$, warm	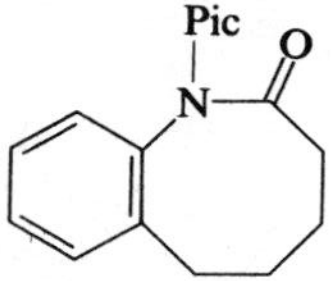 Pic, N, O (structure) (83)	96
(*Z*)	”	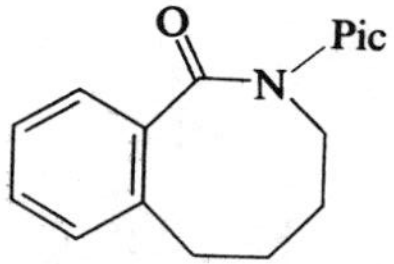 O, N, Pic (structure) (83)	96
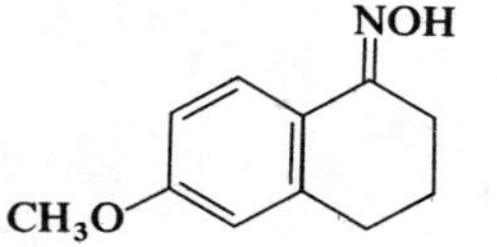 NOH, CH_3O (structure)	PPA, 120°, 10 min	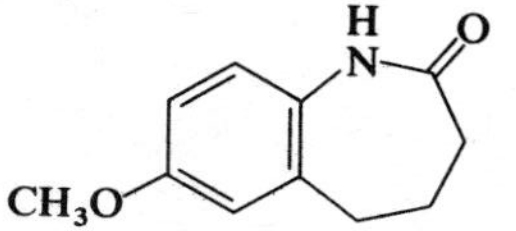 H, N, O, CH_3O (structure) (28) 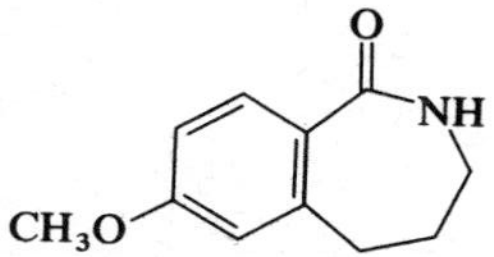 + O, NH, CH_3O (structure) (52)	322, 360

TABLE I. REARRANGEMENTS (*Continued*)

No. of Carbon Atoms	Substrate	Reagent and Conditions	Product(s) and Yield(s) (%)	Refs.
	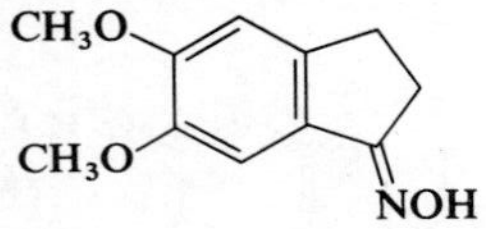 CH_3O, CH_3O, NOH	PPA, 100°, 75 min	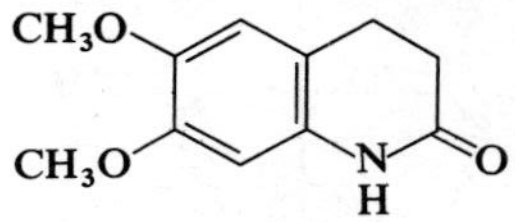 CH_3O, CH_3O, N H, O (50)	319
	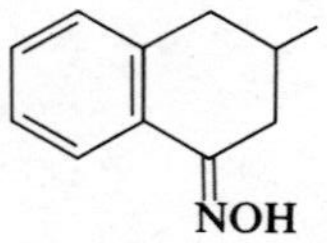 NOH	PPA, 105°, 2 h	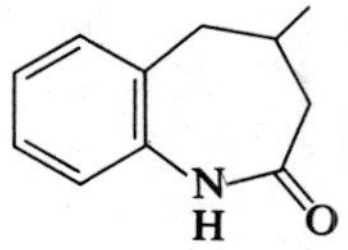 N H, O (90)	361
	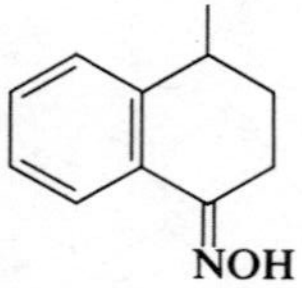 NOH	PPA, 130-140°, 2.5 h	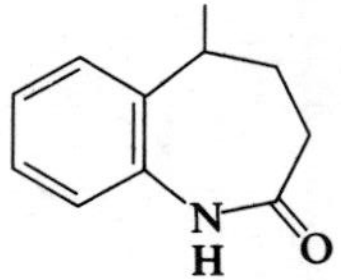 N H, O (70)	361
	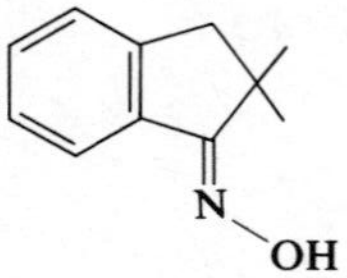 N, OH	PPA, 125–130°, 10 min	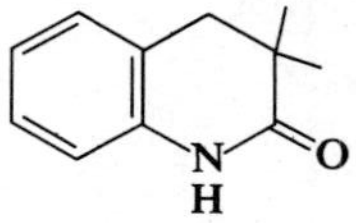 N H, O (97)	362
	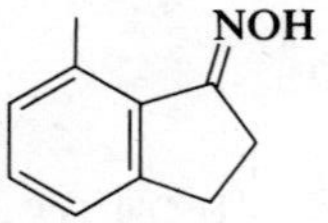 NOH	PPA, 110–120°, 5–10 min	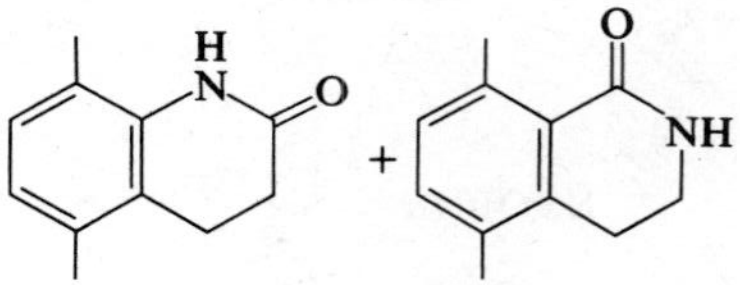 H N, O + O, NH 34:66	70, 71

Reactant		Conditions	Product(s) and Yield(s) (%)	Refs.
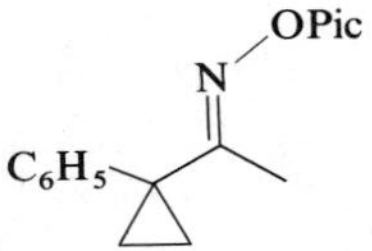 C_6H_5, =N–OPic		1,4-Dichlorobutane, 80°	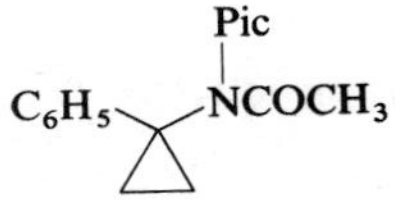 C_6H_5, N(Pic)$COCH_3$ (—)	93
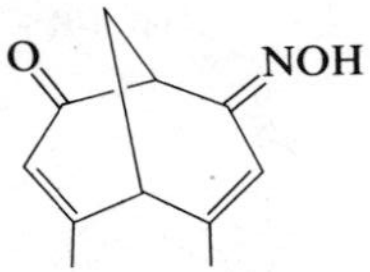 NOH		TsCl, alumina	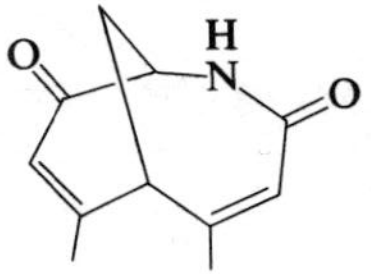 (47)	363
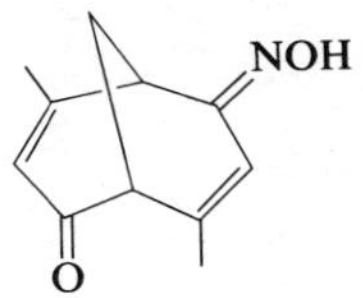 NOH		TsCl, alumina, pyridine	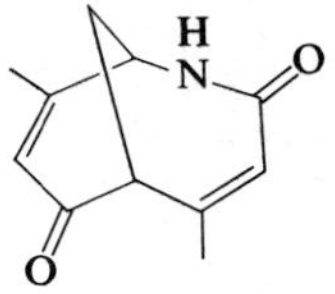 (39)	363
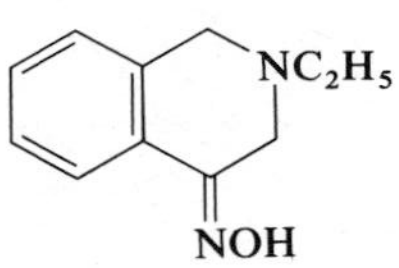 NC_2H_5, NOH		PPA, 100°, 30 min	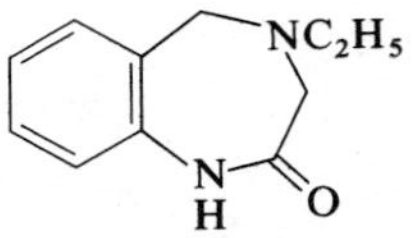 NC_2H_5 (—)	325
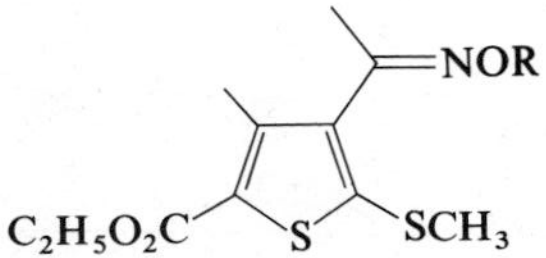 $C_2H_5O_2C$, SCH_3, =NOR	R = H	PCl_5, ether, 1 h	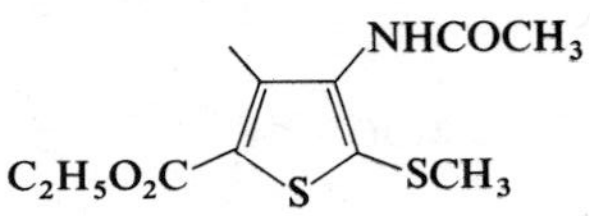 $C_2H_5O_2C$, SCH_3, $NHCOCH_3$ (93)	266
R = $C_6H_4NO_2$-*p*		H_2SO_4, −5°, 10 min	" (32)	266
R = $C_6H_4NO_2$-*p*		CH_3CO_2H, reflux, 1 h	" (81)	266
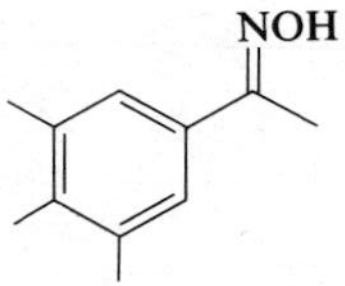 NOH		PCl_5, ether, 5°	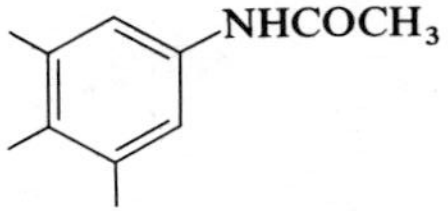 $NHCOCH_3$ (—)	364

TABLE I. REARRANGEMENTS (*Continued*)

No. of Carbon Atoms	Substrate	Reagent and Conditions	Product(s) and Yield(s) (%)	Refs.
	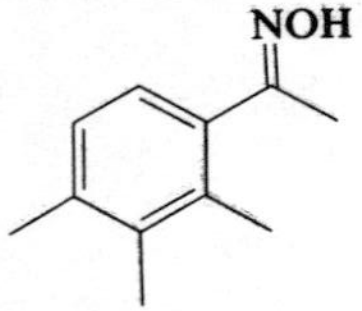 NOH	HCl	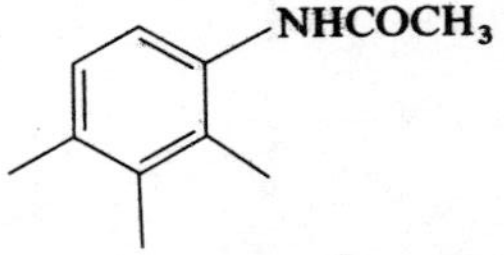 NHCOCH$_3$ (—)	364
		PCl_5	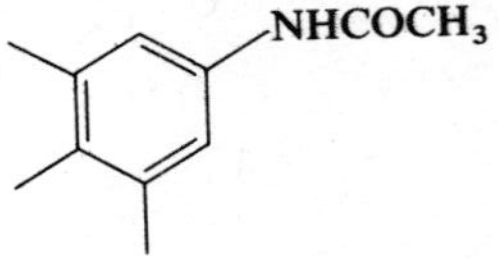 " + NHCOCH$_3$ (—)	364
	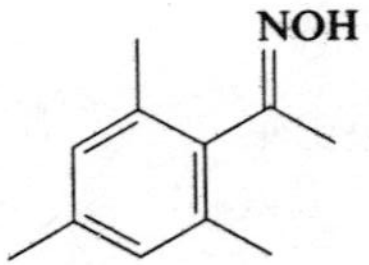 NOH	84.8% H_2SO_4, 80°	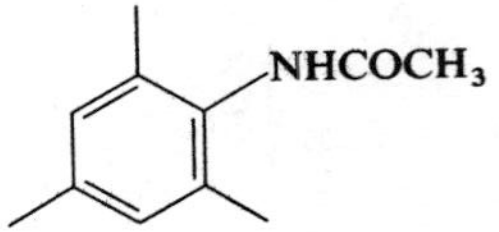 NHCOCH$_3$ (—)	281
	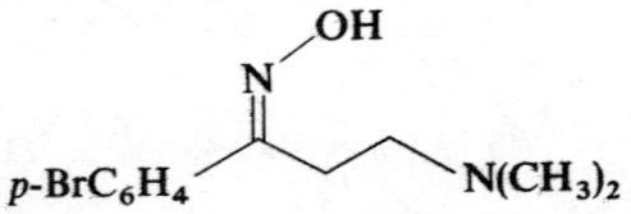 p-BrC$_6$H$_4$C(=NOH)CH$_2$CH$_2$N(CH$_3$)$_2$	PCl_5, ether, reflux, 30 min	p-BrC$_6$H$_4$NHCO(CH$_2$)$_2$N(CH$_3$)$_2$ (98)	114
	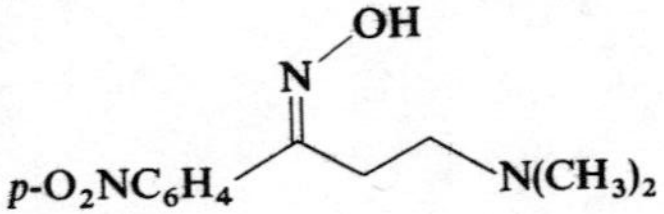 p-O$_2$NC$_6$H$_4$C(=NOH)CH$_2$CH$_2$N(CH$_3$)$_2$	PCl_5, ether, C_6H_6, 40°	p-O$_2$NC$_6$H$_4$NHCO(CH$_2$)$_2$N(CH$_3$)$_2$ (71)	114

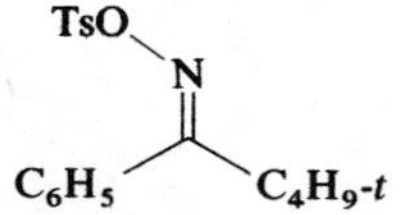 TsO–N=C(C_6H_5)C_4H_9-*t*	$(C_2H_5)_3N$, 80% ethanol, 23°, 48 h	$C_6H_5CONHC_4H_9$-*t* (86)	28
	$FeCl_3$, 1 *N* HCl, dioxane, 15 h	" (75)	28
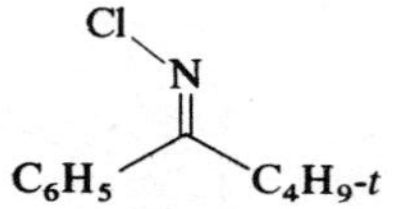 Cl–N=C(C_6H_5)C_4H_9-*t*	$SbCl_5$, CCl_4, 45°	$C_6H_5C{\equiv}\overset{+}{N}C_4H_9\text{-}t\cdot SbCl_6^-$	28
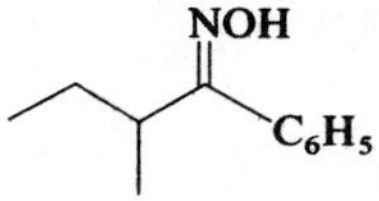 NOH, C_6H_5 (*Z*)	1. TsCl, pyridine 2. $(C_2H_5)_3N$, dioxane, 14 h	$C_6H_5CONHC_4H_9$-*s* (87) I	330
(*E*)	"	$C_6H_5NHCOC_4H_9$-*s* (99) II	330
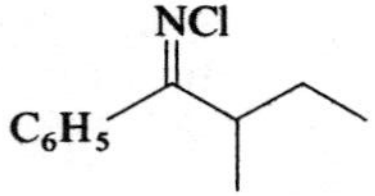 NCl, C_6H_5	$AgBF_4$, 75% dioxane-H_2O, 80°	I (31) + II (11)	120
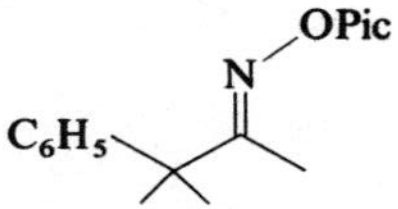 OPic, N, C_6H_5	1,4-Dichlorobutane, 80°	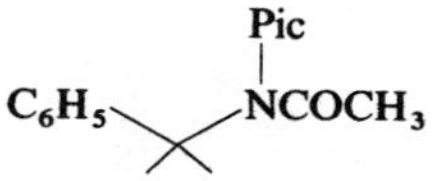 Pic, C_6H_5, $NCOCH_3$ (—)	93
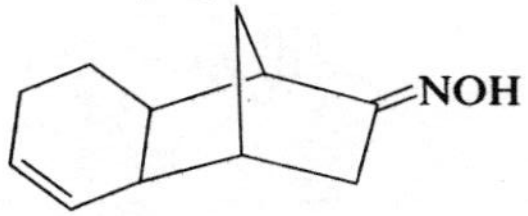 =NOH	$C_6H_5SO_2Cl$, NaOH	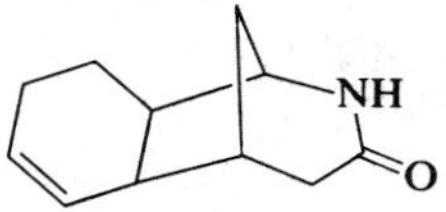 NH, O (31)	265
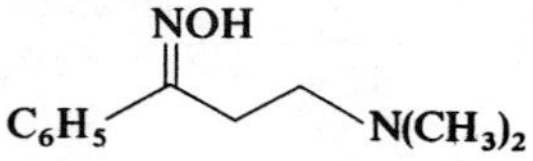 NOH, C_6H_5, $N(CH_3)_2$	PCl_5, ether, 0°, 1 h	$C_6H_5NHCO(CH_2)_2N(CH_3)_2$ (81)	114

TABLE I. REARRANGEMENTS (*Continued*)

No. of Carbon Atoms	Substrate	Reagent and Conditions	Product(s) and Yield(s) (%)	Refs.
	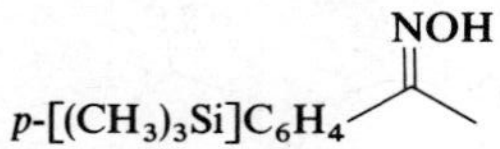 p-[$(CH_3)_3Si$]C_6H_4C(=NOH)CH_3	$SOCl_2$, ether, reflux, 30 min	p-[$(CH_3)_3Si$]$C_6H_4NHCOCH_3$ (—)	365
		PCl_5, ether, reflux, 30 min	" (87)	365
		H_2SO_4	" (73)	365
	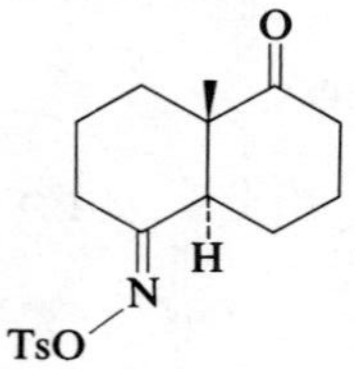 (TsO–N=, H, O)	$(C_2H_5)_3N$, 80% ethanol, 25°	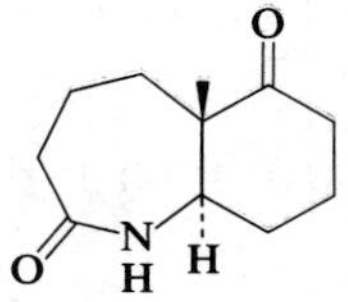 (O, N H, H) I (77)	79
		NaOH, 80% ethanol	I 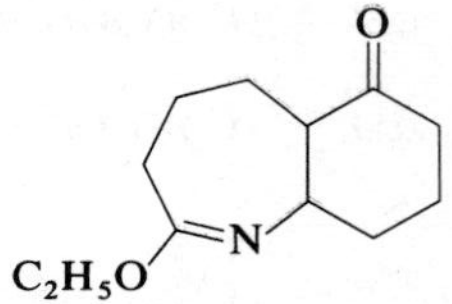 + (O, C_2H_5O, N) (—)	80
	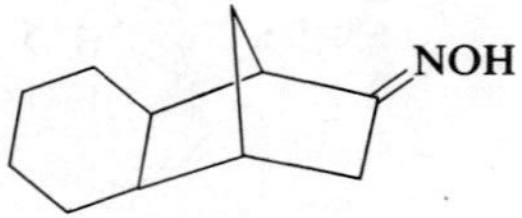 (NOH)	$C_6H_5SO_2Cl$, NaOH	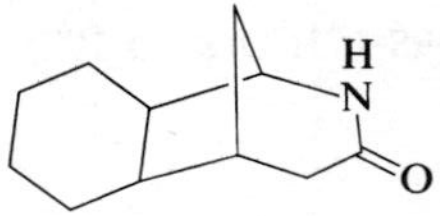 (H, N, O) (44)	265
	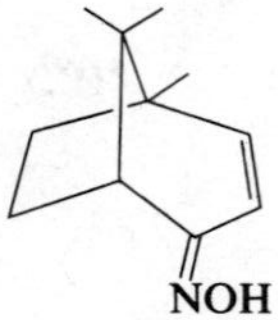 (NOH)	PPA, 125°, 1 h	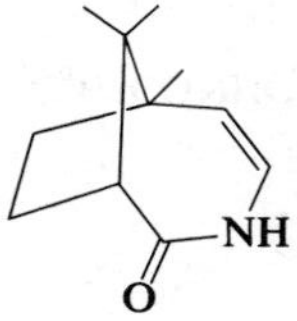 (NH, O) (40–60)	366

Reactant	Conditions	Product(s) and Yield(s) (%)	Refs.
	TsCl, pyridine, 90–95°, 15 min	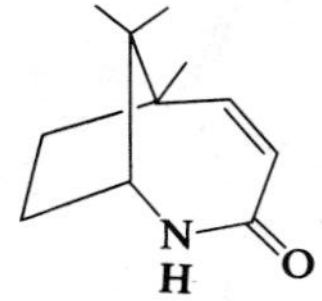 (62)	366
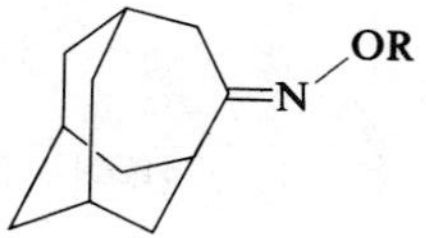 R = H	PPE, $CHCl_3$, 80°	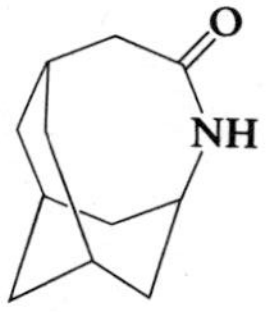 NH (58) I	367
R = H	85% H_2SO_4	I 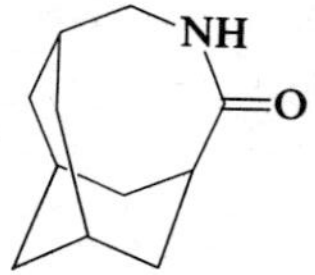 + NH =O II 1:4 (60)	367
R = H	PPE, $CHCl_3$, reflux, 30 min	I (21)	368
R = H	PCl_5, $CHCl_3$, 15 h	I (64)	368
R = H	HCl, CH_3CN, 80°, 2 h	I (56)	368
R = H	TsCl, DMF, 20 h	I (83)	368
R = SO_3H	H_2O, 50 h	I (65)	258
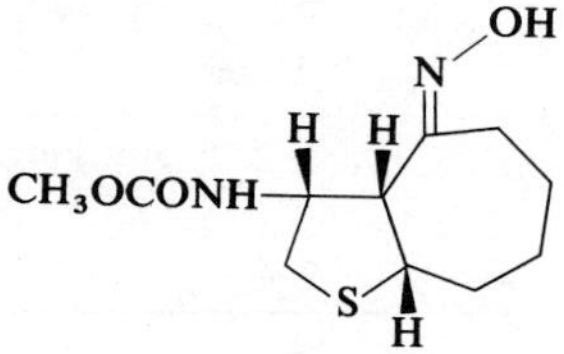 CH_3OCONH (oxime, N–OH)	PPA, 100°, 5 min	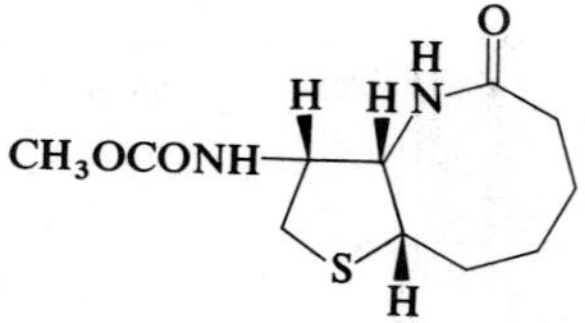 CH_3OCONH (22)	195, 196

TABLE I. REARRANGEMENTS (*Continued*)

No. of Carbon Atoms	Substrate	Reagent and Conditions	Product(s) and Yield(s) (%)	Refs.
	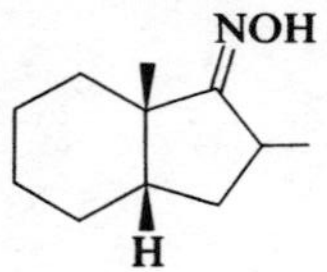 NOH, H	$SOCl_2$, dioxane	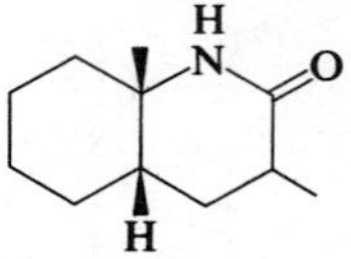 H, N, O, H (41)	311
	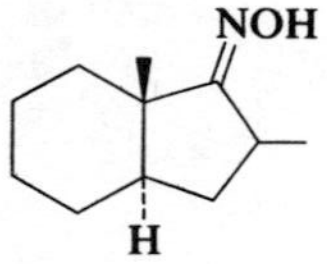 NOH, H	"	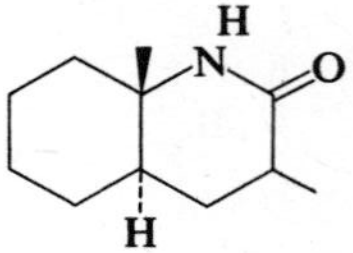 H, N, O, H (40)	311
	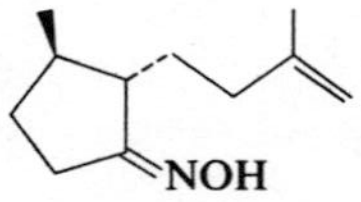 NOH	PCl_5, ether	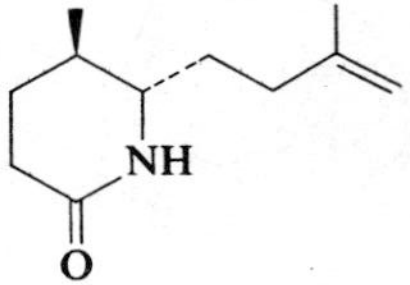 NH, O (68)	295, 369
	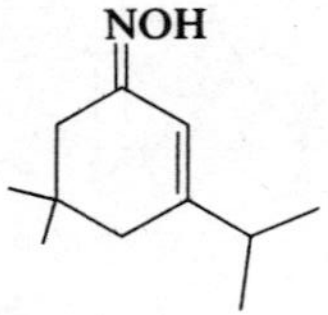 NOH	PPA, 120°	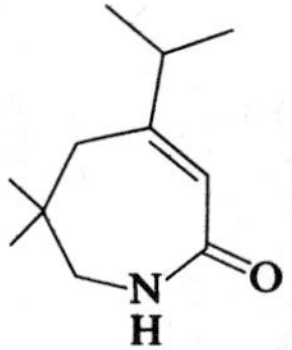 N, H, O (32) 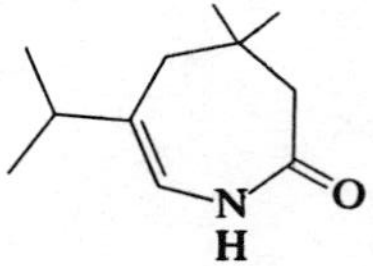 + N, H, O (24)	46

Reactant	Conditions	Product(s) (Yield, %)	Refs.
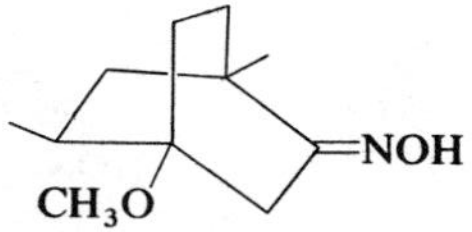 =NOH, CH_3O	$C_6H_5SO_2Cl$, NaOH	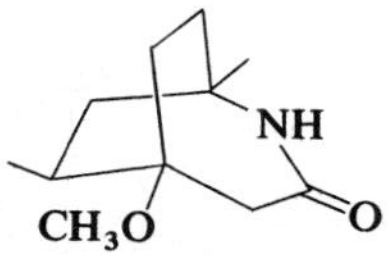 NH, O, CH_3O (23)	297
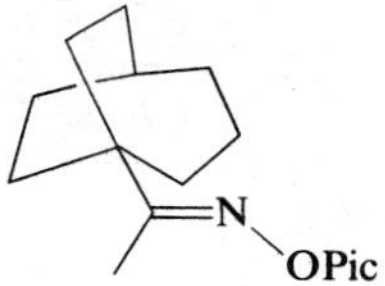 N, OPic	1,4-Dichlorobutane, 80°	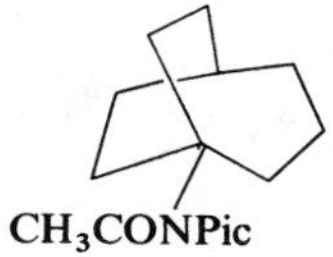 $CH_3CONPic$ (—)	93
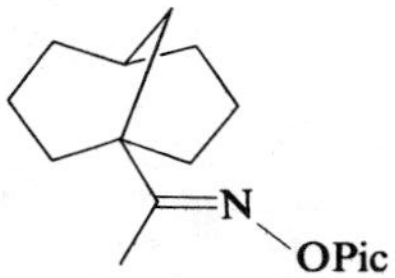 N, OPic	"	 $CH_3CONPic$ (—)	93
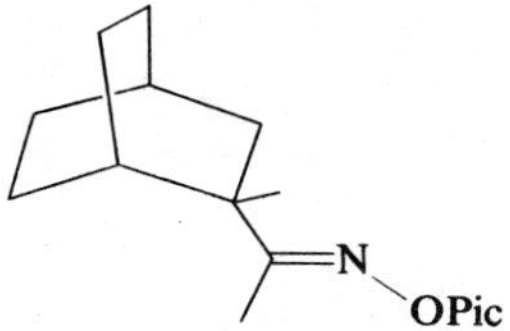 N, OPic	"	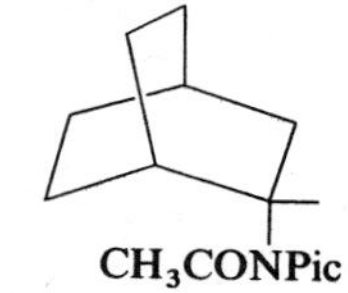 $CH_3CONPic$ (—)	94
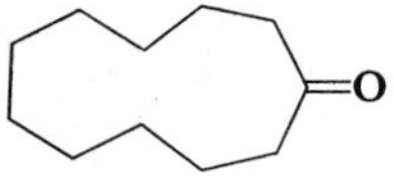 =O	H_2NOSO_3H, HCO_2H, reflux, 4 h	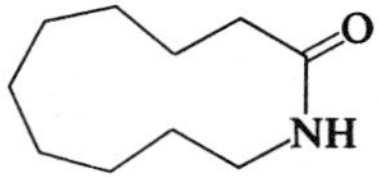 O, NH (77–87)	81
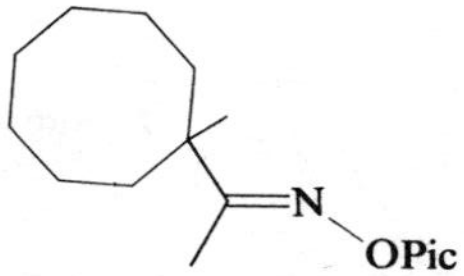 N, OPic	1,4-Dichlorobutane, 80°	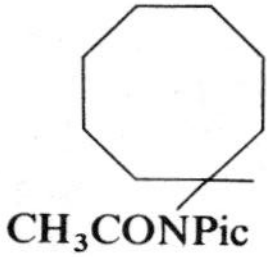 $CH_3CONPic$ (—)	93

TABLE I. REARRANGEMENTS (*Continued*)

No. of Carbon Atoms	Substrate	Reagent and Conditions	Product(s) and Yield(s) (%)	Refs.
	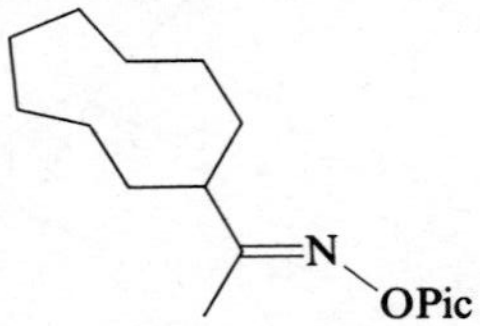 N—OPic	"	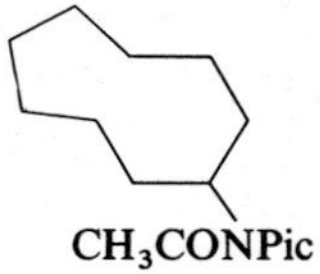 $CH_3CONPic$ (—)	93
	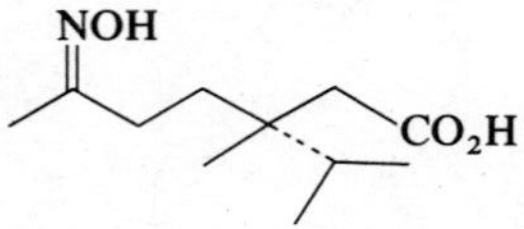 NOH, CO_2H	1. PPA, 135–140° 2. H_2O, reflux, 30 min	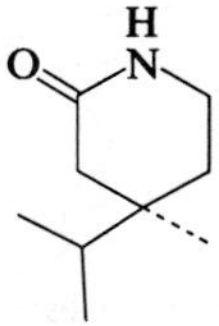 O, H, N (55)	107
	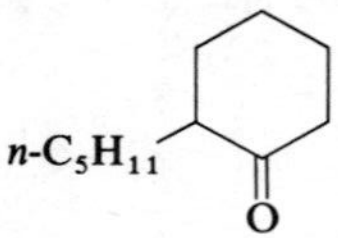 n-C_5H_{11}, O	H_2NOSO_3H, HCO_2H, reflux, 1 h	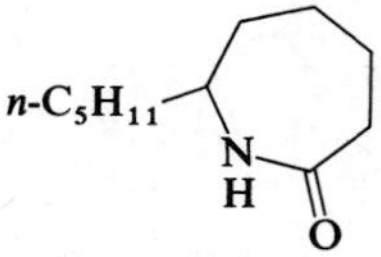 n-C_5H_{11}, N, H, O (82)	370
	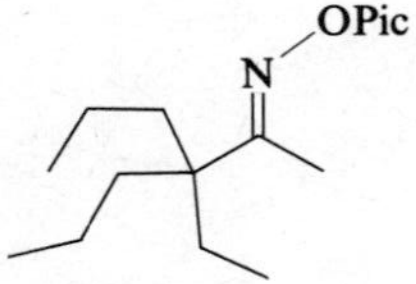 OPic, N	1,4-Dichlorobutane, 80°	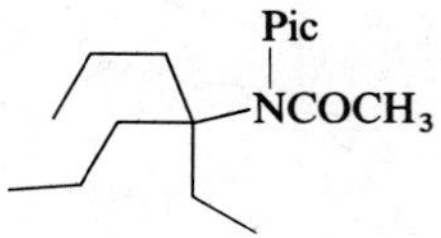 Pic, $NCOCH_3$ (—)	93
	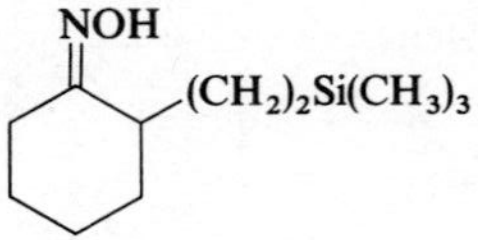 NOH, $(CH_2)_2Si(CH_3)_3$	PCl_5, ether, 4 h	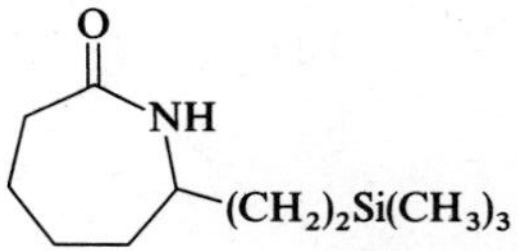 O, NH, $(CH_2)_2Si(CH_3)_3$	371

	Reactant	Conditions	Product(s) and Yield(s) (%)	Refs.
			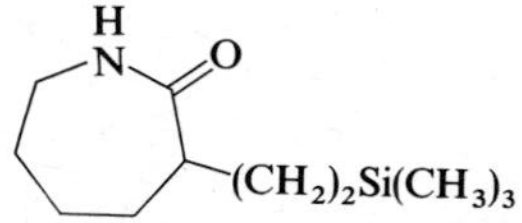 + (CH$_2$)$_2$Si(CH$_3$)$_3$ 3:1 (50)	
C_{12}	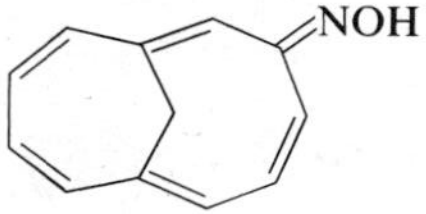 =NOH	1. TsCl, pyridine 2. Dioxane, H_2O, 2,6-lutidine	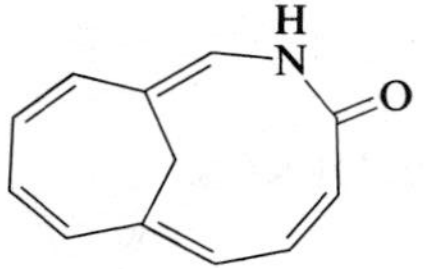 NH, O (71)	372
	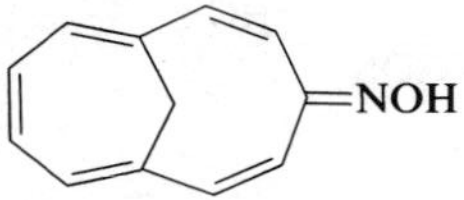 =NOH	TsCl, pyridine, 3 h	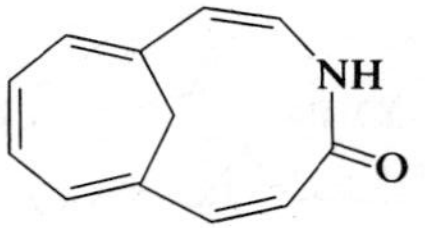 NH, O (70)	372
	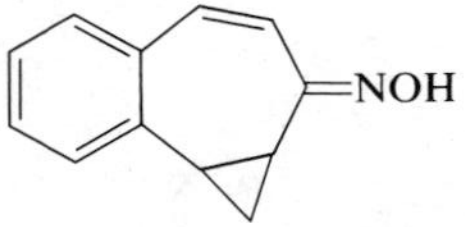 =NOH	TsCl, pyridine	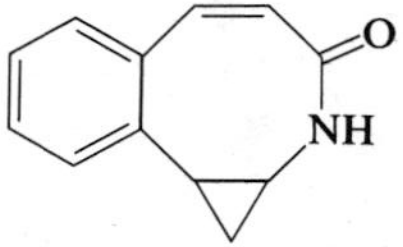 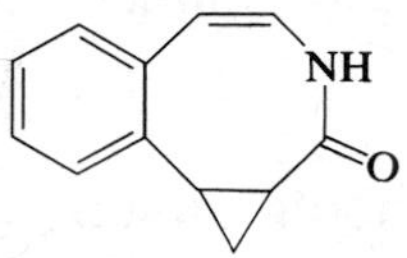 O, NH + NH, O 1:1 (37)	373
	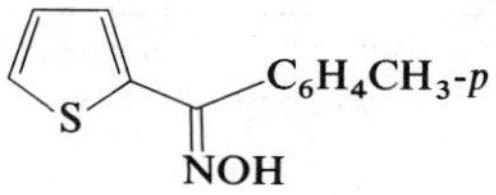 $C_6H_4CH_3$-*p*, NOH, S	PCl_5, ether	 S, $CONHC_6H_4CH_3$-*p* (80)	352
		PPA, 90–115°, 15 min	" (80)	352

TABLE I. REARRANGEMENTS (*Continued*)

No. of Carbon Atoms	Substrate	Reagent and Conditions	Product(s) and Yield(s) (%)	Refs.
	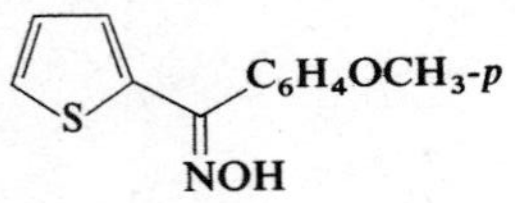 $C_6H_4OCH_3$-*p*, NOH (S)	PCl_5, ether	 (S) $CONHC_6H_4OCH_3$-*p* (88)	352
	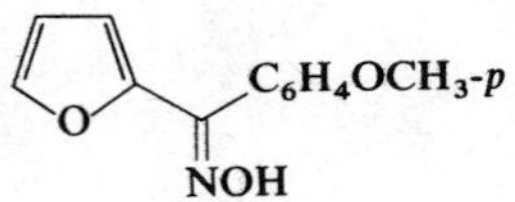 $C_6H_4OCH_3$-*p*, NOH (O)	"	 (O) $CONHC_6H_4OCH_3$-*p* (40)	352
		PPA, 90–115°, 15 min	" (82)	352
		TsCl, pyridine	" (—)	353
	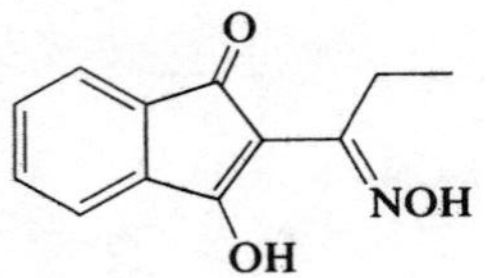 O, NOH, OH	$(CH_3CO)_2O$, cat. pyridine, reflux, 1.5 h	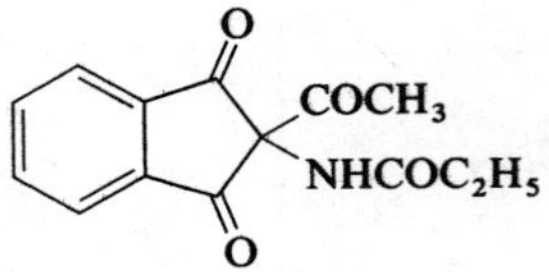 O, $COCH_3$, $NHCOC_2H_5$, O (—)	355
	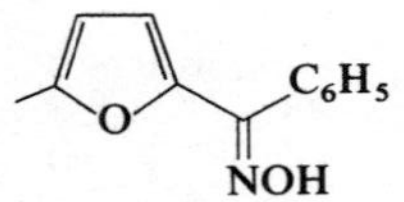 C_6H_5, NOH (O)	PCl_5, ether	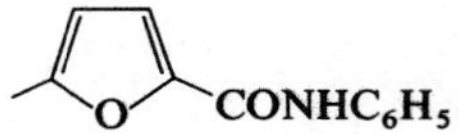 (O) $CONHC_6H_5$ (60)	301
	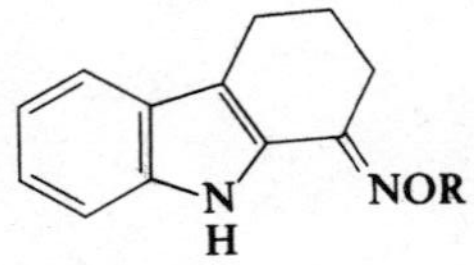 N H, NOR; R = H (*E* or *Z*)	PPA	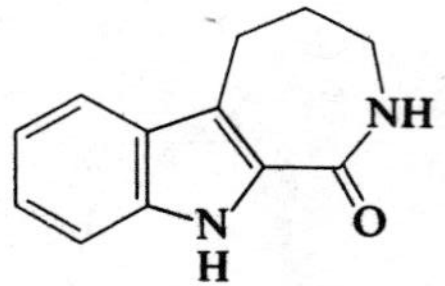 NH, N H, O (73–85)	374
	(*Z*) R = Ts	Alumina	" (81)	374

Reactant	Conditions	Product(s) (% yield)	Ref.
(*E*) R = H	"	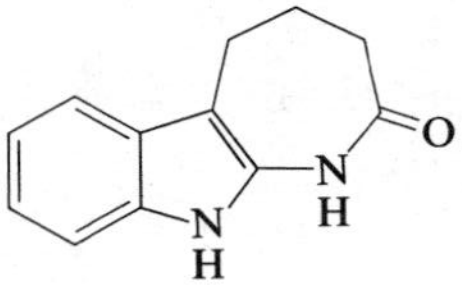 (25)	374
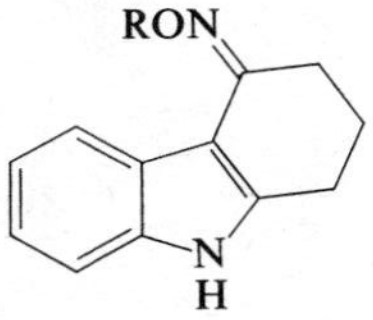 RON=; R = H	PPA, 102–112°, 10 min	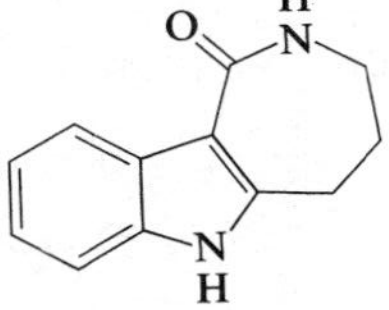 (40)	375
R = Ts	Alumina, C_6H_6	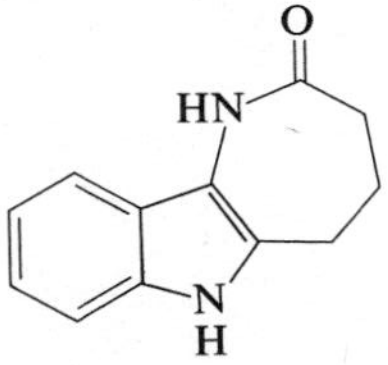 (41)	375
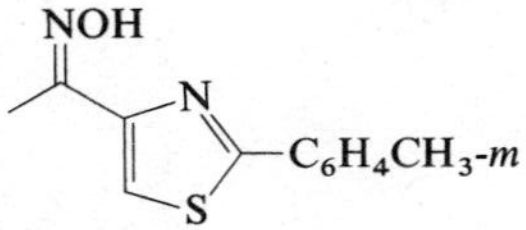 NOH; $C_6H_4CH_3$-*m*	PCl_5, ether	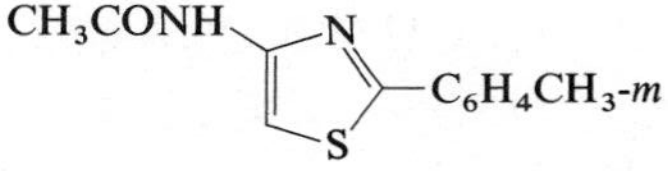 CH_3CONH; $C_6H_4CH_3$-*m* (35)	246
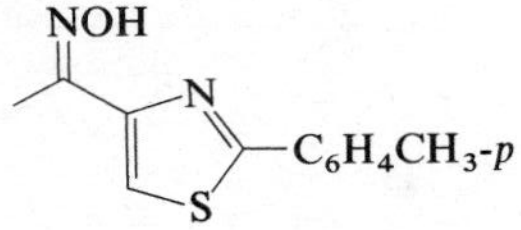 NOH; $C_6H_4CH_3$-*p*	"	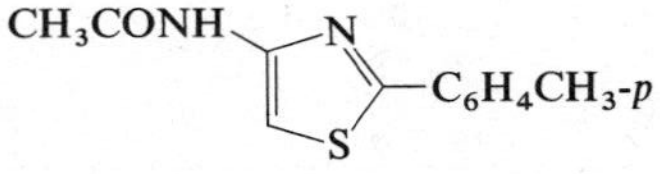 CH_3CONH; $C_6H_4CH_3$-*p* (31)	246
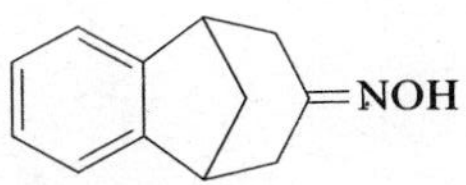 =NOH	PPA, 120°, 30 min	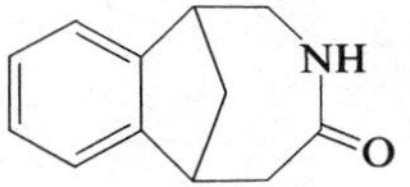 NH; (94)	376

TABLE I. REARRANGEMENTS (*Continued*)

No. of Carbon Atoms	Substrate	Reagent and Conditions	Product(s) and Yield(s) (%)	Refs.
	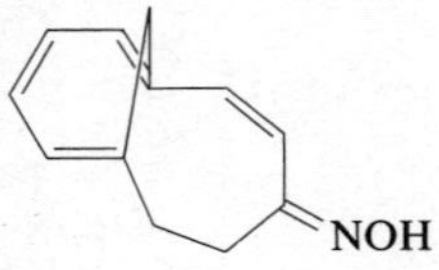 NOH	TsCl, pyridine	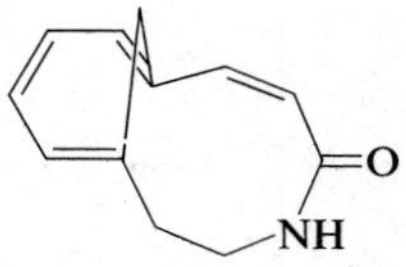 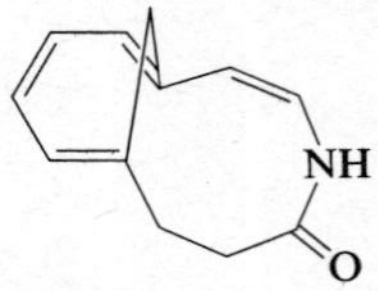 O + NH NH O 5:1 (59)	372
	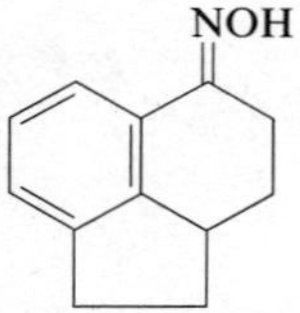 NOH	PPA, 110°, 15 min	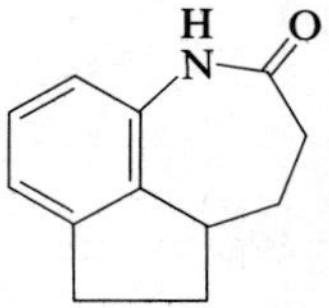 H N O (87)	377
	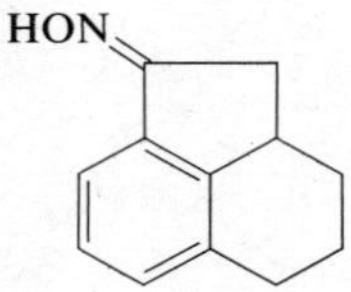 HON	PPA, 120°, 15 min	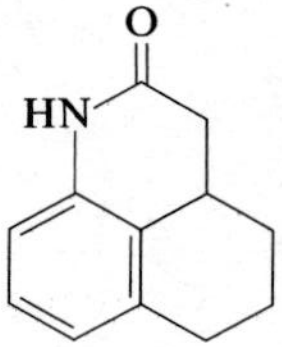 O HN (50)	377
	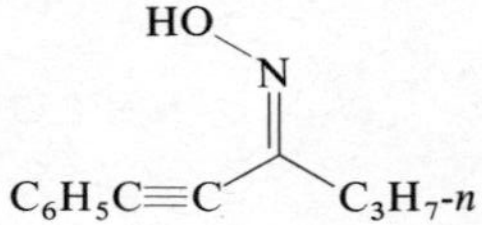 HO N $C_6H_5C{\equiv}C$ C_3H_7-*n*	PCl_5, ether	$C_6H_5CCl{=}CHCONHC_3H_7$-*n* (—)	116

Reactant	Conditions	Product(s) (Yield, %)	Refs.
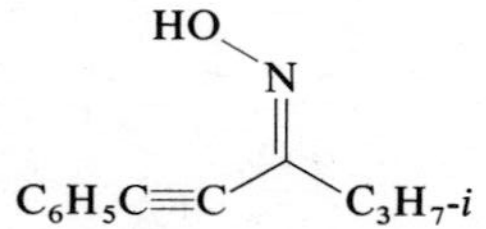 HO N $C_6H_5C\equiv C$ C_3H_7-*i*	"	$C_6H_5CCl{=}CHCONHC_3H_7$-*i* (—)	116
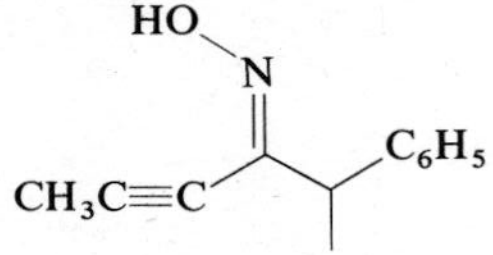 HO N $CH_3C\equiv C$ C_6H_5	"	$CH_3CCl{=}CHCONHCH(CH_3)C_6H_5$ (—)	116
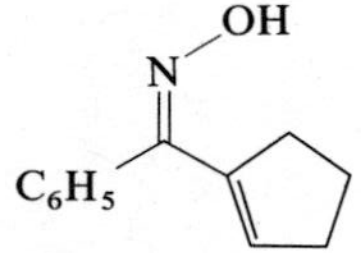 OH N C_6H_5	BF_3, CH_3CO_2H, 0°	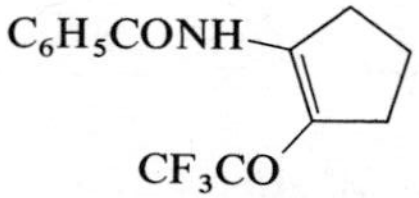 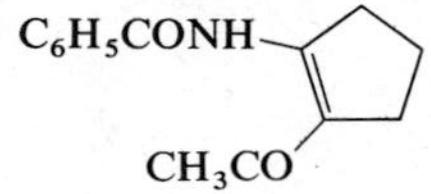 C_6H_5CONH CF_3CO + C_6H_5CONH CH_3CO (25)	288
	PCl_5	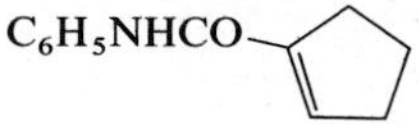 C_6H_5NHCO (—)	288
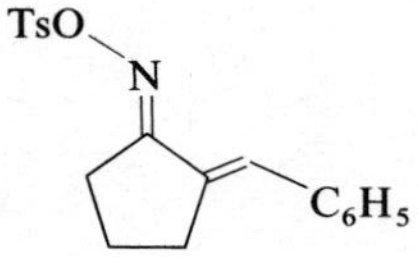 TsO N C_6H_5	CH_3OH, reflux, 7 h	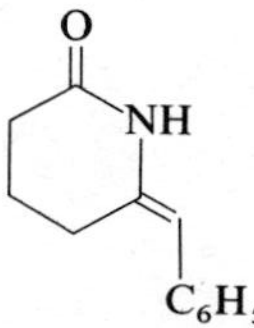 O NH C_6H_5 (4)	31
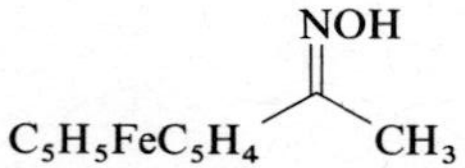 NOH $C_5H_5FeC_5H_4$ CH_3	Cl_3CCN, ether, reflux, 1 h	$C_5H_5FeC_5H_4CONHCH_3$ (74)	232

TABLE I. REARRANGEMENTS (*Continued*)

No. of Carbon Atoms	Substrate	Reagent and Conditions	Product(s) and Yield(s) (%)	Refs.
	3,5-dimethyl-1-phenylpyrazole-4-CH=NOH (C_6H_5)	H_2SO_4, 100°, 1 h	3,5-dimethyl-1-phenylpyrazole-4-$CONH_2$ (C_6H_5) (13)	318
		PPA, 110°, 1 h	" (66)	318
	1-phenylpyrazol-4-yl $C(=NOH)C_2H_5$ (HO–N=, C_2H_5, C_6H_5)		1-phenylpyrazole-4-$CONHC_2H_5$ (C_6H_5) I + 1-phenylpyrazole-4-$NHCOC_2H_5$ (C_6H_5) II	
		PCl_5, $CHCl_3$, 20°, 4 d	I (50) + II (38)	318
		TsCl, pyridine, reflux, 6 h	I (64) + II (9)	318
		H_2SO_4, 100°, 1 h	I (28) + II (2)	318
		PPA, 130°, 30 min	I (50) + II (33)	318
	2-(C_6H_4Cl-*p*)cyclohexanone oxime (HO–N=)	TsCl, pyridine	7-(C_6H_4Cl-*p*)azepan-2-one (O, H–N) (52)	30

Reactant	Conditions	Product(s) (yield %)	Ref.
NOH; N; CH_3	CH_3CO_2H, H_2SO_4	$NHCOCH_3$ (55); N; CH_3	378
C_6H_5; NOH	TsCl, pyridine, 12 h	C_6H_5; HN; O (94)	379
NOH	PPA, 110–120°, 5–10 min	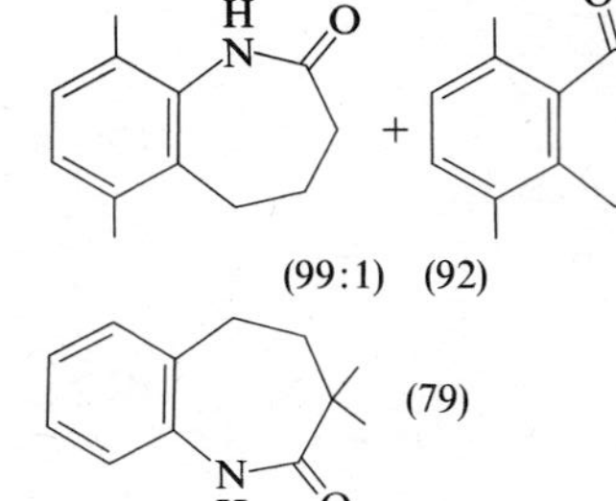 H N O + O H N (99:1) (92)	70
N; OH	PPA, 125–130°, 10 min	(79); N H O	362
NOH; CH_3O	PPA, 120°, 10 min	H N O; CH_3O (34) + O NH; CH_3O (29)	360

TABLE I. REARRANGEMENTS (*Continued*)

No. of Carbon Atoms	Substrate	Reagent and Conditions	Product(s) and Yield(s) (%)	Refs.
	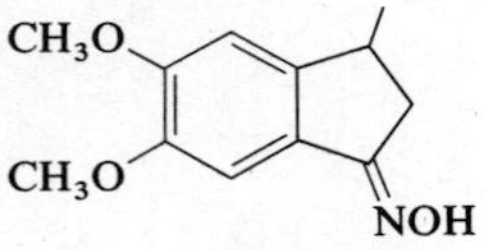 CH_3O, CH_3O, NOH	P_2O_5, CH_3SO_3H	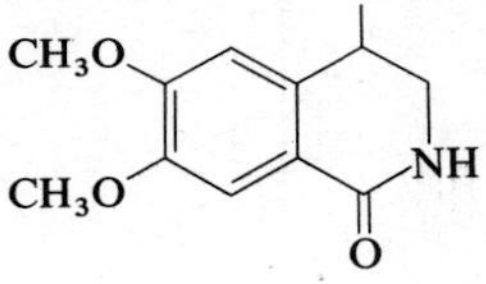 CH_3O, CH_3O, NH, O (72)	380
	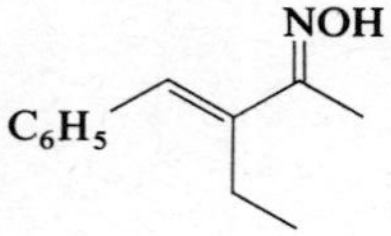 C_6H_5, NOH	PCl_5, ether, 0°, 15 min	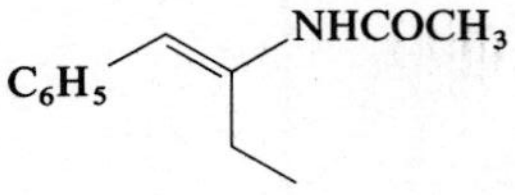 C_6H_5, $NHCOCH_3$ (—)	111
	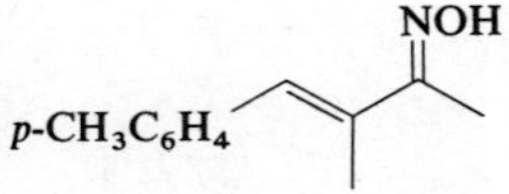 *p*-$CH_3C_6H_4$, NOH	PCl_5, ether, 0–5°, 24 h	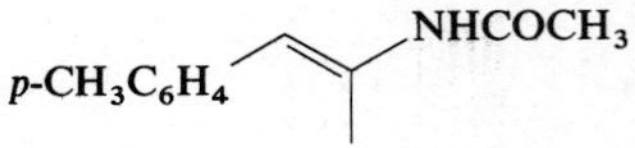 *p*-$CH_3C_6H_4$, $NHCOCH_3$ (—)	111, 112
	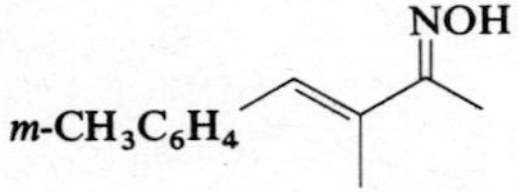 *m*-$CH_3C_6H_4$, NOH	PCl_5, ether, 0°, 15 min	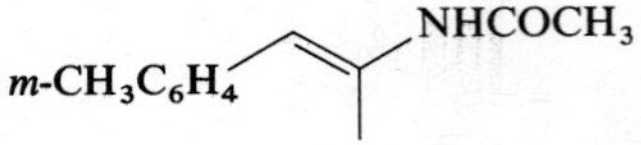 *m*-$CH_3C_6H_4$, $NHCOCH_3$ (—)	111
	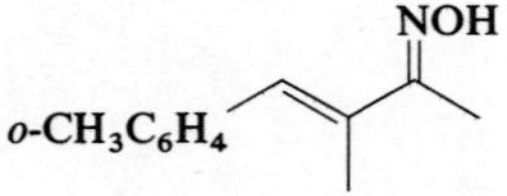 *o*-$CH_3C_6H_4$, NOH	PCl_5, ether, 0–5°, 24 h	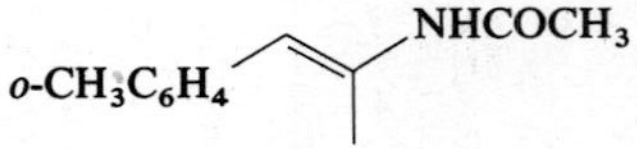 *o*-$CH_3C_6H_4$, $NHCOCH_3$ (—)	112
	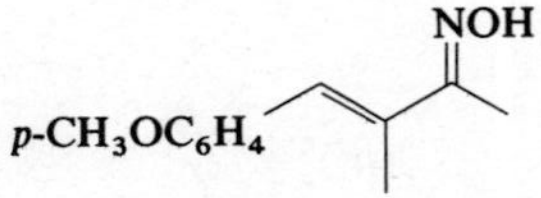 *p*-$CH_3OC_6H_4$, NOH	"	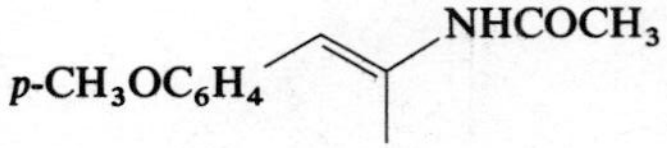 *p*-$CH_3OC_6H_4$, $NHCOCH_3$ (—)	111, 112

Reactant	Conditions	Product(s) and Yield(s) (%)	Refs.
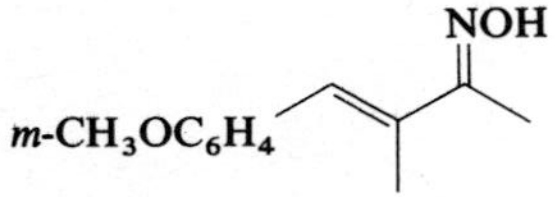 m-$CH_3OC_6H_4CH{=}C(CH_3)C(=NOH)CH_3$	PCl_5, ether, 0°, 15 min	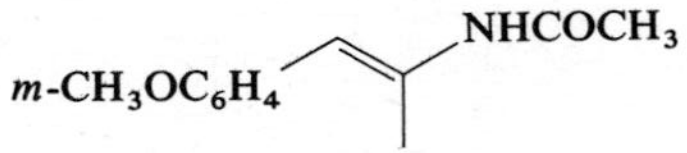 m-$CH_3OC_6H_4CH{=}C(CH_3)NHCOCH_3$ (—)	111
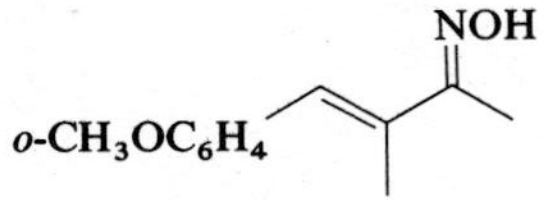 o-$CH_3OC_6H_4CH{=}C(CH_3)C(=NOH)CH_3$	PCl_5, ether, 0–5°, 24 h	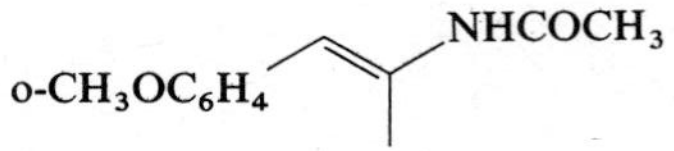 o-$CH_3OC_6H_4CH{=}C(CH_3)NHCOCH_3$ (—)	112
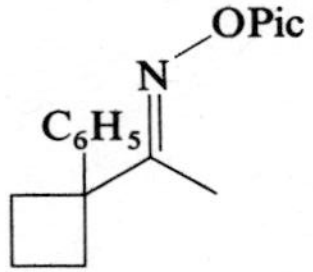 1-C_6H_5-cyclobutyl–C(=N–OPic)CH_3	1,4-Dichlorobutane, 80°	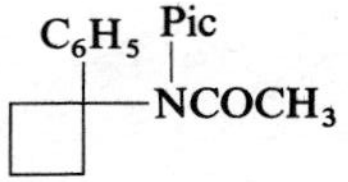 1-C_6H_5-cyclobutyl–N(Pic)$COCH_3$ (—)	93
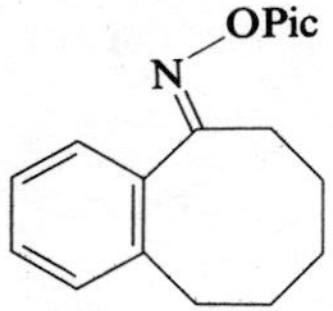 benzocyclooctanone oxime, N–OPic	$C_2H_4Cl_2$, warm	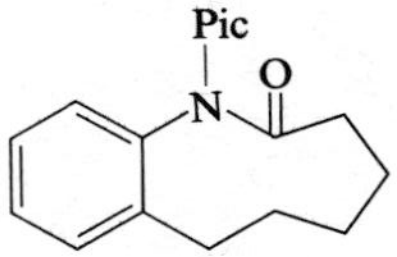 benzo-fused lactam, N-Pic, O (84)	96
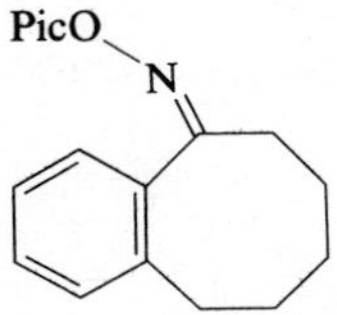 benzocyclooctanone oxime, PicO–N	$C_2H_4Cl_2$, reflux	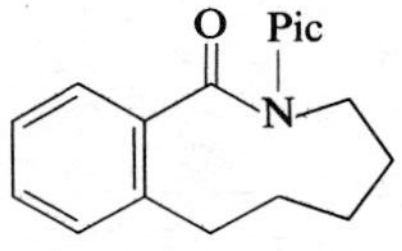 benzo-fused lactam, O, N-Pic (74)	96
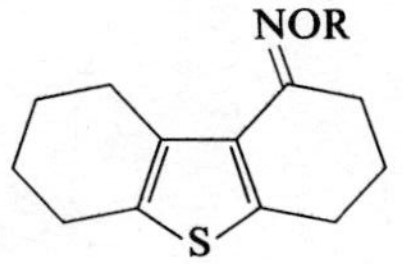 tetrahydrodibenzothiophenone oxime, NOR, R = H	PPA, 120°, 30 min	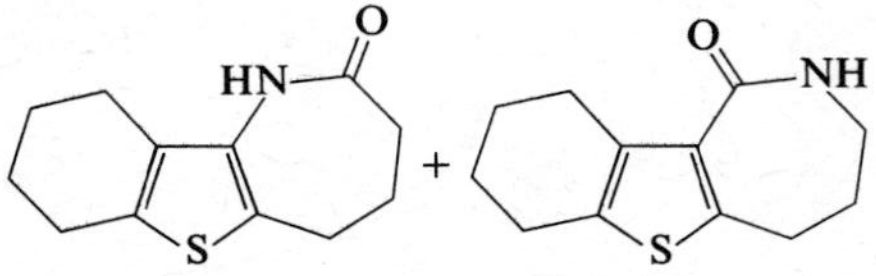 I (HN, O, S) + II (O, NH, S) (83)	381

TABLE I. REARRANGEMENTS (*Continued*)

No. of Carbon Atoms	Substrate	Reagent and Conditions	Product(s) and Yield(s) (%)	Refs.
	$R = SO_2C_6H_5$	NaO_2CCH_3, ethanol, H_2O, reflux, 30 h	I (68)	381
	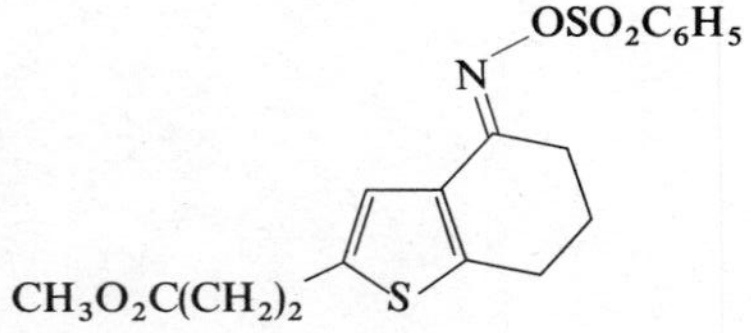 $OSO_2C_6H_5$ N $CH_3O_2C(CH_2)_2$ S	NaO_2CCH_3, ethanol, H_2O, reflux, 4 h	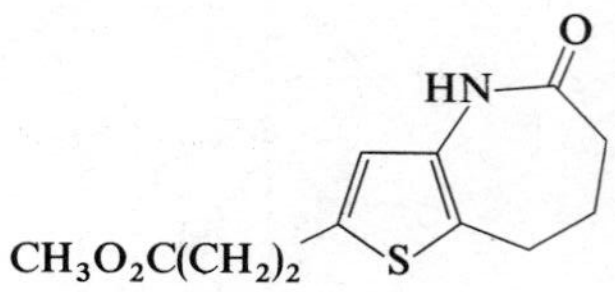 O HN $CH_3O_2C(CH_2)_2$ S (70)	382
	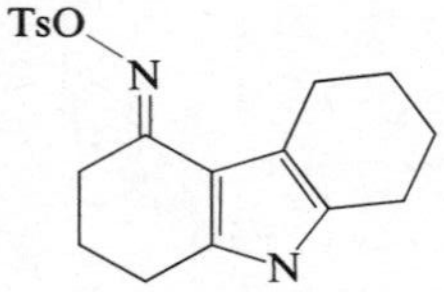 TsO N N	NaO_2CCH_3, ethanol, H_2O, 30 h	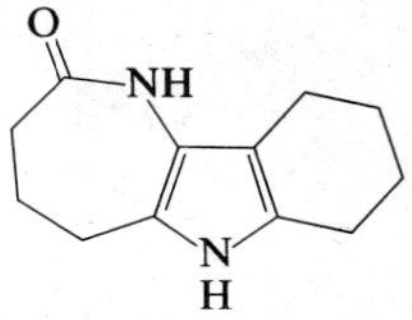 O NH N H (60)	383
	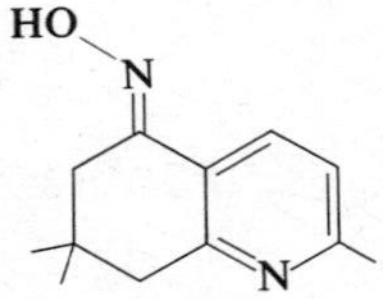 HO N N	PCl_5, ether	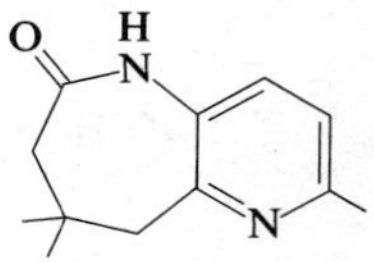 O H N N (15)	384

Reactant	Conditions	Product (Yield, %)	Ref.
CH_3O, NC_2H_5, NOH	PPA, 100°, 30 min	CH_3O, NC_2H_5, N H, O (—)	325
NOH	$C_6H_5SO_2Cl$, NaOH	NH, O (42)	265
HO, N, C_6H_5	1. TsCl, pyridine 2. $(C_2H_5)_3N$, ethanol, H_2O, 14 h	C_6H_5CONH (79)	330
OTs, N, C_6H_5, (−)	80% ethanol, 30°, 6 h	C_6H_5, $NHCOCH_3$, (+) (8)	28
p-$CH_3OC_6H_4C(=NOH)(CH_2)_2N(CH_3)_2$	PCl_5, ether, 0°, 1 h	p-$CH_3OC_6H_4NHCO(CH_2)_2N(CH_3)_2$ (79)	114
NOH	PCl_5, C_6H_6	H, N, O (85)	385

TABLE I. REARRANGEMENTS (*Continued*)

No. of Carbon Atoms	Substrate	Reagent and Conditions	Product(s) and Yield(s) (%)	Refs.
	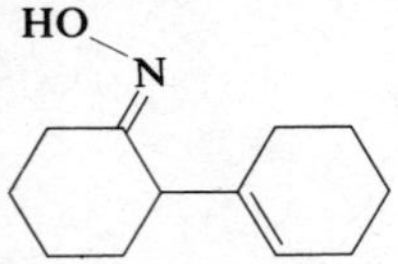 HO–N	PCl_5, ether, 0°	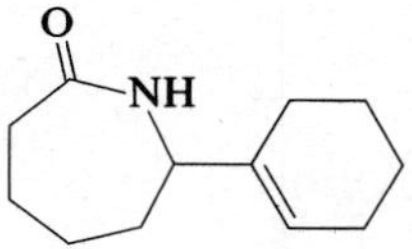 O, NH (25)	386
	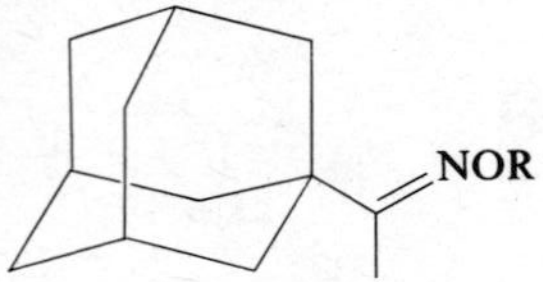 NOR R = H	PPA, 100°, 30 min	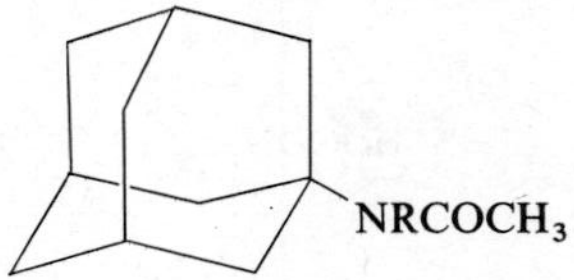 $NRCOCH_3$ (50–80)	61
	R = Pic	1,4-Dichlorobutane, 80°	" (—)	93
	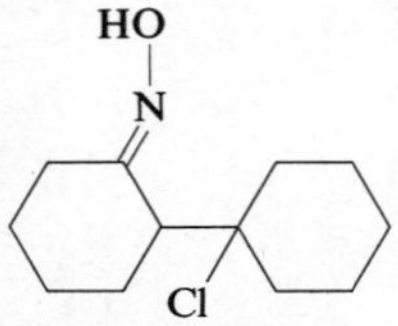 HO–N, Cl	PCl_5, ether, 0°	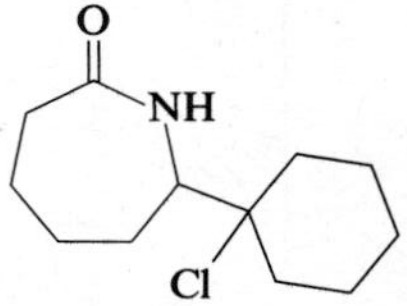 O, NH, Cl (70)	386
	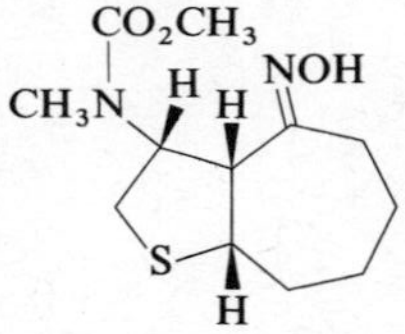 CO_2CH_3, CH_3N, H, H, NOH, S, H (*E*)	PPA, 100°, 15 min	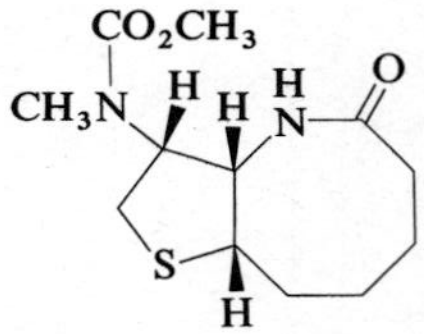 CO_2CH_3, CH_3N, H, H, H, N, O, S, H (—)	196

Reactant	Conditions	Product(s) and Yield(s) (%)	Refs.
(*Z*)	$SOCl_2$, 0°	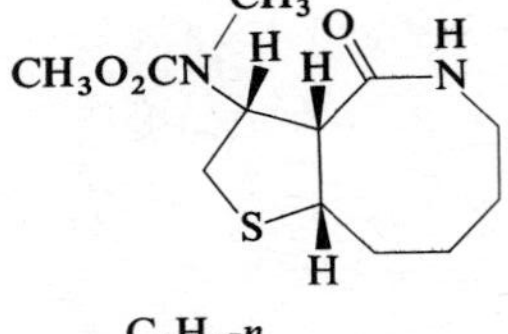 $CH_3O_2CN(CH_3)$, H, H, O, NH, S, H (—)	196
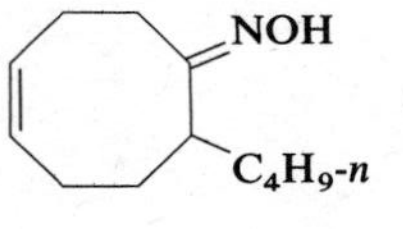 NOH, C_4H_9-*n* (*Z*)	TsCl, pyridine, 18 h	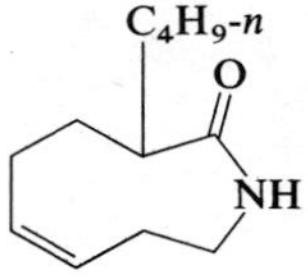 C_4H_9-*n*, O, NH (64)	313
(*E*)	1. TsCl, pyridine, CH_2Cl_2 2. K_2CO_3, THF, H_2O, 15 h	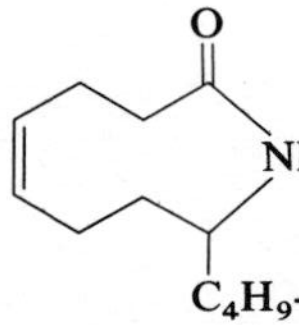 O, NH, C_4H_9-*n* (53)	296
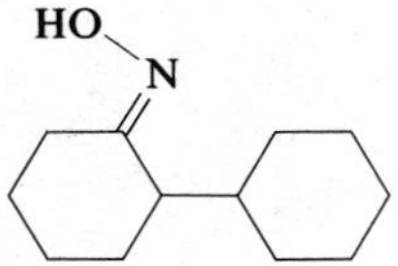 HO–N=	PCl_5, ether, 0°	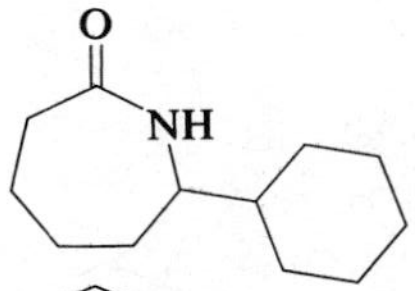 O, NH (90)	386
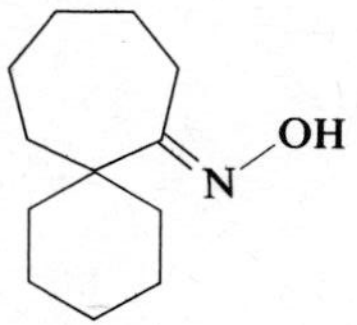 =N–OH	$C_6H_5SO_2Cl$, NaOH, H_2O, acetone	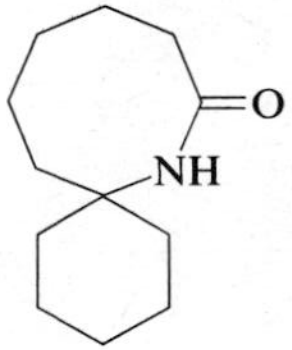 =O, NH (73)	78

TABLE I. REARRANGEMENTS (*Continued*)

No. of Carbon Atoms	Substrate	Reagent and Conditions	Product(s) and Yield(s) (%)	Refs.
	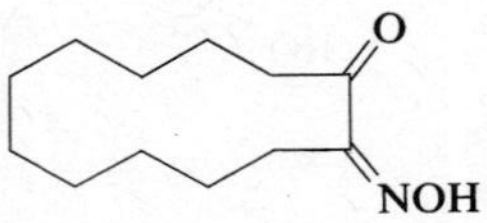 NOH	$COCl_2$	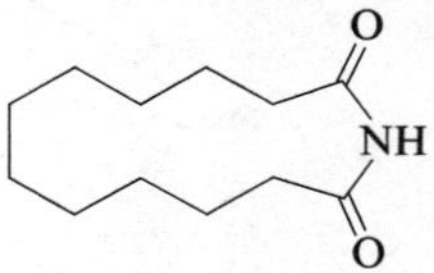 NH (6)	387
	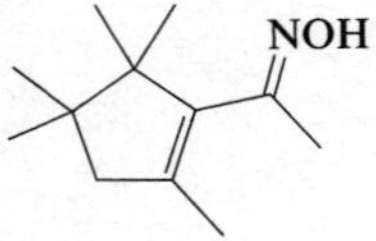 NOH	PCl_5, ether	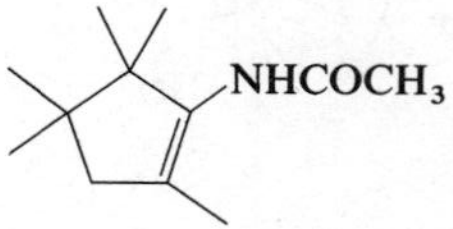 $NHCOCH_3$ (70)	388
	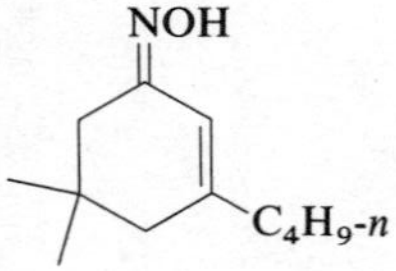 NOH, C_4H_9-*n*	PPA, 120°	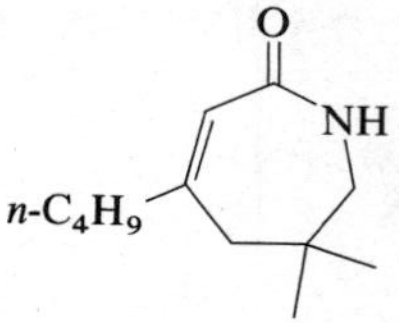 *n*-C_4H_9, NH (38) 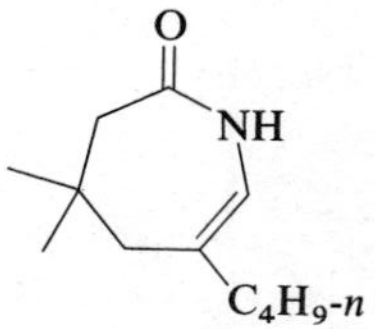 + NH, C_4H_9-*n* (19)	46
	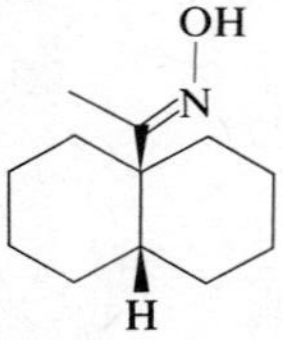 OH, N, H	TsCl, pyridine	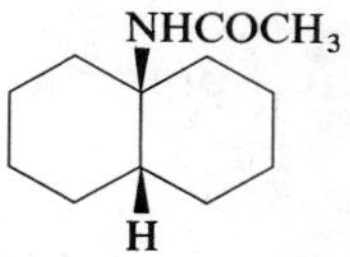 $NHCOCH_3$, H (92)	66, 67

Reactant	Conditions	Product(s) and Yield(s) (%)	Refs.
	PPA, 25°	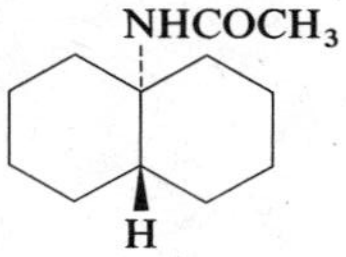 NHCOCH$_3$, H, I (61)	66, 67
	PPA, 125°	I (12) 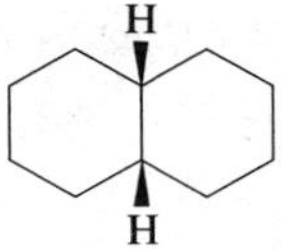 + H, H, II (55)	66, 67
	85–90% H_2SO_4	I (40)	66, 67
	98% H_2SO_4	I (47) + II (34)	66, 67
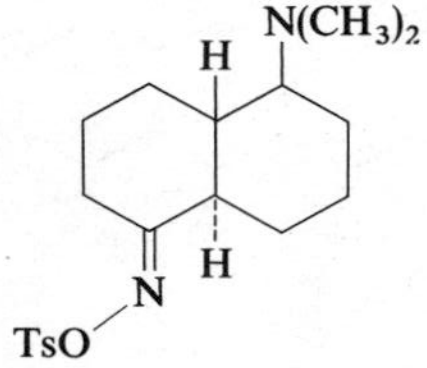 N(CH$_3$)$_2$, H, H, N, TsO	Dioxane, H_2O	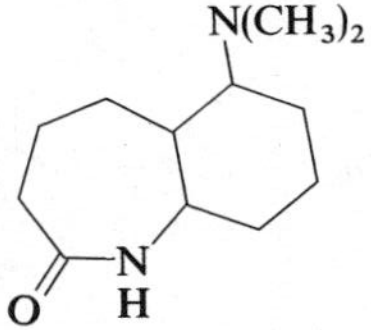 N(CH$_3$)$_2$, O, N, H (100)	200
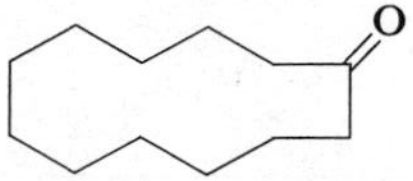 O	H_2NOSO_3H, HCO_2H, reflux, 7 h	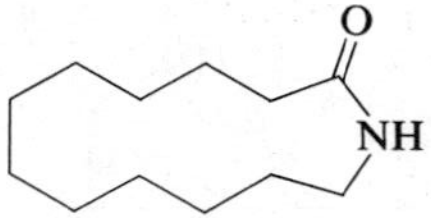 O, NH (87)	81
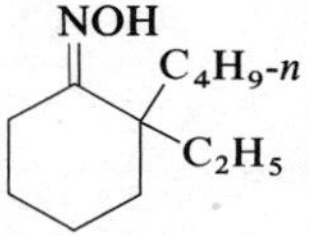 NOH, C_4H_9-*n*, C_2H_5	PPA, 115°, 10 min	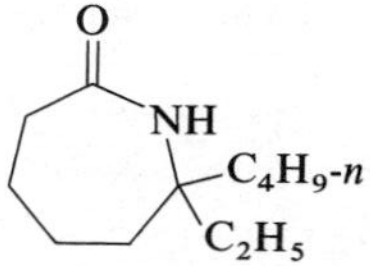 O, NH, C_4H_9-*n*, C_2H_5 (33)	351

TABLE I. REARRANGEMENTS (*Continued*)

No. of Carbon Atoms	Substrate	Reagent and Conditions	Product(s) and Yield(s) (%)	Refs.
	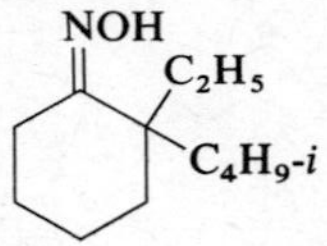 NOH, C_2H_5, C_4H_9-*i*	PPA, 115°, 10 min	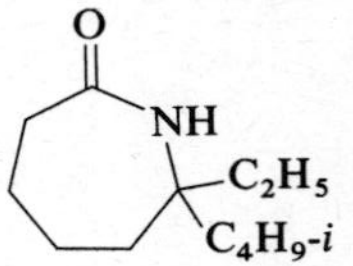 O, NH, C_2H_5, C_4H_9-*i* (17)	351
	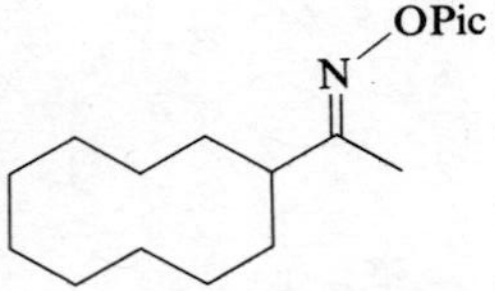 OPic, N	1,4-Dichlorobutane, 80°	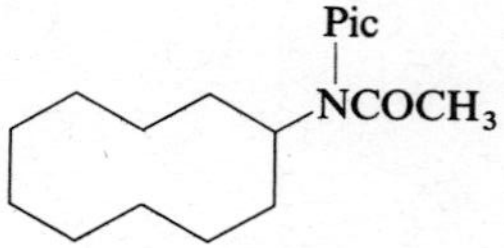 Pic, $NCOCH_3$ (—)	93
	OPic, N, *n*-C_4H_9, C_3H_7-*n*	"	Pic, $NCOCH_3$, *n*-C_4H_9, C_3H_7-*n* (—)	93
	OPic, N, (*n*-C_3H_7)$_3$C	"	Pic, (*n*-C_3H_7)$_3$CNCOCH$_3$ (—)	93
C_{13}	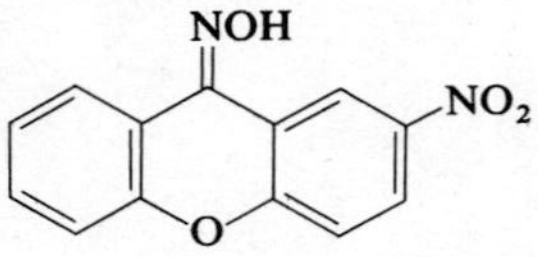 NOH, NO_2, O	PCl_5, ether, 15 h	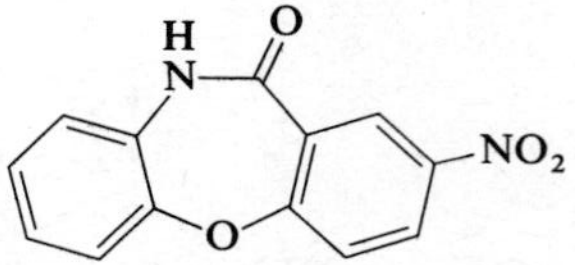 H, N, O, NO_2, O (—)	389

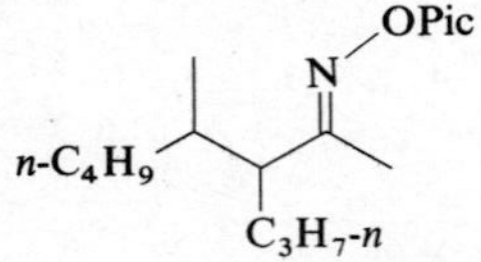

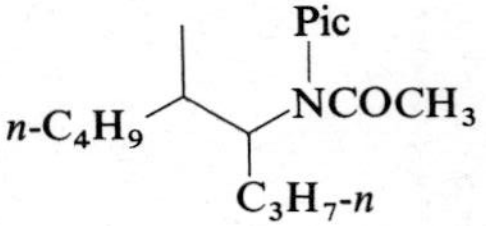

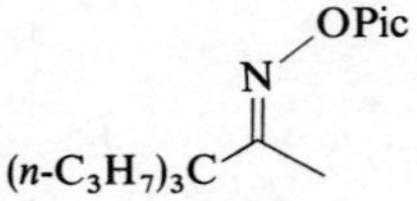

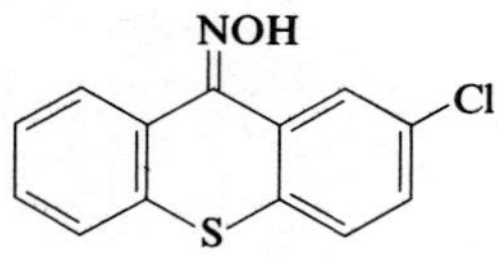

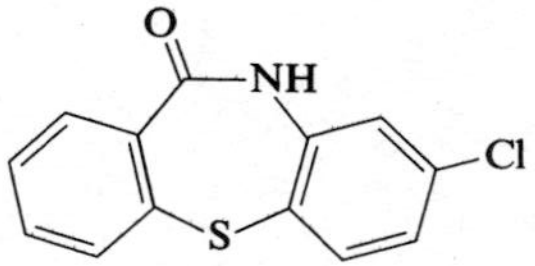

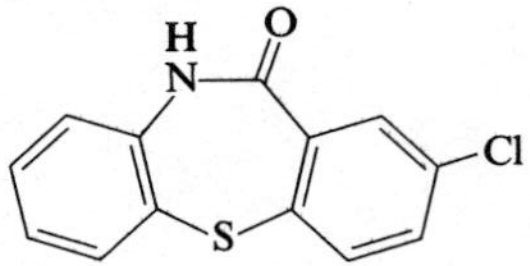

Reactant	Conditions	Product(s) (Yield, %)	Refs.
NOH, Cl, S	"	O, NH, Cl, S	389
		+ H N O, Cl, S (85)	
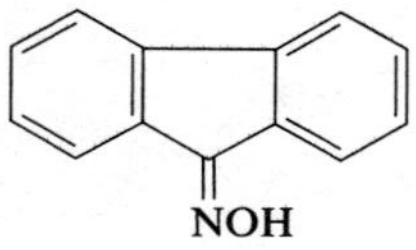 NOH	PPA, 160°, 10 min	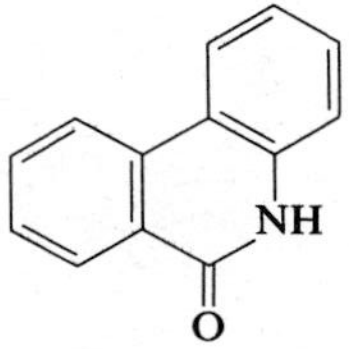 NH, O (95)	390
	PPA, xylene, 150°, 3 h	" (94)	60
	1. $COCl_2$ 2. $AgBF_4$, C_6H_5Cl, 55°	" (87)	391
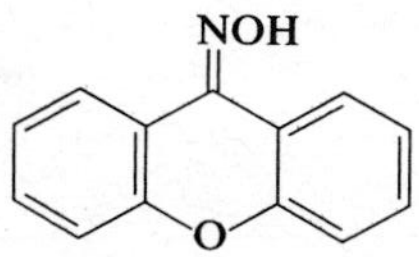 NOH, O	PCl_5, ether, 15 h	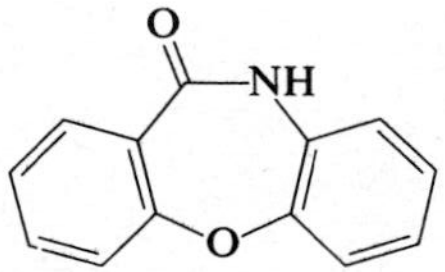 O, NH, O (50)	389
	PPA, 130–160°, 90 min	" (78)	390
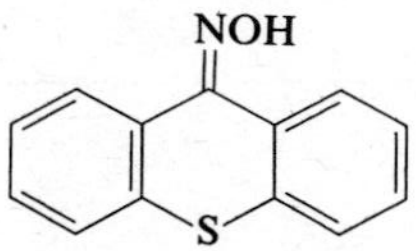 NOH, S	PCl_5, ether, 15 h	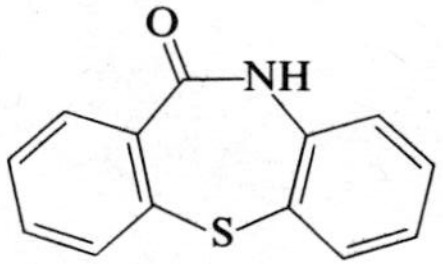 O, NH, S (52)	389
	PPA, 170°, 30 min	" (55)	390

TABLE I. REARRANGEMENTS (*Continued*)

No. of Carbon Atoms	Substrate	Reagent and Conditions	Product(s) and Yield(s) (%)	Refs.
	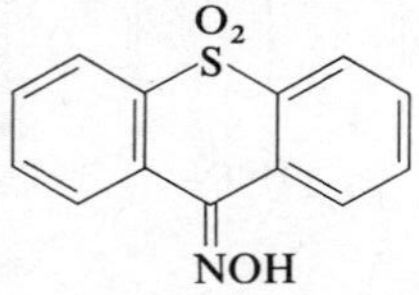 O_2 S; NOH	PPA, 170°, 30 min	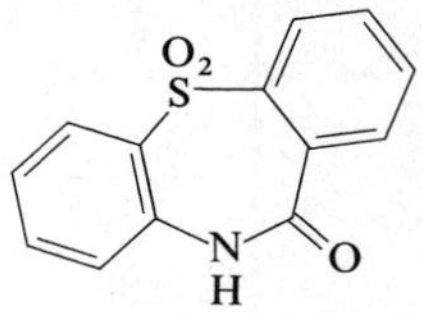 O_2 S; N H; O (87)	390, 392
	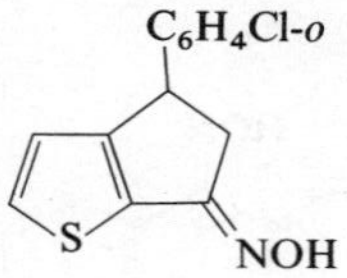 C_6H_4Cl-*o*; S; NOH (*Z*)	PPA, 120°, 2 h	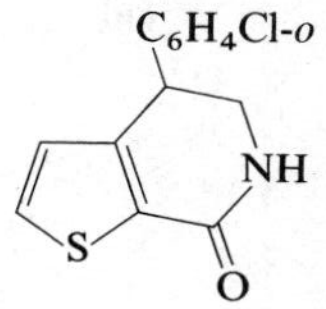 C_6H_4Cl-*o*; S; NH; O (15)	393
	(*E*)	"	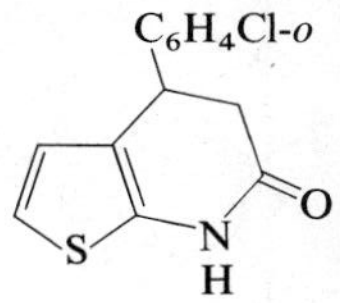 C_6H_4Cl-*o*; S; N H; O (2)	393
	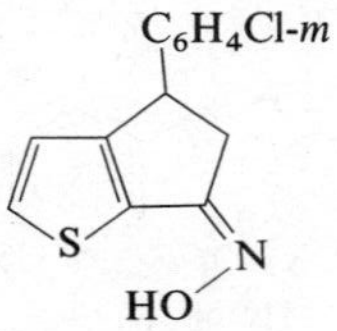 C_6H_4Cl-*m*; S; N; HO	"	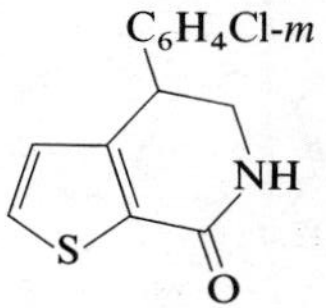 C_6H_4Cl-*m*; S; NH; O (18)	393

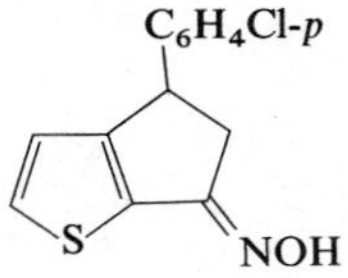 (structure: 4-(C_6H_4Cl-p)-cyclopenta[b]thiophen-6-one oxime, NOH) (Z)	"	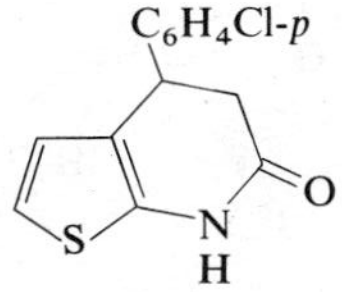 (structure: C_6H_4Cl-p thienoazepinone, N–H, C=O) (15)	393
(E)	"	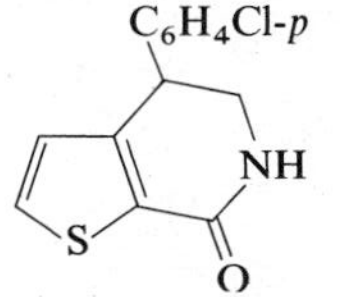 (structure: C_6H_4Cl-p thienoazepinone, NH, C=O) (6)	393
p-$ClC_6H_4C(=NCl)C_6H_5$ (E)	$AgBF_4$, DMF	p-$ClC_6H_4NHCOC_6H_5$ (—)	120
(Z)	"	p-$ClC_6H_4CONH_6H_5$ (—)	120
p-$O_2NC_6H_4C(=NOH)C_6H_5$	$SOCl_2$, C_6H_6	p-$O_2NC_6H_4CONHC_6H_5$ (100)	394
$C_6H_5COC_6H_5$	$H_2NOH \cdot HCl$, CF_3SO_3H, HCO_2H, reflux, 2.5 h	$C_6H_6CONHC_6H_5$ (90)	249
$(C_6H_5)_2C=NOH$	PPA, xylene, 100°, 2 h	" (100)	60
	P_2O_5, CH_3SO_3H, 100°, 1 h	" (90)	73
	P_2O_5, CH_3SO_3H, 50°, 1 h	" (95)	73
	PPE, $CHCl_3$, reflux	" (91)	287
	PPSE, C_6H_6, 20 h	" (94)	72
	HCl, pyridine	" (100)	291
	HCO_2H, reflux, 1 h	" (93)	251
	$C_6H_5SO_2Cl$, alumina	" (90)	77
	$([(CH_3)_2N]_3P)_2O \cdot (BF_4)_2$, CH_3CN, reflux, 4.5 h	" (86)	252
	BCl_3, CH_2Cl_2, − 78°	" (81)	395
	$(CH_3)_3SiI$, $CHCl_3$, 56°, 4 h	" (80)	85
	IF_5, CH_2Cl_2, 3–15°, 2 h	" (74)	253
	HMPA, 225–235°, 10 min	" (54)	282
	HF, 2 h	" (53)	83
	1. $COCl_2$, ether 2. $AgBF_4$, C_6H_5Cl	" (66)	391

TABLE I. REARRANGEMENTS (*Continued*)

No. of Carbon Atoms	Substrate	Reagent and Conditions	Product(s) and Yield(s) (%)	Refs.
		$(C_3H_3N_2)_2CO$, octane	$C_6H_6CONHC_6H_5$ (100)	146
		o-$ClC_6H_4CO_2H$, $(C_6H_5)_3P$, $C_2H_5O_2CN{=}NCO_2C_2H_5$, THF, 18 h	$C_6H_5CON(C_6H_5)COC_6H_4Cl$-*o* (79)	92
		p-$ClC_6H_4CO_2H$, $(C_6H_5)_3P$, $C_2H_5O_2CN{=}NCO_2C_2H_5$, THF, 18 h	$C_6H_5CON(C_6H_5)COC_6H_4Cl$-*p* (76)	92
		m-$O_2NC_6H_4CO_2H$, $(C_6H_5)_3P$, $C_2H_5O_2CN{=}NCO_2C_2H_5$, THF, 18 h	$C_6H_5CON(C_6H_5)COC_6H_4NO_2$-*m* (92)	92
		p-$O_2NC_6H_4CO_2H$, $(C_6H_5)_3P$, $C_2H_5O_2CN{=}NCO_2C_2H_5$, THF, 18 h	$C_6H_5CON(C_6H_5)COC_6H_4NO_2$-*p* (87)	92
		p-$CH_3C_6H_4CO_2H$, $(C_6H_5)_3P$, $C_2H_5O_2CN{=}NCO_2C_2H_5$, THF, 18 h	$C_6H_5CON(C_6H_5)COC_6H_4CH_3$-*p* (74)	92

	p-$CH_3OC_6H_4CO_2H$, $(C_6H_5)_3P$, $C_2H_5O_2CN{=}NCO_2C_2H_5$, THF, 18 h	$C_6H_5CON(C_6H_5)COC_6H_4OCH_3$-$p$ (77)	92
$(C_6H_5)_2C{=}NOTs$	SiO_2, $CHCl_3$, 5°, 30 min	$C_6H_5CONHC_6H_5$ (86)	109
$(C_6H_5)_2C{=}NOSO_2C_6H_2(CH_3)_3$-2,4,6	Al_2O_3, CH_3OH	" (84)	47
$(C_6H_5)_2C{=}NO_2CC_6H_5$	HF, 2 h	" (90)	83
$(C_6H_5)_2C{=}NCl$	$SbCl_5$, CCl_4, 40–50°	$C_6H_5\overset{+}{N}{\equiv}CC_6H_5 \cdot SbCl_6^-$	28
$(C_6H_5)_2C{=}NOCN$	Ethanol, reflux, 30 min	$C_6H_5C(NHCO_2C_2H_5){=}NC_6H_5$ (—)	88
	CCl_4, reflux, 25 min	$C_6H_5C(NCO){=}NC_6H_5$ (—)	88
$(C_6H_5)_2C{=}N{-}O{-}$(1,2,3,4-thiatriazol-5-yl)	CCl_4, reflux,	" (—)	88
$(p\text{-}HOC_6H_4)_2C{=}NOH$	$SOCl_2$, ether	p-$HOC_6H_4NHCOC_6H_4OH$-p (—)	396
2-NH_2-5-Cl-$C_6H_3C(C_6H_5){=}NOCO_2C_2H_5$	Heat at mp	6-Cl-3-C_6H_5-quinazoline-2,4(1H,3H)-dione (50)	397

TABLE I. REARRANGEMENTS (*Continued*)

No. of Carbon Atoms	Substrate	Reagent and Conditions	Product(s) and Yield(s) (%)	Refs.
	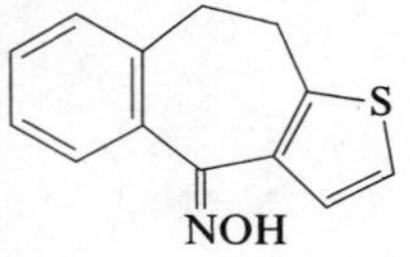 NOH	PCl_5, ether, reflux	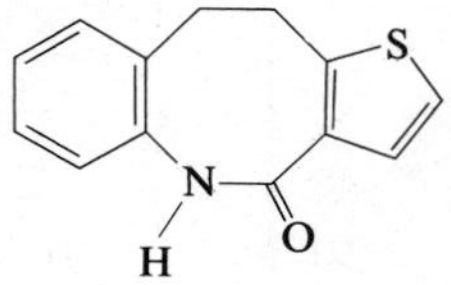 I 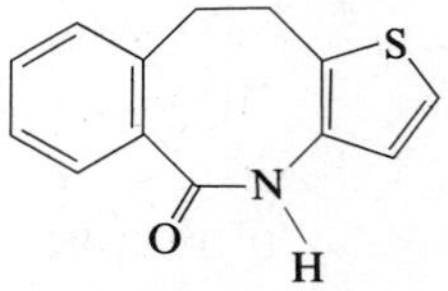 + II 60:40 (—)	398
		PCl_5, $C_6H_5CH_2OH$, reflux	I/II = 60/40 (—)	398
		$POCl_3$, 70°	I/II = 75/25 (—)	398
		H_2SO_4, 125°	I/II = 90/10 (—)	398
		PPA, 90°	I (—)	398
	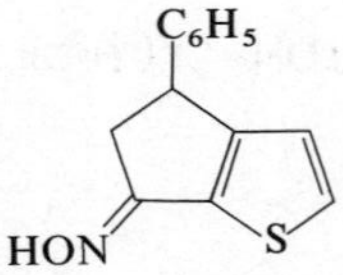 C_6H_5, HON, S (*Z*)	PPA, 120°, 2 h	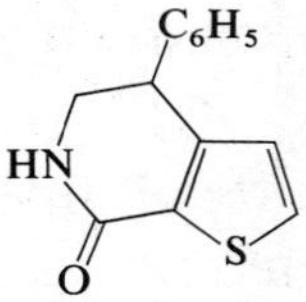 C_6H_5, HN, O, S (16)	393

Oxime	Conditions	Product(s) and Yield(s) (%)	Refs.
(*E*)	"	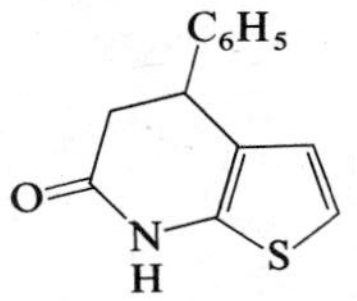 4-C_6H_5-thieno-lactam (10)	393
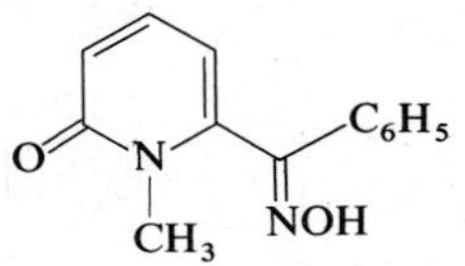 1-CH_3-2-pyridone-6-$C(=NOH)C_6H_5$ (*Z*)	PCl_5, ether, 0°, 5 h	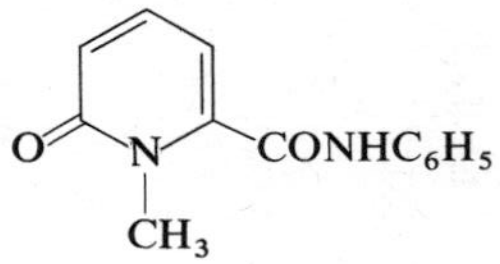 1-CH_3-2-pyridone-6-$CONHC_6H_5$ (86)	293
(*E*)	"	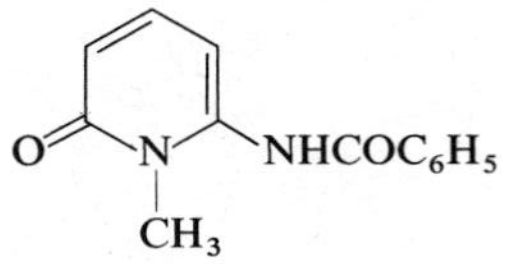 1-CH_3-2-pyridone-6-$NHCOC_6H_5$ (42)	293
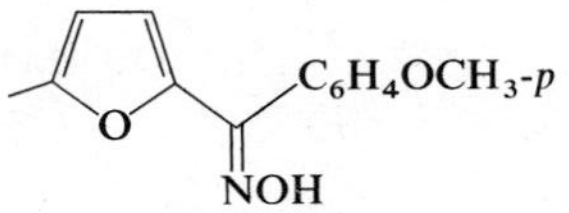 5-methyl-2-furyl-$C(=NOH)C_6H_4OCH_3$-*p*	PCl_5, ether	 5-methyl-2-furyl-$CONHC_6H_4OCH_3$-*p* (20)	301
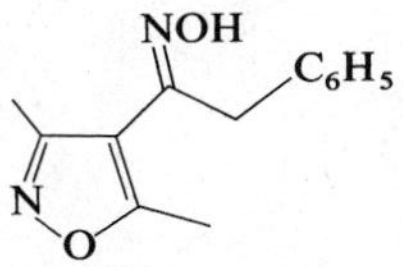 3,5-dimethylisoxazol-4-yl-$C(=NOH)CH_2C_6H_5$	PCl_5, ether, 0°, 20 h	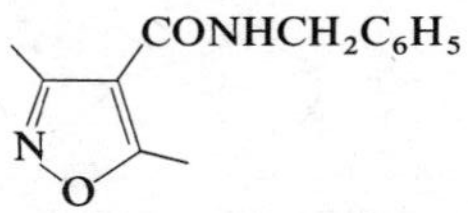 3,5-dimethylisoxazol-4-yl-$CONHCH_2C_6H_5$ (19) 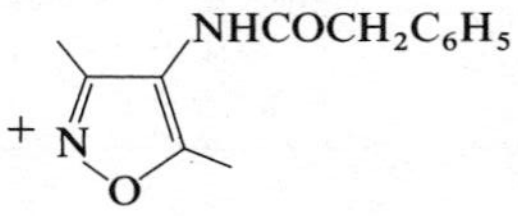 + 3,5-dimethylisoxazol-4-yl-$NHCOCH_2C_6H_5$ (6)	399

TABLE I. REARRANGEMENTS (*Continued*)

No. of Carbon Atoms	Substrate	Reagent and Conditions	Product(s) and Yield(s) (%)	Refs.
	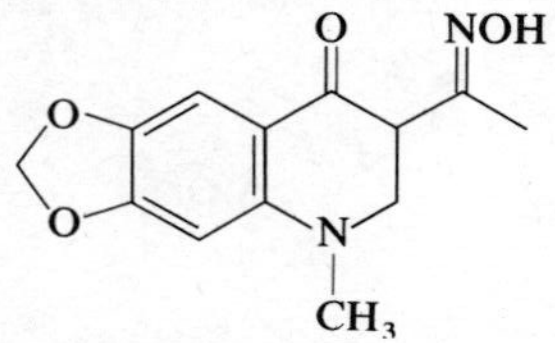 O, NOH; O, O; N; CH_3	HCO_2H, reflux	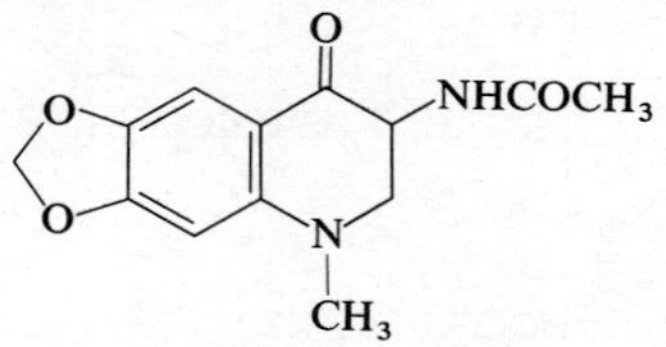 O; $NHCOCH_3$; O, O; N; CH_3	400
	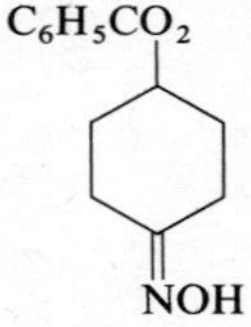 $C_6H_5CO_2$; NOH	PPA, 105–108°, 1 min	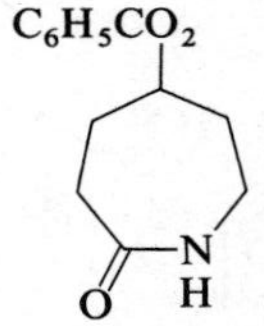 $C_6H_5CO_2$; O, N, H (83)	401
	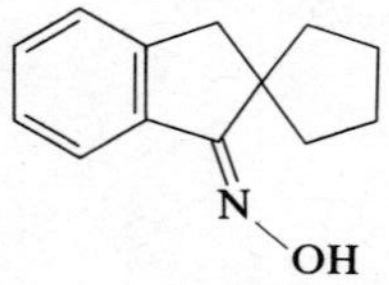 N, OH	PPA, 125–130°, 10 min	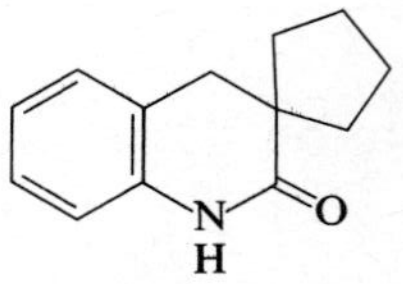 N, H, O (98)	362
	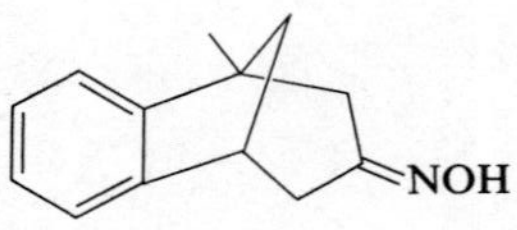 NOH	PPA, 120°, 30 min	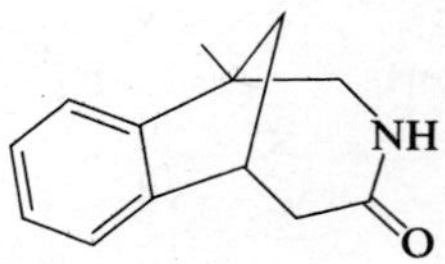 NH, O (43)	376

Reactant	Conditions	Product(s) and Yield(s) (%)	Refs.
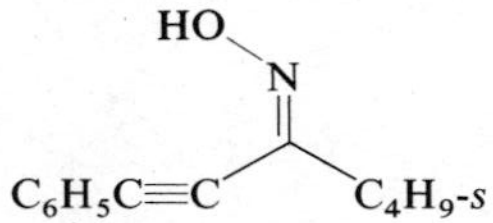 HO–N=C($C_6H_5C{\equiv}C$)(C_4H_9-s)	PCl_5, ether	$C_6H_5CCl{=}CHCONH(C_4H_9$-$s)$	116
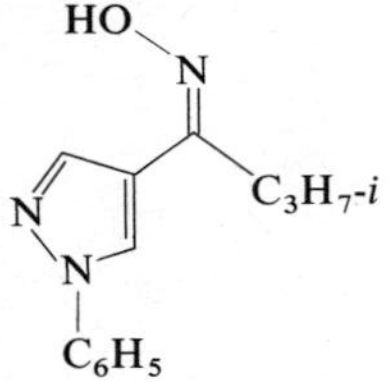 1-C_6H_5-pyrazol-4-yl–C(=N–OH)C_3H_7-i	PCl_5, $CHCl_3$, 20°, 4 d	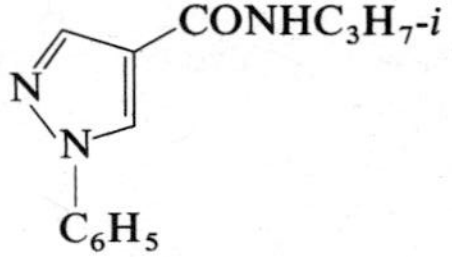 1-C_6H_5-pyrazole-4-$CONHC_3H_7$-i (78)	318
	TsCl, pyridine, reflux, 6 h	" (85)	318
	PPA, 100°, 1 h	" (85)	318
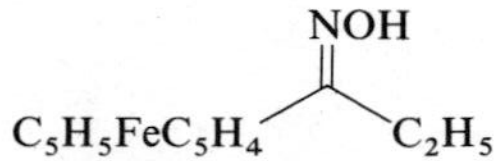 $C_5H_5FeC_5H_4C(=NOH)C_2H_5$	Cl_3CCN, ether, reflux, 4 h	$C_5H_5FeC_5H_4CONHC_2H_5$ (62)	232
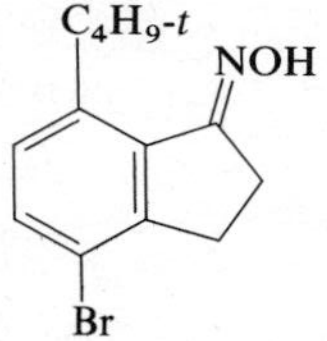 C_4H_9-t, NOH, Br (indanone oxime)	PPA, 130°, 1 min	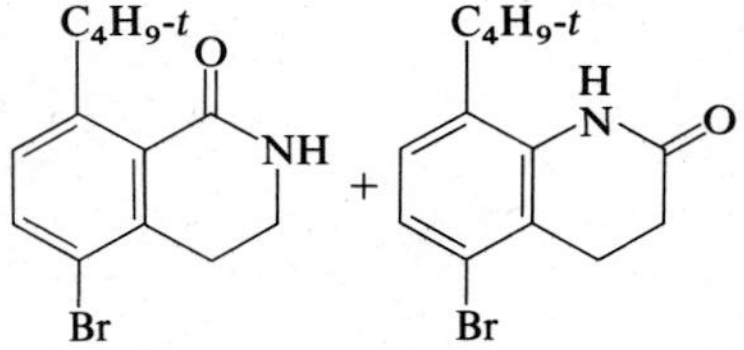 C_4H_9-t, O, NH, Br (13) + C_4H_9-t, H, N, O, Br (3)	68–70
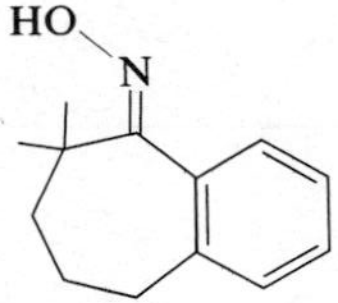 HO, N (benzocycloheptanone oxime)	PPA, 125-130°, 10 min	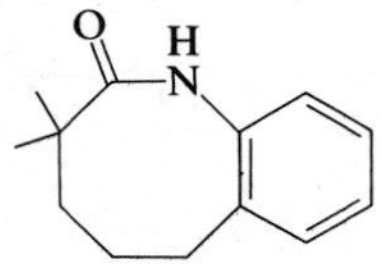 O, H, N (98)	362

TABLE I. REARRANGEMENTS (*Continued*)

No. of Carbon Atoms	Substrate	Reagent and Conditions	Product(s) and Yield(s) (%)	Refs.
	NOH; CH_3O	PPA, 120°, 10 min	H N O; CH_3O (14) + O NH; CH_3O (37)	360
	HO N; $C_6H_4CH_3$-*o*	TsCl, pyridine	O H N $C_6H_4CH_3$-*o* (39)	30
	NOH; $C_6H_4CH_3$-*p*	TsCl, pyridine, 12 h	O H N $C_6H_4CH_3$-*p* (70)	30, 402

Reactant	Conditions	Product (Yield, %)	Refs.
$C_6H_5C(=NOH)C_6H_{11}$ (*Z*)	$SOCl_2$, SO_2 $-70°$, 1 h	$C_6H_{11}NHCOC_6H_5$ (—)	403
(*Z*)	Br_2, SO_2, $-70°$, 1 h	" (—)	403
(*E*)	$SOCl_2$, SO_2 $-70°$, 1 h	$C_6H_{11}CONHC_6H_5$ (—)	403
(*E*)	$SOCl_2$, ether, $-70°$, 1 h	" (—)	403
(*E*)	$C_6H_5SO_2Cl$, SO_2, $-70°$, 1 h	" (—)	403
(*E*)	Br_2, SO_2, $-70°$, 1 h	" (—)	403
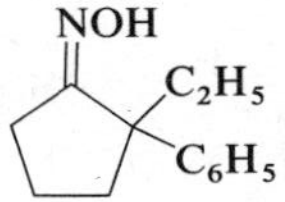 NOH; C_2H_5; C_6H_5	PPA, 115°, 10 min	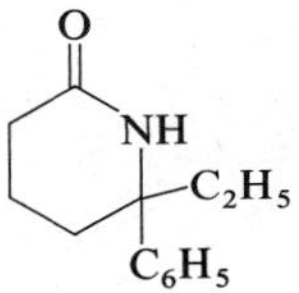 O; NH; C_2H_5; C_6H_5 (55)	351
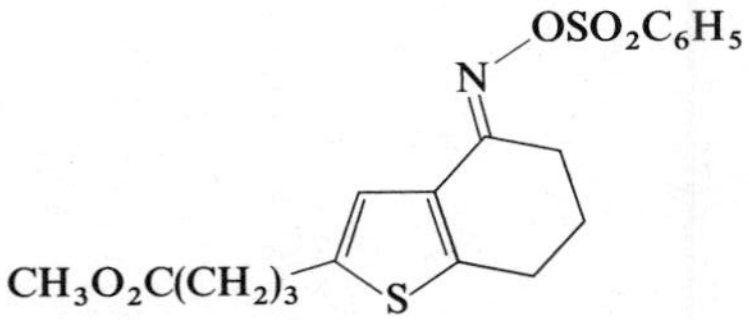 $OSO_2C_6H_5$; N; $CH_3O_2C(CH_2)_3$; S	NaO_2CCH_3, ether, H_2O, reflux, 4 h	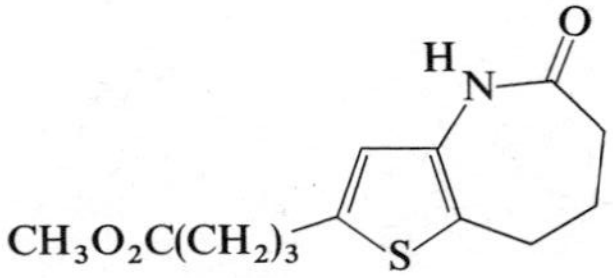 O; H; N; $CH_3O_2C(CH_2)_3$; S (68)	382
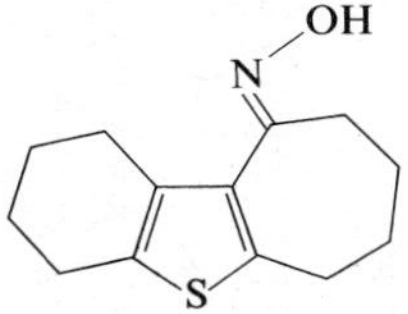 OH; N; S	$C_6H_5SO_2Cl$, pyridine	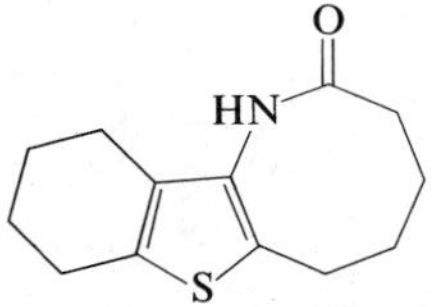 O; HN; S (64)	381
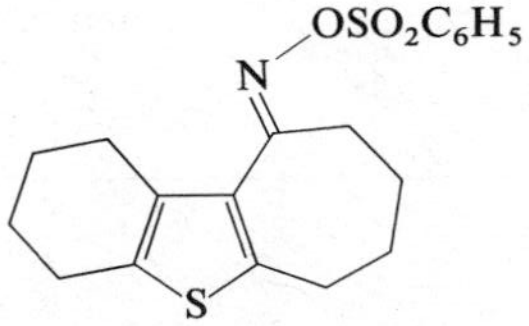 $OSO_2C_6H_5$; N; S	NaO_2CCH_3, ethanol, reflux, 30 h	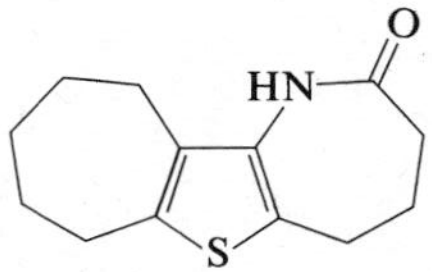 O; HN; S (72)	381

TABLE I. REARRANGEMENTS (*Continued*)

No. of Carbon Atoms	Substrate	Reagent and Conditions	Product(s) and Yield(s) (%)	Refs.
	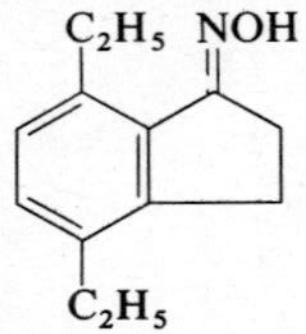 C_2H_5 NOH / C_2H_5	PPA, 120°, 5 min	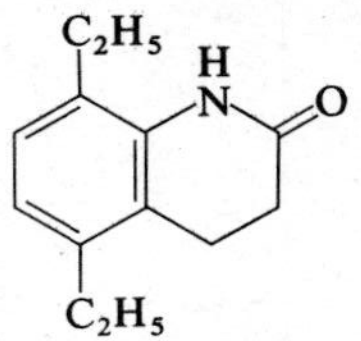 C_2H_5 H N O / C_2H_5 + C_2H_5 O NH / C_2H_5 (—)	71
	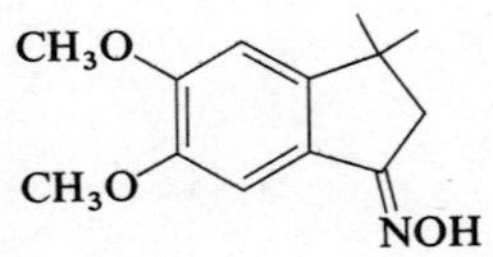 CH_3O / CH_3O NOH	P_2O_5, CH_3SO_3H	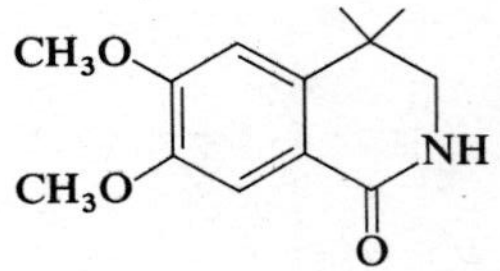 CH_3O / CH_3O NH O (80)	380
	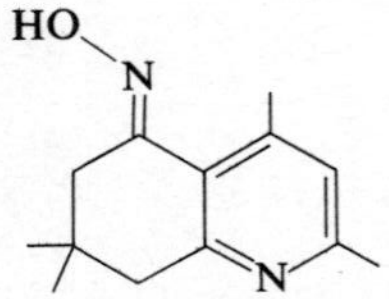 HO N / N	PCl_5, ether	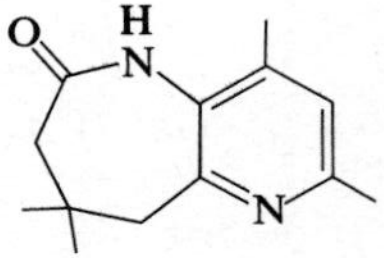 O H N / N (15)	384

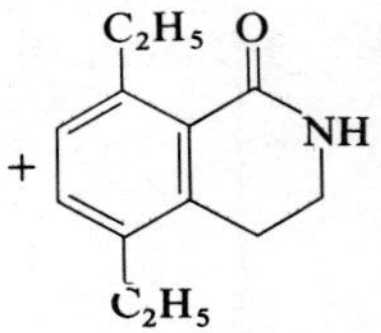

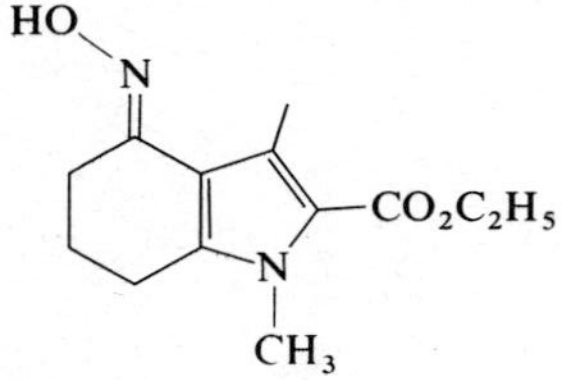

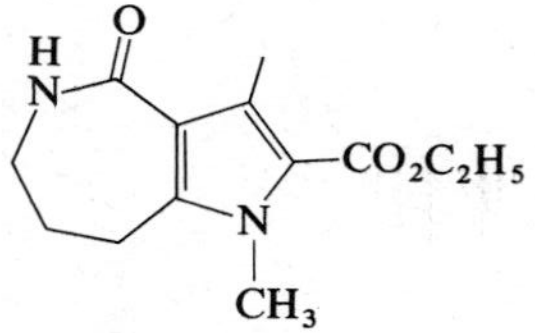

Reactant	Conditions	Product(s) (% Yield)	Refs.
HO–N=; $CO_2C_2H_5$; N–CH_3	(—)	H–N; O; $CO_2C_2H_5$; N–CH_3 (—)	308
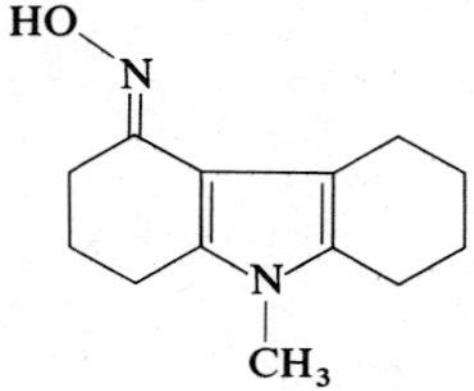 HO–N=; N–CH_3	PPA, 130°, 30 min	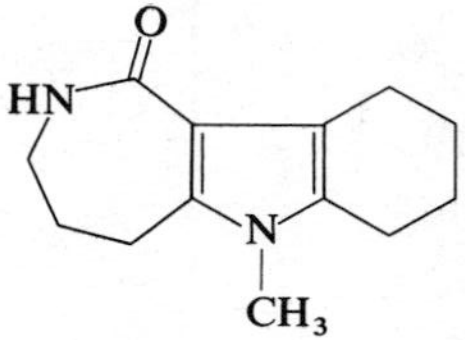 HN; O; N–CH_3 (60)	383
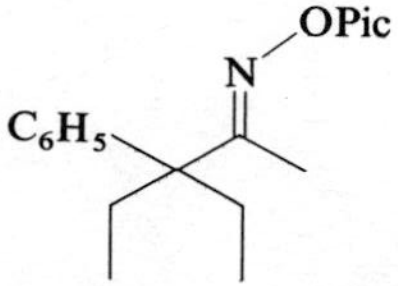 C_6H_5; N–OPic	1,4-Dichlorobutane, 80°	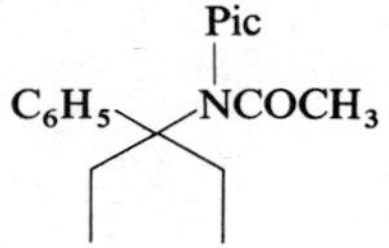 C_6H_5; Pic–$NCOCH_3$ (—)	93
NOH; NOH	PPA	O; H–N; H–N; O + I (—)	42
		HN; O; O; NH + II (—)	

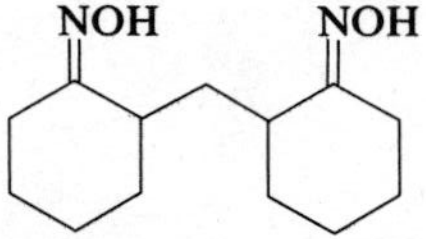

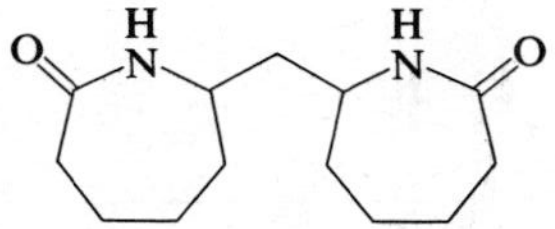

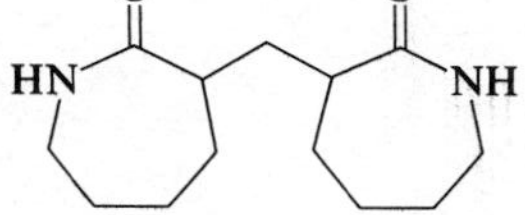

TABLE I. REARRANGEMENTS (*Continued*)

No. of Carbon Atoms	Substrate	Reagent and Conditions	Product(s) and Yield(s) (%)	Refs.
			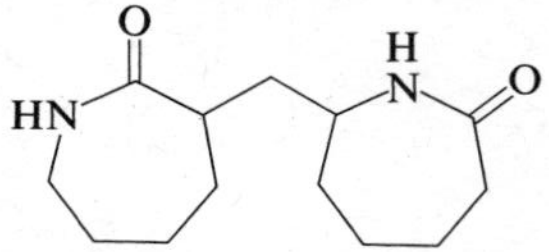 HN, N–H, O, O; III (—)	
		PPA	I + III, I/III = 1/1 (80)	404
		H_2SO_4, $CHCl_3$, reflux, 8 h	" (25)	404
	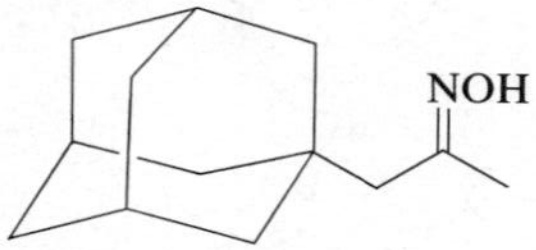 NOH	PPA, 100°, 30 min	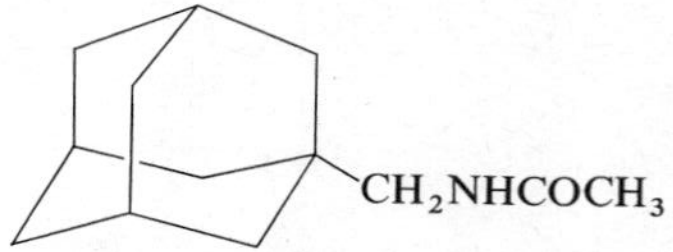 $CH_2NHCOCH_3$ (85)	61
	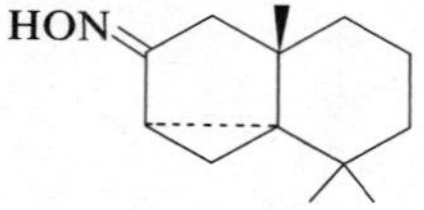 HON (*Z*)	PPA, 90°, 5 min	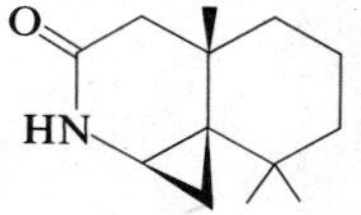 O, HN; I (12) 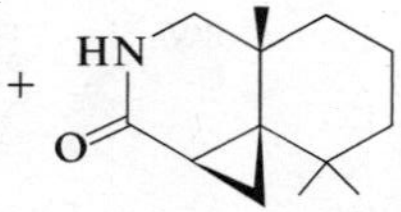 + HN, O; II (40)	405
	(*E*)	"	I (46) + II (10)	405
		$SOCl_2$, dioxane, 40°	II (74)	344
	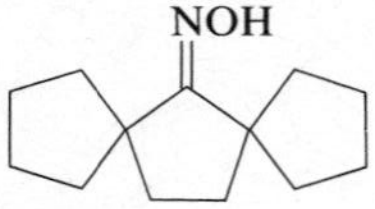 NOH	PPA, 75°, 1 h	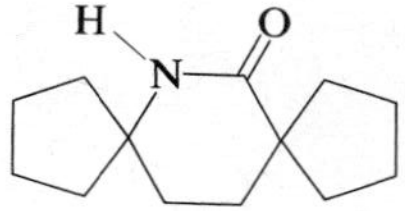 H, N, O; III	406

Formula	Reactant	Conditions	Product(s) (% yield)	Refs.
			+ 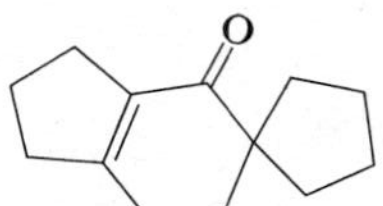 IV (64)	
		PPA, 20°, 15 h	III (24)	406
		$SOCl_2$, 20°, 15 h	III (50)	406
		TsCl, pyridine, 20°, 15 h	III (58)	406
	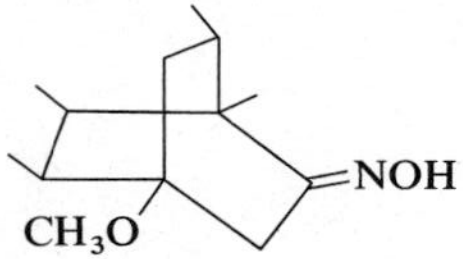 (=NOH, CH_3O)	$C_6H_5SO_2Cl$, NaOH	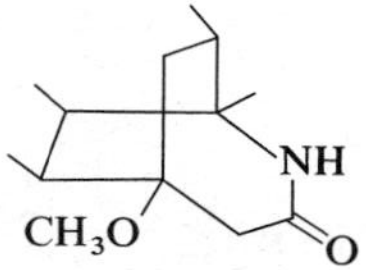 (NH, CH_3O, O) (67)	297
		TsCl, pyridine	" (65)	297
	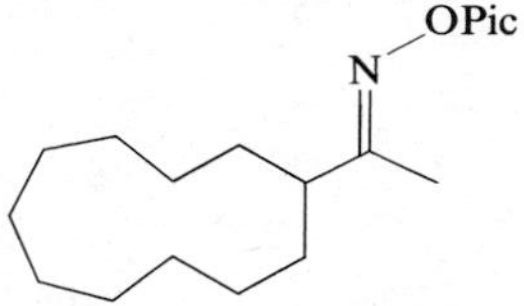 (N–OPic)	1,4-Dichlorobutane, 80°	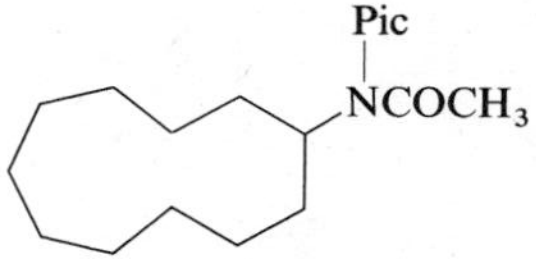 (Pic, $NCOCH_3$) (—)	93
	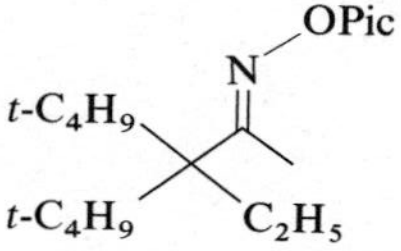 (t-C_4H_9, t-C_4H_9, C_2H_5, N–OPic)	"	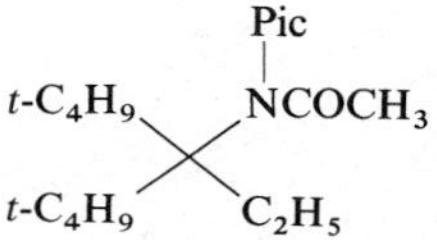 (Pic, $NCOCH_3$, t-C_4H_9, t-C_4H_9, C_2H_5) (—)	93
C_{14}	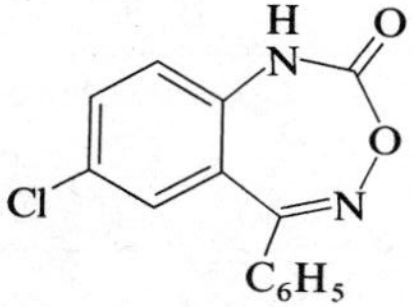 (H, N, O, O, N, Cl, C_6H_5)	Heat at mp	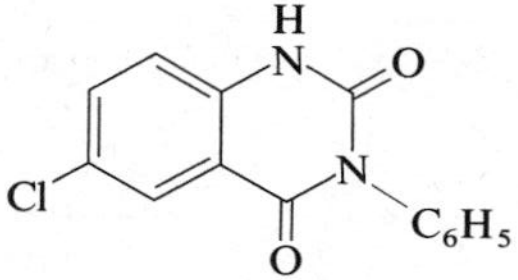 (H, N, O, N, Cl, O, C_6H_5) (54)	397

TABLE I. REARRANGEMENTS (*Continued*)

No. of Carbon Atoms	Substrate	Reagent and Conditions	Product(s) and Yield(s) (%)	Refs.
	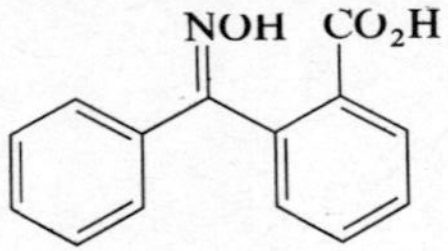 NOH CO_2H	HCO_2H, reflux, 6 h	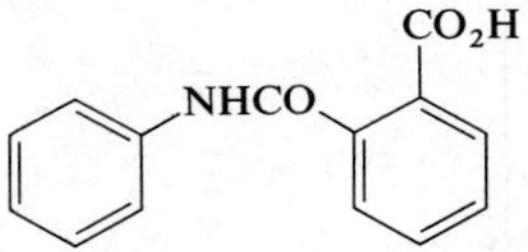 NHCO CO_2H (85)	251
	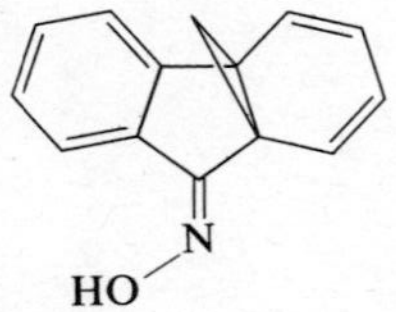 N HO	$C_6H_5SO_2Cl$, NaOH, acetone, H_2O	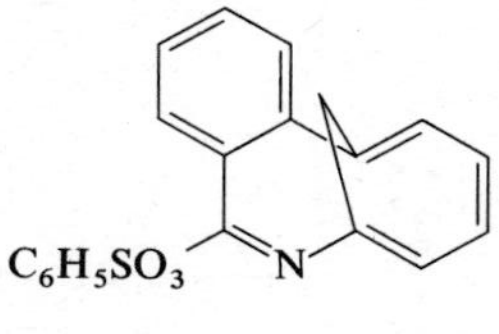 $C_6H_5SO_3$ N (19)	174
	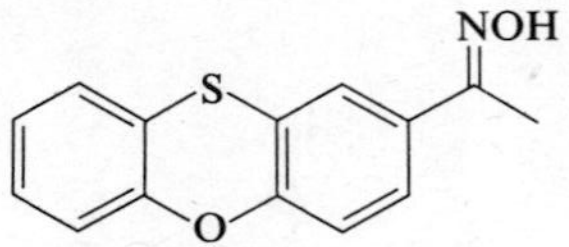 S O NOH	PCl_5, C_6H_6, 10 min	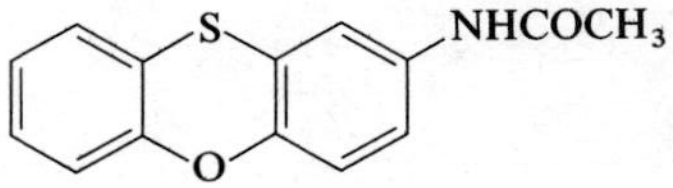 S O $NHCOCH_3$ (—)	407
	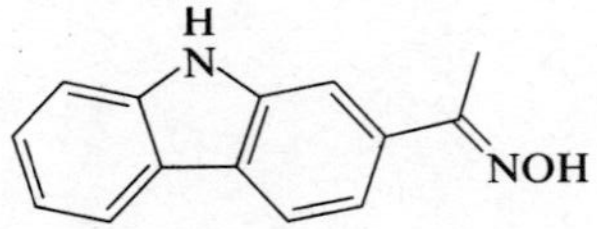 H N NOH	PCl_5, THF, 1 h	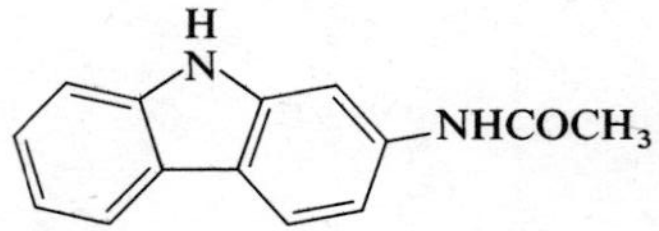 H N $NHCOCH_3$ (51)	408
	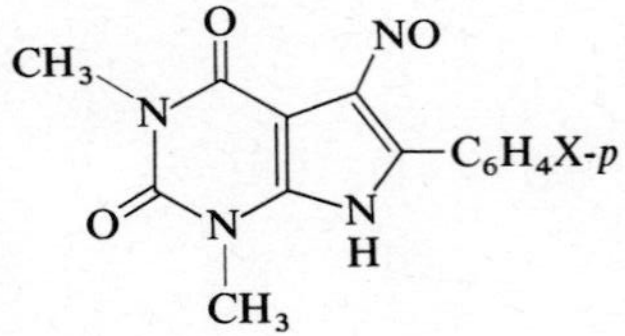 CH_3 N O NO C_6H_4X-*p* N H O N CH_3 X = Cl	TsCl, DMF, reflux, 15 min	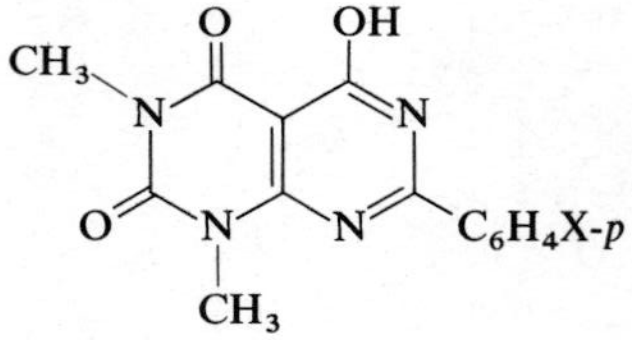 CH_3 N O OH N C_6H_4X-*p* N O N CH_3 (51)	409

X = Br	”	” (40)	409
p-$CH_3C_6H_4C(=NOH)C_6H_5$ (*Z*)	1. Cl_2CO, ether 2. $AgBF_4$, C_6H_5Cl	p-$CH_3C_6H_5CONHC_6H_5$ (79) + I p-$CH_3C_6H_5NHCOC_6H_5$ (11) II	391
(*E*)	”	I (14) + II (82)	391
$C_6H_5C(=NOH)CH_2C_6H_5$	PPSE, C_6H_6, 18 h	$C_6H_5NHCOCH_2C_6H_5$ (92)	72
$C_6H_5CH_2C(=NOSO_2C_6H_2(CH_3)_3)C_6H_5$	Alumina, CH_3OH	$C_6H_5CH_2CONHC_6H_5$ (71)	47
$C_6H_5C(=NOPic)CH_2C_6H_5$	$C_2H_4Cl_2$, warm	$C_6H_5N(Pic)COCH_2C_6H_5$ (88)	97
$C_6H_4CH_3$-*m*; HON; S (*Z*)	PPA, 120°, 2 h	$C_6H_4CH_3$-*m*; HN; O; S (15)	393
(*E*)	”	” (5)	393
$C_6H_4CH_3$-*p*; HON; S (*Z*)	”	$C_6H_4CH_3$-*p*; HN; O; S (15)	393
(*E*)	”	” (6)	393

TABLE I. Rearrangements (*Continued*)

No. of Carbon Atoms	Substrate	Reagent and Conditions	Product(s) and Yield(s) (%)	Refs.
	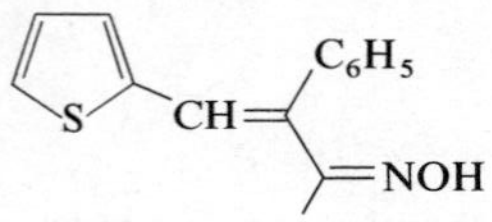 S, CH=, C_6H_5, =NOH	PCl_5, ether, dioxane, 0°, 6 h	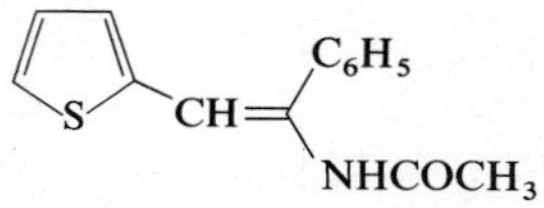 S, CH=, C_6H_5, $NHCOCH_3$ (20)	113
	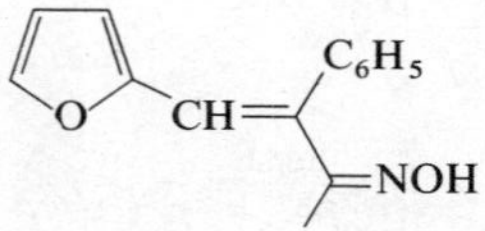 O, CH=, C_6H_5, =NOH	"	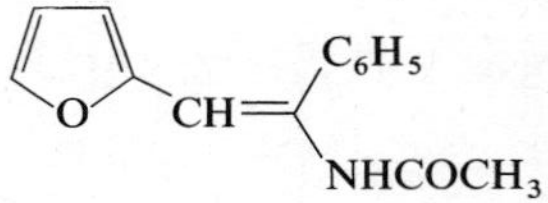 O, CH=, C_6H_5, $NHCOCH_3$ (5)	113
	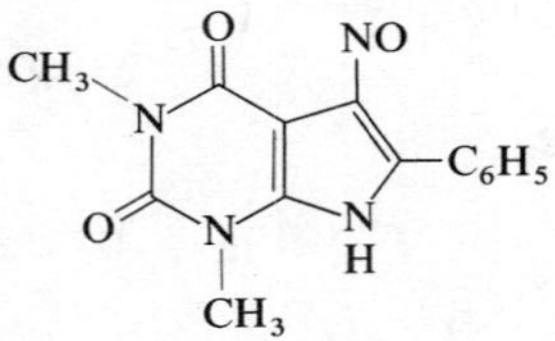 CH_3, N, O, O, NO, N, H, C_6H_5, N, CH_3	TsCl, DMF, reflux, 15 min	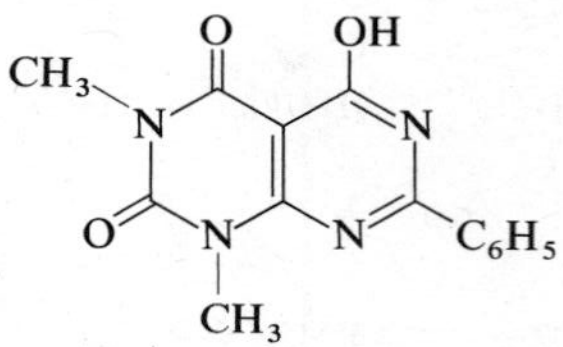 CH_3, N, O, O, OH, N, N, C_6H_5, N, CH_3 (40)	409
	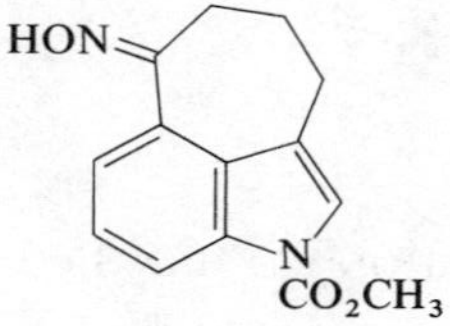 HON=, N, CO_2CH_3	$SOCl_2$	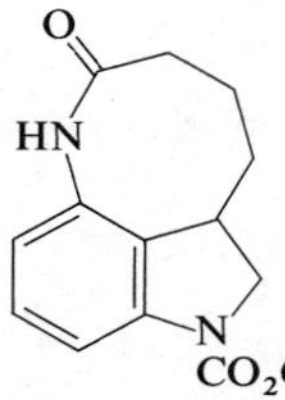 O, HN, N, CO_2CH_3 (50)	410

Reactant	Conditions	Product (Yield, %)	Refs.
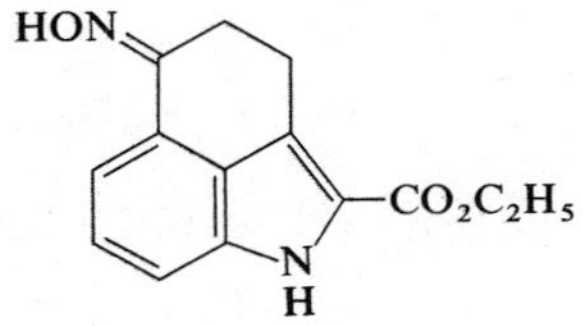 HON, $CO_2C_2H_5$, N H	PPA, 110°, 30 min	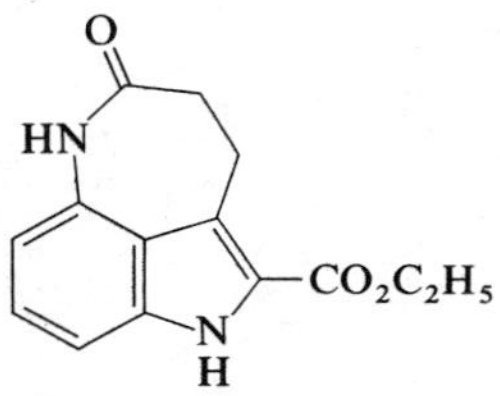 O, HN, $CO_2C_2H_5$, N H (45)	410
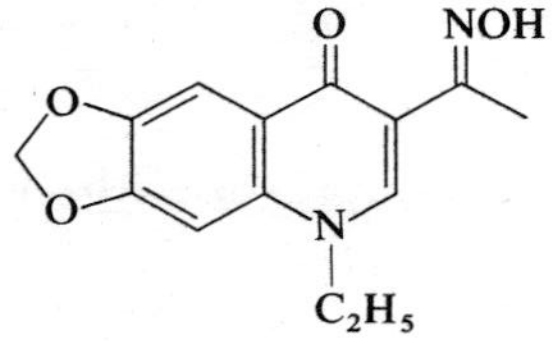 O, NOH, O, O, N, C_2H_5	HCO_2H, reflux	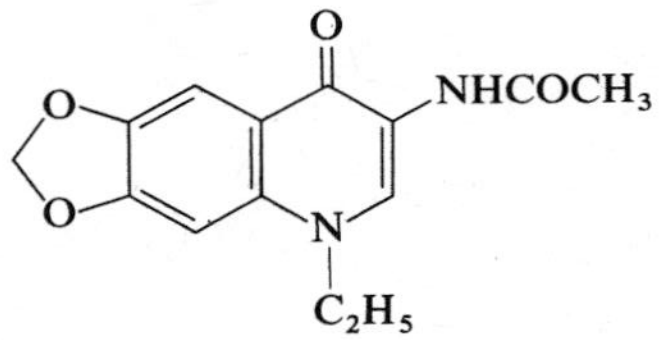 O, $NHCOCH_3$, O, O, N, C_2H_5 (—)	400
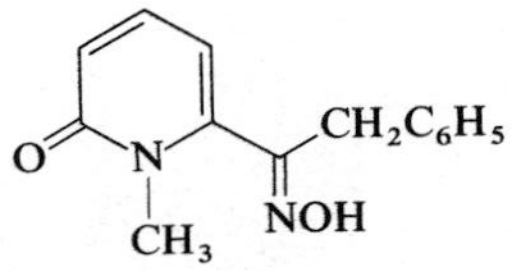 O, N, CH_3, $CH_2C_6H_5$, NOH (*Z*)	PCl_5, ether, 0°, 5 h	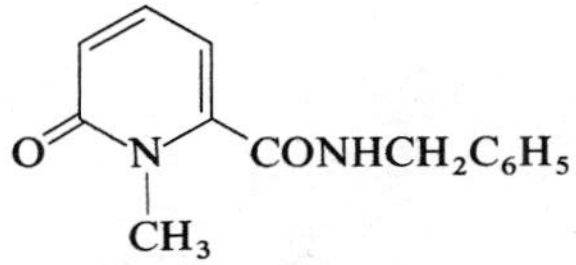 O, N, CH_3, $CONHCH_2C_6H_5$ (84)	293
(*E*)	"	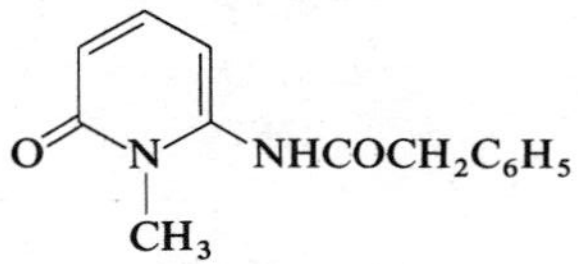 O, N, CH_3, $NHCOCH_2C_6H_5$ (90)	293
$TsNHC({=}NOH)C_6H_5$	$C_6H_5SO_2Cl$, 2 *N* NaOH, 4 h	$TsNHCONHC_6H_5$ (50)	118
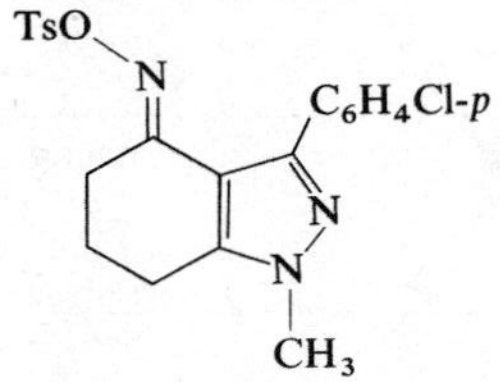 TsO, N, C_6H_4Cl-*p*, N, N, CH_3	NaO_2CCH_3, ethanol, H_2O	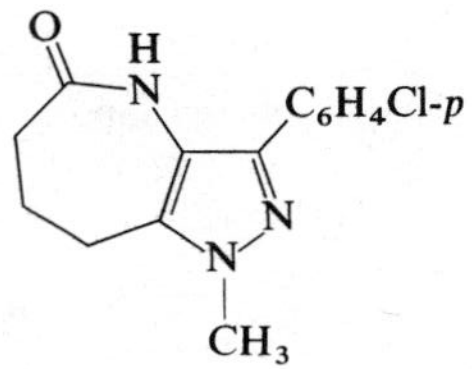 O, H N, C_6H_4Cl-*p*, N, N, CH_3 (20)	411

TABLE I. REARRANGEMENTS (*Continued*)

No. of Carbon Atoms	Substrate	Reagent and Conditions	Product(s) and Yield(s) (%)	Refs.
	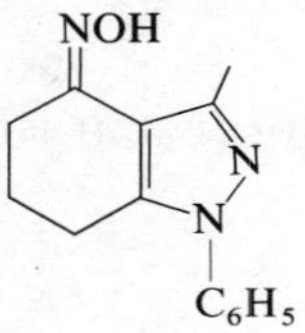 NOH, N, N, C_6H_5	PPA, 140°, 1 h	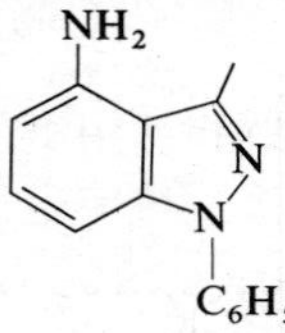 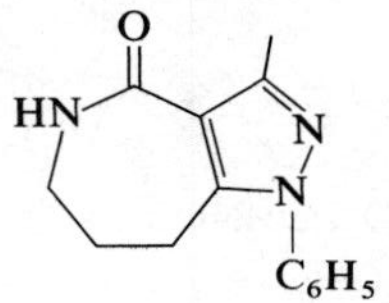 NH_2, N, N, C_6H_5 + O, HN, N, N, C_6H_5 (34)	412
	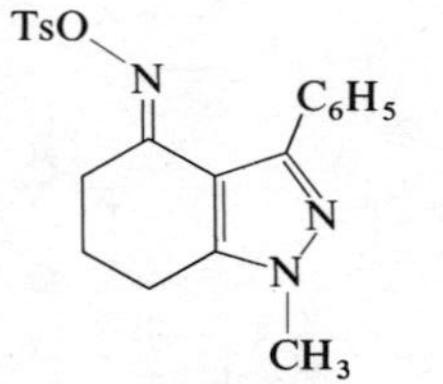 TsO, N, C_6H_5, N, N, CH_3	NaO_2CCH_3, ethanol, H_2O	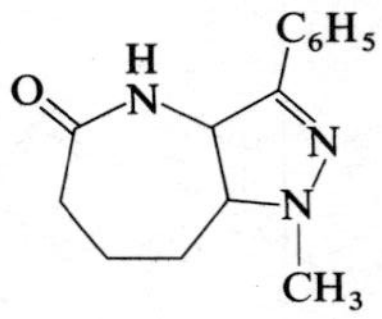 O, H, N, C_6H_5, N, N, CH_3 (64)	411
	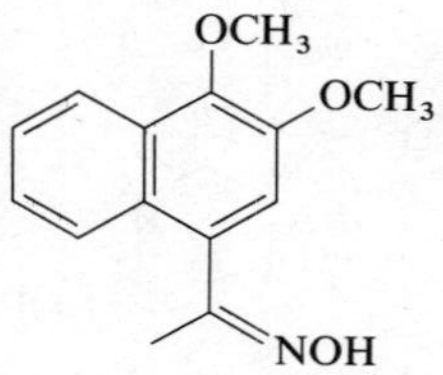 OCH_3, OCH_3, NOH	95% HCO_2H, reflux, 5 h	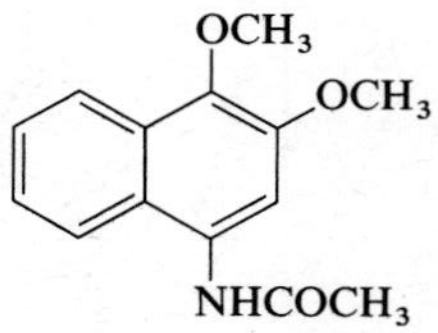 OCH_3, OCH_3, $NHCOCH_3$ (84)	413

Reactant	Conditions	Product(s) and Yield(s) (%)	Refs.
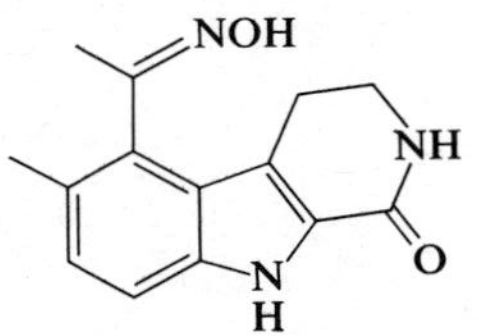 NOH; NH; N H; O	H_2SO_4, CH_3CO_2H, 100°, 1 h	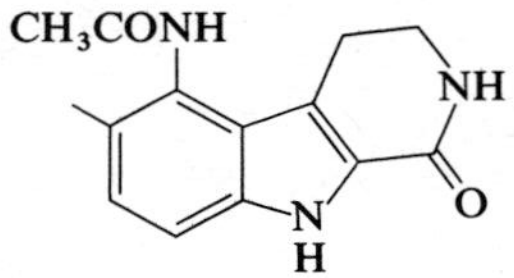 CH_3CONH; NH; N H; O (44)	414
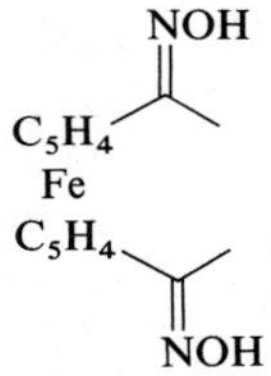 NOH; C_5H_4 Fe C_5H_4; NOH	$C_6H_5SO_2Cl$, NaOH	$C_5H_4COCH_3$ Fe $C_5H_4CONHCH_3$ (15-20)	99
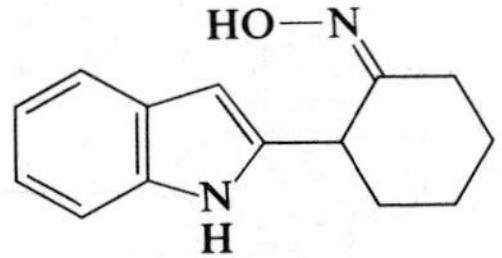 HO—N; N H	PPA, H_2SO_4, 60°, 3 h	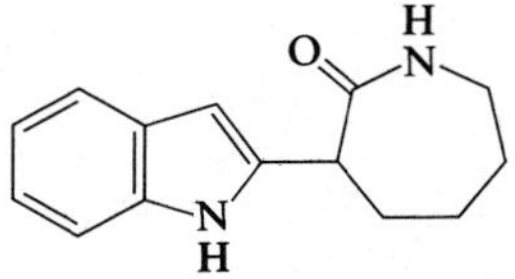 O; H N; N H (35)	415
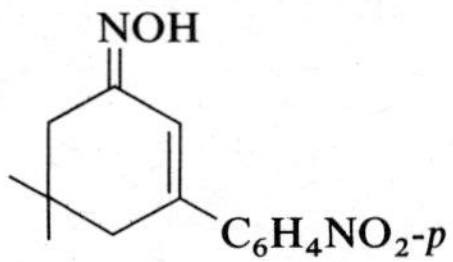 NOH; $C_6H_4NO_2$-*p*	PPA, 120°	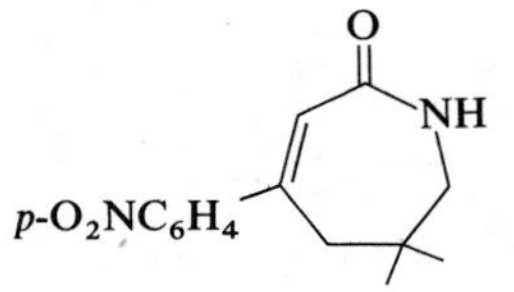 O; NH; *p*-$O_2NC_6H_4$ (14) 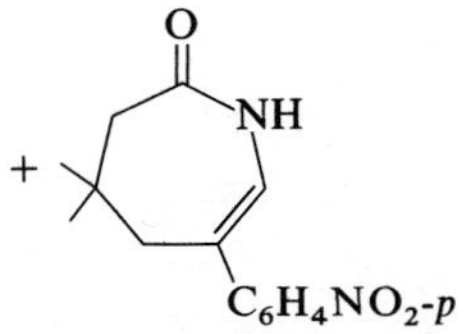 + O; NH; $C_6H_4NO_2$-*p* (26)	46

TABLE I. REARRANGEMENTS (*Continued*)

No. of Carbon Atoms	Substrate	Reagent and Conditions	Product(s) and Yield(s) (%)	Refs.
	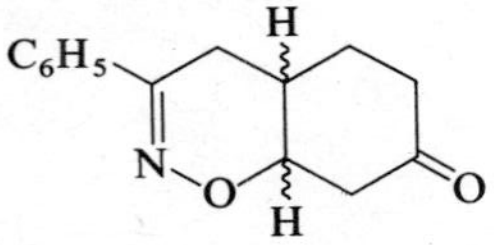 C_6H_5, H, N–O, H, O	CF_3CO_2H, reflux, 5 h	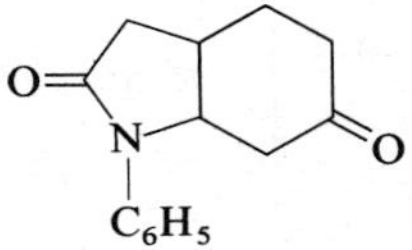 O, N–C_6H_5, O (70)	416
	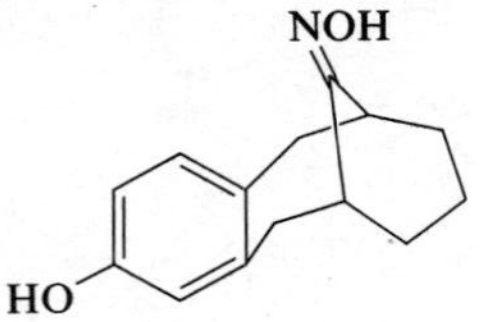 NOH, HO (*Z*)	TsCl, pyridine, 15 h	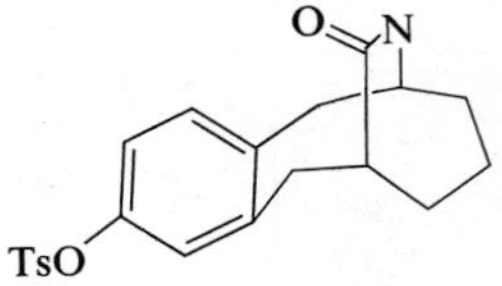 O, N, TsO (70)	417
	(*E*)	"	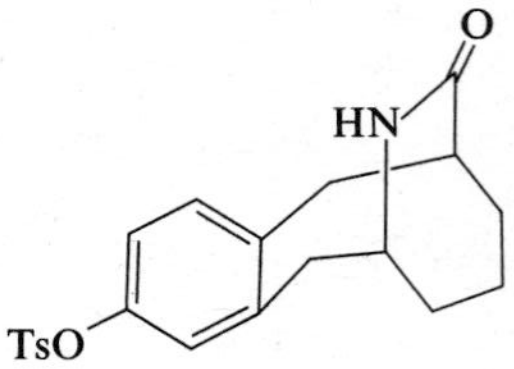 O, HN, TsO (69)	417
	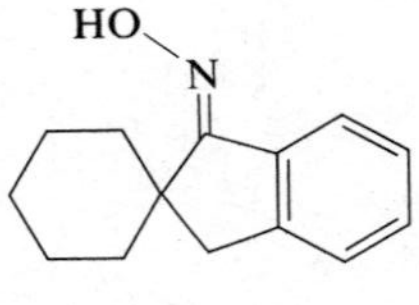 HO–N	PPA, 130°, 10 min	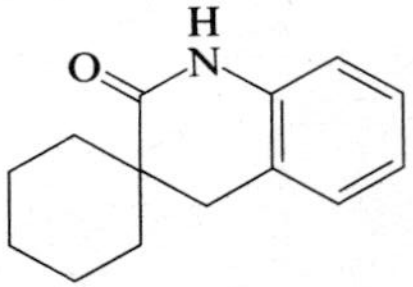 O, H–N (95)	362

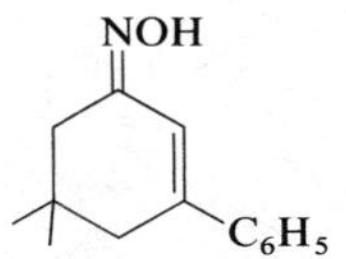 NOH, C_6H_5	PPA, 120°	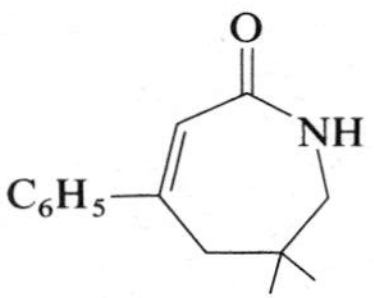 C_6H_5, O, NH (70)	46
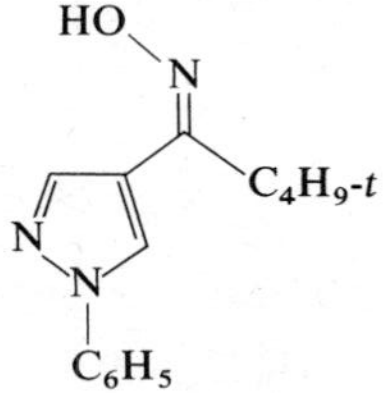 HO, N, C_4H_9-*t*, N, N, C_6H_5	PCl_5, $CHCl_3$, 20°, 4 d	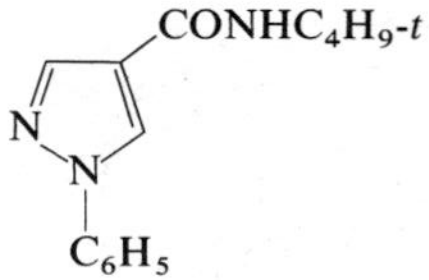 $CONHC_4H_9$-*t*, N, N, C_6H_5 (32)	318
	TsCl, pyridine, 20°, 4 d	" (30)	318
	H_2SO_4, 40–100°, 1 h	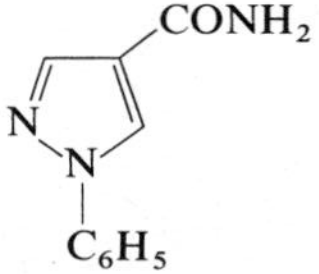 $CONH_2$, N, N, C_6H_5 (87)	318
	PPA, 40–100°, 1 h	" (85)	318
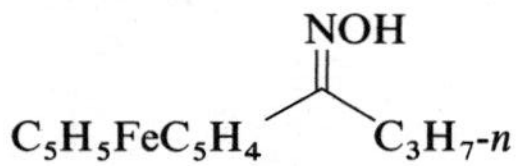 NOH, $C_5H_5FeC_5H_4$, C_3H_7-*n*	Cl_3CCN, ether, reflux, 6 h	$C_5H_5FeC_5H_4CONHC_3H_7$-*n* (—)	232
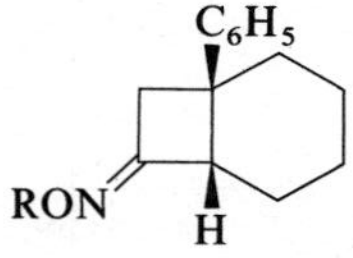 C_6H_5, RON, H (*Z*) R = H	(—)	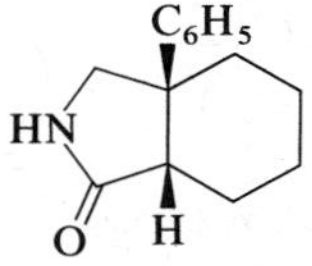 C_6H_5, HN, O, H, I (—)	127

TABLE I. REARRANGEMENTS (*Continued*)

No. of Carbon Atoms	Substrate	Reagent and Conditions	Product(s) and Yield(s) (%)	Refs.
	(*E*) R = H	P_2O_5, CH_3SO_3H, 100°, 1 h	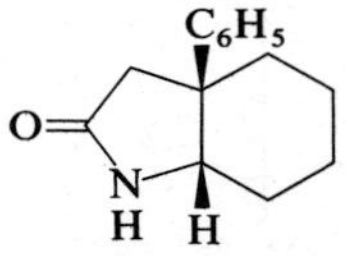 C_6H_5; O=; N; H H; II (93)	126
	(*E*) R = H	PPA, 53°, 2 h	II (27)	126
	(*E*) R = Ts	Al_2O_3, CH_2Cl_2	I/II = 60/40 (—)	126
	(*E*) R = 2,4,6-$(CH_3)_3C_6H_2$	"	I/II = 57/43 (—)	126
	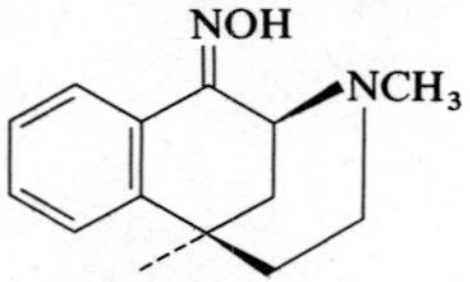 NOH; NCH_3	PPA, 150°, 1 h	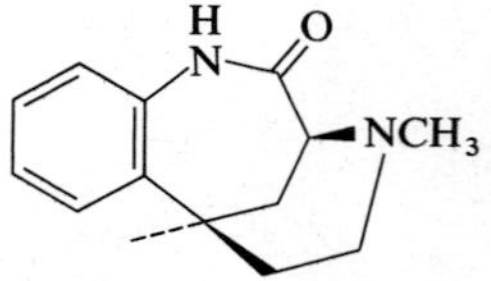 H; N; O; NCH_3 (59)	418
	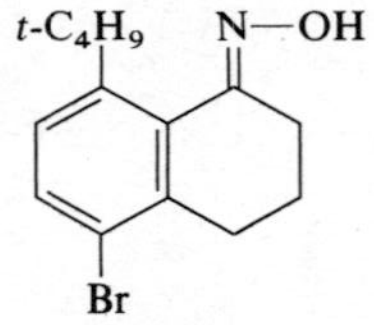 *t*-C_4H_9; N—OH; Br	PPA, 110–120°, 10 min	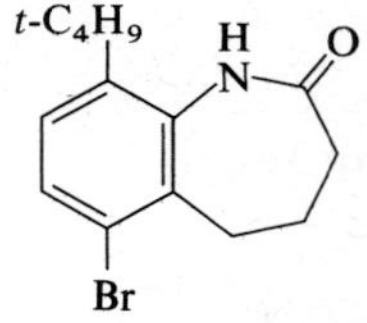 *t*-C_4H_9; H; N; O; Br (97)	70
	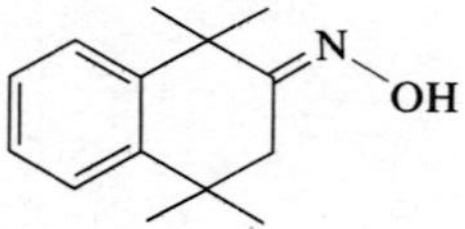 N; OH	PPA, 140°, 20 min	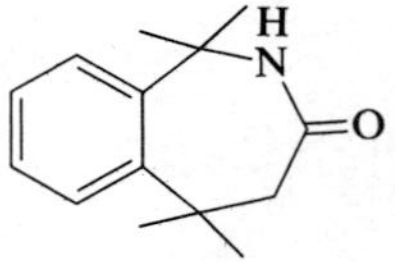 H; N; =O (24)	419, 420

Reactant	Conditions	Product(s) (Yield, %)	Ref.
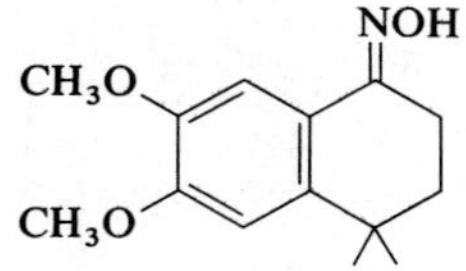 CH_3O, CH_3O, NOH	PPA, 120°, 10 min	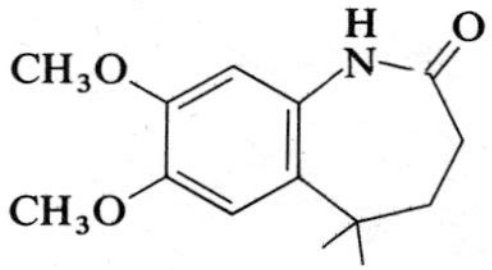 CH_3O, CH_3O, H, N, O (12) 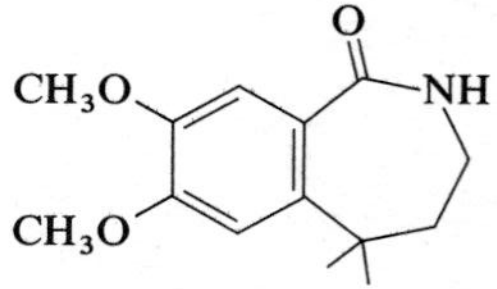 + CH_3O, CH_3O, O, NH (17)	360
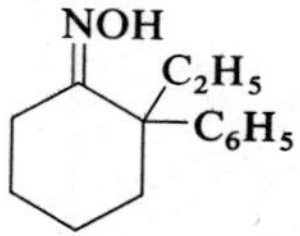 NOH, C_2H_5, C_6H_5	PPA, 115°, 10 min	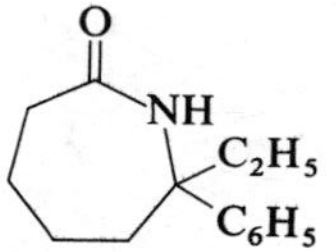 O, NH, C_2H_5, C_6H_5 (62)	351
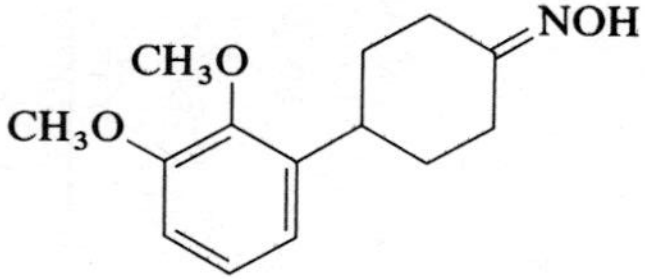 CH_3O, CH_3O, NOH	PPA, 110°, 10 min	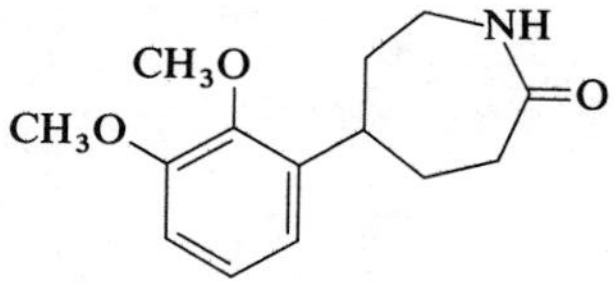 CH_3O, CH_3O, NH, O (43)	421
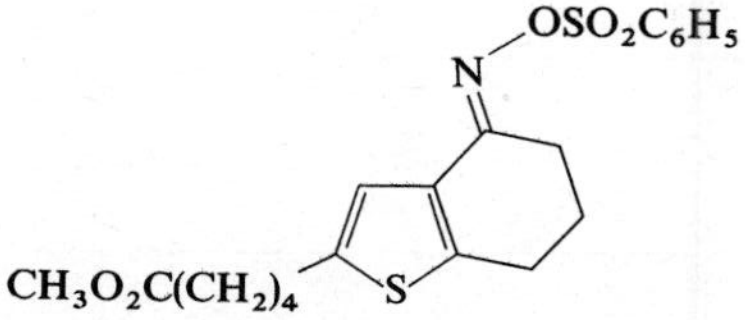 $CH_3O_2C(CH_2)_4$, S, N, $OSO_2C_6H_5$	NaO_2CCH_3, ethanol, H_2O, reflux, 4 h	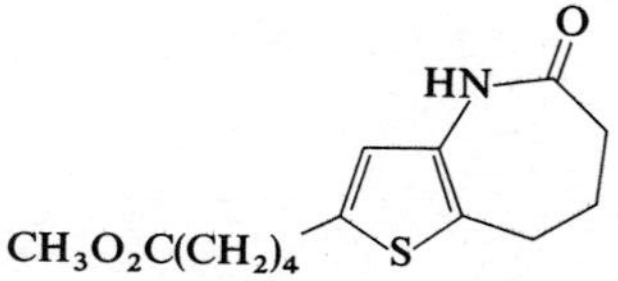 $CH_3O_2C(CH_2)_4$, S, HN, O (67)	382

TABLE I. REARRANGEMENTS (*Continued*)

No. of Carbon Atoms	Substrate	Reagent and Conditions	Product(s) and Yield(s) (%)	Refs.
	[structure: oxime, N–OH]	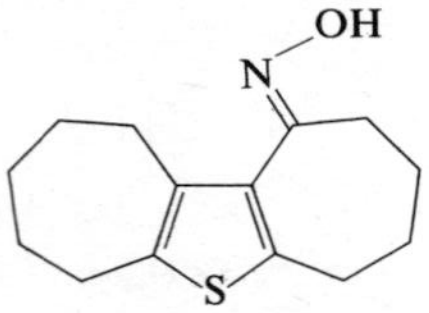 $C_6H_5SO_2Cl$, pyridine	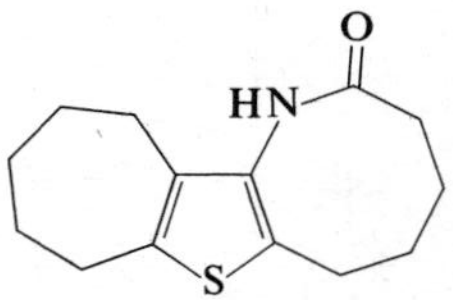 [structure: HN, O, S lactam] (60)	381
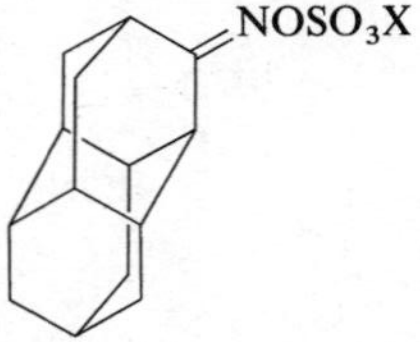	[structure: =NOSO$_3$X] X = H	H_2O, 30 h	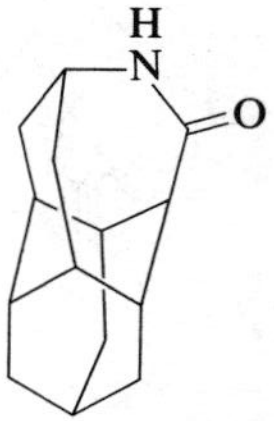 [structure: H N, O lactam] (60)	258
	X = NH_4	200–210°, 2 min	" (85)	258
	[structure: bis-oxime, NOH, NOH]	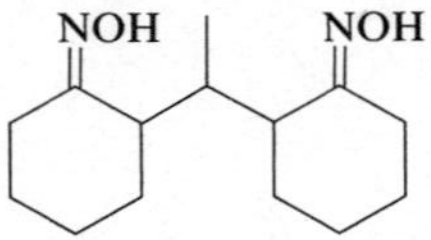 H_2SO_4	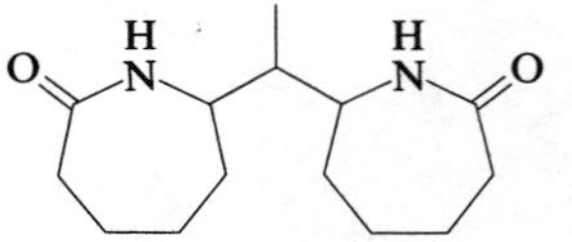 [structure: bis-lactam, O, H N, H N, O] (25)	42
	[structure: H, OH, $COCH_3$]	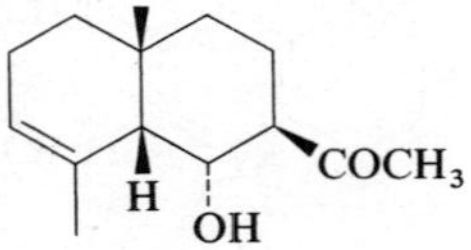 1. $H_2NOSO_2C_6H_2(CH_3)_3$-2,4,6 2. Alumina	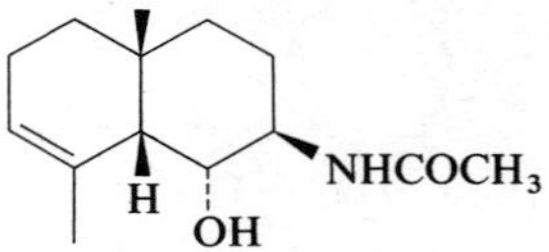 [structure: H, OH, $NHCOCH_3$] (50)	422

Reactant	Conditions	Product(s) and Yield(s) (%)	Refs.
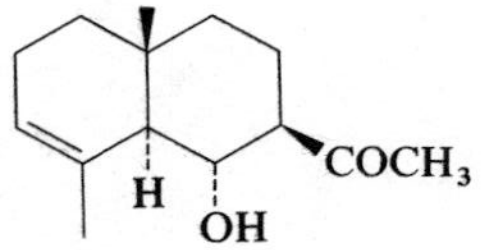 [structure: H, OH, $COCH_3$]	"	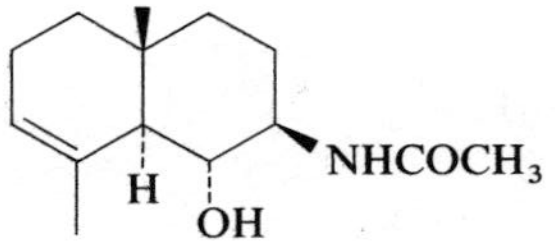 [structure: H, OH, $NHCOCH_3$] (—)	422
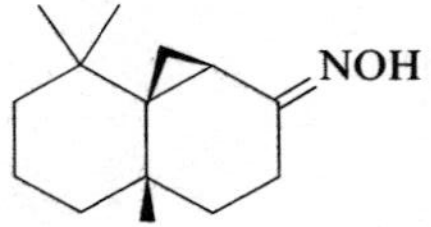 [structure: NOH] (*E*)	PPA, 90°, 10 min	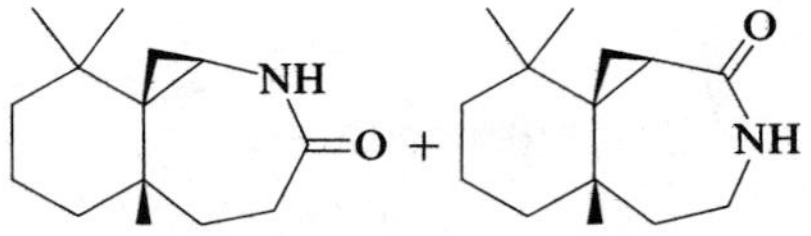 [structure: NH, =O] I (40) + [structure: O, NH] II (7)	405
(*Z*)	"	I (10) + II (41)	405
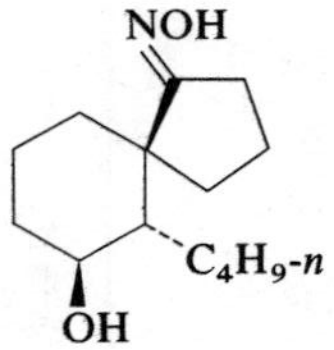 [structure: NOH, C_4H_9-*n*, OH]	TsCl, pyridine, C_6H_6, 25°, 12 h	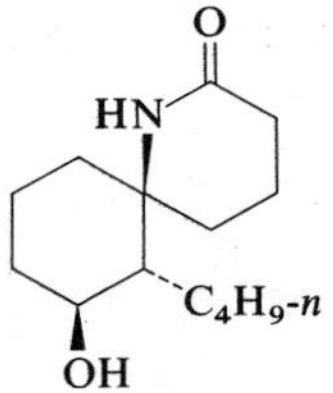 [structure: O, HN, C_4H_9-*n*, OH] (—)	423
	TsCl, pyridine, 0°, 17 h	" (25)	182, 183
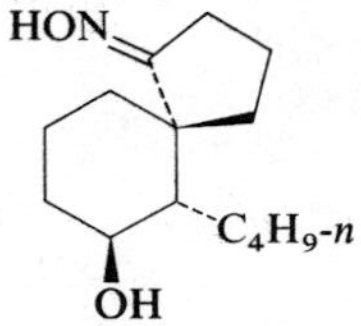 [structure: HON, C_4H_9-*n*, OH]	TsCl, pyridine	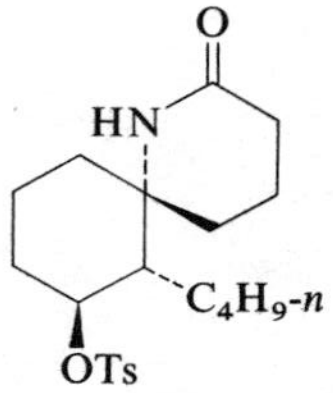 [structure: O, HN, C_4H_9-*n*, OTs] (38)	183, 424
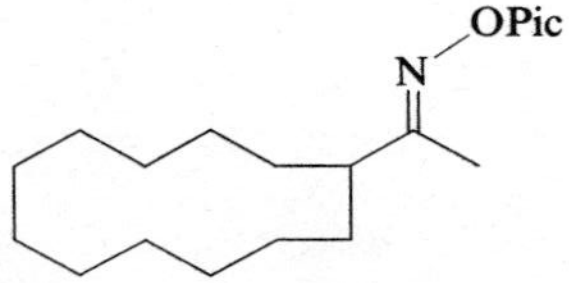 [structure: N, OPic]	1,4-Dichlorobutane, 80°	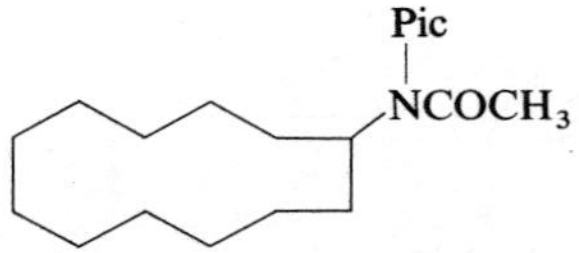 [structure: Pic, $NCOCH_3$] (—)	425

TABLE I. REARRANGEMENTS (*Continued*)

No. of Carbon Atoms	Substrate	Reagent and Conditions	Product(s) and Yield(s) (%)	Refs.
	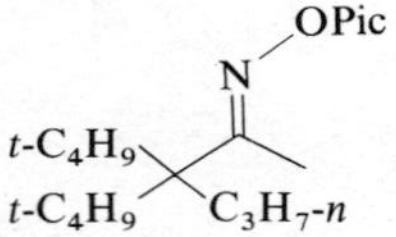 $(t\text{-}C_4H_9)_2C(C_3H_7\text{-}n)C(CH_3){=}NOPic$	1,4-Dichlorobutane, 80°	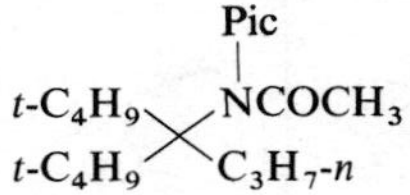 $(t\text{-}C_4H_9)_2C(C_3H_7\text{-}n)N(Pic)COCH_3$ (—)	93
C_{15}	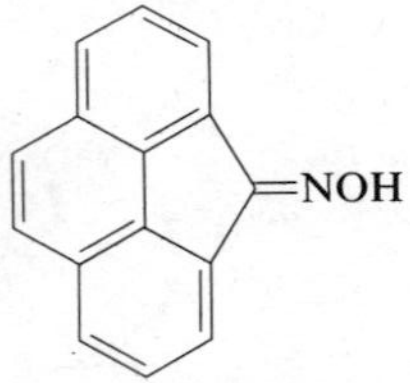 Acenaphthenone-type oxime (=NOH)	PPA, 175–180°, 5 min	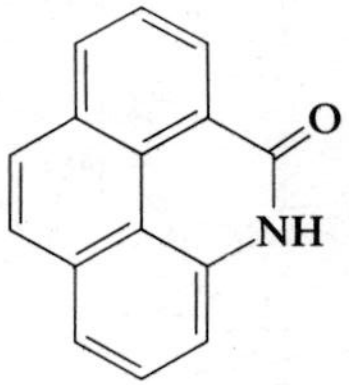 Lactam (C=O, NH) (57)	426
	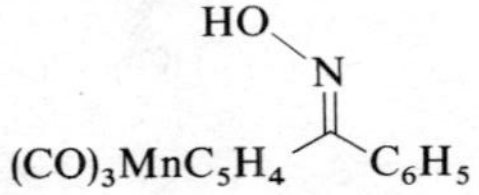 $(CO)_3MnC_5H_4C(C_6H_5){=}NOH$	PCl_5, pyridine, C_6H_6, 0°, 30 min	$(CO)_3MnC_5H_4CONHC_6H_5$ (—)	101, 102
	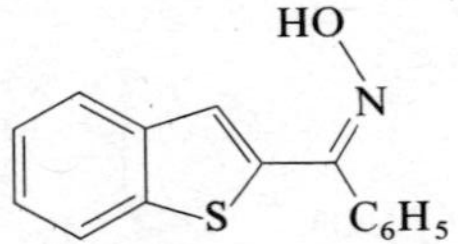 2-Benzothienyl-$C(C_6H_5){=}NOH$	PCl_5, ether, 0°, 24 h	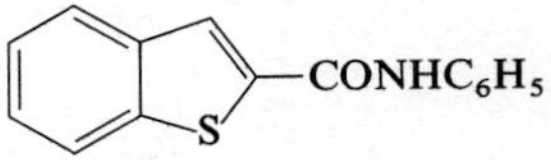 2-Benzothienyl-$CONHC_6H_5$ (100)	427
	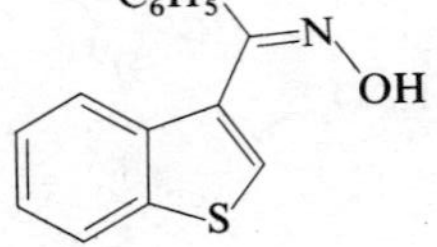 3-Benzothienyl-$C(C_6H_5){=}NOH$	"	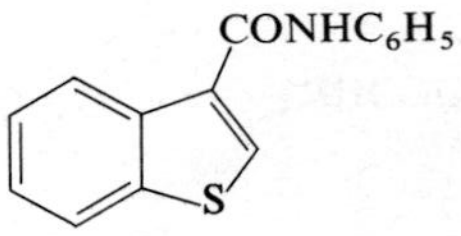 3-Benzothienyl-$CONHC_6H_5$ (100)	427

Substrate	Reagents	Product (Yield %)	Ref.
2-Benzofuranyl ketoxime tosylate: NOTs, C_6H_5	1. TsCl, pyridine 2. CH_3OH	2-Benzofuranyl $-CONHC_6H_5$ (—)	353
HO–N=C(C_6H_5)–C≡CC_6H_5 ($C_6H_5C{\equiv}C$, C_6H_5)	PCl_5, Et_2O	$C_6H_5CCl{=}CHCONHC_6H_5$ (—)	116
3-(COC_6H_5)indole (N–H)	$H_2NOH \cdot HCl$, ethanol reflux, 3 d	3-($NHCOC_6H_5$)indole (N–H) (19)	320
3-[C_6H_5C=N–OH]indole (N–H)	PCl_5, ether, 0°, 24 h	3-($CONHC_6H_5$)indole (N–H) (100)	428
	CH_3CO_2H, reflux, 2 h	3-($NHCOC_6H_5$)indole (N–H) (55)	320
2-[HO–N=C(C_6H_5)]indole (N–H) (*Z*)	PCl_5, ether, 0°, 24 h	2-Indolyl $-CONHC_6H_5$ (N–H) (100)	428

TABLE I. REARRANGEMENTS (*Continued*)

No. of Carbon Atoms	Substrate	Reagent and Conditions	Product(s) and Yield(s) (%)	Refs.
	4-(1-hydroxyiminoethyl)-2-(2-naphthyl)thiazole (NOH)	PCl_5, ether	CH_3CONH-4-[2-(2-naphthyl)thiazole] (37)	246
	4-(1-hydroxyiminoethyl)-2-(1-naphthyl)thiazole (NOH)	"	CH_3CONH-4-[2-(1-naphthyl)thiazole] (34)	246
	1-C_6H_5-1,2,4-triazol-3-yl–C(=NOH)C_6H_5 (HO–N)	PPA, 120°, 2 min	1-C_6H_5-1,2,4-triazole-3-$CONHC_6H_5$ (70)	429
		PCl_5, $CHCl_3$, 0° 2 d	1-C_6H_5-1,2,4-triazole-3-$NHCOC_6H_5$ (20)	321
	C_6H_5C(=NOR)CH=CHC_6H_5 R = H	Cl_3CCN, reflux, 5-6 h	$C_6H_5NHCOCH{=}CHC_6H_5$ (60–70)	100

Reactant	Conditions	Product(s) and Yield(s) (%)	Refs.
R = H	PPSE, CH_2Cl_2, 30 min	" (57)	72
R = Ts	SiO_2, $CHCl_3$, 5°, 30 min	" (90)	109
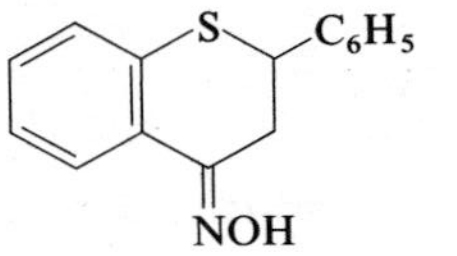 S, C_6H_5, NOH	PPA, 120°, 1 h	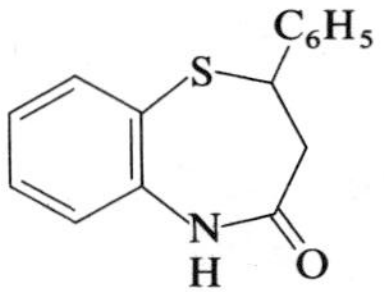 C_6H_5, S, N, H, O (65)	430
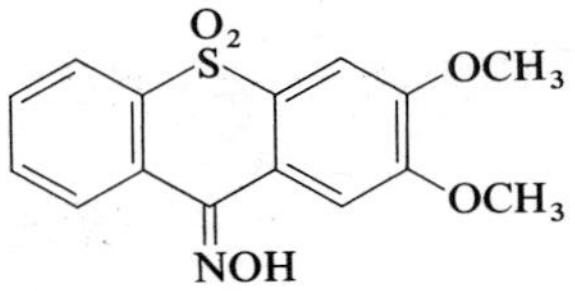 O_2, S, OCH_3, OCH_3, NOH	PPA, 155°, 10 min	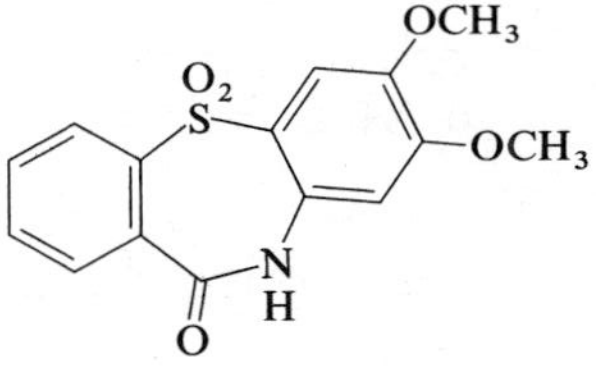 OCH_3, O_2, S, OCH_3, N, H, O (72)	390
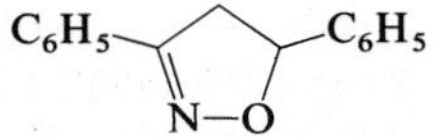 C_6H_5, C_6H_5, N–O	$(CH_3CO)_2O$, BF_3·ether, 3 d	$C_6H_5NHCOCH_2CH(C_6H_5)O_2CCH_3$ + $C_6H_5NHCOCH{=}CHC_6H_5$ (—)	431
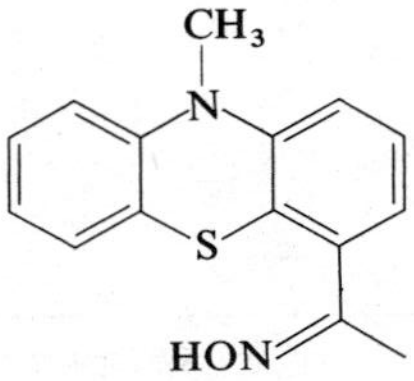 CH_3, N, S, HON	PCl_5, pyridine, C_6H_6	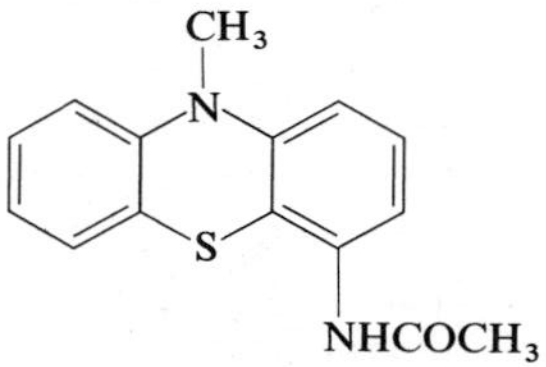 CH_3, N, S, $NHCOCH_3$ (47)	432
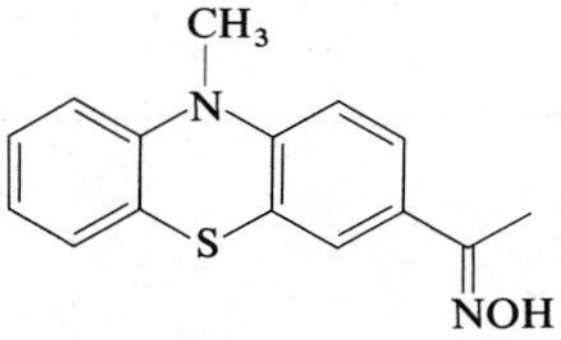 CH_3, N, S, NOH	"	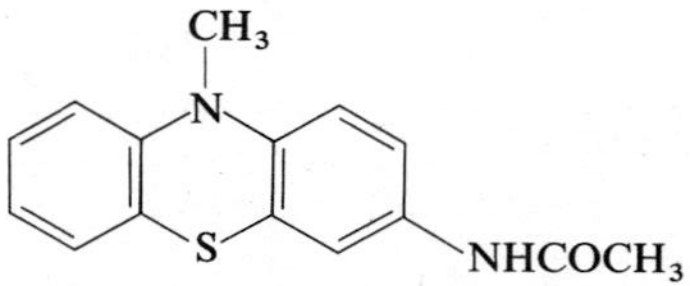 CH_3, N, S, $NHCOCH_3$ (31)	432

TABLE I. Rearrangements (*Continued*)

No. of Carbon Atoms	Substrate	Reagent and Conditions	Product(s) and Yield(s) (%)	Refs.
	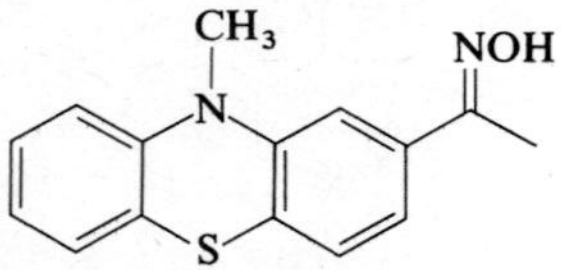 CH_3, N, S, NOH	PCl_5, pyridine, C_6H_6	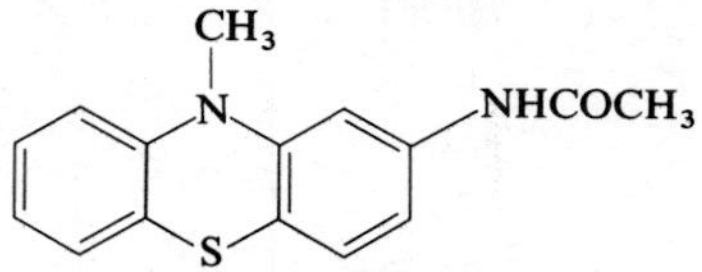 CH_3, N, S, $NHCOCH_3$ (78)	432
	$(p\text{-}CH_3C_6H_4)_2C{=}NOH$	HCO_2H, reflux, 6 h	$p\text{-}CH_3C_6H_4NHCOC_6H_4CH_3\text{-}p$ (95)	251
	$(C_6H_5CH_2)_2C{=}NOH$	PPSE, C_6H_6, 38 h	$C_6H_5CH_2NHCOCH_2C_6H_5$ (76)	72
	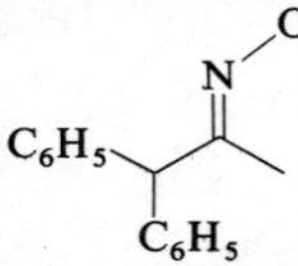 OTs, N, C_6H_5, C_6H_5	$(C_2H_5)_3N$, 80% ethanol, 30°, 60 h	$(C_6H_5)_2CHNHCOCH_3$ (43)	28
	$o\text{-}C_2H_5C_6H_4C({=}NOH)C_6H_5$	PCl_5, C_6H_6	$o\text{-}C_2H_5C_6H_4CONHC_6H_5$ + $o\text{-}C_2H_5C_6H_4NHCOC_6H_5$ (—)	154
	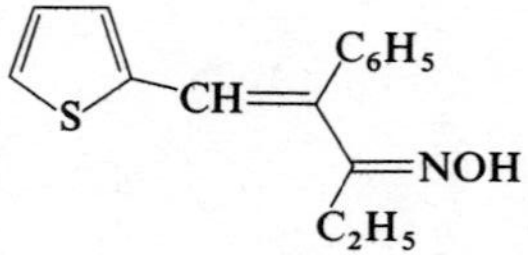 S, CH=, C_6H_5, =NOH, C_2H_5	PCl_5, ether, dioxane, 0°, 6 h	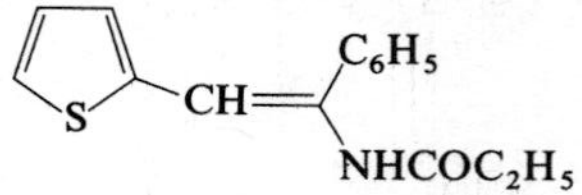 S, CH=, C_6H_5, $NHCOC_2H_5$ (40)	113
	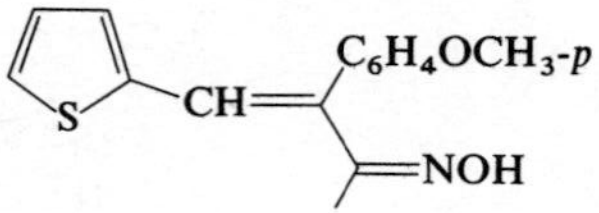 S, CH=, $C_6H_4OCH_3\text{-}p$, =NOH	"	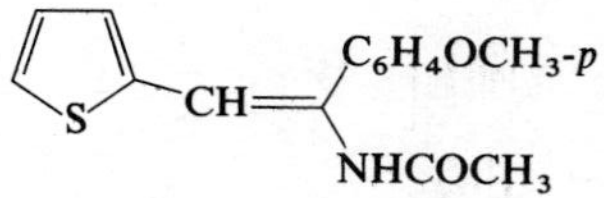 S, CH=, $C_6H_4OCH_3\text{-}p$, $NHCOCH_3$ (—)	113

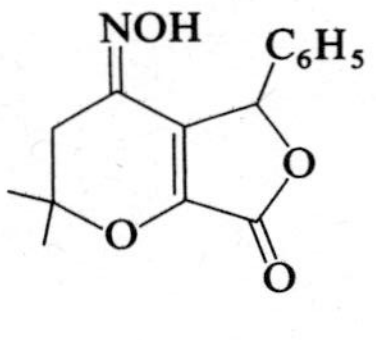	PPA, 110–130°, 20 min	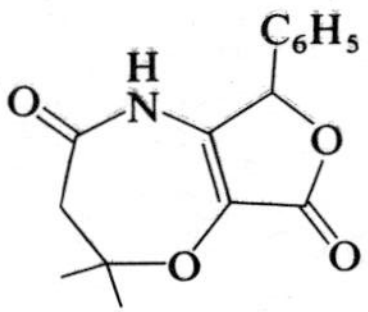	306
		+ 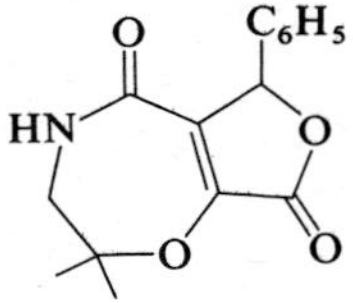 (25)	
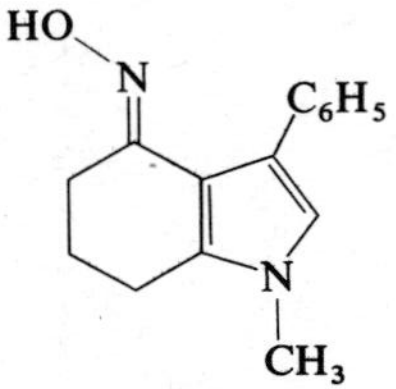	PPA, 130°, 30 min	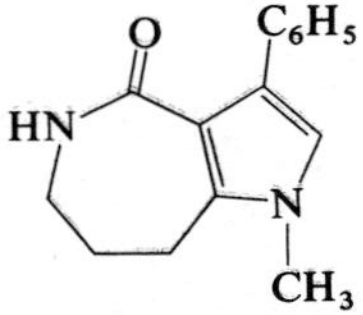 (30)	308
	TsCl, pyridine, 15 h	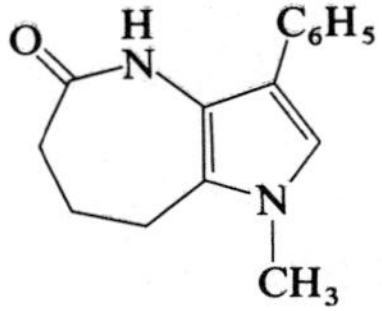 (25)	308
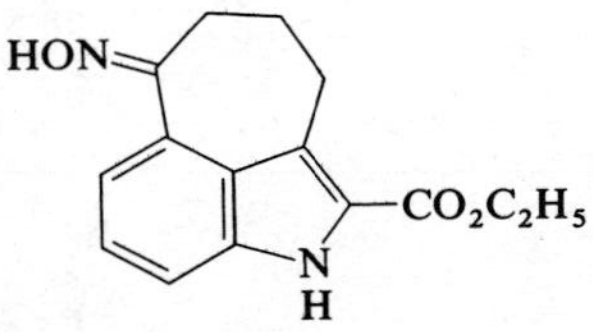	PPA, 110°, 30 min	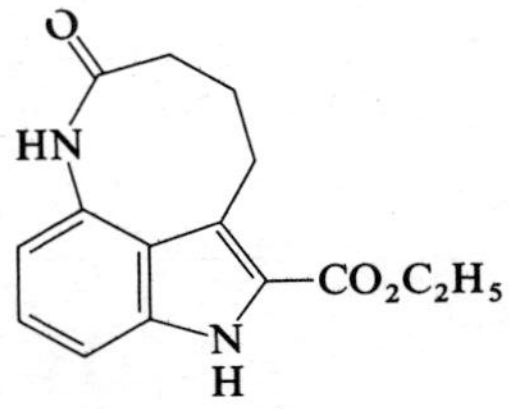 (61)	410

TABLE I. REARRANGEMENTS (*Continued*)

No. of Carbon Atoms	Substrate	Reagent and Conditions	Product(s) and Yield(s) (%)	Refs.
	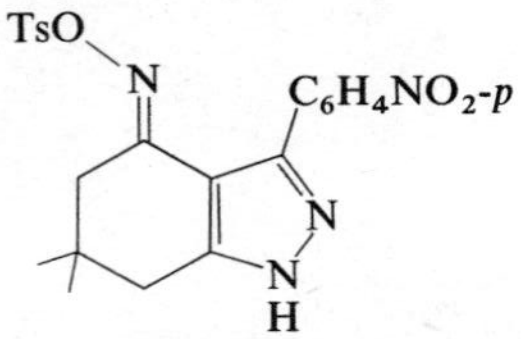 TsO–N=; $C_6H_4NO_2$-*p*; N; N; H	NaO_2CCH_3, ethanol, H_2O	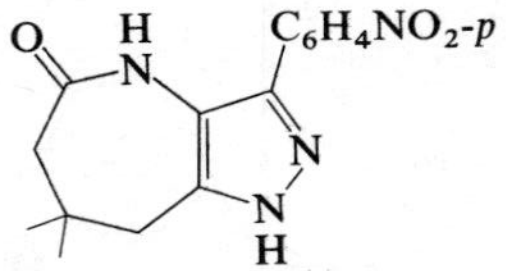 O; H; N; $C_6H_4NO_2$-*p*; N; N; H (20)	411
	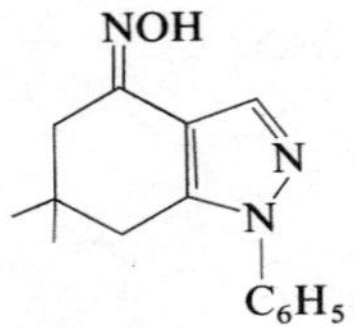 NOH; N; N; C_6H_5	PPA, 130–140°, 1 h	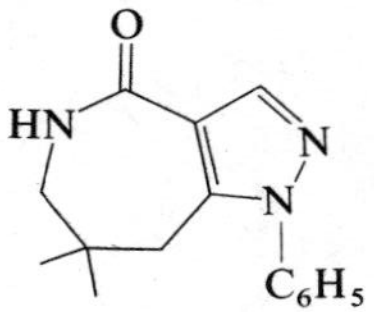 O; HN; N; N; C_6H_5 (100)	412
	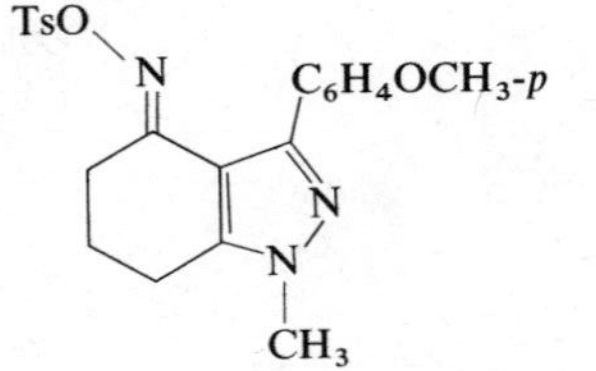 TsO–N=; $C_6H_4OCH_3$-*p*; N; N; CH_3	NaO_2CCH_3, ethanol, H_2O	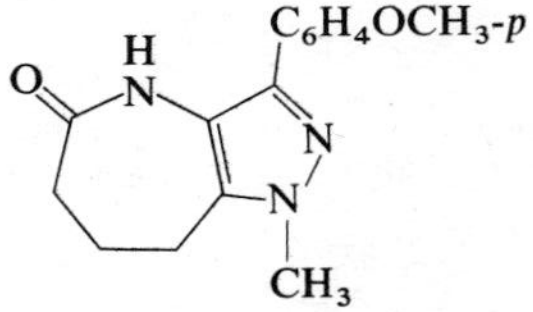 O; H; N; $C_6H_4OCH_3$-*p*; N; N; CH_3 (29)	411
	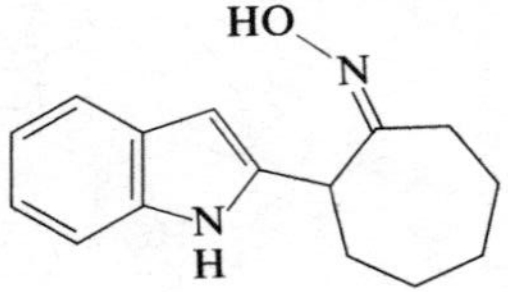 HO–N=; N; H	PPA, H_2SO_4, 60–70°, 3 h	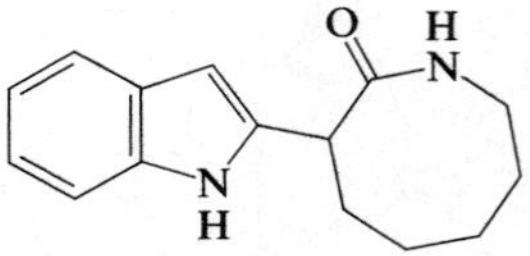 O; H; N; N; H (40)	415

Oxime	Conditions	Product(s) (Yield, %)	Refs.
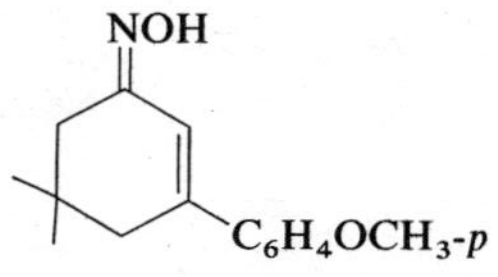 NOH; $C_6H_4OCH_3$-*p*	PPA, 120°	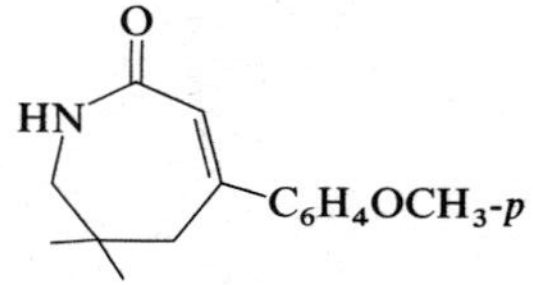 HN; O; $C_6H_4OCH_3$-*p* (90)	46
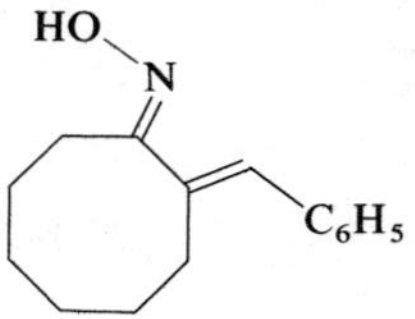 HO–N; C_6H_5	PCl_5, ether, −10°	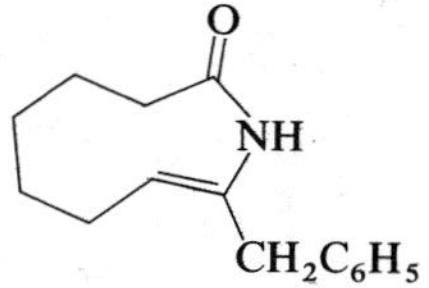 O; NH; $CH_2C_6H_5$ (—)	31
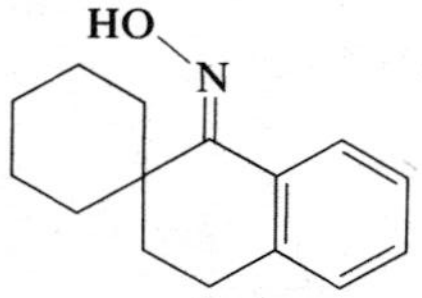 HO–N	PPA, 125–130°, 10 min	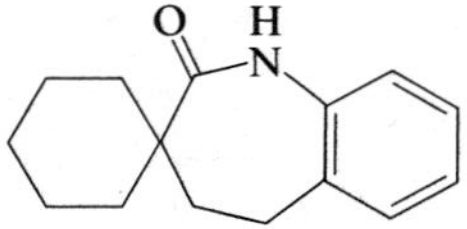 O; H–N (82)	362
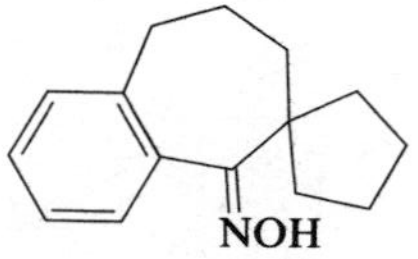 NOH	PPA, 125–130°, 10 min	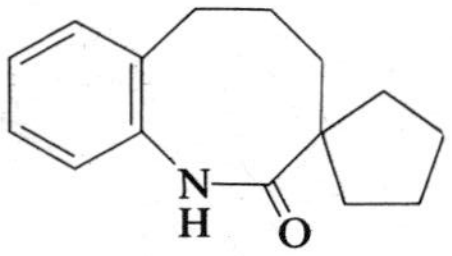 N–H; O (92)	362
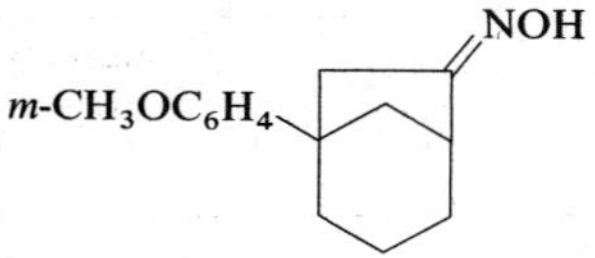 *m*-$CH_3OC_6H_4$; NOH	1. TsCl, DMF, 22 h 2. 10% NaOH	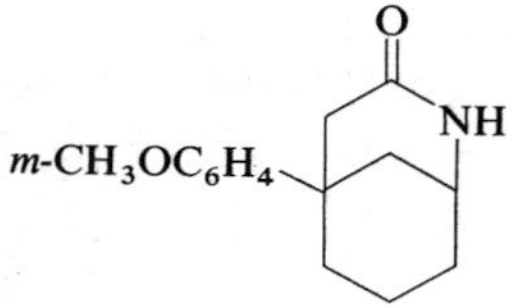 *m*-$CH_3OC_6H_4$; O; NH (8)	433

TABLE I. REARRANGEMENTS (*Continued*)

No. of Carbon Atoms	Substrate	Reagent and Conditions	Product(s) and Yield(s) (%)	Refs.
		PCl_5, ether, 27 h	(26)	434
	$(CH_2)_2N(CH_3)_2$	PPA, 150°, 1 h	$(CH_2)_2N(CH_3)_2$ (—)	435
	CH_3O, CH_3O, NOTs	CH_3CO_2H	CH_3O, CH_3O (90)	436
	HON, OH	TsCl, pyridine	OH (—)	437

Reactant	Conditions	Product(s) and Yield(s) (%)	Refs.
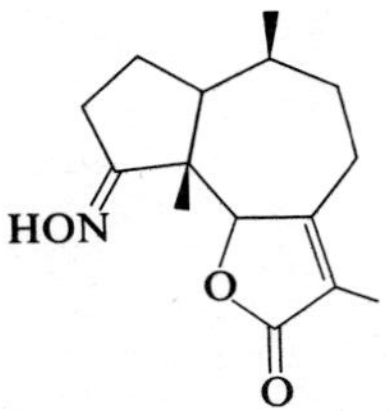 HON= (oxime)	”	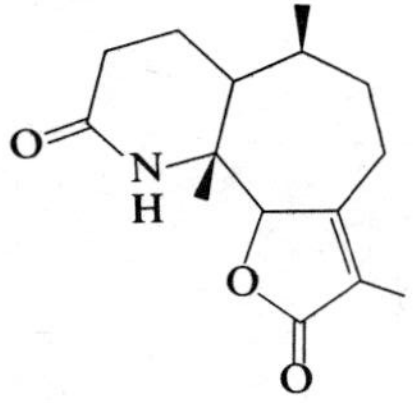 (—)	437
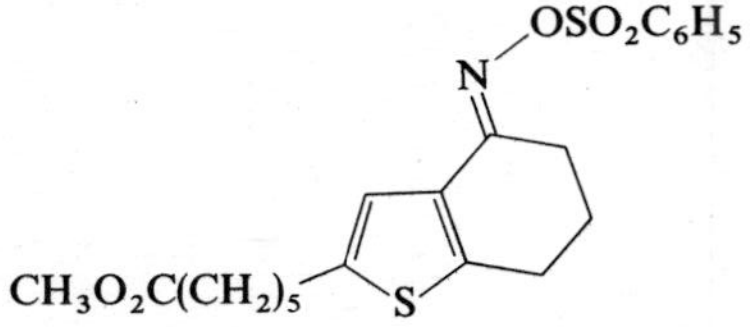 $CH_3O_2C(CH_2)_5$, $N{-}OSO_2C_6H_5$	NaO_2CCH_3, ethanol, H_2O, 0°, 4 h	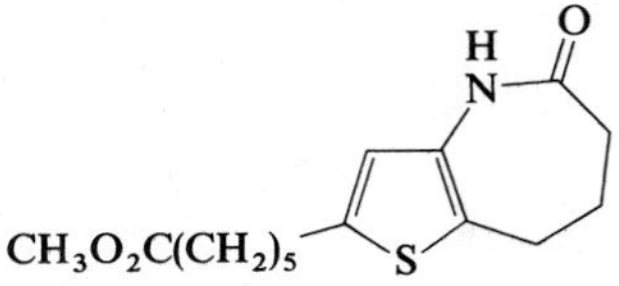 $CH_3O_2C(CH_2)_5$ (68)	382
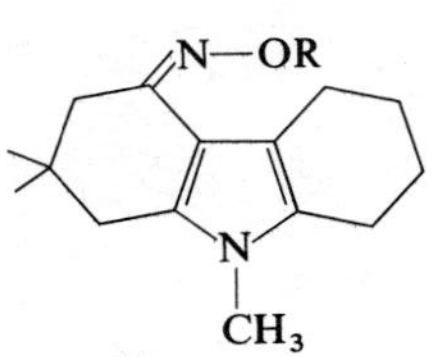 $N{-}OR$, $N{-}CH_3$; R = H	PPA, 130°, 30 min	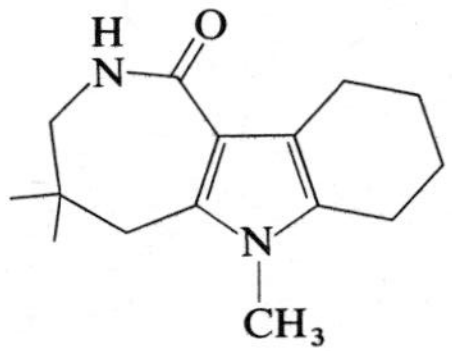 $N{-}CH_3$ (45)	383
R = Ts	NaO_2CCH_3, ethanol, H_2O, 30 h	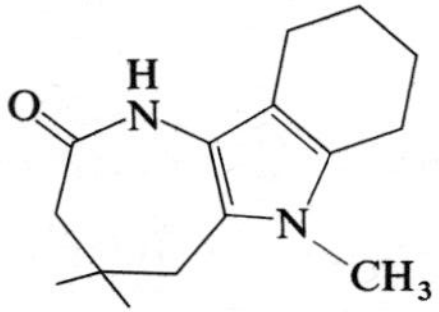 $N{-}CH_3$ (50)	383
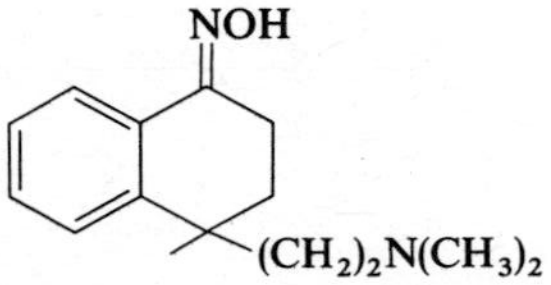 NOH, $(CH_2)_2N(CH_3)_2$	PPA, 150°, 1 h	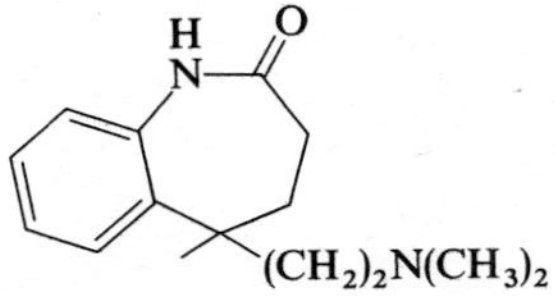 $(CH_2)_2N(CH_3)_2$ (72)	435

TABLE I. REARRANGEMENTS (*Continued*)

No. of Carbon Atoms	Substrate	Reagent and Conditions	Product(s) and Yield(s) (%)	Refs.
	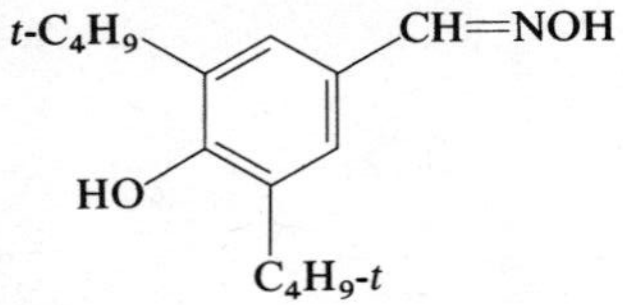 *t*-C_4H_9, CH=NOH, HO, C_4H_9-*t*	H_2SO_4, 20°, 2 h	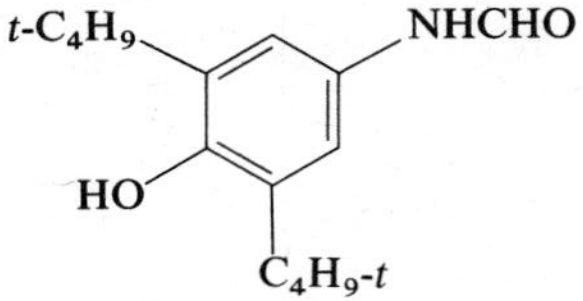 *t*-C_4H_9, NHCHO, HO, C_4H_9-*t* (—)	135
	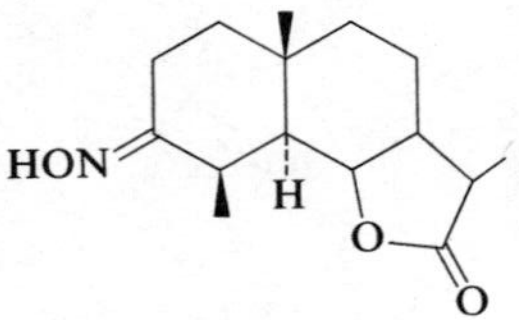 HON, H, O, O (*E*/*Z* = 4/1)	$SOCl_2$, 70°, dioxane	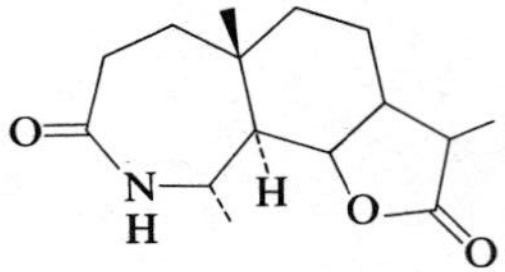 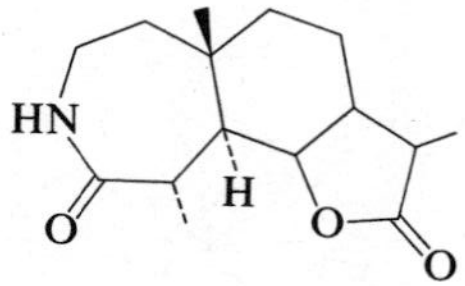 O, N, H, H, O, O; HN, O, H, O, O (30) 2:3	438
	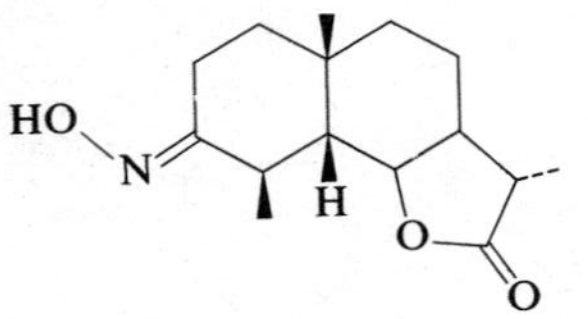 HO, N, H, O, O	TsCl, pyridine, 50°, 1 h	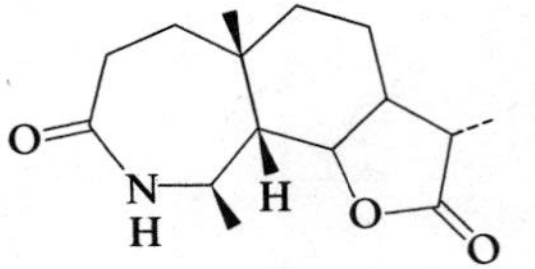 O, N, H, H, O, O (76)	438

Reactant	Conditions	Product(s) and Yield(s) (%)	Refs.
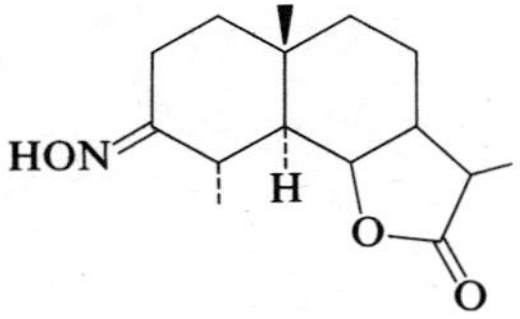 HON, H (E/Z = 5/1)	$SOCl_2$, 70°, dioxane	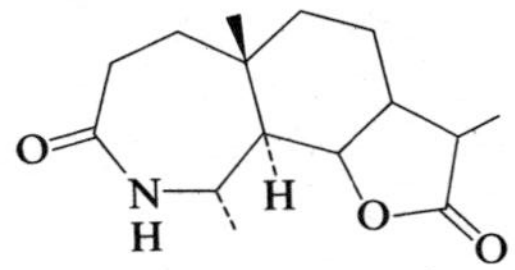 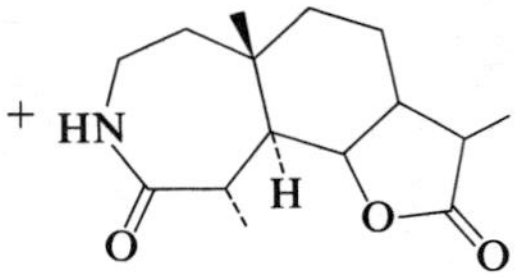 O, N, H, H, O, O + HN, H, O, O (—) 2:1	438
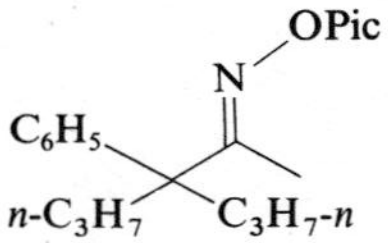 C_6H_5, n-C_3H_7, C_3H_7-n, N, OPic	1,4-Dichlorobutane, 80°	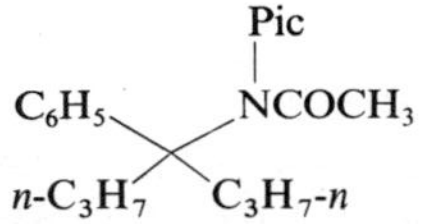 Pic, C_6H_5, NCOCH$_3$, n-C_3H_7, C_3H_7-n (—)	93
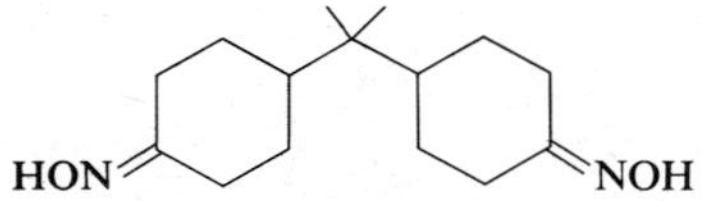 HON, NOH	H_2SO_4	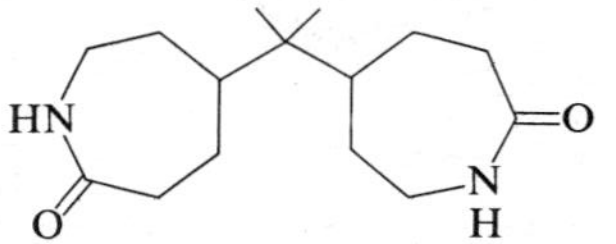 HN, O, O, N, H (89)	333
	PPA	" (83)	333
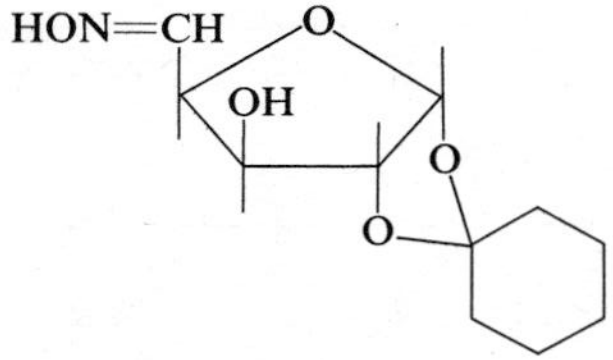 HON=CH, O, OH, O, O	$Cu(O_2CCH_3)_2 \cdot (H_2O)_4$, dioxane, 70°, 5–10 h	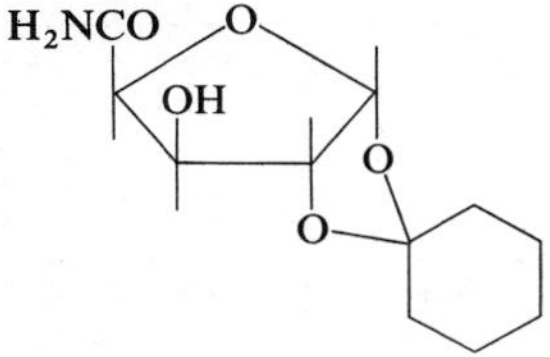 H_2NCO, O, OH, O, O (47)	131

TABLE I. REARRANGEMENTS (*Continued*)

No. of Carbon Atoms	Substrate	Reagent and Conditions	Product(s) and Yield(s) (%)	Refs.
	$(n\text{-}C_4H_9)_3C(=NOPic)CH_3$	1,4-Dichlorobutane, 80°	$(n\text{-}C_4H_9)_3CN(Pic)COCH_3$ (—)	93
C_{16}	$(CO)_3Mn$–(methylcyclopentadienyl)–$C(=NOH)C_6H_5$	PCl_5, pyridine, C_6H_6, 0°, 30 min	$(CO)_3Mn$–(methylcyclopentadienyl)–$CONHC_6H_5$ (—)	101, 281
	$(CO)_3Mn$–(methylcyclopentadienyl)–$C(=NOH)C_6H_5$	"	$(CO)_3Mn$–(methylcyclopentadienyl)–$CONHC_6H_5$ (—)	101, 281
	NOH	PCl_5, ether	O, N, H (75)	373
	HO, N, CH_3C, O, S, N, OH, CCH_3	PPA, 110°, 3 h	CH_3CONH, O, S, $NHCOCH_3$ (94)	439

Reactant	Conditions	Product(s) and Yield(s) (%)	Refs.
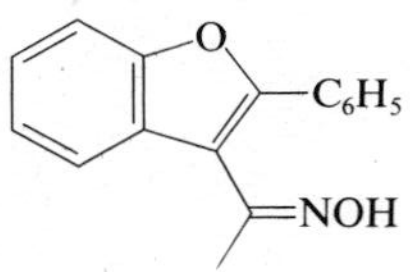 (benzofuran, 2-C_6H_5, 3-C(=NOH)CH_3)	TsCl, pyridine, 0°	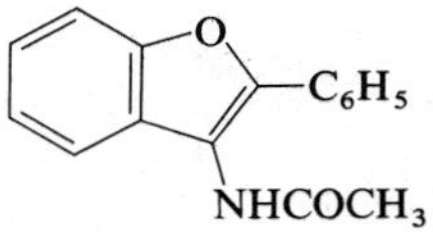 (benzofuran, 2-C_6H_5, 3-$NHCOCH_3$) (—)	440
	PCl_5, ether	" (—)	440
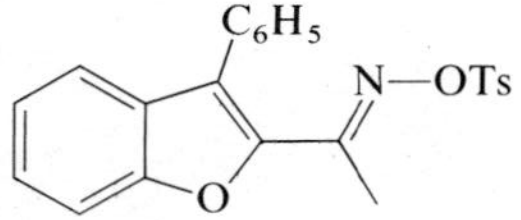 (benzofuran, 3-C_6H_5, 2-C(=N—OTs)CH_3)	CH_3OH, reflux, 6 h	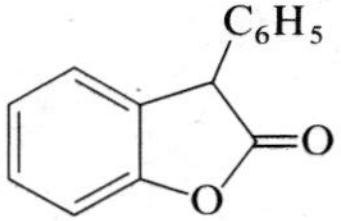 (3-C_6H_5-benzofuran-2(3H)-one, =O) (—)	540
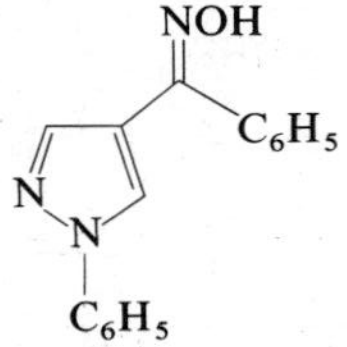 (1-C_6H_5-pyrazol-4-yl)C(=NOH)C_6H_5 (Z)	TsCl, pyridine, 4 d	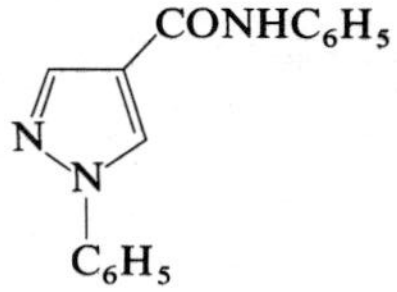 (1-C_6H_5-pyrazole, 4-$CONHC_6H_5$) (78)	318
(Z)	PCl_5, $CHCl_3$, 4 d	" (90)	318
(Z)	PPA, 100°, 1 h	" (93)	318
(Z)	H_2SO_4, 100°, 3 h	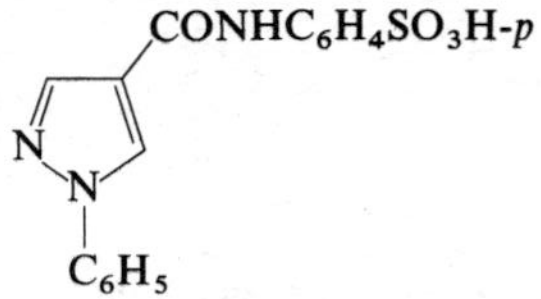 (1-C_6H_5-pyrazole, 4-$CONHC_6H_4SO_3H$-*p*) (84)	318
(E)	TsCl, pyridine	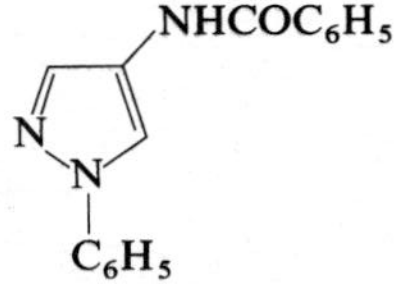 (1-C_6H_5-pyrazole, 4-$NHCOC_6H_5$) (62)	318
(E)	PCl_5, $CHCl_3$, 4 d	" (63)	318
(E)	PPA, 100°, 12 h	" (86)	318

TABLE I. REARRANGEMENTS (*Continued*)

No. of Carbon Atoms	Substrate	Reagent and Conditions	Product(s) and Yield(s) (%)	Refs.
	COC_6H_5 / N / CH_3	$H_2NOH \cdot HCl$, ethanol, reflux, 3 d	$NHCOC_6H_5$ / N / CH_3 (43)	320
	HON C_6H_5 / N / CH_3 (*Z*)	CH_3CO_2H, reflux, 2 h	" (45)	320
	(*Z*)	PCl_5, ether	$CONHC_6H_5$ / N / CH_3 (100)	320, 428
	(*E*)	PCl_5, ether, 24 h	$NHCOC_6H_5$ / Cl / N / CH_3 (40)	320

Reactant	Conditions	Product (Yield, %)	Refs.
HO–N=C(C_6H_5)– (1-CH_3-indol-2-yl)	PCl_5, ether, 0°	3-Cl-indol-2-yl–$CONHC_6H_5$ (—)	428
9-$COCH_3$-carbazol-3-yl–C(=NOH)CH_3	PCl_5, THF, 1 h	9-$COCH_3$-carbazol-3-yl–$NHCOCH_3$ (90)	408
HO–N=C(C_6H_5)CH_2–(indol-2-yl, N–H)	PPA, 2 h	indol-2-yl (N–H)–$CH_2CONHC_6H_5$ (68)	415
HON=C(C_6H_5)–(1-$C_6H_4OCH_3$-p-1,2,4-triazol-3-yl) (Z)	PPA	C_6H_5NHCO–(1-p-$CH_3OC_6H_4$-1,2,4-triazol-3-yl) (—)	321
(E)	PCl_5, $CHCl_3$, 0°, 3 d	C_6H_5CONH–(1-p-$CH_3OC_6H_4$-1,2,4-triazol-3-yl) (20)	321

TABLE I. REARRANGEMENTS (*Continued*)

No. of Carbon Atoms	Substrate	Reagent and Conditions	Product(s) and Yield(s) (%)	Refs.
	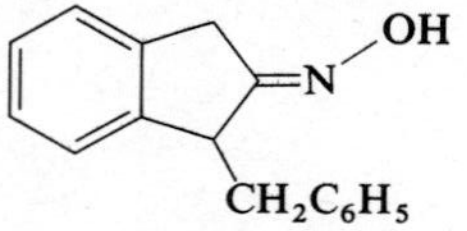 (structure: OH, N, $CH_2C_6H_5$)	PCl_5, $CHCl_3$, 2 h	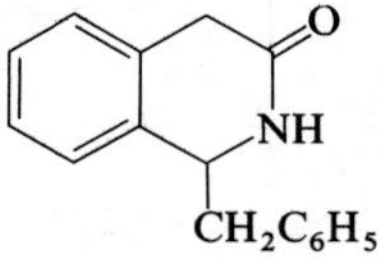 (structure: O, NH, $CH_2C_6H_5$) (37)	441
	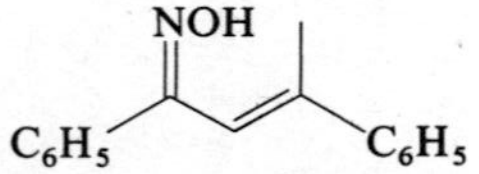 (structure: NOH, C_6H_5, C_6H_5)	PCl_5	$C_6H_5NHCOCH{=}C(CH_3)C_6H_5$ (—)	431
	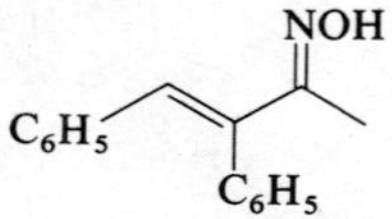 (structure: NOH, C_6H_5, C_6H_5)	PCl_5, ether	$C_6H_5CH{=}C(C_6H_5)NHCOCH_3$ (68)	442
	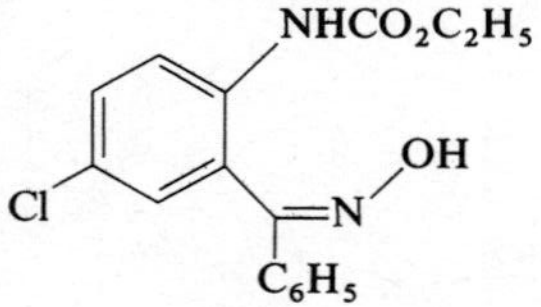 (structure: $NHCO_2C_2H_5$, OH, N, Cl, C_6H_5)	Heat at mp	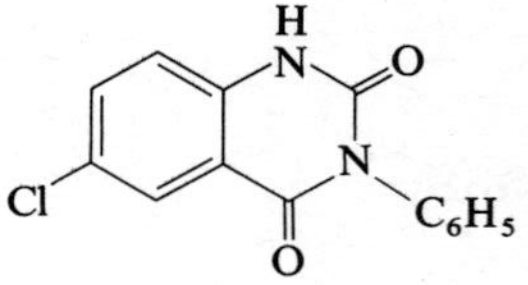 (structure: H, N, O, N, C_6H_5, Cl, O) (55)	397
	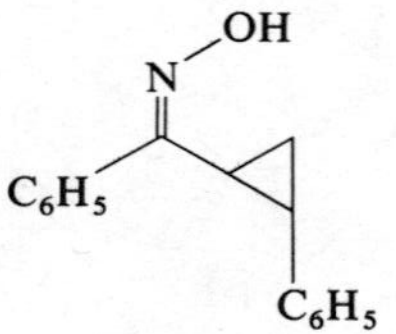 (structure: OH, N, C_6H_5, C_6H_5)	PPSE, CH_2Cl_2, 20 h	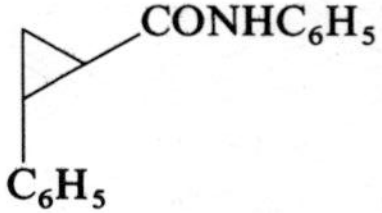 (structure: $CONHC_6H_5$, C_6H_5) (76)	72

Reactant	Conditions	Product(s) and Yield(s) (%)	Refs.
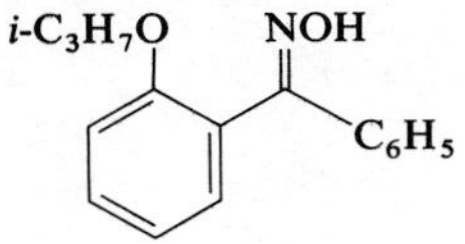 i-C_3H_7O, NOH, C_6H_5 (structure)	PCl_5, C_6H_6	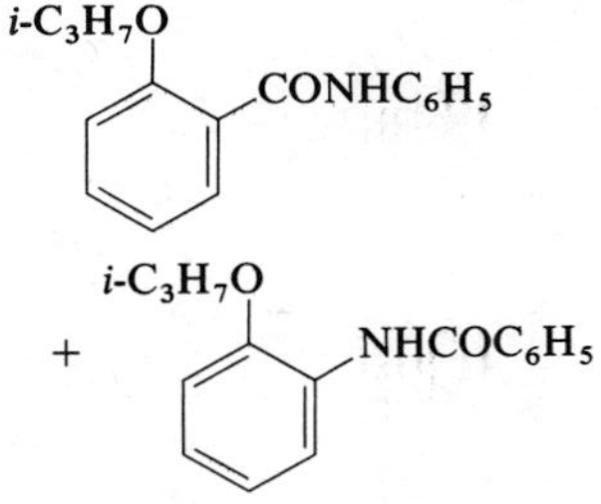 i-C_3H_7O, $CONHC_6H_5$ (structure) + i-C_3H_7O, $NHCOC_6H_5$ (structure)	443
	Oximation temp 60°	13.5:86.5 (75)	
	Oximation temp 118°	28:72 (98)	
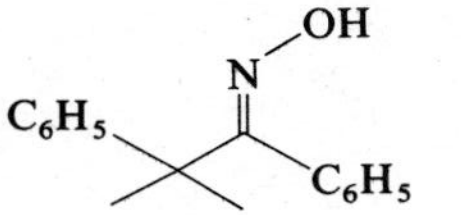 OH, N, C_6H_5, C_6H_5 (structure)	PPA, 130°, 10 min	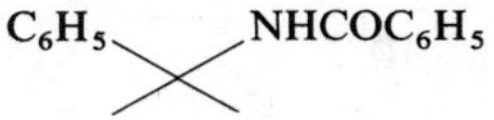 C_6H_5, $NHCOC_6H_5$ (structure) (—)	64, 66
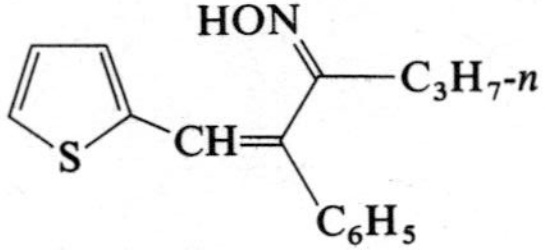 HON, C_3H_7-n, S, CH=, C_6H_5 (structure)	PCl_5, ether, dioxane, 0° 6 h	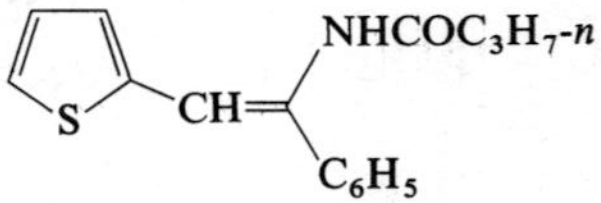 S, CH=, $NHCOC_3H_7$-n, C_6H_5 (structure) (25)	113
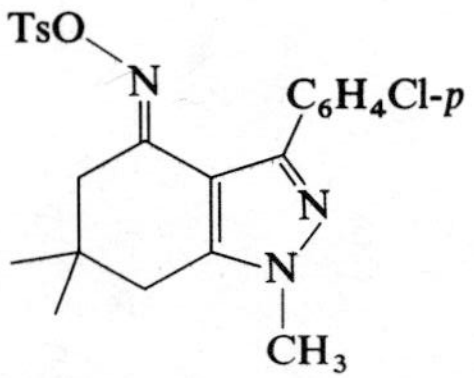 TsO, N, C_6H_4Cl-p, N, N, CH_3 (structure)	NaO_2CCH_3, ethanol, H_2O	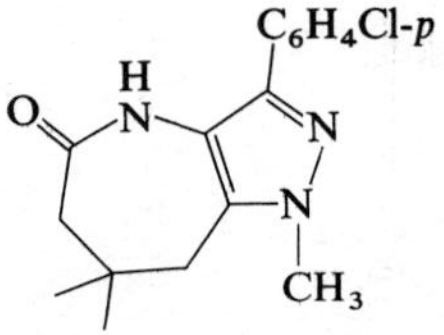 O, H, N, C_6H_4Cl-p, N, N, CH_3 (structure) (23)	411
TsO, N, $C_6H_4OCH_3$-p, N, N, H (structure)	"	O, H, N, $C_6H_4OCH_3$-p, N, N, H (structure) (30)	411

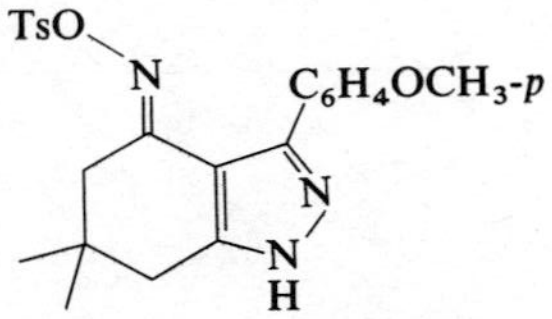

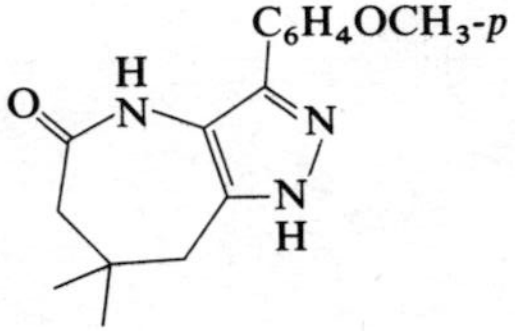

TABLE I. REARRANGEMENTS (*Continued*)

No. of Carbon Atoms	Substrate	Reagent and Conditions	Product(s) and Yield(s) (%)	Refs.
	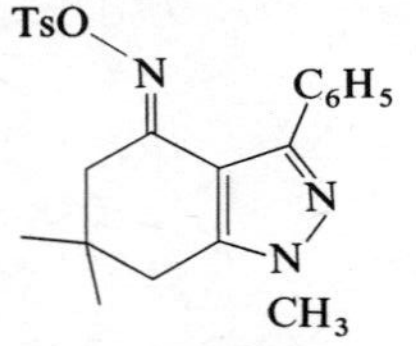 TsO, N, C_6H_5, N, N, CH_3	NaO_2CCH_3, ethanol, H_2O	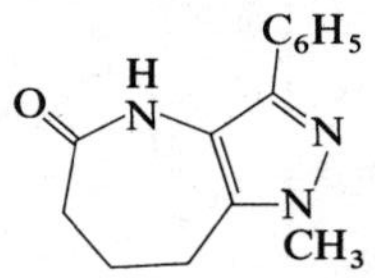 O, H, N, C_6H_5, N, N, CH_3 (61)	411
	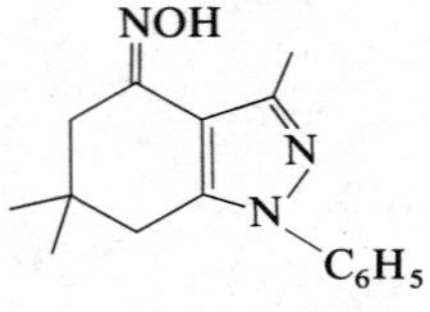 NOH, N, N, C_6H_5	PPA, 130–140°, 1 h	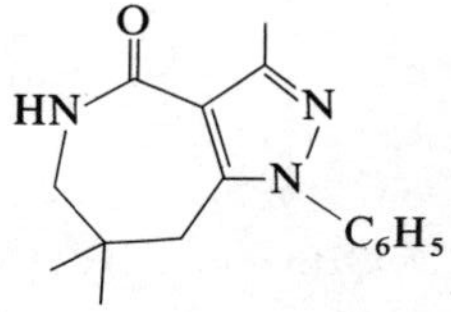 O, HN, N, N, C_6H_5 (34)	412
	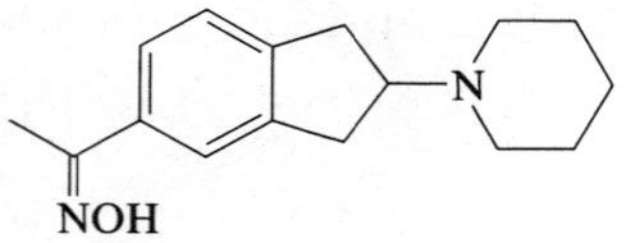 N, NOH	PPA, 100°, 3 h	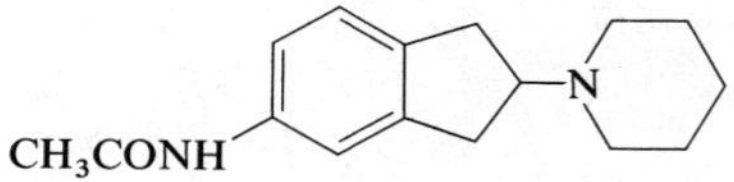 CH_3CONH, N (50)	444
	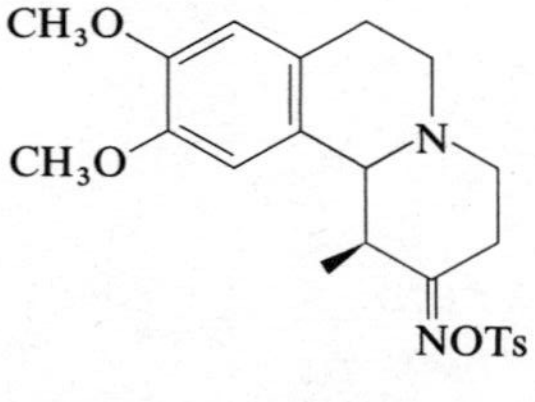 CH_3O, CH_3O, N, NOTs	CH_3CO_2H	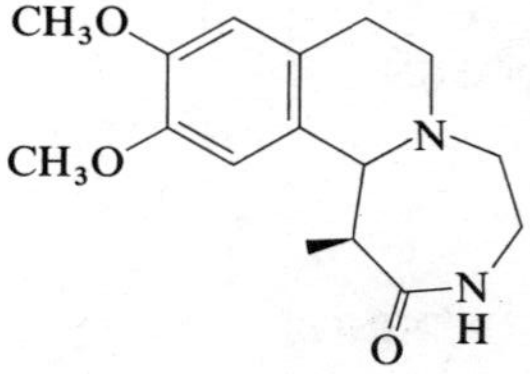 CH_3O, CH_3O, N, N, H, O	436

Reactant	Conditions	Product(s) (Yield, %)	Refs.
		+ CH_3O, CH_3O ... N ... N H ... O (69)	
CH_3O, CH_3O ... N ... N–OTs	"	CH_3O, CH_3O ... N ... N H ... O (67)	436
NOTs ... N–Ts	CH_3OH, reflux, 1 h	O ... N–Ts ... $\overset{+}{N}$ H H ^{-}OTs (93)	317
NOH ... N	PPA, 130°, 1 h	O ... NH ... N (28)	445

TABLE I. REARRANGEMENTS (*Continued*)

No. of Carbon Atoms	Substrate	Reagent and Conditions	Product(s) and Yield(s) (%)	Refs.
	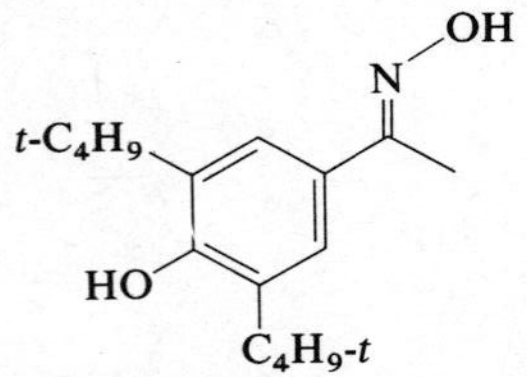 OH, N, t-C_4H_9, HO, C_4H_9-t	H_2SO_4, 2 h	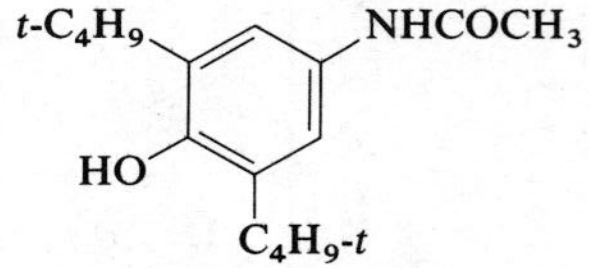 t-C_4H_9, $NHCOCH_3$, HO, C_4H_9-t (—)	135
	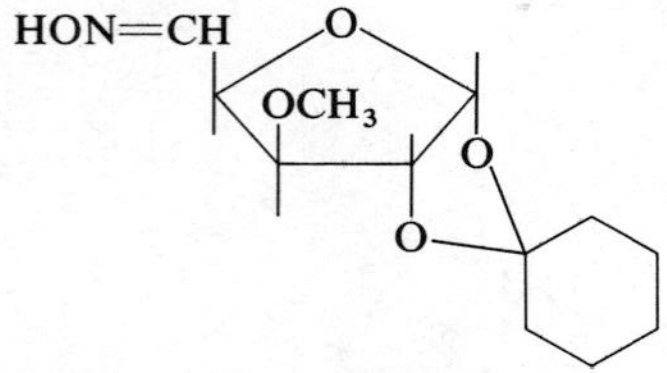 HON=CH, O, OCH_3, O, O	$Cu(O_2CCH_3)_2 \cdot (H_2O)_4$, dioxane, 70°, 5–10 h	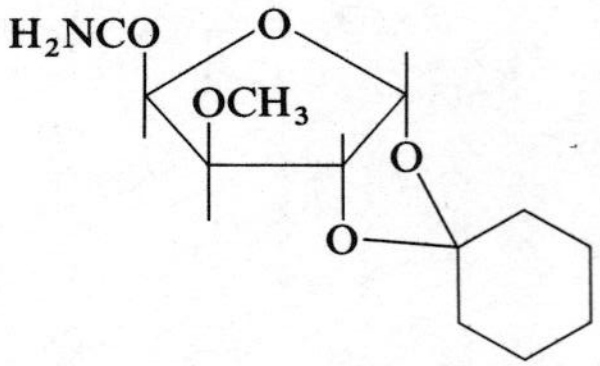 H_2NCO, O, OCH_3, O, O (51)	131
	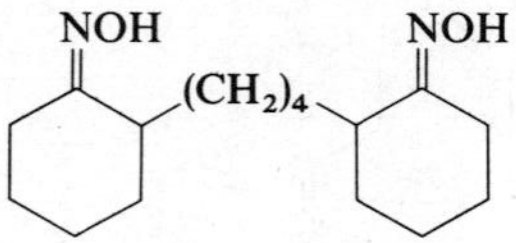 NOH, $(CH_2)_4$, NOH	PPA	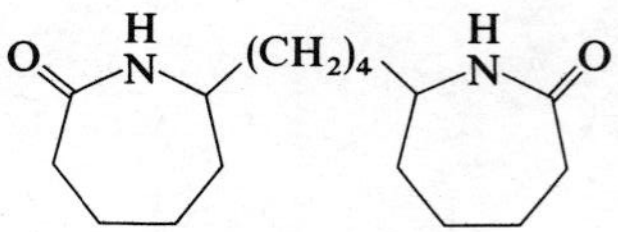 O, H, N, $(CH_2)_4$, H, N, O (65)	42
	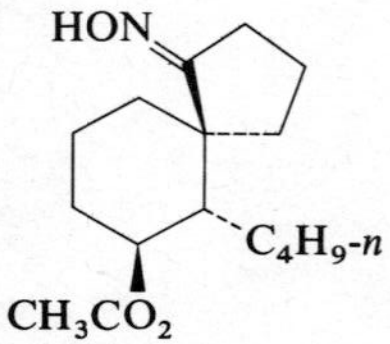 HON, C_4H_9-n, CH_3CO_2	TsCl, pyridine	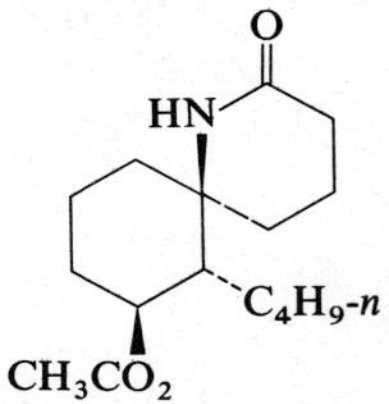 O, HN, C_4H_9-n, CH_3CO_2 (33)	183, 424

	Reactant	Conditions	Product (Yield %)	Refs.
	[structure: cyclic ketoxime HON=, 2-(2-methylallyl), ring $(CH_2)_{10}$]	$POCl_3$, pyridine, 80°	[structure: lactam O=C–NH, $(CH_2)_{10}$ ring, 2-methylallyl] (20)	220
C_{17}	[structure: N-phthalimido $NCH_2CH_2C(=NOH)C_6H_5$]	PCl_5 ether, 30 min	[structure: phthalimido] $N(CH_2)_2CONHC_6H_5$ (63)	309
	[structure: 3-quinolyl$CH_2C(=NOH)C_6H_5$]	PCl_5, ether, 20 h	[structure: 3-quinolyl] $CH_2CONHC_6H_5$ (55)	446
	$C_5H_5FeC_5H_4C(=NOH)C_6H_5$	Cl_3CCN, reflux, 5–6 h	$C_5H_5FeC_5H_4CONHC_6H_5$ (60–70)	100, 232
		$C_6H_5SO_2Cl$, NaOH	" (18)	99
		TsCl, pyridine, reflux, 30 min	" (23)	98
	[structure: phthalide, 3-substituted] $C_6H_5CHC(=NOH)CH_3$	1. TsCl, pyridine 2. PCl_5, BF_3	[structure: phthalide, 3-substituted] $CH(C_6H_5)NHCOCH_3$ (15)	447
	[structure: anthrone oxime (NOH), 10-substituted $C(CH_3)_2NO_2$]	PCl_5, CCl_4, reflux, 2 h	[structure: dibenzazepinone (H–N–C=O), 11-substituted $C(CH_3)_2NO_2$] (62)	142

TABLE I. REARRANGEMENTS (*Continued*)

No. of Carbon Atoms	Substrate	Reagent and Conditions	Product(s) and Yield(s) (%)	Refs.
	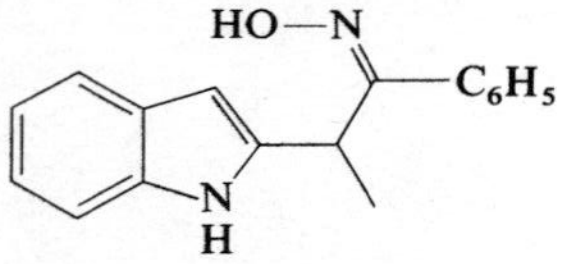 HO—N; C_6H_5	PPA, 75°, 3 h	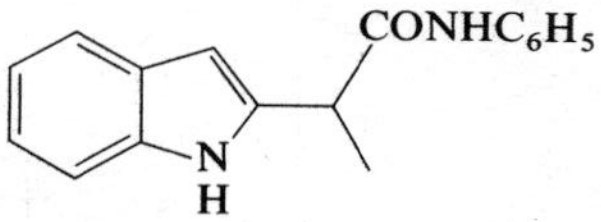 $CONHC_6H_5$; N H (53)	415
	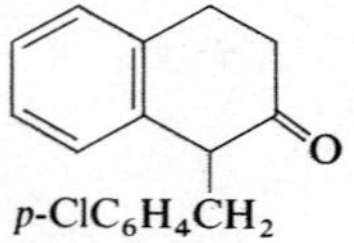 O; p-$ClC_6H_4CH_2$	1. H_2NOH, PPA, 55° 2. 95–110°, 30 min	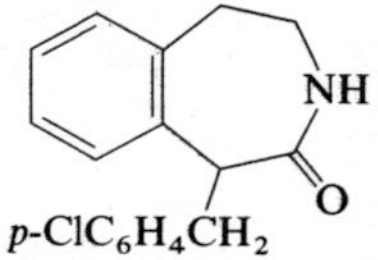 NH (36); O; p-$ClC_6H_4CH_2$	448
	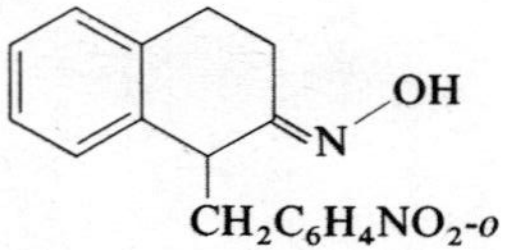 OH; N; $CH_2C_6H_4NO_2$-o	PCl_5, $CHCl_3$, 2 h	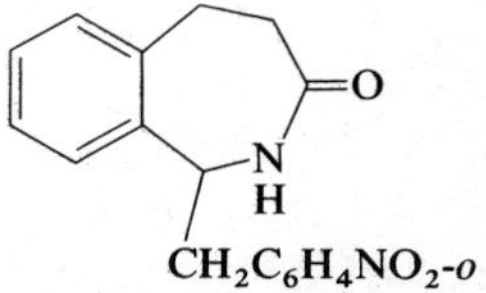 O; N H; $CH_2C_6H_4NO_2$-o (93)	441
	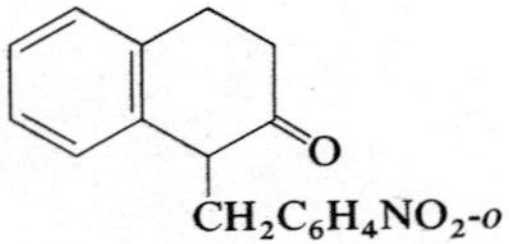 O; $CH_2C_6H_4NO_2$-o	1. H_2NOH, PPA, 55° 2. 95–110°, 30 min	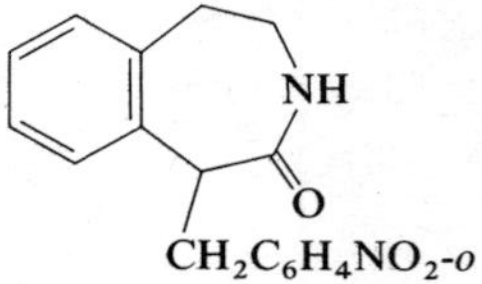 NH; O; $CH_2C_6H_4NO_2$-o (29)	448
	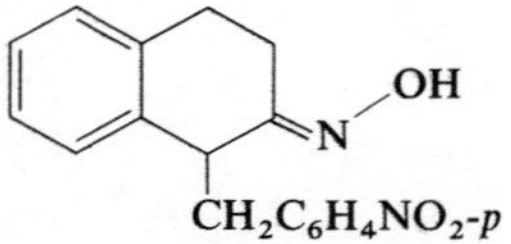 OH; N; $CH_2C_6H_4NO_2$-p	PCl_5, $CHCl_3$, 2 h	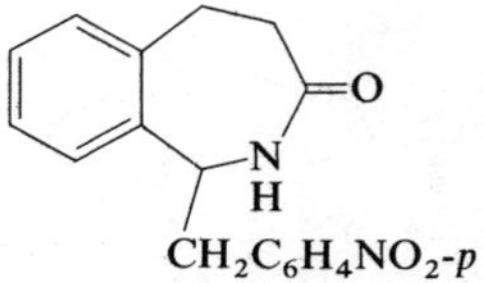 O; N H; $CH_2C_6H_4NO_2$-p (58)	441

Reactant	Conditions	Product(s) and Yield(s) (%)	Refs.
1-benzyl-2-tetralone ($C_6H_5CH_2$)	1. H_2NOH, PPA, 55° 2. 95–110°, 30 min	5-benzyl-1,2,4,5-tetrahydro-3H-3-benzazepin-4-one ($C_6H_5CH_2$, NH, O) (27)	448
1-benzyl-2-tetralone oxime ($C_6H_5CH_2$, =N–OH)	PCl_5, $CHCl_3$, 2 h	1-benzyl-benzazepinone ($C_6H_5CH_2$, NH, =O) (84)	441
p-$CH_3C_6H_4CH{=}C(C_6H_5)C(=NOH)CH_3$	PCl_5, ether	p-$CH_3C_6H_4CH{=}C(C_6H_5)NHCOCH_3$ (20)	442
p-$CH_3OC_6H_4CH{=}C(C_6H_5)C(=NOH)CH_3$	"	p-$CH_3OC_6H_4CH{=}C(C_6H_5)NHCOCH_3$ (20)	442
$C_6H_5CH{=}C(C_6H_5)C(=NOH)C_2H_5$	"	$C_6H_5CH{=}C(C_6H_5)NHCOC_2H_5$ (5)	442
$C_6H_5CH{=}C(C_6H_4OCH_3$-$p)C(=NOH)CH_3$	PCl_5, ether, dioxane, 0°, 6 h	$C_6H_5CH{=}C(C_6H_4OCH_3$-$p)NHCOCH_3$ (50)	113
2-[2-(2-pyridyl)ethyl]-1-tetralone oxime (NOH, N)	PPA, 120–135°, 10 min	3-[2-(2-pyridyl)ethyl]-1-benzazepin-2-one (H, N, O, N) (56)	449

TABLE I. REARRANGEMENTS (*Continued*)

No. of Carbon Atoms	Substrate	Reagent and Conditions	Product(s) and Yield(s) (%)	Refs.
	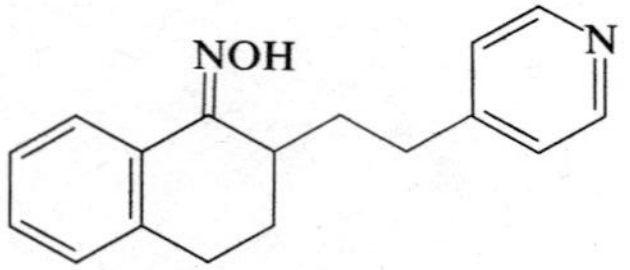 NOH	CH_3CO_2H, $(CH_3CO)_2O$, HCl	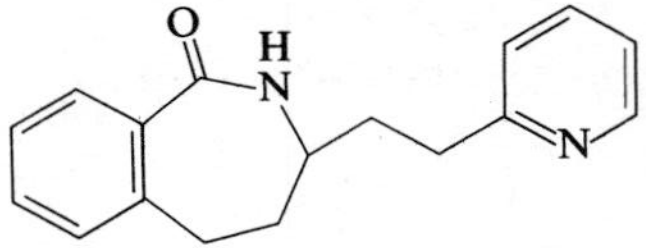 O H N N (10)	449
		PPA, 120–135°, 10 min	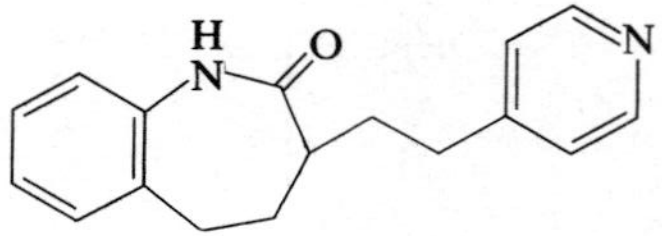 H N O N (72)	449
		CH_3CO_2H, $(CH_3CO)_2O$, HCl	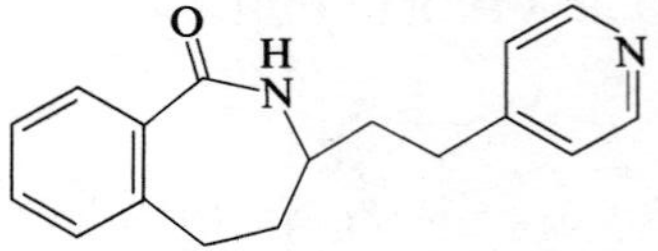 O H N N (56)	449
	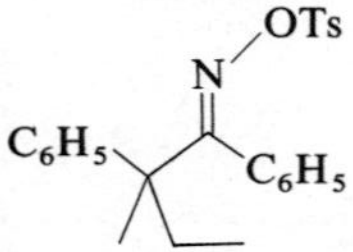 OTs N C_6H_5 C_6H_5	$(C_2H_5)_3N$, 80% ethanol, 30° 12 h	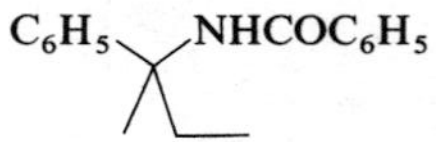 C_6H_5 $NHCOC_6H_5$ (—)	28
		$(C_2H_5)_3N$, 80% CH_3OH, 30°, 12 h	" (4)	450
		$(C_2H_5)_3N$, P_2O_5, $CHCl_3$	" (7)	450
	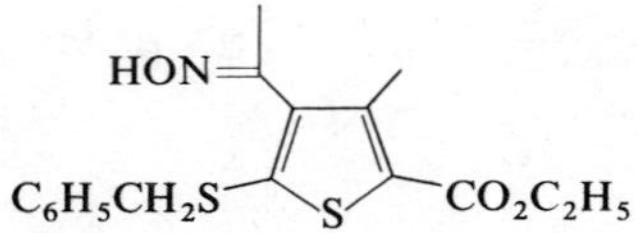 HON $C_6H_5CH_2S$ S $CO_2C_2H_5$	PCl_5, ether, 1 h	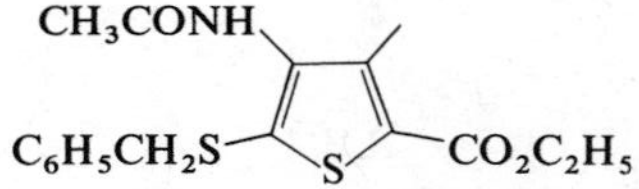 CH_3CONH $C_6H_5CH_2S$ S $CO_2C_2H_5$ (83)	266

Reactant	Conditions	Product(s) and Yield(s) (%)	Refs.
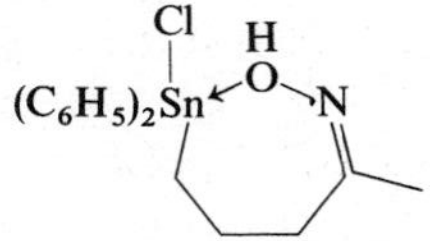 $(C_6H_5)_2Sn(Cl)$←O(H)–N=C(CH$_3$) ring	PCl_5, ether, 90°, 10 min	$(C_6H_5)_2SnCl(CH_2)_3NHCOCH_3$ (24)	103
$[p\text{-}(CH_3)_2NC_6H_4]_2C{=}NOH$	HCO_2H, reflux, 3 h	$p\text{-}(CH_3)_2NC_6H_4NHCOC_6H_4N(CH_3)_2\text{-}p$ (60)	251
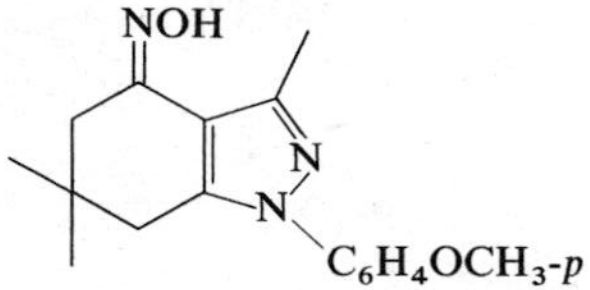 NOH; N; N; $C_6H_4OCH_3\text{-}p$	PPA, 130-140°, 1 h	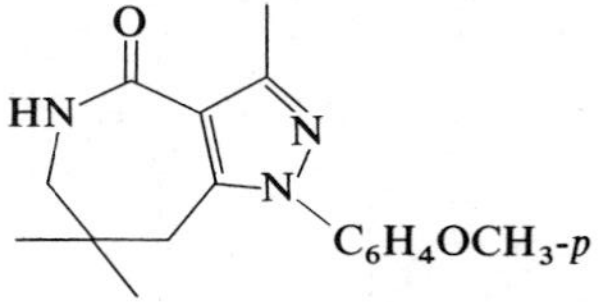 O; HN; N; N; $C_6H_4OCH_3\text{-}p$ (—)	412
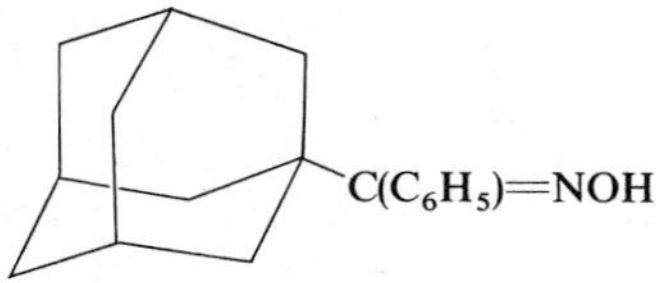 $C(C_6H_5){=}NOH$	PPA, 100°, 30 min	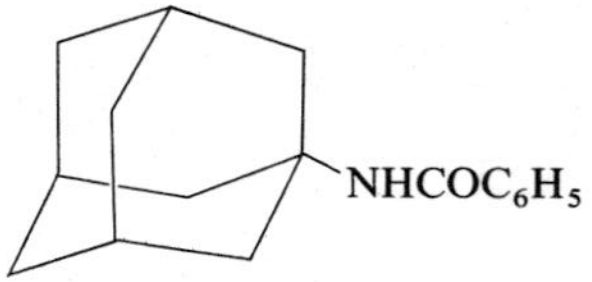 $NHCOC_6H_5$ (93)	61
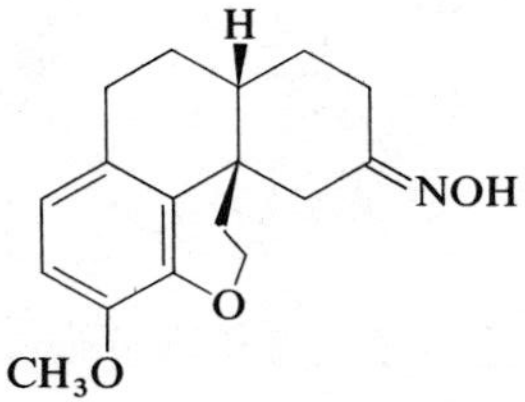 H; NOH; O; CH_3O	"	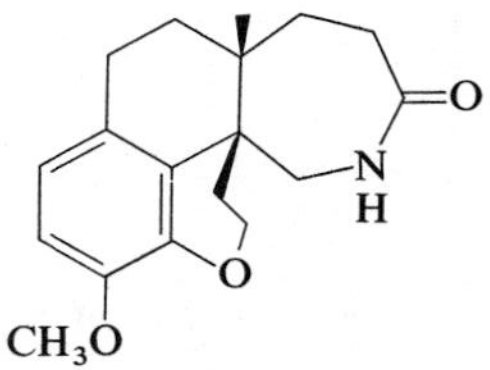 O; N; H; O; CH_3O (54)	451
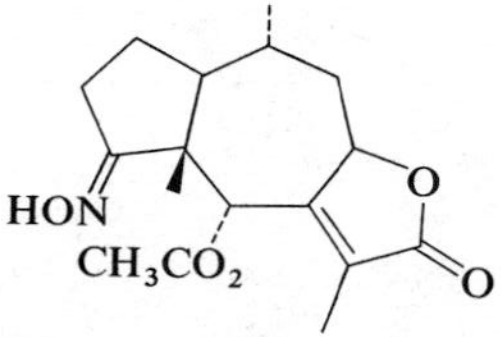 HON; CH_3CO_2; O; O	TsCl, pyridine	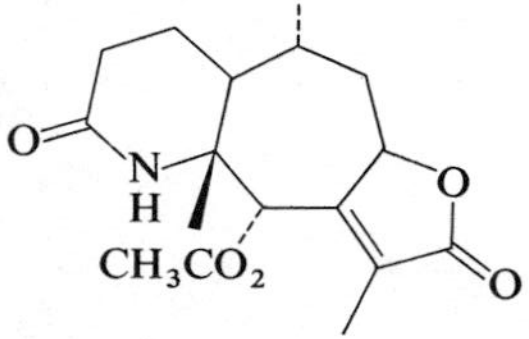 O; N; H; CH_3CO_2; O; O (—)	437

TABLE I. REARRANGEMENTS (*Continued*)

No. of Carbon Atoms	Substrate	Reagent and Conditions	Product(s) and Yield(s) (%)	Refs.
	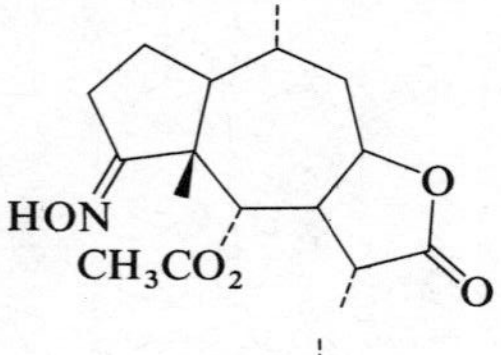 HON, CH_3CO_2, O, O	TsCl, pyridine	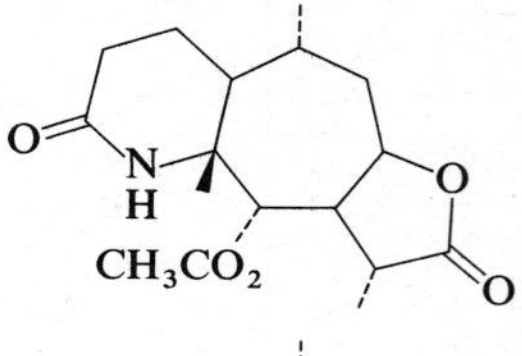 O, N, H, CH_3CO_2, O, O (—)	437
	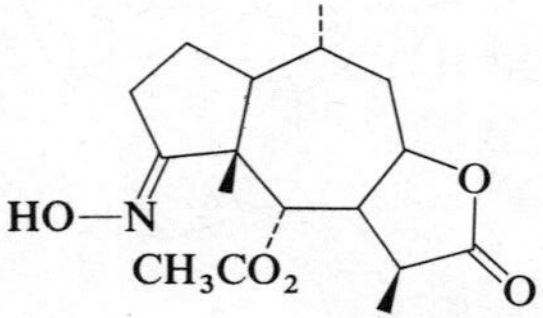 HO—N, CH_3CO_2, O, O	"	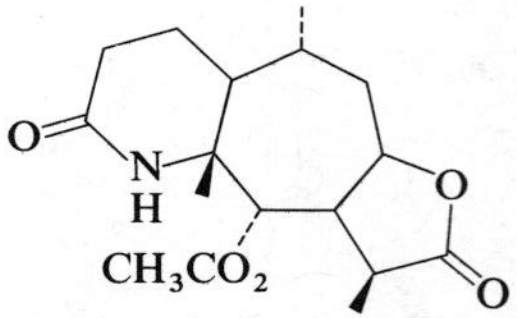 O, N, H, CH_3CO_2, O, O (—)	437
	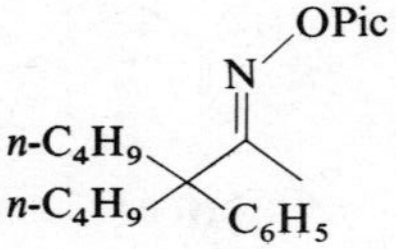 OPic, N, n-C_4H_9, n-C_4H_9, C_6H_5	1,4-Dichlorobutane, 80°	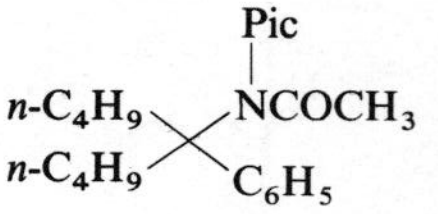 Pic, n-C_4H_9, n-C_4H_9, $NCOCH_3$, C_6H_5 (—)	93
	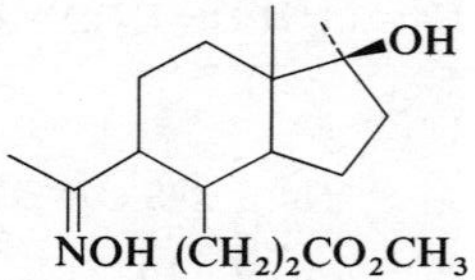 OH, NOH, $(CH_2)_2CO_2CH_3$	TsCl, pyridine, 24 h	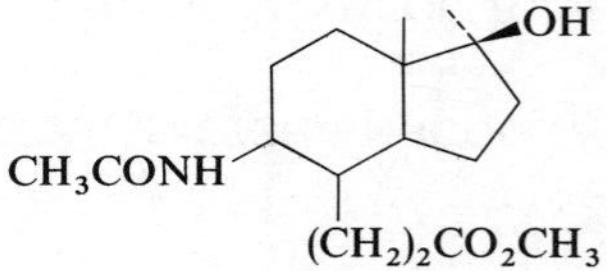 OH, CH_3CONH, $(CH_2)_2CO_2CH_3$ (19)	452

	Reactant	Conditions	Product(s) and Yield(s) (%)	Refs.
C_{18}	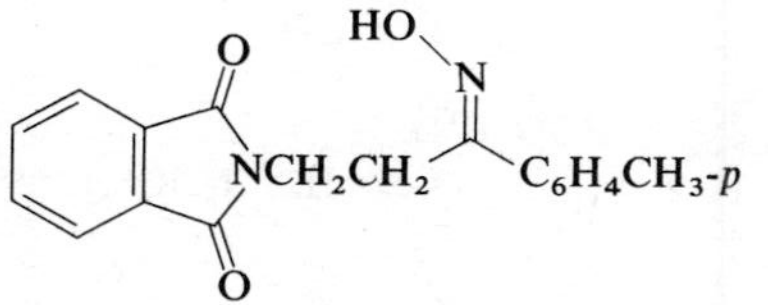 $NCH_2CH_2C(=NOH)C_6H_4CH_3$-*p* (phthalimido)	PCl_5, ether, 30 min	 $NCH_2CH_2CONHC_6H_4CH_3$-*p* (phthalimido) (66)	309
	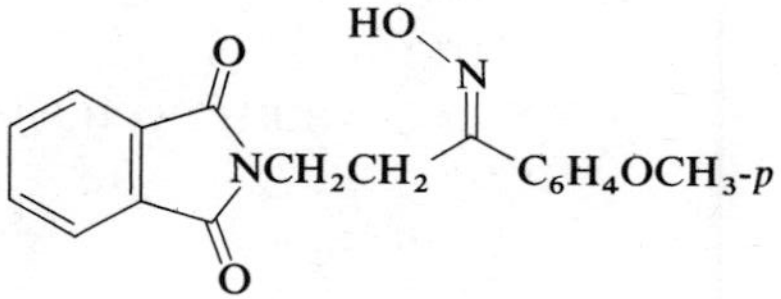 $NCH_2CH_2C(=NOH)C_6H_4OCH_3$-*p* (phthalimido)	"	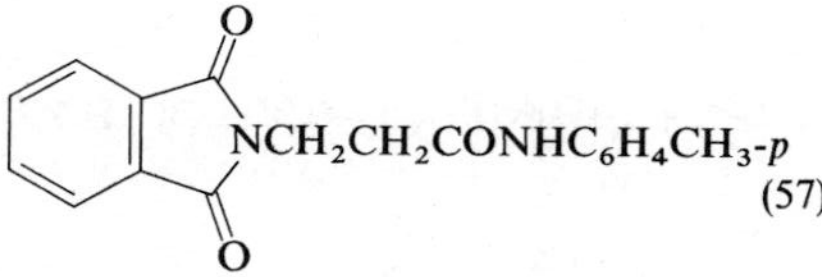 $NCH_2CH_2CONHC_6H_4CH_3$-*p* (phthalimido) (57)	309
	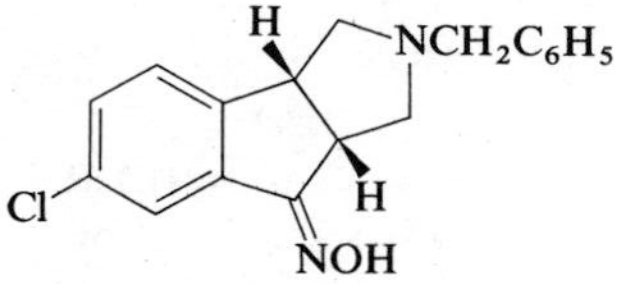 Cl, H, H, $NCH_2C_6H_5$, NOH	PPA, 200°, 1 h	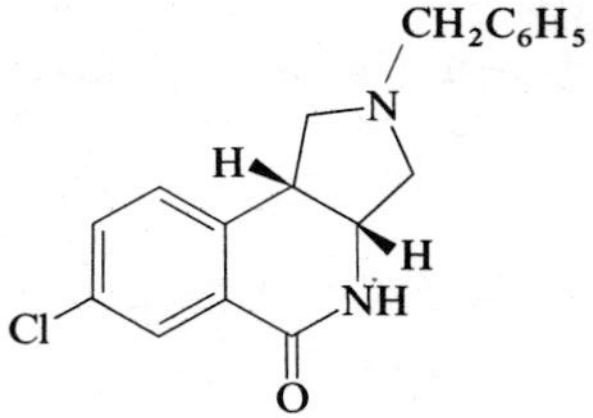 Cl, H, H, NH, O, N–$CH_2C_6H_5$ (63)	453
	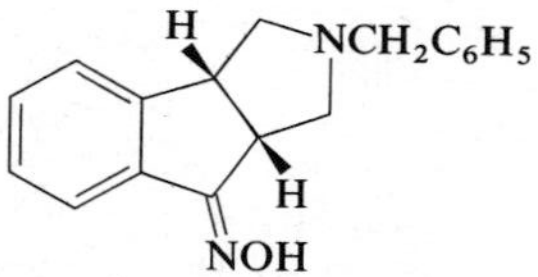 H, H, $NCH_2C_6H_5$, NOH	"	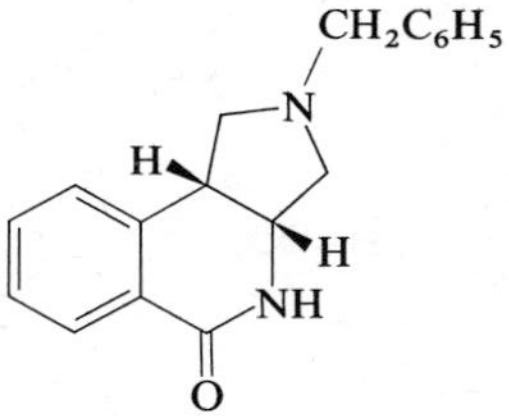 H, H, NH, O, N–$CH_2C_6H_5$ (50)	453
	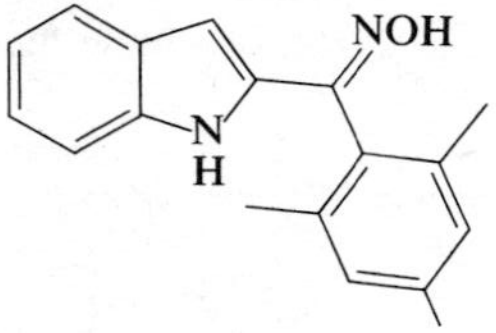 NOH, N–H (indolyl mesityl ketoxime)	PCl_5, ether, 0°	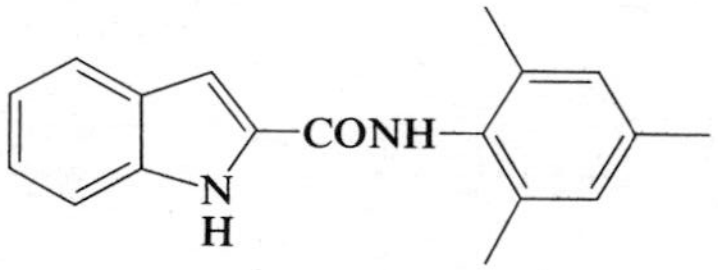 CONH, N–H (—)	428

TABLE I. REARRANGEMENTS (*Continued*)

No. of Carbon Atoms	Substrate	Reagent and Conditions	Product(s) and Yield(s) (%)	Refs.
	HO–N=C(3-indolyl)(2,4,6-trimethylphenyl)	PCl_5, ether	3-indolyl–CONH–(2,4,6-trimethylphenyl) (—)	428
	$C_6H_5CH{=}C(C_6H_5)C({=}NOH)C_3H_7$-*n*	"	$C_6H_5CH{=}C(C_6H_5)NHCOC_3H_7$-*n* (15-30)	442
	p-$CH_3C_6H_4CH{=}C(C_6H_5)C({=}NOH)C_2H_5$	"	*p*-$CH_3C_6H_4CH{=}C(C_6H_5)NHCOC_2H_5$ (25)	442
	p-$CH_3OC_6H_4CH{=}C(C_6H_5)C({=}NOH)C_2H_5$	"	*p*-$CH_3OC_6H_4CH{=}C(C_6H_5)NHCOC_2H_5$ (15)	442
	p-$CH_3OC_6H_4CH{=}C(C_6H_4OCH_3$-*p*$)C({=}NOH)CH_3$	PCl_5, ether, dioxane, 0°, 6 h	*p*-$CH_3OC_6H_4CH{=}C(C_6H_4OCH_3$-*p*$)NHCOCH_3$ (50)	113

Reactant	Conditions	Product (Yield, %)	Refs.
	PCl_5, ether	(15)	384
	$C_6H_5SO_2Cl$, acetone, H_2O, reflux, 6 h	(45)	454
	PPA, 115°, 15 min	(70)	455
	PPA, 100°, 1 h	(24)	451, 456
	PPA, 100°, 1 h	(44)	451

TABLE I. REARRANGEMENTS (*Continued*)

No. of Carbon Atoms	Substrate	Reagent and Conditions	Product(s) and Yield(s) (%)	Refs.
	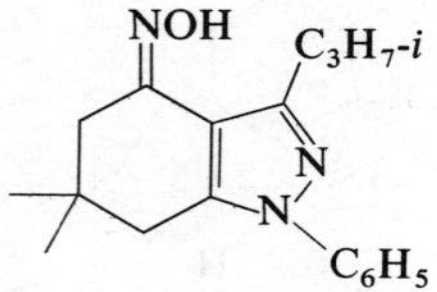 NOH, C_3H_7-*i*, N, N, C_6H_5	PPA, 130-140°, 1 h	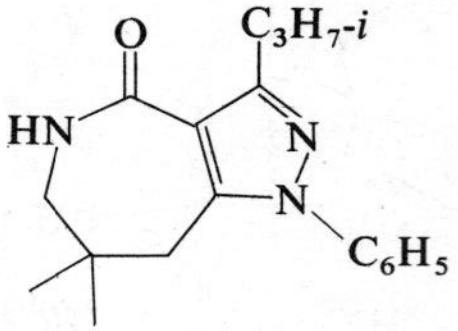 O, C_3H_7-*i*, HN, N, N, C_6H_5 (20)	412
	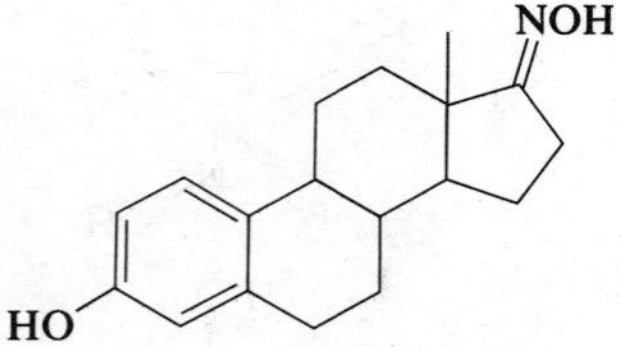 NOH, HO	*p*-$CH_3CONHC_6H_4SO_2Cl$, pyridine, 24 h	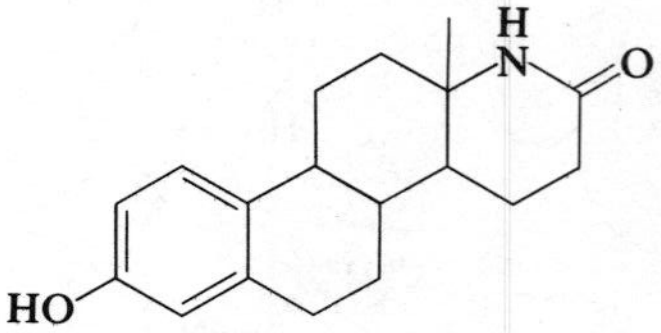 H, N, O, HO (—)	457, 458
		$SOCl_2$	" (—)	458
		$SOCl_2$, dioxane	" (83)	74, 457
		$SOCl_2$, CCl_4	" (76)	75
	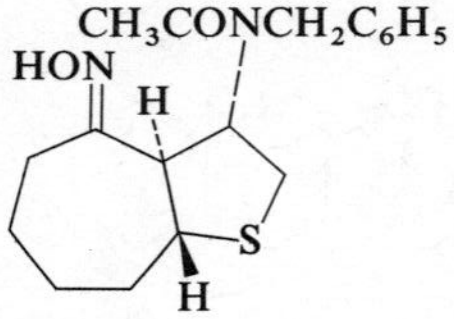 $CH_3CONCH_2C_6H_5$, HON, H, S, H	1. CH_3SO_2Cl, $(C_2H_5)_3N$, THF, 0° 2. CH_3CN, reflux	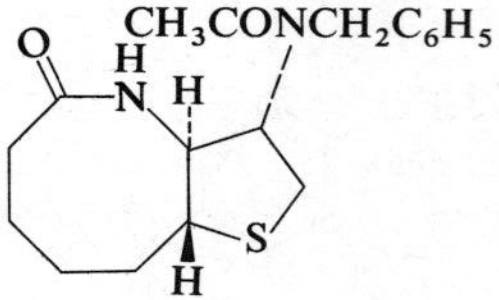 O, $CH_3CONCH_2C_6H_5$, H, N, H, S, H (—)	459

Reactant	Conditions	Product(s) and Yield(s) (%)	Refs.
	1. CH_3SO_2Cl, $(C_2H_5)_3N$, THF, 0° 2. Alumina	(15)	459
	PPA, 25°, 24 h	(51)	196
	PPA, 100°, 1 h	(55)	451
	"	(65)	451

TABLE I. REARRANGEMENTS (*Continued*)

No. of Carbon Atoms	Substrate	Reagent and Conditions	Product(s) and Yield(s) (%)	Refs.
	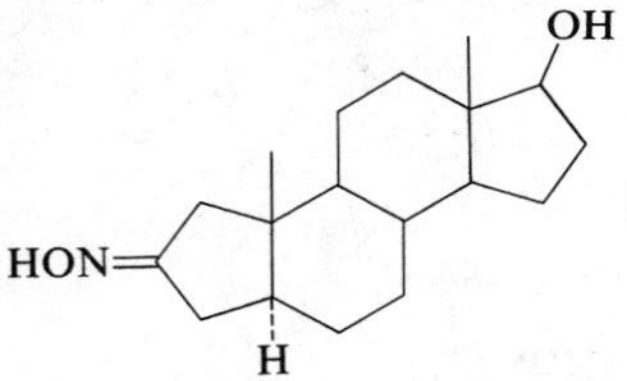 HON= … OH (*Z*)	TsCl, pyridine	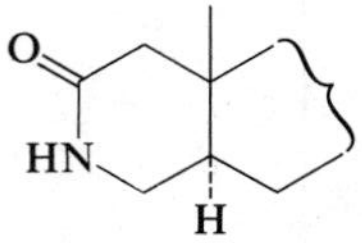 O, HN, H (—)	460
	(*E*)	"	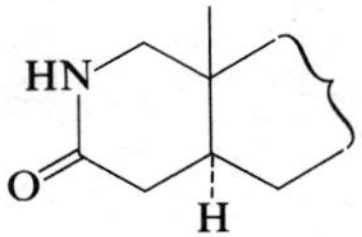 HN, O, H (—)	460
	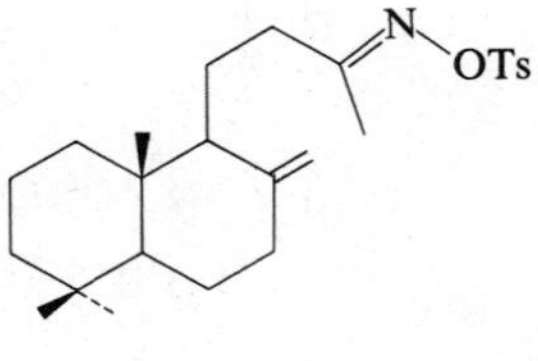 N–OTs (*E*)	7% KOH, dioxane, 1.5 h	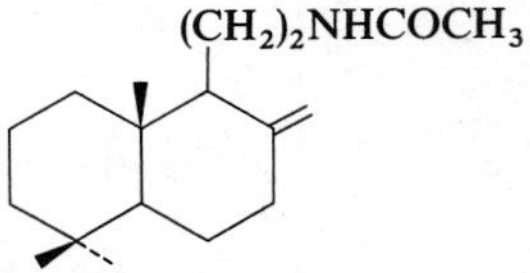 $(CH_2)_2NHCOCH_3$ (100)	461
	(*Z*)	25% KOH, dioxane, 100°, 9 h	" (27)	461
		HCl, CH_3CO_2H, 100°, 20 min	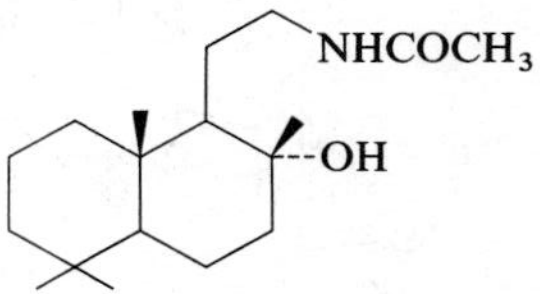 $NHCOCH_3$, OH	461

	Reactant	Conditions	Product(s) (Yield, %)	Refs.
			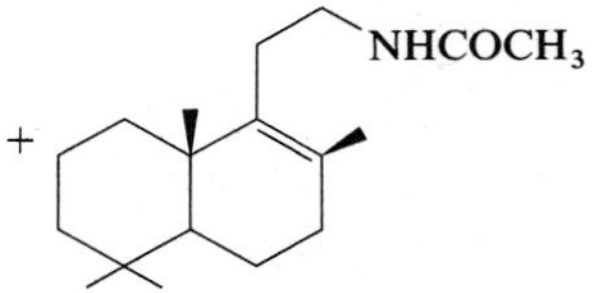 + $NHCOCH_3$ (—)	
	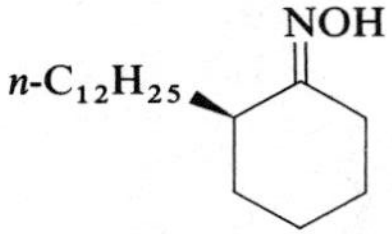 NOH, n-$C_{12}H_{25}$	PPA, 115–120°, 30 min	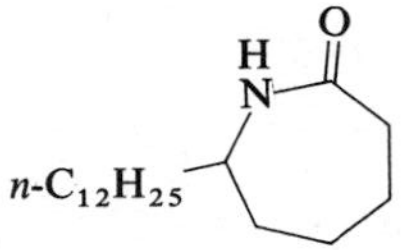 H, N, O, n-$C_{12}H_{25}$ (86)	462
C_{19}	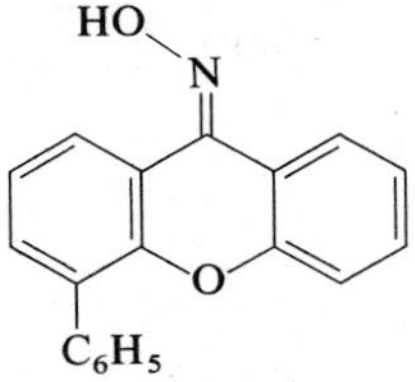 HO, N, O, C_6H_5	PPA, 150–160°, 1 h	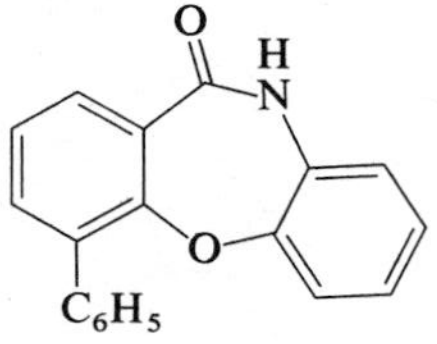 O, H, N, O, C_6H_5 (60)	463
	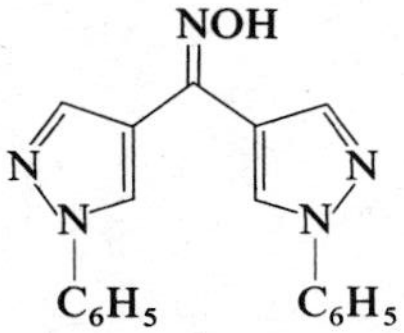 NOH, N, N, N, N, C_6H_5, C_6H_5	TsCl, pyridine, 4 d	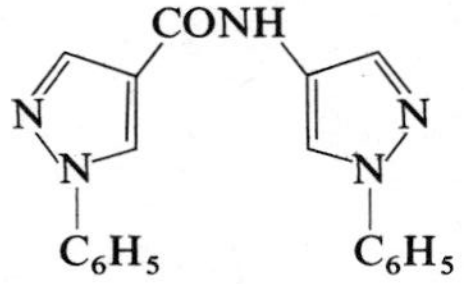 CONH, N, N, N, N, C_6H_5, C_6H_5 (60)	318
		PCl_5, $CHCl_3$, 4 d	" (80)	318
		H_2SO_4, 130°, 30 min	" (79)	318
		PPA, 130°, 30 min	" (92)	318
	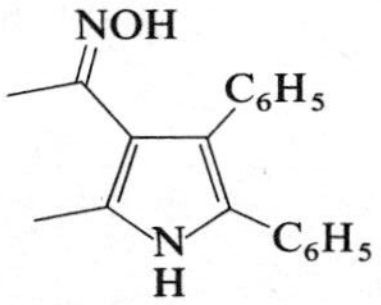 NOH, C_6H_5, N, H, C_6H_5	HCl, ethanol, reflux, 2 h	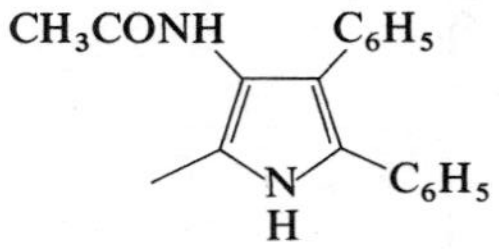 CH_3CONH, C_6H_5, N, H, C_6H_5 (—)	464

TABLE I. REARRANGEMENTS (*Continued*)

No. of Carbon Atoms	Substrate	Reagent and Conditions	Product(s) and Yield(s) (%)	Refs.
	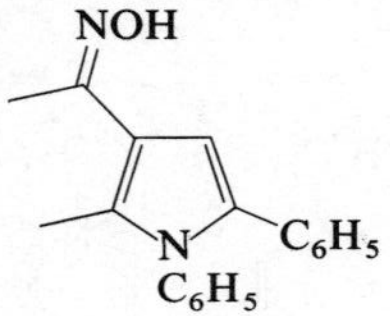 NOH; N; C_6H_5; C_6H_5	HCl, ethanol, reflux, 2 h	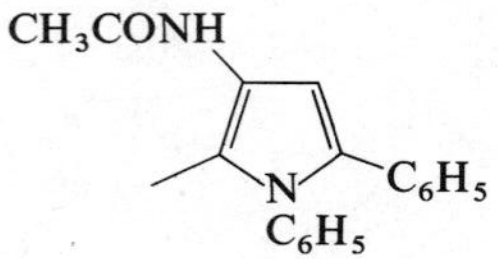 CH_3CONH; N; C_6H_5; C_6H_5 (—)	464
	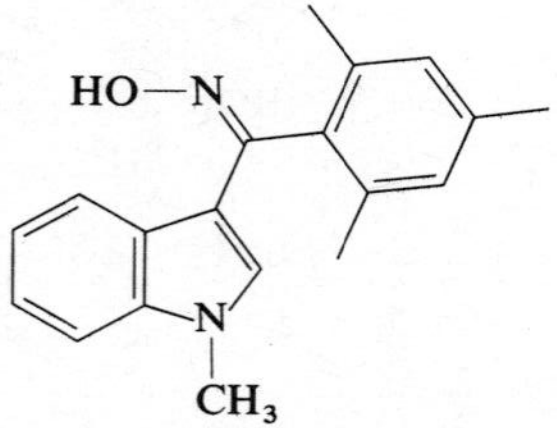 HO—N; N; CH_3	PCl_5, ether, 0°	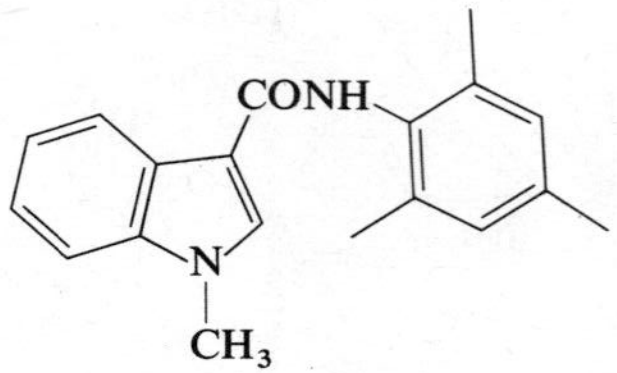 CONH; N; CH_3 (—)	428
	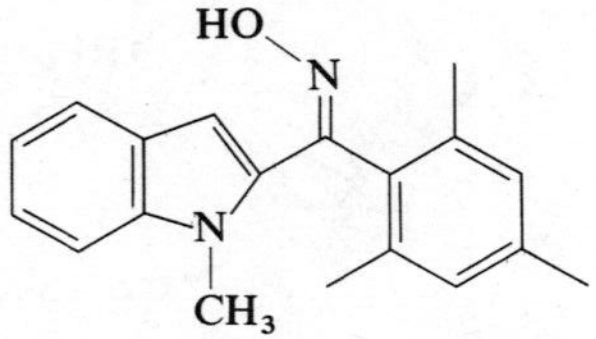 HO; N; N; CH_3	"	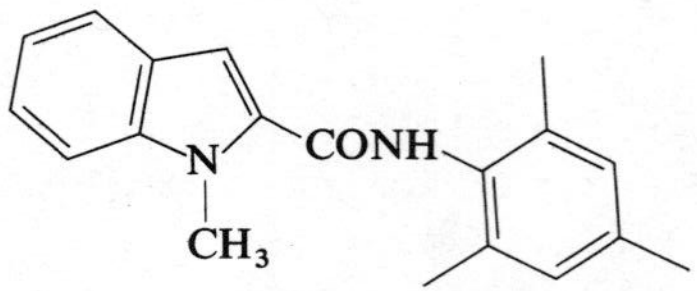 N; CONH; CH_3 (—)	428
	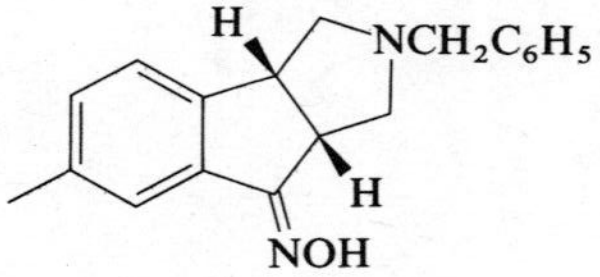 H; $NCH_2C_6H_5$; H; NOH	PPA, 200°, 1 h	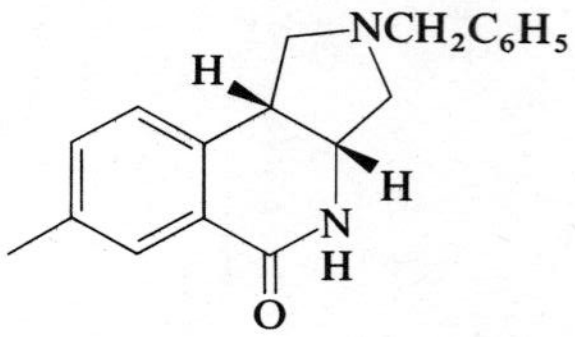 $NCH_2C_6H_5$; H; H; N; H; O (45)	453

Reactant	Conditions	Product(s) and Yield(s) (%)	Refs.
		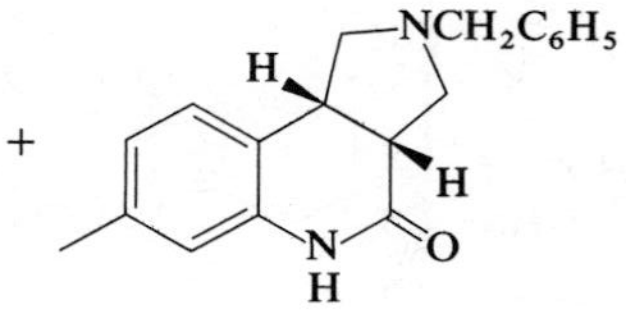 + (7) $NCH_2C_6H_5$, H, H, N H, O	
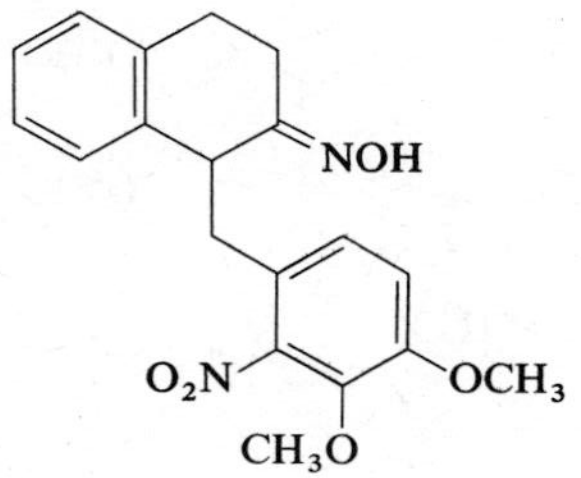 NOH, O_2N, OCH_3, CH_3O	PPA, 70°, 45 min	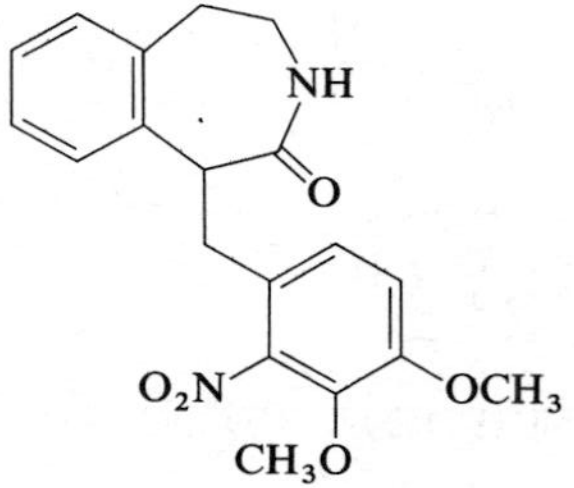 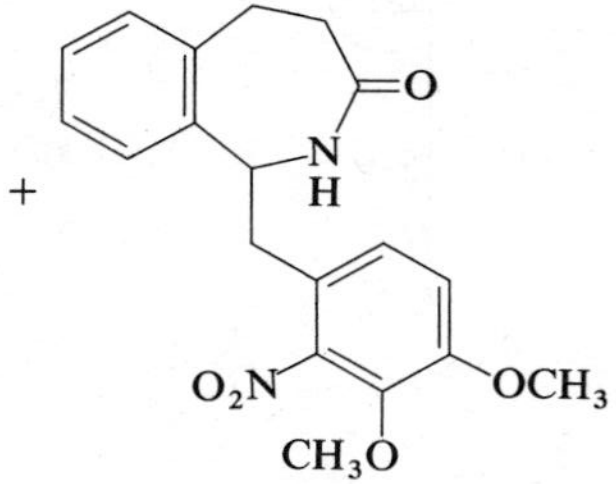 NH, O, O_2N, OCH_3, CH_3O + O, N H, O_2N, OCH_3, CH_3O 3:2 (51)	448
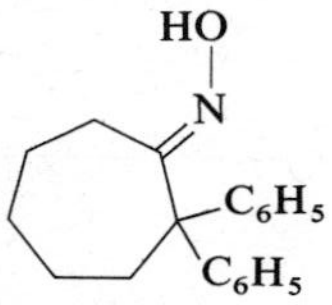 HO–N, C_6H_5, C_6H_5	PPA	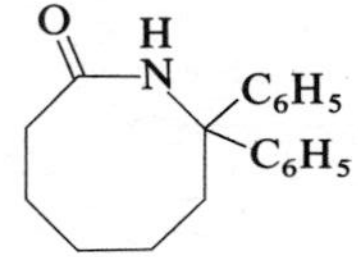 O, H N, C_6H_5, C_6H_5 (12)	165

TABLE I. REARRANGEMENTS (*Continued*)

No. of Carbon Atoms	Substrate	Reagent and Conditions	Product(s) and Yield(s) (%)	Refs.
	p-(i-C_3H_7)$C_6H_4CH=C(C_6H_5)C(=NOH)CH_3$	PCl_5, ether	p-(i-C_3H_7)$C_6H_4CH=C(C_6H_5)NHCOCH_3$ (30)	442
	p-$CH_3C_6H_4CH=C(C_6H_5)C(=NOH)C_3H_7$-$n$	"	p-$CH_3C_6H_4CH=C(C_6H_5)NHCOC_3H_7$-$n$ (5)	442
	p-$CH_3OC_6H_4CH=C(C_6H_5)C(=NOH)C_3H_7$-$n$	"	p-$CH_3OC_6H_4CH=C(C_6H_5)NHCOC_3H_7$-$n$ (25)	442
	NOH; CH_3O, CH_3O; $CH_2C_6H_5$	PPA	H, N, O; CH_3O, CH_3O; $CH_2C_6H_5$ (50) + O, NH; CH_3O, CH_3O; $CH_2C_6H_5$ (50)	465

Reactant	Conditions	Product (Yield, %)	Ref.
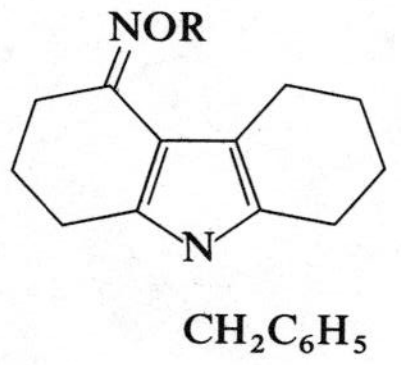 NOR; $CH_2C_6H_5$; R=H	PPA, 130°, 30 min	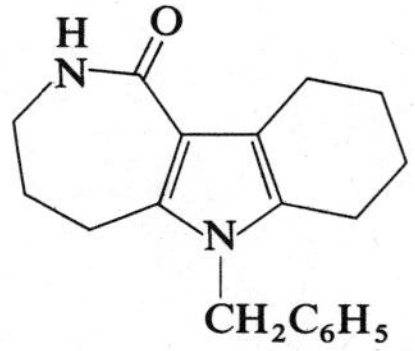 H N O; N—$CH_2C_6H_5$ (49)	383
R = Ts	NaO_2CCH_3, ethanol, H_2O, 30 h	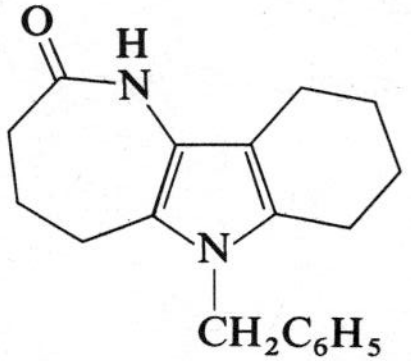 O H N; N—$CH_2C_6H_5$ (34)	383
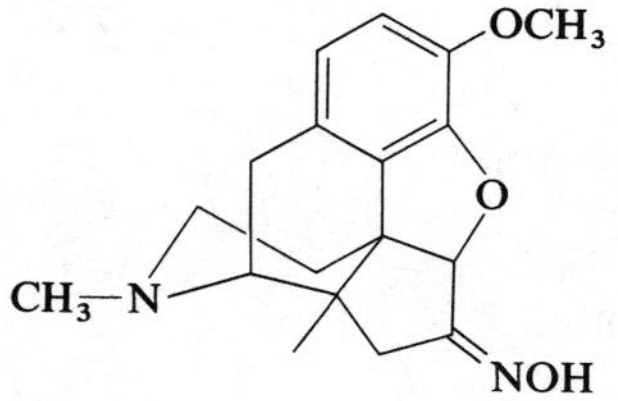 OCH_3; O; CH_3—N; NOH	P_2O_5, CH_3SO_3H, 100°, 1 h	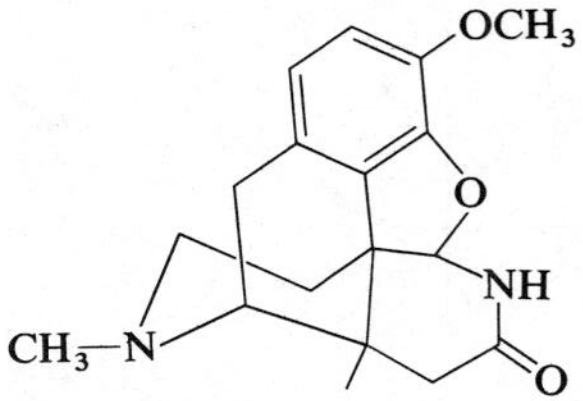 OCH_3; O; NH; CH_3—N; O (94)	466
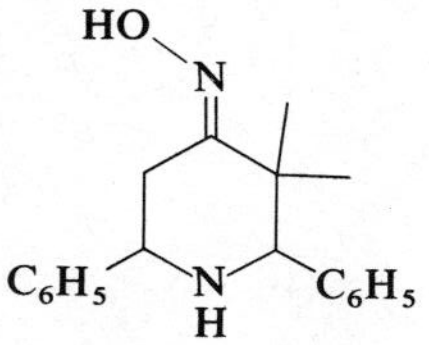 HO N; C_6H_5 N H C_6H_5	PPA, 115°, 15 min	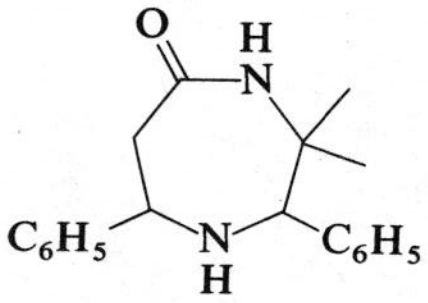 O H N; C_6H_5 N H C_6H_5 (85)	455
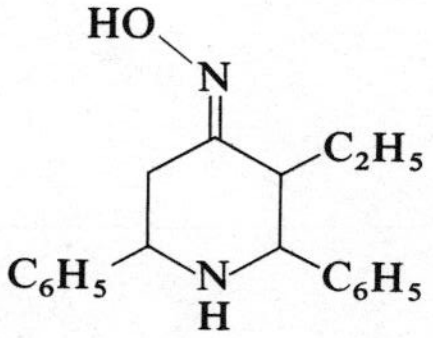 HO N; C_2H_5; C_6H_5 N H C_6H_5	”	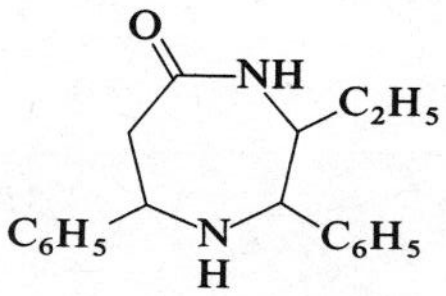 O NH C_2H_5; C_6H_5 N H C_6H_5 (70)	455

TABLE I. REARRANGEMENTS (*Continued*)

No. of Carbon Atoms	Substrate	Reagent and Conditions	Product(s) and Yield(s) (%)	Refs.
	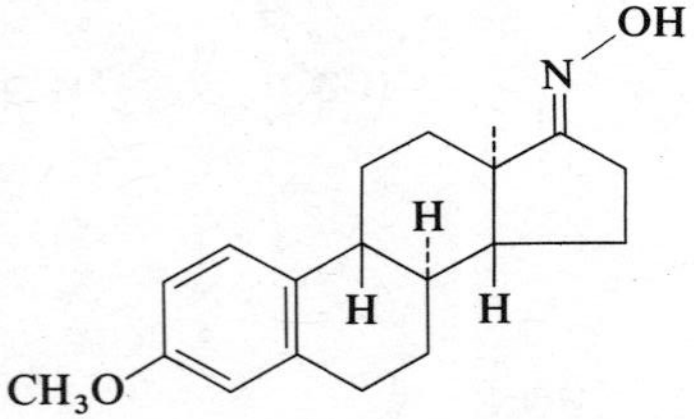 N—OH; H; H; H; CH_3O	$POCl_3$, $CHCl_3$, pyridine, −5°, 90 min	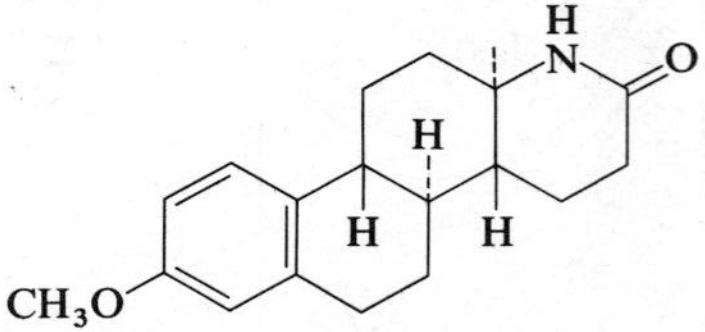 H N O; H; H; H; CH_3O (79)	467
	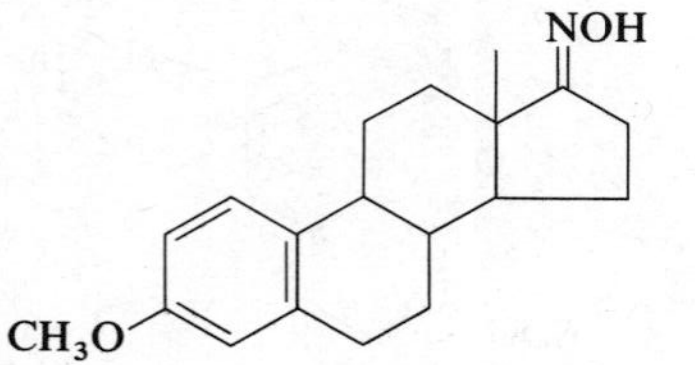 NOH; CH_3O	$SOCl_2$	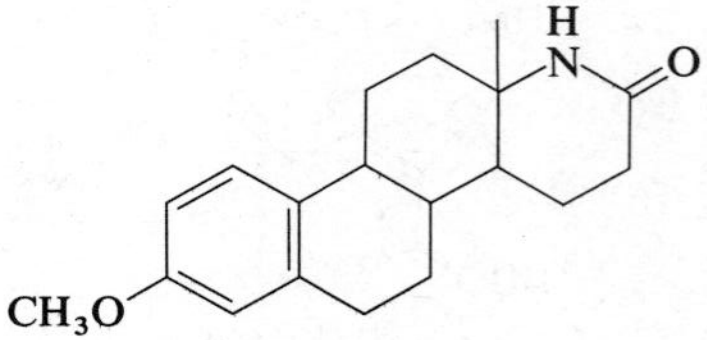 H N O; CH_3O (—)	458
		p-$CH_3CONHC_6H_4SO_2Cl$, pyridine 24 hr	" (—)	457, 458
		DCC, DMSO, CF_3CO_2H, 16 h	" (39)	468
		HN_3	" (38)	469
	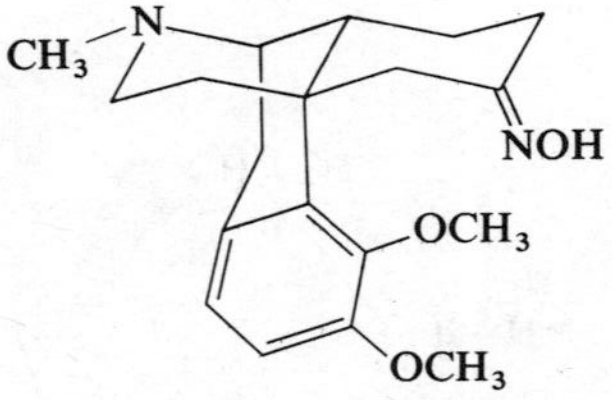 CH_3 N; NOH; OCH_3; OCH_3	PPA, 100°, 1 h	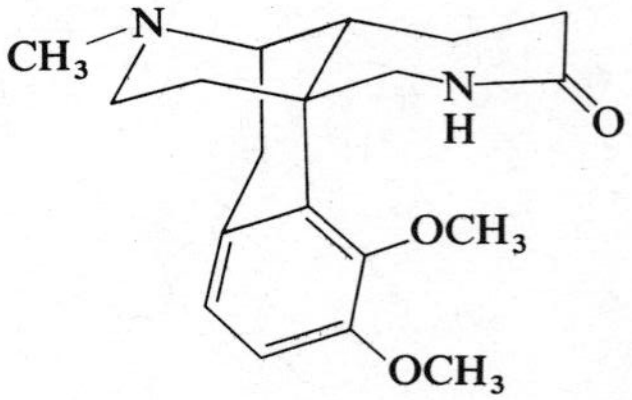 CH_3 N; N H; O; OCH_3; OCH_3 (57)	451

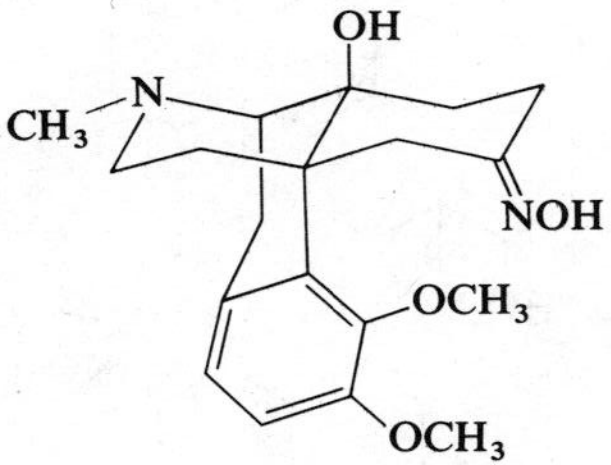	PPA, 100°, 1 h	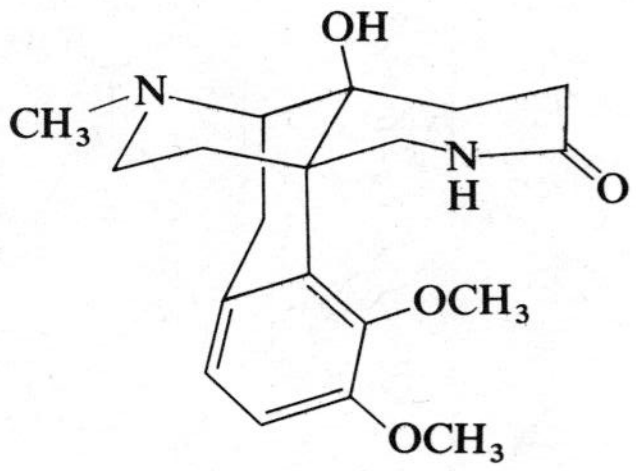 (72)	451
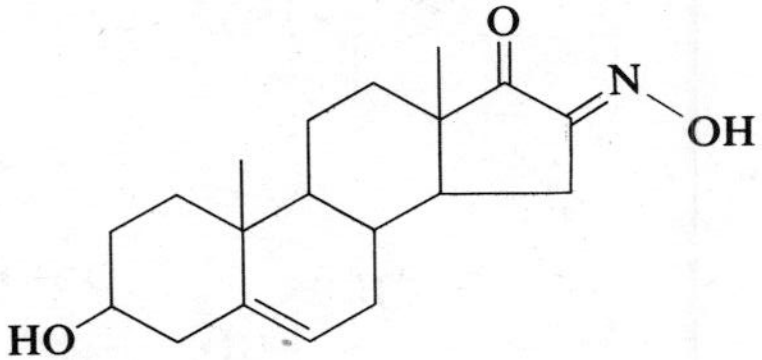	CH_3CO_2H, $(CH_3CO)_2O$	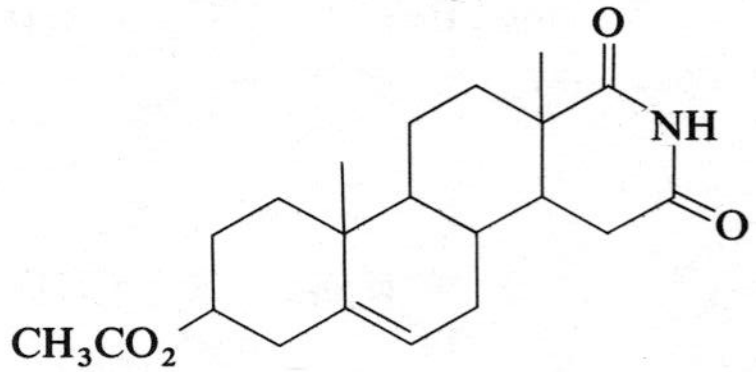 (—)	65
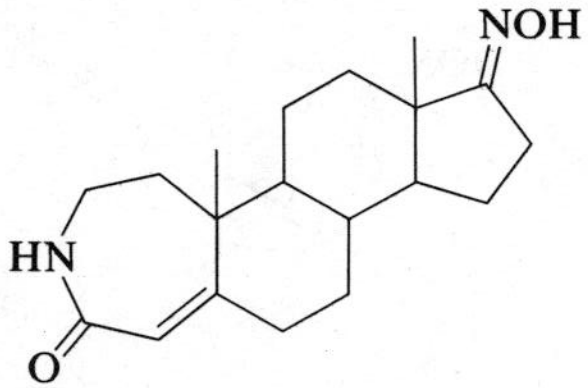	$SOCl_2$, 20 h	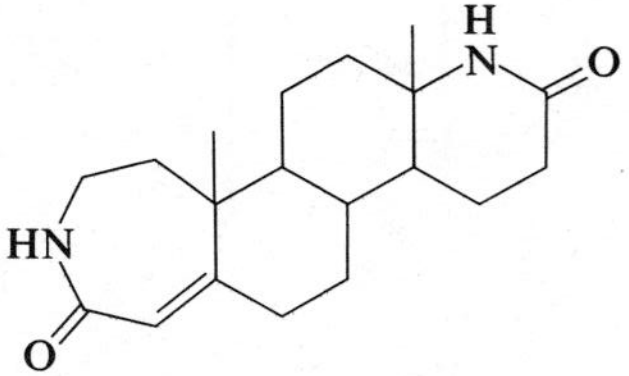 (54)	470
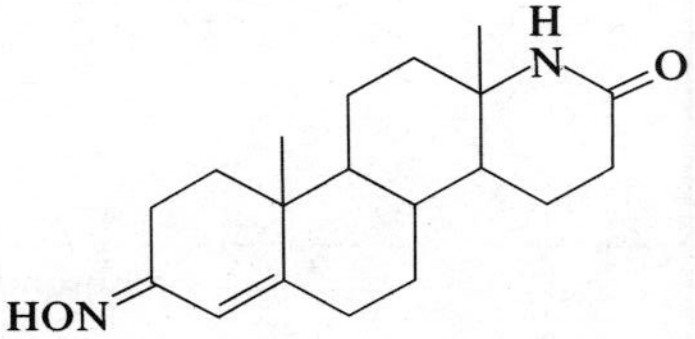	$SOCl_2$, dioxane	" (55)	470, 471

TABLE I. REARRANGEMENTS (*Continued*)

No. of Carbon Atoms	Substrate	Reagent and Conditions	Product(s) and Yield(s) (%)	Refs.
	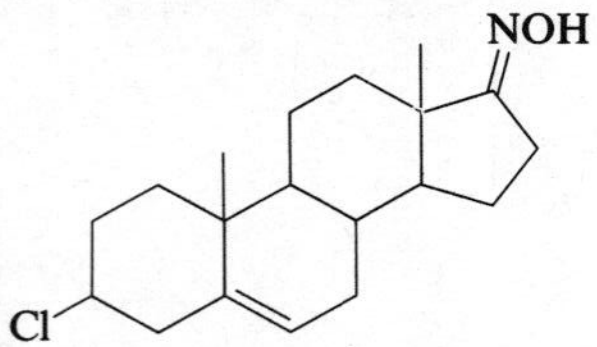 Cl, NOH	$SOCl_2$, dioxane, 15°, 15 min	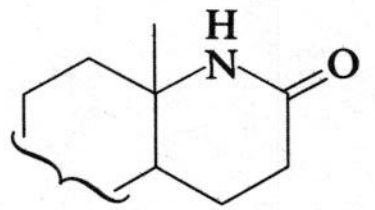 H, N, O (—)	472
	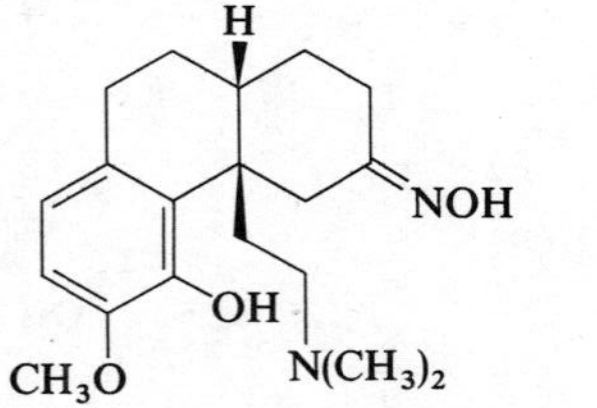 H, NOH, OH, CH_3O, $N(CH_3)_2$	PPA, 100°, 1 h	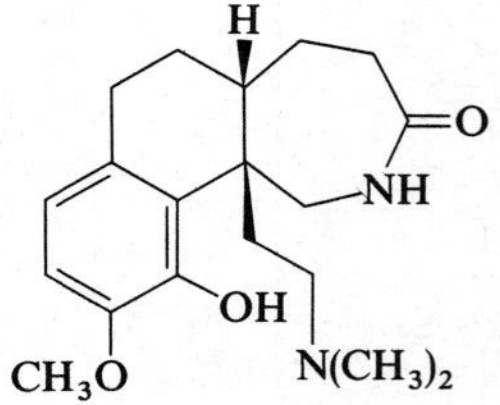 H, O, NH, OH, CH_3O, $N(CH_3)_2$ (34)	451
	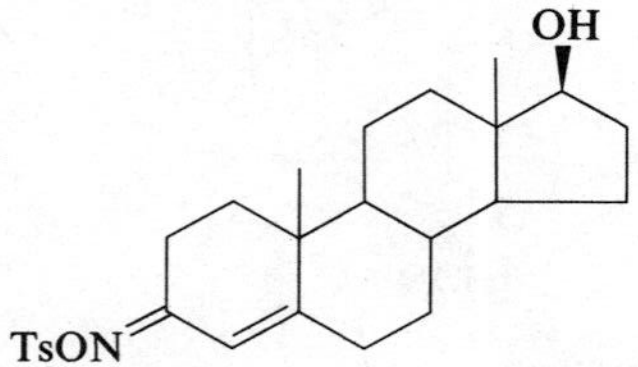 OH, TsON	HCl, CH_3OH, 50°, 30 min	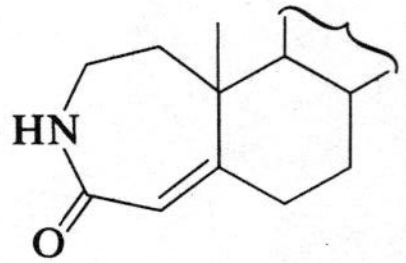 HN, O (85-90)	40
	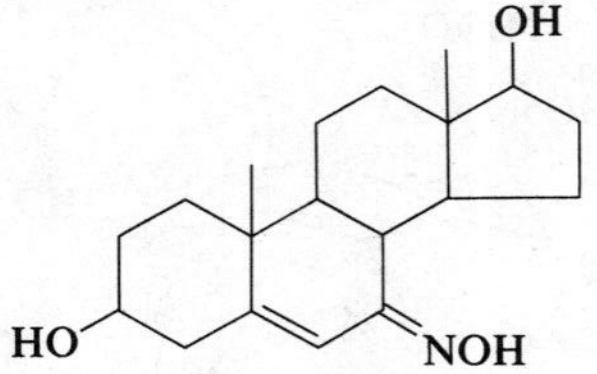 OH, HO, NOH	$SOCl_2$, dioxane, 10-25°, 25 min	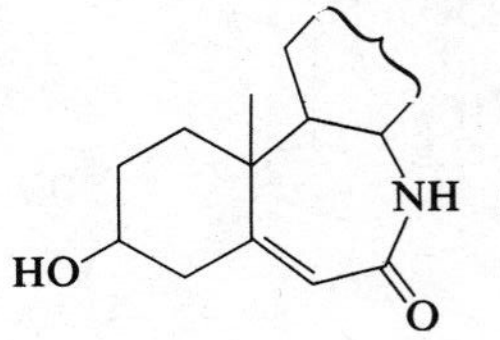 HO, NH, O (—)	473

Reactant	Conditions	Product(s) and Yield(s) (%)	Refs.
CH_3CO_2, CH_3CO_2, C_5H_{11}-*n*	2,4,6-$(CH_3)_3C_6H_2SO_3NH_2$, CH_2Cl_2, 0°, 30 min	HN, CH_3CO_2, CH_3CO_2, C_5H_{11}-*n* (55) + CH_3CO_2, CH_3CO_2, C_5H_{11}-*n* (15)	82
NOH	TsCl, pyridine, 16 h	(53)	474
	HN_3	" (40)	469
HON	$SOCl_2$, ether, −20°, 5 min	(92)	475
HON, OH (*Z*)	TsCl, pyridine	HN (100)	460

TABLE I. REARRANGEMENTS (*Continued*)

No. of Carbon Atoms	Substrate	Reagent and Conditions	Product(s) and Yield(s) (%)	Refs.
	(*E*)	TsCl, pyridine	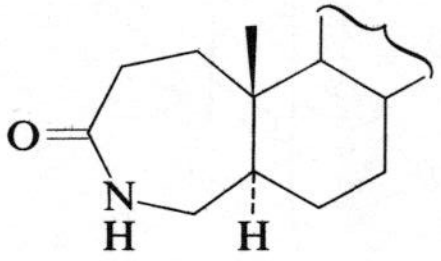 O= N H H (100)	460
	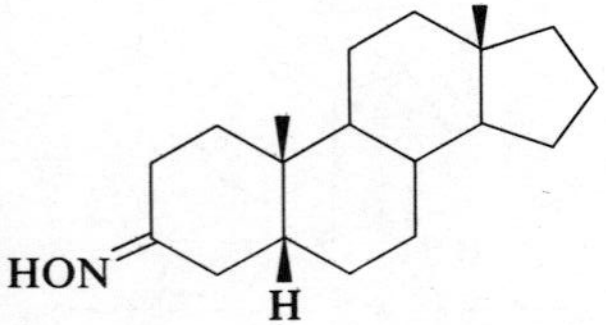 HON H	SOCl$_2$, ether, $-20°$, 10 min	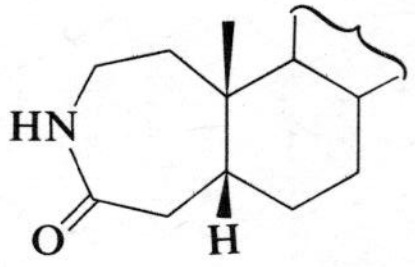 HN O H	475
			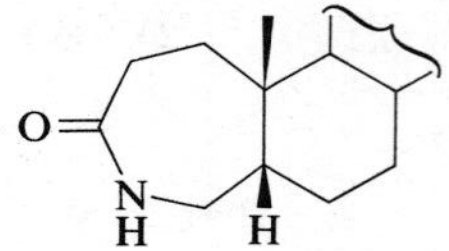 + O= N H H (—)	475
	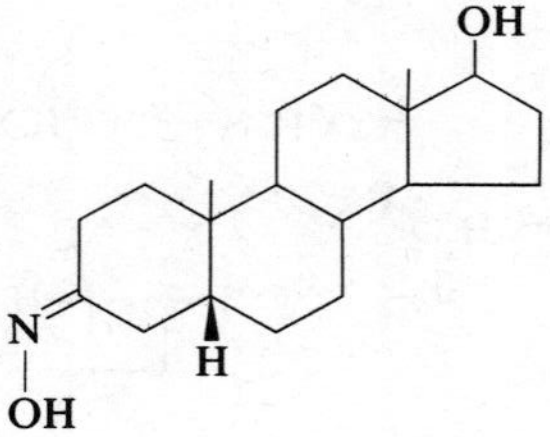 OH (*Z*) N OH H	TsCl, pyridine	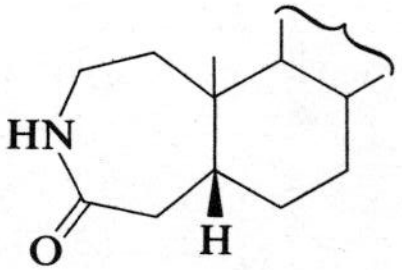 HN O H (100)	460

Substrate	Conditions	Product(s) (Yield %)	Refs.
(*E*) [3-oxime, 5β-H, 16-OH steroid: N–OH, H, OH]	"	[ring-expanded lactam: O=, HN, H] (100)	460
	"	[ring-expanded lactam: HN, O, H] (50)	476
[HO, OH, NOH steroid]	$SOCl_2$, dioxane	[HO, NH_2, CH_2CO_2H] (—)	473, 477
[HO, OH, H, N–OH steroid]	"	[lactam: HO, NH, O, H] (—)	473

TABLE I. REARRANGEMENTS (*Continued*)

No. of Carbon Atoms	Substrate	Reagent and Conditions	Product(s) and Yield(s) (%)	Refs.
	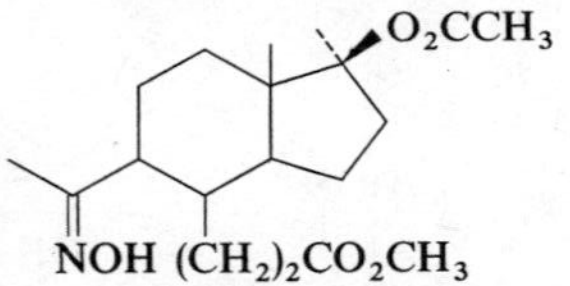 O_2CCH_3; NOH $(CH_2)_2CO_2CH_3$	$POCl_3$, pyridine, 0°	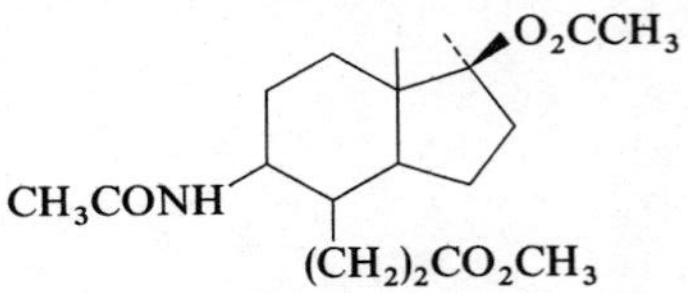 O_2CCH_3; CH_3CONH; $(CH_2)_2CO_2CH_3$ (—)	452
C_{20}	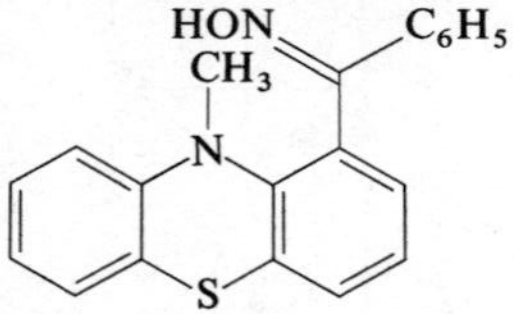 HON; C_6H_5; CH_3; N; S	PCl_5, pyridine, C_6H_6	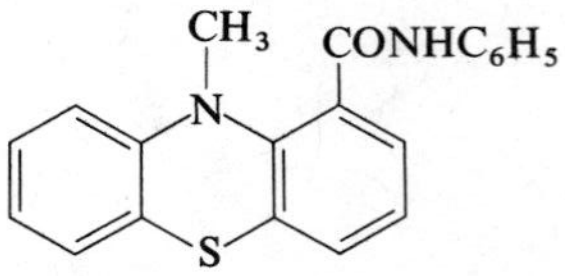 CH_3; $CONHC_6H_5$; N; S (82)	432
	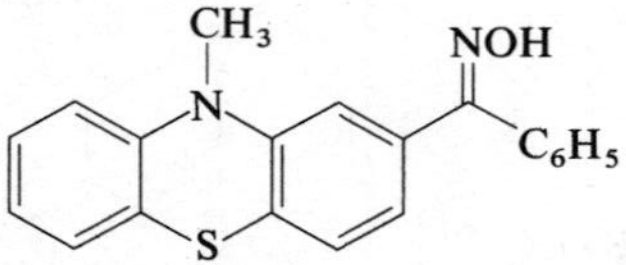 CH_3; N; S; NOH; C_6H_5	"	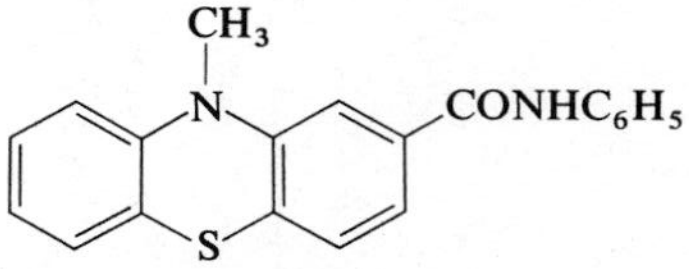 CH_3; N; S; $CONHC_6H_5$ (55)	432
	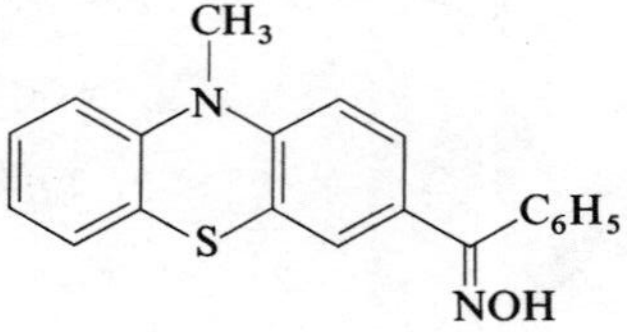 CH_3; N; S; C_6H_5; NOH	"	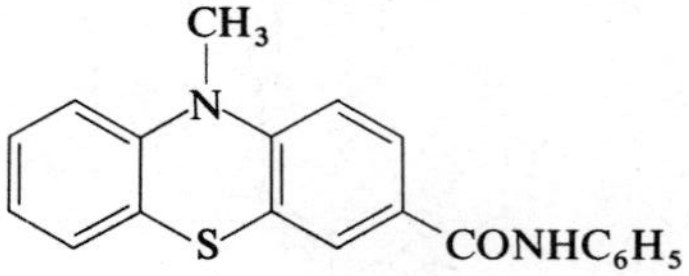 CH_3; N; S; $CONHC_6H_5$ (62)	432

Reactant	Conditions	Product(s) and Yield(s) (%)	Refs.
10-Methylphenothiazin-4-yl phenyl ketone oxime (CH_3, N, S, HON=C–C_6H_5)	"	10-Methyl-4-($CONHC_6H_5$)phenothiazine (65) + 10-Methyl-4-($NHCOC_6H_5$)phenothiazine (6)	432
2-(p-$CH_3C_6H_4$)C_6H_4C(=NOH)C_6H_5	PCl_5, C_6H_6, 3–8 h	2-(p-$CH_3C_6H_4$)$C_6H_4CONHC_6H_5$ + 2-(p-$CH_3C_6H_4$)$C_6H_4NHCOC_6H_5$	443
	Oximation temp 60°	15.5:84.5 (93)	
	Oximation temp 118°	22:78 (98)	
4-Oximino-2-C_6H_5-1-(C_6H_4Br-p)-4,5,6,7-tetrahydroindole (N–OH)	PPA, 150°, 20 min	HN, O, C_6H_5, N–C_6H_4Br-p (pyrrolo-azepinone) (76)	478

TABLE I. REARRANGEMENTS (*Continued*)

No. of Carbon Atoms	Substrate	Reagent and Conditions	Product(s) and Yield(s) (%)	Refs.
	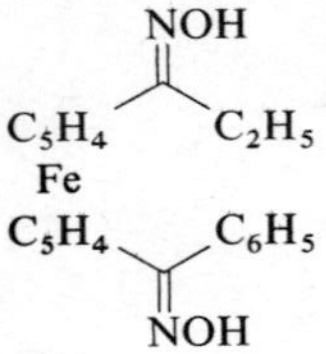 1,1'-Ferrocene bis-oxime: $C_5H_4C(=NOH)C_2H_5$ / Fe / $C_5H_4C(=NOH)C_6H_5$	$C_6H_5SO_2Cl$, NaOH	$C_5H_4COC_2H_5$ / Fe / $C_5H_4CONHC_6H_5$ (15–20)	99
	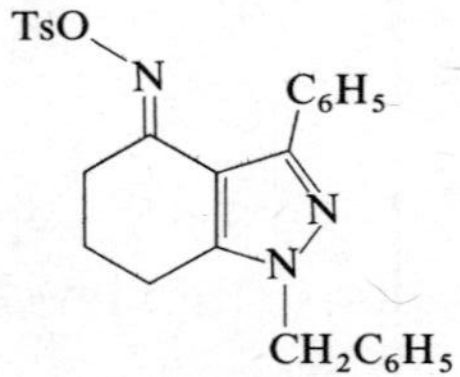 TsO–N= tetrahydroindazolone oxime (C_6H_5, N–$CH_2C_6H_5$)	NaO_2CCH_3, C_2H_5OH, H_2O	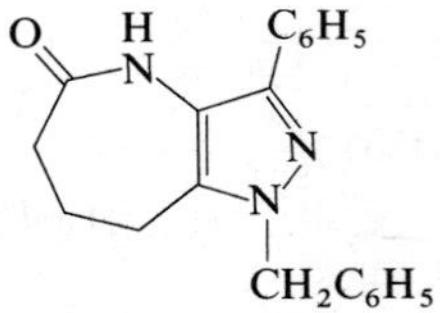 Lactam (O, H, N; C_6H_5; N–$CH_2C_6H_5$) (10)	411
	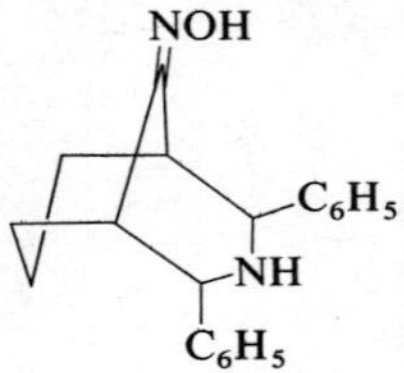 Bicyclic oxime (NOH; C_6H_5, NH, C_6H_5)	PPA, 120–130°, 10–15 min	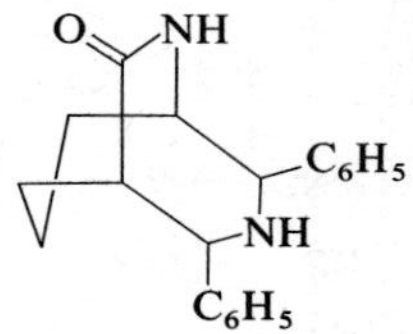 Lactam (O, NH; C_6H_5, NH, C_6H_5) (50)	479
	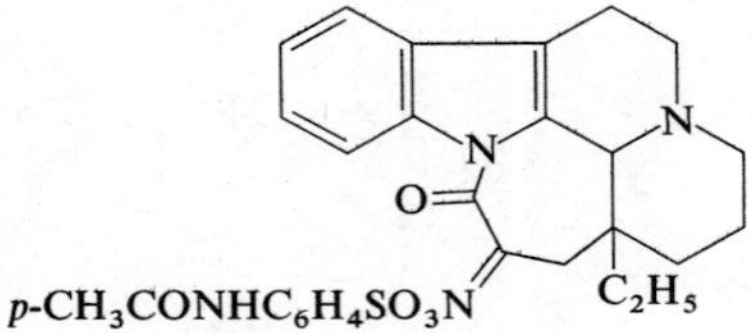 p-$CH_3CONHC_6H_4SO_3N$= indoloquinolizidine ketolactam (N, O, N, C_2H_5)	$SOCl_2$	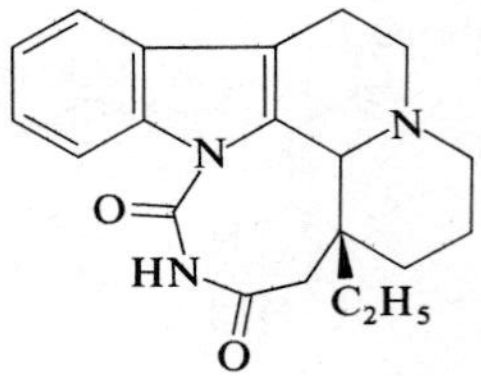 Imide (N, O, HN, O, N, C_2H_5) (71)	480

Reactant	Conditions	Product(s) (Yield, %)	Refs.
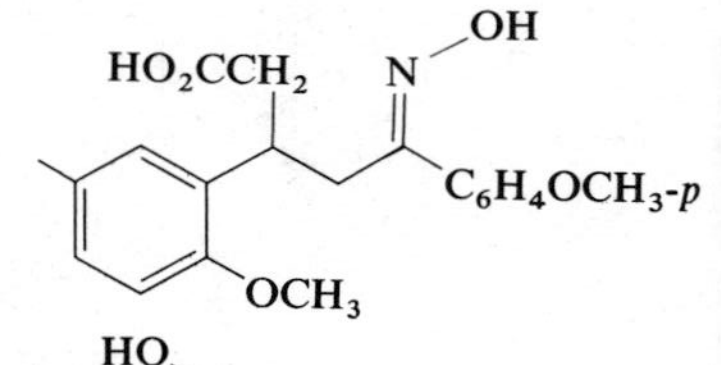 HO_2CCH_2, N—OH, $C_6H_4OCH_3$-*p*, OCH_3	PPA, 100°, 8 h	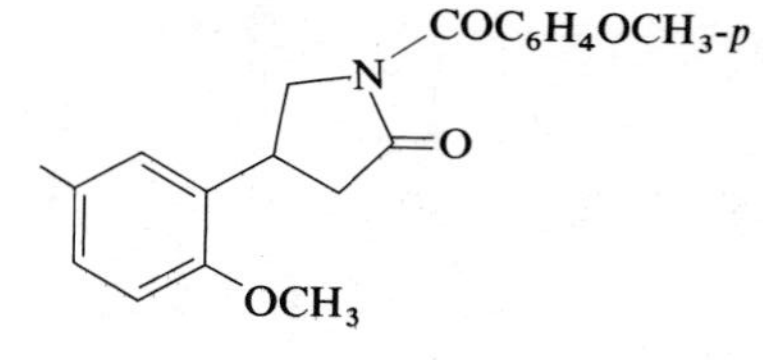 $COC_6H_4OCH_3$-*p*, N, O, OCH_3 (70)	108
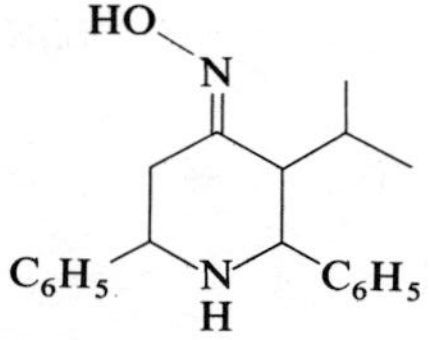 HO—N, C_6H_5, N H, C_6H_5	PPA, 115°, 15 min	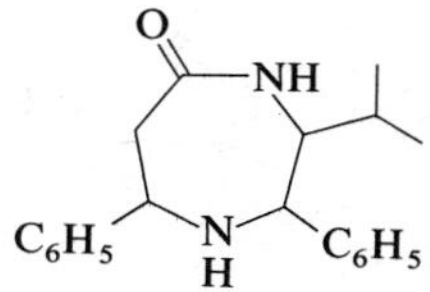 O, NH, C_6H_5, N H, C_6H_5 (80)	455
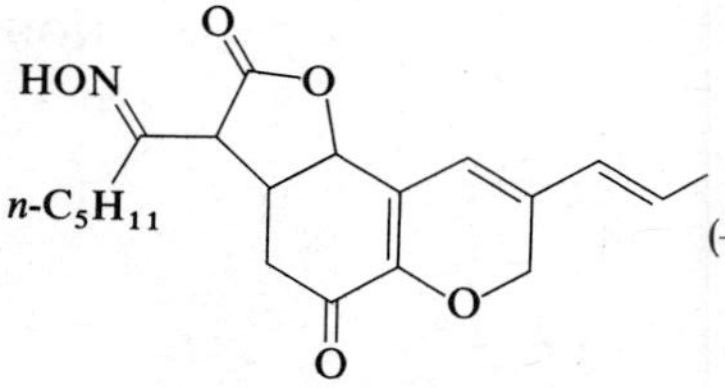 HON, O, O, *n*-C_5H_{11}, O, O (—)		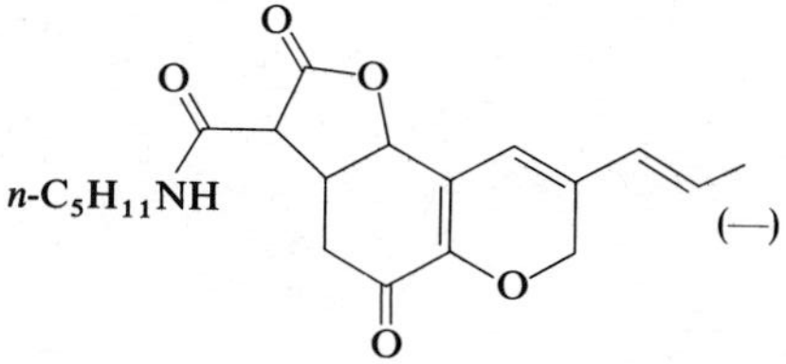 O, O, O, *n*-C_5H_{11}NH, O, O (—)	481
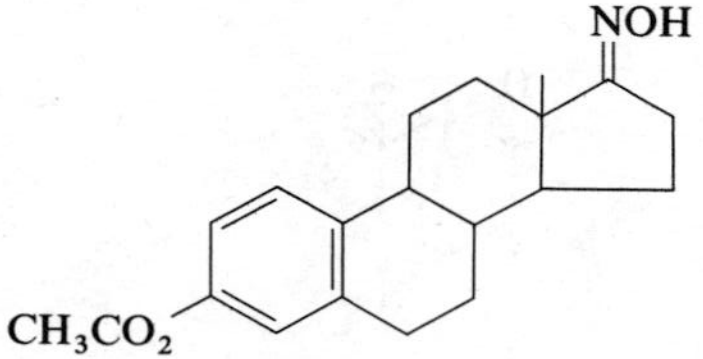 NOH, CH_3CO_2	$SOCl_2$	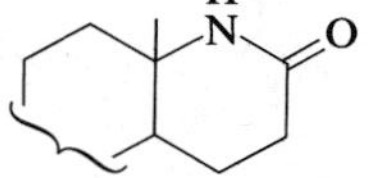 H N, O (—)	458
	p-$CH_3CONHC_6H_4SO_2Cl$, pyridine, 24 h	" (—)	457, 458

TABLE I. REARRANGEMENTS (*Continued*)

No. of Carbon Atoms	Substrate	Reagent and Conditions	Product(s) and Yield(s) (%)	Refs.
	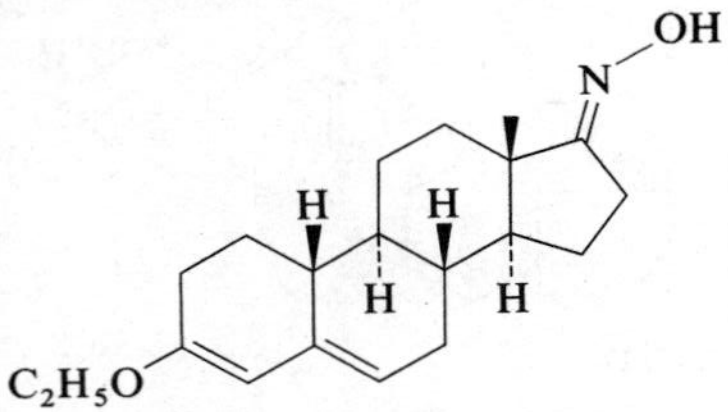 (structure: C_2H_5O, N–OH)	$POCl_3$, pyridine, $CHCl_3$, $-5°$, 90 min	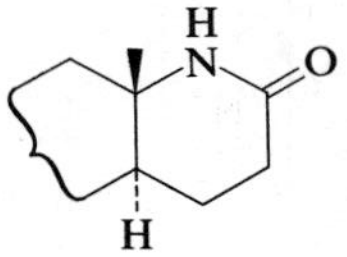 (structure) (56)	467
	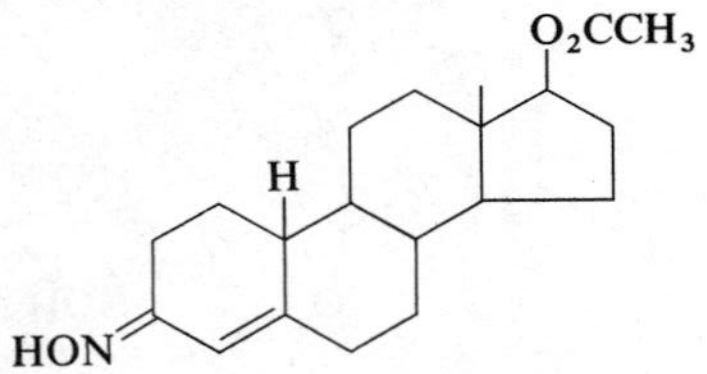 (structure: O_2CCH_3, HON)	TsCl, DMF	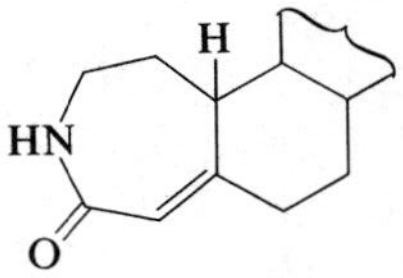 (structure) (20)	57
	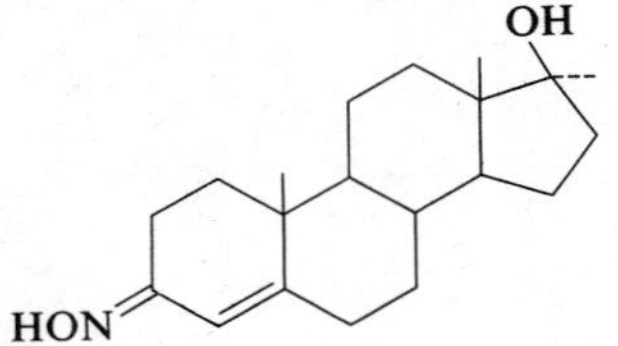 (structure: OH, HON)	$SOCl_2$, dioxane, 10–25°, 1 h	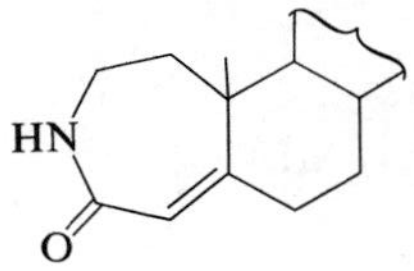 (structure) (97)	43
	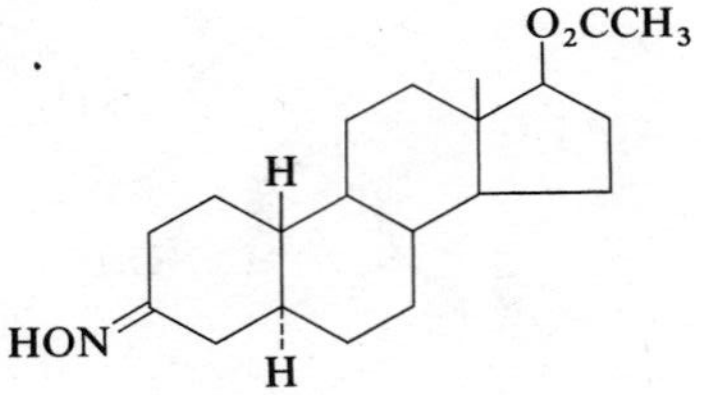 (structure: O_2CCH_3, HON)	$SOCl_2$, dioxane, 10 min	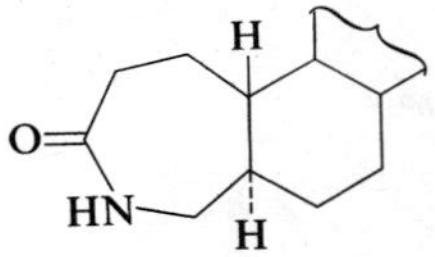 (structure)	482

Formula	Reactant	Conditions	Product(s) (% Yield)	Refs.
			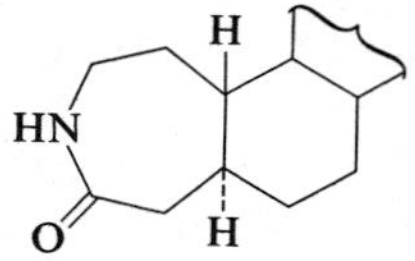 + HN, O, H, H (55)	
	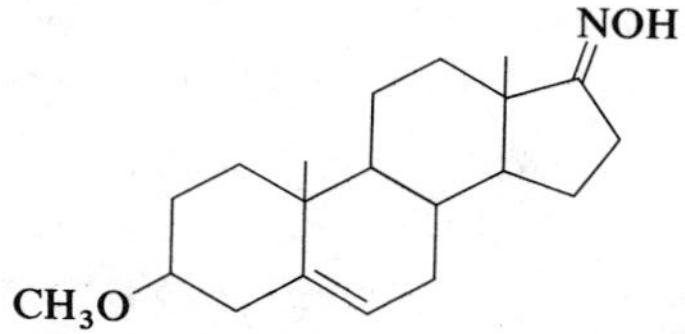 CH_3O, NOH	DCC, DMSO, CF_3CO_2H, 16 h	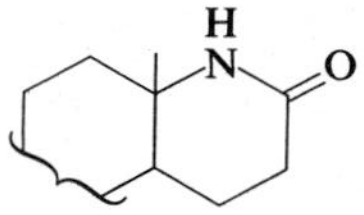 H, N, O (21)	468
	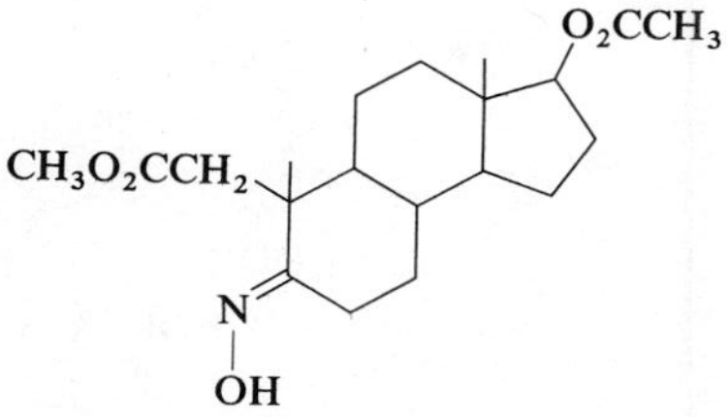 $CH_3O_2CCH_2$, O_2CCH_3, N, OH	$SOCl_2$, dioxane, 10°, 10 min	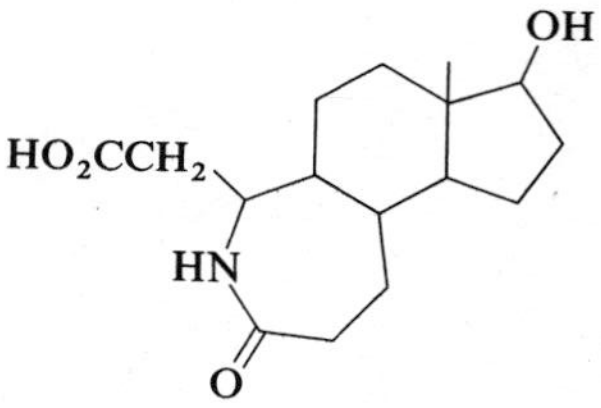 HO_2CCH_2, OH, HN, O (70–80)	483
C_{21}	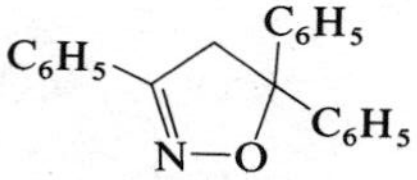 C_6H_5, C_6H_5, C_6H_5, N–O	$(CH_3CO)_2$, $BF_3\cdot$ether, 3 d	$C_6H_5CONHCH{=}C(C_6H_5)_2$ (—)	431
	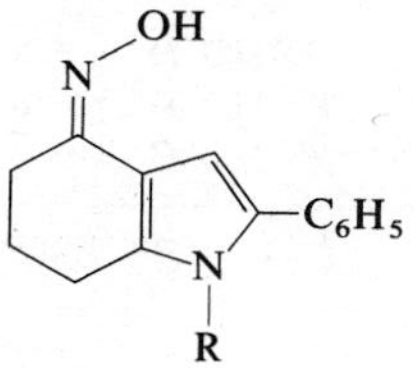 N, OH, C_6H_5, N, R	PPA, 150°, 20 min	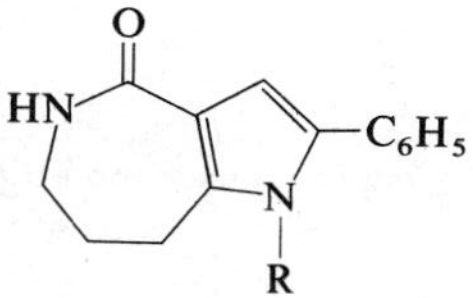 O, HN, C_6H_5, N, R	

No. of Carbon Atoms	Substrate	Reagent and Conditions	Product(s) and Yield(s) (%)	Refs.
	$R = C_6H_4CH_3$-*o*	PPA, 150°, 20 min	(84)	478
	$R = C_6H_4CH_3$-*m*	"	(78)	478
	$R = C_6H_4CH_3$-*p*	"	(85)	478
	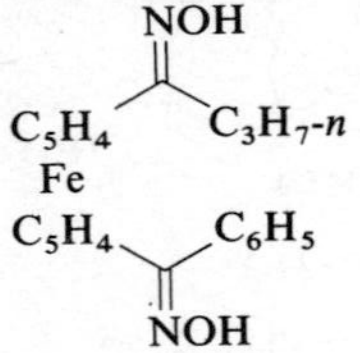 $C_5H_4C(=NOH)C_3H_7$-*n* / Fe / $C_5H_4C(=NOH)C_6H_5$	$C_6H_5SO_2Cl$, NaOH	$C_5H_4COC_3H_7$-*n* / Fe / $C_5H_4CONHC_6H_5$ (15–20)	99
	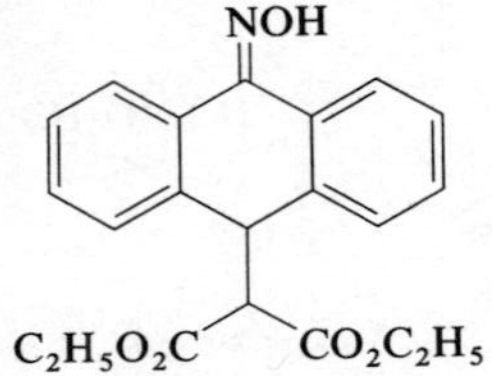 NOH; $C_2H_5O_2C$, $CO_2C_2H_5$	PCl_5, CCl_4, reflux, 2 h	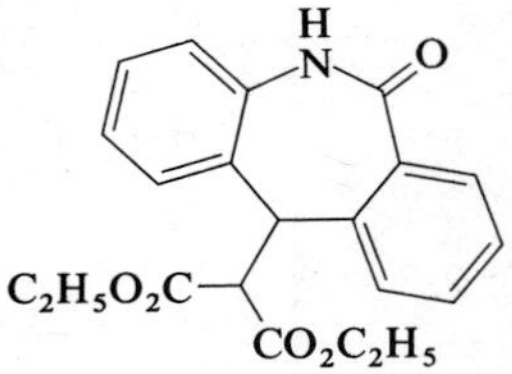 H, N, O; $C_2H_5O_2C$, $CO_2C_2H_5$ (85)	142
	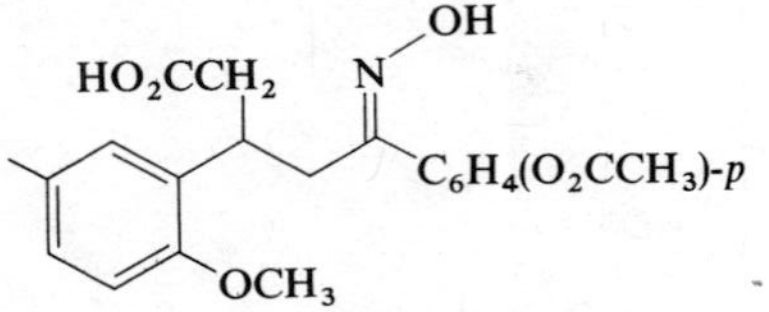 HO_2CCH_2, N–OH, $C_6H_4(O_2CCH_3)$-*p*, OCH_3	PPA, 100°, 8 h	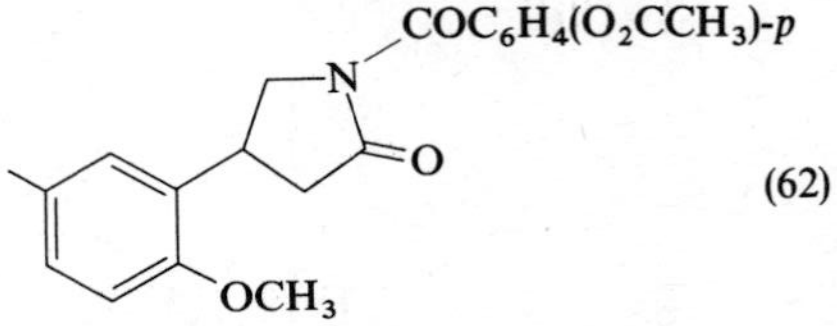 N–$COC_6H_4(O_2CCH_3)$-*p*, O, OCH_3 (62)	108
	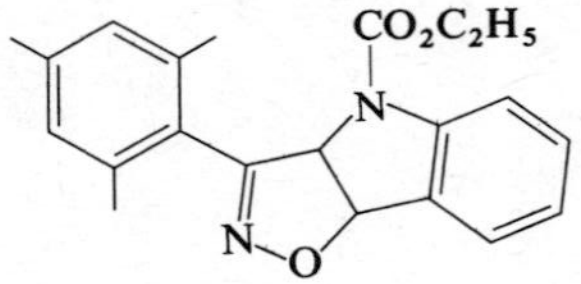 $CO_2C_2H_5$, N, N–O	H_2SO_4, CH_3CO_2H, reflux	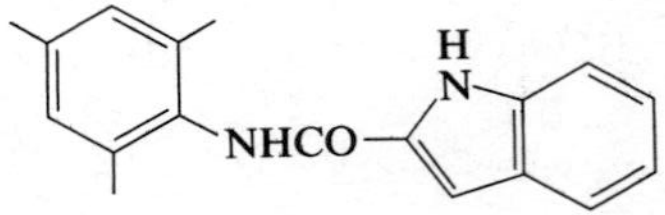 NHCO, H, N (—)	428

Reactant	Conditions	Product(s) and Yield(s) (%)	Refs.
	"	(—)	428
	PPA, 100°, 8 h	(65)	108
	"	(58)	108
	"	(55)	108

TABLE I. REARRANGEMENTS (*Continued*)

No. of Carbon Atoms	Substrate	Reagent and Conditions	Product(s) and Yield(s) (%)	Refs.
	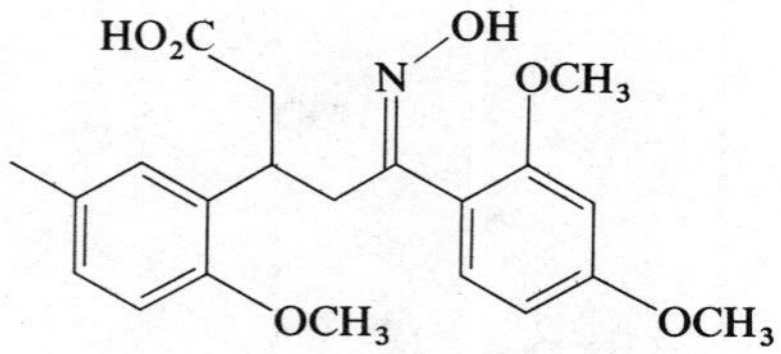 [structure: oxime, HO_2C, OH, N, OCH_3, OCH_3, OCH_3]	PPA, 100°, 8 h	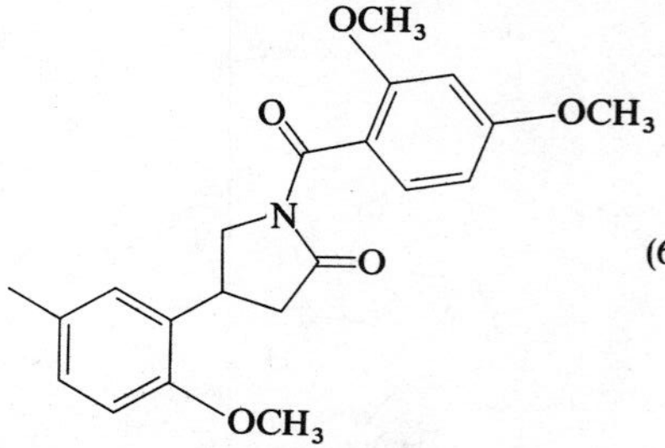 [structure: N-acylpyrrolidinone, OCH_3, OCH_3, O, N, O, OCH_3] (66)	108
	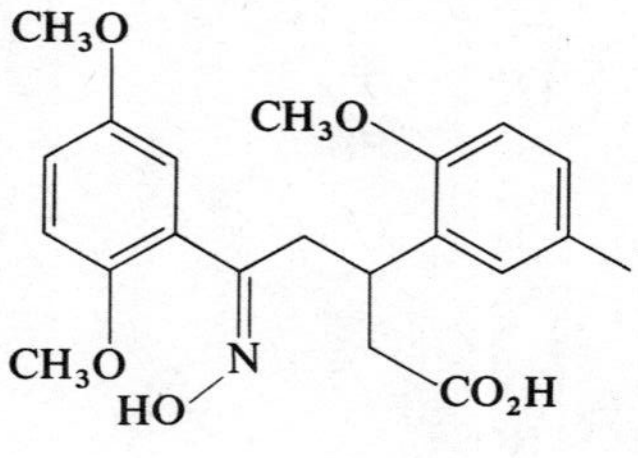 [structure: oxime, CH_3O, CH_3O, CH_3O, N, HO, CO_2H]	"	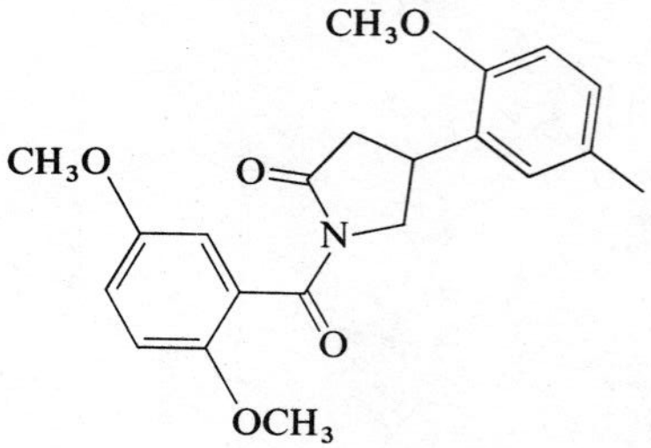 [structure: N-acylpyrrolidinone, CH_3O, CH_3O, O, N, O, OCH_3] (69)	108
	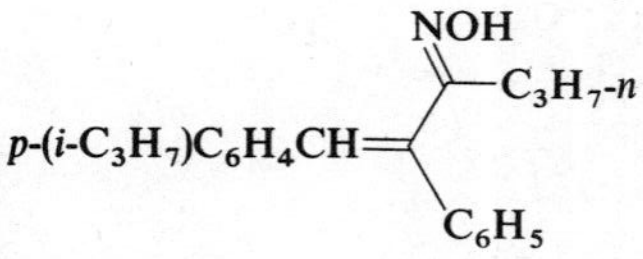 p-(i-C_3H_7)$C_6H_4CH{=}C(C_6H_5)C(=NOH)C_3H_7$-$n$	PCl_5, ether	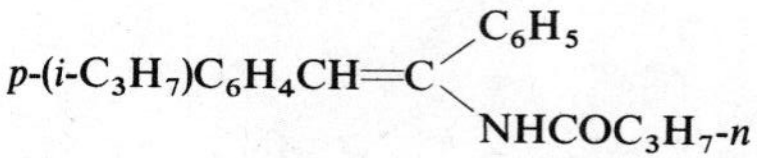 p-(i-C_3H_7)$C_6H_4CH{=}C(C_6H_5)NHCOC_3H_7$-$n$ (—)	442

Reactant	Conditions	Product(s) (Yield, %)	Refs.
	$AgBF_4$, dioxane, H_2O, reflux, 3 h	(43)	484
	PPA, 130°, 10 min	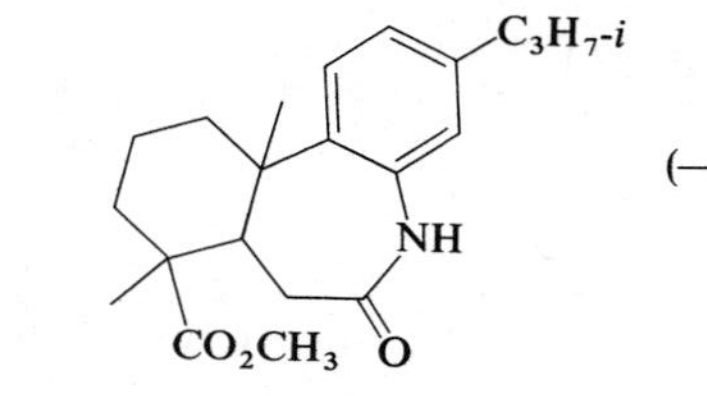 (—)	485
	$SOCl_2$, dioxane	(32)	311
	$SOCl_2$, dioxane, 1 h	(63)	486

TABLE I. REARRANGEMENTS (*Continued*)

No. of Carbon Atoms	Substrate	Reagent and Conditions	Product(s) and Yield(s) (%)	Refs.
	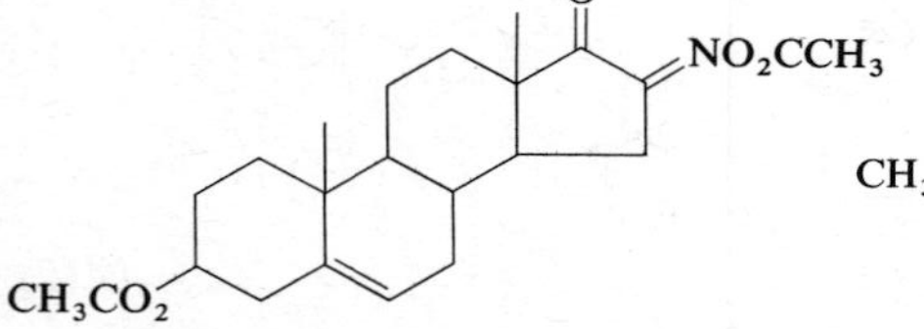 CH_3CO_2, O, NO_2CCH_3	CH_3CO_2H, reflux, 18 h	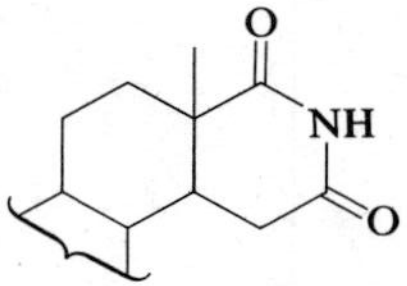 O, NH, O (—)	65
	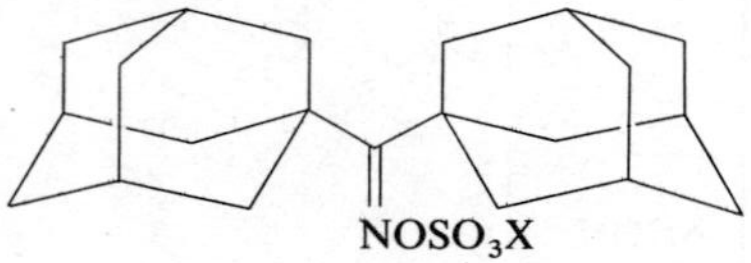 $NOSO_3X$ X = H	H_2O, 18 h	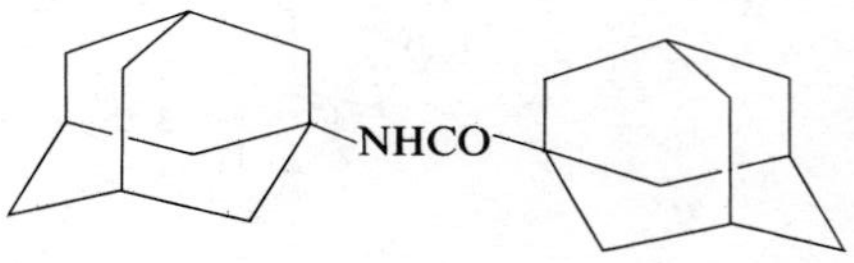 NHCO (32)	258
	X = NH_4	180°, 5 min	" (85)	258
	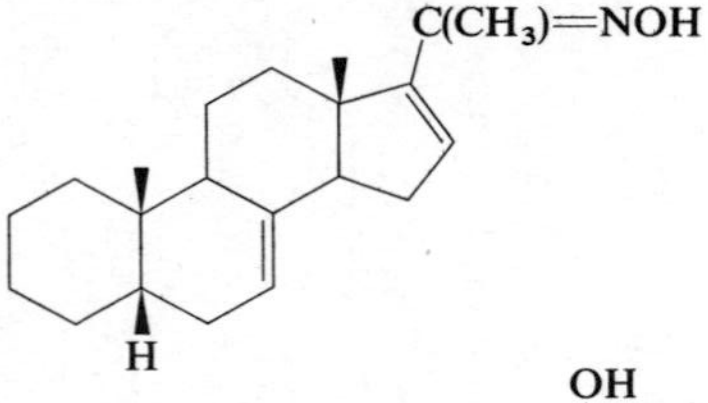 $C(CH_3){=}NOH$, H	p-$CH_3CONHC_6H_4SO_2Cl$, pyridine, 0°, 2 h	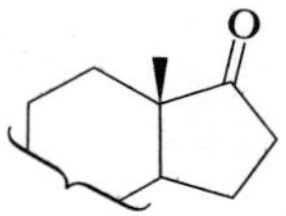 O (49)	115
	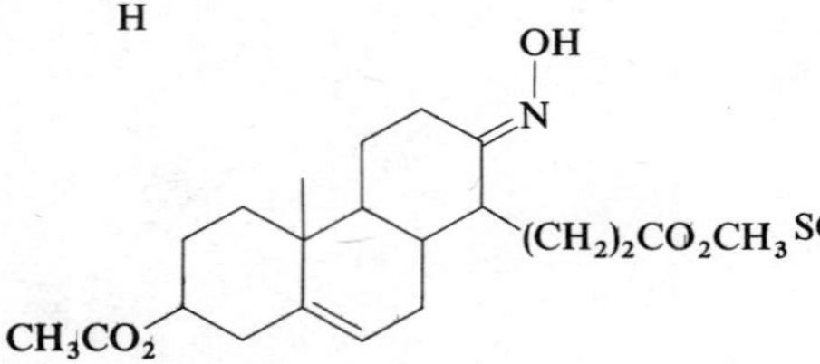 OH, N, $(CH_2)_2CO_2CH_3$, CH_3CO_2	$SOCl_2$, ether, 30 min	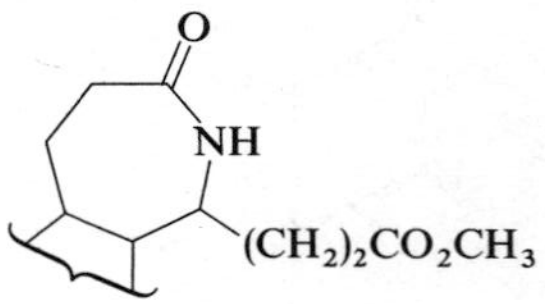 O, NH, $(CH_2)_2CO_2CH_3$ (75)	487

Reactant	Conditions	Product(s) and Yield(s) (%)	Refs.
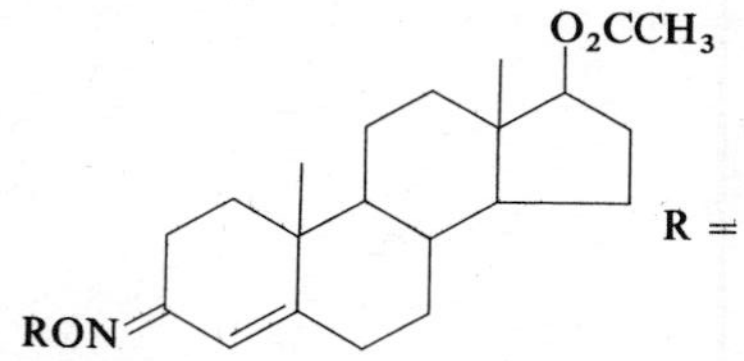 RON (steroid, O_2CCH_3), R = H	TsCl	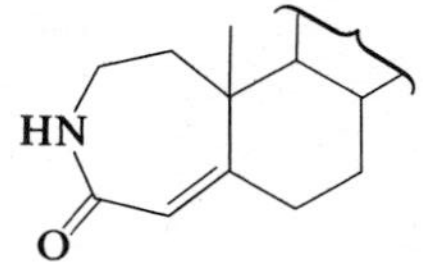 HN, O (25)	392
R = H	$SOCl_2$, dioxane, 1 h	" (50)	488
R = H	BF_3·ether, dioxane, 90°, 3 h	" (17)	488
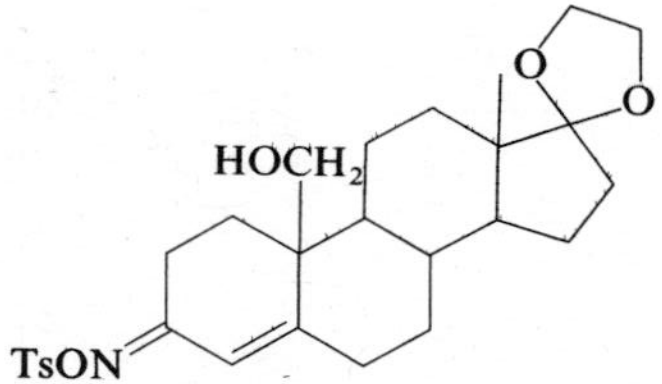 TsON (steroid, $HOCH_2$, ketal)	HCl, CH_3CO_2H, 60°, 30 min	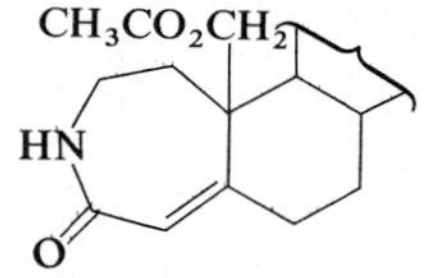 $CH_3CO_2CH_2$, HN, O (85)	40
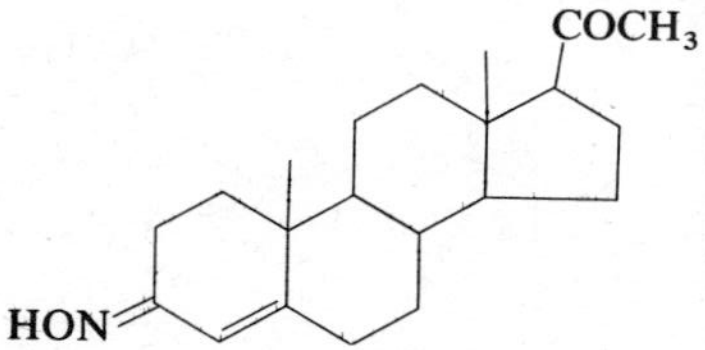 HON (steroid, $COCH_3$)	$SOCl_2$, dioxane, 1 h	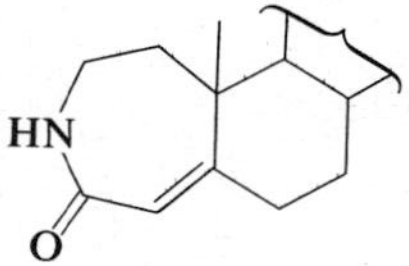 HN, O (58)	489
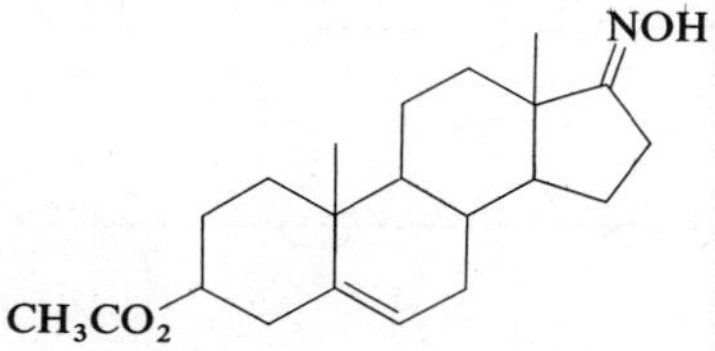 CH_3CO_2 (steroid, NOH)	*p*-$CH_3CONHC_6H_4SO_2Cl$, pyridine	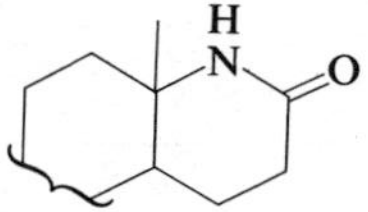 H, N, O (—)	490

TABLE I. Rearrangements (*Continued*)

No. of Carbon Atoms	Substrate	Reagent and Conditions	Product(s) and Yield(s) (%)	Refs.
	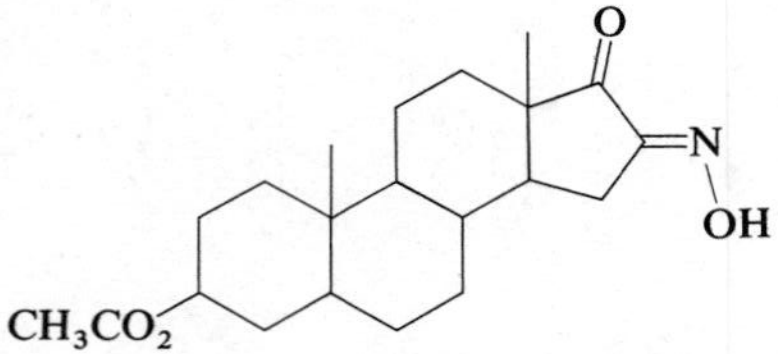 O; =N–OH; CH_3CO_2	1. $(CH_3CO)_2O$ 2. Ethanol, reflux	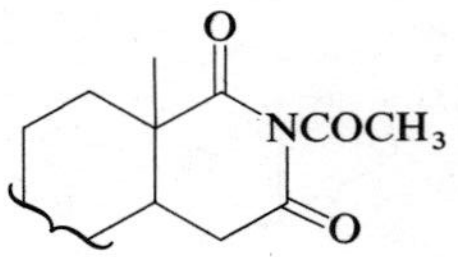 O; $NCOCH_3$; O (—)	491
	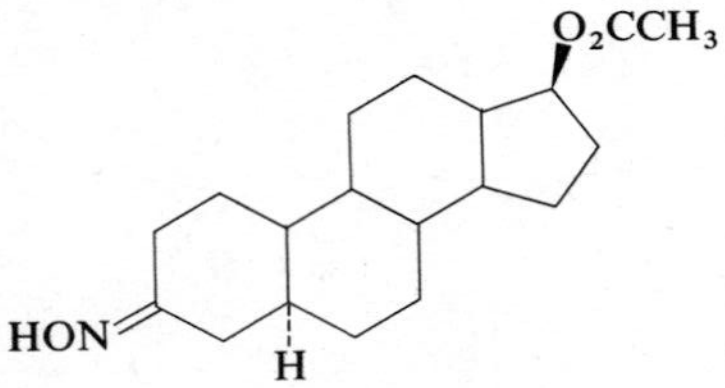 O_2CCH_3; HON=; H	$SOCl_2$, ether, $-20°$	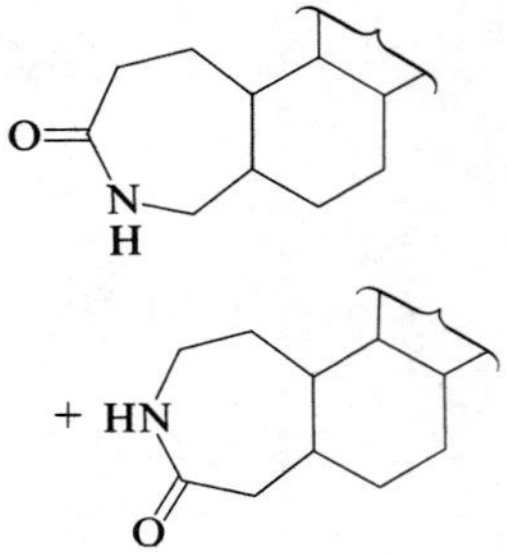 O=; N–H + HN; O (—)	475
	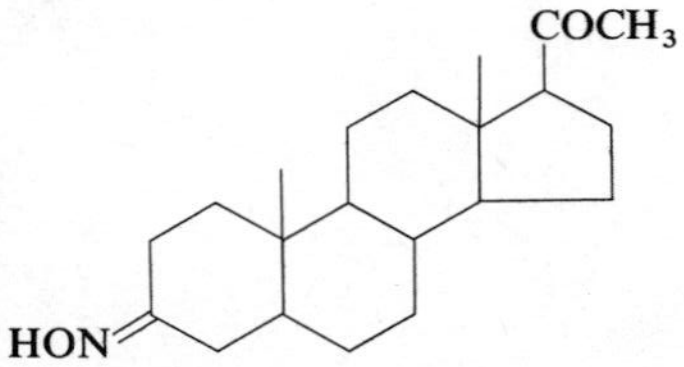 $COCH_3$; HON=	$C_6H_5SO_2Cl$, pyridine, 30 min	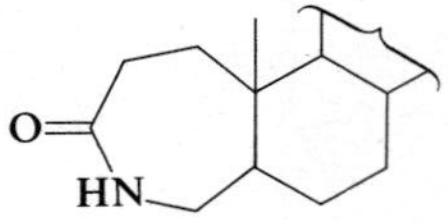 O=; HN (93)	492

	Reactant	Conditions	Product(s) and Yield(s) (%)		Refs.
	C_2H_5; HON	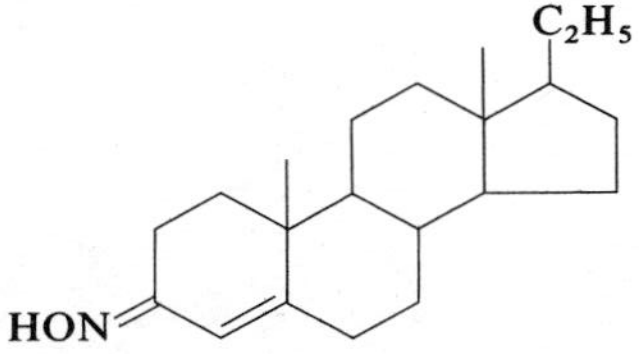 1. $SOCl_2$, 10° 2. KOH, H_2O, 90°	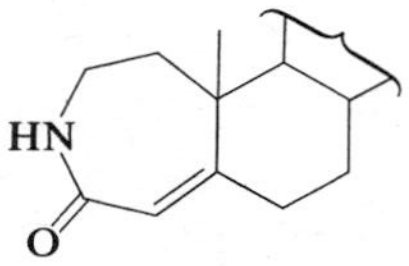 HN; O	(35)	489
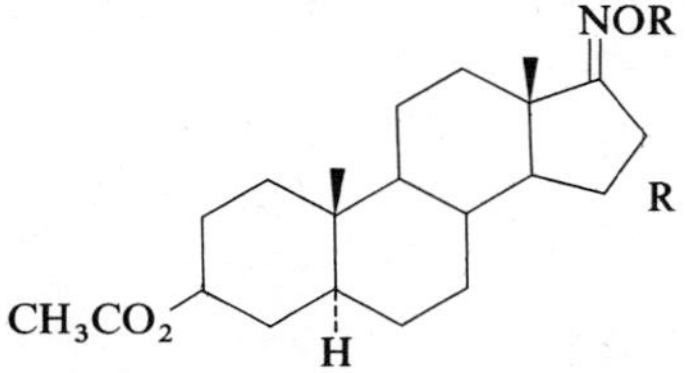	NOR; CH_3CO_2; H; R = H	TsCl, pyridine, reflux, 1.75 h	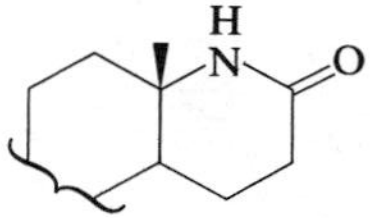 H N; O	(70)	163, 493
	R = H	DCC, DMSO, CF_3CO_2H, 3 d	"	(40)	163
	R = H	HN_3	"	(48)	469
	R = $COCH_3$	BF_3·ether, dioxane, 10–15 min	"	(40)	494
	ICH_2; N-OTs	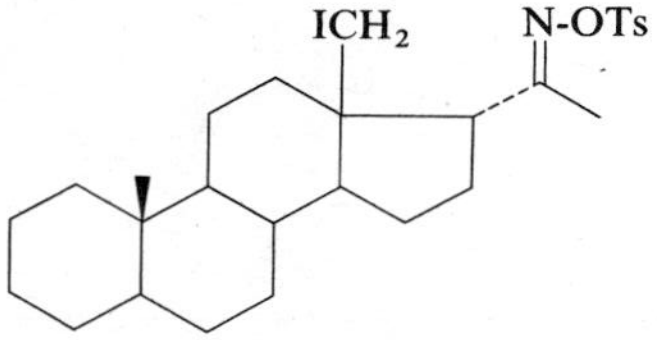 K_2CO_3, 90°	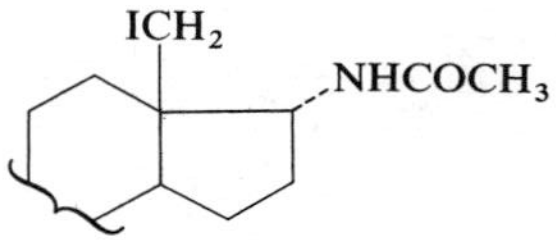 ICH_2; $NHCOCH_3$	(—)	495
C_{22}	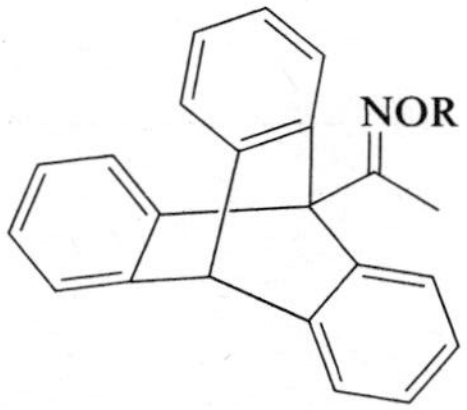 NOR; R = H	PPA, 130–140°	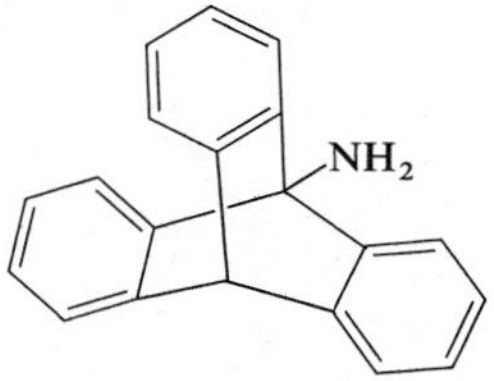 NH_2	(35)	496
	R = Ts	$(C_2H_5)_3N$, 80% ethanol, reflux	"	(73)	496

TABLE I. REARRANGEMENTS (*Continued*)

No. of Carbon Atoms	Substrate	Reagent and Conditions	Product(s) and Yield(s) (%)	Refs.
	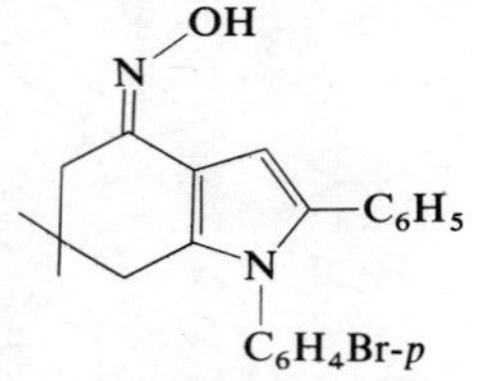 N–OH; C_6H_5; N–C_6H_4Br-*p*	PPA, 150°, 20 min	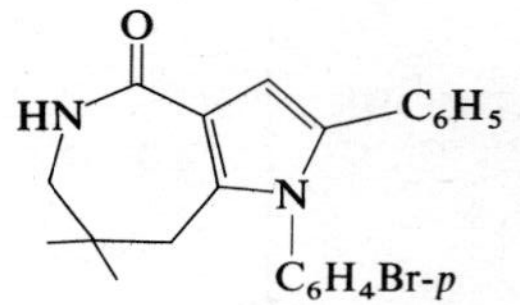 O; HN; C_6H_5; N–C_6H_4Br-*p* (73)	478
	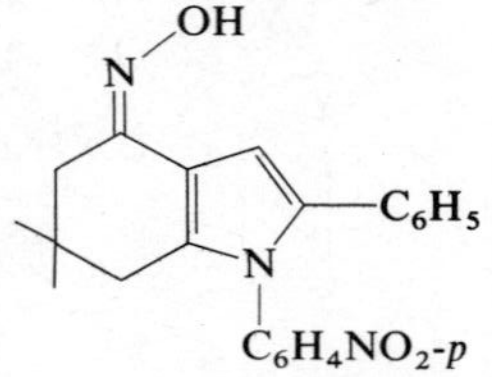 N–OH; C_6H_5; N–$C_6H_4NO_2$-*p*	"	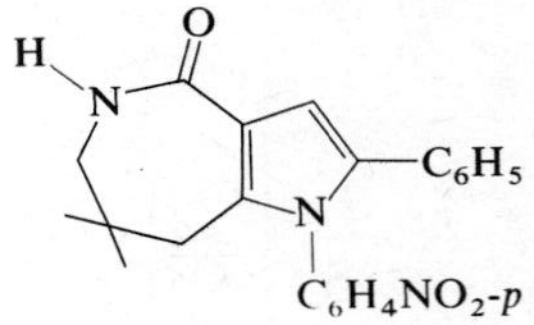 O; H–N; C_6H_5; N–$C_6H_4NO_2$-*p* (40)	478
	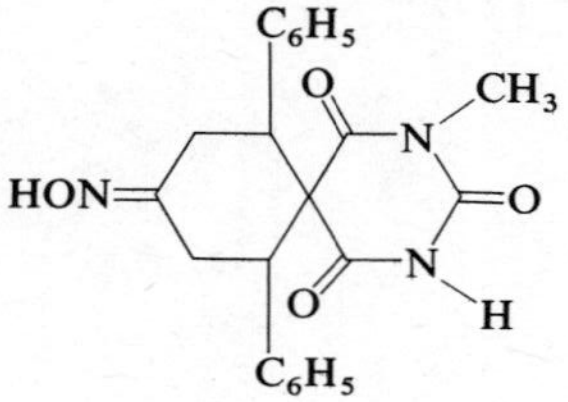 HON=; C_6H_5; O; N–CH_3; =O; O; N–H; C_6H_5	PCl_5, THF, 12 h	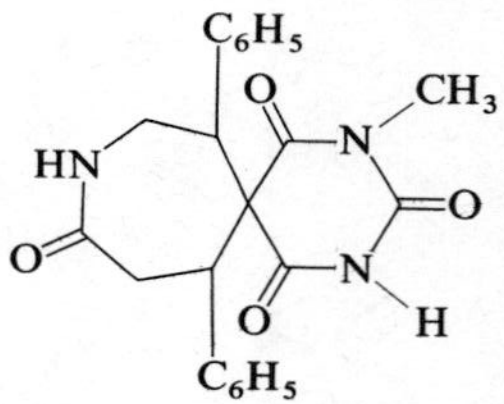 C_6H_5; HN; O; O; N–CH_3; =O; O; N–H; C_6H_5 (57)	497

Reactant	Conditions	Product(s) and Yield(s) (%)	Refs.
	PCl_5, ether, 12 h	" (46)	497
NOH; C_6H_5; N; C_6H_5	PPA, 150°, 20 min	O; HN; C_6H_5; N; C_6H_5 (67)	478
NOH; $CO_2C_2H_5$; $CO_2C_2H_5$	PCl_5, CCl_4, 2 h	O; HN; $CO_2C_2H_5$; $CO_2C_2H_5$ (49)	142
TsO; N; C_6H_5; N; N; $CH_2C_6H_5$	NaO_2CCH_3, ethanol, H_2O	O; H; N; C_6H_5; N; N; $CH_2C_6H_5$ (10)	411
HO; N; C_6H_5; C_6H_5; O; C_4H_9-*t*	PCl_5, ether, 0°	C_6H_5; C_6H_5; O; $CH_2CONHC_4H_9$-*t* (26)	161

TABLE I. REARRANGEMENTS (*Continued*)

No. of Carbon Atoms	Substrate	Reagent and Conditions	Product(s) and Yield(s) (%)	Refs.
	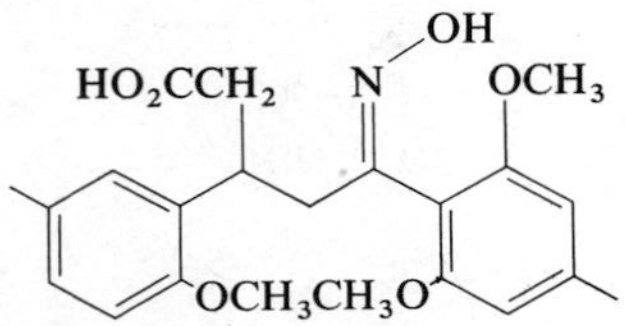 HO_2CCH_2, N–OH, OCH_3, OCH_3CH_3O	PPA, 100°, 8 h	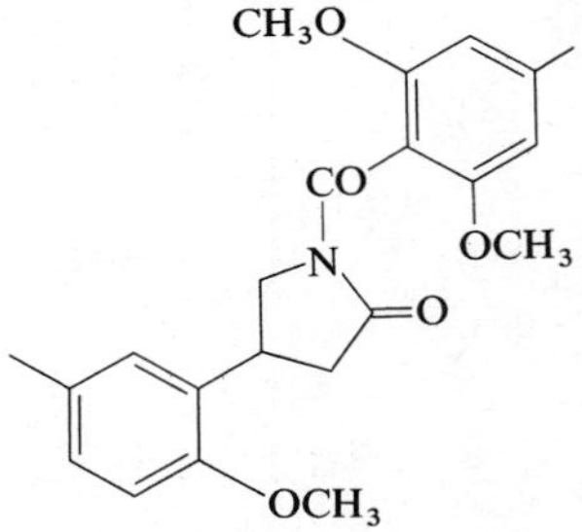 CH_3O, CO, N, OCH_3, O, OCH_3 (60)	108
	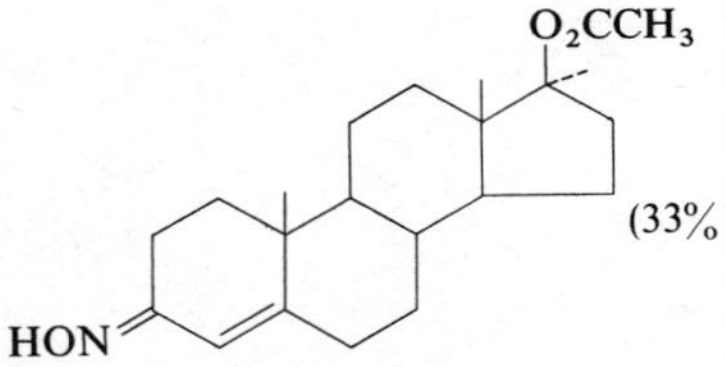 O_2CCH_3 (33% *Z*) HON (*E*)	TsCl, DMF	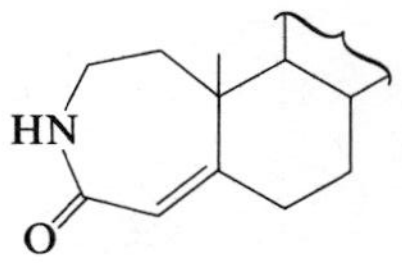 HN, O (33)	57, 392
		$SOCl_2$, dioxane, 10°, 1 h	" (59)	43
	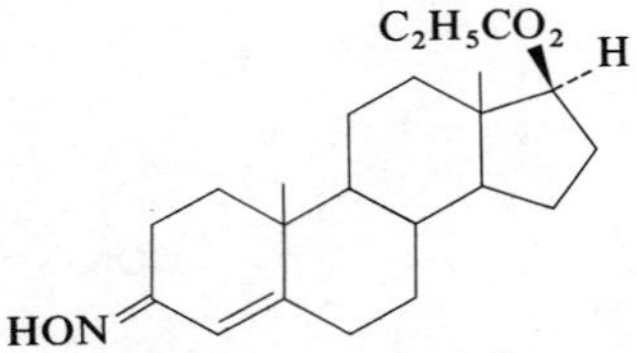 $C_2H_5CO_2$ H HON	$SOCl_2$, dioxane, 1 h	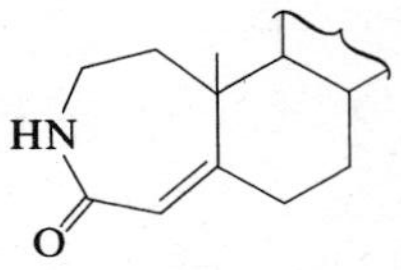 HN, O (50)	43

	Reactant	Conditions	Product(s) and Yield(s) (%)	Refs.
	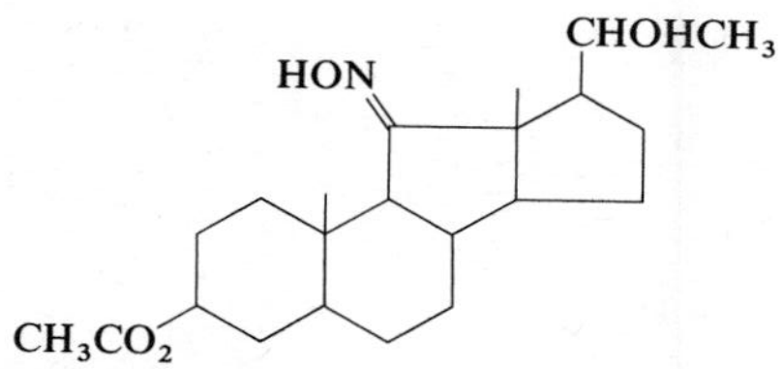 HON, CHOHCH$_3$, CH$_3$CO$_2$	TsCl, pyridine, 0°, 24 h	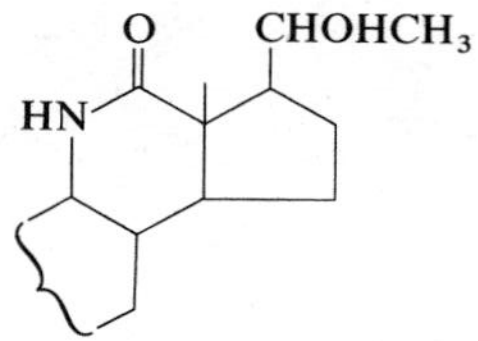 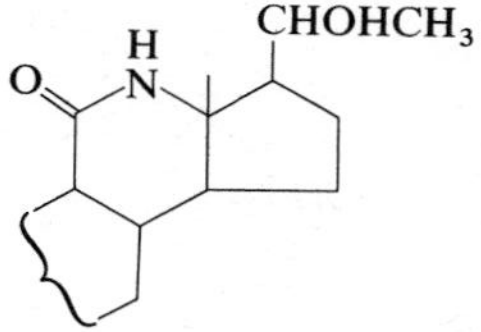 O, CHOHCH$_3$, HN + H, O, N, CHOHCH$_3$ (14)	498
C_{23}	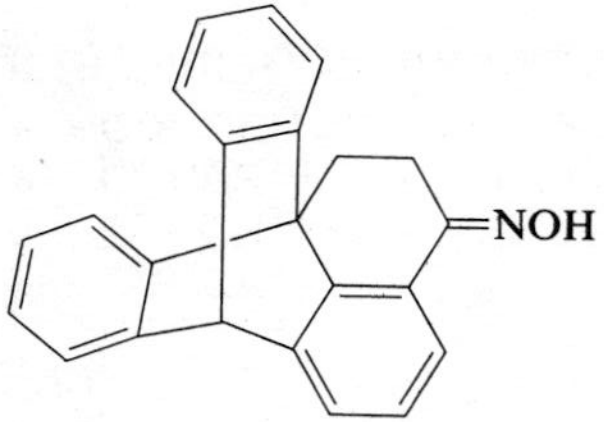 =NOH	PPA, 130°	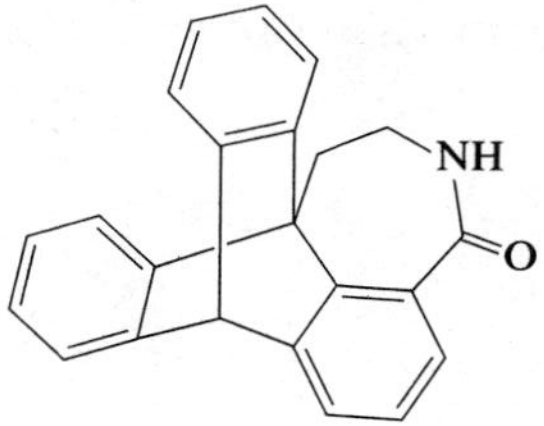 NH, O (90)	496
		$(C_6H_5)_3P$, Cl_2, C_6H_6	" (90)	496
	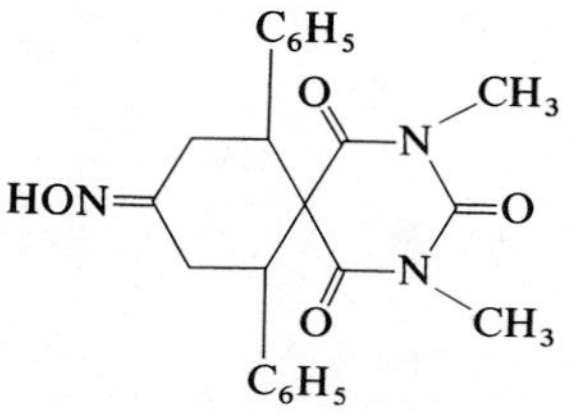 HON=, C_6H_5, O, N, CH$_3$, O, N, CH$_3$, O, C_6H_5	PCl_5, THF, 12 h	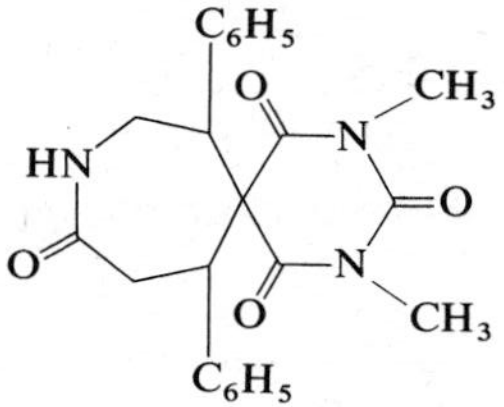 HN, O, C_6H_5, O, N, CH$_3$, O, N, CH$_3$, O, C_6H_5 (44)	497

TABLE I. REARRANGEMENTS (*Continued*)

No. of Carbon Atoms	Substrate	Reagent and Conditions	Product(s) and Yield(s) (%)	Refs.
	[structure: oxime, substituents OH, N, C_6H_5, N–R]	PPA, 150°, 20 min	[structure: HN, O, C_6H_5, N–R]	
	$R = C_6H_4CH_3$-*o*		(80)	478
	$R = C_6H_4CH_3$-*m*		(73)	478
	$R = C_6H_4CH_3$-*p*		(90)	478
	$R = C_6H_4OCH_3$-*p*		(85)	478
	$(C_6H_5)_3Sn(CH_2)_3C({=}NOH)CH_3$	TsCl, pyridine, 0°, 14 h	$(C_6H_5)_3Sn(CH_2)_3NHCOCH_3$ + $(C_6H_5)_3Sn(CH_2)_3CONHCH_3$ (81)	104, 499
	[structure: steroid, substituents =NOH, O, CH_3CO_2]	*p*-$CH_3CONHC_6H_4SO_2Cl$, pyridine, 10°, 2 h	[structure: substituents $NHCOCH_3$, O] (—)	500
	[structure: steroid, substituents NOH, C_5H_9O]	*p*-$CH_3CONHC_6H_4SO_2Cl$, pyridine, 24 h	[structure: lactam, substituents H, N, O] (—)	457, 458

Reactant	Conditions	Product(s) and Yield(s) (%)	Refs.
CO_2CH_3, N, RO, $R = H$	$SOCl_2$	" (—)	458
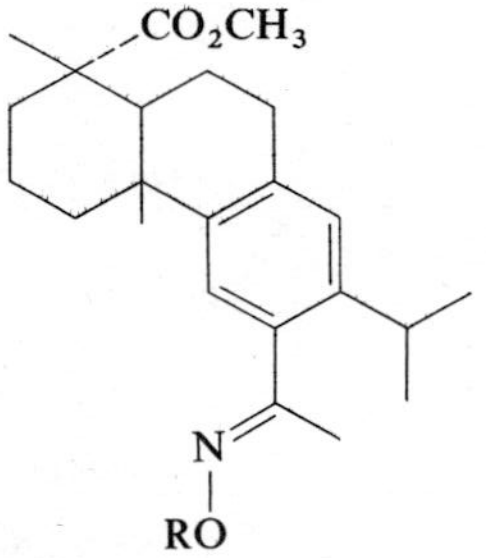	HCl, CH_3CO_2H, 2 h	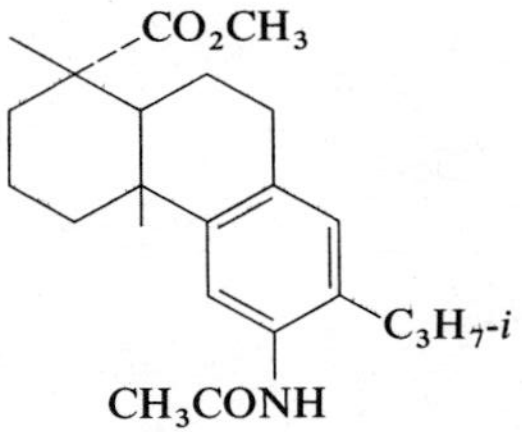 CO_2CH_3, C_3H_7-*i*, CH_3CONH (97)	501
$R = COCH_3$	BF_3·ether, dioxane, 35 min	" (65)	501
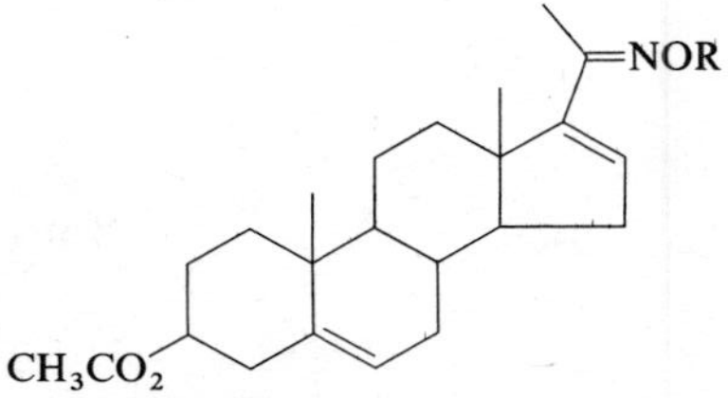 $=NOR$, CH_3CO_2, R = H or CH_3CO—	BF_3·ether, $(CH_3CO)_2O$	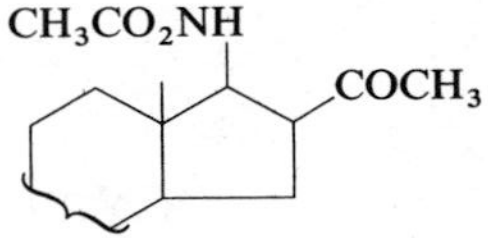 CH_3CO_2NH, $COCH_3$ (72)	502
$R = CH_3CO$—	BF_3·ether, C_6H_6, 3 h	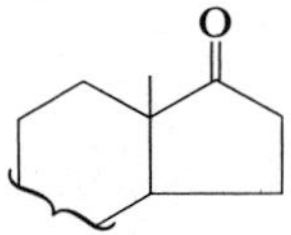 O (26)	502
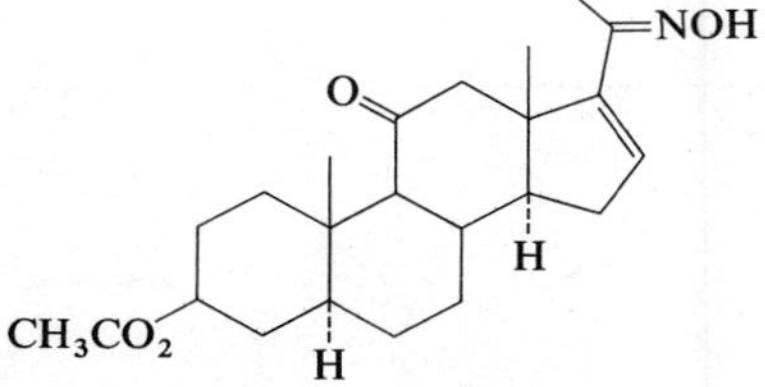 $=NOH$, O, H, CH_3CO_2, H	(—)	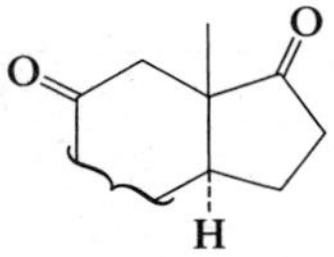 O, O, H (65)	503

TABLE I. REARRANGEMENTS (*Continued*)

No. of Carbon Atoms	Substrate	Reagent and Conditions	Product(s) and Yield(s) (%)	Refs.
	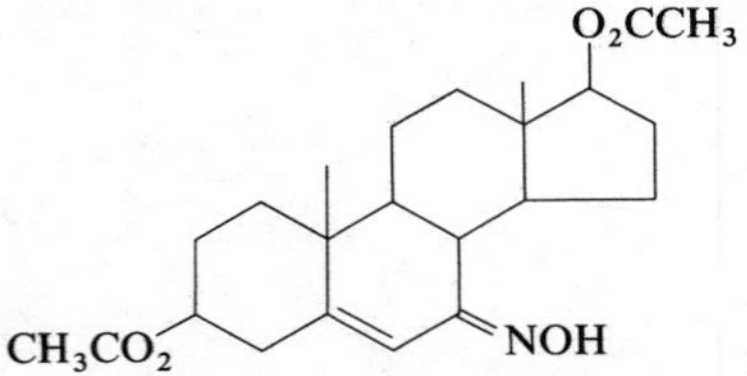 O_2CCH_3, CH_3CO_2, NOH	PPA, 120–130°, 30 min	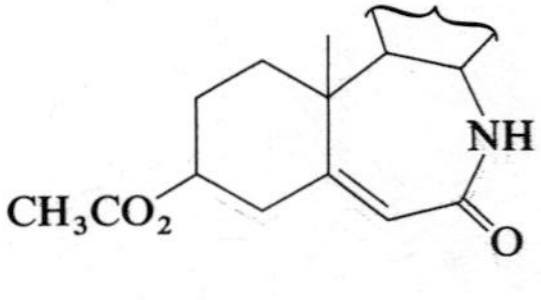 CH_3CO_2, NH, O (—)	473
	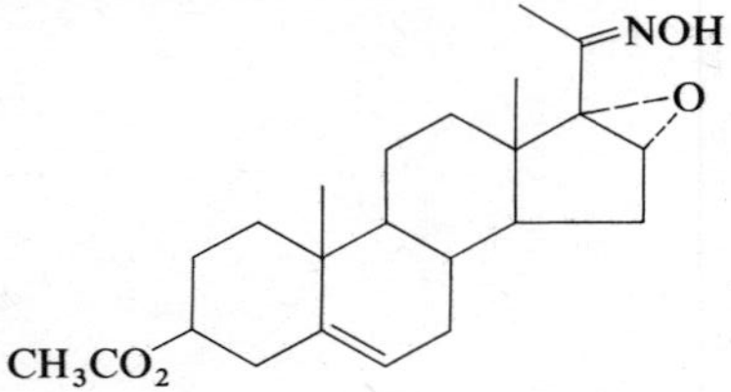 NOH, O, CH_3CO_2	$POCl_3$, pyridine, −10°, 2 h	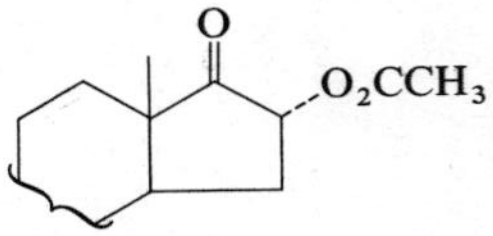 O, O_2CCH_3 (44)	504
	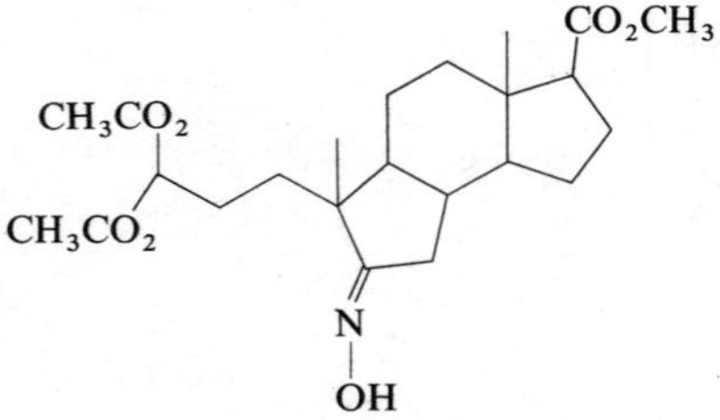 CO_2CH_3, CH_3CO_2, CH_3CO_2, N–OH	$SOCl_2$, ether, −15°, 20 min	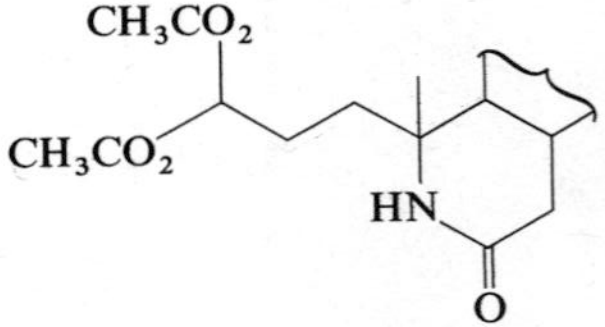 CH_3CO_2, CH_3CO_2, HN, O (—)	505

Reactant	Conditions	Product (Yield, %)	Refs.
$CH_3CO_2CH_2$, HON, H (Z)	TsCl, pyridine, 37°, 3 h	$CH_3CO_2CH_2$, HN, O, H (85)	506
(E)	"	$CH_3CO_2CH_2$, O, N, H, H (87)	506
O_2CCH_3, CH_3CO_2, H, NOH	PPA	NH, O (—)	473
O_2CCH_3, CH_3CO_2, H, NOH (2)	$SOCl_2$, dioxane, 10–25°, 25 min	CH_3CO_2, H, NH, O (—)	473

TABLE I. REARRANGEMENTS (*Continued*)

No. of Carbon Atoms	Substrate	Reagent and Conditions	Product(s) and Yield(s) (%)	Refs.
	(*E*)	$SOCl_2$, −15°, 40 min	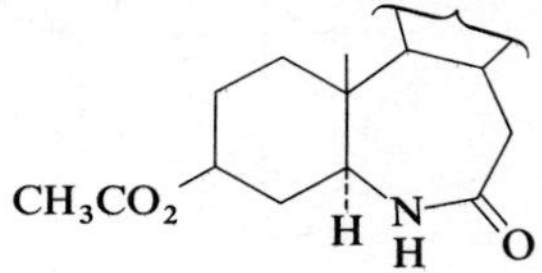 CH_3CO_2 H N H O (73)	477
	(*E*)	$SOCl_2$, dioxane, 25 min	" (60)	473, 477
	(*E*)	$SOCl_2$, acetone, −78°	" (73)	473
	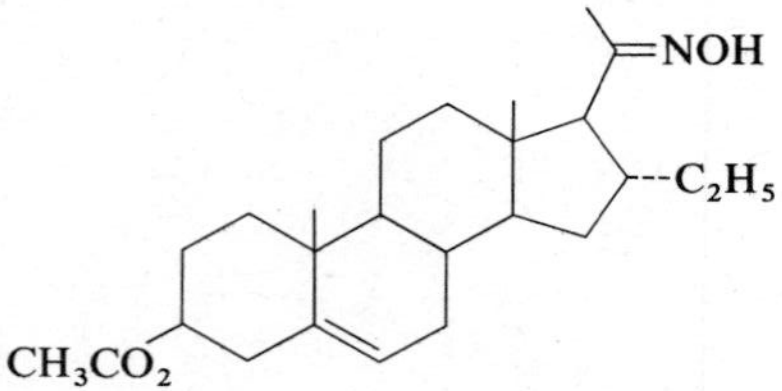 =NOH C_2H_5 CH_3CO_2	TsCl, pyridine, 3 h	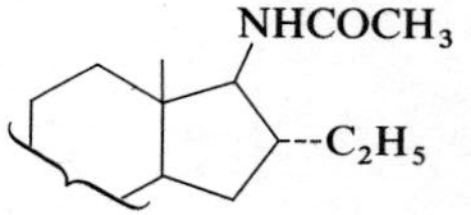 $NHCOCH_3$ C_2H_5 (68)	502
C_{24}	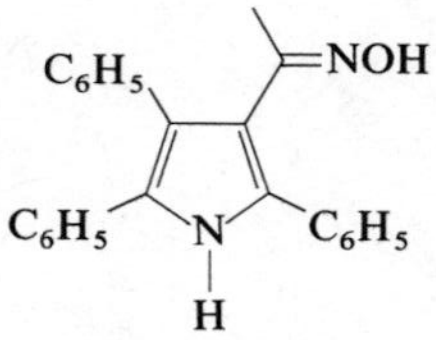 C_6H_5 =NOH C_6H_5 N C_6H_5 H	HCl, ethanol, reflux, 2 h	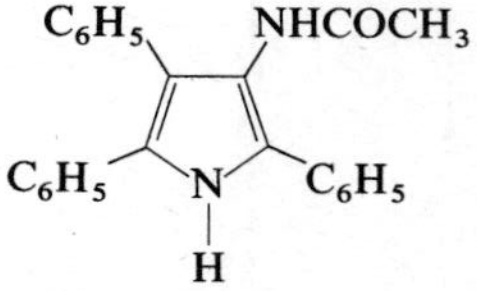 C_6H_5 $NHCOCH_3$ C_6H_5 N C_6H_5 H (—)	464
	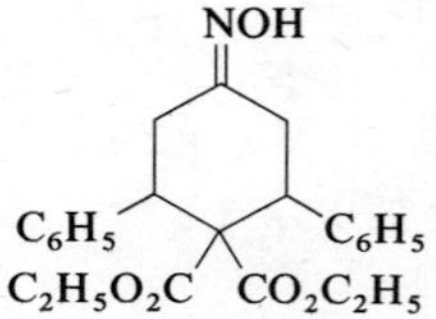 NOH C_6H_5 C_6H_5 $C_2H_5O_2C$ $CO_2C_2H_5$	PPSE, CH_2Cl_2, 12 h	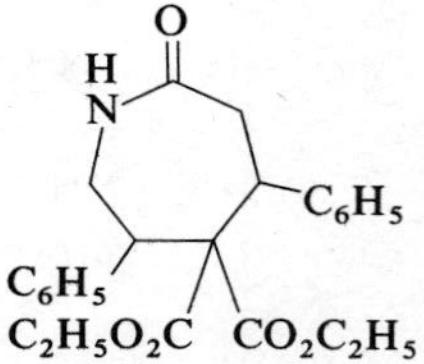 O H N C_6H_5 C_6H_5 $C_2H_5O_2C$ $CO_2C_2H_5$ (63)	72

Reactant	Reagents and Conditions	Product (Yield, %)	Refs.
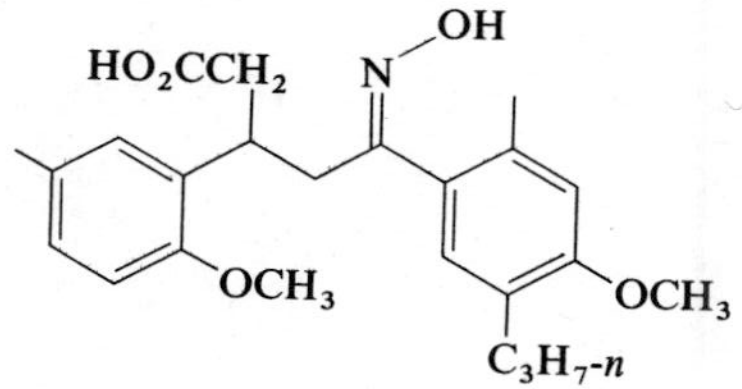	PPA, 100°, 8 h	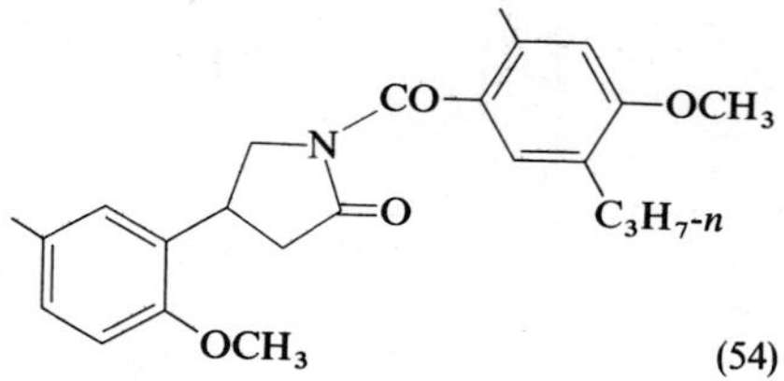 (54)	108
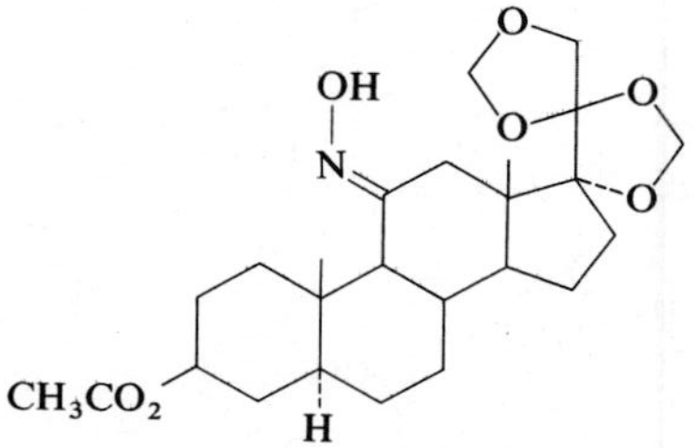	$POCl_3$, pyridine	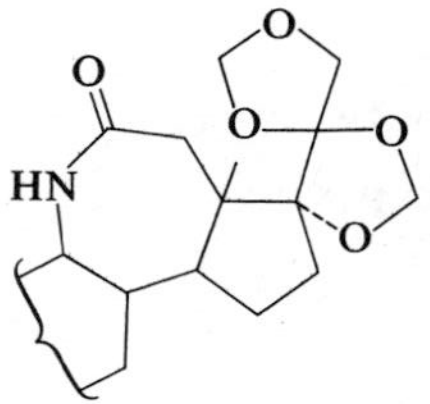 (16)	507
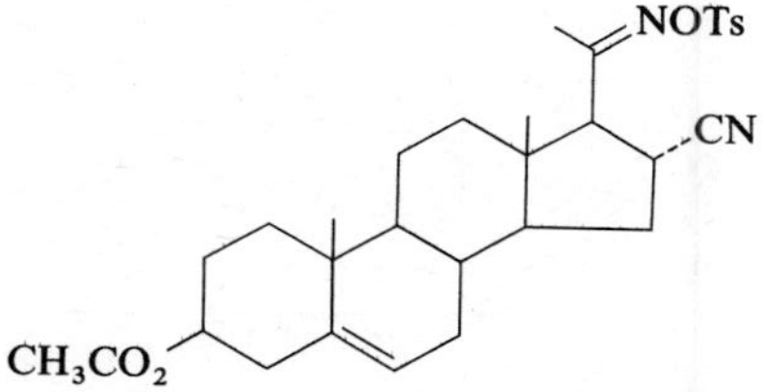	Pyridine, reflux, 10 h	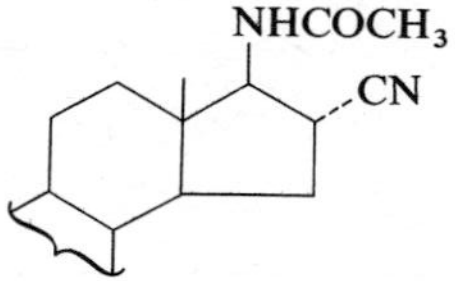 (66)	508
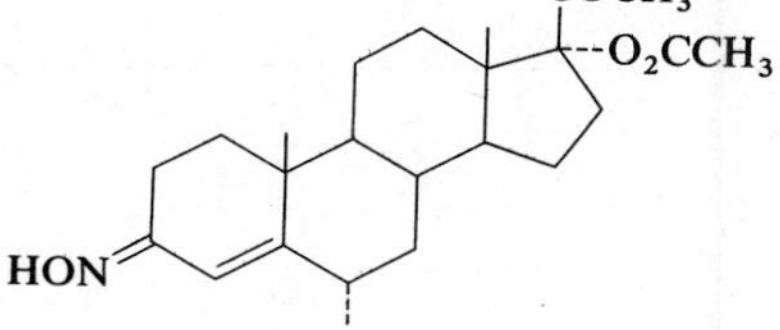	$SOCl_2$, dioxane, 30°	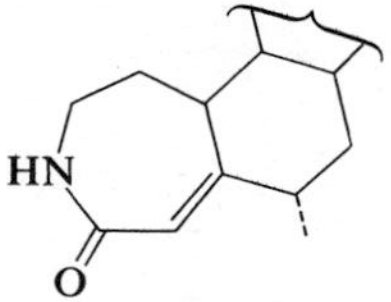 (—)	509

TABLE I. REARRANGEMENTS (*Continued*)

No. of Carbon Atoms	Substrate	Reagent and Conditions	Product(s) and Yield(s) (%)	Refs.
	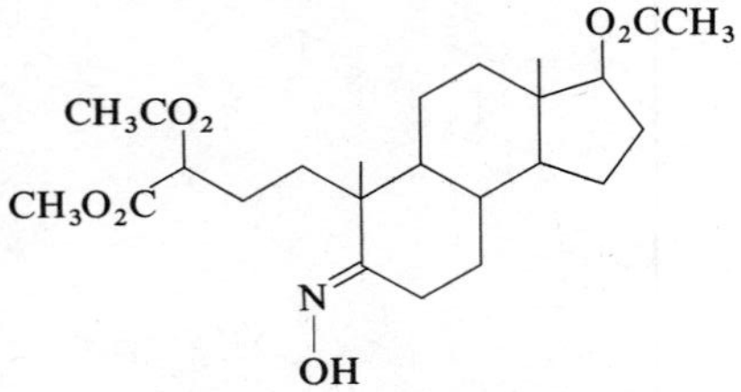 CH_3CO_2, CH_3O_2C, O_2CCH_3, N—OH (structure)	$SOCl_2$, ether, $-10°$, 15 min	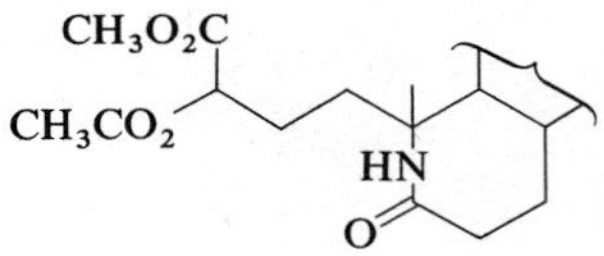 CH_3O_2C, CH_3CO_2, HN, O (structure) (58)	510
	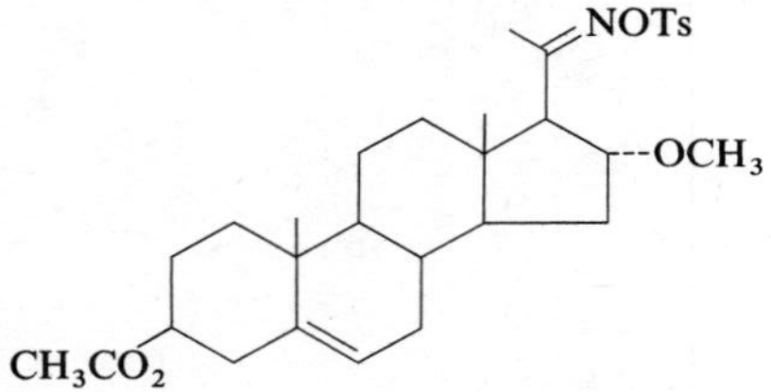 NOTs, OCH_3, CH_3CO_2 (structure)	Pyridine, 90°, 2 h	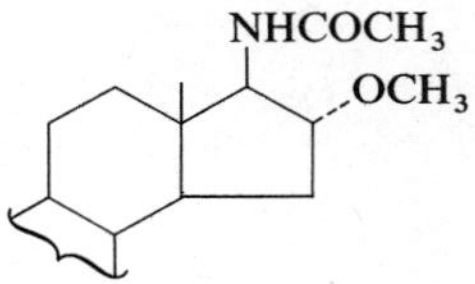 $NHCOCH_3$, OCH_3 (structure) (90)	508
C_{25}	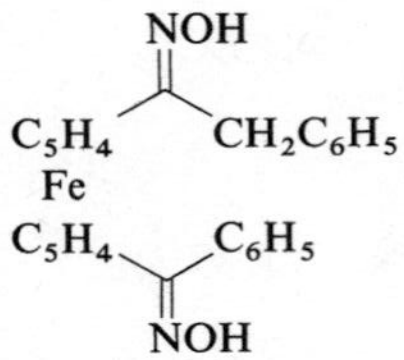 $C_5H_4C(=NOH)CH_2C_6H_5$ Fe $C_5H_4C(=NOH)C_6H_5$	$C_6H_5SO_2Cl$, NaOH	$C_5H_4COCH_2C_6H_5$ Fe $C_5H_4CONHC_6H_5$ (15-20)	99

Reactant	Conditions	Product(s) (Yield, %)		Refs.
C_2H_5, =NOH, C_6H_5, C_6H_5, C_6H_5, N, H	HCl, ethanol, reflux, 2 h	C_6H_5, $NHCOC_2H_5$, C_6H_5, N, C_6H_5, H	(—)	464
OH, HO_2CCH_2, N, $C_6H_4OC_6H_5$-*p*, OCH_3	PPA, 100°, 8 h	$COC_6H_4OC_6H_5$-*p*, N, O, OCH_3	(68)	108
OH, HO_2CCH_2, N, OCH_3, O_2CCH_3, C_3H_7-*n*	"	O, O_2CCH_3, N, O, C_3H_7-*n*, OCH_3	(59)	108
$CH(CH_3)(CH_2)_2CO_2H$, HO, NOH	$SOCl_2$, dioxane	HO, N, H, O	(—)	511

TABLE I. REARRANGEMENTS (*Continued*)

No. of Carbon Atoms	Substrate	Reagent and Conditions	Product(s) and Yield(s) (%)	Refs.
C_{26}	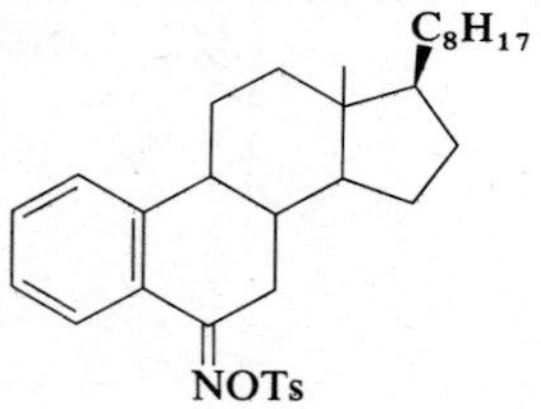 C_8H_{17}; NOTs	Alumina, pet ether, 10 h	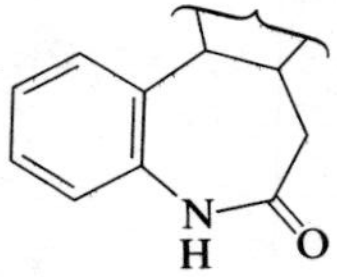 N H; O (100)	512
	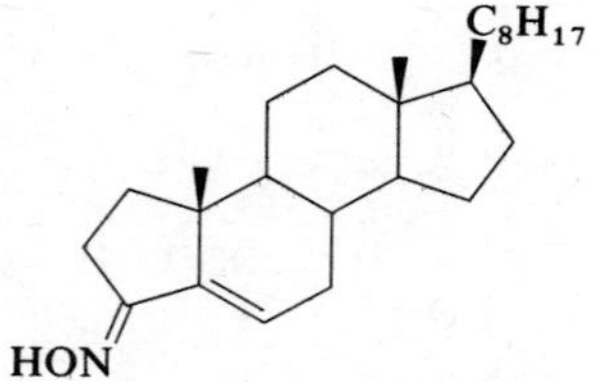 C_8H_{17}; HON	$SOCl_2$, ether, −20°, 5 min	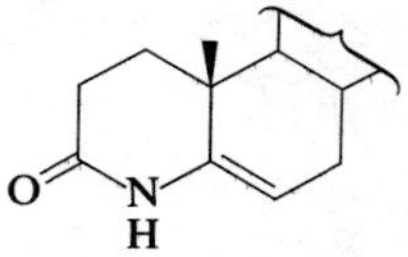 O; N H (12)	513
	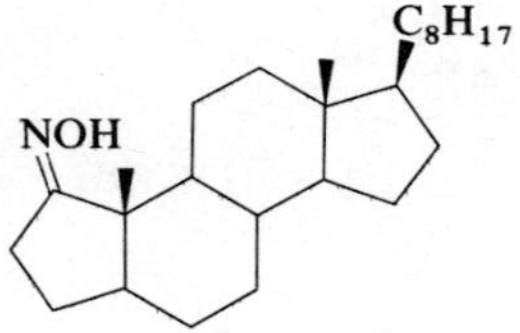 C_8H_{17}; NOH	"	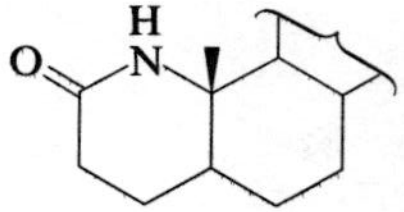 H N; O (35)	164
	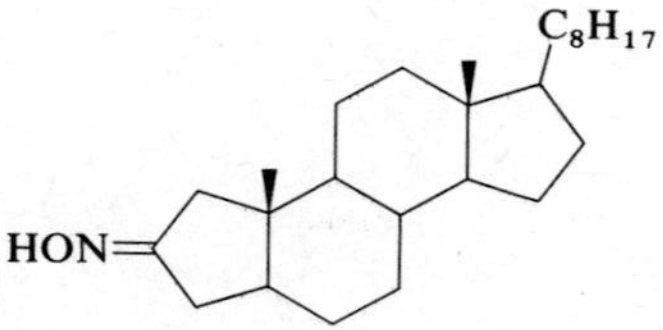 C_8H_{17}; HON	PPA, 120–130°, 40 min	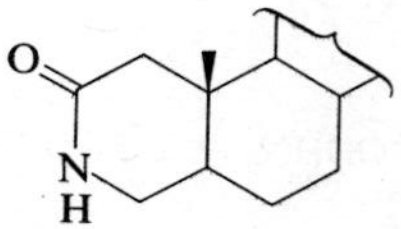 O; N H	514

	Reactant	Conditions	Product(s) and Yield(s) (%)	Refs.
C_{27} (Contd.)	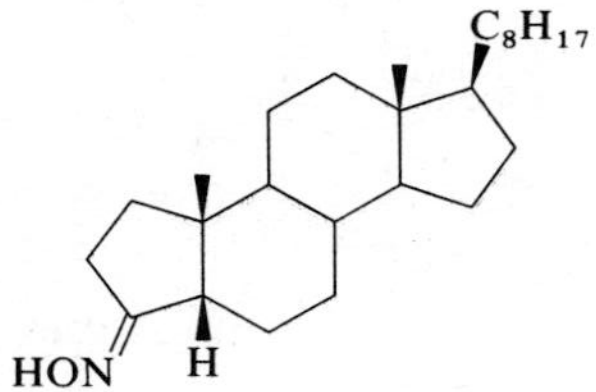 HON, H, C_8H_{17}	$SOCl_2$, ether, $-20°$, 5 min	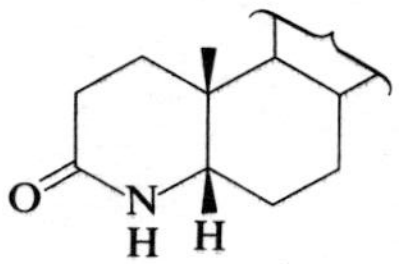 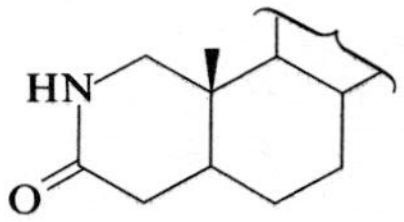 + HN, O (30); O, N, H, H (77)	513
	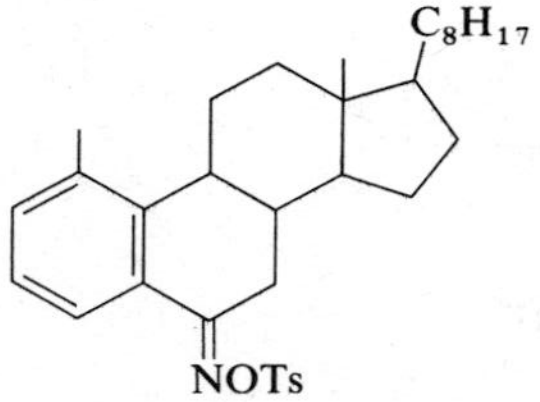 C_8H_{17}, NOTs	Alumina, pet ether, 15 h	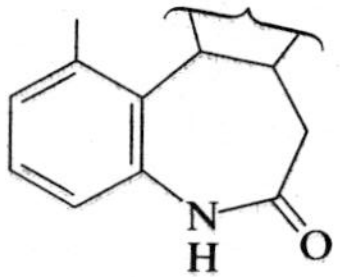 N, H, O (100)	512
	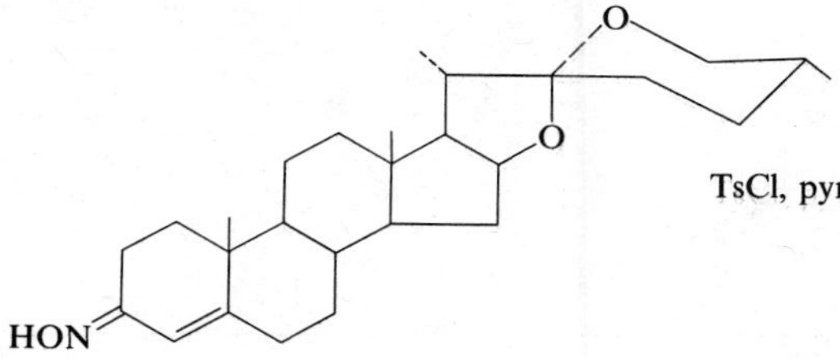 HON, O, O	TsCl, pyridine, 4 h	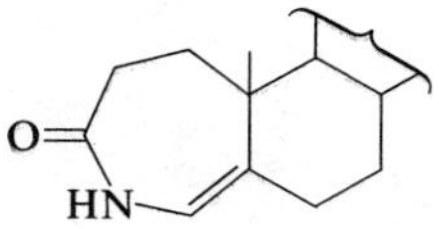 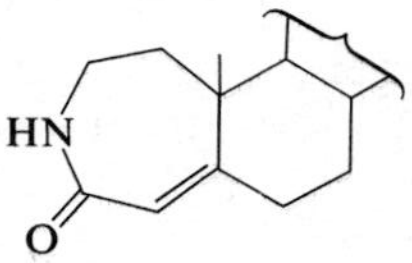 O, HN + HN, O (40)	515

TABLE I. REARRANGEMENTS (*Continued*)

No. of Carbon Atoms	Substrate	Reagent and Conditions	Product(s) and Yield(s) (%)	Refs.
	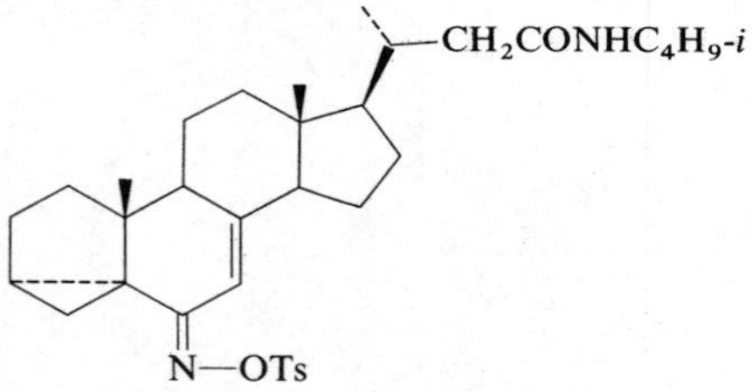 $CH_2CONHC_4H_9$-*i*; N—OTs	TsCl, pyridine, 15 h	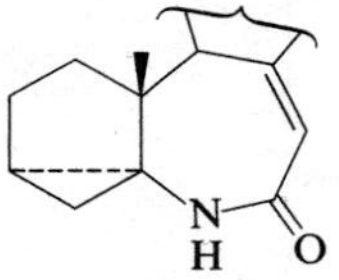 N H O (—)	516
	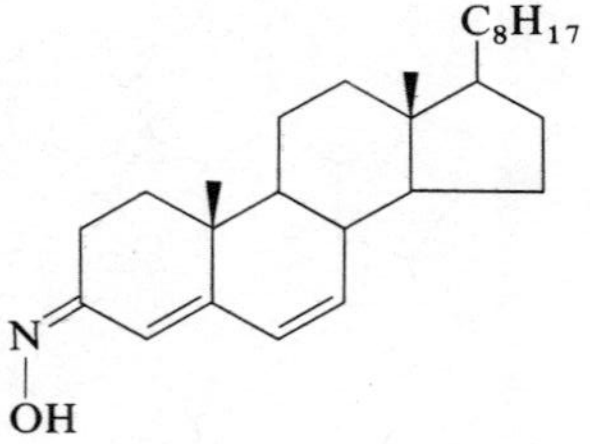 C_8H_{17}; N—OH	TsCl, pyridine, 15°, 15 h	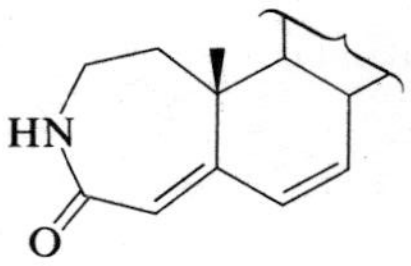 HN O (40)	517
		1. TsCl, pyridine 2. Alumina	" (45)	517
	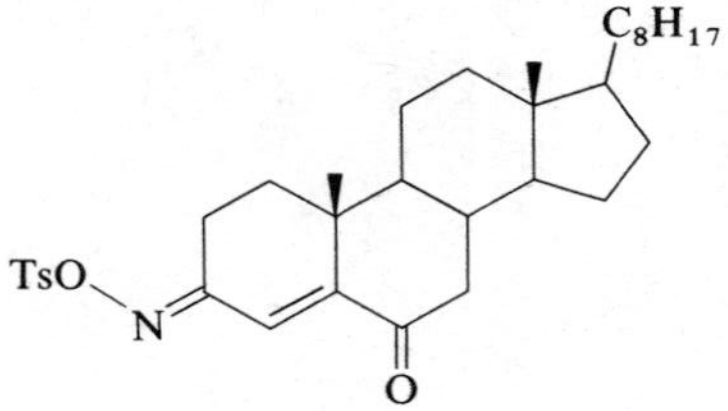 C_8H_{17}; TsO N; O	HCl, CH_3OH	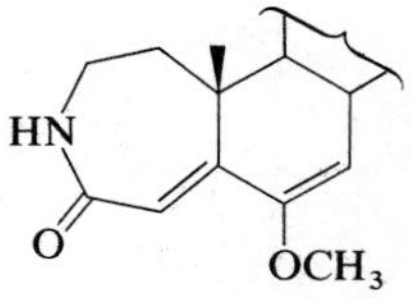 HN O OCH_3 (86)	40

Reactant	Conditions	Product(s) (Yield, %)	Refs.
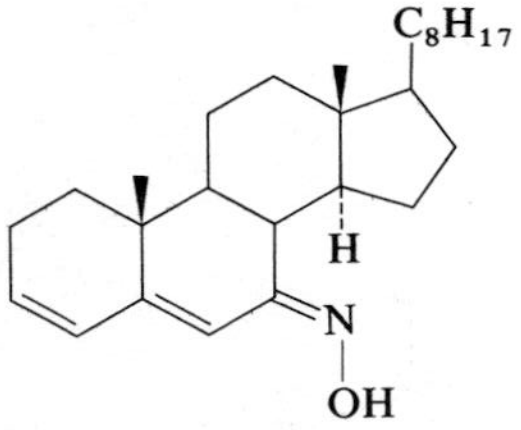 C_8H_{17}, H, N–OH	1. TsCl, pyridine 2. Alumina, 1 h	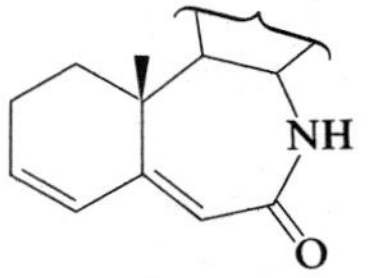 NH, O (78)	518
	$SOCl_2$, $-10°$, 4 *M* KOH	" (24)	518
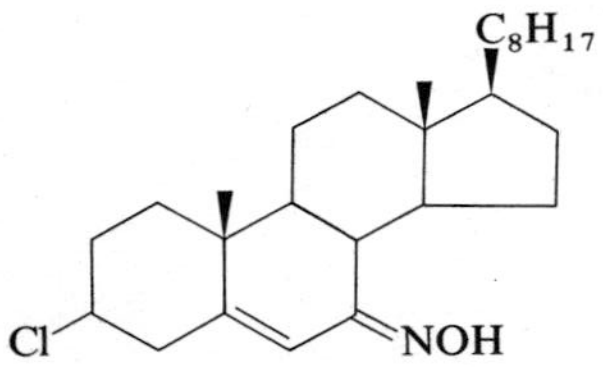 C_8H_{17}, Cl, NOH	1. $SOCl_2$, 0° 2. KOH, 80°	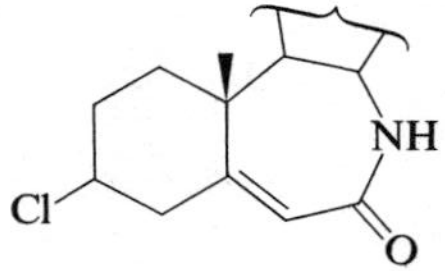 Cl, NH, O (43)	519, 520
	1. TsCl, pyridine 2. Alumina	" (75)	520
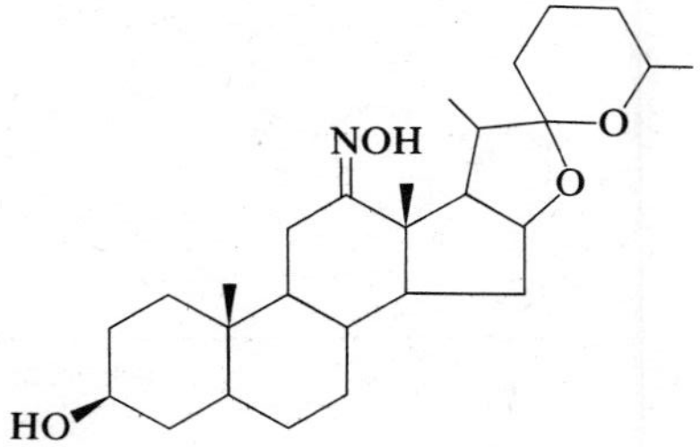 HO, NOH, O, O	(–)	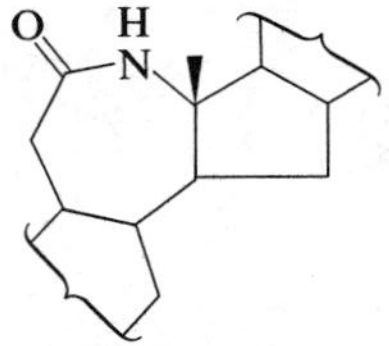 O, H, N (—)	521, 522

TABLE I. REARRANGEMENTS (*Continued*)

No. of Carbon Atoms	Substrate	Reagent and Conditions	Product(s) and Yield(s) (%)	Refs.
	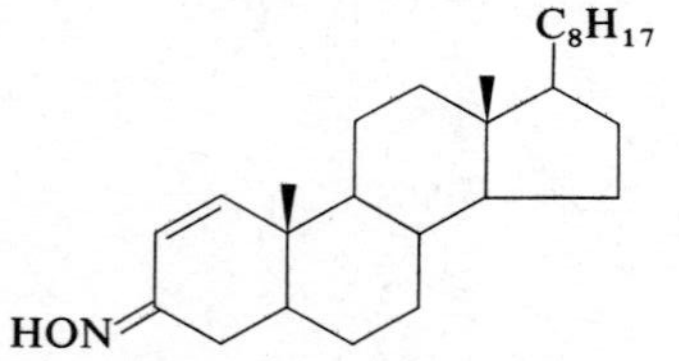 C_8H_{17}, HON (*Z*)	PPA, 125°, 10 min	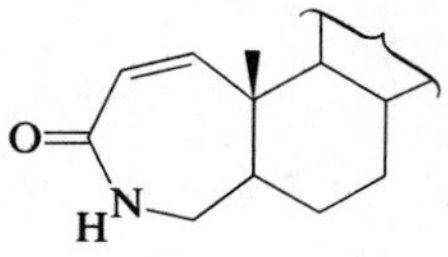 O, HN I (44)	493
	(*Z*)	$SOCl_2$, 10°, 20 min	I (73)	44
	(*E*)	"	I (19) 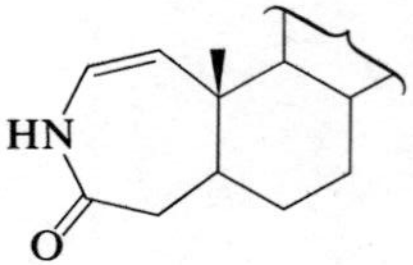 + HN, O II (32)	493
	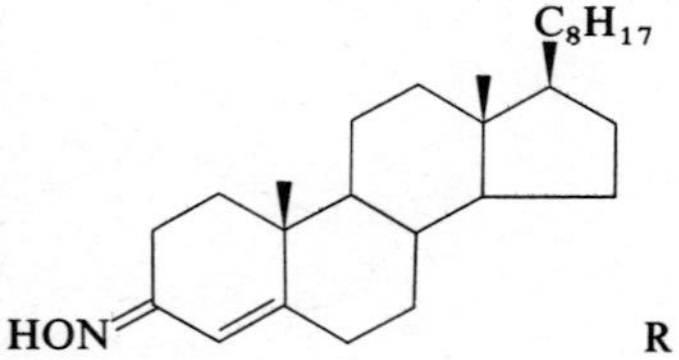 C_8H_{17}, HON, R = H	TsCl	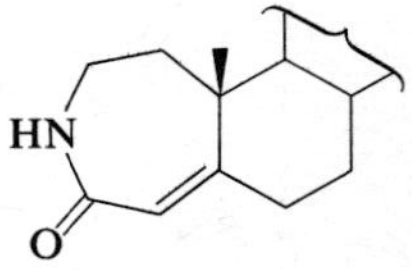 HN, O (60)	294, 392
	R = H	$SOCl_2$, dioxane	" (—)	523
	(*Z*) R = H	$SOCl_2$, ether	" (20)	41, 42
	(*E*) R = H	1. *p*-$CH_3CONHC_6H_4SO_2Cl$ 2. KOH, ethanol, reflux, 1 h	" (11)	42
	(25% *Z*) R = Ts	HCl, CH_3OH, 50°, 30 min	" (87)	40

R = H	PPA, 125–135°, 10 min	" (10) + 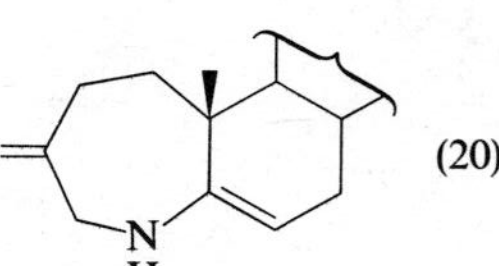 (11)	493
	$SOCl_2$, ether, −20°, 5 min	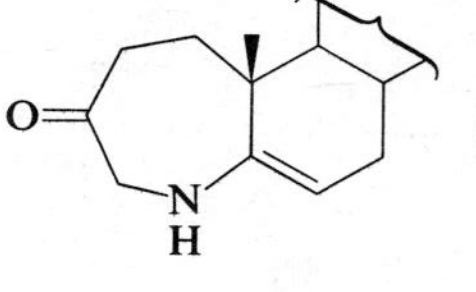 (20)	41
	1. $SOCl_2$, 0° 2. KOH, 90°	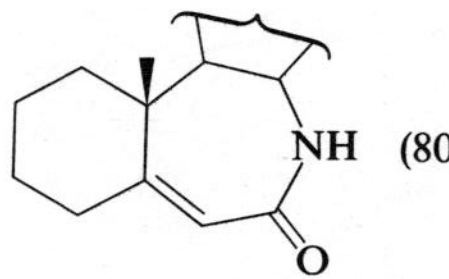 (80)	524
	1. $SOCl_2$, dioxane, 10° 2. $KHCO_3$, H_2O, 20°	" (72)	524
	$SOCl_2$, dioxane, 10–25°, 30 min	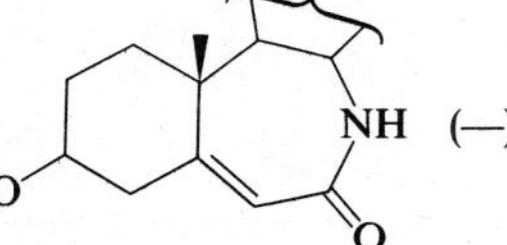 (—)	525

TABLE I. REARRANGEMENTS (*Continued*)

No. of Carbon Atoms	Substrate	Reagent and Conditions	Product(s) and Yield(s) (%)	Refs.
	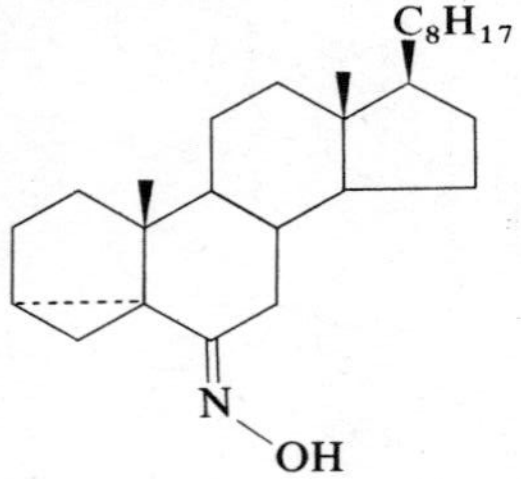 C_8H_{17}; N–OH; R = H	TsCl, pyridine, 15 h	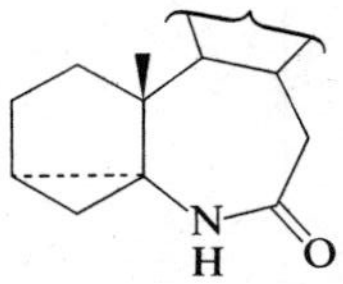 N–H, O (73)	526
	R = H	(—)	" (—)	527
	R = Ts	(—)	" (—)	119
	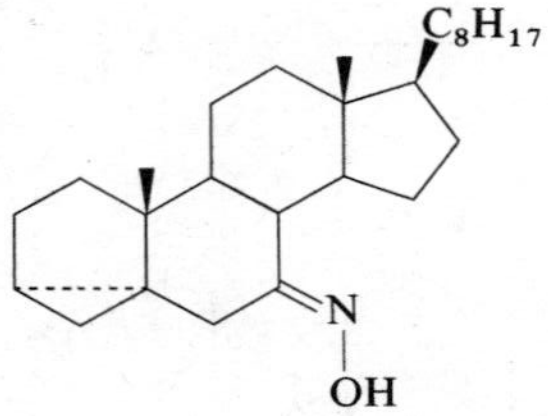 C_8H_{17}; N–OH	$SOCl_2$, dioxane, 10 min	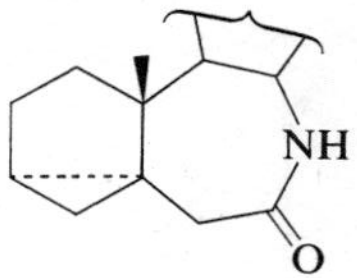 NH, O (66)	528
	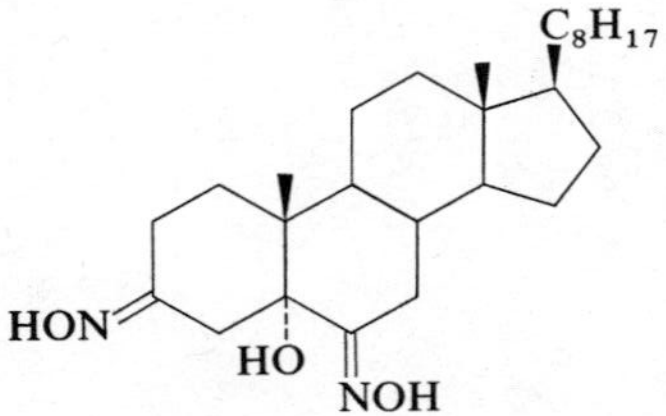 C_8H_{17}; HON; HO; NOH	1. $SOCl_2$, $-20°$ 2. 4 *M* KOH	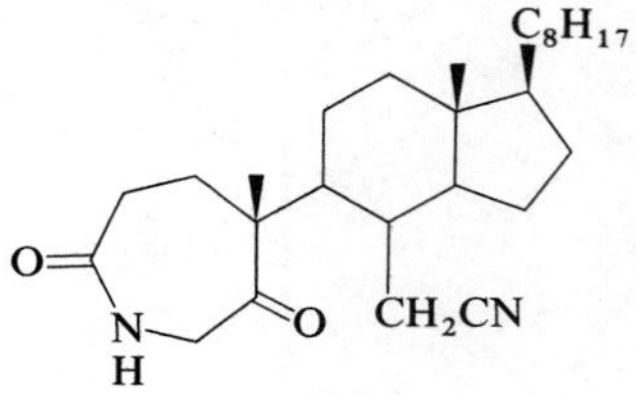 C_8H_{17}; O; N–H; O; CH_2CN (26)	529

Reactant	Conditions	Product(s) and Yield(s) (%)	Refs.
C_8H_{17}; X; NOH; X = Cl	TsCl, pyridine	X; N H; O (50)	530
X = Br	1. $SOCl_2$, 0° 2. 4 *M* KOH	" (72)	526
X = I	"	" (54)	531
X = I	TsCl, pyridine, 15 h	" (50)	531
C_8H_{17}; Cl; H; NOH	TsCl, pyridine	Cl; H; NH; O (65)	520
C_8H_{17}; NOH	1. $SOCl_2$, ether, − 20°, 5 min 2. 4 *M* KOH	O; H N (32)	164
C_8H_{17}; HON	$SOCl_2$, ether, − 20°	O; H N (22) + HN; O (47)	164

TABLE I. REARRANGEMENTS (*Continued*)

No. of Carbon Atoms	Substrate	Reagent and Conditions	Product(s) and Yield(s) (%)	Refs.
	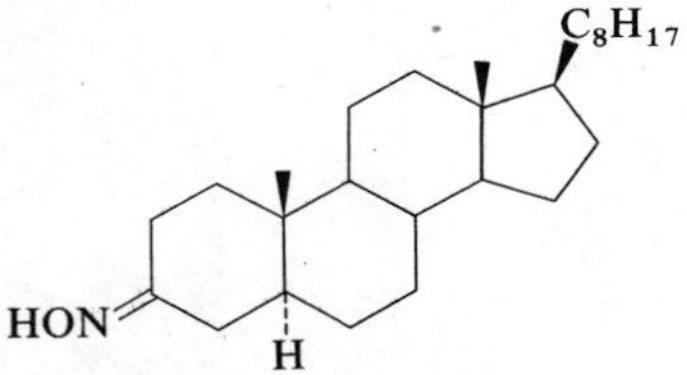 HON= ... H; C_8H_{17}	$SOCl_2$	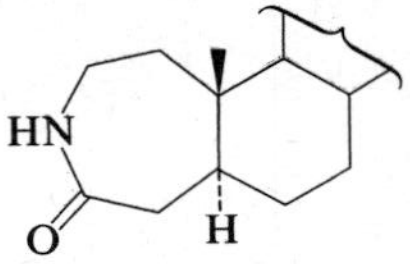 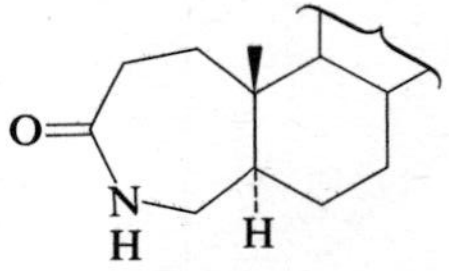 HN, O, H; I (—) + O=, N H, H; II	125
	(*Z*)	$SOCl_2$, dioxane, 40°, 15 min	I (79)	532
		"	I + II (67)	499
		"	II (—)	533
		TsCl, pyridine, 0°	I + II (73)	534
		1. TsCl 2. Alumina	I (85)	77
		$(C_6H_5)_3P$, CCl_4, reflux, 5 h	I (78)	535
		PPA, 130°, 7 h	I (58)	493, 535, 536
	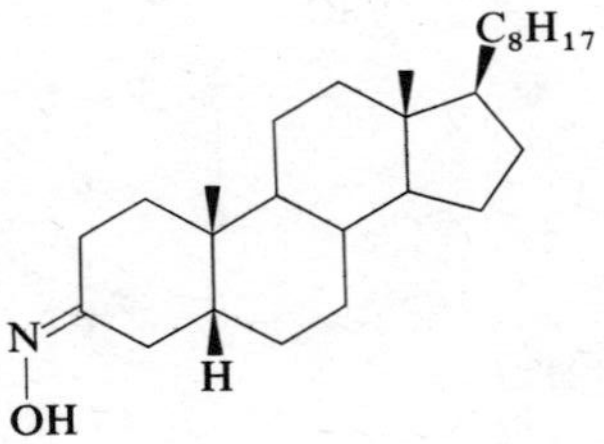 N(—OH)= ... H; C_8H_{17}	$SOCl_2$, dioxane, 40°, 25 min	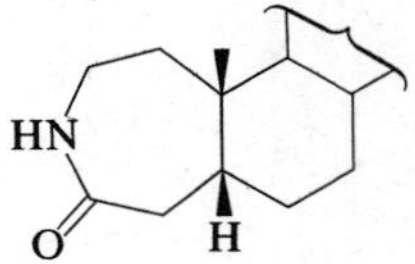 HN, O, H (42)	532

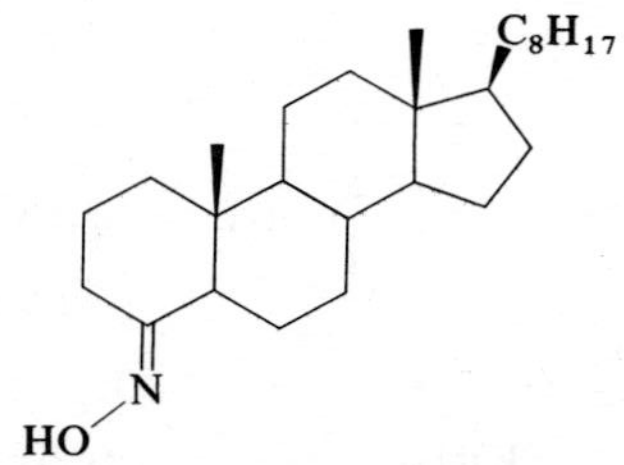	PPA, 120°, 10 min	" (90)	493, 536
	$SOCl_2$, ether, −20°	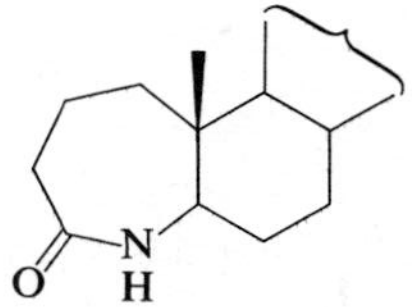 (80)	164
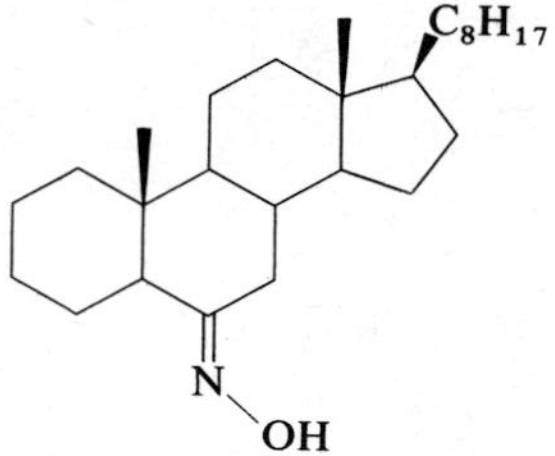	$SOCl_2$, ether, −20°, 10 min	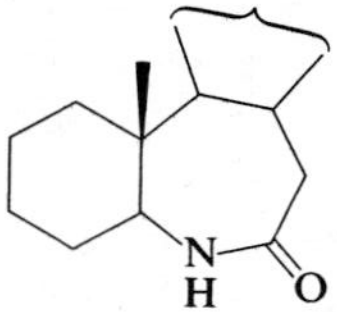 (—)	164
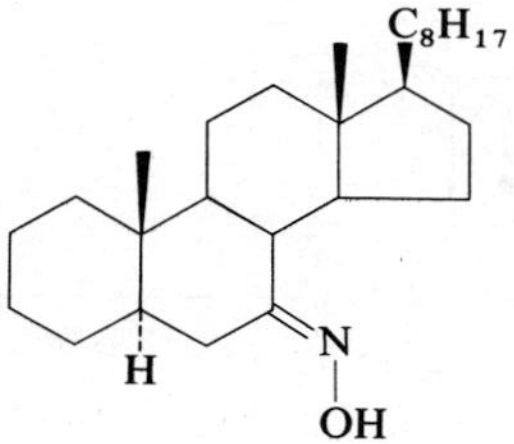	1. $SOCl_2$, 0° 2. KOH, 90°	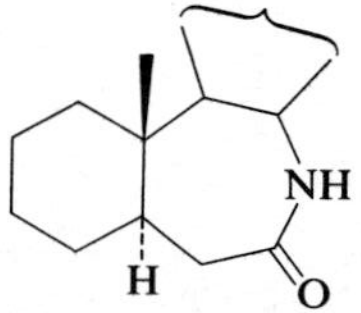 (93)	524

TABLE I. REARRANGEMENTS (*Continued*)

No. of Carbon Atoms	Substrate	Reagent and Conditions	Product(s) and Yield(s) (%)	Refs.
	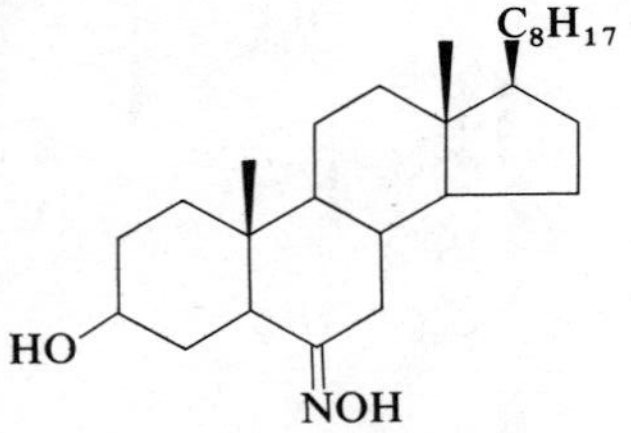 HO, NOH, C_8H_{17}	$SOCl_2$, $-20°$	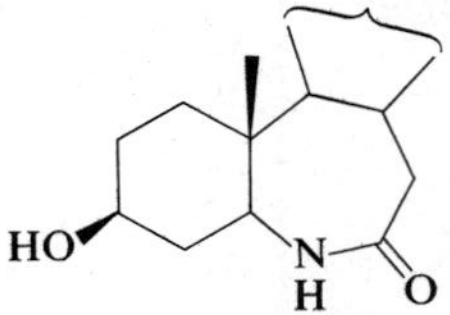 HO, N, H, O (—)	525
	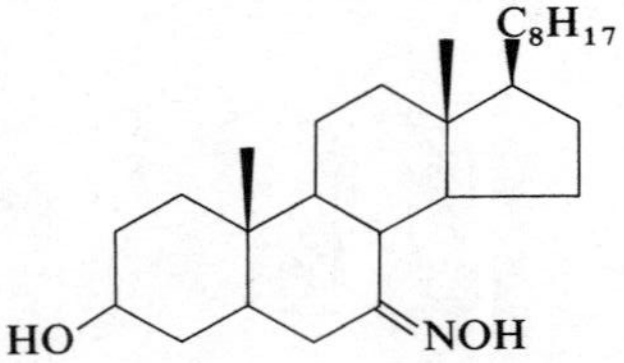 HO, NOH, C_8H_{17}	(–)	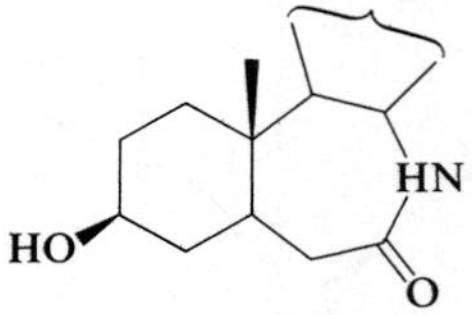 HO, HN, O (—)	525
	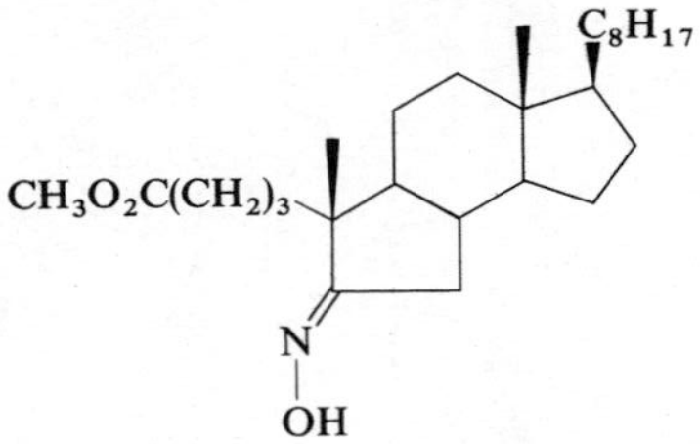 $CH_3O_2C(CH_2)_3$, N, OH, C_8H_{17}	$SOCl_2$, ether, $-10°$	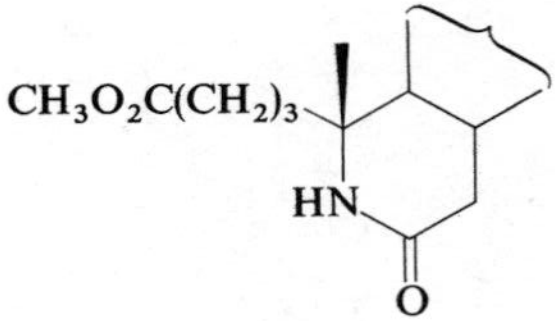 $CH_3O_2C(CH_2)_3$, HN, O (45)	537–539

Reactant	Conditions	Product (Yield %)	Refs.
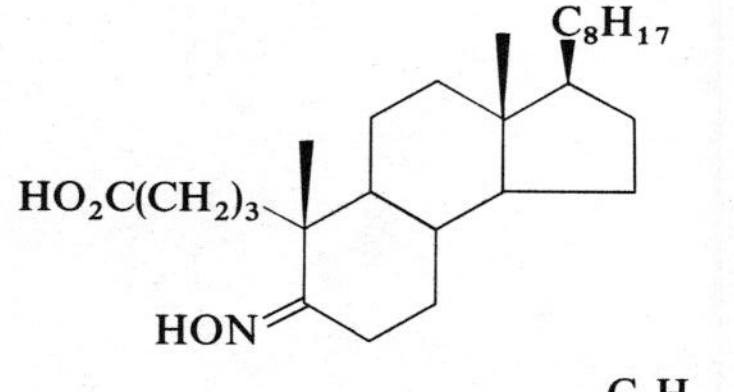 $HO_2C(CH_2)_3$ … C_8H_{17} … HON	$SOCl_2$, $-20°$	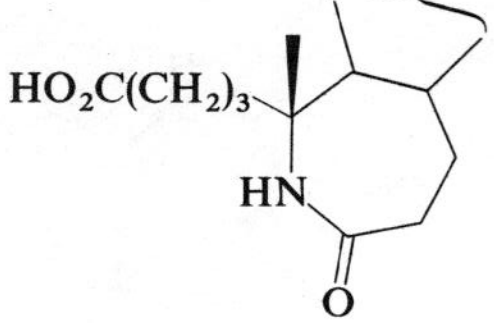 $HO_2C(CH_2)_3$ … HN … O (36)	540
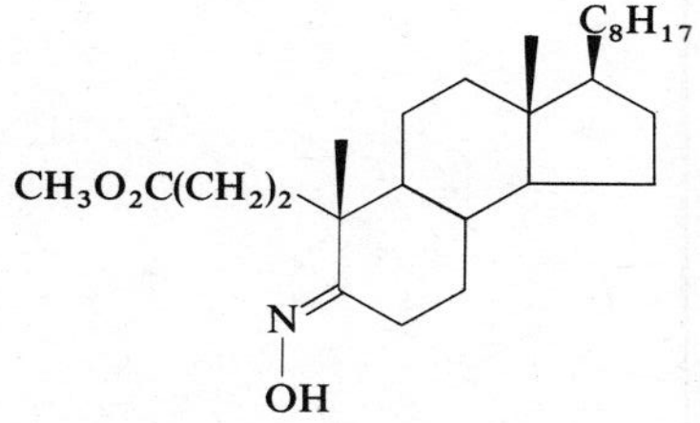 $CH_3O_2C(CH_2)_2$ … C_8H_{17} … N–OH	$SOCl_2$, ether, 0°	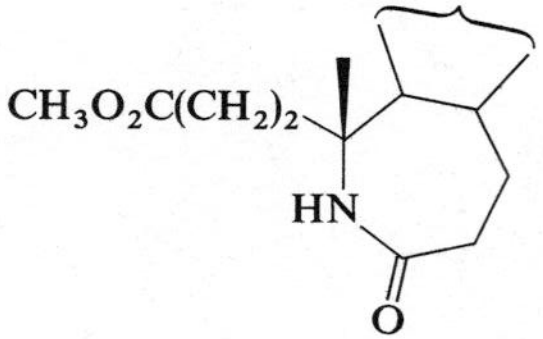 $CH_3O_2C(CH_2)_2$ … HN … O (74)	541, 542
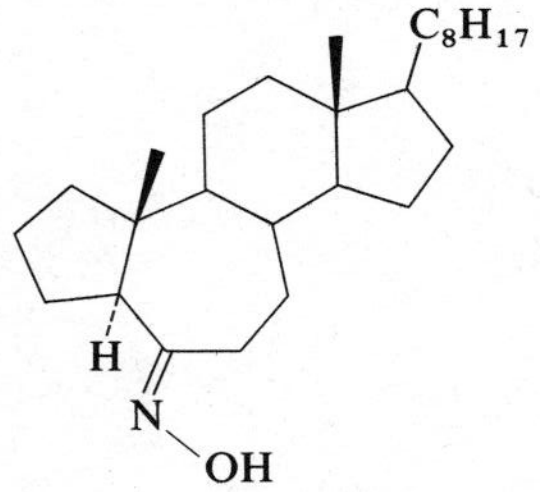 C_8H_{17} … H … N–OH	$SOCl_2$, dioxane, 20 min	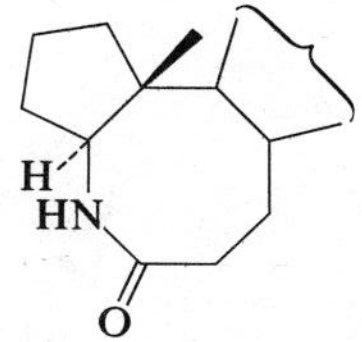 H … HN … O (89)	543
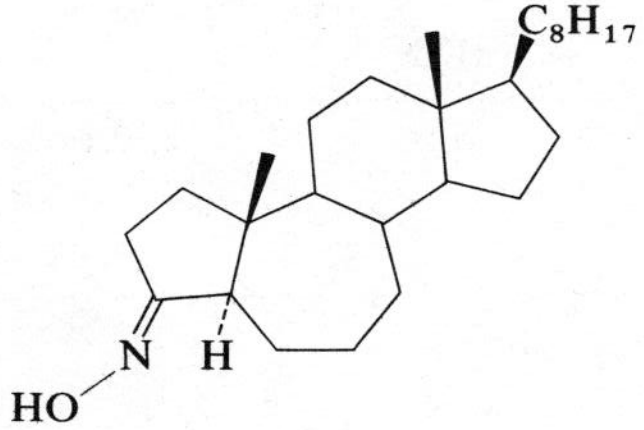 C_8H_{17} … H … N–HO	$SOCl_2$, 20 min	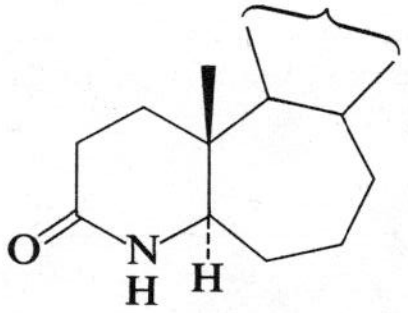 O … N H … H (75)	543

TABLE I. REARRANGEMENTS (*Continued*)

No. of Carbon Atoms	Substrate	Reagent and Conditions	Product(s) and Yield(s) (%)	Refs.
	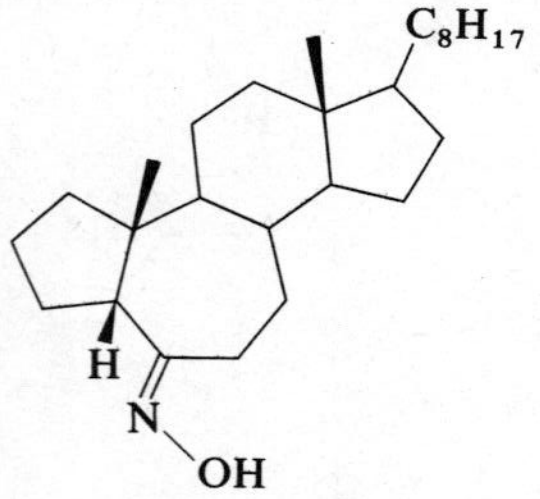 C_8H_{17}; H; N–OH	$SOCl_2$, dioxane, 30 min	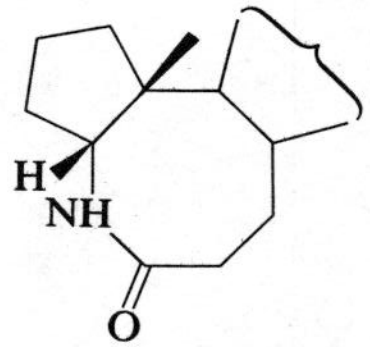 H; NH; O (67)	543
C_{28}	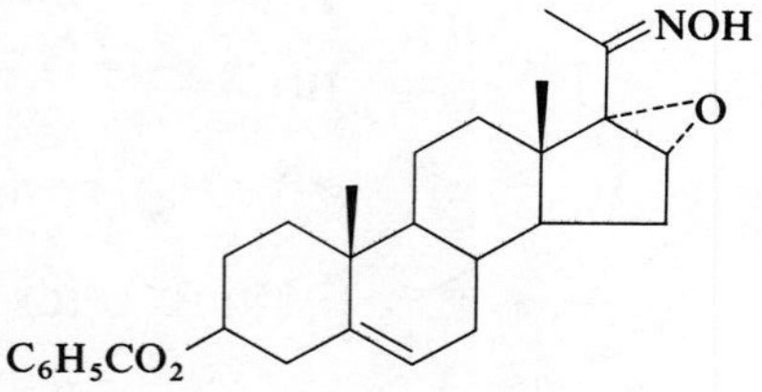 NOH; O; $C_6H_5CO_2$	$POCl_3$, pyridine, $-10°$, 2 h	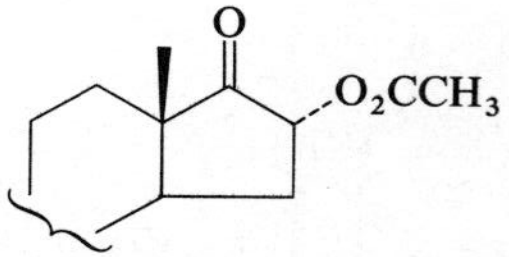 O; O_2CCH_3 (36)	504
	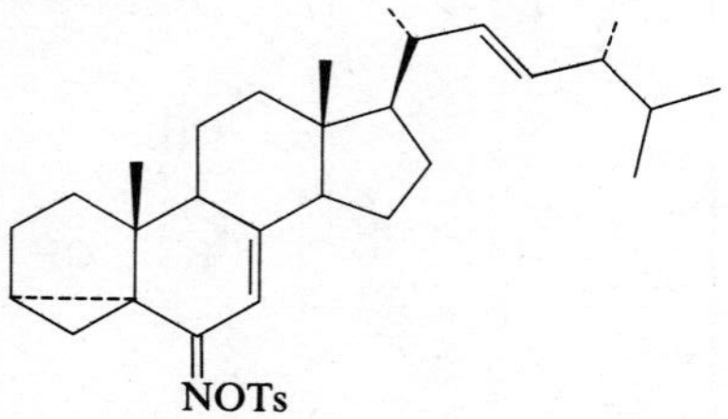 NOTs	TsCl, pyridine, 15 h	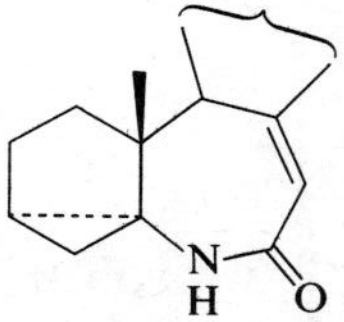 N; H; O (90)	516

Reactant	Conditions	Product(s) and Yield(s) (%)		Refs.
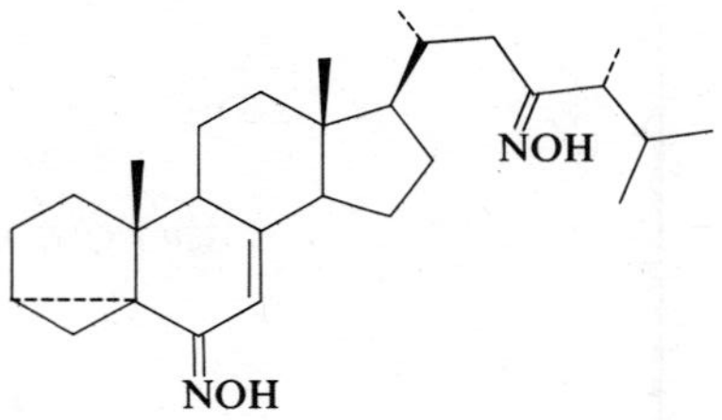 NOH / NOH	"	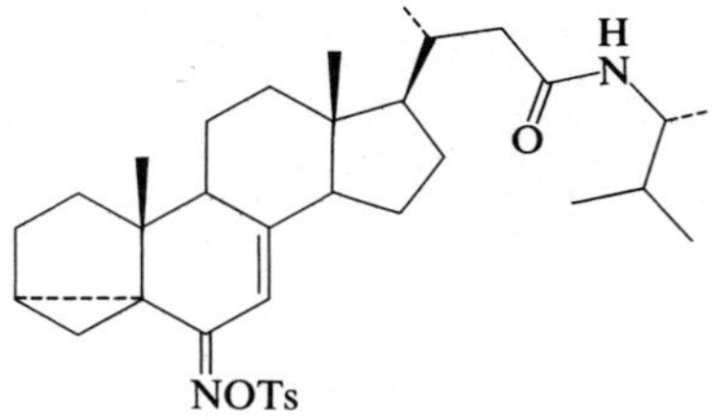 NOTs / O, N, H	(90)	516
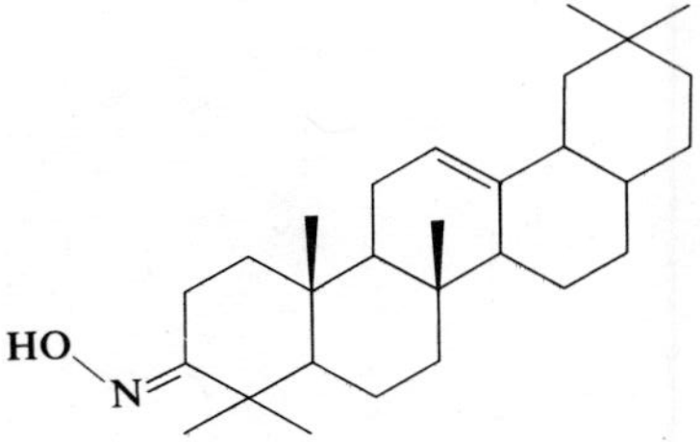 HO, N	TsCl, pyridine, 16 h	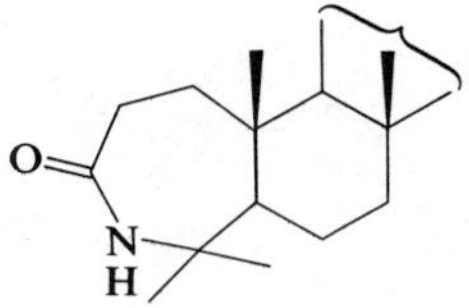 O, N, H	(67)	544
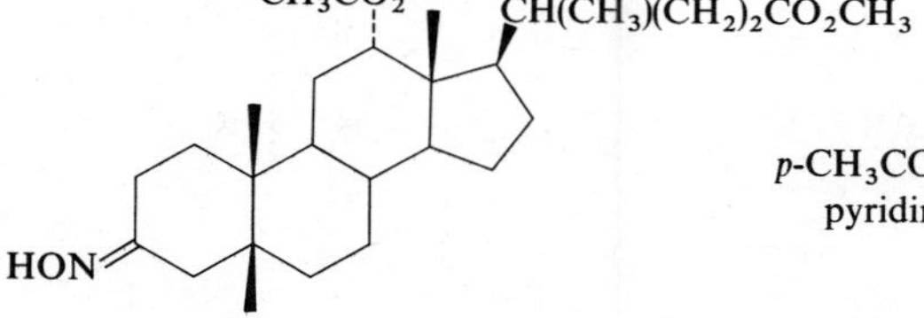 CH_3CO_2, $CH(CH_3)(CH_2)_2CO_2CH_3$, HON	p-$CH_3CONHC_6H_4SO_2Cl$, pyridine	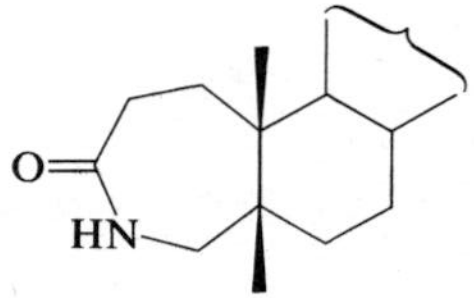 O, HN	(—)	511
	$SOCl_2$, dioxane, 10°	"	(—)	511
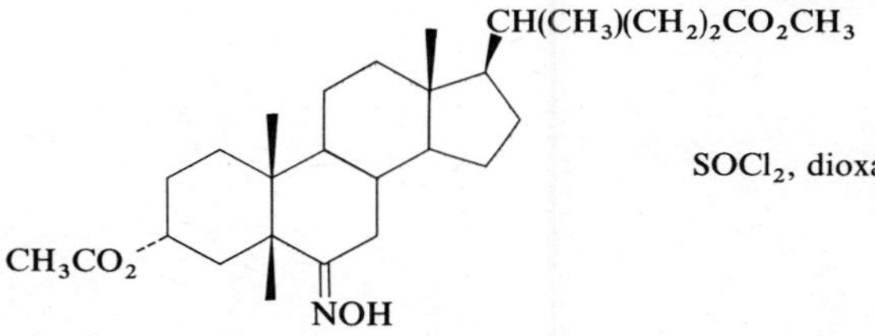 $CH(CH_3)(CH_2)_2CO_2CH_3$, CH_3CO_2, NOH	$SOCl_2$, dioxane, 1 h	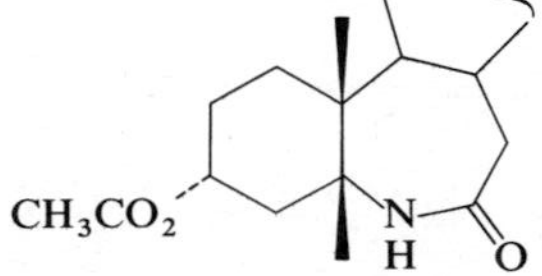 CH_3CO_2, N, H, O	(—)	511

TABLE I. REARRANGEMENTS (*Continued*)

No. of Carbon Atoms	Substrate	Reagent and Conditions	Product(s) and Yield(s) (%)	Refs.
	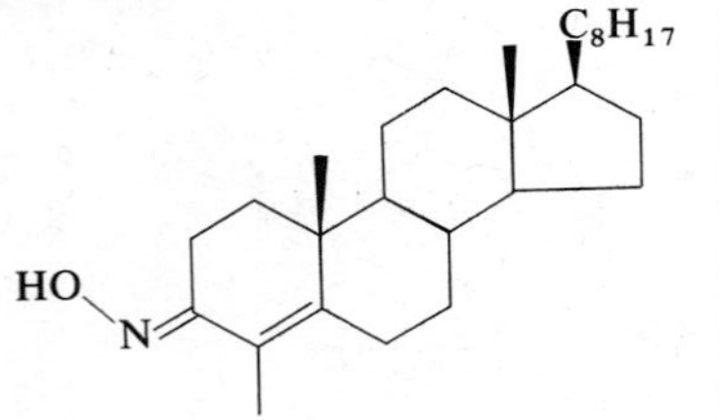 HO–N=; C_8H_{17}	PPA, 100°, 1 h	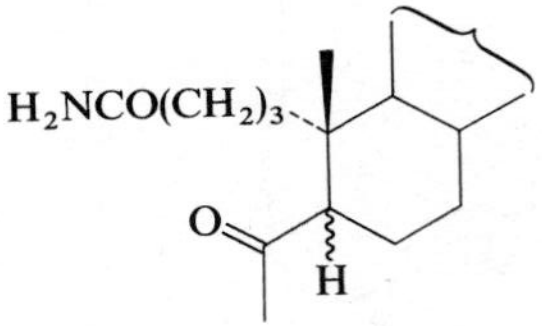 $H_2NCO(CH_2)_3$; O; H (60)	545
	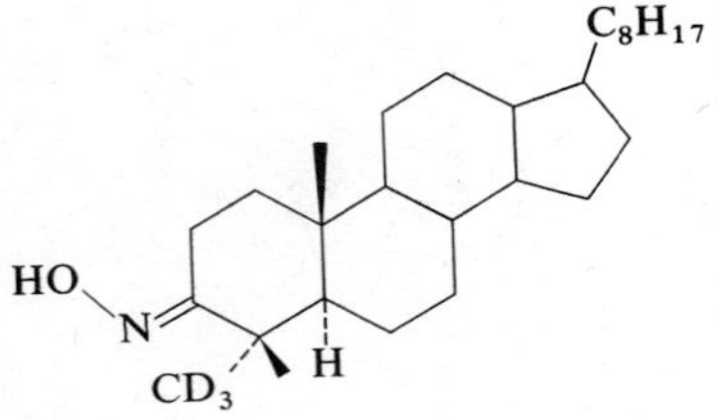 HO–N=; CD_3; H; C_8H_{17}	TsCl, pyridine	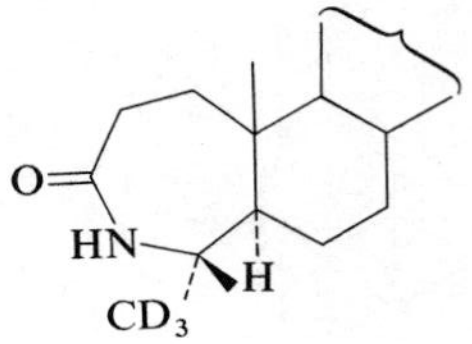 O; HN; CD_3; H (—)	51
	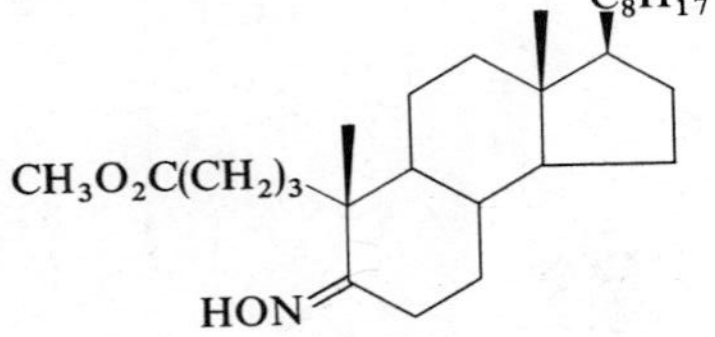 $CH_3O_2C(CH_2)_3$; HON=; C_8H_{17}	(—)	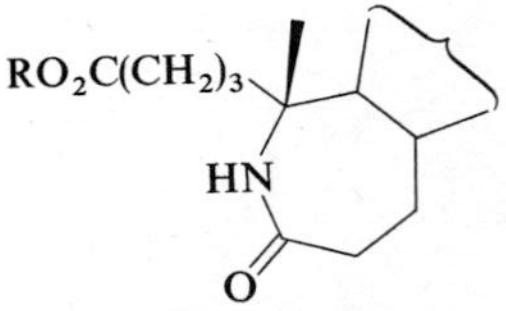 $RO_2C(CH_2)_3$; HN; O (61) R = H, CH_3	540

	Reactant	Conditions	Product (Yield %)	Refs.
C_{29}	$C_6H_5CH_2O$; HO–N	TsCl, pyridine, 60°, 1 h	$C_6H_5CH_2O$; O; N–H (65)	54
	CO_2CH_3; CH_3CO_2; $C_6H_5CH_2O_2C$; N–OH	$SOCl_2$, ether, $-15°$, 20 min	CO_2CH_3; HN; O (—)	505
	HON; O; CH_3CO_2	TsCl, pyridine, 64 h	O; H–N; CH_3CO_2 (52)	546
	HO–N; O; CH_3CO_2	TsCl, pyridine, 21 h	O; H–N; CH_3CO_2 (50)	522

TABLE I. REARRANGEMENTS (*Continued*)

No. of Carbon Atoms	Substrate	Reagent and Conditions	Product(s) and Yield(s) (%)	Refs.
	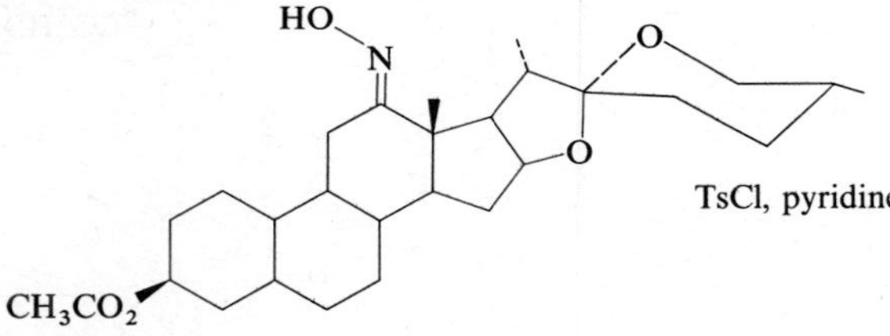 HO–N=; CH_3CO_2	TsCl, pyridine, 100°, 3 h	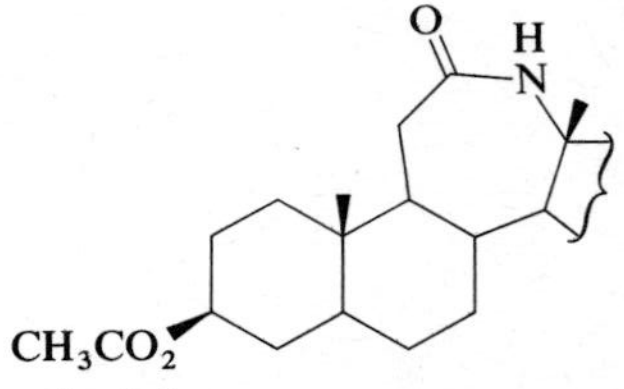 O, H, N; CH_3CO_2 (74)	522, 547
		TsCl, pyridine, 4 d	" (94)	546
	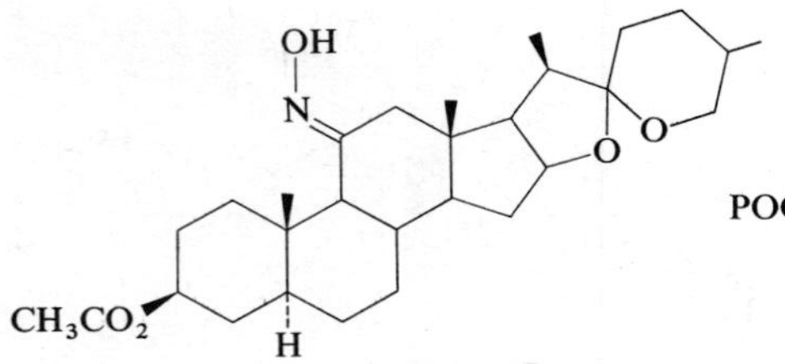 OH–N=; CH_3CO_2; H	$POCl_3$, pyridine	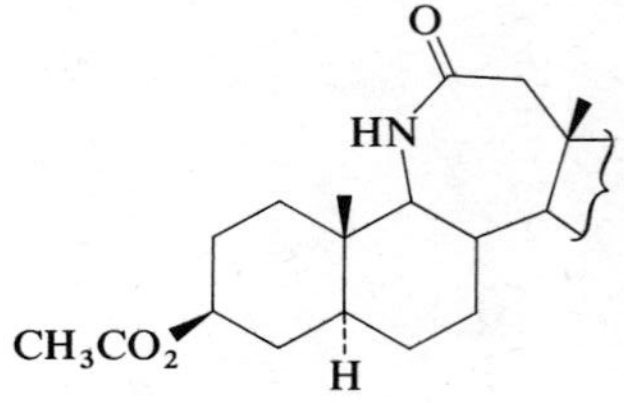 O, HN; CH_3CO_2; H (31)	507
	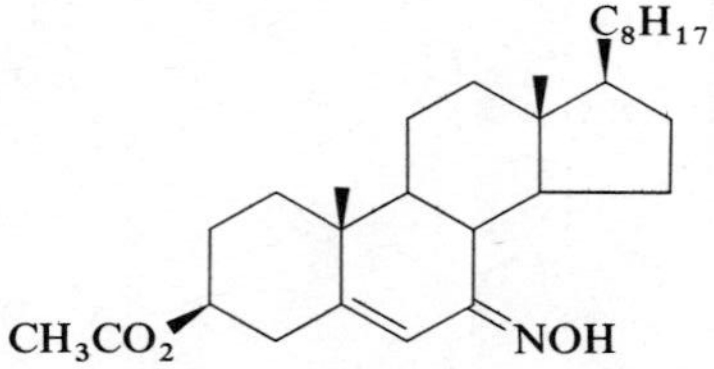 C_8H_{17}; CH_3CO_2; NOH	(—)	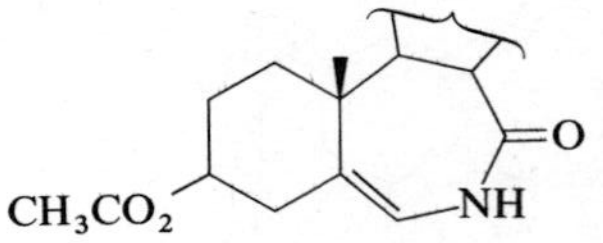 O, NH; CH_3CO_2 (—)	548

Reactant	Conditions	Product(s) and Yield(s) (%)	Refs.
	$SOCl_2$, $-20°$	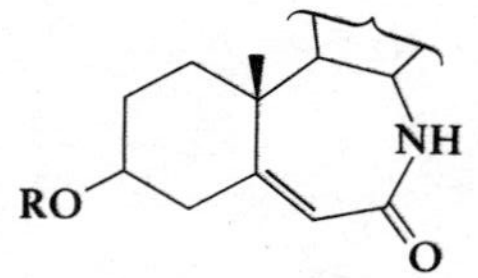 RO, NH, O; I, R = H; II, R = CH_3CO (—)	525
	p-$CH_3CONHC_6H_4SO_2Cl$, pyridine	II (—)	525
	$SOCl_2$, dioxane	II (59)	549
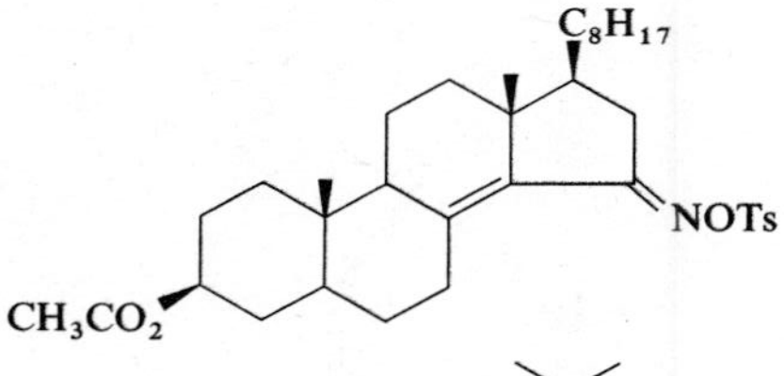 C_8H_{17}, NOTs, CH_3CO_2	HCl, CH_3OH, 90 min	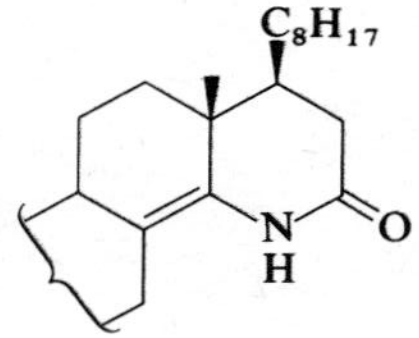 C_8H_{17}, N, H, O (30)	550
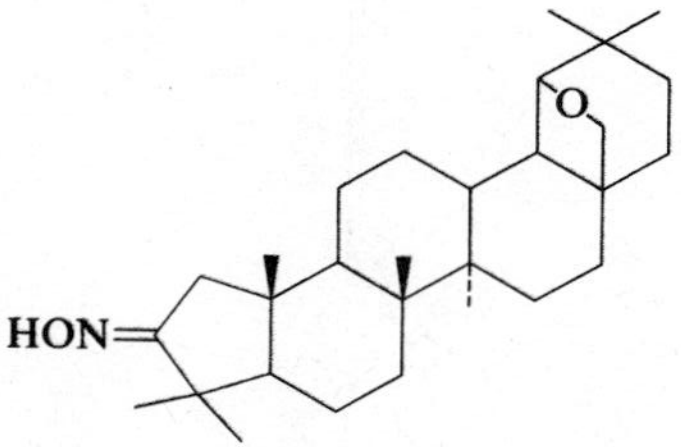 O, HON=	TsCl, pyridine, 3 d	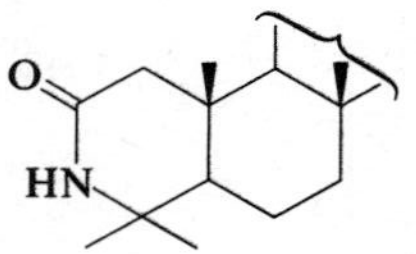 O, HN (59)	50
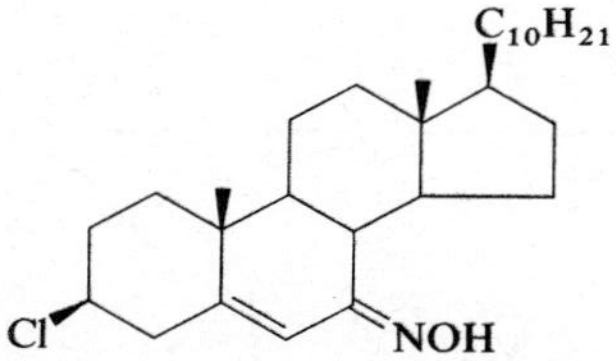 $C_{10}H_{21}$, Cl, NOH	TsCl, ether	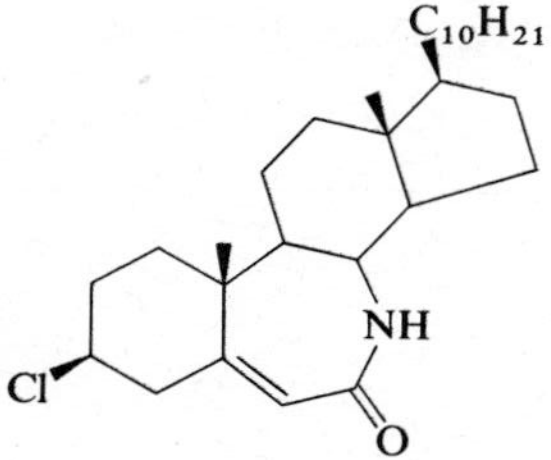 $C_{10}H_{21}$, Cl, NH, O (—)	551

TABLE I. REARRANGEMENTS (*Continued*)

No. of Carbon Atoms	Substrate	Reagent and Conditions	Product(s) and Yield(s) (%)	Refs.
	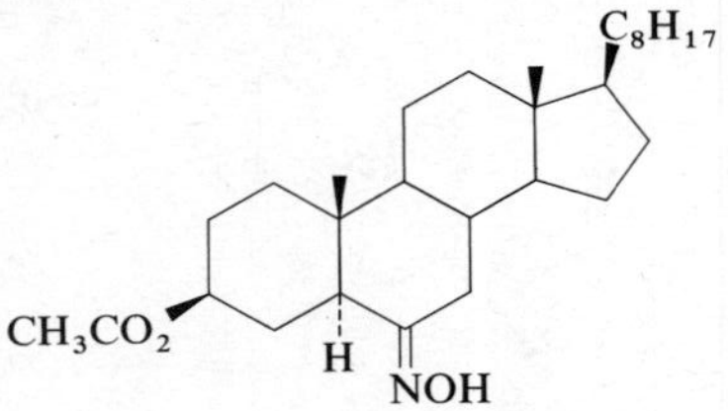 CH_3CO_2, H, NOH, C_8H_{17}	$SOCl_2$, C_6H_6, 20 min	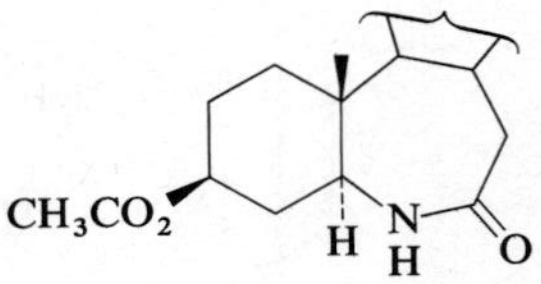 CH_3CO_2, H, N, H, O (70)	525, 552
	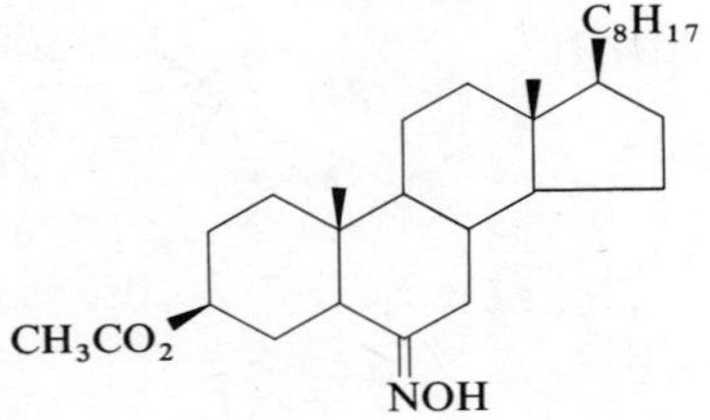 CH_3CO_2, NOH, C_8H_{17}	$SOCl_2$, $-20°$, 5 min	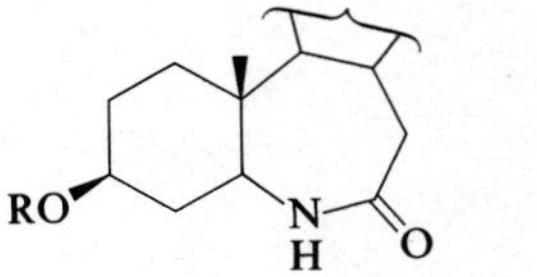 RO, N, H, O I R = H, CH_3CO (—)	525
		(—)	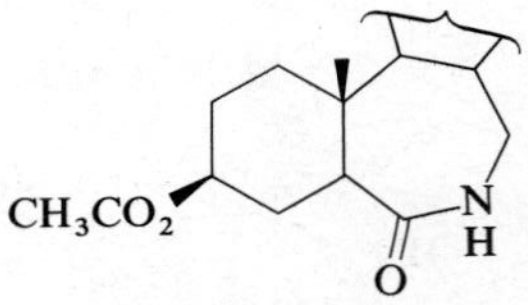 CH_3CO_2, O, N, H + I (R = CH_3CO) (—)	548

	Reactant	Conditions	Product(s) and Yield(s) (%)	Refs.
		$SOCl_2$, −20°	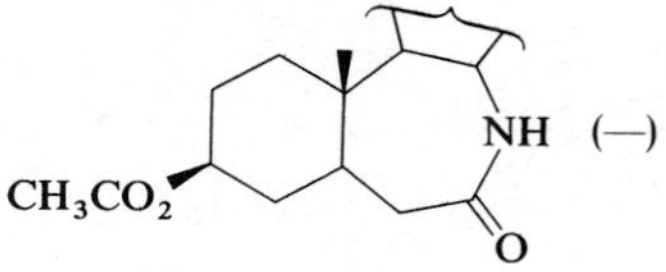 CH_3CO_2 … NH … O (—)	525
		p-$CH_3CONHC_6H_4SO_2Cl$	" (—)	525
	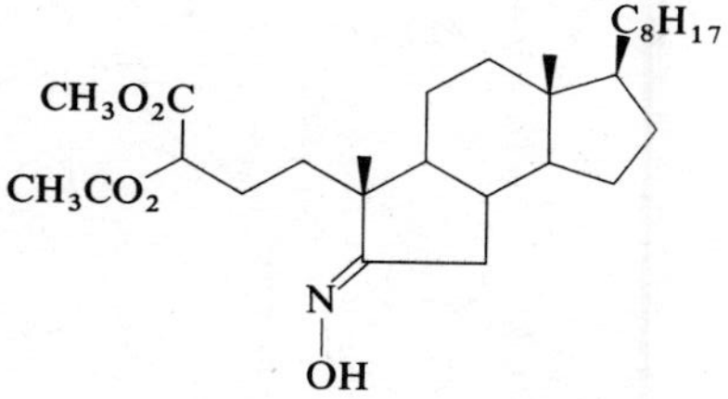 CH_3O_2C, CH_3CO_2, C_8H_{17}, N–OH	$SOCl_2$, ether	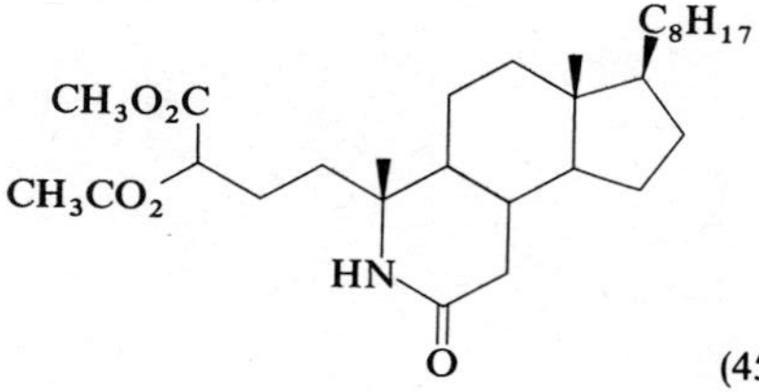 CH_3O_2C, CH_3CO_2, C_8H_{17}, HN, O (45)	356, 553
	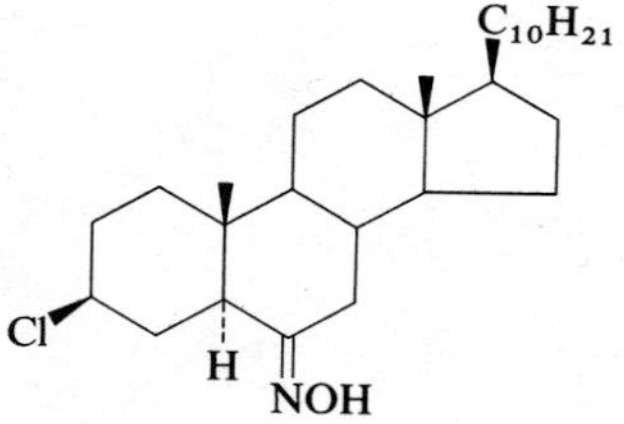 $C_{10}H_{21}$, Cl, H, NOH	1. TsCl 2. Alumina, ether	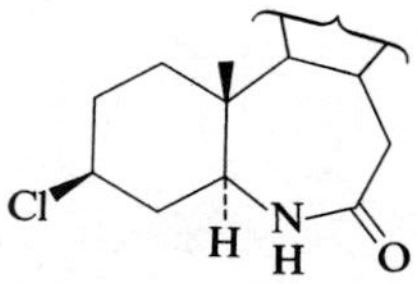 Cl, H, N, H, O (—)	551
C_{30}	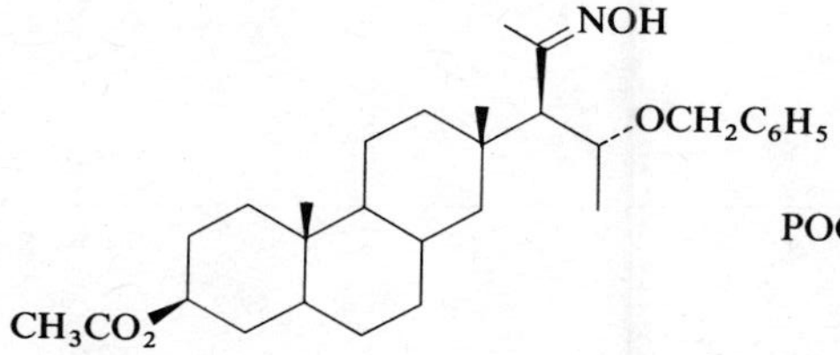 CH_3CO_2, NOH, $OCH_2C_6H_5$	$POCl_3$, pyridine, 0°, 3 h	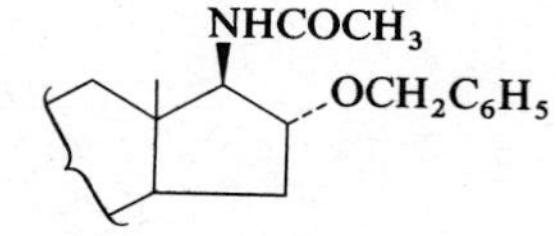 $NHCOCH_3$, $OCH_2C_6H_5$ (—)	554

TABLE I. REARRANGEMENTS (*Continued*)

No. of Carbon Atoms	Substrate	Reagent and Conditions	Product(s) and Yield(s) (%)	Refs.
	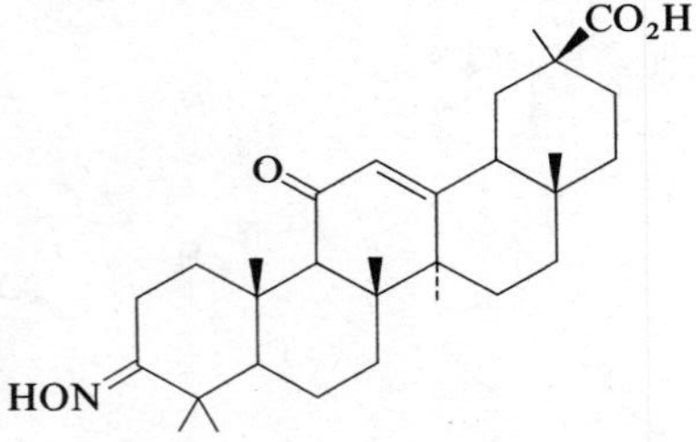 HON, O, CO_2H	$SOCl_2$, dioxane, 1 h	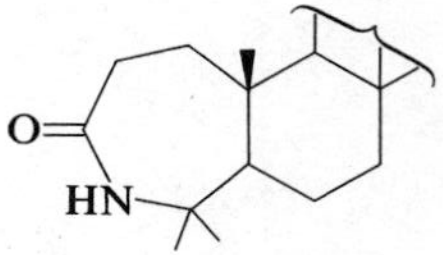 O, HN (35)	555
	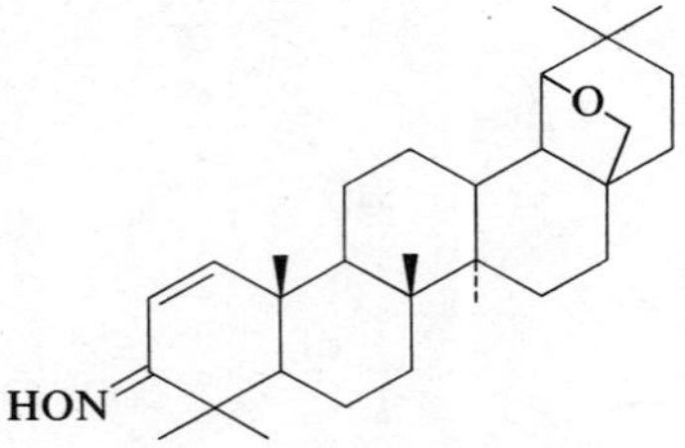 HON, O	TsCl, pyridine, 2 d	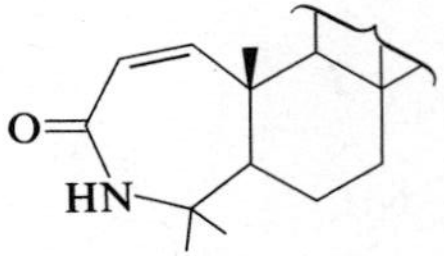 O, HN (57)	50
		$POCl_3$, pyridine, 100°, 2 h	" (36)	50
	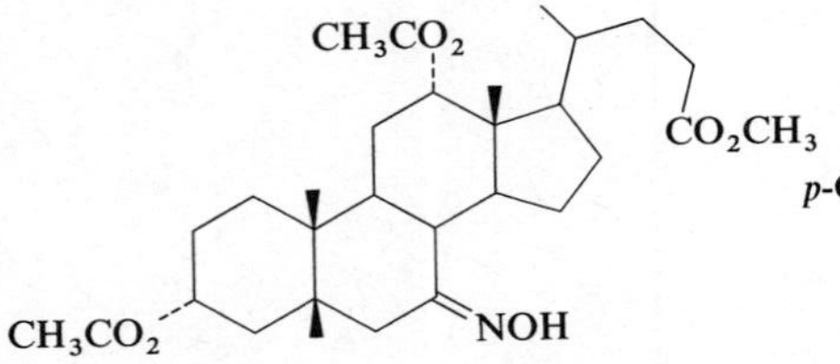 CH_3CO_2, CH_3CO_2, CO_2CH_3, NOH	p-$CH_3CONHC_6H_4SO_2Cl$, pyridine	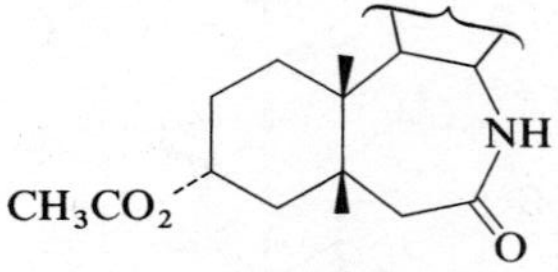 CH_3CO_2, NH, O (—)	511

Reactant	Conditions	Product(s) and Yield(s) (%)	Refs.
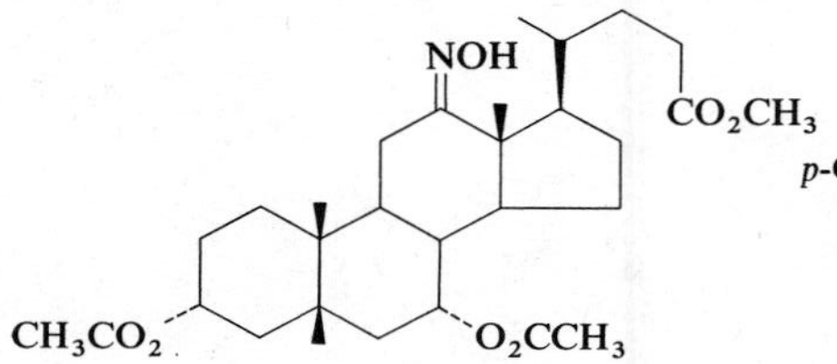 NOH, CO_2CH_3, CH_3CO_2, O_2CCH_3	p-$CH_3CONHC_6H_4SO_2Cl$, pyridine	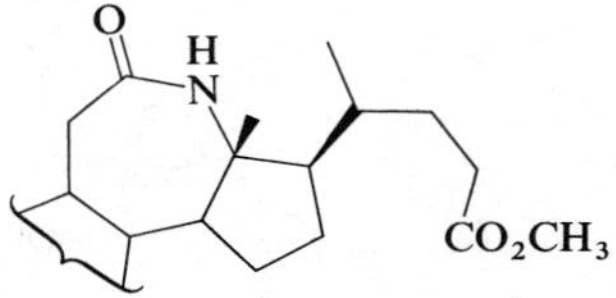 O, H, N, CO_2CH_3 (—)	511
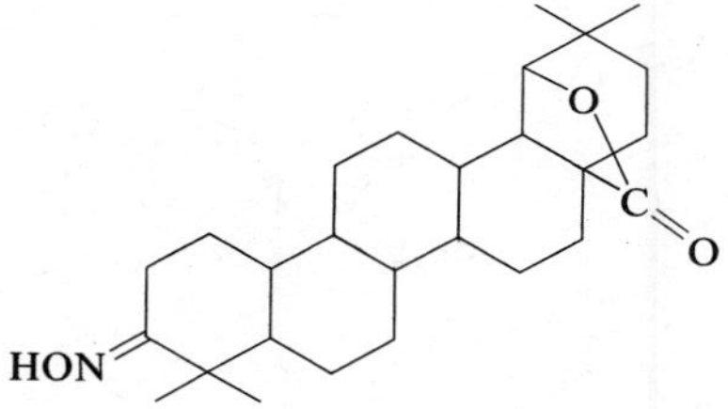 O, C=O, HON	$POCl_3$, pyridine, 35°, 4 h	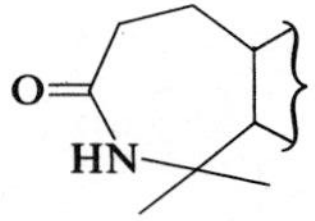 O=, HN (—)	556
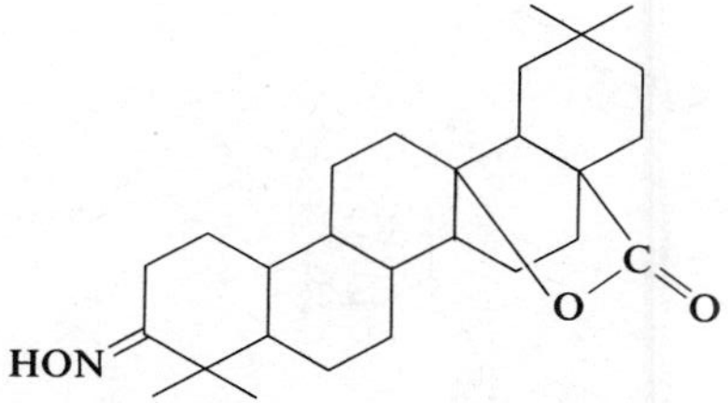 C=O, O, HON	"	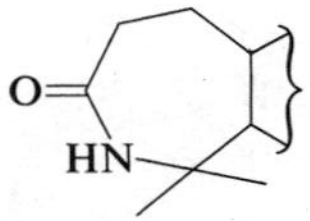 O=, HN (—)	556
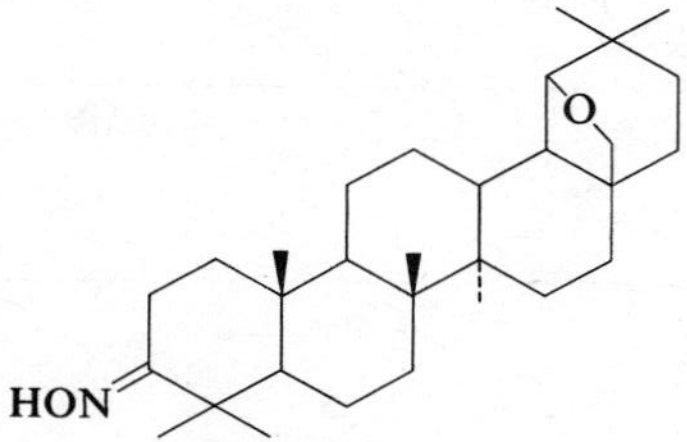 O, HON	TsCl, pyridine, 2 d	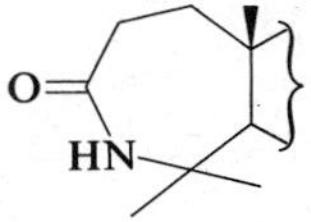 O=, HN (65-75)	50
	$POCl_3$, pyridine, 2 h	" (57)	50

TABLE I. REARRANGEMENTS (*Continued*)

No. of Carbon Atoms	Substrate	Reagent and Conditions	Product(s) and Yield(s) (%)	Refs.
	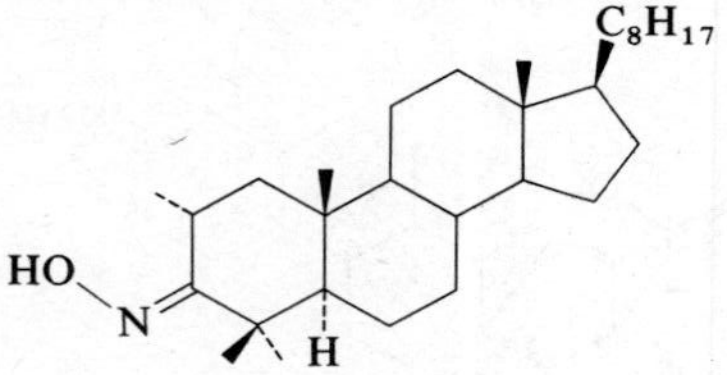 HO–N=, C_8H_{17}, H	$SOCl_2$, dioxane, 15°, 15 min	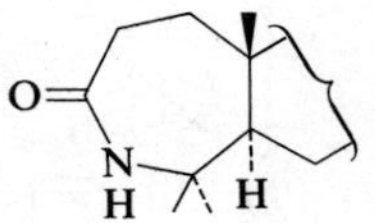 O=, N H, H (—)	557
C_{31}	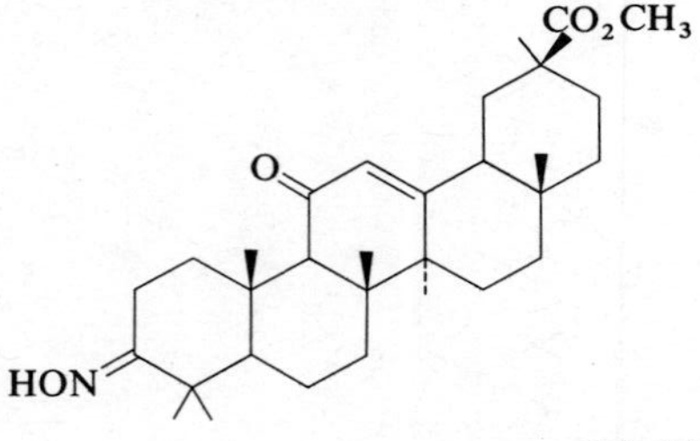 HON=, CO_2CH_3, O	$SOCl_2$, dioxane, 1 h	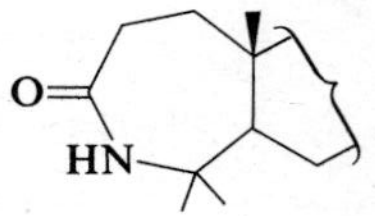 O=, HN (—)	555
	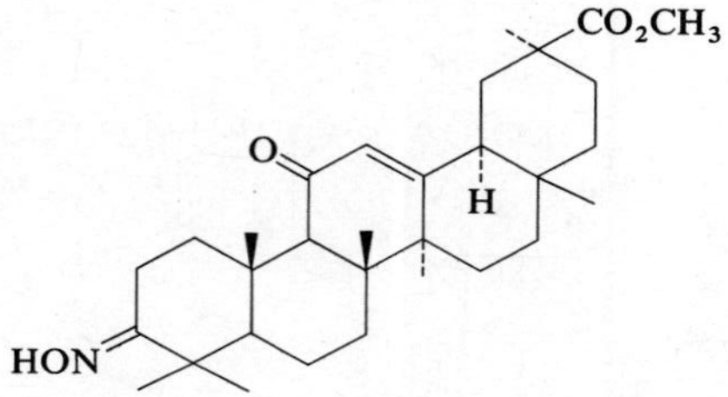 HON=, CO_2CH_3, O, H	TsCl, pyridine	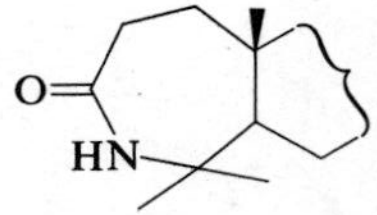 O=, HN (—)	558

Reactant	Conditions	Product(s) and Yield(s) (%)	Refs.
HON, O, H, CO_2CH_3	"	O, HN (—)	558
$C(CH_3){=}NOH$, CH_3CO_2, O, N, H	$POCl_3$, pyridine, 0°, 3 h	$NHCOCH_3$, O, N, H (30)	554
HON, CO_2CH_3	$POCl_3$, pyridine, 35°, 4 h	O, HN (40)	559

TABLE I. REARRANGEMENTS (*Continued*)

No. of Carbon Atoms	Substrate	Reagent and Conditions	Product(s) and Yield(s) (%)	Refs.
	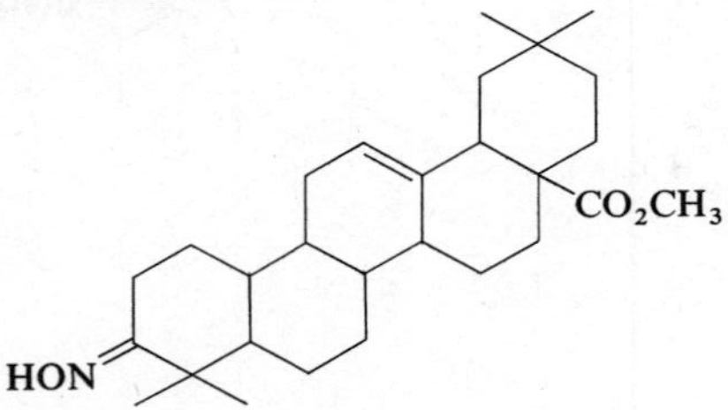 HON, CO_2CH_3	PPA, 130°, 30 min	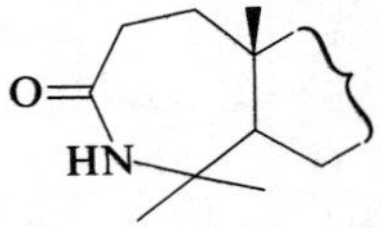 O, HN (—)	556
	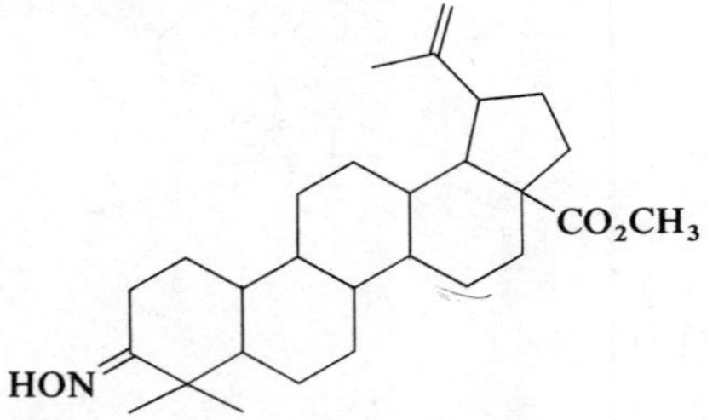 HON, CO_2CH_3	"	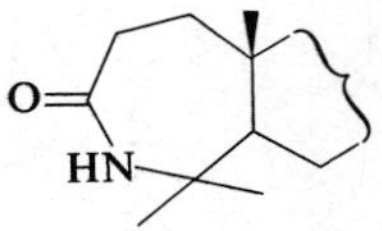 O, HN (—)	556
	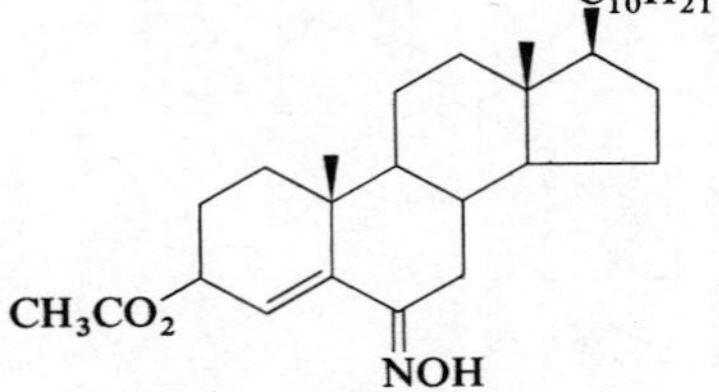 CH_3CO_2, NOH, $C_{10}H_{21}$	1. TsCl 2. Alumina, ether	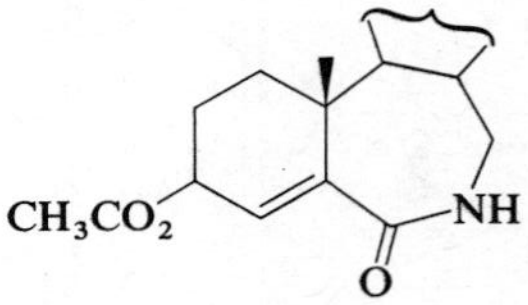 CH_3CO_2, NH, O (—)	551

Reactant	Conditions	Product (Yield, %)	Refs.
$C_{10}H_{21}$; CH_3CO_2; NOH	"	CH_3CO_2; NH; O (—)	551
$C_{10}H_{21}$; CH_3CO_2; H; NOH	"	CH_3CO_2; H; N; H; O (75)	551
HON	$POCl_3$, pyridine, 35°, 4 h	O; HN (—)	559

TABLE I. REARRANGEMENTS (*Continued*)

No. of Carbon Atoms	Substrate	Reagent and Conditions	Product(s) and Yield(s) (%)	Refs.
C_{32}	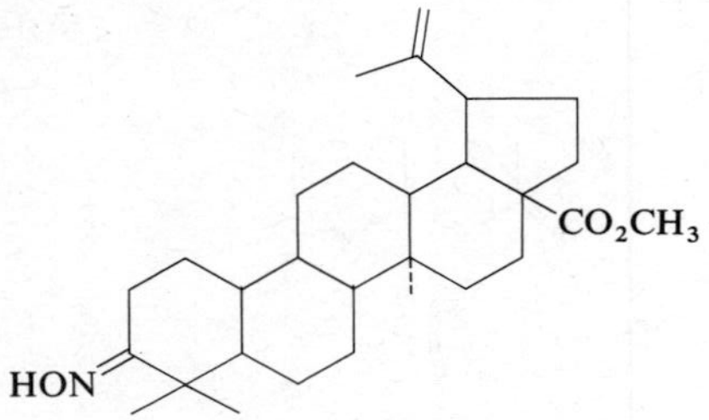 HON, CO_2CH_3	"	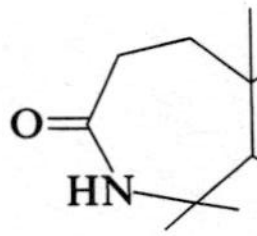 O, HN (—)	559
C_{34}	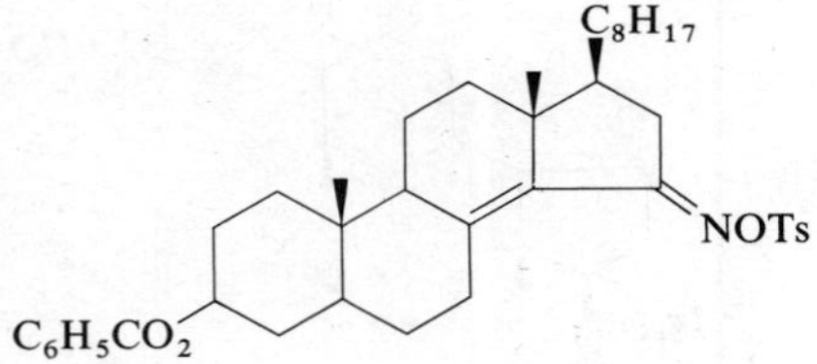 C_8H_{17}, NOTs, $C_6H_5CO_2$	HCl, CH_3OH, 90 min	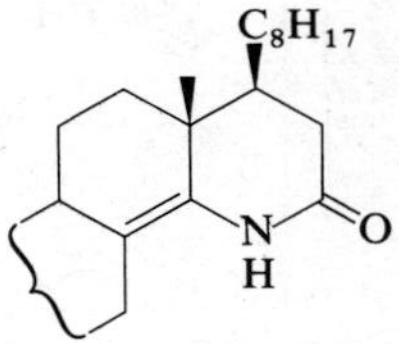 C_8H_{17}, N H, O (56)	550
	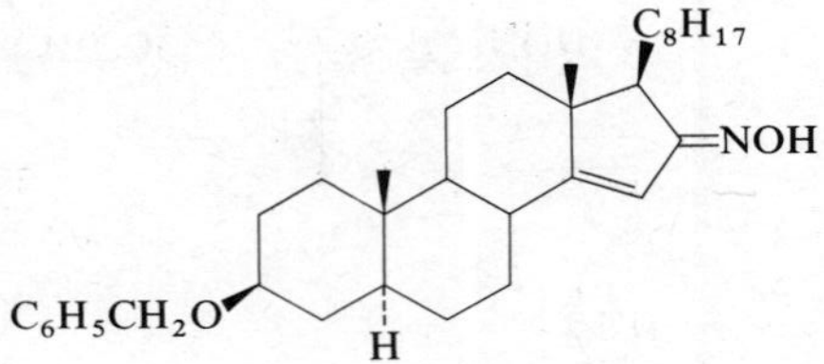 C_8H_{17}, =NOH, $C_6H_5CH_2O$, H	TsCl, pyridine, 3 h	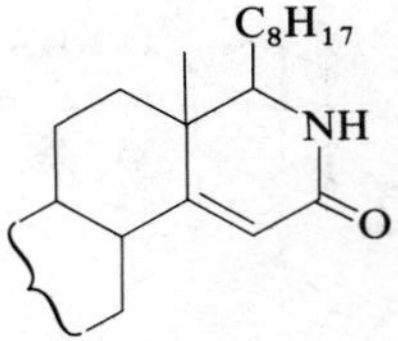 C_8H_{17}, NH, O (—)	560

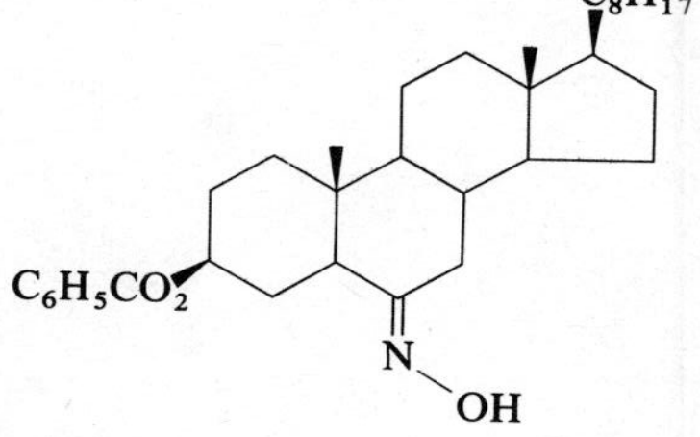 $C_6H_5CO_2$, C_8H_{17}, N–OH	$SOCl_2$, $-20°$	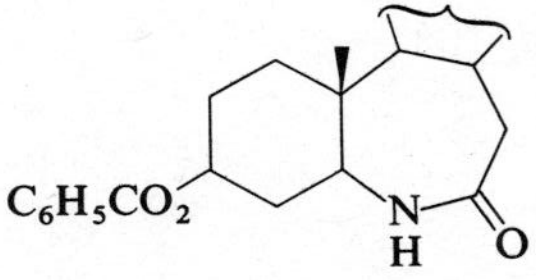 $C_6H_5CO_2$, NH, O; I (—)	525
	(—)	I 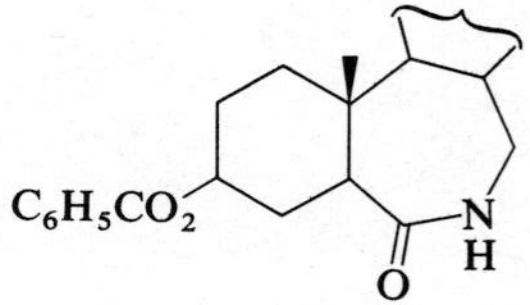 + $C_6H_5CO_2$, NH, O (—)	548
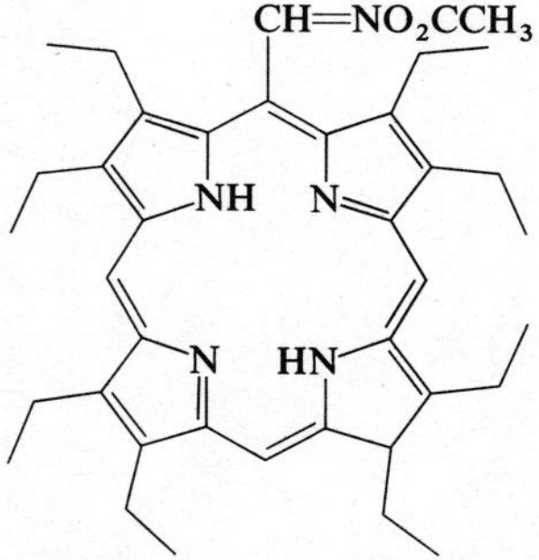 C_{37} $CH{=}NO_2CCH_3$, NH, N, HN, N	$(CH_3CO)_2O$, reflux	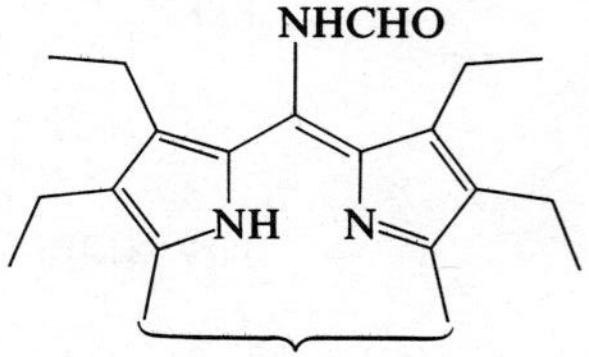 NHCHO, NH, N (28)	134

TABLE II. ELIMINATION-ADDITIONS

No. of Carbon Atoms	Substrate	Reagent and Conditions	Product(s) and Yield(s) (%)	Refs.
C_3	$(CH_3)_2C{=}NOMs$	1. $(C_2H_5)_2AlI$ 2. C_6H_5MgBr 3. DIBAL	$C_6H_5CH(CH_3)NHCH_3$ (92)	86
		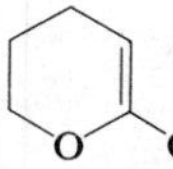 O, $OSi(CH_3)_3$ + $(C_2H_5)_2AlCl$, CH_2Cl_2, −78 to 20°, 1 h	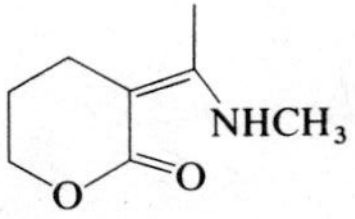 $NHCH_3$, O, O (48)	159
		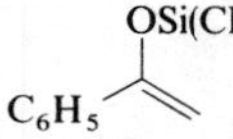 $OSi(CH_3)_3$, C_6H_5 + $(C_2H_5)_2AlCl$, CH_2Cl_2, −78 to 20°, 1 h	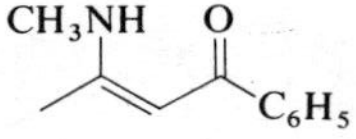 CH_3NH, O, C_6H_5 (65)	159
		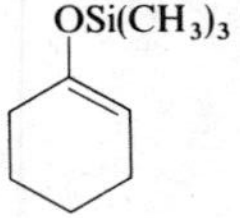 $OSi(CH_3)_3$ + $C_2H_5AlCl_2$ CH_2Cl_2, −78 to 20°, 1 h	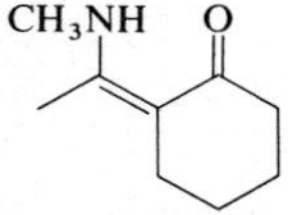 CH_3NH, O (69)	159
C_5	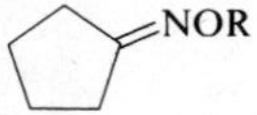 =NOR R = Ts	$(i\text{-}C_4H_9)_2AlSCH_3$, CH_2Cl_2, 40°, 6 min	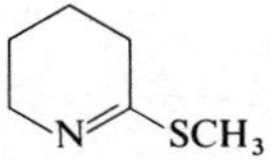 N, SCH_3 (5)	147
		1. $(n\text{-}C_3H_7)_3Al$, CH_2Cl_2, 40° 2. DIBAL, 0°	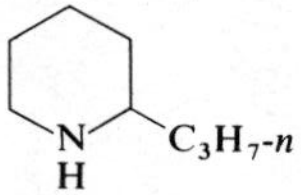 N, H, $C_3H_7\text{-}n$ (58)	147, 153
		1. $(CH_3)_3Al$, CH_2Cl_2, −78° 2. See below	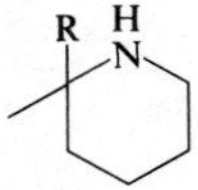 R, H, N	

	Reagents: R = Ms	$HC\equiv CCH_2MgBr$	R = $CH_2C\equiv CH$ (55)	147
	R = Ts	$CH_2=CHCH_2MgBr$	R = $CH_2CH=CH_2$ (51)	147
	$(C_2H_5)_2C=NOR$			
	R = $CO_2C_2H_5$	$(CH_3)_3SiI$, $CDCl_3$	$C_2H_5N=ClC_2H_5$ (97)	86
	R = $COCH_3$	"	" (54)	86
	R = COC_6H_5	"	" (37)	86
	R = Ms	1. $(C_2H_5)_2AlI$ 2. C_6H_5MgBr 3. DIBAL	$C_6H_5CH(C_2H_5)NHC_2H_5$ (63)	86
		n-$C_4H_9CH=C(OCH_3)OSi(CH_3)_3$ + $(C_2H_5)AlCl_2$, CH_2Cl_2, −78 to 20°, 1 h	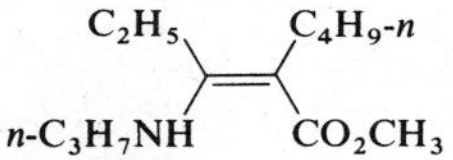 C_2H_5, C_4H_9-n, n-C_3H_7NH, CO_2CH_3 (73)	159
		$CH_2=C(C_6H_{13}$-$n)OSi(CH_3)_3$ + $(C_2H_5)_2AlCl$, CH_2Cl_2, −78 to 20°, 1 h	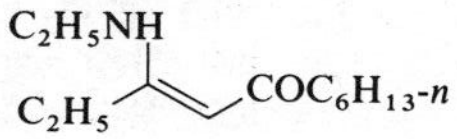 C_2H_5NH, C_2H_5, COC_6H_{13}-n (95)	159
C_6	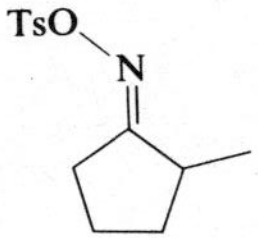 TsO, N	$(i$-$C_4H_9)_2AlSCH_3$, CH_2Cl_2, 40°, 6 min	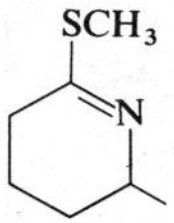 SCH_3, N (46)	147
		1. $(n$-$C_3H_7)_3Al$, CH_2Cl_2, 40° 2. DIBAL	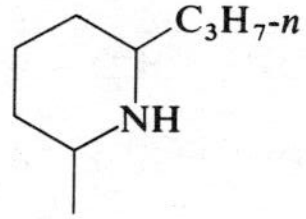 C_3H_7-n, NH (70)	147, 153
	NOR R = Ms	1. 2 $(C_2H_5)_2AlI$, CH_2Cl_2, −78°, 1 h 2. C_6H_5MgBr, −78 to 0°, 1 h 3. DIBAL, 0°, 1 h	H, N, R R = C_6H_5 (81)	86
		1. $(CH_3)_3Al$, CH_2Cl_2, −78 to 0° 2. DIBAL, 0°	R = CH_3 (70)	147, 153

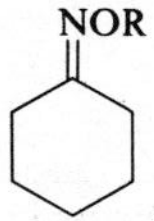

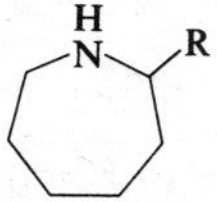

TABLE II. ELIMINATION-ADDITIONS (*Continued*)

No. of Carbon Atoms	Substrate	Reagent and Conditions	Product(s) and Yield(s) (%)	Refs.
		1. $(C_2H_5)_3Al$, CH_2Cl_2, −78 to 0° 2. DIBAL, 0°	R = C_2H_5 (47)	147, 153
		1. $(n\text{-}C_3H_7)_3Al$, −78 to 0° 2. DIBAL, 0°	R = C_3H_7-*n* (64)	147, 153
		1. *n*-C_4H_9MgBr, $C_6H_5CH_3$, −78 to 0° 2. DIBAL	R = C_4H_9-*n* (63)	157
		1. $(i\text{-}C_4H_9)_3Al$, CH_2Cl_2, −78 to 0° 2. DIBAL,	R = C_4H_9-*i* (52)	147, 153
		1. $(C_2H_5)_2AlC{\equiv}CC_4H_9$-*n*, CH_2Cl_2, −78° 2. DIBAL	R = $C{\equiv}CC_4H_9$-*n* (67)	147, 153
	R = Ts	$(i\text{-}C_4H_9)_2AlSR$, CH_2Cl_2, 0°, 1 h	SR R = CH_3 (52)	147
	R = Ms	"	R = CH_3 (58)	147
		"	R = C_2H_5 (62)	147
	R = Ms	1. $(CH_3)_3Al$, CH_2Cl_2, −78° 2. $CH_2{=}CHCH_2MgBr$	H N R R = $CH_2CH{=}CH_2$ (60)	147
		1. CH_3MgBr, $C_6H_5CH_3$, −78° 2. $CH_2{=}CHCH_2MgBr$, 0°, 1 h	R = $CH_2CH{=}CH_2$ (72)	157
		1. $(CH_3)_3Al$, CH_2Cl_2, −78° 2. $HC{\equiv}CCH_2MgBr$	R = $CH_2C{\equiv}CH$ (84)	147
		1. CH_3MgBr, $C_6H_5CH_3$, −78° 2. $HC{\equiv}CCH_2MgBr$, 0°, 1 h	R = $CH_2C{\equiv}CH$ (66)	157

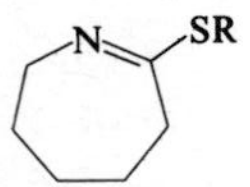

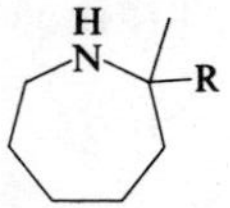

	Reactant	Reagents	Product(s) (Yield %)	Refs.
C_7	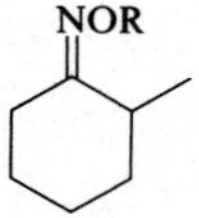 NOR (2-methylcyclohexanone oxime) (*Z*) R = H	1. $(n\text{-}C_3H_7)_3Al$, CH_2Cl_2, $-78°$ 2. $CH_2{=}CHCH_2MgBr$	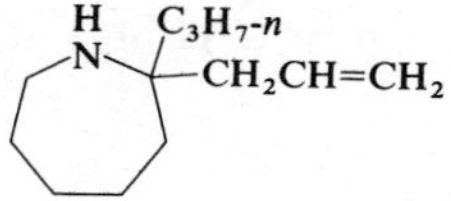 H, N, C_3H_7-*n*, $CH_2CH{=}CH_2$ (63)	147
		$C_2H_5CH{=}CHOSi(CH_3)_3$ + $(C_2H_5)_2AlCl$, CH_2Cl_2, -78 to 20°, 1 h	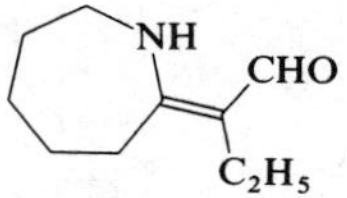 NH, CHO, C_2H_5 (53)	159
		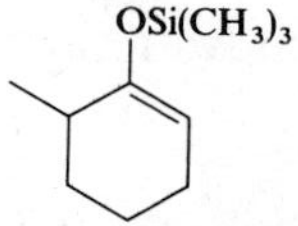 $OSi(CH_3)_3$ + $C_2H_5AlCl_2$, CH_2Cl_2, -78 to 20°, 1 h	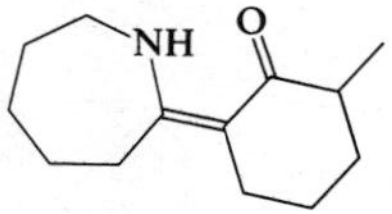 NH, O (82)	159
		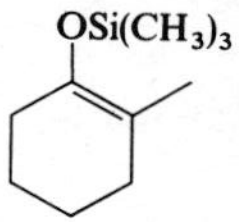 $OSi(CH_3)_3$ + $C_2H_5AlCl_2$, CH_2Cl_2, -78 to 20°, 1 h	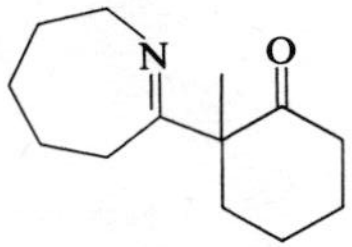 N, O (42)	1 59
		1. MsCl, $(C_2H_5)_3N$ 2. $(i\text{-}C_4H_9)_2AlSCH_3$, CH_2Cl_2, 0°, 1 h	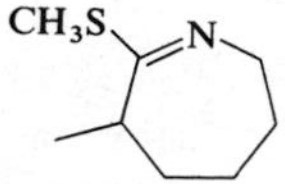 CH_3S, N (67)	147
	(*E*) R = Ms or Ts	$(i\text{-}C_4H_9)_2AlSCH_3$, CH_2Cl_2, 0°, 1 h	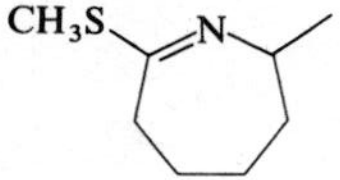 CH_3S, N (66)	147
	(*E*) R = Ms	1. $(n\text{-}C_3H_7)_3Al$, CH_2Cl_2, -78 to 0° 2. DIBAL, 0°	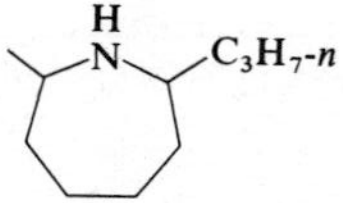 H, N, C_3H_7-*n* (48)	147

TABLE II. ELIMINATION-ADDITIONS (*Continued*)

No. of Carbon Atoms	Substrate	Reagent and Conditions	Product(s) and Yield(s) (%)	Refs.
			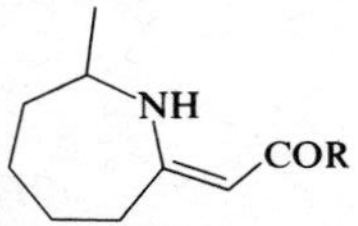 [structure: 2-methyl-7-(acylmethylene)azepane, =CHCOR]	
		$CH_2{=}C(C_6H_{13}\text{-}n)OSi(CH_3)_3$ + $(C_2H_5)_2AlCl$, CH_2Cl_2, −78 to 20°, 1 h	R = C_6H_{13}-*n* (90)	159
		$CH_2{=}C(C_6H_5)OSi(CH_3)_3$ + $(C_2H_5)_2AlCl$, CH_2Cl_2, −78 to 20°, 1 h	R = C_6H_5 (66)	159
	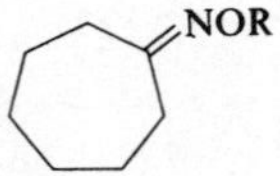 [structure: cycloheptanone oxime, =NOR] R = Ts	CH_3OH, 15 h	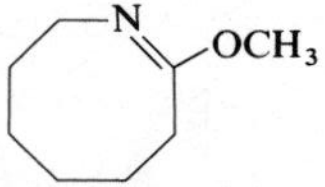 [structure: 2-methoxy-azocine, cyclic imidate with OCH_3] (62)	140
		$(CH_3)_2AlSC_2H_5$, CH_2Cl_2, 0°, 1 h	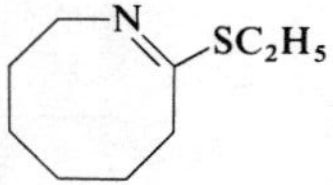 [structure: cyclic thioimidate with SC_2H_5] (90)	147
	R = Ms	1. $(C_3H_7)_3Al$, CH_2Cl_2, −78° 2. DIBAL	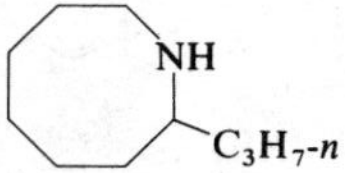 [structure: 2-(C_3H_7-*n*)azocane, NH] (68)	147
		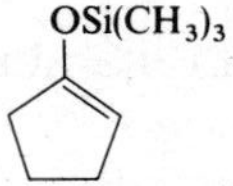 [structure: 1-($OSi(CH_3)_3$)cyclopentene] + $(C_2H_5)_2AlCl$, CH_2Cl_2, −78 to 20°, 1 h	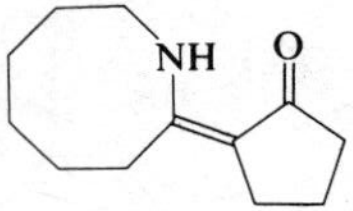 [structure: 2-(azocan-2-ylidene)cyclopentanone, NH, O] (74)	159
	$(i\text{-}C_3H_7)_2C{=}NOH$	CH_3SO_2Cl, $(C_2H_5)_3N$, $C_6H_5CH_3$, −15 to 100°	$(CH_3)_2C{=}C{=}NC_3H_7\text{-}i$ (85)	138

	Substrate	Conditions	Product(s) and Yield(s) (%)	Refs.
	$(n\text{-}C_3H_7)_2C{=}NOH$	1. CH_3SO_2Cl, $(C_2H_5)_3N$ 2. $(CH_3)_3SiCN$, $(C_2H_5)_2AlCl$, CH_2Cl_2, −78 to −20°, 1 h	$n\text{-}C_3H_7N{=}C(CN)C_3H_7\text{-}n$ (90)	147
C_8	$C_6H_5C({=}NOR)CH_3$ R = H	DIBAL, 5 eq, 0°, 2 h	$C_6H_5NHC_2H_5$ (92)	154
	R = Ms	DIBAL, CH_2Cl_2, −78°	" (87)	147, 153
	R = Ms	$(C_3H_3N_2)_2C{=}O$ (1,1′-carbonyldiimidazole), CH_2Cl_2	$C_6H_5N{=}C(CH_3)(N\text{-imidazolyl})$ (77)	146
			$+C_6H_5N{=}C(CH_3)R$	
	R = Ms	$(i\text{-}C_4H_9)_2AlSCH_3$, 0°, 1 h	R = SCH_3 (90)	147
		$(CH_3)_2AlSC_4H_9\text{-}t$, 0°, 1 h	R = $SC_4H_9\text{-}t$ (85)	147
		$(CH_3)_2AlSC_6H_5$, −78°, 4 h	R = SC_6H_5 (88)	147
	R = $CO_2C_2H_5$	1. $(CH_3)_3SiI$ 2. C_2H_5SLi	R = SC_2H_5 (67)	86
	R = Ms	$(CH_3)_2AlSeCH_3$, CH_2Cl_2, 0°, 1.5 h	R = $SeCH_3$ (49)	147
		$(i\text{-}C_4H_9)_2AlSeC_6H_5$, CH_2Cl_2, 0°, 30 min	R = SeC_6H_5 (61)	147
		$(CH_3)_2AlS$-(2-pyridyl), CH_2Cl_2, 0–25°	$C_6H_5N{=}C(CH_3)S$-(2-pyridyl) (68)	147
		$(CH_3)_2AlS(CH_2)_6SAl(CH_3)_2$, 0°, 1 h	$C_6H_5N{=}C(CH_3)S(CH_2)_6SC(CH_3){=}NC_6H_5$ (46)	147
		$(CH_3)_3SiCN$, $(C_2H_5)_2AlCl$, −78 to −20°, 1 h	$C_6H_5N{=}C(CN)CH_3$ (95)	147
	R = $COCH_3$ or $CO_2C_2H_5$	$(CH_3)_3SiI$, $CDCl_3$	$C_6H_5N{=}ClCH_3$ (100)	86
		1. See below 2. DIBAL	$C_6H_5NHCH(CH_3)R$	
	R = Ms	Reagent: $(CH_3)_3Al$, CH_2Cl_2, −78°	R = CH_3 (67)	147, 153
		$(C_2H_5)_2AlC{\equiv}CCH_3$, CH_2Cl_2, −78°	R = $C{\equiv}CCH_3$ (60)	147, 153
		$(C_2H_5)_2AlC{\equiv}CC_4H_9\text{-}n$, CH_2Cl_2, −78°	R = $C{\equiv}CC_4H_9\text{-}n$ (83)	147, 153

TABLE II. ELIMINATION-ADDITIONS (*Continued*)

No. of Carbon Atoms	Substrate	Reagent and Conditions	Product(s) and Yield(s) (%)	Refs.
		$(C_2H_5)_2AlC{\equiv}CC_6H_5$, CH_2Cl_2, −78°	R = $C{\equiv}CC_6H_5$ (67)	147, 153
		CH_3MgBr, $C_6H_5CH_3$, −78°	R = CH_3 (52)	157
		n-C_4H_9MgBr, $C_6H_5CH_3$, −78°	R = C_4H_9-n (55)	157
		n-$C_4H_9C{\equiv}CMgBr$, $C_6H_5CH_3$, −78°	R = $C{\equiv}CC_4H_9$-n (47)	157
		1. $(C_2H_5)_2AlI$, −78° 2. C_6H_5MgBr	R = C_6H_5 (44)	86
	R = $CO_2C_2H_5$	1. $(CH_3)_3SiI$ 2. $CH_3CH{=}CHMgBr$	R = $CH{=}CHCH_3$ (69)	86
		1. $(CH_3)_3SiI$ 2. C_6H_5MgBr	R = C_6H_5 (61)	86
	R = Ms	$(C_2H_5)_2AlCH_2CH{=}CH_2$, CH_2Cl_2	$C_6H_5NHC(CH_3)(CH_2CH{=}CH_2)_2$ (47)	147
		1. $(CH_3)_3Al$, CH_2Cl_2 2. $CH_2{=}CHCH_2MgBr$	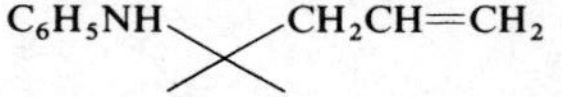 C_6H_5NH–C(CH_3)$_2$–$CH_2CH{=}CH_2$ (54)	147
		1. $(C_2H_5)_2AlC{\equiv}CC_4H_9$-$n$, CH_2Cl_2 2. $CH_2{=}CHCH_2MgBr$	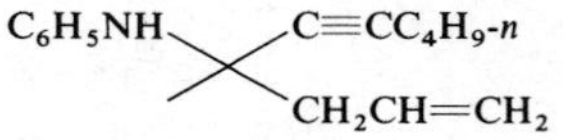 C_6H_5NH–C(–$C{\equiv}CC_4H_9$-n)(–$CH_2CH{=}CH_2$) (74)	147
		1. $(CH_3)_3Al$, CH_2Cl_2 2. $CH_3CH{=}CHCH_2MgBr$	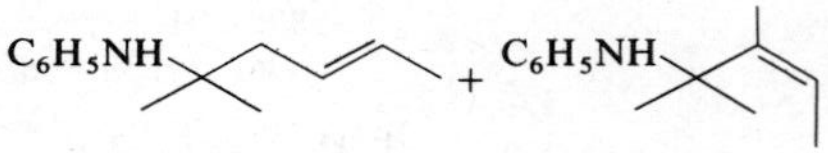 C_6H_5NH[structure] + C_6H_5NH[structure] (56)	147
	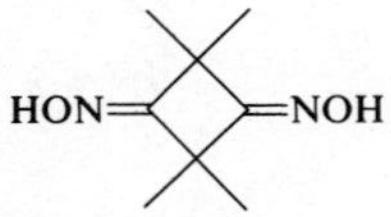 HON=[cyclobutane, tetramethyl]=NOH	$C_6H_5SO_2Cl$, NaOH, H_2O, 14 h	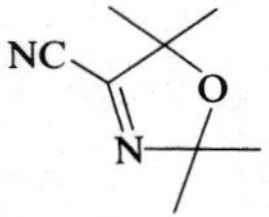 NC-[oxazoline structure] (—)	156
	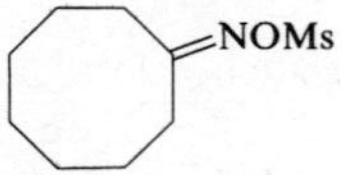 [cyclooctanone]=NOMs	$(CH_3)_3SiCN$, $(C_2H_5)_2AlCl$, −78 to −20°, 1 h	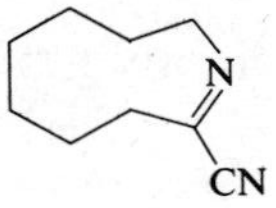 [azacyclononene, N, CN] (48)	147

	Substrate	Conditions	Product(s) and Yield(s) (%)	Refs.
C_9	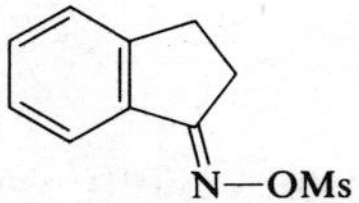 N—OMs	DIBAL, CH_2Cl_2, $-78°$	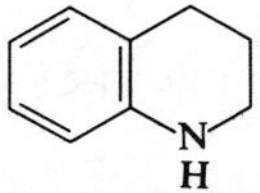 N H (80)	147, 153
	$C_6H_5C({=}NOR)C_2H_5$ R = H	$(C_6H_5)_3P$, CCl_4, CH_3CN, 0°, 15 h	$C_6H_5N{=}CClC_2H_5$ (67)	144
	R = Ms	$(CH_3)_3SiCN$, $(C_2H_5)_2AlCl$ -78 to $-20°$, 1 h	$C_6H_5N{=}C(C_2H_5)CN$ (93)	147
C_{10}	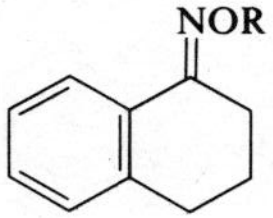 NOR R = H	DIBAL, 5 eq, 0-20°	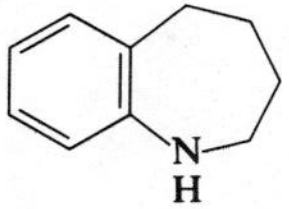 N H (92)	154
	R = Ms	1. $(n\text{-}C_3H_7)_3Al$ 2. DIBAL	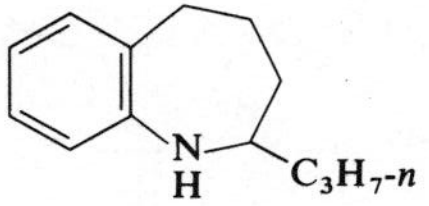 N H C_3H_7-n (88)	147, 153
	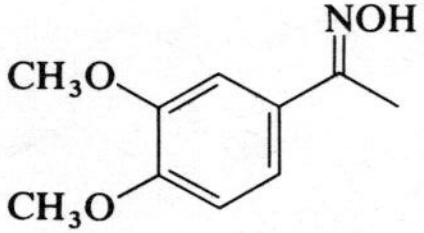 CH_3O CH_3O NOH	DIBAL, 5 eq	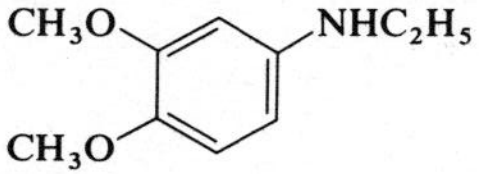 CH_3O CH_3O NHC_2H_5 (74)	154
	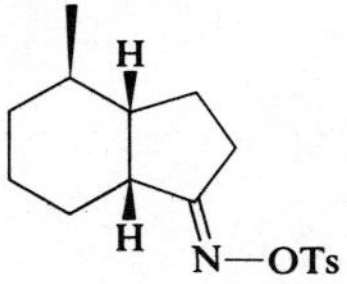 H H N—OTs	$(i\text{-}C_4H_9)_2AlSCH_3$, $C_2H_4Cl_2$, 50°, 6 min	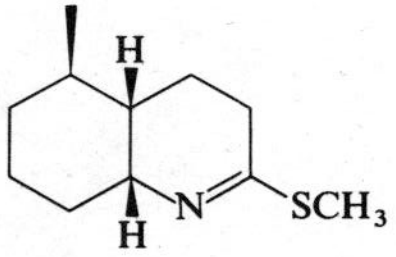 H H N SCH_3 (70)	147
		1. $(CH_3)_3Al$ 2. DIBAL	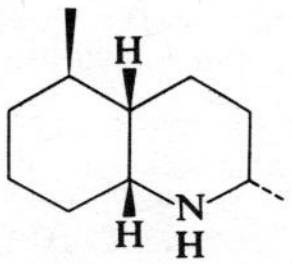 H H N H (57)	147, 153

TABLE II. ELIMINATION-ADDITIONS (*Continued*)

No. of Carbon Atoms	Substrate	Reagent and Conditions	Product(s) and Yield(s) (%)	Refs.
	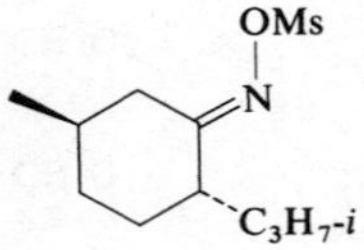 [structure: OMs oxime of 5-methyl-2-isopropylcyclohexanone, C_3H_7-*i*]	1. $(n\text{-}C_3H_7)_3Al$ 2. DIBAL	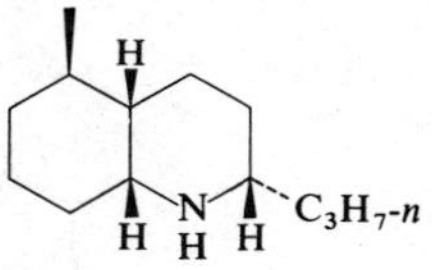 [structure: decahydroquinoline, C_3H_7-*n*] (60)	147, 153
		DIBAL, ether, −78 to 0°, 1 h	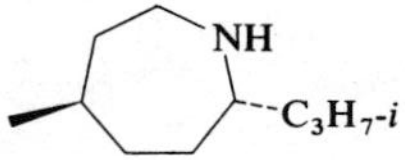 [structure: azepane, NH, C_3H_7-*i*] (82)	147, 153
		1. $(CH_3)_3Al$, $C_6H_5CH_3$, −20 to 0°, 2 h 2. DIBAL, 0°, 1 h	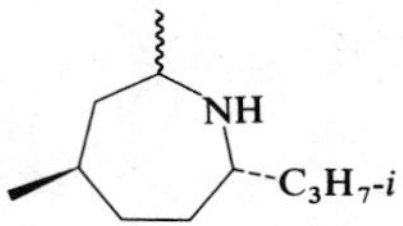 [structure: methyl-substituted azepane, NH, C_3H_7-*i*] (57)	147, 153
C_{11}	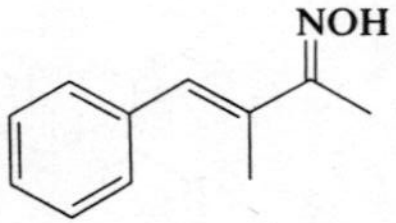 [structure: $C_6H_5CH{=}C(CH_3)C(CH_3){=}NOH$]	PCl_5, ether, 0°, 15 min	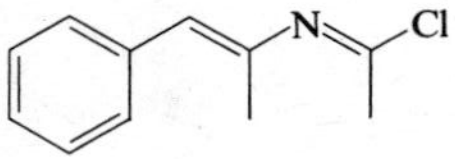 [structure: $C_6H_5CH{=}C(CH_3)N{=}C(Cl)CH_3$] (—)	143
	p-$CH_3C_6H_4C({=}NOH)C_3H_7$-*i*	CH_3SO_2Cl, $(C_2H_5)_3N$, $C_6H_5CH_3$, −15 to 100°	*p*-$CH_3C_6H_4N{=}C{=}C(CH_3)_2$ (83)	138
	$(n\text{-}C_5H_{11})_2C{=}NOH$	DIBAL, 5 eq, 0–20°	*n*-$C_5H_{11}NHCH_2C_5H_{11}$-*n* (85)	154
		1. CH_3SO_2Cl, $(C_2H_5)_3N$ 2. $(CH_3)_3SiCN$, $(C_2H_5)_2AlCl$, −78 to −20°, 1 h	*n*-$C_5H_{11}N{=}C(CN)C_5H_{11}$-*n* (91)	147

	Reactant	Reagents/Conditions	Product(s) and Yield(s) (%)	Refs.
C_{12}	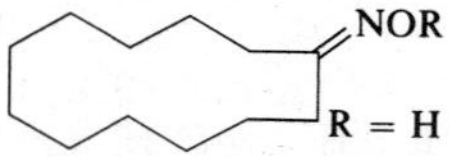 NOR, R = H or Ms	DIBAL, CH_2Cl_2, $-78°$	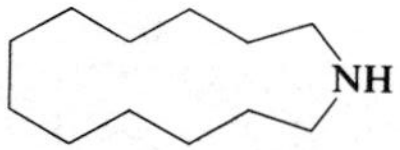 NH (73)	147, 154
	R = Ms	$(CH_3)_3SiI$, $CDCl_3$, 0°	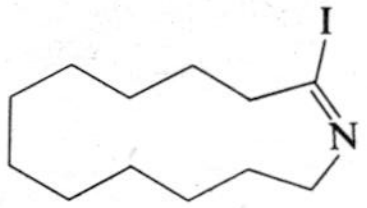 I, N (96)	86
	R = $CO_2C_2H_5$	$(CH_3)_3SiI$, $CDCl_3$, 30°	" (85)	86
	R = Ms	$(CH_3)_3SiCN$, $(C_2H_5)_2AlCl$, -78 to $-20°$, 1 h	R = CN (92)	147
		$(i\text{-}C_4H_9)_2AlSCH_3$, CH_2Cl_2, 0°, 1 h	R = SCH_3 (95)	147
		$(CH_3)_2AlSeCH_3$, CH_2Cl_2, 0°, 1 h	R = $SeCH_3$ (71)	147
		$(i\text{-}C_4H_9)_2AlSeC_6H_5$, 0°, 30 min	R = SeC_6H_5 (57)	147
	R = Ts	$(CH_3)_2AlSCH_2CH{=}CH_2$, CH_2Cl_2, 0°, 1 h	R = $SCH_2CH{=}CH_2$ (80)	147
		$(CH_3)_2AlSC_6H_5$, CH_2Cl_2, 0°, 1 h	R = SC_6H_5 (82)	147
		$(i\text{-}C_4H_9)_2AlSCH_3$, CH_2Cl_2, 0°, 1 h	R = SCH_3 (97)	147
		1. See below 2. DIBAL	R, NH	
	R = Ts	Reagents: $(CH_3)_3Al$, $-78°$, CH_2Cl_2	R = CH_3 (56)	147
		$(C_2H_5)_2AlC{\equiv}CC_6H_5$, $-78°$, CH_2Cl_2	R = $C{\equiv}CC_6H_5$ (71)	147
	R = Ms	$(CH_3)_3Al$, $-78°$, CH_2Cl_2	R = CH_3 (60)	147
		CH_3MgBr, $C_6H_5CH_3$, -78 to 0°, 1 h	R = CH_3 (66)	157
		$n\text{-}C_4H_9MgBr$, $C_6H_5CH_3$, -78 to 0°, 1 h	R = $C_4H_9\text{-}n$ (63)	157
		$n\text{-}C_8H_{17}MgBr$, $C_6H_5CH_3$, -78 to 0°, 1 h	R = C_8H_{17} (68)	157
		1. $(CH_3)_3SiI$ 2. C_6H_5MgBr	R = C_6H_5 (51)	86

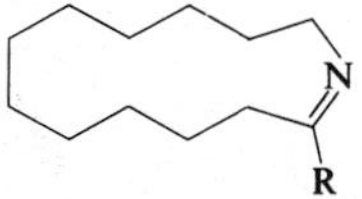

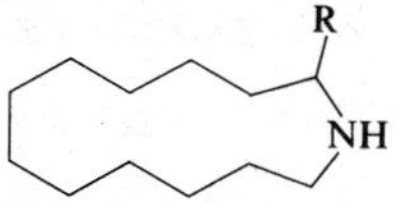

TABLE II. ELIMINATION-ADDITIONS (*Continued*)

No. of Carbon Atoms	Substrate	Reagent and Conditions	Product(s) and Yield(s) (%)	Refs.
			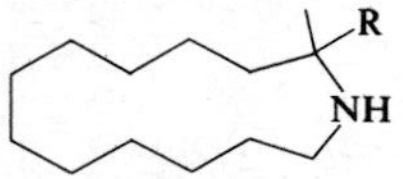 (ring structure, NH, R)	
		1. $(CH_3)_3Al$, −78°, CH_2Cl_2 2. $CH_2{=}CHCH_2MgBr$	R = $CH_2CH{=}CH_2$ (84)	147
		1. CH_3MgBr, $C_6H_5CH_3$, −78 to 0°, 1 h 2. $CH_2{=}CHCH_2MgBr$	R = $CH_2CH{=}CH_2$ (76)	157
		1. $(CH_3)_3Al$ 2. $HC{\equiv}CCH_2MgBr$	R = $CH_2C{\equiv}CH$ (61)	147
		1. CH_3MgBr, $C_6H_5CH_3$ 2. $HC{\equiv}CCH_2MgBr$	R = $CH_2CH{\equiv}CH$ (79)	157
		1. $(C_2H_5)_2AlC{\equiv}CC_6H_5$, CH_2Cl_2 2. $CH_2{=}CHCH_2MgBr$	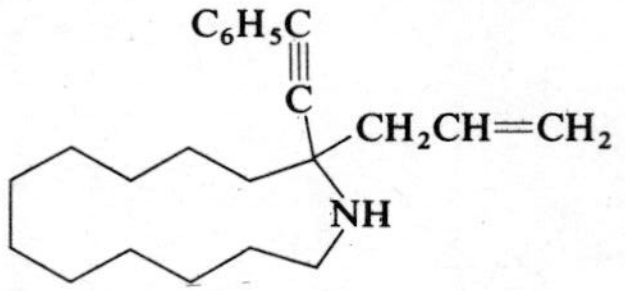 $C_6H_5C{\equiv}C$, $CH_2CH{=}CH_2$, NH (ring structure) (88)	147
		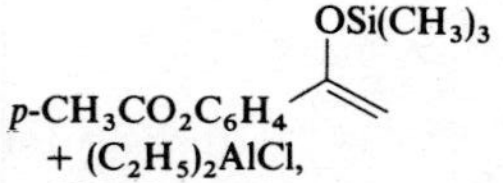 p-$CH_3CO_2C_6H_4C(OSi(CH_3)_3){=}CH_2$ + $(C_2H_5)_2AlCl$, CH_2Cl_2, −78 to 20°, 1 h	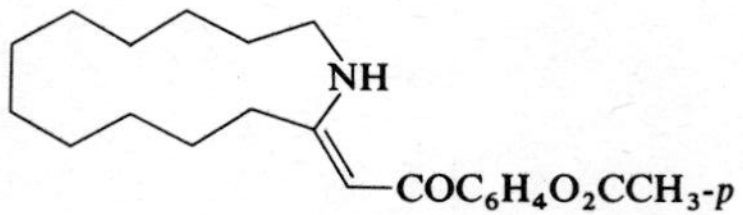 NH (ring structure) $=CHCOC_6H_4O_2CCH_3$-p (—)	159
C_{13}	$(p\text{-}BrC_6H_4)_2C{=}NOH$	PCl_5, C_6H_6	p-$BrC_6H_4N{=}CClC_6H_4Br$-p (80)	141
	$(p\text{-}O_2NC_6H_4)_2C{=}NOH$	"	p-$O_2NC_6H_4N{=}CClC_6H_4NO_2$-$p$ (82)	141
	p-$ClC_6H_4C({=}NOH)C_6H_5$	$(C_6H_5)_3P$, CCl_4, CH_3CN, 0°, 15 h	$C_6H_5N{=}CClC_6H_4Cl$-p (87)	144
	$(C_6H_5)_2C{=}NOH$	$(C_6H_5)_3P$, CCl_4, CH_3CN, 0°, 15 h	$C_6H_5N{=}CClC_6H_5$ (79)	144
		Polymer-$P(C_6H_5)_2$, CCl_4, $C_2H_4Cl_2$, reflux	" (88)	145

	Reactant	Conditions	Product(s) (Yield, %)	Refs.
		1,1'-Carbonyldiimidazole ($(C_3H_3N_2)_2C{=}O$)	$C_6H_5N{=}C(C_6H_5)(N\text{-imidazolyl})$ (95)	146
	$(C_6H_5)_2C{=}NCl$	$AgBF_4$, $NaOC_2H_5$, DME	$C_6H_5N{=}C(OC_2H_5)C_6H_5$ (—)	120
	$C_6H_5SO_2NHC({=}NOH)C_6H_4NO_2$-*m*	PCl_5, ether, 1 h	$C_6H_5SO_2N{=}CClNHC_6H_4NO_2$-*m* (—)	118
	$C_6H_5SO_2NHC({=}NOH)C_6H_4NO_2$-*p*	"	$C_6H_5SO_2N{=}CClNHC_6H_4NO_2$-*p* (—)	118
	$C_6H_5SO_2NHC({=}NOH)C_6H_5$	"	$C_6H_5SO_2N{=}CClNHC_6H_5$ (—)	118
		$SOCl_2$, ether, reflux, 30 min	"	118
	2-Benzylidenecyclohexanone oxime ($HO{-}N{=}C_6H_8{=}CHC_6H_5$)	TsCl, pyridine, reflux	2-(1-Pyridinio)-7-benzylidene-azepine tosylate ($\bar{O}Ts$) (75)	31
	2-n-C_8H_{17}-cyclopentanone O-mesyloxime ($N{-}OMs$)	1. CH_3MgBr, $C_6H_5CH_3$, −78 to 0°, 1 h 2. DIBAL	2-n-C_8H_{17}-6-methylpiperidine (N–H) (36)	157
C_{14}	p-$CH_3C_6H_4SO_2NHC({=}NOH)C_6H_4Br$-*p*	PCl_5, ether, 1 h	p-$CH_3C_6H_4SO_2N{=}CClNHC_6H_4Br$-*p* (81)	118
	p-$CH_3C_6H_4SO_2NHC({=}NOH)C_6H_5$	"	p-$CH_3C_6H_4SO_2N{=}CClNHC_6H_5$ (84)	118

TABLE II. ELIMINATION-ADDITIONS (*Continued*)

No. of Carbon Atoms	Substrate	Reagent and Conditions	Product(s) and Yield(s) (%)	Refs.
	$C_6H_5SO_2NH-C(=NOH)-C_6H_4CH_3$-*p*	"	$C_6H_5SO_2N{=}CClNHC_6H_4CH_3$-*p* (77)	118
C_{15}	$(C_6H_5CH_2)_2C{=}NOH$	DIBAL, 5 eq	$C_6H_5CH_2NH(CH_2)_2C_6H_5$ (71)	154
	p-$CH_3C_6H_4SO_2NH-C(=NOH)-C_6H_4CH_3$-*p*	PCl_5, ether, 1 h	*p*-$CH_3C_6H_4SO_2N{=}CClNHC_6H_4CH_3$-*p* (75)	118
C_{16}	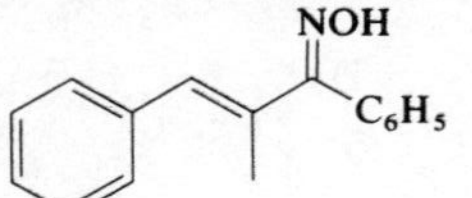 NOH, C_6H_5	"	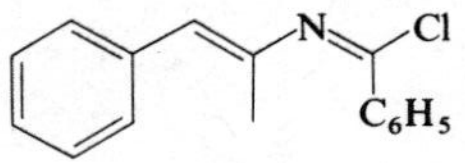 N, Cl, C_6H_5 (—)	143
	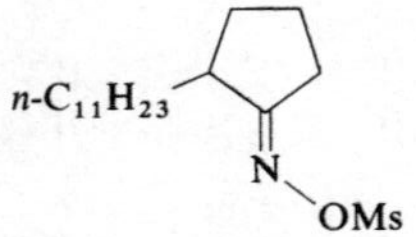 *n*-$C_{11}H_{23}$, N, OMs	$(CH_3)_3Al$, CH_2Cl_2, −78 to 25°, 1 h	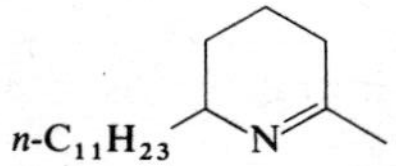 *n*-$C_{11}H_{23}$, N (57)	147, 158
C_{17}	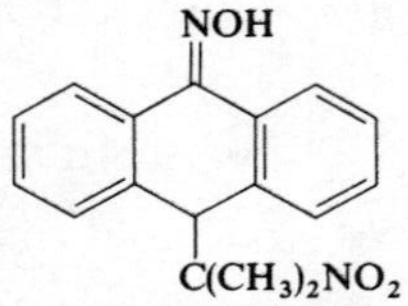 NOH, $C(CH_3)_2NO_2$	PCl_5, CCl_4	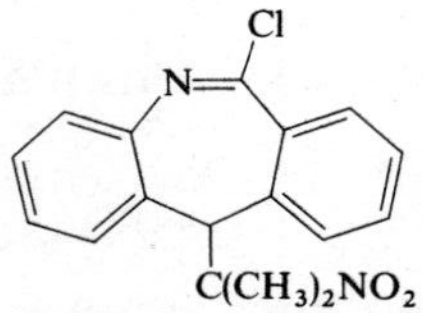 Cl, N, $C(CH_3)_2NO_2$ (—)	142
	$[p\text{-}(CH_3)_2NC_6H_4]_2C{=}NOH$	$C_6H_5SO_2Cl$, KOH, acetone, H_2O, 4 KCN	*p*-$(CH_3)_2NC_6H_4N{=}C(CN)C_6H_4N(CH_3)_2$-*p* 87)	155

C_{18}	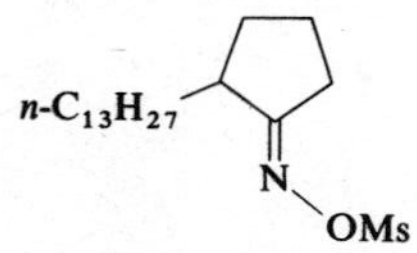 n-$C_{13}H_{27}$ (cyclopentanone oxime, N–OMs)	$(CH_3)_3Al$, CH_2Cl_2, −78 to 25°, 1 h	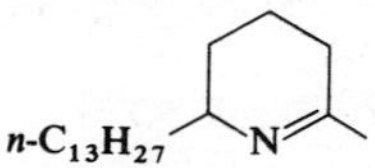 n-$C_{13}H_{27}$ (tetrahydropyridine) (—)	147
C_{21}	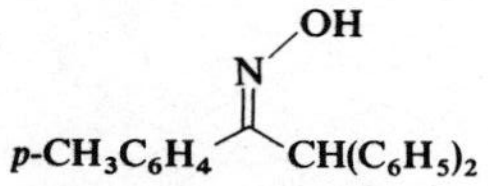 p-$CH_3C_6H_4$C(=NOH)$CH(C_6H_5)_2$	CH_3SO_2Cl, $(C_2H_5)_3N$, $C_6H_5CH_3$, −15 to 100°	$(C_6H_5)_2C{=}C{=}NC_6H_4CH_3$-p (95)	138
	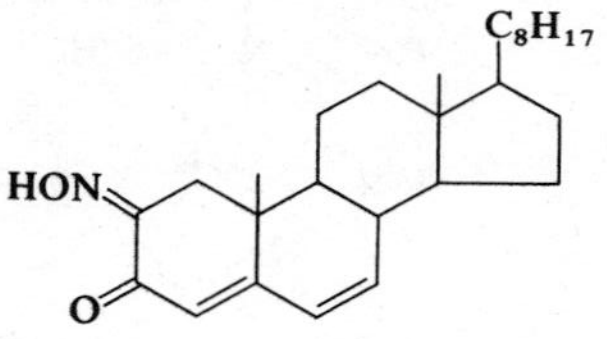 HON, O, C_8H_{17} (steroid)	TsCl, pyridine	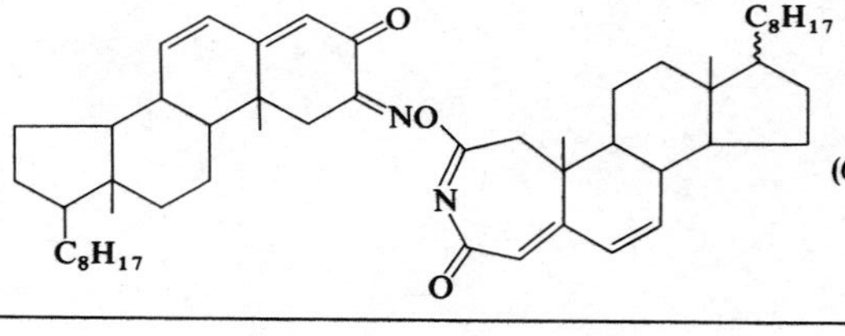 O, NO, N, O, C_8H_{17}, C_8H_{17} (63)	139

TABLE III. KETOXIME FRAGMENTATIONS

No. of Carbon Atoms	Substrate	Reagent and Conditions	Product(s) and Yield(s) (%)	Refs.
C_4	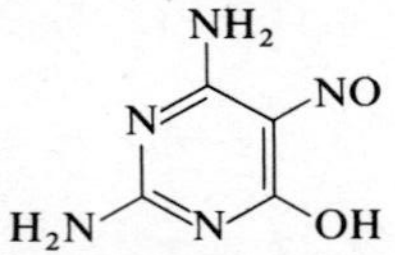 (pyrimidine ring: NH_2, NO, OH, H_2N)	$(CF_3CO)_2O$	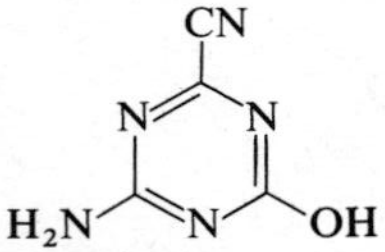 (triazine ring: CN, H_2N, OH) (—)	561
	$CH_3C(NF_2)_2C(=NF)CH_3$	H_2SO_4 or HSO_3F or HNF_2	$CH_3C(NF_2)_3$ (3)	121
C_5	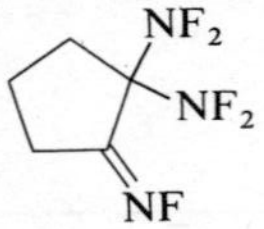 (cyclopentane ring: NF_2, NF_2, =NF)	"	$NC(CH_2)_3C(NF_2)_3$ (15)	121
		BF_3, CH_2Cl_2	" (—)	121
	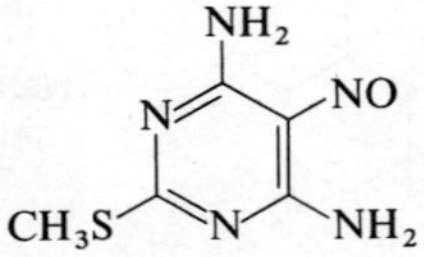 (pyrimidine ring: NH_2, NO, CH_3S, NH_2)	$(CH_3CO)_2O$	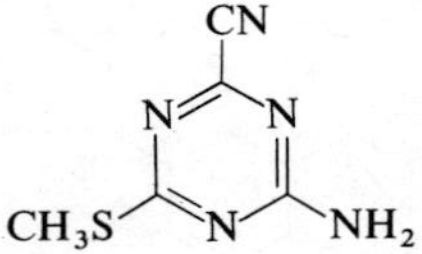 (triazine ring: CN, CH_3S, NH_2) (100)	561
	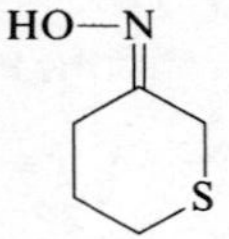 (thiane ring: HO—N=)	$SOCl_2$, C_6H_6	$NC(CH_2)_3SCH_2Cl$ (72)	193
		PCl_5, ether, 15 h	$[NC(CH_2)_3S]_2CH_2$ (85)	193
		TsCl, pyridine, 24 h	" (—)	194
		1. TsCl, pyridine 2. $(C_2H_5)_3N$, ethanol	$C_2H_5OCH_2S(CH_2)_3CN$ (92)	194
	$(CH_3)_2C(NF_2)C(=NF)CH_3$	HNF_2, H_2SO_4	$(CH_3)_2C(NF_2)_2$ (—)	121

	Reactant	Conditions	Product(s) (Yield, %)	Refs.
	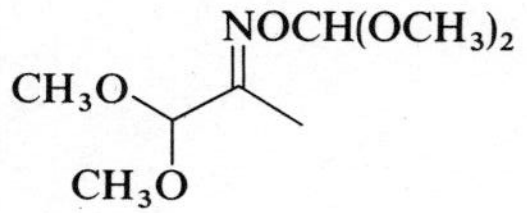 $NOCH(OCH_3)_2$, CH_3O, CH_3O	CH_3SO_3H, $CHCl_3$, 75°, 2 h	$CH_3CO_2CH_3$ (60) + CH_3CN (60) + $HC(OCH_3)_3$ (60) + $CH_3C(OCH_3)_3$ (10)	189
C_6	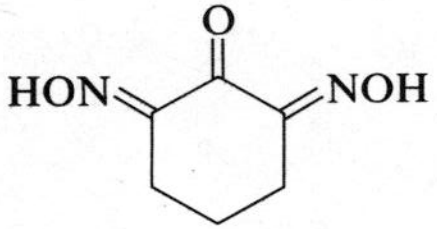 O, HON, NOH	1. $NaOCH_3$, CH_3OH 2. $(CH_3CO)_2O$	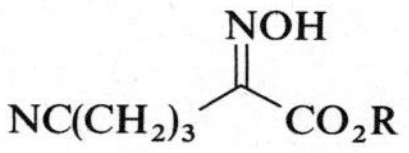 NOH, $NC(CH_2)_3$, CO_2R R = CH_3 (74)	562
		1. $Mg(OCH_3)_2$ 2. $(CH_3CO)_2O$	" R = CH_3 (65)	562
		1. $C_6H_5CH_2\overset{+}{N}(CH_3)_3\bar{O}H$, CH_3OH 2. $(CH_3CO)_2O$	" R = CH_3 (59)	562
		1. $NaOC_2H_5$, ethanol 2. $(CH_3CO)_2O$	" R = C_2H_5 (92)	562
		1. KOH, ethanol 2. $(CH_3CO)_2O$	" R = C_2H_5 (65)	562
		$(CH_3CO)_2O$, NaOH	" R = H (62)	563
		1. $NaOC_3H_7$-*i* 2. $(CH_3CO)_2O$	" R = C_3H_7-*i* (51)	562
		1. $NaOCH_2C_6H_5$ 2. $(CH_3CO)_2O$	" R = $CH_2C_6H_5$ (54)	562
	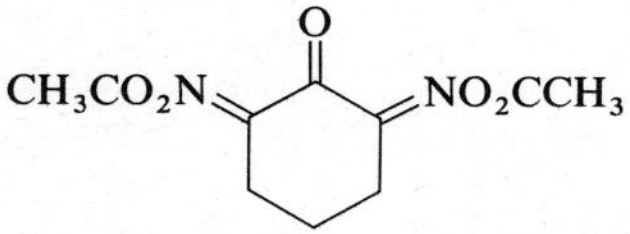 O, CH_3CO_2N, NO_2CCH_3	$NaOC_2H_5$, ethanol	" R = C_2H_5 (88)	564
	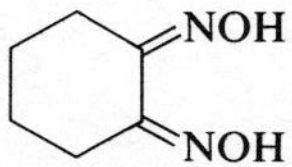 NOH, NOH	SO_3, SO_2, 1 h	$NC(CH_2)_4CO_2H$ (23)	248

TABLE III. KETOXIME FRAGMENTATIONS (*Continued*)

No. of Carbon Atoms	Substrate	Reagent and Conditions	Product(s) and Yield(s) (%)	Refs.
	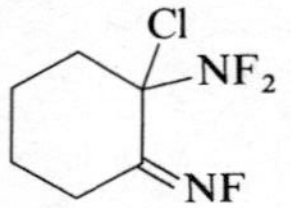 Cl, NF_2, NF (cyclohexane ring)	BF_3, CH_2Cl_2	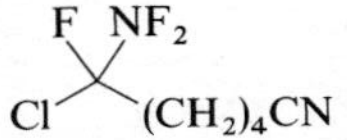 F, NF_2, Cl, $(CH_2)_4CN$ (25)	121, 565
		HNF_2, HSO_3F	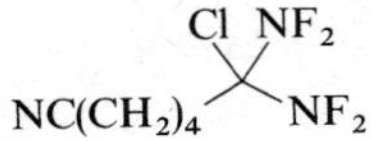 Cl, NF_2, $NC(CH_2)_4$, NF_2 (23)	121
	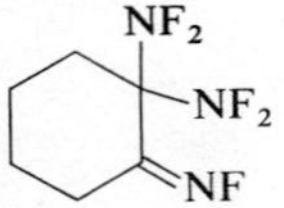 NF_2, NF_2, NF (cyclohexane ring)	H_2SO_4 or HSO_3F or HNF_2	$NC(CH_2)_4C(NF_2)_3$ (11)	121
	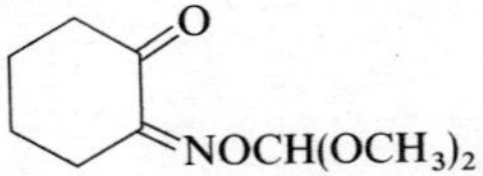 O, $NOCH(OCH_3)_2$ (cyclohexane ring)	(—)	$NC(CH_2)_4CO_2CH_3$ (—)	189
	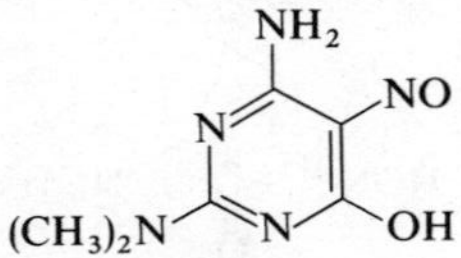 NH_2, NO, N, N, $(CH_3)_2N$, OH (pyrimidine ring)	$(CH_3CO)_2O$	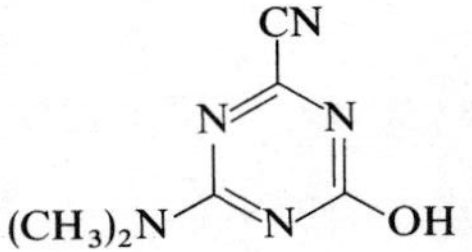 CN, N, N, N, $(CH_3)_2N$, OH (triazine ring) (84)	561
	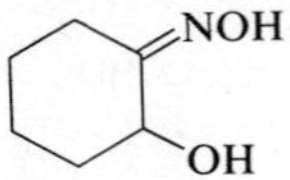 NOH, OH (cyclohexane ring)	C_6H_5NCO, $(C_2H_5)_3N$, C_6H_6, reflux, 1 h	$NC(CH_2)_4CHO$ (72)	440
		$(PNCl_2)_3$	" (75)	566

	Reactant	Conditions	Product(s) (Yield, %)	Ref.
	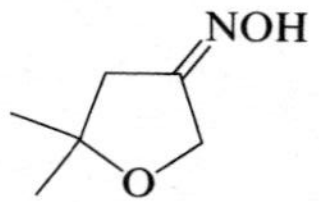 NOH (2,2-dimethyltetrahydrofuran-4-one oxime)	PCl_5, C_6H_6, 10 h	CH_2O (43)	187
	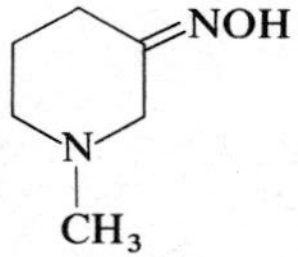 NOH, N–CH_3 (1-methylpiperidin-3-one oxime)	TsCl, KOH, H_2O, 24 h	$NC(CH_2)_3N(CH_3)Ts$ (46)	190
	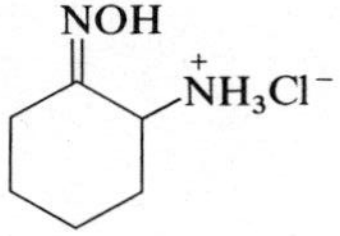 NOH, $\overset{+}{N}H_3Cl^-$ (2-aminocyclohexanone oxime hydrochloride)	H_2SO_4	$NC(CH_2)_4CHO$ (—)	567
	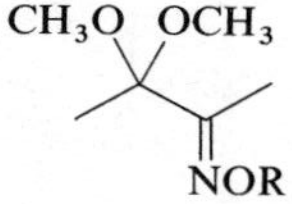 CH_3O OCH_3, NOR; R = H	SO_2, 6 h	CH_3CN (100) + $CH_3CO_2CH_3$ (100) + CH_3OH (100)	189
	R = $CH(OCH_3)_2$	SO_2, 1 h	" (100) + " (97) + " (97)	189
		CH_3SO_3H, CCl_4, 1 h	" (100) + " (97) + " (97)	189
	R = $C(OCH_3)C_2H_5$	"	" (100) + " (50) + " (50) + $C_2H_5CO_2CH_3$ (50) + $CH_3C(OCH_3)_3$ (50)	189
C_7	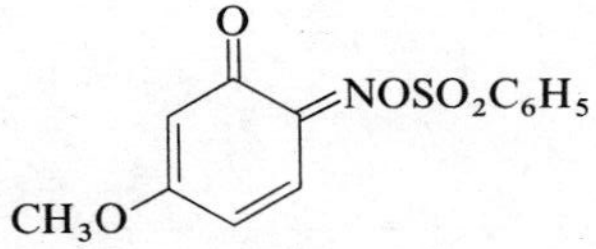 O, $NOSO_2C_6H_5$, CH_3O (4-methoxy-1,2-benzoquinone 2-(O-benzenesulfonyl)oxime)	Pyridine, H_2O	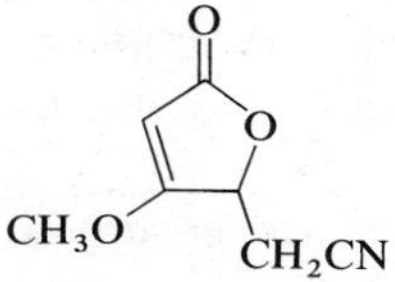 O, O, CH_3O, CH_2CN (4-methoxy-5-cyanomethyl-2(5H)-furanone) (—)	568

TABLE III. KETOXIME FRAGMENTATIONS (*Continued*)

No. of Carbon Atoms	Substrate	Reagent and Conditions	Product(s) and Yield(s) (%)	Refs.
	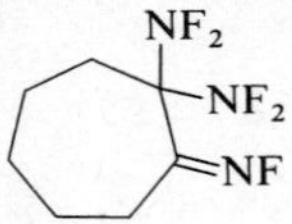 [structure: cycloheptane ring bearing NF_2, NF_2 and =NF]	H_2SO_4 or HSO_3F or HNF_2	$NC(CH_2)_5C(NF_2)_3$ (8)	121
	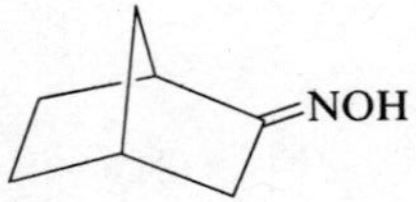 [structure: bicyclic ketoxime, =NOH]	$C_6H_5SO_2Cl$, NaOH	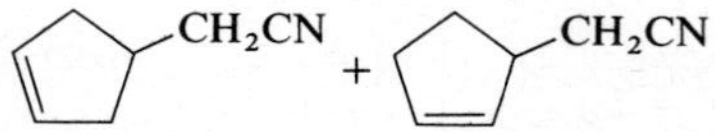 [structure: cyclopentenyl-CH_2CN] (10) + [structure: cyclopentenyl-CH_2CN] (20)	268
	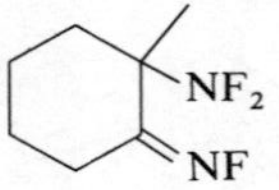 [structure: methylcyclohexane ring bearing NF_2 and =NF]	HNF_2, HSO_3F	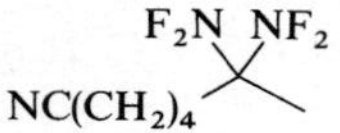 $NC(CH_2)_4C(CH_3)(NF_2)_2$ [structure: F_2N, NF_2] (15)	121
		BF_3, CH_2Cl_2	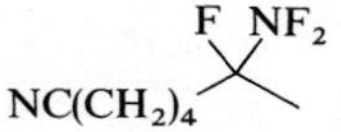 $NC(CH_2)_4C(CH_3)(F)(NF_2)$ [structure: F, NF_2] (—)	121, 565
	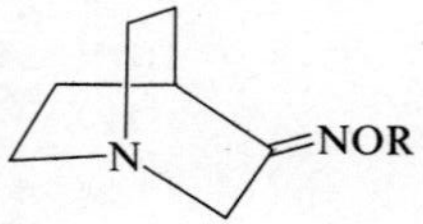 [structure: quinuclidinone oxime, =NOR] R = H	PPA, 130°, 10 min	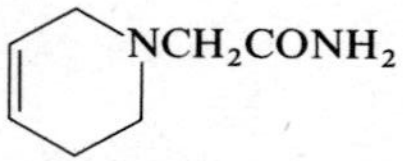 [structure: tetrahydropyridine]-NCH_2CONH_2 (—)	274
		$C_6H_5SO_2Cl$, NaOH, H_2O, 90 min	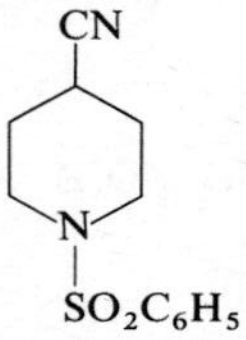 [structure: 4-cyanopiperidine, N-$SO_2C_6H_5$] (40)	273

	TsCl, NaOH, H_2O, 90 min	 [structure: 4-cyano-1-tosylpiperidine, CN / N—Ts] (24)	190, 273
$R = COC_6H_5$	KOH, 80% CH_3OH	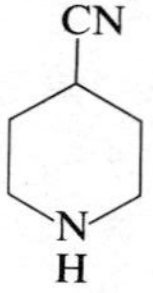 [structure: 4-cyanopiperidine, CN / N—H] (55)	190
n-$C_3H_7COC(=NOH)C_2H_5$	85% H_2SO_4, 120°	$C_2H_5CONH_2$ (19) + n-$C_3H_7CO_2H$ (78)	197, 569
	$C_6H_5SO_2Cl$, NaOH	C_2H_5CN (45) + " (84)	197, 569
$CH_3COC(=NOH)C_4H_9$-n	"	n-C_4H_9CN (82)	197, 569
	PCl_5, ether	" (70)	197, 569
	CH_3COCl, NaOH	" (70–80)	197
	CF_3CO_2H	" (58)	197
	85% H_2SO_4, 120°	n-$C_4H_9CONH_2$ (59)	197, 569
	PPA, 120°, 18 min	" (24) + n-$C_4H_9CO_2H$ (28)	197
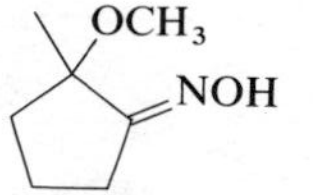 [structure: 2-methyl-2-methoxycyclopentanone oxime, OCH_3, NOH]	TsCl, C_6H_6, H_2O	$CH_3CO(CH_2)_3CN$ (70)	570
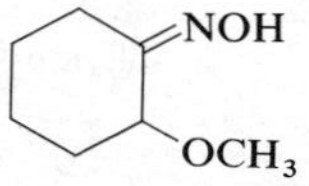 [structure: 2-methoxycyclohexanone oxime, NOH, OCH_3]	PCl_5, ether, 0°	$NC(CH_2)_4CHO$ (64)	571
		$NC(CH_2)_4CHClOCH_3$ (90)	572

TABLE III. KETOXIME FRAGMENTATIONS (*Continued*)

No. of Carbon Atoms	Substrate	Reagent and Conditions	Product(s) and Yield(s) (%)	Refs.
	2,2-dimethylcyclopentanone oxime (NOH)	PCl_5, ether, 0–25°, 24 h	$(CH_3)_2C{=}CH(CH_2)_2CN$ (94)	226
	CH_3O, OCH_3 (2,2-dimethoxycyclopentanone oxime, =NOH)	$C_2H_5C(OCH_3)_2 \cdot BF_4$, 1 h	$NC(CH_2)_3CO_2CH_3$ (90)	189
	$N(CH_3)_2$ (2-dimethylaminocyclopentanone oxime, =NOH)	TsCl, dioxane	$NC(CH_2)_3CHO$ (16)	570
	$(CH_3)_2N-C(CH_3)_2-C(=NOH)CH_3$	TsCl, NaOH, H_2O	CH_3COCH_3 (62)	570
C_8	$C_6H_5C(=N-O-C_6H_3(NO_2)_2)CH_2NH_2$ (2,4-dinitrophenyl, NO_2, NO_2)	80% ethanol, 80°, 1000 h	$C_6H_5CONH_2$ (95)	95
	cyclooctane-1,2-dione mono-oxime (O, NOH) (*Z*)	97% H_3PO_4	$NC(CH_2)_6CO_2H$ (80)	387
	(*E*)	"	" (78)	387

Reactant	Conditions	Product(s) and Yield(s) (%)	Refs.
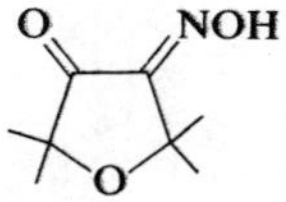 O, NOH (Z)	170°, 5 min	 HO_2C … CO_2H (50)	573
(E)	"	" (50)	573
(Z)	TsCl, 10% NaOH, reflux, 15 min	" (8)	573
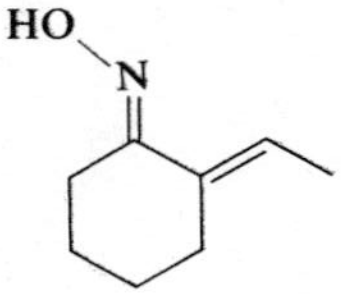 HO, N	TsCl, pyridine	$NC(CH_2)_4COC_2H_5$ (—)	31
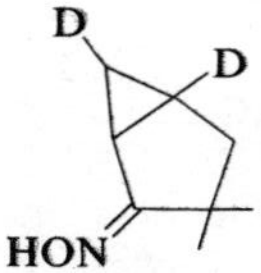 D, D, HON	1. PCl_5, pyridine, 23 h 2. TsOH, $C_6H_5CH_3$, reflux, 2 h	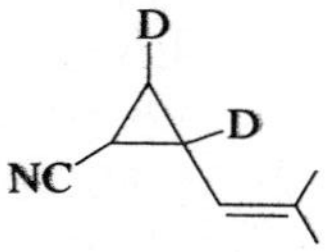 D, D, NC (98)	574
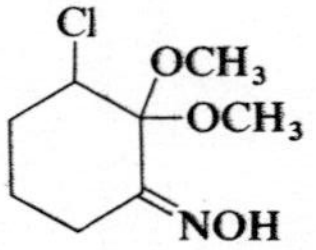 Cl, OCH_3, OCH_3, NOH	$C_2H_5C(OCH_3)_2 \cdot BF_4$, CH_2Cl_2	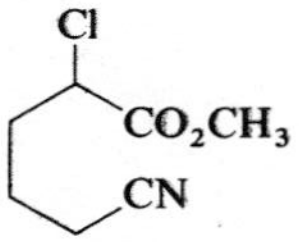 Cl, CO_2CH_3, CN (—)	575
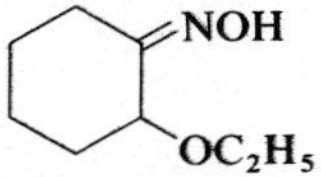 NOH, OC_2H_5	PCl_5, ether, 0°	$NC(CH_2)_4CHO$ (60)	571
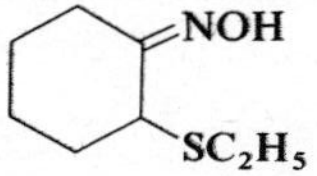 NOH, SC_2H_5	"	" (32)	571

TABLE III. KETOXIME FRAGMENTATIONS (*Continued*)

No. of Carbon Atoms	Substrate	Reagent and Conditions	Product(s) and Yield(s) (%)	Refs.
	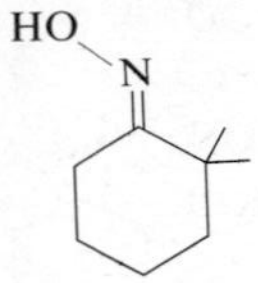 HO N	PCl_5, ether, 24 h	$(CH_3)_2C{=}CH(CH_2)_3CN$ (98) I	226
		$C_6H_5SO_2Cl$, NaOH, acetone, H_2O, reflux, 4 h	I (27) + i-$C_3H_7(CH_2)_5CONH_2$ (54)	226
	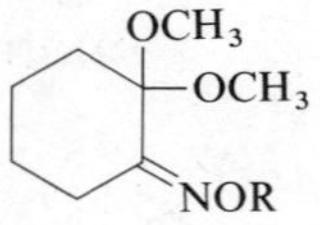 OCH_3 OCH_3 NOR R = H	SO_2, 24 h	$NC(CH_2)_4CO_2CH_3$ (16)	189
		$HC(OCH_3)_3$, SO_2, 24 h	" (97)	188, 189
		$CH_3C(OCH_3)_3$, CH_3SO_3H, C_6H_6, 80°, 6 h	" (96)	189
		$C_2H_5C(OCH_3)_2\cdot BF_4$, CH_2Cl_2, 1 h	" (93)	189
		$SOCl_2$, SO_2, $-10°$, 25 h	" (100)	189
		TsCl, pyridine, 0°, 3 h	" (61)	189
	R = $CH(OCH_3)_2$	CH_3SO_3H, CCl_4, 24 h	" (93)	189
		SO_2, 1 h	" (95)	189
	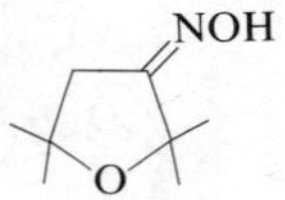 NOH O	PCl_5, C_6H_6, 48 h	$(CH_3)_2C{=}CHCN$ (43)	187
		$SOCl_2$, ether, 18 h	" (—)	187
		TsCl, pyridine, 64 h	" + $(CH_3)_2C(OH)CH_2CN$ (—)	187

	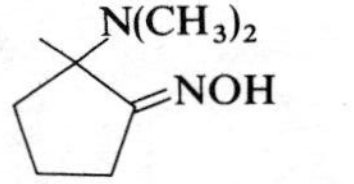 $N(CH_3)_2$, =NOH	TsCl, NaOH, H_2O	$CH_3CO(CH_2)_3CN$ (60)	570
	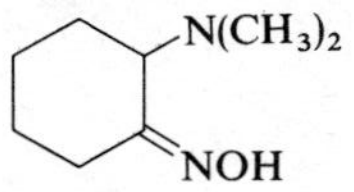 $N(CH_3)_2$, NOH	TsCl, dioxane	$NC(CH_2)_4CHO$ (48)	570
	(*Z*)	TsCl, C_6H_6, H_2O	" (68)	570
	(*E*)	SO_3, SO_2	" (50)	570
	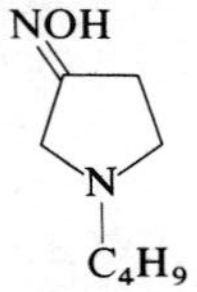 NOH, N, C_4H_9	TsCl, H_2O, <25°	$NC(CH_2)_2N(Ts)C_4H_9$ (94)	190
C_9	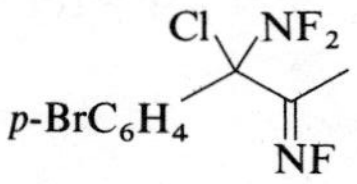 Cl, NF_2, *p*-BrC_6H_4, NF	H_2SO_4, CH_2Cl_2	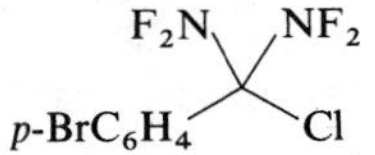 F_2N, NF_2, *p*-BrC_6H_4, Cl (32)	121
		HNF_2, H_2SO_4	" (72)	121
	$C_6H_5COC(=NOH)CH_3$	85% H_2SO_4	$C_6H_5CO_2H$ (84)	197, 569
		$C_6H_5SO_2Cl$, NaOH	" (91)	197, 569
		1. $NaOCH_3$, CH_3OH 2. $(CH_3CO)_2O$	$C_6H_5CO_2CH_3$ (54)	576
		1. $NaOC_2H_5$, ethanol 2. $(CH_3CO)_2O$	$C_6H_5CO_2C_2H_5$ (73)	576
	$CH_3COC(=NOH)C_6H_5$	85% H_2SO_4, 120°	$C_6H_5CONH_2$ (92)	197, 569
		$C_6H_5SO_2Cl$, NaOH	C_6H_5CN (87)	197, 569
		$SOCl_2$, ether	" (88)	197, 569
		$POCl_3$	" (37)	197
	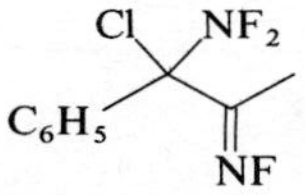 Cl, NF_2, C_6H_5, NF	H_2SO_4, CH_2Cl_2	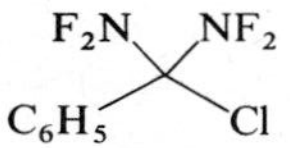 F_2N, NF_2, C_6H_5, Cl (27)	121
		HNF_2 + H_2SO_4 or BF_3	" (60)	121

TABLE III. KETOXIME FRAGMENTATIONS (*Continued*)

No. of Carbon Atoms	Substrate	Reagent and Conditions	Product(s) and Yield(s) (%)	Refs.
	$C_6H_5C(Br)(NF_2)C(=NF)CH_3$	BF_3, CH_2Cl_2	$C_6H_5CCl(F)NF_2$ (68)	121, 565
		H_2SO_4, CH_2Cl_2	$C_6H_5CBr(NF_2)_2$ (28)	121
		HNF_2, H_2SO_4	" (41)	121
		BF_3, CH_2Cl_2	$C_6H_5CBr(F)NF_2$ (—)	121, 565
	$C_6H_5C(NF_2)_2C(=NF)CH_3$	H_2SO_4 or HSO_3F or HNF_2	$C_6H_5C(NF_2)_3$ (17)	121
	$CH_3C(NF_2)_2C(=NF)C_6H_5$	"	$CH_3C(NF_2)_3$ (3)	121
		TsCl, pyridine, -20 to 0°, 2 d	C_8H_7CN (57)	577
		$SOCl_2$	o-$NCC_6H_4CH_2SCH_2Cl$ (81)	193

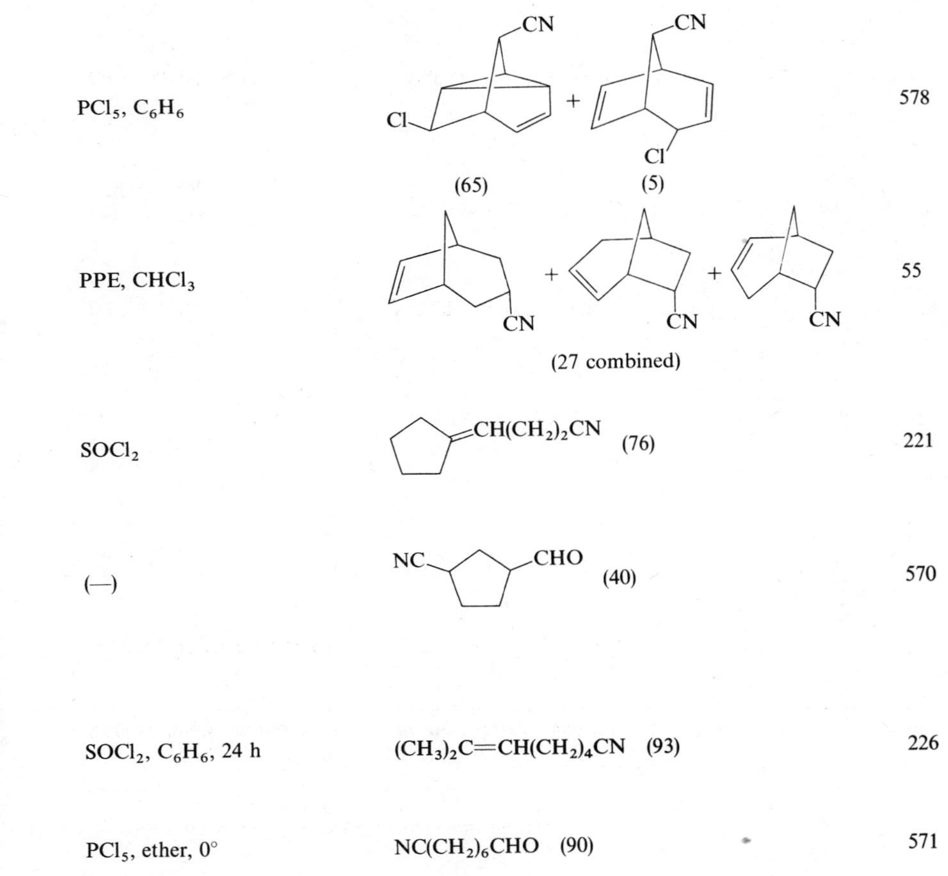

Conditions	Product(s) (Yield, %)	Refs.
PCl_5, C_6H_6	(65) + (5)	578
PPE, $CHCl_3$	(27 combined)	55
$SOCl_2$	$CH(CH_2)_2CN$ (76)	221
(—)	(40)	570
$SOCl_2$, C_6H_6, 24 h	$(CH_3)_2C{=}CH(CH_2)_4CN$ (93)	226
PCl_5, ether, 0°	$NC(CH_2)_6CHO$ (90)	571

TABLE III. KETOXIME FRAGMENTATIONS (*Continued*)

No. of Carbon Atoms	Substrate	Reagent and Conditions	Product(s) and Yield(s) (%)	Refs.
	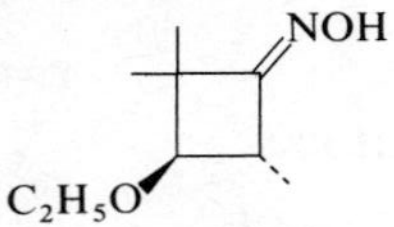 NOH; C_2H_5O	m-$O_2NC_6H_4COCl$, CH_2Cl_2	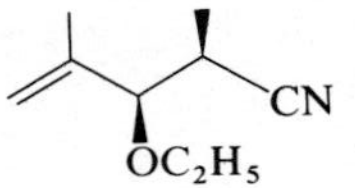 CN (64); OC_2H_5	176
	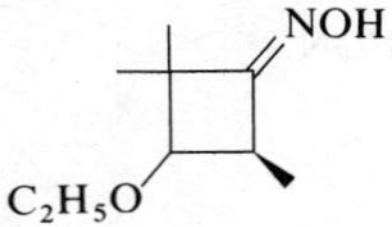 NOH; C_2H_5O	"	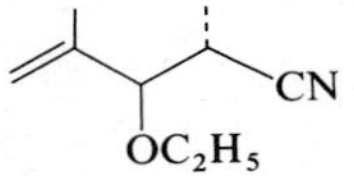 CN (61); OC_2H_5	176
	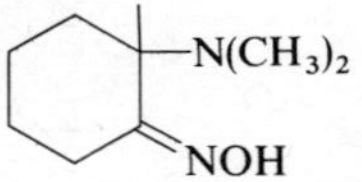 $N(CH_3)_2$; NOH	TsCl, NaOH, H_2O	$CH_3CO(CH_2)_4CN$ (68)	570
	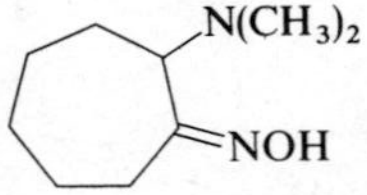 $N(CH_3)_2$; NOH	TsCl, dioxane	$NC(CH_2)_5CHO$ (30)	570
	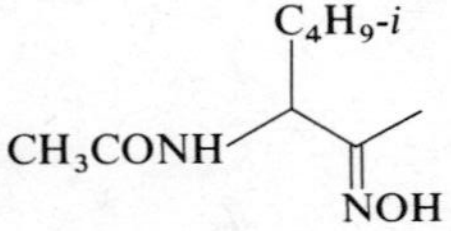 C_4H_9-i; CH_3CONH; NOH	TsCl, $(C_2H_5)_3N$, CH_3CN, -10 to 20°, 2 h	$CH_3CONHCH{=}CHC_3H_7$-i $E/Z = 54/36$ (43)	579
	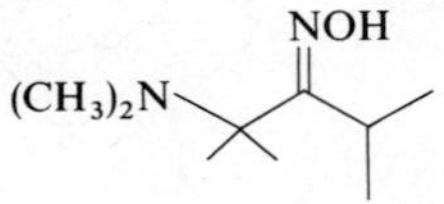 NOH; $(CH_3)_2N$	TsCl, NaOH, H_2O	CH_3COCH_3 (57) + i-C_3H_7CN (36)	570

Formula	Reactant	Conditions	Product(s) (% yield)	Refs.
C_{10}		TsCl, dioxane, reflux, 30 min	(51)	199
		1. TsCl, $C_6H_5CH_3$, reflux 2. 10% NaOH, reflux, 15 min	(32)	580
		"	(88)	580
		TsCl, NaOH, acetone, H_2O	" (49)	203
		$SOCl_2$, SO_2, 70°, 6 h	" (54)	202
		1. $C_6H_5SO_2Cl$, acetone, reflux 2. 10% NaOH, reflux, 15 min	(20)	580
		$SOCl_2$, 70°, 6 h	(98)	202
		$SOCl_2$, DMF, 0°	$-CH_2CN$ (74)	175

TABLE III. KETOXIME FRAGMENTATIONS (*Continued*)

No. of Carbon Atoms	Substrate	Reagent and Conditions	Product(s) and Yield(s) (%)	Refs.
	=NOTs	CH_3OH, reflux, 8 h	(—)	265
	NH_2, NO, C_6H_5, NH_2	$C_6H_5SO_2Cl$, pyridine	CN, C_6H_5, NH_2 (30)	561
		$POCl_3$, pyridine	" (30)	561
		$SOCl_2$	" (30)	561
	$CH_3COC(=NOH)CH_2C_6H_5$	$C_6H_5SO_2Cl$, NaOH	$C_6H_5CH_2CN$ (87)	197, 569
		PCl_5	" (86)	197, 569
		85% H_2SO_4, 120°	$C_6H_5CH_2CONH_2$ (83)	197, 569
	$C_6H_5C(NF_2)_2C(=NF)C_2H_5$	H_2SO_4 or HSO_3F or HNF_2	$C_6H_5C(NF_2)_3$ (17)	121
	F_2N NF_2, p-$CH_3OC_6H_4$, NF	"	p-$CH_3OC_6H_4C(NF_2)_3$ (12)	121
	=$NOSO_2C_6H_5$	THF, H_2O	$HOCH_2$, CH_2OH (—)	329
	$(CH_3)_2C(NF_2)C(=NF)C_6H_5$	HNF_2, H_2SO_4	$CH_3C(NF_2)_3$ (—)	121

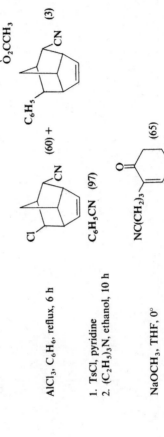

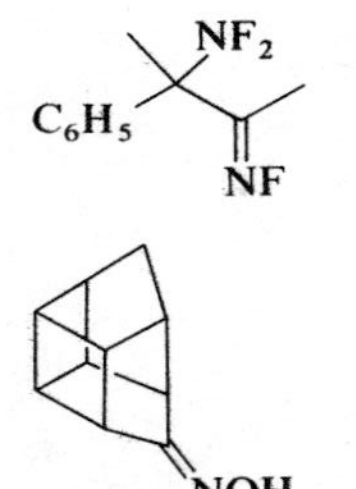	BF_3, CH_2Cl_2	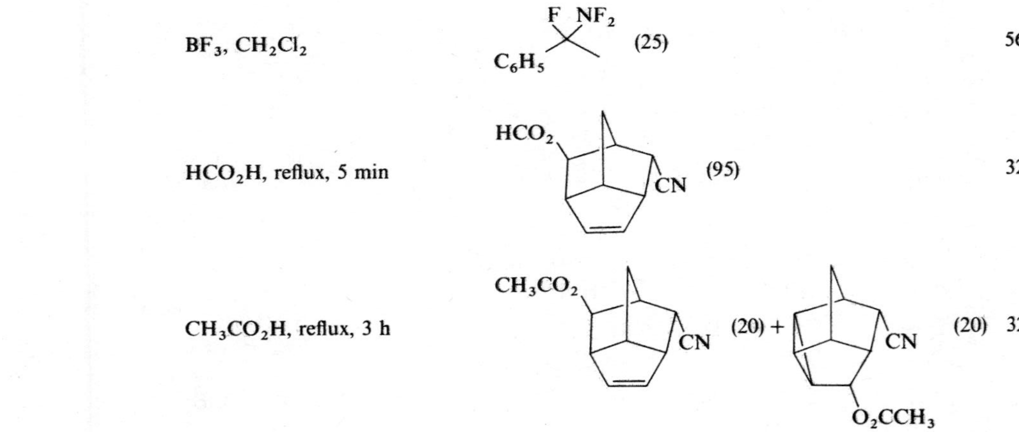 (25)	565
	HCO_2H, reflux, 5 min	(95)	327
	CH_3CO_2H, reflux, 3 h	(20) + (20)	327
	$AlCl_3$, C_6H_6, reflux, 6 h	(60) + (3)	327
$C_6H_5C(=NOH)CH(SCH_3)CH_3$	1. TsCl, pyridine 2. $(C_2H_5)_3N$, ethanol, 10 h	C_6H_5CN (97)	330
(TsO-oxime of octahydronaphthalenone; H, H, O, N, TsO)	$NaOCH_3$, THF, 0°	$NC(CH_2)_3$-cyclohexenone (65)	337
	KOC_4H_9-*t*, THF, 0°	" (85)	79, 200 337
	NaOH, dioxane, H_2O	" (60)	337

TABLE III. KETOXIME FRAGMENTATIONS (*Continued*)

No. of Carbon Atoms	Substrate	Reagent and Conditions	Product(s) and Yield(s) (%)	Refs.
	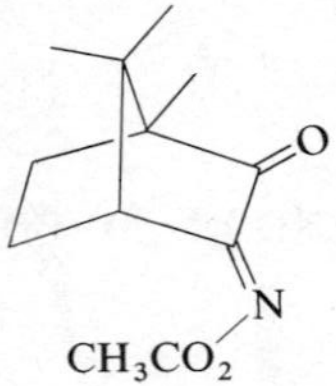 O; N; CH_3CO_2	CH_3CO_2H, H_2O, reflux	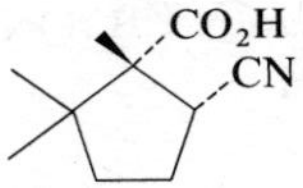 CO_2H; CN (—)	65
		Dilute NaOH	" (—)	65
	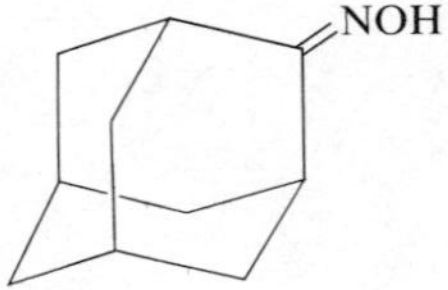 NOH	HCl, CH_3CN	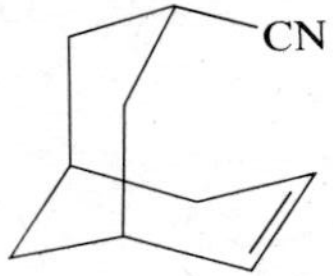 CN (35)	334
	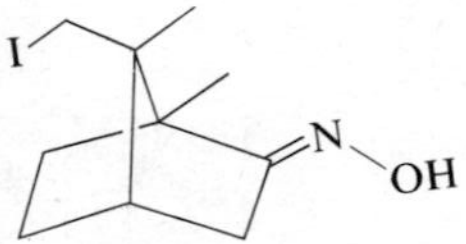 I; N; OH	TsCl, pyridine, 0°	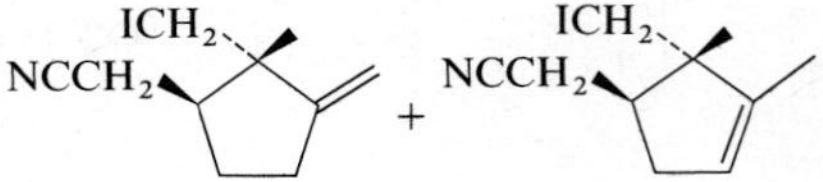 ICH_2; $NCCH_2$ + ICH_2; $NCCH_2$ (81)	169
	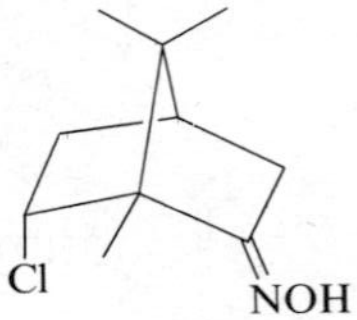 Cl; NOH	65% H_2SO_4	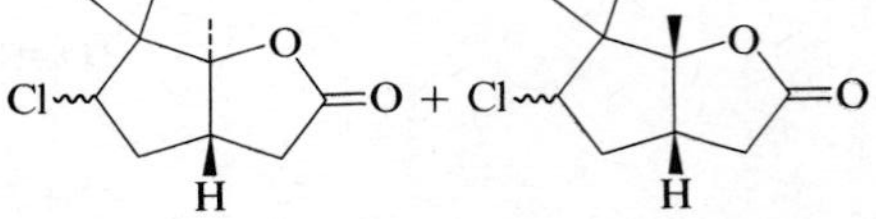 Cl; O; =O; H (7) + Cl; O; =O; H (11)	167

Reactant	Conditions	Product(s) and Yield(s) (%)	Refs.
Br NOH		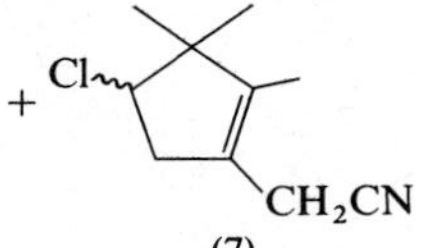 + Cl CH_2CN (7)	
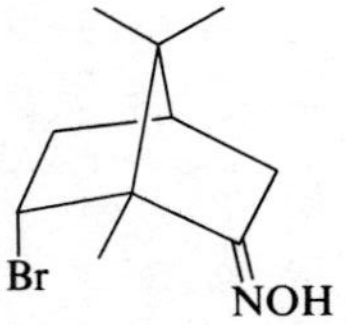	Zn, CH_3CO_2H, H_2O, reflux, 4 h	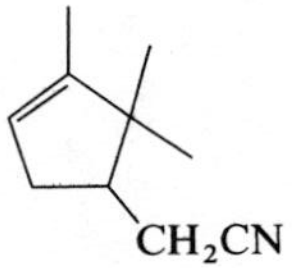 CH_2CN (100)	581
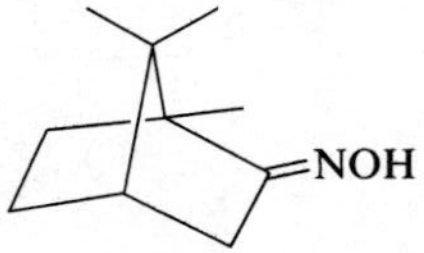 NOH	HCl, 110°, 10 h	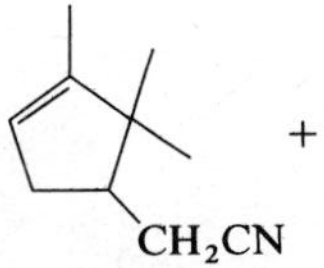 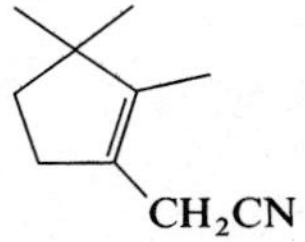 CH_2CN + CH_2CN (—)	168
	TsCl, pyridine, 0°, 20 h	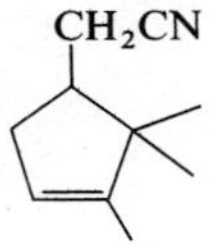 CH_2CN (55) 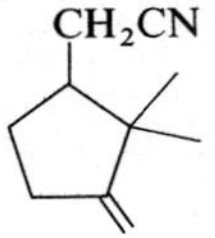 + CH_2CN (45)	170
	240°, 6–7 min	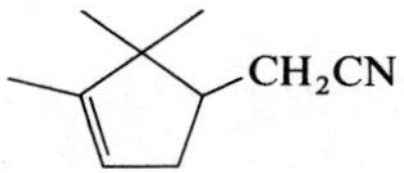 CH_2CN I + CH_2CN (—) 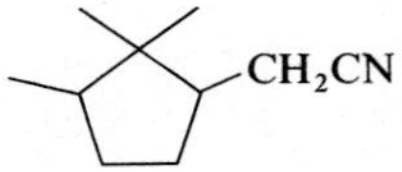	582
	P_2O_5, $C_6H_5CH_3$, reflux, 30 min	I (77)	173
	500°, 1 s	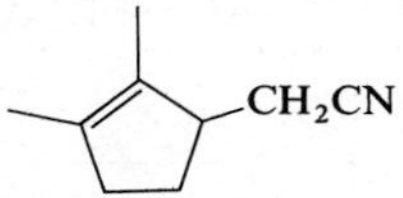 CH_2CN (—)	582

TABLE III. KETOXIME FRAGMENTATIONS (*Continued*)

No. of Carbon Atoms	Substrate	Reagent and Conditions	Product(s) and Yield(s) (%)	Refs.
	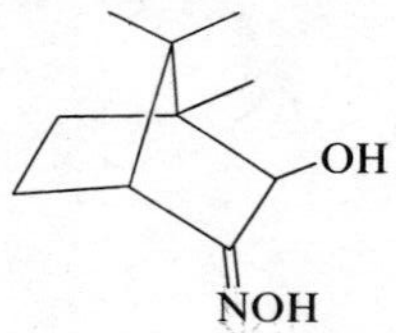 OH; NOH	Dilute H_2SO_4, reflux, 2 min	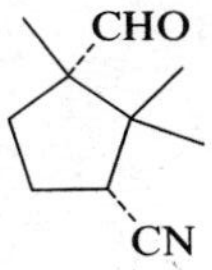 CHO; CN (95)	180
		Cl_2C:	" (85)	178
		CF_3SO_2Cl	" (65)	583
		$(CF_3SO_2)_2O$	" (81)	583
		$(CF_3CO)_2O$	" (79)	583
		C_6H_5NCO, $(C_2H_5)_3N$, C_6H_6, reflux, 1 h	" (73)	440
	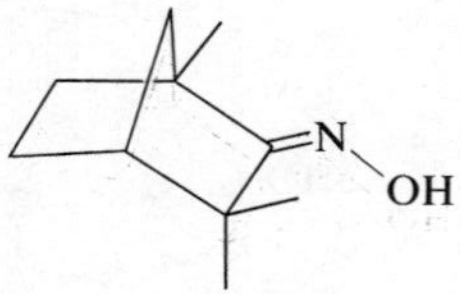 N; OH	TsCl, pyridine	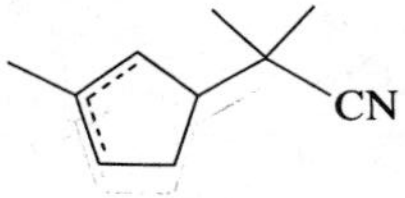 CN (79)	348
		PCl_5, pet ether	" (70)	348
		H_2SO_4	" (38)	348
		P_2O_5, $C_6H_5CH_3$, reflux, 30 min	" (63)	173
	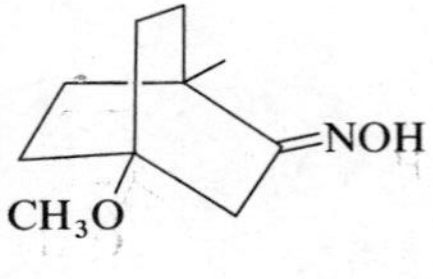 NOH; CH_3O	$C_6H_5SO_2Cl$, NaOH	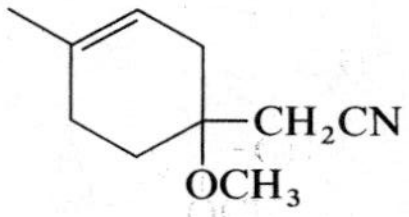 CH_2CN; OCH_3 (52)	297
		TsCl, pyridine	" (60)	297

Reactant	Conditions	Product(s) and Yield(s) (%)	Refs.
NOH, H	$SOCl_2$, dioxane	$(CH_2)_2CN$ (35–40)	311
	PCl_5, ether, 0°	$(CH_2)_2CN$ (—)	343
N, OH	$SOCl_2$	$=CH(CH_2)_3CN$ (7)	221
	PPA, 120–125°, 10 min	$(CH_2)_4CONH_2$ (32)	221
	PCl_5, C_6H_6	$(CH_2)_4CN$ (10)	342
HO, N	$SOCl_2$	$(CH_2)_3CN$ (30)	221
	PCl_5, C_6H_6	" (96)	221
	PPA, 120–125°, 10 min	$(CH_2)_3CONH_2$ (83)	221

TABLE III. KETOXIME FRAGMENTATIONS (*Continued*)

No. of Carbon Atoms	Substrate	Reagent and Conditions	Product(s) and Yield(s) (%)	Refs.
	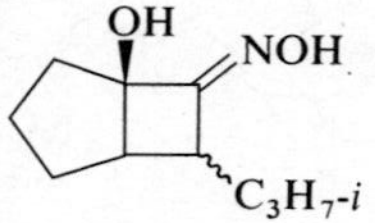 OH, NOH, C_3H_7-*i*	100°, 15 h	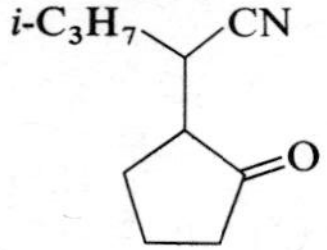 *i*-C_3H_7, CN, O (—)	584
	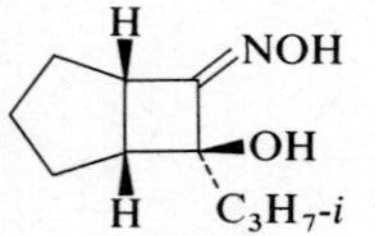 H, NOH, OH, H, C_3H_7-*i*	"	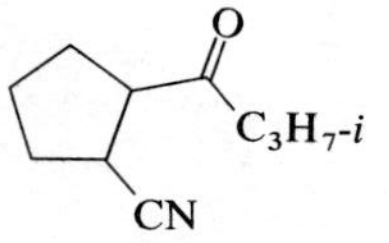 O, C_3H_7-*i*, CN (—)	584
	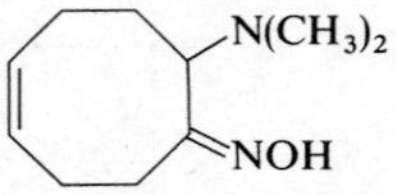 $N(CH_3)_2$, NOH	TsCl, dioxane	$NC(CH_2)_2CH{=}CH(CH_2)_2CHO$ (40)	570
	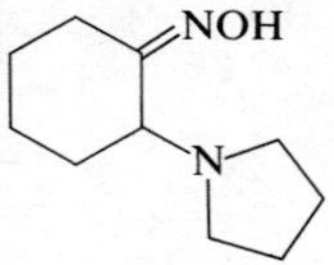 NOH, N	PCl_5, ether, 0°	$NC(CH_2)_4CHO$ (24)	571
	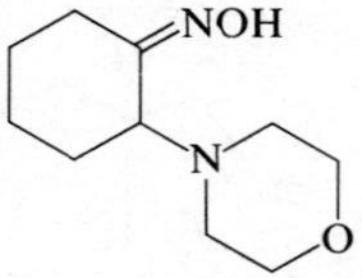 NOH, N, O	"	" (30)	571

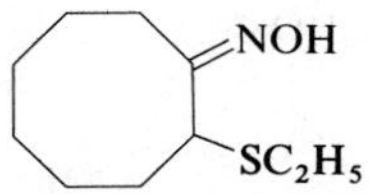 NOH; SC_2H_5	”	$NC(CH_2)_6CHO$ (50)	571
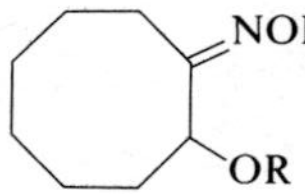 NOH; OR; $R = C_2H_5$	”	” (82)	571
$R = CH_3$	”	” (85)	233
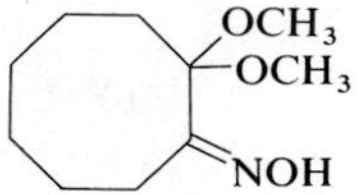 OCH_3; OCH_3; NOH	$C_2H_5C(OCH_3)_2 \cdot BF_4$, 1 h	$NC(CH_2)_6CO_2CH_3$ (95)	189
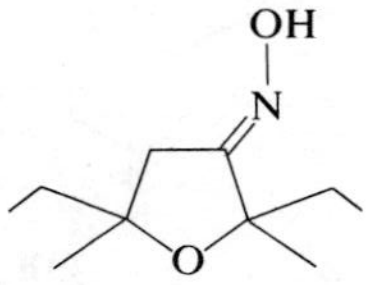 OH; N; O	PCl_5, C_6H_6	$C_2H_5COCH_3$ (54) + “nitriles”	187
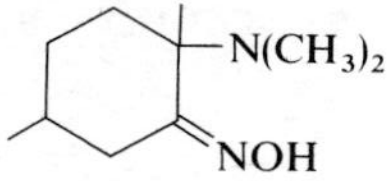 $N(CH_3)_2$; NOH	TsCl, NaOH, H_2O	$CH_3CO(CH_2)_2CH(CH_3)CH_2CN$ (60)	570
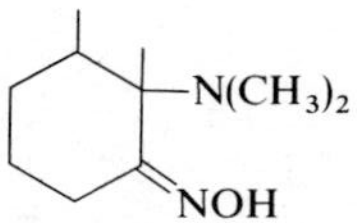 $N(CH_3)_2$; NOH	”	$CH_3COCH(CH_3)(CH_2)_3CN$ (65)	570
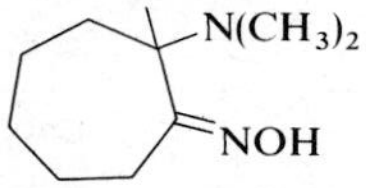 $N(CH_3)_2$; NOH	”	$CH_3CO(CH_2)_5CN$ (43)	570
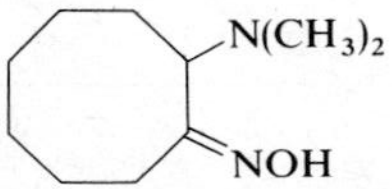 $N(CH_3)_2$; NOH	TsCl, dioxane	$NC(CH_2)_6CHO$ (25)	570

TABLE III. KETOXIME FRAGMENTATIONS (*Continued*)

No. of Carbon Atoms	Substrate	Reagent and Conditions	Product(s) and Yield(s) (%)	Refs.
	RO–N=(2-$CH_2Si(CH_3)_3$-cyclohexylidene) R = H	CH_3SO_2Cl, pyridine	$CH_2{=}CH(CH_2)_4CN$ (52)	204
		PCl_5	" (28)	204
		$POCl_3$	" (48)	204
		P_2O_5	" (73)	204
	R = CH_3CO	$CF_3SO_3Si(CH_3)_3$, CH_2Cl_2, 0°	" (89)	204
	RO–N=(2-methyl-3-$Si(CH_3)_3$-cyclohexylidene) R = H	P_2O_5	(E)-$CH_3CH{=}CH(CH_2)_3CN$ (51)	204
	R = CH_3CO	$CF_3SO_3Si(CH_3)_3$, CH_2Cl_2, 0°	" (94)	204
	CH_3CO_2–N=(2-methyl-3-$Si(CH_3)_3$-cyclohexylidene, cis)	"	(Z)-$CH_3CH{=}CH(CH_2)_3CN$ (90)	204
C_{11}	7-CH_3O-1-OH-2-NO-naphthalene	$C_6H_5SO_2Cl$, NaOH, acetone, H_2O, reflux	2-CN-4-CH_3O-C$_6H_3$-CH=CH-CO_2H (Z) (86)	585

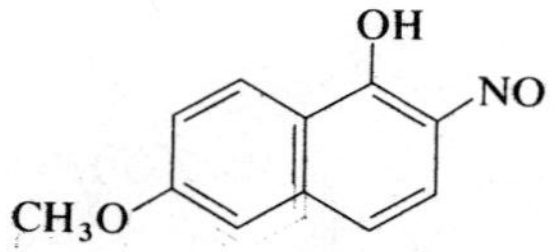 OH, NO, CH_3O (structure)	”	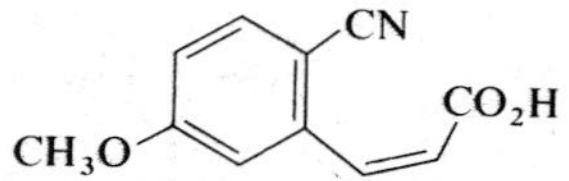 CN, CO_2H, CH_3O (structure) (26)	585
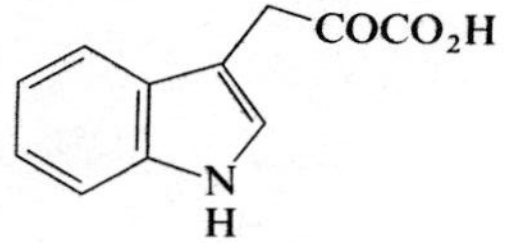 $COCO_2H$, N, H (structure)	$H_2NOH \cdot HCl$, H_2O, reflux, 2.5 h	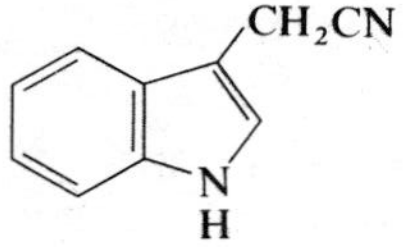 CH_2CN, N, H (structure) (60)	198
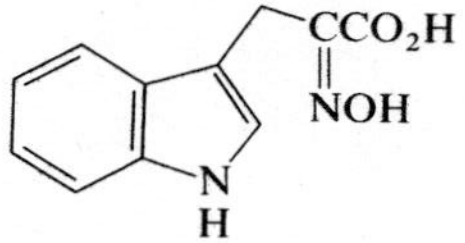 CCO_2H, NOH, N, H (structure)	0.05 M H_2SO_4, reflux, 3 h	” (95)	198
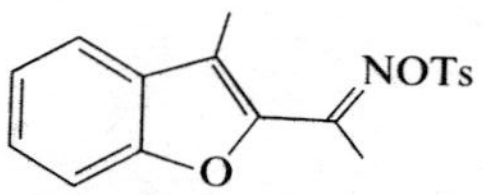 NOTs, O (structure)	CH_3OH, reflux, 5 h	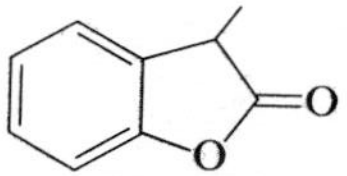 O, =O (structure) (—)	440
$C_6H_5CH{=}C(CH_3)C(CH_3){=}NOH$	PCl_5, ether	$CH_3CN + C_6H_5CH_2COCH_3$ (33)	586
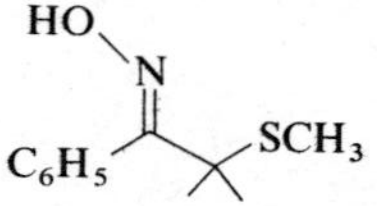 HO, N, C_6H_5, SCH_3 (structure)	1. TsCl, pyridine 2. $(C_2H_5)_3N$, ethanol, 10 h	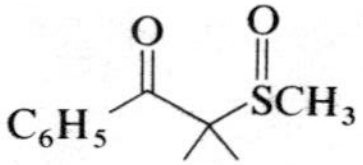 C_6H_5CN (53) 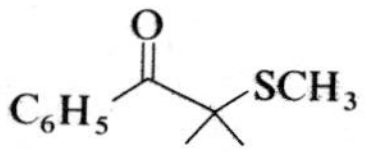 + O, C_6H_5, SCH_3 (structure) (13) + O, O, C_6H_5, SCH_3 (structure) (28)	330
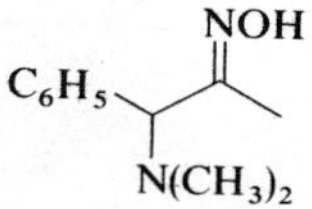 NOH, C_6H_5, $N(CH_3)_2$ (structure)	TsCl, NaOH, H_2O	C_6H_5CHO (56)	570

TABLE III. KETOXIME FRAGMENTATIONS (*Continued*)

No. of Carbon Atoms	Substrate	Reagent and Conditions	Product(s) and Yield(s) (%)	Refs.
	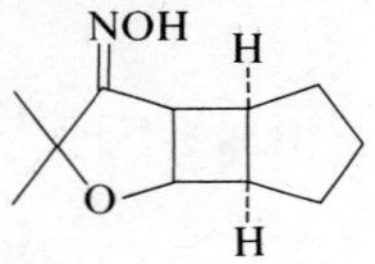 NOH, H, O, H	$SOCl_2$, CCl_4, 0°	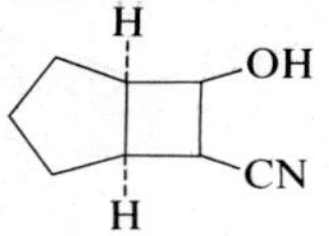 H, OH, CN, H (—)	587
	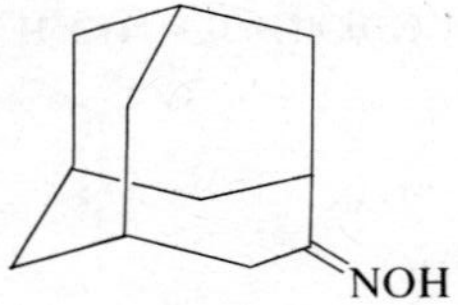 NOH	PPE, $CHCl_3$, reflux, 1 h	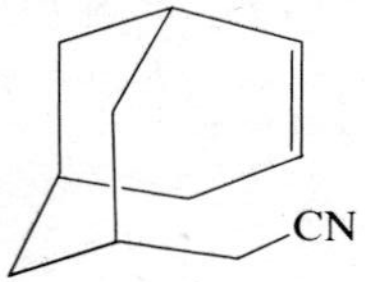 CN (79)	368
	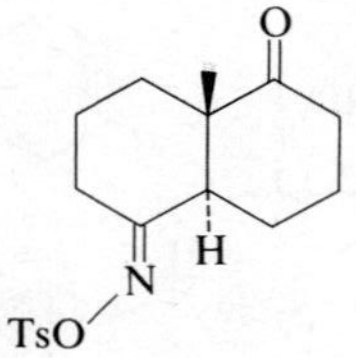 O, H, N, TsO	NaOR, ethanol	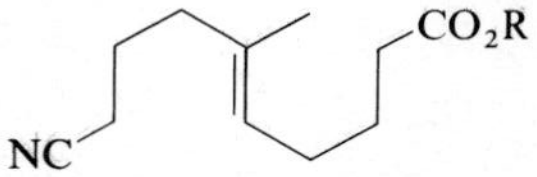 NC, CO_2R (100) R = H or C_2H_5	79, 80
	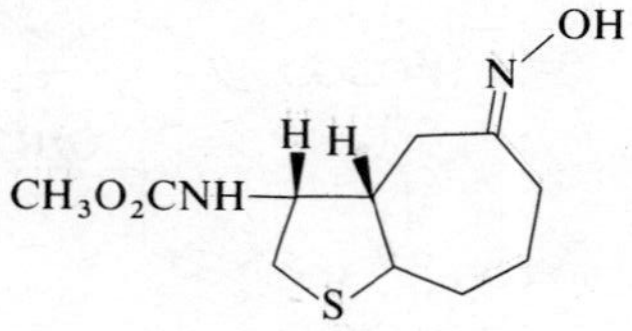 OH, N, H H, CH_3O_2CNH, S	PPA, 100°, 15 min	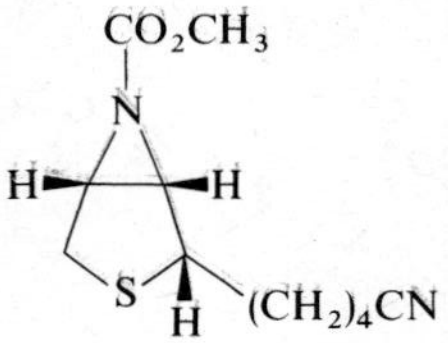 CO_2CH_3, N, H, H, S, H, $(CH_2)_4CN$ (—)	195, 196

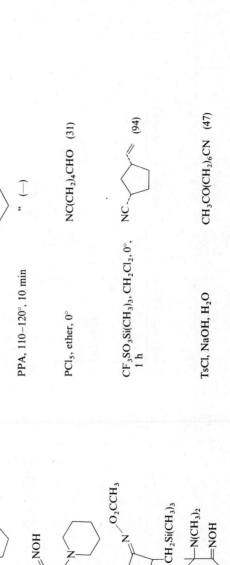

Reactant	Conditions	Product(s) (Yield, %)	Refs.
NOH, H	$SOCl_2$, dioxane	CN + CN (12–17)	311
NOH, CH_3O	$C_6H_5SO_2Cl$, NaOH	CH_2CN, OCH_3 (44)	297
HO, N, O	PCl_5, C_6H_6, 10 h	O (—)	187
	PPA, 110–120°, 10 min	" (—)	187
NOH, N	PCl_5, ether, 0°	$NC(CH_2)_4CHO$ (31)	571
O_2CCH_3, N, $CH_2Si(CH_3)_3$	$CF_3SO_3Si(CH_3)_3$, CH_2Cl_2, 0°, 1 h	NC (94)	204
$N(CH_3)_2$, NOH	TsCl, NaOH, H_2O	$CH_3CO(CH_2)_6CN$ (47)	570

TABLE III. KETOXIME FRAGMENTATIONS (*Continued*)

No. of Carbon Atoms	Substrate	Reagent and Conditions	Product(s) and Yield(s) (%)	Refs.
	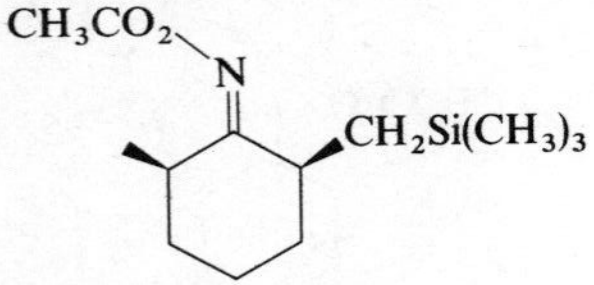 CH_3CO_2 N; $CH_2Si(CH_3)_3$	$CF_3SO_3Si(CH_3)_3$, CH_2Cl_2, 0°, 4 h	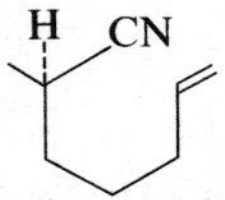 H CN (92)	204
	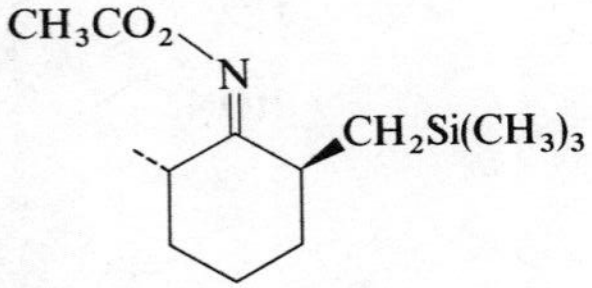 CH_3CO_2 N; $CH_2Si(CH_3)_3$	" 1 h	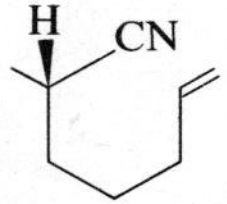 H CN (82)	204
C_{12}	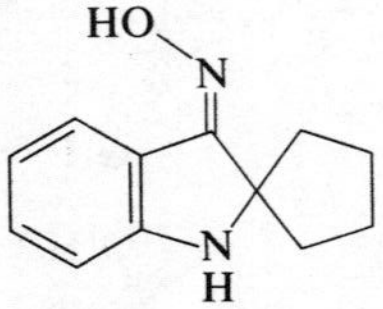 HO N; N H	PPA	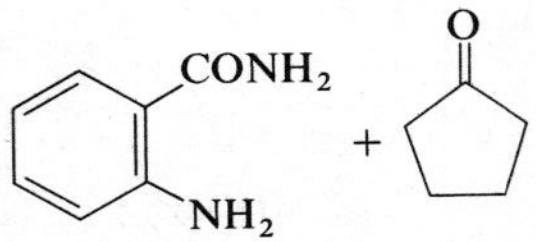 $CONH_2$ NH_2 + O (—)	588
	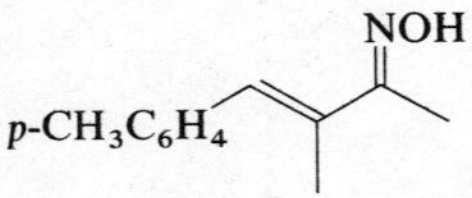 NOH; p-$CH_3C_6H_4$	PCl_5, ether	CH_3CN + p-$CH_3C_6H_4CH_2COCH_3$ (30)	586
	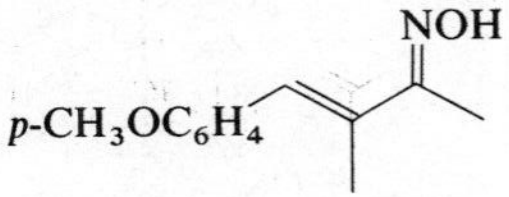 NOH; p-$CH_3OC_6H_4$	"	" + p-$CH_3OC_6H_4CH_2COCH_3$ (42)	586

Reactant	Conditions	Product(s) (Yield, %)	Refs.
HON=C(CH$_3$)C(C$_2$H$_5$)=CHC$_6$H$_5$	"	" + $C_6H_5CH_2COCH_3$ (64)	442
1,1-dimethyl-2-tetralone oxime (NOH)	"	2-(CH$_2$)$_2$CN-isopropenylbenzene (93)	420
7-CH$_3$O-2-SCH$_3$-1-tetralone O-R oxime (NOR), R = H	CH_3SO_2Cl, pyridine, 85°, 4 h	CH$_3$O, CN-substituted benzene with CH$_2$CH=CHSCH$_3$ side chain (57)	192, 589
	$(C_2H_5)_2NCF_2CHFCl$, dioxane, 70°	" (46)	589
R = Ts	Pyridine, reflux, 12 h	" (35)	589
	KOC_4H_9-*t*	" (—)	590
5-C$_6$H$_5$-2,2-dimethyltetrahydrofuran-3-one oxime (OH–N)	PCl_5	$CH_3COCH_3 + C_6H_5CH{=}CHCN$ (—)	187
	PPA	" + $C_6H_5CH{=}CHCONH_2$ (—)	187
CH$_3$CONH–CH(CH$_2$C$_6$H$_5$)–C(CH$_3$)=NOH	TsCl, $(C_2H_5)_3N$, CH_3CN, −10 to 20°, 2 h	$CH_3CONHCH{=}CHC_6H_5$ (*E*/*Z*) = 73/27 (81)	579
C$_6$H$_5$C(CH$_3$)(C$_2$H$_5$)C(CH$_3$)=N–OTs	$(C_2H_5)_3N$, 80% ethanol, 30°, 6 h	C$_6$H$_5$C(CH$_3$)(C$_2$H$_5$)OR, R = H (52), R = C_2H_5 (26)	28

TABLE III. KETOXIME FRAGMENTATIONS (*Continued*)

No. of Carbon Atoms	Substrate	Reagent and Conditions	Product(s) and Yield(s) (%)	Refs.
	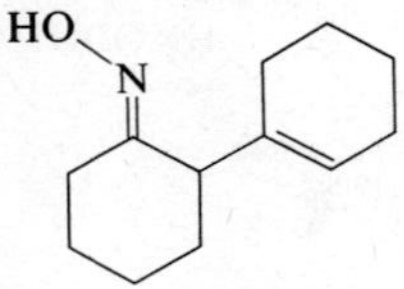 HO N	PCl_5, ether	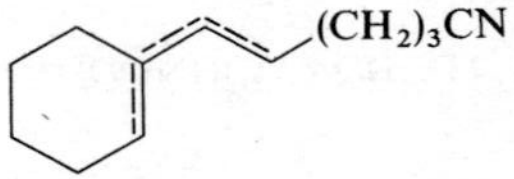 $(CH_2)_3CN$ (—)	386
	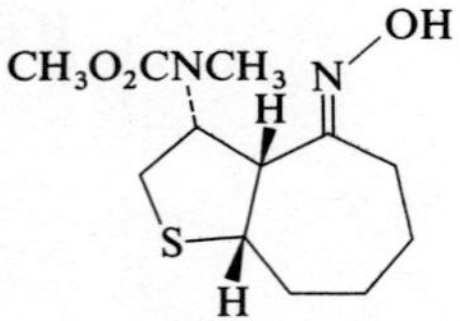 $CH_3O_2CNCH_3$ N OH H S H	$SOCl_2$, CH_2Cl_2, 0°	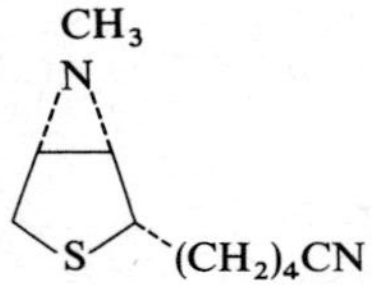 CH_3 N S $(CH_2)_4CN$ (75)	196
	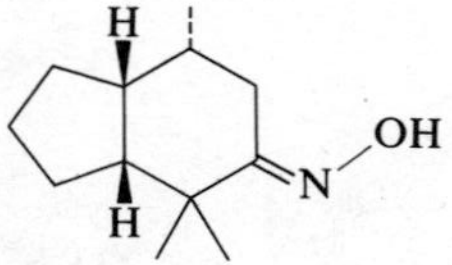 H H N OH	TsCl, pyridine	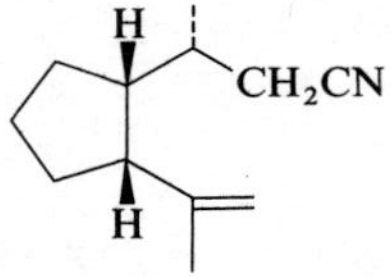 H H CH_2CN (—)	591
	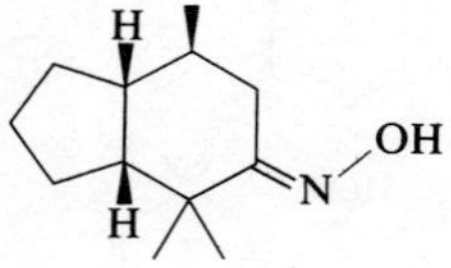 H H N OH	"	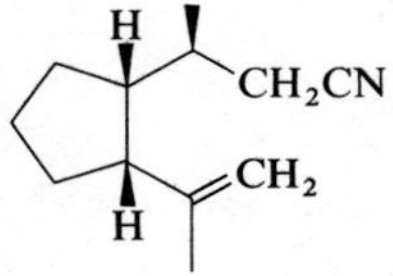 H H CH_2CN CH_2 (—)	592
	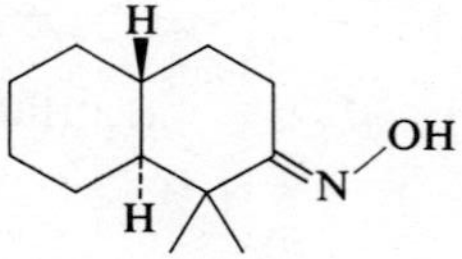 H H N OH	"	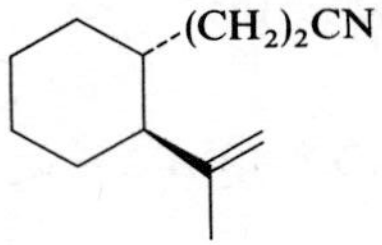 $(CH_2)_2CN$ (66)	160

Reactant	Conditions	Product(s) (%)	Refs.
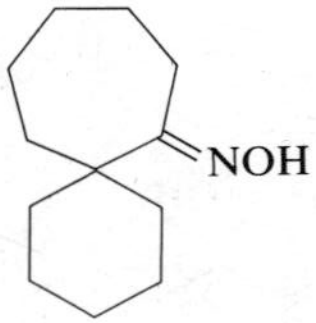 NOH	PCl_5, C_6H_6	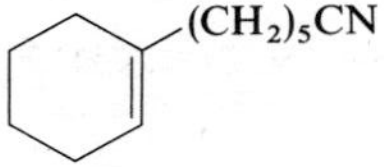 $(CH_2)_5CN$ (92)	78
	P_2O_5, C_6H_6, 70°	" (76)	342
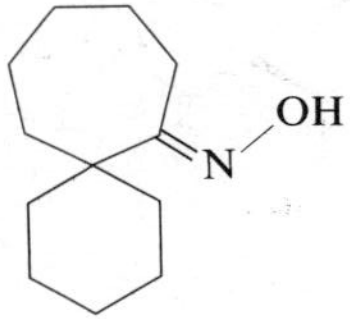 OH, N	PPA, 120°	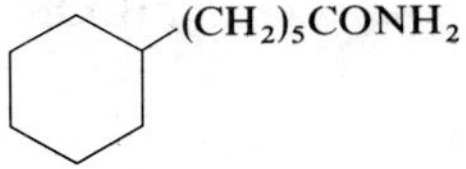 $(CH_2)_5CONH_2$ (46)	78
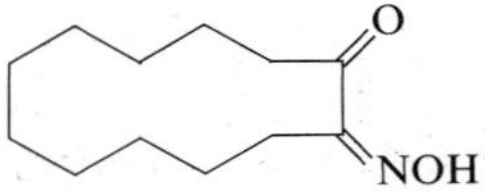 O, NOH	$SOCl_2$	$NC(CH_2)_{10}COCl$ (62)	387
	$SOCl_2$, NH_3	$NC(CH_2)_{10}CONH_2$ (96)	387
	$(CH_3CO)_2O$, H_2SO_4	$NC(CH_2)_{10}CO_2CH_3$ (95)	387
	$COCl_2$, CH_3OH	$NC(CH_2)_{10}CO_2CH_3$ (95) + $H_2NCO(CH_2)_{10}CO_2CH_3$ (15)	387
	H_2SO_4, H_2O	$H_2NCO(CH_2)_{10}CO_2H$ (96)	387
	97% H_3PO_4, H_2O	$NC(CH_2)_{10}CO_2H$ (98)	387
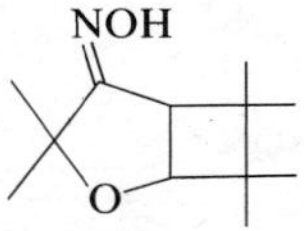 NOH, O	$SOCl_2$, CCl_4, 0°	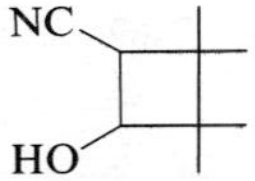 NC, HO (—)	593
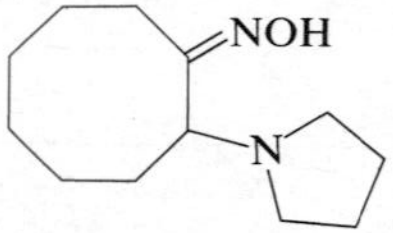 NOH, N	PCl_5, ether, 0°	$NC(CH_2)_6CHO$ (24)	571

TABLE III. Ketoxime Fragmentations (*Continued*)

No. of Carbon Atoms	Substrate	Reagent and Conditions	Product(s) and Yield(s) (%)	Refs.
	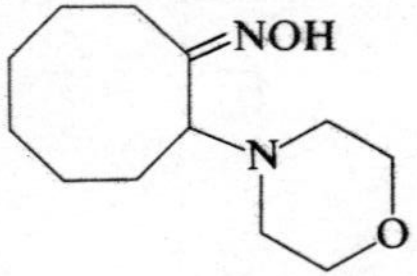 NOH; N; O	"	" (69)	571
	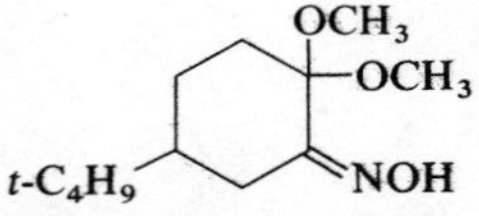 OCH_3; OCH_3; t-C_4H_9; NOH	$C_2H_5C(OCH_3)_2 \cdot BF_4$, CH_2Cl_2, 1 h	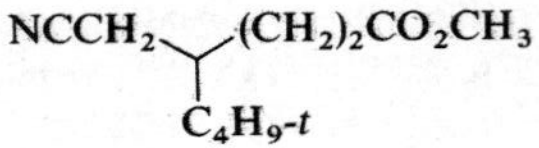 $NCCH_2$ $(CH_2)_2CO_2CH_3$; C_4H_9-t (95)	189
	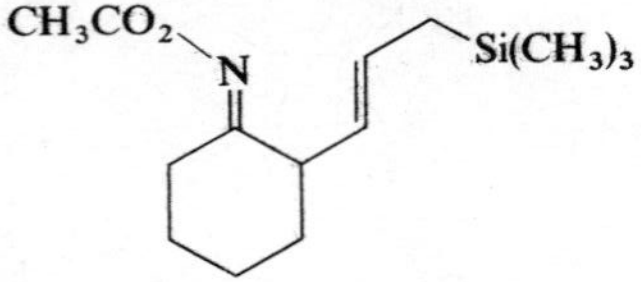 CH_3CO_2; N; $Si(CH_3)_3$	$CF_3SO_3Si(CH_3)_3$, CH_2Cl_2, 12 h	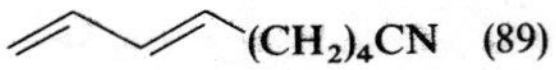 $(CH_2)_4CN$ (89)	204
	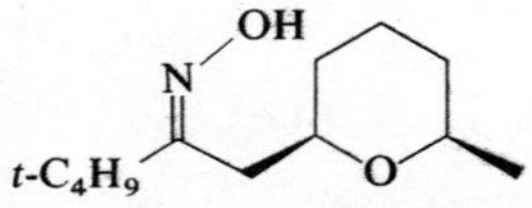 OH; N; t-C_4H_9; O	PCl_5, ether	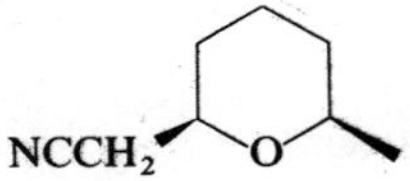 $NCCH_2$; O (69)	161
	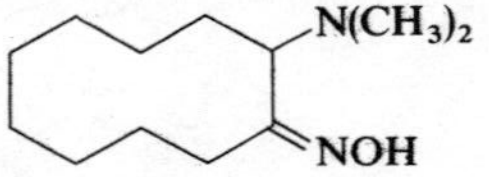 $N(CH_3)_2$; NOH	SO_3, SO_2	$NC(CH_2)_8CHO$ (30)	570
C_{13}	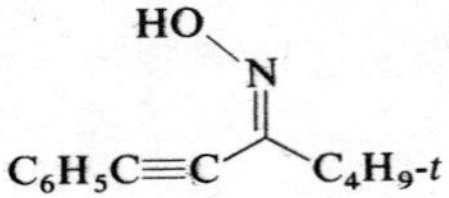 HO; N; $C_6H_5C{\equiv}C$; C_4H_9-t	PCl_5, ether	$C_6H_5C{\equiv}CCN$ (—)	116

Reactant	Conditions	Product(s) (Yield, %)	Refs.
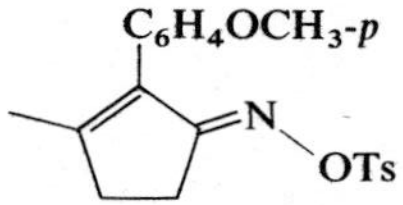 $C_6H_4OCH_3$-*p*; =N–OTs (cyclopentenone oxime tosylate)	80% ethanol, 110°	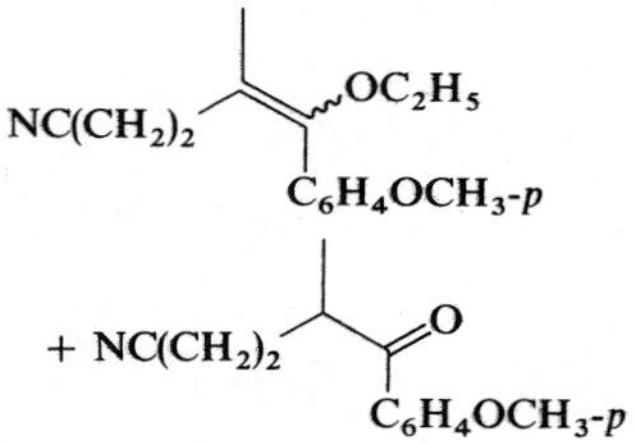 $NC(CH_2)_2$–C(CH_3)=C(OC_2H_5)($C_6H_4OCH_3$-*p*) + $NC(CH_2)_2$CH(CH_3)CO$C_6H_4OCH_3$-*p* (11)	49
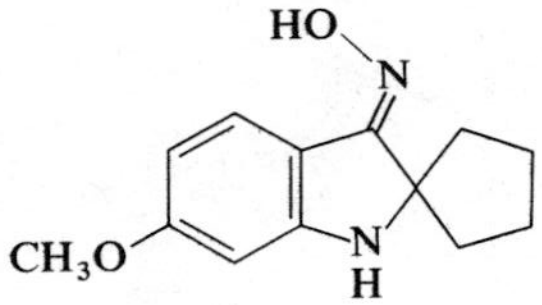 HO–N=; CH_3O; N–H (spiro-cyclopentane indolone oxime)	TsCl, pyridine, reflux, 2 h	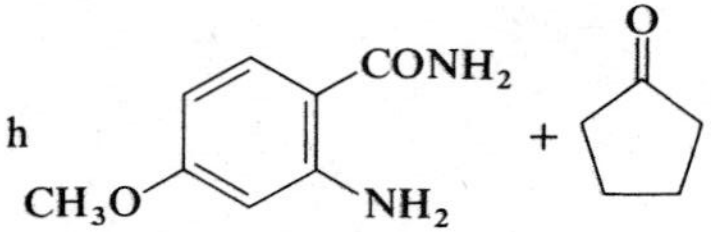 CH_3O–C_6H_3($CONH_2$)(NH_2) + cyclopentanone (—)	588
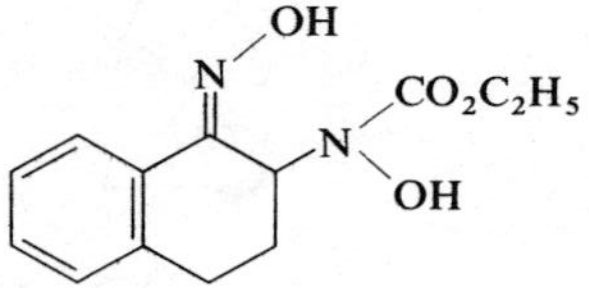 N–OH; N($CO_2C_2H_5$)OH (tetralone oxime)	$NaOC_2H_5$, ethanol, 1 h	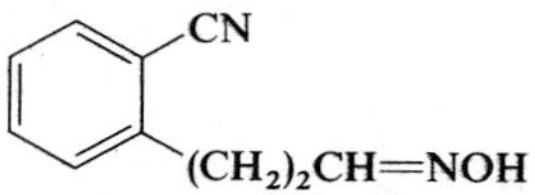 C_6H_4(CN)$(CH_2)_2CH{=}NOH$ (87)	594
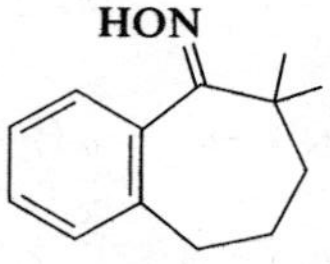 HON= (dimethylbenzocycloheptenone oxime)	PPA	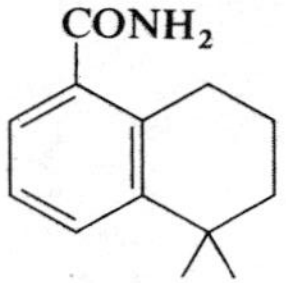 $CONH_2$ (dimethyltetralin carboxamide) (70)	166
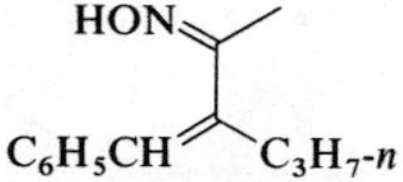 HON=C(CH_3)–C(=CHC_6H_5)C_3H_7-*n*; C_6H_5CH	PCl_5, ether	$CH_3CN + C_6H_5CH_2CO(CH_2)_2CH_3$ (62)	442
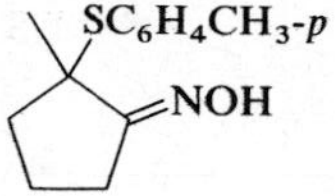 $SC_6H_4CH_3$-*p*; =NOH (2-methylcyclopentanone oxime)	TsCl, C_6H_6, H_2O	$CH_3CO(CH_2)_3CN$ (—)	570

TABLE III. KETOXIME FRAGMENTATIONS (*Continued*)

No. of Carbon Atoms	Substrate	Reagent and Conditions	Product(s) and Yield(s) (%)	Refs.
	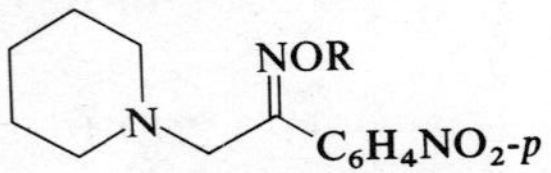 [structure: 1-piperidinyl-CH$_2$–C(=NOR)–$C_6H_4NO_2$-*p*] (*Z*), R = COC_6H_5	80% ethanol, 80°, 200 h	*p*-$O_2NC_6H_4CN$ (40)	95
	(*E*)	80% ethanol, 80°, 1 h	" (99)	95
	(*Z*) R = 2,4-$(O_2N)_2C_6H_3$	" 6 h	" (84)	95
	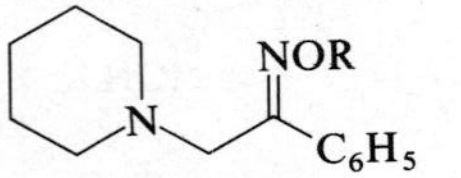 [structure: 1-piperidinyl-CH$_2$–C(=NOR)–C_6H_5]	TsCl, $(C_2H_5)_3N$	C_6H_5CN (91)	595
	R = COC_6H_5	80% ethanol, 70°	" (—)	595
	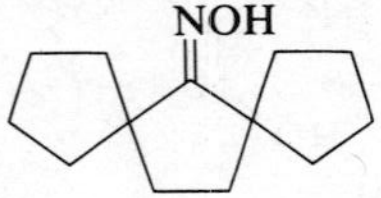 [structure: dispiro ketoxime, NOH]	$SOCl_2$, 15 h	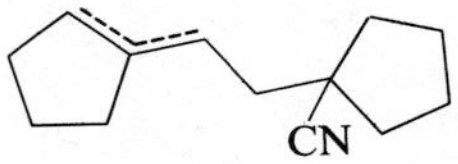 [structure: cyclopentylidene/cyclopentenyl-ethyl cyclopentane carbonitrile, CN] (25)	406
		TsCl, 15 h	" (28)	406
	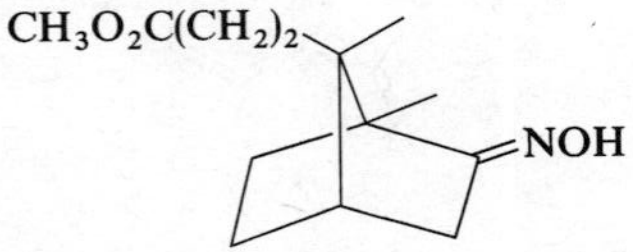 [structure: $CH_3O_2C(CH_2)_2$-substituted bicyclic ketoxime, =NOH]	$(CF_3CO)_2O$, CF_3CO_2H, 24 h	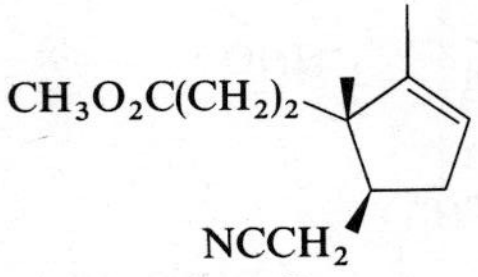 [structure: $CH_3O_2C(CH_2)_2$-substituted methylcyclopentene with $NCCH_2$] I (80)	172
		TsCl, pyridine	I 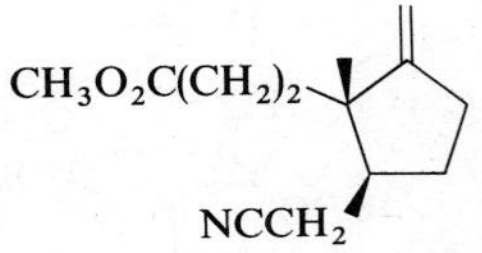 + [structure: $CH_3O_2C(CH_2)_2$-substituted methylenecyclopentane with $NCCH_2$] II; I/II = 3/2 (70)	171, 172

Reactant	Conditions	Product(s) and Yield(s) (%)	Refs.
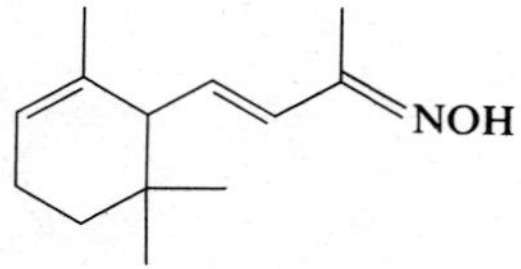 (structure) NOH	p-$CH_3CONHC_6H_4SO_2Cl$, pyridine, 0°	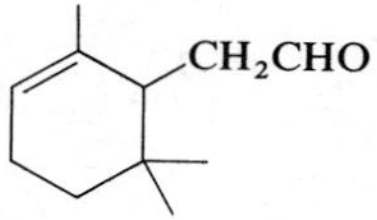 (structure) CH_2CHO (62)	596
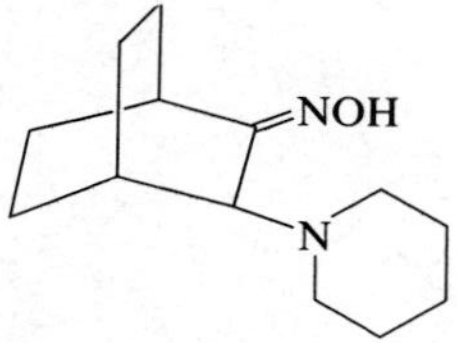 (structure) NOH, N	(—)	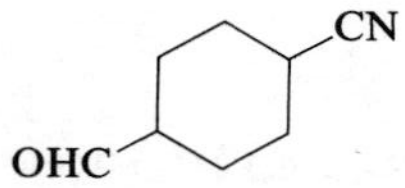 (structure) OHC, CN (35)	570
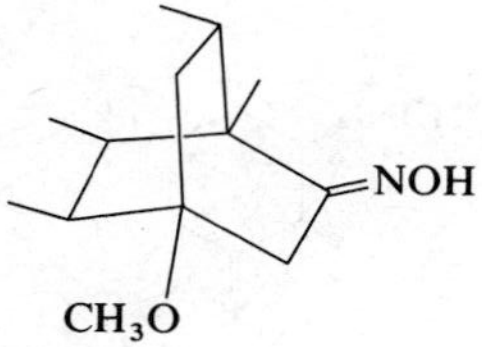 (structure) NOH, CH_3O	$C_6H_5SO_2Cl$, NaOH	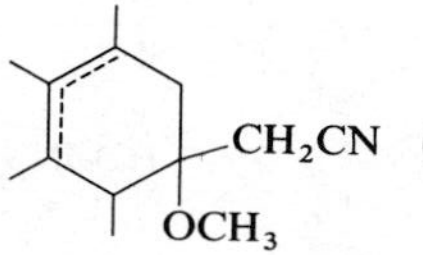 (structure) CH_2CN, OCH_3 (16)	297
	TsCl, pyridine	" (19)	297
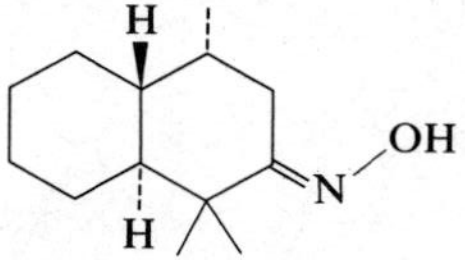 (structure) H, H, N, OH	", reflux	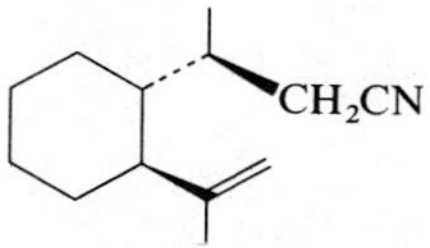 (structure) CH_2CN (74)	160, 592
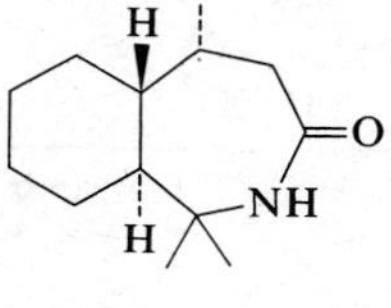 (structure) H, H, O, NH	TsCl, pyridine	" (61)	160

TABLE III. KETOXIME FRAGMENTATIONS (*Continued*)

No. of Carbon Atoms	Substrate	Reagent and Conditions	Product(s) and Yield(s) (%)	Refs.
	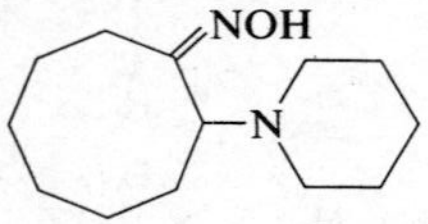 NOH, N	PCl_5, ether, 0°	$NC(CH_2)_6CHO$ (65)	571
	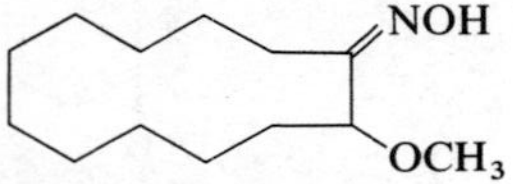 NOH, OCH_3	"	$NC(CH_2)_{10}CHO$ (82)	571
	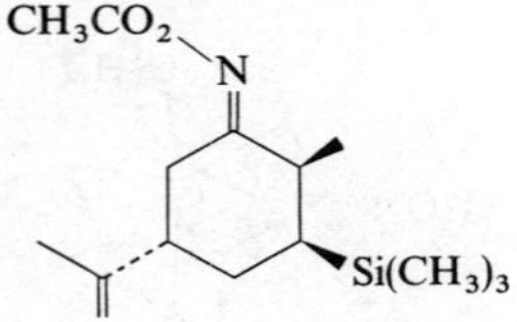 CH_3CO_2, N, $Si(CH_3)_3$	$CF_3SO_3Si(CH_3)_3$, CH_2Cl_2, 0°, 4 h	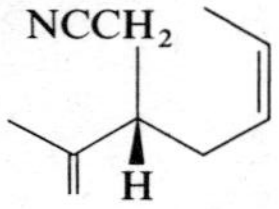 $NCCH_2$, H (94)	204
C_{14}	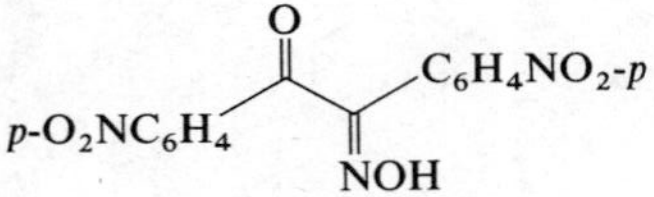 p-$O_2NC_6H_4$, O, $C_6H_4NO_2$-p, NOH	1. $Pd[P(C_6H_5)_3]_4$, 4 eq, CH_3CN, 60° 2. O_2	p-$O_2NC_6H_4CN$ (68) + p-$O_2NC_6H_4CHO$ (6)	597
	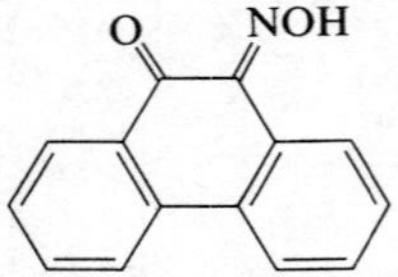 O, NOH	$SOCl_2$, SO_2, 70°, 6 h	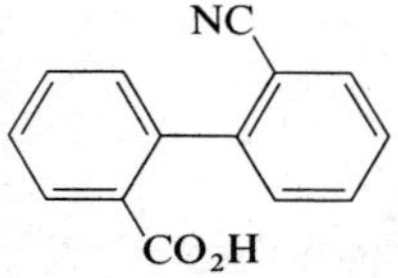 NC, CO_2H (60) 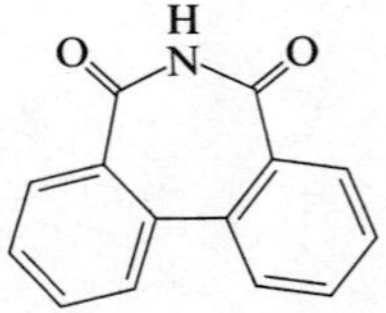 + O, H, N, O (11)	202

Reactant	Conditions	Product(s) (Yield, %)	Refs.
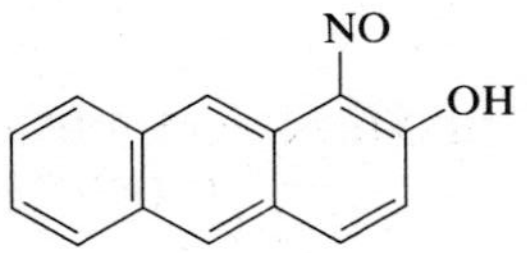 NO, OH	TsCl, pyridine, acetone	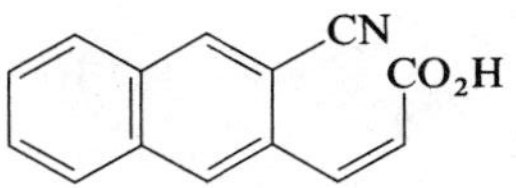 CN, CO_2H (70 – 80)	598
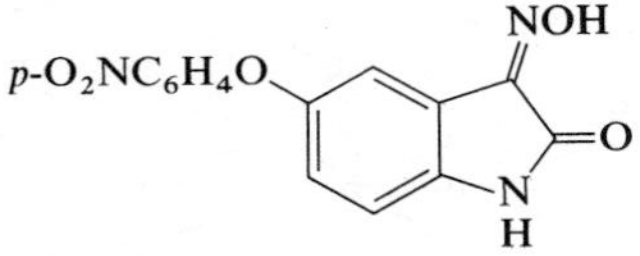 p-$O_2NC_6H_4O$, NOH, O, N, H	TsCl, pyridine, NaOH	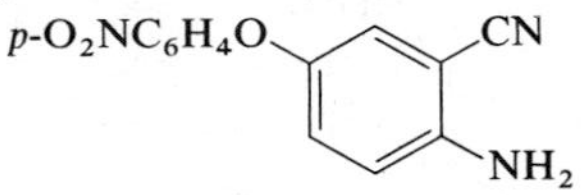 p-$O_2NC_6H_4O$, CN, NH_2 (80)	599
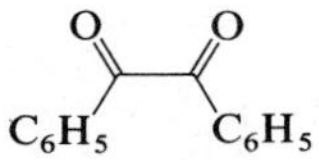 O, O, C_6H_5, C_6H_5	H_2NOSO_3H, HCO_2H, reflux	$C_6H_5CO_2H$ (90) + C_6H_5CN + $C_6H_5CONH_2$ (71)	600
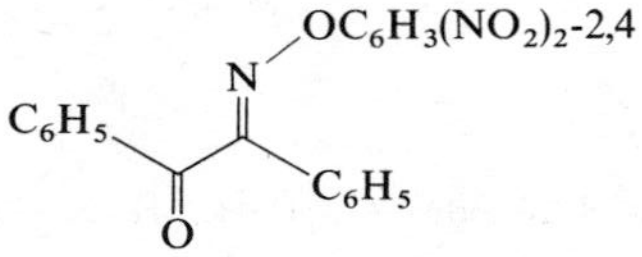 $OC_6H_3(NO_2)_2$-2,4, N, C_6H_5, C_6H_5, O	NH_3 (liq)	$C_6H_5CONH_2$ + C_6H_5CN (—)	601
	NaOH, dioxane, H_2O	$C_6H_5CO_2H$ + " (—)	601
	NaOR, ROH, dioxane, R = CH_3, i-C_3H_7, t-C_4H_9	$C_6H_5CO_2R$ + " (—)	601
$C_6H_5COC(=NOH)C_6H_5$ (E)	PPA	$C_6H_5CO_2H$ + C_6H_5CN + $C_6H_5CONH_2$	
	25°, 420 min	(96) (87) (3)	179, 188, 197
	65°, 90 min	(98) (80) (15)	179, 188, 197
	120°, 15 min	(98) (0) (92)	179, 188, 197
	$HC(OC_2H_5)_3$, SO_2, reflux	$C_6H_5CO_2C_2H_5$ (98) + C_6H_5CN (95)	188
	CF_3CO_2H	$C_6H_5CO_2H$ (88) + C_6H_5CN (94)	197, 569
	1. $NaOC_2H_5$, ethanol 2. $(CH_3CO)_2O$	$C_6H_5CO_2C_2H_5$ + C_6H_5CN (75) (81)	576

TABLE III. KETOXIME FRAGMENTATIONS (*Continued*)

No. of Carbon Atoms	Substrate	Reagent and Conditions	Product(s) and Yield(s) (%)	Refs.
	(*Z*)	1. $Pd[P(C_6H_5)_3]_4$, 4 eq, CH_3CN, 60° 2. O_2	C_6H_5CN (85) + C_6H_5CHO (43)	597
	(*E*)	"	" (45) + " (11)	597
	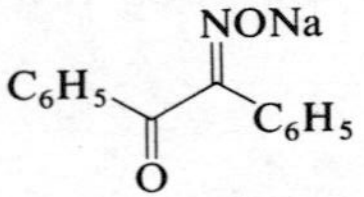 $C_6H_5COC(=NONa)C_6H_5$	$FClO_3$	C_6H_5CN 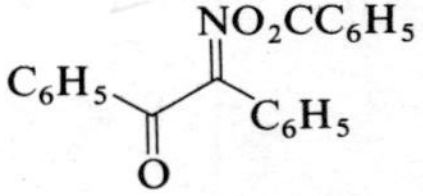 + $C_6H_5COC(=NO_2CC_6H_5)C_6H_5$ (36)	602
	p-$ClC_6H_4CHOHC(=NOH)C_6H_4Cl$-*p*		*p*-ClC_6H_4CHO + *p*-ClC_6H_4CN	
		CF_3SO_3Cl	(66) (62)	583
		$(CF_3SO_2)_2O$	(80) (70)	583
		$(CF_3CO)_2O$	(76) (70)	583
	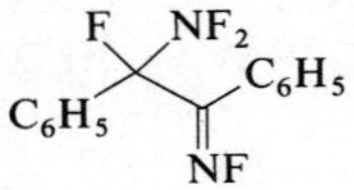 $C_6H_5CF(NF_2)C(=NF)C_6H_5$	$HNF_2 + HSO_3F$	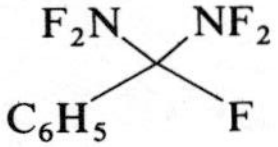 $C_6H_5CF(NF_2)_2$ (23)	121
	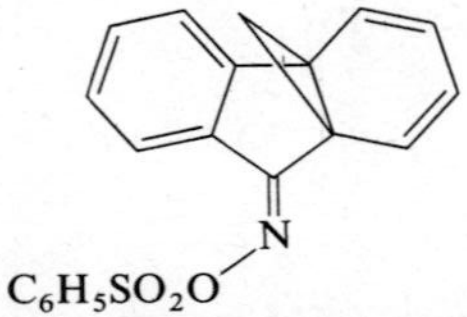 (structure; =N, $C_6H_5SO_2O$)	Reflux, acetone, H_2O	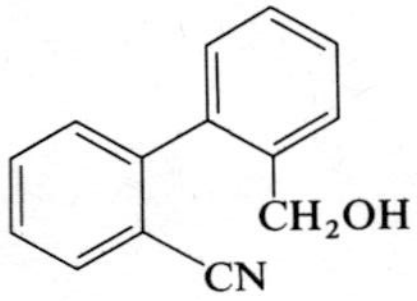 (structure; CH_2OH, CN) (31)	174
	$C_6H_5CH_2C(=NOH)C_6H_5$ (*E*)	1. $Pd[P(C_6H_5)_3]_4$, 4 eq, CH_3CN, 60° 2. O_2	C_6H_5CN (58)	597

Reactant	Conditions	Product(s) and Yield(s) (%)		Refs.
$C_6H_5CHOHC(=NOH)C_6H_5$		C_6H_5CN +	C_6H_5CHO	
	CF_3SO_2Cl	(77)	(74)	583
	$(CF_3SO_2)_2O$	(78)	(72)	583
	$(CF_3CO)_2O$	(75)	(71)	583
	C_6H_5NCO, $(C_2H_5)_3N$, C_6H_6, reflux, 1 h	(70)	(72)	440
	1. $Pd[P(C_6H_5)_3]_4$, 4 eq, CH_3CN, 60° 2. O_2	(60)	(26)	597
	PPA, 25°, 8 h	(26)	(33)	179
	Cl_2C:	(76)	(80)	178
$C_6H_5C(=NOH)CH_2NHC_6H_4Cl$-*p*	PCl_5, ether	C_6H_5CN + 1-(C_6H_4Cl-*p*)-4-C_6H_5-imidazoline 3-oxide (—)		603
6-*t*-C_4H_9-1-nitroso-2-naphthol	TsCl, NaOH, acetone, 55–60°	4-*t*-C_4H_9-2-(CH=CHCO$_2$H)benzonitrile (56)		604
1-CH_3-2-(*p*-ClC_6H_4)-3,3-dimethyl-5-(hydroxyimino)piperidin-4-one	$N_2H_4\cdot H_2O$, KOH, 190°, 3 h	1-CH_3-5-(*p*-ClC_6H_4)-4,4-dimethyl-3-X-pyrrolidin-2-one (—), X = H, OH		605

TABLE III. KETOXIME FRAGMENTATIONS (*Continued*)

No. of Carbon Atoms	Substrate	Reagent and Conditions	Product(s) and Yield(s) (%)	Refs.
	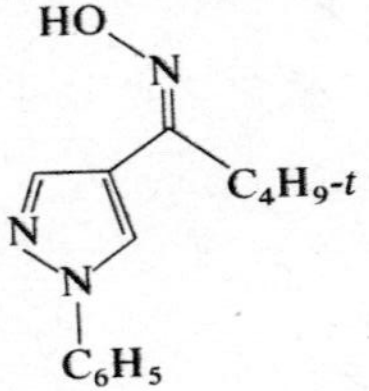 HO–N=C(C_4H_9-*t*)–(1-C_6H_5-pyrazol-4-yl)	PCl_5, $CHCl_3$, 4 d	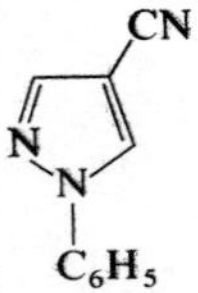 4-CN-1-C_6H_5-pyrazole (50)	318
		TsCl, pyridine, 4 d	" (58)	318
	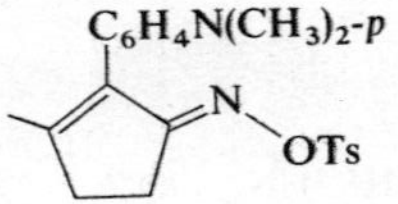 $C_6H_4N(CH_3)_2$-*p*, N–OTs (cyclopentenone oxime tosylate)	80% ethanol, 80°	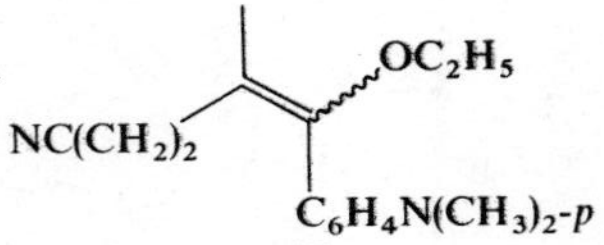 $NC(CH_2)_2$, OC_2H_5, $C_6H_4N(CH_3)_2$-*p* (63) 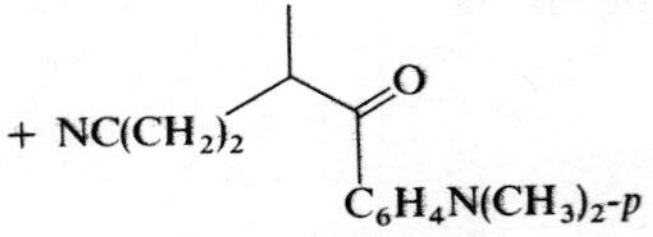 + $NC(CH_2)_2$, O, $C_6H_4N(CH_3)_2$-*p* (10)	49
	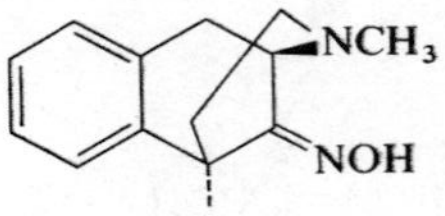 NCH_3, NOH	PCl_5, CH_2Cl_2, 5 min	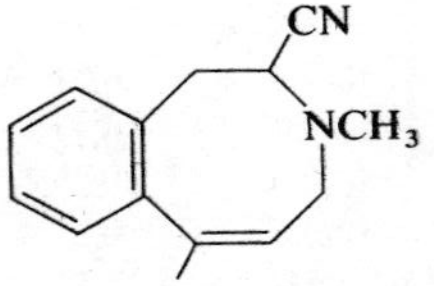 CN, NCH_3 (59)	606

Reactant	Conditions	Product(s) and Yield(s) (%)	Refs.
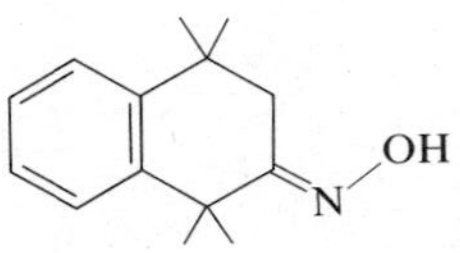 OH, N	PCl_5, C_6H_6	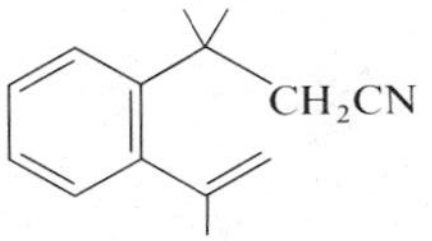 CH_2CN (94)	420
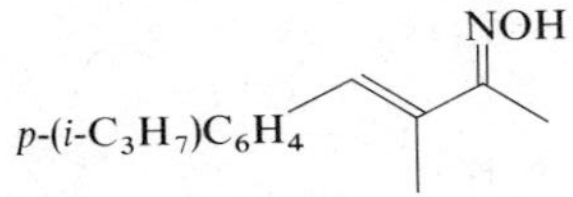 p-(i-C_3H_7)C_6H_4, NOH	PCl_5, ether	CH_3CN 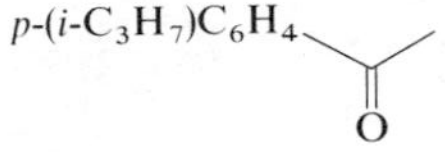 + p-(i-C_3H_7)C_6H_4, O (21)	586
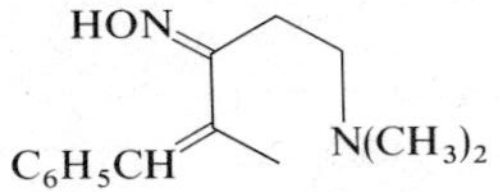 HON, C_6H_5CH, $N(CH_3)_2$	PCl_5, dioxane, 12 h	$C_6H_5CH_2COCH_3$ + $(CH_3)_2N(CH_2)_2CN$ (—)	442
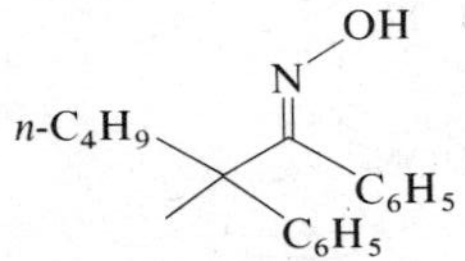 n-C_4H_9, N, OH, C_6H_5, C_6H_5	PPA, 25°, 2 h or 110°, 10 min	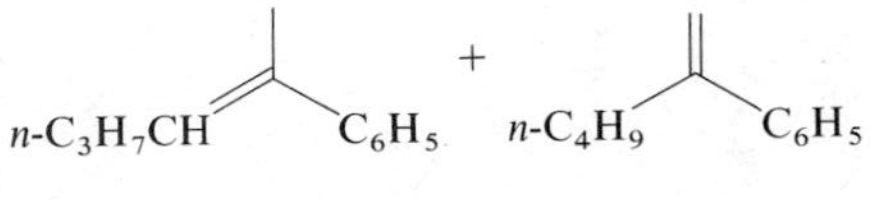 n-C_3H_7CH, C_6H_5 + n-C_4H_9, C_6H_5 (98)	162
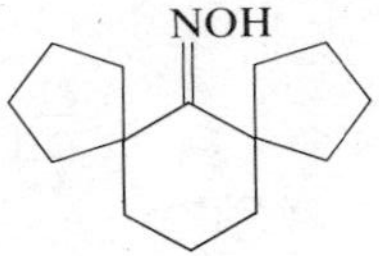 NOH	$SOCl_2$, 15 h	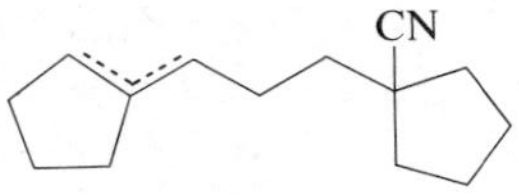 CN (79)	406
	TsCl, 15 h	" (73)	406
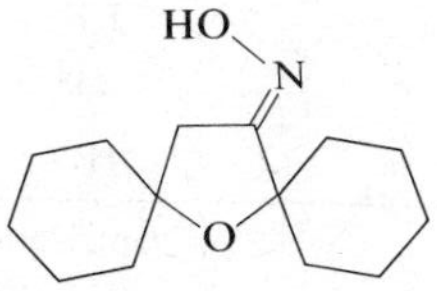 HO, N, O	PCl_5, C_6H_6	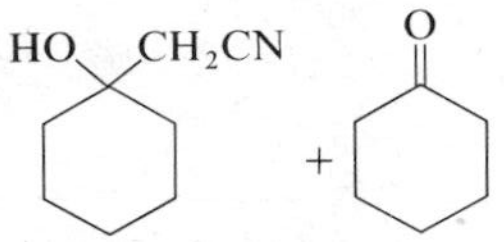 HO, CH_2CN + O (45)	187

TABLE III. KETOXIME FRAGMENTATIONS (*Continued*)

No. of Carbon Atoms	Substrate	Reagent and Conditions	Product(s) and Yield(s) (%)	Refs.
	OH, N, C_4H_9-*n*, HO	TsCl, pyridine, 0°, 17 h	$(CH_2)_3CN$, CHO, C_4H_9-*n* (23)	182, 183
	NOH, OC_2H_5	PCl_5, ether, 0°	$NC(CH_2)_{10}CHO$ (—)	571
	NOH, SC_2H_5	"		571
	OCH_3, OCH_3, NOH	$C_2H_5C(OCH_3)_2 \cdot BF_4$, 1 h	$NC(CH_2)_{10}CO_2CH_3$ (100)	189
	$N(CH_3)_2$, NOH	SO_3, SO_2	$NC(CH_2)_{10}CHO$ (35)	570
	CH_3CO_2, N, C_5H_{11}-*n*, $Si(CH_3)_3$	$CF_3SO_3Si(CH_3)_3$, CH_2Cl_2, 0°	*n*-$C_5H_{11}$$(CH_2)_3CN$ (93)	607

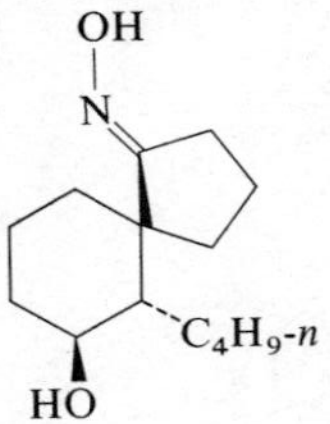

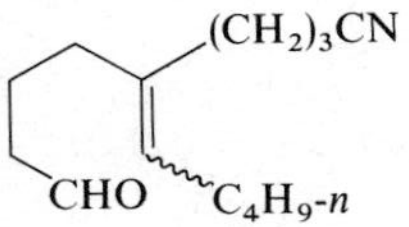

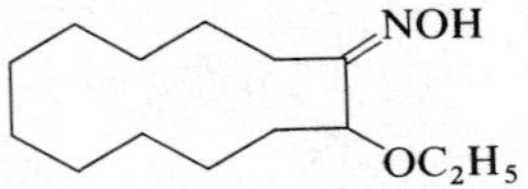

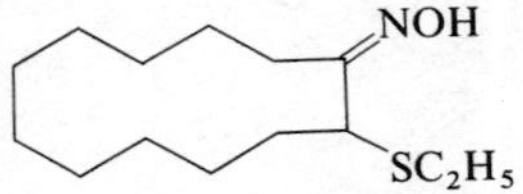

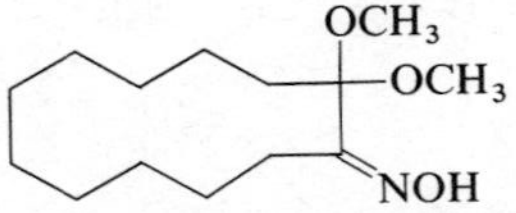

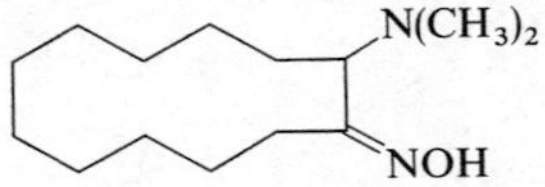

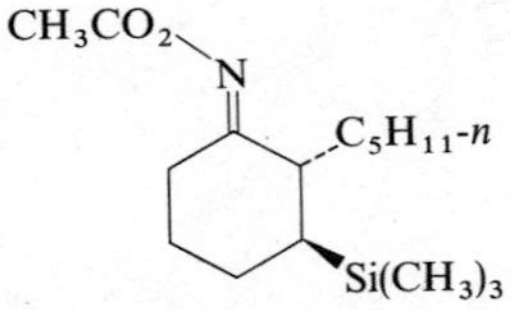

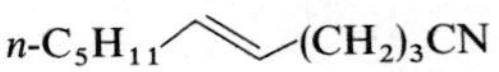

	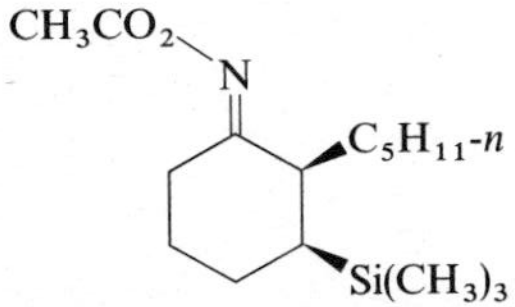	"	 (88)	607
C_{15}	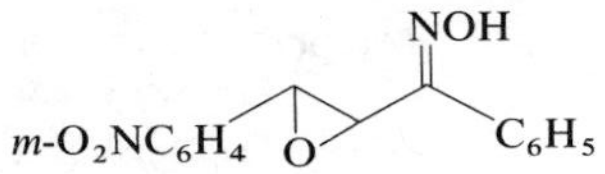	PCl_5, C_6H_6	C_6H_5CN + m-$O_2NC_6H_4CHOHCHO$ (—)	187
	$C_6H_5CH_2COC(=NOH)C_6H_5$	$C_6H_5SO_2Cl$, NaOH	C_6H_5CN (77) + $C_6H_5CH_2CO_2H$ (74)	197, 569
		$(CH_3CO)_2O$, $NaOC_2H_5$, ethanol	C_6H_5CN (52) + $C_6H_5CH_2CO_2C_2H_5$ (43)	576
		85% H_2SO_4	$C_6H_5CONH_2$ (61) + $C_6H_5CH_2CO_2H$ (68)	197, 569
	$C_6H_5CH_2C(=NOH)COC_6H_5$	$C_6H_5SO_2Cl$, NaOH	$C_6H_5CH_2CN$ (68) + $C_6H_5CO_2H$ (74)	197
		85% H_2SO_4, 120°	$C_6H_5CH_2CONH_2$ (47) + $C_6H_5CO_2H$ (86)	197
	p-$ClC_6H_4CH(C_6H_5)C(=NOH)CH_3$	PCl_5, ether, 0°, 5 min	CH_3CN + $C_6H_5CHClC_6H_4Cl$-p (38)	608
	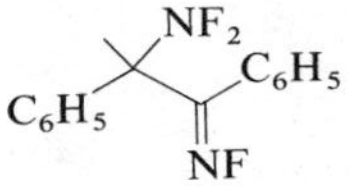	HNF_2, H_2SO_4	$C_6H_5COCH_3$ (—)	121
	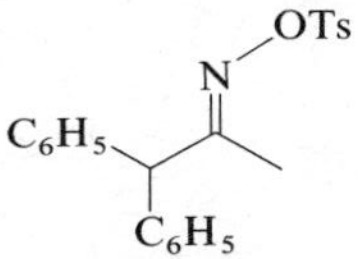	$(C_2H_5)_3N$, 80% ethanol, 30°, 60 h	CH_3CN + $(C_6H_5)_2CHOR$ R = H (16) R = C_2H_5 (38)	28
	$C_6H_5CH(OCH_3)C(=NOH)C_6H_5$	TsCl, pyridine, 16 h	C_6H_5CHO + C_6H_5CN (—)	187
		PCl_5, C_6H_6	" " (—)	187
		PPA, 100–120°, 10 min	" + $C_6H_5CONH_2$ (—)	187
	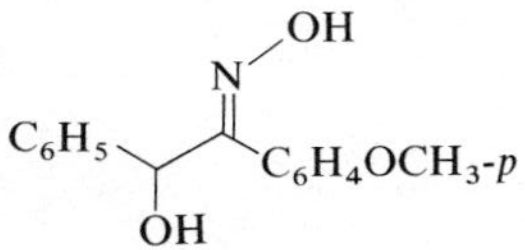		C_6H_5CHO + p-$CH_3OC_6H_4CN$	

TABLE III. KETOXIME FRAGMENTATIONS (*Continued*)

No. of Carbon Atoms	Substrate	Reagent and Conditions	Product(s) and Yield(s) (%)		Refs.
		CF_3SO_2Cl	(70)	(65)	583
		$(CF_3SO_2)_2O$	(76)	(70)	583
		$(CF_3CO)_2O$	(69)	(62)	583
		Cl_2C:	(74)	(65)	178
	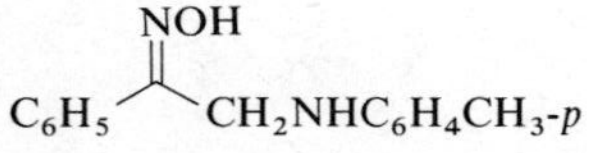 $C_6H_5C(=NOH)CH_2NHC_6H_4CH_3$-*p*	PCl_5, ether, 0°	C_6H_5CN 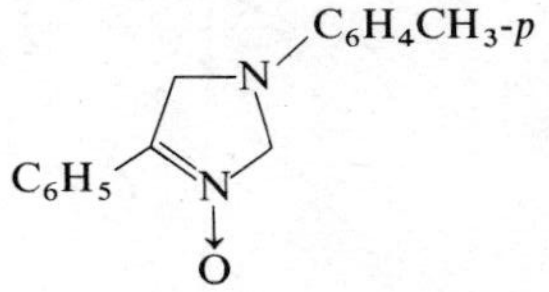 + C_6H_5, N→O, N–$C_6H_4CH_3$-*p*	(—)	603
	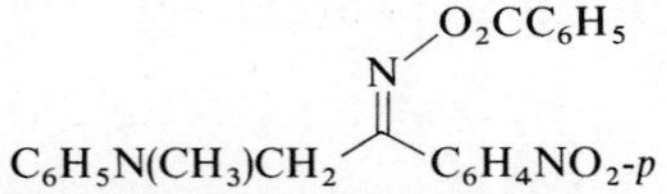 $C_6H_5N(CH_3)CH_2C(=NO_2CC_6H_5)C_6H_4NO_2$-*p*	80% ethancl, 80°, 50 h	*p*-$O_2NC_6H_4CN$ (85)		95
	$C_6H_5N(CH_3)CH_2C(=NOR)C_6H_5$	80% ethanol, 80°	C_6H_5CN		
	R = $COC_6H_3(NO_2)_2$-3,5		(90)		95
	R = $C_6H_3(NO_2)_2$-2,4		(81)		95
	R = $COC_6H_4NO_2$-*p*		(91)		95
	R = COC_6H_5		(88)		95
	R = $COC_6H_4OCH_3$-*p*		(84)		95
	N–OH	PCl_5, C_6H_6	$(CH_2)_2CN$	(96)	420

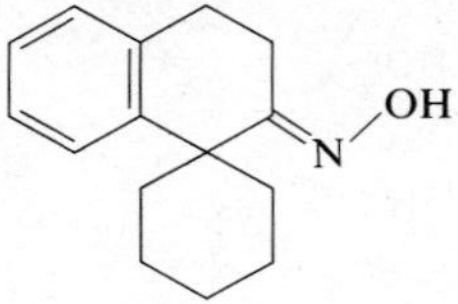

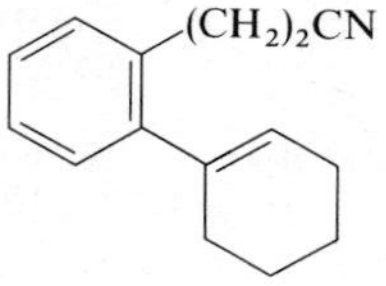

Formula	Reactant	Conditions	Product(s) and Yield(s) (%)	Refs.
	[structure]	TsCl, pyridine	[structure] OCH_3, CH_2CN, C_6H_5 (77)	297
	[structure] $CO_2C_4H_9$-*i*	$NaOC_2H_5$, ethanol, 1 h	[structure] CN, $(CH_2)_2CH{=}NOH$ (89)	594
	[structure] NOH, $(CH_2)_2N(CH_3)_2$	(—)	[structure] $(CH_2)_2CN$, $CH_2N(CH_3)_2$ (64)	606
	[structure] CH_3CO_2, $CH_2C_6H_5$, $Si(CH_3)_3$	$CF_3SO_3Si(CH_3)_3$, CH_2Cl_2, 0°, 4 h	$C_6H_5CH_2CH{=}CH(CH_2)_2CN$ (93)	204
C_{16}	[structure] (dimer of isatin 3-oxime)	TsCl, dioxane, reflux, 30 min	[structure] H_2N, NC, CN, NH_2 (14)	199
	[structure] R = O	"	[structure] NC, CN, H_2N, NH_2 (54)	199
	R = S		(68)	199

TABLE III. KETOXIME FRAGMENTATIONS (*Continued*)

No. of Carbon Atoms	Substrate	Reagent and Conditions	Product(s) and Yield(s) (%)	Refs.
	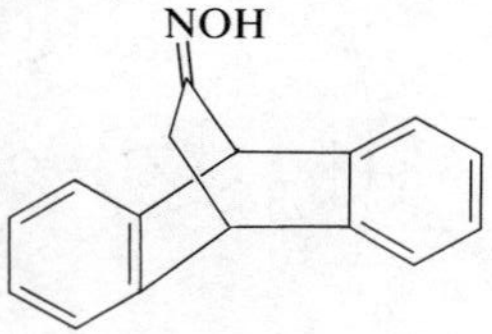 NOH	$C_6H_5SO_2Cl$, NaOH, reflux, 2 h	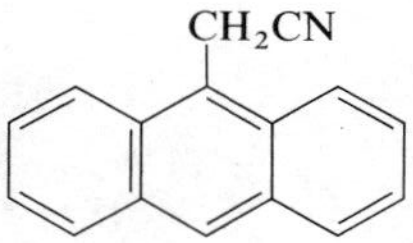 CH_2CN (79)	609
		$POCl_3$, PCl_5, 0°	" (94)	609
	p-$CH_3C_6H_4COCOC_6H_4CH_3$-p	H_2NOSO_3H, HCO_2H, reflux	p-$CH_3C_6H_4CO_2H$ (87) + p-$CH_3C_6H_4CN$ + p-$CH_3C_6H_4CONH_2$ (53)	600
	p-$CH_3OC_6H_4COCOC_6H_4OCH_3$-$p$	"	p-$CH_3OC_6H_4CO_2H$ (59) + p-$CH_3OC_6H_4CN$ + p-$CH_3OC_6H_4CONH_2$ (55) + p-$CH_3OC_6H_4NH_2$ (28)	600
	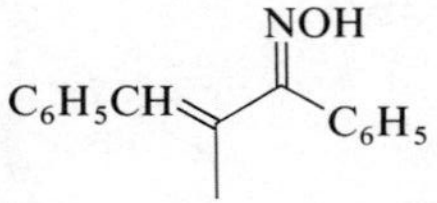 C_6H_5CH= NOH C_6H_5	PCl_5, ether	$C_6H_5CH_2COCH_3$, C_6H_5CN (—)	442
	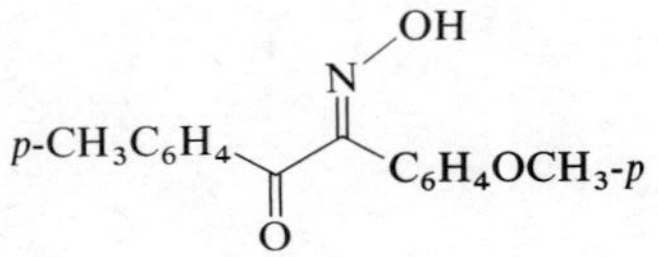 p-$CH_3C_6H_4$ O N OH $C_6H_4OCH_3$-p	1. $Pd[P(C_6H_5)_3]_4$, 5 eq, CH_3CN 2. 220°	p-$CH_3OC_6H_4CN$ (49)	597
	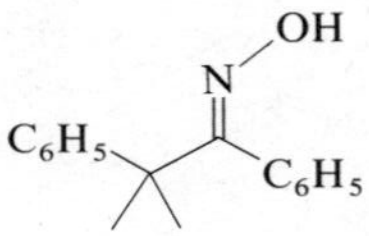 C_6H_5 N OH C_6H_5	PPA, 130°, 10 min	$C_6H_5CONH_2$ (—)	64

HO–N=C(C_6H_5)–C(C_6H_5)(OCH_3)$_2$ [HO, N, C_6H_5, C_6H_5, CH_3O, OCH_3]	SO_2, 75°, 48 h	C_6H_5CN (100) + $C_6H_5CO_2CH_3$ (100)	189
p-$CH_3OC_6H_4$CH(C_6H_5)C(=N–OH)CH_3 [OH, N, p-$CH_3OC_6H_4$, C_6H_5]	PCl_5, ether, 0°, 5 min	$C_6H_5CHClC_6H_4CH_3$-p (64)	608
p-$CH_3OC_6H_4$CH(OH)C(=N–OH)$C_6H_4OCH_3$-p [OH, N, p-$CH_3OC_6H_4$, $C_6H_4OCH_3$-p, OH]		p-$CH_3OC_6H_4CHO$ + p-$CH_3OC_6H_4CN$	
	CF_3SO_2Cl	(79) (69)	583
	$(CF_3SO_2)_2O$	(81) (72)	583
	$(CF_3CO)_2O$	(72) (68)	583
[NHCOCH$_3$, =NOH, N, COCH$_3$]	TsCl, $(C_2H_5)_3N$, CH_3CN, −10 to 20°, 2 h	[CH=CHNHCOCH$_3$, N, COCH$_3$] (54)	579
[O, N, NOH, $(CH_2)_2$, CHC$_6$H$_5$]	PCl_5, dioxane, 12 h	$C_6H_5CH_2COCH_3$ + [O, N]$(CH_2)_2CN$ (—)	442
[OH, N, t-C_4H_9, O, C_6H_5]	PCl_5, ether, 0°	[$NCCH_2$, O, C_6H_5] (—)	161

TABLE III. KETOXIME FRAGMENTATIONS (*Continued*)

No. of Carbon Atoms	Substrate	Reagent and Conditions	Product(s) and Yield(s) (%)	Refs.
	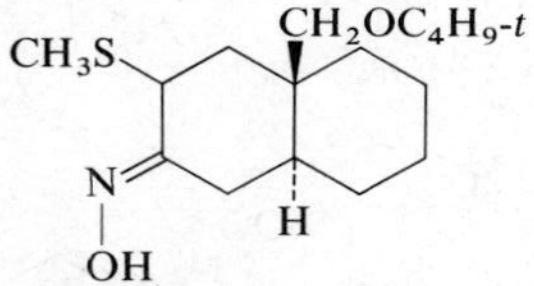 CH_3S, $CH_2OC_4H_9$-*t*, N–OH, H	CH_3SO_2Cl, pyridine, reflux, 1.5 h	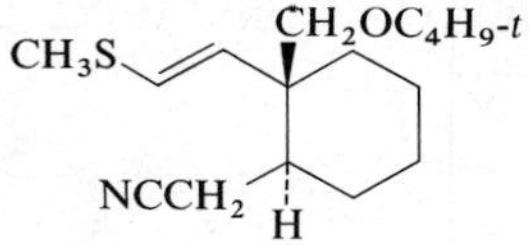 CH_3S, $CH_2OC_4H_9$-*t*, $NCCH_2$, H (45)	610, 611
	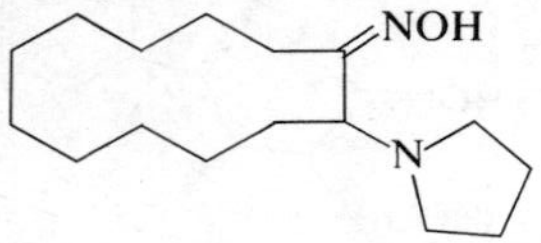 NOH, N	PCl_5, ether, 0°	$NC(CH_2)_{10}CHO$ (21)	571
	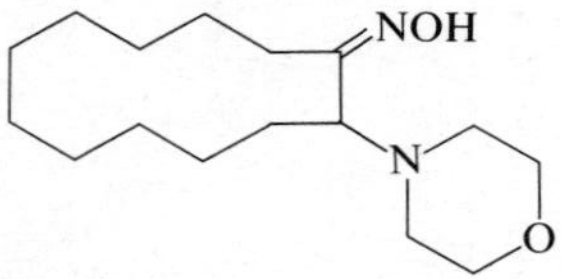 NOH, N, O	"	" (84)	571
	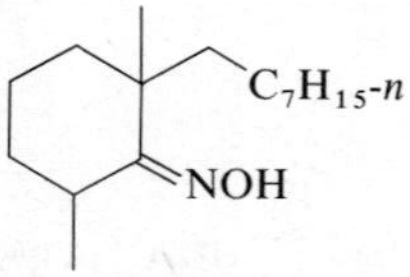 C_7H_{15}-*n*, NOH	TsCl, pyridine	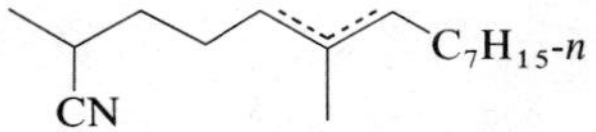 CN, C_7H_{15}-*n* (90)	612
	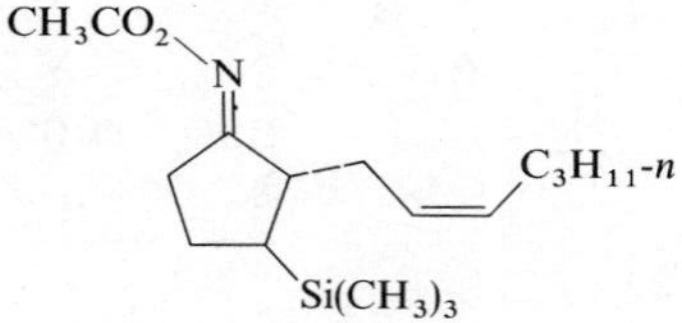 CH_3CO_2, N, C_3H_{11}-*n*, $Si(CH_3)_3$	$CF_3SO_3Si(CH_3)_3$, CH_2Cl_2, 0°, 2.5 h	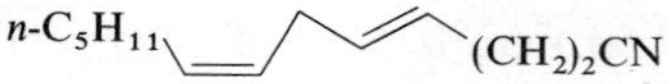 *n*-C_5H_{11}, $(CH_2)_2CN$ (79)	607

Formula	Reactant	Conditions	Product (Yield, %)	Refs.
	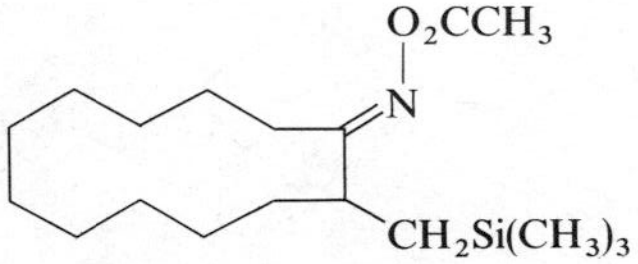 O_2CCH_3, N, $CH_2Si(CH_3)_3$	"	$CH_2{=}CH(CH_2)_{10}CN$ (81)	204
C_{17}	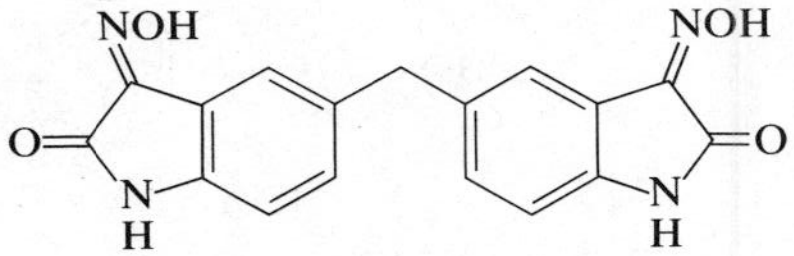 NOH, O, N, H, NOH, O, N, H	TsCl, dioxane, reflux, 30 min	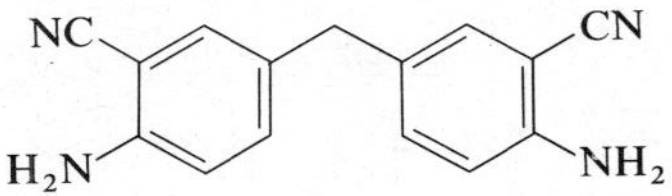 NC, H_2N, CN, NH_2 (28)	199
	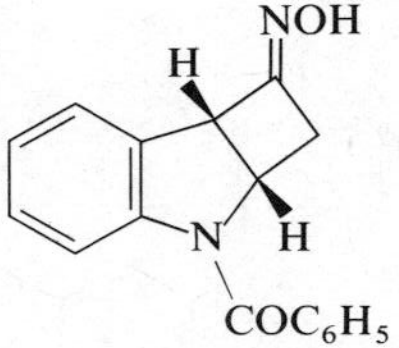 NOH, H, N, H, COC_6H_5	$SOCl_2$, DMF, 0°	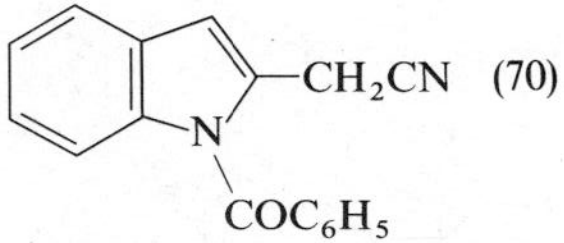 CH_2CN, N, COC_6H_5 (70)	175
	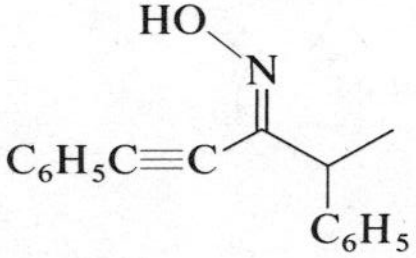 HO, N, $C_6H_5C{\equiv}C$, C_6H_5	PCl_5, ether	$C_6H_5C{\equiv}CCN$ (—)	116
	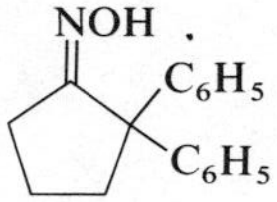 NOH, C_6H_5, C_6H_5	PCl_5, ether, 24 h	$(C_6H_5)_2C{=}CH(CH_2)_2CN$ (96)	165, 613
		PPA, 110–120°, 10 min	" (95)	165
		$SOCl_2$, C_6H_6, 24 h	" (95)	165
	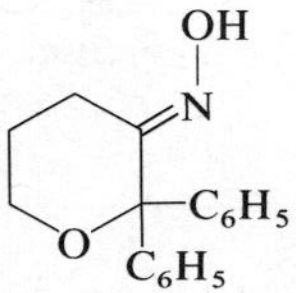 OH, N, O, C_6H_5, C_6H_5	PCl_5, $CHCl_3$, 10 h	$C_6H_5COC_6H_5$ (95)	187

TABLE III. KETOXIME FRAGMENTATIONS (*Continued*)

No. of Carbon Atoms	Substrate	Reagent and Conditions	Product(s) and Yield(s) (%)	Refs.
		PPA, 100–120°, 10 min	" (27)	187
		TsCl, pyridine	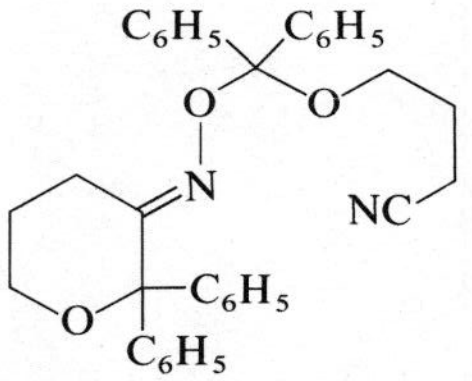 [structure: C_6H_5, C_6H_5, O, O, N, NC, C_6H_5, C_6H_5, O] (95)	187
	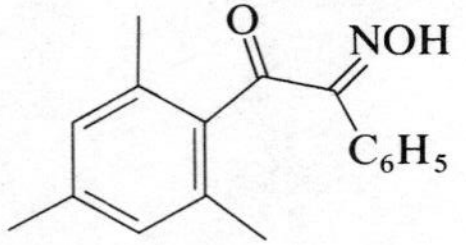 [structure: O, NOH, C_6H_5]	$C_6H_5SO_2Cl$, NaOH	C_6H_5CN 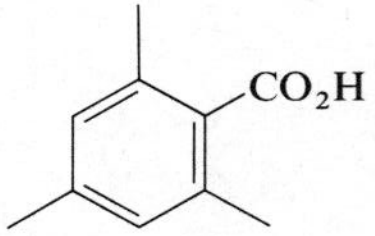 + [structure: CO_2H] (—)	58
	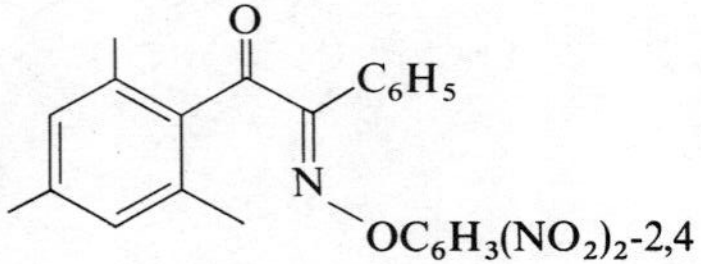 [structure: O, C_6H_5, N, $OC_6H_3(NO_2)_2$-2,4]	NaOR, ROH, dioxane R = H, CH_3, i-C_3H_7, t-C_4H_9	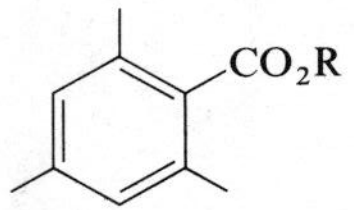 [structure: CO_2R] (—)	601
	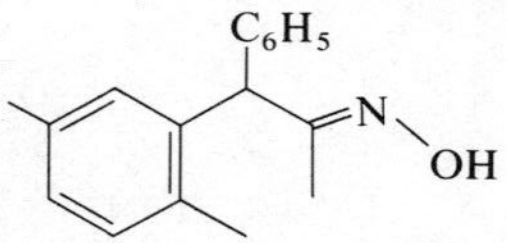 [structure: C_6H_5, N, OH]	PCl_5, ether, 0°, 5 min	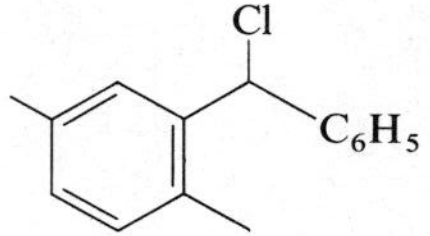 [structure: Cl, C_6H_5] (63)	608
	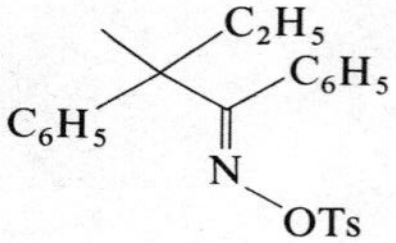 [structure: C_2H_5, C_6H_5, C_6H_5, N, OTs]	$(C_2H_5)_3N$, 80% ethanol, 30°	C_6H_5CN (71) 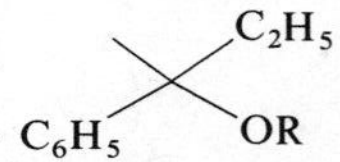 + [structure: C_6H_5, C_2H_5, OR]	28

			$R = H$ (47) + $R = C_2H_5$ (33)	
	NOH, C_6H_5CH, N (piperidino)	PCl_5, dioxane, 12 h	$C_6H_5CH_2COCH_3$ + N-$(CH_2)_2CN$ (piperidine) (—)	442
	OCH_3, CH_3O, CH_3O, CH_3, N, O, N–OH	$N_2H_4 \cdot H_2O$, KOH, 190°, 3 h	OCH_3, CH_3O, CH_3O, CH_3, N, O, OH (—)	605
	NOH, N (piperidino)	PCl_5, ether, 0°	$NC(CH_2)_{10}CHO$ (51)	571
C_{18}	NOH, N, H, COC_6H_5	$SOCl_2$, DMF, 0°	CH_2CN, N, COC_6H_5 (64)	175
	OH, N, C_6H_5, C_6H_5	PCl_5, ether, 24 h	$(C_6H_5)_2C{=}CH(CH_2)_3CN$ (96)	165

TABLE III. KETOXIME FRAGMENTATIONS (*Continued*)

No. of Carbon Atoms	Substrate	Reagent and Conditions	Product(s) and Yield(s) (%)	Refs.
	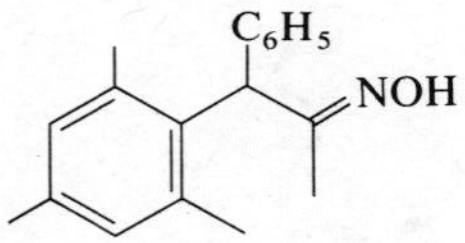	$SOCl_2$, C_6H_6, 18 h	" (99)	165
	C_6H_5, NOH	PCl_5, ether, 0°, 5 min	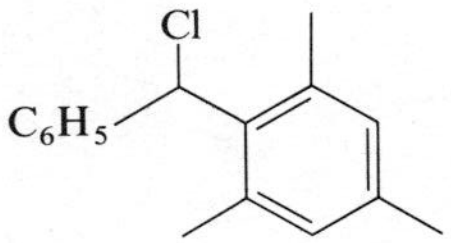 Cl, C_6H_5 (65)	608
	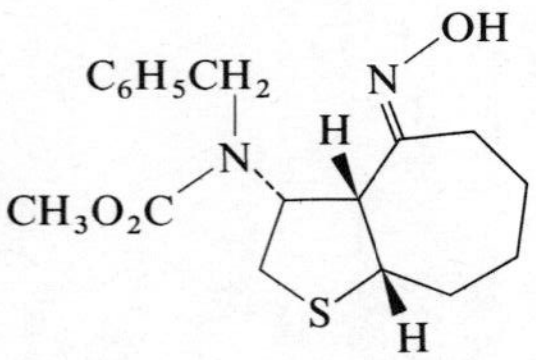 $C_6H_5CH_2$, N, CH_3O_2C, H, N, OH, S, H	$SOCl_2$, CH_2Cl_2, 0°	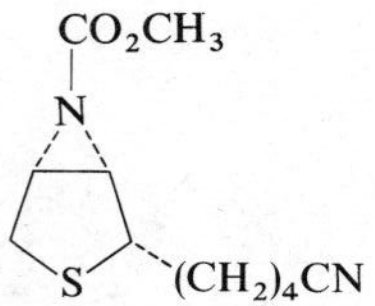 CO_2CH_3, N, S, $(CH_2)_4CN$ (—)	196
C_{19}	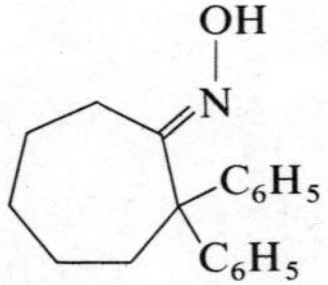 OH, N, C_6H_5, C_6H_5	$SOCl_2$, C_6H_6, 24 h	$(C_6H_5)_2C{=}CH(CH_2)_4CN$ (98)	165
		PPA, 125–135°, 10 min	$(C_6H_5)_2CH(CH_2)_5CONH_2$ (67)	165
	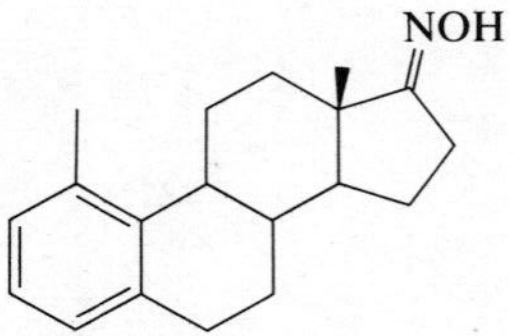 NOH	DCC, CF_3CO_2H, DMSO, C_6H_6, 3 h	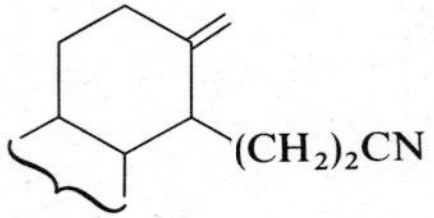 $(CH_2)_2CN$ (80)	614

Reagents / Conditions	Structures	Yield	Ref.
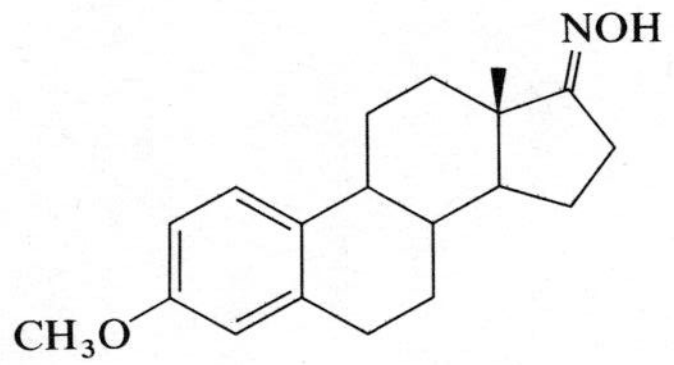 DCC, CF_3CO_2H, DMSO, 16 h	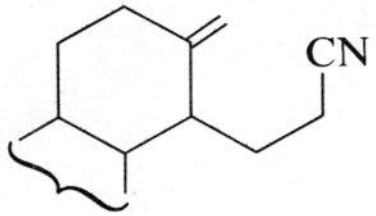 CN	(41)	468
HN_3	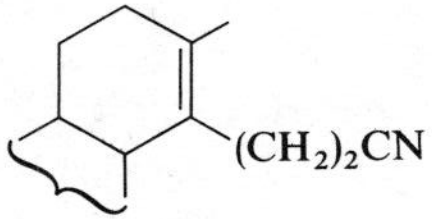 $(CH_2)_2CN$	(9)	469
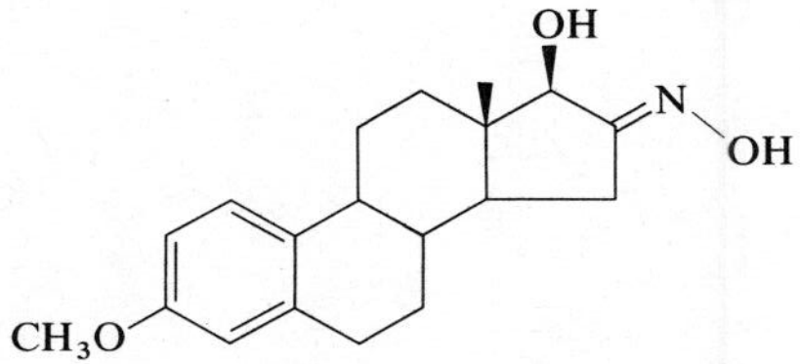 TsCl, pyridine, 3 h	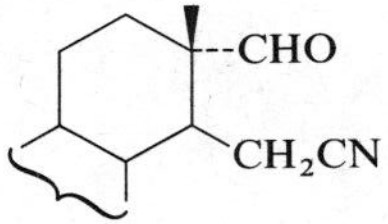 CHO; CH_2CN	(98)	181
KOH, $HO(CH_2)_2OH$, reflux	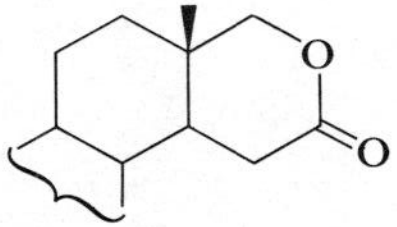 O; O	(—)	181
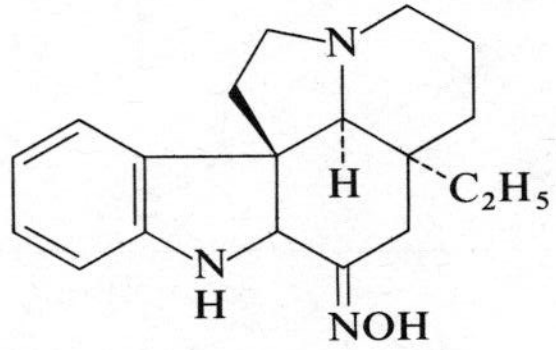 $POCl_3$, 6 h	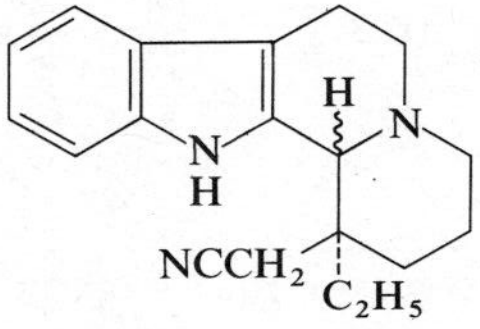 H; N; N H; $NCCH_2$; C_2H_5	(55)	191

TABLE III. KETOXIME FRAGMENTATIONS (*Continued*)

No. of Carbon Atoms	Substrate	Reagent and Conditions	Product(s) and Yield(s) (%)	Refs.
	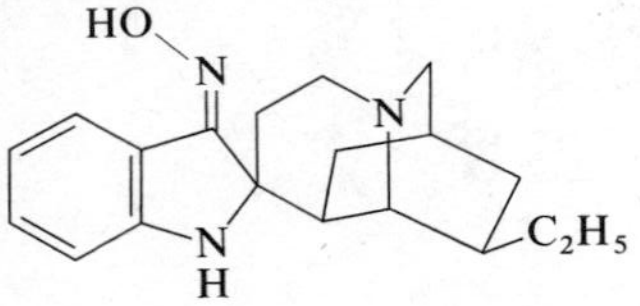 [structure: HO–N=, N, H, C_2H_5]	TsCl, reflux, 2 h	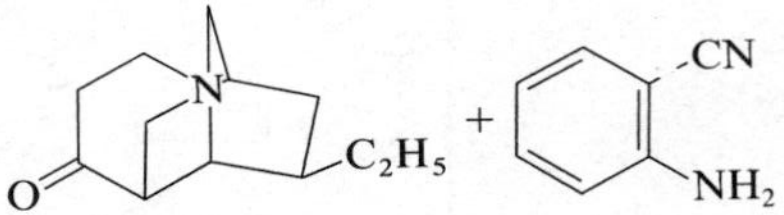 [structures: O, N, C_2H_5 + CN, NH_2] (33)	588
	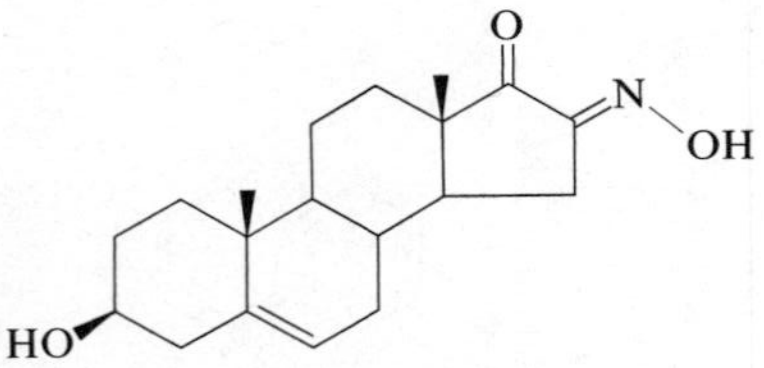 [structure: O, N–OH, HO]	$(CH_3CO)_2O$	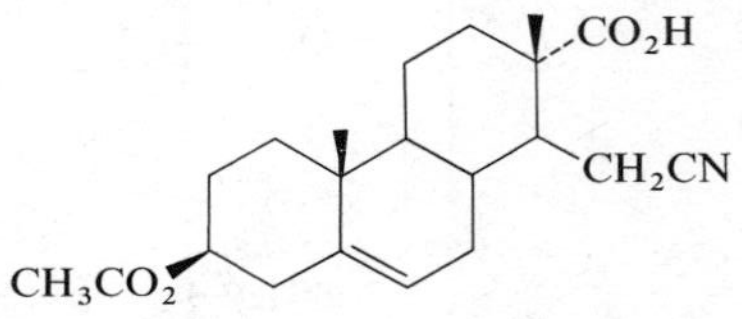 [structure: CO_2H, CH_2CN, CH_3CO_2] (—)	615, 616
	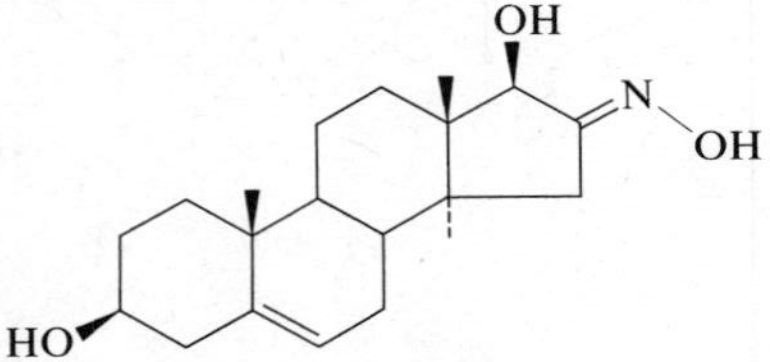 [structure: OH, N–OH, HO]	TsCl, pyridine, 0°, 12 h	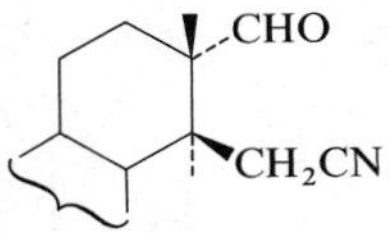 [structure: CHO, CH_2CN] (95)	180
	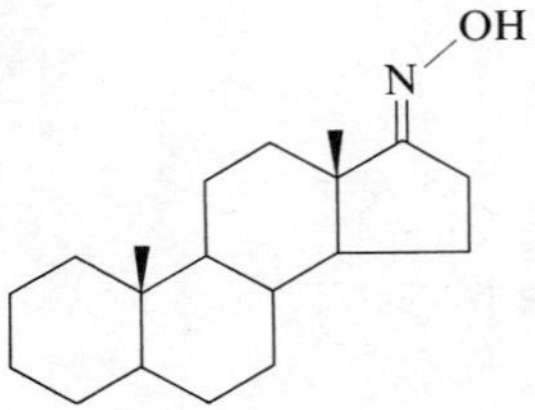 [structure: N–OH]	TsCl, pyridine, 16 h	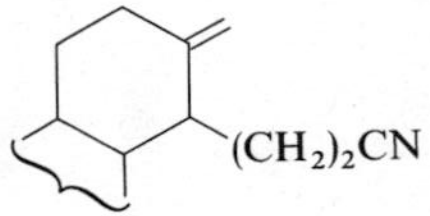 [structure: $(CH_2)_2CN$] (53)	474

Substrate	Conditions	Product(s) and Yield(s) (%)	Refs.
C_{20} NOTs, C_2H_5	HCl	$NCCH_2$, C_2H_5, N, H (26)	480
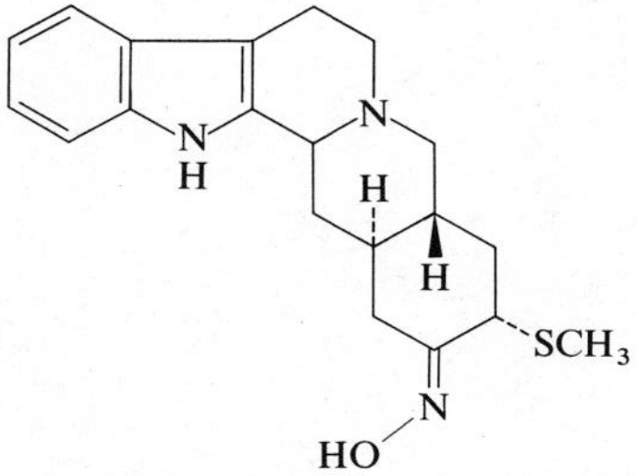 N, H, H, H, SCH_3, N, HO	1. $SOCl_2$, ether 2. Dil. NH_3, H_2O	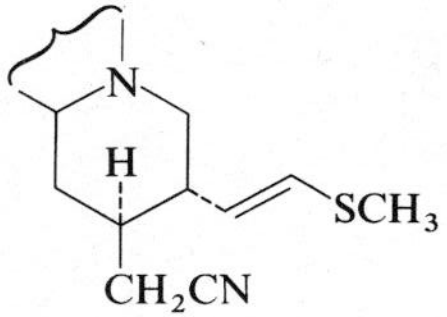 N, H, SCH_3, CH_2CN (57)	617
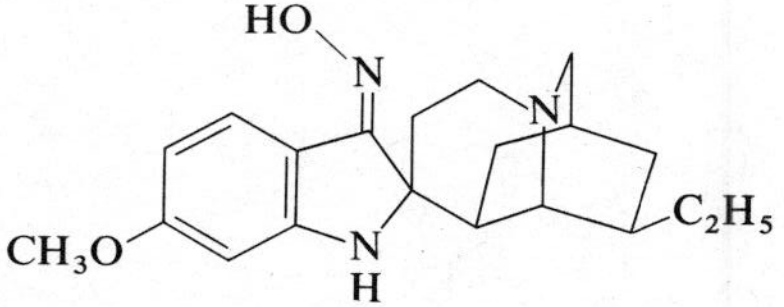 HO, N, N, CH_3O, N, H, C_2H_5	TsCl, pyridine, reflux, 2 h	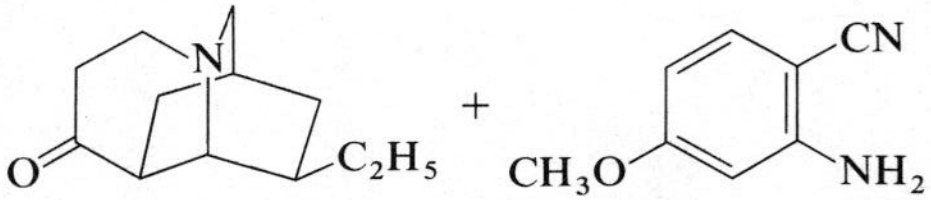 O, N, C_2H_5 + CH_3O, CN, NH_2 (40)	588
HO, N, N, CH_3O, N, H, C_2H_5	"	" + CH_3O, CN, NH_2 (32)	588
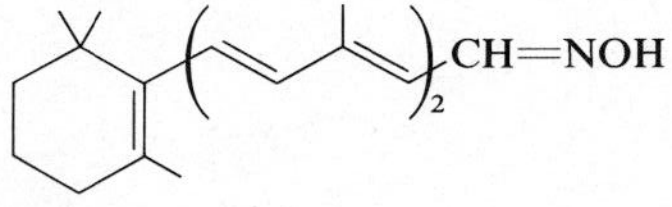 ( )$_2$CH=NOH	$(CF_3CO)_2O$, pyridine, 0–25° 3 h	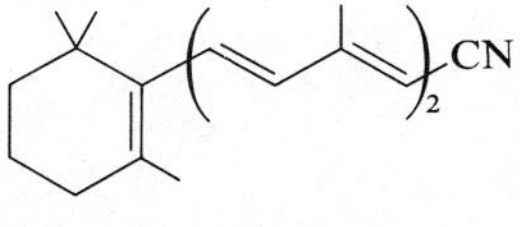 ( )$_2$CN (80)	618

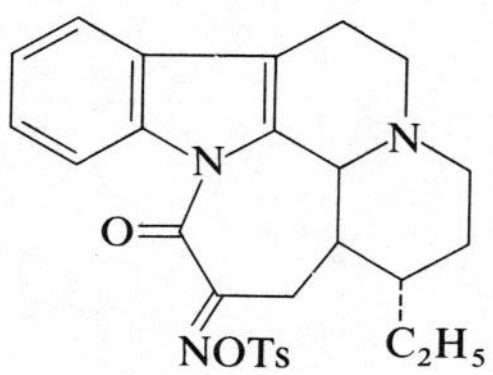

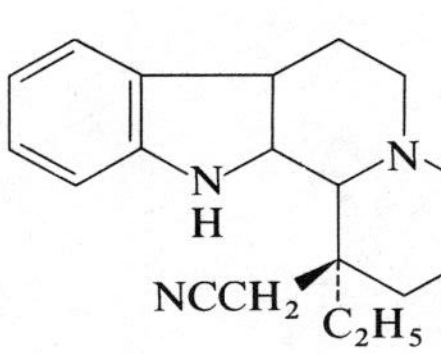

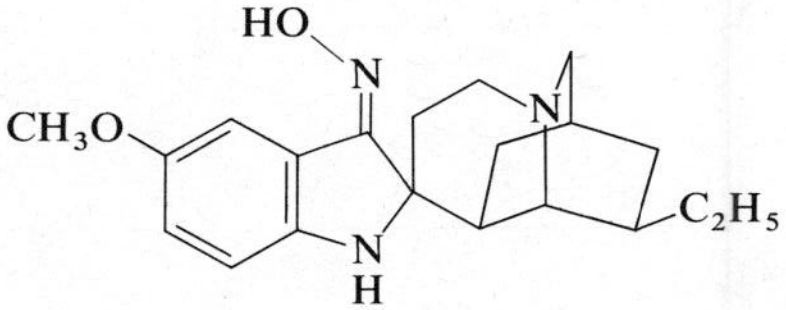

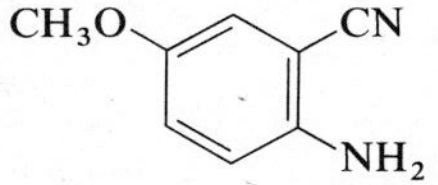

TABLE III. KETOXIME FRAGMENTATIONS (*Continued*)

No. of Carbon Atoms	Substrate	Reagent and Conditions	Product(s) and Yield(s) (%)	Refs.
	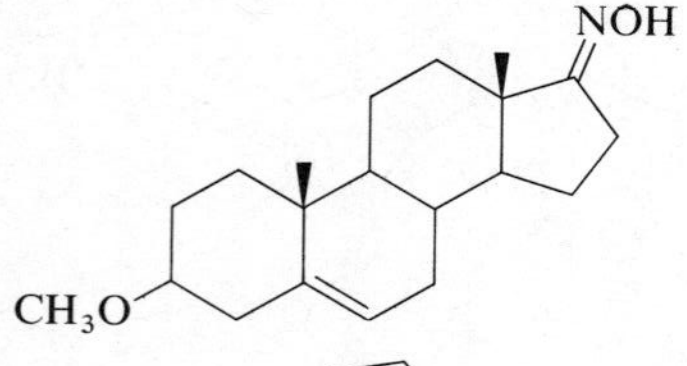 NOH; CH_3O	DCC, CF_3CO_2H, DMSO, 16 h	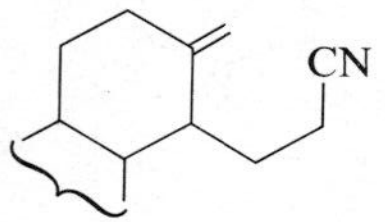 CN (61)	468
C_{21}	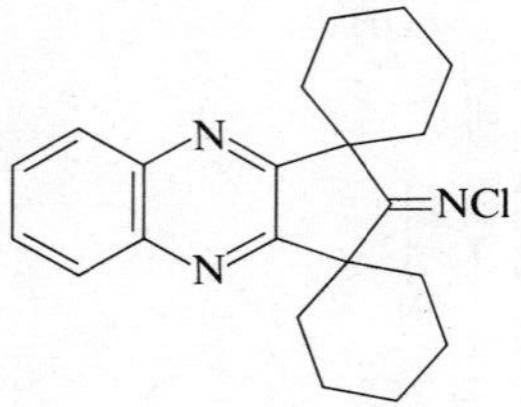 N; N; =NCl	$AgBF_4$, dioxane, H_2O, reflux, 3 h	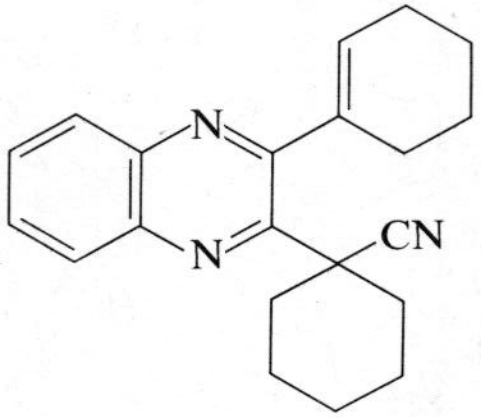 N; N; CN (12)	484
	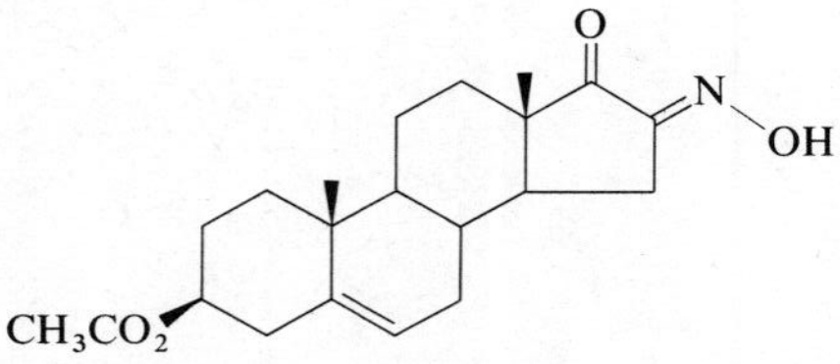 O; N; OH; CH_3CO_2	CH_3CO_2H, reflux, 1 h	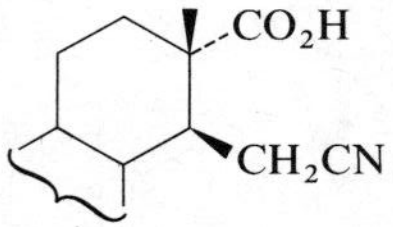 CO_2H; CH_2CN (—)	65
		$SOCl_2$, 1 h	" (—)	65

Reactant	Conditions	Product(s) and Yield(s) (%)	Refs.
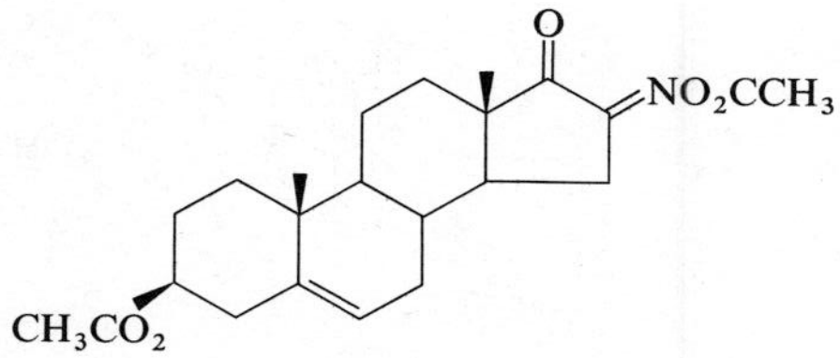 CH_3CO_2 … NO_2CCH_3 (steroid)	CH_3OH, 24 h	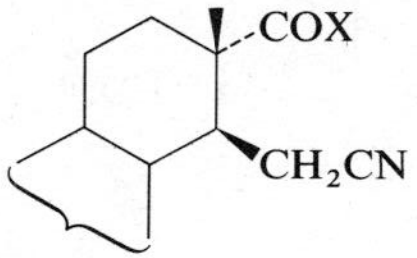 COX, CH_2CN; X = OCH_3 (—)	57
	KOC_4H_9-*t*	" X = OC_4H_9-*t* (—)	57
	$NaSC_4H_9$-*n*	" X = SC_4H_9-*n* (75)	57
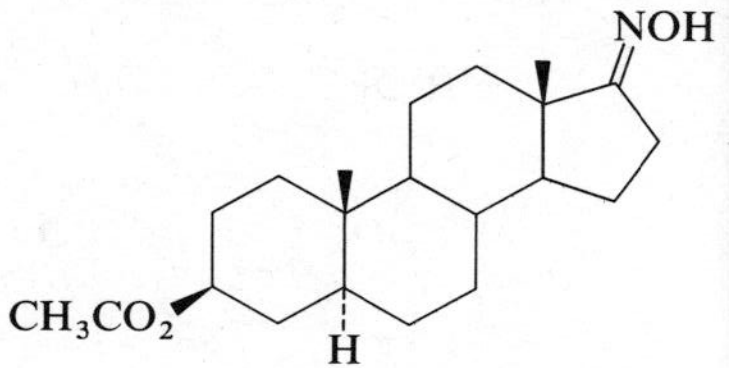 NOH, CH_3CO_2, H (steroid)	TsCl, pyridine, reflux, 1.75 h	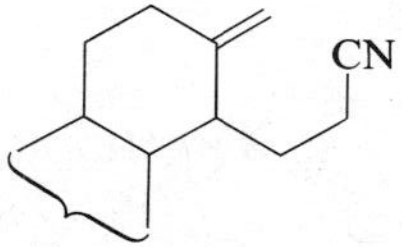 CN (21)	163
NO_2CCH_3, CH_3CO_2 (steroid)	DCC, CF_3CO_2H, DMSO, 3 d	" (52)	163
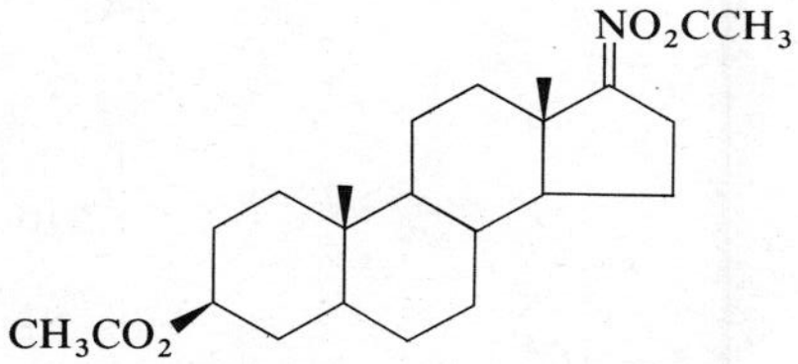	BF_3·ether, dioxane, 10–15 min	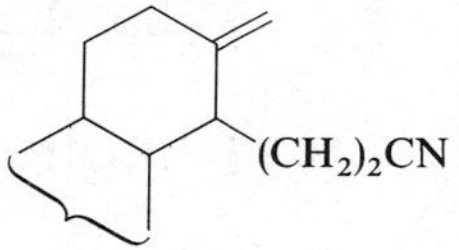 $(CH_2)_2CN$ (50)	494
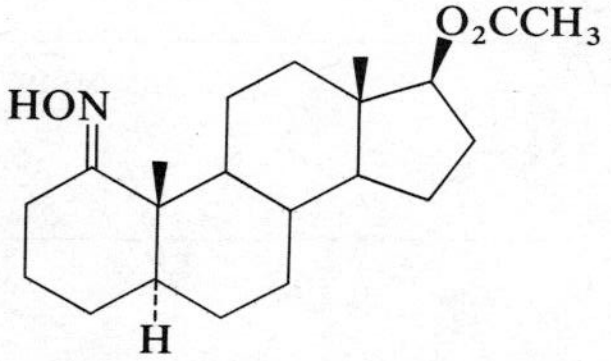 HON, O_2CCH_3, H (steroid)	TsCl, pyridine, reflux, 2.5 h	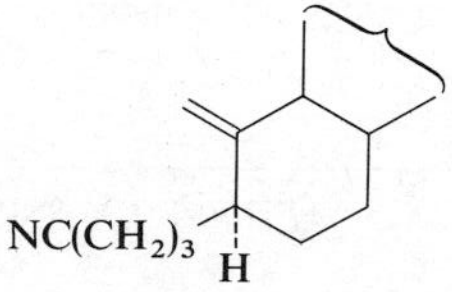 $NC(CH_2)_3$, H (90)	163

TABLE III. KETOXIME FRAGMENTATIONS (*Continued*)

No. of Carbon Atoms	Substrate	Reagent and Conditions	Product(s) and Yield(s) (%)	Refs.
	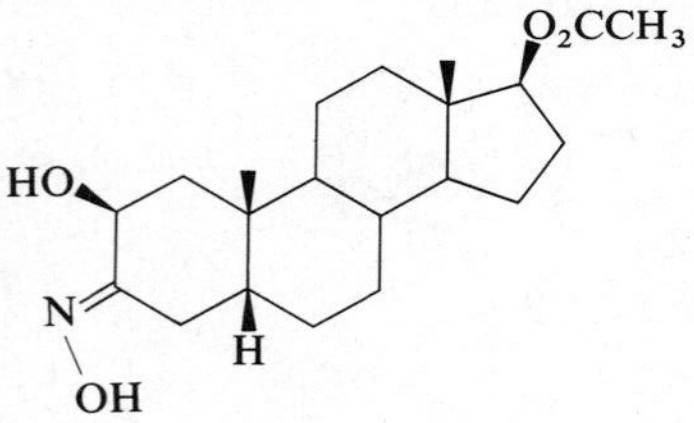 HO, N—OH, H, O_2CCH_3	$SOCl_2$, NaOH, H_2O	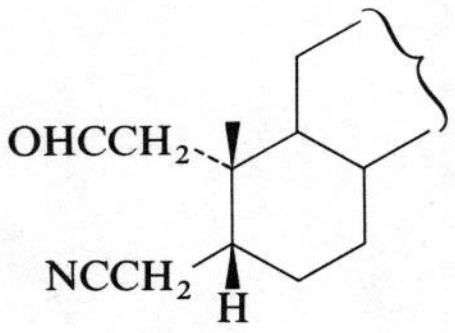 $OHCCH_2$, $NCCH_2$, H (>70)	619
C_{22}	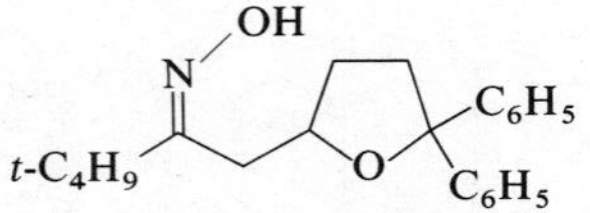 t-C_4H_9, N—OH, O, C_6H_5, C_6H_5	PCl_5, ether, 0°	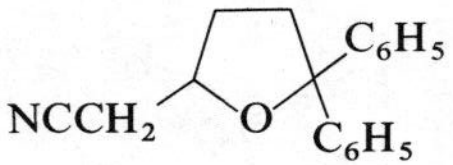 $NCCH_2$, O, C_6H_5, C_6H_5 (69)	161
C_{24}	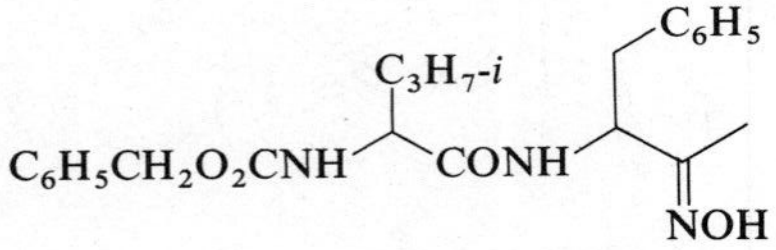 $C_6H_5CH_2O_2CNH$, C_3H_7-i, CONH, C_6H_5, NOH	TsCl, $(C_2H_5)_3N$, −10 to 20°, 2 h	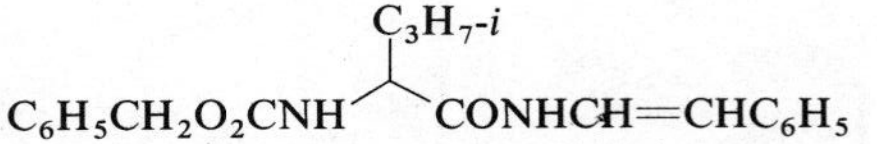 $C_6H_5CH_2O_2CNH$, C_3H_7-i, $CONHCH{=}CHC_6H_5$ (65)	579
C_{27}	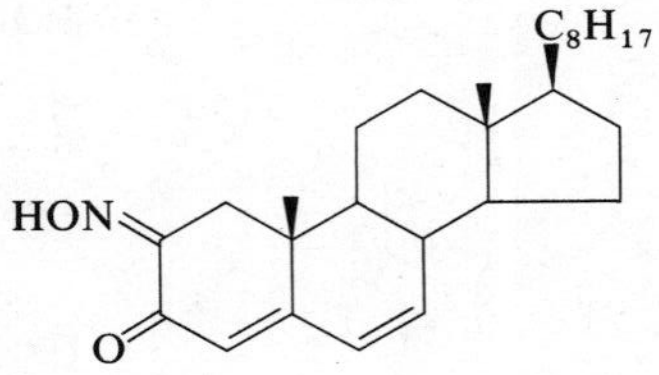 HON, O, C_8H_{17}	$SOCl_2$	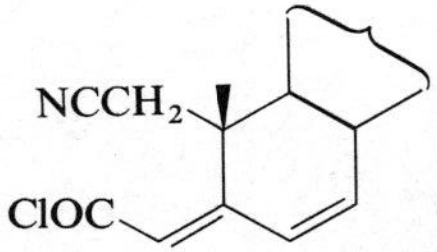 $NCCH_2$, ClOC (96)	139

Reactant	Conditions	Product(s) and Yield(s) (%)	Refs.
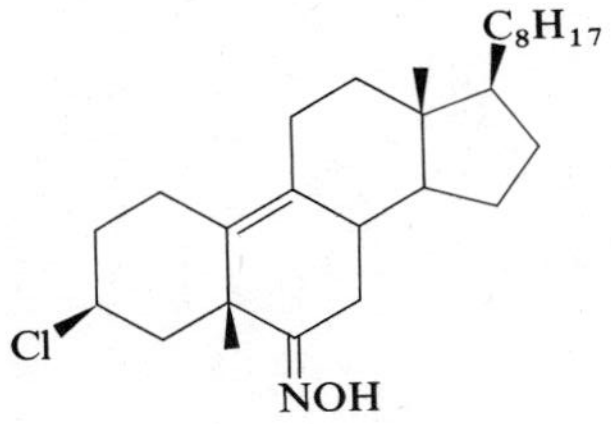 [structure: 3-Cl steroid, 6-NOH, C_8H_{17}]	$SOCl_2$, 4 *M* KOH	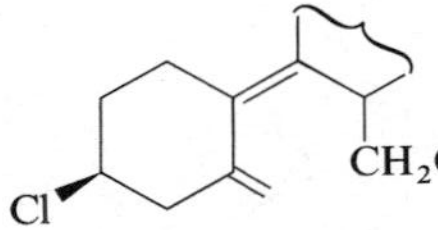 [structure] CH_2CN (31)	620
	TsCl, pyridine	" (—)	620
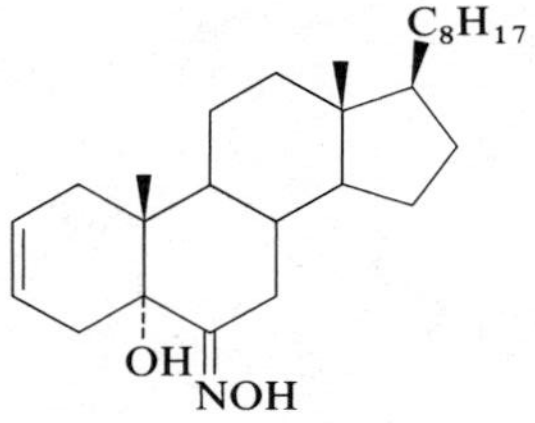 [structure: 5-OH steroid, 6-NOH, C_8H_{17}]	$SOCl_2$, −20°, 4 *M* KOH	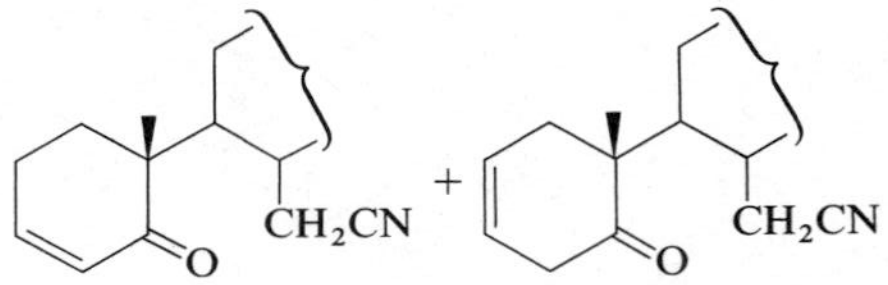 [structure] CH_2CN (37) + [structure] CH_2CN (22)	529
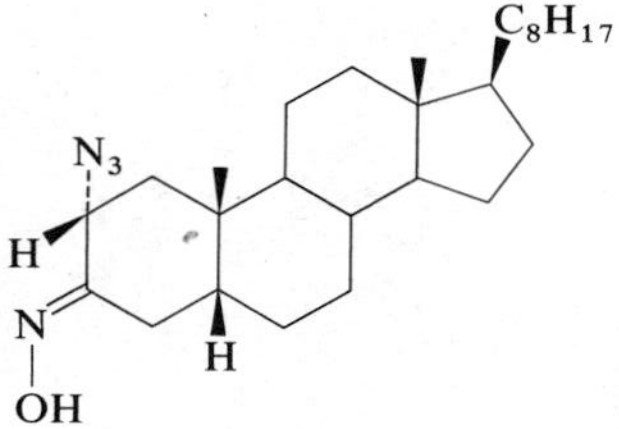 [structure: N_3, H, N–OH steroid, H, C_8H_{17}]	$POCl_3$, pyridine, 110°, 20 min	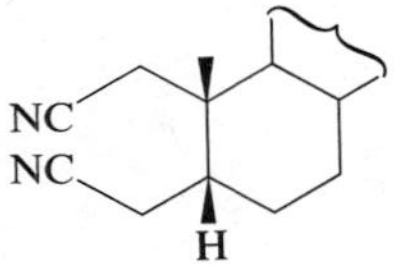 [structure] NC NC H (94)	621
	TsCl, pyridine, 110°, 30 min	" (85)	621
	TsCl, alumina	" (22)	621
	$SOCl_2$, 0°, 15 min	" (70)	621
	$SOCl_2$, pyridine, 60 min	" (73)	621
	PCl_5, pyridine, 110°, 20 min	" (70)	621
	CH_3SO_2Cl, pyridine, 100°, 30 min	" (65)	621

TABLE III. KETOXIME FRAGMENTATIONS (*Continued*)

No. of Carbon Atoms	Substrate	Reagent and Conditions	Product(s) and Yield(s) (%)	Refs.
	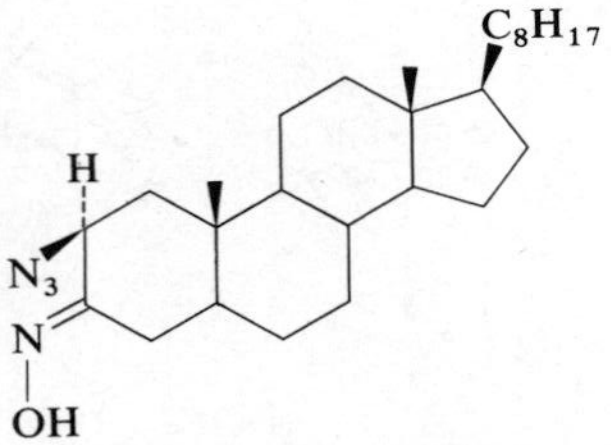 C_8H_{17}; H; N_3; N—OH	P_2O_5, C_6H_6, 80°, 20 min	" (39)	621
		$SOCl_2$, 0°, 15 min	" (64)	621
	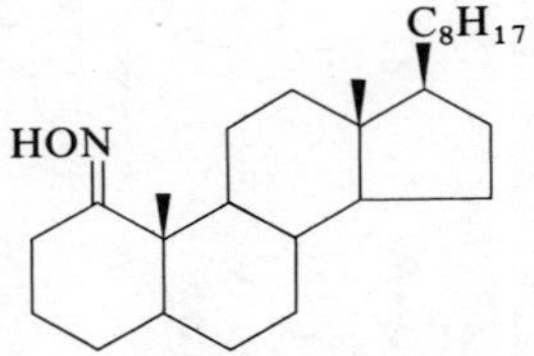 C_8H_{17}; HON	$SOCl_2$, ether, −20°, 5 min	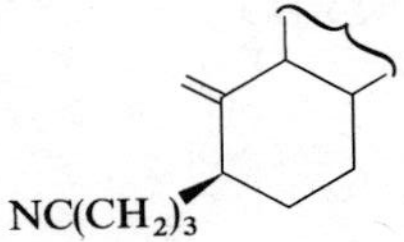 $NC(CH_2)_3$ (35)	164
	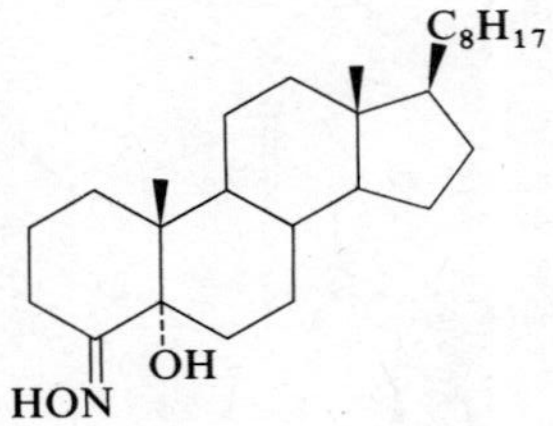 C_8H_{17}; HON; OH	"	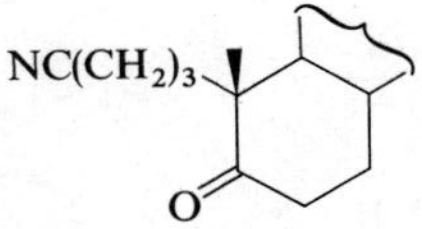 $NC(CH_2)_3$; O (94)	177
		Cl_2C:	" (86)	178

Reactant	Conditions	Product(s) and Yield(s) (%)	Refs.
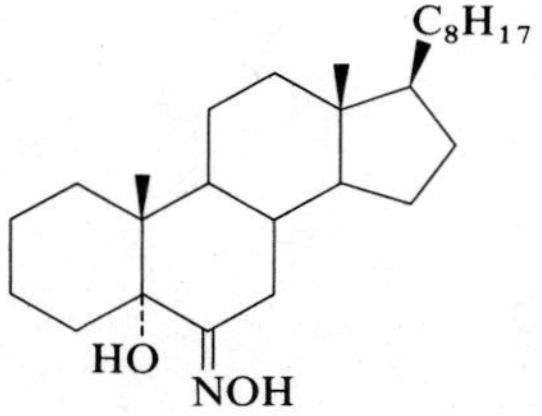 [structure: C_8H_{17}, HO, NOH]	$SOCl_2$, ether, $-20°$, 5 min	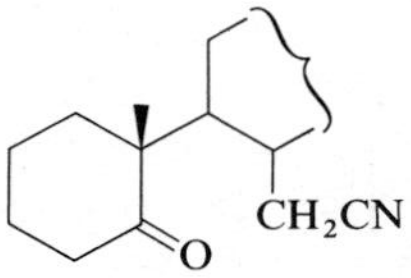 [structure: CH_2CN, O] (90)	177
	$(CH_3CO)_2O$, pyridine, heat	" (—)	622
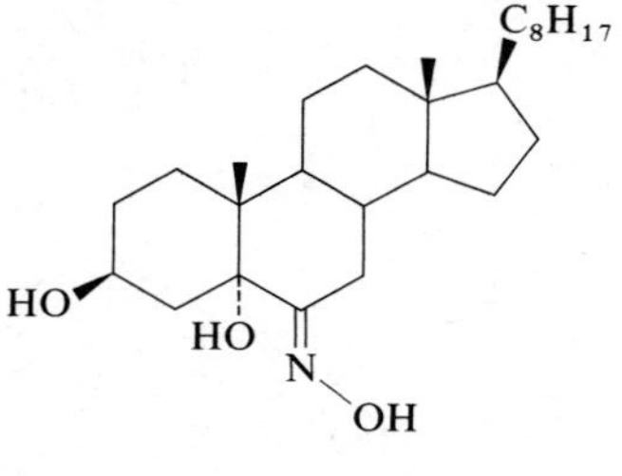 [structure: C_8H_{17}, HO, HO, N–OH]	$(PNCl_2)_3$, pyridine, THF	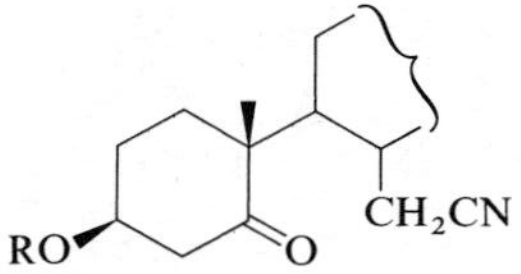 [structure: RO, CH_2CN, O] R = H (85)	566
	$(CH_3CO)_2$, pyridine	" R = CH_3CO (50)	622
	$SOCl_2$, 4 *M* KOH	" R = H 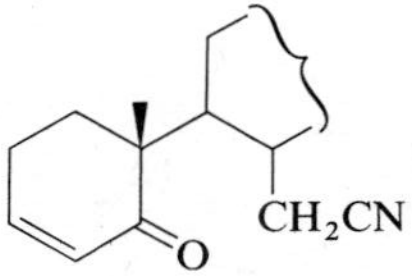 (—) + [structure: CH_2CN, O] (—)	529
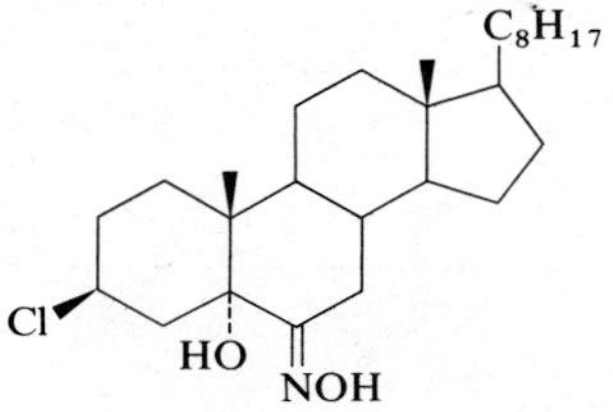 [structure: C_8H_{17}, Cl, HO, NOH]	"	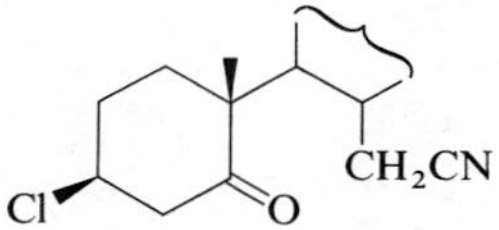 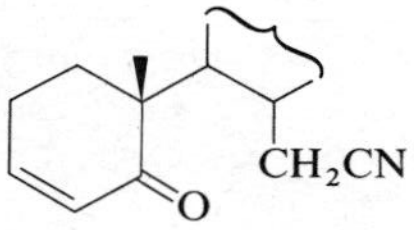 [structure: Cl, CH_2CN, O] + [structure: CH_2CN, O] (—)	529

TABLE III. KETOXIME FRAGMENTATIONS (*Continued*)

No. of Carbon Atoms	Substrate	Reagent and Conditions	Product(s) and Yield(s) (%)	Refs.
	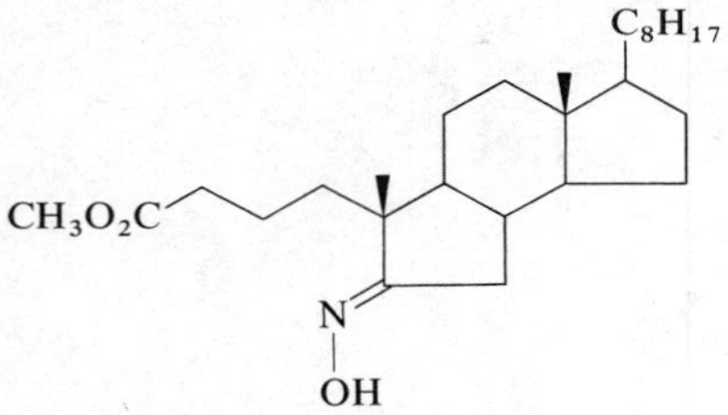 CH_3O_2C ... C_8H_{17}, N—OH	$SOCl_2$, ether, −10°	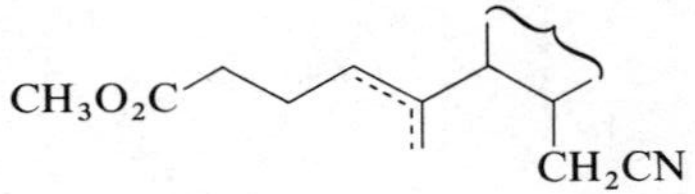 CH_3O_2C ... CH_2CN (—)	537
C_{28}	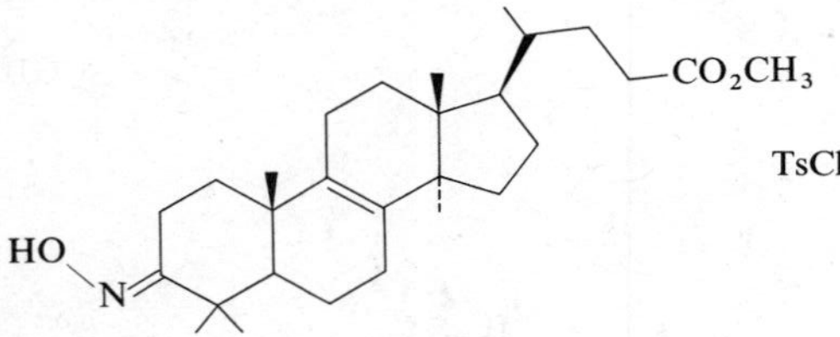 HO—N= ... CO_2CH_3	TsCl, pyridine, reflux, 3 h	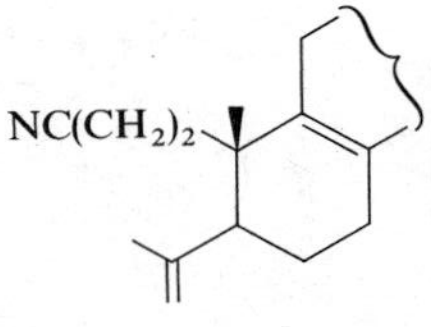 $NC(CH_2)_2$ ... (—)	623
	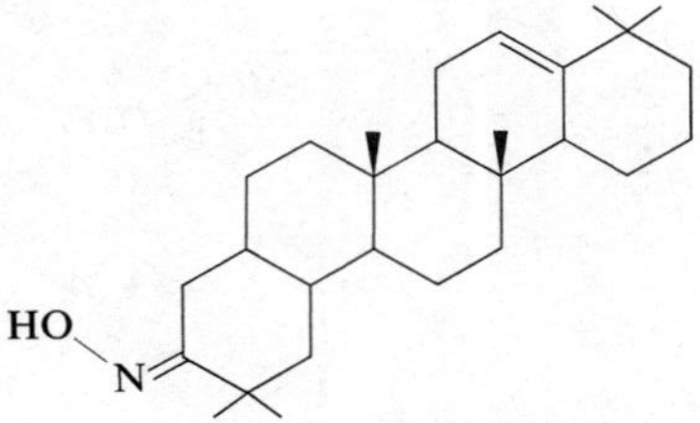 HO—N= ...	TsCl, pyridine, 16 h	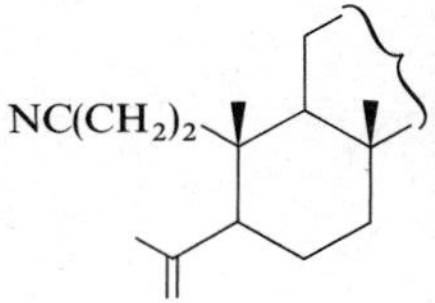 $NC(CH_2)_2$ ... (19)	559

	Reactant	Conditions	Product(s) (Yield, %)	Refs.
C_{29}	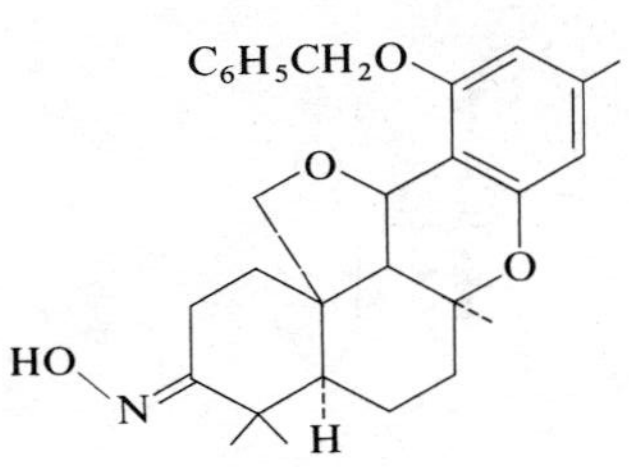 $C_6H_5CH_2O$; HO–N=	TsCl, pyridine, reflux, 3 h	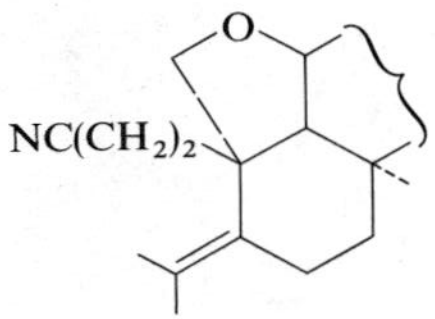 $NC(CH_2)_2$ (69)	54
	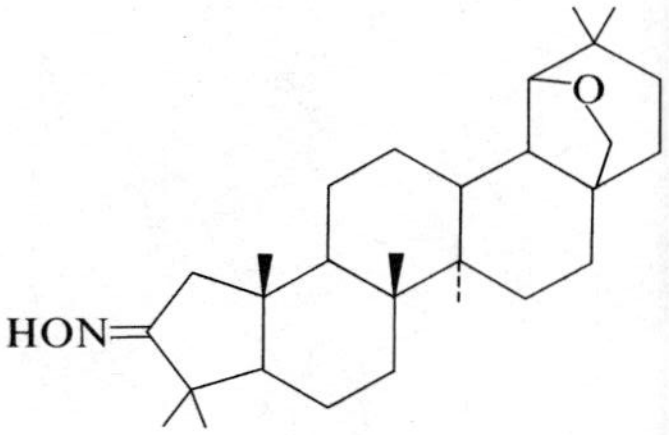 HON=	TsCl, pyridine, 3 d	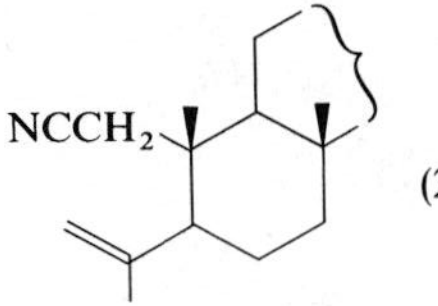 $NCCH_2$ (21)	50
		$POCl_3$, pyridine, 100°, 3 h	" (53)	50
	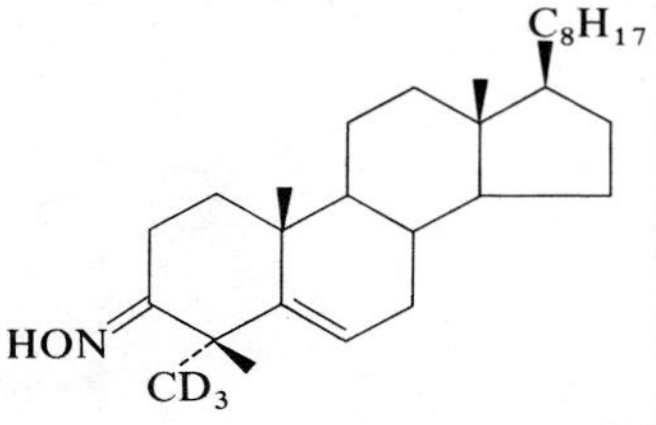 C_8H_{17}; HON=; CD_3	TsCl, pyridine	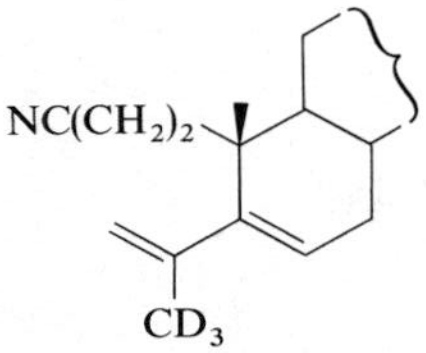 $NC(CH_2)_2$; CD_3 (—)	53
	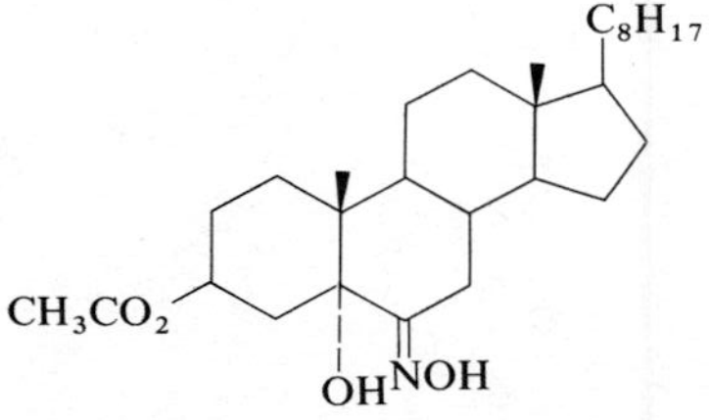 C_8H_{17}; CH_3CO_2; OH; NOH	C_6H_5NCO, $(C_2H_5)_3N$, C_6H_6, reflux, 1 h	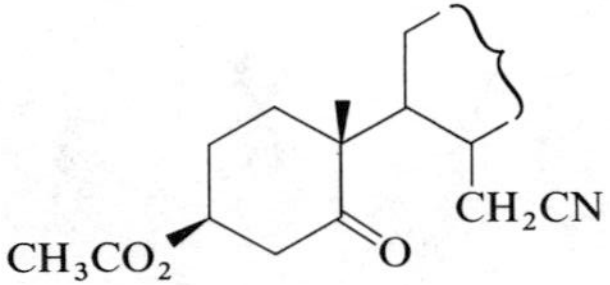 CH_3CO_2; CH_2CN (83)	440

TABLE III. KETOXIME FRAGMENTATIONS (*Continued*)

No. of Carbon Atoms	Substrate	Reagent and Conditions	Product(s) and Yield(s) (%)	Refs.
		Cl_2C:	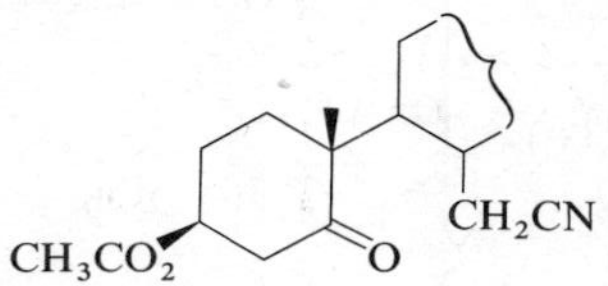 CH_3CO_2 … CH_2CN (85)	178
		$(CH_3CO)_2O$, pyridine	" (—)	622
		1. $(CH_3CO)_2O$ 2. CH_3OH, reflux, 2 h	" (100)	622
		$SOCl_2$, $-20°$, 4 *M* KOH	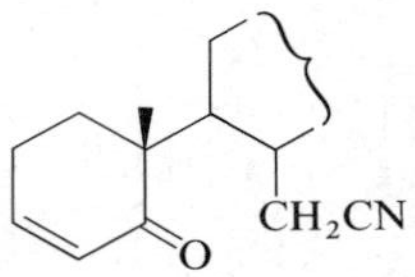 CH_2CN ("major")	529
	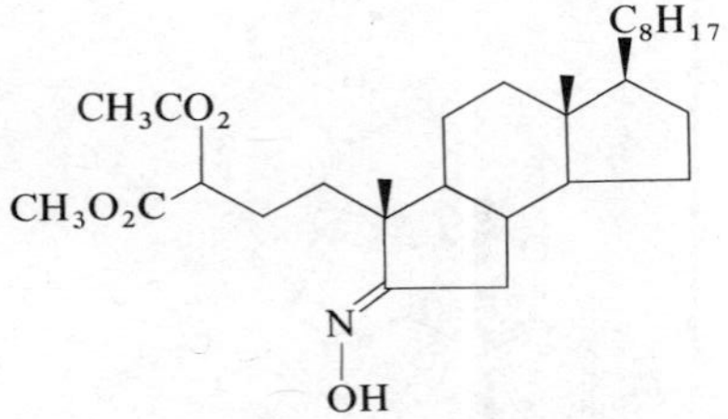 CH_3CO_2, CH_3O_2C, C_8H_{17}, N–OH	$SOCl_2$, ether, $-10°$	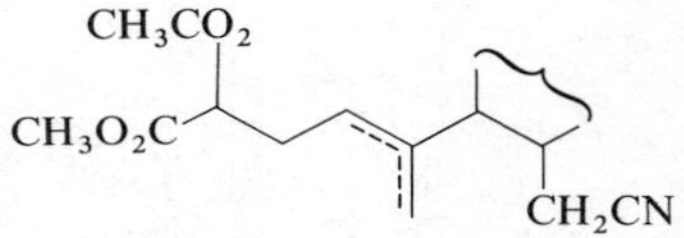 CH_3CO_2, CH_3O_2C, CH_2CN (—)	537

	Substrate	Conditions	Product(s) (Yield, %)	Refs.
	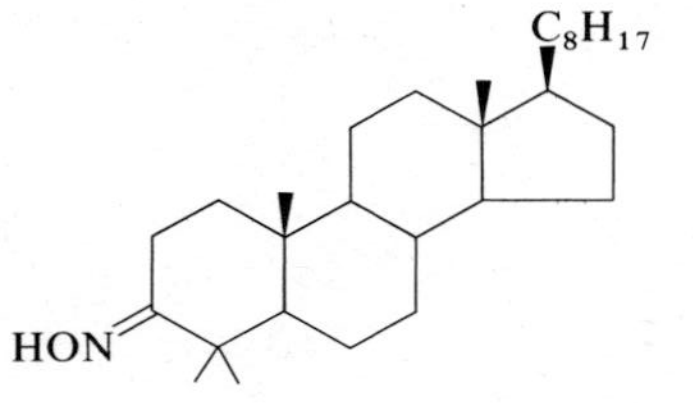 HON, C_8H_{17}	TsCl, pyridine, reflux	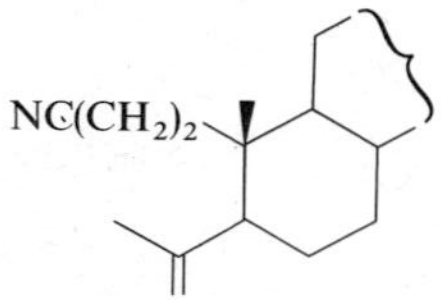 $NC(CH_2)_2$ (85)	493, 557
	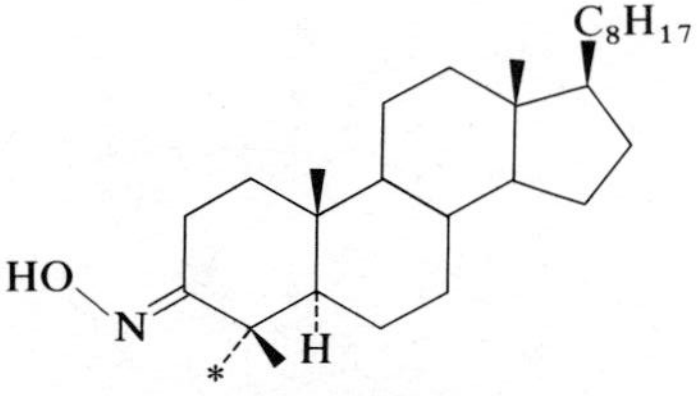 HO, N, H, $* = CD_3$ or $^{14}CH_3$, C_8H_{17}	TsCl, pyridine, $-10°$	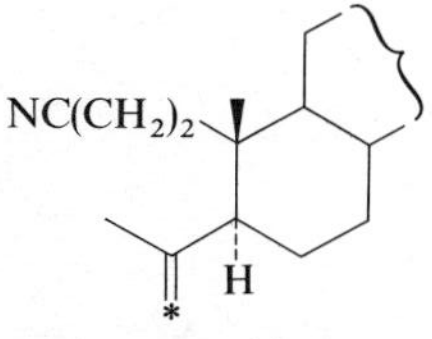 $NC(CH_2)_2$, H, $* = CD_2$ or $^{14}CH_2$ (—)	51, 52
C_{30}	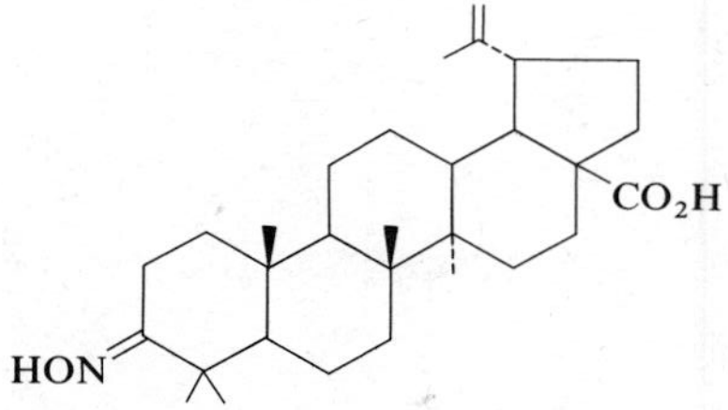 HON, CO_2H	TsCl, pyridine, reflux, 5 h 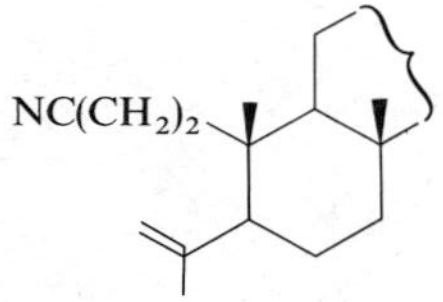	$NC(CH_2)_2$ (65)	624

TABLE III. KETOXIME FRAGMENTATIONS (*Continued*)

No. of Carbon Atoms	Substrate	Reagent and Conditions	Product(s) and Yield(s) (%)	Refs.
	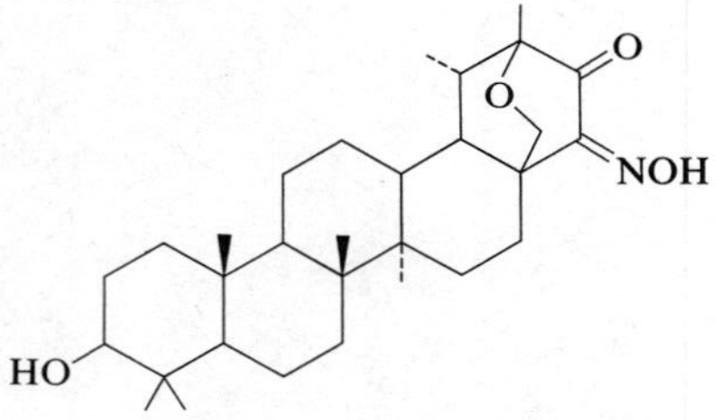 HO ... NOH	$(CH_3CO)_2O$, pyridine, reflux, 2 h	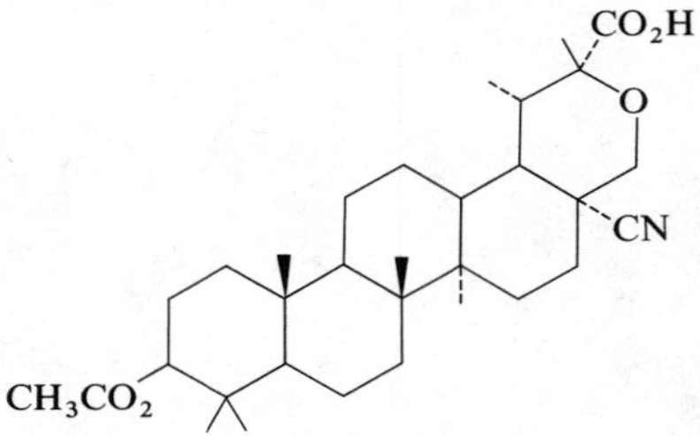 CH_3CO_2 ... CO_2H ... CN (45)	186
	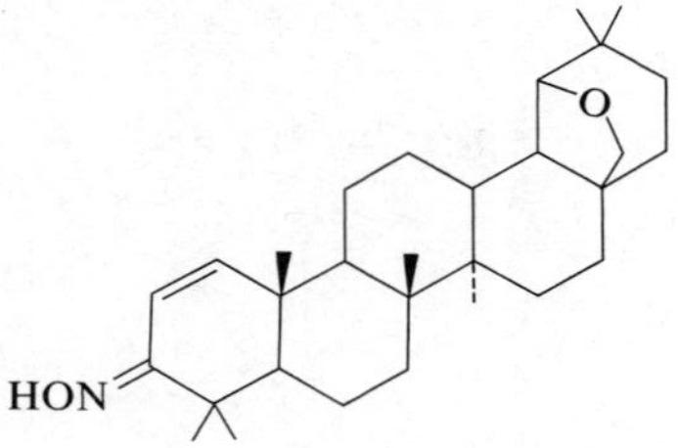 HON	TsCl, pyridine, 2 d	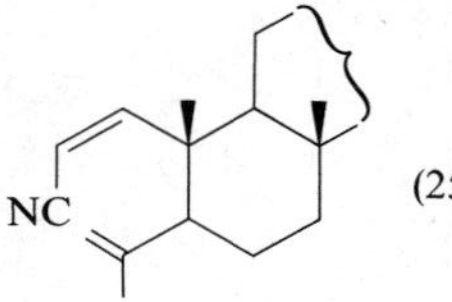 NC (25)	50
		$POCl_3$, pyridine, 100°, 2 h	" (23)	50
	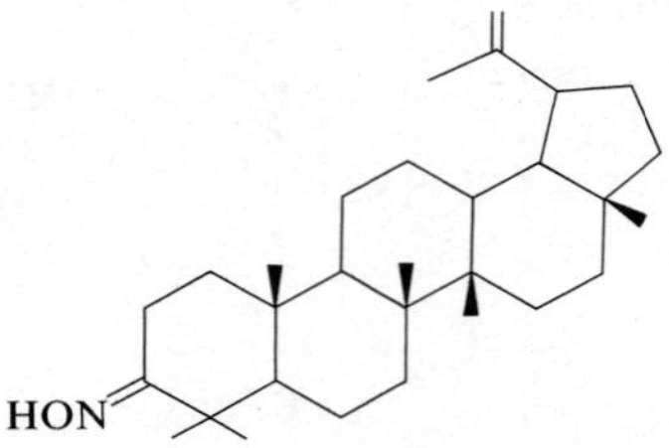 HON	TsCl, pyridine, 120°, 3 h	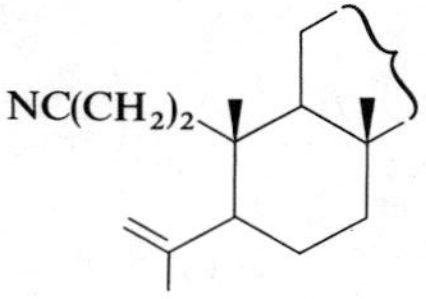 $NC(CH_2)_2$ (60)	559

Substrate	Conditions	Product (Yield, %)		Refs.
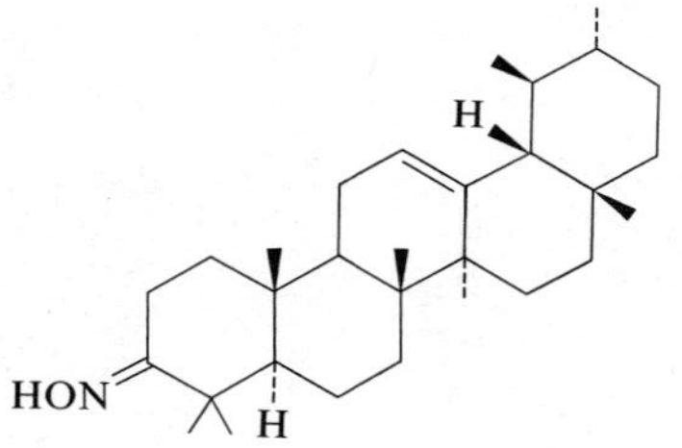 HON, H	TsCl, pyridine, reflux	$NC(CH_2)_2$, H	(—)	625
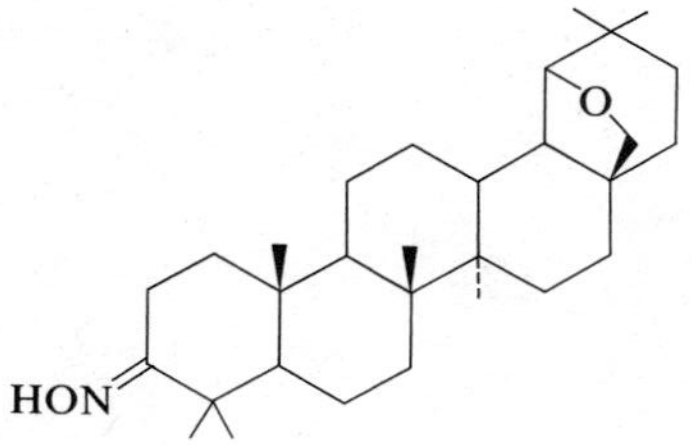 HON, O	TsCl, pyridine, 2 d	$NC(CH_2)_2$	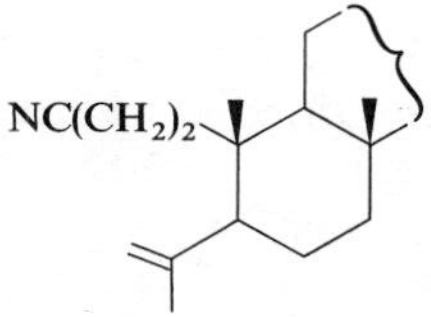 (20–25)	50
	TsCl, pyridine, reflux, 2–4 h	”	(84–90)	50
	$POCl_3$, pyridine, 2 h	”	(30)	50
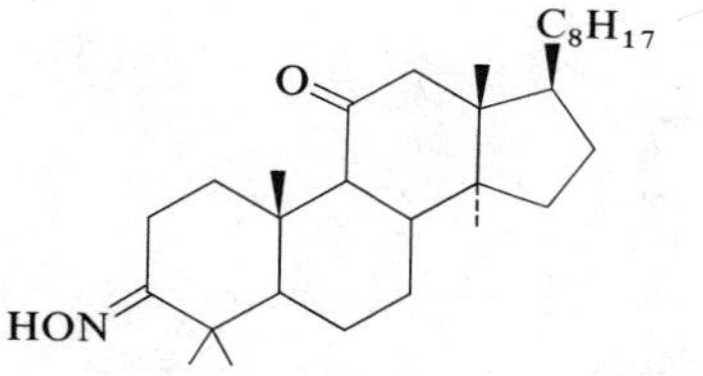 HON, O, C_8H_{17}	TsCl, pyridine, reflux, 14 h	$NC(CH_2)_2$	(78)	412
HO–N, H, C_8H_{17}	TsCl, pyridine	$NC(CH_2)_2$, H	(—)	626

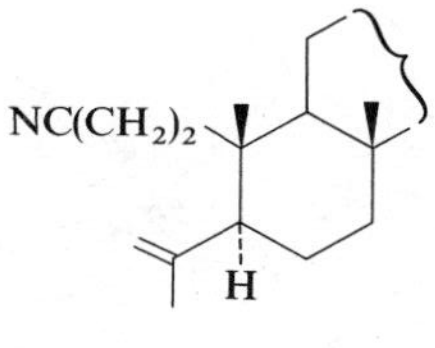

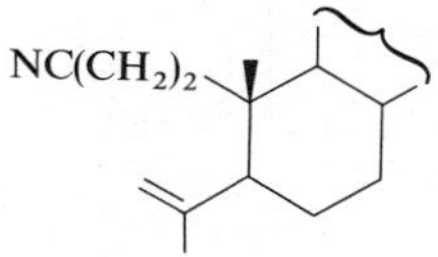

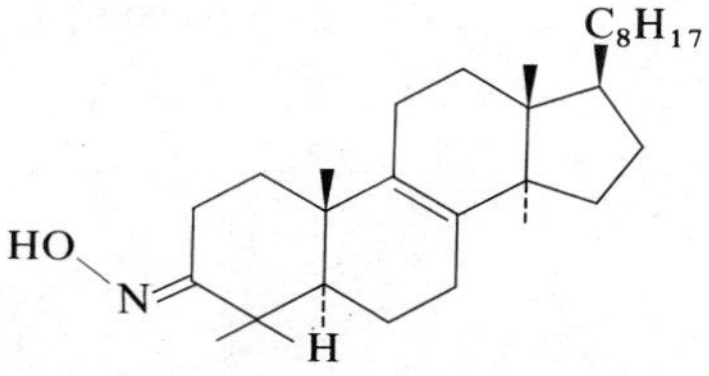

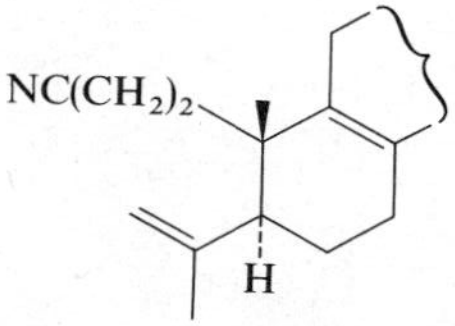

TABLE III. Ketoxime Fragmentations (*Continued*)

No. of Carbon Atoms	Substrate	Reagent and Conditions	Product(s) and Yield(s) (%)	Refs.
C_{31}	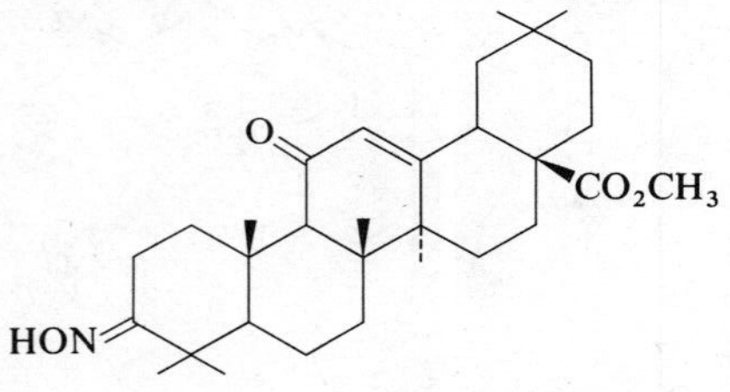 HON=, CO_2CH_3	TsCl, pyridine, 120°, 3 h	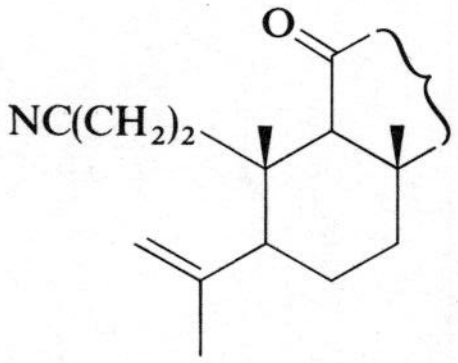 $NC(CH_2)_2$ (76)	627
	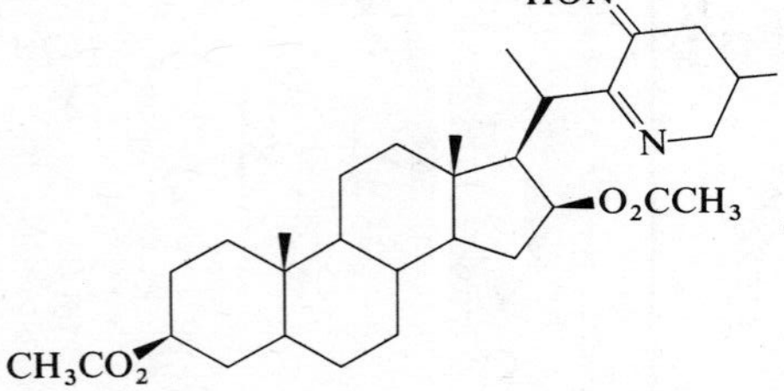 HON, N, O_2CCH_3, CH_3CO_2	TsCl, pyridine, 12 h	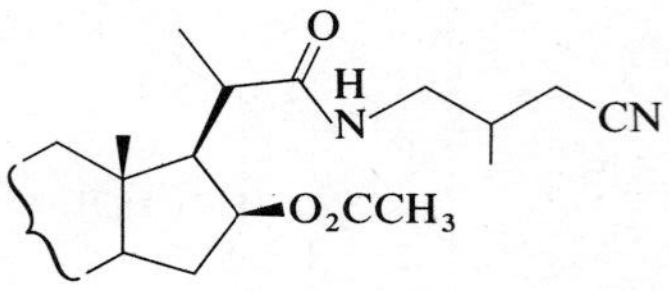 O, H, N, CN, O_2CCH_3 (97)	201
	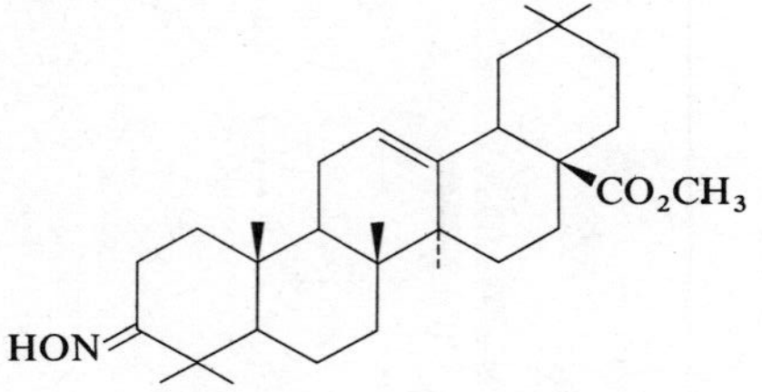 HON=, CO_2CH_3	$POCl_3$, pyridine, 35°, 4 h	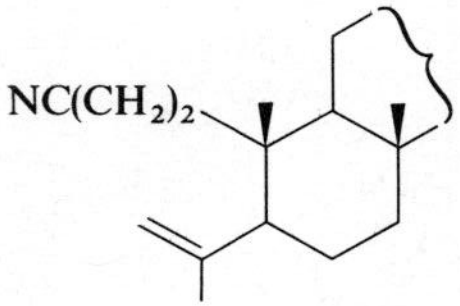 $NC(CH_2)_2$ (10)	559

Reactant	Conditions	Product (Yield, %)	Refs.
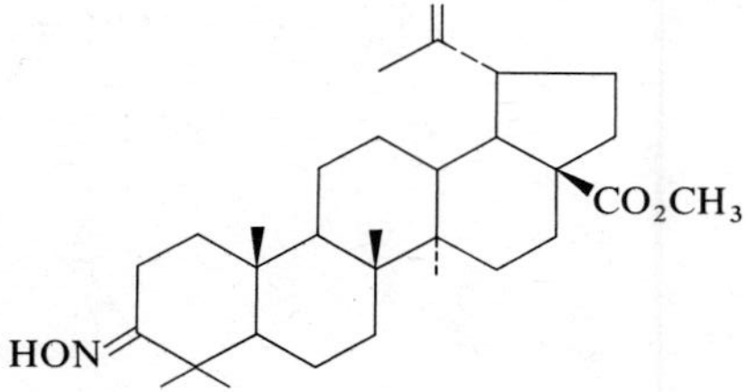	TsCl, pyridine, reflux, 5 h	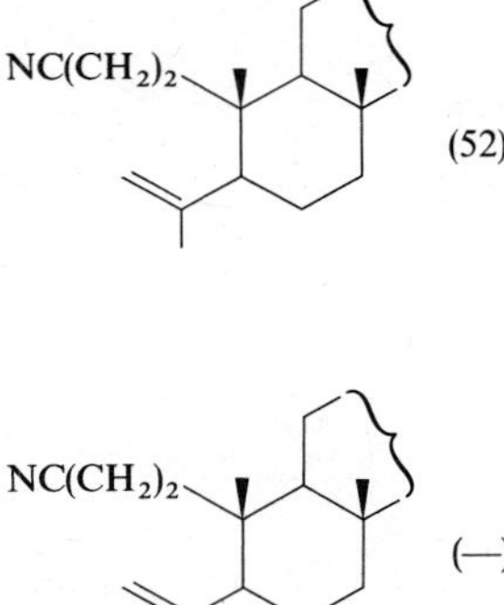 (52)	559, 624
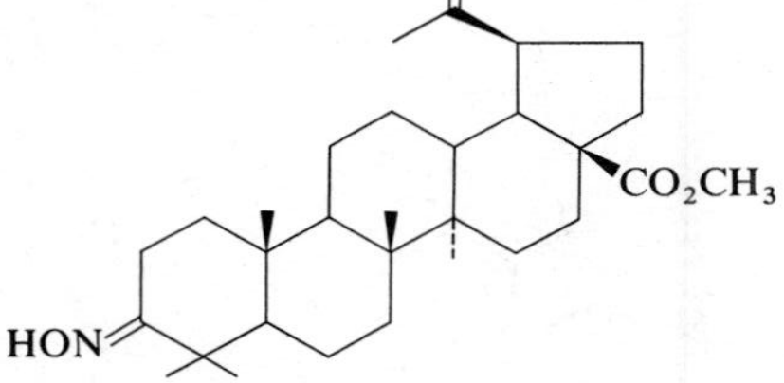	$POCl_3$, 35°, 4 h	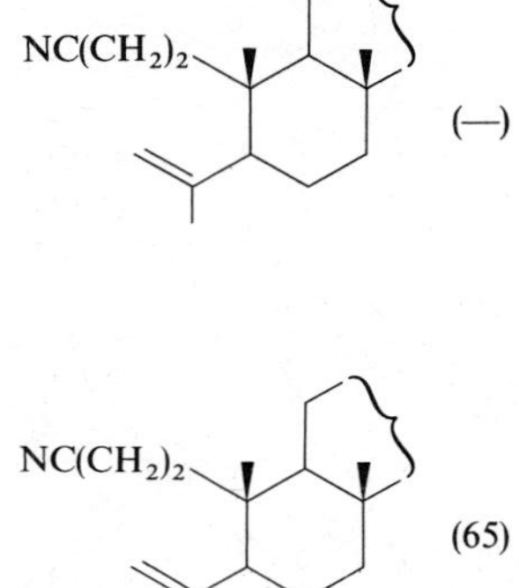 (—)	559
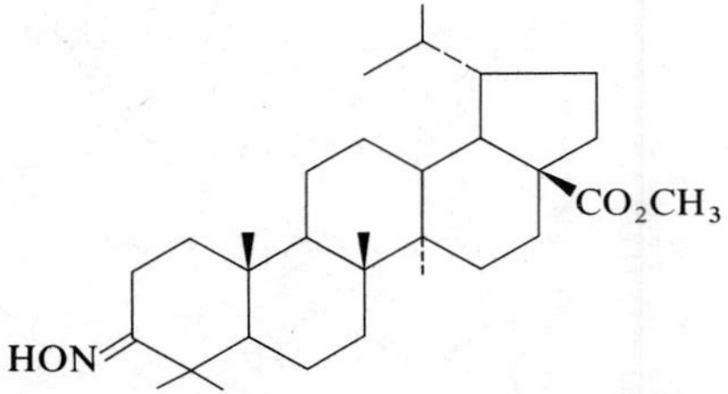	TsCl, pyridine, reflux, 5 h	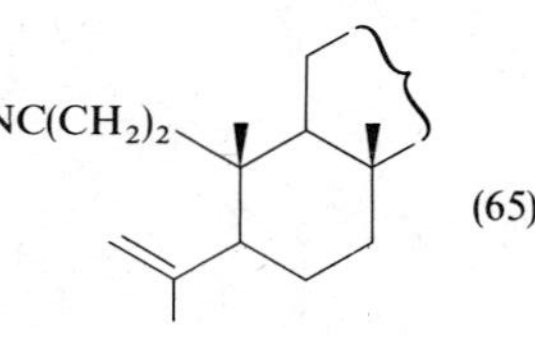 (65)	624
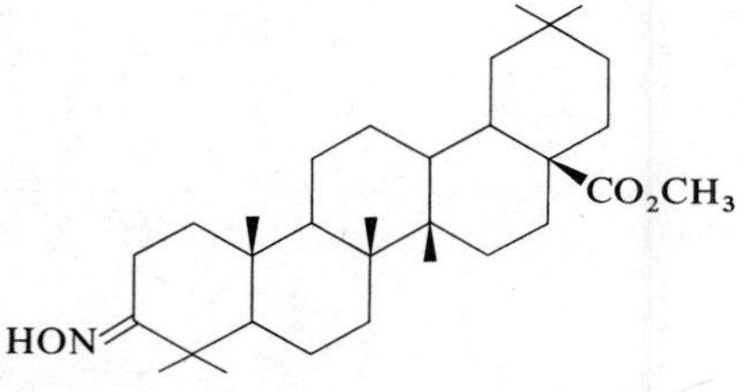	TsCl, pyridine, 120°, 3 h	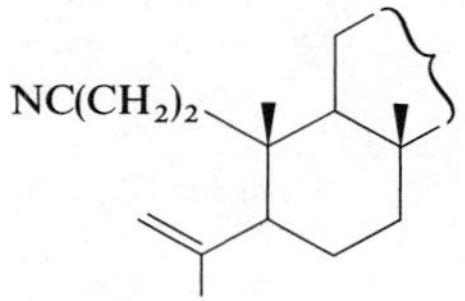 (60)	559

TABLE III. KETOXIME FRAGMENTATIONS (*Continued*)

No. of Carbon Atoms	Substrate	Reagent and Conditions	Product(s) and Yield(s) (%)	Refs.
	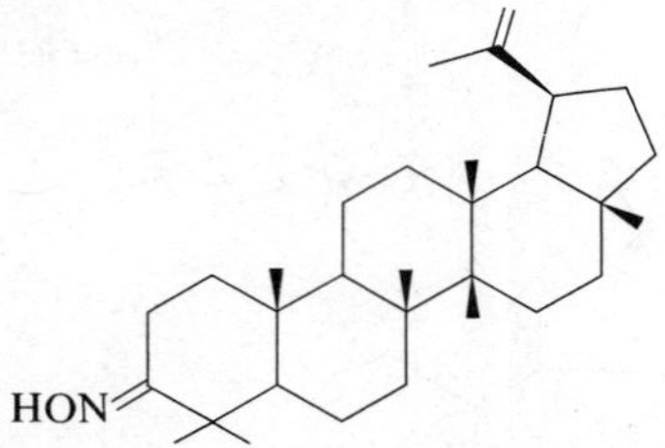 HON= (structure)	$POCl_3$, pyridine, 35°, 4 h	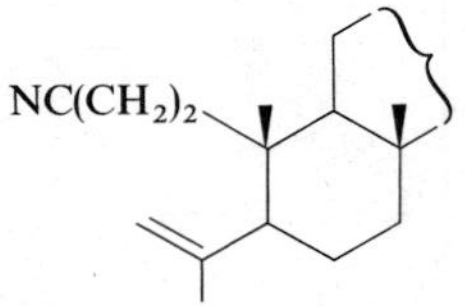 $NC(CH_2)_2$ (structure) (—)	559
	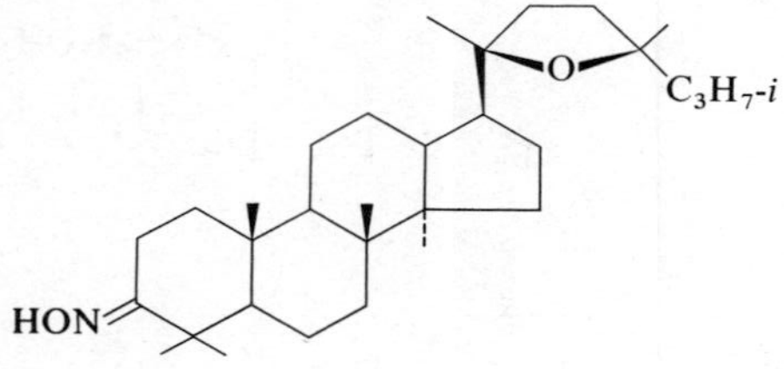 HON= (structure), C_3H_7-*i*	TsCl, pyridine	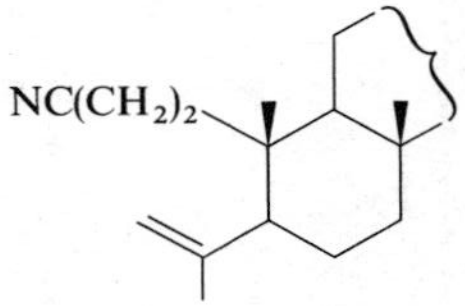 $NC(CH_2)_2$ (structure) (—)	628
C_{32}	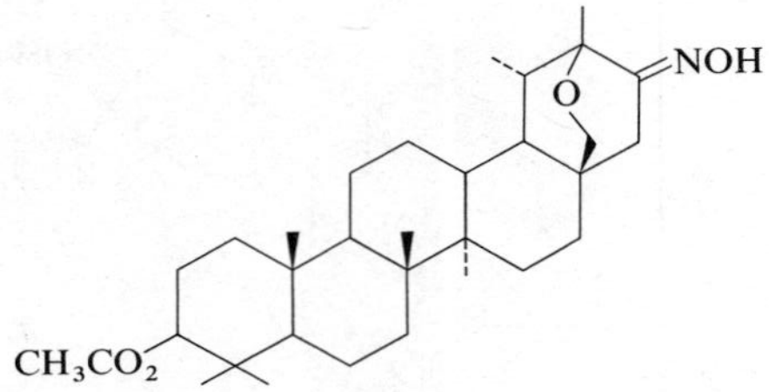 CH_3CO_2 (structure) =NOH	$(CH_3CO)_2O$, pyridine, 4 h	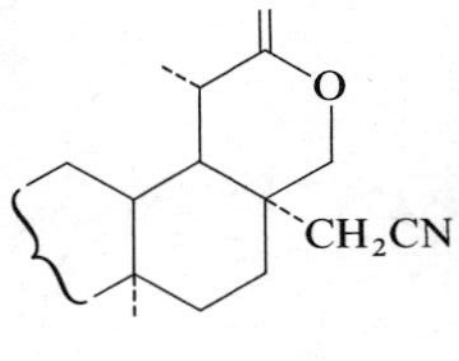 (structure) CH_2CN (93)	186

	Reactant	Reagents	Product	Ref.
	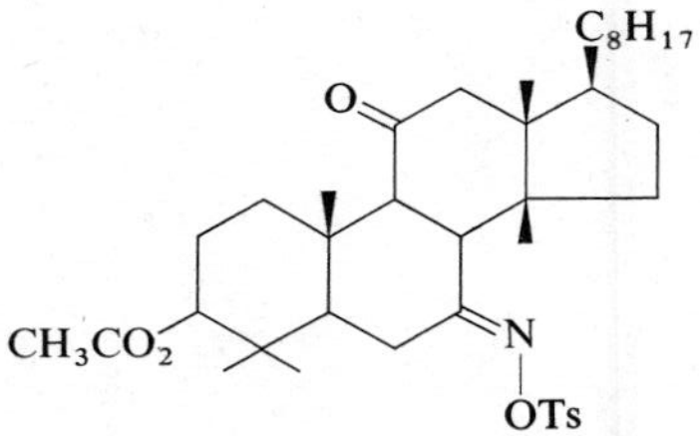 CH_3CO_2, C_8H_{17}, O, =N–OTs	$NaOC_4H_9$-*t*, *t*-C_4H_9OH	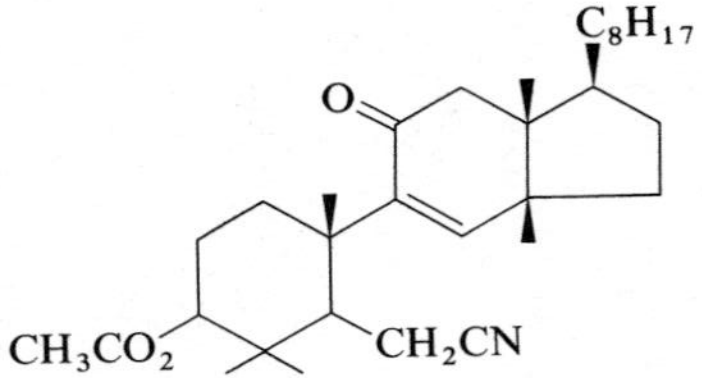 CH_3CO_2, C_8H_{17}, O, CH_2CN (29)	629
C_{33}	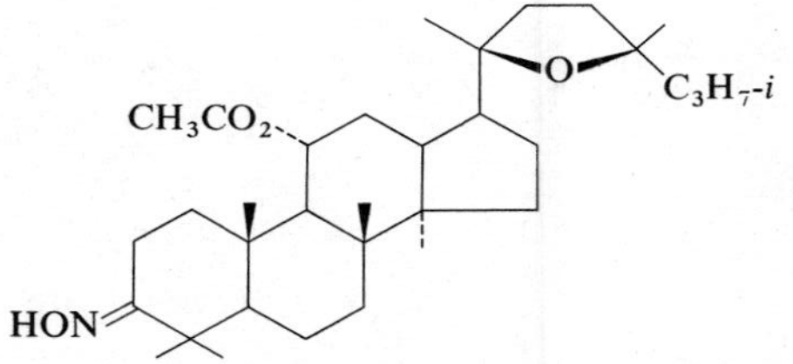 HON=, CH_3CO_2, O, C_3H_7-*i*	TsCl, pyridine	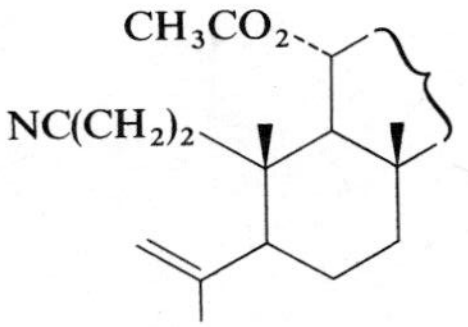 CH_3CO_2, $NC(CH_2)_2$ (—)	630

TABLE IV. ALDOXIME FRAGMENTATIONS

No. of Carbon Atoms	Substrate	Reagent and Conditions	Product(s) and Yield(s) (%)	Refs.
C_1	CH_2O	H_2NOSO_3H	HCN (—)	631, 632
C_2	$CH_3CH{=}NOH$	$(C_6H_5)_3P$, CCl_4, $(C_2H_5)_3N$, 60°, 2.5 h	CH_3CN (86)	633
		$(CH_3)_2N^+{=}CCl_2\ Cl^-$, $CHCl_3$, 2 h	" (92)	289
		$TiCl_4$, dioxane	" (81)	234
		SeO_2, DMF, 2 h	" (70)	634
	$HOCH_2CH{=}NOH$	H_2NOSO_3H, H_2O	$HOCH_2CN$ (60)	632
C_3	$C_2H_5CH{=}NOH$	"	C_2H_5CN (80)	632
		$TiCl_4$, THF	" (89)	234
	$CH_3CHOHCH{=}NOH$	SeO_2, $CHCl_3$, 2 h	$CH_3CHOHCN$ (71)	634
C_4	$CH_3CH{=}CHCH{=}NOH$	"	$CH_3CH{=}CHCN$ (75)	634
	n-$C_3H_7CH{=}NOH$	H_2NOSO_3H, H_2O	$CH_3(CH_2)_2CN$ (90)	632
		$(C_6H_5O)_2PHO$, CCl_4, $(C_2H_5)_3N$, 4 h	" (40)	635
		P_2I_4	" (40)	636
		$HC(OC_2H_5)_3$, H^+	" (93)	188
		$CH_3C(OC_2H_5)_3$, H^+	" (94)	188
		$CH_3C(OC_2H_5)_3$, SO_2, H^+	" (96)	188
		HCO_2Na, HCO_2H, reflux, 1 h	" (30)	637
		$TiCl_4$, THF	" (80)	234
		SeO_2, DMF, 2 h	" (80)	634
		SeO_2, $CHCl_3$	" (71–80)	634
		Cyanuric chloride	" (64)	638
	i-$C_3H_7CH{=}NOH$	$TiCl_4$, THF	$(CH_3)_2CHCN$ (85)	234

	Reactant	Reagents	Product (Yield, %)	Refs.
C_5	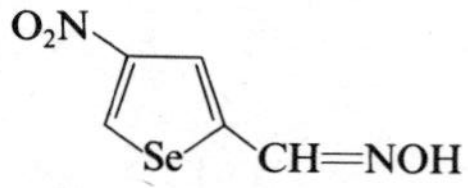 O_2N, Se, CH=NOH	PCl_5	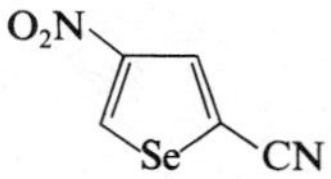 O_2N, Se, CN (70)	639
	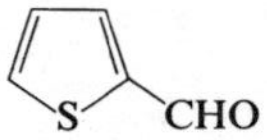 S, CHO	$H_2NOH \cdot HCl$, pyridine, $C_6H_5CH_3$, reflux	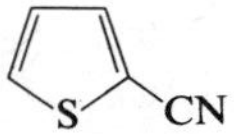 S, CN (78)	640
		$C_2H_5NO_2$, HCl, pyridine, reflux, 1 h	" (88)	641
	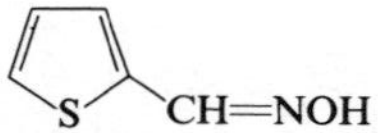 S, CH=NOH	Cl_3CCOCl, $(C_2H_5)_3N$, CH_2Cl_2	" (91)	642
		$TiCl_4$, dioxane	" (96)	234
		$CuSO_4 \cdot 5H_2O$, DCC, $(C_2H_5)_3N$, CH_2Cl_2	" (93)	643
	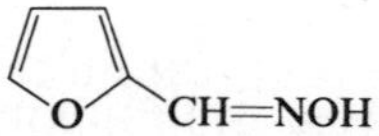 O, CH=NOH	C_6H_5OSOCl, pyridine, ether, 5–25°, 15 h	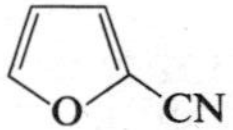 O, CN (92)	644
		$(CH_3)_3N \cdot SO_2$	" (76)	645
		$ClSO_2F$, $(C_2H_5)_3N$, CH_2Cl_2, 8 h	" (70)	646
		$(PNCl_2)_3$, $(C_2H_5)_3N$	" (70)	647
		P_2I_4	" (89)	636
		$(C_6H_5O)_2PHO$, $(C_2H_5)_3N$, CCl_4, 4 h	" (66)	635
		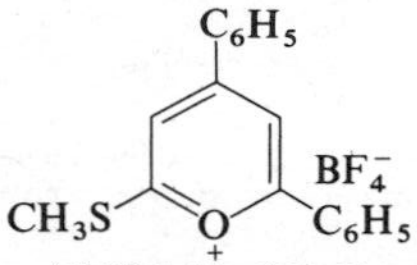 C_6H_5, CH_3S, $\overset{+}{O}$, BF_4^-, C_6H_5; $(C_2H_5)_3N$, CH_2Cl_2, 30 min	" (75)	648
		Cl_3CCOCl, $(C_2H_5)_3N$, CH_2Cl_2	" (75)	642
		SeO_2, ethanol, $CHCl_3$, 2 h	" (77)	634

TABLE IV. Aldoxime Fragmentations (*Continued*)

No. of Carbon Atoms	Substrate	Reagent and Conditions	Product(s) and Yield(s) (%)	Refs.
	C_4H_3NH–2-$CH{=}NOH$ (pyrrole)	H_2NOSO_3H, H_2O	C_4H_3NH–2-CN (pyrrole) (88)	632
	$HON{=}CH(CH_2)_3CH{=}NOH$	$(CF_3CO)_2O$, pyridine, 0–25°, 3 h	$NC(CH_2)_3CN$ (76)	618
		SeO_2, $CHCl_3$, 2 h	" (72)	634
	n-C_4H_9CHO	$H_2NOH\cdot HCl$, HCO_2H, reflux, 30 min	$CH_3(CH_2)_3CN$ (77)	236
	n-$C_4H_9CH{=}NOH$	$TiCl_4$, THF	" (87)	234
	$HO(CH_2)_4CH{=}NOH$	SeO_2, $CHCl_3$, 2 h	$HO(CH_2)_4CN$ (79)	634
	i-C_4H_9CHO	$H_2NOH\cdot HCl$, HCO_2H, reflux, 30 min	$(CH_3)_2CHCH_2CN$ (88)	236
	t-$C_4H_9CH{=}NOH$	$(C_3H_3N_2)_2C{=}O$ (imidazole), CH_2Cl_2	*t*-C_4H_9CN (95)	649
C_6	C_5H_4N–2-$CH{=}NOH$ (pyridine)	$(PNCl_2)_3$, $(C_2H_5)_3N$	C_5H_4N–2-CN (pyridine) (95)	647
		H_2NOSO_3H, H_2O	" (75)	632
	(*E*)	Cyanuric chloride, CH_2Cl_2	" (63)	638
	C_5H_4N–3-CHO (pyridine)	$H_2NOH\cdot HCl$, NaO_2CCH_3, reflux, 15 h	C_5H_4N–3-CN (pyridine) (—)	650

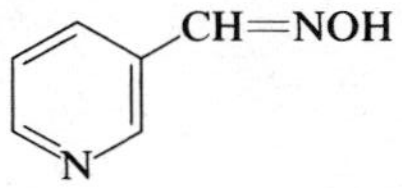 CH=NOH, N	Cl_3CCOCl, $(C_2H_5)_3N$, CH_2Cl_2	" (85)	642
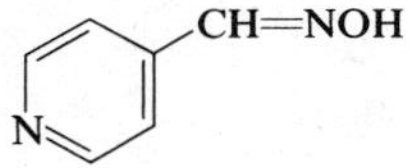 CH=NOH, N	H_2NOSO_3H, H_2O	" (76)	632
	"	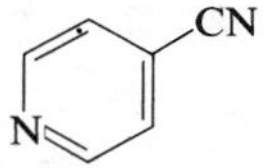 CN, N (70)	632
	$(CF_3CO)_2O$, pyridine, 0–25°, 3 h	" (99)	618
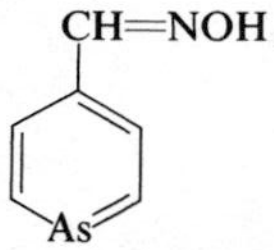 CH=NOH, As	CH_3CO_2H	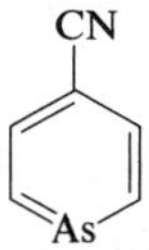 CN, As (—)	651
$HC{\equiv}CC(CH_3){=}CHCH{=}NOH$	$(CH_3)_2N^+{=}CHCl\ Cl^-$, CH_3CN, 0°	$HC{\equiv}CC(CH_3){=}CHCN$ (83)	652
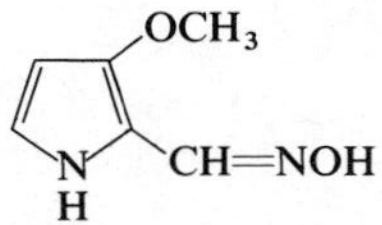 OCH_3, CH=NOH, N, H	H_2NOSO_3H, H_2O	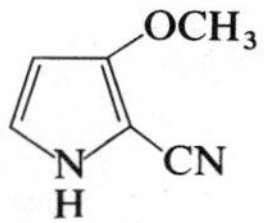 OCH_3, CN, N, H (60)	632
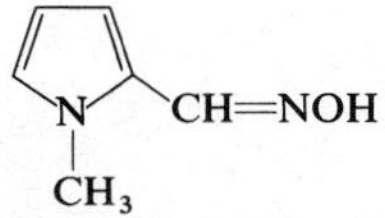 CH=NOH, N, CH_3	Cyanuric chloride, dioxane	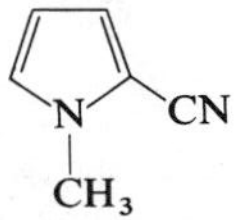 CN, N, CH_3 (70)	638
n-$C_5H_{11}CHO$	$H_2NOH \cdot HCl$, HCO_2H, reflux, 30 min	n-$C_5H_{11}CN$ (95)	236
n-$C_5H_{11}CH{=}NOH$	$ClSO_2NCO$, $(C_2H_5)_3N$, 8 h	" (75)	653
	$ClSO_2F$, $(C_2H_5)_3N$, CH_2Cl_2, 8 h	" (92)	646
	$(CH_3)_3N \cdot SO_2$	" (81)	645
	$(CF_3CO)_2O$, pyridine, 0–25°	" (94)	618

TABLE IV. ALDOXIME FRAGMENTATIONS (*Continued*)

No. of Carbon Atoms	Substrate	Reagent and Conditions	Product(s) and Yield(s) (%)	Refs.
		$(CF_3SO_2)_2O$, $(C_2H_5)_3N$, CH_2Cl_2, −78 to 25°, 2 h	n-$C_5H_{11}CN$ (94)	172
		$CH_3C{\equiv}NC_2H_5 \cdot BF_4$, 25–80°, CH_3CN, 8 h	" (88)	654
		2-CH_3S-4-C_6H_5-6-C_6H_5-pyrylium BF_4^-, $(C_2H_5)_3N$, CH_2Cl_2, 30 min	" (80)	648
		$HCSN(CH_3)_2$, CH_3I, reflux, 18 h	" (80)	655
		2,4-$(O_2N)_2C_6H_3F$, KOC_4H_9-t CH_3CN	" (74)	655
	n-$C_6H_{13}CH{=}NOC(OC_2H_5)_2C_2H_5$	BF_3, HgO, ether, reflux, 3 h	" (67)	137
C_7	2,6-$Cl_2C_6H_3CH{=}NOH$	H_2NOSO_3H, H_2O	2,6-$Cl_2C_6H_3CN$ (68)	632
		$CuSO_4 \cdot H_2O$, DCC, $(C_2H_5)_3N$, CH_2Cl_2	" (99)	643
		$CH_3C{\equiv}NC_2H_5 \cdot BF_4$, CH_3CN, 25–80°, 8.5 h	" (74)	654

Reactant	Reagent(s)	Product (Yield, %)	Refs.
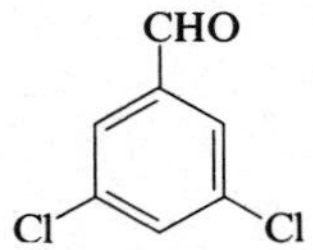 CHO, Cl, Cl	$H_2NOH \cdot HCl$, NaO_2CCH_3, reflux, 15 h	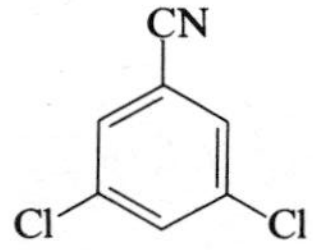 CN, Cl, Cl (34)	650
o-ClC_6H_4CHO	$H_2NOH \cdot HCl$, HCO_2H, reflux, 30 min	o-ClC_6H_4CN (94)	236
	H_2NOSO_3H, H_2O	" (85)	632
o-$ClC_6H_4CH{=}NOH$	Cl_3CCOCl, $(C_2H_5)_3N$, CH_2Cl_2	" (88)	642
p-ClC_6H_4CHO	$H_2NOH \cdot HCl$, HCO_2H, reflux, 30 min	p-ClC_6H_4CN (97)	236, 640
	$C_2H_5NO_2$, HCl, pyridine, reflux, 1 h	" (98)	641
p-$ClC_6H_4CH{=}NOH$	C_6H_5OSOCl, pyridine, ether, 5–25°, 15 h	" (96)	644
	p-ClC_6H_4OSOCl	" (42–61)	644
	$SOCl_2$, CCl_4, 12 h	" (98)	75
	P_2I_4	" (85)	636
(*Z*)	$(PNCl_2)_3$, $(C_2H_5)_3N$	" (78)	647
(*E*)	"	" (76)	647
	$CH_3C{\equiv}CN(C_2H_5)_2$, CH_3CN	" (77)	656
(*Z*)	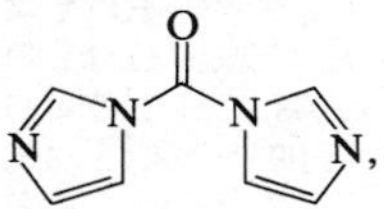 O, N, N, N, N, CH_2Cl_2	" (98)	649
(*E*)	"	" (95)	649
	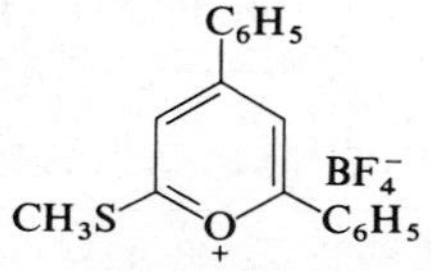 C_6H_5, BF_4^-, CH_3S, O^+, C_6H_5, $(C_2H_5)_3N$, CH_2Cl_2, 30 min	" (79)	648
	$CuSO_4 \cdot 5\ H_2O$, DCC, $(C_2H_5)_3N$, CH_2Cl_2	" (97)	643

TABLE IV. Aldoxime Fragmentations (*Continued*)

No. of Carbon Atoms	Substrate	Reagent and Conditions	Product(s) and Yield(s) (%)	Refs.
		$(CH_3)_2N^+{=}CCl_2\ Cl^-$, $CHCl_3$, 2 h	*p*-ClC_6H_4CN (97)	289
		Cl_3CCOCl, $(C_2H_5)_3N$, CH_2Cl_2	" (95)	642
		$ClCO_2C_6H_5$, pyridine, ether, 0–25°, 10 h	" (90)	657
		HCO_2Na, HCO_2H, reflux, 1 h	" (97)	637
		SeO_2, $CHCl_3$, ethanol, 2 h	" (75–84)	634
	o-BrC_6H_4CHO	$H_2NOH{\cdot}HCl$, HCO_2H, reflux, 30 min	*o*-BrC_6H_4CN (95)	236
		1. 2,4-$(O_2N)_2C_6H_3ONH_2$, HCl, ethanol 2. KOH, reflux, 3 h	" (89)	658
	p-$BrC_6H_4CH{=}NOH$	$SOCl_2$, CCl_4, 12 h	*p*-BrC_6H_4CN (98)	75
		DCC, DMSO, $(CF_3CO)_2O$, 2.5 h	" (84)	468
		HCO_2Na, HCO_2H, reflux, 1 h	" (84)	637
	o-$O_2NC_6H_4CHO$	$H_2NOH{\cdot}HCl$, HCO_2H, reflux, 30 min	*o*-$O_2NC_6H_4CN$ (94)	236
	o-$O_2NC_6H_4CH{=}NOH$	$CuSO_4{\cdot}5H_2O$, DCC, $(C_2H_5)_3N$, CH_2Cl_2	" (99)	643
		HCO_2Na, HCO_2H, reflux, 1 h	" (83)	637
	m-$O_2NC_6H_4CHO$	$CF_3CO_2NHCOCF_3$, pyridine, C_6H_6, reflux, 2 h	" (79)	659
	m-$O_2NC_6H_4CH{=}NOH$	$(CF_3SO_2)_2O$, $(C_2H_5)_3N$, CH_2Cl_2, −78 to 25°, 2 h	*m*-$O_2NC_6H_4CN$ (88)	172
		$(CF_3CO)_2O$, pyridine, 0–25°, 3 h	" (99)	618
		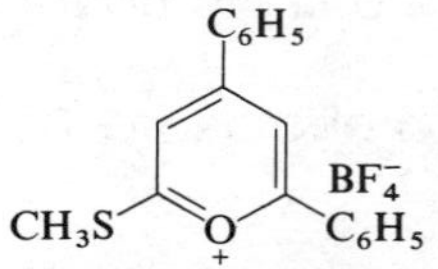 [structure: C_6H_5; CH_3S; O^+; BF_4^-; C_6H_5], $(C_2H_5)_3N$, CH_2Cl_2, 30 min	" (72)	648

Starting material	Reagents and conditions	Product (yield, %)	Ref.
	$CuSO_4 \cdot 5H_2O$, DCC, $(C_2H_5)_3N$, CH_2Cl_2	" (99)	643
	$Cu(O_2CCH_3)_2H_2O$, CH_3CN, reflux, 5 h	" (85)	660
	HCO_2Na, HCO_2H, reflux, 1 h	" (90)	637
p-$O_2NC_6H_4CHO$	$H_2NOH \cdot HCl$, pyridine $C_6H_5CH_3$, reflux	*p*-$O_2NC_6H_4CN$ (75)	640
	1. 2,4-$(O_2N)_2C_6H_3ONH_2$, HCl, ethanol 2. KOH, reflux, 3 h	" (94)	658
p-$O_2NC_6H_4CH{=}NOH$	H_2NOSO_3H, H_2O	" (95)	632
	$(C_6H_5O)_2PHO$, CCl_4, $(C_2H_5)_3N$, 4 h	" (85)	635
	P_2I_4	" (67)	636
	$SOCl_2$, CCl_4, 12 h	" (99)	75
	p-ClC_6H_4OSOCl, CH_2Cl_2	" (70)	661
	$CuSO_4 \cdot 5H_2O$, DCC, $(C_2H_5)_3N$, CH_2Cl_2	" (99)	643
	Cl_3CCOCl, $(C_2H_5)_3N$, CH_2Cl_2	" (86)	642
	$(CF_3CO)_2O$, pyridine, 0–25°, 3 h	" (99)	618
	$ClCO_2C_6H_5$, pyridine, 0–25°, 10 h	" (90)	657
	$(C_3H_3N_2)_2C{=}O$, CH_2Cl_2, reflux, 6 h	" (99)	649
	$(CH_3)_2N^+{=}CCl_2\ Cl^-$, $CHCl_3$, 2 h	" (98)	289
	$CH_3C{\equiv}CN(C_2H_5)_2$, CH_3CN	" (80)	656
	Cyanuric chloride	" (92)	638
	HCO_2Na, HCO_2H, reflux, 1 h	" (90)	637
	$TiCl_4$, dioxane	" (90)	234
	SeO_2, $CHCl_3$, ethanol, 2 h	" (70–82)	634

TABLE IV. ALDOXIME FRAGMENTATIONS (*Continued*)

No. of Carbon Atoms	Substrate	Reagent and Conditions	Product(s) and Yield(s) (%)	Refs.
	4-O_2N-C_6H_4-CH=N-O-C(C_2H_5)(OC_2H_5)$_2$ [structure: C_2H_5O, OC_2H_5, C_2H_5, O, N, H, O_2N]	BF_3, HgO, ether, reflux, 3 h	p-$O_2NC_6H_4CN$ (80)	137
	C_6H_5CHO	$H_2NOH \cdot HCl$, HCO_2H, reflux, 30 min	C_6H_5CN (99)	236, 640
		1. 2,4-$(O_2N)_2C_6H_3ONH_2$, HCl, ethanol 2. KOH, reflux, 3 h	" (84)	658
		$C_2H_5NO_2$, HCl, pyridine, reflux, 1 h	" (80)	641
	$C_6H_5CH{=}NOH$	H_2NOSO_3H, H_2O	C_6H_5CN (85)	632
		$(CH_3)_3N \cdot SO_2$	" (84)	645
		$SOCl_2$, CCl_4, 12 h	" (89)	75
		C_6H_5OSOCl, pyridine, ether, 5–25°, 15 h	" (92–95)	644
		$ClSO_2F$, $(C_2H_5)_3N$, CH_2Cl_2, 8 h	" (97)	646
		$(CF_3SO_2)_2O$, $(C_2H_5)_3N$, CH_2Cl_2, −78 to 25°	" (90)	172
		$ClSO_2NCO$, $(C_2H_5)_3N$, CH_2Cl_2, 8 h	" (79)	653
		PPE, $CHCl_3$, reflux, 5 min	" (80)	287
		$(C_6H_5)_3P$, CCl_4, 60°, 2.5 h	" (80)	633
		$(C_6H_5)_2P$-polymer, CCl_4, $C_2H_4Cl_2$, reflux	" (76)	145

Reagents and Conditions	Product (Yield %)	Refs.
$(C_6H_5O)_2PHO$, CCl_4, $(C_2H_5)_3N$, 4 h	" (88)	635
PI_3, $(C_2H_5)_3N$, CH_2Cl_2, 15 min	" (53)	662
$(C_2H_5O)_3PI_2$, $(C_2H_5)_3N$, CH_2Cl_2, 0–25°, 2 h	" (61)	663
$(C_2H_5O)_3PO$, SO_3, $C_2H_4Cl_2$	" (—)	255
$(CF_3CO)_2O$, pyridine, 0–25°, 3 h	" (89–92)	618
Cl_3CCOCl, $(C_2H_5)_3N$, CH_2Cl_2	" (85)	642
$CuSO_4 \cdot 5H_2O$, DCC, $(C_2H_5)_3N$, CH_2Cl_2	" (98)	643
$ClCO_2C_6H_5$, pyridine, ether, 0–25°, 10 h	" (85)	657
$ClC(=NC_6H_5)OC_2H_5$, C_6H_6, 0°	" (78)	664
$Cl_2C=NC_6H_5$, C_6H_6	" (75)	664
$(CH_3)_2N^+=CHCl\ Cl^-$, CH_3CN, 0°	" (95)	652
1. $(CH_3)_2N^+=CCl_2Cl^-$ CH_2Cl_2, 0° 2. Reflux, 2 h	" (97)	289
$HCSN(CH_3)_2$, CH_3I, reflux, 18 h	" (65)	655
2,4-$(O_2N)_2C_6H_3F$, KOC_4H_9-*t*, CH_3CN, 80°, 30 min	" (100)	655
$CH_3C\equiv CN(C_2H_5)_2$, CH_3CN	" (21–69)	654, 656
Cyanuric chloride, CH_2Cl_2	" (82)	638
$(PNCl_2)_3$, $(C_2H_5)_3N$	" (73)	647
2-CH_3S-4,6-$(C_6H_5)_2$-pyrylium BF_4^-, $(C_2H_5)_3N$, CH_2Cl_2, 30 min	" (85)	648

TABLE IV. ALDOXIME FRAGMENTATIONS (*Continued*)

No. of Carbon Atoms	Substrate	Reagent and Conditions	Product(s) and Yield(s) (%)	Refs.
		Cl_3CCN, reflux, 30 min	C_6H_5CN (81)	665
		$C_6H_5CH_2N^+(C_2H_5)_3\ Cl^-$, NaOH, H_2O, $CHCl_3$	" (51)	666
		HCO_2Na, HCO_2H, reflux, 1 h	" (89)	637
		$([(CH_3)_2N]_3P)_2O\cdot(BF_4)_2$, DMF, reflux	" (90)	252
		$TiCl_4$, dioxane	" (97)	234
		$Cu(O_2CCH_3)_2\cdot H_2O$, CH_3CN, reflux, 2 h	" (90)	660
		$HC(OC_2H_5)_3$, H^+, SO_2	" (90)	188
		$CH_3C(OC_6H_5)_3$, H^+, SO_2	" (60)	188
	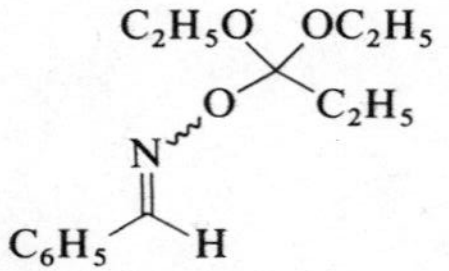 $C_6H_5CH=N\sim OC(OC_2H_5)_2C_2H_5$	BF_3, HgO, ether, reflux, 2 h	" (50–60)	137
	o-HOC_6H_4CHO	$CF_3CO_2NHCOCF_3$, pyridine, C_6H_6, reflux, 2 h	o-HOC_6H_4CN (53–56)	659
		$C_2H_5NO_2$, HCl, pyridine, reflux, 1 h	" (80)	641
	o-$HOC_6H_4CH=NOH$	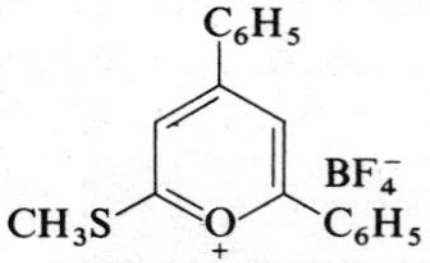 2-CH_3S-4,6-$(C_6H_5)_2$-pyrylium BF_4^-, $(C_2H_5)_3N$, CH_2Cl_2, 30 min	" (86)	648
		HCO_2Na, HCO_2H, reflux, 1 h	" (87)	637
	m-$HOC_6H_4CH=O$	H_2NOSO_3H, H_2O	m-HOC_6H_4CN (93)	632

m-$HOC_6H_4CH{=}NOH$	$CuSO_4 \cdot 5H_2O$, DCC, $(C_2H_5)_3N$, CH_2Cl_2	" (70)	643
p-$HOC_6H_4CH{=}O$	H_2NOSO_3H, H_2O	p-HOC_6H_4CN (80)	632
p-$HOC_6H_4CH{=}NOH$	$CuSO_4 \cdot 5H_2O$, DCC, $(C_2H_5)_3N$, CH_2Cl_2	" (—)	643
o-$H_2NC_6H_4CH{=}NOH$	2-CH_3S-4,6-$(C_6H_5)_2$-pyrylium BF_4^-	o-$H_2NC_6H_4CN$ (76)	648
p-$H_2NC_6H_4CH{=}NOH$	$CuSO_4 \cdot 5H_2O$, DCC, $(C_2H_5)_3N$, CH_2Cl_2	p-$H_2NC_6H_4CN$ (85)	643
$C_6H_{11}CH{=}NOH$	$ClSO_2NCO$, $(C_2H_5)_3N$, CH_2Cl_2, 8 h	$C_6H_{11}CN$ (86)	653
	$CuSO_4 \cdot 5H_2O$, DCC, $(C_2H_5)_3N$, CH_2Cl_2	" (88)	643
	SeO_2, $CHCl_3$, ethanol, 2 h	" (74–82)	634
n-$C_6H_{13}CH{=}NOH$	$ClSO_2F$, $(C_2H_5)_3N$, CH_3Cl_2, 8 h	n-$C_6H_{13}CN$ (73)	646
$C_2H_5O_2C(CH_2)_3CH{=}NOH$	$HC(OC_2H_5)_3$, H^+, ether	$C_2H_5O_2C(CH_2)_3CN$ (93)	188
n-$C_6H_{13}CHO$	$CF_3CO_2NHCOCF_3$, pyridine, C_6H_6, reflux, 2 h	n-$C_6H_{13}CN$ (72)	659
	2,4-$(O_2N)_2C_6H_3ONH_2$, HCl, ethanol, $(C_2H_5)_3N$, reflux, 5 min	" (91)	658
n-$C_6H_{13}CH{=}NOH$	C_6H_5OSOCl, pyridine, ether, 0–25°, 15 h	" (90)	644
	p-ClC_6H_4OSOCl, ether	" (43)	661
	$(CH_3)_3N \cdot SO_2$	" (83)	645
	$(PNCl_2)_3$, $(C_2H_5)_3N$	" (93)	647
	HCO_2Na, HCO_2H, reflux, 1 h	" (42)	637
	$HC(OC_2H_5)_3$, H^+	" (95)	188
	SeO_2, $CHCl_3$, ethanol or DMF, 2 h	" (84)	634
	SeO_2, $CHCl_3$, reflux	" (71–84)	235

TABLE IV. Aldoxime Fragmentations (*Continued*)

No. of Carbon Atoms	Substrate	Reagent and Conditions	Product(s) and Yield(s) (%)	Refs.
C_8	O_2N, CH=NOH, O, O (methylenedioxybenzene ring)	$CuSO_4 \cdot 5H_2O$, DCC, $(C_2H_5)_3N$, CH_2Cl_2	O_2N, CN, O, O (methylenedioxybenzene ring) (98)	643
	p-NCC_6H_4CH=NOH	HCO_2Na, HCO_2H, reflux, 1 h	*p*-NCC_6H_4CN (96)	637
	o-HON=CHC_6H_4CH=NOH	$CuSO_4 \cdot 5H_2O$, DCC, $(C_2H_5)_3N$, CH_2Cl_2	*o*-NCC_6H_4CN (55)	643
	p-HON=CHC_6H_4CH=NOH	$(CF_3CO)_2O$, 0–25°, 3 h	*p*-NCC_6H_4CN (98)	618
		$CuSO_4 \cdot 5H_2O$, DCC, $(C_2H_5)_3N$, CH_2Cl_2	" (95)	643
	CH=NOH, O_2N, NO_2, CH_3O (benzene ring)	"	CN, O_2N, NO_2, CH_3O (benzene ring) (97)	643
	NO_2, CH_3O, CH=NOH, O_2N (benzene ring)	"	NO_2, CH_3O, CN, O_2N (benzene ring) (85)	643
	O, O, CHO (methylenedioxybenzene ring)	$H_2NOH \cdot HCl$, CH_3CO_2Na	O, O, CN (methylenedioxybenzene ring) (50)	650

Reactant	Reagents	Product (Yield, %)	Refs.
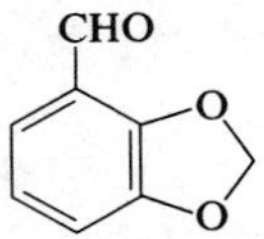 CHO	1. 2,4-$(O_2N)_2C_6H_3ONH_2$, HCl, ethanol 2. KOH, reflux, 3 h	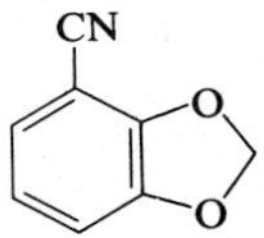 CN (81)	658
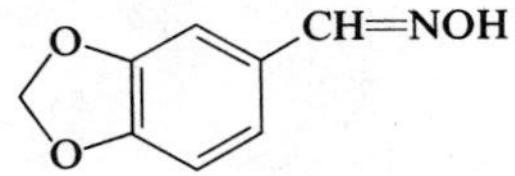 CH=NOH	$CuSO_4 \cdot 5H_2O$, DCC, $(C_2H_5)_3N$, CH_2Cl_2	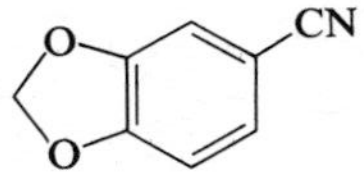 CN (99)	643
$C_6H_5CH_2CH{=}NOH$	HCO_2Na, HCO_2H, reflux, 1 h	" (83)	637
	P_2I_4	$C_6H_5CH_2CN$ (37)	636
	PI_3, $(C_2H_5)_3N$, CH_2Cl_2, 15 min	" (83)	662
	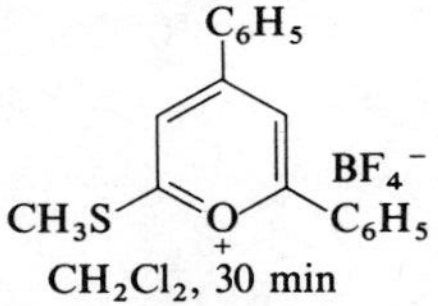 C_6H_5, BF_4^-, CH_3S, C_6H_5, $(C_2H_5)_3N$, CH_2Cl_2, 30 min	" (84)	648
	$(CH_3)_3SiI$, $[(CH_3)_3Si]_2NH$, $CHCl_3$, 56°, 4 h	" (84)	85
p-$CH_3C_6H_4CHO$	$H_2NOH \cdot HCl$, HCO_2H, reflux, 30 min	*p*-$CH_3C_6H_4CN$ (98)	236, 640
p-$CH_3C_6H_4CH{=}NOH$	$SOCl_2$, CCl_4, 12 h	" (87)	75
	$(CF_3SO_2)_2O$, $(C_2H_5)_3N$, CH_2Cl_2, −78 to 25°, 2 h	" (89)	172
	$ClSO_2NCO$, $(C_2H_5)_3N$, CH_2Cl_2, 8 h	" (82)	653
	$ClSO_2F$, $(C_2H_5)_3N$, CH_2Cl_2, 8 h	" (81)	646
	C_6H_5OSOCl, pyridine, ether, 5–25°, 15 h	" (90)	644
	$(CH_3)_3N \cdot SO_2$	" (84)	645
	$CuSO_4 \cdot 5H_2O$, DCC, $(C_2H_5)_3N$, CH_2Cl_2	" (95)	643

TABLE IV. ALDOXIME FRAGMENTATIONS (*Continued*)

No. of Carbon Atoms	Substrate	Reagent and Conditions	Product(s) and Yield(s) (%)	Refs.
		Cl_3CCOCl, $(C_2H_5)_3N$, CH_2Cl_2	p-$CH_3C_6H_4CN$ (85)	642
		$HCSN(CH_3)_2$, CH_3I, reflux, 18 h	" (85)	655
		2,4-$(O_2N)_2C_6H_3F$, KOC_4H_9-t, CH_3CN	" (94)	655
		$C_6H_5O_2CCl$, pyridine, ether, 0–25°, 10 h	" (90)	657
		$CH_3C{\equiv}CNC_2H_5{\cdot}BF_4$, CH_3CN, 25–80°, 8.5 h	" (78)	654
		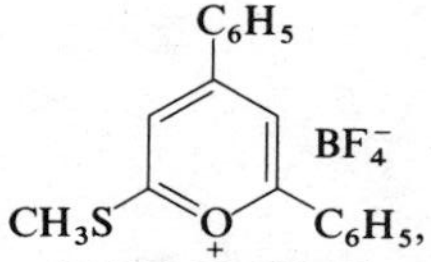 C_6H_5; CH_3S; $\overset{+}{O}$; C_6H_5; BF_4^-, $(C_2H_5)_3N$, CH_2Cl_2, 30 min	" (89)	648
		Cl_3CCN, reflux, 5 h	" (94)	665
		$Cu(O_2CCH_3)_2{\cdot}H_2O$, CH_3CN, reflux, 4 h	" (98)	660
	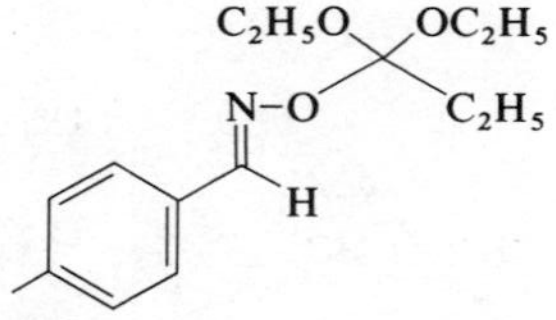 C_2H_5O, OC_2H_5; N–O; C_2H_5; H (*Z*)	BF_3, HgO, ether, reflux, 3 h	" (80)	137
	(*E*)	BF_3, HgO, ether, reflux, 3 h	" (23)	137
		$TiCl_4$, ether, reflux, 3 h	" (75)	137
		$ZnCl_2$, ether, reflux, 14 h	" (60)	137
	o-$CH_3OC_6H_4CHO$	$C_2H_5NO_2$, HCl, pyridine, reflux, 1 h	o-$CH_3OC_6H_4CN$ (90)	641

p-$CH_3OC_6H_4CHO$	$H_2NOH \cdot HCl$, HCO_2H, reflux, 30 min	p-$CH_3OC_6H_4CN$ (98)	236, 640, 650
	1. 2,4-$(O_2N)_2C_6H_3ONH_2$, HCl, ethanol 2. KOH, reflux, 3 h	" (91)	658
	$CF_3CO_2NHCOCF_3$, pyridine, C_6H_6, reflux, 2 h	" (74)	659
p-$CH_3OC_6H_4CH{=}NOH$	$SOCl_2$, CCl_4, 12 h	" (88)	75
	C_6H_5OSOCl, pyridine, ether, 5–25°, 15 h	" (98)	644
	$ClSO_2NCO$, $(C_2H_5)_3N$, CH_2Cl_2, 8 h	" (83)	653
	$ClSO_2F$, $(C_2H_5)_3N$, CH_2Cl_2, 8 h	" (85)	646
	$(CH_3)_3N \cdot SO_2$	" (79)	645
	$(C_6H_5O)_2PHO$, $(C_2H_5)_3N$, CCl_4, 4 h	" (85)	635
	P_2I_4	" (83)	636
	$CuSO_4 \cdot 5H_2O$, DCC, $(C_2H_5)_3N$, CH_2Cl_2	" (97)	643
	$C_6H_5O_2CCl$, pyridine, ether, 0–25°, 10 h	" (85)	657
	$(CH_3)_2N^+{=}CCl_2\ Cl^-$, reflux, 3 h	" (96)	289
	Cl_3CCOCl, $(C_2H_5)_3N$, CH_2Cl_2	" (94)	642
	$(CF_3CO)_2O$, pyridine, 0–25°, 2 h	" (99)	618
	$CH_3C{\equiv}CN(C_2H_5)_2$, CH_3CN	" (73)	654, 656
	[structure: C_6H_5; BF_4^-; CH_3S; O^+; C_6H_5], $(C_2H_5)_3N$, CH_2Cl_2, 30 min	" (82)	648
	$Cu(O_2CCH_3)_2 \cdot H_2O$, CH_3CN, reflux, 30 min	" (98)	660
	$HC(OC_2H_5)_3$, SO_2, H^+	" (90)	188
	$CH_3C(OC_2H_5)_3$, SO_2, H^+	" (90)	188
	HCO_2Na, HCO_2H, reflux, 1 h	" (81)	637
	$TiCl_4$, dioxane	" (90)	234
	SeO_2, $CHCl_3$, ethanol	" (76–89)	634

TABLE IV. ALDOXIME FRAGMENTATIONS (*Continued*)

No. of Carbon Atoms	Substrate	Reagent and Conditions	Product(s) and Yield(s) (%)	Refs.
	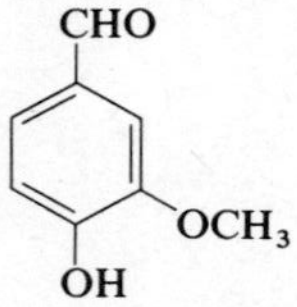 4-hydroxy-3-methoxybenzaldehyde (CHO, OCH_3, OH)	$H_2NOH\cdot HCl$, CH_3CO_2Na	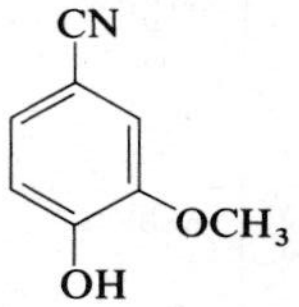 4-hydroxy-3-methoxybenzonitrile (CN, OCH_3, OH) (80)	650
	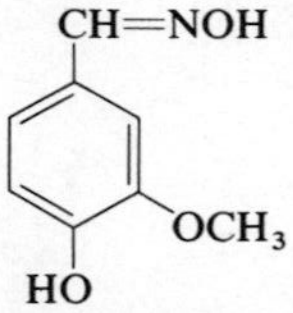 4-hydroxy-3-methoxybenzaldoxime (CH=NOH, OCH_3, HO)	HCO_2Na, HCO_2H, reflux, 1 h	" (95)	637
	$CH_3CH{=}C{=}CHC(CH_3)_2CH{=}NOH$	$(CH_3)_2N^+{=}CHCl\ Cl^-$, CH_3CN, 0°	$CH_3CH{=}C{=}CHC(CH_3)_2CN$ (90)	652
	$C_2H_5O_2C(CH_2)_4CH{=}NOH$	$HC(OC_2H_5)_3$, H^+, $CHCl_3$	$C_2H_5O_2C(CH_2)_4CN$ (96)	188
	$n\text{-}C_7H_{15}CHO$	$H_2NOH\cdot HCl$, pyridine, $C_6H_5CH_3$, reflux	$n\text{-}C_7H_{15}CN$ (83)	640, 667
	$n\text{-}C_7H_{15}CH{=}NOH$	P_2I_4	" (64)	636
		Cl_3CCOCl, $(C_2H_5)_3N$, CH_2Cl_2	" (92)	642
		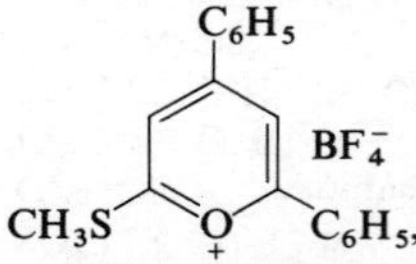 2-(CH_3S)-4,6-di(C_6H_5)pyrylium BF_4^-, $(C_2H_5)_3N$, CH_2Cl_2, 30 min	" (86)	648
		$Cu(O_2CCH_3)_2\cdot H_2O$, CH_3CN, reflux	" (85)	660
		SeO_2, $CHCl_3$, reflux	" (59–74)	235
	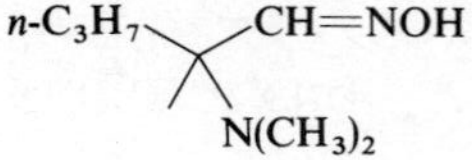 $n\text{-}C_3H_7C(CH_3)(N(CH_3)_2)CH{=}NOH$	TsCl, NaOH, H_2O	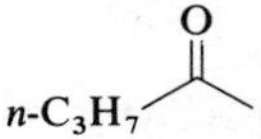 $n\text{-}C_3H_7COCH_3$ (67)	570

	Reactant	Reagents/Conditions	Product (Yield %)	Refs.
C_9	indole-3-carboxaldehyde (3-CHO, 1-NH indole)	$CF_3CO_2NHCOCF_3$, pyridine, C_6H_6, reflux	indole-3-carbonitrile (3-CN, 1-NH indole) (82)	659
		$C_2H_5NO_2$, HCl, pyridine, reflux, 1 h	" (70)	641
	indole-3-carboxaldehyde oxime (3-CH=NOH, 1-NH indole)	$(PNCl_2)_3$, $(C_2H_5)_3N$	" (98)	647
		$CuSO_4 \cdot 5H_2O$, DCC, $(C_2H_5)_3N$, CH_2Cl_2	" (91)	643
		SeO_2, $CHCl_3$, ethanol, 2 h	" (82)	634
	$C_6H_5CH{=}CHCHO$	$H_2NOH \cdot HCl$, pyridine, $C_6H_5CH_3$, reflux	$C_6H_5CH{=}CHCN$ (92)	640
		$CF_3CO_2NHCOCF_3$, pyridine, C_6H_6, reflux, 2 h	" (88)	659
		$C_2H_5NO_2$, HCl, pyridine, reflux, 1 h	" (85)	641
	$C_6H_5CH{=}CHCH{=}NOH$	$(CF_3SO_2)_2O$, $(C_2H_5)_3N$, CH_2Cl_2, −78 to 25°, 2 h	" (85–93)	172
		C_6H_5OSOCl, pyridine, ether, 5−25°, 15 h	" (98)	644
		$ClSO_2F$, $(C_2H_5)_3N$, CH_2Cl_2, 8 h	" (88)	646
		$(CH_3)_3N \cdot SO_2$	" (74)	645
		P_2I_4	$C_6H_5CH{=}CHCN$ (68)	636
		$(C_6H_5)_3P$, CCl_4, $(C_2H_5)_3N$, 60°, 2.5 h	" (88)	633
		$(C_6H_5O)_2PHO$, CCl_4, $(C_2H_5)_3N$, 4 h	" (95)	635
		$(PNCl_2)_3$, $(C_2H_5)_3N$	" (98)	647
		$HCSN(CH_3)_2$, CH_3I, reflux, 18 h	" (72)	655
		2,4-$(O_2N)_2C_6H_3F$, KOC_4H_9-*t*, CH_3CN, 80°, 30 min	" (71)	655
		Cl_3CCOCl, $(C_2H_5)_3N$, CH_2Cl_2	" (87)	642
		$CH_3C{\equiv}NC_2H_5 \cdot BF_4$, CH_3CN, 25–80°, 8.5 h	" (56)	654

TABLE IV. ALDOXIME FRAGMENTATIONS (*Continued*)

No. of Carbon Atoms	Substrate	Reagent and Conditions	Product(s) and Yield(s) (%)	Refs.
		C_6H_5, BF_4^-, CH_3S, $\overset{+}{O}$, C_6H_5 (pyrylium salt), $(C_2H_5)_3N$, CH_2Cl_2, 30 min	$C_6H_5CH{=}CHCN$ (80)	648
		$Cu(O_2CCH_3)_2 \cdot H_2O$, CH_3CN, reflux	" (98)	660
		Cl_3CCN, reflux, 30 min	" (92)	665
		SeO_2, $CHCl_3$, ethanol, 2 h	" (81)	634
	$C_6H_5(CH_2)_2CH{=}NOH$	$(CF_3SO_2)_2O$, $(C_2H_5)_3N$, CH_2Cl_2, -78 to $25°$	$C_6H_5(CH_2)_2CN$ (92)	172
		P_2I_4	" (37)	636
		Cl_3CCOCl, $(C_2H_5)_3N$, CH_2Cl_2	" (85)	642
		$CuSO_4 \cdot 5H_2O$, DCC, $(C_2H_5)_3N$, CH_2Cl_2	" (93)	643
		$HC(OC_2H_5)_3$, H^+	" (93)	188
		$Cu(O_2CCH_3)_2 \cdot H_2O$, CH_3CN, reflux	" (98)	660
	CHO, CH_3O, OCH_3 (benzene ring)	$C_2H_5NO_2$, HCl, pyridine, reflux, 1 h	CN, CH_3O, OCH_3 (benzene ring) (92)	641
	$CH{=}NOH$, CH_3O, OCH_3 (benzene ring)	$CuSO_4 \cdot 5H_2O$, DCC, $(C_2H_5)_3N$, CH_2Cl_2	" (98)	643
	CH_3O, CHO, CH_3O (benzene ring)	$CF_3CO_2NHCOCF_3$, pyridine, C_6H_6, reflux, 2 h	CH_3O, CN, CH_3O (benzene ring) (87)	659

	Reactant	Reagents and Conditions	Product (Yield, %)	Refs.
	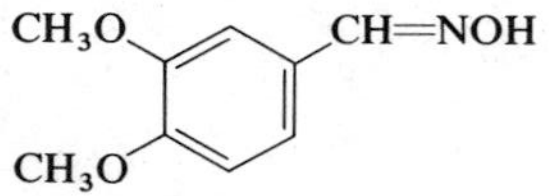 CH_3O, CH_3O [structure] $CH{=}NOH$	p-ClC_6H_4OSOCl, ether	" (61)	661
		$HCSN(CH_3)_2$, CH_3I, reflux, 18 h	" (86)	655
		2,4-$(O_2N)_2C_6H_3F$, KOC_4H_9-t, CH_3CN	" (96)	655
		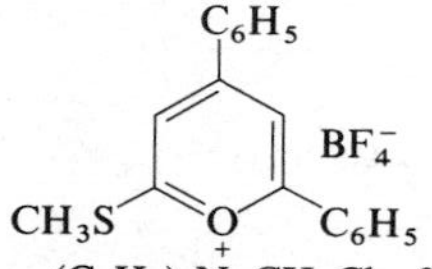 [structure] C_6H_5, BF_4^-, CH_3S, O^+, C_6H_5; $(C_2H_5)_3N$, CH_2Cl_2, 30 min	" (93)	648
		Cl_3CCN, reflux, 30 min	" (75)	665
	p-$(CH_3)_2NC_6H_4CHO$	$H_2NOH\cdot HCl$, pyridine, $C_6H_5CH_3$, reflux	p-$(CH_3)_2NC_6H_4CN$ (86)	640
		$C_2H_5NO_2$, HCl, pyridine, reflux, 1 h	" (85)	641
	p-$(CH_3)_2NC_6H_4CH{=}NOH$	P_2I_4	" (62)	636
		$CuSO_4\cdot 5H_2O$, DCC, $(C_2H_5)_3N$, CH_2Cl_2	" (97)	643
		Cl_3CCOCl, $(C_2H_5)_3N$, CH_2Cl_2	" (94)	642
		$(CH_3)_2N^+{=}CCl_2\ Cl^-$, $CHCl_3$, 2 h	" (82)	289
		HCO_2Na, HCO_2H, reflux, 1 h	" (84)	637
		$TiCl_4$, dioxane	" (84)	234
		SeO_2, $CHCl_3$, ethanol, 2 h	" (82)	634
	$(CH_3)_2C{=}C{=}CHC(CH_3)_2CH{=}NOH$	$(CH_3)_2N^+{=}CHCl\ Cl^-$, CH_3CN, 0°	$(CH_3)_2C{=}C{=}CHC(CH_3)_2CN$ (95)	652
	$(CH_3)_3SiC{\equiv}CC(CH_3){=}CHCH{=}NOH$	"	$(CH_3)_3SiC{\equiv}CC(CH_3){=}CHCN$ (85)	652
	n-$C_8H_{17}CH{=}NOH$	$(CH_3)_3SiI$, $[(CH_3)_3Si]_2NH$, $CHCl_3$, 56°, 4 h	n-$C_8H_{17}CN$ (88)	85
C_{10}	$C_6H_5CH{=}CHCOCH{=}NOH$	1. C_2H_5ONa, ethanol 2. $(CH_3CO)_2O$	$C_6H_5CH{=}CHCO_2C_2H_5$ (—)	576
	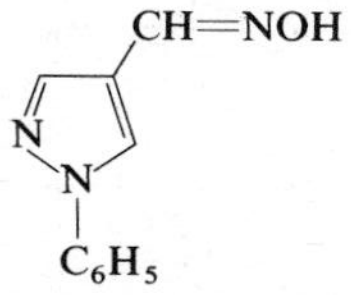 [structure] $CH{=}NOH$, N, N, C_6H_5	TsCl, pyridine, 4 d	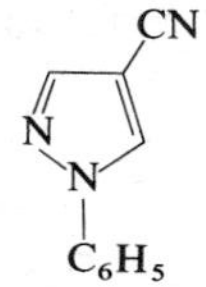 [structure] CN, N, N, C_6H_5 (87–95)	318

TABLE IV. ALDOXIME FRAGMENTATIONS (*Continued*)

No. of Carbon Atoms	Substrate	Reagent and Conditions	Product(s) and Yield(s) (%)	Refs.
	CH=NOH	$CuSO_4 \cdot 5H_2O$, DCC, $(C_2H_5)_3N$, CH_2Cl_2	CN (92)	643
	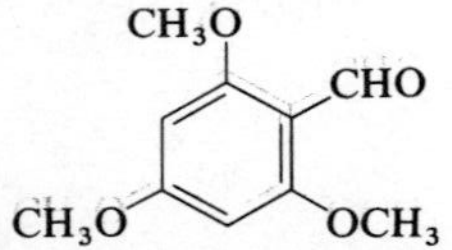 CH_3O, CHO, CH_3O, OCH_3	$C_2H_5NO_2$, HCl, pyridine, reflux, 1 h	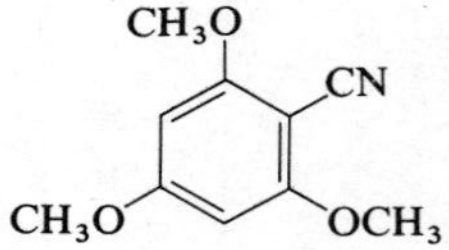 CH_3O, CN, CH_3O, OCH_3 (91)	641
	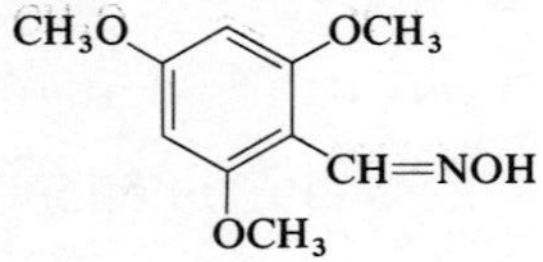 CH_3O, OCH_3, CH=NOH, OCH_3	Cyanuric chloride, CH_2Cl_2	" (—)	638
		$CuSO_4 \cdot 5H_2O$, DCC, $(C_2H_5)_3N$, CH_2Cl_2	" (92)	643
	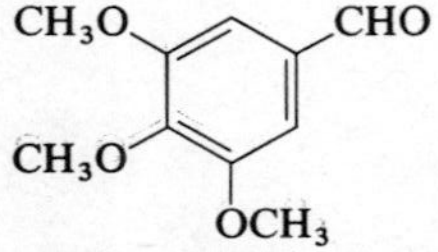 CH_3O, CHO, CH_3O, OCH_3	$C_2H_5NO_2$, HCl, pyridine, reflux, 1 h	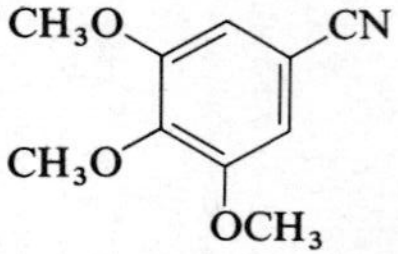 CH_3O, CN, CH_3O, OCH_3 (93)	641
	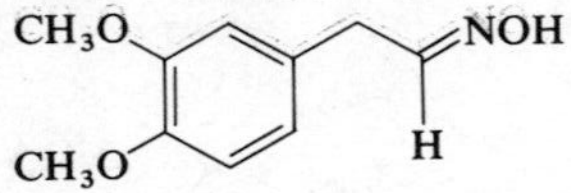 CH_3O, CH_3O, NOH, H	$(CH_3)_3SiI$, $[(CH_3)_3Si]_2NH$, $CHCl_3$, 56°, 4 h	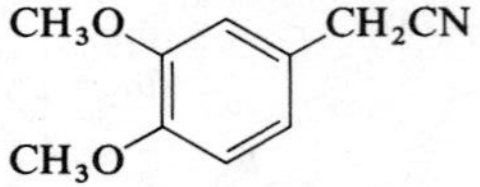 CH_3O, CH_2CN, CH_3O (—)	85, 668
	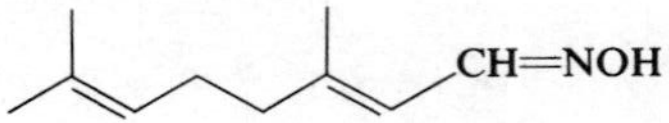 CH=NOH	$(PNCl_2)_3$, $(C_2H_5)_3N$	 CN (98)	647

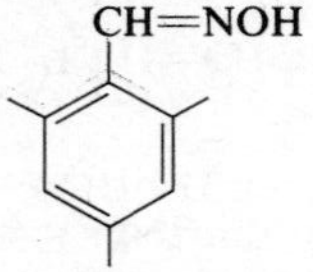

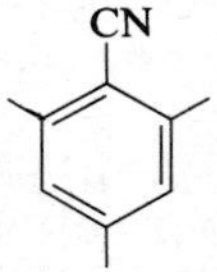

	Reactant	Reagents	Product (Yield, %)	Refs.
	$n\text{-}C_9H_{19}CHO$	$H_2NOH \cdot HCl$, pyridine, $C_6H_5CH_3$, reflux	$n\text{-}C_9H_{19}CN$ (52)	640
	$n\text{-}C_9H_{19}CH{=}NOH$	Cl_3CCOCl, $(C_2H_5)_3N$, CH_2Cl_2	" (90)	642
		$Cu(O_2CCH_3)_2 \cdot H_2O$, CH_3CN, reflux	" (98)	642
C_{11}	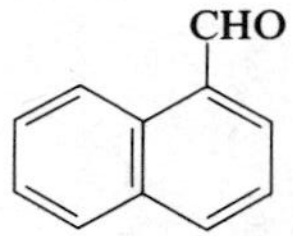 CHO	$H_2NOH \cdot HCl$, pyridine, $C_6H_5CH_3$, reflux	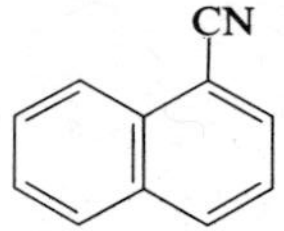 CN (85)	640
	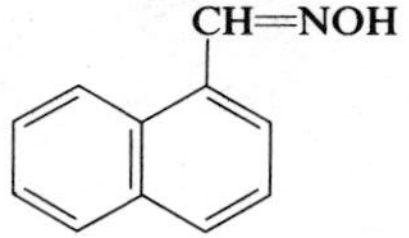 CH=NOH	$ClSO_2F$, $(C_2H_5)_3N$, CH_2Cl_2, 8 h	" (95)	646
		Cl_3CCOCl, $(C_2H_5)_3N$, CH_2Cl_2	" (93)	642
		HCO_2Na, HCO_2H, reflux, 1 h	" (93)	637
		Cl_3CCN, reflux, 30 min	" (95)	665
	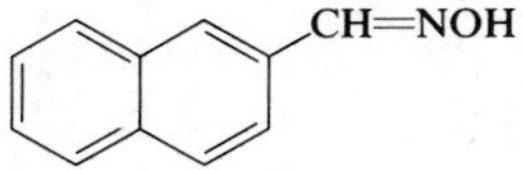 CH=NOH	$CuSO_4 \cdot 5H_2O$, DCC, $(C_2H_5)_3N$, CH_2Cl_2	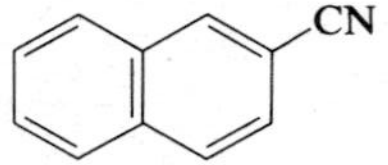 CN (100)	643
		SeO_2, $CHCl_3$, ethanol, 2 h	" (84)	634
	$C_5H_5FeC_5H_4CH{=}NOH$	Cl_3CCN	$C_5H_5FeC_5H_4CN$ (100)	232
	$p\text{-}(C_2H_5)_2NC_6H_4CH{=}NOH$	$CuSO_4 \cdot 5H_2O$, DCC, $(C_2H_5)_3N$, CH_2Cl_2	$p\text{-}(C_2H_5)_2NC_6H_4CN$ (99)	643
	$CH_2{=}CH(CH_2)_8CHO$	1. 2,4-$(O_2N)_2C_6H_3ONH_2$, HCl, ethanol 2. $(C_2H_5)_3N$, reflux, 5 min	$CH_2{=}CH(CH_2)_8CN$ (—)	658
	$n\text{-}C_{10}H_{21}CH{=}NOH$	$(PNCl_2)_3$, $(C_2H_5)_3N$	$n\text{-}C_{10}H_{21}CN$ (95)	647
		PI_3, $(C_2H_5)_3N$, CH_2Cl_2, 15 s	" (85)	662
C_{12}	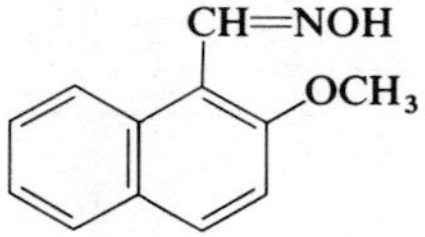 CH=NOH, OCH_3	$CuSO_4 \cdot 5H_2O$, DCC, $(C_2H_5)_3N$, CH_2Cl_2	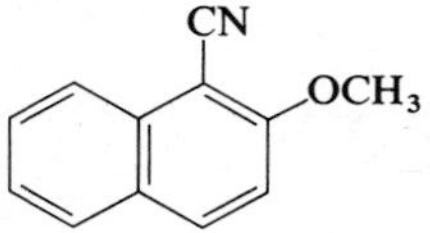 CN, OCH_3 (98)	643

TABLE IV. ALDOXIME FRAGMENTATIONS (*Continued*)

No. of Carbon Atoms	Substrate	Reagent and Conditions	Product(s) and Yield(s) (%)	Refs.
	1-C_6H_5-3,5-dimethylpyrazole-4-CH=NOH (structure)	TsCl, pyridine, 4 d	1-C_6H_5-3,5-dimethylpyrazole-4-CN (structure) (82)	318
		PCl_5, $CHCl_3$, 4 d	" (80)	318
	(cyclohexylidene)C=CH–C(CH_3)$_2$–CH=NOH (structure)	$(CH_3)_2N^+{=}CHCl\ Cl^-$, CH_3CN, 0°	(cyclohexylidene)C=CH–C(CH_3)$_2$–CN (structure) (92)	652
	n-$C_{11}H_{23}CH{=}NOH$	$CuSO_4\cdot 5H_2O$, DCC, $(C_2H_5)_3N$, CH_2Cl_2	n-$C_{11}H_{23}CN$ (95)	643
C_{13}	p-$C_6H_5C_6H_4CH{=}NOH$	$(PNCl_2)_3$, $(C_2H_5)_3N$	p-$C_6H_5C_6H_4CN$ (69)	647
C_{14}	HON=CH–C_6H_4–N=N–C_6H_4–CH=NOH (meta, meta; structure)	$CuSO_4\cdot 5H_2O$, DCC, $(C_2H_5)_3N$, CH_2Cl_2	NC–C_6H_4–N=N–C_6H_4–CN (meta, meta; structure) (96)	643
	10-CH_3-phenothiazine-3-CH=NOH (structure)	"	10-CH_3-phenothiazine-3-CN (structure) (95)	643
	p-$C_6H_5CH_2OC_6H_4CH{=}NOH$	"	p-$C_6H_5CH_2OC_6H_4CN$ (97)	643
	$(C_6H_5)_2CHCH{=}NOH$	$(CF_3SO_2)_2O$, $(C_2H_5)_3N$, CH_2Cl_2, − 78 to 25°, 2 h	" (93)	172
	n-$C_{13}H_{27}CHO$	$H_2NOH\cdot HCl$, reflux, 6 h	n-$C_{13}H_{27}CN$ (90)	667

C_{15}	9-anthryl–CH=NOH	$CuSO_4 \cdot 5H_2O$, DCC, $(C_2H_5)_3N$, CH_2Cl_2	9-anthryl–CN	(99)	643
	3-Br-4-$C_6H_5CH_2O$-5-OCH_3-C_6H_2–CH=NOH	”	3-Br-4-$C_6H_5CH_2O$-5-OCH_3-C_6H_2–CN	(99)	643
	3-$C_6H_5CH_2O$-4-CH_3O-C_6H_3–CH=NOH	”	3-$C_6H_5CH_2O$-4-CH_3O-C_6H_3–CN	(99)	643
	3-CH_3O-4-$C_6H_5CH_2O$-C_6H_3–CH=NOH	”	3-CH_3O-4-$C_6H_5CH_2O$-C_6H_3–CN	(99)	643
	$(CH_3)_2N$–C_6H_4–N=N–C_6H_4–CH=NOH	”	$(CH_3)_2N$–C_6H_4–N=N–C_6H_4–CN	(91)	643
	HON=CH–(OH, O,O-cyclohexylidene furanose)	(–)	CN–(OH, O,O-cyclohexylidene furanose)	(—)	131
C_{16}	3-CH_3O-4-n-$C_8H_{17}O$-C_6H_3–CH=NOH	$CuSO_4 \cdot 5H_2O$, DCC, $(C_2H_5)_3N$, CH_2Cl_2	3-CH_3O-4-n-$C_8H_{17}O$-C_6H_3–CN	(93)	643

TABLE IV. ALDOXIME FRAGMENTATIONS (*Continued*)

No. of Carbon Atoms	Substrate	Reagent and Conditions	Product(s) and Yield(s) (%)	Refs.
	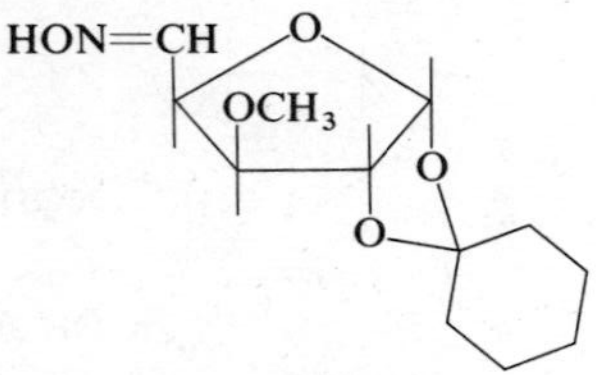 HON=CH (furanose, OCH₃, cyclohexylidene acetal)	(—)	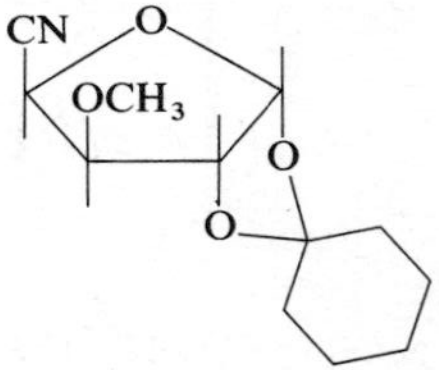 CN (furanose, OCH₃, cyclohexylidene acetal) (—)	131
C_{18}	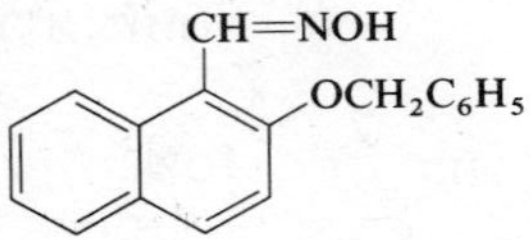 1-(CH=NOH)-2-($OCH_2C_6H_5$)naphthalene	$CuSO_4 \cdot 5H_2O$, DCC, $(C_2H_5)_3N$, CH_2Cl_2	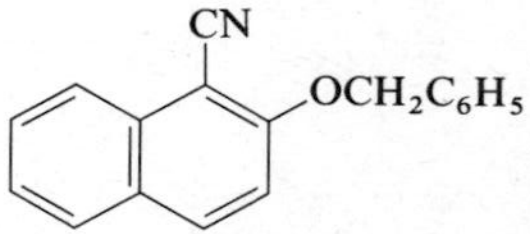 1-CN-2-($OCH_2C_6H_5$)naphthalene (99)	643
	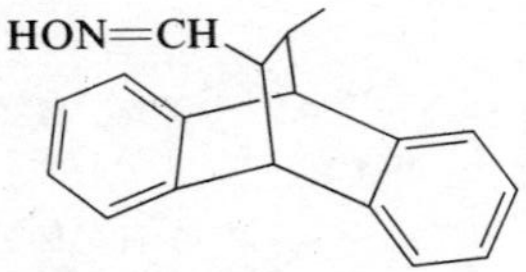 HON=CH (dibenzobicyclic)	"	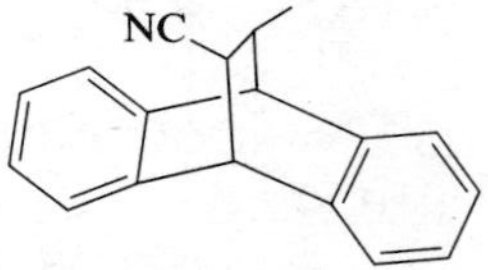 NC (dibenzobicyclic) (94)	643
	$[n\text{-}C_{17}H_{35}CHO]_3$	$H_2NOH \cdot HCl$, H_2O, reflux, 6 h	$n\text{-}C_{17}H_{35}CN$ (94)	667
C_{20}	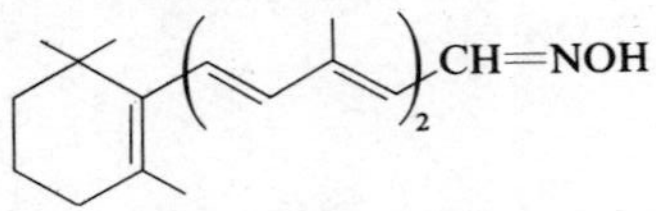 (trimethylcyclohexenyl)(CH=CH–C(CH₃)=CH)₂CH=NOH	SeO_2, $CHCl_3$, ethanol, 2 h	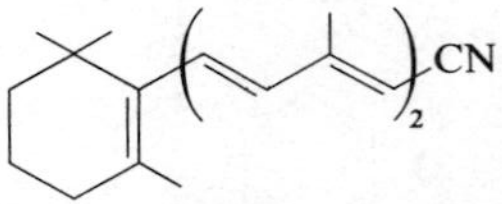 (trimethylcyclohexenyl)(CH=CH–C(CH₃)=CH)₂CN (79)	634
C_{21}	$p\text{-}[(C_6H_5CH_2)_2N]C_6H_4CH{=}NOH$	$CuSO_4 \cdot 5H_2O$, DCC, $(C_2H_5)_3N$, CH_2Cl_2	$p\text{-}[(C_6H_5CH_2)_2N]C_6H_4CN$ (97)	643

	Reactant	Conditions	Product (Yield, %)	Refs.
	HON=CH, $COCH_3$	$POCl_3$, pyridine	NC, $COCH_3$ (—)	669
	HO, HON=CH	$(CH_3CO)_2O$	HO, NC (—)	669
	HO, HON=CH	"	HO, NC (—)	669
C_{32}	C_8H_{17}, CH=NOH, CH_3CO_2, H, OH	CH_3SO_2Cl, pyridine, 0°, 15 h	C_8H_{17}, CN (85)	670

TABLE V. REARRANGEMENT-CYCLIZATIONS

No. of Carbon Atoms	Substrate	Reagent and Conditions	Product(s) and Yield(s) (%)	Refs.
C_6	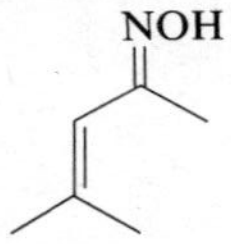 NOH	H_2SO_4	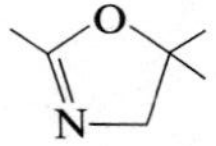 O, N (80)	206
		PPA	" (8)	62
C_7	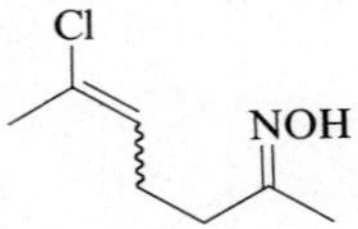 Cl, NOH	P_2O_5, CCl_4, reflux	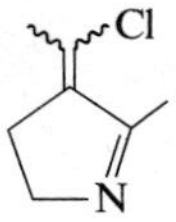 Cl, N (50)	217, 219
	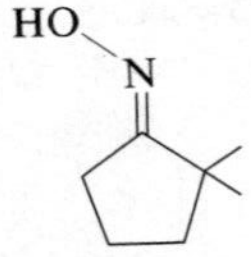 HO, N	1. PPA, 125–130°, 12 min 2. NaOH, ice	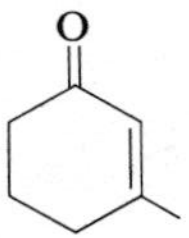 O (32)	226
C_8	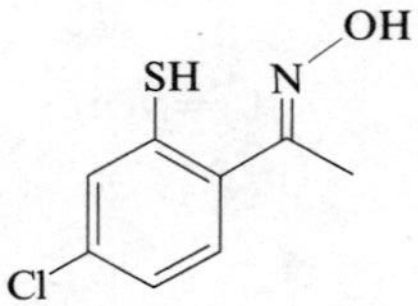 SH, N, OH, Cl	PPA, 125°, 1.5 h	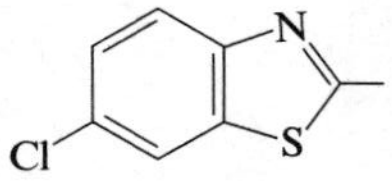 N, S, Cl (70)	208
	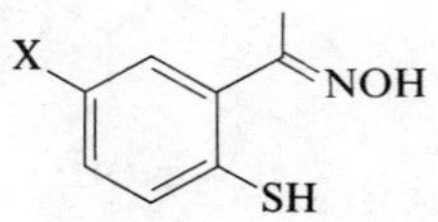 X, NOH, SH X=Cl	"	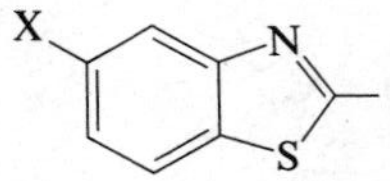 X, N, S (70)	208

$X=NO_2$	"	" (30)	208
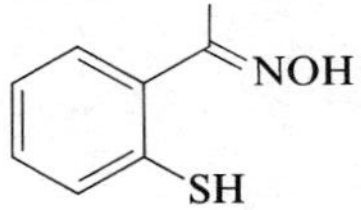 NOH, SH	"	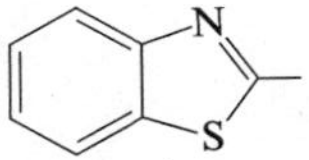 N, S (—)	208
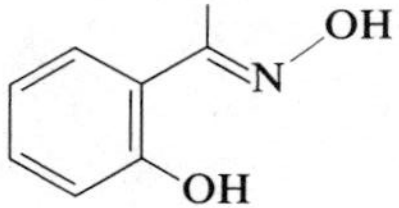 OH, N, OH	$Cl_3CCONCO$	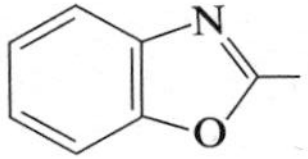 N, O (15)	671
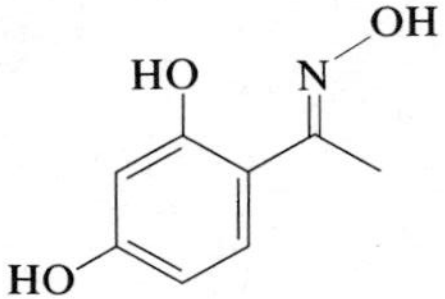 HO, N, OH, HO	$POCl_3$, DMAC, CH_3CN, 30°, 30 min	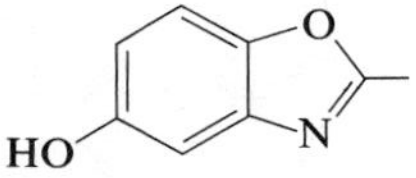 O, N, HO (83)	207
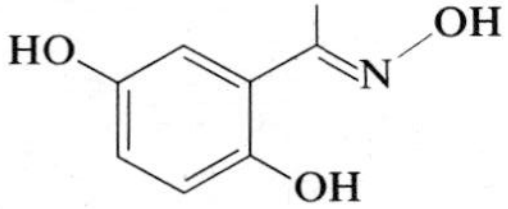 HO, OH, N, OH	"	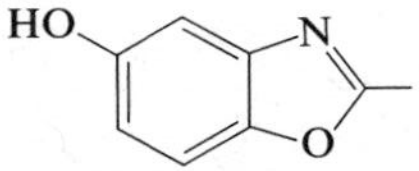 HO, N, O (68)	207
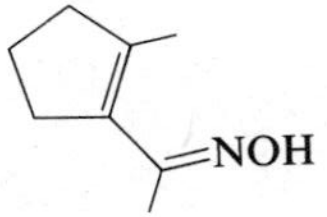 NOH	HCl, CH_3CO_2H, $(CH_3CO)_2O$	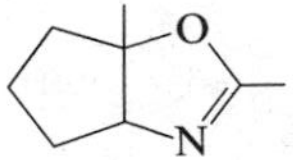 O, N (—)	15, 205
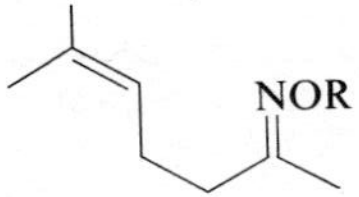 NOR R=H	P_2O_5, CCl_4, reflux	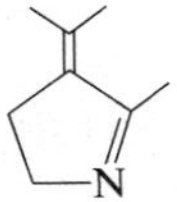 N (73)	216, 217
	PPSE, CCl_4, reflux	" (74)	217
(*E*) R=Ms	$SnCl_4$, CH_2Cl_2, −20 to 0°	" (88)	218
	$(C_2H_5)_2AlCl$, CH_2Cl_2	" (57)	218

TABLE V. REARRANGEMENT-CYCLIZATIONS (*Continued*)

No. of Carbon Atoms	Substrate	Reagent and Conditions	Product(s) and Yield(s) (%)	Refs.
C_9	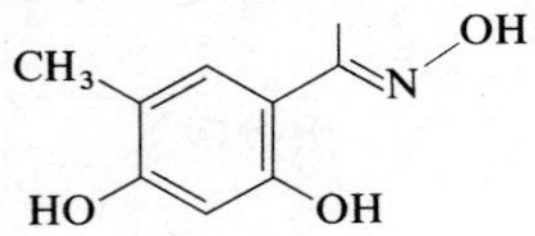 CH_3, HO, OH, N, OH	$POCl_3$, DMAC, CH_3CN, 30°, 30 min	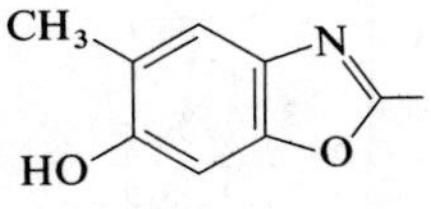 CH_3, HO, N, O (82)	207
	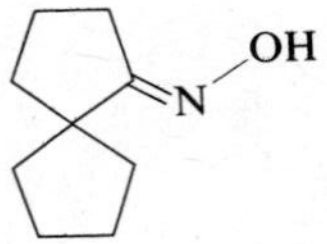 N, OH	PPA, 120–130°, 10 min	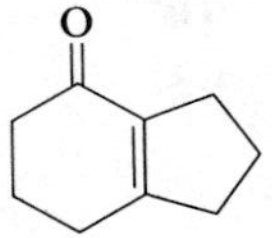 O (98)	221
	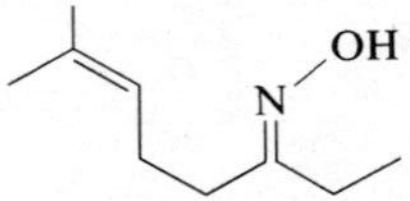 N, OH	P_2O_5, CCl_4, reflux	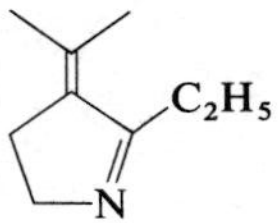 C_2H_5, N	217, 219
	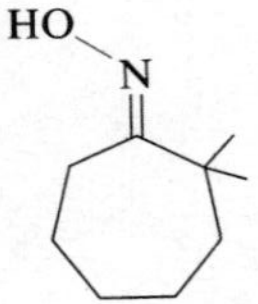 HO, N	1. PPA, 125–130°, 10 min 2. NaOH, ice	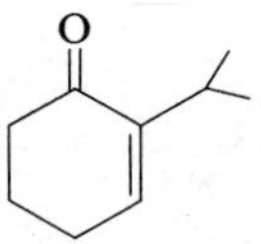 O (30)	226
C_{10}	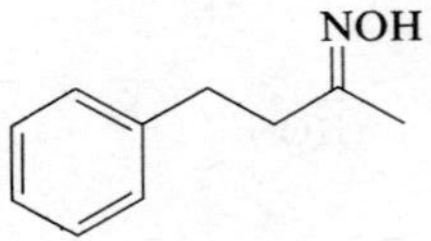 NOH	P_2O_5, $POCl_3$, SO_2, 70°, 12 h	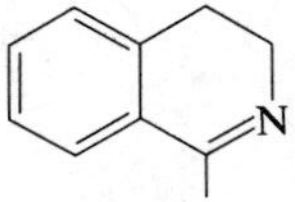 N (45)	215

Reactant	Conditions	Product(s) and Yield(s) (%)	Refs.
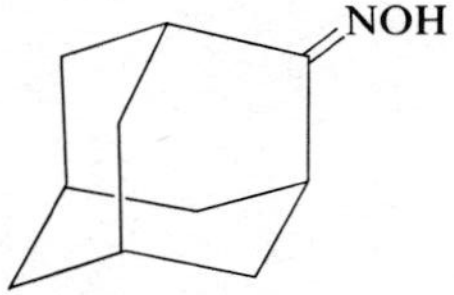 NOH	48% HBr, 140°, 2 h	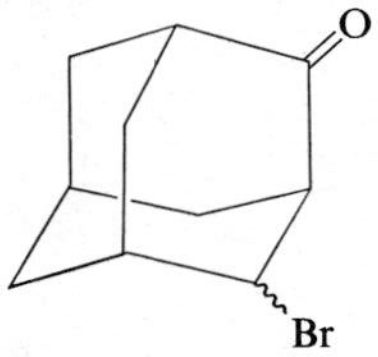 O, Br (56)	335
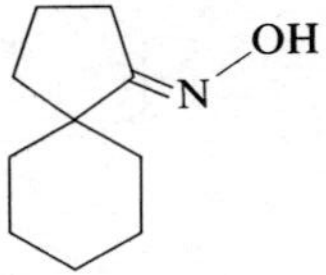 OH, N	PPA, 125–130°, 10 min	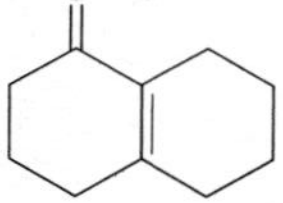 O (94)	221
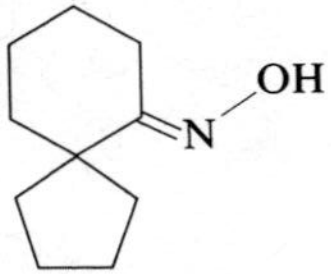 OH, N	PPA, 120–125°, 10 min	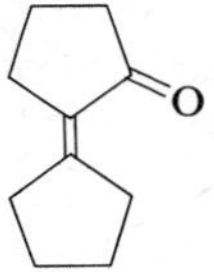 O (56)	221
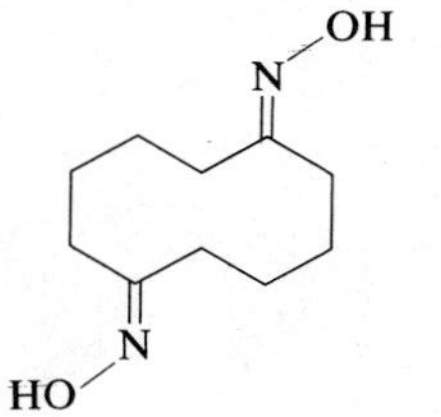 OH, N, N, HO	$SOCl_2$, SO_2	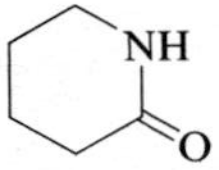 NH, O (33)	210
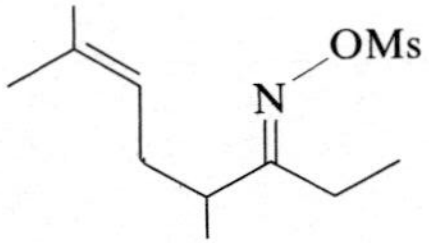 OMs, N	$SnCl_4$, CH_2Cl_2, −20 to 0°	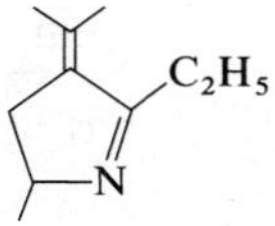 C_2H_5, N (80)	218

TABLE V. REARRANGEMENT-CYCLIZATIONS (*Continued*)

No. of Carbon Atoms	Substrate	Reagent and Conditions	Product(s) and Yield(s) (%)	Refs.
C_{11}	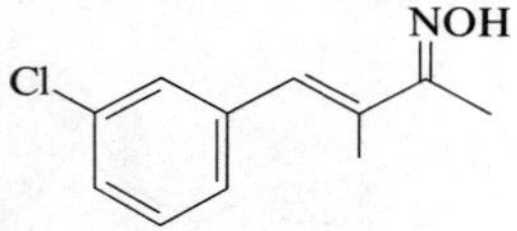 Cl NOH	1. PCl_5, 0°, 15 min 2. P_2O_5, decalin, reflux, 30 min	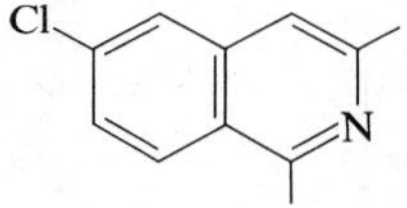 Cl N (48)	213
	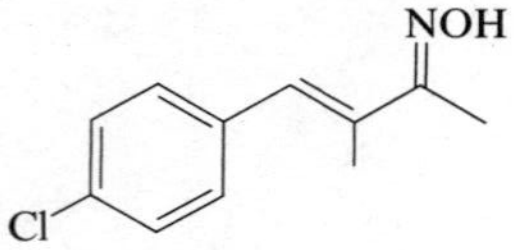 NOH Cl	"	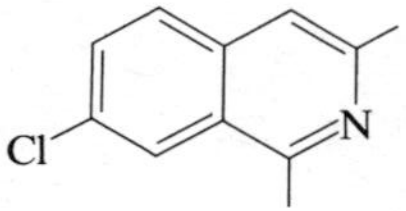 Cl N (31)	213
	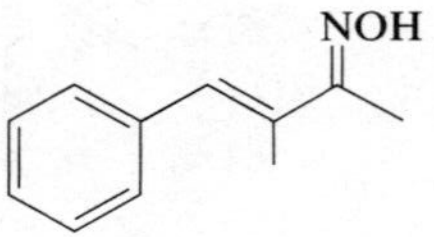 NOH	"	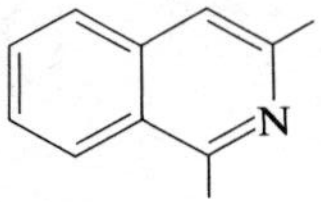 N (65)	213
	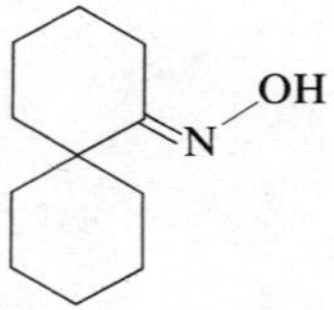 OH N	PPA, 120–130°, 10 min	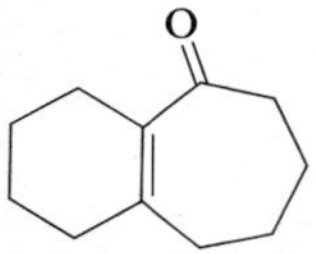 O (10)	221
	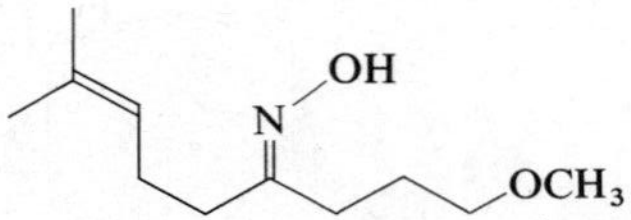 OH N OCH_3	P_2O_5, CCl_4, reflux	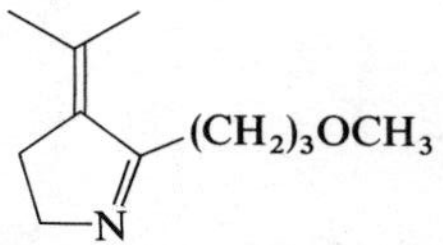 $(CH_2)_3OCH_3$ N (<10)	217

	Reactant	Conditions	Product(s) and Yield(s) (%)	Refs.
	OMs oxime	1. $(C_2H_5)_2AlCl$, CH_2Cl_2 2. DIBAL	NH (73)	218
C_{12}	NOH	1. PCl_5, 0°, 15 min 2. P_2O_5, decalin, reflux, 30 min	N (32)	213
	C_6H_5 NOH	P_2O_5, CCl_4, reflux	C_6H_5 N (50)	217, 219
		PPSE, CCl_4, reflux	" (59)	217
	NOH	1. PCl_5, 0°, 15 min 2. P_2O_5, decalin, reflux, 30 min	N (45)	213
	NOH	"	N (58)	213

TABLE V. REARRANGEMENT-CYCLIZATIONS (*Continued*)

No. of Carbon Atoms	Substrate	Reagent and Conditions	Product(s) and Yield(s) (%)	Refs.
	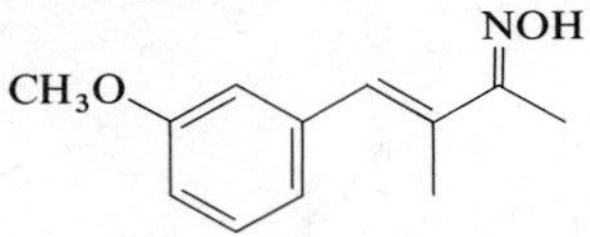 CH_3O, NOH	"	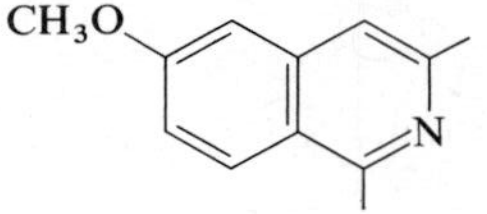 CH_3O, N (41)	213
	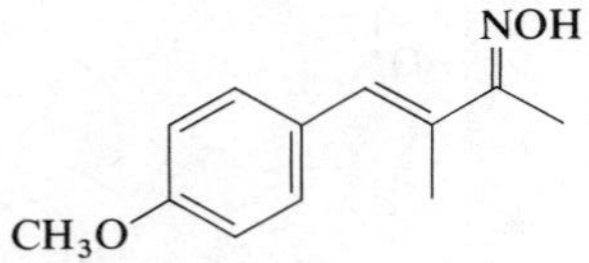 NOH, CH_3O	"	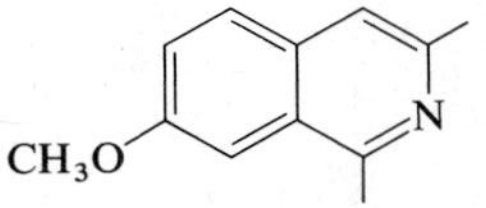 N, CH_3O (6)	213
	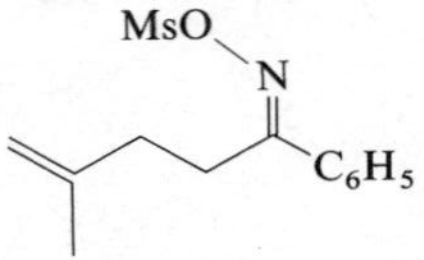 MsO, N, C_6H_5	1. $SnCl_4$, CH_2Cl_2, −20 to 0° 2. DIBAL	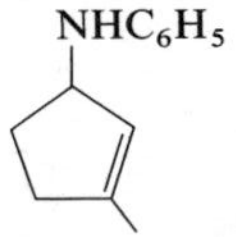 NHC_6H_5 (65)	218
	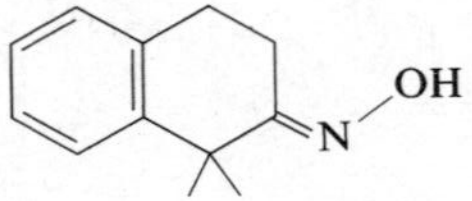 N, OH	PPA,125-130°, 10 min	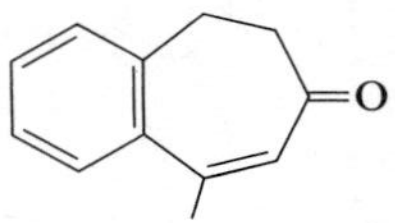 O (71)	420
	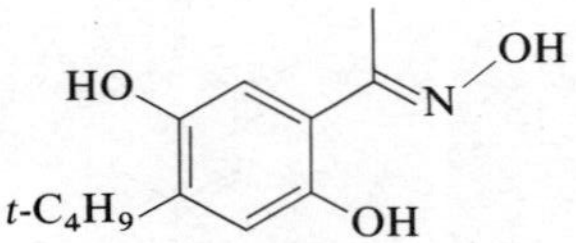 HO, N, OH, t-C_4H_9, OH	$POCl_3$, DMAC, 30°, 30 min	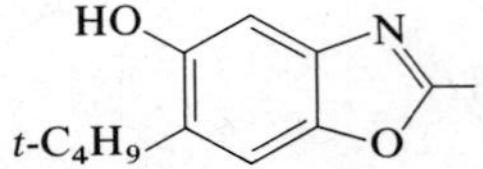 HO, N, t-C_4H_9, O (88)	207

No. of C Atoms	Reactant	Conditions	Product(s) and Yield(s) (%)	Refs.
	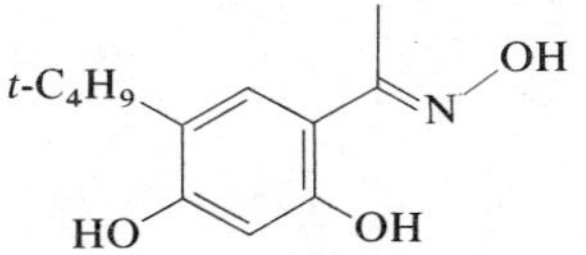 t-C_4H_9, HO, OH, N, OH	$POCl_3$, DMAC, 30°, 30 min	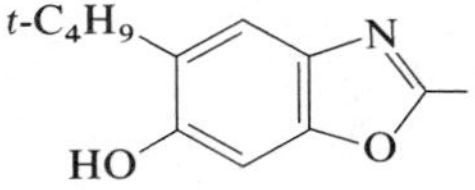 t-C_4H_9, HO, N, O (88)	207
	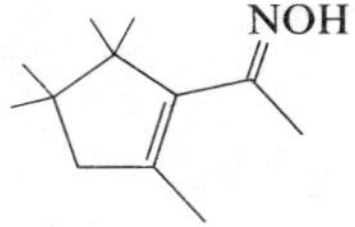 NOH	H_2SO_4	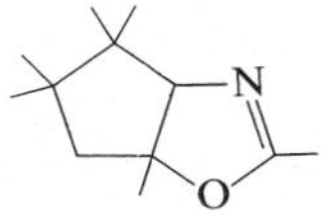 N, O (60–70)	388
	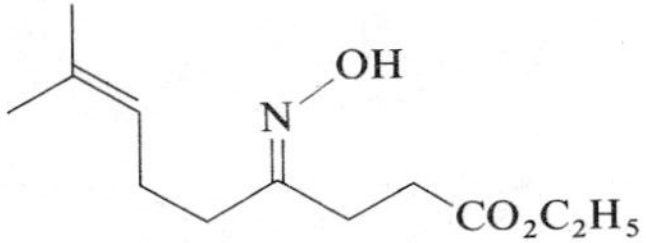 N, OH, $CO_2C_2H_5$	PPSE, CCl_4, reflux	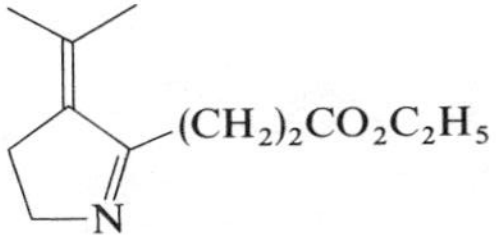 $(CH_2)_2CO_2C_2H_5$, N (<10)	217
C_{13}	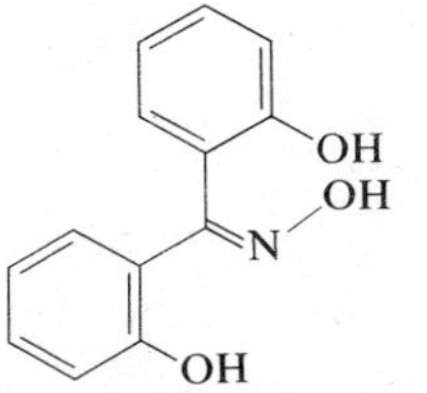 OH, OH, N, OH	$Cl_3CCONCO$	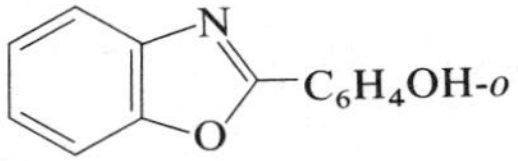 N, O, C_6H_4OH-o (26)	671
	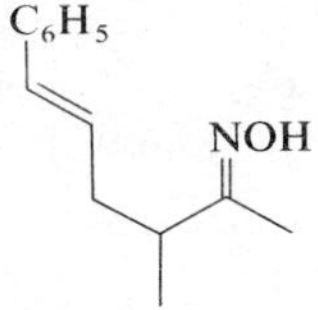 C_6H_5, NOH	P_2O_5, CCl_4, reflux	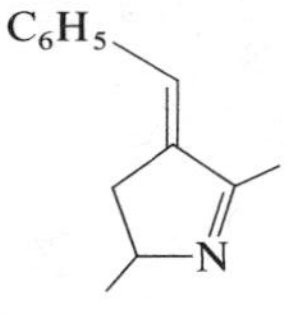 C_6H_5, N (82)	217, 219
	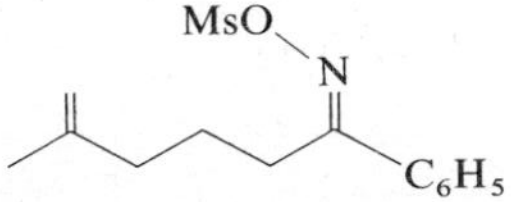 MsO, N, C_6H_5	1. $SnCl_4$, CH_2Cl_2, −20 to 0° 2. DIBAL	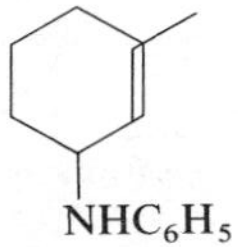 NHC_6H_5 (65)	218

TABLE V. REARRANGEMENT-CYCLIZATIONS (*Continued*)

No. of Carbon Atoms	Substrate	Reagent and Conditions	Product(s) and Yield(s) (%)	Refs.
		1. $(C_2H_5)_2AlCl$, CH_2Cl_2 2. DIBAL	" (31)	218
		1. $(CH_3)_3Al$, CH_2Cl_2 2. DIBAL	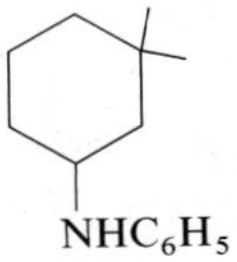 NHC_6H_5 (63)	218
	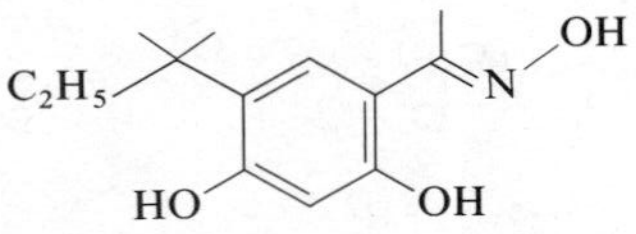 C_2H_5, HO, OH, N, OH	$POCl_3$, DMAC, CH_3CN, 30°, 30 min	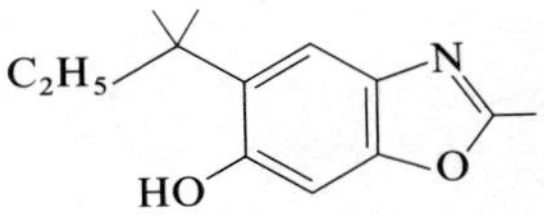 C_2H_5, HO, N, O (62)	207
	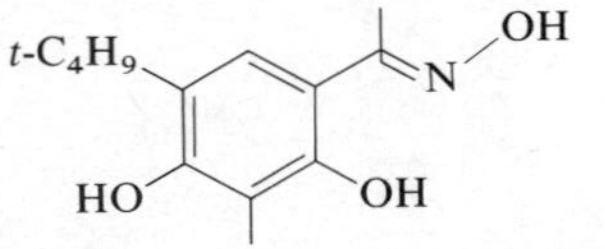 t-C_4H_9, HO, OH, N, OH	"	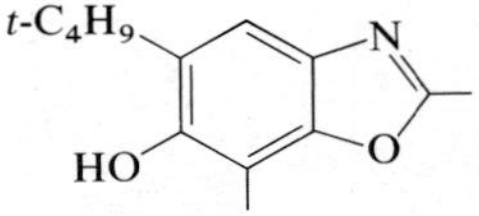 t-C_4H_9, HO, N, O (85)	207
	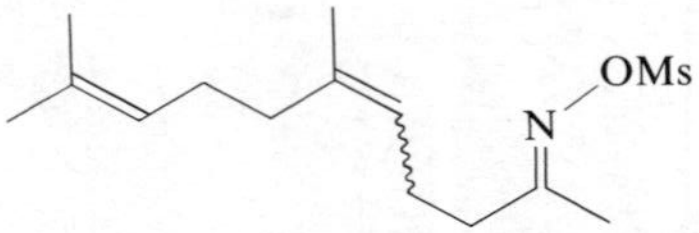 N, OMs	$SnCl_4$, CH_2Cl_2, −20 to 0°	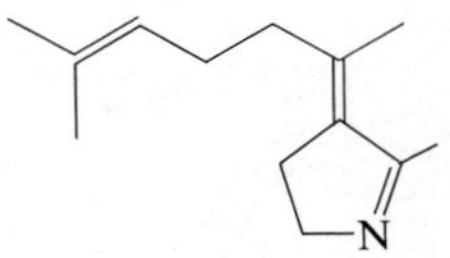 N (74)	218

	Reactant	Conditions	Product(s) and Yield(s) (%)	Refs.
C_{14}	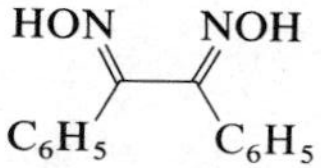 HON, NOH, C_6H_5, C_6H_5	PPA, 120°, 12 min	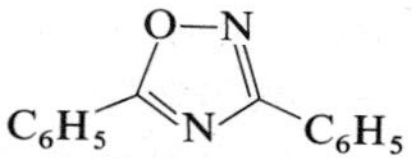 O–N, N, C_6H_5, C_6H_5 (99)	179
	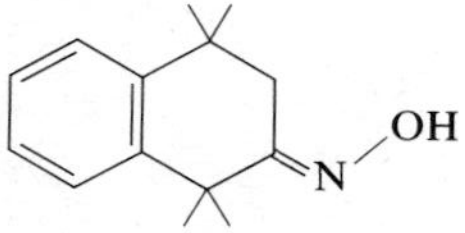 OH, N	PPA, 125–130°, 10 min	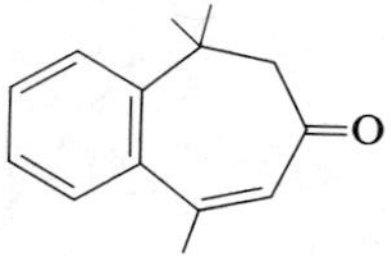 O (72)	420
	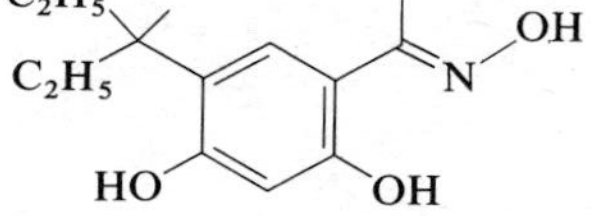 C_2H_5, C_2H_5, OH, N, HO, OH	$POCl_3$, DMAC, CH_3CN, 30°, 30 min	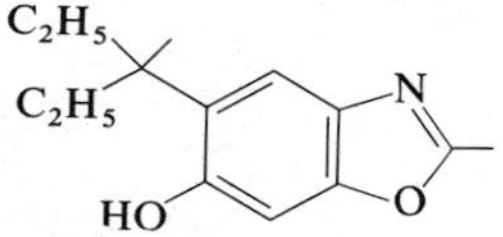 C_2H_5, C_2H_5, N, O, HO (87)	207
	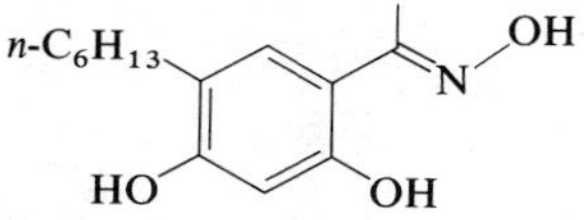 n-C_6H_{13}, OH, N, HO, OH	"	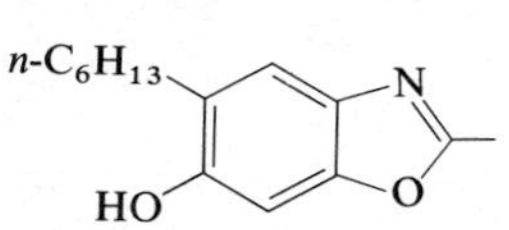 n-C_6H_{13}, N, O, HO (68)	207
C_{15}	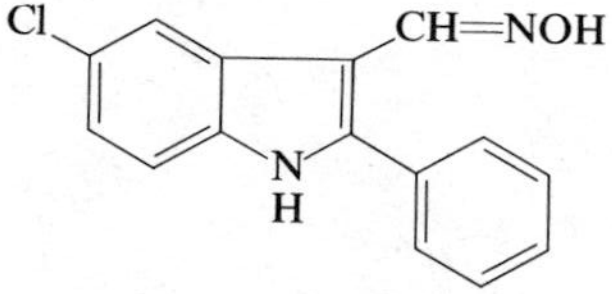 Cl, CH=NOH, N, H	H_2SO_4, ethanol, reflux, 1 h	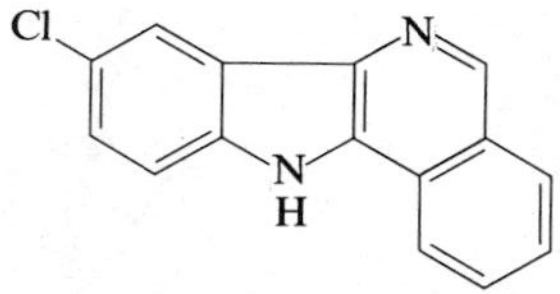 Cl, N, N, H (–)	212
	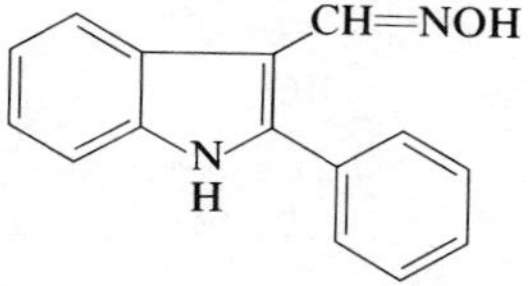 CH=NOH, N, H	"	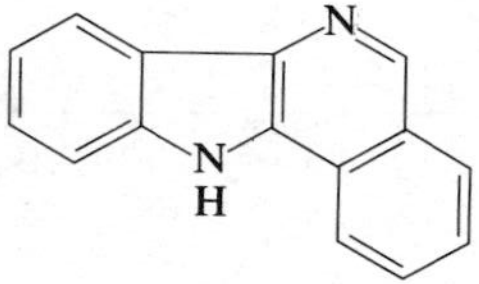 N, N, H (–)	212

TABLE V. REARRANGEMENT-CYCLIZATIONS (*Continued*)

No. of Carbon Atoms	Substrate	Reagent and Conditions	Product(s) and Yield(s) (%)	Refs.
	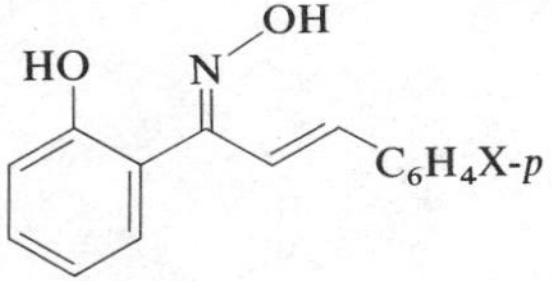 HO, N–OH, C_6H_4X-*p*	$SOCl_2$, ether, 0°	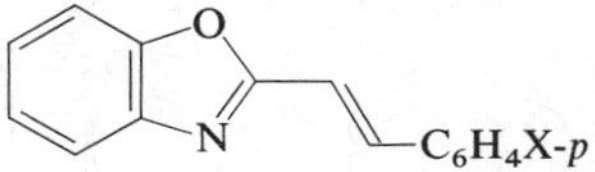 O, N, C_6H_4X-*p*	
	X=Cl		(52)	672
	X=Br		(54)	672
	X=I		(42)	672
	$X=NO_2$		(58)	672
	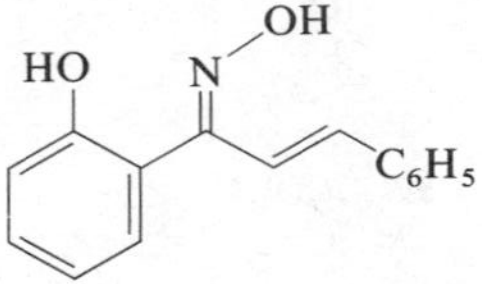 HO, N–OH, C_6H_5	"	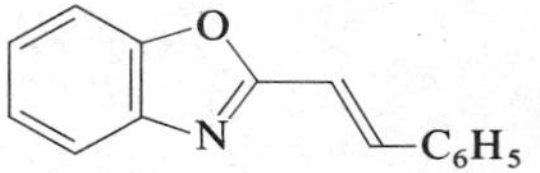 O, N, C_6H_5 (54)	673
	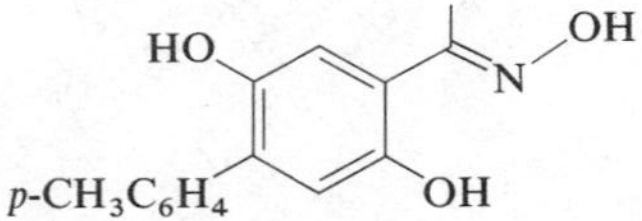 HO, N–OH, OH, *p*-$CH_3C_6H_4$	$POCl_3$, DMAC, CH_3CN, 30 °, 30 min	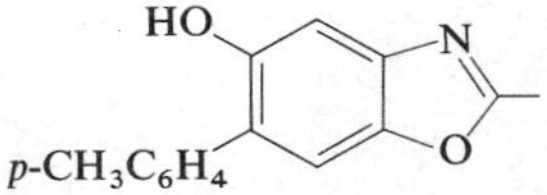 HO, N, O, *p*-$CH_3C_6H_4$ (92)	207
	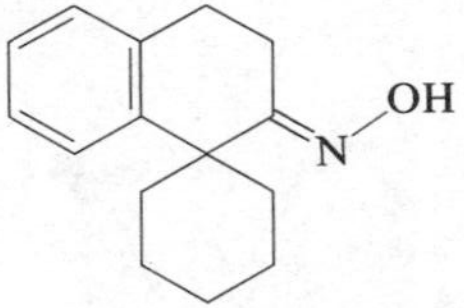 N–OH	PPA, 125–130°, 10 min	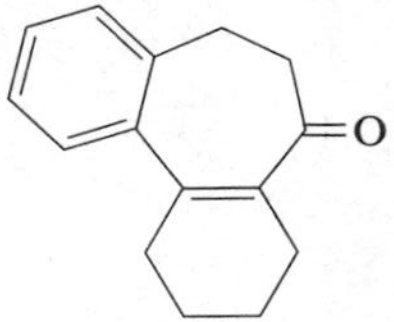 O (71)	420

	Reactant	Conditions	Product(s) and Yield(s) (%)	Refs.
	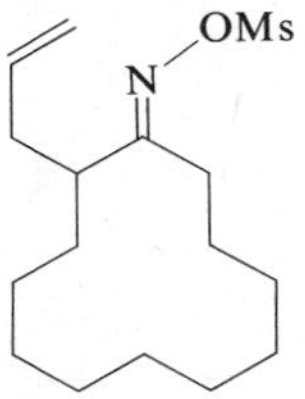 N–OMs	1. $CF_3SO_3Si(CH_3)_3$, $CDCl_3$, 1 h 2. DIBAL	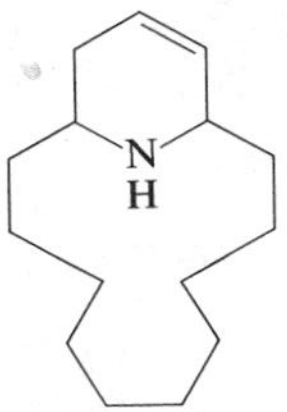 N H (77)	674
C_{16}	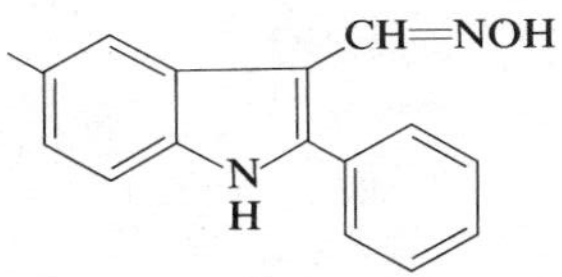 CH=NOH, N H	H_2SO_4, ethanol, reflux, 1 h	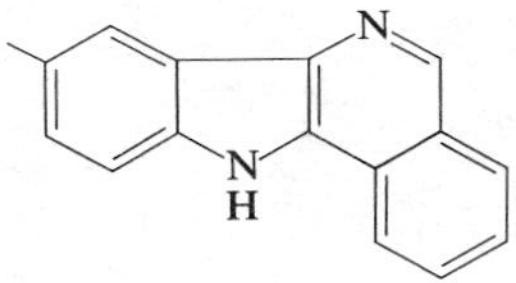 N, N H (–)	212
	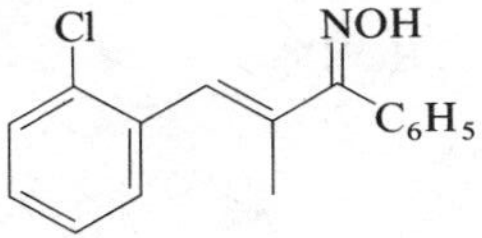 Cl, NOH, C_6H_5	1. PCl_5, decalin, 0° 2. P_2O_5, reflux	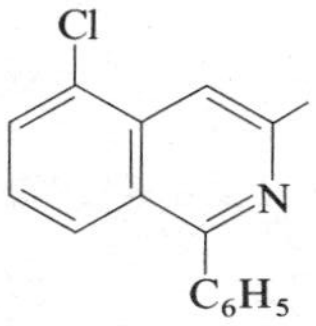 Cl, N, C_6H_5 (63)	214
	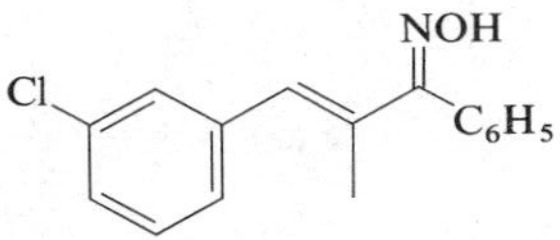 Cl, NOH, C_6H_5	"	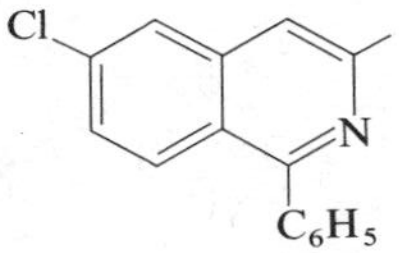 Cl, N, C_6H_5 (45)	214
	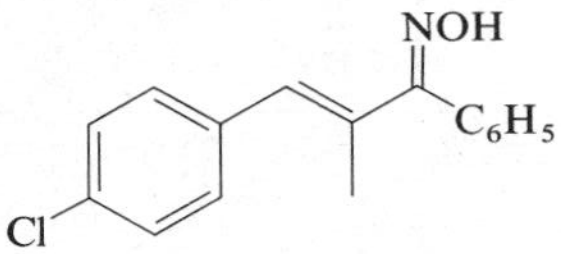 Cl, NOH, C_6H_5	1. PCl_5, decalin, 0° 2. P_2O_5, reflux	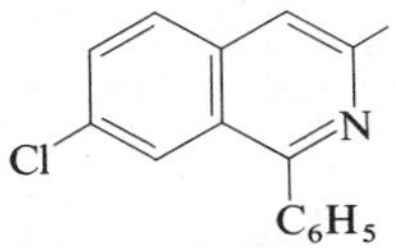 Cl, N, C_6H_5 (30)	214

TABLE V. REARRANGEMENT-CYCLIZATIONS (*Continued*)

No. of Carbon Atoms	Substrate	Reagent and Conditions	Product(s) and Yield(s) (%)	Refs.
	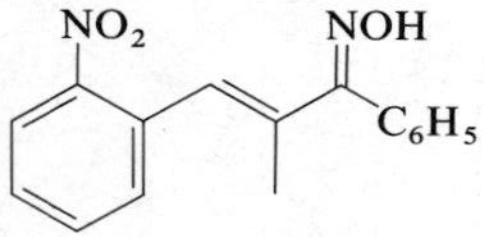 NO_2, NOH, C_6H_5	"	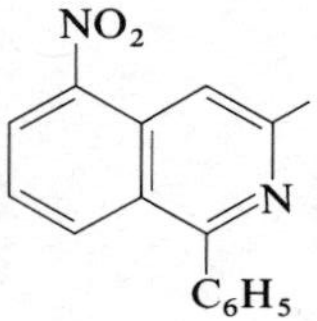 NO_2, N, C_6H_5 (25)	214
	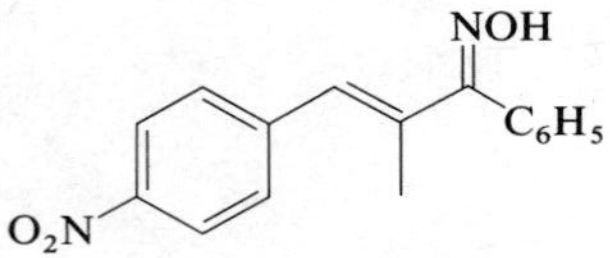 NOH, C_6H_5, O_2N	"	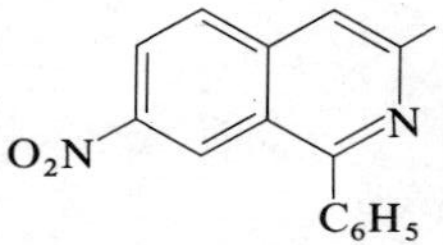 O_2N, N, C_6H_5 (36)	214
	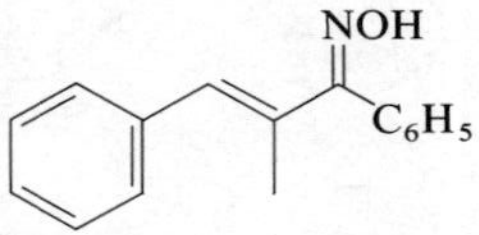 NOH, C_6H_5	"	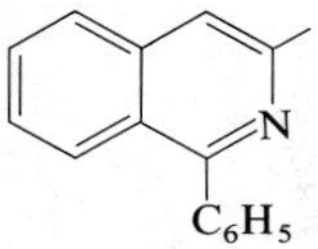 N, C_6H_5 (67)	214
	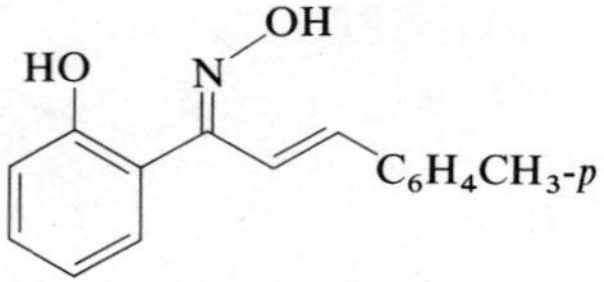 HO, N, OH, $C_6H_4CH_3$-*p*	$SOCl_2$, ether, 0°	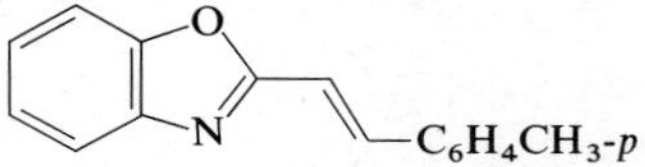 O, N, $C_6H_4CH_3$-*p* (42)	672
	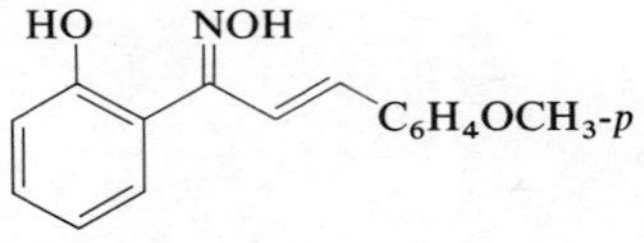 HO, NOH, $C_6H_4OCH_3$-*p*	"	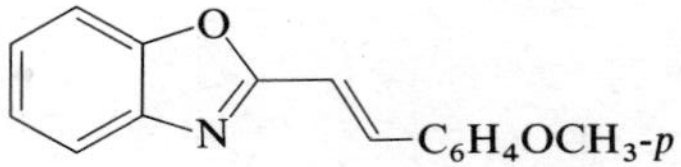 O, N, $C_6H_4OCH_3$-*p* (45)	672

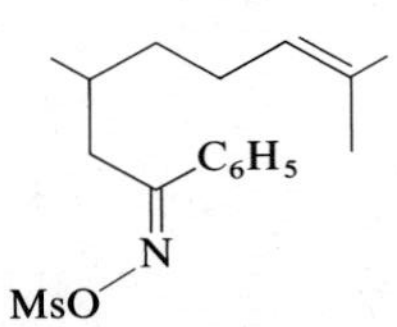	$SnCl_4$, CH_2Cl_2, 0–20°	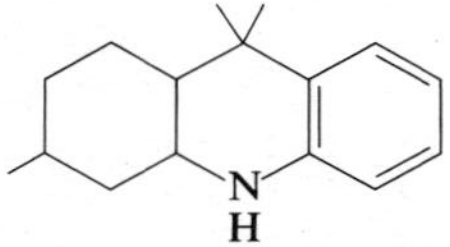 (51)	218
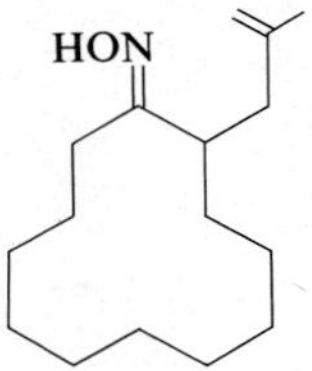	$POCl_3$, pyridine, 80°	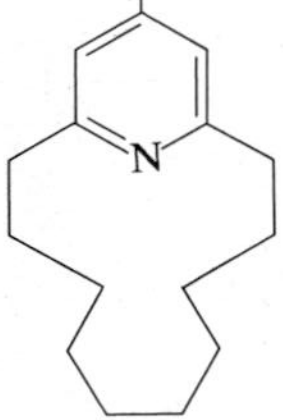 (25)	220
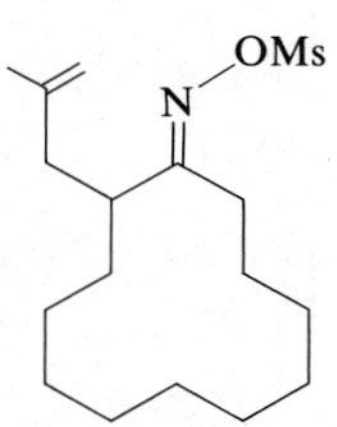	1. $CF_3SO_3Si(CH_3)_3$, $CDCl_3$, 1 h 2. DIBAL	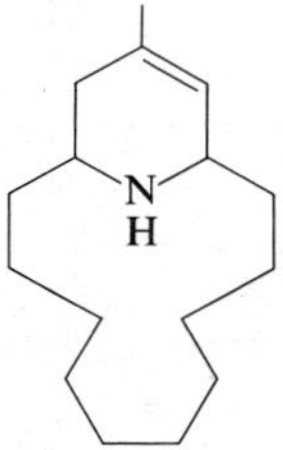 (87)	674
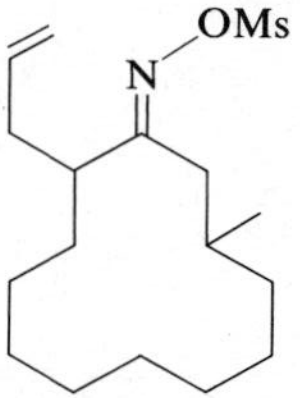	$CF_3SO_3Si(CH_3)_3$	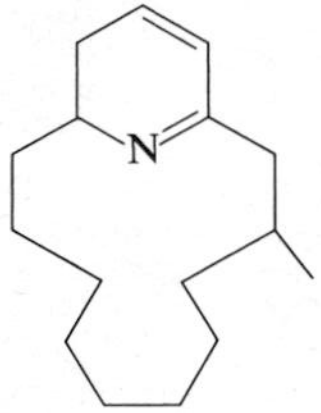 (80)	218

TABLE V. REARRANGEMENT-CYCLIZATIONS (*Continued*)

No. of Carbon Atoms	Substrate	Reagent and Conditions	Product(s) and Yield(s) (%)	Refs.
	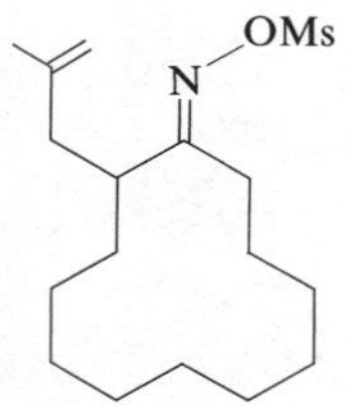 OMs, N	1. $(CH_3)_3Al$, CH_2Cl_2 2. DIBAL	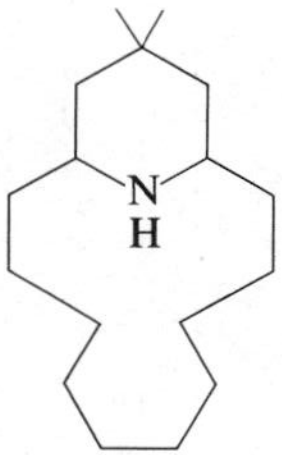 N H (59)	218
C_{17}	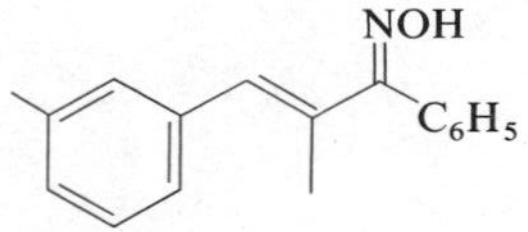 NOH, C_6H_5	1. PCl_5, decalin, 0° 2. P_2O_5, reflux	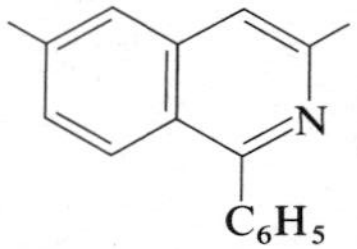 N, C_6H_5 (70)	214
	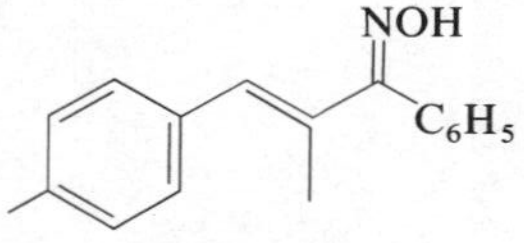 NOH, C_6H_5	"	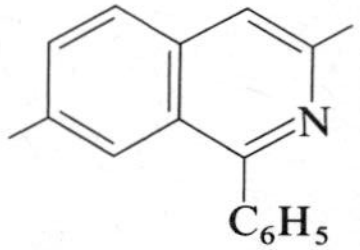 N, C_6H_5 (64)	214
	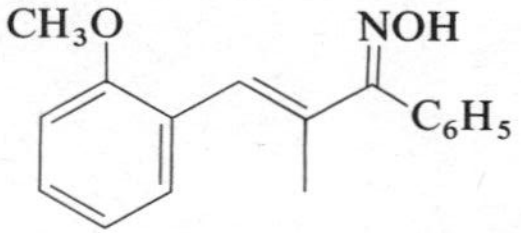 CH_3O, NOH, C_6H_5	"	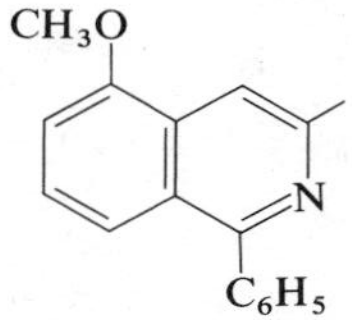 CH_3O, N, C_6H_5 (46)	214

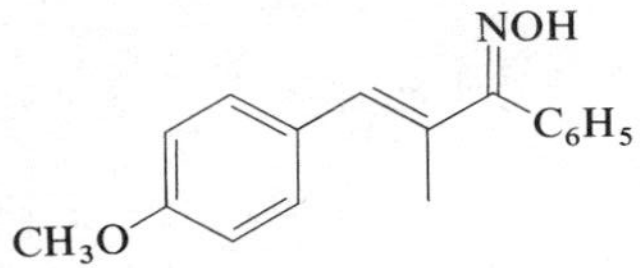

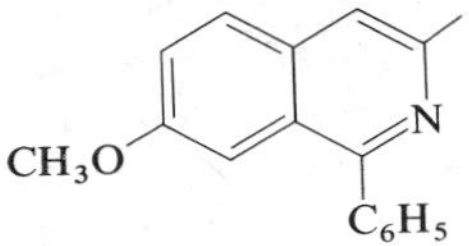

	Reactant	Conditions	Product(s) (Yield %)	Refs.
	CH_3O; NOH; C_6H_5	"	CH_3O; N; C_6H_5 (4)	214
	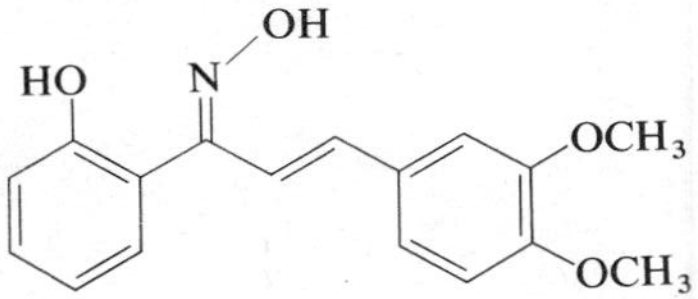 HO; N; OH; OCH_3; OCH_3	$SOCl_2$, ether, 0° 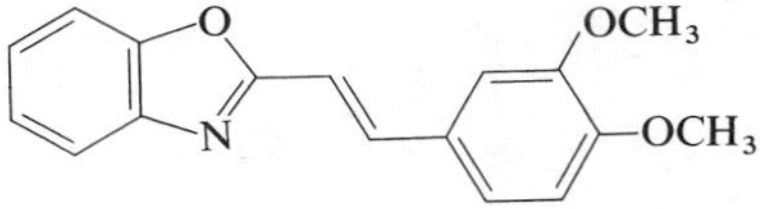	O; N; OCH_3; OCH_3 (36)	672
	OH; N; OH; $C_6H_4OC_2H_5$-*p*	"	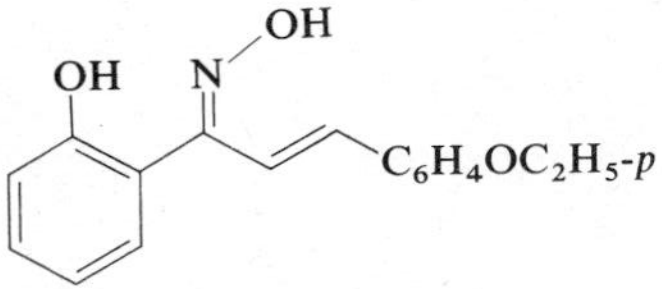 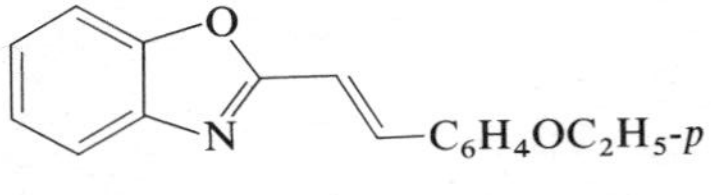 O; N; $C_6H_4OC_2H_5$-*p* (40)	672
C_{18} 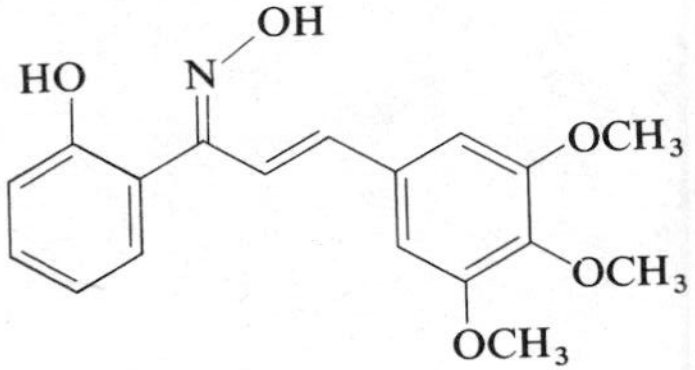 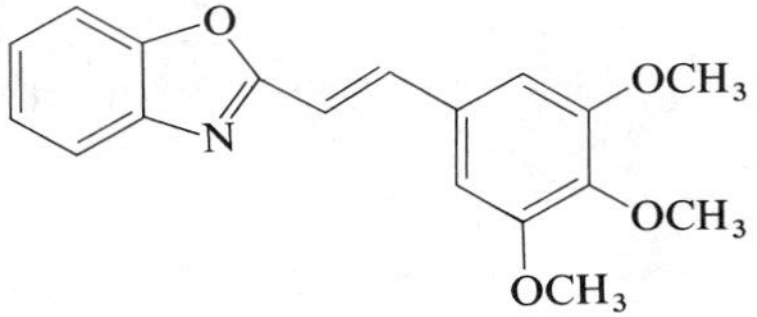	HO; N; OH; OCH_3; OCH_3; OCH_3	"	O; N; OCH_3; OCH_3; OCH_3 (–)	672
	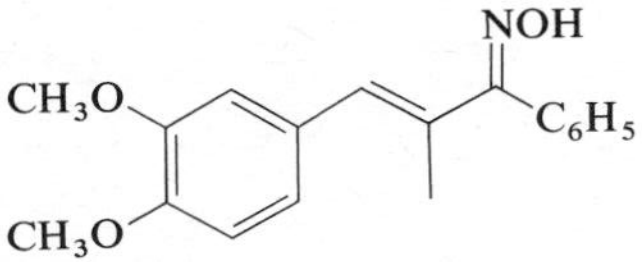 CH_3O; CH_3O; NOH; C_6H_5	1. PCl_5, decalin, 0° 2. P_2O_5, reflux 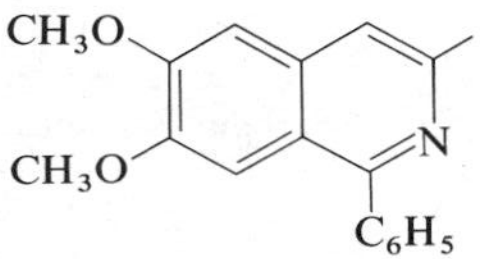	CH_3O; CH_3O; N; C_6H_5 (7)	214

TABLE V. REARRANGEMENT-CYCLIZATIONS (*Continued*)

No. of Carbon Atoms	Substrate	Reagent and Conditions	Product(s) and Yield(s) (%)	Refs.
	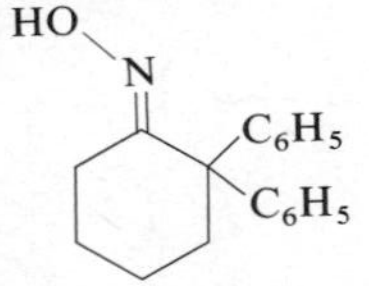 HO–N=; C_6H_5; C_6H_5	PPA, 120–125°, 10 min	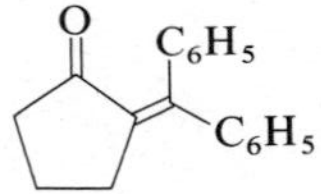 O; C_6H_5; C_6H_5 (15)	165
		1. PPA, 120–125°, 10 min 2. Neutralize 3. HCl	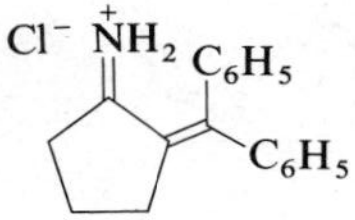 Cl^- $\overset{+}{N}H_2$; C_6H_5; C_6H_5 (87)	165
C_{19}	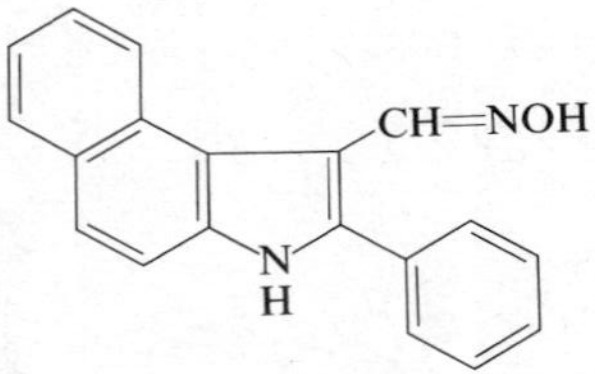 CH=NOH; N; H	H_2SO_4, ethanol, reflux, 1 h	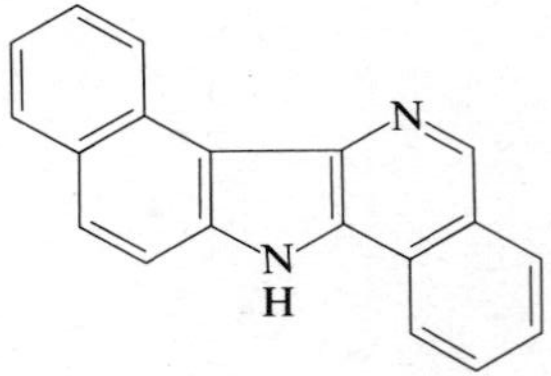 N; N; H (–)	212
	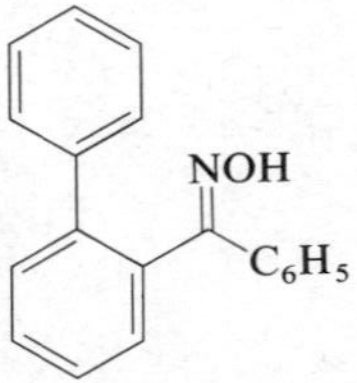 NOH; C_6H_5	P_2O_5, $POCl_3$, SO_2, 70°, 3 h	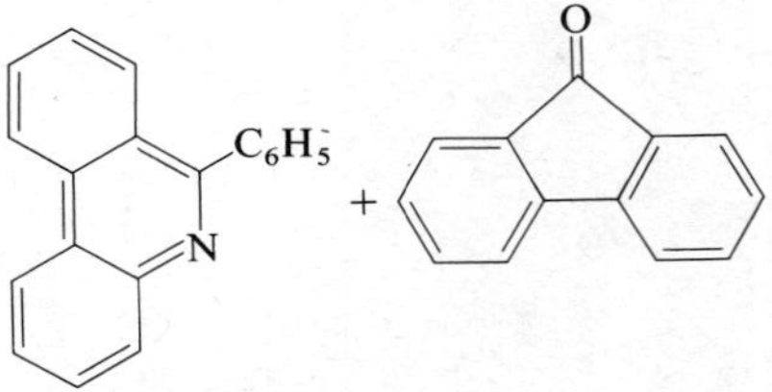 C_6H_5; N (29) + O (75)	215

	Cl; NH; CH_3O_2C; CO_2CH_3; N; $OP^+(C_6H_5)_3Cl^-$; C_6H_5	$(C_6H_5)_3P$, CCl_4	CH_3O_2C; CO_2CH_3; N; C_6H_5; N; Cl + N; CO_2CH_3; CO_2CH_3; NHC_6H_5; Cl (—)	675
	n-$C_{13}H_{27}$; N; OMs	1. $(C_2H_5)_2AlCl$, CH_2Cl_2 2. DIBAL	n-$C_{13}H_{27}$; N; H (53)	218
C_{24}	n-$C_{16}H_{33}O$; OH; N; OH	$POCl_3$, DMAC, 30°, 30 min	n-$C_{16}H_{13}O$; N; O (76)	207
C_{29}	N; H; C_6H_5; N; $N(C_6H_5)_2$; N; H; O	PPA, 85°, 15 min	C_6H_5; N; N; N; H; O (80)	209
C_{35}	$C_6H_5CH_2$; H; N; C_6H_5; N; $N(C_6H_5)_2$; N; H; O	PPA, 65°, 10 min	C_6H_5; N; N; $CH_2C_6H_5$; N; H; O (52)	209

REFERENCES

[1] E. Beckmann, *Chem. Ber.*, **89**, 988 (1886).

[2] A. H. Blatt, *Chem. Rev.*, **12**, 215 (1933).

[3] B. Jones, *Chem. Rev.*, **35**, 335 (1944).

[4] F. Moller, in *Methoden der Organischen Chemie*, E. Muller, Ed., Thieme Verlag, Stuttgart, 1957, Vol. 11, Part 1, p. 892.

[5] L. G. Donaruma and W. Z. Heldt, *Org. React.*, **11**, 1 (1960).

[6] A. L. J. Beckwith, in *The Chemistry of Amides*, J. Zabicky, Ed., Interscience, New York, 1970, p. 131.

[7] C. G. McCarty, in *The Chemistry of the Carbon–Nitrogen Double Bond*, S. Patai, Ed., Interscience, New York, 1970, p. 408.

[8] P. A. S. Smith, in *Molecular Rearrangements*, P. De Mayo, Ed., Interscience, New York, 1963, p. 457.

[9] H. Mukamal, *Nuova Chim.*, **47**, 79 (1971) [*C.A.*, **75**, 75559w (1971)].

[10] G. Hornke, H. Krauch and W. Kunz, *Chem.-Ztg.*, **89**, 525 (1965).

[11] O. Wallach, *Justus Liebigs Ann. Chem.*, **309**, 1 (1889).

[12] R. T. Conley and S. Ghosh, *Mech. Mol. Migr.*, **4**, 197 (1971).

[13] C. Goldschmidt, *Chem. Ber.*, **28**, 818 (1895).

[14] O. Wallach, *Justus Liebigs Ann. Chem.*, **319**, 1901 (1877).

[15] W. H. Perkin, Jr., *J. Chem. Soc.*, **1890**, 204.

[16] J. H. Armin and P. De Mayo, *Tetrahedron Lett.*, **1963**, 1585.

[17] A. Padwa, *Chem. Rev.*, **77**, 37 (1977).

[18] J. E. Blackwood, C. L. Gladys, K. L. Loening, A. E, Petrarca, and J. E. Rush, *J. Am. Chem. Soc.*, **90**, 509 (1968).

[19] J. W. Schulenberg and S. Archer, *Org. React.*, **14**, 1 (1965).

[20] B. J. Gregory, R. B. Moodie, and K. Schofield, *Chem. Commun.*, **1968**, 1380.

[21] D. E. Pearson, J. F. Baxter, and J. C. Martin, *J. Org. Chem.*, **17**, 1511 (1952).

[22] R. E. Gawley, *J. Org. Chem.*, **46**, 4595 (1981).

[23] B. J. Gregory, R. B. Moodie, and K. Schofield, *J. Chem. Soc. D*, **1969**, 645.

[24] Y. Yukawa and T. Ando, *J. Chem. Soc. D*, **1971**, 1601.

[25] W. Z. Heldt, *J. Org. Chem.*, **26**, 1695 (1961).

[26] L. I. Krimen and D. J. Cota, *Org. React.*, **17**, 213 (1969).

[27] A. I. Meyers, J. Schneller, and N. K. Ralhan, *J. Org. Chem.*, **28**, 2944 (1963).

[28] C. A. Grob, H. P. Fischer, and W. Raudenbusch, *Helv. Chim. Acta*, **47**, 1003 (1964).

[29] K. K. Kelly and J. S. Matthews, *J. Org. Chem.*, **36**, 2159 (1971).

[30] A. C. Huitric and S. D. Nelson, Jr., *J. Org. Chem.*, **34**, 1230 (1969).

[31] T. Sato, H. Wakatsuka, and K. Amano, *Tetrahedron*, **27**, 5381 (1971).

[32] J. M. Riego, A. Costa, P. Deya, J. V. Sinisterra, and J. M. Marinas, *React. Kinet. Catal. Lett.*, **19**, 61 (1982).

[33] H. P. Fischer, *Tetrahedron Lett.*, **1968**, 285.

[34] I. I. Kukhtenko, *Zh. Org. Khim.*, **7**, 333 (1971); *J. Org. Chem. USSR* (*Engl. transl.*), **7**, 327 (1971).

[35] A. Maquestiau, Y. Van Haverbeke, C. De Meyer, C. Duthoit, P. Meyrant, and R. Flammang, *Nouv. J. Chim.*, **3**, 517 (1979).

[36] A. Maquestiau, Y. Van Haverbeke, R. Flammang, and P. Meyrant, *Org. Mass Spectrom.*, **15**, 80 (1980).

[37] G. R. Krow and S. Szczepanski, *Tetrahedron Lett.*, **21**, 4593 (1980).

[38] G. R. Krow and S. Szczepanski, *J. Org. Chem.*, **47**, 1153 (1982).

[39] N. Wakabayashi, R. M. Waters, and M. W. Law, *Org. Prep. Proced. Int.*, **6**, 203 (1974).

[40] K. Oka and S. Hara, *J. Org. Chem.*, **43**, 3790 (1978).

[41] C. W. Shoppee, G. Kruger, and R. N. Mirrington, *J. Chem. Soc.*, **1962**, 1050.

[42] J. Kondelikova, J. Kralicek, and V. Kubanek, *Collect. Czech. Chem. Commun.*, **38**, 3773 (1973).

[43] R. H. Mazur, *J. Org. Chem.*, **28**, 248 (1963).

[44] C. W. Shoppee, R. Lack, R. N. Mirrington, and L. R. Smith, *J. Chem. Soc.*, **1965**, 5868.

[45] R. H. Mazur, *J. Org. Chem.*, **26**, 1289 (1961).
[46] G. I. Hutchison, R. H. Prager, and A. D. Ward, *Aust. J. Chem.*, **33**, 2477 (1980).
[47] Y. Tamura, H. Fujiwara, K. Sumoto, M. Ikeda, and Y. Kita, *Synthesis*, **1973**, 215.
[48] C. A. Grob and P. Wenk, *Tetrahedron Lett.*, **1976**, 4191.
[49] C. A. Grob and P. Wenk, *Tetrahedron Lett.*, **1976**, 4195.
[50] J. Klinot and A. Vystrcil, *Collect. Czech. Chem. Commun.*, **27**, 377 (1962).
[51] G. P. Moss and S. A. Nicolaidis, *J. Chem. Soc. D*, **1969**, 1077.
[52] A. Savva and V. F. Martynov. *Zh. Obshch. Khim.*, **44**, 1655 (1974); *J. Gen. Chem. USSR* (*Engl. transl.*), **44**, 1628 (1974).
[53] A. Savva and T. E. Ryzhkina, *Zh. Org. Khim.*, **10**, 1997 (1974); *J. Org. Chem. USSR* (*Engl. transl.*), **10**, 2013 (1974).
[54] I. Seki, T. Shindo, A. Naito, and K. Nakagawa, *Annu. Rep. Sankyo Res. Lab.*, **28**, 53 (1976) [*C.A.*, **87**, 608a].
[55] T. Sasaki, S. Eguchi, and O. Hiroaki, *J. Org. Chem.*, **41**, 1803 (1976).
[56] H. P. Fischer and C. A. Grob, *Tetrahedron Lett.*, **26**, 22 (1960).
[57] F. Kohen, *Chem. Ind.* (*London*), **1966**, 1378.
[58] J. P. Freeman, *J. Org. Chem.*, **26**, 3507 (1961).
[59] D. E. Pearson and R. M. Stone, *J. Am. Chem. Soc.*, **83**, 1715 (1961).
[60] A. Guy, J. P. Guette, and G. Lang, *Synthesis*, **1980**, 222.
[61] G. Y. Stepanova, V. M. Dikolenko, E. I. Dikolenko, and F. Jethwa, *Zh. Org. Khim.*, **10**, 1455 (1974); *J. Org. Chem. USSR* (*Engl. transl.*), **10**, 1464 (1974).
[62] N. Thoai and J. Wiemann, *Bull. Soc. Chim. Fr.*, **1965**, 2474.
[63] R. T. Conley and T. M. Tencza, *Tetrahedron Lett.*, **1963**, 1781.
[64] R. T. Conley, *J. Org. Chem.*, **28**, 278 (1963).
[65] A. Hassner, W. A. Wentworth, and J. H. Pomerantz, *J. Org. Chem.*, **28**, 304 (1963).
[66] R. K. Hill, R. T. Conley, and O. T. Chortyk, *J. Am. Chem. Soc.*, **87**, 5646 (1965).
[67] R. K. Hill and O. T. Chortyk, *J. Am. Chem. Soc.*, **84**, 1064 (1962).
[68] P. T. Lansbury and J. G. Colson, *J. Am. Chem. Soc.*, **84**, 4167 (1962).
[69] P. T. Lansbury, J. G. Colson, and N. R. Mancuso, *J. Am. Chem. Soc.*, **86**, 5225 (1964).
[70] P. T. Lansbury and N. R. Mancuso, *Tetrahedron Lett.*, **1965**, 2445.
[71] P. T. Lansbury and N. R. Mancuso, *J. Am. Chem. Soc.*, **88**, 1205 (1966).
[72] T. Imamoto, H. Yokoyama, and M. Yokoyama, *Tetrahedron Lett.*, **22**, 1803 (1981).
[73] P. E. Eaton, G. R. Carlson, and J. T. Lee, *J. Org. Chem.*, **38**, 4071 (1973).
[74] B. M. Regan and F. N. Hayes, *J. Am. Chem. Soc.*, **78**, 639 (1956).
[75] R. N. Butler and D. A. O'Donoghue, *J. Chem. Res.* (*S*), **1983**, 18.
[76] M. Gates and S. P. Malchick, *J. Am. Chem. Soc.*, **79**, 5546 (1957).
[77] J. C. Craig and A. R. Naik, *J. Am. Chem. Soc.*, **84**, 3410 (1962).
[78] R. T. Conley and M. C. Annis, *J. Org. Chem.*, **27**, 1961 (1962).
[79] W. Eisele, C. A. Grob, E. Renk, and H. Von Tschammer, *Helv. Chim. Acta*, **51**, 816 (1968).
[80] C. A. Grob and H. Von Tschammer, *Helv. Chim. Acta*, **51**, 1083 (1968).
[81] G. A. Olah and A. P. Fung, *Synthesis*, **1979**, 537.
[82] W. Bartmann, G. Beck, J. Knolle, and R. H. Rupp, *Tetrahedron Lett.*, **23**, 3647 (1982).
[83] K. D. Kopple and J. J. Katz, *J. Org. Chem.*, **24**, 1975 (1959).
[84] R. M. Waters, N. Wakabayashi, and E. S. Fields, *Org. Prep. Proced. Int.*, **6**, 53 (1974).
[85] M. E. Jung and Z. Long-Mei, *Tetrahedron Lett.*, **24**, 4533 (1983).
[86] Y. Ishida, S. Sasatani, K. Maruoka, and H. Yamamoto, *Tetrahedron Lett.*, **24**, 3255 (1983).
[87] F. I. Luknitskii and B. A. Vovsi, *Dokl. Akad. Nauk SSSR*, **182**, 350 (1968); *Proc. Acad. Sci. USSR, Chem. Sec.* (*Engl. transl.*), **182**, 812 (1968).
[88] A. Holm, C. Christophersen, T. Ottersen, H. Hope, and A. Christensen, *Acta Chem. Scand., Ser. B*, **31**, 687 (1977).
[89] Y. Iwakura, K. Uno, K. Haga, and K. Nakamura, *J. Polym. Sci., Polym. Chem. Ed.*, **11**, 367 (1973).
[90] M. Rothe and R. Timler, *Chem. Ber.*, **95**, 783 (1962).
[91] G. I. Glover, R. B. Smith, and H. Rapoport, *J. Am. Chem. Soc.*, **87**, 2003 (1965).

[92] S. Bittner and S. Grinberg, *J. Chem. Soc., Perkin Trans. 1*, **1976**, 1708.
[93] H. Langhals and C. Ruechardt, *Chem. Ber.*, **114**, 3831 (1981).
[94] H. Langhals, G. Range, E. Wistuba, and C. Ruechardt, *Chem. Ber.*, **114**, 3818 (1981).
[95] H. P. Fischer, C. A. Grob, and E. Renk, *Helv. Chim. Acta*, **45**, 2539 (1962).
[96] R. Huisgen, J. Witte, and I. Ugi, *Chem. Ber.*, **90**, 1844 (1957).
[97] R. Huisgen, J. Witte, and W. Jira, *Chem. Ber.*, **90**, 1850 (1957).
[98] N. Weliky and E. S. Gould, *J. Am. Chem. Soc.*, **79**, 2742 (1957).
[99] H. Patin, *Tetrahedron Lett.*, **1974**, 2893.
[100] V. I. Boev and A. V. Dombrovskii, *Zh. Org. Khim.*, **13**, 1125 (1977); *J. Org. Chem. USSR* (*Engl. transl.*), **13**, 1035 (1977).
[101] E. Cuingnet and M. Adalberon, *C. R. Hebd. Seances Acad. Sci.*, **257**, 713 (1963).
[102] E. Cuingnet and M. Tarterat-Adalberon, *Bull. Soc. Chim. Fr.*, **1965**, 3728.
[103] S. Z. Abbas and R. C. Poller, *J. Organomet. Chem.*, **104**, 187 (1976).
[104] S. Z. Abbas and R. C. Poller, *J. Organomet. Chem.*, **55**, C9 (1973).
[105] L. I. Zakharkin, V. N. Kalinin, and V. V. Gedymin, *Tetrahedron*, **27**, 1317 (1971).
[106] L. I. Zakharkin and I. V. Pisareva, *Izv. Akad. Nauk SSSR, Ser. Khim.*, **1979**, 1886; *Bull. Acad. Sci. USSR, Div. Chem. Sci.* (*Engl. transl.*), **1979**, 1750.
[107] L. M. Jackman, R. L. Webb, and H. C. Yick, *J. Org. Chem.*, **47**, 1824 (1982).
[108] D. N. Chaturvedi and R. A. Kulkarni, *J. Indian Chem. Soc.*, **55**, 161 (1978).
[109] A. Costa, R. Mestres, and J. M. Riego, *Synth. Commun.*, **12**, 1003 (1982).
[110] H. B. Kagan, N. Langlois and T. P. Dang, *J. Organomet. Chem.*, **90**, 353 (1975).
[111] W. Zielinski, *Pol. J. Chem.*, **52**, 2233 (1978).
[112] S. Goszczynski, D. Rusinska-Roszak, and M. Lozynski, *Pol. J. Chem.*, **53**, 849 (1979).
[113] B. Unterhalt and H. J. Reinhold, *Arch. Pharm.* (*Weinheim, Ger.*), **308**, 346 (1975).
[114] F. L. Scott, R. J. MacConaill, and J. C. Riordan, *J. Chem. Soc. C*, **1967**, 44.
[115] J. S. Dixon, I. Midgley, and C. Djerassi, *J. Am. Chem. Soc.*, **99**, 3432 (1977).
[116] Z. Hamlet and M. Rampersad, *J. Chem. Soc. D*, **1970**, 1230.
[117] T. E. Stevens, *J. Org. Chem.*, **28**, 2436 (1963).
[118] E. A. Abrazhanova and L. F. Pronskii, *Zh. Org. Khim.*, **9**, 780 (1973); *J. Org. Chem. USSR* (*Engl. transl.*), **9**, 803 (1973).
[119] M. S. Ahmad, Shafiullah, and M. Mushfiq, *Tetrahedron Lett.*, **1970**, 2739.
[120] R. N. Loeppley and M. Rotman, *J. Org. Chem.*, **32**, 4010 (1967).
[121] T. E. Stevens, *J. Org. Chem.*, **34**, 2451 (1969).
[122] E. O. Beckmann, *Chem. Ber.*, **23**, 3331 (1890).
[123] G. Zinner, *Chem. -Ztg.*, **102**, 58 (1978).
[124] D. H. R. Barton, M. J. Day, R. H. Hesse, and M. M. Pechet, *J. Chem. Soc. D*, **1971**, 945.
[125] D. H. R. Barton, M. J. Day, R. H. Hesse, and M. M. Pechet, *J. Chem. Soc., Perkin Trans. 2*, **1975**, 1764.
[126] P. W. Jeffs, G. Molina, N. A. Cortese, P. R. Hauck, and J. Wolfram, *J. Org. Chem.*, **47**, 3876 (1982).
[127] P. W. Jeffs and G. Molina, *J. Chem. Soc., Chem. Commun.*, **1973**, 3.
[128] J. B. Chattopadhyaya and A. V. Rama Rao, *Tetrahedron*, **30**, 2899 (1974).
[129] J. P. De Keersmaeker and F. Fontyn, *Ind. Chim. Belge*, **32**, 1087 (1967).
[130] G. N. Dorofeenko and Y. I. Ryabukhin, *Zh. Obshch. Khim.*, **48**, 1668 (1978); *J. Gen. Chem. USSR* (*Engl. transl.*), **48**, 1528 (1978).
[131] Y. A. Zhdanov, Y. E. Alekseev, and T. P. Sudareva, *Dokl. Akad. Nauk SSSR*, **238**, 580 (1978); *Proc. Acad. Sci. USSR, Chem. Sec.* (*Engl. transl.*), **238**, 35 (1978).
[132] A. J. Leusink, T. G. Meerbeek, and J. G. Noltes, *Recl. Trav. Chim. Pays-Bas*, **95**, 123 (1976).
[133] A. J. Leusink, T. G. Meerbeek, and J. G. Noltes, *Recl. Trav. Chim. Pays-Bas*, **96**, 142 (1977).
[134] P. S. Clezy, C. L. Lim, and J. S. Shannon, *Aust. J. Chem.*, **27**, 1103 (1974).
[135] E. P. Ivakhnenko, V. P. Panov, and O. Y. Okhlobystin, *Dokl. Akad. Nauk SSSR*, **243**, 659 (1978); *Proc. Acad. Sci. USSR, Chem. Sec.* (*Engl. transl.*), **243**, 547 (1978).

[136] E. M. Burgess, H. R. Penton, Jr., and E. A. Taylor, *J. Org. Chem.*, **38**, 26 (1973).

[137] T. Mukaiyama, K. Tonooka, and K. Inoue, *J. Org. Chem.*, **26**, 2202 (1961).

[138] S. Huenig and W. Rehder, *Angew. Chem., Int. Ed. Engl.*, **7**, 304 (1968).

[139] M. P. Cava and Q. A. Ahmed, *J. Org. Chem.*, **33**, 2440 (1968).

[140] W. Z. Heldt, *J. Am. Chem. Soc.*, **80**, 5880 (1958).

[141] B. Greenberg and J. G. Aston, *J. Org. Chem.*, **25**, 1894 (1960).

[142] R. H. Williams and H. R. Snyder, *J. Org. Chem.*, **38**, 809 (1973).

[143] W. Zielinski, *Pol. J. Chem.*, **54**, 745 (1980).

[144] R. Appel and K. Warning, *Chem. Ber.*, **108**, 1437 (1975).

[145] C. R. Harrison, P. Hodge, and W. J. Rogers, *Synthesis*, **1977**, 41.

[146] H. G. Foley and D. R. Dalton, *Synth. Commun.*, **4**, 251 (1974).

[147] K. Maruoka, T. Miyazaki, M. Ando, Y. Matsumura, S. Sakane, K. Hattori, and H. Yamamoto, *J. Am. Chem. Soc.*, **105**, 2831 (1983).

[148] G. A. Smith and R. E. Gawley, unpublished results.

[149] M. Harfenist and E. Magnien, *J. Am. Chem. Soc.*, **80**, 6080 (1958).

[150] M. N. Rerick, C. H. Trottier, R. A. Daignault, and J. D. De Foe, *Tetrahedron Lett.*, **1963**, 629.

[151] S. H. Graham and A. J. S. Williams, *Tetrahedron*, **21**, 3263 (1965).

[152] Y. Girault, M. Decouzon, and M. Azzaro, *Bull. Soc. Chim. Fr.*, **1975**, 385.

[153] K. Hattori, Y. Matsumura, T. Miyazaki, K. Maruoka, and H. Yamamoto, *J. Am. Chem. Soc.*, **103**, 7368 (1981).

[154] S. Sasatani, T. Miyazaki, K. Maruoka, and H. Yamamoto, *Tetrahedron Lett.*, **24**, 4711 (1983).

[155] R. H. Poirier, R. D. Morin, R. W. Pfeil, A. E. Bearse, D. N. Kramer, and F. M. Miller, *J. Org. Chem.*, **27**, 1547 (1962).

[156] H. K. Hall, *J. Org. Chem.*, **29**, 3139 (1964).

[157] K. Hattori, K. Maruoka, and H. Yamamoto, *Tetrahedron Lett.*, **23**, 3395 (1982).

[158] Y. Matsumura, K. Maruoka, and H. Yamamoto, *Tetrahedron Lett.*, **23**, 1929 (1982).

[159] Y. Matsumura, J. Fujiwara, K. Maruoka, and H. Yamamoto, *J. Am. Chem. Soc.*, **105**, 6312 (1983).

[160] J. A. Marshall, N. H. Andersen, and J. W. Schlicher, *J. Org. Chem.*, **35**, 858 (1970).

[161] D. Seebach, M. Pohmakotr, C. Schregenberger, B. Weidmann, R. S. Mali, and S. Pohmakotr, *Helv. Chim. Acta*, **65**, 419 (1982).

[162] R. M. Palmere, R. T. Conley, and J. L. Rabinowitz, *J. Org. Chem.*, **37**, 4095 (1972).

[163] J. C. Chapman and J. T. Pinhey, *Aust. J. Chem.*, **27**, 2421 (1974).

[164] C. W. Shoppee, R. E. Lack, and S. K. Roy, *J. Chem. Soc.*, **1963**, 3767.

[165] R. T. Conley and B. E. Nowak, *J. Org. Chem.*, **27**, 1965 (1962).

[166] B. Amit and A. Hassner, *Synthesis*, **1978**, 932.

[167] C. H. Brieskorn and E. Hemmer, *Arch. Pharm.* (*Weinheim, Ger.*), **310**, 65 (1977).

[168] N. G. Kozlov and T. Pehk, *Zh. Org. Khim.*, **18**, 1118 (1982); *J. Org. Chem. USSR* (*Engl. transl.*), **18**, 968 (1982).

[169] A. S. Narula and S. P. Sethi, *Tetrahedron Lett.*, **25**, 685 (1984).

[170] G. E. Gream, D. Wege, and M. Mular, *Aust. J. Chem.*, **27**, 567 (1974).

[171] R. V. Stevens and F. C. A. Gaeta, *J. Am. Chem. Soc.*, **99**, 6105 (1977).

[172] R. V. Stevens, F. C. A. Gaeta, and D. S. Lawrence, *J. Am. Chem. Soc.*, **105**, 7713 (1983).

[173] M. Nazir, Naeemuddin, I. Ahmed, M. K. Bhatty, and Karimullah, *Pak. J. Sci. Ind. Res.*, **10**, 13 (1967).

[174] R. J. Hunadi and G. K. Helmkamp, *J. Org. Chem.*, **46**, 2880 (1981).

[175] M. Ikeda, T. Uno, K. Homma, K. Ono, and Y. Tamura, *Synth. Commun.*, **10**, 437 (1980).

[176] G. Frater, U. Mueller, and W. Guenther, *Tetrahedron Lett.*, **25**, 1133 (1984).

[177] C. W. Shoppee and S. K. Roy, *J. Chem. Soc.*, **1963**, 3774.

[178] J. N. Shah, Y. P. Mehta, and G. M. Shah, *J. Org. Chem.*, **43**, 2078 (1978).

[179] R. T. Conley and F. A. Mikulski, *J. Org. Chem.*, **24**, 97 (1959).

[180] D. Miljkovic, J. Petrovic, M. Stajic, and M. Miljkovic, *J. Org. Chem.*, **38**, 3585 (1973).

[181] D. Miljkovic and J. Petrovic, *J. Org. Chem.*, **42**, 2101 (1977).

[182] T. Ibuka, Y. Mitsui, K. Hayashi, H. Minakata, and Y. Inubushi, *Tetrahedron Lett.*, **22**, 4425 (1981).

[183] T. Ibuka, H. Minakata, Y. Mitsui, K. Hayashi, T. Taga, and Y. Inubushi, *Chem. Pharm, Bull.*, **30**, 2840 (1982).

[184] H. Suginome and C.-M. Shea, *Synthesis*, **1980**, 229.

[185] M. Boes and W. Fleischhacker, *Justus Liebigs Ann. Chem.*, **1982**, 112.

[186] E. Rihova and A. Vystrcil, *Collect. Czech. Chem. Commun.*, **34**, 240 (1969).

[187] R. K. Hill, *J. Org. Chem.*, **27**, 29 (1962).

[188] M. M. Rogic, J. F. Van Peppen, K. P. Klein, and T. R. Demmin, *J. Org. Chem.*, **39**, 3424 (1974).

[189] K. P. Klein, T. R. Demmin, B. C. Oxenrider, M. M. Rogic, and M. T. Tetenbaum, *J. Org. Chem.*, **44**, 275 (1979).

[190] C. A. Grob, H. P. Fischer, H. Link, and E. Renk, *Helv. Chim. Acta*, **46**, 1190 (1963).

[191] G. Hugel, B. Gourdier, J. Levy, and J. Le Men, *Tetrahedron Lett.*, **1974**, 1597.

[192] R. L. Autrey and P. W. Scullard, *J. Am. Chem. Soc.*, **90**, 4924 (1968).

[193] R. K. Hill and D. A. Cullison, *J. Am. Chem. Soc.*, **95**, 2923 (1973).

[194] C. A. Grob and J. Ide, *Helv. Chim. Acta*, **57**, 2562 (1974).

[195] P. N. Confalone, E. D. Lollar, G. Pizzolato, and M. R. Uskokovic, *J. Am. Chem. Soc.*, **100**, 6291 (1978).

[196] P. N. Confalone, G. Pizzolato, D. Lollar-Confalone, and M. R. Uskokovic, *J. Am. Chem. Soc.*, **102**, 1954 (1980).

[197] A. F. Ferris, *J. Org. Chem.*, **25**, 12 (1960).

[198] A. Ahmad and I. D. Spenser, *Can. J. Chem.*, **38**, 1625 (1960).

[199] E. L. Zaitseva, A. N. Flerova, R. M. Gitina, L. N. Kurkovskaya, E. N. Teleshov, A. N. Pravednikov, E. S. Botvinnik, N. N. Shmagina, and E. L. Gefter, *Zh. Org. Khim.*, **12**, 1987 (1976); *J. Org. Chem. USSR* (*Engl. transl.*), **12**, 1939 (1976).

[200] K. G. Artz and C. A. Grob, *Helv. Chim. Acta*, **51**, 807 (1968).

[201] Y. Sato and N. Ikekawa, *J. Org. Chem.*, **26**, 5058 (1961).

[202] D. Murakami and N. Tokura, *Bull. Chem. Soc. Jpn.*, **31**, 1044 (1958).

[203] L. Edwards, M. Gouterman, and C. B. Rose, *J. Am. Chem. Soc.*, **98**, 7638 (1976).

[204] H. Nishiyama, K. Sakuta, N. Osaka, and K. Itoh, *Tetrahedron Lett.*, **24**, 4021 (1983).

[205] T. Wagner-Jauregg and M. Roth, *Chem. Ber.*, **93**, 3036 (1960).

[206] J. Wiemann and Y. Dubois, *Bull. Soc. Chim. Fr.*, **1961**, 1873.

[207] S. Fujita, K. Koyama, and Y. Inagaki, *Synthesis*, **1982**, 68.

[208] K. Clarke, C. G. Hughes, and R. M. Scrowston, *J. Chem. Soc., Perkin Trans. 1*, **1973**, 356.

[209] G. Kollenz, *Justus Liebigs Ann. Chem.*, **762**, 23 (1972).

[210] N. Tokura, R. Tada, and K. Suzuki, *Bull. Chem. Soc. Jpn.*, **32**, 654 (1959).

[211] W. M. Whaley and T. R. Govindachari, *Org. React.*, **6**, 74 (1951).

[212] S. P. Hiremath, J. S. Biradar, and M. G. Purohit, *Indian J. Chem., Sect. B*, **21**, 249 (1982).

[213] W. Zielinski, *Synthesis*, **1980**, 70.

[214] W. Zielinski, *Pol. J. Chem.*, **54**, 2209 (1980).

[215] R. Tada, H. Sakuraba, and N. Tokura, *Bull. Chem. Soc. Jpn.*, **31**, 1003 (1958).

[216] R. E. Gawley, E. J. Termine, and K. D. Onan, *J. Chem. Soc., Chem. Commun.*, **1981**, 568.

[217] R. E. Gawley and E. J. Termine, *J. Org. Chem.*, **49**, 1946 (1984).

[218] S. Sakane, Y. Matsumura, Y. Yamamura, Y. Ishida, K. Maruoka, and H. Yamamoto, *J. Am. Chem. Soc.*, **105**, 672 (1983).

[219] R. E. Gawley and E. J. Termine, *Tetrahedron Lett.*, **23**, 307 (1982).

[220] P. Dubs and R. Stuessi, *J. Chem. Soc., Chem. Commun.*, **1976**, 1021.

[221] R. K. Hill and R. T. Conley, *J. Am. Chem. Soc.*, **82**, 645 (1960).

[222] H. Wolff, *Org. React.*, **3**, 307 (1946).

[223] G. I. Koldobskii, G. F. Tereshchenko, E. S. Gerasimova, and L. I. Bagal, *Russ. Chem. Rev.* (*Engl. transl.*), **40**, 835 (1971).

[224] P. A. S. Smith and J. P. Horowitz, *J. Am. Chem. Soc.*, **72**, 3718 (1950).

[225] E. Oliveros, M. Riviere, and A. Lattes, *Nouv. J. Chim.*, **3**, 739 (1979).
[226] R. T. Conley and B. E. Nowak, *J. Org. Chem.*, **27**, 3196 (1962).
[227] M. Cunningham, L. S. Ng Lim, and G. Just, *Can. J. Chem.*, **49**, 2891 (1971).
[228] W. G. Kofron and M.-K. Yeh, *J. Org. Chem.*, **41**, 439 (1976).
[229] M. E. Jung, P. A. Blair, and J. A. Lowe, *Tetrahedron Lett.*, **1976**, 1439.
[230] R. E. Gawley and T. Nagy, *Tetrahedron Lett.*, **25**, 263 (1984).
[231] D. W. Burrows and R. H. Eastman, *J. Am. Chem. Soc.*, **79**, 3756 (1957).
[232] V. I. Boev and A. V. Dombrovskii, *Zh. Obshch. Khim.*, **47**, 1892 (1977); *J. Gen. Chem. USSR* (*Engl. transl.*), **47**, 1728 (1977).
[233] M. Ohno, N. Naruse, and I. Terasawa, *Org. Synth.*, Coll. Vol. **5**, 266 (1973).
[234] W. Lehnert, *Tetrahedron Lett.*, **1971**, 559.
[235] G. Sosnovsky and J. A. Krogh, *Synthesis*, **1978**, 703.
[236] G. A. Olah and T. Keumi, *Synthesis*, **1979**, 112.
[237] A. Striegler, *J. Prakt. Chem.*, **15**, 1 (1961).
[238] A. F. McKay, E. J. Tarlton, S. I. Petri, P. R. Steyermark, and M. A. Mosley, *J. Am. Chem. Soc.*, **80**, 1510 (1958).
[239] J. D. Butler and T. C. Poles, *J. Chem. Soc., Perkin Trans. 2*, **1973**, 41.
[240] J. D. Butler and T. C. Poles, *J. Chem. Soc., Perkin Trans. 2*, **1973**, 1262.
[241] J. T. Doi, W. K. Musker, D. L. De Leeuw, and A. S. Hirschon, *J. Org. Chem.*, **46**, 1239 (1981).
[242] A. Buzas and J. Teste, *Bull. Soc. Chim. Fr.*, **1960**, 359.
[243] O. Meth-Cohn and B. Narine, *Synthesis*, **1980**, 133.
[244] S. Gelin, *J. Heterocycl. Chem.*, **18**, 535 (1981).
[245] Y. Tamura, Y. Kita, and M. Terashima, *Chem. Pharm. Bull.*, **19**, 529 (1971).
[246] A. Benko and I. Rotaru, *Monatsh. Chem.*, **106**, 1027 (1975).
[247] O. D. Strizhakov, E. N. Zil'berman, and S. V. Svetozarskii, *Zh. Obshch. Khim.*, **35**, 628 (1965); *J. Gen. Chem. USSR* (*Engl. transl.*), **631** (1965).
[248] N. Tokura, R. Tada, and K. Yokoyama, *Bull. Chem. Soc. Jpn.*, **34**, 1812 (1961).
[249] I. Ganboa and C. Palomo, *Synth. Commun.*, **13**, 941 (1983).
[250] I. Sakai, N. Kawabe, and M. Ohno, *Bull. Chem. Soc. Jpn.*, **52**, 3381 (1979).
[251] T. Van Es, *J. Chem. Soc.*, **1965**, 3881.
[252] I. J. Galpin, P. F. Gordon, R. Ramage, and W. D. Thorpe, *Tetrahedron*, **32**, 2417 (1976).
[253] T. E. Stevens, *J. Org. Chem.*, **26**, 2531 (1961).
[254] R. Tada and N. Tokura, *Bull. Chem. Soc. Jpn.*, **31**, 387 (1958).
[255] A. F. Turbak, *Ind. Eng. Chem., Prod. Res. Dev.*, **7**, 189 (1968).
[256] Y. Izumi, S. Sato, and K. Urabe, *Chem. Lett.*, **1983**, 1649.
[257] A. Costa, P. M. Deya, J. V. Sinisterra, and J. M. Marinas, *Can. J. Chem.*, **58**, 1266 (1980).
[258] E. F. Novoselov, S. D. Isaev, A. G. Yurchenko, L. Vodicka, and J. Triska, *Zh. Org. Khim.*, **17**, 2558 (1981); *J. Org. Chem. USSR* (*Engl. transl.*), **17**, 2284 (1981).
[259] K. Fukui, M. Uchida, and M. Masaki, *Bull. Chem. Soc. Jpn.*, **46**, 3168 (1973).
[260] N. Tokura, T. Kawahara, and S. Ikeda, *Bull. Chem. Soc. Jpn.*, **37**, 138 (1964).
[261] M. Masaki, J. Kita, K. Fukui, and S. Matsunami, *Tetrahedron Lett.*, **1974**, 191.
[262] J. Wiemann, N. Van Thoai, and P. Ham, *Bull. Soc. Chim. Fr.*, **1967**, 3920.
[263] S. W. Shalaby, S. Sifniades, K. P. Klein, and D. Sheehan, *Polym. Prep., Am. Chem. Soc., Div. Polym. Chem.*, **15**, 429 (1974).
[264] K. Aparajithan, A. C. Thompson, and J. Sam, *J. Heterocycl. Chem.*, **3**, 466 (1966).
[265] H. K. Hall, *J. Am. Chem. Soc.*, **82**, 1209 (1960).
[266] K. Clarke, W. R. Fox, and R. M. Scrowston, *J. Chem. Soc., Perkin Trans. 1*, **1980**, 1029.
[267] I. M. Zalesskaya, A. N. Blakitnyi, E. P. Saenko, Y. A. Fialkov, and L. M. Yagupol'skii, *Zh. Org. Khim.*, **16**, 1194 (1980); *J. Org. Chem. USSR* (*Engl. transl.*), **16**, 1031 (1980).
[268] G. H. Schmid and P. H. Fitzgerald, *Can. J. Chem.*, **46**, 3758 (1968).
[269] B. L. Fox and J. E. Reboulet, *J. Org. Chem.*, **33**, 3639 (1968).
[270] R. C. Elderfield and E. T. Losin, *J. Org. Chem.*, **26**, 1703 (1961).
[271] A. P. Cowling, J. Mann, and A. A. Usmani, *J. Chem. Soc., Perkin Trans. 1*, **1981**, 2116.

[272] C. Santelli, *Tetrahedron Lett.*, **21**, 2893 (1980).

[273] M. V. Rubtsov, E. E. Mikhlina, V. Y. Vorob'eva, and A. D. Yanina, *Zh. Obshch. Khim.*, **34**, 2222 (1964); *J. Gen. Chem. USSR (Engl. transl.)*, **34**, 2232 1964).

[274] E. E. Mikhlina, V. Y. Vorob'eva, V. I. Schedchenko, and M. V. Rubtsov, *Zh. Org. Khim.*, **1**, 1336 (1965); *J. Org. Chem. USSR (Engl. transl.)*, **1**, 1352 (1965).

[275] G. G. Lyle and R. M. Barrera, *J. Org. Chem.*, **29**, 3311 (1964).

[276] M.-K. Yeh, *J. Chin. Chem. Soc. (Taipei)*, **25**, 83 (1978) [*C.A.*, **90** 22385q (1979)].

[277] D. D. Ridley and G. W. Simpson, *Aust. J. Chem.*, **34**, 569 (1981).

[278] N. Thoai, N. Ngoc-Chieu, and C. Beaute, *Bull. Soc. Chim. Fr.*, **1970**, 3656.

[279] U. R. Joshi and P. A. Limaye, *Indian J. Chem., Sect. B*, **21**, 1122 (1982).

[280] N. H. P. Smith, *J. Chem. Soc.*, **1961**, 4209.

[281] B. J. Gregory, R. B. Moodie, and K. Schofield, *J. Chem. Soc. B*, **1970**, 338.

[282] R. S. Monson and B. M. Broline, *Can. J. Chem.*, **51**, 942 (1973).

[283] V. M. Gurav and U. K. Jagwani, *Marathwada Univ. J. Sci., Nat. Sci.*, **14**, 7 (1975) [*C.A.*, **89**, 108503t (1978)].

[284] V. M. Gurav and U. K. Jagwani, *J. Indian Chem. Soc.*, **56**, 325 (1979).

[285] V. M. Gurav, *Acta Cienc. Indica*, **4**, 25 (1978) [*C.A.*, **90**, 6041u (1979)].

[286] B. P. Fabrichnyi, I. F. Shalavina, and Y. L. Gol'dfarb, *Zh. Org. Khim.*, **5**, 361 (1965); *J. Org. Chem. USSR (Engl. transl.)*, **5**, 346 (1969).

[287] Y. Kanaoka, O. Yonemitsu, E. Sato, and Y. Ban, *Chem. Pharm. Bull.*, **16**, 280 (1968).

[288] C. Beaute, N. Thoai, and J. Wiemann, *Bull. Soc. Chim. Fr.*, **1971**, 3327.

[289] V. P. Kukhar and V. I. Pasternak, *Synthesis*, **1974**, 563.

[290] S.-G. Kim, T. Kawakami, T. Ando, and Y. Yukawa, *Bull. Chem. Soc. Jpn.*, **52**, 1115 (1979).

[291] R. Royer, P. Demerseman, G. Colin, and A. Cheutin, *Bull. Soc. Chim. Fr.*, **1968**, 4090.

[292] A. P. Stoll and F. Troxler, *Helv. Chim. Acta*, **51**, 1864 (1968).

[293] H. Weber, *Arch. Pharm. (Weinheim, Ger.)*, **310**, 20 (1977).

[294] E. H. Billett, I. Fleming, and S. W. Hanson, *J. Chem. Soc., Perkin Trans. 2*, **1973**, 1661.

[295] R. T. LaLonde, N. Muhammad, C. F. Wong, and E. R. Sturiale, *J. Org. Chem.*, **45**, 3664 (1980).

[296] S. R. Wilson and R. A. Sawicki, *J. Org. Chem.*, **44**, 330 (1979).

[297] K. Morita and Z. Suzuki, *J. Org. Chem.*, **31**, 233 (1966).

[298] K. B. Becker and C. A. Gabutti, *Tetrahedron Lett.*, **23**, 1883 (1982).

[299] A. Guggisberg, U. Kramer, C. Heidelberger, R. Charubala, E. Stephanou, M. Hesse, and H. Schmid, *Helv. Chim. Acta*, **61**, 1050 (1978).

[300] M. W. Majchrzak, A. Kotelko, R. Guryn, J. B. Lambert, and S. M. Wharry, *Tetrahedron*, **37**, 1075 (1981).

[301] C. Tsuchiya, *Nippon Kagaku Zasshi*, **82**, 1549 (1961) [*C.A.*, **59** 2752g (1963)].

[302] R. E. Lyle and D. A. Walsh, *Org. Prep. Proced. Int.*, **5**, 299 (1973).

[303] A. Jossang-Yanagida and C. Gansser, *J. Heterocycl. Chem.*, **15**, 249 (1978).

[304] S. S. Stradling, D. Hornick, J. Lee, and J. Riley, *J. Chem. Educ.*, **60**, 502 (1983).

[305] B. P. Fabrichnyi, I. F. Shalavina, and Y. L. Gol'dfarb, *Zh. Org. Khim.*, **3**, 2079 (1967); *J. Org. Chem. USSR (Engl. transl.)*, **3**, 2027 (1967).

[306] M. M. Payard, J. M. Paris, J. M. Couquelet, and J. D. Couquelet, *Bull. Soc. Chim. Fr.*, **1979**, 299.

[307] Y. Tamura, Y. Kita, and J. Uraoka, *Chem. Pharm. Bull.*, **20**, 876 (1972).

[308] V. Bardakos and W. Sucrow, *Chem. Ber.*, **111**, 1780 (1978).

[309] H. Gnichtel and K. Hirte, *Chem. Ber.*, **108**, 3380 (1975).

[310] T. Duong, R. H. Prager, J. M. Tippett, D. A. Ward, and D. I. B. Ken, *Aust. J. Chem.*, **29**, 2667 (1976).

[311] E. J. Moriconi and M. A. Stemniski, *J. Org. Chem.*, **37**, 2035 (1972).

[312] A. Kuemin, E. Maverick, P. Seiler, N. Vanier, L. Damm, R. Hobi, J. D. Dunitz, and A. Eschenmoser, *Helv. Chim. Acta*, **63**, 1158 (1980).

[313] S. R. Wilson and R. A. Sawicki, *J. Heterocycl. Chem.*, **19**, 81 (1982).

[314] A. Zabza, C. Wawrzenczyk, and H. Kuczynski, *Bull. Acad. Pol. Sci., Ser. Sci. Chim.*, **22**, 855 (1974).

[315] A. Zabza, C. Wawrzenczyk, and H. Kuczynski, *Bull. Acad. Pol. Sci., Ser. Sci. Chim.*, **20**, 623 (1972).

[316] H. P. Fischer and C. A. Grob, *Helv. Chim. Acta*, **51**, 153 (1968).

[317] E. G. Rozantsev, A. V. Chudinov, and V. D. Sholle, *Izv. Akad. Nauk SSSR, Ser. Khim.*, **1980**, 2114; *Bull. Acad. Sci. USSR, Div. Chem. Sci.* (*Engl. transl.*), **1980**, 1510.

[318] I. L. Finar and H. E. Saunders, *J. Chem. Soc. C*, **1969**, 1495.

[319] V. Joshi and M. I. Hari, *Indian J. Chem., Sect. B*, **22**, 65 (1983).

[320] F. M. Albini, R. Oberti, and P. Caramella, *J. Chem. Res.* (*S*), **1983**, 4.

[321] M. Ruccia, N. Vivona, G. Cusmano, and G. Macaluso, *J. Chem. Soc., Perkin Trans. 2*, **1977**, 589.

[322] J. Huet, M. Sado-Odeye, M. Martin, P. Guibet, Ph. Linee, P. Lacroix, P. Quiniou, and J. Laurent, *Eur. J. Med. Chem.—Chim. Ther.*, **9**, 376 (1974).

[323] M. D. Nair and J. A. Desai, *Indian J. Chem., Sect. B*, **19**, 765 (1980).

[324] H. Boehme and R. Malcherek, *Arch. Pharm.* (*Weinheim, Ger.*), **312**, 648 (1979).

[325] V. K. Venugopal, N. Rao, M. F. Rahman, U. T. Bhalerao, and G. Thyagarajan, *Indian J. Chem., Sect. B*, **20**, 156 (1981).

[326] D. Huckle, I. M. Lockart, and M. Wright, *J. Chem. Soc., Perkin Trans. 1*, **1972**, 2425.

[327] K. Hirao, H. Miura, and O. Yonemitsu, *Heterocycles*, **7**, 857 (1977).

[328] H. J. Goelz, J. M. Muchowski, and M. L. Maddox, *Angew. Chem., Int. Ed. Engl.*, **17**, 855 (1978).

[329] W. J. Lipa, H. T. Crawford, P. C. Radlick, and G. K. Helmkamp, *J. Org. Chem.*, **43**, 3813 (1978).

[330] C. A. Grob and J. Ide, *Helv. Chim. Acta*, **57**, 2571 (1974).

[331] M. Schaefer-Ridder, A. Wagner, M. Schwamborn, H. Schreiner, E. Devrout, and E. Vogel, *Angew. Chem., Int. Ed. Engl.*, **17**, 853 (1978).

[332] P. Kovacic, K. W. Field, and T. A. Wnuk, *J. Chem. Eng. Data*, **16**, 141 (1971).

[333] J. Kralicek, J. Kondelikova, and V. Kubanek, *Collect. Czech. Chem. Commun.*, **39**, 249 (1974).

[334] J. G. Korsloot and V. G. Keizer, *Tetrahedron Lett.*, **1969**, 3517.

[335] J. Triska, L. Vodicka, and J. Hlavaty, *Collect. Czech. Chem. Commun.*, **44**, 1448 (1979).

[336] V. P. Arya, R. Prasad, and V. Sadhale, *Indian J. Chem.*, **13**, 1267 (1975).

[337] W. Eisele, C. A. Grob, and E. Renk, *Tetrahedron Lett.*, **1963**, 75.

[338] F. T. Bond, J. E. Stemke, and D. W. Powell, *Synth. Commun.*, **5**, 427 (1975).

[339] M. Tichy, P. Malon, I. Fric, and K. Blaha, *Collect. Czech. Chem. Commun.*, **44**, 2653 (1979).

[340] A. Zabza, H. Kuczynski, Z. Chabudzinski, and D. Zedzik-Hibner, *Bull. Acad. Pol. Sci., Ser. Sci. Chim.*, **20**, 841 (1972).

[341] R. R. Rao, K. Chakraborti, S. Bhattacharya, and A. Rao, *Indian J. Chem., Sect. B*, **22**, 1122 (1983).

[342] R. Lukes and J. Hofman, *Collect. Czech. Chem. Commun.*, **26**, 523 (1961).

[343] G. Di Maio and V. Permutti, *Tetrahedron*, **22**, 2059 (1966).

[344] H. Erdtman and S. Thoren, *Acta Chem. Scand.*, **24**, 87 (1970).

[345] W. D. Lloyd and G. W. Hedrick, *Ind. Eng. Chem., Prod. Res. Dev.*, **2**, 143 (1963).

[346] A. Zabza, C. Wawrzenczyk, and H. Kuczynski, *Bull. Acad. Pol. Sci., Ser. Sci. Chim.*, **20**, 521 (1972).

[347] A. Zabza, C. Wawrzenczyk, and H. Kuczynski, *Bull. Acad. Pol. Sci., Ser. Sci. Chim.*, **20**, 631 (1972).

[348] R. W. Cottingham, *J. Org. Chem.*, **25**, 1473 (1960).

[349] H. P. Fischer and C. A. Grob, *Helv. Chim. Acta*, **47**, 564 (1964).

[350] A. Zabza, H. Kuczynski, Z. Chabudzinski, and G. Piotrowska, *Bull. Acad. Pol. Sci., Ser. Sci. Chim.*, **21**, 1 (1973).

[351] P. Nedenskov, W. Taub, and D. Ginsburg, *Acta Chem. Scand.*, **12**, 1405 (1958).

[352] C. Tsuchiya, *Nippon Kagaku Zasshi*, **82**, 1395 (1961) [*C.A.*, **59**, 2751a (1963)].

[353] J. N. Chatterjea and K. R. R. P. Singh, *J. Indian Chem. Soc.*, **59**, 527 (1982).
[354] M. D. Nair, V. Sudarsanam and J. A. Desai, *Indian J. Chem., Sect. B*, **21**, 1027 (1982).
[355] L. Geita, I. Dalberga, A. Grinvalde, and E. Liepins, *Zh. Org. Khim.*, **13**, 1461 (1977); *J. Org. Chem. USSR* (*Engl. transl.*), **13**, 1346 (1977).
[356] R. M. Coates and E. F. Johnson, *J. Am. Chem. Soc.*, **93**, 4016 (1971).
[357] T. Toda, S. Ryu, Y. Hagiwara, and T. Nozoe, *Bull. Chem. Soc. Jpn.*, **48**, 82 (1975).
[358] B. Rouot and G. Leclerc, *Bull. Soc. Chim. Fr.*, **1979**, 520.
[359] B. Unterhalt and H. J. Reinhold, *Arch. Pharm.* (*Weinheim, Ger.*), **308**, 41 (1975).
[360] Y. Sawa, T. Kato, T. Masuda, M. Hori, and H. Fujimura, *Chem. Pharm. Bull.*, **23**, 1917 (1975).
[361] P. B. Terent'ev, O. E. Vendrova, V. M. Dem'yanovich, L. D. Solov'eva, and V. M. Potapov, *Khim. Geterotsikl. Soedin.*, **1983**, 1236 [*C.A.*, **100**, 85577z (1984)].
[362] R. T. Conley and L. J. Frainier, *J. Org. Chem.*, **27**, 3844 (1962).
[363] J. M. Mellor, R. Pathirana, and J. H. A. Stibbard, *J. Chem. Soc., Perkin Trans. 1*, **1983**, 2541.
[364] N. P. Buu-Hoi, P. Jacquignon, and O. Roussel, *Bull. Soc. Chim. Fr.*, **1964**, 867.
[365] R. G. Neville, *J. Org. Chem.*, **24**, 870 (1959).
[366] I. Fleming and R. B. Woodward, *J. Chem. Soc., Perkin Trans. 2*, **1973**, 1653.
[367] T. Sasaki, S. Eguchi, and T. Toru, *J. Org. Chem.*, **36**, 2454 (1971).
[368] T. Sasaki, S. Eguchi, and M. Mizutani, *J. Org. Chem.*, **37**, 3961 (1972).
[369] R. T. LaLonde, N. Muhammad, and C. F. Wong, *J. Org. Chem.*, **42**, 2113 (1977).
[370] P. Duhamel and M. Kotera, *J. Chem. Res.* (*S*), **1982**, 276.
[371] N. S. Fedotov, I. G. Rybalka, A. V. Kisin, G. I. Nikishin, and V. F. Mironov, *Zh. Obshch. Khim.*, **50**, 617 (1980); *J. Org. Chem. USSR* (*Engl. transl.*), **50**, 501 (1980).
[372] L. A. Paquette, H. C. Berk, and S. V. Ley, *J. Org. Chem.*, **40**, 902 (1975).
[373] L. A. Paquette, G. D. Ewing, S. V. Ley, H. C. Berk, and S. G. Traynor, *J. Org. Chem.*, **43**, 4712 (1978).
[374] J. B. Hester, Jr., *J. Org. Chem.*, **35**, 875 (1970).
[375] J. B. Hester, Jr., *J. Org. Chem.*, **32**, 3804 (1967).
[376] P. H. Mazzocchi, E. W. Kordoski, and R. Rosenthal, *J. Heterocycl. Chem.*, **19**, 941 (1982).
[377] D. D. Evans, J. Weale, and D. J. Weyell, *Aust. J. Chem.*, **26**, 1333 (1973).
[378] A. S. Bailey, C. J. Barnes, and P. A. Wilkinson, *J. Chem. Soc., Perkin Trans. 2*, **1974**, 1321.
[379] E. Wenkert and B. F. Barnett, *J. Am. Chem. Soc.*, **82**, 4671 (1960).
[380] E. E. Smissman, J. R. Reid, D. A. Walsh, and R. T. Borchardt, *J. Med. Chem.*, **19**, 127 (1976).
[381] B. P. Fabrichnyi, I. F. Shalavina, Y. L. Gol'dfarb, and S. M. Kostrova, *Zh. Org. Khim.*, **10**, 1956 (1974); *J. Org. Chem. USSR* (*Engl. transl.*), **10**, 1966 (1974).
[382] Y. L. Gol'dfarb, B. P. Fabrichnyi, I. F. Shalavina, and S. M. Kostrova, *Zh. Org. Khim.*, **11**, 2400 (1975); *J. Org. Chem. USSR* (*Engl. transl.*), **11**, 2449 (1975).
[383] V. Bardakos and W. Sucrow, *Chem. Ber.*, **111**, 853 (1978).
[384] A. Maquestiau, Y. Van Haverbeke, J. J. Vanden Eynde, and N. De Pauw, *Bull. Soc. Chim. Belge*, **89**, 45 (1980).
[385] E. V. Dehmlow and H. P. Joswig, *Justus Liebigs Ann. Chem.*, **1975**, 916.
[386] K. K. Kelly and J. S. Matthews, *Tetrahedron*, **26**, 1555 (1970).
[387] M. Kataoka and M. Ohno, *Bull. Chem. Soc. Jpn.*, **46**, 3474 (1973).
[388] J. Wiemann and P. Ham, *Bull. Soc. Chim. Fr.*, **1961**, 1005.
[389] K. Nagarajan, C. L. Kulkarni, and A. Venkateswarlu, *Indian J. Chem.*, **12**, 247 (1974).
[390] P. Catsoulacos, *Chim. Ther.*, **5**, 401 (1970).
[391] P. Beak and J. A. Barron, *J. Org. Chem.*, **38**, 2771 (1973).
[392] P. Catsoulacos, *Bull. Soc. Chim. Fr.*, **1973**, 2136.
[393] J. P. Maffrand, R. Boigegrain, J. Courregelongue, G. Ferrand, and D. Frehel, *J. Heterocycl. Chem.*, **18**, 727 (1981).
[394] T. Ogata, S. Tanaka, H. Yoshida, and S. Inokawa, *Bull. Chem. Soc. Jpn.*, **44**, 2853 (1971).
[395] W. Gerrard, M. F. Lappert, and J. W. Wallis, *J. Chem. Soc.*, **1960**, 2141.
[396] L. Makaruk, H. Polanska, and B. Wazynska, *Polym. Prep. Am. Chem. Soc., Div. Polym. Chem.*, **24**, 260 (1983).
[397] T. S. Sulkowski and S. J. Childress, *J. Org. Chem.*, **27**, 4424 (1962).

[398] J. M. Bastian, A. Ebnoether, and E. Jucker, *Helv. Chim. Acta*, **54**, 283 (1971).
[399] H. Yamanaka, M. Shiraiwa, T. Sakamoto, and S. Konno, *Chem. Pharm. Bull.*, **29**, 3548 (1981).
[400] M. Pesson, P. De Lajudie, and M. Antoine, *C. R. Hebd. Seances Acad. Sci., Ser. C*, **273**, 907 (1971).
[401] Z. Majer, M. Kajtar, M. Tichy, and K. Blaha, *Collect. Czech. Chem. Commun.*, **47**, 950 (1982).
[402] R. H. Prager, J. M. Tippett, and A. D. Ward, *Aust. J. Chem.*, **31**, 1989 (1978).
[403] R. Tada, Y. Masubuchi, and N. Tokura, *Bull. Chem. Soc. Jpn.*, **34**, 209 (1961).
[404] J. Kondelikova, J. Kralicek, J. Smolikova, and K. Blaha, *Collect. Czech. Chem. Commun.*, **38**, 523 (1973).
[405] H. Takahashi and M. Ito, *Bull. Chem. Soc. Jpn.*, **56**, 2050 (1983).
[406] J. Brugidou, H. Christol, and M. F. Feldmann, *Bull. Soc. Chim. Fr.*, **1974**, 1005.
[407] J. P. Coic and G. Saint Ruf, *J. Heterocycl. Chem.*, **15**, 1367 (1978).
[408] J. B. Kyziol and A. Lyzniak, *Tetrahedron*, **36**, 3017 (1980).
[409] F. Yoneda, M. Higuchi, K. Senga, M. Kanahori, and S. Nishigaki, *Chem. Pharm. Bull.*, **21**, 473 (1973).
[410] T. Nagasaka and S. Ohki, *Chem. Pharm. Bull.*, **25**, 3023 (1977).
[411] V. Bardakos and W. Sucrow, *Chem. Ber.*, **109**, 1898 (1976).
[412] K. F. Cohen, R. Kazlauskas, and J. T. Pinhey, *J. Chem. Soc., Perkin Trans. 2*, **1973**, 2076.
[413] S. Archer, P. Osei-Gyimah, and S. Silbering, *J. Med. Chem.*, **23**, 516 (1980).
[414] M. Y. Uritskaya, O. S. Anisimova, and L. N. Yakhontov, *Khim. Geterotsikl. Soedin.*, **1979**, 80 [*C.A.*, **90**, 152048t (1979)].
[415] P. Rosenmund, D. Sauer, and W. Trommer, *Chem. Ber.*, **103**, 496 (1970).
[416] G. M. Iskander and V. S. Gulta, *J. Chem. Soc., Perkin Trans. 1*, **1982**, 1891.
[417] P. C. Belanger, J. Scheigetz, and R. N. Young, *Can. J. Chem.*, **61**, 2177 (1983).
[418] D. Carr, B. Iddon, H. Suschitzky, and R. T. Parfitt, *J. Chem. Soc., Perkin Trans. 1*, **1980**, 2374.
[419] H. A. Bruson, F. W. Grant, and E. Bobko, *J. Am. Chem. Soc.*, **80**, 3633 (1958).
[420] R. T. Conley and R. J. Lange, *J. Org. Chem.*, **28**, 210 (1963).
[421] K. Ito and Y. Oka, *Chem. Pharm. Bull.*, **28**, 2862 (1980).
[422] H. Nagano, Y. Ishikawa, Y. Matsuo, and M. Shiota, *Chem. Lett.*, **1982**, 1947.
[423] E. J. Corey, J. F. Arnett, and G. N. Widiger, *J. Am. Chem. Soc.*, **97**, 430 (1975).
[424] T. Ibuka, H. Minakata, Y. Mitsui, E. Tabushi, T. Taga, and Y. Inubushi, *Chem. Lett.*, **1981**, 1409.
[425] Z. Kafka, L. Vodicka, and V. Galik, *Chem. Prum.*, **29**, 596 (1979).
[426] T. Kimura, M. Minabe, and K. Suzuki, *J. Org. Chem.*, **43**, 1247 (1978).
[427] P. Caramella, G. Cellerino, P. Grunanger, F. M. Albini, and M. R. R. Cellerino, *Tetrahedron*, **34**, 3545 (1978).
[428] P. Caramella, C. A. Coda, A. Corsaro, D. Del Monte, and A. F. Marinone, *Tetrahedron*, **38**, 173 (1982).
[429] E. J. Browne and J. B. Polya, *J. Chem. Soc. C*, **1968**, 824.
[430] A. Levai, *Acta Chim. Acad. Sci. Hung.*, **107**, 361 (1981).
[431] G. Cainelli, S. Morrocchi, and A. Quilico, *Tetrahedron Lett.*, **1963**, 1959.
[432] G. Cauquil, E. Casadevall, and A. Casadevall, *Bull. Soc. Chim. Fr.*, **1962**, 608.
[433] M. E. Rogers and E. L. May, *J. Med. Chem.*, **17**, 1328 (1974).
[434] H. Alper and E. C. H. Keung, *J. Heterocycl. Chem.*, **10**, 637 (1973).
[435] D. Carr, B. Iddon, H. Suschitzky, and R. T. Parfitt, *J. Chem. Soc., Perkin Trans. 1*, **1980**, 2380.
[436] A. Buzas, F. Cossais, and J. P. Jacquet, *Bull. Soc. Chim. Fr.*, **1972**, 4397.
[437] A. Velez and J. Romo-A, *Bol. Inst. Quim. Univ. Nac. Auton. Mex.*, **20**, 49 (1968) [*C.A.*, **71**, 22207r (1969)].
[438] H. Ogura, H. Takayanagi, and C. Miyahara, *J. Org. Chem.*, **37**, 519 (1972).
[439] K. Niume, K. Nakamichi, F. Toda, K. Uno, M. Hasegawa, and Y. Iwakura, *J. Polym. Sci., Polym. Chem. Ed.*, **18**, 2163 (1980).
[440] J. N. Chatterjea and K. R. R. P. Singh, *Indian J. Chem., Sect. B*, **20**, 1053 (1981).
[441] B. L. Jensen and D. P. Michaud, *J. Heterocycl. Chem.*, **15**, 321 (1978).
[442] B. Unterhalt, *Arch. Pharm. (Weinheim, Ger.)*, **300**, 748 (1967).

[443] P. A. S. Smith and E. P. Antoniades, *Tetrahedron*, **9**, 210 (1960).

[444] J. E. Sundeen, J. A. Reid, J. A. Osband, and F. P. Hauck, *J. Med. Chem.*, **20**, 1478 (1977).

[445] H. Reinshagen and A. Stuetz, *Monatsh. Chem.*, **110**, 567 (1979).

[446] S. Konno, M. Shiraiwa, and H. Yamanaka, *Chem. Pharm. Bull.*, **29**, 3554 (1981).

[447] G. I. Hutchison, P. A. Marshall, R. H. Prager, J. M. Tippett, and A. D. Ward, *Aust. J. Chem.*, **33**, 2699 (1980).

[448] D. Berney and K. Schuh, *Helv. Chim. Acta*, **59**, 2059 (1976).

[449] L. Bauer and R. E. Hewiston, *J. Org. Chem.*, **27**, 3982 (1962).

[450] H. P. Fischer, *Helv. Chim. Acta*, **48**, 1279 (1965).

[451] I. Seki, *Chem. Pharm. Bull.*, **18**, 1269 (1970).

[452] W. J. Rodewald and B. M. Jagodzinska, *Pol. J. Chem.*, **54**, 911 (1980).

[453] R. Achini, *Helv. Chim. Acta*, **64**, 2203 (1981).

[454] G. G. Lyle and E. T. Pelosi, *J. Am. Chem. Soc.*, **88**, 5276 (1966).

[455] V. Baliah, M. Lakshmanan, and K. Pandiarajan, *Indian J. Chem., Sect. B*, **16**, 72 (1978).

[456] R. Bognar, S. Makleit, L. Radics, and I. Seki, *Org. Prep. Proced. Int.*, **5**, 49 (1973).

[457] B. Matkovics, B. Tarodi, and L. Balaspiri, *Acta Chim. Acad. Sci. Hung.*, **80**, 79 (1974).

[458] B. Matkovics, G. Gondos, and B. Tarodi, *Acta Chim. Acad. Sci. Hung.*, **57**, 119 (1968).

[459] H. J. Monteiro, *An. Acad. Bras. Cienc.*, **52**, 493 (1980). [*C.A.*, **94**, 208767d (1980)].

[460] K. Oka and S. Hara, *Chem. Ind. (London)*, **1969**, 168.

[461] P. K. Grant, J. S. Prasad, and D. D. Rowan, *Aust. J. Chem.*, **36**, 1197 (1983).

[462] Y. V. Tanchuk and S. I. Kotenko, *Zh. Org. Khim.*, **17**, 758 (1981); *J. Org. Chem. USSR (Engl. transl.)*, **17**, 661 (1981).

[463] A. T. Troshchenko and T. P. Lobanova, *Zh. Org. Khim.*, **3**, 501 (1967); *J. Org. Chem. USSR (Engl. transl.)*, **3**, 480 (1967).

[464] V. Sprio, P. Madonia, and R. Caronia, *Ann. Chim. (Rome)*, **49**, 169 (1959) [*C.A.*, **53**, 16105i (1959)].

[465] J. Henin and J. Gardent, *Bull. Soc. Chim. Fr.*, **1977**, 89.

[466] M. Boeş, W. Fleischhacker, and M. Kratzel, *J. Heterocycl. Chem.*, **19**, 1113 (1982).

[467] R. J. Marshall, I. McIndewar, J. A. M. Peters, N. P. Van Vliet, and, F. J. Zeelen, *Eur. J. Med. Chem.—Chim. Ther.*, **19**, 43 (1984).

[468] A. H. Fenselau, E. H. Hamamura, and J. G. Moffatt, *J. Org. Chem.*, **35**, 3546 (1970).

[469] A. Cervantes, P. Crabbe, J. Iriarte, and G. Rosenkranz, *J. Org. Chem.*, **33**, 4294 (1968).

[470] C. Camoutsis and P. Catsoulacos, *J. Heterocycl. Chem.*, **20**, 1093 (1983).

[471] H. Singh and V. V. Parashar, *Tetrahedron Lett.*, **1966**, 983.

[472] R. T. Blickenstaff and E. L. Foster, *J. Org. Chem.*, **26**, 5029 (1961).

[473] B. Matkovics and Z. Tegyey, *Acta Chim. Acad. Sci. Hung.*, **80**, 211 (1974).

[474] D. C. Mammato and G. A. Eadon, *J. Org. Chem.*, **40**, 1784 (1975).

[475] C. W. Shoppee and G. Kruger, *J. Chem. Soc.*, **1961**, 3641.

[476] K. Oka and S. Hara, *Tetrahedron Lett.*, **1969**, 1193.

[477] B. Matkovics and Z. Tegyey, *Acta Chim. Acad. Sci. Hung.*, **60**, 413 (1969).

[478] C. Dagher, R. Hanna, P. B. Terent'ev, Y. G. Boundel, B. I. Maksimov, and N. S. Kulikov, *J. Heterocycl. Chem.*, **20**, 989 (1983).

[479] V. Baliah and R. Jeyaraman, *Indian J. Chem., Sect. B*, **15**, 796 (1977).

[480] D. Cartier, J. Levy, and J. Le Men, *Bull. Soc. Chim. Fr.*, **1976**, 1961.

[481] M. Ohashi, S. Yamamura, A. Terahara, and K. Nakanishi, *Bull. Chem. Soc. Jpn.*, **33**, 1630 (1960).

[482] W. Klyne, D. N. Kirk, J. Tilley, and H. Suginome, *Tetrahedron*, **36**, 543 (1980).

[483] S. D. Levine, *J. Org. Chem.*, **35**, 1064 (1970).

[484] S. Mohr, *Chem. Ber.*, **114**, 2146 (1981).

[485] H. Erdtman and L. Malmborg, *Acta Chem. Scand.*, **24**, 2252 (1970).

[486] H. Singh, S. K. Gupta, S. Padmanabhan, D. Paul, and T. R. Bhardwaj, *Indian J. Chem., Sect. B*, **15**, 101 (1977).

[487] W. J. Rodewald and J. W. Morzycki, *Pol. J. Chem.*, **52**, 2101 (1978).

[488] P. Catsoulacos, *Chem. Chron.*, **1**, 147 (1972).

[489] C. W. Shoppee, R. E. Lack, and B. C. Newman, *J. Chem. Soc.*, **1964**, 3388.

[490] R. Anliker, M. Muller, M. Perelman, J. Wohlfahrt, and H. Heusser, *Helv. Chim. Acta*, **42**, 1071 (1959).

[491] R. D. Heard, M. T. Ryan, and H. I. Bolker, *J. Org. Chem.*, **24**, 172 (1959).

[492] H. R. Nace and A. C. Watterson, *J. Org. Chem.*, **31**, 2109 (1966).

[493] M. Kobayashi, Y. Shimizu, and H. Mitsuhashi, *Chem. Pharm. Bull.*, **17**, 1255 (1969).

[494] P. Catsoulacos and L. Boutis, *Chim. Ther.*, **8**, 215 (1973).

[495] P. Choay, C. Monneret, and H. Q. Khuong, *Tetrahedron*, **34**, 1529 (1978).

[496] I. I. Brunovlenskaya, A. B. Alekseeva, and V. R. Skvarchenko, *Zh. Org. Khim.*, **16**, 2141 (1980); *J. Org. Chem. USSR* (*Engl. transl.*), **16**, 1824 (1980).

[497] H. H. Otto and J. Triepel, *Justus Liebigs Ann. Chem.*, **1978**, 1809.

[498] H. Mitsuhashi and K. Tomimoto, *Chem. Pharm. Bull.*, **19**, 1974 (1971).

[499] H. Ogura, H. Takayanagi, and K. Furuhata, *J. Chem. Soc., Perkin Trans. 2*, **1976**, 665.

[500] E. S. Rothman and M. E. Wall, *J. Org. Chem.*, **25**, 1396 (1960).

[501] P. Catsoulacos, *Chim. Ther.*, **6**, 449 (1971).

[502] J. Romo and A. Romo de Vivar, *J. Am. Chem. Soc.*, **81**, 3446 (1959).

[503] D. G. Patterson, C. Djerassi, Y. Yuh, and N. L. Allinger, *J. Org. Chem.*, **42**, 2365 (1977).

[504] L. I. Klimova and N. N. Suvorov, *Zh. Obshch. Khim.*, **33**, 3011 (1963); *J. Gen. Chem. USSR* (*Engl. transl.*), **33**, 2937 (1963).

[505] W. J. Rodewald and J. R. Jaszczynski, *Tetrahedron Lett.*, **1976**, 2977.

[506] K. Oka and S. Hara, *J. Am. Chem. Soc.*, **99**, 3859 (1977).

[507] J. A. Zderic and J. Iriarte, *J. Org. Chem.*, **27**, 1756 (1962).

[508] M. Kobayashi and H. Mitsuhashi, *Chem. Pharm. Bull.*, **21**, 1069 (1973).

[509] A. P. Shroff and C. J. Shaw, *Anal. Chem.*, **43**, 455 (1971).

[510] W. J. Rodewald and A. Zaworska, *Pol. J. Chem.*, **54**, 1147 (1980).

[511] B. Matkovics, G. Balazs, and L. Balaspiri, *Acta Phys. Chem.*, **21**, 181 (1975).

[512] I. Shafiullah, *Acta Chim. Acad. Sci. Hung.*, **101**, 319 (1979).

[513] C. W. Shoppee, R. W. Killick, and G. Kruger, *J. Chem. Soc.*, **1962**, 2275.

[514] N. J. Doorenbos and R. E. Havranek, *J. Org. Chem.*, **30**, 2474 (1965).

[515] M. P. Irismetov, G. Y. Tsvetkova, and M. I. Goryaev, *Zh. Obshch. Khim.*, **47**, 941 (1977); *J. Gen. Chem. USSR* (*Engl. transl.*), **47**, 857 (1977).

[516] D. H. R. Barton, J. P. Poyser, and P. G. Sammes, *J. Chem. Soc., Perkin Trans. 1*, **1972**, 53.

[517] M. S. Ahmad, A. H. Siddiqui, and S. C. Logani, *Aust. J. Chem.*, **22**, 271 (1969).

[518] M. S. Ahmad and A. H. Siddiqui, *Aust. J. Chem.*, **21**, 1371 (1968).

[519] M. S. Ahmad, Shafiullah, and Islamuddin, *Indian J. Chem.*, **12**, 1323 (1974).

[520] A. H. Siddiqui, M. H. Baig, and C. Lakshmi, *Aust. J. Chem.*, **30**, 2271 (1977).

[521] J. A. Zderic, H. Carpio, D. C. Limon, and A. Ruiz, *J. Org. Chem.*, **26**, 2842 (1961).

[522] R. H. Mazur, *J. Am. Chem. Soc.*, **81**, 1454 (1959).

[523] V. Dave, J. B. Stothers, and E. W. Warnhoff, *Can. J. Chem.*, **57**, 1557 (1979).

[524] C. W. Shoppee, M. I. Akhtar, and R. E. Lack, *J. Chem. Soc.*, **1964**, 3392.

[525] B. Matkovics, Z. Tegyey, M. Resch, F. Sirokman, and E. Boga, *Acta Chim. Acad. Sci. Hung.*, **66**, 333 (1970).

[526] M. S. Ahmad and M. Mushfig, *Aust. J. Chem.*, **24**, 213 (1971).

[527] S. Stiver and P. Yates, *J. Chem. Soc., Chem. Commun.*, **1983**, 50.

[528] H. Suginome and C.-M. Shea, *J. Chem. Soc., Perkin Trans. 1*, **1980**, 2268.

[529] M. S. Ahmad, N. K. Pillai, and Z. H. Chaudhry, *Aust. J. Chem.*, **27**, 1537 (1974).

[530] M. S. Ahmad, Shafiullah, and A. H. Siddiqi, *Indian J. Chem.*, **7**, 1167 (1969).

[531] M. S. Ahmad and Shafiullah, *Indian J. Chem.*, **10**, 1136 (1972).

[532] C. W. Shoppee and J. C. P. Sly, *J. Chem. Soc.*, **1958**, 3458.

[533] T. Sasaki, S. Eguchi, and T. Toru, *J. Chem. Soc. D*, **1970**, 1239.

[534] V. Dave, J. B. Stothers, and E. W. Warnhoff, *Can. J. Chem.*, **58**, 2666 (1980).

[535] J. C. Kim, S.-K. Choi, W.-W. Park, and Y.-T. Lee, *Daehan Hwahak Hwoejee* (*J. Kor. Chem. Soc.*), **22**, 218 (1979) [*C.A.*, **91**, 141085x (1979)].

[536] N. J. Doorenbos and M. T. Wu, *J. Org. Chem.*, **26**, 2548 (1961).

[537] W. J. Rodewald and B. Achmatowicz, *Rocz. Chem.*, **46**, 203 (1972).
[538] W. J. Rodewald and J. Wicha, *Bull. Acad. Pol. Sci.*, **12**, 95 (1964).
[539] W. J. Rodewald and B. Achmatowicz, *Rocz. Chem.*, **45**, 1501 (1971).
[540] M. S. Ahmad and F. Waris, *Indian J. Chem., Sect. B*, **15**, 1141 (1977).
[541] W. J. Rodewald and K. Olejniczak, *Rocz. Chem.*, **50**, 1089 (1976).
[542] W. J. Rodewald and J. Wicha, *Bull. Acad. Pol. Sci., Ser. Sci. Chim.*, **11**, 437 (1963).
[543] H. Suginome, T. Yabe, and E. Osawa, *J. Chem. Soc., Perkin Trans. 1*, **1982**, 931.
[544] G. H. Whitham, *J. Chem. Soc.*, **1960**, 2016.
[545] M. Kobayashi and H. Mitsuhashi, *Chem. Pharm. Bull.*, **20**, 1567 (1972).
[546] P. Bladon and W. McMeekin, *J. Chem. Soc.*, **1961**, 3504.
[547] P. Bladon and W. McMeekin, *Chem. Ind.* (*London*), **1960**, 1307.
[548] B. Matkovics, G. Gondos, and Z. Tegyey, *Acta Chim. Acad. Sci. Hung.*, **53**, 417 (1967).
[549] H. Singh and S. Padmanabhan, *Tetrahedron Lett.*, **1967**, 3689.
[550] D. H. R. Barton, X. Lusinchi, A. Martinez Menendez, and P. Milliet, *Tetrahedron*, **39**, 2201 (1983).
[551] M. S. Ahmad, I. A. Ansari, and G. Moinuddin, *Indian J. Chem., Sect. B*, **20**, 602 (1981).
[552] L. Knof, *Justus Liebigs Ann. Chem.*, **642**, 194 (1961).
[553] W. J. Rodewald and B. Achmatowicz, *Tetrahedron*, **27**, 5467 (1971).
[554] N. N. Suvorov, L. I. Klimova, and L. M. Morozovskaya, *Zh. Obshch. Khim.*, **32**, 3308 (1962); *J. Gen. Chem. USSR* (*Engl. transl.*), **32**, 3250 (1962).
[555] S. C. Puri, K. L. Dhar, and C. K. Atal, *Indian J. Chem., Sect. B*, **15**, 917 (1977).
[556] K. L. Rao, S. K. Ramraj, and T. Sundararamaiah, *J. Indian Chem. Soc.*, **57**, 833 (1980).
[557] H. Suginome and Y. Takahash, *J. Chem. Soc., Perkin Trans. 1*, **1979**, 2920.
[558] G. A. Tolstikov and M. I. Goryaev, *Zh. Org. Khim.*, **2**, 1718 (1966); *J. Org. Chem. USSR* (*Engl. transl.*), **2**, 1694 (1966).
[559] T. Sundararamaiah, S. K. Ramraj, K. L. Rao, and V. V. Bai, *J. Indian Chem. Soc.*, **53**, 664 (1976).
[560] K. Tsuda and R. Hayatsu, *J. Am. Chem. Soc.*, **78**, 4107 (1956).
[561] E. C. Taylor, C. W. Jefford, and C. C. Cheng, *J. Am. Chem. Soc.*, **83**, 1261 (1961).
[562] A. F. Ferris, G. S. Johnson, F. E. Gould, and H. Stange, *J. Org. Chem.*, **25**, 1302 (1960).
[563] A. F. Ferris, G. S. Johnson, F. E. Gould, and H. K. Latourette, *J. Org. Chem.*, **25**, 492 (1960).
[564] A. F. Ferris, G. S. Johnson, and F. E. Gould, *J. Org. Chem.*, **25**, 496 (1960).
[565] T. E. Stevens, *Tetrahedron Lett.*, **1967**, 3017.
[566] G. Rosini and A. Medici, *Synthesis*, **1975**, 665.
[567] S. L. Reid and D. B. Sharp, *J. Org. Chem.*, **26**, 2567 (1961).
[568] H. Gaskin, D. L. Swallow, P. J. Taylor, and M. J. Rix, *J. Chem. Soc., Chem. Commun.*, **1972**, 547.
[569] A. F. Ferris, *J. Org. Chem.*, **24**, 580 (1959).
[570] K. Lunkwitz, W. Pritzkow, and G. Schmid, *J. Prakt. Chem.*, **37**, 319 (1968).
[571] M. Ohno, N. Naruse, S. Torimitsui, and J. Teresawa, *J. Am. Chem. Soc.*, **88**, 3168 (1966).
[572] M. Ohno and I. Terasawa, *J. Am. Chem. Soc.*, **88**, 5683 (1966).
[573] L. L. Rodina, L. V. Koroleva, and I. K. Korobitsyna, *Zh. Org. Khim.*, **6**, 2336 (1970); *J. Org. Chem. USSR* (*Engl. transl.*), **6**, 2345 (1970).
[574] W. Von E. Doering and Y. Yamashita, *J. Am. Chem. Soc.*, **105**, 5368 (1983).
[575] B. C. Oxenrider and M. M. Rogic, *J. Org. Chem.*, **47**, 2629 (1982).
[576] A. F. Ferris, G. S. Johnson, and F. E. Gould, *J. Org. Chem.*, **25**, 1813 (1960).
[577] T. A. Antkowiak, D. C. Sanders, G. B. Trimitsis, J. B. Press, and H. Shechter, *J. Am. Chem. Soc.*, **94**, 5366 (1972).
[578] G. Mehta and K. S. Rao, *Indian J. Chem., Sect. B*, **21**, 981 (1982).
[579] U. Redeker, N. Engel, and W. Steglich, *Tetrahedron Lett.*, **22**, 4263 (1981).
[580] A. Stankevicius and A. N. Kost, *Zh. Org. Khim.*, **6**, 1022 (1970); *J. Org. Chem. USSR* (*Engl. transl.*), **6**, 1026 (1970).
[581] C. H. Brieskorn and E. Hemmer, *Chem. Ber.*, **109**, 1418 (1976).
[582] T. Sato and H. Obase, *Tetrahedron Lett.*, **1967**, 1633.

[583] G. A. Olah, Y. D. Vankar, and A. L. Berrier, *Synthesis*, **1980**, 45.
[584] P. R. Brook and D. E. Kitson, *J. Chem. Soc., Chem. Commun.*, **1978**, 87.
[585] G. Dudenas, A. Stankevicius, A. N. Kost, and J. Sulekiene, *Zh. Org. Khim.*, **13**, 2185 (1977); *J. Org. Chem. USSR* (*Engl. transl.*), **13**, 2035 (1977).
[586] B. Unterhalt, *Arch. Pharm.* (*Weinheim, Ger.*), **299**, 608 (1966).
[587] S. W. Baldwin and M. T. Crimmins, *Tetrahedron Lett.*, **1978**, 4197.
[588] M. F. Bartlett, D. F. Dickel, and W. I. Taylor, *J. Am. Chem. Soc.*, **80**, 126 (1958).
[589] R. L. Autrey and P. W. Scullard, *J. Am. Chem. Soc.*, **87**, 3284 (1965).
[590] C. A. Grob and P. W. Schiess, *Angew. Chem., Int. Ed. Engl.*, **6**, 1 (1967).
[591] J. A. Marshall, N. H. Andersen, and P. C. Johnson, *J. Am. Chem. Soc.*, **89**, 2748 (1967).
[592] J. A. Marshall and N. H. Andersen, *Tetrahedron Lett.*, **1967**, 1219.
[593] S. W. Baldwin and J. M. Wilkinson, *Tetrahedron Lett.*, **1979**, 2657.
[594] N. I. Kiseleva, Y. A. Baskakov, M. I. Faddeeva, and Y. G. Putsykin, *Zh. Org. Khim.*, **17**, 343 (1981); *J. Org. Chem. USSR* (*Engl. transl.*), **171**, 289 (1981).
[595] Y. L. Chow, *J. Am. Chem. Soc.*, **87**, 4642 (1965).
[596] C. F. Seidel and A. Storni, *Chimia*, **13**, 63 (1959).
[597] K. Maeda, I. Moritani, T. Hosokawa, and S. Murahashi, *J. Chem. Soc., Chem. Commun.*, **1975**, 689.
[598] V. N. Kopranenkov, A. M. Vorotnikov, and E. A. Luk'yanets, *Zh. Obshch. Khim.*, **49**, 2783 (1979); *J. Gen. Chem. USSR* (*Engl. transl.*), **49**, 2467 (1979).
[599] A. V. Serbin, A. N. Flerova, V. N. Yarosh, N. N. Voznesenskaya, E. N. Teleshov, and A. N. Pravednikov, *Vysokomol. Soedin., Ser. A*, **25**, 1204; *Macromol. Comp., Ser. A* (*Engl. transl.*), **25**, 1394 (1983).
[600] K. N. Carter and J. E. Hulse, III, *J. Org. Chem.*, **47**, 2208 (1982).
[601] M. Green and S. C. Pearson, *J. Chem. Soc. B*, **1969**, 593.
[602] J. P. Freeman, *J. Am. Chem. Soc.*, **82**, 3869 (1960).
[603] T. Kiersznicki and A. Rajca, *Pol. J. Chem.*, **53**, 1147 (1979).
[604] V. N. Kopranenkov, E. A. Tarkhanova, and E. A. Luk'yanets, *Zh. Org. Khim.*, **15**, 642 (1979); *J. Org. Chem. USSR* (*Engl. transl.*), **15**, 570 (1979).
[605] J. Bosch, M. Moral, and M. Rubiralta, *Heterocycles*, **20**, 509 (1983).
[606] B. Iddon, D. Price, H. Suschitzky, and D. I. C. Scopes, *J. Chem. Soc., Perkin Trans. 1*, **1983,** 2583.
[607] H. Nishiyama, K. Sakuta, and K. Itoh, *Tetrahedron Lett.*, **25**, 223 (1984).
[608] A. Hassner and E. G. Nash, *Tetrahedron Lett.*, **1965**, 525.
[609] S. Wawzonek and J. V. Hallum, *J. Org. Chem.*, **24**, 364 (1959).
[610] P. A. Grieco, K. Hiroi, J. J. Reap, and J. A. Noguez, *J. Org. Chem.*, **40**, 1450 (1975).
[611] P. A. Grieco and K. Hiroi, *Tetrahedron Lett.*, **1973**, 1831.
[612] P. J. Kocienski and J. M. Ansell, *J. Org. Chem.*, **42**, 1102 (1977).
[613] S. H. Graham and A. J. S. Williams, *J. Chem. Soc.*, **1959**, 4066.
[614] O. Ribeiro, S. T. Hadfield, A. F. Clayton, C. W. Vose, and M. M. Coombs, *J. Chem. Soc., Perkin Trans. 1*, **1983**, 87.
[615] A. Hassner and I. H. Pomerantz, *J. Org. Chem.*, **27**, 1760 (1962).
[616] F. H. Stodola, E. C. Kendall, and B. F. McKenzie, *J. Org. Chem.*, **6**, 843 (1941).
[617] R. L. Autrey and P. W. Scullard, *J. Am. Chem. Soc.*, **90**, 4917 (1968).
[618] A. Carotti, F. Campagna, and R. Ballini, *Synthesis*, **1979**, 56.
[619] J. K. Paisley and L. Weiler, *Tetrahedron Lett.*, **1972**, 261.
[620] Shafiullah and M. A. Ghaffari, *J. Indian Chem. Soc.*, **57**, 762 (1980).
[621] T. T. Takahashi, K. Nomura, and J. Y. Satoh, *J. Chem. Soc., Chem. Commun.*, **1983,** 1441.
[622] C. R. Narayanan and M. S. Parkar, *Chem. Ind.* (*London*), **1974**, 163.
[623] M. Fryberg, L. Avruch, A. C. Oehlschlager, and A. M. Unrau, *Can. J. Biochem.*, **53**, 881 (1975).
[624] J. Klinot, V. Sumanova, and A. Vystrcil, *Collect. Czech. Chem. Commun.*, **37**, 603 (1972).
[625] L. Mangoni and M. Belardini, *Tetrahedron Lett.*, **1963**, 921.
[626] G. P. Moss and S. A. Nicolaidis, *J. Chem. Soc. D*, 1969, 1072.

[627] K. K. Purushothaman and A. Sarada, *Indian J. Chem., Sect. B*, **14**, 635 (1976).
[628] T. Hirata, R. Ideo, and T. Suga, *Chem. Lett.*, **1977**, 283.
[629] E. S. Olson and J. H. Richards, *J. Org. Chem.*, **33**, 434 (1968).
[630] T. Hirata, R. Ideo, and T. Suga, *Chem. Lett.*, **1977**, 711.
[631] L. Cambi and G. Bargigia, *Chim. Ind. (Milan)*, **47**, 517 (1965).
[632] C. Fizet and J. Streith, *Tetrahedron Lett.*, **1974**, 3187.
[633] R. Appel, R. Kleinstuck, and K.-D. Ziehn, *Chem. Ber.*, **104**, 2025 (1971).
[634] G. Sosnovsky, J. A. Krogh, and S. G. Umhoefer, *Synthesis*, **1979**, 722.
[635] P. J. Foley, *J. Org. Chem.*, **34**, 2805 (1969).
[636] H. Suzuki, T. Fuchita, A. Iwasa, and T. Mishina, *Synthesis*, **1978**, 905.
[637] T. Van Es, *J. Chem. Soc.*, **1965**, 1564.
[638] J. K. Chakrabarti and T. M. Hotten, *J. Chem. Soc., Chem. Commun.*, **1972**, 1226.
[639] G. Ah-Kow, C. Paulmier, and P. Pastour, *Bull. Soc. Chim. Fr.*, **1976**, 151.
[640] A. Saednya, *Synthesis*, **1982**, 190.
[641] D. Dauzonne, P. Demerseman, and R. Royer, *Synthesis*, **1981**, 739.
[642] A. Saednya, *Synthesis*, **1983**, 748.
[643] E. Vowinkel and J. Bartel, *Chem. Ber.*, **107**, 1221 (1974).
[644] J. G. Krause and S. Shaikh, *Synthesis*, **1975**, 502.
[645] G. A. Olah and Y. D. Vankar, *Synthesis*, **1978**, 702.
[646] G. A. Olah, S. C. Narang, and A. Garcia-Luna, *Synthesis*, **1980**, 659.
[647] G. Rosini, G. Baccolini, and S. Cacchi, *J. Org. Chem.*, **38**, 1060 (1973).
[648] P. Molina, M. Alajarin, and M. J. Vilaplana, *Synthesis*, **1982**, 1016.
[649] H. G. Foley and D. R. Dalton, *J. Chem. Soc., Chem. Commun.*, **1973**, 628.
[650] J. H. Hunt, *Chem. Ind. (London)*, **1961**, 1873.
[651] G. Maerkle, J. B. Rampal, and V. Schoeberl, *Tetrahedron Lett.*, **1979**, 3141.
[652] J.-P. Dulcere, *Tetrahedron Lett.*, **22**, 1599 (1981).
[653] G. A. Olah, Y. D. Vankar, and A. Garcia-Luna, *Synthesis*, **1979**, 227.
[654] T.-L. Ho, *Synthesis*, **1975**, 401.
[655] T.-L. Ho and C. M. Wong, *Synth. Commun.*, **1975**, 299.
[656] C. Bernhart and C.-G. Wermuth, *Synthesis*, **1977**, 338.
[657] J. M. Prokipcak and P. A. Forte, *Can. J. Chem.*, **49**, 1321 (1971).
[658] M. J. Miller and G. M. Loudon, *J. Org. Chem.*, **40**, 126 (1975).
[659] J. H. Pomeroy and C. A. Craig, *J. Am. Chem. Soc.*, **81**, 6340 (1959).
[660] O. Attanasi, P. Palma, and F. Serra-Zanetti, *Synthesis*, **1983**, 741.
[661] D. L. J. Clive, *J. Chem. Soc. D*, **1970**, 1014.
[662] J. N. Denis and A. Krief, *J. Chem. Soc., Chem. Commun.*, **1980**, 544.
[663] D. Cooper and S. Trippett, *Tetrahedron Lett.*, **1979**, 1725.
[664] T. Mukaiyama, T. Fujisawa, and O. Mitsunobu, *Bull. Chem. Soc. Jpn.*, **35**, 1104 (1962).
[665] T.-L. Ho and C. M. Wong, *J. Org. Chem.*, **38**, 2241 (1973).
[666] T. Saraie, T. Ishiguro., K. Kawashima, and K. Morita, *Tetrahedron Lett.*, **1973**, 2121.
[667] J. A. Findlay and C. S. Tang, *Can. J. Chem.*, **45**, 1014 (1967).
[668] M. E. Jury and S. J. Miller, *J. Am. Chem. Soc.*, **103**, 1984 (1981).
[669] X. Lusinchi, *Tetrahedron Lett.*, **1967**, 177.
[670] T. J. Bentley and J. F. McChie, *Tetrahedron Lett.*, **1965**, 2497.
[671] G. Stokker, *J. Org. Chem.*, **48**, 2613 (1983).
[672] Z. Witczak and M. Krolikowska, *Pol. J. Chem.*, **53**, 1033 (1979).
[673] Z. Witczak and M. Krolikowska, *Pol. J. Chem.*, **52**, 2479 (1978).
[674] S. Sakane, K. Maruoka, and H. Yamamoto, *Tetrahedron Lett.*, **24**, 943 (1983).
[675] J. B. Hester, Jr., *J. Org. Chem.*, **39**, 2137 (1974).

CHAPTER 2

THE PERSULFATE OXIDATION OF PHENOLS AND ARYLAMINES (THE ELBS AND THE BOYLAND–SIMS OXIDATIONS)

E. J. Behrman

The Ohio State University, Columbus, Ohio

CONTENTS

ACKNOWLEDGMENTS

I thank the following: the Editorial Board of Organic Reactions for their advice, and in particular Drs. Robert Joyce and Ralph Hirschmann for the meticulous care that they devoted to my typescript, the late Dr. Burnett M. Pitt for introducing me to the Elbs oxidation, and Prof. J. O. Edwards, to whom this paper is dedicated, for invaluable help over the years.

THE ELBS OXIDATION: INTRODUCTION

A phenolate anion reacts with persulfate ion in alkaline solution to yield a product in which a sulfate group enters the ring *para* or *ortho* to the phenolic group. *Para* substitution predominates. Subsequent acid-catalyzed hydrolysis yields the dihydric phenol.

$$C_6H_5O^- + S_2O_8^{2-} \longrightarrow p\text{-}{}^-O_3SOC_6H_4O^- + o\text{-}{}^-O_3SOC_6H_4O^-$$

$$\xrightarrow[H_2O]{H^+} p\text{-}HOC_6H_4OH + o\text{-}C_6H_4(OH)_2$$

The reaction was discovered by Karl Elbs* in 1893[1] and named the *Elbs persulfate oxidation.*[2] The reaction is generally applicable to *ortho-*, *meta-*, and

* A brief obituary of Karl Elbs appears in *Ber.*, **66**, 74 (1933). For further bibliographical material, see J. S. Fruton, *A Bio-Bibliography for the History of the Biochemical Sciences Since 1800*, American Philosophical Society, Philadelphia, 1982, p. 191.

para-substituted phenols with isomer distributions as shown:

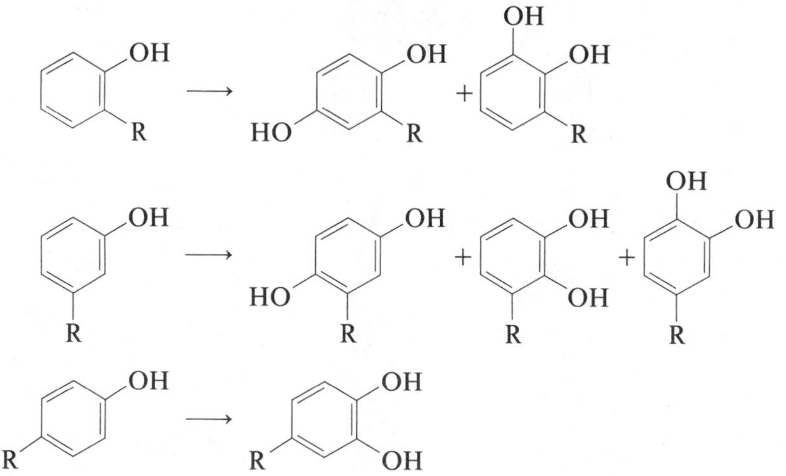

The yields are not very high, particularly from *para*-substituted phenols, but the major contaminant is usually unchanged starting material that can be separated easily from the intermediate sulfate ester by solvent extraction. Other generally oxidizable groups such as an aldehyde or a double bond are usually not affected under the reaction conditions. The reaction was last thoroughly reviewed in 1951.[3] Subsequent partial reviews include Refs. 4–8. T. R. Seshadri [see W. Baker and S. Rangaswami, *Biograph. Memoirs Fell. Roy. Soc.*, **25**, 505 (1979) and Fruton, *loc. cit.*, p. 661] has made major contributions to the development of the Elbs oxidation. Nearly 30% of the references in this review are due to him and his colleagues.

MECHANISM

Studies of the kinetics of the reaction[9–11] reveal a first-order dependence on both persulfate and phenol and a positive salt effect. The relationship between pH and reaction rate shows that the phenolate ion is the reactive species. Allyl acetate, a reagent that reacts rapidly with sulfate radical ions, has no effect on either the rate of disappearance of persulfate or the rate of product formation (for *o*-nitrophenol as substrate). These data and the substituent effects discussed below make it clear that the reaction proceeds via electrophilic attack of the persulfate ion on the phenolate ion.

The observed ionic strength effect is consistent with a reaction between two ions of the same charge. The phenolate ion is, of course, much more susceptible to electrophilic attack than is the undissociated phenol. Accordingly, the pH at which a maximum rate is achieved is dependent on the pK_a. The effect of a series of substituents on the reaction rate has been reported,[10] and representative second-order rate constants are given in Table A. Again, the substituent effects are in the expected direction for electrophilic attack by the persulfate ion. The

TABLE A. Representative Rate Constants for Reaction of Phenols and Persulfate Ion[a]

Substituent	Rate Constant $10^2 k_2$ ($L\,mol^{-1}\,s^{-1}$)
H	1.93
o-NO_2	0.15
m-NO_2	0.32
o-CN	0.24
m-CN	0.38
o-CHO	0.53
m-CHO	0.73
o-CO_2^-	4.08
m-CO_2^-	0.55
o-Cl	1.60
m-Cl	0.58
o-Br	1.56
m-Br	0.64
o-I	2.12
m-I	0.66
o-F	1.61
m-F	1.25
o-CH_3	8.42
m-CH_3	2.40
o-C_4H_9-*t*	16.40
m-C_4H_9-*t*	1.63
o-OCH_3	31.70
m-OCH_3	4.25

[a] Conditions: 30°, 1.7 *M* KOH, Ref. 10.

question of whether initial attack is at carbon or at oxygen followed by rearrangement has not been definitely settled. This point is discussed in further detail below.

Electrophilic attack by the persulfate ion on each of the three resonance forms of the phenolate anion will give rise to products **1**, **2**, and **3**:

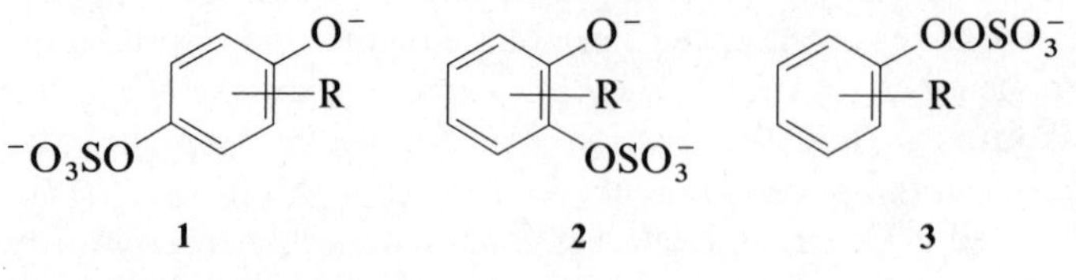

O^-, R, ^-O_3SO **1** O^-, R, OSO_3^- **2** $OOSO_3^-$, R **3**

O^-, OSO_3^-, R, + **4** O^-, R, +, ^-O_3SO **5**

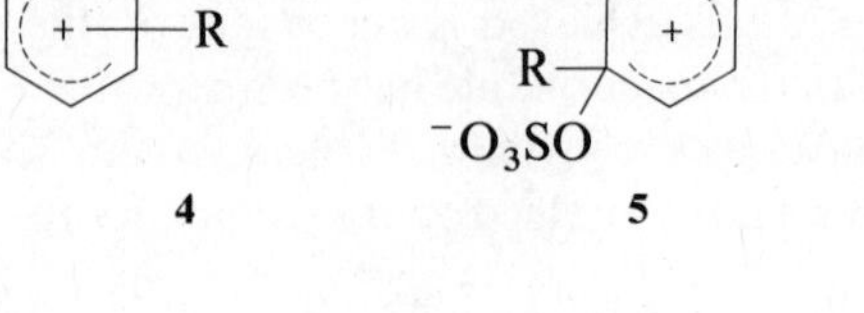

There are, in addition, possibilities for electrophilic attack by persulfate at two sets of *ipso* positions to give **4** and **5**.

The final product could arise by direct attack at either carbon or oxygen (or an *ipso* position) followed by rearrangement. The evidence bearing on the site of initial attack rests principally on the kinetic effects of substitutents in the *ortho* and *meta* positions. The overall correlation with Hammett substituent constants is much better for the assumption of attack at oxygen (or the oxygen *ipso* position) than for attack at carbon. Thus the order of reactivity of each pair of *ortho*- and *meta*-substituted phenols (except for carboxylate) in Table A is consistent with what would be expected on the assumption of attack at oxygen. This conclusion is based on the fact that the substituent with the more negative Hammett sigma constant should give the higher rate and that the relative rates should be reversed depending on whether rate-limiting attack at carbon or oxygen is assumed.[10] The principal contrary evidence[10] originally adduced to support attack at carbon came from a comparison of the rates of oxidation of 2,4- and 2,6-disubstituted phenols. Less reliance must be placed on these data because the yields from *para*-substituted phenols are too low to provide a basis for reliable conclusions.[12] This *caveat* also applies to arguments in support of attack at oxygen based on the relative reactivity of *p*-fluorophenol.[13] The preponderance of the evidence appears to support attack at the phenolic oxygen followed by inter- or intramolecular rearrangement.

Although *para*-substituted phenols react at normal rates, the yields of *o*-sulfate are typically very low, and little unchanged starting material is generally recovered. These yields can be substantially increased by increasing the persulfate:phenol ratio,[14] in contrast to the result for *ortho*- and *meta*-substituted phenols, where increasing the persulfate:phenol ratio usually decreases the yield of *p*-sulfate. These facts argue for a pathway in which reaction with 1 mol of persulfate leads to an intermediate that neither gives the normal Elbs product nor reverts to starting material. It is possible to view structures **3**, **4**, and **5** as intermediates of this sort. An *ipso* intermediate of type **5** has also been implicated in the reaction of persulfate with 2,4,6-trichlorophenol by the formation of substantial quantities of chloride ion[12] and in the formation of 2,5-dihydroxy-3-iodo-4-methoxybenzoic acid by persulfate oxidation of 2-hydroxy-3,5-diiodo-4-methoxybenzoic acid.[15]

SCOPE AND LIMITATIONS

Phenols substituted with a wide variety of functional groups are successfully oxidized by the Elbs procedure in spite of the large redox potential of persulfate ion. This is due to the fact that, although persulfate ion is capable of oxidizing many substituents, these reactions do not take place at significant rates under the typical conditions of the Elbs oxidation. Alcohols, aldehydes, and olefins are essentially inert to the action of persulfate at room temperature and below in aqueous alkali. Some oxidative cleavage of the double bond of stilbenes is reported,[16] but coumarins, which react as the *o*-hydroxycinnamic acid dianions, generally give good yields (Table IV). Some functional groups, however, undergo reaction with persulfate more rapidly than do typical Elbs substrates. Among these are thiol groups, which are oxidized by persulfate to disulfides.[17] Aliphatic amines also appear to be oxidized sufficiently rapidly (to unknown products) to suggest effective competition with the Elbs oxidation.[18] *p*-Nitrosophenol is oxidized to *p*-nitrophenol without any observable formation of the *o*-sulfate.[19] For summaries of the reactions of persulfate with a variety of organic substrates, see Refs. 8 and 20–22.

Another source of interference is the instability of some substrates under alkaline conditions, for example, 4-methoxycoumarins.[23,24] When this instability is due to reaction of the phenolate anion with oxygen, the difficulty can be circumvented by working in an inert atmosphere. Thus 1,3,5-trihydroxybenzene is successfully oxidized under nitrogen,[25] and the yield of 5,8-dihydroxyflavone is increased 15% by purging the system with nitrogen.[26] On the other hand, certain quinones react rapidly with hydroxyl ion. The oxidation of 5-hydroxy-1,4-naphthoquinone (juglone) is unsuccessful because of this fact.[27] Hydroquinone can be oxidized to quinhydrone by persulfate.[28]

There are a number of other examples in which the substrate is recovered unchanged (Table XV). It is not known why these compounds fail to react, but perhaps increasing the persulfate:substrate ratio might be beneficial.[14]

Isomer Distribution

There are only a few reports on the isomer distribution in the Elbs persulfate oxidation.[11,29–32] Ratios of *para* to *ortho* isomers are reported for seven subtrates based on isolated yields.[29] Table B lists those subtrates for which ratios have been determined by methods that do not depend on the isolation procedure. The *para*:*ortho* ratio is reported to increase with decreasing ionic strength.[11]

Ratios of *para* to *ortho* isomer determined by isolation are in the range 6–22 and are thus not widely different from the three examples determined by analytical methods, with the following notable exception. In the oxidation of two *meta*-substituted phenols, the 3,4-dihydroxy isomer could not be found.[29] It has since been shown that this isomer is indeed formed in substantial proportion.[30] Failure to detect it earlier was simply a result of the isolation scheme.

TABLE B. ISOMER DISTRIBUTION

Substrate	Products (Relative Yield)	Method	Ref.
Phenol	Hydroquinone (5.9–2.3), catechol (1)[a]	GLC	11
m-Hydroxybenzoic acid	2,5-Dihydroxybenzoic acid (8), 3,4-dihydroxybenzoic acid (3), 2,3-dihydroxybenzoic acid (1)	GLC, HPLC	30
2-Pyridone	2,5-Dihydroxypyridine (11.5), 2,3-dihydroxypyridine (1)	Colorimetric	31

[a] The *para*:*ortho* ratio is a function of pH, ionic strength, temperature, and ratio of reactants.

Byproducts

Substantial quantities of apparently polymeric material of the humic acid type have been noted as products in the persulfate oxidation of phenols. For example, a 40% yield of a dark brown amorphous material is produced in the oxidation of *m*-hydroxybenzaldehyde.[33] Similar observations have been noted incidently throughout the literature, but the products have not been well characterized. The only extensive studies show that dihydric phenols, aminophenols, and even monohydric phenols such as phenol itself, *o*-cresol, and salicylic acid all give rise to substantial quantities of "humic acids" when oxidized at persulfate:phenol ratios greater than 1.[34,35] It should be recalled that the yield of the Elbs product (the sulfate ester) generally increases as the persulfate: phenol ratio is decreased.[9] In addition to the humic acids, biphenyls have been detected as byproducts,[36] especially with activated phenols. These presumably arise from radical coupling reactions, and their formation might possibly be prevented by the inclusion of radical trapping agents such as allyl alcohol. The biphenyls, however, do not appear to be formed in large quantities.

COMPARISON WITH OTHER METHODS

Many methods exist for the synthesis of hydroquinones. A superb and extensive summary can be found in Wedemeyer's volume of Houben-Weyl.[37] There are also less detailed treatments.[38–42] The methods can be divided into those that involve replacement of some substituent other than hydrogen and those in which hydrogen is replaced (direct methods). The first group includes alkali fusion of halophenols and phenolsulfonic acids, the Dakin oxidation of phenolic aldehydes and the related Baeyer–Villiger oxidation of hydroxyacetophenones, diazotization and hydrolysis of aminophenols, the Bucherer reaction, and hydrolysis of halophenols via Grignard reagents. The direct methods, which include the Elbs oxidation, usually offer the considerable advantage of fewer steps from starting material to product. Direct methods other than the Elbs oxidation include the use of Fenton's reagent, oxidation by hydrogen peroxide or peracids, electrochemical oxidation of phenols, reduction of quinones available by oxidation of phenols with Fremy's salt, and three promising newer methods: (1) benzeneselenic anhydride oxidation of phenols to *o*-quinones;[43] (2) an *ortho*-hydroxylation procedure[44] using copper(I) chloride and oxygen in acetonitrile at 0–50° with yields of 70–90%; and (3) a *para*-hydroxylation method [45] involving alkylation with cyclopentadiene, isomerization, and finally oxidation with hydrogen peroxide in acetonitrile. These latter two methods have not yet been tested for generality.

The Elbs oxidation remains a useful procedure, in spite of its generally moderate yields, because of the simplicity of the process and the fact that the conditions for the synthesis are compatible with a number of sensitive functional groups that might not survive other procedures. It offers the unique advantage that the sulfate ester is produced on the way to the hydroquinone.

This means that the water solubility of the sulfate ester can be used to advantage in separating the product from both unchanged starting materials and from byproducts. Beyond this, the production of the unsymmetrical alkali-stable hydroquinone sulfate ester can be used to synthetic advantage as it allows distinction to be made between the hydroxyl groups in what otherwise might be a symmetrical molecule.[2]

EXPERIMENTAL CONDITIONS

Most of the studies reported in the literature deal with isolated yields. There are few consistent trends to be derived from these data. In a few reactions, yields have been determined by an analytical method that is not subject to the variables of a particular isolation method (Table B). The major factors that have been shown to influence yield are pH, ratio of reactants, free-radical traps, metal ion chelators, and, perhaps, ionic strength. In addition, small effects have been noted by varying the temperature.

pH and Nature of the Base

The stoichiometry of the reaction is

$$R(H)OH + S_2O_8^{2-} + 2OH^- \longrightarrow {}^-OROSO_3^- + SO_4^{2-} + 2H_2O$$

One mole of alkali is needed to ionize the phenol and a second mole to neutralize the proton displaced from the ring. A quantity of base less than this requirement will decrease the yield correspondingly; the addition of alkali in excess of this requirement appears to offer no advantage and may actually decrease the yield as the result of an ionic strength effect.[11] The identity of the base may influence the yield. Tetramethylammonium hydroxide is reported to give better yields than either potassium hydroxide or sodium hydroxide in some reactions.[46] Tetraethylammonium hydroxide gives much better yields than does tetramethylammonium hydroxide in the oxidation of 5-hydroxyflavone.[26] Sodium hydroxide is reported to give better yields than potassium hydroxide.[47]

Ratio of Reactants

With 2-pyridone as substrate, substrate:persulfate ratios in the range 10–20 give yields of 85–90%; a 1:1 ratio gives a yield of only about 55%, and excess persulfate sharply decreases the yield.[9] With phenol as substrate, however, excess substrate appears to lower the yield slightly.[11] The reason for this apparent discrepancy is not known. The former result is probably usually true for *ortho*- and *meta*-substituted phenols since the monosulfate formed as the initial product can undergo further attack by persulfate to yield a disulfate.[32,48,49] However, for four *p*-substituted phenols, increased yields are obtained with a substrate:persulfate ratio of 0.3.[14]

Under typical synthetic conditions, the phenol will usually be the cost-limiting component, so that it will seldom prove practicable to use excess phenol. The presence of excess phenol during the initial stages of the reaction, at least, can be achieved by adding a solution of the persulfate slowly to the solution of phenol. Alternatively, one can add potassium persulfate as a solid to the solution of phenol in alkali and take advantage of the fact that this salt dissolves slowly relative to the sodium and ammonium salts. However, improved yields can sometimes be obtained with the more concentrated solutions of persulfate made possible by using the more soluble ammonium salt.[50] At 20°, saturated solutions of potassium, sodium, and ammonium persulfates in water are 0.17, 2.3, and 2.5 *M* respectively.[51]

Nature and Position of Ring Substituents

Reactions of a set of *ortho*-, *meta*-, and *para*-substituted phenols with persulfate, using a colorimetric method to determine the dihydric phenol produced following acid hydrolysis, show that the *ortho*- and *meta*-substituted phenols give yields in the range 60–75% regardless of the nature of the substituent.[12] The yields from the *para*-substituted phenols are, however, in only the 15–20% range.

Free-Radical Traps and Metal Ion Chelators

The yield of 2-phenylhydroquinone from *o*-phenylphenol is increased by about 9% on the addition of allylbenzene to the reaction mixture and by about 5% by the addition of ethylenediaminetetraacetic acid (EDTA).[50] Allyl alcohol decreases the formation of dark-colored materials during the oxidation of guaiacol and *o*-*tert*-butylphenol.[10] By contrast, allyl acetate has no effect on yield in the oxidation of *o*-nitrophenol.[9] This result may reflect a difference between activated and deactivated phenols.

Temperature and Ionic Strength

Temperature variation over the range 0–70° appears to affect the yield only slightly, perhaps by a decrease of 3–5% as temperature increases. The oxidation of water by persulfate may be a competing reaction at the upper end of this temperature range, especially for phenols that react very slowly.[52]

A significant drop in yield with an increase in ionic strength amounting to about 15% for an increase from 0.05 to 0.3 *M* is reported,[11] but this effect has been questioned.[9] These studies should be repeated, using a wider range of substrates.

Effects of Ferrous Ion

It has been recommended that ferrous ion be added to reaction mixtures.[53] This addition may have been rationalized on the assumption that the Elbs oxidation is a free-radical process with ferrous ion serving as initiator. However, it has since been shown that the reaction does not involve free-radical intermediates and further that the addition of ferrous ion does not affect the rate of the reaction.[9,10] Indeed, there is evidence that metal ions reduce the yield, presumably by the promotion of competing free-radical processes leading to other organic products and also by direct consumption of persulfate.[47] See, however, Ref. 53a.

Solvents Other Than Water

Although Elbs oxidations are usually carried out in aqueous solution, pyridine[54] or 1,4-dioxane[10] can be used as cosolvents to aid in solubilizing certain phenols. Kinetic studies in aqueous mixtures of ethanol, *tert*-butyl alcohol, and acetonitrile give linear plots of log k versus $1/D$ (where D is the dielectric constant) with a negative slope.[55] There exists the possibility, as yet unexplored, for carrying out the Elbs oxidation in pure organic solvents since persulfate can be solubilized by crown ethers and quaternary ammonium salts.[56]

Conditions for the Hydrolysis of Aryl Sulfates

Rate constants for the acid-catalyzed hydrolysis of a variety of aryl sulfates vary by a factor of about 10 from the *p*-nitrophenyl sulfates (highest) to the *p*-methoxyphenyl sulfates (lowest).[57] Hydrolysis takes place with cleavage of the sulfur–oxygen bond.[58]

EXPERIMENTAL PROCEDURES

Phenylhydroquinone.[50] A solution of 17.0 g (0.1 mol) of *o*-phenylphenol, 0.5 g (0.0017 mol) of EDTA, 34.0 g (0.85 mol) of sodium hydroxide, and 2.4 g (0.02 mol) of allylbenzene in 180 mL of distilled water was prepared. This solution was cooled to 5° and kept under nitrogen while a solution of 22.8 g (0.1 mol) of ammonium persulfate in 100 mL of distilled water was added over a period of 1 hour. The resulting solution was kept at 5° for an additional 4 hours. Then 100 mL of methylene chloride was added and the mixture acidified with 2 *N* HCl to pH 1. A small amount of tar was filtered off on glass wool and the filter washed with water. The filtrate was extracted with methylene chloride, after which the aqueous phase was treated with 25 mL of concentrated HCl and hydrolyzed on a steam bath for 1 hour under nitrogen. The solution was cooled and extracted with methylene chloride (4 × 150 mL). Evaporation of the methylene chloride left the product, which crystallized to give 7.9 g of brown crystals of 95% purity (38.4% yield of pure product).

The yield dropped to 29% in the absence of allylbenzene and to 25% in the absence of both allylbenzene and EDTA.

2,5-Dihydroxypyridine (5-Hydroxy-2-pyridone).[31] In this procedure, 38 g (0.4 mol) of 2-pyridone and 80 g (2 mol) of sodium hydroxide were dissolved in 1.5 L of water. The solution was cooled to 5°, and then 135 g (0.5 mol) of potassium persulfate was added all at once. The mixture was stirred for 20 hours while the temperature was allowed to rise to 20°. The reaction mixture was filtered, cooled, and brought to pH 0.75 with concentrated sulfuric acid. The mixture was hydrolyzed at 100° for 30 minutes. The hydrolysate was cooled to 5°, brought to pH 6.5 with 10 *N* sodium hydroxide under nitrogen, and evaporated to dryness *in vacuo*. After a final drying over phosphorus pentoxide, the salt cake was thoroughly extracted with 2-propanol in a Soxhlet apparatus. The

2-propanol extract was decolorized with charcoal and then concentrated until crystals began to form. After standing overnight at −10°, the solution deposited 19 g (42%) of crude 2,5-dihydroxypyridine, which, after two recrystallizations from ethanol, gave 8 g of nearly colorless crystals that darkened at 230° and decomposed at 250–260° without melting. UV (water) nm max (ε): 230.5 (7390), 320 (5620).

The original procedure recommended the addition of 2 g of ferrous sulfate. As discussed earlier, ferrous ion offers no advantage and indeed merely decreases the yield by reduction of persulfate.

Hydrogen Cytosine-5-sulfate Monohydrate and 5-Hydroxycytosine.[59] To a solution of 2 g (0.018 mol) of cytosine in 100 mL of 1.0 *N* KOH was added 7.3 g (0.027 mol) of potassium persulfate. The solution was stirred at 25° for 18 hours. The pale yellow solution was acidified by the addition of 9 mL of concentrated HCl with cooling. Hydrogen cytosine-5-sulfate precipitated from the solution. It was washed with cold water, acetone, and ether to give 3.3 g (89%) of crude material. One recrystallization from 45 mL of water gave 2.6 g (70%) of pure product as the monohydrate. UV (water, pH 6.8) nm max (ε): 277 (5400).

A 3-g sample of hydrogen cytosine-5-sulfate monohydrate (0.014 mol) and 7 mL of 6 *N* HCl was heated in a boiling water bath for 15 minutes. Cooling produced 2 g (85%) of 5-hydroxycytosine hydrochloride. This material was dissolved in 30 mL of warm water and the pH adjusted to 7 with 4 *N* potassium hydroxide. The precipitate of 5-hydroxycytosine (1.2 g, 77%) was washed with water, acetone, and ether. UV (water, pH 6.8) nm max (ε): 288 (5000).

5,8-Dihydroxy-3-ethoxy-7,3′,4′,5′-tetramethoxyflavone.[60] To a solution of 1.4 g (0.0035 mol) of 5-hydroxy-3-ethoxy-7,3′,4′,5′-tetramethoxyflavone in 20 mL of pyridine was gradually added a solution of 1 g of potassium hydroxide in 250 mL of water. To this mixture was added during 2 hours a solution of 1.4 g (0.0052 mol) of potassium persulfate in 50 mL of water. After 24 hours, the solution was acidified, filtered, and then extracted three times with ether to remove unchanged starting material. Concentrated HCl (25 mL) and sodium sulfate (2 g) were then added to the aqueous phase, and the mixture was heated on a boiling water bath for 30 minutes. A yellow precipitate of product separated. This was combined with some further material obtained by ether extraction to yield 0.7 g (48%) of product. Recrystallization from ethanol gave deep yellow short needles, mp 190–192°.

5,8-Dihydroxy-2-methyl-4′,5′-dihydro[furano-3′2′:6,7-chromone](8-Hydroxydihydronorvisnagin).[61] To a solution of 1 g of dihydronorvisnagin (5-hydroxy-2-methyl-4′,5′-dihydro[furano-3′,2′:6,7-chromone]) in 20 mL of pyridine and 18 mL of 10% aqueous tetramethylammonium hydroxide was added 2.2 g of potassium persulfate dissolved in 150 mL of water during 4 hours. The reaction mixture was kept for 20 hours under nitrogen at 15–20°. The deep red solution was then acidified to pH 2 (Congo red) and filtered to remove 0.4 g of a brown precipitate. The filtrate was extracted twice with ether, the ether evaporated, and the residue combined with the brown precipitate. The combined residues were extracted with chloroform. On chromatographic purification of the chloroform solution on alumina, 0.3 g of starting material was recovered.

The aqueous filtrate was treated with 2 g of sodium sulfite, 30 mL of concentrated HCl, heated for 30 minutes at 90°, and cooled. Extraction with ether (5 × 30 mL) gave 0.45 g (42%) of 8-hydroxydihydronorvisnagin, which, following treatment with Norit, crystallized from methanol containing sulfurous acid in deep yellow thin plates, mp 260–262°.

THE BOYLAND–SIMS OXIDATION: INTRODUCTION

By analogy with the Elbs persulfate oxidation of phenols, it might be expected that aromatic amines would react with persulfate to give *p*-aminoaryl sulfates. Although the Elbs reaction had been known since 1893, it was not until 60 years later that Boyland et al.[62] reported the extension of this reaction to aromatic amines. In accordance with expectations, aminoaryl sulfates were indeed the major products of the reaction, but, unexpectedly, the substitution took place exclusively *ortho* to the amino group rather than predominantly in the *para* position as in the phenol oxidation. *Para* substitution takes place only if the *ortho* positions are occupied by substituents other than hydrogen. Boyland and Sims explored the preparative aspects of this reaction in a series of papers.[16,62–65] It seems appropriate to name the reaction the Boyland–Sims oxidation.[66] Primary, secondary, and tertiary aromatic amines are all converted to the corresponding *o*-aminoaryl sulfates under conditions similar to those used for the Elbs oxidation, that is, room temperature or below, aqueous alkali, and equimolar quantities of amine and persulfate.

MECHANISM

The mechanistic evidence favors a polar rather than a free-radical reaction involving electrophilic displacement by the peroxide oxygen on the unprotonated amine.[66] In particular, radical traps have no effect on either the rate or extent of product formation. However, a single electron transfer mechanism is possible provided that the radicals are confined to a solvent cage.[66a] The rate law, like that for the Elbs persulfate oxidation, is $v = k[S_2O_8^{2-}]$ [amine]. Selected rate constants are given in Tables C–E. The exclusive *ortho* orientation of the entering sulfate group could, in principle, arise from attack at the (1) *ortho* carbon atom assisted by interaction with the amino group (**6**), (2) nitrogen atom followed by rearrangement (**7**), or (3) *ipso* carbon atom followed by rearrangement (**8**).

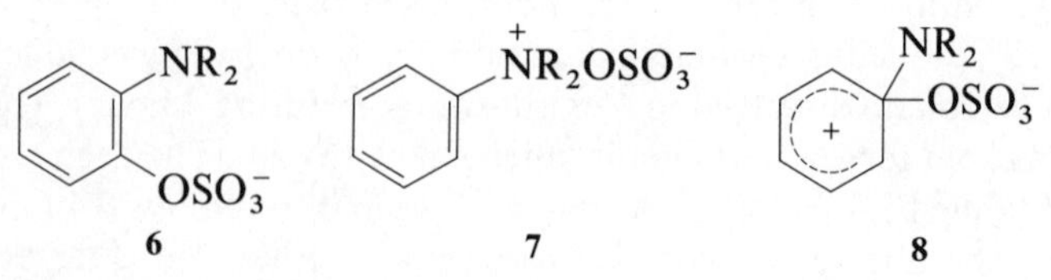

NR_2, OSO_3^- — **6** $\overset{+}{N}R_2OSO_3^-$ — **7** NR_2, OSO_3^- — **8**

TABLE C. SELECTED RATE CONSTANTS FOR OXIDATION OF PRIMARY ANILINES BY PERSULFATE[a]

Substituent	Rate constant $10^3\,k_2$ (L mol^{-1} s^{-1}) *o*	*m*	*p*
—H	12	12	12
—OCH_3	56	17	165
—CH_3	27	18	32
—F	4	5	17
—Cl	3	5	15
—CO_2^-	3	3	4
—NO_2	0.15	1	0.3

[a] Conditions: 30°, pH 7, 20% aqueous ethanol (v/v).[67]

TABLE D. SELECTED RATE CONSTANTS FOR OXIDATION OF PRIMARY, SECONDARY, AND TERTIARY ANILINES BY PERSULFATE.[a]

Substrate	Rate Constant, $10^3\,k_2$ (L mol^{-1} s^{-1})
Aniline	5
N-Methylaniline	70
N,*N*-Dimethylaniline	28

[a] Conditions: 30°, pH 7, 50% aqueous ethanol (v/v).[68] The data in Ref. 69 suggest that rate constants obtained in 50% ethanol can be converted to those expected in 20% ethanol by multiplying by a factor of about 2.

TABLE E. RATE Constants for OXIDATION OF SUBSTITUTED *N*,*N*-DIMETHYLANILINES BY PERSULFATE.

Substituent	Rate Constant, $10^3\,k_2$ (L mol^{-1} s^{-1})	Ref.
4-OCH_3	91.5	70
3-OCH_3	18	70
4-CH_3	38.5	70
3-CH_3	24	70
4-Cl	18	70
2-CH_3-4-Cl	1.8	71
2-CH_3-3-Cl	1	71
2,4-$(CH_3)_2$	2.5	71
2,3-$(CH_3)_2$	2.1	71

[a]Conditions: 30°, pH 7,[70] or 0.1 *M* KOH, 50% aqueous ethanol (v/v).[71]

The observed effects of substituents on the rate of reaction eliminate rate-limiting attack at the *ortho* carbon atom[66–71] for primary and tertiary anilines. The choice between the other two possibilities appears to have been solved for tertiary anilines and may be generally applicable. Intermediate **7**, R = methyl, was synthesized and was shown under the reaction conditions not to rearrange to the *o*-sulfate but rather to hydrolyze as shown.[72] Inasmuch as the substituent effects for a series of tertiary anilines are the same as those for primary anilines, *ipso* attack followed by rearrangement seems the most likely alternative.[70,71]

$$C_6H_5\overset{+}{N}(CH_3)_2OSO_3^- \xrightarrow{H_2O} C_6H_5N(CH_3)_2\!\rightarrow\!O + H_2SO_4$$

When a substrate such as 2,6-dimethylaniline is oxidized, the *para* sulfate is formed. This might occur either by direct attack of persulfate at the *para* carbon or by an intermolecular rearrangement. Kinetic measurements of the oxidation of 2,6- and 2,4-disubstituted anilines would be revealing. Since no *p*-substituted products are ordinarily formed, direct attack at the *para* position must be slow compared with the rate of *ortho* substitution. If, however, the rate of oxidation of the 2,6 isomer is approximately equal to that of the 2,4 isomer (as the limited data in Ref. 73 suggest), we might assume an intermolecular rearrangement.

SCOPE AND LIMITATIONS

The Boyland–Sims oxidation of aromatic amines is not as well represented in the literature as the Elbs oxidation of phenols. Consequently, the scope and limitations of the reaction are less well known. The general limitations on the Elbs oxidation apply since the two reactions are usually carried out under similar conditions. The principal possible difference in the reaction conditions (although most Boyland–Sims reactions have been run in dilute alkali) is due to the widely different values for the pK_a of typical aromatic amines compared with phenols: phenol has a pK_a of 10, while aniline has a pK_a of 4.6. Thus, while the Elbs oxidation of phenol should be run above pH 11, the Boyland–Sims oxidation of aniline can be carried out at neutrality because the reactive species are the phenolate anion and the uncharged amine, respectively. Therefore, the Boyland–Sims reaction can be performed in the presence of alkali-sensitive functional groups, in contrast to the Elbs oxidation.

Overall yields in the Boyland–Sims oxidation appear to be lower than those in the Elbs oxidation. Thus 2-pyridone gives a yield of 85% of 2,5-dihydroxypyridine,[9] in contrast to a yield of 55% of 2-amino-3-hydroxypyridine from 2-aminopyridine.[66] It is reasonable to attribute this difference to the more facile formation of condensation polymers of the humic acid type from aromatic amines.

Isomer Distribution

When there is a free *ortho* position, *ortho* substitution is generally exclusive, although small quantities of *para*-substituted products have been detected in the oxidations of three related anilines: anthranilic acid, *o*-aminoacetophenone, and kynurenine.[64] 3-Methylindole is attacked at all free positions in the benzene ring and so may be reacting by a different mechanism.[74] The ratio of the two possible *ortho* isomers for *meta*-substituted anilines has not been studied to any extent; however, the 6-sulfate (the least sterically hindered) is the major product in the oxidation of 3-methylaniline and 3-chloroaniline.[63]

Byproducts

In addition to sulfation of the ring, competitive oxidation reactions occur at the nitrogen atom. A number of these byproducts have been isolated under typical Boyland–Sims conditions; others, under more acidic conditions and include imines, quinones, and their condensation products. The final stages of condensation are a humic acid-like polymer. Typical structures are shown in Table F. While the structures of these products are well established, the mechanisms by which they are formed are not well understood. Kinetic investigations of the formation of these colored products have been carried out by monitoring the increase in absorbance in the vicinity of 400 nm.[75–81] Mechanistic schemes for the reaction based on these studies include the following points: (1) there is some free-radical involvement as judged by the inhibitory effects of allyl acetate and allyl alcohol;[81] (2) the rates are first order in both amine and persulfate, but the derived second-order "constants" are a function of initial concentrations;[81] (3) electron-withdrawing substituents generally increase the rate, although there is conflicting evidence on this point,[79,80] especially by the fact that the protonated amine is unreactive[78,81a]; and (4) the quantity of polymer increases with increasing persulfate:amine ratio.[81] Resolution of some of the conflicting evidence may lie in the interpretation of the kinetic data. The method measures the formation of both the imine and quinone intermediates as well as the condensation products. The reactions leading to these products probably have different electronic requirements. Attack by persulfate at the amine nitrogen is probably accelerated by electron-donating substituents on the ring, while the condensation reactions leading to polymer formation could be dominated by the electrophilicity of the quinones formed in the initial reactions. It is not clear at what stage free radicals are involved, but it must be remembered that allyl acetate has no effect on the rate of disappearance of persulfate so that radical involvement must follow any steps requiring persulfate.

Only brown amorphous material is formed from the persulfate oxidation of 2-aminofluorene and 4-amino-4′-fluorobiphenyl, while 2-aminoanthracene, 2-aminoanthraquinone, 2-aminochrysene, and 4-aminoazobenzene all fail to react appreciably.[16] Likewise, 4-dimethylaminostilbene is not attacked, except with some cleavage of the double bond to yield 4-dimethylaminobenzoic acid.[16]

A tentative overall view of the reactions between persulfate and aromatic amines can be seen as a partitioning of products due to competition between

TABLE F. SOME BYPRODUCTS FROM THE REACTION BETWEEN PERSULFATE AND ARYL AMINES

Substrate	Byproducts	Yield (%)	Ref.
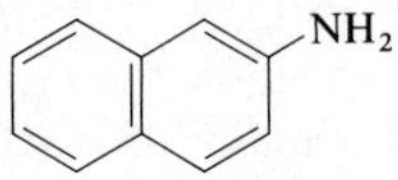 NH_2	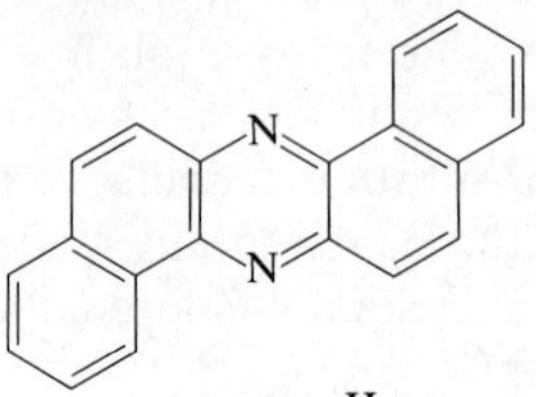 N, N	(—)	62
	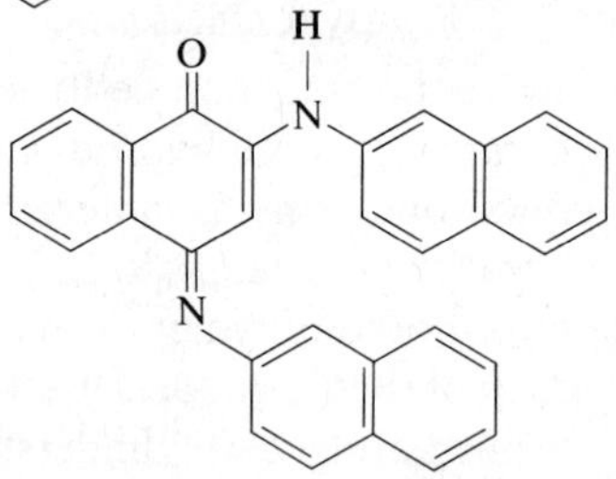 O, H, N, N	(—)	
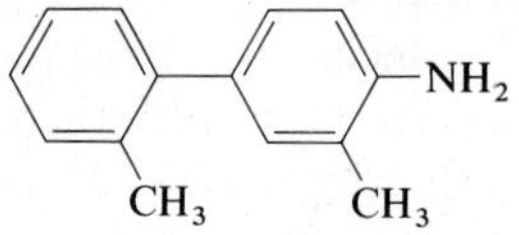 NH_2, CH_3, CH_3	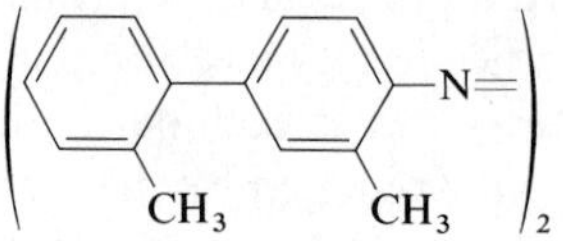 $(\ldots N{=})_2$, CH_3, CH_3	(26),	65
	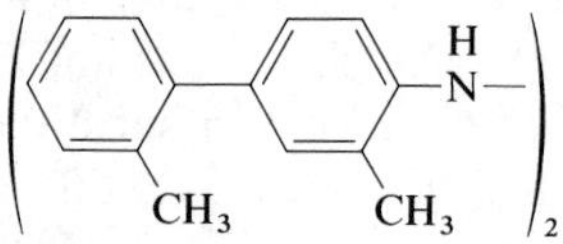 $(\ldots NH{-})_2$, CH_3, CH_3	(6)	
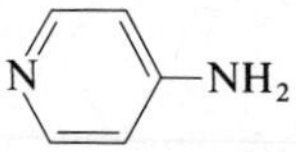 N, NH_2	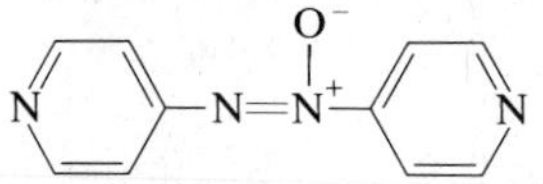 N, $N{=}N^{+}(O^{-})$, N	(20)	65
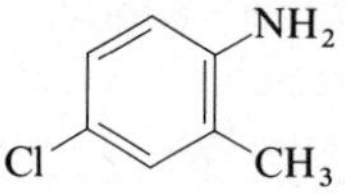 NH_2, Cl, CH_3	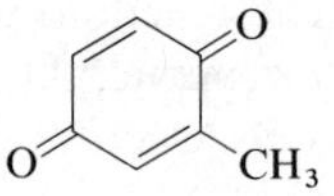 O, O, CH_3	(—)	75
	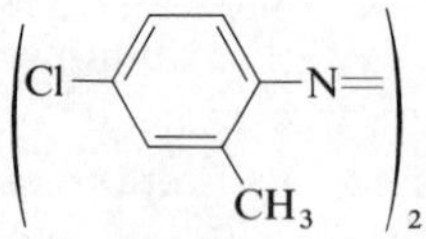 $(Cl \ldots N{=})_2$, CH_3	(—)	
	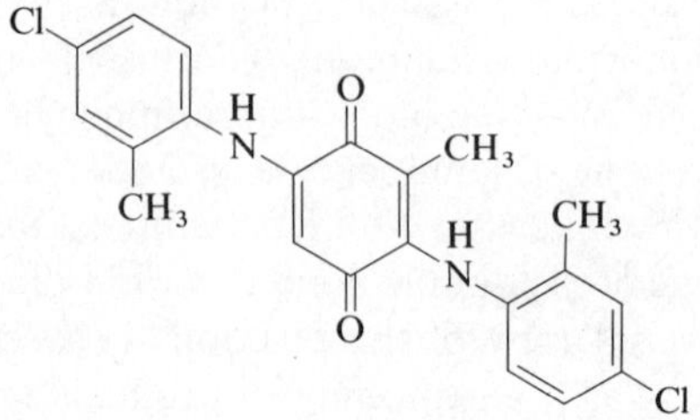 Cl, H, N, O, CH_3, CH_3, H, N, CH_3, O, Cl	(—)	

attack at the *ipso* carbon atom leading to ring sulfation (the Boyland–Sims oxidation) and attack at nitrogen leading eventually to polymeric products.

COMPARISON WITH OTHER METHODS

The principal alternative methods for synthesis of aminophenols are the reduction of nitrophenols and diazophenols. Summaries of methods for the synthesis of aminophenols are to be found in Refs. 37, 38, 40, 41, 82, and 83. Monoperphosphoric acid reacts with some aromatic amines (only a few have been looked at) in the presence of carbonyl compounds and acid to give aminophenols and their *O*-phosphate esters.[83a] The advantage of the Boyland–Sims oxidation lies in the mild conditions under which it can be run. While the yields for some compounds are respectable, generally better yields are often obtained by these other methods.

EXPERIMENTAL CONDITIONS

Experimental conditions have been varied in only a few studies. The only factors known to influence the yield are the ratio of reactants and pH. The yield of sulfate ester drops markedly as the amine:persulfate ratio is decreased.[66] Concomitantly, the yield of polymeric product increases.[78] The yield of sulfate ester falls as the pH is increased,[66] but this might conceivably be an ionic strength effect. Considering the stoichiometry of the reaction, however, in the absence of other data, it would appear prudent to use no more than 1 equivalent of base.

$$HO^- + S_2O_8^{2-} + R(H)NH_2 \longrightarrow H_2O + SO_4^{2-} + R\begin{cases} NH_2 \\ OSO_3^- \end{cases}$$

The reaction can be carried out in water or in acetone–water mixtures. Other water–solvent mixtures can be used as well, but the rate of reaction decreases with decreasing dielectric constant.[69] Temperature does not appear to influence the yield over the range 30–50°.[66]

EXPERIMENTAL PROCEDURES

It has been the usual practice, in contrast to the Elbs oxidation, to isolate the intermediate sulfate and subsequently hydrolyze it to the aminophenol.

The general procedure is as follows.[63] The amine (5 g) in water (250 mL) is brought into solution by the addition of acetone, or, in the case of amines containing an acidic group, by the addition of 2 *N* sodium or potassium hydroxide. Sodium or potassium hydroxide (2 *N*, 20% molar excess) is added, followed by 1 molar equivalent of persulfate in aqueous solution during 8 hours with continuous stirring. The mixture is kept overnight, evaporated to 200 mL under reduced pressure, and filtered. The solution is washed with ether and further treated according to the nature of the amine. Toluidine *o*-sulfates can be extracted from the ether-washed solution with butanol, aminobenzoic acid *o*-sulfates can be isolated by extracting the dried reaction mixture with methanol, and some sulfates can be crystallized directly from the ether-extracted reaction mixture (diphenylamine *o*-sulfate). The aminophenol is then formed by acid-catalyzed hydrolysis of the sulfate.

Comparative experiments in which the aminophenol is isolated directly have not been reported.

***o*-Dimethylaminophenyl Hydrogen Sulfate and *o*-Dimethylaminophenol.**[62] *N,N*-Dimethylaniline (5 g, 0.04 mol) in a mixture of 250 mL of water, 400 mL of acetone, and 30 mL of 2 *N* potassium hydroxide (0.06 mol) were mixed with a saturated aqueous solution of 11.2 g of potassium persulfate (0.04 mol) and the mixture stirred for 8 hours at room temperature. The mixture was kept overnight, filtered, concentrated to 250 mL, washed with ether (3 × 150 mL), and then evaporated to dryness under reduced pressure. The residue was extracted with hot 95% ethanol (3 × 50 mL). The combined extracts were diluted with 1.5 L of ether, yielding *o*-dimethylaminophenyl potassium sulfate (4.2 g, 40%), which was recrystallized from 95% ethanol.

The potassium salt (0.46 g) was dissolved in 2 mL of water and treated with 2 mL of concentrated HCl to yield *o*-dimethylaminophenyl hydrogen sulfate (0.31 g, 82%). Recrystallization from aqueous ethanol gave prisms, mp 217–219° (dec.).

The hydrogen sulfate (0.4 g) was heated at 100° with concentrated HCl (5 mL) for 1 hour, and the solution was then cooled to near 0° and partially neutralized with 2 *N* NaOH. *o*-Dimethylaminophenol (0.21 g, 83%) separated as needles, mp 44–45°, which was raised to 46° by crystallization from aqueous ethanol.

TABULAR SURVEY

The literature has been searched through mid-1984. Some oxidations of phenols and aromatic amines may have been missed because they were not indexed if incidental to the principal theme of the reference. In addition to Chemical Abstracts, the ISI citation index was found very valuable.

In each table, entries are arranged in order of increasing number of carbon atoms and, within each carbon-number group, in order of increasing number of hydrogen atoms. Yields are given in parentheses, and conversions based on recovered starting material are in brackets. Yields marked with a double dagger (‡) are for the phenol sulfate ester; those marked with an asterisk (*) were determined by a chromatographic or colorimetric procedure. A dash in parentheses (—) in the yield column indicates that no yield was reported. Since most reactions have been carried out under similar conditions, no details are given in the tables. Table XV lists a number of unsuccessful oxidations together with comments.

TABLE I. Persulfate Oxidation of Phenols

No. of	Phenol	Product(s) and Yield(s) (%)	Refs.
C_6	2,6-Dichlorophenol	2,6-Dichlorohydroquinone (50) [100]	84
	2,6-Dibromophenol	2,6-Dibromohydroquinone (50) [100]	84
	2-Fluorophenol	Fluorohydroquinone (40)	85
	2-Chlorophenol	Chlorohydroquinone (50) [62]	2
		" (25), 3-chloro-1,2-dihydroxybenzene (3)	29
	3-Chlorophenol	" (20) [36]	84
	4-Chlorophenol	4-Chloro-1,2-dihydroxybenzene (25)	14
		" (—)	86
	2-Bromophenol	Bromohydroquinone (20) [33]	84
	4-Bromophenol	4-Bromo-1,2-dihydroxybenzene (—)	86
	2-Iodophenol	Iodohydroquinone (14)	87
	2-Nitrophenol	Nitrohydroquinone (30–40) [60–80]	1
		" (20)[‡]	88
		" (78)[*]	9
		" (38)	89
		" (30)	90
		" (—)	91
		" (13), 1,2-dihydroxy-3-nitrobenzene (2)	29
	3-Nitrophenol	" (20) [67]	84
		" (13)[‡]	88
	4-Nitrophenol	1,2-Dihydroxy-4-nitrobenzene (10)[‡]	88
		" (7)	14
		" (—)	86
	Phenol	Hydroquinone (18) [34]	2
		" (—)	29, 92
		" (42), catechol (8)[a]	11
		" (—)[‡]	93
	1,3,4-Trihydroxybenzene	1,2,4,5-Tetrahydroxybenzene (11)	25
	4-Hydroxyphenylarsonic acid	3,4-Dihydroxyphenylarsonic acid (—)	94

TABLE I. PERSULFATE OXIDATION OF PHENOLS (*Continued*)

No. of Carbon Atoms	Phenol	Product(s) and Yield(s) (%)	Refs.
C_7	2-Cyanophenol	Cyanohydroquinone (19)	95
		" (—)	96
	3-Trifluoromethylphenol	(Trifluoromethyl)hydroquinone (6)	85
	2-Hydroxybenzaldehyde	2,5-Dihydroxybenzaldehyde (25) [33]	2, 97
		" (34)	50
		" (32), 2,3-dihydroxybenzaldehyde (4)	29
		" (20)	98
	3-Hydroxybenzaldehyde	" (19)	33
		" (16), 3,4-dihydroxybenzaldehyde (0.7)	29
	2,4-Dihydroxybenzaldehyde	2,4,5-Trihydroxybenzaldehyde (8)	99
	2-Hydroxybenzoic acid	2,5-Dihydroxybenzoic acid (60)	100
		" (40)–60)	101
		" (47)	102
		" (30 – 50)[40 – 70]	32
		" (34), (—)‡	98
		" (65)‡	103
		" (—)	104
		" (—), 2,3-dihydroxybenzoic acid (—) 6:1	29
	2-Hydroxybenzoic acid (carboxyl–^{14}C)	" (carboxyl–^{14}C) (46)	105
	3-Hydroxybenzoic acid	" (10), " (0.7)	29
		" (—)‡	106
		" (—)*, 3,4-dihydroxybenzoic acid (—), 2,3-dihydroxybenzoic acid (—) 8:3:1	30
	4-Hydroxybenzoic acid	3,4-Dihydroxybenzoic acid (0.6) [2]	2
		" (—)‡	106
		" (—)	86
		" (50)	14
	2,4-Dihydroxybenzoic acid	2,4,5-Trihydroxybenzoic acid (6)	107
	3,5-Dihydroxybenzoic acid	2,3,5-Trihydroxybenzoic acid (18)	32

	Substrate	Product	Ref.
	2,3-Methylenedioxyphenol	2,3-Methylenedioxyhydroquinone (20) [32]	108
	2-Methyl-3-nitrophenol	2-Methyl-3-nitrohydroquinone (29)	109
	2-Methylphenol	Methylhydroquinone (—)	92
	3-Methylphenol	" (—)	92
	4-Methylphenol	3,4-Dihydroxytoluene (14)	14
		" (9) [11]	2
		" (—)	86
C_8	OH, HO_2C, O, CH_2, O	OH, HO_2C, O, CH_2, O, OH (5) [9]	108
	OH, I, CO_2H, CH_3O, I	OH, I, CO_2H, CH_3O, OH (—)	15
	3-Cyano-2-methylphenol	3-Cyano-2-methylhydroquinone (23)	109
	OH, Br, $COCH_3$	OH, Br, $COCH_3$, OH (—)	110
	OH, I, CO_2H, CH_3O	OH, I, CO_2H, CH_3O, OH (19)	15

TABLE I. PERSULFATE OXIDATION OF PHENOLS (*Continued*)

No. of Carbon Atoms	Phenol	Product(s) and Yield(s) (%)	Refs.
	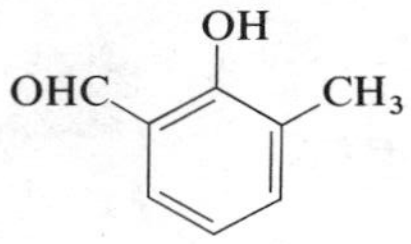 OH, OHC, CH_3	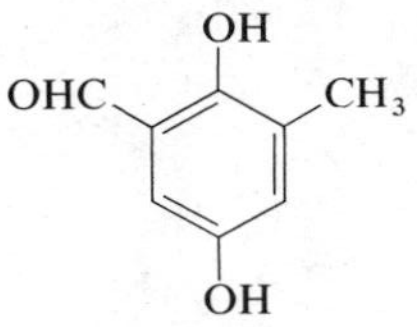 OH, OHC, CH_3, OH (13)	111
	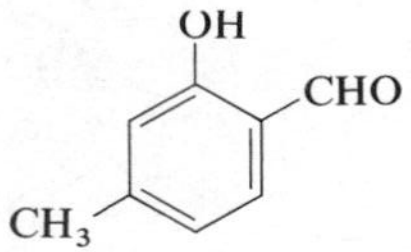 OH, CHO, CH_3	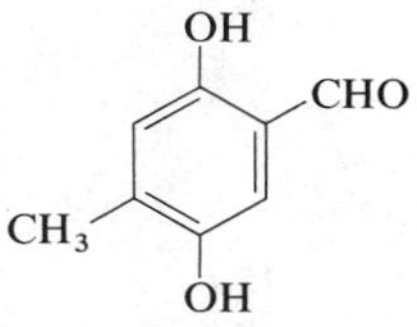 OH, CHO, CH_3, OH (15)	111
	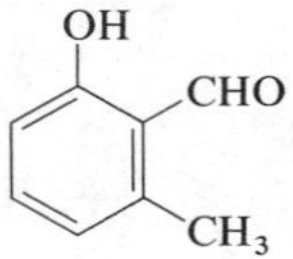 OH, CHO, CH_3	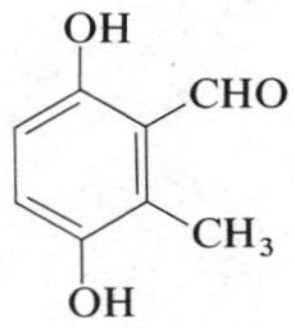 OH, CHO, CH_3, OH (17)	111
	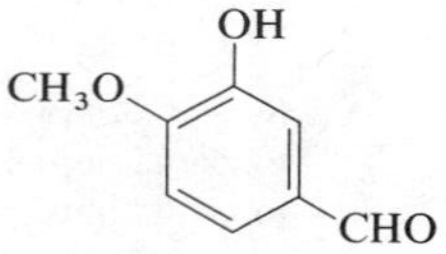 OH, CH_3O, CHO	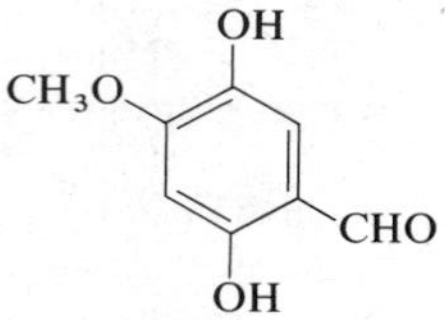 OH, CH_3O, CHO, OH (3) [4]	112

Substrate	Product	Yield	Refs.
CH_3, OH, CO_2H	CH_3, OH, CO_2H, OH	(—)	104, 113, 114
	”	(24) [44]	115
OH, CO_2H, CH_3	OH, CO_2H, CH_3, OH	(71)	116
	”	(33)	117
	”	(—)	104, 113
OH, CO_2H, CH_3	HO, OH, CO_2H, CH_3	(—)	86
OH, CH_3, CO_2H	OH, CH_3, CO_2H, OH	(—)	118
OH, CO_2H, CH_3	OH, CO_2H, CH_3, OH	(26) [36]	117

TABLE I. Persulfate Oxidation of Phenols (*Continued*)

No. of Carbon Atoms	Phenol	Product(s) and Yield(s) (%)	Refs.
		OH, CO_2H, CH_3, OH (40)	119
	OHC, OH, OCH_3	OHC, OH, OCH_3, OH (10) [14]	36
		" (14)	120
	OH, OCH_3, CHO	HO, OH, OCH_3, CHO (2) [4]	2
	OH, OCH_3, OHC	OH, OCH_3, OHC, OH (36)	107

Substrate	Product		Reference
OH, CHO, CH_3O	OH, CHO, CH_3O, OH	(18)	107
OH, $COCH_3$, HO	OH, $COCH_3$, HO, OH	(18)	121
	”	(30)	122
	”	(17)	123, 124
OH, $COCH_3$, OH	OH, HO, $COCH_3$, OH	(29)	36
OH, CHO, HO, OCH_3	OH, CHO, HO, OCH_3, OH	(8)	125
OH, HO_2C, OCH_3	OH, HO_2C, OCH_3, OH	(14)	32

TABLE I. PERSULFATE OXIDATION OF PHENOLS (*Continued*)

No. of Carbon Atoms	Phenol	Product(s) and Yield(s) (%)	Refs.
	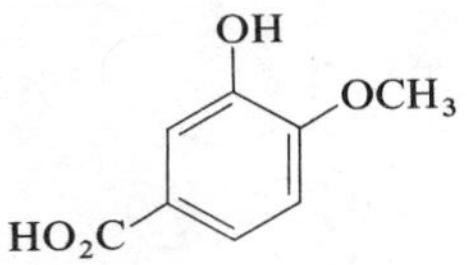 OH, OCH_3, HO_2C	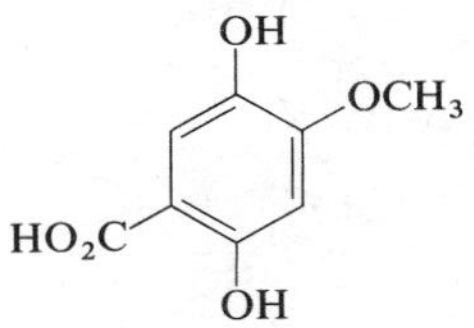 OH, OCH_3, HO_2C, OH (32)	107
	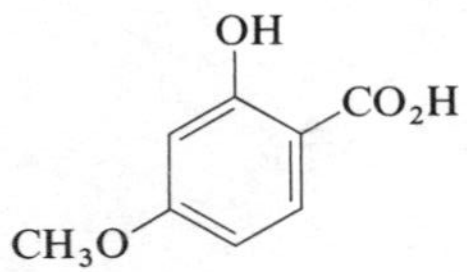 OH, CO_2H, CH_3O	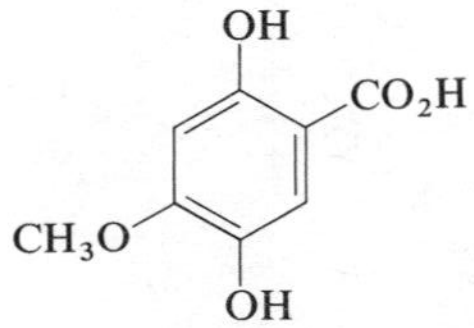 OH, CO_2H, CH_3O, OH (32)	107
		" (27)	126
	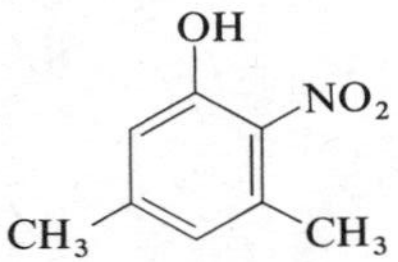 OH, NO_2, CH_3, CH_3	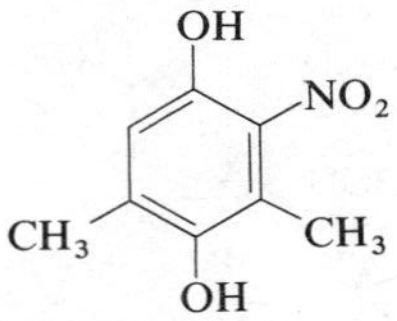 OH, NO_2, CH_3, CH_3, OH (50)	109
	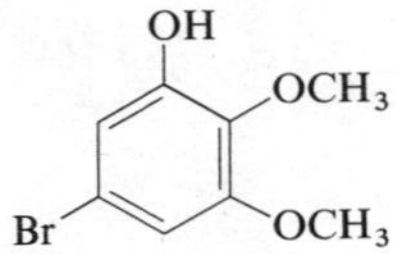 OH, OCH_3, Br, OCH_3	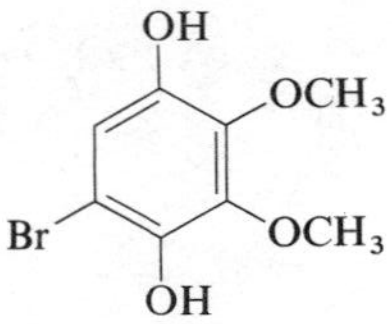 OH, OCH_3, Br, OCH_3, OH (41)	126a

	OH, CH_3, CH_3	OH, CH_3, CH_3, OH (36) [51]	2
		” (—)	127
	OH, CH_3, CH_3	OH, CH_3, CH_3, OH (42) [56]	2
	OH, CH_3, CH_3	OH, CH_3, CH_3, OH (30) [40]	2
		” (35)	128
		” (—)	129
	OH, OCH_3, OCH_3	OH, OCH_3, OCH_3, OH (18)	108
C_9	OH, HO_2C, CHO, OCH_3	OH, HO_2C, CHO, OCH_3, OH (48)	120

TABLE I. PERSULFATE OXIDATION OF PHENOLS (*Continued*)

No. of Carbon Atoms	Phenol	Product(s) and Yield(s) (%)	Refs.
	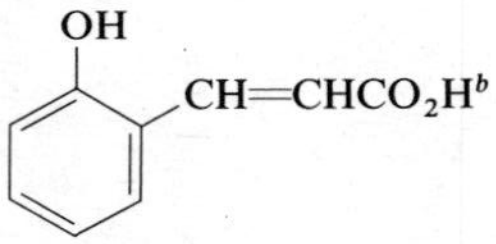 OH, CH=CHCO$_2$H[b]	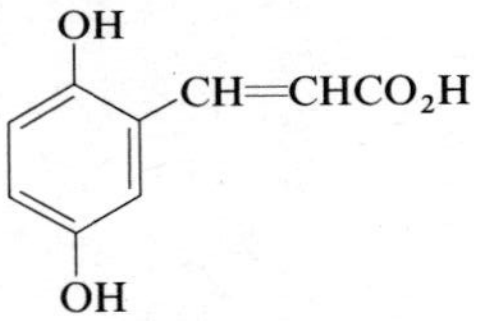 OH, CH=CHCO$_2$H, OH (23) [36]	130
	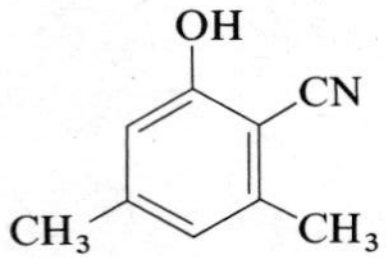 OH, CN, CH$_3$, CH$_3$	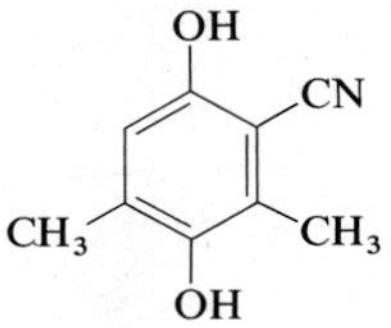 OH, CN, CH$_3$, CH$_3$, OH (52)	109
	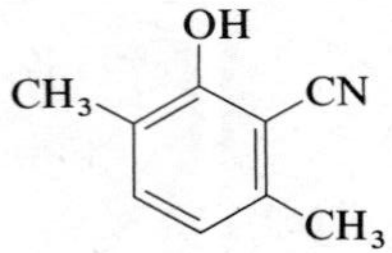 OH, CN, CH$_3$, CH$_3$	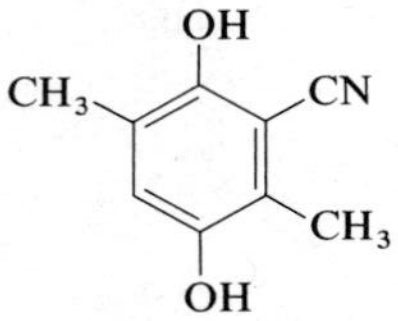 OH, CN, CH$_3$, CH$_3$, OH (36)	109
	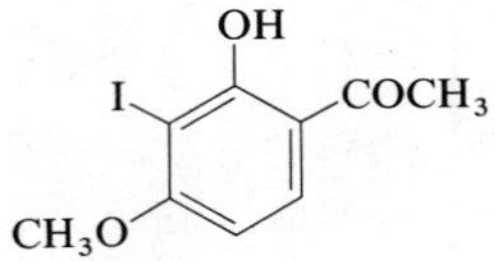 OH, I, COCH$_3$, CH$_3$O	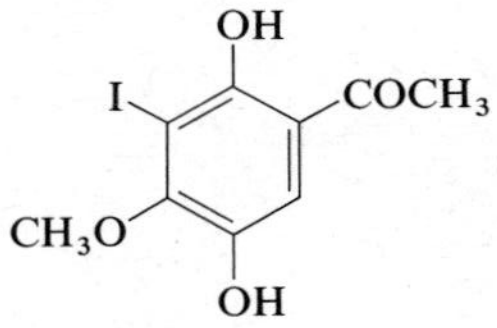 OH, I, COCH$_3$, CH$_3$O, OH (21)	15

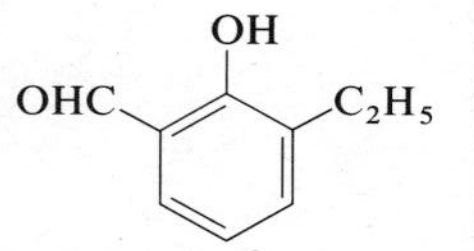

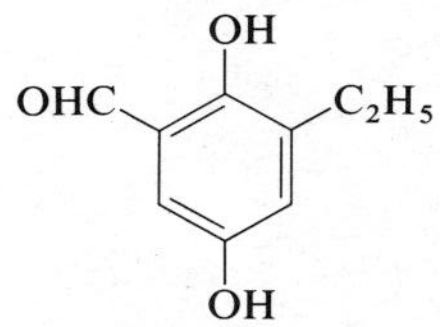

OH, OHC, C_2H_5	OH, OHC, C_2H_5, OH	(23)	111
OH, CHO, CH_3, CH_3	OH, CHO, CH_3, CH_3, OH	(—)	131
OH, HO_2C, C_2H_5	OH, HO_2C, C_2H_5, OH	(—)	113
OH, HO_2C, CH_3, CH_3	OH, HO_2C, CH_3, CH_3, OH	(—)	113
OH, CHO, CH_3O, CH_3	OH, CHO, CH_3O, CH_3, OH	(24)	132

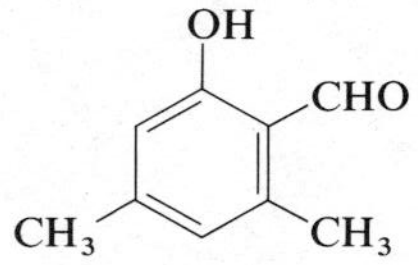

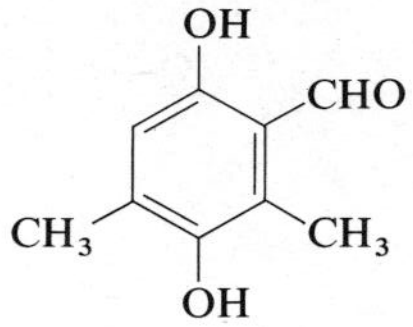

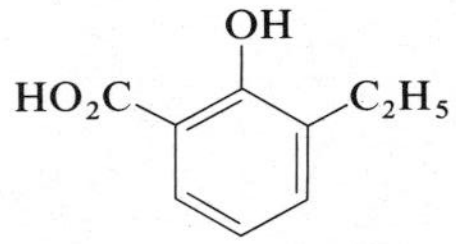

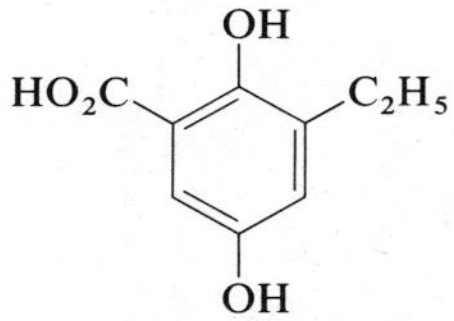

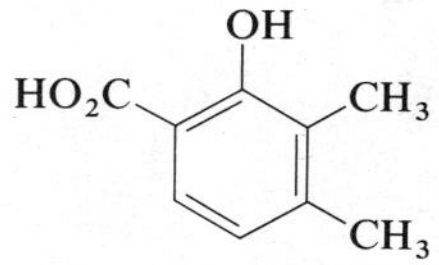

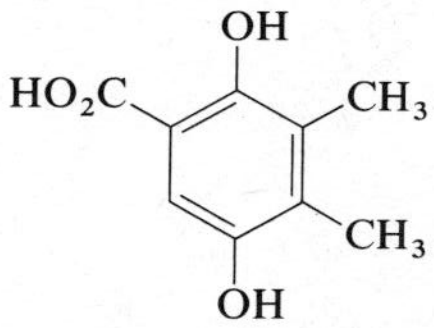

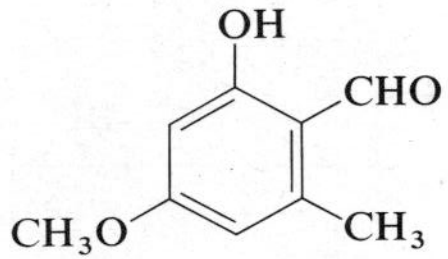

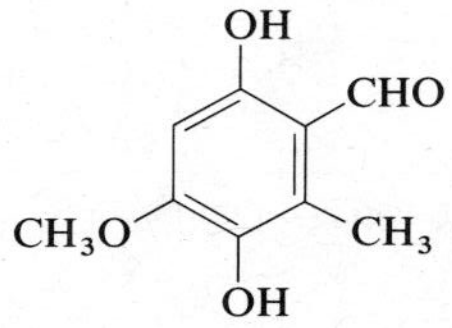

TABLE I. PERSULFATE OXIDATION OF PHENOLS (*Continued*)

No. of Carbon Atoms	Phenol	Product(s) and Yield(s) (%)	Refs.
	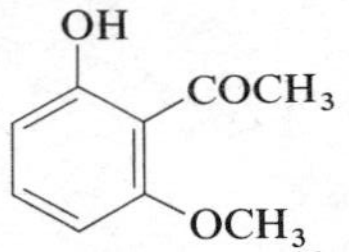 OH, $COCH_3$, OCH_3	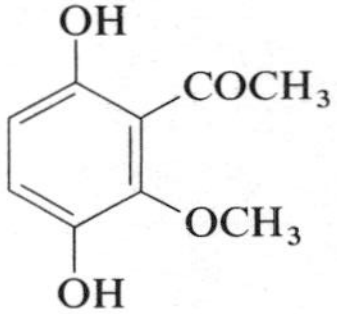 OH, $COCH_3$, OCH_3, OH (33)	133
		" (28)	134
	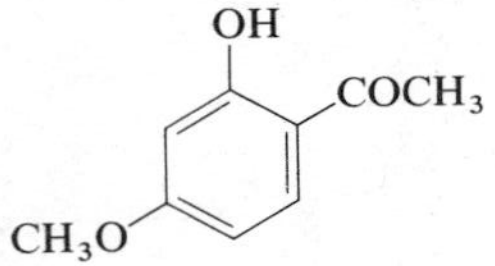 OH, $COCH_3$, CH_3O	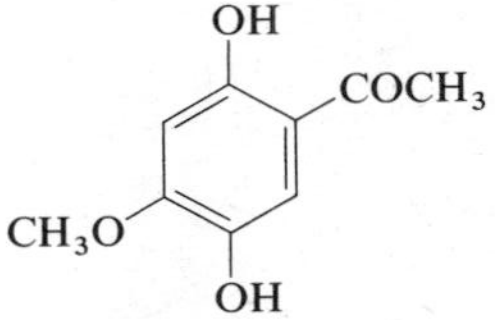 OH, $COCH_3$, CH_3O, OH (18)	135
		" (—)	122
	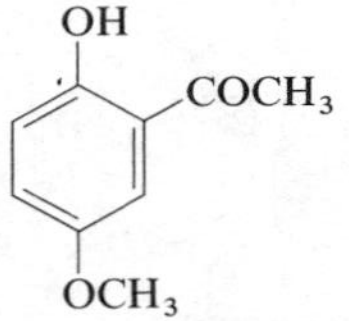 OH, $COCH_3$, OCH_3	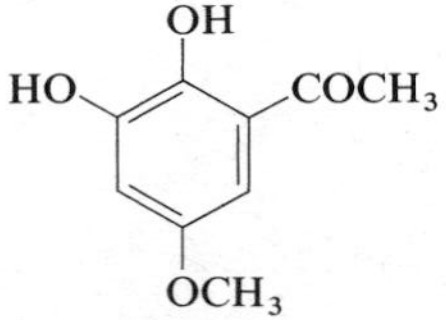 OH, HO, $COCH_3$, OCH_3 (1)	36
	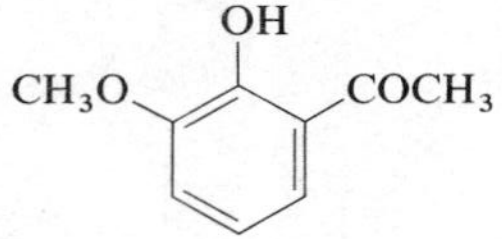 OH, CH_3O, $COCH_3$	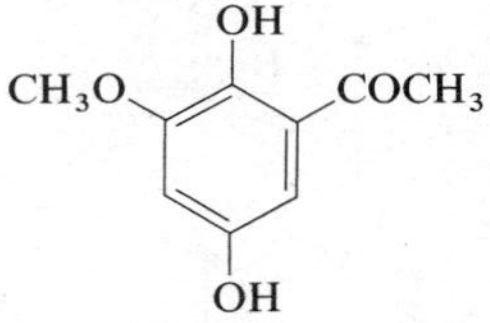 OH, CH_3O, $COCH_3$, OH (2) [3]	36
		" (26) [29]	136

OH, CHO, CH_3O, OCH_3	OH, CHO, CH_3O, OCH_3, OH (8)	99
OH, CO_2H, C_2H_5O	OH, CO_2H, C_2H_5O, OH (15)	107
OH, CO_2H, CH_3O, CH_3	OH, CO_2H, CH_3O, CH_3, OH (23)‡	132
OH, CH_3O, CO_2H, CH_3O	OH, CH_3O, CO_2H, CH_3O, OH (25)	108
OH, CO_2CH_3, HO, CH_3	OH, CO_2CH_3, HO, OH (18)	132

TABLE I. PERSULFATE OXIDATION OF PHENOLS (*Continued*)

No. of Carbon Atoms	Phenol	Product(s) and Yield(s) (%)	Refs.
	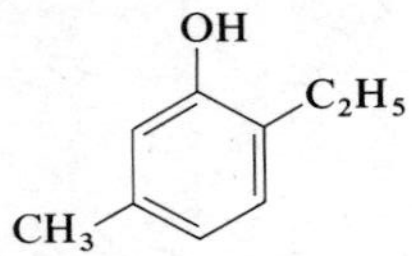 2-C_2H_5-5-CH_3-phenol (OH, C_2H_5, CH_3)	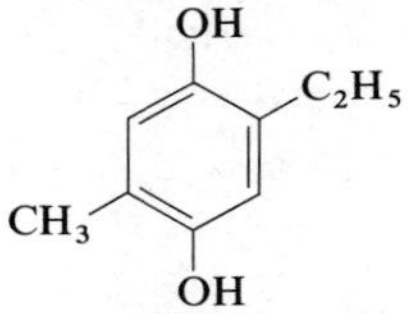 2-C_2H_5-5-CH_3-hydroquinone (OH, C_2H_5, CH_3, OH) (33) [66]	137
	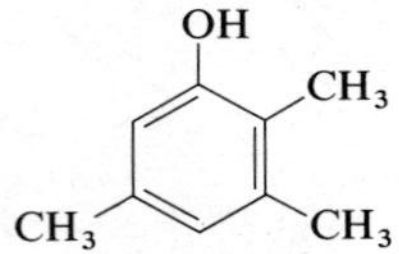 2,3,5-(CH_3)$_3$-phenol (OH, CH_3, CH_3, CH_3)	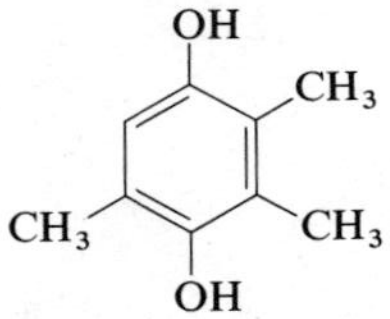 2,3,5-(CH_3)$_3$-hydroquinone (OH, CH_3, CH_3, CH_3, OH) (—)	138
	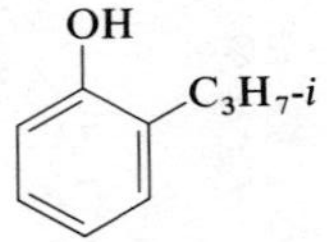 2-C_3H_7-*i*-phenol (OH, C_3H_7-*i*)	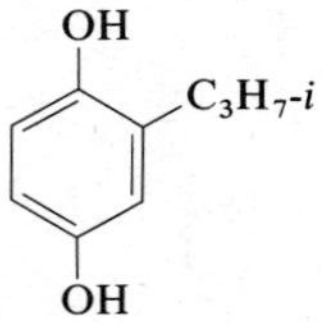 2-C_3H_7-*i*-hydroquinone (OH, C_3H_7-*i*, OH) (33)	128
		" (11)	139
	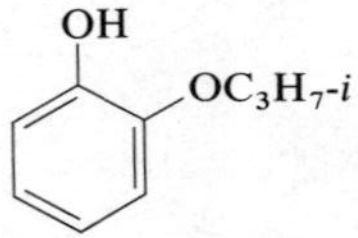 2-OC_3H_7-*i*-phenol (OH, OC_3H_7-*i*)	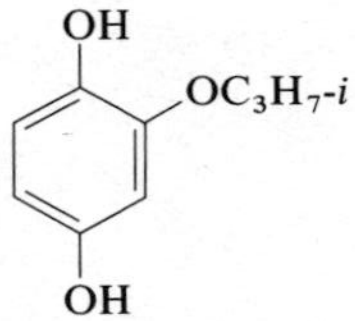 2-OC_3H_7-*i*-hydroquinone (OH, OC_3H_7-*i*, OH) (36)	128

C_{10}	$CH_2{=}CHCH_2$, OH, CN	$CH_2{=}CHCH_2$, OH, CN, OH	(13)	95
	OHC, OH, $CH{=}CHCH_3$	OHC, OH, $CH{=}CHCH_3$, OH	(15)	111
	$CH_2{=}CHCH_2$, OH, O, CH_2, O	$CH_2{=}CHCH_2$, OH, O, CH_2, O, OH	(4) [6]	108
	HO, OH, O, $COCH_3$	HO, OH, O, $COCH_3$, HO	(28)	140
	OHC, OH, C_3H_7-*n*	OHC, OH, C_3H_7-*n*, OH	(20)	111

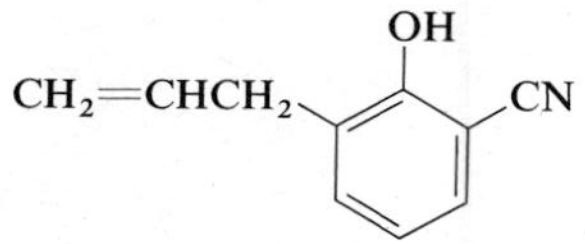

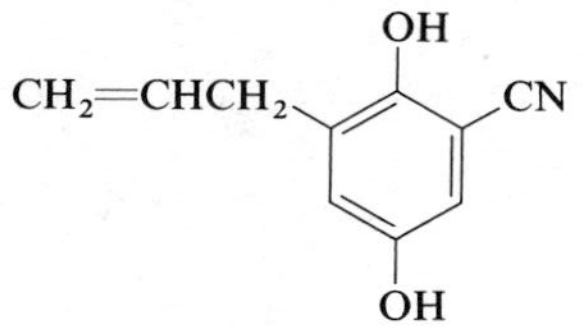

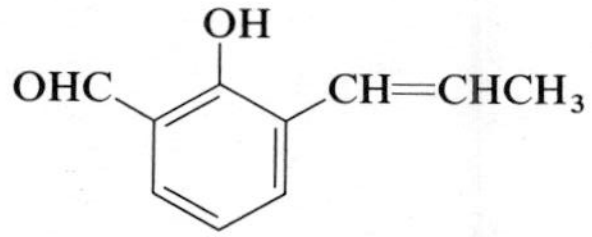

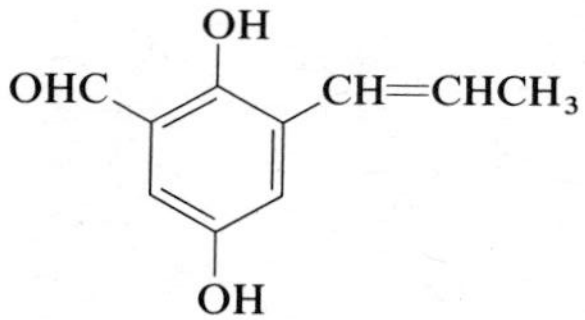

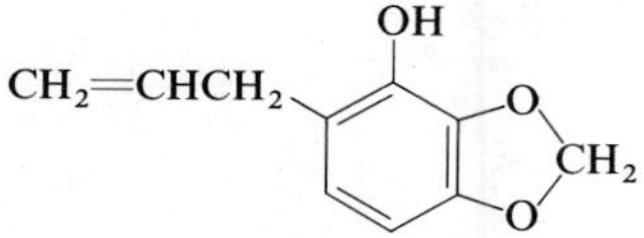

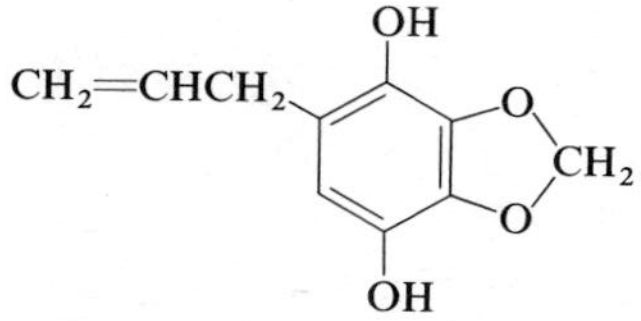

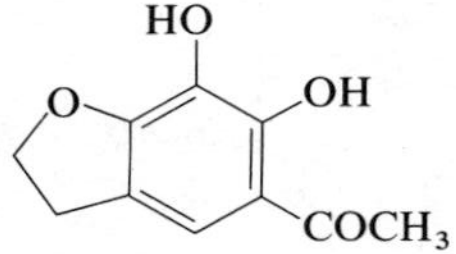

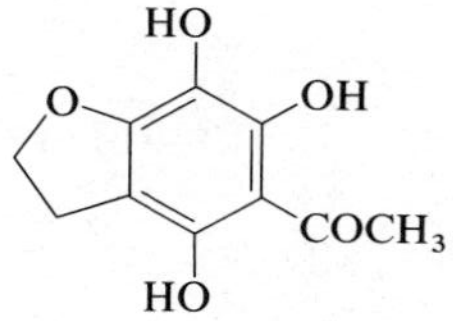

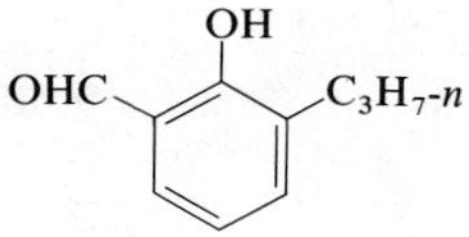

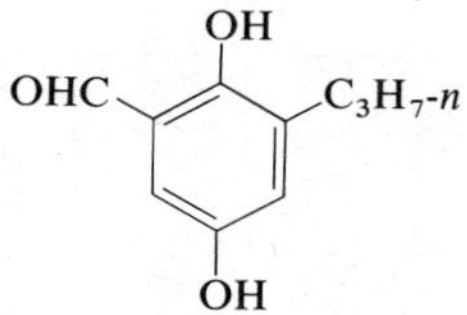

TABLE I. Persulfate Oxidation of Phenols (*Continued*)

No. of Carbon Atoms	Phenol	Product(s) and Yield(s) (%)	Refs.
	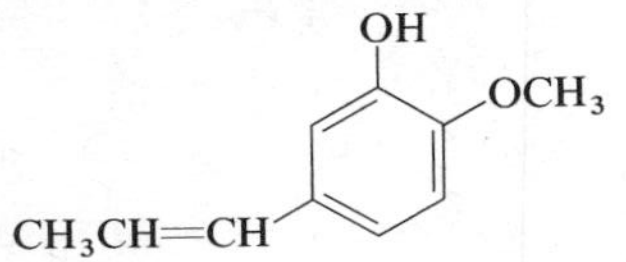 2-OH, 1-OCH_3, 4-($CH_3CH{=}CH$) benzene (OH, OCH_3, $CH_3CH{=}CH$)	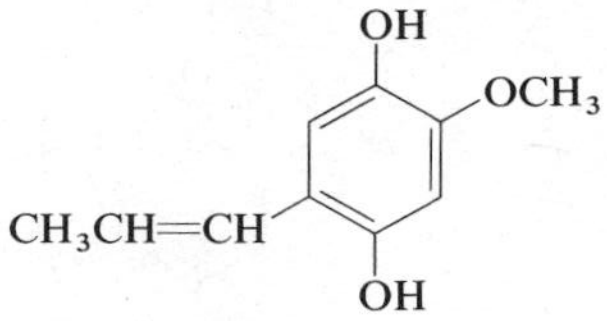 OH, OCH_3, $CH_3CH{=}CH$, OH (4.5) [9]	141
	"	" (—)	142
	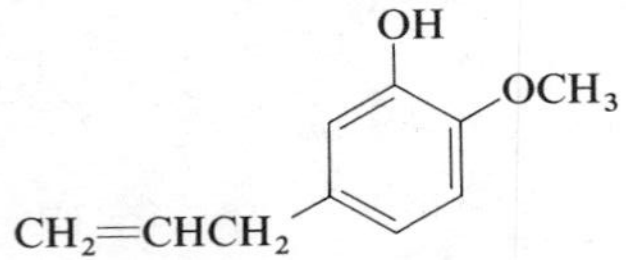 OH, OCH_3, $CH_2{=}CHCH_2$	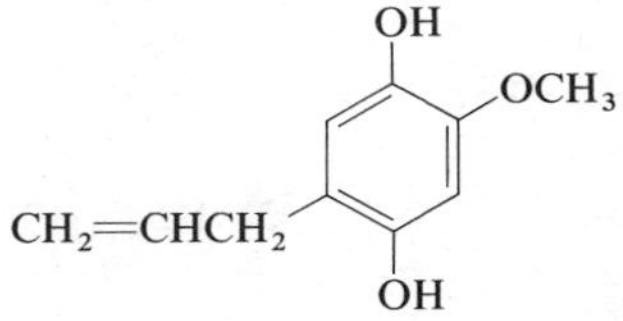 OH, OCH_3, $CH_2{=}CHCH_2$, OH (13)	95
	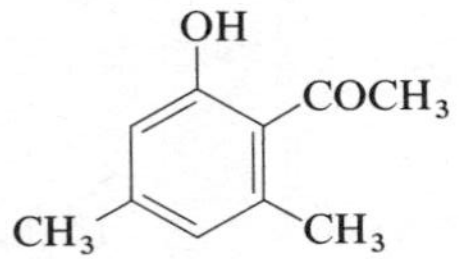 OH, $COCH_3$, CH_3, CH_3	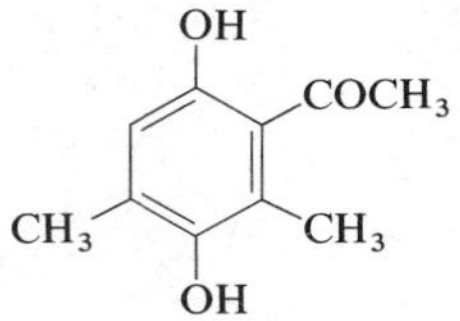 OH, $COCH_3$, CH_3, CH_3, OH (34)	109
	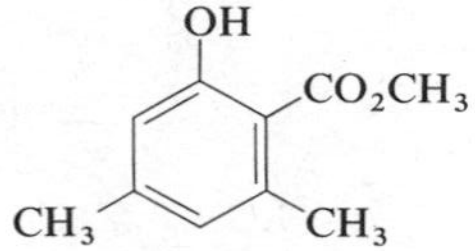 OH, CO_2CH_3, CH_3, CH_3	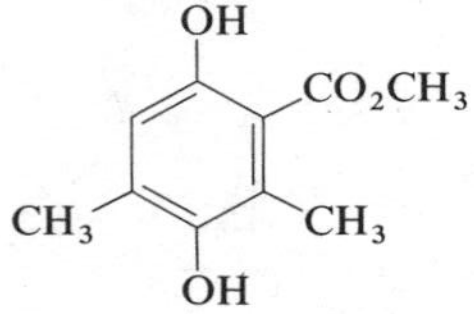 OH, CO_2CH_3, CH_3, CH_3, OH (31)	109

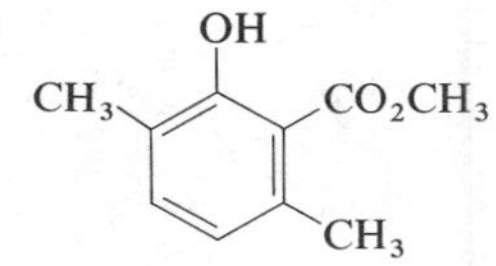 OH, CH_3, CO_2CH_3, CH_3	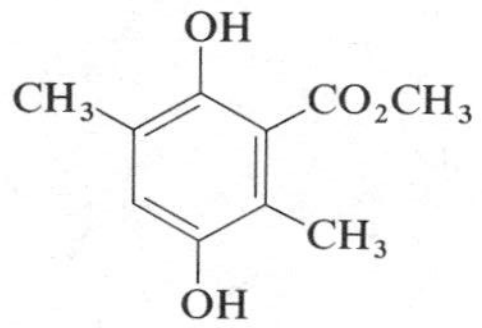 OH, CH_3, CO_2CH_3, CH_3, OH	(29)	109
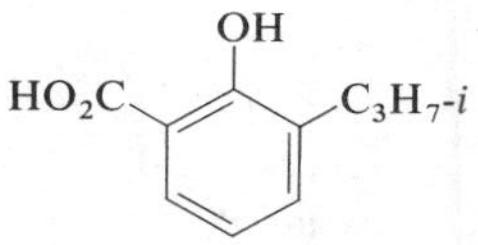 OH, HO_2C, C_3H_7-*i*	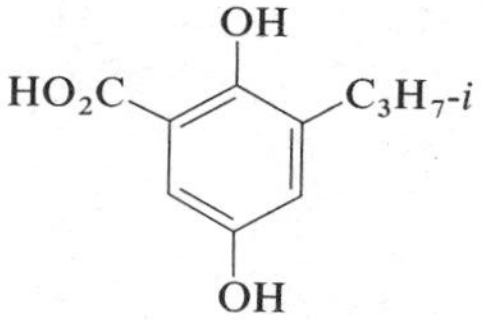 OH, HO_2C, C_3H_7-*i*, OH	(—)	118
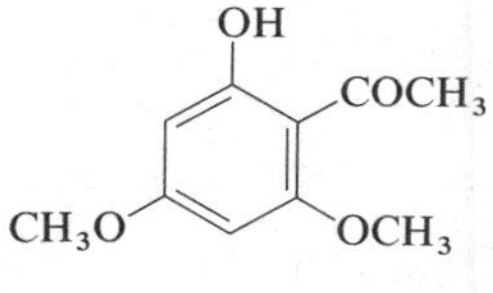 OH, $COCH_3$, CH_3O, OCH_3	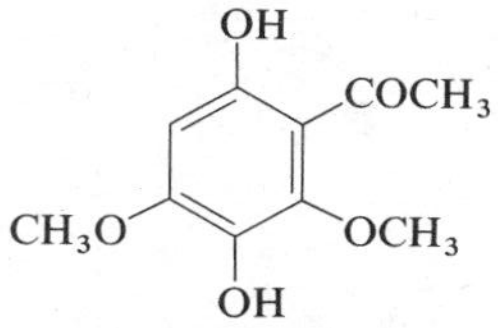 OH, $COCH_3$, CH_3O, OCH_3, OH	(37)	143
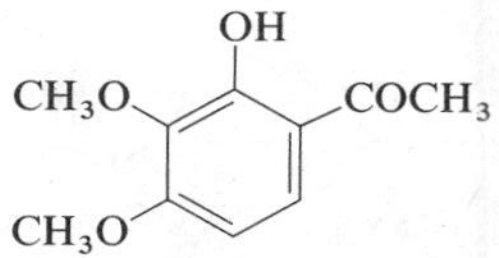 OH, CH_3O, $COCH_3$, CH_3O	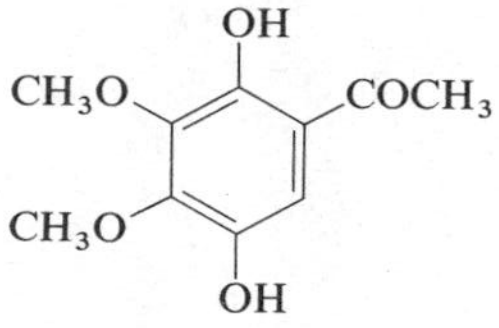 OH, CH_3O, $COCH_3$, CH_3O, OH	(31) [36]	47
	"	(10)	53
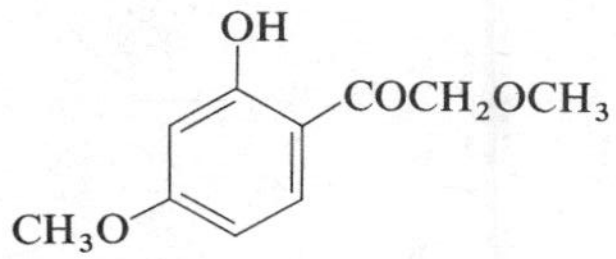 OH, $COCH_2OCH_3$, CH_3O	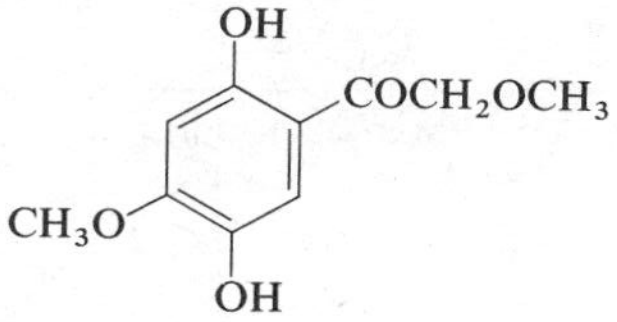 OH, $COCH_2OCH_3$, CH_3O, OH	(15) [17]	144

TABLE I. PERSULFATE OXIDATION OF PHENOLS (*Continued*)

No. of Carbon Atoms	Phenol	Product(s) and Yield(s) (%)	Refs.
	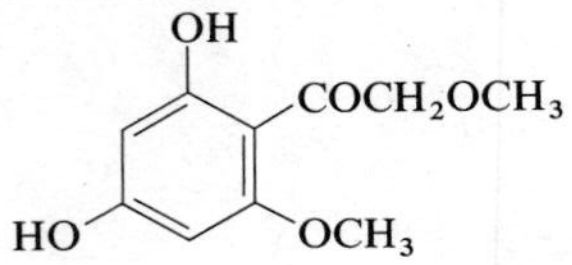 OH, $COCH_2OCH_3$, HO, OCH_3	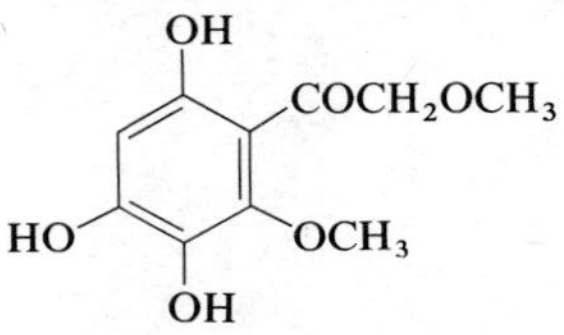 OH, $COCH_2OCH_3$, HO, OCH_3, OH (11) [13]	145
	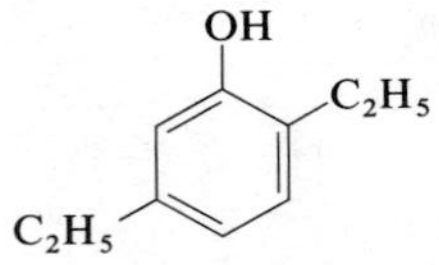 OH, C_2H_5, C_2H_5	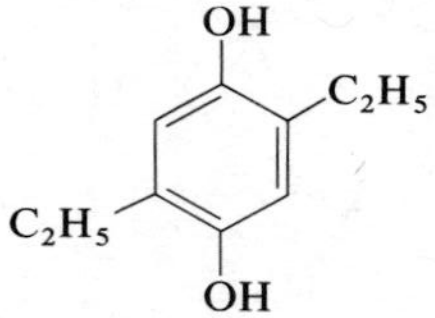 OH, C_2H_5, C_2H_5, OH (49) [117?]	137
	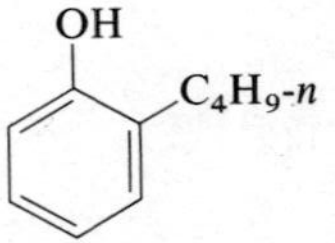 OH, C_4H_9-*n*	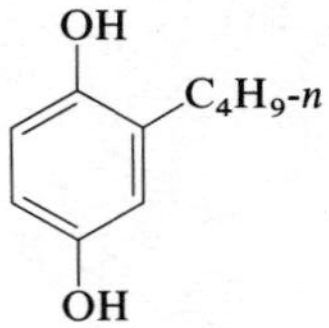 OH, C_4H_9-*n*, OH (—)	111
	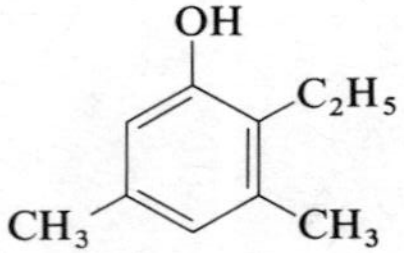 OH, C_2H_5, CH_3, CH_3	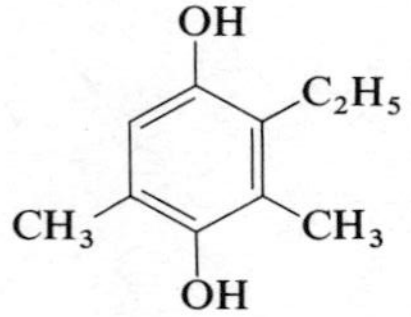 OH, C_2H_5, CH_3, CH_3, OH (—)	138

	Reactant	Product	Yield (%)	Ref.
C_{11}	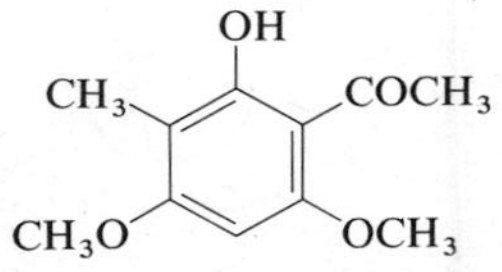 OH, CH_3, $COCH_3$, CH_3O, OCH_3	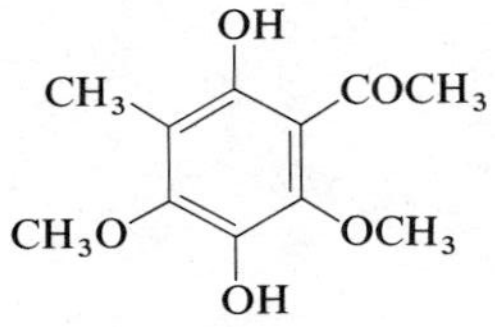 OH, CH_3, $COCH_3$, CH_3O, OCH_3, OH	(34.5)	146
	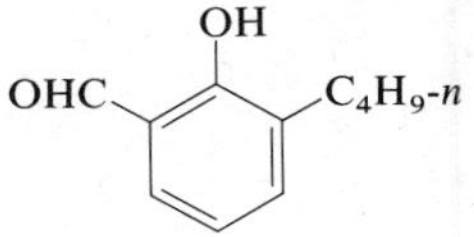 OH, OHC, C_4H_9-*n*	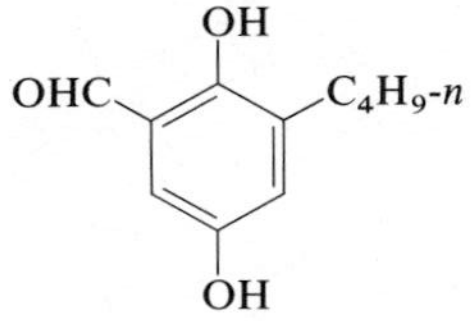 OH, OHC, C_4H_9-*n*, OH	(—)	111
	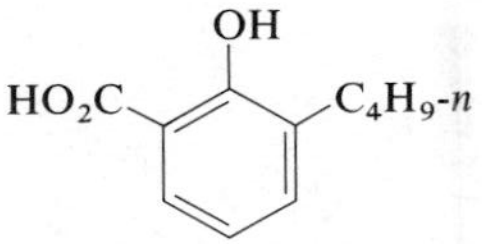 OH, HO_2C, C_4H_9-*n*	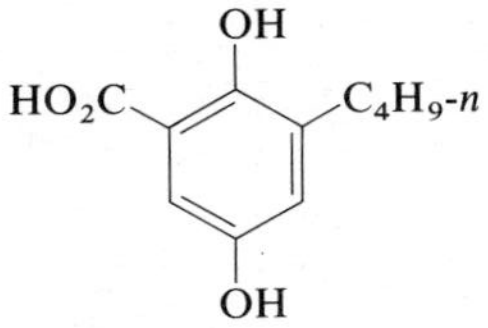 OH, HO_2C, C_4H_9-*n*, OH	(16)	111
	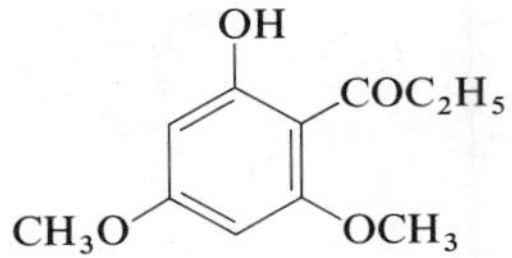 OH, COC_2H_5, CH_3O, OCH_3	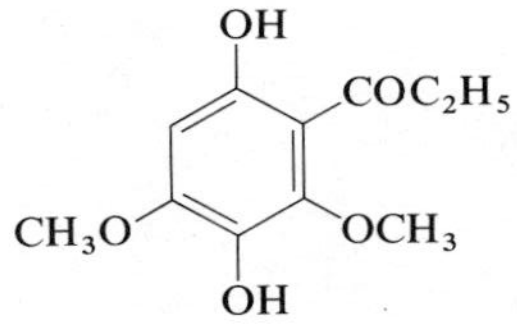 OH, COC_2H_5, CH_3O, OCH_3, OH	(31) [37]	147
	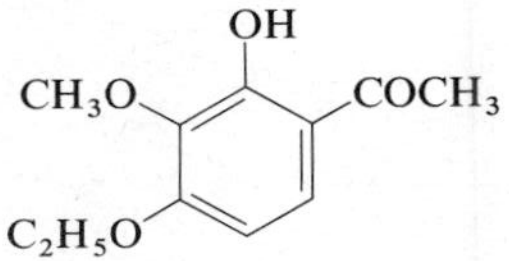 OH, CH_3O, $COCH_3$, C_2H_5O	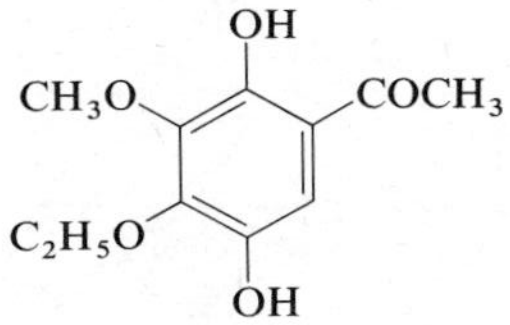 OH, CH_3O, $COCH_3$, C_2H_5O, OH	(40)[c]	148

TABLE I. PERSULFATE OXIDATION OF PHENOLS (*Continued*)

No. of Carbon Atoms	Phenol	Product(s) and Yield(s) (%)	Refs.
	2-$CO_2C_2H_5$-3-CH_3-5-CH_3O-phenol (OH; $CO_2C_2H_5$; CH_3; CH_3O)	2-$CO_2C_2H_5$-3-CH_3-5-CH_3O-1,4-dihydroxybenzene (37)	132
	2-$COCH_3$-3-OCH_3-5,6-di-CH_3O-phenol (OH; $COCH_3$; OCH_3; CH_3O; CH_3O)	2-$COCH_3$-3-OCH_3-5,6-di-CH_3O-1,4-dihydroxybenzene (29)	47
	2-$COCH_2OCH_3$-3-OCH_3-5-CH_3O-phenol (OH; $COCH_2OCH_3$; OCH_3; CH_3O)	2-$COCH_2OCH_3$-3-OCH_3-5-CH_3O-1,4-dihydroxybenzene (23) [30]	149
C_{12}	2-C_6H_5-phenol (OH; C_6H_5)	2-C_6H_5-1,4-dihydroxybenzene (38)	50

Substrate	Product	Yield (%)	Ref.
OH, OHC, C_5H_{11}-*i*	OH, OHC, C_5H_{11}-*i*, OH	(—)	111
OH, $COCH_3$, C_2H_5O, OC_2H_5	OH, $COCH_3$, C_2H_5O, OC_2H_5, OH	(21)	150
OH, $COCH_2OCH_3$, C_2H_5O, OCH_3	OH, $COCH_2OCH_3$, C_2H_5O, OCH_3, OH	(23)	151
OH, CH_3O, $COCH_2OCH_3$, CH_3O, OCH_3	OH, CH_3O, $COCH_2OCH_3$, CH_3O, OCH_3, OH	(14) [16]	152
OH, C_6H_{13}-*n*	OH, C_6H_{13}-*n*, OH	(—)	111

TABLE I. PERSULFATE OXIDATION OF PHENOLS (*Continued*)

No. of Carbon Atoms	Phenol	Product(s) and Yield(s) (%)	Refs.
	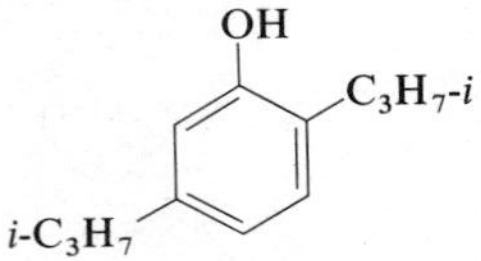 OH, C_3H_7-*i*, *i*-C_3H_7	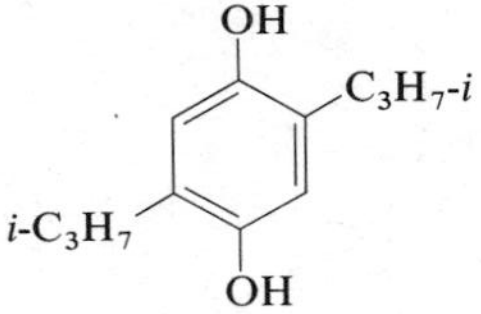 OH, C_3H_7-*i*, *i*-C_3H_7, OH (26.5)	139
	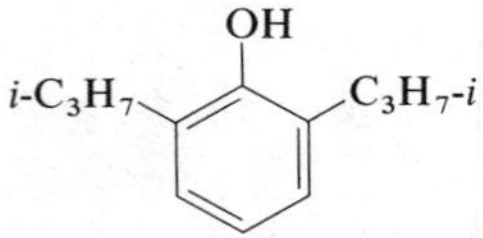 OH, *i*-C_3H_7, C_3H_7-*i*	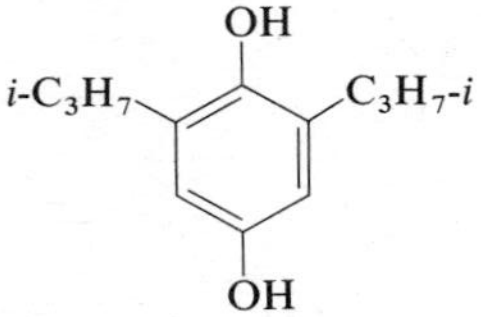 OH, *i*-C_3H_7, C_3H_7-*i*, OH (33)	128
C_{13}	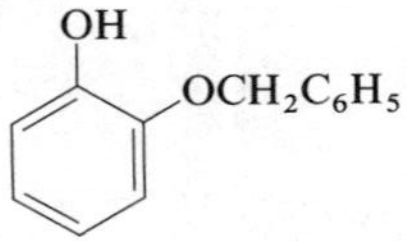 OH, $OCH_2C_6H_5$	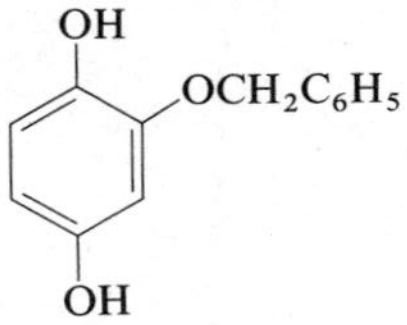 OH, $OCH_2C_6H_5$, OH (33)[‡]	128
	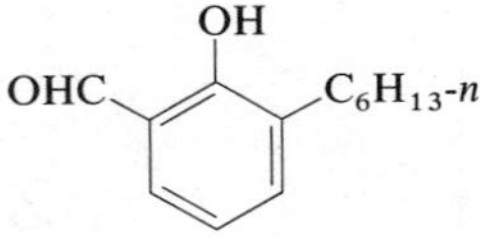 OH, OHC, C_6H_{13}-*n*	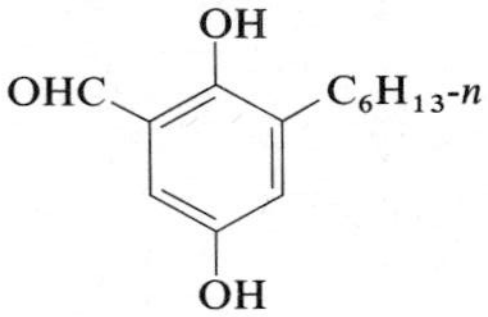 OH, OHC, C_6H_{13}-*n*, OH (—)	111

	Starting material	Product	Yield [ref]	Ref.
	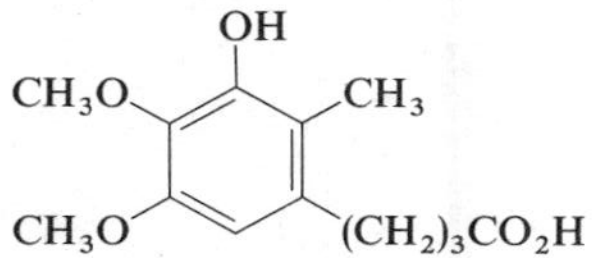 OH, CH_3O, CH_3, CH_3O, $(CH_2)_3CO_2H$	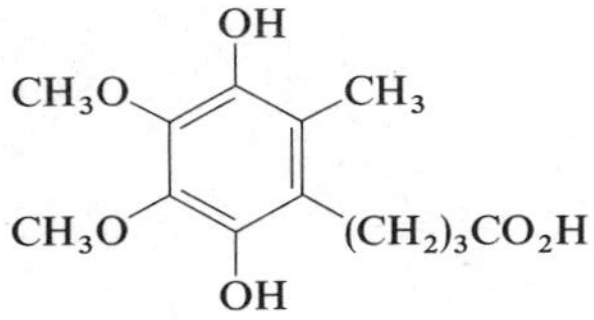 OH, CH_3O, CH_3, CH_3O, $(CH_2)_3CO_2H$, OH	(—)	152a
C_{14}	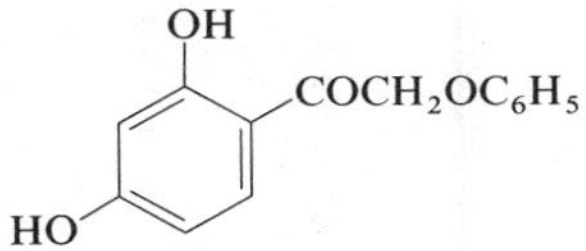 OH, $COCH_2OC_6H_5$, HO	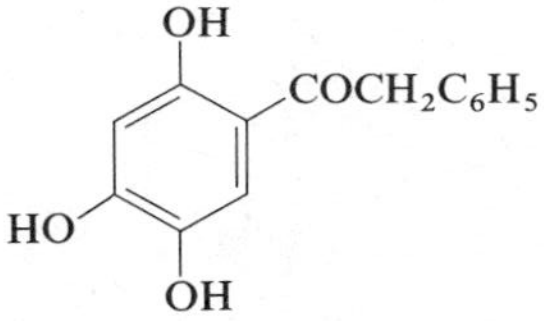 OH, $COCH_2C_6H_5$, HO, OH	(—)	153
	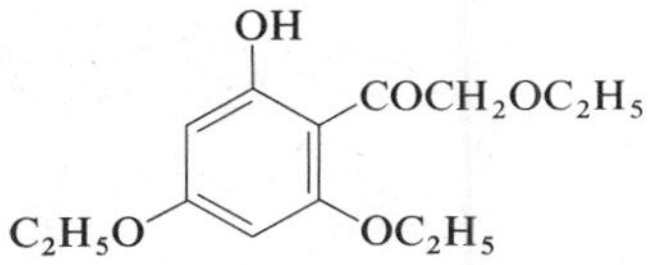 OH, $COCH_2OC_2H_5$, C_2H_5O, OC_2H_5	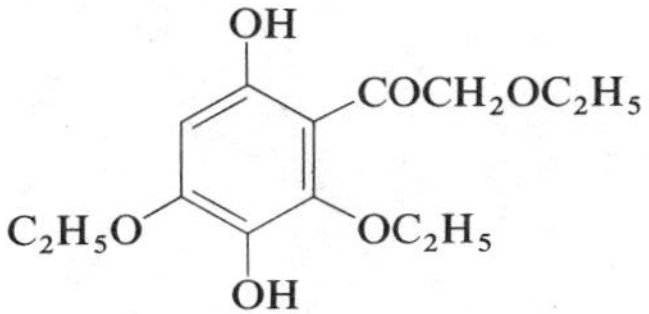 OH, $COCH_2OC_2H_5$, C_2H_5O, OC_2H_5, OH	(21) [30]	154
C_{15}	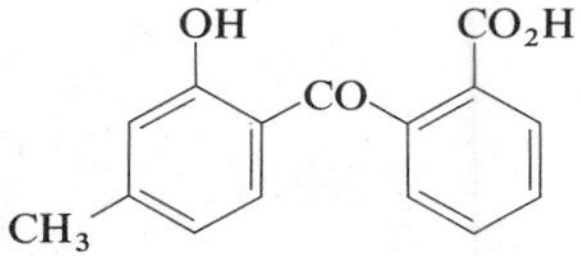 OH, CO_2H, CO, CH_3	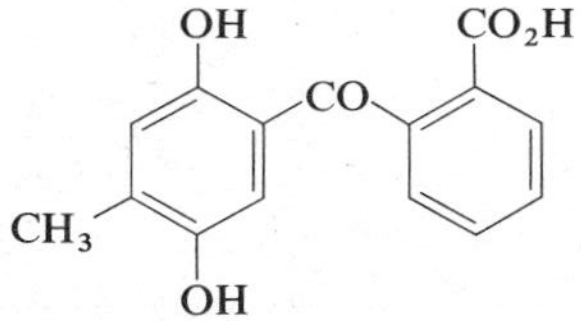 OH, CO_2H, CO, CH_3, OH	(19) [24]	155
	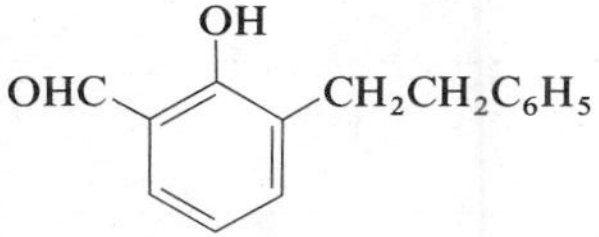 OH, OHC, $CH_2CH_2C_6H_5$	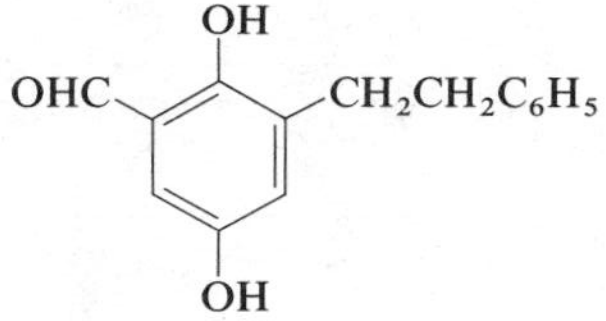 OH, OHC, $CH_2CH_2C_6H_5$, OH	(16)	111

TABLE I. PERSULFATE OXIDATION OF PHENOLS (*Continued*)

No. of Carbon Atoms	Phenol	Product(s) and Yield(s) (%)	Refs.
	2-OH, 4-($C_6H_5CH_2O$)-$C_6H_3COCH_3$ (structure: OH, $COCH_3$, $C_6H_5CH_2O$)	2,5-(OH)$_2$-4-($C_6H_5CH_2O$)-$C_6H_2COCH_3$ (structure: OH, $COCH_3$, $C_6H_5CH_2O$, OH) (23)	122
		" (25)	156
	2-OH, 4-(CH_3O)-$C_6H_3COCH_2C_6H_5$ (structure: OH, $COCH_2C_6H_5$, CH_3O)	2,5-(OH)$_2$-4-(CH_3O)-$C_6H_2COCH_2C_6H_5$ (structure: OH, $COCH_2C_6H_5$, CH_3O, OH) (—)	157
	2-OH, 6-($OCH_2C_6H_5$)-$C_6H_3COCH_3$ (structure: OH, $COCH_3$, $OCH_2C_6H_5$)	2,5-(OH)$_2$-6-($OCH_2C_6H_5$)-$C_6H_2COCH_3$ (structure: OH, $COCH_3$, $OCH_2C_6H_5$, OH) (49) [83]	36
	2-OH, 3,4-(CH_3O)$_2$-$C_6H_2COC_6H_5$ (structure: CH_3O, OH, COC_6H_5, CH_3O)	2,5-(OH)$_2$-3,4-(CH_3O)$_2$-$C_6HCOC_6H_5$ (structure: CH_3O, OH, COC_6H_5, CH_3O, OH) (10)	53

C_{16}	OH, $COCH{=}CHC_6H_5$, CH_3O	OH, $COCH{=}CHC_6H_5$, CH_3O, OH	(24)	158
	HO_2C, OH, CO, CH_3, CH_3O	HO_2C, OH, CO, CH_3, CH_3O, OH	(47)	155
	CH_3O, OH, HO_2C, CO, CH_3	CH_3O, OH, HO_2C, CO, CH_3, OH	(9)	159
	CH_3O, OH, $COCH_2$, HO, OCH_3	CH_3O, OH, $COCH_2$, HO, OCH_3, OH	(19)	160
	OH, $COCH_3$, $C_6H_5CH_2O$, OCH_3	OH, $COCH_3$, $C_6H_5CH_2O$, OCH_3, OH	(12) [23]	161
		”	(—)	162

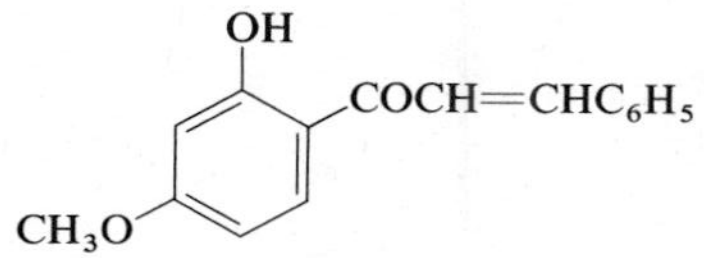

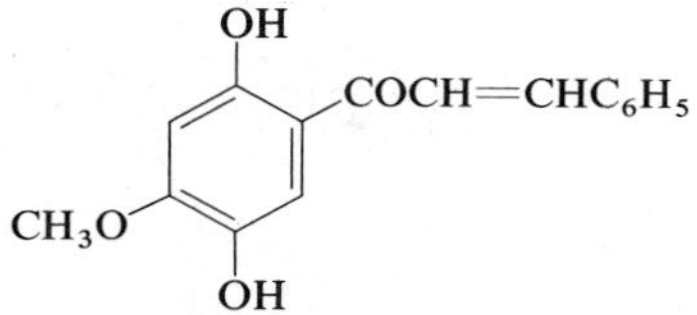

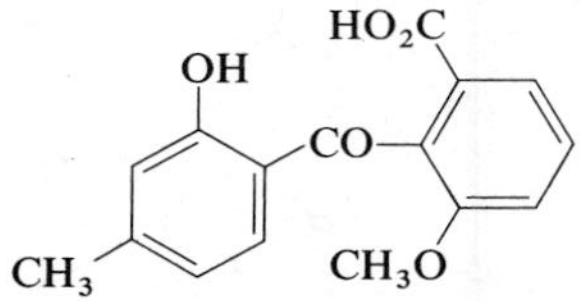

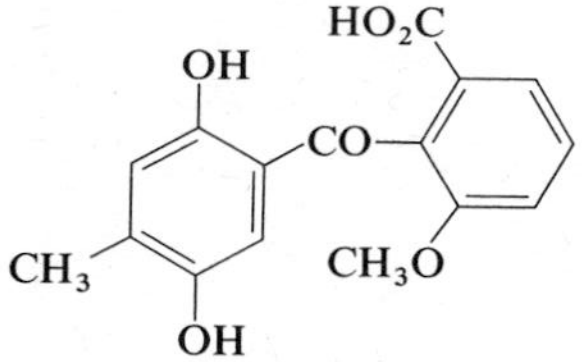

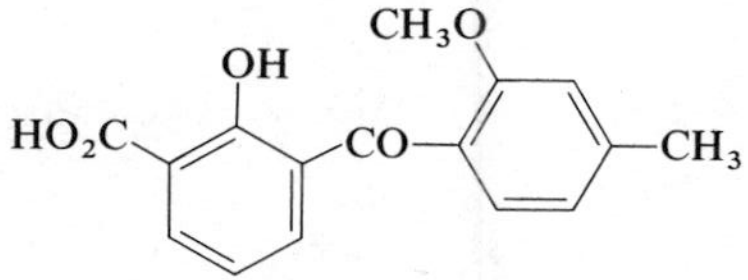

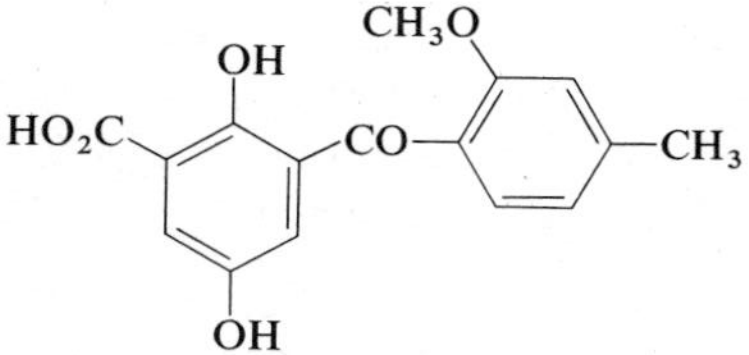

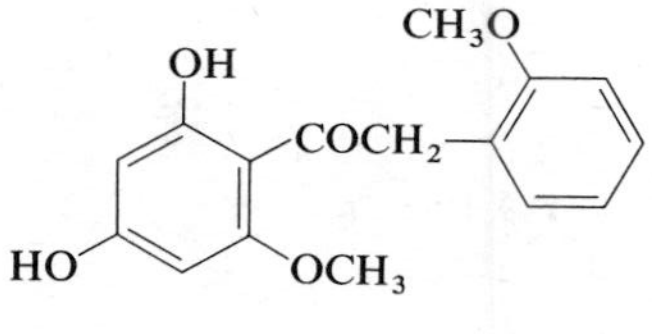

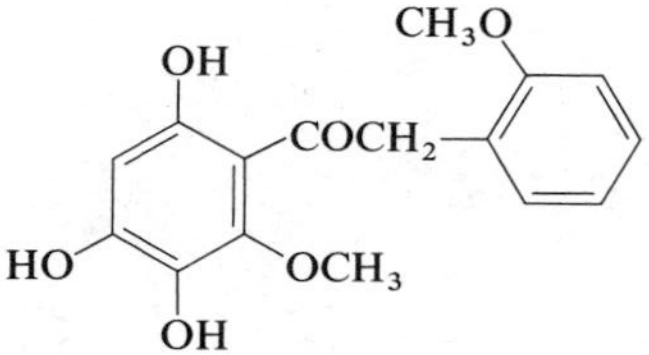

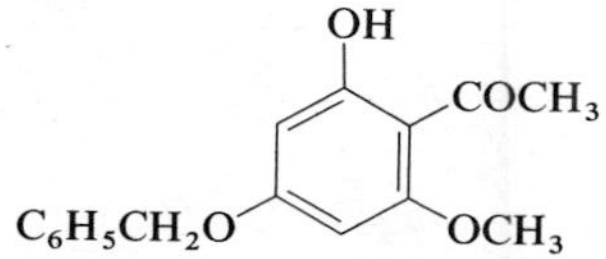

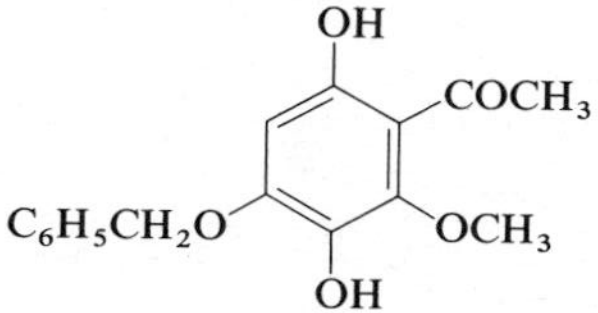

TABLE I. PERSULFATE OXIDATION OF PHENOLS (*Continued*)

No. of Carbon Atoms	Phenol	Product(s) and Yield(s) (%)	Refs.
	OH, $COCH_2C_6H_5$, CH_3O, OCH_3	OH, $COCH_2C_6H_5$, CH_3O, OCH_3, OH (—)	163
	OH, $COCH_2C_6H_4OCH_3$-4, HO, OCH_3	OH, $COCH_2C_6H_4OCH_3$-4, HO, OCH_3, OH (10)	164
	OH, $COCH_2C_6H_4OCH_3$-4, CH_3O	OH, $COCH_2C_6H_4OCH_3$-4, CH_3O, OH (21)	165
C_{17}	OH, $COCH{=}CHC_6H_5$, CH_3O, OCH_3	OH, $COCH{=}CHC_6H_5$, CH_3O, OCH_3, OH (19)	158

Substrate	Product	Yield (%)	Ref.
OH; HO_2C; OCH_3; CO; CH_3; CH_3O	OH; HO_2C; OCH_3; CO; CH_3; CH_3O; OH	(42)	155
OH; HO_2C; CH_3; CO; CH_3O; OCH_3	OH; HO_2C; CH_3; CO; CH_3O; OCH_3; OH	(11)	159
OH; HO_2C; CO; OCH_3; CH_3; CH_3O	OH; HO_2C; CO; OCH_3; CH_3; CH_3O; OH	(81)	166
OH; $COCH_2C_6H_4OCH_3$-2; CH_3O; OCH_3	OH; $COCH_2C_6H_4OCH_3$-2; CH_3O; OCH_3; OH	(9.5)	160
OH; CH_3O; $COC_6H_4OC_2H_5$-4; CH_3O	OH; CH_3O; $COC_6H_4OC_2H_5$-4; CH_3O; OH	(24)	167

TABLE I. PERSULFATE OXIDATION OF PHENOLS (*Continued*)

No. of Carbon Atoms	Phenol	Product(s) and Yield(s) (%)	Refs.
C_{18}	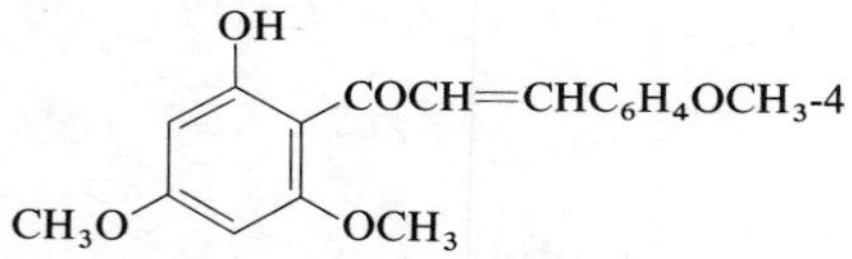 OH, $COCH{=}CHC_6H_4OCH_3$-4, CH_3O, OCH_3	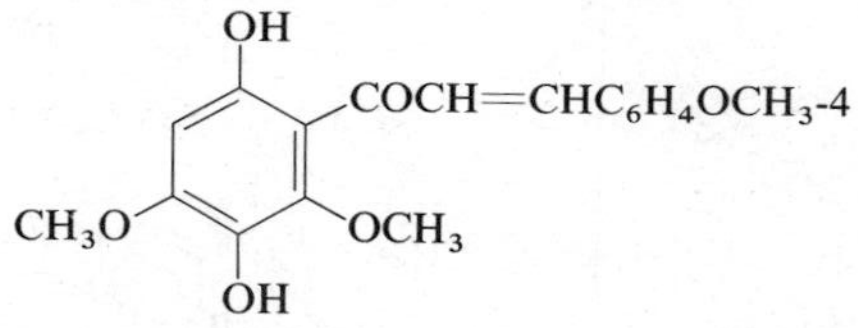 OH, $COCH{=}CHC_6H_4OCH_3$-4, CH_3O, OCH_3, OH (28)	168
	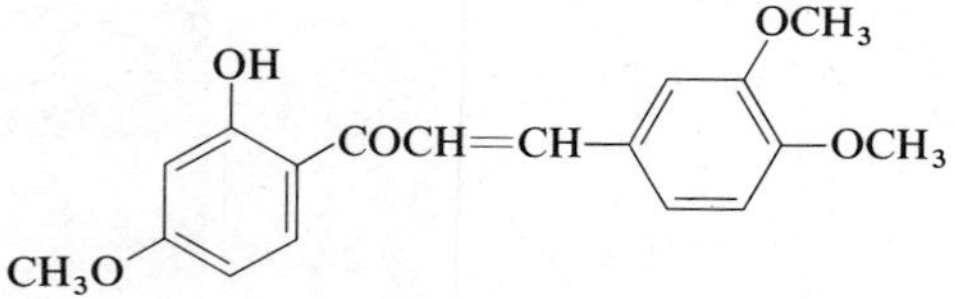 OH, $COCH{=}CH$–, OCH_3, OCH_3, CH_3O	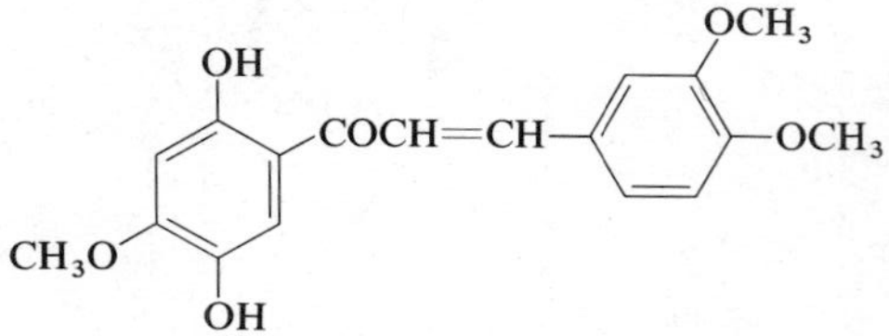 OH, $COCH{=}CH$–, OCH_3, OCH_3, CH_3O, OH (19)	122
	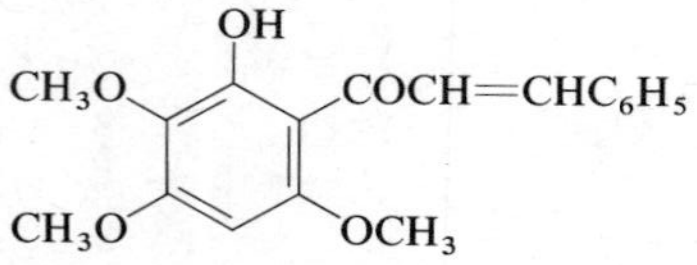 OH, CH_3O, $COCH{=}CHC_6H_5$, CH_3O, OCH_3	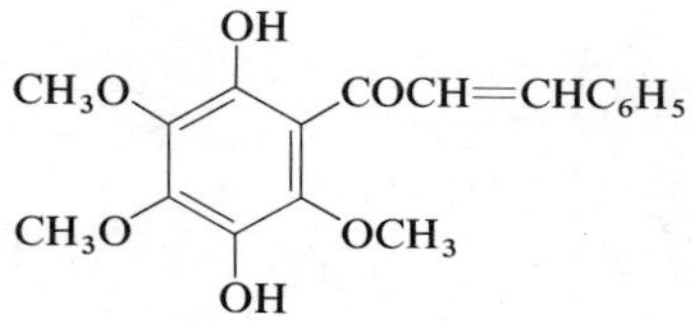 OH, CH_3O, $COCH{=}CHC_6H_5$, CH_3O, OCH_3, OH (19)	169
	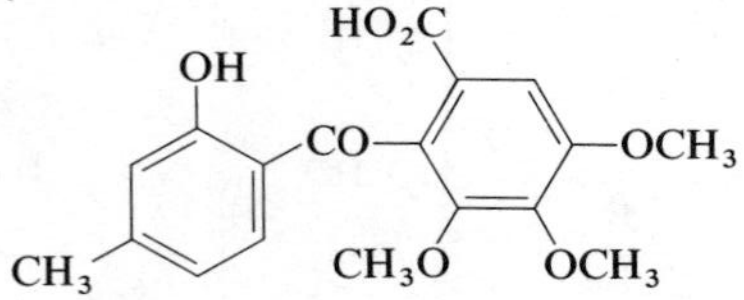 HO_2C, OH, CO, OCH_3, CH_3, CH_3O, OCH_3	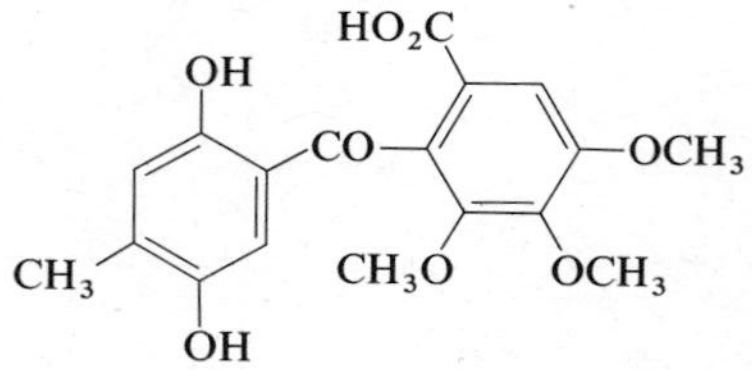 HO_2C, OH, CO, OCH_3, CH_3, CH_3O, OCH_3, OH (38)	159

	Reactant	Product	Yield (%)	Ref.
	OH, CH_3O, $COCH_2OCH_3$, $C_6H_5CH_2O$, OCH_3	OH, CH_3O, $COCH_2OCH_3$, $C_6H_5CH_2O$, OCH_3, OH	(10)	170
C_{19}	OH, CH_3O, CH_3, CH_3O, $(CH_2)_9CO_2H$	OH, CH_3O, CH_3, CH_3O, $(CH_2)_9CO_2H$, OH	(—)	152a

[a] This reference reports variations in overall yield and the (*ortho:para*) ratio as functions of ionic strength, pH, and temperature.
[b] See also Table IV, first entry.
[c] The product was isolated as the trimethoxy derivative following treatment with methyl iodide.

Note added in proof: The oxidation of 3,4-dimethylphenol to 4,5-dimethylcatechol in 16% yield and, more significantly, the conversion of mesitol (although in minute yield) to the only example of a *meta*-substitution product has been reported: R.G.R. Bacon, *Sci. Proc. R. Dublin Soc.*, **27**, 177 (1956).

TABLE II. PERSULFATE OXIDATION OF NAPHTHOLS

No. of Carbon Atoms	Naphthol	Product(s) and Yield(s) (%)	Refs.
C_{10}	1-Naphthol	1,4-Dihydroxynaphthalene (17) [20]	171
		" (41)	172
		" (15)	128
		" (—)‡	93
	2-Naphthol	1,2-Dihydroxynaphthalene (12)	172
		" (—)	171
C_{11}	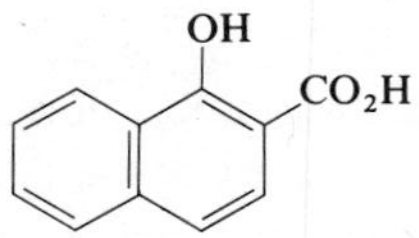 OH, CO_2H	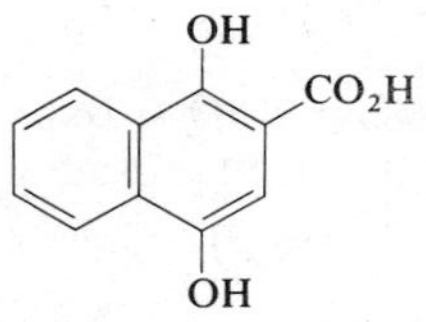 OH, CO_2H, OH (24) [35]	171
		" (—)	118
	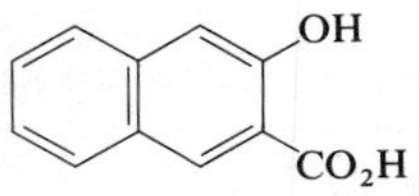 OH, CO_2H	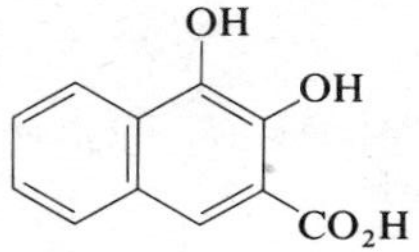 OH, OH, CO_2H (16) [33]	171
C_{12}	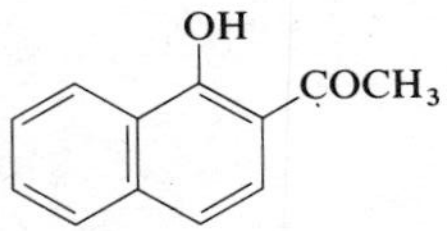 OH, $COCH_3$	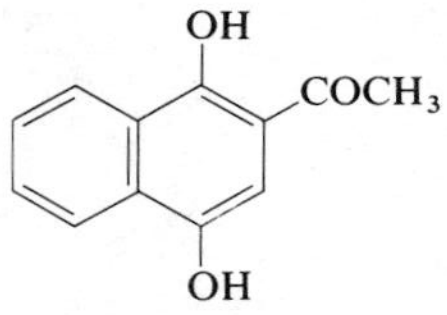 OH, $COCH_3$, OH (24) [67]	171
C_{15}	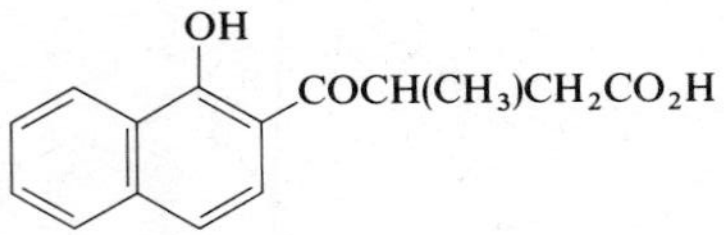 OH, $COCH(CH_3)CH_2CO_2H$	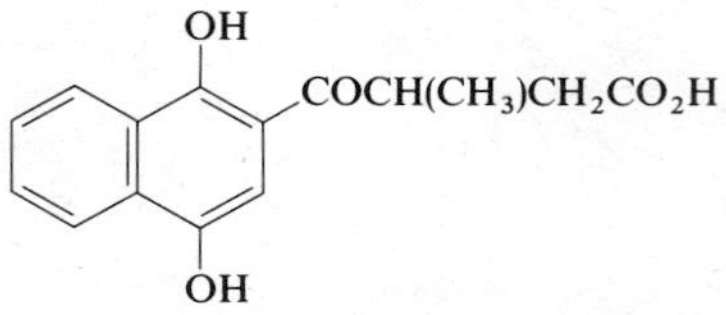 OH, $COCH(CH_3)CH_2CO_2H$, OH (48)	173

TABLE III. PERSULFATE OXIDATION OF HYDROXYQUINONES

No. of Carbon Atoms	Hydroxyquinones	Product(s) and Yield(s) (%)	Ref.
C_{10}	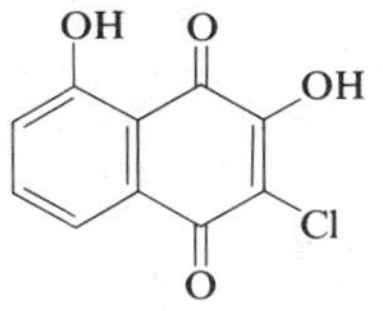 OH, O, OH, Cl, O	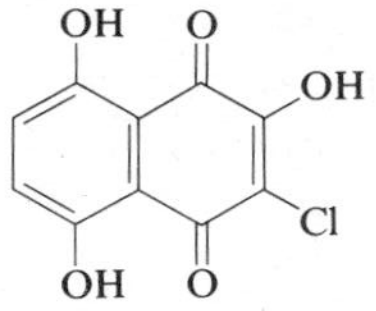 OH, O, OH, Cl, OH, O (—)	27
	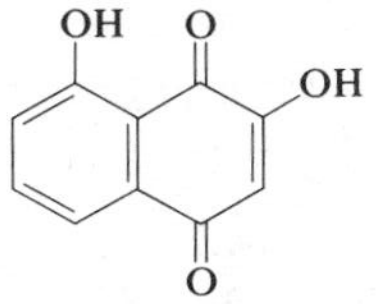 OH, O, OH, O	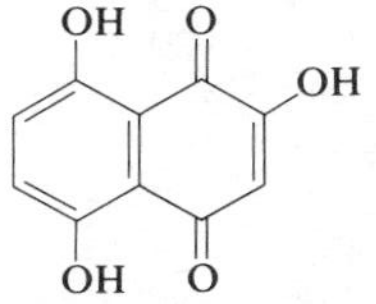 OH, O, OH, OH, O (0.5)	27
C_{14}	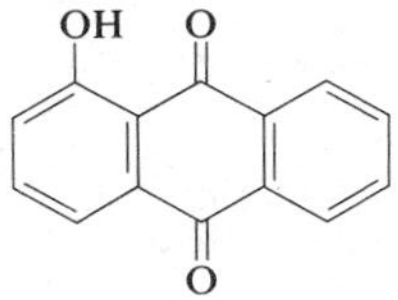 OH, O, O	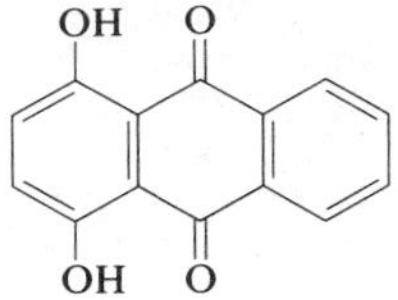 OH, O, OH, O (0.3)	27
C_{16}	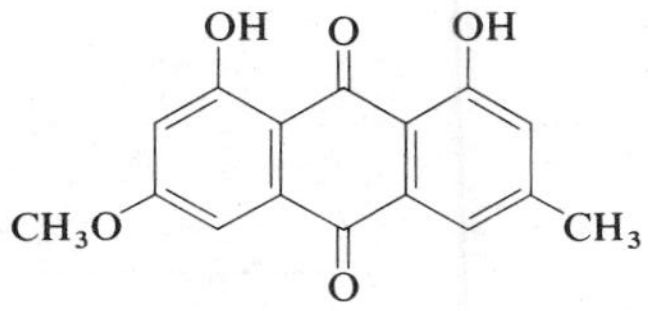 OH, O, OH, CH_3O, CH_3, O	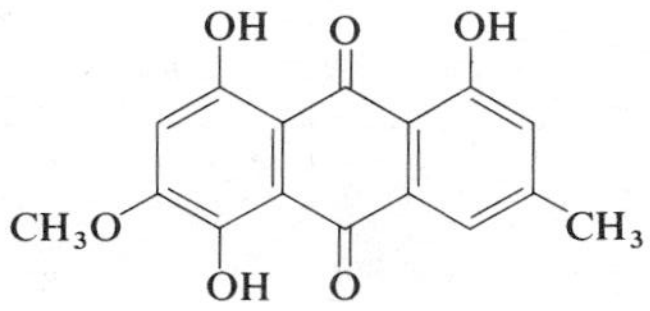 OH, O, OH, CH_3O, CH_3, OH, O (1.5) [7]	174
C_{21}	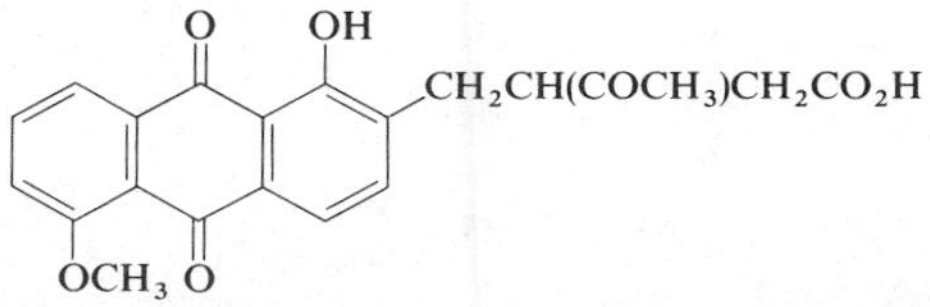 O, OH, $CH_2CH(COCH_3)CH_2CO_2H$, OCH_3, O	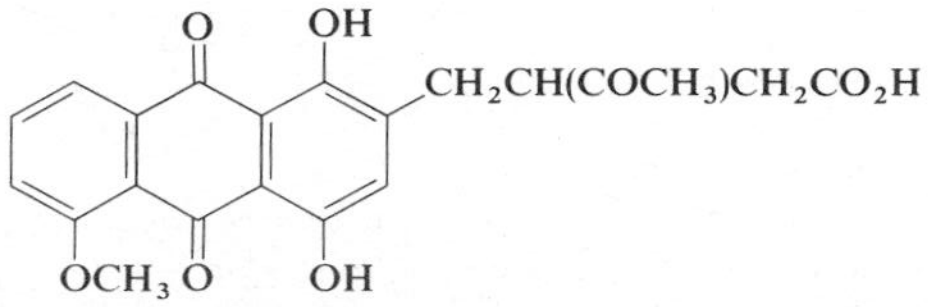 O, OH, $CH_2CH(COCH_3)CH_2CO_2H$, OCH_3, O, OH (42) [68]	175

TABLE IV. Persulfate Oxidation of Coumarins

No. of Carbon Atoms	Coumarin	Product(s) and Yield(s) (%)	Refs.
C_9	Coumarin	6-Hydroxycoumarin (15)	176
		" (11)	177
		" (27) [39]	178
	7-Hydroxycoumarin	6,7-Dihydroxycoumarin (10)	179
	3,4-Dihydrocoumarin	6-Hydroxy-3,4-dihydrocoumarin (53)	50
		" (12)	180
C_{10}	3-Carboxy-5-nitrocoumarin	3-Carboxy-6-hydroxy-5-nitrocoumarin (18)	181
	3-Carboxy-7-nitrocoumarin	3-Carboxy-6-hydroxy-7-nitrocoumarin (12)	181
	7-Formylcoumarin	6-Hydroxy-7-formylcoumarin (23)	182
	3-Carboxycoumarin	3-Carboxy-6-hydroxycoumarin (—)	181
	4-Methylcoumarin	6-Hydroxy-4-methylcoumarin (42) [45]	183
	7-Hydroxy-4-methylcoumarin	6,7-Dihydroxy-4-methylcoumarin (10)	179
	7-Methoxycoumarin	6-Hydroxy-7-methoxycoumarin (—)	177, 184
		" (25)	179
	8-Methoxycoumarin	6-Hydroxy-8-methoxycoumarin (—)	185
	Methylcoumarin-7-sulfonate	Methyl-6-hydroxycoumarin-7-sulfonate (23) [32]	186
C_{11}	[structure]	[structure, HO] (32–37)	187
	4-Methyl-7-formylcoumarin	4-Methyl-6-hydroxy-7-formylcoumarin (28)	182

	HO_2C, OH, CH_3	OH, HO_2C, OH, CH_3 (9) [17]	188
	NO_2, OCH_3, CH_3	NO_2, HO, OCH_3, CH_3 (27)	189
	4,6-Dimethylcoumarin	4,6-Dimethyl-8-hydroxycoumarin (3) [6]	188
	5-Methoxy-4-methylcoumarin	6-Hydroxy-5-methoxy-4-methylcoumarin (45)	188
	7-Methoxy-4-methylcoumarin	6-Hydroxy-7-methoxy-4-methylcoumarin (22) [31]	179
		" (17) [25]	183
		" (32) [38]	178
	5,7-Dimethoxycoumarin	5,7-Dimethoxy-6-hydroxycoumarin (9) [12]	188
		" (—)	178
	7,8-Dimethoxycoumarin	7,8-Dimethoxy-6-hydroxycoumarin (35)	178
		" (28)	179
		" (19)	190
		" (10)	53
	4,7-Dimethoxycoumarin	2,5-Dihydroxy-4-methoxyacetophenone (—)	23
C_{12}	7-Allyloxycoumarin	7-Allyloxy-6-hydroxycoumarin (37)	187
	8-Ethoxy-7-methoxycoumarin	8-Ethoxy-6-hydroxy-7-methoxycoumarin (8)	190
	CH_3, CH_3O, CH_3	CH_3, HO, CH_3O, CH_3 (40) [54]	183
		" (—)	178

TABLE IV. Persulfate Oxidation of Coumarins (*Continued*)

No. of Carbon Atoms	Coumarin	Product(s) and Yield(s) (%)	Refs.
	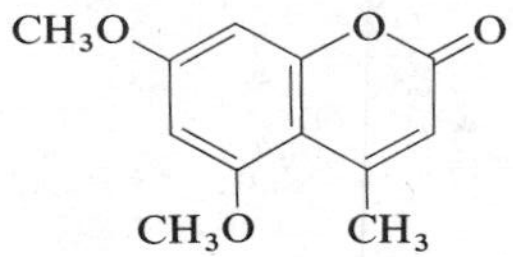 CH_3O, O, O, CH_3O, CH_3	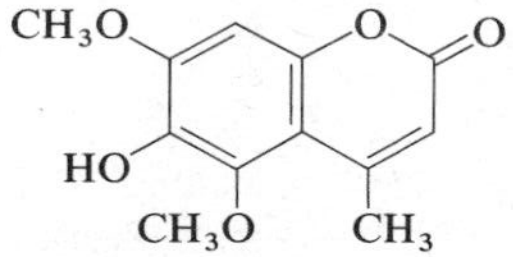 CH_3O, O, O, HO, CH_3O, CH_3 (20)	179
		" (25) [34]	183
		" (25)	178
	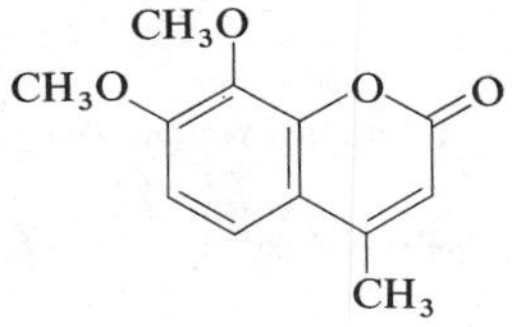 CH_3O, CH_3O, O, O, CH_3	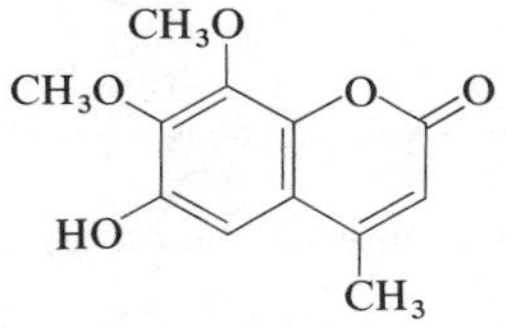 CH_3O, CH_3O, O, O, HO, CH_3 (30)	179
		" (25) [32]	183
	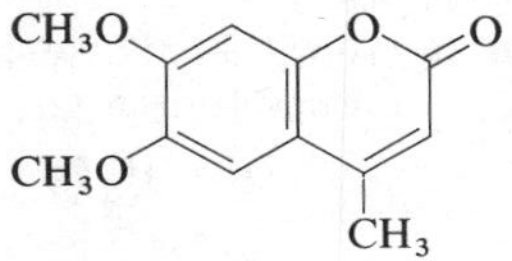 CH_3O, O, O, CH_3O, CH_3	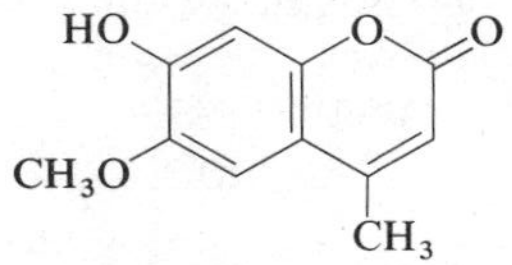 HO, O, O, CH_3O, CH_3 (?) (23) [26]	178
	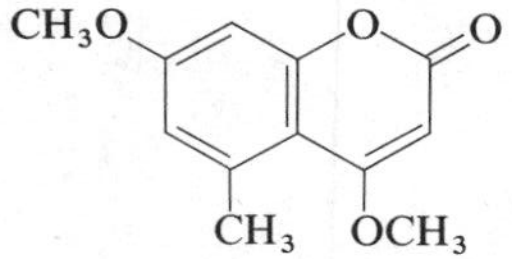 CH_3O, O, O, CH_3, OCH_3	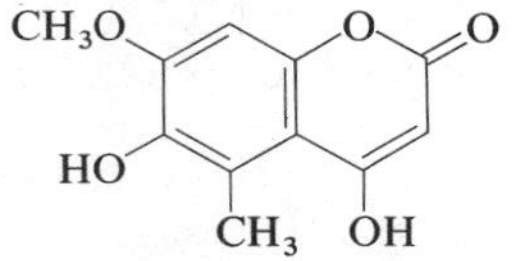 CH_3O, O, O, HO, CH_3, OH (35) [47],	
		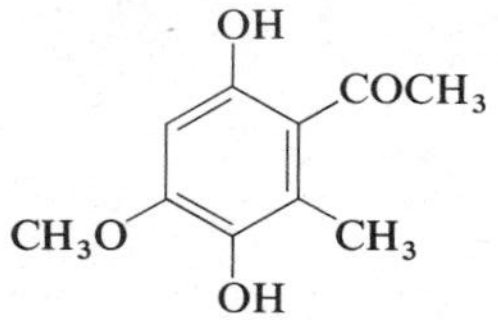 OH, $COCH_3$, CH_3O, CH_3, OH (22) [30]	24

	Substrate	Product (yield)	Ref.
	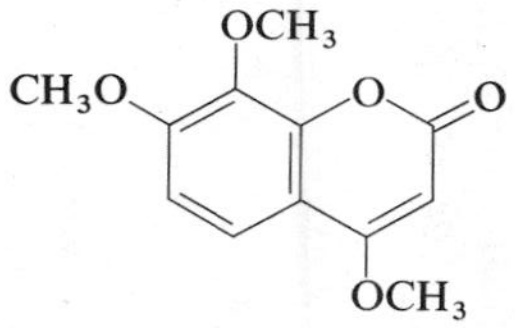 CH_3O, OCH_3, O, O, OCH_3	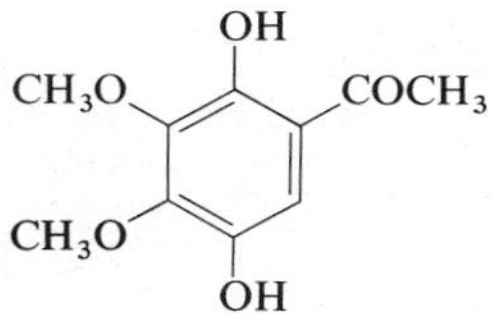 OH, CH_3O, $COCH_3$, CH_3O, OH (—)	23
C_{13}	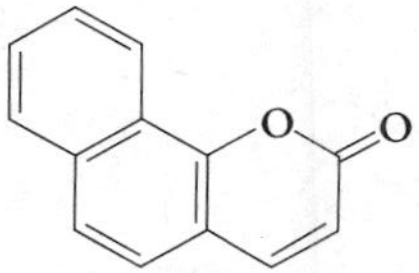 O, O	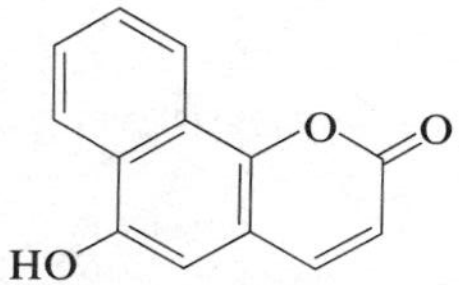 HO, O, O (66)	191
	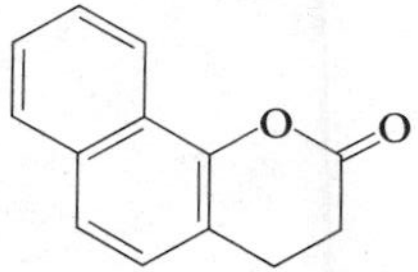 O, O	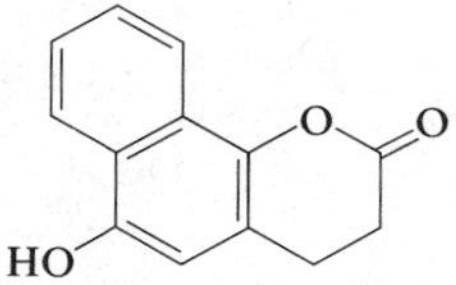 HO, O, O (41)	191
	7,8-Diethoxycoumarin	7,8-Diethoxy-6-hydroxycoumarin (—)	190
C_{14}	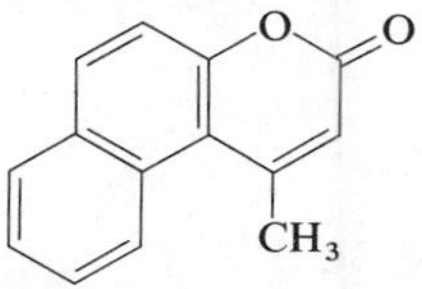 O, O, CH_3	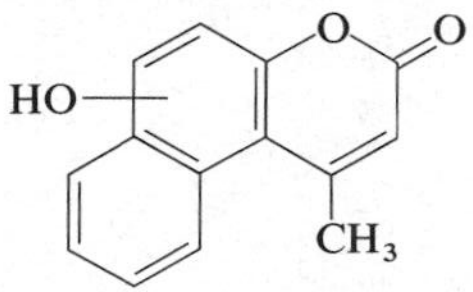 HO, O, O, CH_3 (5) [7]	171
	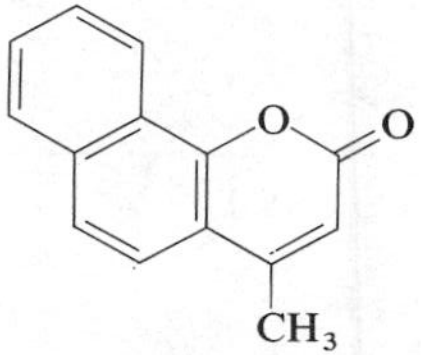 O, O, CH_3	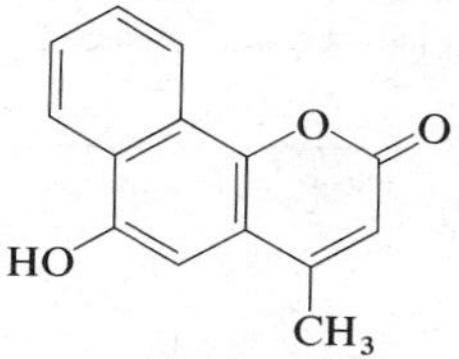 HO, O, O, CH_3 (37) [40]	171
		" (10)	178

TABLE IV. PERSULFATE OXIDATION OF COUMARINS (*Continued*)

No. of Carbon Atoms	Coumarin	Product(s) and Yield(s) (%)	Refs.
C_{15}	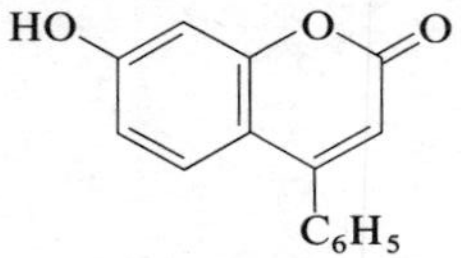 HO, O, O, C_6H_5	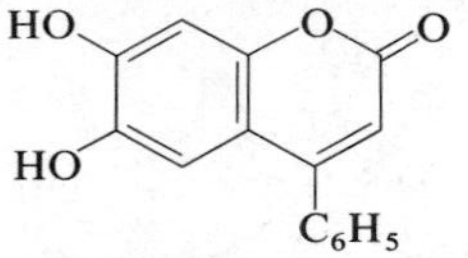 HO, HO, O, O, C_6H_5 (16)	192
	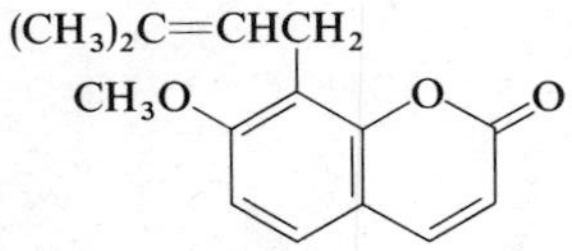 $(CH_3)_2C{=}CHCH_2$, CH_3O, O, O	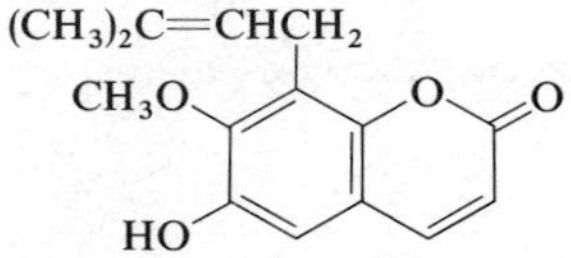 $(CH_3)_2C{=}CHCH_2$, CH_3O, HO, O, O (16)	193
C_{16}	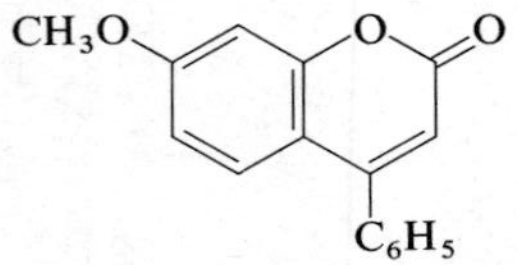 CH_3O, O, O, C_6H_5	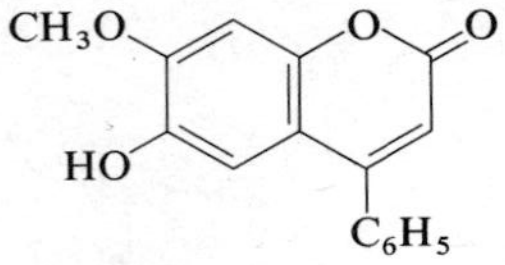 CH_3O, HO, O, O, C_6H_5 (51)	192
	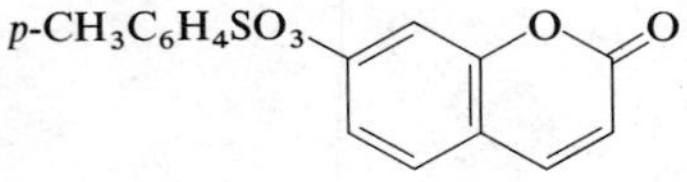 p-$CH_3C_6H_4SO_3$, O, O	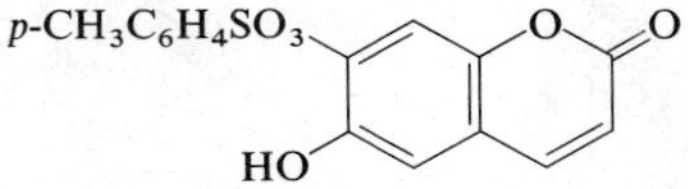 p-$CH_3C_6H_4SO_3$, HO, O, O (5) [8]	194
C_{17}	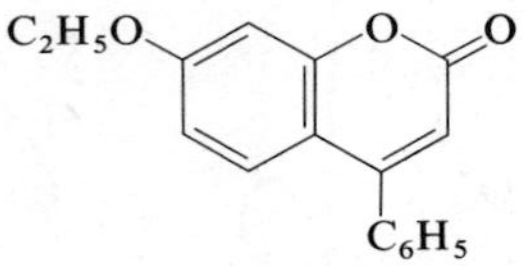 C_2H_5O, O, O, C_6H_5	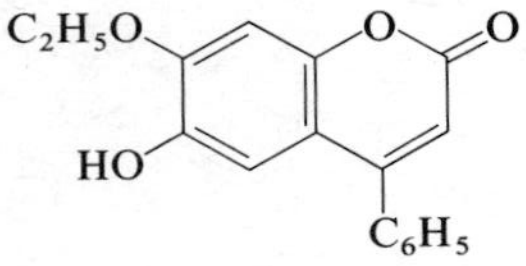 C_2H_5O, HO, O, O, C_6H_5 (23)	192

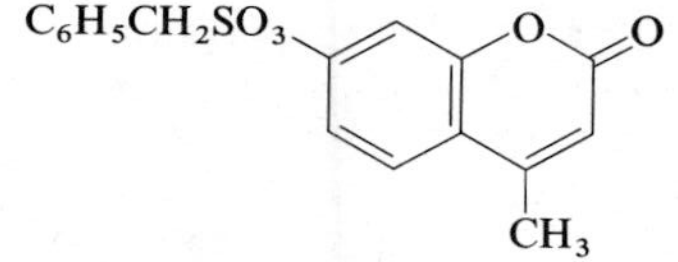

	$C_6H_5CH_2SO_3$, O, O, CH_3	$C_6H_5CH_2SO_3$, O, O, HO, CH_3	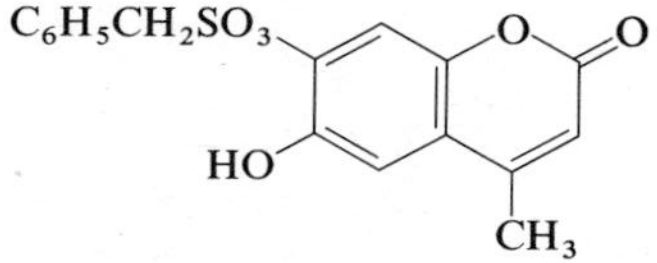 (14) [20]	195
	CH_3O, CH_3O, O, O, C_6H_5	CH_3O, CH_3O, O, O, HO, C_6H_5	(47)	196
	CH_3O, O, O, CH_3O, C_6H_5	CH_3O, O, O, HO, CH_3O, C_6H_5	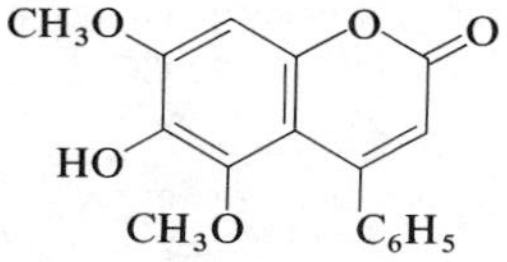 (28)	196
C_{18}	CH_3CO, $C_6H_5CH_2O$, O, O	CH_3CO, $C_6H_5CH_2O$, O, O, HO	(28)	197
	CH_3, O, O, $C_6H_5CH_2SO_3$, CH_3	CH_3, O, O, HO, $C_6H_5CH_2SO_3$, CH_3	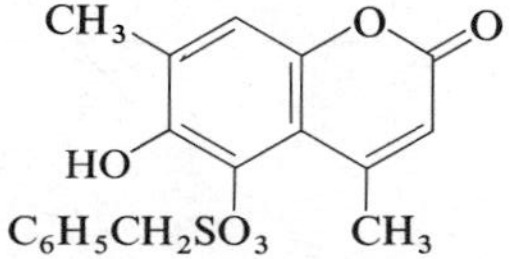 (28)	195
C_{22}	$C_6H_5CH_2O$, O, O, C_6H_5	$C_6H_5CH_2O$, O, O, HO, C_6H_5	(13)	192

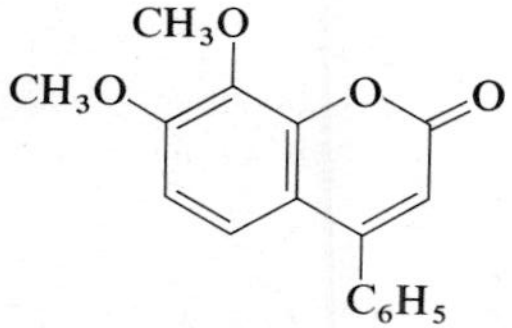

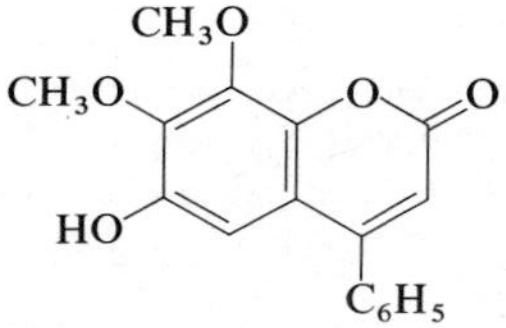

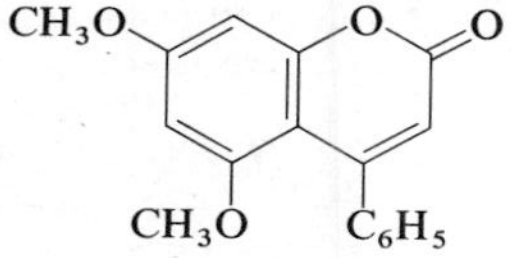

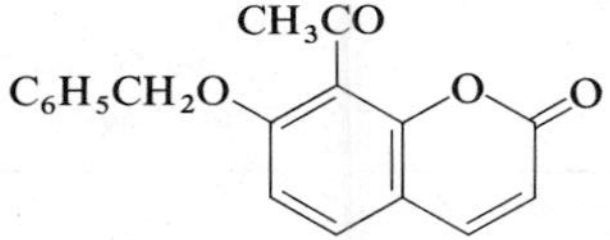

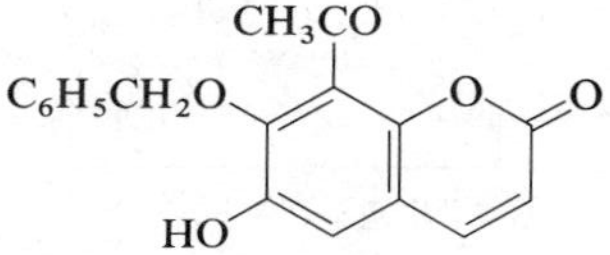

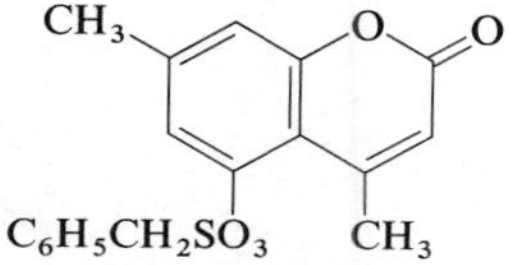

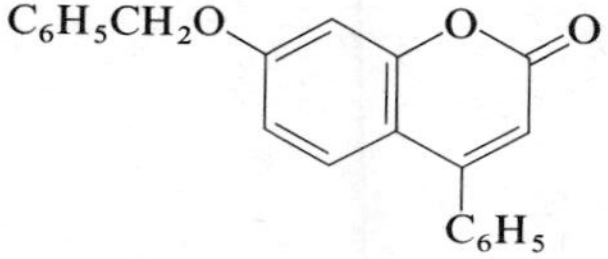

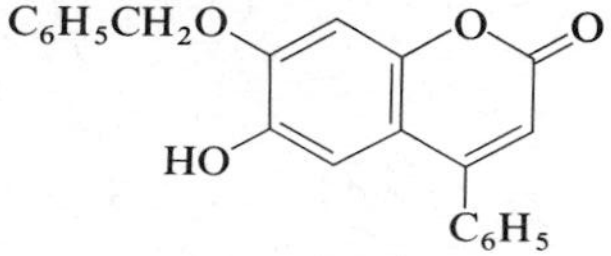

TABLE V. PERSULFATE OXIDATION OF FLAVONES[a]

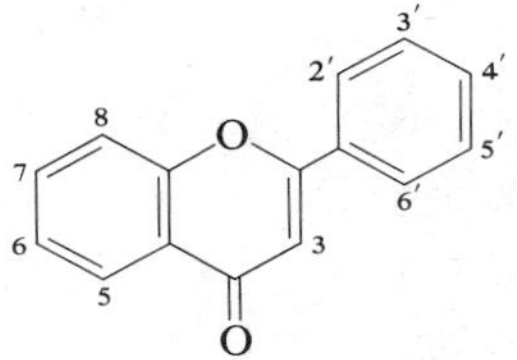

(Flavone ring numbering: 2–8, 2′–6′)

No. of Carbon Atoms	Flavone	Product(s) and Yield(s) (%)	Refs.
C_{15}	5-Hydroxy-6-chloroflavone	5,8-Dihydroxy-6-chloroflavone (24) [35]	26
	5-Hydroxyflavone	5,8-Dihydroxyflavone (42)	198
		" (50)	26
	8-Hydroxyflavone	" (13)	199
	7-Hydroxyflavone	7,8-Dihydroxyflavone (12) [20], 9‡ [16]‡	200
	5,7-Dihydroxyflavone	5,7,8-Trihydroxyflavone (42)	201
C_{16}	5-Hydroxy-3-methoxyflavone	5,8-Dihydroxy-3-methoxyflavone (24) [30]	202
	5-Hydroxy-6-methoxyflavone	5,8-Dihydroxy-6-methoxyflavone (7)	203
	5-Hydroxy-7-methoxyflavone	5,8-Dihydroxy-7-methoxyflavone (42)	201
	5-Hydroxy-4′-methoxyflavone	5,8-Dihydroxy-4′-methoxyflavone (10) [43], 5,6,8-trihydroxy-4′-methoxyflavone (—)	48
	7-Hydroxy-3-methoxyflavone	7,8-Dihydroxy-3-methoxyflavone (20)	200
	8-Hydroxy-4′-methoxyflavone	5,8-Dihydroxy-4′-methoxyflavone (3)	48
	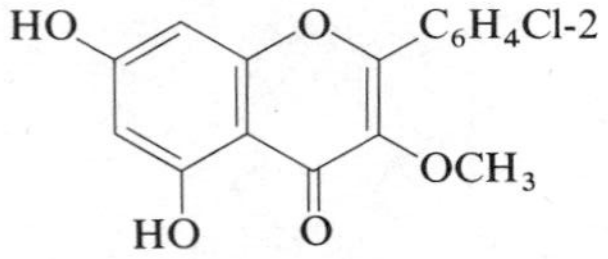 HO, HO, O, O, C_6H_4Cl-2, OCH_3	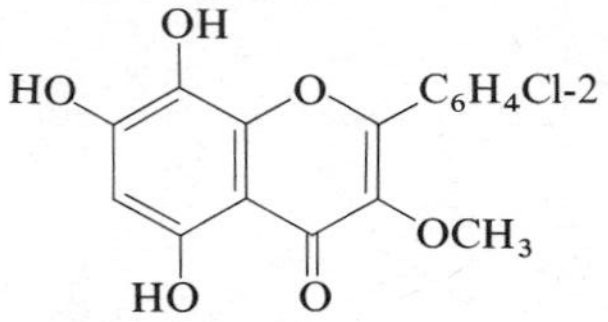 OH, HO, HO, O, O, C_6H_4Cl-2, OCH_3 (—)	203a

	Substrate	Product (Yield, %)	Ref.
C_{17}	HO; HO; O; C_6H_5; OCH_3; O (7-OH, 5-OH, 3-OCH_3, 2-C_6H_5 chromone)	HO; HO; HO; O; C_6H_5; OCH_3; O (47)	204
		" (34)	205
	HO; HO; O; $C_6H_4OCH_3$-4; O	HO; HO; HO; O; $C_6H_4OCH_3$-4; O (33)	206
	HO; OH; O; Cl; OCH_3; OCH_3; O	OH; HO; OH; O; Cl; OCH_3; OCH_3; O (28)	207
	CH_3O; HO; HO; O; Cl; OCH_3; O	OH; CH_3O; HO; HO; O; Cl; OCH_3; O (—)	203a
	HO; O; $C_6H_4OCH_3$-4; OCH_3; O	HO; HO; O; $C_6H_4OCH_3$-4; OCH_3; O (—)	208

TABLE V. Persulfate Oxidation of Flavones (*Continued*)

No. of Carbon Atoms	Flavone	Product(s) and Yield(s) (%)	Refs.
	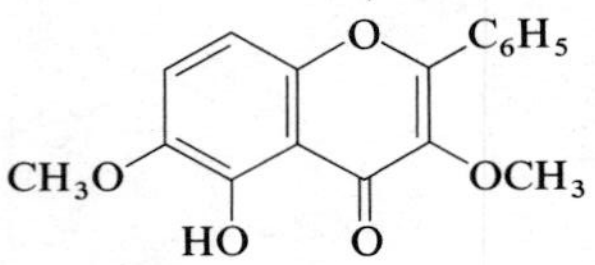 CH_3O, HO, O, C_6H_5, OCH_3, O	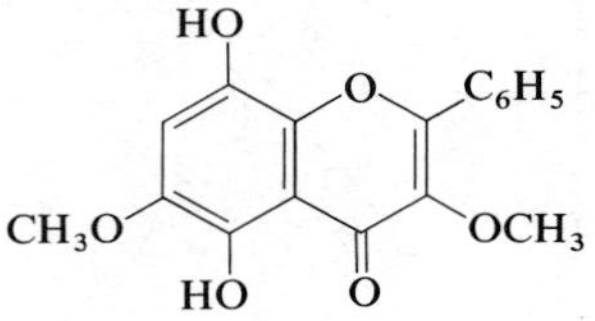 HO, O, C_6H_5, CH_3O, OCH_3, HO, O (—)	209
		" (9)	210
	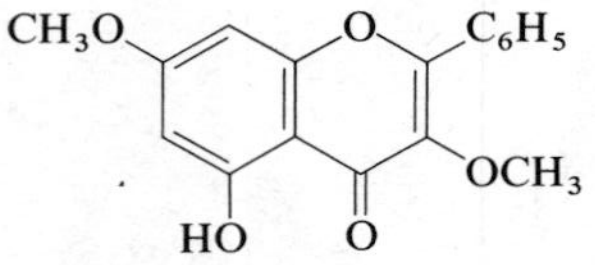 CH_3O, O, C_6H_5, OCH_3, HO, O	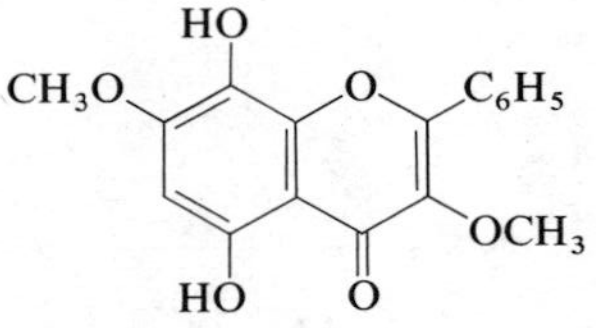 HO, CH_3O, O, C_6H_5, OCH_3, HO, O (52)	204
	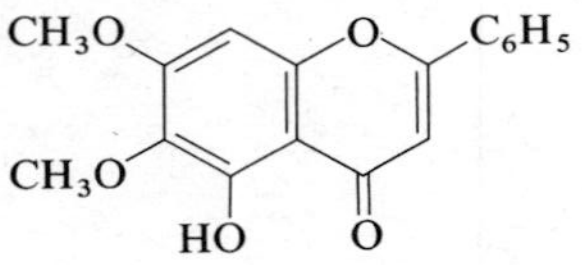 CH_3O, CH_3O, O, C_6H_5, HO, O	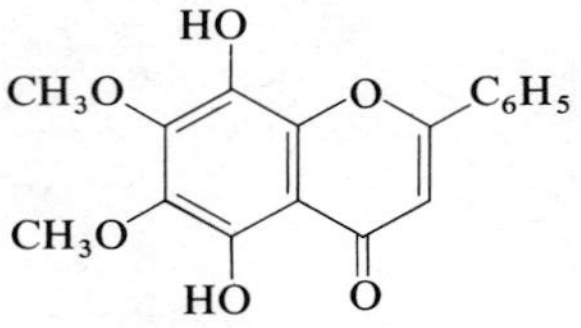 HO, CH_3O, O, C_6H_5, CH_3O, HO, O (33)	211
	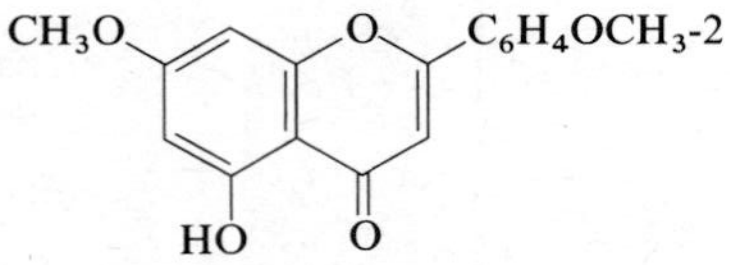 CH_3O, O, $C_6H_4OCH_3$-2, HO, O	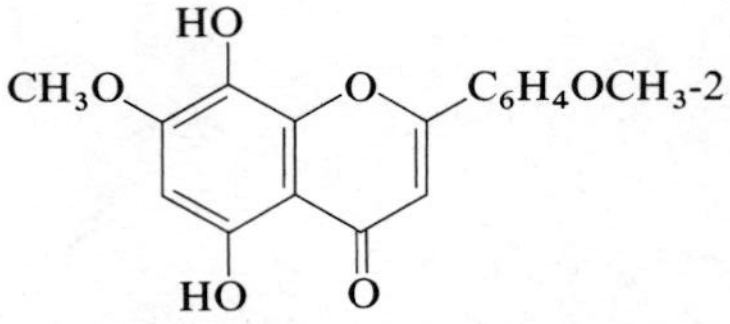 HO, CH_3O, O, $C_6H_4OCH_3$-2, HO, O (26)	212

CH_3O, $C_6H_4OCH_3$-4, HO, O	HO, CH_3O, $C_6H_4OCH_3$-4, HO, O	(28)	206
CH_3O, OCH_3, HO, O	CH_3O, OCH_3, HO, HO, O	(9)	199
CH_3O, OCH_3, HO, O	”	(9)	199
HO, $C_6H_4OCH_3$-4, OCH_3, O	HO, HO, $C_6H_4OCH_3$-4, OCH_3, O	(9)	213
HO, CH_3O, OCH_3, O	HO, CH_3O, OH, OCH_3, O	(25)	214

TABLE V. Persulfate Oxidation of Flavones (*Continued*)

No. of Carbon Atoms	Flavone	Product(s) and Yield(s) (%)	Refs.
C_{18}	HO, HO, O, O, $C_6H_4OCH_3$-2, OCH_3	HO, HO, HO, O, O, $C_6H_4OCH_3$-2, OCH_3 (38)	214
	CH_3O, HO, O, O, $C_6H_4OCH_3$-4, OCH_3	HO, CH_3O, HO, O, O, $C_6H_4OCH_3$-4, OCH_3 (48)	215
		" (6)	216
	CH_3O, CH_3O, HO, O, O, $C_6H_4OCH_3$-4	HO, CH_3O, CH_3O, HO, O, O, $C_6H_4OCH_3$-4 (29)	211
	CH_3O, HO, O, O, CH_3O, OCH_3	HO, CH_3O, HO, O, O, CH_3O, OCH_3 (—)	212

		(38)	206
		(44) [53]	217
		(10)	213
		(10)[b]	199

TABLE V. Persulfate Oxidation of Flavones (*Continued*)

No. of Carbon Atoms	Flavone	Product(s) and Yield(s) (%)	Refs.
	HO, OCH_3, CH_3O, O, OCH_3, O	HO, OCH_3, CH_3O, O, OH, OCH_3, O (16)	218
	CH_3O, OCH_3, HO, O, OCH_3, HO, O	HO, CH_3O, OCH_3, HO, O, OCH_3, HO, O (56)	214
	CH_3O, OCH_3, HO, O, OCH_3, HO, O	CH_3O, OCH_3, HO, HO, O, OCH_3, HO, O (43)	204
		" (32)	219

C_{19}	C_2H_5O, O, $C_6H_4OCH_3$-4 , OCH_3, HO, O	HO, C_2H_5O, O, $C_6H_4OCH_3$-4 OCH_3, HO, O	(—)	220
	CH_3O, OCH_3, O, OCH_3 OCH_3, HO, O	CH_3O, OCH_3, HO, O, OCH_3 OCH_3, HO, O	(37)	221
	CH_3O, OCH_3, CH_3O, O, OCH_3 HO, O	CH_3O, OCH_3, HO, CH_3O, O, OCH_3 HO, O	(38)	206
	CH_3O, OCH_3, CH_3O, O OCH_3, HO, O	CH_3O, OCH_3, HO, CH_3O, O OCH_3, HO, O	(43)	204

TABLE V. PERSULFATE OXIDATION OF FLAVONES (*Continued*)

No. of Carbon Atoms	Flavone	Product(s) and Yield(s) (%)	Refs.
		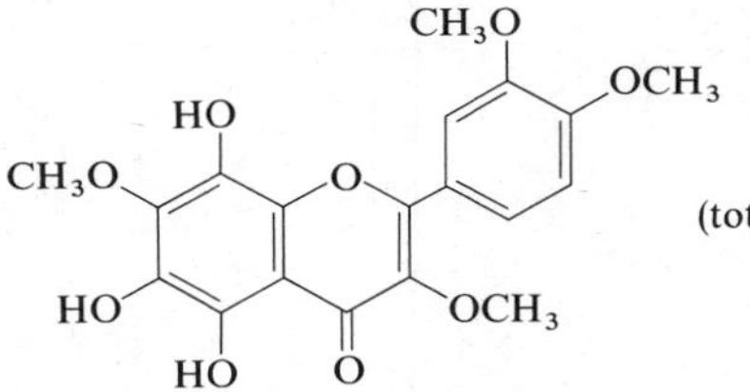 " (—), CH_3O, HO, HO, HO, O, CH_3O, OCH_3, OCH_3 (—) (total 55)[c]	49
	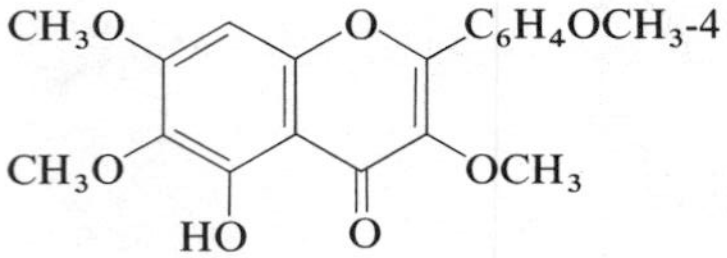 CH_3O, CH_3O, HO, O, $C_6H_4OCH_3$-4, OCH_3	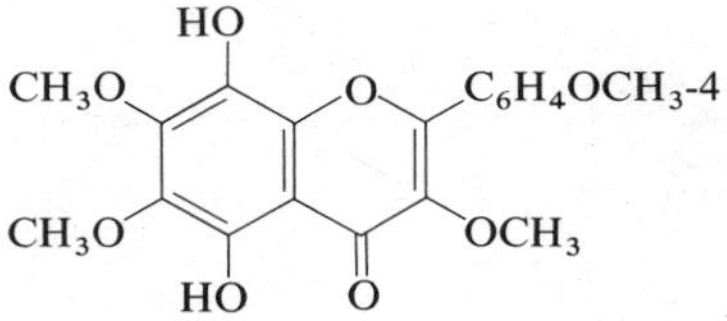 HO, CH_3O, CH_3O, HO, O, $C_6H_4OCH_3$-4, OCH_3 (38)	222
	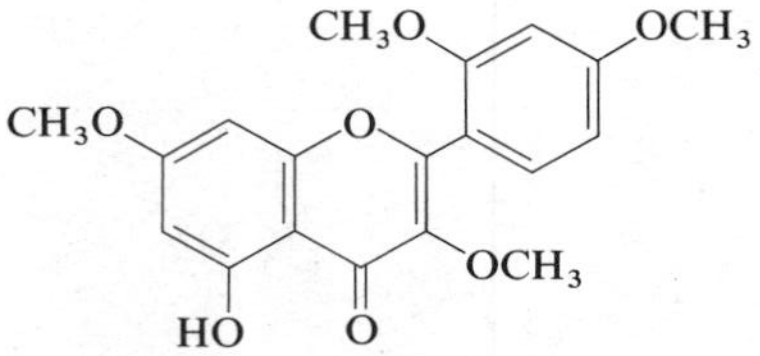 CH_3O, HO, O, CH_3O, OCH_3, OCH_3	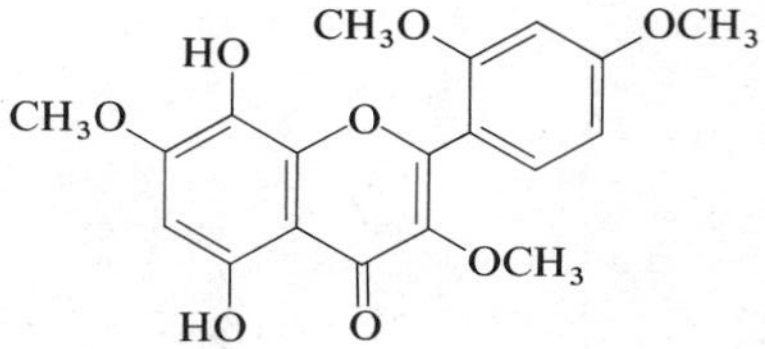 HO, CH_3O, HO, O, CH_3O, OCH_3, OCH_3 (12)	223
	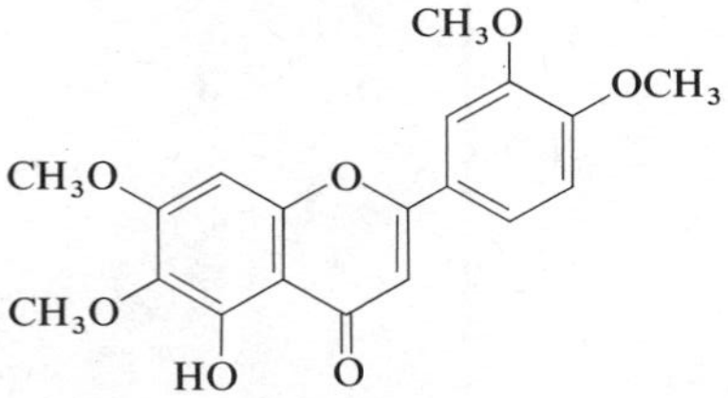 CH_3O, CH_3O, HO, O, CH_3O, OCH_3	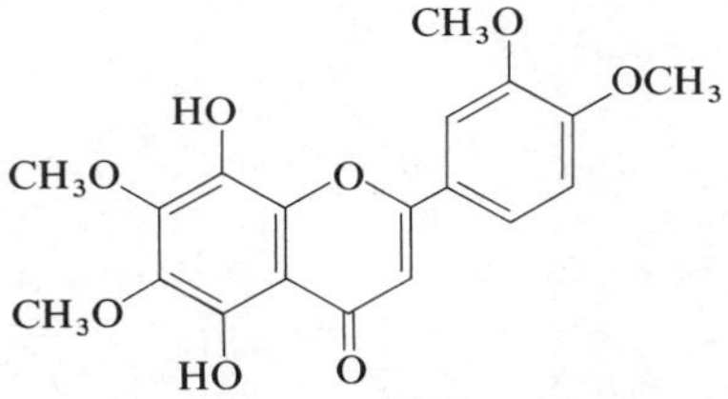 HO, CH_3O, CH_3O, HO, O, CH_3O, OCH_3 (38)	224

	Structure	Structure	Ref.	Page
C_{20}	HO, O, CH_3O, OCH_3, OCH_3, OCH_3, O	HO, HO, O, CH_3O, OCH_3, OCH_3, OCH_3, O	(14)	213
	HO, O, CH_3O, OCH_3, OCH_3, OCH_3, HO, O	HO, HO, O, CH_3O, OCH_3, OCH_3, OCH_3, HO, O	(53)	215
	CH_3O, CH_3O, O, CH_3O, OCH_3, OCH_3, HO, O	HO, CH_3O, CH_3O, O, CH_3O, OCH_3, OCH_3, HO, O	(43) [62]	222
		"	(—)	225
	CH_3O, O, CH_3O, OCH_3, OCH_3, OCH_3, HO, O	HO, CH_3O, O, CH_3O, OCH_3, OCH_3, OCH_3, HO, O	(48)	215

TABLE V. PERSULFATE OXIDATION OF FLAVONES (*Continued*)

No. of Carbon Atoms	Flavone	Product(s) and Yield(s) (%)	Refs.
	CH_3O, OCH_3, CH_3O, OH, OCH_3, CH_3O, O	CH_3O, HO, OCH_3, CH_3O, OH, OCH_3, CH_3O, O (29)	226
C_{21}	C_2H_5O, $C_6H_4OC_2H_5$-4, OC_2H_5, HO, O	HO, C_2H_5O, $C_6H_4OC_2H_5$-4, OC_2H_5, HO, O (—)	227
	C_2H_5O, OC_2H_5, CH_3O, OCH_3, HO, O	C_2H_5O, OC_2H_5, HO, CH_3O, OCH_3, HO, O (19)	218

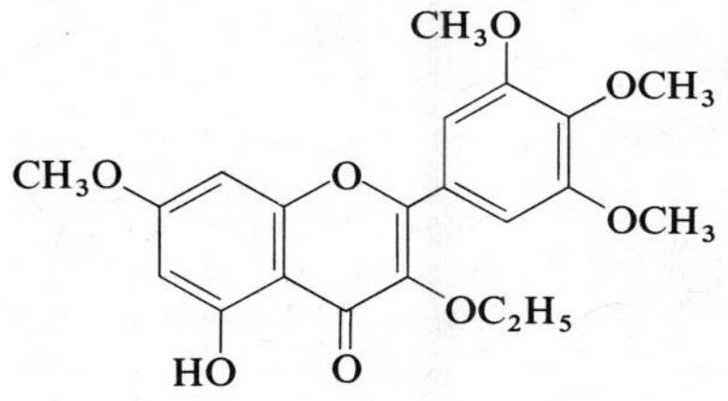

Formula	Starting material	Product	Ref.	Yield
	CH_3O, OCH_3, OCH_3, CH_3O, OC_2H_5, HO, O	CH_3O, OCH_3, HO, OCH_3, CH_3O, OC_2H_5, HO, O	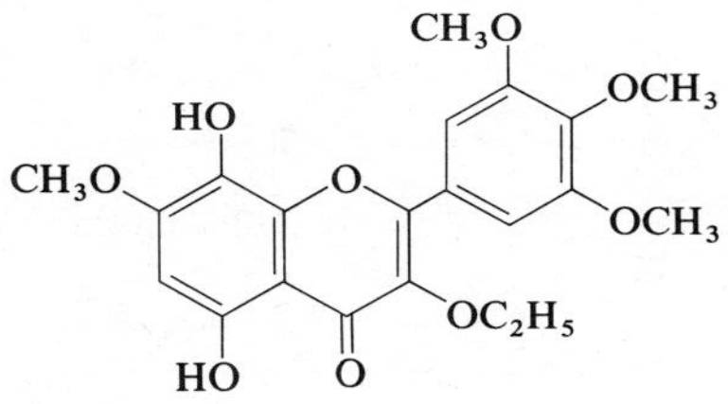 (48)	60
C_{22}	$C_6H_5CH_2O$, C_6H_5, HO, O	HO, $C_6H_5CH_2O$, C_6H_5, HO, O	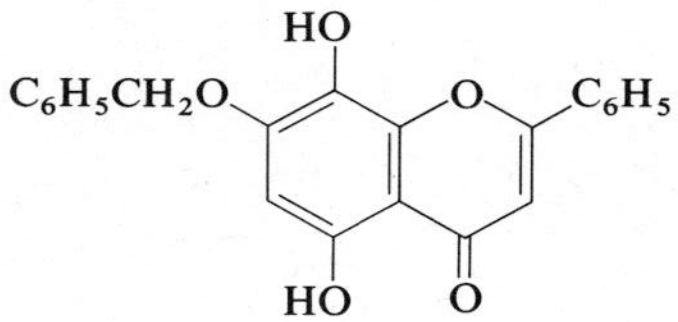 (19) [48]	228
C_{23}	HO, Cl, $OCH_2C_6H_5$, OCH_3, OH, O	OH, HO, Cl, $OCH_2C_6H_5$, OCH_3, OH, O	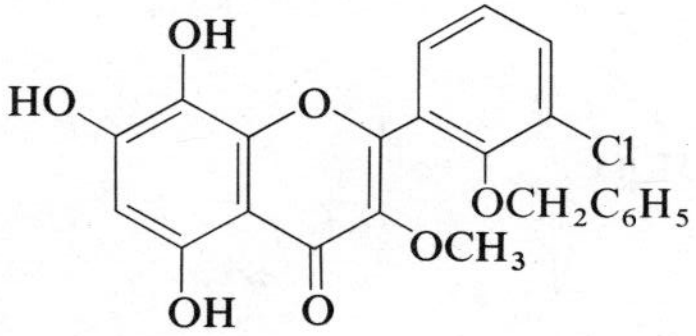 (20)	229
C_{24}	$C_6H_4OCH_3$-4, $COC_6H_4OCH_3$-4, HO, O	HO, $C_6H_4OCH_3$-4, $COC_6H_4OCH_3$-4, HO, O	(4),	48
		OH, $C_6H_4OCH_3$-4, HO, $COC_6H_4OCH_3$-4, OH, O	(—)	

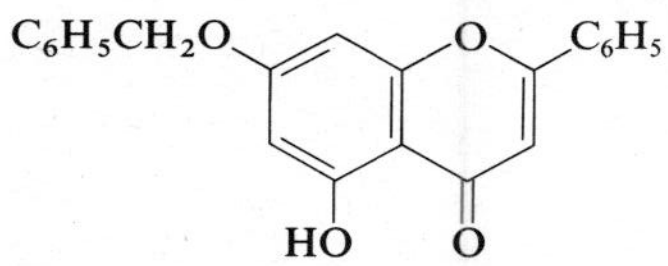

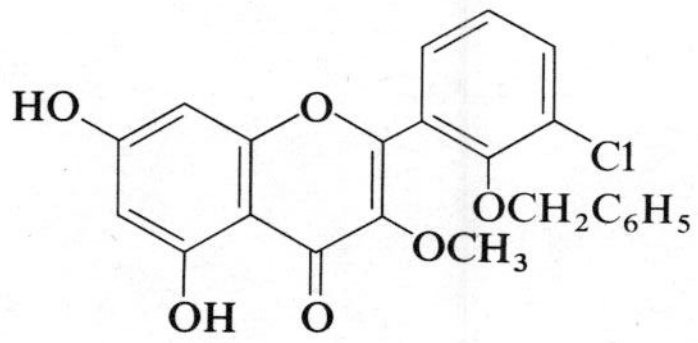

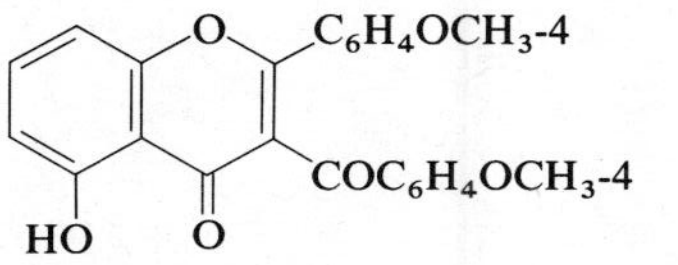

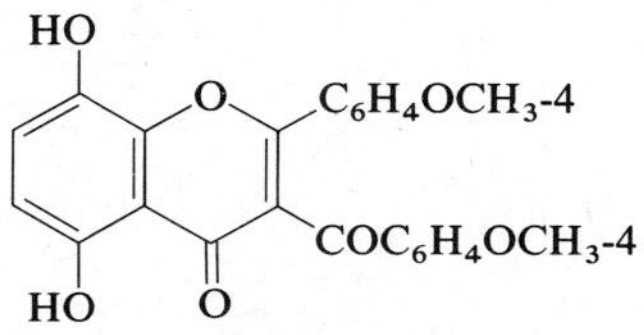

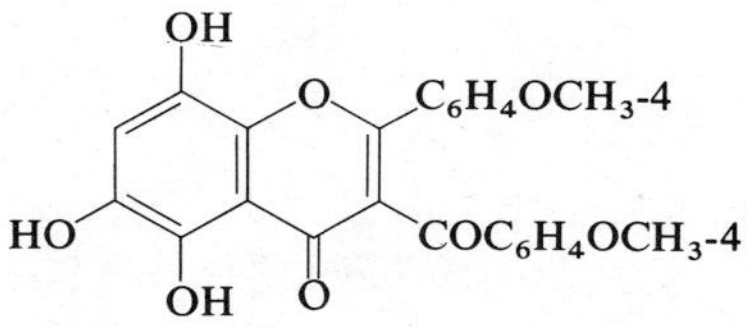

TABLE V. PERSULFATE OXIDATION OF FLAVONES (*Continued*)

No. of Carbon Atoms	Flavone	Product(s) and Yield(s) (%)	Refs.
	$C_6H_5CH_2O$, $C_6H_4OCH_3$-4, OCH_3, HO, O	HO, $C_6H_5CH_2O$, $C_6H_4OCH_3$-4, OCH_3, HO, O (—)	220
		" (9)	230
	HO, OCH_3, $OCH_2C_6H_5$, OCH_3, OH, O	OH, HO, OCH_3, $OCH_2C_6H_5$, OCH_3, OH, O (15)	219
C_{36}	$C_6H_5CH_2O$, $OCH_2C_6H_5$, $OCH_2C_6H_5$, HO, O	HO, $C_6H_5CH_2O$, $OCH_2C_6H_5$, $OCH_2C_6H_5$, HO, O (—)	231

[a] See Refs. 54 and 232 for discussions of the Elbs oxidation of flavones and related compounds.

[b] Attempts to oxidize the 5-hydroxy isomer were unsuccessful.

[c] The proportion of the 6,8-disubstituted product increased with increasing ratio of persulfate to flavone.

TABLE VI. Persulfate Oxidation of Flavanones

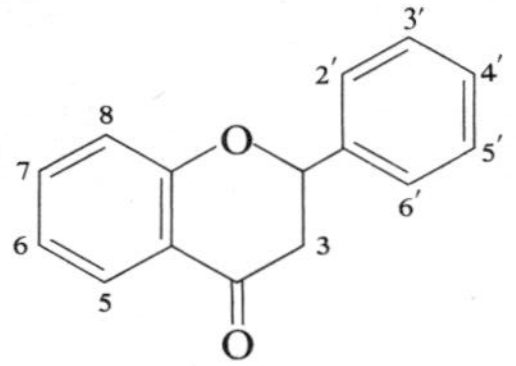

No. of Carbon Atoms	Flavanone	Product(s) and Yield(s) (%)	Refs.
C_{15}	5-Hydroxyflavanone	5,8-Dihydroxyflavanone (14)	233
C_{16}	5-Hydroxy-7-methoxyflavanone	5,8-Dihydroxy-7-methoxyflavanone (19)[a]	234
C_{17}	5-Hydroxy-7,4′-dimethoxyflavanone	5,8-Dihydroxy-7,4′-dimethoxyflavanone (19)[a]	234

[a] The original structural assignments have been corrected.[235]

TABLE VII. Persulfate Oxidation of Isoflavones

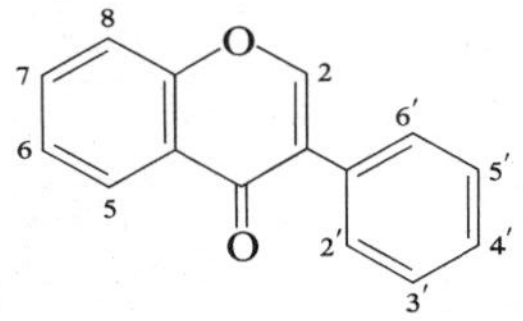

No. of Carbon Atoms	Isoflavone	Product(s) and Yield(s) (%)	Refs.
C_{15}	5,7-Dihydroxyisoflavone	5,7,8-Trihydroxyisoflavone (28)	236
C_{16}	7-Hydroxy-2-methylisoflavone	7,8-Dihydroxy-2-methylisoflavone (5)	236
	5,7-Dihydroxy-2-methylisoflavone	5,7,8-Trihydroxy-2-methylisoflavone (27)	236

TABLE VIII. Persulfate Oxidation of Chromones

No. of Carbon Atoms	Chromone	Product(s) and Yield(s) (%)	Refs.
C_{11}	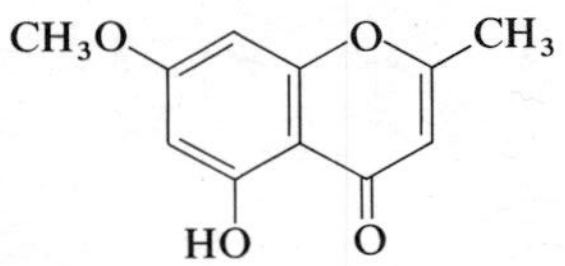 CH_3O, CH_3, HO, O	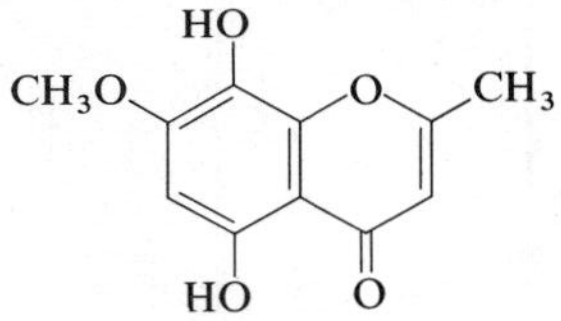 HO, CH_3O, CH_3, HO, O (20)[a]	46
	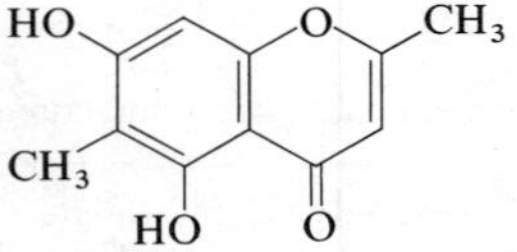 HO, CH_3, CH_3, HO, O	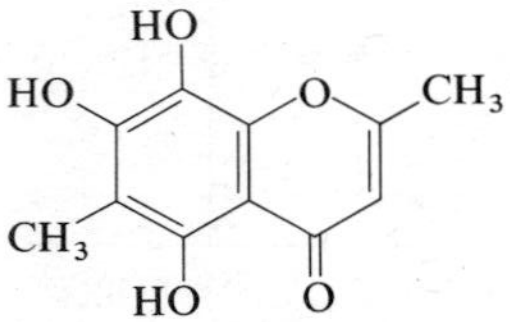 HO, HO, CH_3, CH_3, HO, O (10)[a]	46
	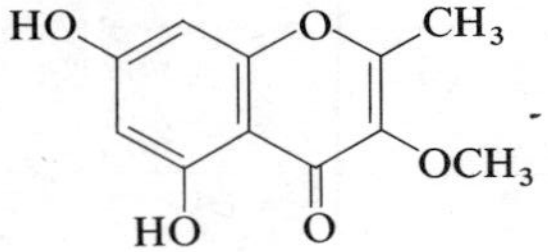 HO, CH_3, OCH_3, HO, O	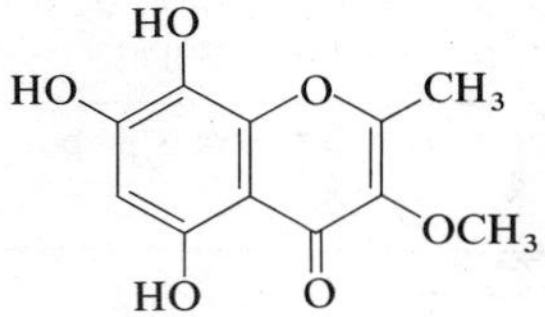 HO, HO, CH_3, OCH_3, HO, O (32) [35]	237
C_{12}	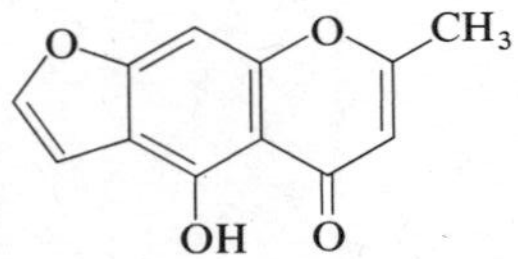 CH_3, OH, O	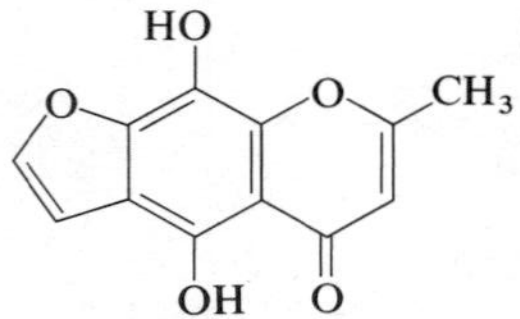 HO, CH_3, OH, O (11)	238

	Substrate	Product	Ref.
		(42) [60]	61
		(11)	46
		(37) [41]	237
C_{14}		(10)	239
		" (14), the quinone (9)	240

TABLE VIII. PERSULFATE OXIDATION OF CHROMONES (*Continued*)

No. of Carbon Atoms	Chromone	Product(s) and Yield(s) (%)	Refs.
	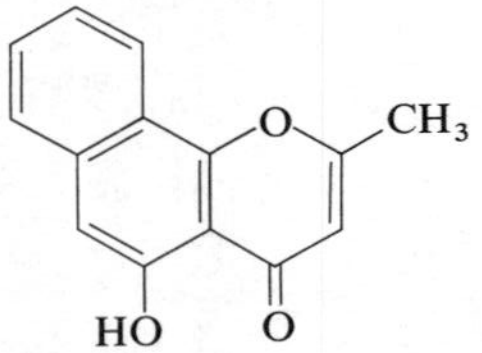 O, CH_3, HO, O	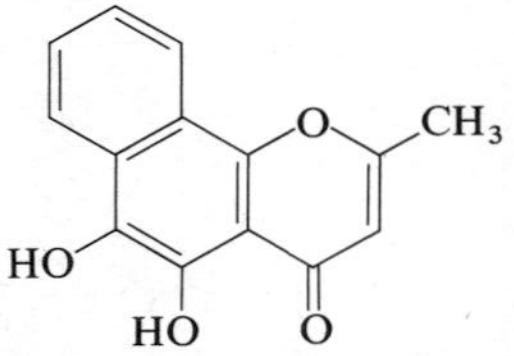 O, CH_3, HO, HO, O (5), the quinone (9)	240
	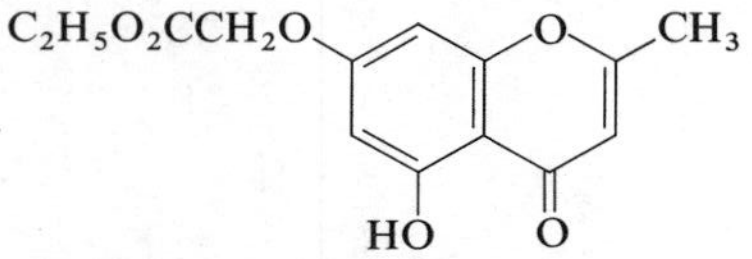 $C_2H_5O_2CCH_2O$, O, CH_3, HO, O	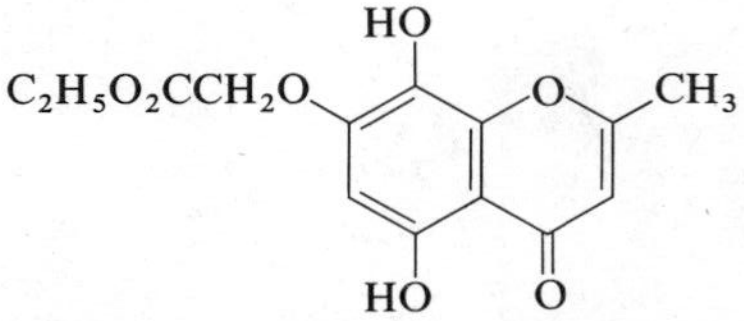 HO, $C_2H_5O_2CCH_2O$, O, CH_3, HO, O (—)	241

[a] Tetramethylammonium hydroxide was used as the base and gave better yields than either NaOH or KOH.

TABLE IX. PERSULFATE OXIDATION OF XANTHONES

No. of Carbon Atoms	Xanthone	Product(s) and Yield(s) (%)	Refs.
C_{13}	1-Hydroxyxanthone	1,4-Dihydroxyxanthone (47)	242
C_{14}	1-Hydroxy-3-methylxanthone	1,4-Dihydroxy-3-methylxanthone (21)	243
	1-Hydroxy-3-methoxyxanthone	1,4-Dihydroxy-3-methoxyxanthone (17)	243
	1-Hydroxy-7-methoxyxanthone	1,4-Dihydroxy-7-methoxyxanthone (38)[a]	244
C_{15}	1-Hydroxy-3,5-dimethylxanthone	1,4-Dihydroxy-3,5-dimethylxanthone (17)	243
	1-Hydroxy-3,6-dimethylxanthone	1,4-Dihydroxy-3,6-dimethylxanthone (15)	243
	1-Hydroxy-6-methoxy-3-methylxanthone	1,4-Dihydroxy-6-methoxy-3-methylxanthone (5)	243
	1-Hydroxy-7-methoxy-3-methylxanthone	1,4-Dihydroxy-7-methoxy-3-methylxanthone (4)	243
		" (24)	245
	1-Hydroxy-8-methoxy-3-methylxanthone	1,4-Dihydroxy-8-methoxy-3-methylxanthone (—)	246
C_{16}	1-Hydroxy-6-methoxy-3,8-dimethylxanthone	1,4-Dihyroxy-6-methoxy-3,8-dimethylxanthone (9)	247

[a] The corresponding dihydroxy compound gave very poor yields.

TABLE X. PERSULFATE OXIDATION OF PYRIDINES, INDOLES, AND QUINOLINES

No. of Carbon Atoms	Heterocycle	Product(s) and Yield(s) (%)	Refs.
C_5	2-Hydroxypyridine	2,5-Dihydroxypyridine (18), 2,3-dihydroxypyridine (1.6)	31
		" (85)[*,a]	9
		" (47)	248
		" (32–34)	249
		" (28,[‡] 23)	249a
		" (—)	250
	3-Hydroxypyridine	" (11), 2,3-dihydroxypyridine (trace), 3,4-dihydroxypyridine (trace)	31
	4-Hydroxypyridine	3,4-Dihydroxypyridine (—)	12
		" (8)	251
	2-Aminopyridine	2-Amino-3-hydroxypyridine (12)[‡]	65
		" (55)[*]	66
	4-Aminopyridine	4-Amino-3-hydroxypyridine (16),[‡] 4,4′-azoxypyridine (22)	65
C_6	2-Hydroxy-3-carboxypyridine	2,5-Dihydroxy-3-carboxypyridine (47, 52[‡])	252
	2-Hydroxy-3-methylpyridine	2,5-Dihydroxy-3-methylpyridine (—)	250
	2-Hydroxy-4-methylpyridine	2,5-Dihydroxy-4-methylpyridine (—)	250
	2-Hydroxy-6-methylpyridine	2,5-Dihydroxy-6-methylpyridine (—)	250
		" (17)	253
C_7	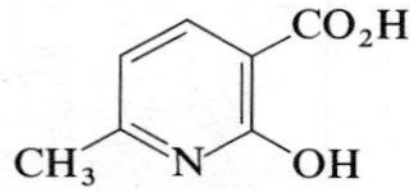 CO_2H, CH_3, N, OH	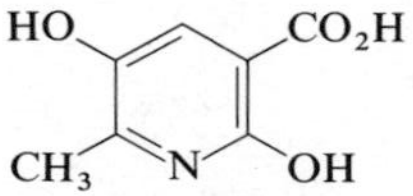 HO, CO_2H, CH_3, N, OH (49, 54[‡])	254
	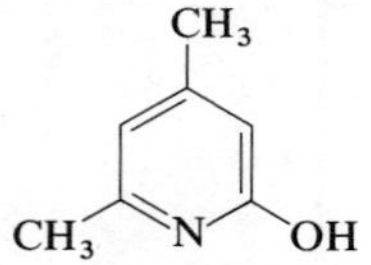 CH_3, CH_3, N, OH	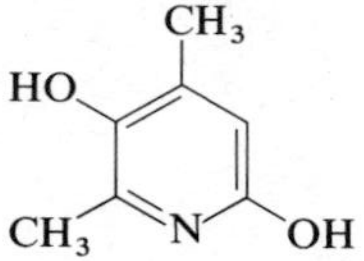 CH_3, HO, CH_3, N, OH (3)	253

	Reactant	Product (Yield, %)	Refs.
	CH_3, HO, CH_3, N	" (4)	253
	OCH_3, O, N, OH, CH_3	OCH_3, HO, O, N, OH, CH_3 (16)	255
C_8	Indole	3-Hydroxyindole (18)[‡]	64
	CO_2H, CO_2H, CH_3, N, OH	CO_2H, HO, CO_2H, CH_3, N, OH (75–84)	53a
C_9	3-Methylindole	Mixture of 4-, 5-, 6-, and 7-hydroxy-3-methylindoles (—)	74
	8-Hydroxyquinoline	5,8-Dihydroxyquinoline (18)[‡]	188
C_{10}	$CH_2OC_2H_5$, CO_2H, CH_3, N, OH	$CH_2OC_2H_5$, HO, CO_2H, CH_3, N, OH (73)	256

[a] This reference gives non-isolated yield data as functions of reactant ratios and $[OH^-]$.

TABLE XI. Persulfate Oxidation of Pyrimidines

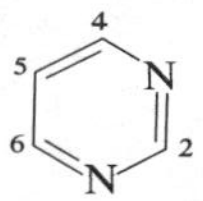

(Pyrimidine ring numbering: 2, 4, 5, 6)

No. of Carbon Atoms	Pyrimidine	Product(s) and Yield(s) (%)	Refs.
C_4	2,4-Dihydroxypyrimidine	2,4,5-Trihydroxypyrimidine (72,[‡] 14)	59
		" (—)	257
	4-Amino-2-hydroxypyrimidine	4-Amino-2,5-dihydroxypyrimidine (87,[‡] 70)	59
		" (—[‡], —)	257
	2-Amino-4-hydroxypyrimidine	2-Amino-4,5-dihydroxypyrimidine (48[‡])[a]	258
	6-Amino-4-hydroxypyrimidine	6-Amino-4,5-dihydroxypyrimidine (21,[‡] 14)	259
	2-Amino-4,6-dihydroxypyrimidine	2-Amino-4,5,6-trihydroxypyrimidine (—)	257
	2,4-Diaminopyrimidine	2,4-Diamino-5-hydroxypyrimidine (73,[‡] 47)	258
	4,6-Diaminopyrimidine	4,6-Diamino-5-hydroxypyrimidine (—,[‡] —)	257
	2,4-Diamino-6-hydroxypyrimidine	2,4-Diamino-5,6-dihydroxypyrimidine (61,[‡] 49)	260
		" (—[‡], —)[b]	257
	4,6-Diamino-2-hydroxypyrimidine	4,6-Diamino-2,5-dihydroxypyrimidine (—[‡])	257
	2,4,6-Triaminopyrimidine	2,4,6-Triamino-5-hydroxypyrimidine (—,[‡] —)	257
C_5	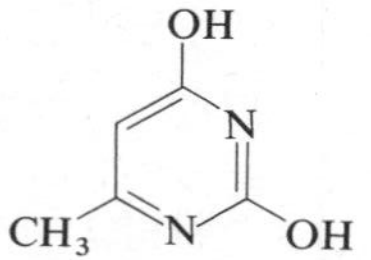 (structure: OH, CH_3, N, N, OH)	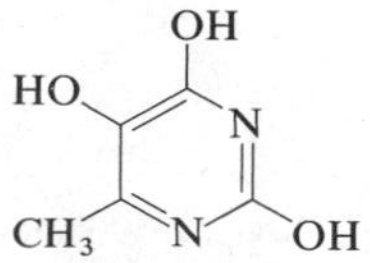 (structure: OH, HO, CH_3, N, N, OH) (—)	257
	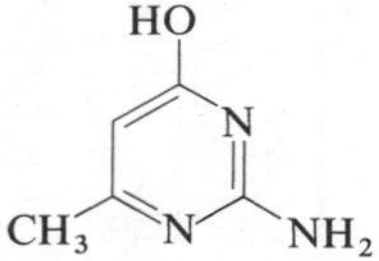 (structure: HO, CH_3, N, N, NH_2)	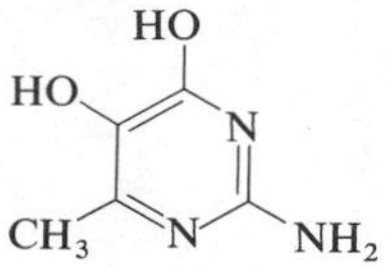 (structure: HO, HO, CH_3, N, N, NH_2) (66,[‡] 54)	258

	Substrate	Product	Yield	Ref.
	NH$_2$ / N / HO N CH$_3$	NH$_2$ / HO N / HO N CH$_3$	(90,‡ 68)	257
	NH$_2$ / N / CH$_3$ N OH	NH$_2$ / HO N / CH$_3$ N OH	(—,‡ —)	257
	NH$_2$ / N / CH$_3$ N NH$_2$	NH$_2$ / HO N / CH$_3$ N NH$_2$	(81,‡ 58)	258
		"	(—,‡ —)	261
	NH$_2$ / N / N NHCH$_3$	NH$_2$ / HO N / N NHCH$_3$	(—)‡	257
C_6	CH$_3$ / N / CH$_3$ N OH	CH$_3$ / HO N / CH$_3$ N OH	(41,‡ 28)	258
		"	(43.5,‡ 34)	257

TABLE XI. PERSULFATE OXIDATION OF PYRIMIDINES (*Continued*)

No. of Carbon Atoms	Pyrimidine	Product(s) and Yield(s) (%)	Refs.
	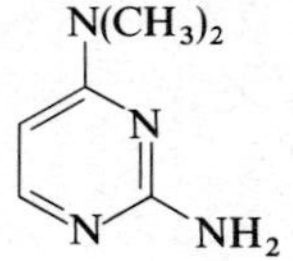 $N(CH_3)_2$, N, N, NH_2	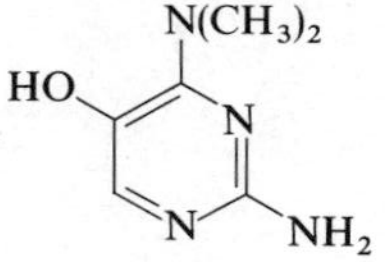 $N(CH_3)_2$, HO, N, N, NH_2 (—‡, —)	257
	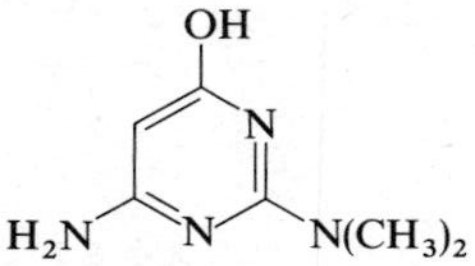 OH, N, H_2N, N, $N(CH_3)_2$	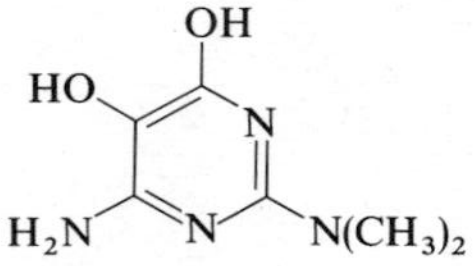 OH, HO, N, H_2N, N, $N(CH_3)_2$ (—)	257
C_8	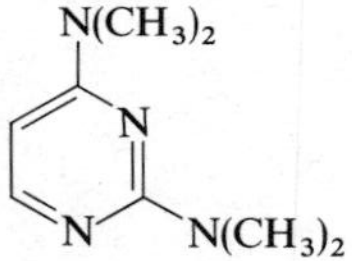 $N(CH_3)_2$, N, N, $N(CH_3)_2$	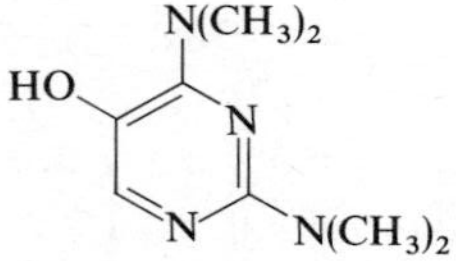 $N(CH_3)_2$, HO, N, N, $N(CH_3)_2$ (—‡, —)	257
C_{10}	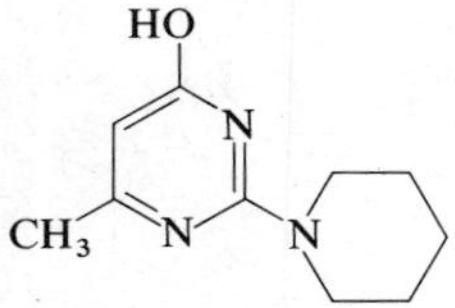 HO, N, CH_3, N, N	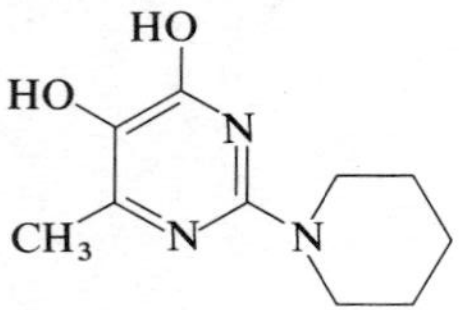 HO, HO, N, CH_3, N, N (26‡, 26)	258

[a] Seven other pyrimidines were reported to react, but the products were not characterized.
[b] Hydrolysis of the sulfate must be carried out at room temperature; at reflux the product is 2-amino-4,5,6-trihydroxypyrimidine.

TABLE XII. PERSULFATE OXIDATION OF PRIMARY ANILINES

No. of Carbon Atoms	Aniline	Product(s) and Yield(s) (%)	Refs.
C_6	3-Chloroaniline	3-Chloro-6-hydroxyaniline (—)[‡]	63
	4-Chloroaniline	4-Chloro-2-hydroxyaniline (—)	262
	4-Bromoaniline	4-Bromo-2-hydroxyaniline (21)[‡]	63
		" (15)	263
	2-Nitroaniline	6-Hydroxy-2-nitroaniline (12)[‡]	63
	Aniline	2-Hydroxyaniline (16)[‡]	62
	4-Aminophenylsulfonic acid	4-Amino-3-hydroxyphenylsulfonic acid (33)[‡]	63
	4-Aminophenylsulfonamide	4-Amino-3-hydroxyphenylsulfonamide (23)[‡]	16
C_7	2-Aminobenzoic acid	2-Amino-3-hydroxybenzoic acid (19)[‡]	63
		" (—)[‡], 2-amino-5-hydroxybenzoic acid (—)[‡]	64
		" (—)[‡]	264
	2-Aminobenzoic acid-1,2-^{14}C	2-Amino-3-hydroxybenzoic acid-1,2-^{14}C (10–15), 2-amino-5-hydroxybenzoic acid-1,2-^{14}C (—)	265
	4-Aminobenzoic acid	4-Amino-3-hydroxybenzoic acid (21)[‡]	63
	2-Methylaniline	6-Hydroxy-2-methylaniline (17)[‡]	63
	3-Methylaniline	6-Hydroxy-3-methylaniline (23)[‡]	63
	4-Methylaniline	2-Hydroxy-4-methylaniline (28)[‡]	63
C_8	2-Aminoacetophenone	2-Amino-3-hydroxyacetophenone (—)[‡], 2-amino-5-hydroxyacetophenone (—)[‡]	266
	3,4-Dimethylaniline	3,4-Dimethyl-6-hydroxyaniline (21)[‡]	16
	2,5-Dimethylaniline	2,5-Dimethyl-6-hydroxyaniline (21)[‡]	65
	2,6-Dimethylaniline	2,6-Dimethyl-4-hydroxyaniline (18)[‡]	65
C_{10}	[structure: benzene ring with NH_2 and $COCH_2CH(NH_2)CO_2H$]	[structure: benzene ring with OSO_3H, NH_2 and $COCH_2CH(NH_2)CO_2H$] (—)	64

TABLE XII. PERSULFATE OXIDATION OF PRIMARY ANILINES (*Continued*)

No. of Carbon Atoms	Aniline	Product(s) and Yield(s) (%)	Refs.
C_{12}	4-Amino-4′-nitrobiphenyl		267
		" (—)	268
	4-Aminobiphenyl	4-Amino-3-hydroxybiphenyl (2.5)[‡]	63
	4-Aminoazobenzene	4-Amino-3-hydroxyazobenzene (4)[‡]	269
C_{14}	$2\text{-}CH_3C_6H_4$–C_6H_3(NH_2)(CH_3)	$2\text{-}CH_3C_6H_4$–C_6H_2(OSO_3H)(NH_2)(CH_3) (22), [$2\text{-}CH_3C_6H_4$–C_6H_3(CH_3)–N]$=_2$ (26), [$2\text{-}CH_3C_6H_4$–C_6H_3(CH_3)–NH]$-_2$ (6)	65

TABLE XIII. PERSULFATE OXIDATION OF SECONDARY AND TERTIARY ANILINES AND PHENOXAZINE

No. of Carbon Atoms	Aniline	Product(s) and Yield(s) (%)	Refs.
C_7	*N*-Methylaniline	2-Hydroxy-*N*-methylaniline (11)‡	65
C_8	*N*,*N*-Dimethylaniline	2-Hydroxy-*N*,*N*-dimethylaniline (40)‡	62
C_{12}	Diphenylamine	*N*-(2-Hydroxyphenyl)aniline (17)‡ *N*-Phenyl-*p*-benzoquinoneimine (—)	65 270
	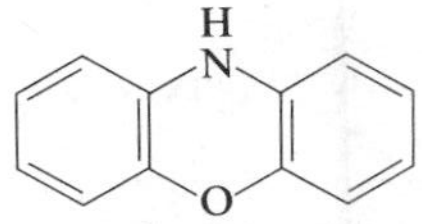 H, N, O	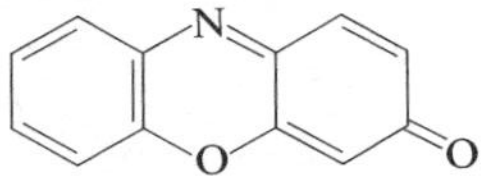 N, O, O (—)	271
C_{13}	4-(*N*-Methylamino)azobenzene	3-Hydroxy-4-(*N*-methylamino)azobenzene (21)‡	269
C_{14}	4-(*N*,*N*-Dimethylamino)biphenyl 4-(*N*,*N*-Dimethylamino)azobenzene	3-Hydroxy-4-(*N*,*N*-dimethylamino)biphenyl (21)‡ 3-Hydroxy-4-(*N*,*N*-dimethylamino)azobenzene (3) [40]‡ " (16)‡	16 16 269
C_{16}	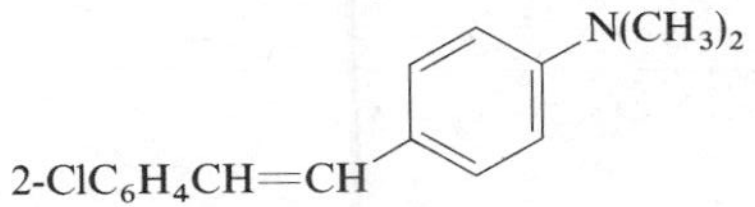 $2\text{-}ClC_6H_4CH{=}CH\text{-}C_6H_4\text{-}N(CH_3)_2$	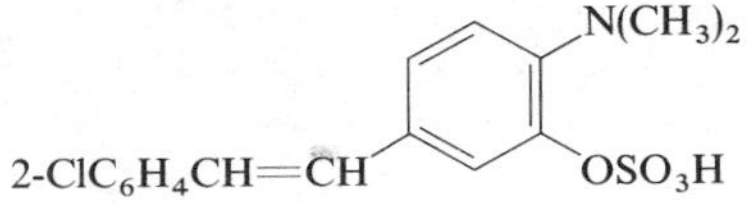 $2\text{-}ClC_6H_4CH{=}CH\text{-}C_6H_3(OSO_3H)\text{-}N(CH_3)_2$ (15) [30]‡	16
C_{17}	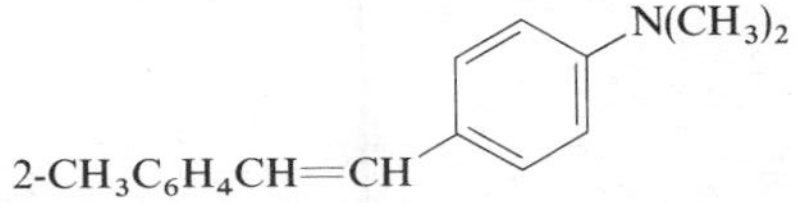 $2\text{-}CH_3C_6H_4CH{=}CH\text{-}C_6H_4\text{-}N(CH_3)_2$	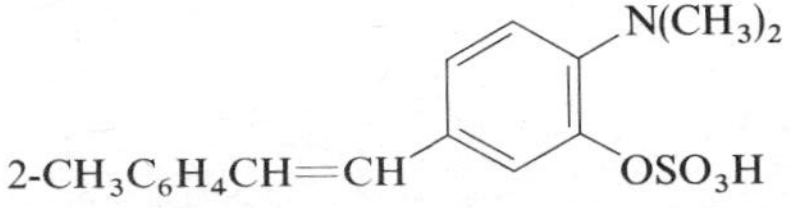 $2\text{-}CH_3C_6H_4CH{=}CH\text{-}C_6H_3(OSO_3H)\text{-}N(CH_3)_2$ (24) [48]‡	16

TABLE XIV. PERSULFATE OXIDATION OF NAPHTHYLAMINES

No. of Carbon Atoms	Naphthylamine	Product(s) and Yield(s) (%)	Refs.
C_{10}	2-Naphthlaminesulfonate	1-Hydroxy-2-napthylaminesulfonate (18)‡	272
		" (6)‡	273
	1-Aminonaphthalene	1-Amino-2-hydroxynaphthalene (18)‡	62
	2-Aminonaphthalene	2-Amino-1-hydroxynaphthalene (45)‡	62
		" (24)‡[a]	272
C_{11}	2-(*N*-Methylamino)naphthalene	1-Hydroxy-2-(*N*-methylamino)naphthalene (5)‡	274
C_{12}	1-(*N*,*N*-Dimethylamino)naphthalene	2-Hydroxy-1-(*N*,*N*-dimethylamino)naphthalene (19)‡	16
	2-(*N*,*N*-Dimethylamino)naphthalene	1-Hydroxy-2-(*N*,*N*-dimethylamino)naphthalene (24)‡	16

[a]An alternative isolation is described involving precipitation of the product as its cetyl pyridinium salt. 2, 2′-Azonaphthalene was also formed in 5% yield.

TABLE XV. Unsuccessful Oxidations

Substrate	Comment	Refs.
Phenols		
2-CH_2CO_2H	Poor yield	275
2-$COCH_2C_6H_5(OCH_3)_2$-3,5	Unsuccessful	276
Anthraquinones		
1,2-$(OH)_2$	Starting material recovered unchanged	277
1,3-$(OH)_2$	"	"
1-OH-2-OCH_3	"	"
1-OH-3-OCH_3	"	"
Coumarins		
4-CH_3-5-OH	A complex, high-melting product	188
4-CH_3-6-OCH_3	"	178, 188
4,7-$(CH_3)_2$-5-OH	"	188
4-CH_3-6-OTs[a]	Starting material recovered unchanged	195
4-CH_3-5,7-$(OTs)_2$[a]	"	"
4-CH_3-7,8-$(OTs)_2$[a]	"	"
Flavones		
5,7,4′-$(OH)_3$	Small yield of impure product	206
4′-OH-3,7-$(OCH_3)_2$	Minute yield of impure product	213
5,7-$(OH)_2$-2′,4′-$(OCH_3)_2$	Unsatisfactory; successful if the 7 position is methylated	212
5-OH-3′,4′,5′-$(OCH_3)_3$	Does not undergo oxidation	199
4′-OH-3,5,7-$(OCH_3)_3$	Minute yield of impure product	213
4′-OH-3,7,3′-$(OCH_3)_3$	"	"
4′-OH-3,5,7,3′-$(OCH_3)_4$	"	"
5-OH-3,7,8,3′,4′-$(OCH_3)_5$	Not successful	200, 222
Chromones		
2-CH_3-5,7-$(OH)_2$	Unsatisfactory yield	237
Xanthones		
1,3-$(OH)_2$	Unsuccessful; satisfactory if 1-OH-3-OCH_3	243
4,6-$(OH)_2$	Some other reaction takes place; satisfactory if 4-OCH_3-6-OH	244
1,6-$(OH)_2$-3-CH_3	Unsuccessful; satisfactory if 1-OH-6-OCH_3	243
Pyrimidines		
2-OH	Did not give the desired product	278
4,6-$(OH)_2$	Unstable in alkali	257
2,4,6-$(OH)_3$	The product was alloxan	"
2,4-$(OH)_2$-6-NH_2	Unsatisfactory elemental analysis	"
2-OH-4-CH_3	Unsuccessful	"
4-OH-6-CH_3	"	"
4,6-$(OH)_2$-2-CH_3	Unstable in alkali	"
2-NH_2-4-CH_3	Unsuccessful	"
4-NH_2-2-OCH_3	"	"
4-NH_2-6-OCH_3	"	"
4-OH-2-SCH_3	Hydrolysis of the methylthio group	"
2,4-$(NH_2)_2$-6-SCH_3	Unsuccessful	"
4,6-$(NH_2)_2$-2-SCH_3	"	"
2-NH_2-4,6-$(CH_3)_2$	"	"
4-OH-6-CH_3-2-SCH_3	Hydrolysis of the methylthio group	"

[a] Ts is the *p*-toluenesulfonyl group.

REFERENCES

[1] K. Elbs, *J. Prakt. Chem.*, **48**, 179 (1893).

[2] W. Baker and N. C. Brown, *J. Chem. Soc.*, **1948**, 2303.

[3] S. M. Sethna, *Chem. Rev.*, **49**, 91 (1951).

[4] K.-F. Wedemeyer, in *Methoden der Organischen Chemie* (*Houben-Weyl*), vol. 6/1c, Pt. 1, G. Thieme, Stuttgart, 1976, pp. 43–54.

[5] E. J. Behrman and J. O. Edwards, in *Prog. Phys. Org. Chem.*, A. Streitwieser, Jr. and R. W. Taft, Eds., Vol. 4, Wiley-Interscience, New York, 1967, p. 108.

[6] G. Sosnovsky and D. J. Rawlinson, in *Organic Peroxides*, D. Swern, Ed., Vol. 2, Wiley-Interscience, New York, 1971, p. 319.

[7] R. Curci and J. O. Edwards, in *Organic Peroxides*, D. Swern, Ed., Vol. 1, Wiley-Interscience, New York, 1970, p. 228.

[8] G. J. Buist, in *Comprehensive Chemical Kinetics*, C. H. Bamford and C. F. H. Tipper, Eds., Vol. 6, Elsevier, Amsterdam, 1972, p. 456 ff.

[9] E. J. Behrman and P. P. Walker, *J. Am. Chem. Soc.*, **84**, 3454 (1962).

[10] E. J. Behrman, *J. Am. Chem. Soc.*, **85**, 3478 (1963).

[11] Y. Ogata and T. Akada, *Tetrahedron*, **26**, 5945 (1970).

[12] E. J. Behrman and M. N. D. Goswami, *Anal. Chem.*, **36**, 2189 (1964).

[13] C. Srinivasan and S. Rajagopal, *Curr. Sci.*, **46**, 669 (1977) [*C.A.*, **88**, 5971u (1978)].

[14] K. B. Rao and N. V. S. Rao, *J. Sci. Ind. Res.*, **14B**, 130 (1955) [*C.A.*, **49**, 10877b (1955)].

[15] M. V. Shah and S. Sethna, *J. Chem. Soc.*, **1959**, 2676.

[16] P. Sims, *J. Chem. Soc.*, **1958**, 44.

[17] I. M. Kolthoff and I. K. Miller, *J. Am. Chem. Soc.*, **73**, 5118 (1951); *ibid.*, **74**, 4419 (1952).

[18] E. J. Behrman and J. E. McIsaac, Jr., in *Mechanisms of Reactions of Sulfur Compounds*, Vol. 2, Intra-Science Research Foundation, Santa Monica, California, 1968, pp. 193–218.

[19] Y. Ogata and T. Akada, *Tetrahedron*, **28**, 15 (1972).

[20] D. A. House, *Chem. Rev.*, **62**, 185 (1962).

[21] W. K. Wilmarth and A. Haim, in *Peroxide Reaction Mechanisms*, J. O. Edwards, Ed., Interscience, New York, 1962, p. 175.

[22] F. Minisci, A. Citterio, and C. Giordano, *Acct. Chem. Res.*, **16**, 27 (1983).

[23] N. J. Desai and S. Sethna, *J. Org. Chem.*, **22**, 388 (1957).

[24] P. Venturella, A. Bellino, and M. L. Marino, *Gazz. Chim. Ital.*, **113**, 819 (1983) [*C.A.*, **101**, 110587h (1983)].

[25] A. Jensen and M. A. Ragan, *Tetrahedron Lett.*, **1978**, 847.

[26] J. H. Looker, J. R. Edman, and C. A. Kingsbury, *J. Org. Chem.*, **49**, 645 (1984).

[27] D. B. Bruce and R. H. Thomson, *J. Chem. Soc.*, **1955**, 1089.

[28] K. C. Khulbe and S. P. Srivastava, *Curr. Sci.*, **33**, 270 (1964) [*C.A.*, **61**, 1789b (1964)].

[29] J. Forrest and V. Petrow, *J. Chem. Soc.*, **1950**, 2340.

[30] E. J. Behrman, *Biochem. J.*, **201**, 677 (1982); E. J. M. Pennings and G. M. J. Van Kempen, *ibid.*, **201**, 677 (1982).

[31] E. J. Behrman and B. M. Pitt, *J. Am. Chem. Soc.*, **80**, 3717 (1958).

[32] R. U. Schock, Jr. and D. L. Tabern, *J. Org. Chem.*, **16**, 1772 (1951).

[33] H. H. Hodgson and H. G. Beard, *J. Chem. Soc.*, **1927**, 2339.

[34] W. Eller, *Justus Liebigs Ann. Chem.*, **431**, 133 (1923).

[35] W. Eller and K. Koch, *Ber.*, **53**, 1469 (1920).

[36] W. Baker, N. C. Brown, and J. A. Scott, *J. Chem. Soc.*, **1939**, 1922.

[37] K.-F. Wedemeyer, in *Methoden der Organischen Chemie* (*Houben-Weyl*), E. Müller, Ed., Vol. 6/1c,Pts. 1 and 2, G. Thieme, Stuttgart, 1976.

[38] C. A. Buehler and D. E. Pearson, *Survey of Organic Syntheses*, Wiley-Interscience, New York, 1970, p. 246; *ibid.*, Vol. 2, 1977, p. 270.

[39] M. Lj. Mihailović and Ž. Čeković, in *Chemistry of the Hydroxyl Group*, S. Patai, Ed., Interscience, New York, 1971, Chapter 10.

[40] J. H. P. Tyman, in *Rodd's Chemistry of Carbon Compounds*, M. F. Ansell, Ed., Vol. 3A, 2nd ed. suppl., Elsevier, Amsterdam, 1983, Chapter 4.

[41] A. R. Forrester and J. L. Wardell, in *Rodd's Chemistry of Carbon Compounds*, S. Coffey, Ed., Vol. 3A, 2nd ed., Elsevier, Amsterdam, 1971, Chapter 4.

[42] T. Matsuura and K. Omura, *Synthesis*, **1974**, 173.

[43] D. H. R. Barton, A. G. Brewster, S. V. Ley, C. M. Read, and M. N. Rosenfeld, *J. Chem. Soc., Perkin Trans. 1*, **1981**, 1473.

[44] P. Capdevielle and M. Maumy, *Tetrahedron Lett.*, **23**, 1573, 1577 (1982).

[45] D. V. Rao and F. A. Stuber, *Tetrahedron Lett.*, **22**, 2337 (1981); *J. Org. Chem.*, **50**, 1722 (1985).

[46] S. K. Mukerjee, T. R. Rajagopalan, and T. R. Seshadri, *J. Sci. Ind. Res.*, **16B**, 58 (1957) [*C.A.*, **51**, 13856a (1957)].

[47] W. Baker, *J. Chem. Soc.*, **1941**, 667.

[48] W. Baker, G. F. Flemons, and R. Winter, *J. Chem. Soc.*, **1949**, 1560.

[49] K. V. Rao and J. A. Owoyale, *J. Heterocycl. Chem.*, **13**, 361 (1976).

[50] H. Bader and E. G. Jahngen, Jr., U.S. Patent 3,652,597 (1972) [*C.A.*, **77**, 5175k (1972)].

[51] J. Balej and A. Regner, *Coll. Czech. Chem. Commun.*, **25**, 1685 (1960); *ibid.*, **28**, 254 (1963); *ibid.*, **31**, 361 (1966).

[52] E. J. Behrman and J. O. Edwards, *Rev. Inorg. Chem.*, **2**, 179 (1980).

[53] G. Bargellini, *Gazz. Chim. Ital.*, **46** (I), 249 (1916) [*C.A.*, **11**, 1139 (1917)].

[53a] E. Matsumura, U.S. Pat. 3,155,673 (1964); Fr. Pat. 1,315,280 (1963) [*C.A.* **59**, 6368g (1963)].

[54] T. R. Seshadri, *Proc. Indian Acad. Sci., Sect. A*, **28**, 1 (1948) [*C.A.*, **43**, 3823f (1949)].

[55] G. P. Panigrahi and R. Panda, *Indian J. Chem.*, **15A**, 1070 (1977).

[56] J. K. Rasmussen and H. K. Smith II, *Makromol. Chem.*, **182**, 701 (1981); *J. Am. Chem. Soc.*, **103**, 730 (1981); E. V. Dehmlow, B. Vehre, and J. K. Makrandi, *Z. Naturforsch., Teil B*, **40**, 1583 (1985)

[57] G. N. Burkhardt, W. G. K. Ford, and E. Singleton, *J. Chem. Soc.*, **1936**, 17; E. J. Fendler and J. H. Fendler, *J. Org. Chem.*, **33**, 3852 (1968).

[58] A. I. Brodskii and N. A. Vysotskaya, *J. Gen. Chem. USSR*, **32**, 2241 (1962).

[59] R. C. Moschel and E. J. Behrman, *J. Org. Chem.*, **39**, 1983 (1974).

[60] P. R. Rao and T. R. Seshadri, *Proc. Indian Acad. Sci., Sect. A.*, **27**, 209 (1948) [*C.A.*, **44**, 1496a (1950)].

[61] R. Aneja, S. K. Mukerjee, and T. R. Seshadri, *Chem. Ber.*, **93**, 297 (1960).

[62] E. Boyland, D. Manson, and P. Sims, *J. Chem. Soc.*, **1953**, 3623.

[63] E. Boyland and P. Sims, *J. Chem. Soc.*, **1954**, 980.

[64] E. Boyland, P. Sims, and D. C. Williams, *Biochem. J.*, **62**, 546 (1956).

[65] E. Boyland and P. Sims, *J. Chem. Soc.*, **1958**, 4198.

[66] E. J. Behrman, *J. Am. Chem. Soc.*, **89**, 2424 (1967).

[66a] M. Chanon and M. L. Tobe, *Angew. Chem. Internat. Ed.*, **21**, 1 (1982); R. E. Ball, A. Chako, J. O. Edwards, and G. Levey, *Inorg. Chim. Acta*, **99**, 49 (1985).

[67] N. Venkatasubramanian and A. Sabesan. *Can. J. Chem.*, **47**, 3710 (1969).

[68] A. Sabesan and N. Venkatasubramanian, *Aust. J. Chem.*, **24**, 1633 (1971).

[69] A. Sabesan and N. Venkatasubramanian, *Indian J. Chem.*, **10**, 1092 (1972).

[70] C. Srinivasan, S. Perumal, and N. Arumugam, *Indian J. Chem.*, **19A**, 160 (1980); *J. Chem. Soc. Perkin Trans. 2*, 1985, 1855

[71] E. J. Behrman and D. M. Behrman, *J. Org. Chem.*, **43**, 4551 (1978).

[72] J. T. Edward and J. Whiting, *Can. J. Chem.*, **49**, 3502 (1971).

[73] T. K. Krishnamurthi and N. Venkatasubramanian, *Indian J. Chem.*, **16A**, 28 (1978).

[74] R. A. Heacock and M. E. Mahon, *Can. J. Biochem.*, **43**, 1985 (1965).

[75] V. K. Gupta and S. P. Srivastava, *Indian Chem. Manuf.*, **18**, 1 (1980) [*C.A.*, **94**, 173951e (1980)].

[76] Rajeev and S. P. Srivastava, *Indian J. Chem.*, **14B**, 141 (1976).

[77] Rajeev, R. C. Gupta, and S. P. Srivastava, *Indian J. Chem.*, **12**, 656 (1974).

[78] R. C. Gupta, L. D. Sharma, and S. P. Srivastava, *Z. Phys. Chem.* (*Leipzig*), **255**, 317 (1974).

[79] S. P. Srivastava, R. C. Gupta, and A. K. Shukla, *Indian J. Chem.*, **15A**, 605 (1977).

[80] S. P. Srivastava and V. K. Gupta, *Natl. Acad. Sci. Lett.* (*India*), **3**, 25 (1980) [*C.A.*, **93**, 238332e (1980)].
[81] R. C. Gupta and S. P. Srivastava, *Indian J. Chem.*, **10**, 706 (1972).
[81a] V. K. Gupta, *React. Kinet. Catal. Lett.*, **27**, 207 (1985) [*C.A.*, **103**, 122763s (1985)].
[82] R. Schrötter and F. Möller, in *Methoden der Organischen Chemie* (*Houben-Weyl*), E. Müller, Ed., Vol. 11/1, G. Thieme, Stuttgart, 1957, p. 341.
[83] M. S. Gibson, in *Chemistry of the Amino Group*, S. Patai, Ed., Interscience, New York, 1968, Chapter 2.
[83a] E. Boyland and D. Manson, *J. Chem. Soc.*, **1957**, 4689.
[84] K. B. Rao and N. V. S. Rao, *J. Sci. Ind. Res.*, **15B**, 14 (1956) [*C.A.*, **51**, 10416a (1957)].
[85] A. E. Feiring and W. A. Sheppard, *J. Org. Chem.*, **40**, 2543 (1975).
[86] German Patent 81298, from P. Friedlaender, Ed., *Fortschritte der Theerfarbenfabrikation*, Pt. 4, 1894—1897, Springer, Berlin, 1899, p. 121.
[87] J. M. Blatchly, J. F. W. McOmie, and J. B. Searle, *J. Chem. Soc.*, *C*, **1969**, 1350.
[88] J. N. Smith, *J. Chem. Soc.*, **1951**, 2861.
[89] J. R. Fyson, G. I. P. Levenson, and G. Wallace, *J. Photogr. Sci.*, **24**, 57 (1976) [*C.A.*, **85**, 184788x (1976)].
[90] M. J. Astle and S. P. Stephenson, *J. Am. Chem. Soc.*, **65**, 2399 (1943).
[91] W. Ruske, *Justus Liebigs Ann. Chem.*, **610**, 156 (1957).
[92] German Patent 81068, from P. Friedlaender, Ed., *Fortschritte der Theerfarbenfabrikation*, Pt. 4, 1894–1897, Springer, Berlin, 1899, p. 126.
[93] A. Johnson and M. L. Rahman, *J. Soc. Dyers Colour.*, **74**, 291 (1958) [*C.A.*, **52**, 10583d (1958)].
[94] German Patent 271,892; from *J. Soc. Chem. Ind.*, **33**, 567 (1914).
[95] E. Seebeck, *Helv. Chim. Acta*, **30**, 149 (1947).
[96] P. R. Hammond, *J. Chem. Soc.*, **1964**, 471.
[97] O. Neubauer and L. Flatow, *Z. Physiol. Chem.* **52**, 375 (1907).
[98] J. R. Velasco, *Ann. R. Soc. Esp. Fis. Quim*, **32**, 358 (1934) [*C.A.*, **28**, 7127[7] (1934)].
[99] L. Ponniah and T. R. Seshadri, *Proc. Indian Acad. Sci.*, *Sect. A*, **37**, 544 (1953) [*C.A.*, **48**, 11402d (1954)].
[100] C. Graebe and E. Martz, *Justus Liebigs Ann. Chem.*, **340**, 213 (1905).
[101] J. M. Sehgal and T. R. Seshadri, *Proc. Indian Acad. Sci.*, *Sect. A*, **36**, 355 (1952) [*C.A.*, **48**, 2055e (1954)].
[102] F. Mauthner, *J. Prakt. Chem.* **156**, 150 (1940).
[103] B. D. Astill, D. W. Fassett, and R. L. Roudabush, *Biochem. J.*, **90**, 194 (1964).
[104] German Patent 81297, from P. Friedlaender, Ed., *Fortschritte der Theerfarbenfabrikation*, Pt. 4, 1894–1897, Springer, Berlin, 1899, p. 127.
[105] S. W. Tanenbaum and E. W. Bassett, *J. Biol. Chem.*, **234**, 1861 (1959).
[106] E. J. M Pennings and G. M. J. Van Kempen, *Biochem. J.*, **191**, 133 (1980).
[107] S. Rajagopalan, T. R. Seshadri, and S. Varadarajan, *Proc. Indian Acad. Sci. Sect. A*, **30**, 265 (1949) [*C.A.*, **46**, 472g (1952)].
[108] W. Baker and R. I. Savage, *J. Chem. Soc.*, **1938**, 1602.
[109] M. F. Ansell, B. W. Nash, and D. A. Wilson, *J. Chem. Soc.*, **1963**, 3028.
[110] D. B. Harper and R. L. Wain, *Ann. Appl. Biol.*, **67**, 395 (1971).
[111] J. Renz, *Helv. Chim. Acta*, **30**, 124 (1947).
[112] J. Daly, L. Horner, and B. Witkop, *J. Am. Chem. Soc.*, **83**, 4787 (1961).
[113] D. J. Hopper and P. J. Chapman, *Biochem. J.*, **122**, 19 (1971).
[114] E. Bernatek and I. Bø, *Acta Chem. Scand.*, **13**, 337 (1959).
[115] I. W. J. Still and D. J. Snodin, *Can. J. Chem.*, **50**, 1276 (1972).
[116] K. E. Schulte and G. Rücker, *Arch. Pharm.* (*Weinheim, Ger.*), **297**, 182 (1964) [*C.A.*, **60**, 13173a (1964)].
[117] H. Inouye, *Pharm. Bull.*, **2**, 359 (1954).
[118] D. J. Hopper, P. J. Chapman, and S. Dagley, *Biochem. J.*, **122**, 29 (1971).

[119] K. Yamamoto, H. Irie, and S. Uyeo, *Yakugaku Zasshi*, **91**, 257 (1971) [*C.A.*, **74**, 125509g (1971)].

[120] J. R. Merchant, R. M. Naik, and A. J. Mountwalla, *J. Chem. Soc.* **1957**, 4142.

[121] G. Bargellini, *Gazz. Chim. Ital.*, **43 (I)**, 164 (1913) [*C.A.*, **7**, 1725 (1913)].

[122] K. R. Laumas, S. Neelakantan, and T. R. Seshadri, *Proc. Indian Acad. Sci., Sect. A*, **46**, 343 (1957) [*C.A.*, **52**, 14569i (1958)].

[123] M. K. Seikel and T. A. Geissman, *J. Am. Chem. Soc.*, **72**, 5720 (1950).

[124] M. K. Seikel, A. L. Haines, and H. D. Thompson, *J. Am. Chem. Soc.*, **77**, 1196 (1955).

[125] L. Ponniah and T. R. Seshadri, *Proc. Indian Acad. Sci., Sect. A*, **38**, 77 (1953) [*C.A.*, **49**, 8270b (1955)].

[126] P. Yates and G. H. Stout, *J. Am. Chem. Soc.*, **80**, 1691 (1958).

[126a] M. Iinuma, T. Tanaka, and S. Matsuura, *Chem. Pharm. Bull.*, **32**, 3354 (1984).

[127] F. Bergel, A. M. Copping, A. Jacob, A. R. Todd, and T. S. Work, *J. Chem. Soc.*, **1938**, 1383.

[128] M. H. Balba and J. E. Casida, *J. Agric. Food Chem.*, **16**, 561 (1968).

[129] H. M. Van Dort and H. J. Geursen, *Recl. Trav. Chim. Pays-Bas*, **86**, 520 (1967).

[130] I. H. Updegraff and H. G. Cassidy, *J. Am. Chem. Soc.*, **71**, 407 (1949).

[131] E. R. Catlin and C. H. Hassal, *J. Chem. Soc., C*, **1971**, 460.

[132] K. Aghoramurthy and T. R. Seshadri, *Proc. Indian Acad. Sci., Sect. A*, **35**, 327 (1952) [*C.A.*, **47**, 8743b (1953)].

[133] W. Baker, *J. Chem. Soc.*, **1939**, 959.

[134] K. Wallenfels, *Ber.*, **75**, 785 (1942).

[135] G. Bargellini and S. Aureli, *Gazz. Chim. Ital.*, **41 (II)**, 590 (1911) [*C.A.*, **6**, 992 (1912)].

[136] T. H. Simpson, *J. Org. Chem.*, **28**, 2107 (1963).

[137] K. Kitahonoki, *Chem. Pharm. Bull.* **7**, 114 (1959).

[138] British Patent 543863 (1940); [*C.A.*, **36**, 6172^4 (1942)].

[139] V. A. Bogolyubskii, *J. Gen. Chem. USSR*, **32**, 862 (1962).

[140] L. R. Row, C. Rukmini, and G. S. R. S. Rao, *Indian J. Chem.* **5**, 105 (1967).

[141] T. R. Seshadri and T. R. Thiruvengadam, *Proc. Indian Acad. Sci., Sect. A*, **32**, 110 (1950) [*C.A.*, **46**, 501g (1952)].

[142] S.-I. Fujita, R. Suemitsu, and Y. Fujita, *Yakugaku Zasshi*, **90**, 1367 (1970) [*C.A.*, **74**, 34553a (1971)].

[143] V. D. N. Sastri and T. R. Seshadri, *Proc. Indian Acad. Sci., Sect. A*, **23**, 262 (1946) [*C.A.*, **41**, 449a (1947)].

[144] L. R. Row and T. R. Seshadri, *Proc. Indian Acad. Sci., Sect. A*, **21**, 155 (1945) [*C.A.*, **39**, 4325^3 (1945)].

[145] B. F. Anderson, L. H. Briggs, T. Cebalo, and M. A. Trotman, *J. Chem. Soc.*, **1964**, 1026.

[146] W. Riedl and E. Leucht, *Chem. Ber.*, **91**, 2784 (1958).

[147] S. K. Mukerjee, T. R. Seshadri, and S. Varadarajan, *Proc. Indian Acad. Sci., Sect A*, **35**, 82 (1952) [*C.A.*, **47**, 8067g (1953)].

[148] W. D. Ollis, B. T. Redman, I. O. Sutherland, and O. R. Gottlieb, *Phytochem.*, **17**, 1379 (1978).

[149] L. R. Row and T. R. Seshadri, *Proc. Indian Acad. Sci., Sect. A*, **23**, 23 (1946) [*C.A.*, **40**, 5050^8 (1946)].

[150] L. R. Row, V. D. N. Sastri, T. R. Seshadri, and T. R. Thiruvengadam, *Proc. Indian Acad. Sci., Sect. A*, **28**, 189 (1948) [*C.A.*, **43**, 5397d (1949)].

[151] S. Rajagopalan and T. R. Seshadri, *Proc. Indian Acad. Sci., Sect. A*, **28**, 31 (1948) [*C.A.*, **43**, 4265b (1949)].

[152] V. V. S. Murti, L. R. Row, and T. R. Seshadri, *Proc. Indian Acad. Sci., Sect. A*, **24**, 233 (1946) [*C.A.*, **41**, 2417e (1947)].

[152a] K. Okamoto, E. Mizuta, K. Kamiya, and I. Imada, *Chem. Pharm. Bull*, **33**, 3756 (1985).

[153] K. Aghoramurthy, T. R. Seshadri, and G. B. Venkatasubramanian, *J. Sci. Ind. Res.*, **15B**, 11 (1956) [*C.A.*, **51**, 11334h (1957)].

[154] L. R. Row and T. R. Seshadri, *Proc. Indian Acad. Sci., Sect. A*, **23**, 140 (1946) [*C.A.*, **41**, 121b (1947)].

[155] S. Neelakantan, T. R. Rajagopalan, and T. R. Seshadri, *Proc. Indian Acad. Sci., Sect. A*, **49**, 234 (1959) [*C.A.*, **54**, 1465f (1960)].

[156] N. Adityachaudhury, C. L. Kirtaniya, and B. Mukherjee, *Tetrahedron*, **27**, 2111 (1971).

[157] A. Ballio and F. Pocchiari, *Gazz. Chim. Ital.*, **79**, 913 (1949) [*C.A.*, **44**, 7840f (1950)].

[158] S. Rajagopalan and T. R. Seshadri, *Proc. Indian Acad. Sci., Sect. A*, **27**, 85 (1948) [*C.A.*, **43**, 640e (1949)].

[159] K. Chandrasenan, S. Neelakantan, and T. R. Seshadri, *J. Indian Chem. Soc.*, **38**, 907 (1961).

[160] S. K. Arora, A. C. Jain, and T. R. Seshadri, *Tetrahedron*, **18**, 559 (1962).

[161] V. V. S. Murti and T. R. Seshadri, *Proc. Indian Acad. Sci., Sect. A*, **29**, 1 (1949) [*C.A.*, **44**, 3985b (1950)].

[162] L. Farkas and J. Strelisky, *Tetrahedron Lett.*, **1970**, 187.

[163] S. K. Arora, A. C. Jain, and T. R. Seshadri, *J. Sci. Ind. Res.*, **18B**, 349 (1959) [*C.A.*, **54**, 7701h (1960)].

[164] S. K. Arora, A. C. Jain, and T. R. Seshadri, *Indian J. Chem.*, **4**, 430 (1966).

[165] S. K. Arora, A. C. Jain, and T. R. Seshadri, *J. Indian Chem. Soc.*, **38**, 61 (1961).

[166] K. Chandrasenan, S. Neelakantan, and T. R. Seshadri, *Proc. Indian Acad. Sci., Sect. A*, **51**, 296 (1960) [*C.A.*, **55**, 8363g (1961)].

[167] W. D. Ollis, B. T. Redman, R. J. Roberts, I. O. Sutherland, O. R. Gottlieb, and M. T. Magalhães, *Phytochemistry*, **17**, 1383 (1978).

[168] N. Narasimhachari, V. D. N. Sastri, and T. R. Seshadri, *Proc. Indian Acad. Sci., Sect. A*, **29**, 404 (1949) [*C.A.*, **44**, 3491i (1950)].

[169] K. V. Rao and T. R. Seshadri, *Proc. Indian Acad. Sci., Sect. A*, **27**, 375 (1948) [*C.A.*, **43**, 4264a (1949)].

[170] V. D. N. Sastri and T. R. Seshadri, *Proc. Indian Acad. Sci., Sect. A*, **24**, 238 (1946) [*C.A.*, **41**, 2417g (1947)].

[171] R. B. Desai and S. Sethna, *J. Indian Chem. Soc.*, **28**, 213 (1951).

[172] Y. Nagase and U. Matsumoto, *Yakugaku Zasshi*, **74**, 9 (1954) [*C.A.*, **49**, 1681c (1955)].

[173] R. D. Gleim, S. Trenbeath, F. Suzuki, and C. J. Sih, *J. Chem. Soc., Chem. Commun.*, **1978**, 242.

[174] K.-E. Stensiö and C. A. Wachtmeister, *Acta Chem. Scand.*, **23**, 144 (1969).

[175] F. Suzuki, S. Trenbeath, R. D. Gleim, and C. J. Sih, *J. Org. Chem.*, **43**, 4159 (1978).

[176] V. D. N. Sastri, N. Narasimhachari, P. Rajagopalan, T. R. Seshadri, and T. R. Thiruvengadam, *Proc. Indian Acad. Sci., Sect. A*. **37**, 681 (1953) [*C.A.*, **48**, 8227f (1954)].

[177] G. Bargellini and L. Monti, *Gazz. Chim. Ital.*, **45** (I), 90 (1915) [*C.A.*, **9**, 2239 (1915)].

[178] M. D. Bhavsar and R. D. Desai, *Indian J. Pharm.*, **13**, 200 (1951).

[179] P. L. Sawhney, T. R. Seshadri, and T. R. Thiruvengadam, *Proc. Indian Acad. Sci., Sect. A*, **33**, 11 (1951) [*C.A.*, **46**, 4536b (1952)].

[180] G. L. Schmir, L. A. Cohen, and B. Witkop, *J. Am. Chem. Soc.*, **81**, 2228 (1959).

[181] M. Ichikawa and H. Ichibagase, *Yakugaku Zasshi*, **80**, 1354, (1960) [*C.A.*, **55**, 5482c (1961)].

[182] K. M. Jainamma and S. Sethna, *J. Indian Chem. Soc.*, **50**, 790 (1973).

[183] R. J. Parikh and S. Sethna, *J. Indian Chem. Soc.*, **27**, 369 (1950).

[184] R. S. Thakur, S. C. Bagadia, and M. L. Sharma, *Experientia*, **34**, 158 (1978).

[185] F. Mauthner, *J. Prakt. Chem.*, **152**, 23 (1939).

[186] R. D. Desai and P. R. Desai, *J. Indian Chem. Soc.*, **40**, 456 (1963).

[187] T. R. Seshadri and M. S. Sood, *J. Indian Chem. Soc.*, **39**, 539 (1962).

[188] V. J. Dalvi, R. B. Desai, and S. Sethna, *J. Indian Chem. Soc.*, **28**, 366 (1951).

[189] R. M. Naik and V. M. Thakor, *J. Org. Chem.*, **22**, 1240 (1957).

[190] F. Wessely and E. Demmer, *Ber.*, **62**, 120 (1929).

[191] A. K. Das Gupta, R. M. Chatterje, and T. P. Bhowmic, *Tetrahedron*, **25**, 4207 (1969).

[192] V. K. Ahluwalia, A. C. Mehta, and T. R. Seshadri, *Proc. Indian Acad. Sci., Sect. A*, **45**, 15 (1957) [*C.A.*, **51**, 11334i (1957)].

[193] S. K. Koul, K. L. Dhar, and R. S. Thakur, *Indian J. Chem.*, **17B**, 396 (1979).

[194] M. G. Patel and S. Sethna, *J. Indian Chem. Soc.*, **39**, 511 (1962).

[195] M. D. Bhavsar and R. D. Desai, *J. Indian Chem. Soc.*, **31**, 141 (1954).

[196] V. K. Ahluwalia, C. L. Rustogi, and T. R. Seshadri, *Proc. Indian Acad. Sci., Sect. A*, **49**, 104 (1959) [*C.A.*, **53**, 18022a (1959)].

[197] V. K. Ahluwalia, V. N. Gupta, C. L. Rustagi, and T. R. Seshadri, *J. Sci. Ind. Res.*, **19B**, 345 (1960) [*C.A.*, **55**, 5478g (1961)].

[198] S. Rajagopalan, K. V. Rao, and T. R. Seshadri, *Proc. Indian Acad. Sci., Sect. A*, **25**, 432 (1947) [*C.A.*, **43**, 637h (1949)].

[199] V. K. Ahluwalia, D. S. Gupta, V. V. S. Murti, and T. R. Seshadri, *Proc. Indian Acad. Sci., Sect. A*, **38**, 480 (1953) [*C.A.*, **49**, 1713d (1955)].

[200] N. Narasimhachari, L. R. Row, and T. R. Seshadri, *Proc. Indian Acad. Sci., Sect. A*, **27**, 37 (1948) [*C.A.*, **43**, 640a (1949)].

[201] K. V. Rao, K. V. Rao, and T. R. Seshadri, *Proc. Indian Acad. Sci., Sect. A*, **25**, 427 (1947) [*C.A.*, **43**, 636i (1949)].

[202] T. R. Seshadri, S. Varadarajan, and V. Venkateswarlu, *Proc. Indian Acad. Sci., Sect. A*, **32**, 250 (1950) [*C.A.*, **46**, 1541b (1952)].

[203] J. E. Gowan, S. P. M. Riogh, G. J. MacMahon, S. Ó'Clérigh, E. M. Philbin, and T. S. Wheeler, *Tetrahedron*, **2**, 116 (1958).

[203a] A. C. Jain, S. K. Gupta, and P. K. Bambah, *Indian J. Chem.*, **23B**, 1002 (1984).

[204] K. V. Rao and T. R. Seshadri, *Proc. Indian Acad. Sci., Sect. A*, **25**, 417 (1947) [*C.A.*, **43**, 636a (1949)].

[205] H. Wagner, G. Maurer, L. Farkas, R. Hänsel, and D. Ohlendorf, *Chem. Ber.*, **104**, 2381 (1971).

[206] K. V. Rao, T. R. Seshadri, and N. Viswanadham, *Proc. Indian Acad. Sci., Sect. A*, **29**, 72 (1949) [*C.A.*, **44**, 3985g (1950)].

[207] A. L. Tőkés, R. Bognár, and E. K. Cservenyák, *Acta Chim. Acad. Sci. Hung.*, **99**, 337 (1979) [*C.A.*, **91**, 157555m (1979)].

[208] V. K. Ahluwalia, G. P. Sachdev, and T. R. Seshadri, *Indian J. Chem.*, **4**, 268 (1966).

[209] K. J. Balakrishna and T. R. Seshadri, *Proc. Indian Acad. Sci., Sect. A*, **27**, 91 (1948) [*C.A.*, **42**, 5454b (1948)].

[210] A. C. Jain, S. K. Mathur, and T. R. Seshadri, *Indian J. Chem.*, **3**, 351 (1965).

[211] V. V. S. Murti, K. V. Rao, and T. R. Seshadri, *Proc. Indian Acad. Sci., Sect. A*, **26**, 182 (1947) [*C.A.*, **42**, 5908i (1948)].

[212] S. R. Gupta and T. R. Seshadri, *Proc. Indian Acad. Sci., Sect. A*, **37**, 611 (1953) [*C.A.*, **48**, 8227b (1954)].

[213] L. R. Row, T. R. Seshadri, and T. R. Thiruvengadam, *Proc. Indian Acad. Sci., Sect. A*, **28**, 98 (1948) [*C.A.*, **44**, 3984b (1950)].

[214] A. C. Jain, T. R. Seshadri, and T. R. Thiruvengadam, *Proc. Indian Acad. Sci., Sect. A*, **36**, 217 (1952) [*C.A.*, **48**, 2701b (1954)].

[215] K. V. Rao and T. R. Seshadri, *Proc. Indian Acad. Sci., Sect. A*, **25**, 444 (1947) [*C.A.*, **43**, 638c (1949)].

[216] M. Krishnamurti, T. R. Seshadri, and P. R. Shankaran, *Indian J. Chem.*, **5**, 137 (1967).

[217] V. Ramanathan and K. Venkataraman, *Proc. Indian Acad. Sci., Sect. A*, **39**, 90 (1954), [*C.A.*, **49**, 6243i (1955)].

[218] N. K. Anand, S. R. Gupta, A. C. Jain, S. K. Mathur, K. S. Pankajamani, and T. R. Seshadri, *J. Sci. Ind. Res.*, **21B**, 322 (1962) [*C.A.*, **57**, 13712f (1962)].

[219] H. Wagner, R. Rüger, G. Maurer, and L. Farkas, *Chem. Ber.*, **110**, 737 (1977).

[220] K. J. Balakrishna and T. R. Seshadri, *Proc. Indian Acad. Sci., Sect. A*, **26**, 214 (1947) [*C.A.*, **43**, 227a (1949)].

[221] V. R. Ahluwalia, S. K. Mukerjee, and T. R. Seshadri, *J. Chem. Soc.*, **1954**, 3988.

[222] S. Rajagopalan, K. V. Rao, and T. R. Seshadri, *Proc. Indian Acad. Sci., Sect. A*, **26**, 18(1947) [*C.A.*, **43**, 639f (1949)].

[223] V. P. Pathak and R. N. Khanna, *Indian J. Chem.*, **21 B**, 891 (1982).

[224] V. V. S. Murti and T. R. Seshadri, *Proc. Indian Acad. Sci., Sect. A*, **27**, 217 (1948) [*C.A.*, **43**, 4263h (1949)].

[225] Z. Čekan and V. Herout, *Collect. Czech. Chem. Commun.*, **21**, 79 (1956).

[226] K. J. Balakrishna, N. P. Rao, and T. R. Seshadri, *Proc. Indian Acad. Sci., Sect. A*, **33**, 151 (1951) [*C. A.*, **46**, 4537c (1952)].

[227] K. J. Balakrishna and T. R. Seshadri, *Proc. Indian Acad. Sci., Sect. A*, **26**, 72 (1947) [*C.A.*, **42**, 6812a (1948)].

[228] K. V. Rao, K. V. Rao, and T. R. Seshadri, *Proc. Indian Acad. Sci., Sect. A*, **26**, 13 (1947) [*C.A.*, **43**, 638h (1949)].

[229] A. L. Tőkés and R. Bognár, *Acta Chim. Acad. Sci. Hung.*, **107**, 365 (1981) [*C.A.*, **96**, 35016u (1982)].

[230] S. C. Bhrara, A. C. Jain, and T. R. Seshadri, *Indian J. Chem.*, **3**, 68 (1965).

[231] K. J. Balakrishna and T. R. Seshadri, *Proc. Indian Acad. Sci., Sect. A*, **26**, 234 (1947) [*C.A.*, **42**, 7296a (1948)].

[232] T. R. Seshadri, *Proc. Indian Acad. Sci., Sect. A*, **30**, 333 (1949); *Experientia, Suppl. 2*, 258 (1955) [*C.A.*, **44**, 9960b (1950)].

[233] H. G. Krishnamurty and T. R. Seshadri, *J. Sci. Ind. Res.*, **18B**, 151 (1959) [*C.A.*, **54**, 12130c (1960)].

[234] I. Dass, D. Rajagopalan, and T. R. Seshadri, *J. Sci. Ind. Res*, **14B**, 335 (1955) [*C.A.*, **50**, 12037a (1956)].

[235] J. Chopin, *C. R. Hebd. Seances Acad. Sci.*, **243**, 588 (1956); T. R. Seshadri, in *The Chemistry of Flavanoid Compounds*, T. A. Geissman, Ed., MacMillan, New York, 1962, pp. 176–177.

[236] N. Narasimhachari, L. R. Row, and T. R. Seshadri, *Proc. Indian Acad. Sci., Sect. A*, **35**, 46, (1952) [*C.A.*, **47**, 6414b (1953)].

[237] D. K. Chakravorty, S. K. Mukerjee, V. V. S. Murti, and T. R. Seshadri, *Proc. Indian Acad. Sci., Sect. A*, **35**, 34 (1952) [*C.A.*, **47**, 6413g (1953)].

[238] S. K. Mukerjee and T. R. Seshadri, *Proc. Indian Acad. Sci., Sect. A*, **35**, 323 (1952) [*C.A.*, **47**, 8742h (1953)].

[239] S. Fukushima, A. Ueno, and Y. Akahori, *Chem. Pharm. Bull.*, **12**, 307 (1964).

[240] A. Ueno, *Chem. Pharm. Bull.*, **14**, 121 (1966).

[241] V. V. S. Murti and T. R. Seshadri, *Proc. Indian Acad. Sci., Sect. A*, **30**, 107 (1949) [*C.A.*, **44**, 5875i (1950)].

[242] K. S. Pankajamani and T. R. Seshadri, *J. Sci. Ind. Res.*, **13B**, 396 (1954) [*C.A.*, **49**, 11639a (1955)].

[243] V. V. Kane, A. B. Kulkarni, and R. C. Shah, *J. Sci. Ind. Res.*, **18B**, 75 (1959) [*C.A.*, 54, 521e (1960)].

[244] K. V. Rao and T. R. Seshadri, *Proc. Indian Acad. Sci., Sect. A*, **26**, 288 (1947) [*C.A.*, **42**, 3393i (1948)].

[245] J. C. Roberts, *J. Chem. Soc.*, **1960**, 785.

[246] V. K. Ahluwalia and T. R. Seshadri, *Proc. Indian Acad. Sci., Sect. A*, **44**, 1 (1956) [*C.A.*, **51**, 5059a (1957)].

[247] V. Jayalakshmi, S. Neelakantan, and T. R. Seshadri, *Indian J. Chem.*, **5**, 180 (1967).

[248] H. Möhrle and H. Weber, *Tetrahedron*, **26**, 3779 (1970).

[249] C. Herdeis and A. Dimmerling, *Arch. Pharm.* (*Weinheim, Ger.*), **317**, 304, (1984).

[249a] P. Nantka-Namirski and A. Rykowski, *Acta Polon. Pharm.*, **31**, 433 (1974) [*C.A.*, **82**, 139899m (1975)].

[250] P. Ashworth, *Tetrahedron*, **32**, 261 (1976).

[251] C. Houghton and R. B. Cain, *Biochem. J.*, **130**, 879 (1972).

[252] P. Nantka-Namirski and A. Rykowski, *Acta Polon. Pharm.*, (Engl. Transl.) **29**, 129 (1972) [*C.A.*, **77**, 114205r (1972)].

[253] H. Loth and B. Hempel, *Arch. Pharm.* (*Weinheim, Ger.*), **305**, 724 (1972).

[254] P. Nantka-Namirski and A. Rykowski, *Acta Polon. Pharm.*, (Engl. transl.), **29**, 233 (1972) [*C.A.*, **78**, 29568n (1973)].

[255] G. A. Swan, *Experienta*, **40**, 687 (1984); *J. Chem. Soc. Perkin Trans. 1*, **1985**, 1757.

[256] E. Matsumura, Japanese Patent 75-10874 (1975) [*C.A.* **85**, 192566p (1976)].

[257] D. T. Hurst, *Aust. J. Chem.*, **36**, 1285 (1983).

[258] R. Hull, *J.* Chem. Soc., **1956**, 2033.

[259] J. H. Chesterfield, D. T. Hurst, J. F. W. McOmie, and M. S. Tute, *J. Chem. Soc.*, **1964**, 1001.
[260] S. I. Zav'yalov and G. V. Pokhvisneva, *Bull. Acad. Sci. USSR, Div. Chem. Sci.*, **1973**, 2308 [*C.A.*, **80**, 37064b (1974)].
[261] N. Oda, Y. Kanie, and I. Ito, *Yakugaku Zasshi*, **93**, 817 (1973) [*C.A.*, **79**, 66309v (1973)].
[262] J. T. Warren, R. Allen, and D. E. Carter, *Drug Metab. Disposition*, **6**, 38 (1978) [*C.A.*, **88**, 182350m (1978)].
[263] K. L. Rinehart, Jr., W. Sobiczewski, J. F. Honegger, R. M. Enanoza, T. R. Witty, V. J. Lee, and L. S. Shield, *Bioorg. Chem.*, **6**, 341 (1977).
[264] J. K. Yeh and R. R. Brown, *Biochem. Med.*, **14**, 12 (1975).
[265] E. L. May, R. C. Millican, and A. H. Mehler, *J.* Org. Chem., **27**, 2274 (1962).
[266] C. E. Dalgliesh, *Biochem. J.*, **61**, 334 (1955).
[267] S. Laham and G. V. Rao, *Chemosphere*, **4**, 317 (1975) [*C.A.*, **84**, 43567s (1976)].
[268] L. Bradshaw and D. B. Clayson, *Nature*, **176**, 974 (1955).
[269] M. Ishidate and A. Hanaki, *Chem. Pharm. Bull.*, **10**, 358 (1962).
[270] N. Ram and K. S. Sidhu, *Indian J. Chem.*, **16A**, 195 (1978).
[271] B. V. Pandav, N. Ram, and K. S. Sidhu, *Indian J. Chem.*, **17A**, 40 (1979).
[272] D. Manson, *Biochem. J.*, **119**, 541 (1970).
[273] E. Boyland, D. Manson, and S. F. D. Orr, *Biochem. J.*, **65**, 417 (1957).
[274] E. Boyland and D. Manson, *Biochem. J.*, **99**, 189 (1966).
[275] G. Leaf and A. Neuberger, *Biochem. J.*, **43**, 606 (1948).
[276] M. Krishnamurti and T. R. Seshadri, *Proc. Indian Acad. Sci., Sect. A*, **39**, 144(1954) [*C.A.*, **45**, 5455h (1955)].
[277] A. C. Jain and T. R. Seshadri, *J. Sci. Ind. Res.* **15B**, 61 (1956) [*C.A.*, **50**, 14684d (1956)].
[278] D. T. Hurst, J. F. W. McOmie, and J. B. Searle, *J. Chem. Soc.*, **1965**, 7116.

CHAPTER 3

FLUORINATION WITH DIETHYLAMINOSULFUR TRIFLUORIDE AND RELATED AMINOFLUOROSULFURANES

MILOŠ HUDLICKÝ

Department of Chemistry,
Virginia Polytechnic Institute and State University,
Blacksburg, Virginia

CONTENTS

ACKNOWLEDGMENTS

The author wishes to express his sincere thanks to the editor of the chapter, Dr. E. Ciganek, E. I. du Pont de Nemours and Company, Wilmington, Delaware, for his efficient help with the preparation of the manuscript, for the procurement of the computer literature search of *Chemical Abstracts* and *Science Citation Indexes*, for the reprints of the patent literature, and for his very thorough editorial work. Thanks are also due to Miss A. V. M. Lomascolo for the artwork and to the secretaries, Ms. A. L. Courtney and Ms. C. F. Walters (and the word processor at Virginia Polytechnic Institute and State University) for the typing of the manuscript. The author also acknowledges with gratitude the critical comments made by Dr. W. J. Middleton.

INTRODUCTION

Among the many organic fluorine compounds containing nitrogen and sulfur, dialkylaminotrifluorosulfuranes and bis(dialkylamino)difluorosulfuranes have become very useful fluorinating agents. The first representative of this group, dimethylaminosulfur trifluoride (**1a**) was prepared in 1964,[1,2] followed in 1970 by diethylaminosulfur trifluoride (**1b**),[3] which became popular under the acronym DAST.* Both compounds were synthesized by treatment of the appropriate dialkylaminotrimethylsilane with sulfur tetrafluoride:

$$R_2NSi(CH_3)_3 + SF_4 \longrightarrow \underset{\substack{\mathbf{1a},\ R = CH_3 \\ \mathbf{1b},\ R = C_2H_5}}{R_2NSF_3} + SiF(CH_3)_3$$

Subsequently, bis(dialkylamino)sulfur difluorides (**2**) were prepared by the reaction of dialkylaminosulfur trifluorides with dialkylaminotrimethylsilanes:[4–6]

$$\underset{\mathbf{1}}{R_2NSF_3} + R_2NSi(CH_3)_3 \longrightarrow \underset{\mathbf{2}}{(R_2N)_2SF_2} + SiF(CH_3)_3$$

Both classes of compounds were found to convert alcohols into fluorides, and aldehydes and ketones into geminal difluorides:

$$R^1OH + R_2NSF_3 \longrightarrow [R^1OSF_2NR_2] \longrightarrow R^1F + R_2NSOF + HF$$

$$R^1COR^2 + R_2NSF_3 \longrightarrow R^1CF_2R^2 + R_2NSOF$$

Further uses include the preparation of acyl and sulfonyl fluorides from carboxylic and sulfonic acids and the conversion of sulfoxides into α-fluoroalkyl sulfides.

Later, the reaction of bis(dialkylamino)difluorosulfuranes with dialkylaminotrimethylsilanes was shown to give tris(dialkylamino)sulfonium difluorotri-

* DAST is indexed in *Chemical Abstracts* under sulfur, (*N*-ethylethanaminato)trifluoro-.

methylsilicates (TASF) (**3**):[7]

$$\underset{\mathbf{2}}{(R_2N)_2SF_2} + R_2NSi(CH_3)_3 \longrightarrow \underset{\mathbf{3}}{(R_2N)_3\overset{+}{S}\,(CH_3)_3\overset{-}{Si}F_2}$$

These salts no longer react with alcohols or carbonyl compounds, but they convert even relatively unreactive halides, such as primary chlorides, into fluorides under very mild conditions. Furthermore, they are readily soluble in organic solvents and are useful sources of "naked," unsolvated fluoride ion.

This chapter deals with the preparation of all three classes of reagents and their use as fluorinating agents. The literature is covered exhaustively until the end of 1984; some papers beyond that date are included.

PREPARATION OF AMINOFLUOROSULFURANES

The reaction by which dimethylaminosulfur trifluoride[1,2] and diethylaminosulfur trifluoride[3] (DAST) were first prepared is still used as the main method for their synthesis. It involves the treatment of dialkylaminotrimethylsilanes with sulfur tetrafluoride.[5,8,9,]

$$(C_2H_5)_2NSi(CH_3)_3 + SF_4 \xrightarrow[-65^\circ \text{ to } -60^\circ;\ \text{room temp.}]{CCl_3F} \underset{\mathbf{(84\%)}}{(C_2H_5)_2NSF_3} + (CH_3)_3SiF$$

Other dialkylamines used for the preparation of dialkylaminosulfur trifluorides are diisopropylamine,[5] pyrrolidine,[5] piperidine,[8,10] morpholine,[8] and *N*-ethylaniline.[8] The yields and physical constants of the products are listed in Table A.

When dialkylaminosulfur trifluorides are treated with dialkylaminotrimethylsilanes, bis(dialkylamino)sulfur difluorides result.[4–6]

$$(CH_3)_2NSF_3 + (C_2H_5)_2NSi(CH_3)_3 \xrightarrow[-78^\circ \text{ to } 25^\circ]{CCl_3F} \underset{\mathbf{(92\%)}}{(C_2H_5)_2NSF_2N(CH_3)_2} + (CH_3)_3SiF$$

Combinations of various dialkylamines linked to sulfur in bis(dialkylamino)sulfur difluorides are tabulated in Table B, together with the yields and physical constants of the products.

Reaction of sulfur tetrafluoride with 3 moles of a dialkylaminotrimethylsilane forms an ionic tris(dialkylamino)sulfonium difluorotrimethylsilicate (**3a**)[7].

$$3\,(CH_3)_2NSi(CH_3)_3 + SF_4 \xrightarrow[-78^\circ \text{ to room temp., 3 d}]{(C_2H_5)_2O} \underset{\mathbf{3a}\ (92\%)}{[(CH_3)_2N]_3\overset{+}{S}\,(CH_3)_3\overset{-}{Si}F_2}$$

TABLE A. AMINOFLUOROSULFURANES AS FLUORINATING AGENTS
DIALKYLAMINOSULFUR TRIFLUORIDES[a]

Compound	Yield (%)	bp (mm)	n_D^{20}	Refs.[b]
$(CH_3)_2NSF_3$[c,d]	(75)[2]	117.5° (760)[1]		**1, 2**
		117.8° (760)[2]		
	(60)	24–25° (12)	1.4018	8
	(82)	49–49.5° (33)		**5, 9**
		117.5° (760)		11
$(C_2H_5)_2NSF_3$ (DAST)[e]	(70)			3
	(70, 90)[4]	43–44° (12)	1.4125	**4**[f], 8
	(80–84)	46–47° (10)		**5, 12**
				9, 13
		30–32° (3)		11
$(i\text{-}C_3H_7)_2NSF_3$	(99)			5[g]
	(—)			14
Pyrrolidino-NSF_3	(83)	54–55° (15)		5
	(76)	23–24° (0.3); mp −18°		15
Morpholino-NSF_3	(98)	33–34° (0.07)	1.4534	**4**[f]
	(70)	" "	1.4538	**16**[h]
	(55)	55–56° (0.8)	1.4540	**16**[i]
	(98)	41–42° (0.5)	1.4536	**4**[j]
	(68)	" "	1.4538	8
Piperidino-NSF_3	(65)	75–77° (12)	1.4515	8
	(97)	" "	1.4517	**4**[j]
	(96)	" "	1.4520	**4**[f]
	(68)	" "	1.4525	**16**[i]
	(65)	76° (12)	1.4520	**16**[h]
	(60)	43° (0.1)		**10**[k]
$C_6H_5N(C_2H_5)SF_3$	(60)	56–57° (0.3)	1.4930	8[l]

[a] The dialkylaminosulfur trifluorides were prepared from dialkylaminotrimethylsilanes and sulfur tetrafluoride unless stated otherwise in footnotes.

[b] References in boldface numbers contain specific preparative procedures.

[c] The density of dimethylaminosulfur trifluoride is 1.3648.[11]

[d] The melting point of dimethylaminosulfur trifluoride is −79°.[1,2]

[e] The density of DAST as calculated from weights and volumes[5,9] is 1.29; as determined at room temperature it is 1.22–1.23[17], 1.220; 1.238.[11]

[f] The compound was prepared from $(R_2N)_2SO$ and SF_4.

[g] The compound decomposed at 60° (2 mm).

[h] The compound was prepared from $p\text{-}CH_3C_6H_4S(O)NR_2$ and SF_4.

[i] The compound was prepared from $R_2NS(O)OC_2H_5$ and SF_4.

[j] The compound was prepared from $R_2NS(O)F$ and SF_4.

[k] The compound decomposed violently at 138°.

[l] The compound decomposed slowly at 20°.

TABLE B. Aminofluorosulfuranes as Fluorinating Agents Bis(dialkylamino)sulfur Difluorides[a]

Compound	Yield (%)	mp	Refs.[b]
$[(CH_3)_2N]_2SF_2$	(60)	64–65.5°	**5, 6**
$(CH_3)_2NSF_2NC_4H_8$[c]	(—)	—	6
$(CH_3)_2NSF_2N(C_2H_5)_2$	(92)	Liquid—not distilled	5, **6**
$(CH_3)_2NSF_2N(C_5H_{10})$[c]	(99)	25–26°	5
$(OC_4H_8N)_2SF_2$[c]	—[d]	101–102°	4
	(98)	"	**16**[e]
$(C_2H_5)_2NSF_2NC_4H_8O$[c]	—[d]	Viscous liquid	4
$[(C_2H_5)_2N]_2SF_2$	(92)	Liquid—not distilled	5, **6**
$OC_4H_8NSF_2NC_5H_{10}$[c]	—[d]	58–59°	4
	(80)	—	**16**[e]
$(C_2H_5)_2NSF_2NC_5H_{10}$[c]	(—)	—	6
$n\text{-}C_4H_9N(C_2H_5)SF_2N(CH_3)C_2H_5$	( )	—	6
$(C_5H_{10}N)_2SF_2$[c]	(—)	—	6
	—[d]	104–105°	4
	(97)	105–106°	16[e]
$[(n\text{-}C_4H_9)_2N]_2SF_2$	(—)	—	6
$[(C_6H_{13})_2N]_2SF_2$	(—)	—	6

[a] The bis(dialkylamino)sulfur difluorides were prepared from dialkylaminotrimethylsilanes and dialkylaminosulfur trifluorides unless stated otherwise in the footnotes.

[b] References in boldface numbers indicate specific preparative procedures described therein.

[c] C_4H_8N is pyrrolidino, OC_4H_8N is morpholino, and $C_5H_{10}N$ is piperidino.

[d] Yields were "close to theoretical."

[e] The compound was prepared from the dialkylamide of sulfurous acid and dialkylaminosulfur trifluoride.

TABLE C. Aminofluorosulfuranes as Fluorinating Agents Tris(dialkylamino)sulfonium Difluorotrimethylsilicates[7]

Compound	Yield (%)	mp
$[(CH_3)_2N]_3\overset{+}{S}\ (CH_3)_3\overset{-}{Si}F_2$ (TASF)	(86)	55–72°[a]
	(99)	61–67°[a] (dec.)
	(71–78)[18]	98–101°
$[(CH_3)_2N]_2\overset{+}{S}NC_4H_8\ (CH_3)_3\overset{-}{Si}F_2$[b]	(78)	40–45°
$[(CH_3)_2N]_2\overset{+}{S}N(C_2H_5)_2\ (CH_3)_3\overset{-}{Si}F_2$	(>51)	
$(CH_3)_2N\overset{+}{S}(NC_4H_8)_2\ (CH_3)_3\overset{-}{Si}F_2$[b]	(90)	49–50°
$(C_4H_8N)_3\overset{+}{S}\ (CH_3)_3\overset{-}{Si}F_2$[b]	(75)	54–57°
$[(C_2H_5)_2N]_3\overset{+}{S}\ (CH_3)_3\overset{-}{Si}F_2$	(98)	90–95°
$(C_5H_{10}N)_3\overset{+}{S}\ (CH_3)_3\overset{-}{Si}F_2$[c]	(89)	87–90°
$(4\text{-}CH_3C_5H_9N)_3\overset{+}{S}\ (CH_3)_3\overset{-}{Si}F_2$[d]	(89)	73–75°
$[n\text{-}C_{18}H_{37}(CH_3)N]_3\overset{+}{S}\ (CH_3)_3\overset{-}{Si}F_2$	(96)	40–60°

[a] The compound was prepared by two different methods.[7] Almost quantitative yield and mp 58–62° are reported in Ref. 19.

[b] C_4H_8N is pyrrolidino.

[c] $C_5H_{10}N$ is piperidino.

[d] $4\text{-}CH_3C_5H_9N$ is 4-methylpiperidino.

Salts **3a** containing different dialkylamines are synthesized from dialkylaminotrimethylsilanes and dialkylaminosulfur trifluorides (Table C).[7]

$$2\ (CH_3)_2NSi(CH_3)_3 + \text{(pyrrolidino)}NSF_3 \xrightarrow[-60^\circ \text{ to room temp., 3 d}]{(C_2H_5)_2O} \text{(pyrrolidino)}N\overset{+}{S}[N(CH_3)_2]_2\ (CH_3)_3\overset{-}{Si}F_2$$

(78%)

Aminofluorosulfuranes of all three types that are reported in the literature but as yet not used for fluorinations are listed in Table D. Table E contains ^1H and ^{19}F NMR data for the most frequently used reagents. Other methods for preparing aminofluorosulfuranes are shown in Scheme 1.[4,16]

$R_2NSOF \xrightarrow[(97\text{–}98\%)]{2\ SF_4}$

$R_2NSO_2C_2H_5 \xrightarrow[(55\text{–}68\%)]{2\ SF_4}$

$R_2NSONR_2 \xrightarrow[(90\text{–}98\%)]{3.3\ SF_4}$

$R_2NSOC_6H_4CH_3\text{-}p \xrightarrow[(65\text{–}70\%)]{1.5\ SF_4}$

$$\xrightarrow[12\ h]{20\text{–}25^\circ,\ \text{autoclave}} R_2NSF_3$$

$$R_2NSONR_2 + R_2NSF_3 \xrightarrow[25^\circ,\ 4\ h;\ 60^\circ,\ 30\ min]{C_6H_6} R_2NSF_2NR_2 + R_2NSOF$$

(97–98%) **(60–67%)**

$R_2 = (C_2H_5)_2$, $(CH_2)_5$ or $O(CH_2CH_2)_2$

Scheme 1

Experimental details illustrating the preparations of aminofluorosulfuranes by a variety of methods are given in "Experimental Procedures."

MECHANISM

Replacement of Hydroxy Groups by Fluorine

The mechanism of the replacement of a hydroxy group by fluorine by the action of dialkylaminosulfur trifluorides and bis(dialkylamino)sulfur difluorides resembles to a certain extent that of sulfur tetrafluoride.[28] The first step is the nucleophilic displacement of fluorine on sulfur by the oxygen of the hydroxy compound accompanied by elimination of hydrogen fluoride:

$$ROH + F_3SNR_2 \longrightarrow \underset{\mathbf{4}}{ROSF_2NR_2} + HF$$

TABLE D. AMINOFLUOROSULFURANES NOT REPORTED AS FLUORINATING AGENTS

No. of Carbon Atoms	Compound (yield %)	mp, bp (mm)	Refs.
C_2	 NSF_3 (—)	Decomposes at room temp	15
	$(CH_3)_2N\overset{+}{S}F_2\overset{-}{B}F_4$ (80)	mp 140–142°	20
C_4	 NSF_3 (68)	mp −20°; bp 18° (0.01); unstable at room temp	15
	$OC_4H_8N\overset{+}{S}F_2\overset{-}{B}F_4$[a] (85)	mp 104–106°	20
	$(C_2H_5)_2N\overset{+}{S}F_2\overset{-}{B}F_4$ (78)	mp 74–76°	20
C_5	$(CH_3)_2NSF_2CF(CF_3)_2$ (89)	bp 42–45° (14)	21
	$OC_4H_8NSF{=}NCFO$[a] (≈100)	bp 115–116° (0.03)	22
	$C_5H_{10}N\overset{+}{S}F_2\overset{-}{B}F_4$[b] (80)	mp 92–94°	20
	$(C_2H_5)_2NSF{=}NCFO$ (≈100)	bp 96–98° (0.4)	22
C_6	$OC_4H_8NSF_2CFClCF_3$[a] (74)	bp 51–52° (0.1)	23
	$OC_4H_8NSF_2C_2F_5$[a] (77)	bp 46–48° (0.13)	24
	$C_5H_{10}NSF{=}NCFO$[b] (≈100)	bp 113–114° (0.08)	22
C_7	$OC_4H_8NSF_2CF(CF_3)_2$[a] (96)	bp 43° (0.05)	21
C_8	$(CH_3)_2NSF_2(CF_2)_2C(CF_3)_3$ (—)	Not isolated[c]	25
	 NSF_3 (76)	mp 17.5°; bp 46° (0.001)	15
C_9	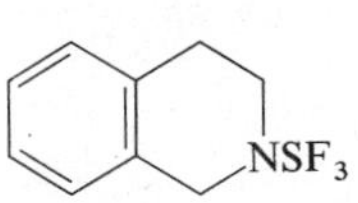 NSF_3 (—)	Decomposes explosively at room temp	15
C_{11}	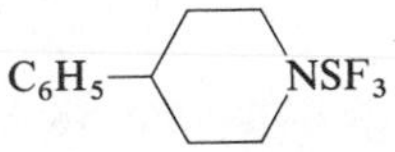 C_6H_5— NSF_3 (—)	mp 68° (dec.), bp 112 (0.03)	15
C_{12}	$(OC_4H_8N)_3\overset{+}{S}\,\overset{-}{B}F_4$[a] (95)	mp 226–227° (dec.)	20
C_{14}	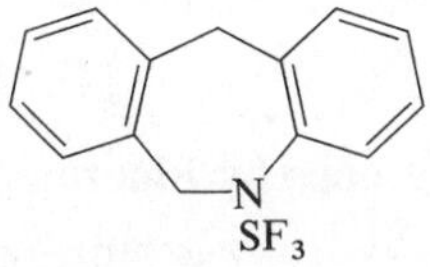 N SF_3 (83)	mp 76° (dec.)	15

[a] OC_4H_8N is morpholino.
[b] $C_5H_{10}N$ is piperidino.
[c] The product reacts with water to give $(CH_3)_2NS(O)(CF_2)_2C(CF_3)_3$ in 50% yield.

TABLE E. ^1H AND ^{19}F NMR SHIFTS OF AMINOFLUOROSULFURANES. AXIAL FLUORINE (F*a*), EQUATORIAL (F*e*)

Compound	Temperature	^1H NMR, δ ppm downfield from $(CH_3)_4Si$	^{19}F NMR, ppm downfield from $FCCl_3$ F*a*	F*e*	Refs.
$(CH_3)_2NSF_3$[a]	25°	3.15	42 (broad)		2
	−68°	3.06 (dt, $J_{HFa} = 5.0$ Hz, $J_{HFe} = 8.2$ Hz)	59.4 (d of heptets, $J_{FaFe} = 58$ Hz, 2F)	20.2 (t of heptets, 1F)	26
$(C_2H_5)_2NSF_3$	25°	1.27 (t, $J_{HH} = 6.9$ Hz), 3.35 (q)	46.4 (broad)		3,10
	−68°	1.25 (t, $J_{HH} = 7.0$ Hz), 3.49 (qdt, $J_{HFa} = 3.0$ Hz, $J_{HFe} = 6.1$ Hz)	54.0 (d of multiplets, $J_{FaFe} = 62$ Hz, 2F)	34.4 (t of quintets, 1F)	26,27
$C_5H_{10}NSF_3$[b]	20°		41.7 (broad)		10
	−86°		55.6	22.2	10
$(i\text{-}C_3H_7)_2NSF_3$	−68°		61.9 (d of multiplets, $J_{FaFe} = 53$ Hz, 2F)	18.3 (t of multiplets, 1F)	14
$[(CH_3)_2N]_2SF_2$[c]		2.80 (s)	6.9 (s)		5,6
$[(C_2H_5)_2N]_2SF_2$		1.36 (t, $J_{HH} = 7.5$ Hz), 3.43 (q)	9.7		5,6
$(CH_3)_2NSF_2N(C_2H_5)_2$		1.35 (t, $J_{HH} = 7.5$ Hz), 2.90 (s), 3.44 (q)	10.9,[5] 10.0[6]		5,6
$(CH_3)_2NSF_2NC_5H_{10}$[b]			5.9		5

[a] In the trigonal–bipyramidal structure two fluorine atoms are axial, one is equatorial, and the dimethylamino group is also equatorial.[2,3]

[b] $C_5H_{10}N$ is piperidino.

[c] Single crystal X-ray diffraction showed that the compound has a trigonal–bipyramidal structure with the fluorine atoms axial and the diethylamino groups equatorial.[27a]

Although intermediate **4** has not been isolated, its temporary existence is inferred from ^{19}F NMR spectroscopy of a mixture of equivalent amounts of DAST and 1,2:5,6-di-*O*-isopropylidene-α-D-glucofuranose. In dichloromethane, the signal of DAST at 40 ppm disappears, and a new five-line signal at 59 ppm ($J = 2$ Hz) could be caused by intermediate **5**.[29] Compounds of type **5** have been prepared by the reaction of dialkylaminosulfur trifluorides with trimethylsilyl polyfluoroalkyl ethers (p. 540).

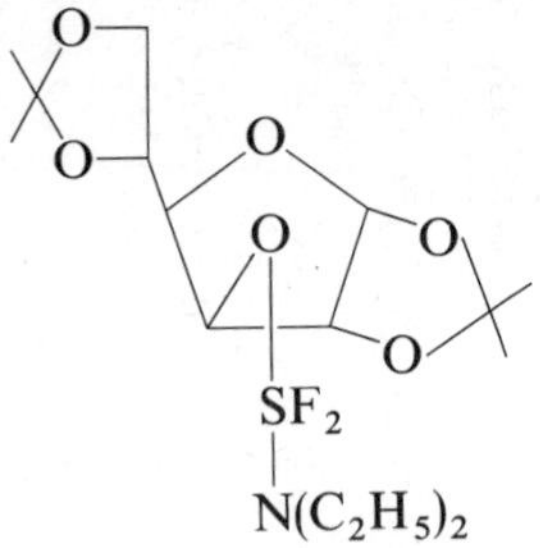

SF_2
$N(C_2H_5)_2$
5

A more reliable proof of such an intermediate is the isolation of a mixture of two compounds **6**, diastereomeric on sulfur, from the partial hydrolysis of the product obtained on treatment of 6α,9α-difluoro-16α-methyl-11β,17,21,-trihydroxy-1,4-pregnadiene-3,20-dione 21-trimethylacetate with piperidinosulfur trifluoride.[30]

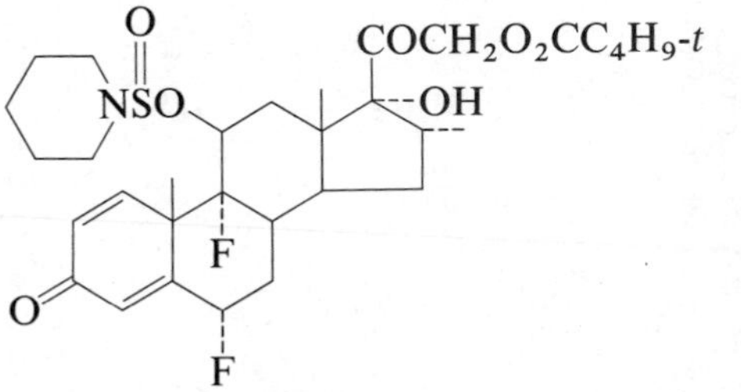

$COCH_2O_2CC_4H_9$-*t*
NSO
OH
F
F
6

The fate of intermediate **4** depends mainly on its structure. With simple alcohols, **4** is converted into the alkyl fluoride by reaction with fluoride ion:

$$\underset{\mathbf{4}}{ROSF_2NR_2} + F^- \longrightarrow RF + FSONR_2 + F^-$$

Occurrence of carbocation-type rearrangements accompanying the conversion of some alcohols into alkyl fluorides implies a mixed S_N1 and S_N2 mechanism. On the other hand, the displacement of many chiral hydroxy groups by fluorine occurs with almost complete inversion of configuration,[31] thus pointing to an S_N2 reaction. No stereo-randomization is reported in many reactions involving carbohydrates[29] and steroids.[32] Complete retention of configuration is observed in case of neighboring group participation.[17]

In view of these results, intramolecular transfer of fluorine from sulfur to carbon in intermediate **4** can be ruled out in most cases. This is in contrast to fluorinations of hydroxy compounds by sulfur tetrafluoride.[28] There are, however, a few examples where intramolecular fluorine transfer may be suspected. In the fluorination of methyl α-D-mannopyranoside with an excess of DAST, fluorine replaces not only the hydroxy group on C-6 but also the one on C-4. It is impossible to limit the reaction to monofluorination at C-6. This result can be best interpreted by assuming intramolecular transfer of fluorine onto C-6 in the intermediate **7**.[33]

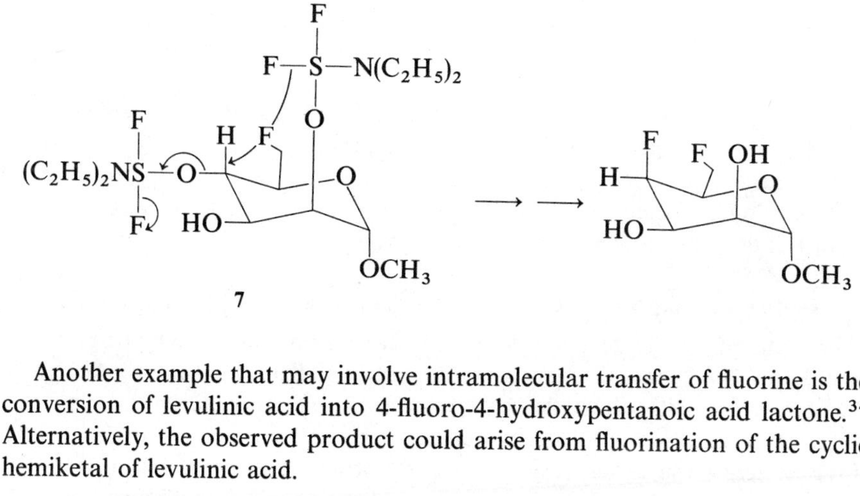

F
F—S—N(C_2H_5)$_2$
O
F
H F
(C_2H_5)$_2$NS—O
-O
F HO
OCH_3
7
→ →
F
F OH
H
-O
HO
OCH_3

Another example that may involve intramolecular transfer of fluorine is the conversion of levulinic acid into 4-fluoro-4-hydroxypentanoic acid lactone.[34] Alternatively, the observed product could arise from fluorination of the cyclic hemiketal of levulinic acid.

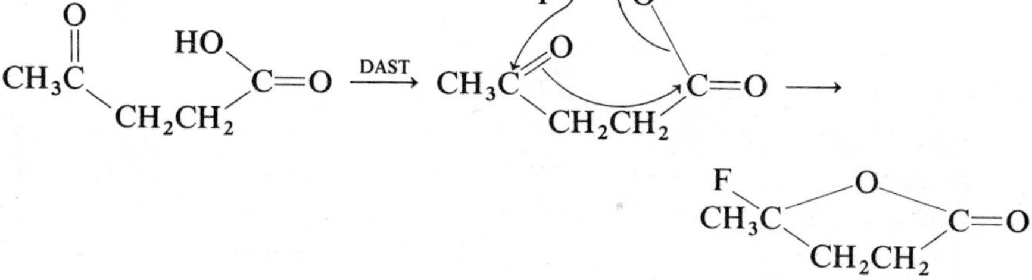

$CH_3C(=O)CH_2CH_2C(=O)OH$ $\xrightarrow{\text{DAST}}$ $CH_3C(=O)CH_2CH_2C(=O)O$—$S(F)_2N(C_2H_5)_2$ →

$CH_3C(F)$—O—$C(=O)$ (ring with CH_2CH_2)

The conversion of allylic alcohols into the isomeric allyl fluorides may occur by either an S_Ni' mechanism involving intramolecular fluorine transfer, or an S_N2' mechanism.[35]

$$\mathrm{F_2S(N(C_2H_5)_2)\text{–}O\text{–}CH(RCH{=}CH)P(O)(OC_2H_5)_2 \xrightarrow{S_Ni'} RCHF\text{–}CH{=}CH\text{–}P(O)(OC_2H_5)_2}$$

$$\mathrm{F_2S(N(C_2H_5)_2)\text{–}O\text{–}CH(RCH{=}CH)P(O)(OC_2H_5)_2 + F^- \xrightarrow{S_N2'} RCHF\text{–}CH{=}CH\text{–}P(O)(OC_2H_5)_2}$$

From the experimental evidence it can be concluded that the replacement of hydroxy groups by fluorine by means of aminofluorosulfuranes proceeds by several mechanisms depending mainly on the structure of the hydroxy compounds. Similar mechanisms can be assumed for the reaction of bis(dialkylamino)sulfur difluorides with hydroxy compounds.

Reaction of Dialkylaminosulfur Trifluorides with Aldehydes and Ketones

In contrast to the reaction of aminofluorosulfuranes with hydroxy compounds, no sulfur- and nitrogen-containing intermediates have been intercepted in the reaction with aldehydes and ketones. It has been suggested that the initial step is addition of hydrogen fluoride, formed from the reagent and traces of water, across the carbonyl group. The resulting α-fluoro alcohol then reacts with dialkyaminosulfur trifluoride or bis(dialkylamino)sulfur difluoride in the way shown for alcohols and affords intermediate **8**.[5]

$$\mathrm{RR^1CO \xrightarrow{HF} RR^1CFOH \xrightarrow[-HF]{R_2NSF_3} \underset{\mathbf{8}}{RR^1CFOSF_2NR_2} \longrightarrow RR^1CF_2 + FSONR_2}$$

Experimental support for the intermediacy of α-fluoroalcohols is the isolation of bis(α-fluoroalkyl) ethers and bis(α-fluoroalkyl) acetals from the reaction of trichloroacetaldehyde with DAST:[36]

$$\mathrm{CCl_3CHO \xrightarrow{DAST} CCl_3CHFOH} \begin{cases} \xrightarrow{\mathrm{CCl_3CHFOH}} \mathrm{(CCl_3CHF)_2O} & (27\%) \\ \xrightarrow{\mathrm{CCl_3CHO}} \mathrm{CCl_3CH(OCHFCCl_3)_2} & (32\%) \end{cases}$$

With monochloroacetaldehyde, the α-fluoroethyl alcohol is less stable and reacts with DAST to form 1-chloro-2,2-difluoroethane in preference to ether formation:[36]

$$ClCH_2CHO \xrightarrow{HF} ClCH_2CHFOH \begin{cases} \xrightarrow{ClCH_2CHFOH} (ClCH_2CHF)_2O \quad (15\%) \\ \xrightarrow{DAST} ClCH_2CHF_2 \quad (28\%) \end{cases}$$

The decomposition of intermediate **8**, formed from the α-fluoroalcohol and dialkylaminosulfur trifluoride, may occur either by intra- or intermolecular transfer of fluoride ion by an S_N2 or S_N1 mechanism. Intervention of fluorocarbocations in substrates with α-hydrogens sometimes results in loss of a proton with formation of a vinyl fluoride.[37,38] Such a reaction is enhanced and may even predominate when polar solvents and especially catalysts such as fuming sulfuric acid are used.[37] The effect of the conditions on the outcome of the reaction is shown in the fluorination of 2-methylcyclohexanone.[37]

DAST, 25°, 40 h; $OSF_2N(C_2H_5)_2$; F; $F\ \bar{O}SF_2N(C_2H_5)_2$; F^-; F; F; A; $-H^+$; F; +; F; B; C

Solvent	A	B	C
CH_2Cl_2	(60%)	(30%)	(10%)
Glyme	(51%)	(36%)	(13%)
Diglyme, H_2SO_4, SO_3	(24%)	(65%)	(11%)
Glyme, H_2SO_4, SO_3	(22%)	(67%)	(11%)

The vinyl fluorides are not formed by elimination of hydrogen fluoride from the geminal difluorides since the latter are stable under these conditions. In certain cases, the fluorocarbocations may undergo a Wagner–Meerwein rearrangement prior to proton loss or fluoride addition. An example is the reaction of pivalaldehyde with DAST.[5] In the nonpolar solvent fluorotrichloromethane,

1,1-difluoro-2,2-dimethylpropane is obtained in 78% yield. When the more polar diglyme is used as the solvent, the 1,1-difluoro-2,2-dimethylpropane is accompanied by two products of a rearrangement:

$$t\text{-}C_4H_9CHO \xrightarrow{HF} t\text{-}C_4H_9CHFOH \xrightarrow[-HF]{DAST}$$

$$t\text{-}C_4H_9CHFOSF_2N(C_2H_5)_2 \begin{cases} \xrightarrow{F^-} t\text{-}C_4H_9CHF_2 + \bar{O}SF_2N(C_2H_5)_2 \quad 24\% \\ \longrightarrow t\text{-}C_4H_9\overset{+}{C}HF\ \bar{O}SF_2N(C_2H_5)_2 \longrightarrow \end{cases}$$

$$\begin{matrix}(CH_3)_2\overset{+}{C}CHFCH_3 \\ \bar{O}SF_2N(C_2H_5)_2\end{matrix} \begin{cases} \xrightarrow{F^-} (CH_3)_2CFCHFCH_3 \quad (31\%) \\ \xrightarrow{-H^+} CH_2{=}C(CH_3)CHFCH_3 \quad (26\%) \end{cases}$$

By comparison, vinyl fluoride formation and cationic rearrangements are not observed in SF_4 fluorinations.

An oxidation occurs in the reaction of DAST with certain β-ketoesters[38a,b] and β-diketones[38b,c] to give α,β-difluoro-α,β-unsaturated esters and ketones, respectively. A possible mechanism involves reaction of DAST with the enol form of the substrate followed by intramolecular fluorine transfer to generate the α-fluorinated species. These intermediates, which have not yet been isolated, react with a second molecule of DAST to give the observed products.

$$CH_3COCH_2CO_2C_2H_5 \xrightarrow[-HF]{DAST} \left[CH_3\overset{O\text{-}SFNR_2\cdots F}{C}{=}CHCO_2C_2H_5 \right] \longrightarrow$$

$$CH_3COCHFCO_2C_2H_5 \xrightarrow[-HF]{DAST} CH_3CF{=}CFCO_2C_2H_5$$

(*cis*:*trans* 1:1; 40–60%)

Conversion of Carboxylic Acids into Acyl Fluorides

The formation of acyl fluorides from carboxylic acids proceeds by a mechanism analogous to that of the reaction of alcohols with dialkylaminosulfur trifluorides. Conversion of a carboxy group into a trifluoromethyl group is reported in only one case,[9] and the replacement of the carbonyl oxygen probably occurs by a mechanism analogous to that for aldehydes or ketones.

Conversion of Halo Compounds into Fluorides

Replacement of reactive halogens by fluorine using aminofluorosulfuranes is a metathetical exchange of halogens as proved by isolation of morpholinosulfur chlorodifluoride from the reaction of morpholinosulfur trifluoride with arylsulfinyl chlorides and benzotrichloride.[39]

$$\mathrm{ArSOCl + O(CH_2CH_2)_2NSF_3 \longrightarrow ArSOF + O(CH_2CH_2)_2NSClF_2}$$

Conversion of Sulfoxides into α-Fluorosulfides

The reaction of DAST with dialkyl or aryl alkyl sulfoxides having at least one α-hydrogen atom gives α-fluorosulfides. The proposed mechanism resembles that of the Pummerer rearrangement.[40]

$$\mathrm{RCH_2S(=O)R^1 + (C_2H_5)_2NSF_3 \xrightarrow{-F^-} [R{-}CH(H){-}\overset{+}{S}(R^1){-}O{-}SF_2{-}N(C_2H_5)_2] \longrightarrow}$$

$$\mathrm{RCH{=}\overset{+}{S}R^1 + F^- \longrightarrow RCHFSR^1 + (C_2H_5)_2NSOF}$$

$$\mathrm{+\ (C_2H_5)_2\overset{+}{N}H{-}SOF}$$

STEREOCHEMISTRY

The steric course of the replacement of hydroxy groups by fluorine by means of aminofluorosulfuranes is not straightforward. On one hand, partial or complete skeletal rearrangement during the fluorination can be best accounted for by assuming ionic or ion-pair mechanisms.[5] Dehydration, which is sometimes very extensive, can also be explained by the intermediate formation of carbocations since it is frequently accompanied by skeletal rearrangements.[5,32,41] In chiral compounds, such reactions should cause partial or complete racemization; however racemization during the reactions of chiral hydroxy compounds with DAST and its analogs has not been explicitly reported. On the contrary, the replacements of hydroxy groups by fluorine are claimed to occur with complete inversion or complete retention of configuration. Treatment of (+)-(*S*)-2-octanol with DAST gives, in addition to octenes, (−)-(*R*)-2-fluorooctane of 97.6% optical purity.[31] Complete inversions have further been reported in the preparation of dimethyl fluoromalate[42] and in the syntheses of fluorinated carbohydrates,[29,33,43-46] steroids,[32,47,48] and other natural products.

On the other hand, complete retention of configuration is observed in many reactions of hydroxy compounds with aminofluorosulfuranes. In most examples of retention, participation of a neighboring group is probably responsible for this result. An example is the conversion of 3-hydroxy-Δ^5-steroids into 3-fluoro-Δ^5-steroids.[32] Thus 3β-cholestanol gives 3α-fluorocholestane, whereas cholesterol affords 3β-fluorocholest-5-ene.[32]

DAST

$(C_2H_5)_2NSF_2O$

$(C_2H_5)_2NSF_2O^-$

F

HO

A similar result is obtained in the conversion of vitamins D into fluorovitamins D, where the configuration may be preserved by temporary formation of a six-membered ring with the participation of the carbonyl oxygen of the acetate group.[49]

DAST

AcO OH

O + O

CH_3

$^-OSF_2N(C_2H_5)_2$

AcO F

Another example of retention of configuration as a result of neighboring group participation is the reaction of aminofluorosulfuranes with 11β-hydroxysteroids with different substituents in position 9. When the 9α substituent is hydrogen or fluorine, piperidinosulfur trifluoride or DAST cause dehydration to 9,11-(**9**) or 11,12-unsaturated steroids (**10**), respectively.[30] When the 9α substituent is chlorine, the 11β-hydroxy group is replaced by fluorine with retention of configuration because of the intermediate formation of a chloronium ion.[30]

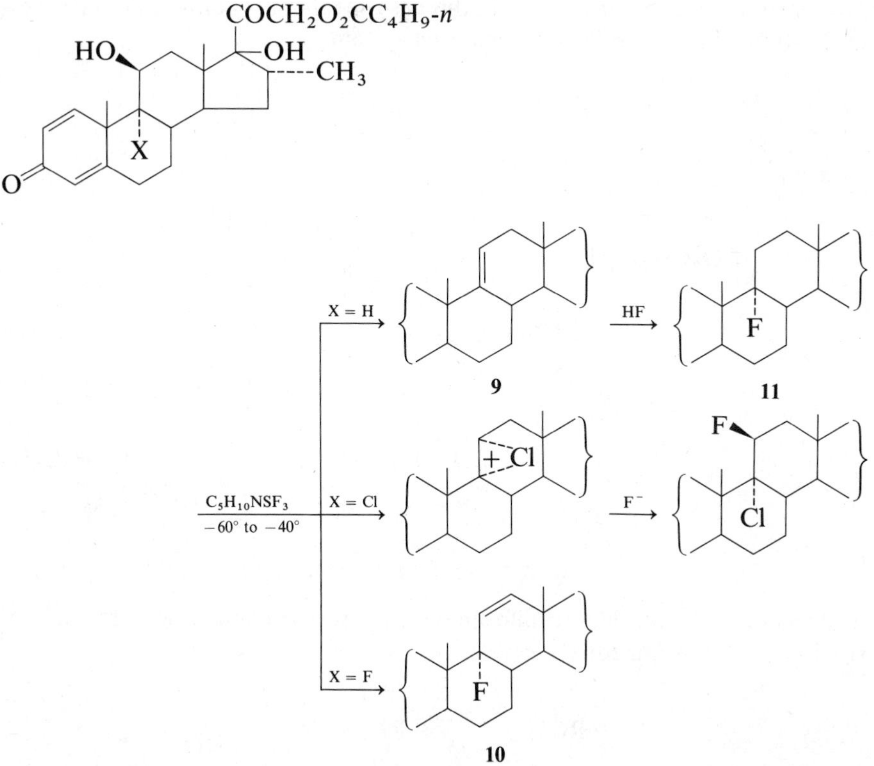

In 11-α-hydroxy steroids with hydrogen in 9α position, the 9α-fluoro derivatives **11** are obtained in low yield, probably resulting from the addition of hydrogen fluoride across the 9,11 double bond in the 9,11 dehydrated products **9**.[48]

It can be inferred that in compounds in which substituents in the vicinity of hydroxy groups are capable of forming short-lived cyclic intermediates, the replacement of hydroxyl by fluorine takes place with retention of configuration.

There are a few examples of retention of configuration in compounds in which neighboring group participation cannot be invoked. Treatment of 6α-hydroxy-5α-androstan-17-one with DAST affords 6α-fluoro-5α-androstan-17-one:[48]

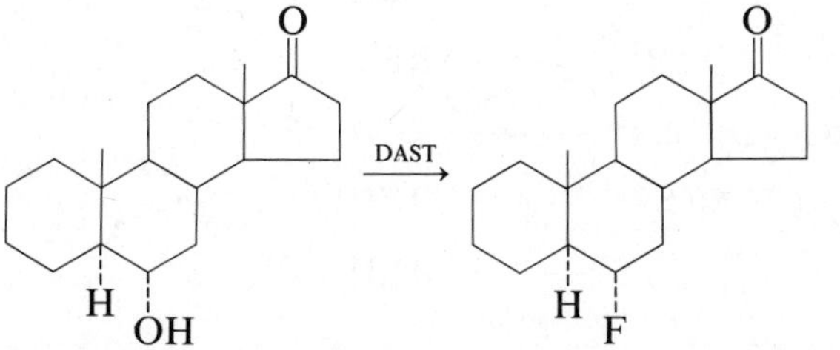

(86%)

The hydroxy group in position 7 of 10β-benzyloxycarbonyl-2-oxo-7α-hydroxy-1β,8β-dimethylgibbane-1α,4aα-carbolactone is replaced by fluorine with retention of configuration;[50] in this case, inversion would lead to an excessively strained *trans*-bridged bicyclic ring system.

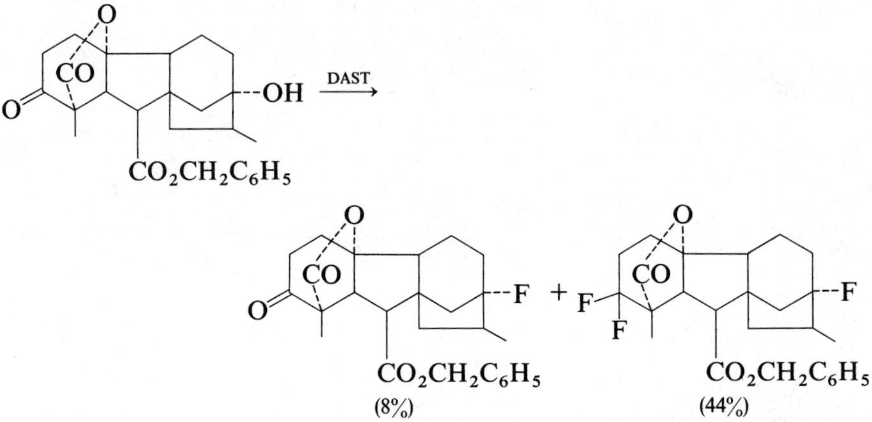

DAST

$CO_2CH_2C_6H_5$ (8%) + $CO_2CH_2C_6H_5$ (44%)

SCOPE AND LIMITATIONS

Reactions of aminofluorosulfuranes with organic compounds are summarized in the following equations:

$$\mathrm{ROH} \xrightarrow[(\mathrm{R_2N})_2\mathrm{SF_2}]{\mathrm{R_2NSF_3\ or}} \mathrm{RF}$$

$$\mathrm{RCOCH_2R^1} \longrightarrow \mathrm{RCF_2CH_2R^1 + RCF{=}CHR^1}$$

$$\mathrm{RCO_2H} \longrightarrow \mathrm{RCOF\ (+RCF_3)^*}$$

$$\mathrm{RSO_3H} \longrightarrow \mathrm{RSO_2F}$$

$$\mathrm{R_2P(O)OH} \longrightarrow \mathrm{R_2P(O)F}$$

$$\mathrm{RCOCl} \longrightarrow \mathrm{RCOF}$$

$$\mathrm{RSOCl} \longrightarrow \mathrm{RSOF}$$

$$\mathrm{RSO_2Cl} \longrightarrow \mathrm{RSO_2F}$$

$$\mathrm{R_3SiCl} \longrightarrow \mathrm{R_3SiF}$$

$$\mathrm{ROSiR^1{}_3} \longrightarrow \mathrm{RF}$$

$$\mathrm{RCH(OR^1)SR^2} \longrightarrow \mathrm{RCH(OR^1)F}$$

$$\mathrm{RCH_2S(O)R^1} \longrightarrow \mathrm{RCHFSR^1}$$

$$\mathrm{(C_6H_5)_3P\ or\ (C_6H_5)_3PS} \longrightarrow \mathrm{(C_6H_5)_3PF_2}$$

* Only one example of the conversion of a carboxy into a trifluoromethyl group is recorded.[9]

Dialkylaminosulfur trifluorides also add to poly- and perfluorinated alkenes (Eq. 1)[21]

$$CF_2{=}CFCF_3 + R_2NSF_3 \xrightarrow{CsF} (CF_3)_2CFSF_2NR_2 \qquad \text{(Eq. 1)}$$

and bis(dialkylamino)sulfur difluorides react with trimethylsilyl isocyanates according to Eq. 2:

$$(R_2N)_2SF_2 + (CH_3)_3SiNCO \longrightarrow (R_2N)_2S{=}NCOF + (CH_3)_3SiF \quad \text{(Eq. 2)}$$

These reactions of aminofluorosulfuranes are not discussed in this chapter, which is limited to transformations that introduce fluorine into organic compounds in place of oxygen, sulfur, or halogen. However, dehydrations that take place during reactions of hydroxy compounds with aminofluorosulfuranes are mentioned.

Apart from differences in the reactivity between dialkylaminotrifluorosulfuranes and the less reactive bis(dialkylamino)difluorosulfuranes,[4] there are small differences among individual aminofluorosulfuranes as to the results of their reactions. Thus, piperidinosulfur trifluoride gives consistently higher yields than DAST in fluorinations of 3-(1-hydroxyethyl)-4-benzyloxycarbonylmethyl-2-azetidinone[51] and causes more dehydration than DAST in the reaction with 9α-fluoro-16β-methyl-12,17α,21-trihydroxypregna-1,4-diene-3,20-dione 17α,21-dipropionate.[52] In the preparation of bis(α-fluoro)alkyl ethers by fluorination of α-haloaldehydes with dimethylaminosulfur, diethylaminosulfur, and morpholinosulfur trifluorides, slight differences in the ratios of diastereomers are observed.[36]

Reactions with Alcohols

The principal application of DAST and other dialkylaminosulfur trifluorides and bis(dialkylamino)sulfur difluorides is in the conversion of alcohols into monofluorides. In this field aminofluorosulfuranes are superior to sulfur tetrafluoride, which requires elevated temperatures and reacts with only relatively acidic hydroxy compounds. DAST and its analogs react with all types of hydroxy compounds at temperatures well below room temperature, sometimes at −78°. Primary, secondary, tertiary, allylic, and benzylic alcohols are converted into fluorides in high yields.[5,9] Lewis-acid catalysis of these reactions has not been observed, in contrast to fluorinations with sulfur tetrafluoride. Carbocation rearrangements occur, although to a lesser extent than with other fluorinating agents.[5] Thus isobutyl alcohol gives 49% of isobutyl fluoride and only 21% of *tert*-butyl fluoride. However, both borneol and isoborneol (*endo*-bornanol and *exo*-bornanol) rearrange to the same 3-fluoro-2,2,3-trimethylbicyclo[2.2.1]heptane (72–74%) accompanied by 17–18% of camphene.[5]

A pinacol rearrangement occurs when DAST reacts with *trans*-7,8-dihydroxy-7,8,9,10-tetrahydrobenzo[*a*]pyrene (**12**). The product, 7-(difluoromethyl)-8,9-dihydro-7*H*-cyclopenta[*a*]pyrene (**13**), suffers another rearrangement to 7-fluorobenzo[*a*]pyrene (**14**) on treatment with 2,3-dichloro-5,6-dicyano-1,4-benzoquinone (DDQ).[53]

HO, OH, **12** —DAST, CH_2Cl_2, −78°, 30 min, 0°→ CHF_2, **13** (40%) —DDQ→ F, **14** (80%)

Allylic rearrangements occur in the reactions of allylic alcohols with aminofluorosulfuranes.[5,9,35] Rearrangement of crotyl alcohol and isocrotyl alcohol to isocrotyl and crotyl fluorides, respectively, is slightly affected by solvents, with more polar solvents such as diglyme causing more rearrangement than nonpolar solvents such as isooctane.[5]

$$CH_3CH{=}CHCH_2OH \xrightarrow{DAST} CH_3CH{=}CHCH_2F + CH_3CHFCH{=}CH_2$$

isooctane	(36%)	(64%)
diglyme	(28%)	(72%)

$$CH_3CHOHCH{=}CH_2 \xrightarrow{DAST} \text{''} + \text{''}$$

isooctane	(9%)	(91%)
diglyme	(22%)	(78%)

Skeletal rearrangements are observed in the reaction of DAST with cholest-5-en-3β,19-diol 3-acetate (**15**) and 6β-hydroxymethyl-19-norcholest-5(10)-en-3β-ol 3-acetate (**16**).[41]

A similar rearrangement involving a ring size change occurs in 17β-ethoxycarbonylmethyl-1β,3β-dihydroxyandrost-5-ene 3-acetate (**17**) to afford derivative **18** with fused five- and seven-membered rings.[54]

HO AcO 15 DAST CH_2Cl_2 AcO F (10% from **15**, 23% from **16**) AcO (31–32%) AcO (12–14%) AcO OH 16

$CO_2C_2H_5$ OH AcO 17 DAST + AcO AcO + AcO F **18** (80%)

An interesting rearrangement takes place with some saccharide derivatives in which a free hydroxylic group is adjacent to an acetal oxygen and to an azido group.[54a,b] With the participation of the neighboring groups, either the replacement of the hydroxy group occurs with retention of configuration, or an exchange of the neighboring substituents takes place with inversion on both chiral centers involved. If the acetal oxygen participates, the hydroxy group is replaced by fluorine (path a), or an ether of a glycosyl fluoride is formed (path b). If azide nitrogen participates, the fluorine replaces the azide group (path c).

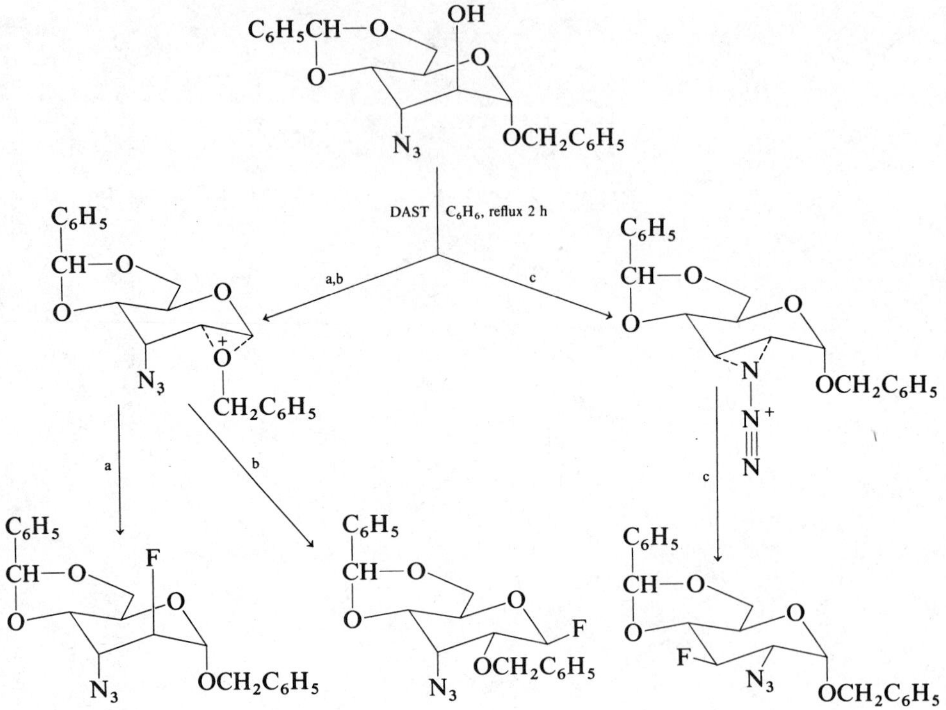

A similar rearrangement occurs when *N*,*N*-dibenzyl-L-serine benzyl ester (**19**) is treated with DAST. *N*,*N*-Dibenzyl-α-fluoro-β-alanine benzyl ester (**21**) is obtained by an intramolecular nucleophilic displacement via an aziridine intermediate. Compound **21** is also obtained by treatment of *N*,*N*-dibenzyl-D-isoserine benzyl ester (**20**) with DAST.[55]

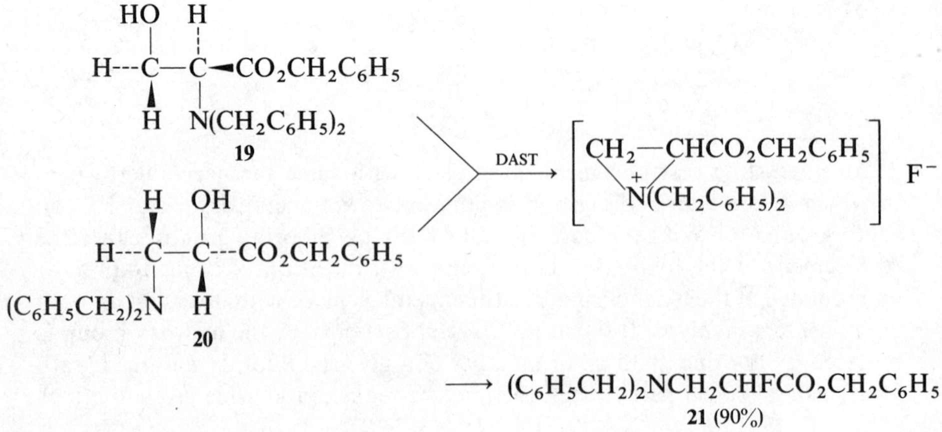

$$\longrightarrow (C_6H_5CH_2)_2NCH_2CHFCO_2CH_2C_6H_5$$

21 (90%)

On the other hand, fluorination of derivatives of the methyl homolog of serine occurs with retention of configuration: threonine derivatives afford *threo* products, and allothreonine derivatives afford *erythro* products.[55]

Dehydration to olefins, which sometimes accompanies the reaction of alcohols with aminofluorosulfuranes, is seldom as extensive as with other agents capable of replacing hydroxy groups by fluorine.[5] Dehydration accompanied by Wagner–Meerwein rearrangement occurs during the fluorination of testosterone; 18-nor-17-methyl-4,13-androstadien-3-one (**23**) is isolated in addition to 4-androsten-17α-fluoro-3-one (**22**).[32]

DAST →

22 (40%) + **23** (32%)

In a few cases, dehydration occurs to the exclusion of fluorination; thus 9α-fluoro-11-hydroxysteroids give 9α-fluoro-Δ^{11}-steroids.[30,52,56]

Intermolecular dehydration to form ethers in addition to fluorides is observed in the reaction of DAST with benzhydryl alcohols.[57]

$$(C_6H_5)_2CHOH \xrightarrow[CH_2Cl_2,\ -30°,\ 30\ min]{DAST} \underset{(40\%)}{(C_6H_5)_2CHF} + \underset{(44\%)}{[(C_6H_5)_2CH]_2O}$$

Reaction of polyhydroxy compounds, such as saccharides, with DAST replaces one or two hydroxy groups by fluorine. More fluorine atoms are not introduced even when up to 6 equivalents of DAST are used.[43,58]

Halogenated alcohols are converted into halofluorides without replacement of halogen by fluorine.[5,9,59]

The high degree of reactivity of DAST and its analogs toward hydroxy groups provides a method for selective conversion of hydroxy ketones into fluoroketones.[32,47,48,60] Many hydroxyketosteroids are transformed into fluoroketosteroids by treatment with DAST at room temperature.[32,47,48]

Examples of fluorinations of hydroxy aldehydes are lacking. Judging from the ease of replacement of aldehydic oxygen by fluorines, it can be anticipated that selective replacement of a hydroxy group in hydroxy aldehydes will be very difficult, if not impossible. Hydroxy acids, treated with an excess of DAST, form fluoroacid fluorides, which on workup in an aqueous medium give fluoroacids.[61] Hydroxy esters[5,9,42,62,63] and hydroxy amides [64–66] are converted into fluoroesters and fluoroamides, respectively, since DAST and its analogs do not react with the carbonyl group of carboxylic acid derivatives.

Reactions with Aldehydes

Reaction of aldehydes with dialkylaminosulfur trifluorides (use of a bis(dialkylamino)sulfur difluoride is reported in one example[4]) affords geminal difluoro compounds in moderate to high yields. The reactions are usually carried out in aprotic solvents, most frequently in dichloromethane, using 1 mol or rarely an excess of the reagent at temperatures ranging from room temperature to 80°.

Aliphatic,[5,8] aromatic,[5,8] and heterocyclic[60,67,68] aldehydes are converted into 1,1-difluoro compounds even in the presence of other functional groups except hydroxy groups. Such groups in hydroxy aldehydes and hydroxy ketones must be protected since they react preferentially with the fluorinating agents. Hydroxy groups in saccharides can be protected by formation of acetonides.[69,70] Aldedyde oxygen reacts in preference to ketone oxygen. Many steroidal ketoaldehydes are selectively fluorinated at the aldehyde group.[71,72] In cephalosporins, DAST replaces the aldehydic oxygen by two fluorines without affecting the ester and the amide groups, which are inert toward this reagent.[67]

Deviations from the regular reaction course are rare. Occurrence of a rearrangement with pivalaldehyde and the formation of ethers from polyhaloacetaldehydes are mentioned in the mechanism section.

Reactions with Ketones

The reaction of ketones with DAST (thus far, other aminofluorosulfuranes have not been used with ketones) parallels that of aldehydes. It is usually carried out in solvents at temperatures ranging from 25 to 80°, using 1 mol or a small excess of the reagent (in rare cases up to 3 mol). Yields as high as 98% are reported. The formation of vinyl fluorides as a possible side reaction is discussed in the section entitled "Mechanism."

If a vinyl fluoride is the desired product, its yield can be maximized by carrying out the fluorination in glyme in the presence of fuming sulfuric acid;[37] any difluoride formed can be converted into the vinyl fluoride with alumina.[73]

$C_6H_5CO_2$–(cyclohexanone) $\xrightarrow[\text{25°, 48 h}]{\text{DAST, glyme; } H_2SO_4,\ SO_3}$ $C_6H_5CO_2$–(1-fluorocyclohexene) (49%) + $C_6H_5CO_2$–(1,1-difluorocyclohexane) (10%)

As may be expected, the hydroxy group is replaced preferentially in hydroxy ketones.[50]

In ketoaldehydes, DAST reacts preferentially with the aldehyde group[71,72] even with 8 equivalents of DAST.[71] In keto esters[74–77a] and keto amides[38,78] only the ketone carbonyl oxygen is replaced with two fluorine atoms since esters and amides do not react with aminofluorosulfuranes.

A number of β-ketoesters[38a,b] and β-diketones[38b,c] react with DAST to give α,β-difluoro-α,β-unsaturated esters or ketones in moderate to good yields. A possible mechanism for this oxidative fluorination has been discussed. The scope of the reaction has not yet been determined.

$$CH_3COCH_2COCH_3 \xrightarrow[N\text{-methylpyrrolidinone}]{\text{DAST, } -70^\circ \text{ to } 25^\circ} CH_3CF{=}CFCOCH_3$$
(*cis*:*trans* 1:1; 40–60%)

Reactions with Epoxides (Oxiranes)

Epoxides (oxiranes) react with diethylaminosulfur trifluoride in ways depending on their structures. Cyclopentene oxide and cyclohexene oxide give mixtures of *cis*-difluorides and bis(α-fluoro) ethers. Styrene oxide affords a mixture of 1,1-difluoro- and 1,2-difluoro-2-phenylethane, and both *cis*- and *trans*-stilbene oxides give, albeit in very poor yields, mixtures of *meso*- and racemic 1,2-difluoro-1,2-diphenylethanes and 1,1-difluoro-2,2-diphenylethane. Cyclooctene oxide and cyclohexene sulfide do not react appreciably under the conditions used (equimolar ratio, no solvent, 50–80°).[78a]

$$\text{RCH—CHR (epoxide, bridged by O)} \xrightarrow{\text{DAST}} \text{RCHF—CHFR} + (\text{RCHF—CHR})_2\text{O}$$

Reactions with Carboxylic and Other Acids

Dialkylaminosulfur trifluorides react with carboxylic acids in ether, dichloromethane, or benzene at 0–20° to give acyl fluorides in yields of 70–96%.[8,9] Conversion of a carboxy group into a trifluoromethyl group is reported in only one instance when benzoic acid is heated with DAST in the presence of sodium fluoride at 80° for 20 hours.[9] Equally exceptional is replacement by two fluorines of the oxygen in a carbonyl group of some lactones (p. 538).[78b]

The fact that aminofluorosulfuranes do not convert carboxy groups into trifluoromethyl groups under mild conditions differentiates these reagents from sulfur tetrafluoride and makes them useful for the replacement of alcoholic hydroxy groups by fluorine in the presence of carboxy groups in the same molecule. An example is the conversion of mandelic acid into α-fluorophenylacetic acid in 68% yield.[61] The reaction requires 2 moles of the dialkylaminosulfur trifluoride since both alcoholic and carboxylic hydroxy groups are replaced by fluorine. Subsequent treatment of the reaction mixture with water transforms the fluoroacyl fluoride into the fluorinated acid.[61]

A rather exceptional example of the generation of a trifluoromethyl group by means of DAST under mild conditions is the conversion of tetraethylthiuram disulfide, a derivative of a dithiocarbamic acid, into diethyltrifluoromethylamine.[8]

$$[(C_2H_5)_2NC({=}S)S]_2 \xrightarrow{\text{DAST}} (C_2H_5)_2NCF_3 \quad (70\%)$$

Like carboxylic acids, sulfonic and phosphonic acids are transformed into acid fluorides by dialkylaminosulfur trifluorides: *p*-toluenesulfonic acid gives *p*-toluenesulfonyl fluoride at 80°,[79] and dibenzylphosphonic acid affords the fluoride at 20°.[8]

Reactions with Lactones

Certain α-hydroxylactones react with DAST under surprisingly mild conditions to give α-fluorolactones with inversion of configuration as well as products in which the carbonyl group has been replaced by two fluorine atoms. Lactones with hydrogen or fluorine in the α-position do not react with DAST. A possible mechanism is shown below.[78b]

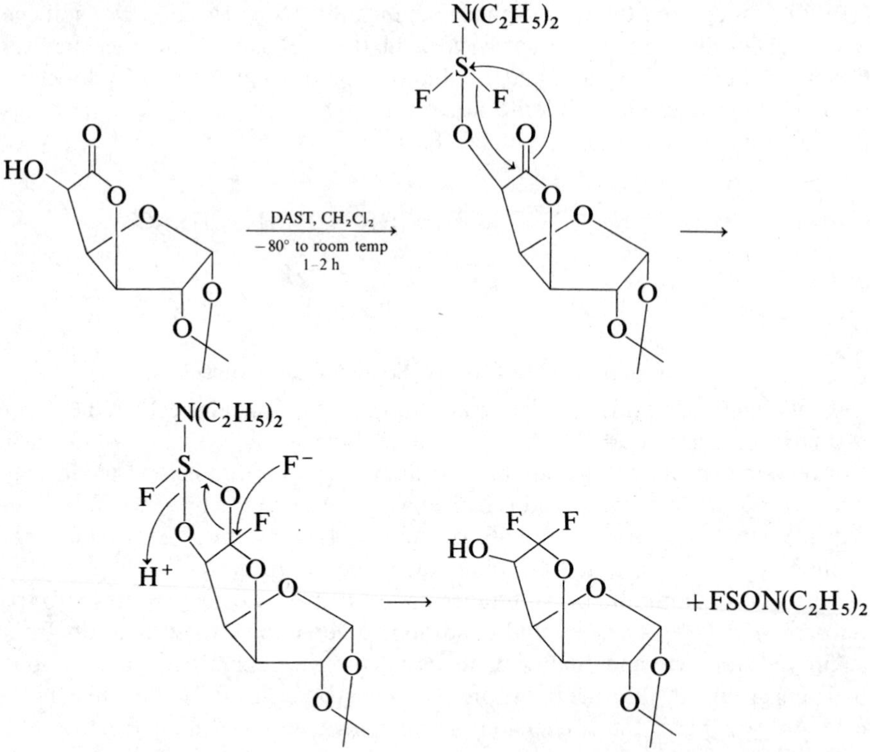

DAST, CH_2Cl_2
−80° to room temp
1–2 h

$N(C_2H_5)_2$

$+ FSON(C_2H_5)_2$

(45%)

Reactions with Halides and Sulfonates

Other halogens, if sufficiently reactive, can be replaced by fluorine using dialkylaminosulfur trifluorides. With a few exceptions, only iodides, allylic and benzylic bromides, and chlorides of carboxylic, sulfinic, sulfonic, and phosphonic acids are fluorinated at temperatures of 20–60°. Trimethylsilyl fluoride[39] and thionyl fluoride[39] can be prepared from the corresponding chloro compounds. Examples of the fluorination of relatively unreactive halo compounds are the conversion of diethyltrichloromethylamine into diethyltrifluoromethylamine by pyrrolidinosulfur trifluoride[39] and the formation of *tert*-butyliminosulfur difluoride from the corresponding dichloride.[39]

$$t\text{-}C_4H_9CN{=}SCl_2 \xrightarrow[60^\circ]{R_2NSF_3} t\text{-}C_4H_9CN{=}SF_2 \qquad (62\%)$$

Tris(dialkylamino)sulfonium difluorotrimethylsilicates, on the other hand, are much more reactive; thus TASF* fluorinates primary chlorides and converts deuteriochloroform into deuteriodichlorofluoromethane.[7] The trifluoromethanesulfonate of 2,3:4,5-di-*O*-isopropylidene-D-fructopyranose (**24**, R = CF_3SO_3) reacts with TASF in refluxing tetrahydrofuran overnight to give the deoxyfluorosugar **24** (R = F) in 80% yield. Direct fluorination of compound **24** (R = OH) with DAST was unsuccessful.[80]

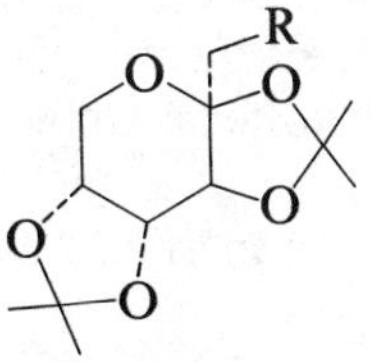

24

Another compound capable of replacing a reactive halogen or a trifluoromethanesulfonyloxy group by fluorine is tris(dimethylamino)sulfonium trifluoromethoxide ($TAS^+CF_3O^-$), prepared by treatment of tris(dimethylamino)-sulfonium difluorotrimethylsilicate with carbonyl fluoride.[80a] However, the reaction is not clear-cut since trifluoromethyl ethers are formed concomitantly with fluorides.[80b]

$$[(CH_3)_2N]_3S^+(CH_3)_3SiF_2^- + COF_2 \longrightarrow [(CH_3)_2N]_3S^+CF_3O^- \ (TAS^+CF_3O^-)$$

$$R^1R^2CHX \xrightarrow{TAS^+CF_3O^-} R^1R^2CHF + R^1R^2CHOCF_3$$
$$X = Br, CF_3SO_3$$

Reactions with Sulfoxides

DAST can be used for an unusual general transformation of dialkyl and alkyl aryl sulfoxides that contain at least one α-hydrogen atom into α-fluoroalkyl sulfides.[40]

$$(CH_3)_2SO \xrightarrow[CHCl_3,\ 25^\circ,\ 16\ h]{DAST} CH_3SCH_2F \qquad (>83\%)$$

This reaction, whose mechanism resembles that of the Pummerer rearrangement, is strongly catalyzed by anhydrous zinc iodide.

* Tris(dimethylamino)sulfonium difluorotrimethylsilicate

Reactions with Hemithioacetals

Another unusual reaction of DAST is the replacement of a phenylthio group in saccharide hemithioacetals by fluorine. The reaction requires the presence of *N*-bromosuccinimide and takes place at 0–25°. The transformation seems to be stereospecific; its mechanism is not well understood.[81]

DAST, NBS, 0-25° → (91%)

Reactions with Alkyl Silyl Ethers

Dialkylaminosulfur trifluorides react readily with trimethylsilyl ethers of α,α,ω-trihydroperfluoroalcohols to form stable polyfluoroalkoxydialkylaminodifluorosulfuranes that decompose at higher temperatures to dialkylaminosulfinyl fluorides and α,α,ω-trihydroperfluoroalkanes.[82]

$$HCF_2CF_2CH_2OSi(CH_3)_3 + O(CH_2CH_2)_2NSF_3 \xrightarrow[\text{0-5°, 10 min; 20°, 30 min}]{(C_2H_5)_2O}$$

$$HCF_2CF_2CH_2OSF_2N(CH_2CH_2)_2O \xrightarrow[\text{0.05 mm}]{90°}$$

$$\underset{(85\%)}{HCF_2CF_2CH_2F} + \underset{(90\%)}{O(CH_2CH_2)_2NS(O)F}$$

The exceptional stability of the primary reaction products is evidently due to the presence of strongly electron-withdrawing substituents that do not favor the formation of a carbocation.

Trimethylsilyl ethers of cyanohydrins react with DAST to give α-fluoronitriles in yields that are higher than those obtained from the cyanohydrins themselves.[83]

Reactions with Phosphorus Compounds

Dialkylaminosulfur trifluorides are used for the preparation of triphenylphosphine difluoride (difluorotriphenylphosphorane) from both triphenylphosphine and triphenylphosphine sulfide.[8]

$$(C_6H_5)_3P \xrightarrow{R_2NSF_3} (C_6H_5)_3PF_2 \quad (93\%)$$

$$(C_6H_5)_3PS \xrightarrow{R_2NSF_3} (C_6H_5)_3PF_2 \quad (82\%)$$

Applications of aminofluorosulfuranes in the fluorination of various classes of compounds are collected in Table F, which should be helpful in the location of fluorination in a specific area of chemistry.

TABLE F. LIST OF REFERENCES BY SUBSTRATE

Carbohydrates	13, 29, 33, 43, 44, 45, 46, 54a,b, 58, 63, 69, 70, 80, 80b, 81, 84–90
Steroids	30, 32, 37, 41, 47, 48, 49, 52, 54, 56, 71, 72, 77, 77a, 91–109
Terpenes	50, 62, 74, 110–118
Amino Acids	55, 76, 119–122
Heterocycles	38, 60, 64, 65, 66, 67, 68, 78, 121, 123
Antibiotics	67, 68, 124–130

Side Reactions

Fluorinations with aminofluorosulfuranes are sometimes accompanied by rearrangements, loss of water, or loss of hydrogen fluoride. Several types of rearrangements have been discussed in the preceding sections.

The majority of the nonfluorinating side reactions are dehydrations to olefins. These accompany conversion of alcohols into fluorides[5,9] and may become the only reaction with certain hydroxy compounds. Exclusive dehydration is observed in a number of 9α-fluoro-11β-hydroxy steroids, which always give nonfluorinated products with an 11,12 double bond.[30,52,56] It is interesting that the 9α-chloro-11β-hydroxy analogs are converted into 9α-chloro-11β-fluoro steroids,[95] whereas the halogen-free compounds are dehydrated to $\Delta^{9(11)}$ steroids.[30] When intermediate **25** of these reactions carries a hydrogen on C-9, elimination of a proton via a carbocation gives the $\Delta^{9(11)}$ product **26**. If C-9 bears chlorine, a chloronium ion **27** bridging C-9 and C-11 is formed; attack by fluoride leads to the 11β-fluoride **28** with retention of the original configuration. With fluorine on C-9, formation of a positive charge is disfavored, and the $C_5H_{10}NSF_2O$ group is eliminated via a six-membered transition state to give the Δ^{11} steroid **29**.[30]

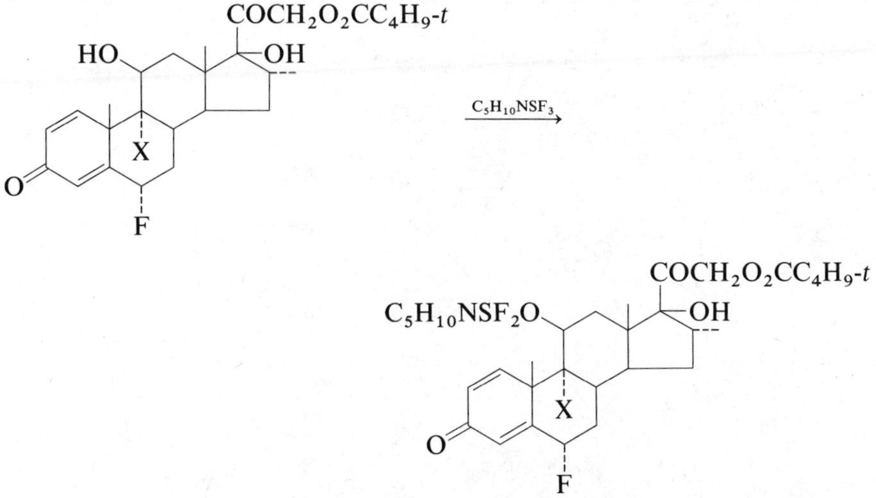

$COCH_2O_2CC_4H_9$-*t*
HO OH
X
O
F

$\xrightarrow{C_5H_{10}NSF_3}$

$COCH_2O_2CC_4H_9$-*t*
$C_5H_{10}NSF_2O$ OH
X
O
F

25

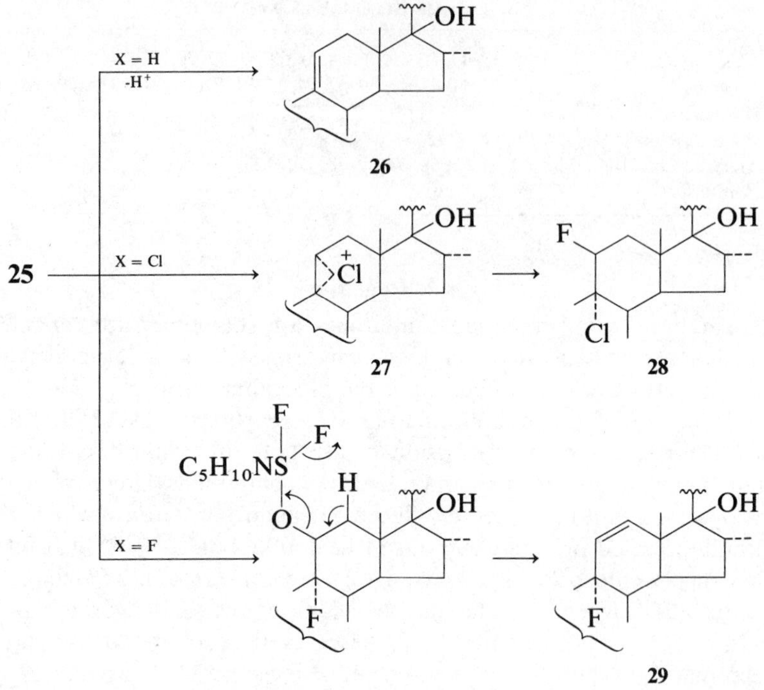

X = H
-H⁺

26

25

X = Cl

27

28

$C_5H_{10}NS$

X = F

29

Regarding other dehydrations, the reaction of DAST in the presence of pyridine with two diastereomeric saccharides should be mentioned. Whereas 1,2:5,6-di-*O*-isopropylidene-α-D-allofuranose (**30**) gives the corresponding fluorodeoxysugar **31** with inversion of configuration, 1,2:5,6-di-*O*-isopropylidene-α-D-glucofuranose (**32**) undergoes exclusive dehydration to product **33**.[29]

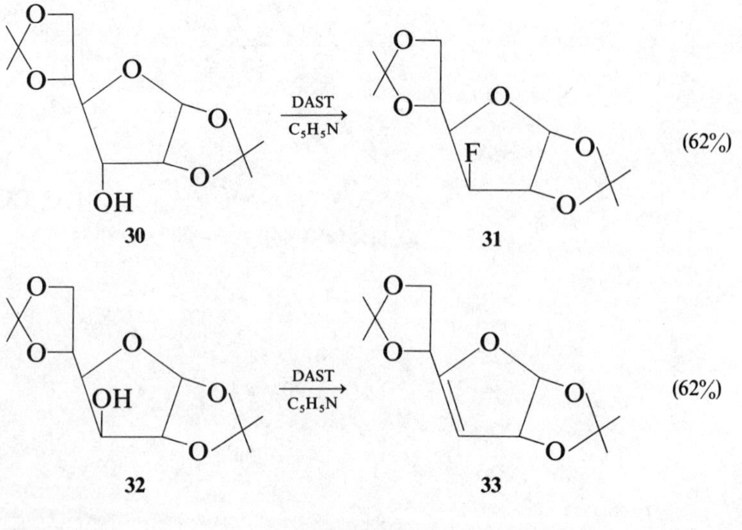

DAST / C_5H_5N (62%)

30 31

DAST / C_5H_5N (62%)

32 33

With several benzhydryl alcohols, dehydrations result in the formation of ethers by intermolecular displacement of the R_2NSF_2O group.[57] Intramolecular displacement of the R_2NSF_2O group in the reaction of aminofluorosulfuranes with alcohols containing nucleophilic substituents such as trialkylsilyloxy groups,[116] *tert*-butoxycarbonylamino groups,[126] or benzyloxycarbonylamino groups[128] gives rise to ethers and oxazolidinones, respectively.

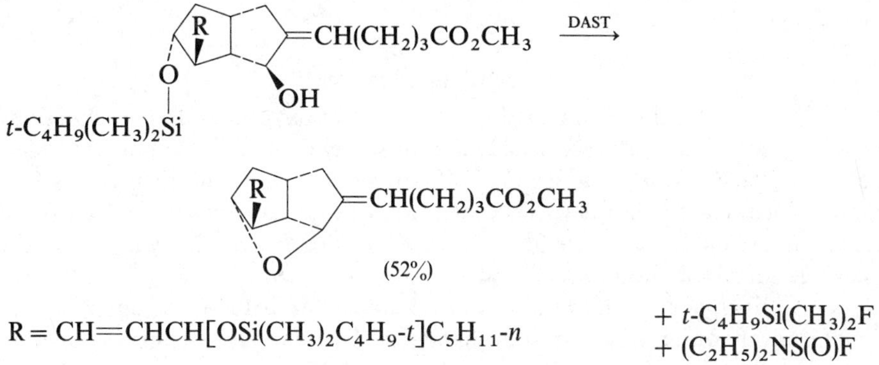

$=CH(CH_2)_3CO_2CH_3$ $\xrightarrow{\text{DAST}}$ OH

$t\text{-}C_4H_9(CH_3)_2Si$

$=CH(CH_2)_3CO_2CH_3$ (52%)

$R = CH{=}CHCH[OSi(CH_3)_2C_4H_9\text{-}t]C_5H_{11}\text{-}n$ + $t\text{-}C_4H_9Si(CH_3)_2F$ + $(C_2H_5)_2NS(O)F$

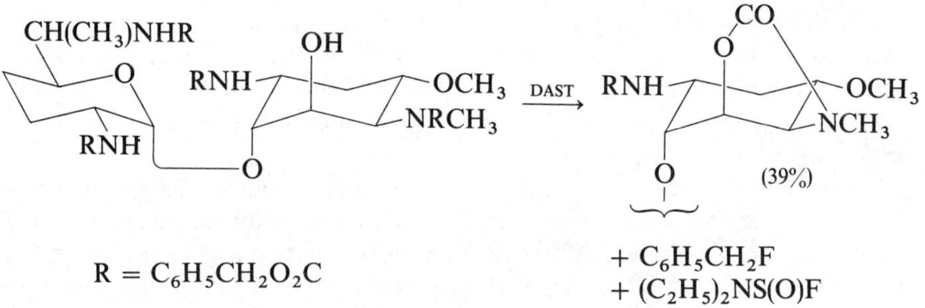

$CH(CH_3)NHR$ O RNH OH RNH OCH_3 $NRCH_3$ O $\xrightarrow{\text{DAST}}$ RNH CO O OCH_3 NCH_3 O (39%)

$R = C_6H_5CH_2O_2C$ + $C_6H_5CH_2F$ + $(C_2H_5)_2NS(O)F$

Vinyl fluoride formation is a side reaction in the fluorination of ketones, but not aldehydes, with dialkylaminosulfur trifluorides.[37,38,78] The effect of solvent polarity and the presence of fuming sulfuric acid on the product distribution has been discussed previously.

Unsuccessful Fluorinations

It is difficult to list reactions in which aminofluorosulfuranes failed to accomplish fluorinations because even if such failures are reported, they may escape indexing and a literature search. A few reported failures discovered in reading papers containing successful fluorinations are included in Tables I–III, VIII, and X. However, reports on unsuccessful fluorinations should not discourage chemists from attempting such reactions since they may succeed under different conditions.

COMPARISON WITH OTHER FLUORINATING AGENTS

Aminofluorosulfuranes are important additions to the host of reagents that convert alcohols into monofluorides and carbonyl compounds into geminal difluorides. They compare favorably with other fluorinating reagents used for this purpose.[131–135]

Hydrogen Fluoride

Hydrogen fluoride or its mixtures with amines (Olah's reagent, 70% hydrogen fluoride and 30% pyridine)[136] can be used with alcohols, but side reactions such as rearrangements and polymerization often occur in the strongly acidic medium.

Sulfur Tetrafluoride

Sulfur tetrafluoride dominated fluorinations of oxygen-containing functions for 25 years. It is useful for conversion of some alcohols into monofluorides, aldehydes and ketones into geminal difluorides, and acids and their derivatives into trigeminal trifluorides. The last application remains unchallenged by aminofluorosulfuranes since they react only with free carboxylic acids and convert them only into acyl fluorides.

One advantage of the liquid dialkylaminosulfur trifluorides over the gas sulfur tetrafluoride is their ease of handling. They are also definitely superior to sulfur tetrafluoride in reactions with alcohols. Sulfur tetrafluoride reacts at room temperature or slightly elevated temperatures only with very acidic alcohols such as nitro alcohols and polyfluoro alcohols, whereas normal alcohols require temperatures above 100°. DAST and related compounds convert primary, secondary, and tertiary alcohols into fluorides at temperatures as low as −78°.[5,9]

Sulfur tetrafluoride and aminofluorosulfuranes are comparable for the preparation of geminal difluorides from aldehydes and ketones, except that the former does not cause vinyl fluoride formation or cationic rearrangements. In contrast to sulfur tetrafluoride, dialkylaminosulfur trifluorides selectively convert hydroxy acids into fluorinated carboxylic acids,[61] and keto esters into geminal difluoro esters.[75]

Phenyltrifluorosulfurane

Phenyltrifluorosulfurane ($C_6H_5SF_3$) is used for the conversion of carbonyl compounds into geminal difluorides and carboxylic acids into trifluoromethyl compounds. However, the reaction temperatures are usually at least 100°,[137] although conversion of an aldehyde into a geminal difluoride at −25° has been reported.[67]

Fluoroalkylamino Reagents (FAR)

These reagents are prepared by addition of secondary amines to polyfluoroalkenes and perfluoroalkenes and are represented mainly by 2-chloro-1,1,2-trifluoroethyldiethylamine (Yarovenko–Raksha reagent)[138] and hexafluoroisopropyldiethylamine (Ishikawa reagent).[139] These α,α-difluoroamines replace only hydroxy groups in alcohols and carboxylic acids by fluorine but do not normally react with aldehydes and ketones.[131] In this respect, they are superior to sulfur tetrafluoride since they react with most alcohols at room temperature.

They are also more selective than aminofluorosulfuranes since they do not attack carbonyl groups. However, aminofluorosulfuranes can be used for selective replacement of hydroxy groups in hydroxy ketones as well.[50,61,95] The disadvantages of the α,α-difluoroamines compared with aminofluorosulfuranes are their lower storage stability and the greater difficulty in isolating products. Whereas the byproducts of fluorinations with aminofluorosulfuranes remain in the aqueous phase during workup, the byproducts of fluorination with α,α-difluoroamines (carboxamides) must be separated by distillation. Also, side reactions are more frequent with α,α-difluoroamines than with aminofluorosulfuranes,[110,113,131] although in some cases α,α-difluoroamines give higher yields than aminofluorosulfuranes.[31,41]

Selenium Tetrafluoride and Molybdenum Hexafluoride

These reagents convert carbonyl compounds into geminal difluorides in good yields under very mild conditions.[140,141] Like the aminofluorosulfuranes, they have the advantage over sulfur tetrafluoride of not requiring pressure equipment. It is therefore suprising that they are used so infrequently. Mixtures of selenium tetrafluoride with anhydrous hydrogen fluoride are used for the conversion of alcohols into monofluorides.[136]

Fluorophosphoranes

Fluorophosphoranes containing one to four fluorine atoms can replace alcoholic hydroxyl by fluorine, but usually only at temperatures well above 150°. A modified procedure, treatment of 2-tosyloxyoctane with methyltributylfluorophosphorane at 25°, or of trimethylsilyloxyoctane with phenyltetrafluorophosphorane at 80°, gives 2-fluorooctane in only 15 and 14% yields, respectively. 2-Chloro-1,1,2-trifluoroethyldiethylamine and DAST convert (+)-(*S*)-2-octanol into (−)-(*R*)-2-fluorooctane in 44 and 23% yields, respectively; the rest are octenes.[31]

Alkali and Tetraalkylammonium Fluorides

These salts displace the sulfonyloxy groups of alkyl *p*-toluenesulfonates, methanesulfonates, and trifluoromethanesulfonates by fluorine in good yields.[131] The reaction requires higher temperatures than does the fluorination of the free alcohols with aminofluorosulfuranes. Ester formation and fluorination can be carried out in one step.[142]

EXPERIMENTAL CONDITIONS

Fluorinations with aminofluorosulfuranes require no special apparatus. They may be carried out in glass equipment, which, however, may be superficially etched by the reaction byproducts. Polyethylene or polytetrafluoroethylene reaction vessels are therefore preferable.

Fluorinations with DAST and its analogs are most frequently carried out in dichloromethane, chloroform, fluorotrichloromethane, carbon tetrachloride, hexane, isooctane, benzene, toluene, ether, glyme, diglyme, or triglyme. It is essential that the solvents be anhydrous. On a few occasions the reagent has served as the solvent. Exclusion of atmospheric moisture is required, especially when the reactions are carried out at low temperatures. Use of an inert atmosphere such as nitrogen or argon is advisable.

Most fluorinations of alcohols are started at $-78°$ and finished by allowing the mixtures to warm to room temperature. Aldehydes and ketones are usually fluorinated between room temperature and 80°. Higher temperatures cannot be used because DAST decomposes above 85°.[143]

The reaction times vary over a wide range. Some fluorinations, especially of alcohols, take place almost instantaneously; others require days and even weeks. Most reactions are, however, completed within hours.

Moisture-sensitive products are isolated by distillation of the crude reaction mixtures. In the majority of cases, the reaction mixture is decomposed by pouring it into water, over ice, or into a solution of sodium bicarbonate. The treatment with ice is advisable, especially when a large excess of DAST or its analogs is used, since the decomposition of the unreacted reagent is very violent. Water hydrolyzes fluorine-containing byproducts to water-soluble acidic compounds. Neutralization of the reaction mixture, usually with sodium bicarbonate, is therefore necessary. After washing of the crude product or its solution with water, the product is isolated and purified by standard procedures.

Dimethylaminosulfur trifluoride, DAST, and TASF are commercially available from Aldrich Chemical Company and Alfa Products, Inc.; DAST is also available from Pfalz and Bauer Research Chemicals. They can be used without purification. If it is necessary to distill DAST, it should be kept in mind that a violent and even explosive decomposition may take place at temperatures above 50°. Consequently, pressures of 10 mm or less are required to ensure safe distillation. It is recommended that the distillation of DAST and its analogs be carried out behind a shield and in a closed hood.

EXPERIMENTAL PROCEDURES

Preparation of Reagents

Diethylaminotrimethylsilane $(C_2H_5)_2NSi(CH_3)_3$.[3] Trimethylchlorosilane (54 g. 0.5 mol) was added dropwise to 80 g (1.1 mol) of neat diethylamine at room temperature.[3] Separation of the salt by suction filtration and distillation of the filtrate afforded 75.6 g (70%) of diethylaminotrimethylsilane, bp 126.1–126.4° (750 mm), n_D^{20} 1.4112.[144]

Dialkylaminotrimethylsilanes.[145] A secondary amine (0.1 mol) was added dropwise to 0.1 mol of ethylmagnesium bromide in 100 mL of ether. The solution was allowed to stand for 1 hour at room temperature and then added during 30 minutes to a vigorously stirred mixture of 10.8 g (0.1 mol) of trimethylchlorosilane in 50 mL of ether. The reaction mixture was heated under reflux for 2 hours and allowed to stand for 15 hours at room temperature. The magnesium halides were removed by suction filtration, and the dialkylaminotrimethylsilanes were obtained by distillation. All operations were carried out under nitrogen or argon. Yields and boiling points (pressure in millimeters) of the products were as follows: $(CH_3)_2N$, 65%, 85–86° (760); $(C_2H_5)_2N$, 72%, 125–126° (760); $(C_3H_7)_2N$, 75%, 67.0° (26); C_4H_8N, 55%, 141–142° (760); $C_5H_{10}N$, 63%, 162° (760); OC_4H_8N, 38%, 57–60° (20).[145]

Dimethylaminosulfur Trifluoride.[5,9] A solution of 40 g (0.34 mol) of dimethylaminotrimethylsilane in 50 mL of fluorotrichloromethane was added dropwise to a solution of 20 mL (0.36 mol, measured at -78°) of sulfur tetrafluoride in 100 mL of fluorotrichloromethane at −65° to −60°. The reaction mixture was warmed to room temperature and distilled to give 37 g (82%) of dimethylaminosulfur trifluoride as a pale yellow liquid, bp 49–49.5° (33 mm), mp-79°.

Diethylaminosulfur Trifluoride (DAST). A detailed procedure is published in *Organic Syntheses.*[12] This compound decomposes violently at temperatures above 50°.[143,146,147] In this procedure, DAST containing ^{18}F is obtained by treatment of DAST with $H^{18}F$.[85,148]

Piperidinosulfur Trifluoride

From N-Piperidyltrimethylsilane.[3,10] A stainless-steel autoclave fitted with a magnetic stirrer was charged with 24 g (0.153 mol) of *N*-piperidyltrimethylsilane and was cooled in liquid nitrogen. Commercial (90–96%) sulfur tetrafluoride (27 g, 0.25 mol) was *slowly* condensed in the autoclave (with rapid condensation an exothermic reaction set in). After slow warming, the contents of the autoclave were magnetically stirred at room temperature for 10 hours. The unreacted sulfur tetrafluoride and trimethylfluorosilane were removed at 10 mm, and the residue was distilled at 43° (0.1 mm) to give piperidinosulfur trifluoride in 60% yield.

Caution: Piperidinosulfur trifluoride decomposes violently at 138°.

From N-p-Toluenesulfinylpiperidine.[16,149] A solution of 13.6 g (0.16 mol) of piperidine in 100 mL of anhydrous ether was added with stirring to a solution of 11.3 g (0.08 mol) of *p*-toluenesulfinyl chloride in 100 mL of anhydrous ether at −40°. After 1 hour, water was added to dissolve the salt, and the ether layer was separated, dried, and evaporated to dryness. The residue, 10.4 g (72% yield), mp 45–52°, was recrystallized from light petroleum to give *N-p*-toluenesulfinylpiperidine, mp 59–60°. A steel autoclave was charged with 22.3 g (0.1 mol) of *N-p*-toluenesulfinylpiperidine, cooled in a dry ice–acetone bath evacuated, and 16.2 g (0.15 mol) of sulfur tetrafluoride was condensed in it. The mixture was allowed to stand at 25° for 12 hours and then was fractionated to give *p*-toluenesulfinyl fluoride, bp 48–49° (0.5 mm), and 11.2 g (65%) of piperidinosulfur trifluoride, bp 76° (12 mm).

Bis(dimethylamino)sulfur Difluoride.[5,6] Dimethylaminotrimethylsilane (29.3 g, 0.25 mol) was added dropwise to a solution of 33.2 g (0.25 mol) of dimethylaminosulfur trifluoride in 100 mL of fluorotrichloromethane cooled to −78°. The reaction mixture was warmed to 25° and filtered under nitrogen to remove a small amount of a solid. The filtrate was evaporated to dryness under reduced pressure to give 23.5 g (60%) of bis(dimethylamino)sulfur difluoride as a white crystalline solid, mp 64–65°.

Diethylaminodimethylaminosulfur Difluoride.[6] Dimethylaminotrimethylsilane (11.7 g, 0.1 mol) was added dropwise to a solution of 16.1 g (0.1 mol) of diethylaminosulfur trifluoride in 50 mL of fluorotrichloromethane at 25°. The mixture was stirred for 1 hour, after which time two liquid phases were formed. The solvent and fluorotrimethylsilane were removed by distillation at 25° (0.5 mm) to leave 11.1 g (92%) of diethylaminodimethylaminosulfur difluoride as a light yellow liquid.

Bis(piperidino)sulfur Difluoride from the Dipiperidide of Sulfurous Acid.[16,150] With the exclusion of air, a solution of 5.3 g (0.045 mol) of thionyl chloride in 25 mL of dry petroleum ether was added dropwise with cooling to a solution of 15 g (0.18 mol) of piperidine in 100 mL of dry petroleum ether (bp <50°). The reaction mixture was filtered rapidly with suction, the salt was washed with petroleum ether, and the filtrate was evaporated on a water bath. The residue crystallized after cooling in a desiccator over phosphorus pentoxide. The crystals were spread over a porous plate and then recrystallized from dry ether to give the dipiperidide of sulfurous acid, mp 46°.

Piperidinosulfur trifluoride (8.65 g, 0.05 mol) was added dropwise to a stirred solution of 10.8 g (0.05 mol) of the dipiperidide of sulfurous acid in 30 mL of benzene. The mixture was heated at 25° for 4 hours and at 60° for 30 minutes. The solvents and piperidinosulfinyl fluoride were distilled into a receiver cooled with liquid nitrogen. The residue (11.9 g, 97%) was bis(piperidino)sulfur difluoride, mp 104–105°.

Tris(dimethylamino)sulfonium Difluorotrimethylsilicate (TASF).[19] A detailed procedure is described in *Organic Syntheses.*[18]

Tris(pyrrolidino)sulfonium Difluorotrimethylsilicate.[7] *N*-Pyrrolidyltrimethylsilane (47.3 g, 0.33 mol) was added dropwise to a solution of 5.5 mL (0.1 mol) of sulfur tetrafluoride in 100 mL of ether cooled to −78°. The reaction mixture was slowly warmed to room temperature and then stirred for 16 hours. The solid was collected under nitrogen and dried in a vacuum desiccator over P_2O_5 to give 29.8 g (75%) of tris(pyrrolidino)sulfonium difluorotrimethylsilicate, mp 54–57°.

Conversion of Alcohols into Fluorides

1-Fluorooctane.[9] A solution of 13.0 g (0.1 mol) of 1-octanol in 25 mL of dichloromethane was added dropwise to a solution of 16.1 g (0.1 mol) of diethylaminosulfur trifluoride in 60 mL of dichloromethane cooled to $-70°$ to $-65°$. The reaction mixture was warmed to 25°, 50 mL of water was added, and the lower organic layer was separated and dried with anhydrous magnesium sulfate and distilled to give 12.0 g (90%) of 1-fluorooctane as a colorless liquid, bp 42–43° (20 mm). ^{19}F NMR (CCl_3F): 218.8 ppm (tt, $J = 49/25$ Hz).

1-Bromo-2-fluoroethane.[5] Ethylene bromohydrin (31.3 g, 0.25 mol) was added dropwise to a solution of 33 g (0.25 mol) of dimethylaminosulfur trifluoride in 150 mL of diglyme cooled to $-50°$. The reaction mixture was warmed to room temperature, and 50 mL of the most volatile portion was removed by distillation at reduced pressure. The distillate was diluted with water, and the organic layer was separated, washed with a 5% solution of sodium bicarbonate, dried with anhydrous magnesium sulfate, and redistilled to give 22.2 g (70%) of 1-bromo-2-fluoroethane as a colorless liquid, bp 71–72°.

α-Fluorobenzeneacetic Acid.[61] A solution of 1.4 g (9.2 mmol) of mandelic acid (α-hydroxybenzeneacetic acid) in 3 mL of dichloromethane was slowly added to a stirred solution of 3.0 g (19 mmol, 2 equiv) of diethylaminosulfur trifluoride in 6 mL of dichloromethane contained in a polyethylene bottle under nitrogen at $-78°$. After the addition had been completed, the solution was allowed to warm to room temperature. The reaction mixture was stirred for several hours with 50 mL of cold water, and the organic layer was washed with two 10-mL portions of water, dried with anhydrous magnesium sulfate, and evaporated under reduced pressure. From the remaining yellow oil an off-white solid was sublimed at room temperature at 0.1 mm. Recrystallization from 95% ethanol gave 0.98 g (68%) of α-fluorobenzeneacetic acid, mp 74–76°.

9α-Chloro-11β-fluoro-17,21-dihydroxy-16β-methyl-1,4-pregnadiene-3,20-dione Dipropionate.[30] Piperidinosulfur trifluoride (0.3 mL, 0.39 g, 2.4 mmol) was added dropwise at $-40°$ to a solution of 0.53 g (1 mmol) of 9α-chloro-11β,17,21-trihydroxy-16β-methyl-1,4-pregnadiene-3,20-dione 17,21-dipropionate in 25 mL of dichloromethane (freshly filtered through basic aluminum oxide). After 2.5 hours at $-40°$, 1.2 mL of water was added, and the mixture was warmed to room temperature and neutralized with a solution of sodium bicarbonate. The aqueous phase was extracted with dichloromethane, and the extracts were washed with water, dried, and evaporated under reduced pressure. Chromatography of the residue over 15 g of silica gel and elution with hexane:ethyl acetate (3:1) gave 470 mg (90%) of 9α-chloro-11β-fluoro-17,21-dihydroxy-16β-methyl-1,4-pregnadiene-3,20-dione dipropionate, mp 160–162° (diethyl ether–diisopropyl ether), $\alpha_D + 82°$; high-resolution mass spectrum m/z 523 ($M^+ + H$), 503 ($M^+ - F$), 487 ($M^+ - Cl$), 486 ($M^+ - HCl$).

Methyl 3,6-Dideoxy-3,6-difluoro-β-D-allopyranoside.[43] Diethylaminosulfur trifluoride (7.5 mL, 9.7 g, 60 mmol) was added to a suspension of 1.94 g

(10 mmol) of methyl β-D-glucopyranoside in 40 mL of anhydrous dichloromethane at $-40°$ under nitrogen. The cooling bath was removed, and the mixture was stirred overnight at room temperature, cooled to $-20°$, quenched by addition of 40 mL of methanol, and concentrated under reduced pressure. Chromatography on silica gel and elution with hexane:ethyl acetate (1:4) afforded 1.01 g (51%) of methyl 3,6-dideoxy-3,6-difluoro-β-D-allopyranoside, mp 129–130°; $[\alpha]_D$ $-47.0°$ ($c = 1.03$, C_2H_5OH). ^{19}F NMR (^1H decoupled): 217.6 ppm (s, F-3); 234.6 ppm (s, F-6).

Conversion of Carbonyl Compounds into Geminal Difluorides or Vinyl Fluorides

1,1-Difluoro-3-methylbutane.[5] Isovaleraldehyde (1.72 g, 0.02 mol) was slowly added to a solution of 2.5 mL (3.2 g, 0.02 mol) of diethylaminosulfur trifluoride in 10 mL of fluorotrichloromethane at 25°. The reaction mixture was stirred for 30 minutes, and mixed with 25 mL of water; the lower organic layer was separated, washed with water, dried with anhydrous magnesium sulfate, and distilled to give 1.73 g (80%) of 1,1-difluoro-3-methylbutane as a colorless liquid of unspecified boiling point. Anal: Calcd. for $C_5H_{10}F_2$: F, 35.1%. Found: F, 35.1%.

1-Cyclohexyl-1,1,2-trifluoroethane.[151] A solution of 2.5 g (17 mmol) of cyclohexyl fluoromethyl ketone in 10 mL of anhydrous benzene was added, under nitrogen and with stirring, to 3.1 mL (4.0 g, 26 mmol) of DAST. The mixture was stirred for 17 hours at 50°. After the mixture had been cooled to 0°, 10 mL of water was slowly added, causing an exothermic reaction. The mixture was washed with a solution of sodium bicarbonate until neutral. The aqueous phase was extracted with ether, the organic solution was dried and evaporated at reduced pressure, and the residue was bulb-to-bulb distilled to give 2.4 g (86%) of 1-cyclohexyl-1,1,2-trifluoroethane, bp 149°. ^{19}F NMR ($CDCl_3$): 115 ppm (quintet, 2F; $J_{FF} = 14.8$ Hz, $J_{HF} = 12.7$ Hz); 235.5 ppm (ttm, 1F; $J_{HF} = 46$ Hz).

3,3-Difluoro-1,3-dihydro-1-methyl-2*H*-indol-2-one.[75] A mixture of 8.06 g (0.05 mol) of 1-methylisatin and 12.6 mL (1.61 g, 0.1 mol) of diethylaminosulfur trifluoride was warmed gently to 60° and held at that temperature for 15 minutes, during which time the solid dissolved. An exothermic reaction required cooling to maintain the temperature at 60°. The reaction mixture was cooled and poured over ice, and the solid that formed was collected on a filter, washed with water, dried in air, and recrystallized from heptane to give 8.70 g (95%) of 3,3-difluoro-1,3-dihydro-1-methyl-2*H*-indol-2-one as yellow crystals, mp 90–92°. ^{19}F NMR ($CDCl_3$): 112.8 ppm (m).

4-Fluoro-3-cyclohexenyl Benzoate (Conversion of a Ketone into a Vinyl Fluoride).[37] In a Teflon® bottle, 4 g (0.018 mol) of 4-ketocyclohexyl benzoate was dissolved in 25 mL of glyme. With magnetic stirring, 0.5 g of 20% fuming sulfuric acid was added under nitrogen, the mixture was stirred for 5 minutes, and 5.6 g (0.035 mol) of diethylaminosulfur trifluoride was added. Stirring was continued at room temperature for 48 hours. The reaction mixture was then poured into aqueous sodium bicarbonate. The product was extracted with dichloromethane, and the extract was washed with water and brine, dried with anhydrous magnesium sulfate, filtered, and evaporated under vacuum to yield 5.2 g of a yellow oil. Fractionation in a spinning band column at 76° (1 mm) gave 2.4 g of a partly crystalline mixture containing 2 g (49%) of 4-fluoro-3-cyclohexenyl benzoate and 0.4 g (10%) of 4,4-difluorocyclohexyl benzoate. ^{19}F NMR: =CF 102.33 ppm; CF_2 93.32, 95.84, 100.28, and 102.94 ppm.

Conversion of Carboxylic Acids into Acyl Fluorides

Benzoyl Fluoride.[8] A solution of 0.01 mol of dimethylamino-, diethylamino-, piperidino-, or morpholinosulfur trifluoride in 10 mL of ether was added dropwise with stirring to a solution of 1.22 g (0.01 mol) of benzoic acid in 30 mL of ether cooled in an ice bath. The mixture was stirred at 20° for 15 minutes, the ether was removed by distillation, and the residue was distilled to give 10.5 g (85%) of benzoyl fluoride, bp 43–44° (12 mm), n_D^{20} 1.4960.

Conversion of Halides and Sulfonates into Fluorides

Ethyl Fluoroformate.[39] Diethylamino-, piperidino-, or morpholinosulfur trifluoride (0.02 mol) was added dropwise to 2.17 g (0.02 mol) of stirred and cooled ethyl chloroformate. The mixture was stirred for 15–20 minutes at 20° and then at 60° until the gas evolution stopped (about 30 minutes). After cooling to 20° the reaction mixture was fractionated to give 0.94 g (51%) of ethyl fluoroformate, bp 56–57° (755 mm), n_D^{20} 1.3370.

Allyl Fluoride.[7] Allyl bromide (4.4 mL, 6.1 g, 0.05 mol) was added to a stirred solution of 7.1 g (0.2 mol) of tris(pyrrolidino)sulfonium difluorotrimethylsilicate in 5 mL of acetonitrile, and the reaction mixture was stirred at room temperature for 2 hours. The liquid portion was distilled off into a dry ice or liquid nitrogen trap under reduced pressure to give a solid residue, which, after recrystallization from acetone–ether, afforded 5.0 g (78%) of tris(pyrrolidino)sulfonium bromide as colorless crystals, mp 85–88°. The distillate was fractionated through a low-temperature microcolumn to give 1.0 g (83%) of allyl fluoride as a colorless liquid, bp −3 to 0°. ^{19}F NMR (CCl_3F): 216.7 ppm (td, $J = 47/14$ Hz).

1-Deoxy-1-fluoro-2,3:4,5-di-*O*-isopropylidene-D-fructopyranose.[80] Tris(dimethylamino)sulfonium difluorotrimethylsilicate (16.5 g, 0.06 mol) was transferred in a dry box into a flask capped with a rubber septum. A solution of 21.56 g (0.055 mol) of 2,3:4,5-di-*O*-isopropylidene-1-trifluoromethanesulfonyloxy-D-fructopyranose in 50 mL of anhydrous tetrahydrofuran was added by means of a syringe. The septum was replaced by a reflux condenser and the mixture was heated under reflux overnight under nitrogen. It was then poured into water and the mixture extracted with ether. The ether extract was dried and concentrated under reduced pressure, and the residue was chromatographed on silica gel. Elution with a mixture of hexane:ethyl acetate (2:1) afforded 1-deoxy-1-fluoro-2,3:4,5-di-*O*-isopropylidene-D-fructopyranose as a colorless syrup that on Kugelrohr distillation gave 11.58 g (80%) of the pure product, bp 85–95° (0.1 mm); $[\alpha]_D$ −19.2° (c = 0.63, $CHCl_3$). ^{19}F NMR (acetone-d_6): 230.1 ppm (t, J_{HF} = 48 Hz).

Miscellaneous Fluorinations

1-Fluorohexadecyl 4-Methoxyphenyl Sulfide (Conversion of a Sulfoxide into an α-Fluoroalkyl Sulfide).[40] To a stirred solution of 3.80 g (10 mmol) of 1-hexadecyl 4-methoxyphenyl sulfoxide and 0.096 g (0.3 mmol) of zinc iodide in 20 mL of chloroform was added 3.22 g (20 mmol) of diethylaminosulfur trifluoride under nitrogen, and the dark mixture was stirred at room temperature for 16 hours. Treatment of the reaction mixture with an ice-cold solution of sodium bicarbonate gave 3.58 g (94%) of 1-fluorohexadecyl 4-methoxyphenyl sulfide as a pale yellow solid, mp 40–42°. ^1H NMR ($CHCl_3$): 5.54 ppm (dt, J_{HF} = 54.6 Hz, J_{HH} = 6.5 Hz).

4-*O*-(β-2′,6′-Dideoxy-3′-*O*-methyl-4′-*O*-dimethyl-*tert*-butylsilylglucopyranosyl)-2,6-dideoxy-3-*O*-methyl-α-D-glucopyranosyl Fluoride (Conversion of a Hemithioacetal into an α-Fluoroether).[81] To a solution of 56 mg (0.11 mmol) of 4-*O*-(β-2′,6′-dideoxy-3′-methyl-4′-dimethyl-*tert*-butylsilylglucopyranosyl)-2,6-dideoxy-3-methyl-α-phenylthioglucopyranoside in 20 mL of dichloromethane at −15° was added 0.02 mL (0.16 mmol) of diethylaminosulfur trifluoride, followed by 25 mg (0.14 mmol) of *N*-bromosuccinimide. After 15 minutes at −15° the reaction mixture was poured into 5 mL of a saturated sodium bicarbonate solution and extracted with three 10-mL portions of ether. The ether extracts were washed with 5 mL of brine and dried with anhydrous magnesium sulfate. Evaporation of the solvent, followed by flash column chromatography on silica using mixtures of ether—petroleum ether, furnished 40 mg (85%) of a 5:1 (α:β) anomeric mixture of 4-*O*-(β-2′,6′-dideoxy-3′-*O*-methyl-4′-*O*-dimethyl-*tert*-butylsilylglucopyranosyl)-2,6-dideoxy-3-*O*-methyl-α-D-glucopyranosyl fluorides, R_F 0.24 (ether: petroleum ether 3:7). High-resolution mass spectrum *m/z* calculated for $C_{18}H_{39}FO_6Si$: 398.6018. Found: 398.6020.

TABULAR SURVEY

Fluorinations with DAST and other aminofluorosulfuranes are listed by substrate type in Tables I–XI. Reactions in which these reagents cause dehydration rather than fluorination are in Table XII. The computer search of *Chemical Abstracts and Science Citation Index* covers the literature to the end of 1984, and some later papers have also been included. About 50 compounds reported in patents without indication of experimental conditions and yields are not included in the tables.

The reactants are arranged in order of increasing number of carbons and further according to the number of hydrogens and next elements in the molecules.

However, slight deviations occur when similar derivatives containing the same number of carbon atoms are compiled into general schemes. The listing of molar equivalents of DAST and other fluorinating reagents is not systematic. Where nothing is mentioned, either 1 or approximately 1 equivalent per mole was used, or the amounts were not explicitly listed in the papers. If no mention is made of a solvent, the reagents were used neat. A dash (—) means that yields were not reported. Where a reaction has been reported in more than one publication, the conditions producing the highest yields are given and the reference to that paper is listed first.

The following abbreviations are used in the tables:

Ac	acetyl
ax	axial
C_4H_8N	pyrrolidyl
$C_5H_{10}N$	piperidyl
DAST	diethylaminosulfur trifluoride
diglyme	diethylene glycol dimethyl ether
eq	equatorial
equiv.	equivalent
ether	diethyl ether
glyme	1,2-dimethoxyethane
NBS	*N*-bromosuccinimide
OC_4H_8N	morpholinyl
TASF	tris(dimethylamino)sulfonium difluorotrimethylsilicate
THF	tetrahydrofuran
triglyme	triethylene glycol dimethyl ether
AcO*, AcO**	asterisks used to distinguish AcO groups at different carbons

TABLE I. ALCOHOLS

Reactant	Reagent (Molar Equiv.)	Conditions	Product(s) and Yield(s) (%)	Refs.
C_1				
CH_3OH	^{18}DAST	—	$CH_3{}^{18}F$ (20)	148
C_2				
$BrCH_2CH_2OH$	$(CH_3)_2NSF_3$	Diglyme, $-50°$ to room temp	$BrCH_2CH_2F$ (70)	5
$ClCH_2CH_2OH$	DAST	Diglyme, $-78°$ to $-50°$ to room temp	$ClCH_2CH_2F$ (69)	5
C_2H_5OH	^{18}DAST	CH_2Cl_2	$C_2H_5{}^{18}F$ (22)	148
$HOCH_2CH_2OH$	DAST (2)	Diglyme, $-78°$ to room temp	FCH_2CH_2F (70),[5] (54)[9]	5, 9
	^{18}DAST	CH_2Cl_2	$^{18}FCH_2CH_2OH$ (12)	148
C_4				
$HO_2CCHOHCHBrCO_2H$	DAST		Results uncertain	17
$HO_2CCHOHCHOHCO_2H$	"		" "	17
3-hydroxy-γ-butyrolactone (structure, OH)	"		3-fluoro-γ-butyrolactone (structure, F) (>85)	54
$CH_3CH{=}CHCH_2OH$			$CH_3CH{=}CHCH_2F$ + $CH_3CHFCH{=}CH_2$	
	"	Isooctane, $-78°$ to $-50°$, warm to 0°	(36) (64)	5, 9
	"	Diglyme, ", "	(28) (72)	5, 9
	$(CH_3)_2NSF_2N(C_2H_5)_2$	Isooctane, ", "	(57) (9)	6
	"	Diglyme, $-78°$ to 25°	(57) (15)	6
$CH_3CHOHCH{=}CH_2$	DAST	Isooctane, $-78°$ to $-50°$, warm to 25°	(9) (91)	5, 9
		Diglyme, ", "	(22) (78)	5, 9
i-C_4H_9OH	"	$-78°$ to $-50°$, warm to room temp	i-C_4H_9F (49) + t-C_4H_9F (21)	5

Substrate	Reagent	Conditions	Product(s) and Yield(s) (%)	Refs.
C_5				
$(CH_3)_2COHC{\equiv}CH$	"	Diglyme, −78°	$(CH_3)_2CFC{\equiv}CH$ (75)	9
$CH_3CHOHCO_2C_2H_5$	"	CH_2Cl_2, −78° to room temp	$CH_3CHFCO_2C_2H_5$ (78)	5, 9
$(CH_3)_2COHC_2H_5$	"	Diglyme, −78°	$(CH_3)_2CFC_2H_5$ (88)	5, 9
C_6				
$HOCH_2C{\equiv}CCO_2C_2H_5$	"	CH_2Cl_2, ", 45 min, room temp, 200 min	$FCH_2C{\equiv}CCO_2C_2H_5$ (59)	62
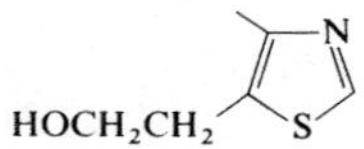 5-($HOCH_2CH_2$)-4-methylthiazole (structure)	"	$CHCl_3$, 0° to room temp, 30 min	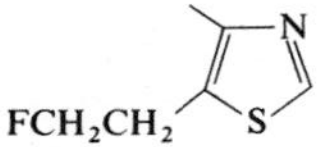 5-(FCH_2CH_2)-4-methylthiazole (structure) (55)	123
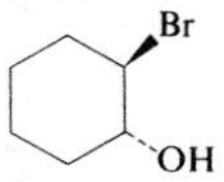 trans-2-Bromocyclohexanol (structure: Br, OH)	"	CH_2Cl_2, −78° to room temp, 1.5 h	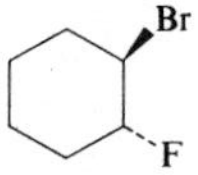 1-Bromo-2-fluorocyclohexane (structure: Br, F) (67)	151a
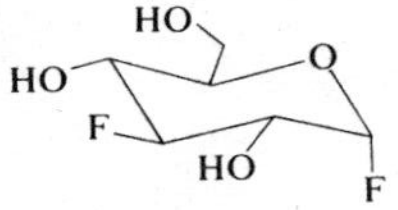 Pyranose (structure: HO, HO, F, HO, F)	DAST (6)	Room temp, 5 d	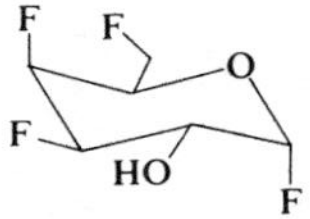 Pyranose (structure: F, F, F, HO, F) (48)	44
(S)-$CH_3O_2CCHOHCH_2CO_2CH_3$	"	$CHCl_3$, 0° to room temp, 30 min	(R)-$CH_3O_2CCHFCH_2CO_2CH_3$ (85)	42
Cyclohexanol	$[(CH_3)_2N]_2SF_2$	CH_2Cl_2, −78° to room temp	Fluorocyclohexane (−) + cyclohexene (−) "equal parts"	5, 6
Cyclohexen-3-ol	DAST	−30° to −50° to room temp, 40 min	3-Fluorocyclohexene (60)[a]	17
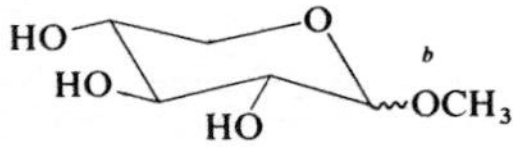 Methyl pyranoside (structure: HO, HO, HO, ~OCH_3 [b])			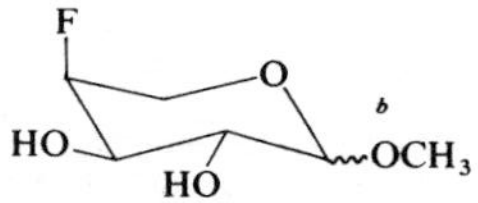 Methyl 4-fluoropyranoside (structure: F, HO, HO, ~OCH_3 [b])	
ax	DAST (4)	CH_2Cl_2, −40° to room temp	ax (19)	43
eq	"	", ", "	eq (52)	43
$Br(CH_2)_6OH$	"	", −78° to 25°	$Br(CH_2)_6F$ (53)	59

TABLE 1. ALCOHOLS (*Continued*)

Reactant					Reagent (Molar Equiv.)	Conditions	Product(s) and Yield(s) (%)					Refs.
C_7												
2,6-$Cl_2C_6H_3CH_2OH$					DAST	CH_2Cl_2, −70° to room temp	2,6-$Cl_2C_6H_3CH_2F$ (50-80)					152
p-$O_2NC_6H_4CH_2OH$					"	", 10°, 45 min to room temp	*p*-$O_2NC_6H_4CH_2F$ (90-95)[c]					153
$C_6H_5CH_2OH$					"	CCl_3F, −78° to room temp	$C_6H_5CH_2F$ (75)					5
					$(CH_3)_2NSF_3$	CH_2Cl_2, "	" (≈100)					5, 9
					$[(CH_3)_2N]_2SF_2$	", "	" (91)					5, 6
R^1	R^2	R^3	R^4	R^5			R^1	R^2	R^3	R^4	R^5	
OCH_3 (ax)	OH (eq)	OH (eq)	F (eq)	OH	DAST (4)	CH_2Cl_2, −40° to room temp	OCH_3 (ax)	OH (eq)	OH (eq)	F (eq)	F (71)	43
"	"	"	N_3 (eq)	"	" "	", ", room temp overnight	"	"	"	N_3 (eq)	" (68)	43
OCH_3 (eq)	"	"	OH (eq)	N_3	" "	", ", ", 3 h	OCH_3 (eq)	"	F (ax)	OH (eq)	N_3 (28)	43
OCH_3 (ax)	OH (ax)	"	"	"	" "	", ", ", 1 h	OCH_3 (ax)	OH (ax)	OH (eq)	F (ax)	" (56)	43
"	"	"	"	OH	" (2)	−10°, 2 h, room temp	"	"	"	"	F (72)	58
					" (5)	CH_2Cl_2, −40°, room temp, 2 h	"	"	"	"	" (80)	33
OCH_3 (eq)	OH (eq)	"	OH (ax)	"	"	", ", ", 45 min	OCH_3 (eq)	OH (eq)	"	OH (ax)	" (15)	84
OCH_3 (ax)	"	"	OH (eq)	"	" (6)	", ", ", 1 h	OCH_3 (ax)	"	"	OH (eq)	" (70)	33
					" (2)	room temp overnight	{ "	"	"	"	" (9) }	58
							{ "	"	"	F (ax)	" (60) }	
					" (6)	", overnight	"	"	"	"	" (60)	45
					" (5)	CH_2Cl_2, −40° to room temp, 72 h	"	"	"	"	" (40)	43

Reactant	Reagent	Conditions	Product(s) (Yield, %)	Refs.
OCH_3 (eq) ” ” ” ”	” (2)	−10°, 2 h, room temp overnight	{ OCH_3 (eq) ” ” ” ” (8); ” ” ” OH (eq) ” (—); ” ” F (ax) ” ” (32) }	58
	” (5)	CH_2Cl_2, −40° room temp, overnight	” ” OH (eq) ” ” (60)	43
	” (6)	”, ”, ”, ”	” ” F (ax) ” ” (51)	43
$CH_2{=}CHCHOHP(O)(OC_2H_5)_2$	DAST	CH_2Cl_2, 0°	$FCH_2CH{=}CHP(O)(OC_2H_5)_2$ (≈100)	35
C_8				
3,4-$Cl_2C_6H_3CHOHCN$[d]	”	”, 5° to room temp, 10–30 min	3,4-$Cl_2C_6H_3CHFCN$ (100)	83
2,6-$Cl_2C_6H_3CHOHCN$[d]	”	”, ”, ”, ”,	2,6-$Cl_2C_6H_3CHFCN$ (86)	83
m-$FC_6H_4CHOHCN$[d]	”	”, ”, ”, ”,	*m*-FC_6H_4CHFCN (47)	83
p-$O_2NC_6H_4CHOHCN$[d]	”	”, ”, ”, ”,	*p*-$O_2NC_6H_4CHFCN$ (50)	83
$C_6H_5CHOHCN$[d]	”	”, ”, ”, ”,	C_6H_5CHFCN (64)	83
3,5-$Br_2C_6H_3CH_2CH_2OH$	”	”, −78° to room temp, 1 h	3,5-$Br_2C_6H_3CH_2CH_2F$ (60)	154
$C_6H_5CHOHCO_2H$	”	”, ”, ”	$C_6H_5CHFCO_2H$ (68)	61
$C_6H_5CH_2CH_2OH$	”	”, −78° to 40°	$C_6H_5CH_2CH_2F$ (60)	5
Cyclooctanol	”	CCl_3F, −78° to room temp	Fluorocyclooctane (70) + cyclooctene (30)	5,9,155
$Br(CH_2)_8OH$	”	CH_2Cl_2, −78° to 25°	$Br(CH_2)_8F$ (61)	59
$CH_3CH{=}CHCHOHP(O)(OC_2H_5)_2$	”	”, 0°	$CH_3CHFCH{=}CHP(O)(OC_2H_5)_2$ (≈100)	35
(+)-(*S*)-*n*-$C_6H_{13}CHOHCH_3$	”	”, −60° to room temp, overnight	(−)-(*R*)-*n*-$C_6H_{13}CHFCH_3$ (23)	31
n-$C_8H_{17}OH$	”	”, −70° to −65° to 25°	*n*-$C_8H_{17}F$ (90)	5,9
C_9				
$C_6H_5COH(CH_3)CN$[d]	”	”, 5° to room temp, 10-30 min	$C_6H_5CF(CH_3)CN$ (50)	83
m-$CH_3OC_6H_4CHOHCN$[d]	”	”, ”, ”, ”	*m*-$CH_3OC_6H_4CHFCN$ (48)	83
R-substituted 1-indanol (OH); R = H, 4-F or 5-F	”	”, −70° to room temp	R-substituted 1-fluoroindane (F) (50–80)	152

TABLE I. ALCOHOLS (*Continued*)

Reactant	Reagent (Molar Equiv.)	Conditions	Product(s) and Yield(s) (%)	Refs.
o-$FC_6H_4C(CH_3)_2OH$	DAST	Glyme, −78°, 60°, 7 d	o-$FC_6H_4C(CH_3)_2F$ (40)	156
$C_6H_5CHOHCH(NH_2)CO_2H$	"	HF, −40°, 2 h	$C_6H_5CHFCH(NH_2)CO_2H$ (—)[e]	120
$C_6H_5C(CH_3)_2OH$	"	CH_2Cl_2, −70° to room temp	$C_6H_5C(CH_3)_2F$ (50–80)	152
2,6-$(CH_3)_2C_6H_3CH_2OH$	"	", "	2,6-$(CH_3)_2C_6H_3CH_2F$ (50–80)	152
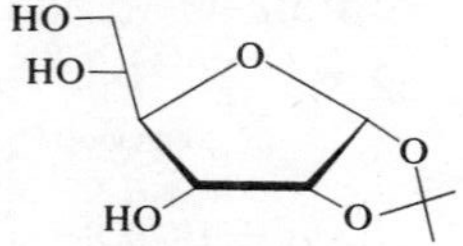 HO— HO— O HO O O (structure)	" (4)	", −40°, room temp, 90 min	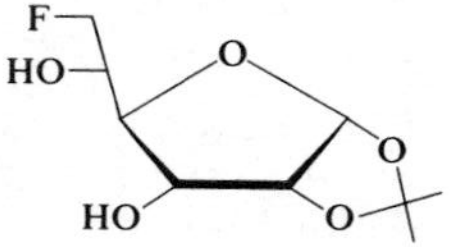 F— HO— O HO O O (structure) (70)	43
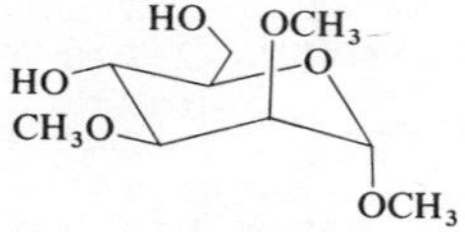 HO OCH_3 HO O CH_3O OCH_3 (structure)	" (3)	", ", ", 1 h	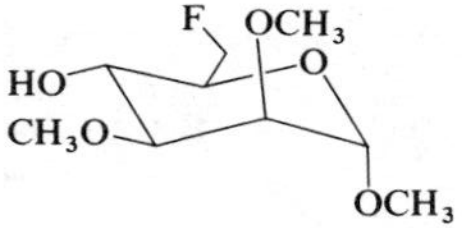 F OCH_3 HO O CH_3O OCH_3 (structure) (60)	43
C_{10}				
m-$CH_3CO_2C_6H_4CHOHCN$[d]	"	", 5° to room temp, 10–30 min	m-$CH_3CO_2C_6H_4CHFCN$ (67)	83
$C_6H_5COH(C_2H_5)CN$[d]	"	", ", ", "	$C_6H_5CF(C_2H_5)CN$ (64)	83
3,4-$(CH_3O)_2C_6H_3CHOHCN$	"	CH_2Cl_2, 5° to room temp, 10–30 min	3,4-$(CH_3O)_2C_6H_3CHFCN$ (43)	83
"[d]	"	", ", ", "	" (82)	83
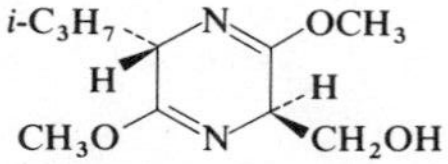 i-C_3H_7 N OCH_3 H H CH_3O N CH_2OH (structure)	"	"	No reaction	121
$HC{\equiv}C(CH_2)_8OH$	"	", −78° to 25°	$HC{\equiv}C(CH_2)_8F$ (82)	59

Substrate	Reagent (equiv)	Conditions	Product(s) and Yield(s) (%)	Refs.
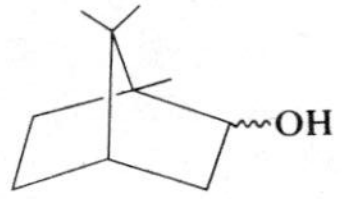 [structure: borneol-type bicyclic alcohol, ~OH] *endo*	"	CCl_3F, $-78°$ to room temp	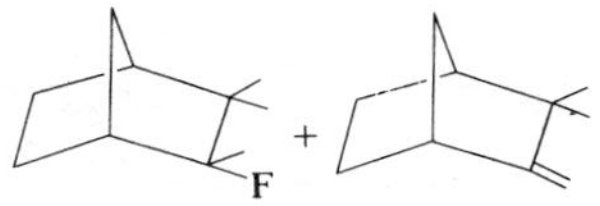 [structure: fluoride, F] + [structure: methylene bicyclic] (72) (17)	5, 9
exo			(74) (18)	5, 9
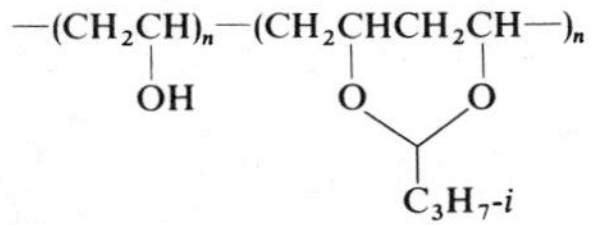 $—(CH_2CH)_n—(CH_2CHCH_2CH—)_n$ with OH; O–CH(C_3H_7-*i*)–O acetal	"	Glyme, 0° to room temp	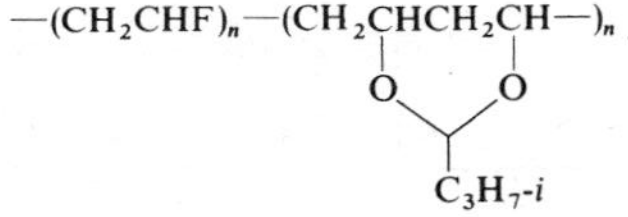 $—(CH_2CHF)_n—(CH_2CHCH_2CH—)_n$ with O–CH(C_3H_7-*i*)–O acetal (—)	9
[structure: HO, HO-CH$_2$, O-isopropylidene sugar, OCH_3]	DAST (3)	CH_2Cl_2, $-40°$ to room temp; room temp, 15 min; then CH_3OH	[structure: R^1, R^2, isopropylidene sugar, OCH_3] R^1 / R^2: {OH (eq), F (14); OH (eq), OCH_3 (12)}	43
	" (3)	", ", room temp 24 h	F (ax), F (23)	43
[structure: 4,6-O-isopropylidene glucoside, HO, HO, OCH_3]	" (4.5)	CH_2Cl_2, $-40°$, room temp, overnight	[structure: 4,6-O-isopropylidene, F, HO, OCH_3] (45)	43
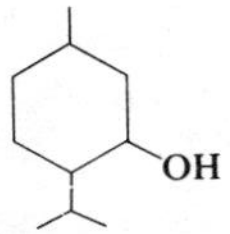 [structure: menthol, OH] (—)	"	CCl_3F, $-78°$ to room temp	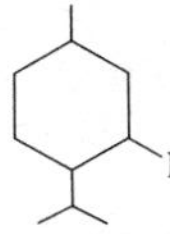 [structure: menthyl fluoride, F] (50)	5, 9
$Br(CH_2)_{10}OH$	"	CH_2Cl_2,"	$Br(CH_2)_{10}F$ (82)	59

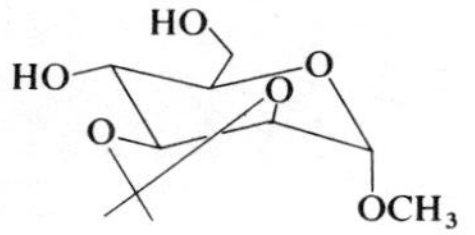

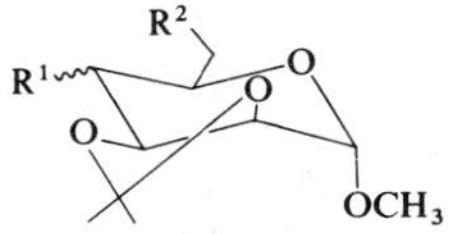

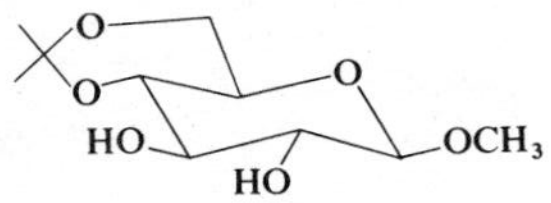

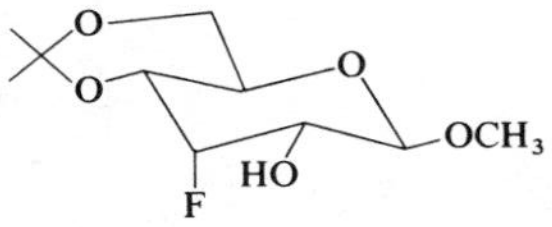

TABLE I. ALCOHOLS (*Continued*)

Reactant	Reagent (Molar Equiv.)	Conditions	Product(s) and Yield(s) (%)	Refs.
C_{11}				
$XC_6H_4CHOHP(O)(OC_2H_5)_2$	DAST	CH_2Cl_2, 0°	$XC_6H_4CHFP(O)(OC_2H_5)_2$	
X = H			X = H (≈100)	35
X = *m*-Cl			X = *m*-Cl (≈100)	35
X = *p*-Cl			X = *p*-Cl (≈100)	35
C_{12}				
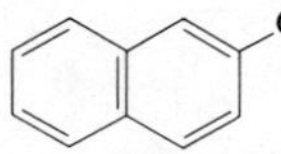 2-Naphthyl-$CHOHCN$[d]	"	", 5° to room temp 10–30 min	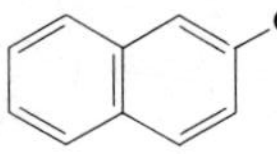 2-Naphthyl-$CHFCN$ (46)	83
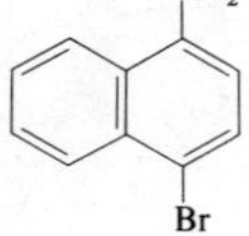 4-Bromo-1-naphthyl-CH_2CH_2OH	"[f]	CH_2Cl_2, −50° to room temp	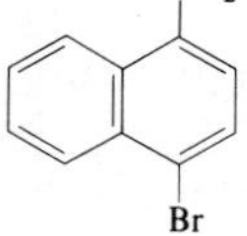 4-Bromo-1-naphthyl-CH_2CH_2F	157
Pyranoside (6-HO, 4-R^3, 3-R^2, 2-HO, 1-OR^1): R^1 / R^2 / R^3			6-Fluoro pyranoside (4-R^3, 3-R^2, 2-HO, 1-OR^1): R^1 / R^2 / R^3	
p-$O_2NC_6H_4$ (eq) / OH (eq) / OH (eq)	" (2)	CH_2Cl_2, −40°, room temp, 35 min	*p*-$O_2NC_6H_4$ (eq) / OH (eq) / OH (eq) (55)	43
	" (6)	", ", ", overnight	" / F (ax) / " (78)	43
C_6H_5(eq) / " / "	" (5)	", ", ", 25 min	C_6H_5(eq) / OH (eq) / " (29)	43
	" (5)	", ", ", overnight	" / F (ax) / " (70)	43
C_6H_5(ax) / " / "	DAST (6)	CH_2Cl_2, −40°, room temp, 2 h	C_6H_5(ax) / OH (eq) / OH (eq) (58)	43
	" (6)	", ", ", 5 d	" / " / F (ax) (38)	43

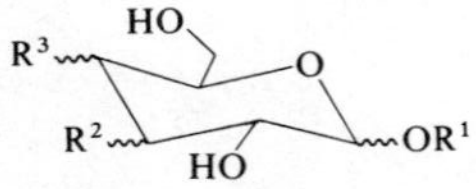

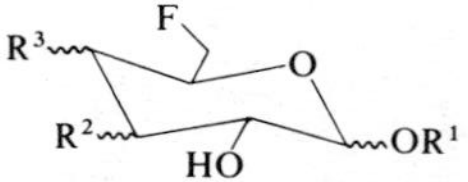

Substrate	Reagent	Conditions	Product(s) and Yield(s) (%)	Refs.
p-$XC_6H_4CROHP(O)(OC_2H_5)_2$			p-$XC_6H_4CRFP(O)(OC_2H_5)_2$	
X = H, R = CH_3	"	CH_2Cl_2, 0°	X = H, R = CH_3 (≈100)	35
X = CH_3, R = H	"	" "	X = CH_3, R = H (≈100)	35
X = Cl, R = CH_3	"	" "	X = Cl, R = CH_3 (≈100)	35
$(CH_3)_2C{=}CH(CH_2)_2(HOCH_2)C{=}CH(CH_2OAc)$	$(CH_3)_2NSF_2$-$N(C_2H_5)_3$	−65°, 10 min to room temp	$(CH_3)_2C{=}CH(CH_2)_2(FCH_2)C{=}CH(CH_2OAc)$ (55)	62
[furanose diacetonide, R^1, R^2]			[furanose diacetonide, R]	
R^1 = H, R^2 = OH	DAST	CH_2Cl_2, C_5H_5N (2.5 equiv), 0° to room temp	R = F (97)[g]	29
R^1 = OH, R^2 = H	^{18}DAST	—	R = ^{18}F (—)	85
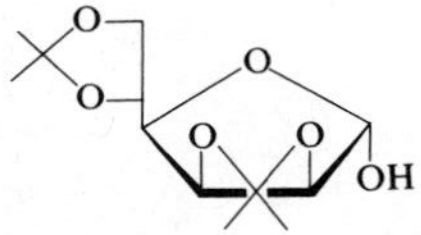 [furanose diacetonide, anomeric OH]	DAST	THF, −30°, room temp, 20 min	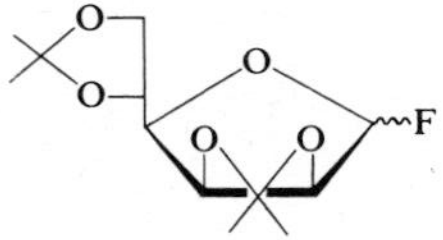 [glycosyl fluoride] α (87) β (13)	90
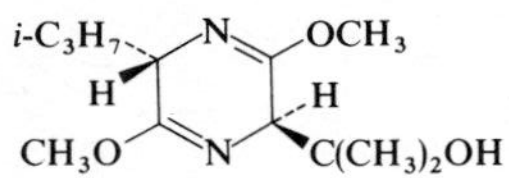 [i-C_3H_7, H, CH_3O, OCH_3, H, $C(CH_3)_2OH$ dihydropyrazine]	"	CH_2Cl_2, −70° to room temp, 1 h	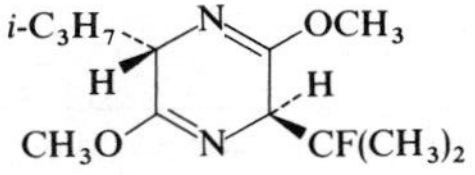 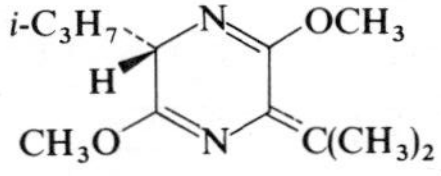 [i-C_3H_7, H, CH_3O, OCH_3, H, $CF(CH_3)_2$ dihydropyrazine] (—) + [i-C_3H_7, H, CH_3O, OCH_3, $=C(CH_3)_2$ dihydropyrazine] (—)	121

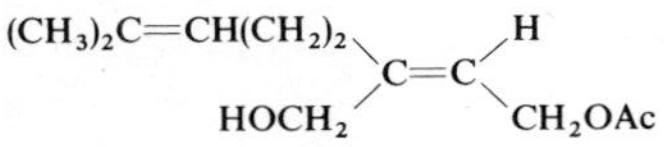

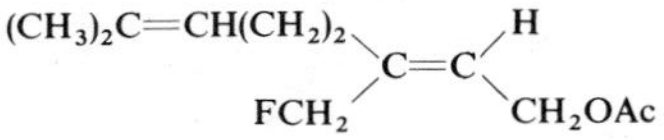

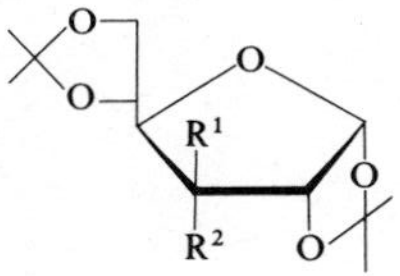

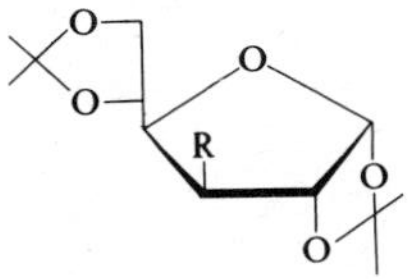

TABLE I. ALCOHOLS (*Continued*)

Reactant	Reagent (Molar Equiv.)	Conditions	Product(s) and Yield(s) (%)	Refs.
[structure: di-O-isopropylidene sugar, CH₂OH]	DAST	Diglyme, −10° to room temp, 60°, 2 h	[structure: corresponding CH₂F compound] (73)	89
[structure: di-O-isopropylidene sugar, CH₂OH]	"	(—)	No reaction	80
$t\text{-}C_4H_9CO_2$, HO, HO, HO, OCH_3 [pyranoside structure]	"	CH_2Cl_2, −40° to room temp, overnight	$t\text{-}C_4H_9CO_2$, HO, F, HO, OCH_3 [pyranoside structure] (35)	43
C_{13}				
[structure: 9-fluorenol, H, OH]	"	", −30°, 30 min	[structure: 9-fluorofluorene, H, F] (48) + [structure: bis(9-fluorenyl) ether, O, 2] (13)	57
$(C_6H_5)_2CHOH$	"	CH_2Cl_2, −30°, 30 min	$(C_6H_5)_2CHF$ (40) + $[(C_6H_5)_2CH]_2O$ (44)	57
$C_6H_5CH{=}CHCHOHP(O)(OC_2H_5)_2$	"	", 0°	$C_6H_5CHFCH{=}CHP(O)(OC_2H_5)_2$ ($\approx$100)	35
C_{14}				
$C_6H_5CHOHCOC_6H_5$	"	", −78° to room temp	$C_6H_5CHFCOC_6H_5$ (86)	61

Substrate	Reagent	Conditions	Product(s) and Yield(s) (%)	Refs.
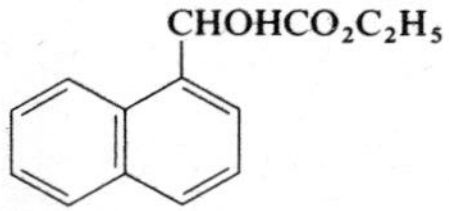 $CHOHCO_2C_2H_5$ (1-naphthyl)	"	CCl_3F, $-78°$ to $50°$ to room temp	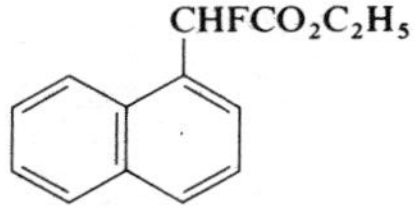 $CHFCO_2C_2H_5$ (1-naphthyl) (60)	5
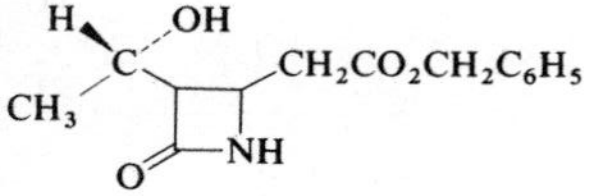 H, OH, C, CH_3, $CH_2CO_2CH_2C_6H_5$, O, NH	"	CH_2Cl_2, $-110°$ or $-78°$ to room temp	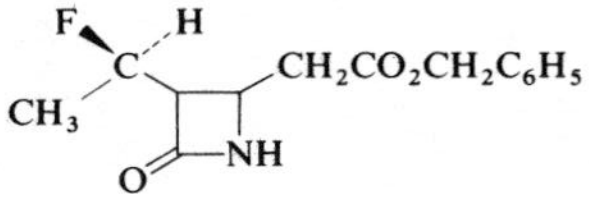 F, H, C, CH_3, $CH_2CO_2CH_2C_6H_5$, O, NH **I** (20–40)[h] 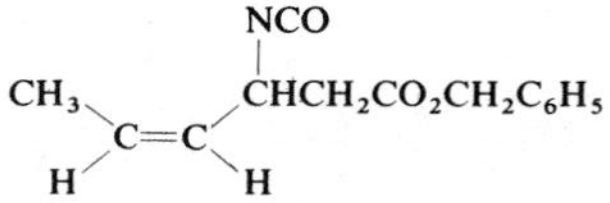 + CH_3, H, C=C, H, NCO, $CHCH_2CO_2CH_2C_6H_5$ **II** (7–26)[i]	51
	$C_5H_{10}NSF_3$	", ", "	**I** (10-45) + **II** (5–19)[i]	51
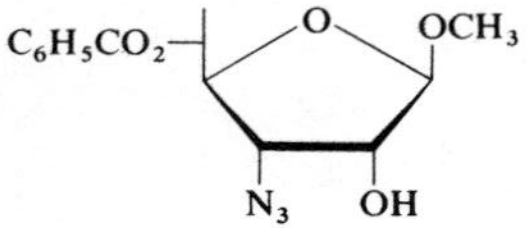 $C_6H_5CO_2$, O, OCH_3, N_3, OH	DAST (4)	$C_6H_5CH_3$, $-10°$, $60°$, 2 h	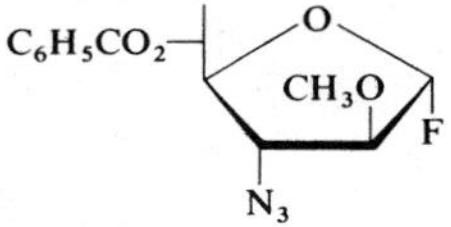 $C_6H_5CO_2$, O, CH_3O, F, N_3 (50)	54b
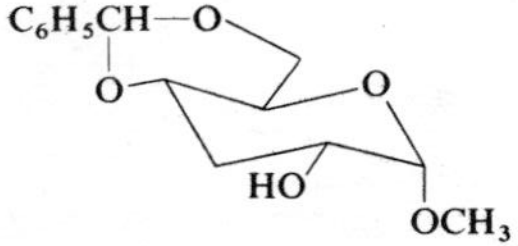 C_6H_5CH—O, O, O, HO, OCH_3	DAST (4.5)	C_6H_6, C_5H_5N, $0°$, room temp, 30 min, $60°$, 3 h	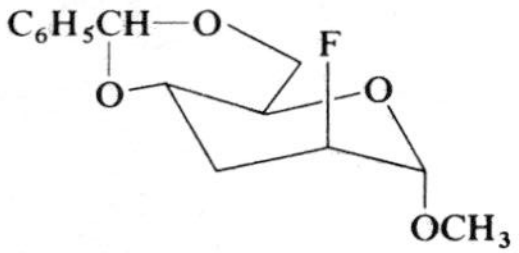 C_6H_5CH—O, O, F, O, OCH_3 (78) 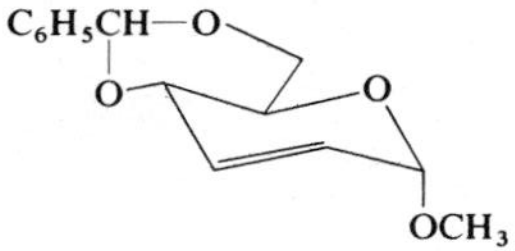 + C_6H_5CH—O, O, O, OCH_3 (6)	157a

TABLE I. ALCOHOLS (*Continued*)

Reactant	Reagent (Molar Equiv.)	Conditions	Product(s) and Yield(s) (%)	Refs.
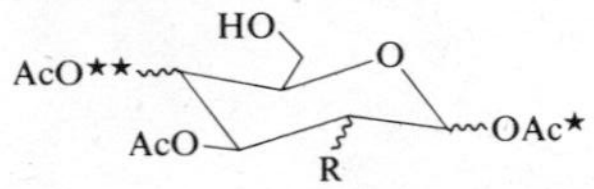			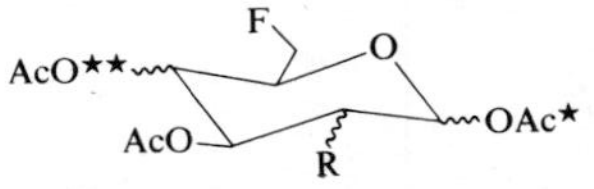	
AcO★ / R / AcO★★			AcO★ / R / AcO★★	
(eq) / AcO (eq) / (eq)	DAST	Diglyme, −10°, 10–15 min, 60°, 3 h	(eq) / AcO (eq) / (eq) (85)	63
" / AcNH (eq) / "	"	", ", ", 25°, 2 h	" / AcNH (eq) / " (85)	63
(ax, eq) / AcNH (ax) / "	"	", ", ", ", 1.5 h	(ax, eq) / AcNH (ax) / " (70)	63
", " / AcNH (eq) / (ax)	"	", ", ", ", "	", " / AcNH (eq) / (ax) (68)	63
(ax) / " / (eq)	"	", 0°, 1 h, 40°, 1 h	(ax) / " / (eq) (80)	63
2,4,6-$(CH_3)_3C_6H_2CHOHP(O)(OC_2H_5)_2$	"	CH_2Cl_2, 0°	2,4,6-$(CH_3)_3C_6H_2CHFP(O)(OC_2H_5)_2$ (≈100)	35
C_{15}				
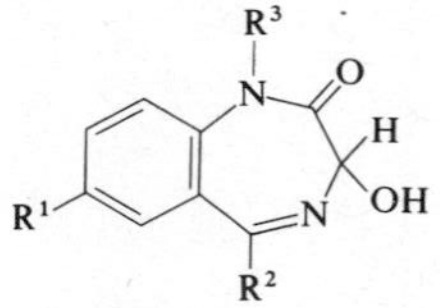			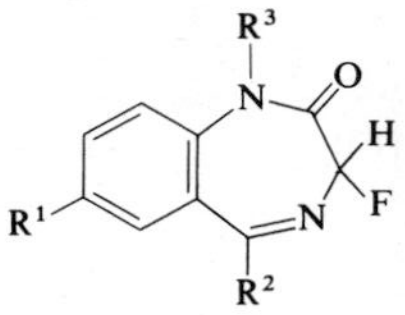	
R^1 / R^2 / R^3				
H / C_6H_5 / H	DAST (7)	CH_2Cl_2, −70° to −10°, 25 min	(88)	64
				64–66
Br / " / "	" (2.5)	", ", "	(88)	64–66
Cl / " / "	" (7)	", ", "	(82)	64–66
NO_2 / " / "	" (2.5)	", ", "	(83)	64–66
Cl / o-ClC_6H_4 / "	" (7)	", ", "	(53)	64–66
Br / o-FC_6H_4 / "	" (7)	", ", "	(74)	64–66
Cl / " / "	" (6)	", ", "	(98)	64–66

C_{16}					
Br *o*-FC_6H_4 CH_3	"	(7)	", ", "	(92)	64
Cl " "	"	(2.5)	", −70° to 5°, 45 min	(83)	64–66
Cl C_6H_5 "	"	(2.5)	", ", "	(90)	64–66
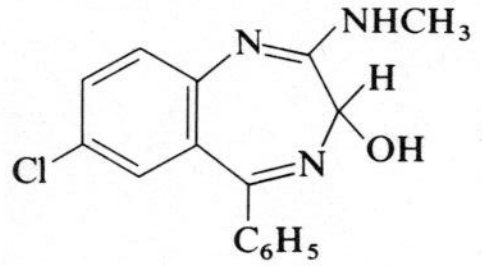 Cl, N, $NHCH_3$, H, OH, N, C_6H_5	DAST	(7)	CH_2Cl_2, −70° to −10° 25 min	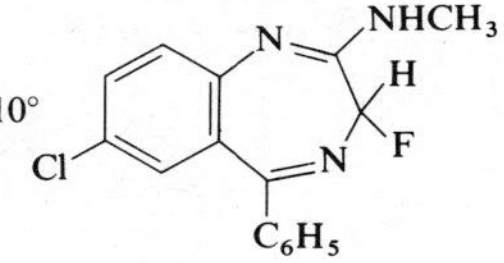 Cl, N, $NHCH_3$, H, F, N, C_6H_5 (82)	64
$(CH_2)_2OH$, R				$(CH_2)_2F$, R	
R: H			", −78° to room temp, 30 h	(68)	157, 158
Br	"		", ", overnight	(90)	157, 158
C_{17}					
Cl, N, N, N, H, OH, N, C_6H_5	"		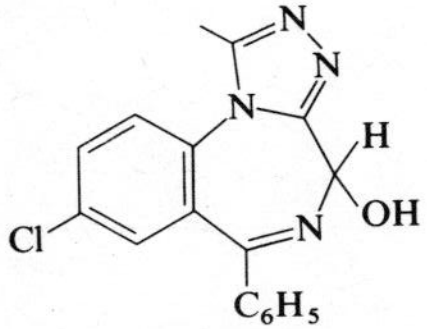 ", −78° to −20°	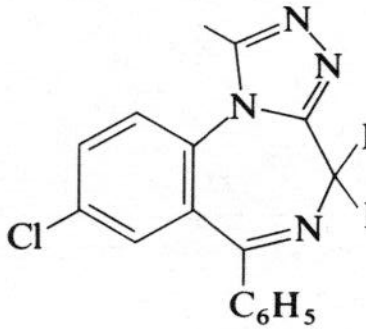 Cl, N, N, N, H, F, N, C_6H_5 (85)	64–66
C_2H_5, O, N, Cl, H, OH, N, C_6H_5	"		", −70° to 5°	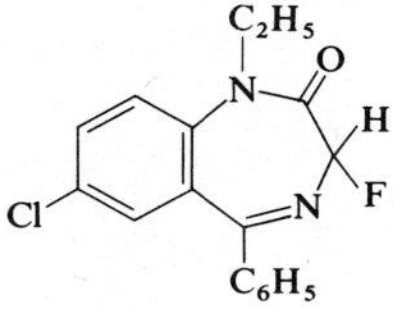 C_2H_5, O, N, Cl, H, F, N, C_6H_5 (90)	64–66

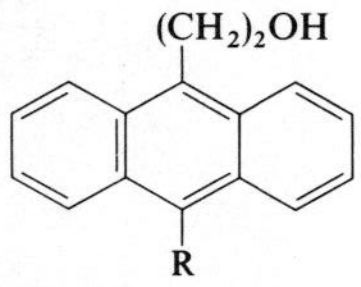

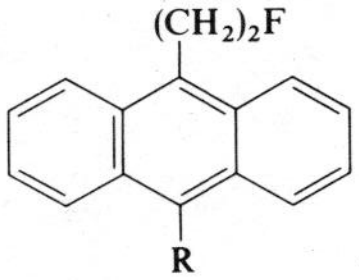

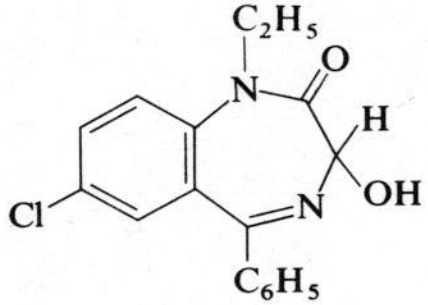

TABLE I. ALCOHOLS (*Continued*)

Reactant	Reagent (Molar Equiv.)	Conditions	Product(s) and Yield(s) (%)	Refs.
9-$(CH_2)_2OH$-10-methylanthracene	DAST	CH_2Cl_2, −78° to room temp	9-$(CH_2)_2F$-10-methylanthracene (76)	158
p-$HOC_6H_4CH(C_2H_5)CH(CH_2OH)C_6H_4OH$-$p$	"		No reaction	159
i-C_3H_7, H, OCH_3, H, CH_3O, $COH(CH_3)C_6H_5$ (dihydropyrazine)	"	", −70° to room temp, 1 h	i-C_3H_7, H, OCH_3, H, CH_3O, $CF(CH_3)C_6H_5$ (dihydropyrazine) (96)	121
THP-$O(CH_2)_2CH{=}CHCH_2CH{=}CH(CH_2)_4CHROH$	"		THP-$O(CH_2)_2CH{=}CHCH_2CH{=}CH(CH_2)_4CHRF$	
R = H	$(CH_3)_2NSF_3$(2) + $(CH_3)_2NSi$-$(CH_3)_3$	CCl_3F, −78°, 45 min room temp, 1 h	(59)	59
R = T	DAST	CH_2Cl_2, −78° to 25°, 20 min	(96)	59
C_{18}				
$C_6H_5CH_2CONH$-cephem S-oxide, CH_2OH, $CO_2CH_2CCl_3$	"	", −78°, few h	$C_6H_5CH_2CONH$-cephem S-oxide, CH_2F, $CO_2CH_2CCl_3$ (—)	68

Reactant	Reagent	Conditions	Product(s) and Yield(s) (%)	Refs.
p-HOC_6H_4–CH(H)–C(H)(C_6H_4OH-p)$CHROH$			p-HOC_6H_4–CH(H)–C(H)(C_6H_4OH-p)$CHRF$	
R = H	"	THF, ice–salt cooling, to room temp, 7 h	(76)	160
R = T	"	", −10° to room temp, 37 h	(10)[k]	161
C_{19}				
$(C_6H_5)_3COH$	"	CH_2Cl_2, −30°, 30 min	$(C_6H_5)_3CF$ (85)	57
(dihydrodiol, HO, OH)	"	", −68°, 30 min, room temp, 2 h p-$CH_3C_6H_4SO_3H$, reflux 4 h	(F) (—)	162
(diol, HO, OH)	" (1.5)	", ", room temp overnight, reflux 30 min	(F) (22)	162
(R^1, OH; R^2, OH) *cis* and/or *trans*			(R^1, R^2)	
R^1 R^2: CH_3 H	DAST	CH_2Cl_2, room temp, 2 h	R^1 R^2: CH_3 F (53) + R^1 R^2: CH_3 H (11)	163
H CH_3	"	", ", ",	F CH_3 (48) + H CH_3 (14)	163

TABLE I. ALCOHOLS (*Continued*)

Reactant	Reagent (Molar Equiv.)	Conditions	Product(s) and Yield(s) (%)	Refs.
$(C_6H_5)_2P(O)CHOHC_6H_5$	DAST	CH_2Cl_2, 0°	$(C_6H_5)_2P(O)CHFC_6H_5$ (≈100)	35
$[(C_6H_5)_3\overset{+}{P}CH_2OH]\overset{-}{B}F_4$	"	", 0° to room temp; room temp, 1 h	$[(C_6H_5)_3\overset{+}{P}CH_2F]\ \overset{-}{B}F_4$ (88)	164
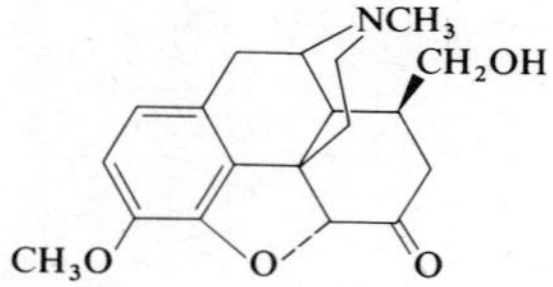 [structure: CH_3O, NCH_3, CH_2OH, O, O]	" (2)	", 0°, 15 min	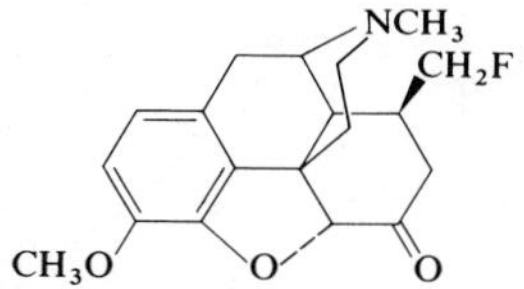 [structure: CH_3O, NCH_3, CH_2F, O, O] (31)	60
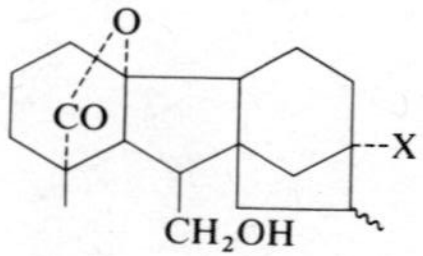 [structure: O, CO, X, CH_2OH]			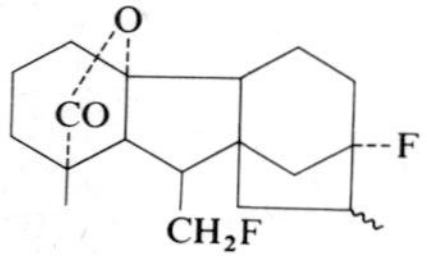 [structure: O, CO, F, CH_2F]	
X = F	"	CH_2Cl_2, 0°, 30 min	(83)	113
X = OH	"	", ", "	(77)	113
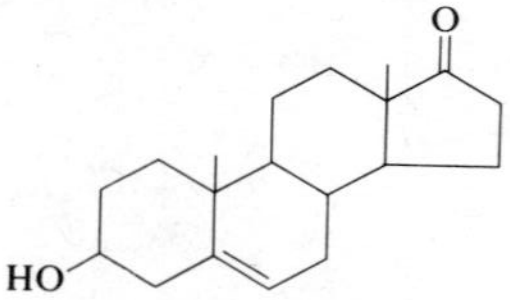 [structure: HO, O]	"	", −78° to room temp	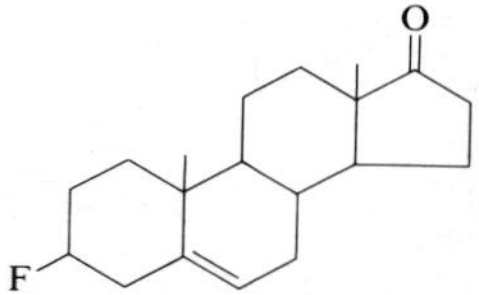 [structure: F, O] (81)	32
	"	", 20°, 30 min,	" (91)	47, 48
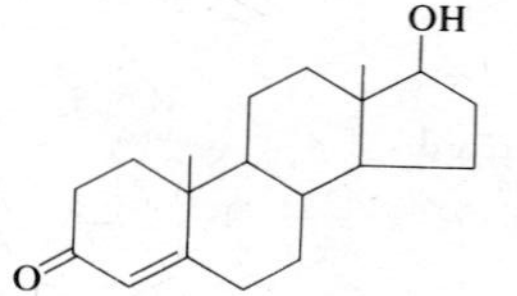 [structure: O, OH]	"	CH_2Cl_2, −78° to room temp	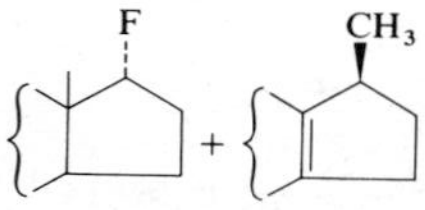 [structure: F] (40) + [structure: CH_3] (32)	32

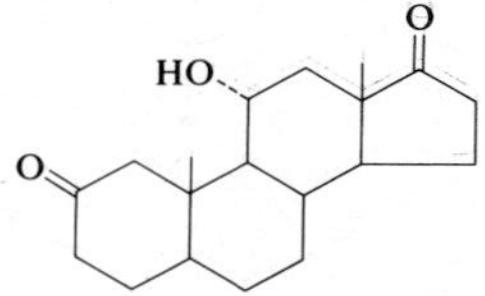	"	", 20°, 30 min	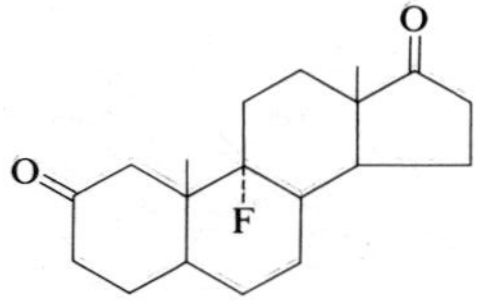 (16) + 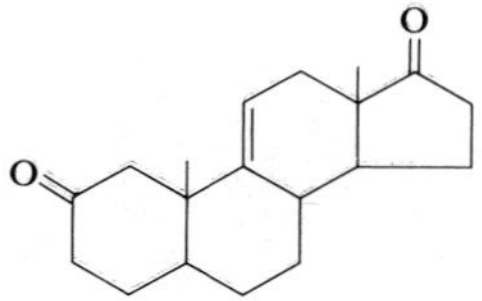 (67)	47, 48
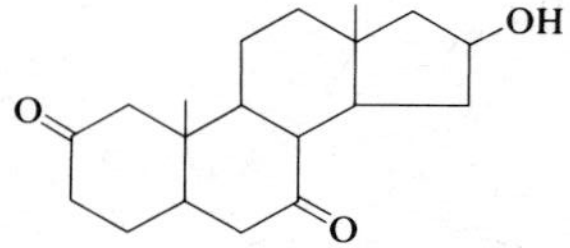	DAST (2)	CH_2Cl_2, 20°, 30 min	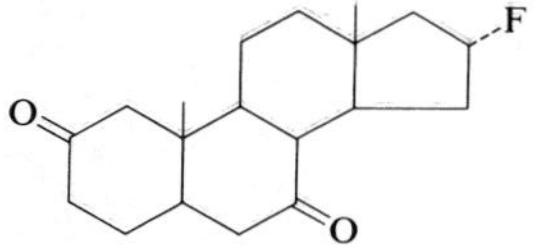 (72)	47, 48
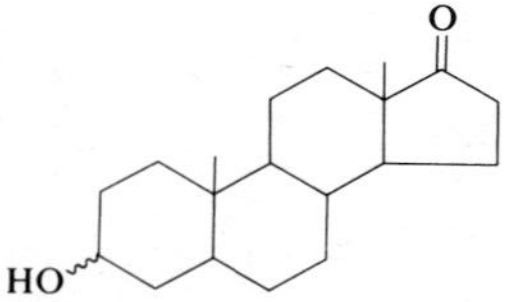 α-OH	"	CH_2Cl_2, −78° to room temp	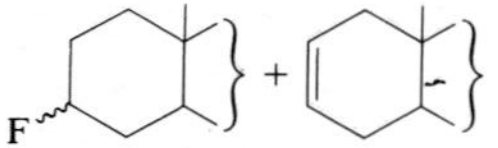 β-F (35) (—)	32
α-OH	"	", 20°, 30 min	β-F (14) (83)	47, 48
β-OH	"	", ", ",	α-F (47) (44)	47, 48
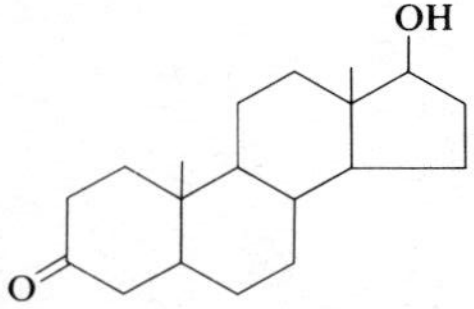	"	", ", ",	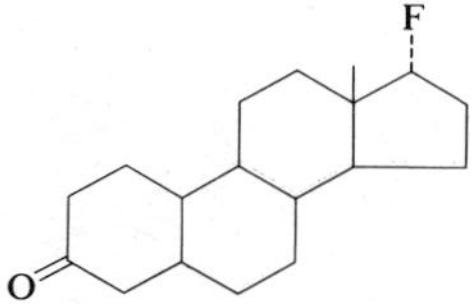 (39)	47, 48

TABLE I. ALCOHOLS (*Continued*)

Reactant	Reagent (Molar Equiv.)	Conditions	Product(s) and Yield(s) (%)	Refs.
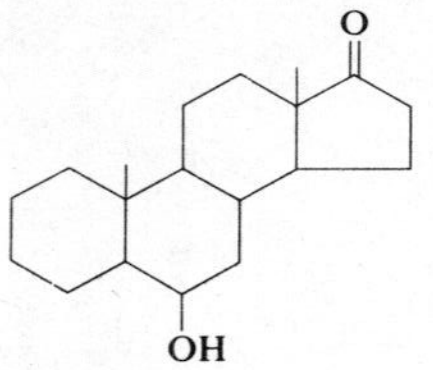 OH, O	DAST	CH_2Cl_2, 20°, 30 min	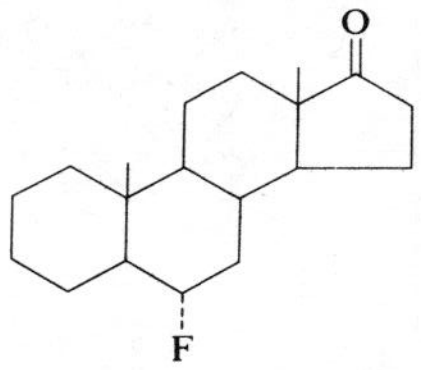 O, F (86)	47, 48
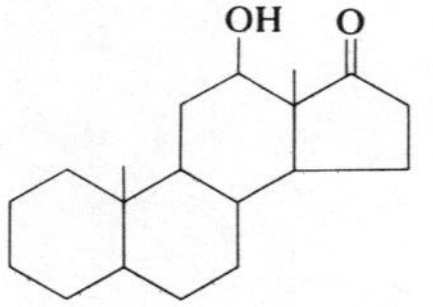 OH, O	"	", ", ",	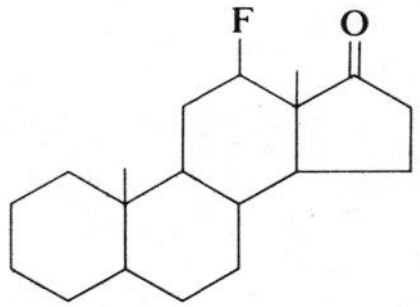 F, O (77)	47, 48
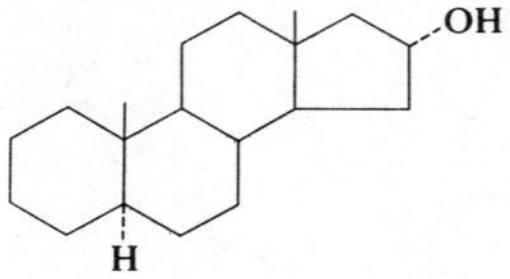 OH, H	" (2)	", ", "	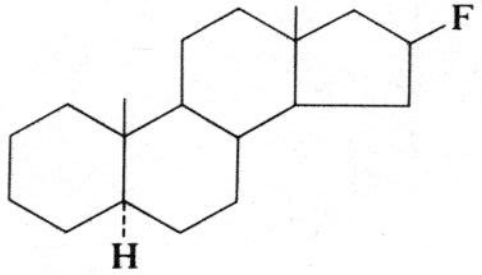 F, H (88)	47, 48
C_{20}				
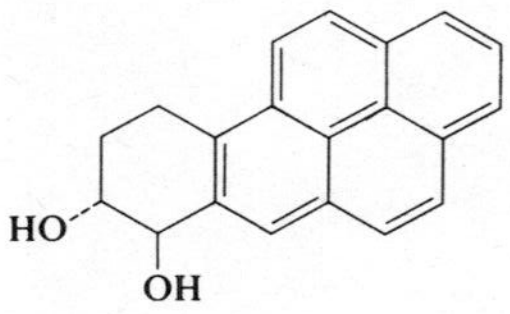 HO, OH	DAST (3)	CH_2Cl_2, −78°, 30 min; warm to 0°	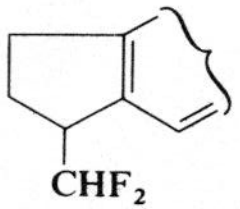 CHF_2 (40)	53
	" (3)	", ", ", DDQ[l], C_6H_6 reflux 16 h	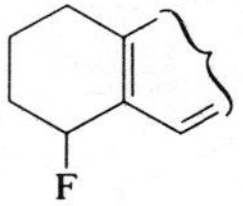 F (60)	53

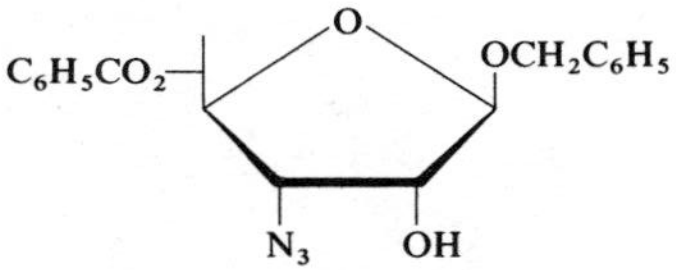	" (8)	$C_6H_5CH_3$, −10°, 60°, 2 h	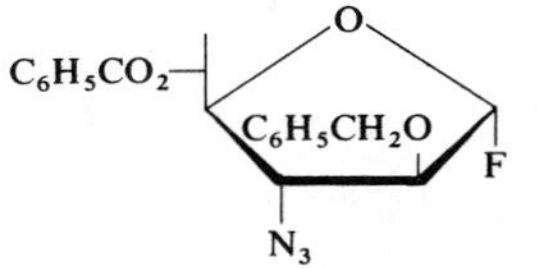 (79)	54b
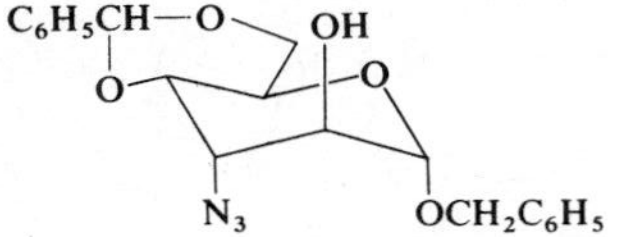	" (3)	C_6H_6, reflux, 2 h	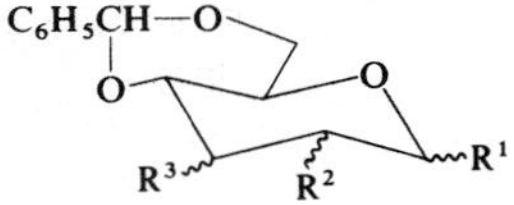 R^1 = ax-$OCH_2C_6H_5$, R^2 = ax-F, R_3 = ax-N_3 (40) eq-F, eq-$OCH_2C_6H_5$, ax-N_3 (15) ax-$OCH_2C_6H_5$, eq-N_3, eq-F (40)	54a
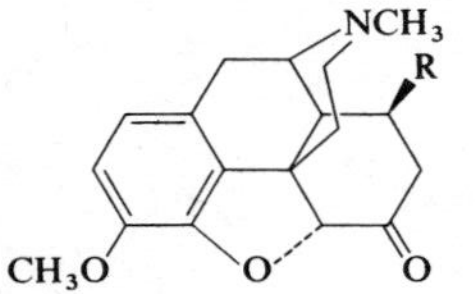 R			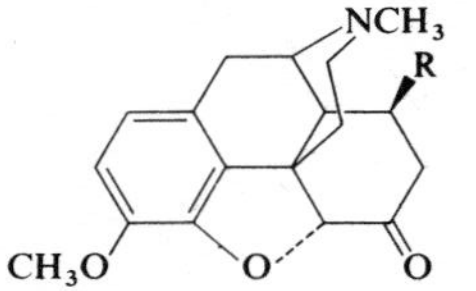 R	
CH_2CH_2OH	" (3)	", 0°, 30 min	CH_2CH_2F (28)	60
SCH_2CH_2OH	" (3)	", 0°, 30 min	SCH_2CH_2F (35)	60
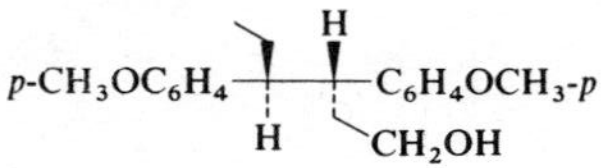	"	CH_2Cl_2, −70° to room temp, 3 h	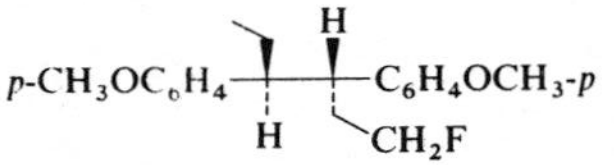 (45)	160
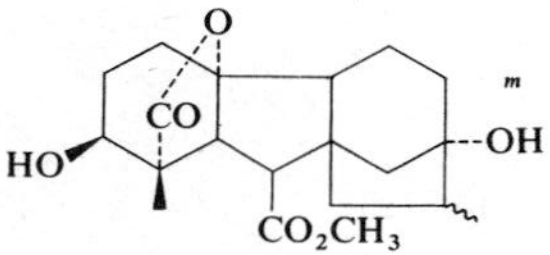 [m]	" (2)	", 0°, 45 min	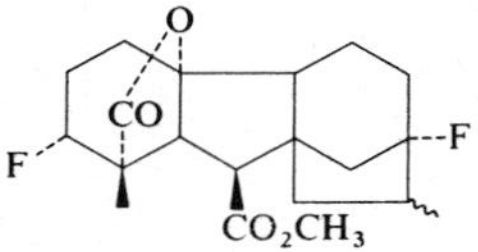 (2)[m] + 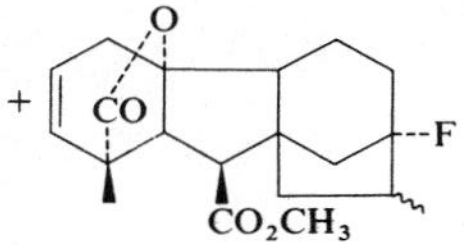 (26)[m]	113

TABLE I. ALCOHOLS (*Continued*)

Reactant	Reagent (Molar Equiv.)	Conditions	Product(s) and Yield(s) (%)	Refs.
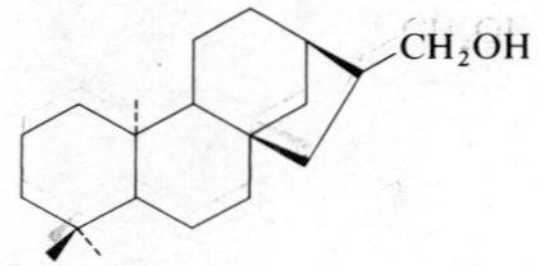 C_{21} CH$_2$OH	DAST (2)	room temp, 7 h	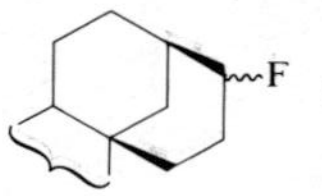 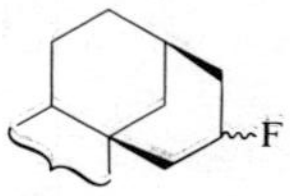 F or F (41)	110
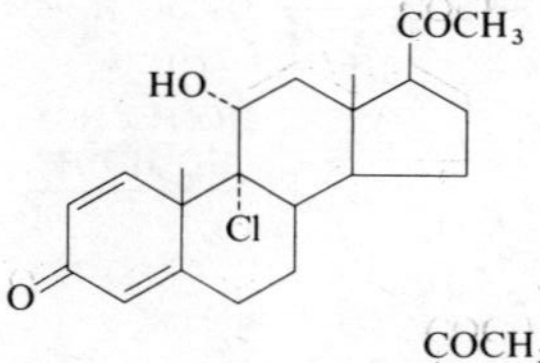 HO, COCH$_3$, Cl, O	$C_5H_{10}NSF_3$	CH_2Cl_2, −60°, 5 h	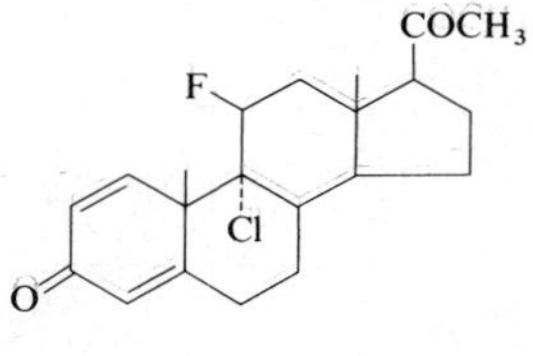 F, COCH$_3$, Cl, O (—)	95
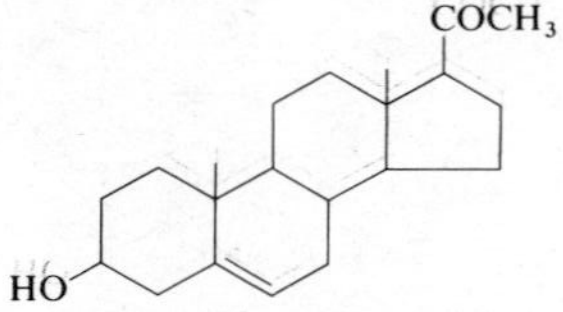 COCH$_3$, HO	DAST	CH_2Cl_2, −78° to room temp	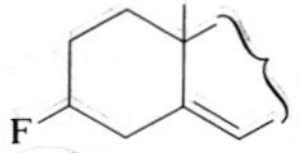 F (82)	32
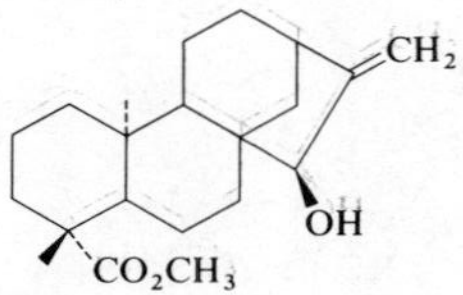 CH$_2$, OH, CO$_2$CH$_3$	"	"; 0°, 5 min	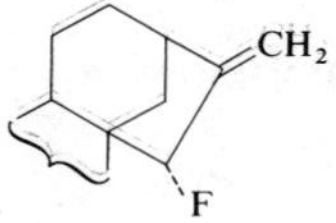 CH$_2$, F (80)	114
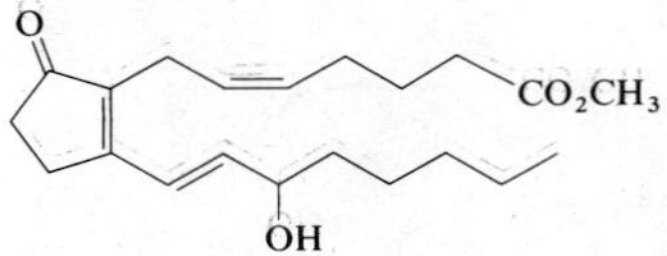 O, CO$_2$CH$_3$, OH	$OC_4H_8NSF_3$	CH_2Cl_2, −78°, 1.5 h to −10°	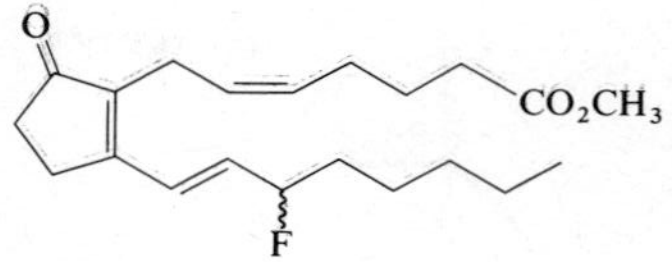 O, CO$_2$CH$_3$, F (80)	111

Substrate	Reagent	Conditions	Product(s) and Yield(s) (%)	Refs.
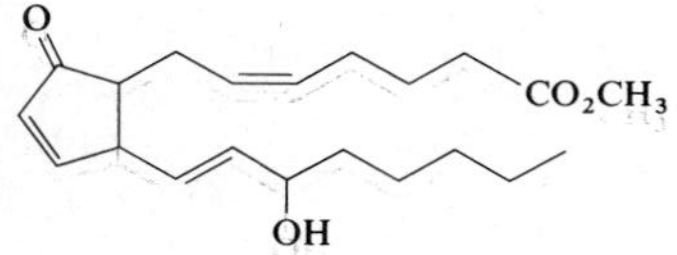 (structure: OH, CO_2CH_3)	"	", ", ", "	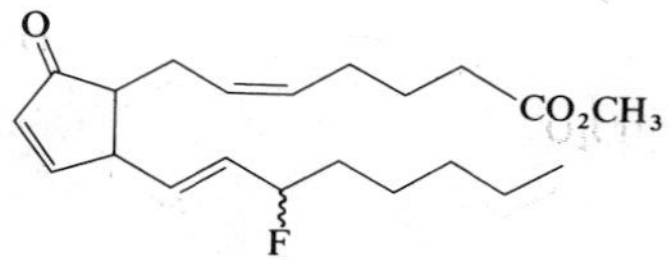 (structure: F, CO_2CH_3) (75)	111
C_{23}				
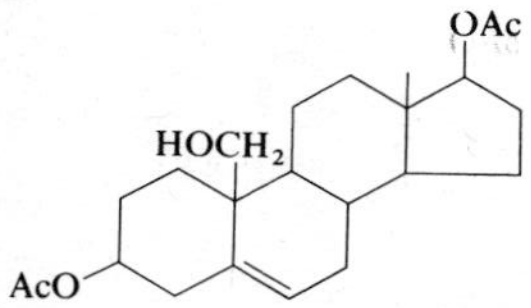 (structure: $HOCH_2$, AcO, OAc)	DAST	60°, 5 h	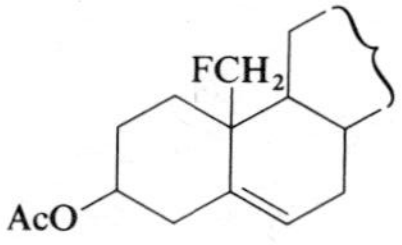 (structure: FCH_2, AcO) (—)	98
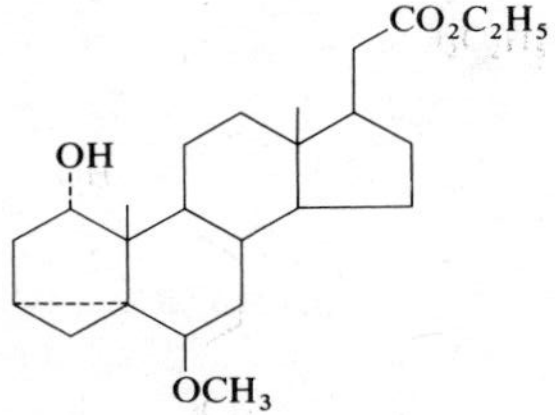 (structure: OH, $CO_2C_2H_5$, OCH_3)	"	CH_2Cl_2, −78°	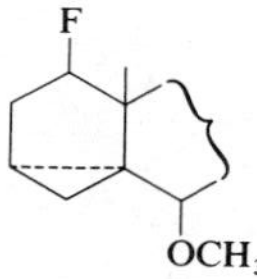 (structure: F, OCH_3) (~80)[n]	54
C_{24-25}				
$RCHOHCHCO_2CH_2C_6H_5$ / $N(CH_2C_6H_5)_2$ R: H; CH_3, *threo*; CH_3, *erythro*	"	THF, room temp, 50 min	$RCHCHFCO_2CH_2C_6H_5$ ($N(CH_2C_6H_5)_2$) + $RCHFCHCO_2CH_2C_6H_5$ ($N(CH_2C_6H_5)_2$) R: H (90) / (0); CH_3, *threo* (60) / CH_3, *threo* (26); CH_3, *erythro* (90) / (0)	55
C_{25}				
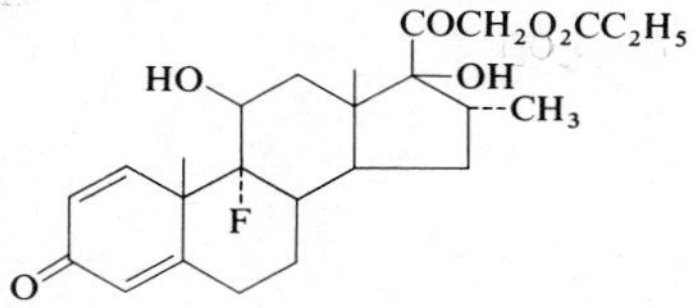 (structure: HO, F, O, $COCH_2O_2CC_2H_5$, OH, CH_3)	$C_5H_{10}NSF_3$	CH_2Cl_2, dioxane −30°, 3.5 h	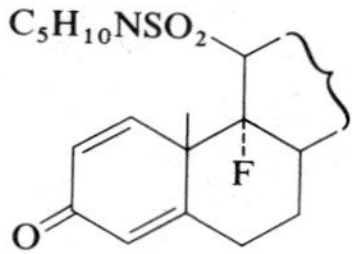 (structure: $C_5H_{10}NSO_2$, F, O) (77)	30

TABLE I. ALCOHOLS (*Continued*)

Reactant	Reagent (Molar Equiv.)	Conditions	Product(s) and Yield(s) (%)	Refs.
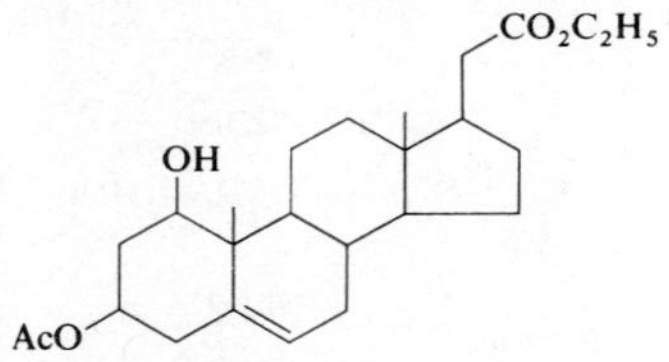 AcO, OH, $CO_2C_2H_5$	DAST	−78° to 0°	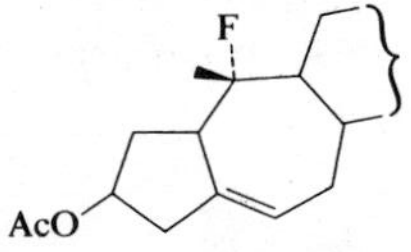 AcO, F (~80)	54
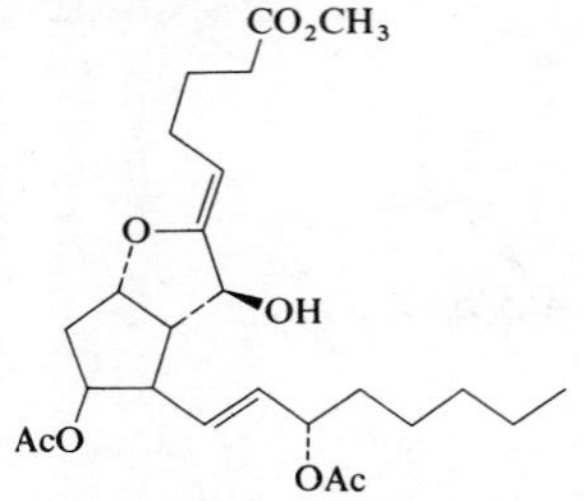 CO_2CH_3, AcO, OH, OAc	"	CH_2Cl_2, −40°, −20°, 13 h	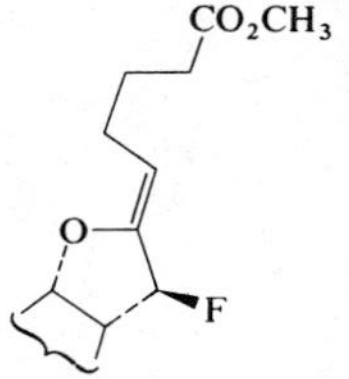 CO_2CH_3, F (32) 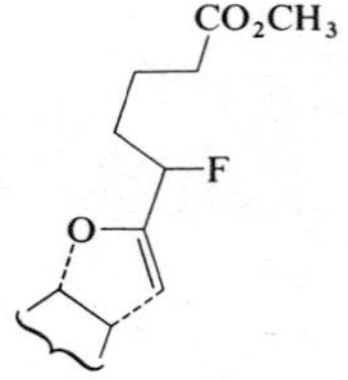 + CO_2CH_3, F (22)	116
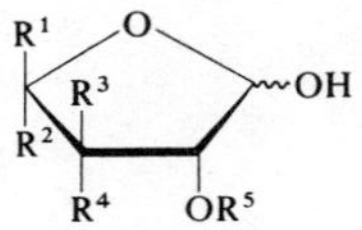 R^1, R^2, R^3, R^4, OR^5, OH		THF, −30°, room temp, 20 min	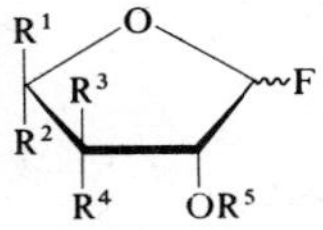 R^1, R^2, R^3, R^4, OR^5, F	

R^1	R^2	R^3	R^4	R^5	$\alpha:\beta$	Refs.
$C_6H_5CO_2$	H	H	$C_6H_5CO_2$	C_6H_5CO	(90) 42:58	90
$C_6H_5CH_2O$	H	H	$C_6H_5CH_2O$	$C_6H_5CH_2$	(99) 9:91[e]	90
H	$C_6H_5CH_2O$	$C_6H_5CH_2O$	H	$C_6H_5CH_2$	(95) 91:9	90

C_{26}

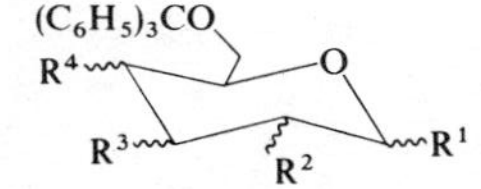

Substrate R¹	R²	R³	R⁴	Reagent	Conditions	Product R¹	R²	R³	R⁴	(Yield)	Refs.
OCH_3 (eq)	OH (eq)	OH (eq)	OH (eq)	DAST (4.5)	CH_2Cl_2, −40° room temp overnight	OCH_3 (eq)	OH (eq)	F (ax)	OH (eq)	(50)	43
OCH_3 (ax)	"	"	"	" (3.3)	", ", ", 72 h	OCH_3 (ax)	"	OH (eq)	F (ax)	(42)	43, 58
"	OH (ax)	"	"	" (3)	", ", ", 2 h	"	OH (ax)	"	"	(40)	43

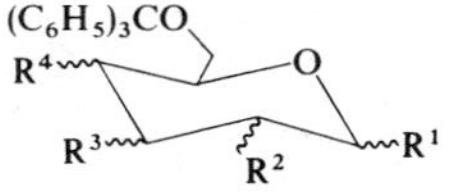

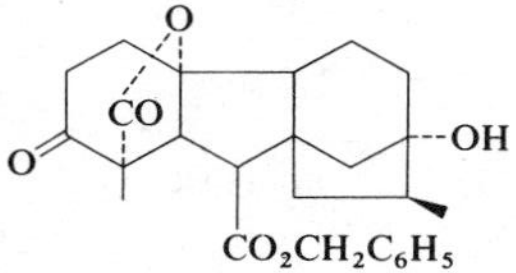

Substrate ring: $(C_6H_5)_3CO$-substituted pyranoside bearing R^1, R^2, R^3, R^4.

Substrate	Reagent	Conditions	Product(s) and Yield(s) (%)	Refs.
Gibberellin-type lactone ketone with --OH and $CO_2CH_2C_6H_5$	"	", 0°, 8 h	Fluoride (--F), $CO_2CH_2C_6H_5$ (8) + trifluoride (F, F, --F), $CO_2CH_2C_6H_5$ (44) + vinyl fluoride (F, --F), $CO_2CH_2C_6H_5$ (9)	50

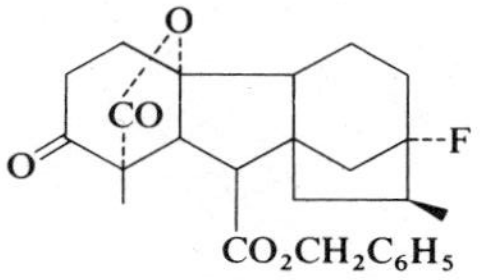

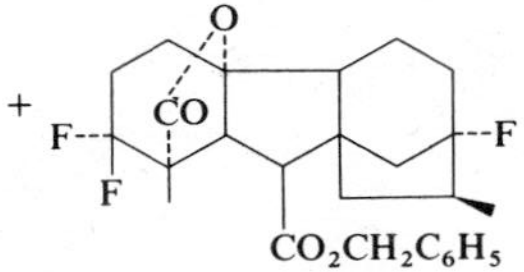

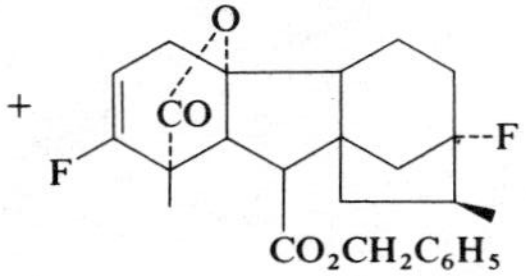

TABLE I. ALCOHOLS (*Continued*)

Reactant	Reagent (Molar Equiv.)	Conditions	Product(s) and Yield(s) (%)	Refs.
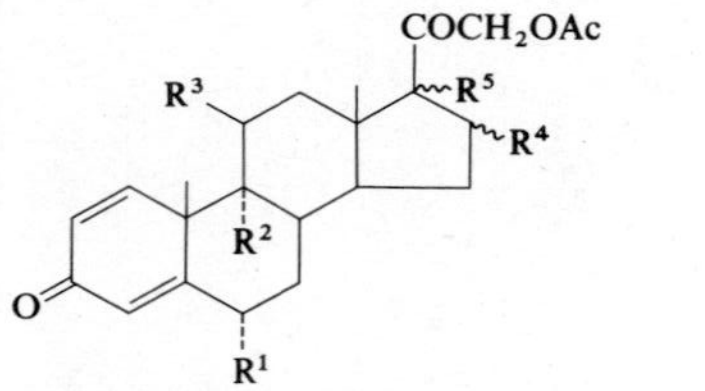 [steroid structure: 3-oxo-1,4-diene; R^1 at C-6, R^2 at C-9, R^3 at C-11, R^4 and R^5 at C-16/C-17, $COCH_2OAc$ at C-17]			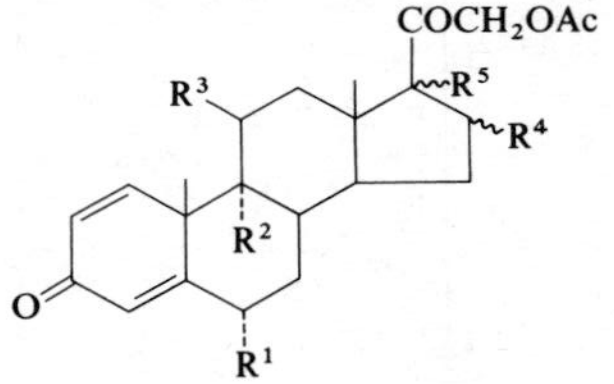 [steroid structure: 3-oxo-1,4-diene; R^1 at C-6, R^2 at C-9, R^3 at C-11, R^4 and R^5 at C-16/C-17, $COCH_2OAc$ at C-17]	
R^1 R^2 R^3 R^4 R^5			R^1 R^2 R^3 R^4 R^5	
F Cl OH α-CH_3 β-OAc	DAST	Dioxane, 4°, 2 h	F Cl F α-CH_3 β-OAc (—)	95
" F " α-$OC(CH_3)_2O$-α	$C_5H_{10}NSF_3$	", CH_2Cl_2, −10°	" F $C_5H_{10}NSO_2$ α-$OC(CH_3)_2O$-α (56)	30
C_{27}				
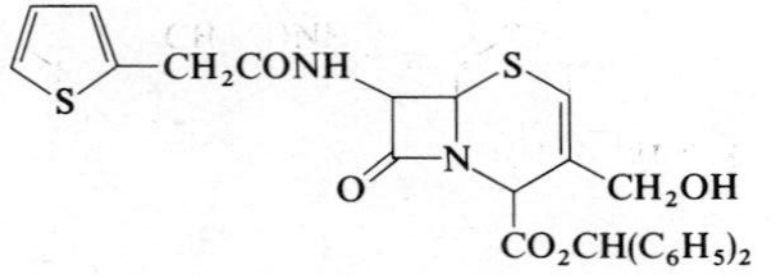 [2-thienyl-CH_2CONH-cephem structure with CH_2OH and $CO_2CH(C_6H_5)_2$]	DAST	CH_2Cl_2 −78°, 0.5 h	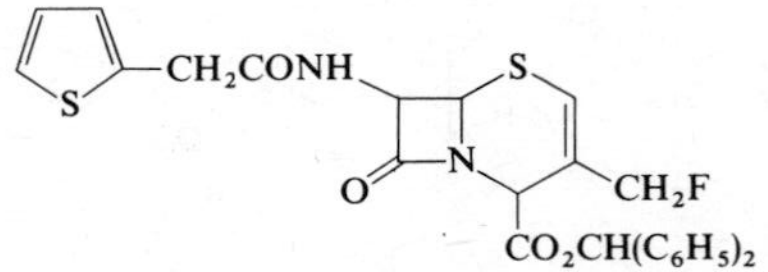 [2-thienyl-CH_2CONH-cephem structure with CH_2F and $CO_2CH(C_6H_5)_2$] (—)	68
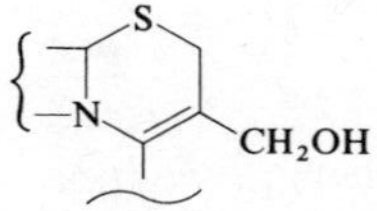 [cephem partial structure: S, N, CH_2OH]	"	", ", "	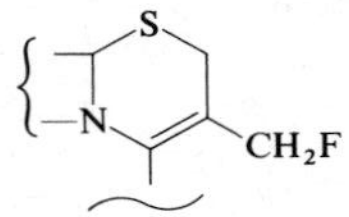 [cephem partial structure: S, N, CH_2F] (—)	9, 68
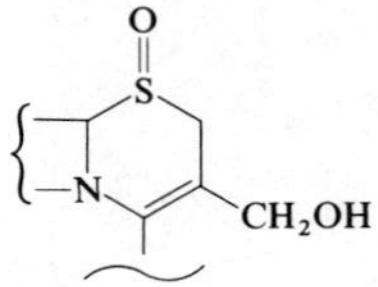 [cephem sulfoxide partial structure: S=O, N, CH_2OH]	"	", ", "	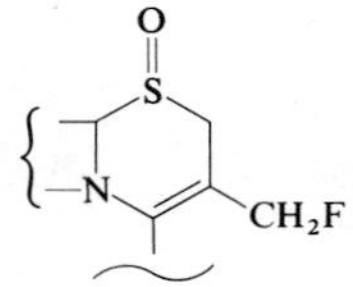 [cephem sulfoxide partial structure: S=O, N, CH_2F] (6)	9, 68

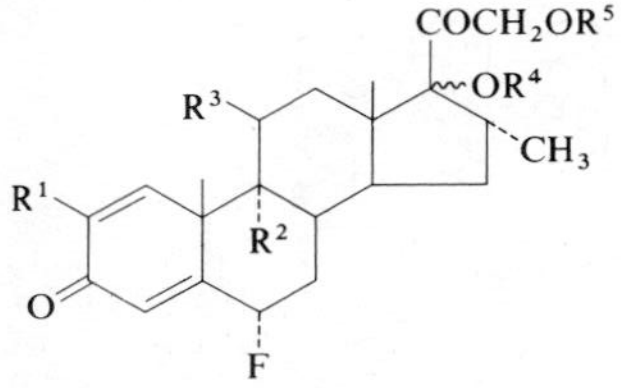			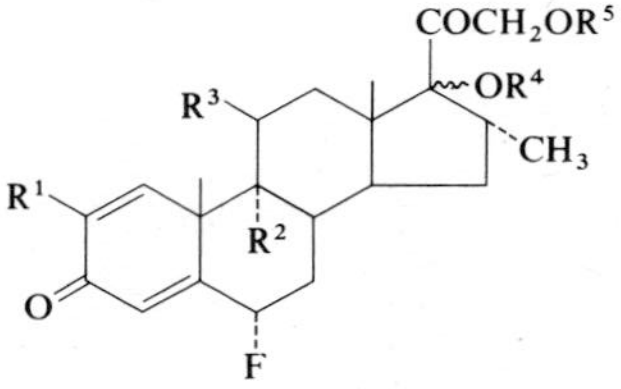	
$i\text{-}C_3H_7CHOHCH(N(CH_2C_6H_5)_2)CO_2CH_2C_6H_5$ *erythro* *threo*	"	THF, room temp, 50 min	$i\text{-}C_3H_7CH(N(CH_2C_6H_5)_2)CHFCO_2CH_2C_6H_5$ *erythro* (90) *threo* (90)	55
$COCH_2OR^5$, OR^4, CH_3, R^1, R^2, R^3, O, F; R¹ R² R³ R⁴ R⁵			$COCH_2OR^5$, OR^4, CH_3, R^1, R^2, R^3, O, F; R¹ R² R³ R⁴ R⁵	
Cl Cl OH $C(C_2H_5)(OC_2H_5)$	$C_5H_{10}NSF_3$	Dioxane, CH_2Cl_2, 4°, 2 h	Cl Cl F $C(C_2H_5)(OC_2H_5)$ (—)	95
" " " OH $COC_4H_9\text{-}t$	"	CH_2Cl_2, −40°, 2.5 h	" " " OH $COC_4H_9\text{-}t$ (92)	30, 95
H " " " $COC_4H_9\text{-}n$	"	Dioxane, 4–6°, 2 h	H " " " $COC_4H_9\text{-}n$ (—)	95
" F " " $COC_4H_9\text{-}t$	"	", −30°, 3.5 h	" F $C_5H_{10}NSO_2$ " $COC_4H_9\text{-}t^p$ (94)	30, 95
HO	DAST	CH_2Cl_2, −78° to room temp	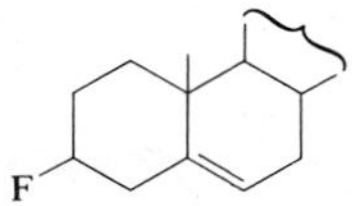 F (95)	32
	$OC_4H_8NSF_3$	" , 5°, 10 min	" (90)	164a
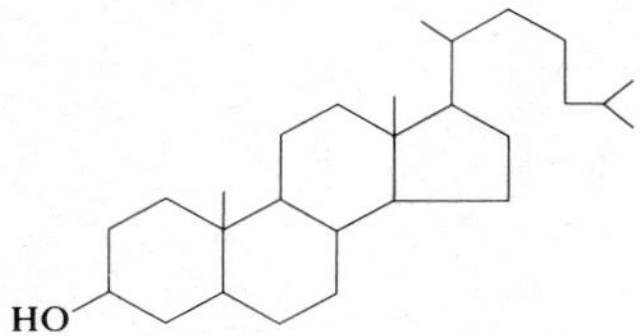 HO	"	", ", "	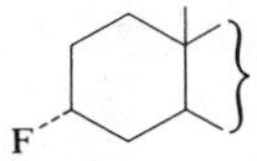 F (43) 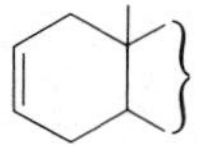 + (32)	32

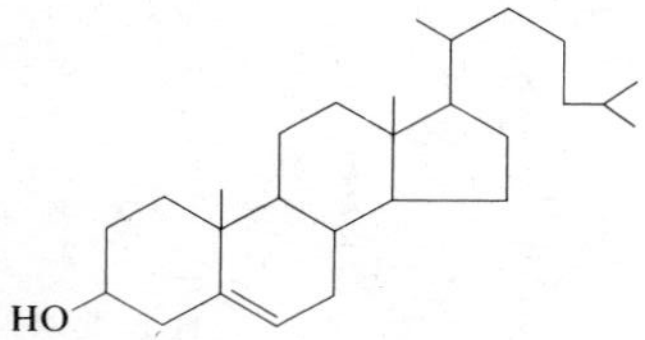

TABLE I. ALCOHOLS (*Continued*)

Reactant	Reagent (Molar Equiv.)	Conditions	Product(s) and Yield(s) (%)	Refs.
C_{28}				
$C_6H_5CO_2$, OH, $C_6H_5CO_2$, $C_6H_5CO_2$, O, OCH_3	DAST (1.5)	CH_2Cl_2, −25°, room temp overnight	$C_6H_5CO_2$, F, $C_6H_5CO_2$, $C_6H_5CO_2$, O, OCH_3 (41)	33
$COCH_2O_2CC_2H_5$, HO, $O_2CC_2H_5$, R^3, R^2, Cl, O, R^1	$C_5H_{10}NSF_3$	", −40°, 2.5 h	F, R^2, Cl, O, R^1	30, 95
R^1 R^2 R^3: H H β-CH_3			(88)	
F Cl α-CH_3			(95)	
R^3, R^4, R^1, R^2, OCH_3			R^3, R^4, R^1, R^2, OCH_3	

Reactant	Fluorinating Agent	Conditions	Product(s) and Yield(s) (%)	Refs.
R¹ R² R³ R⁴			R¹ R² R³ R⁴	
OH H H H	DAST	CH_2Cl_2, −78°	F H H H (86)	91, 108
H OH " "	"	", −78° to 20°, 1 h	H F[q] " " (24)	108
" H OH "	" (3)	", −78° to room temp, 15 min	" H F[q] " (91)	108
" " H OH[r]	"	", −78° to 20°, 20 min	" " H F[s] (85)	108
C_{29}				
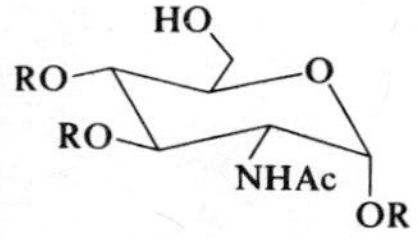 RO, RO, HO, O, NHAc, OR; R=$C_6H_5CH_2$	"	CH_2Cl_2, 0°, 1 h; 25°, 1 h	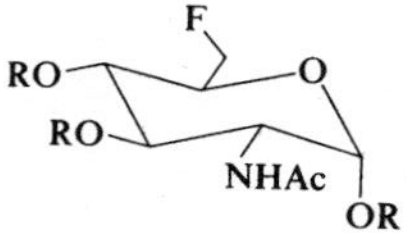 RO, RO, F, O, NHAc, OR (92)	63
R¹, R², R³, R⁴, R⁵ (structure)			R¹, R², R³, R⁴, R⁵ (structure)	
R¹ R² R³ R⁴ R⁵			R¹ R² R³ R⁴ R⁵	
H CH_2 α-OH β-OAc H, H	"	CH_2Cl_2, −78°, 5 min	H CH_2 F[q] β-OAc H, H (42)	94, 100
OH " " " "	"	", −70°, 3 min	F " α-F " " (35-42)	49, 94
" " H " "	"	", −78°, 5 min to room temp, 15 min	F " H " " (59)	94, 102, 103
" H, H β-OAc H CH_2	"	", −78°, to room temp	F H, H β-OAc H CH_2 (—)	94

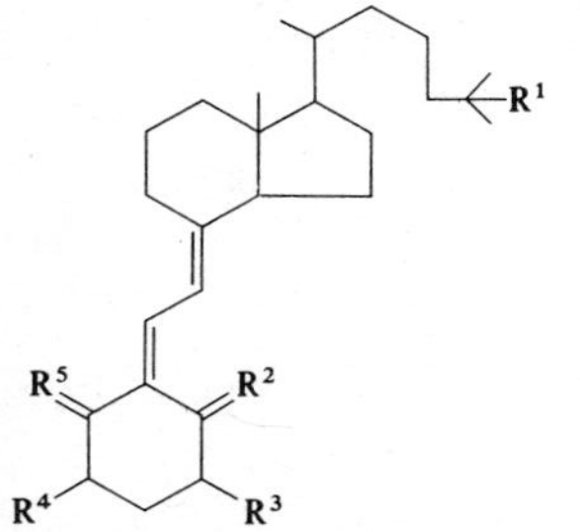

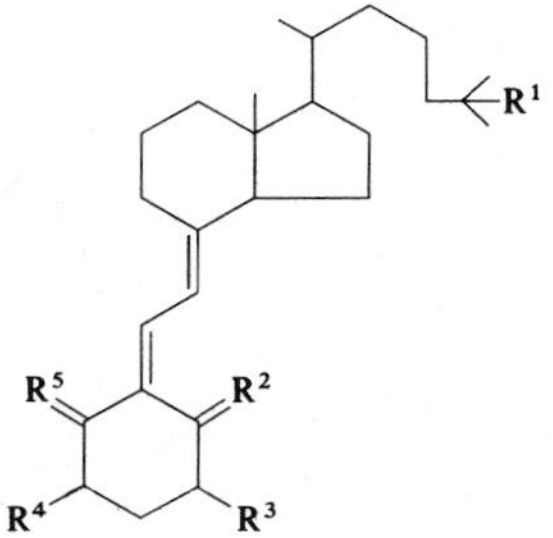

TABLE I. ALCOHOLS (*Continued*)

Reactant	Reagent (Molar Equiv.)	Conditions	Product(s) and Yield(s) (%)	Refs.
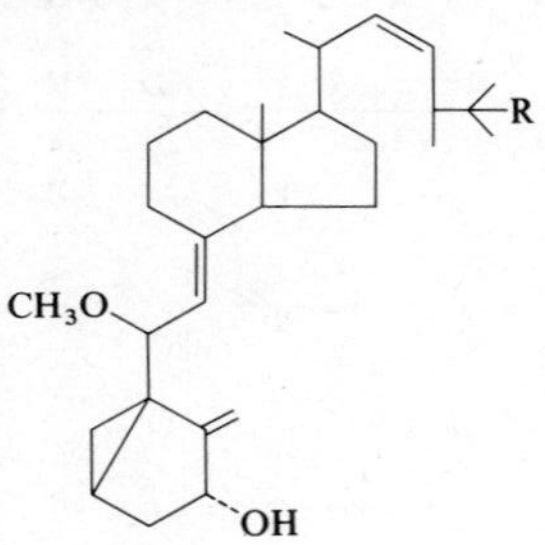 CH_3O … OH; R = H	DAST	CH_2Cl_2, −78°, 15 min to room temp	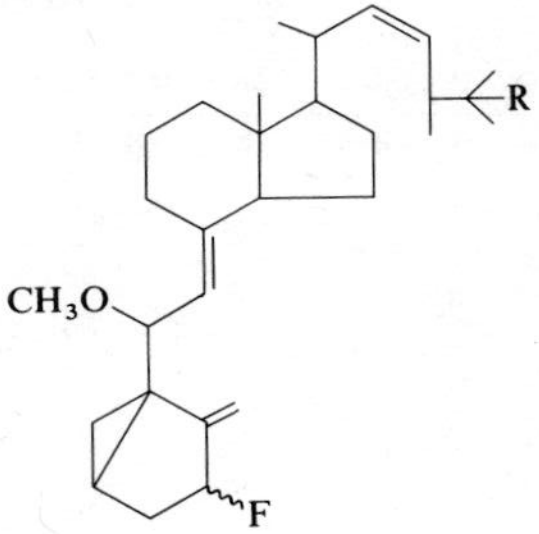 CH_3O … F; R = H(—)	49, 94
R = OH	"	", −70°, 3 min	R = F(75)	49, 94
AcO … OH	"		F (76)	93
HO … OH, *t*	"	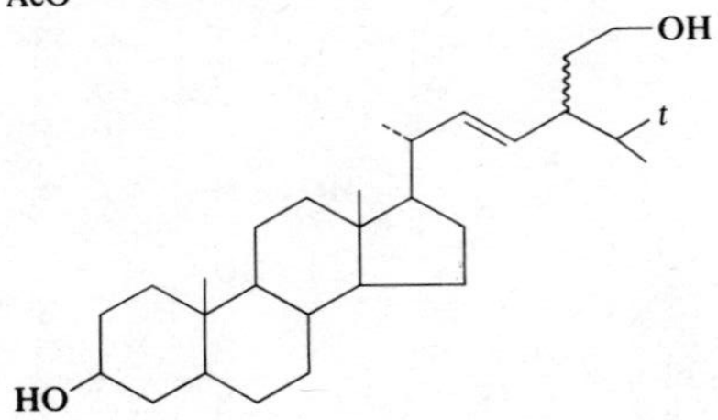 ", −78° to 25°	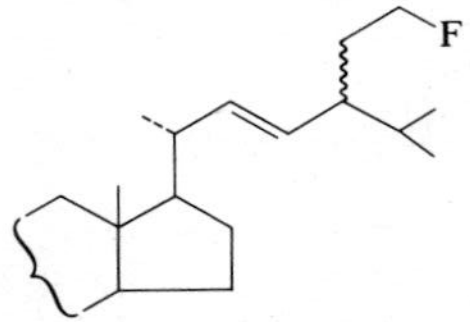 F (63)[f]	104

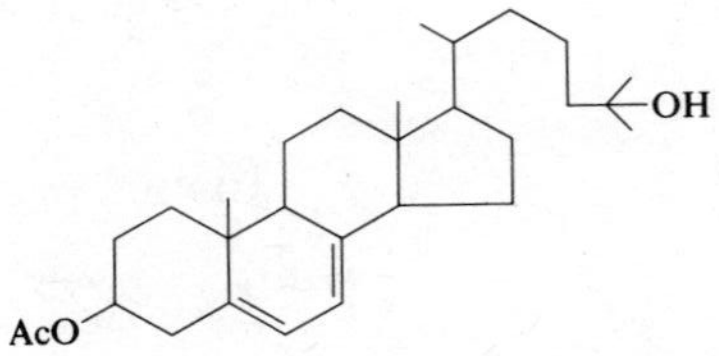

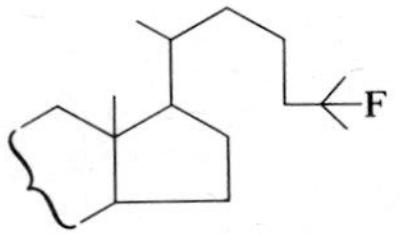

Substrate	Reagent	Conditions	Product(s) (Yield, %)	Ref.
AcO ... OH **A** or HO ... AcO **B**	"	CH_2Cl_2, 3°, 24 h	AcO **D** + AcO ... F **C** + AcO **E**	41

	C	D	E
From **A**	(23)	(32)	(12)
From **B**	(10)	(31)	(14)

Substrate	Reagent	Conditions	Product(s) (Yield, %)	Ref.
AcO ... OH	"	CH_2Cl_2, −78°	F (92)	91
OH ... O O ... OCH_3	" (5)	", −78° to 20°, 30 min	F ... O O (90)	108

TABLE I. ALCOHOLS (*Continued*)

Reactant	Reagent (Molar Equiv.)	Conditions	Product(s) and Yield(s) (%)	Refs.
C_{30}				
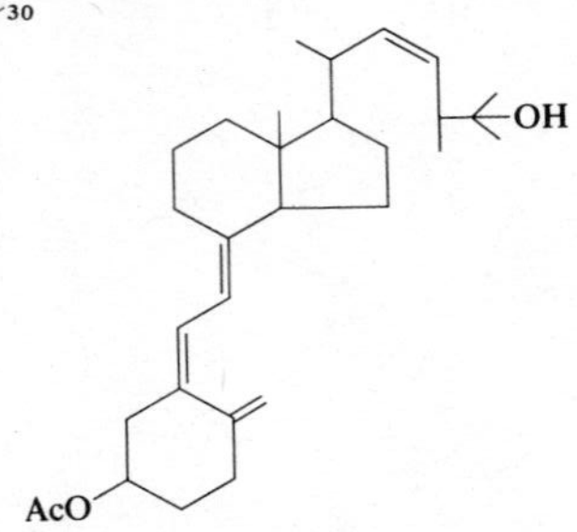 [structure] OH, AcO	DAST	CH_2Cl_2 ", −78° to room temp	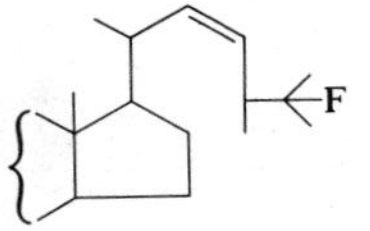 [structure] F (—)	94
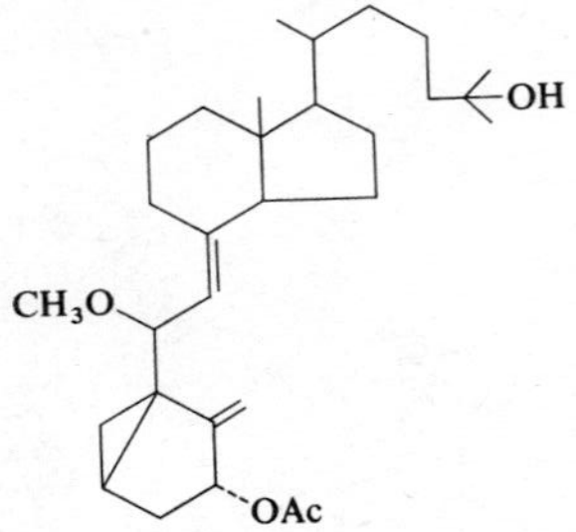 [structure] OH, CH_3O, OAc	"	", ", "	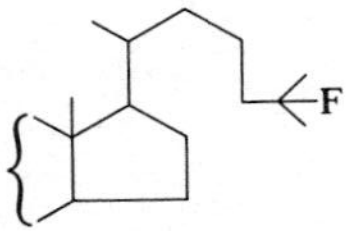 [structure] F (66)	94

Reactant	Fluorinating Agent (Equiv.)	Conditions	Product(s) and Yield(s) (%)	Refs.
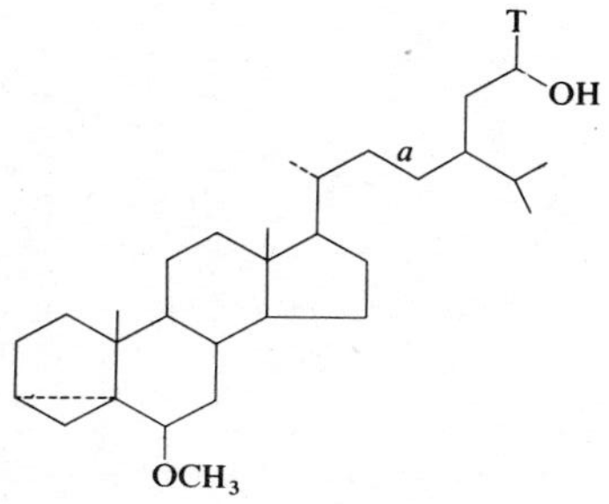 (structure with T, OH, bond *a*, OCH_3)			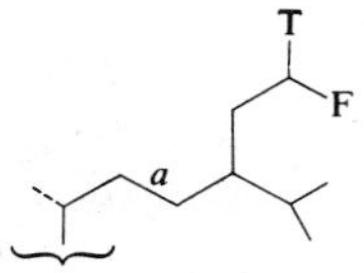 (structure with T, F, bond *a*)	
a: Double bond	$(CH_3)_2NSF_2$-$N(C_2H_5)_2$, (2.5)	CH_2Cl_2, −78° to 20°, 1 h	(88)	109
a: Single bond	DAST (2)	", −5°, 0°–20° overnight	(56)	109
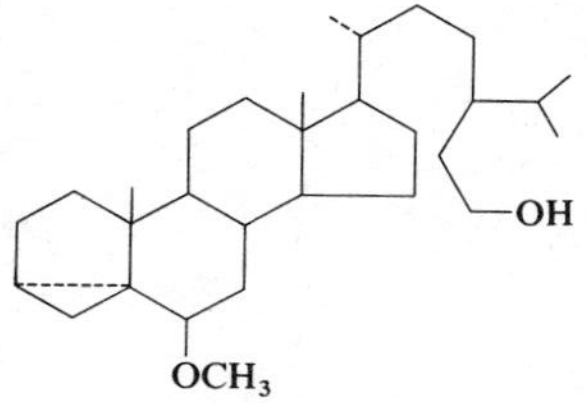 (structure with OH, OCH_3)	"	", −78° to 25°	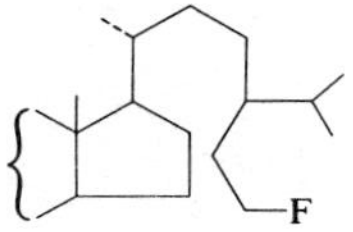 (structure with F) (63)	104

TABLE I. ALCOHOLS (*Continued*)

Reactant	Reagent (Molar Equiv.)	Conditions	Product(s) and Yield(s) (%)	Refs.
C_{31}				
[structure] R¹ R² R³ R⁴ R⁵ R⁶			[structure] R¹ R² R³ R⁴ R⁵ R⁶	
OH H CH_2 α-OAc β-OAc H, H	DAST	CCl_4, 10°, 10 min	F H CH_2 α-OAc β-OAc H, H (>5)	94
" " " β-OAc α-OAc "	"	CCl_3F, −78°	" " " β-OAc α-OAc " (>44)[u]	91
" OAc " H " "	"	CH_2Cl_2, ", 2 min	" OAc " H " " (>18)	94, 101
OAc H " β-OH β-OAc "	"	", ", to room temp	OAc H " F[q] β-OAc " (35)	94
OH " H, H β-OAc α-OAc CH_2	"	", ", 15 min	F H H,H β-OAc α-OAc CH_2 (>18)	94
OAc " " " α-OH "	"	", ", 5 min	OAc " " " F[q] " (>30)	94
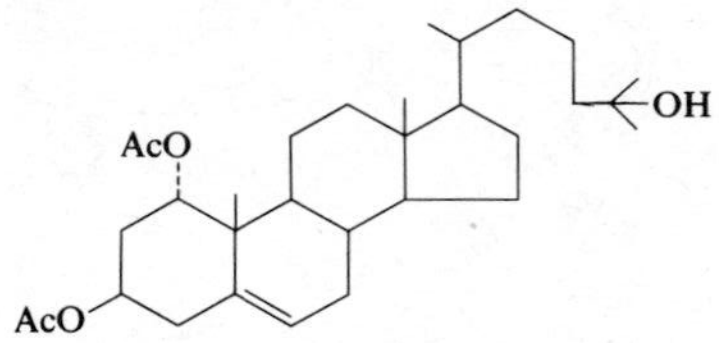 [structure: AcO, AcO, OH]	"	CCl_3F, −78°	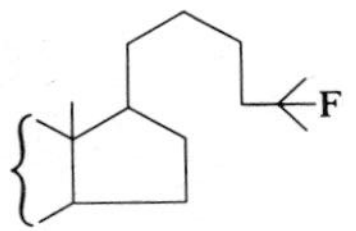 [structure: F] (97)	91

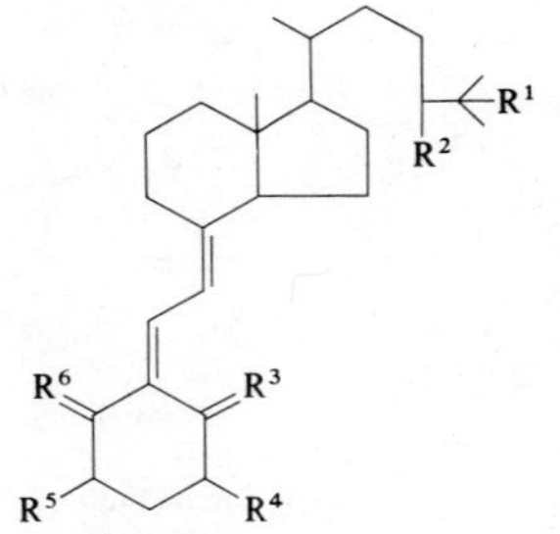

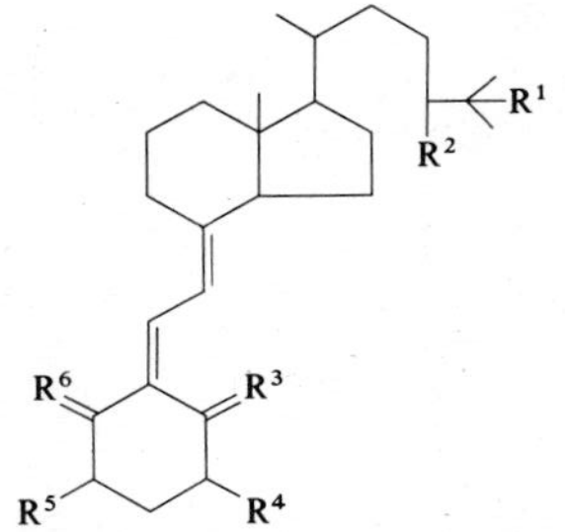

Substrate	Reagent	Conditions	Product(s) and Yield(s) (%)	Refs.
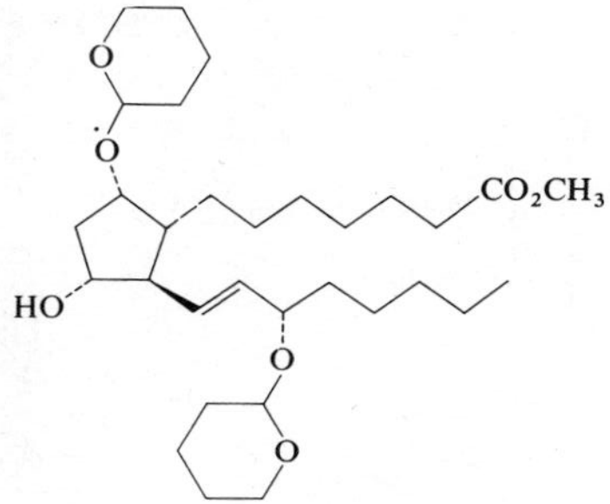 [structure]	$OC_4H_8NSF_3$	CH_2Cl_2, −78°, 3 h	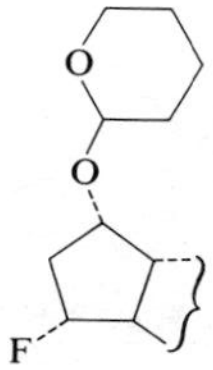 [structure] (>40)	111
C_{32} 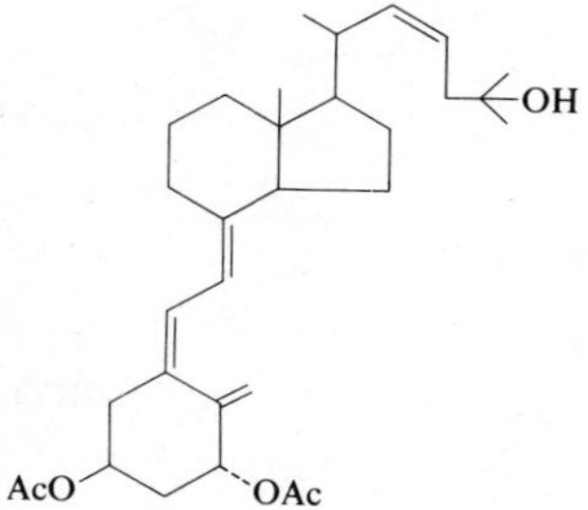 [structure]	DAST	", −78° to room temp	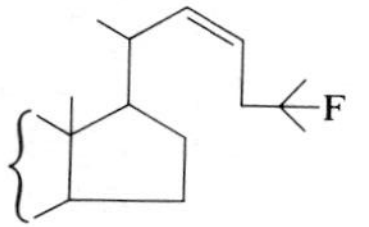 [structure] (—)	94
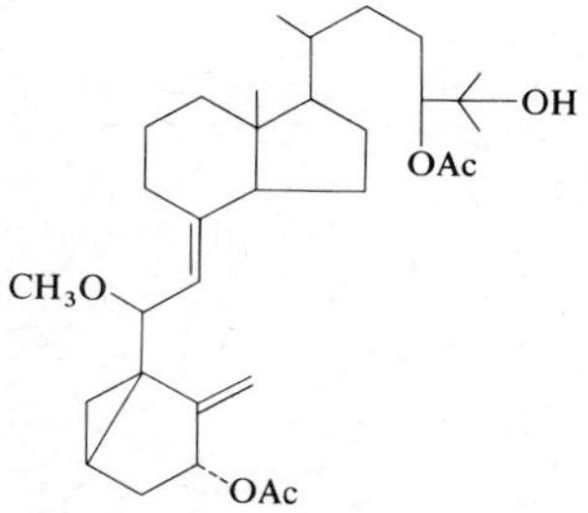 [structure]	"	CH_2Cl_2, −78°	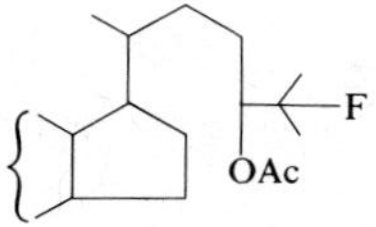 [structure] (33)	94

TABLE I. ALCOHOLS (*Continued*)

Reactant	Reagent (Molar Equiv.)	Conditions	Product(s) and Yield(s) (%)	Refs.
C_{33}				
$C_6H_5CHCONH$, t-$C_4H_9O_2CNH$, $(C_6H_5)_2CHO_2C$, OH	$C_5H_{10}NSF_3$, (3)	CCl_4, room temp	$(C_6H_5)_2CHO_2C$, F (3)	130
C_{34}				
RO, RO, RO, RO, OH; R: C_6H_5CO, $C_6H_5CH_2$	DAST	THF, $-30°$, room temp, 20 min	RO, RO, RO, RO, F; α (8), β (84); α (12), β (88)	90
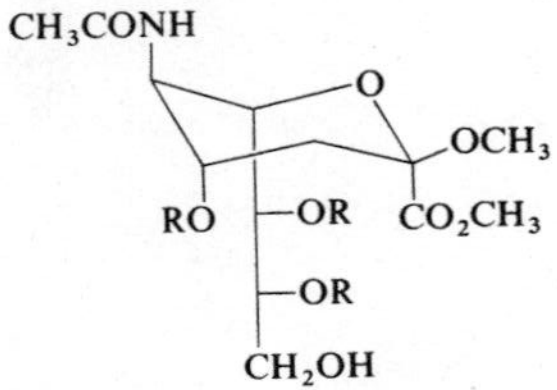 CH_3CONH, OCH_3, RO, OR, CO_2CH_3, OR, CH_2OH; $R = C_6H_5CH_2$	"	CH_2Cl_2, $-10°$ to $0°$, 3 h	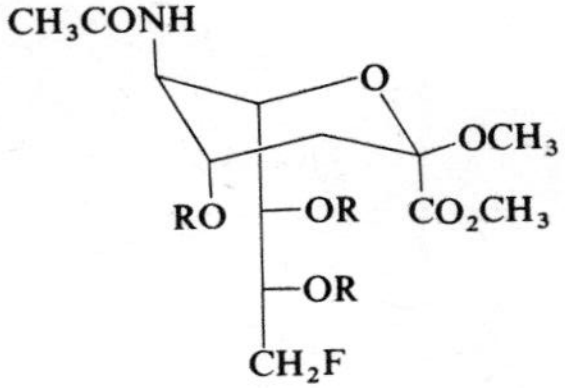 CH_3CONH, OCH_3, RO, OR, CO_2CH_3, OR, CH_2F (59)	88

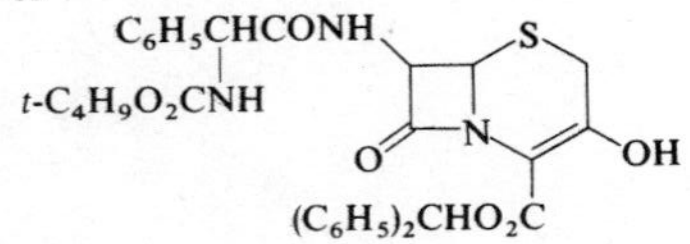

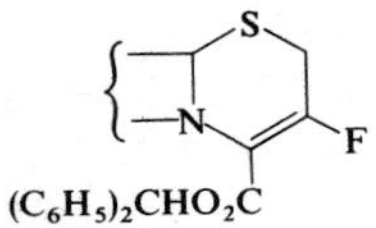

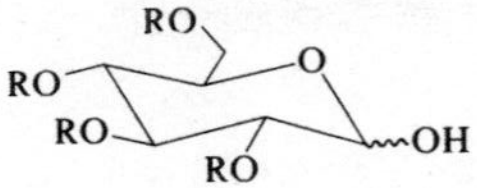

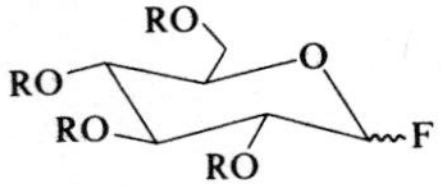

Reactant	Fluorinating Agent	Conditions	Product(s) and Yield(s) (%)	Refs.
C_{35-36}				
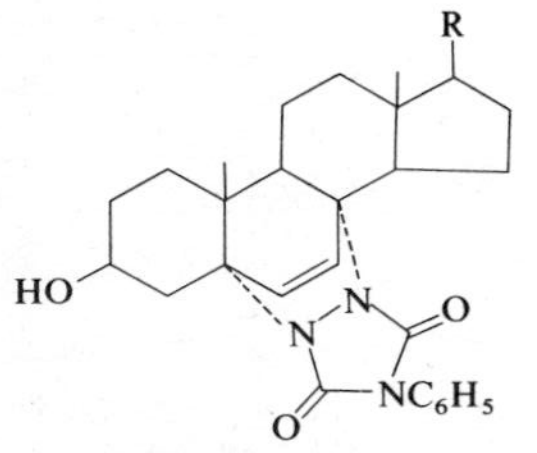 HO, R; N, N, O, O, NC_6H_5 $R = CH(CH_3)(CH_2)_3C_3H_7$-*i* $R = CH(CH_3)CH{=}CHCH(CH_3)C_3H_7$-*i*	$OC_4H_8NSF_3$	", 5°, 20 min	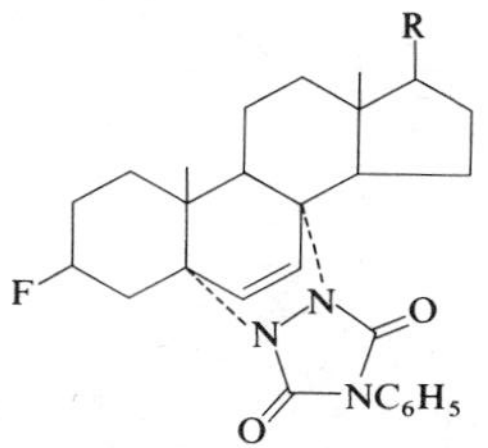 F, R; N, N, O, O, NC_6H_5 (65) (75)	164a
C_{39}				
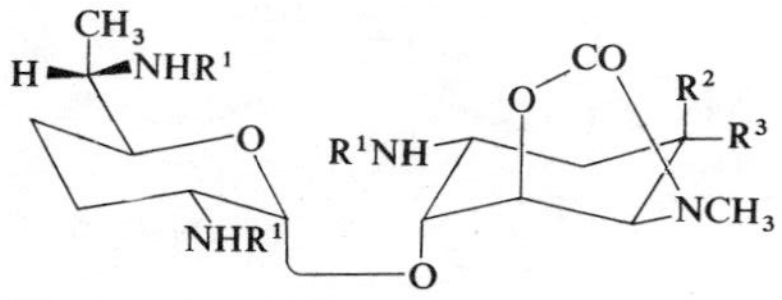 CH_3, H, NHR^1, O, NHR^1, O, R^1NH, O, CO, R^2, R^3, NCH_3 R^1 / R^2 / R^3: $C_6H_5CH_2O_2C$ / H / OH; " / OH / H	DAST	CH_2Cl_2, −15°, 15 min, room temp, 30 min	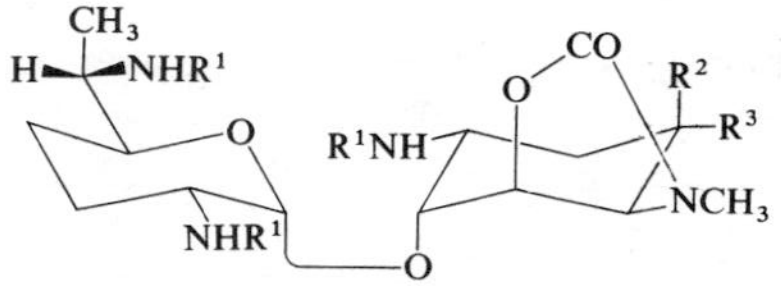 CH_3, H, NHR^1, O, NHR^1, O, R^1NH, O, CO, R^2, R^3, NCH_3 R^1 / R^2 / R^3: $C_6H_5CH_2O_2C$ / F / H (59); " / H / F (68)	125, 129
C_{41}				
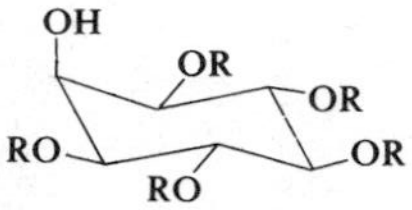 OH, OR, OR, RO, RO, OR $R = C_6H_5CH_2$	"	$C_6H_5CH_3$, 70–80° [v]	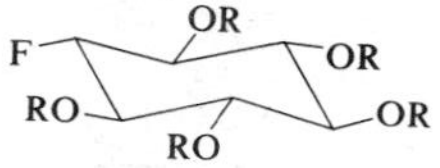 F, OR, OR, RO, RO, OR (86)	46
C_{42}				
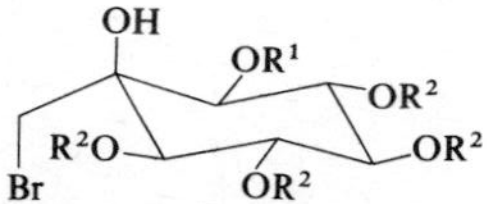 OH, OR^1, OR^2, R^2O, OR^2, OR^2, Br $R^1 = C_6H_5CO$, $R^2 = C_6H_5CH_2$	" (2)	", 60–65° 40 min	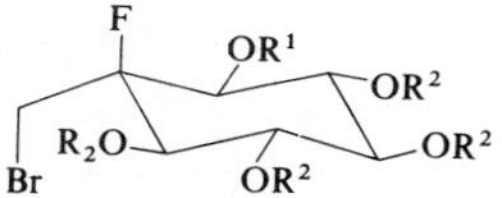 F, OR^1, OR^2, R_2O, OR^2, OR^2, Br (83)	164b

TABLE I. ALCOHOLS (*Continued*)

Reactant	Reagent (Molar Equiv.)	Conditions	Product(s) and Yield(s) (%)	Refs.
C_{47}				
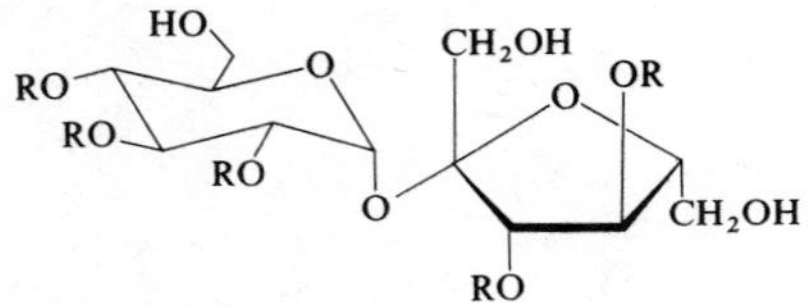 HO, RO, RO, RO, CH_2OH, OR, CH_2OH, RO R = C_6H_5CO	DAST	Diglyme, −5°, 90°, 1 h	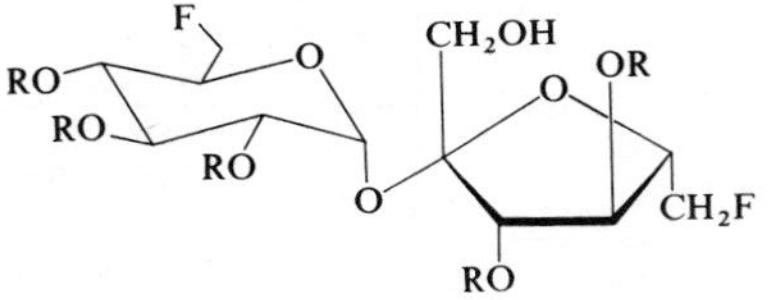 F, RO, RO, RO, CH_2OH, OR, CH_2F, RO (28)	87
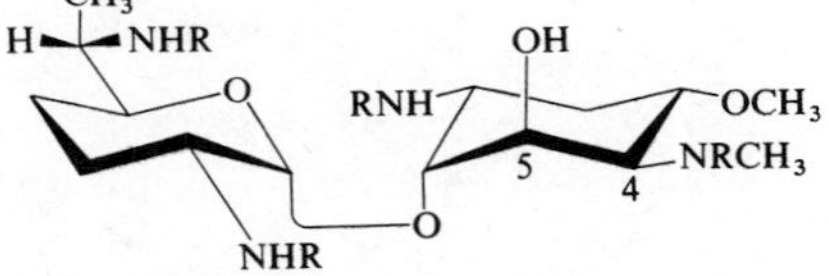 CH_3, H, NHR, RNH, OH, OCH_3, $NRCH_3$, 5, 4, NHR R = $C_6H_5CH_2O_2C$	"	CH_2Cl_2, −15°, 15 min, room temp, 30 min	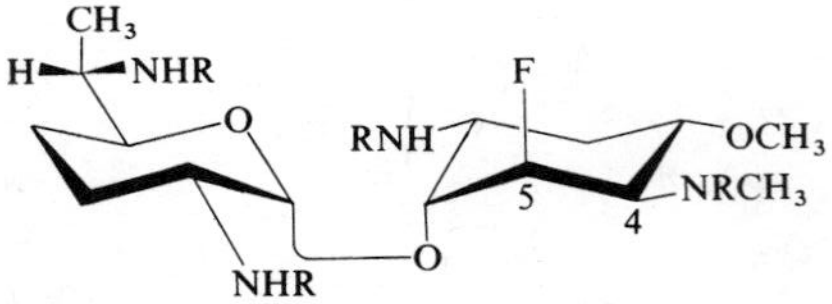 CH_3, H, NHR, RNH, F, OCH_3, $NRCH_3$, 5, 4, NHR (31)[w]	128
C_{48}				
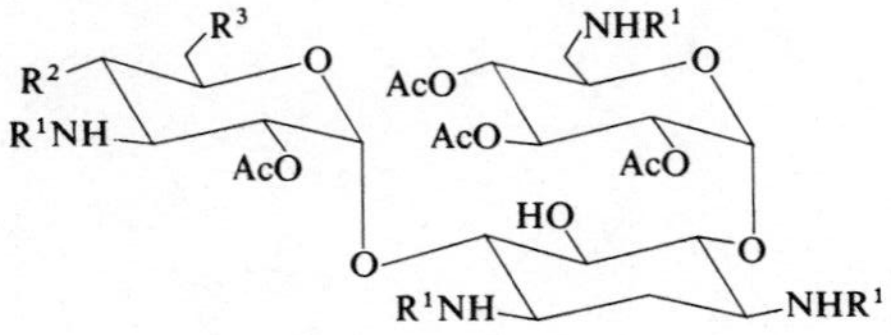 R^3, R^2, R^1NH, AcO, AcO, AcO, NHR^1, AcO, HO, R^1NH, NHR^1 R^1 / R^2 / R^3 t-$C_4H_9O_2C$ / AcO / OH " / OH / AcO	"	", −70°	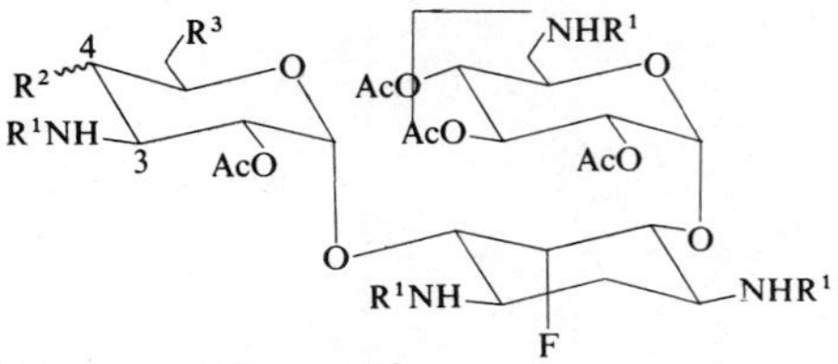 4, R^3, R^2, R^1NH, 3, AcO, AcO, AcO, NHR^1, AcO, R^1NH, F, NHR^1 R^1 / R^2 / R^3 t-$C_4H_9O_2C$ / AcO(eq) / F (79) " / OH (ax) / AcO (—)[x]	126

C_{59}

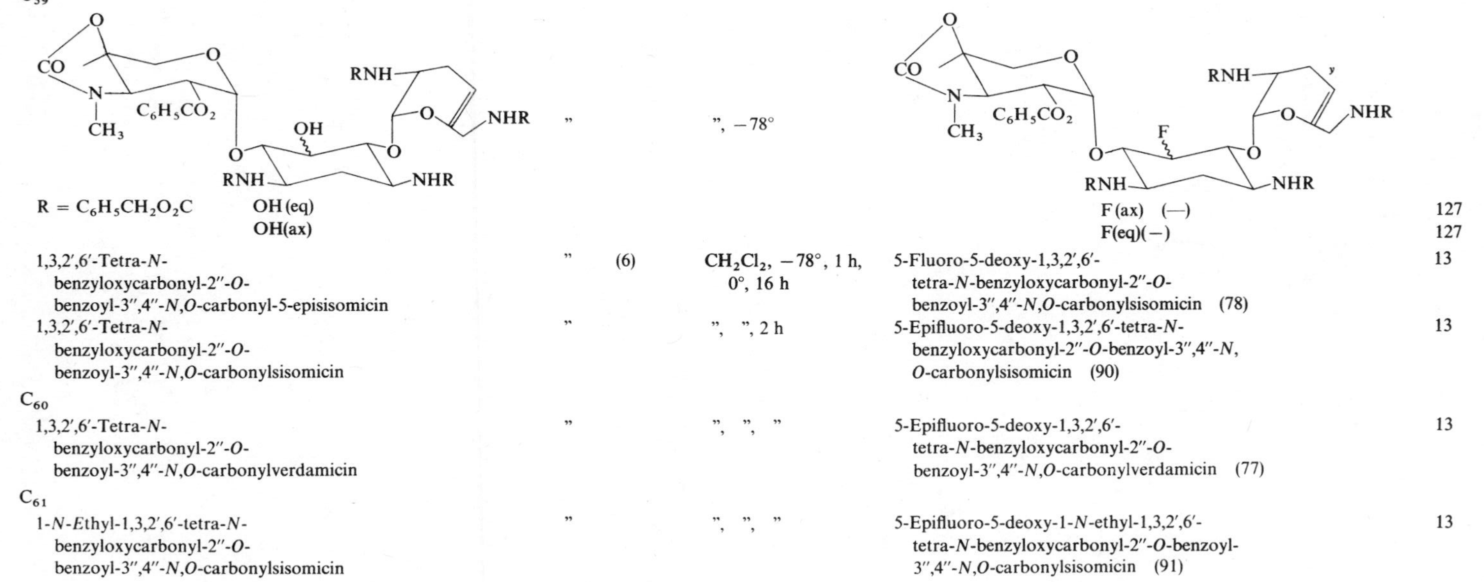

$R = C_6H_5CH_2O_2C$ OH (eq)	”		”, −78°	F (ax) (—)	127
OH(ax)				F(eq)(−)	127
1,3,2′,6′-Tetra-*N*-benzyloxycarbonyl-2″-*O*-benzoyl-3″,4″-*N*,*O*-carbonyl-5-episisomicin	”	(6)	CH_2Cl_2, −78°, 1 h, 0°, 16 h	5-Fluoro-5-deoxy-1,3,2′,6′-tetra-*N*-benzyloxycarbonyl-2″-*O*-benzoyl-3″,4″-*N*,*O*-carbonylsisomicin (78)	13
1,3,2′,6′-Tetra-*N*-benzyloxycarbonyl-2″-*O*-benzoyl-3″,4″-*N*,*O*-carbonylsisomicin	”		”, ”, 2 h	5-Epifluoro-5-deoxy-1,3,2′,6′-tetra-*N*-benzyloxycarbonyl-2″-*O*-benzoyl-3″,4″-*N*,*O*-carbonylsisomicin (90)	13
C_{60}					
1,3,2′,6′-Tetra-*N*-benzyloxycarbonyl-2″-*O*-benzoyl-3″,4″-*N*,*O*-carbonylverdamicin	”		”, ”, ”	5-Epifluoro-5-deoxy-1,3,2′,6′-tetra-*N*-benzyloxycarbonyl-2″-*O*-benzoyl-3″,4″-*N*,*O*-carbonylverdamicin (77)	13
C_{61}					
1-*N*-*E*thyl-1,3,2′,6′-tetra-*N*-benzyloxycarbonyl-2″-*O*-benzoyl-3″,4″-*N*,*O*-carbonylsisomicin	”		”, ”, ”	5-Epifluoro-5-deoxy-1-*N*-ethyl-1,3,2′,6′-tetra-*N*-benzyloxycarbonyl-2″-*O*-benzoyl-3″,4″-*N*,*O*-carbonylsisomicin (91)	13

TABLE I. ALCOHOLS (*Continued*)

Reactant	Reagent (Molar Equiv.)	Conditions	Product(s) and Yield(s) (%)	Refs.
C_{75} 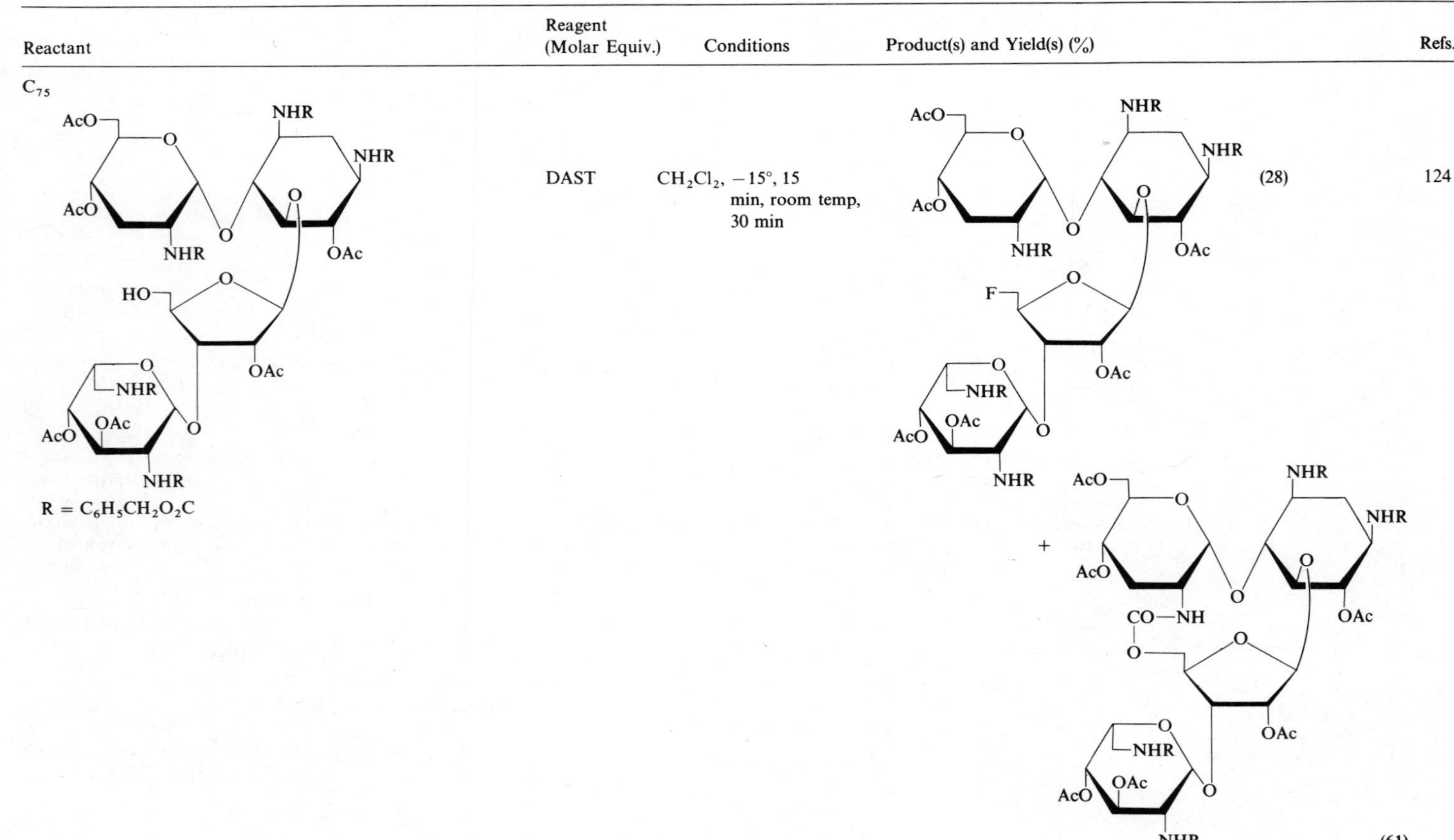 $R = C_6H_5CH_2O_2C$	DAST	CH_2Cl_2, $-15°$, 15 min, room temp, 30 min	(28) + (61)	124

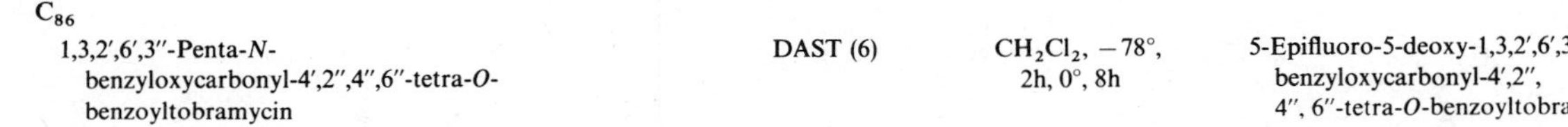

Substrate	Reagent	Conditions	Product(s) (Yield, %)	Refs.
C_{86}				
1,3,2′,6′,3″-Penta-*N*-benzyloxycarbonyl-4′,2″,4″,6″-tetra-*O*-benzoyltobramycin	DAST (6)	CH_2Cl_2, −78°, 2h, 0°, 8h	5-Epifluoro-5-deoxy-1,3,2′,6′,3″penta-*N*-benzyloxycarbonyl-4′,2″, 4″, 6″-tetra-*O*-benzoyltobramycin (95)	13

[a] The yield is of the crude product, which decomposed at room temperature within hours.

[b] The sugar had the *L* configuration.

[c] The yield of the recrystallized product was 67%.

[d] The reaction was carried out with the cyanohydrin trimethylsilyl ether prepared from the aldehyde and trimethylsilyl cyanide.

[e] Equal yields (30%) of the two diastereomers were obtained, and 40% of the starting material was recovered.

[f] The procedure of ref. 5 was used.

[g] The product was 90% pure.

[h] The yields were increased from 20–25% to 28% and 40% by addition of 2 molar equivalents of pyridine or potassium fluoride, respectively.

[i] In some experiments a mixture of both *cis* and *trans* isomers was obtained.

[j] The ratio of diastereomers (3*R*, 6*S*, 3′*S*) : (3*R*, 6*S*, 3′*R*) was 12:1.

[k] The radiochemical purity was 92%.

[l] DDQ is 2,3-dichloro-5,6-dicyano-1,4-benzoquinone. Without the isolation of the intermediate difluoride in pure form, the overall yield was 60%.

[m] The substrate and products were mixtures of 8 epimers (α and β).

[n] The β-fluoro compound was obtained from both the α-hydroxy and the β-hydroxy compounds.

[o] The effects of six solvents on the α:β ratio are tabulated in the paper.

[p] The product was an 85:15 mixture of isomers on sulfur.

[q] The stereochemistry of the fluorine atom was not determined.

[r] The substrate consisted of 90% of the 20-(*S*) epimer.

[s] The product was a mixture of epimers at C-20.

[t] The substrate and the product were mixtures of two epimers at C-24.

[u] The $\triangle^{24(25)}$ and $\triangle^{25(26)}$ dehydration products were isolated in 4% yield.

[v] No reaction took place at 0–25°.

[w] The C-4, C-5 oxazolidinone derivative was isolated in 39% yield; starting material was recovered in 13% yield.

[x] The C-3, C-4 oxazolidinone derivative was isolated in unspecified yield.

[y] Deprotection *in situ* gave 5-epifluoro- or 5-fluoro-5-deoxysisomicin.

TABLE II. ALDEHYDES

No. of Carbon Atoms	Reactant	Reagent (Molar Equiv.)	Conditions	Product(s) and Yield(s) (%)	Refs.
C_2	CBr_3CHO	DAST	CCl_4, 25–60°, 1 h; 40°,8 h	$(CBr_3CHF)_2O$[a] (20)	36, 165
	CCl_3CHO	$(CH_3)_2NSF_3$	35–40°, 4 h	$(CCl_3CHF)_2O$[a] (62)	36
		$OC_4H_8NSF_3$	36–43°, 5 h	" (80)	36
		DAST	$C_6H_5CH_3$, 28–43°, 12.5 h,	" (78)	36, 165
		"	25 to 40°	" (27) + $CCl_3CH(OCHFCCl_3)_2$ (32)	36
	CF_3CHO	"	Diglyme, 60°, 2 h; 60–64°, 6 h	$(CF_3CHF)_2O$[a] (65)	36, 165
	$CHCl_2CHO$	"	CCl_4, −10° to −1°; warm to room temp, 2 h	$(CHCl_2CHF)_2O$[a] (67)	36
	CH_2ClCHO	"	", −26° to −11°; warm to room temp	CH_2ClCHF_2 (28) + $(CH_2ClCHF)_2O$[a] (15)	36
		"	", −20°, 2 h, to 0°	$(CH_2ClCHF)_2O$ (37)	165, 36
C_3	$CH_3N(CHO)_2$	" (2.5)	0°, 80°, 30 min	No reaction	143
	C_2H_5CHO	"	CCl_3F, 25°, 30 min	$C_2H_5CHF_2$ (95)	5
C_5	i-C_4H_9CHO	"	", ", "	i-$C_4H_9CHF_2$ (80)	5, 9
	t-C_4H_9CHO	"	", ", 1 h	t-$C_4H_9CHF_2$ (78)[b]	5
			Diglyme, 25°, 1 h	t-$C_4H_9CHF_2$ (24)	5
C_7	p-XC_6H_4CHO			p-$XC_6H_4CHF_2$	
	X=H	"	CH_2Cl_2, 25–35°, 2 h	X=H (75)	5, 9
		R_2NSF_3[c]	60°, 15 min	" (63)	8
	X=Br	"	", "	X=Br (71)	8
	X=Cl	"	", "	X=Cl (70)	8
		$(OC_4H_8N)_2SF_2$	C_6H_6, reflux, 1.5 h	" (61)	4
	X=NO_2	R_2NSF_3[c]	60°, 15 min	X=NO_2 (75)	8

	Reactant	Reagent	Conditions	Product(s) and Yield(s) (%)	Refs.
	n-$C_6H_{13}CHO$	"	Cooling, then 80°, 30 min	n-$C_6H_{13}CHF_2$ (65)	8
C_9	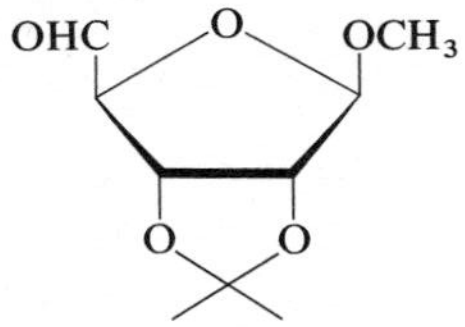 OHC, O, OCH_3 (furanoside, O,O-isopropylidene)	DAST	CH_2Cl_2, room temp, 16 h	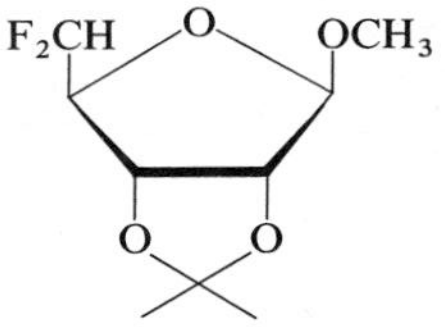 F_2CH, O, OCH_3 (furanoside, O,O-isopropylidene) (45)	70
C_{11}	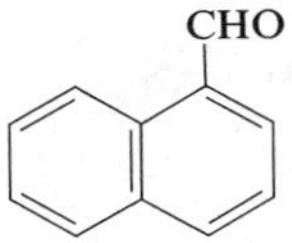 CHO (1-naphthaldehyde)	"	", 25°, 18 h	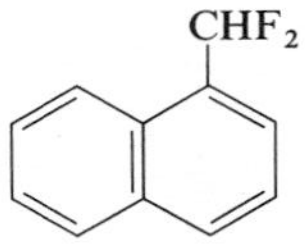 CHF_2 (1-naphthyl) (72)[d]	5
C_{12}	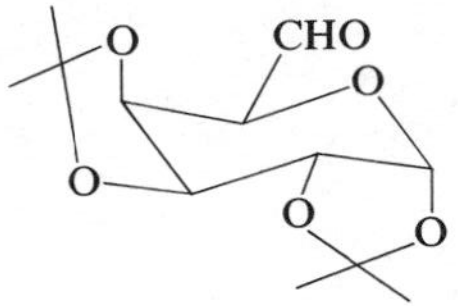 CHO (di-O-isopropylidene pyranose)			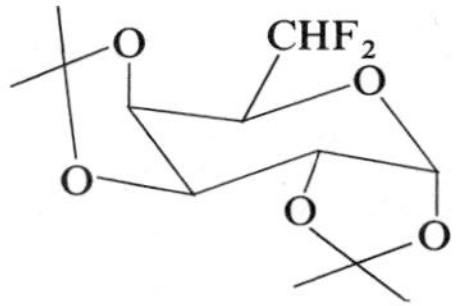 CHF_2 (di-O-isopropylidene pyranose)	
	D	" (2.5)	", room temp, 16 h	(46)	70
	L	" (2)	", 22°, 24 h	(62)	69
C_{18}	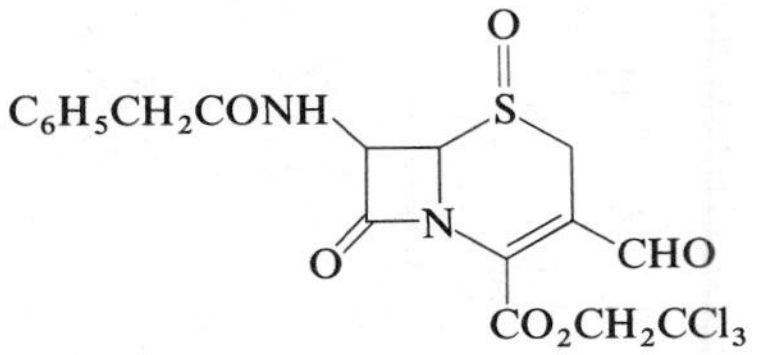 $C_6H_5CH_2CONH$, O=S, N, O, CHO, $CO_2CH_2CCl_3$ (cephem sulfoxide)	"	", 27°, 1 h	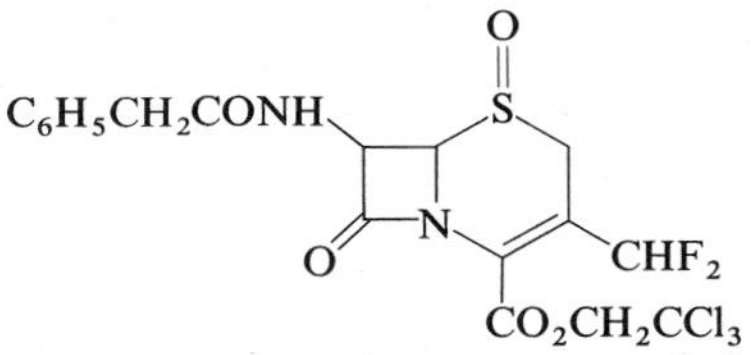 $C_6H_5CH_2CONH$, O=S, N, O, CHF_2, $CO_2CH_2CCl_3$ (cephem sulfoxide) (—)	68

TABLE II. ALDEHYDES (*Continued*)

No. of Carbon Atoms	Reactant	Reagent (Molar Equiv.)	Conditions	Product(s) and Yield(s) (%)	Refs.
C_{19}	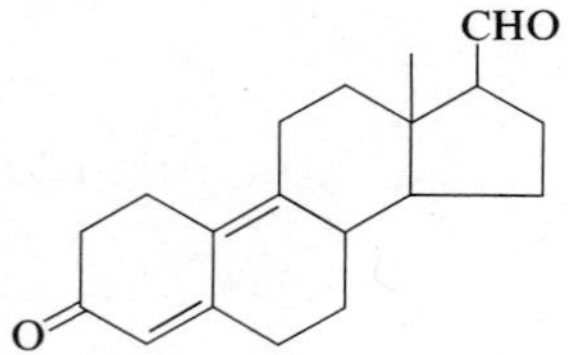 CHO	DAST	CH_2Cl_2, 0°, 20–25°, several h	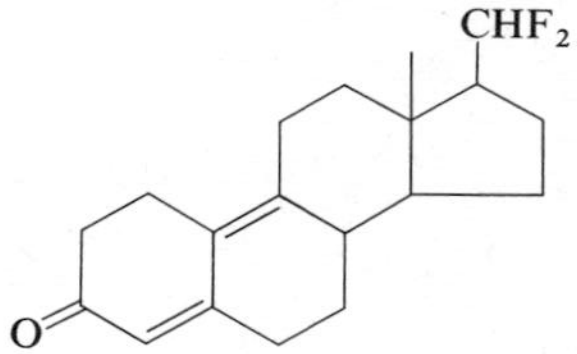 CHF_2 (—)	72
	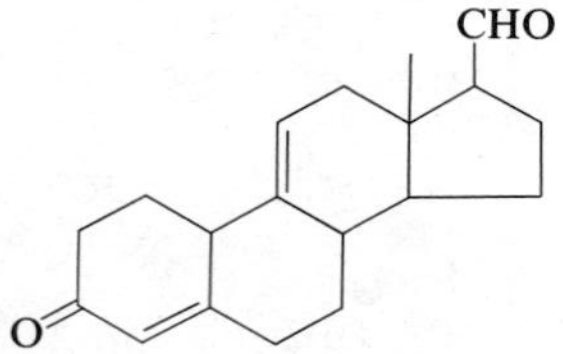 CHO	"	", ", ", "	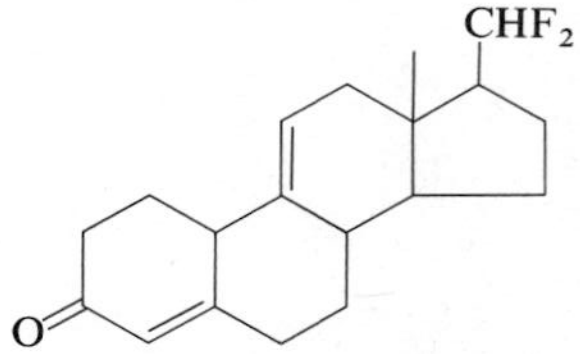 CHF_2 (—)	72
	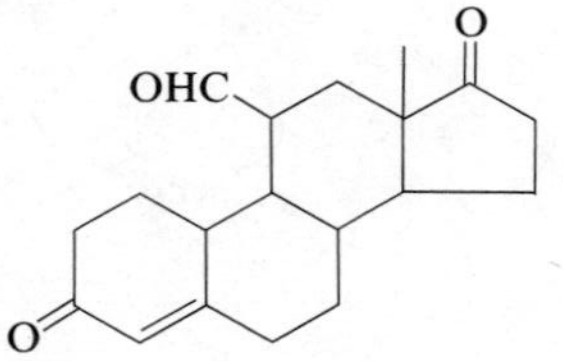 OHC, O	" (5)	", 20–25°, 24 h	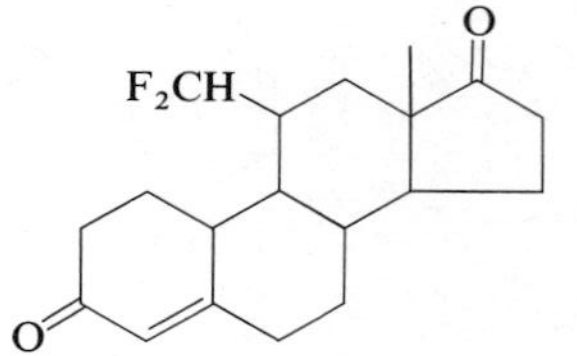 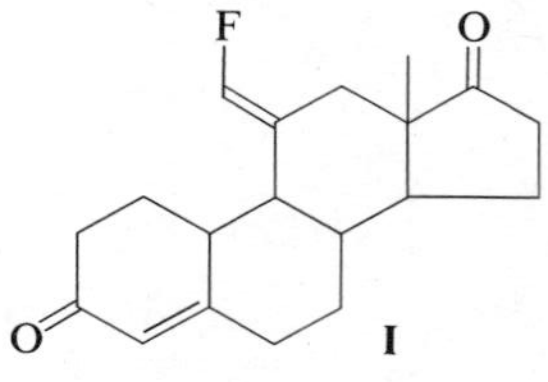 F_2CH, O (—) + F, O, **I** (—)	165a

	Reactant	Reagent	Conditions	Product(s) and Yield(s) (%)	Refs.
		" (5)	CH_3CN, ", "	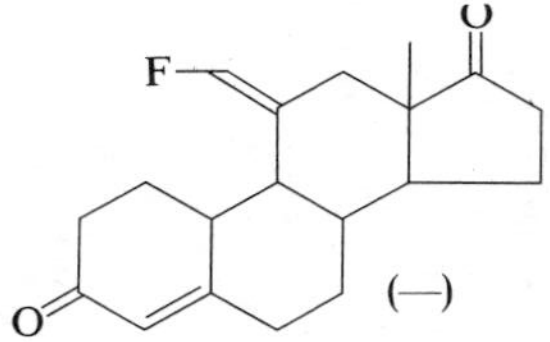 F ; O ; O (—) + **I** (—)	165a
	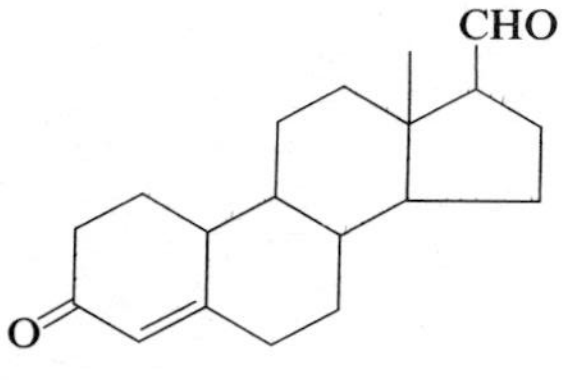 CHO ; O	"	CH_2Cl_2, 0°, 20–25°, several h	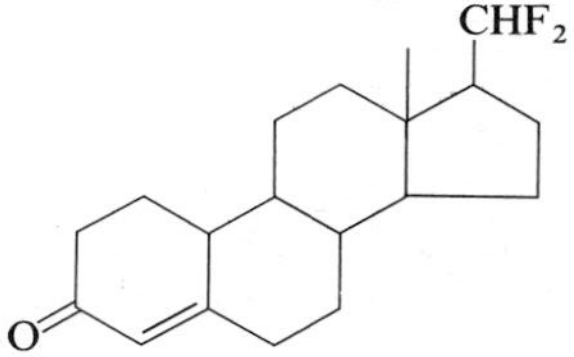 CHF_2 ; O (—)	72
C_{20}	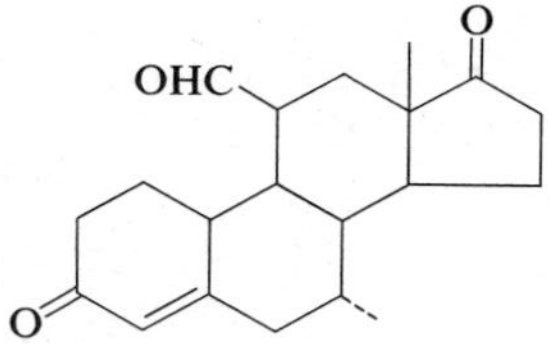 OHC ; O ; O	" (5)	", 20 h	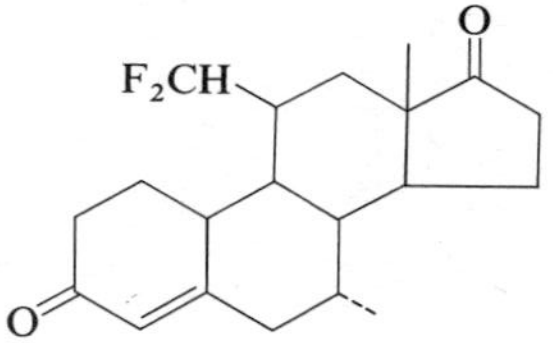 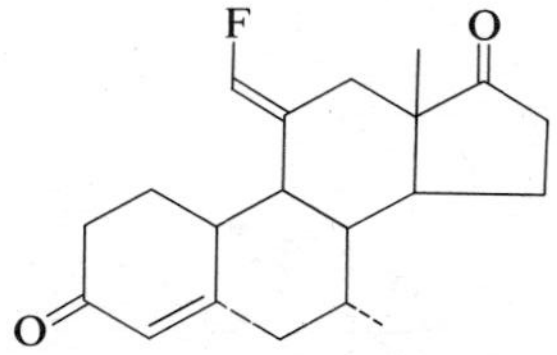 F_2CH ; O ; O (—) + F ; O ; O (—)	165a

TABLE II. ALDEHYDES (*Continued*)

No. of Carbon Atoms	Reactant	Reagent (Molar Equiv.)	Conditions	Product(s) and Yield(s) (%)	Refs.
	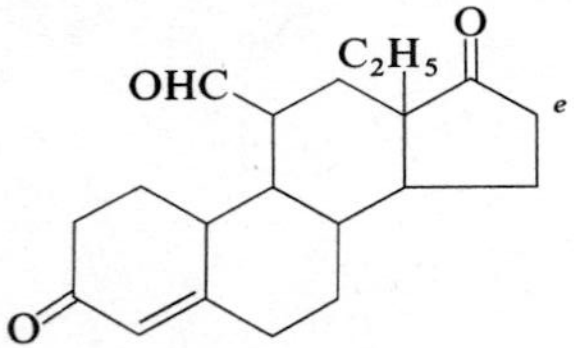 OHC, C_2H_5, O [e]	DAST (5)	CH_3CN, 20–25°, 24 h	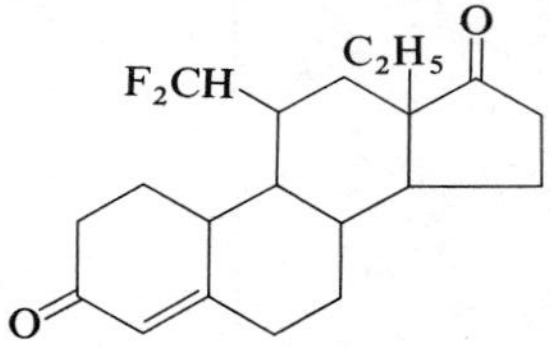 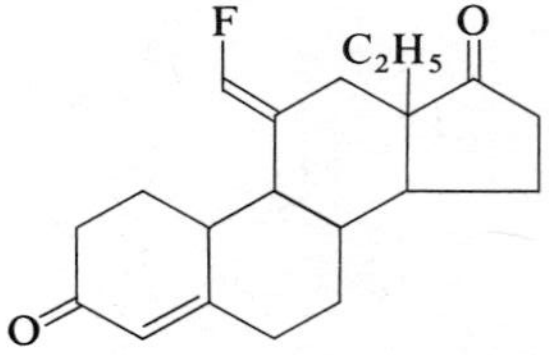 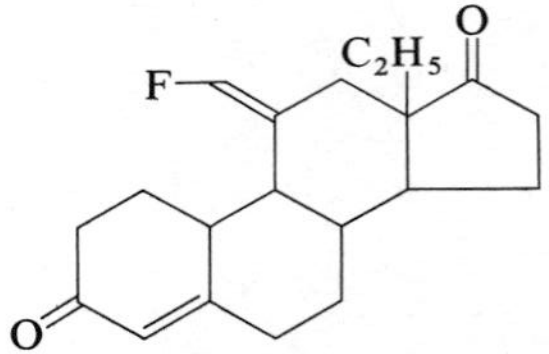 F_2CH, C_2H_5, O (—) + F, C_2H_5, O (—) + F, C_2H_5, O (—)	165a
	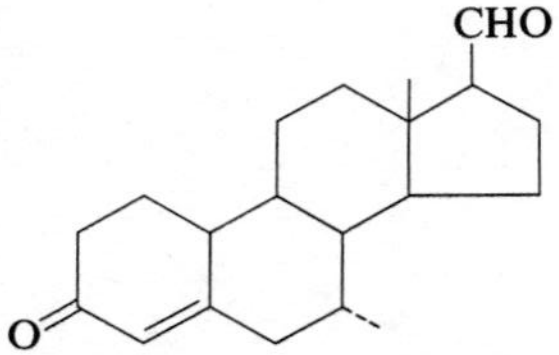 CHO	"	CH_2Cl_2, 0°, 20–25°, several h	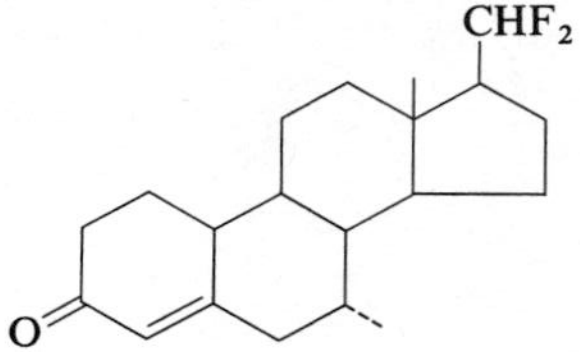 CHF_2 (—)	72

Formula	Reactant	Reagent	Conditions	Product (Yield, %)	Refs.
	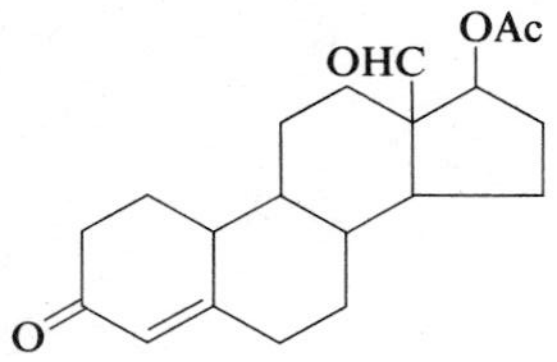 OHC, OAc, O	"	Room temp, 60°, 1 h	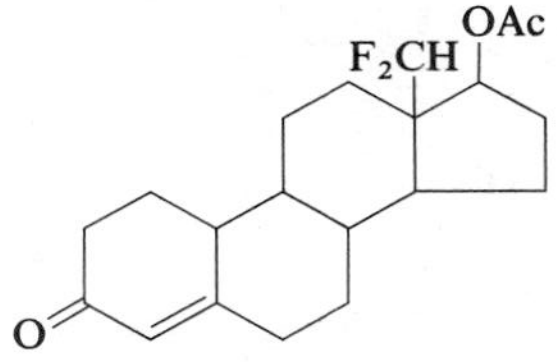 F_2CH, OAc, O (—)	71
	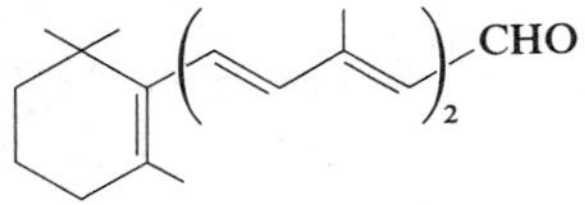 CHO, 2	"	80°, 30 min	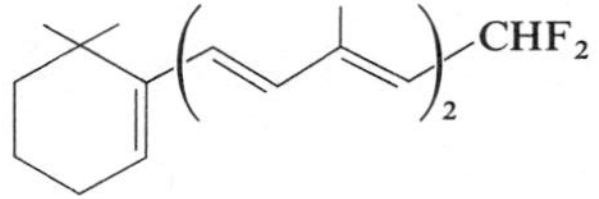 CHF_2, 2 (56)	115
C_{21}	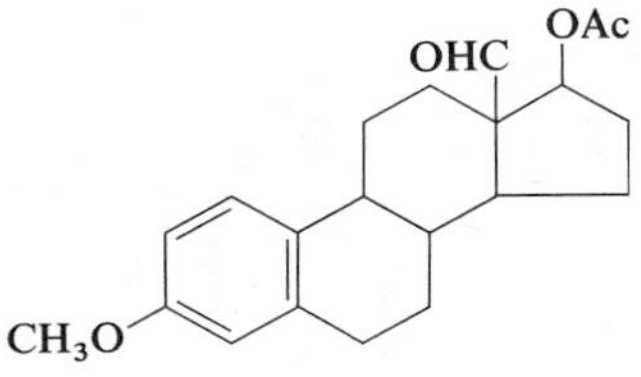 OHC, OAc, CH_3O	$C_5H_{10}NSF_3$	50°, 90 min	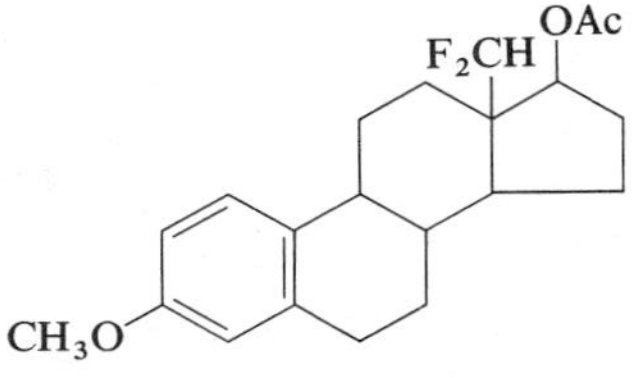 F_2CH, OAc, CH_3O (—)	71, 96
C_{22}	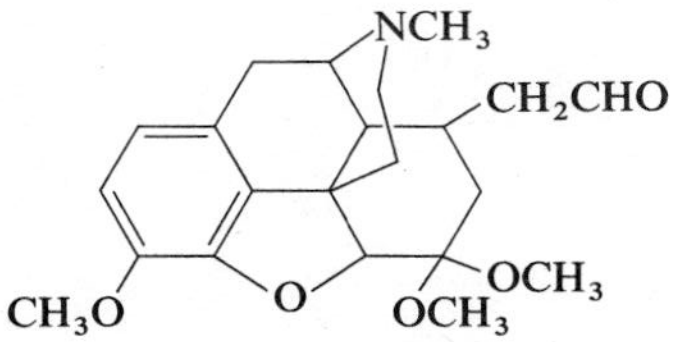 NCH_3, CH_2CHO, CH_3O, O, OCH_3, OCH_3	DAST	$C_6H_5CH_3$, 50°, 3 h	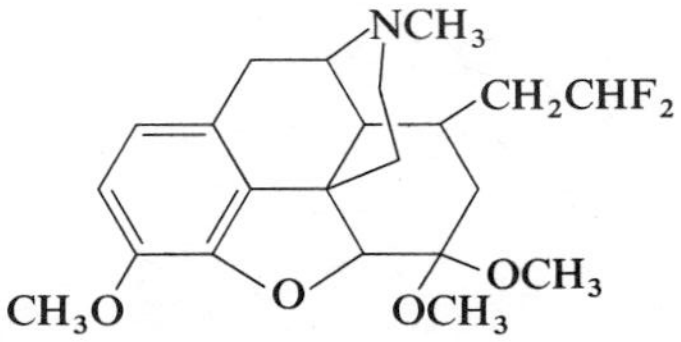 NCH_3, CH_2CHF_2, CH_3O, O, OCH_3, OCH_3 (30)	60
C_{23}	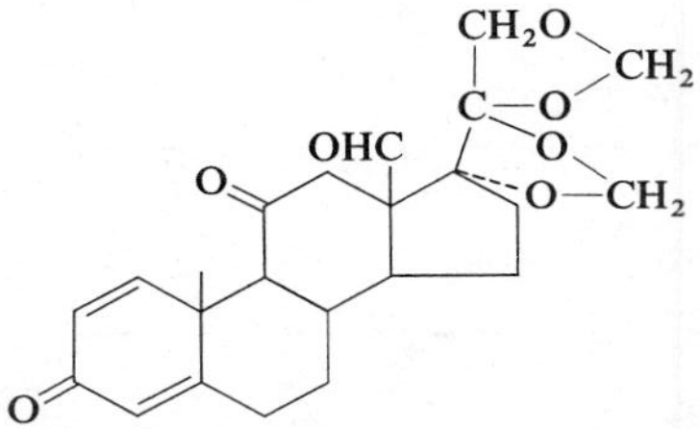 CH_2O, CH_2, C—O, O, OHC, O—CH_2, O, O	" (8)	55°, 3 h	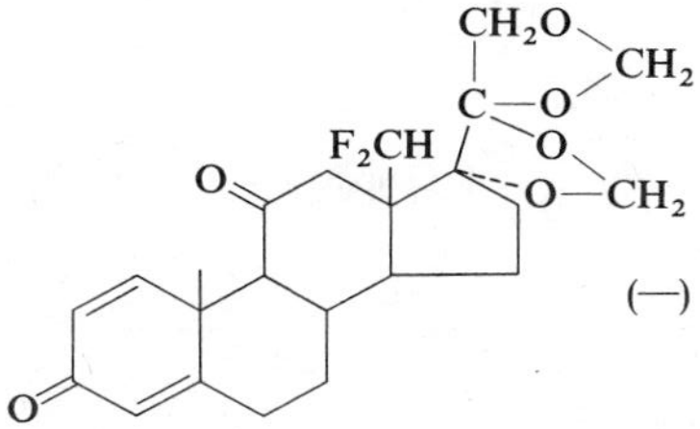 CH_2O, CH_2, C—O, O, F_2CH, O—CH_2, O, O (—)	71

TABLE II. ALDEHYDES (*Continued*)

No. of Carbon Atoms	Reactant	Reagent (Molar Equiv.)	Conditions	Product(s) and Yield(s) (%)	Refs.
	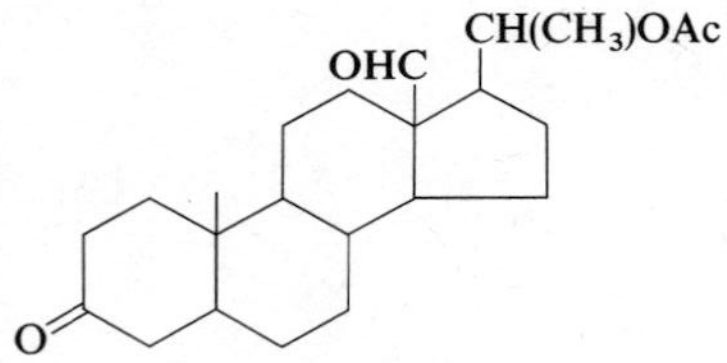 OHC, CH(CH$_3$)OAc, O	$C_5H_{10}NSF_3$	50°, 90 min	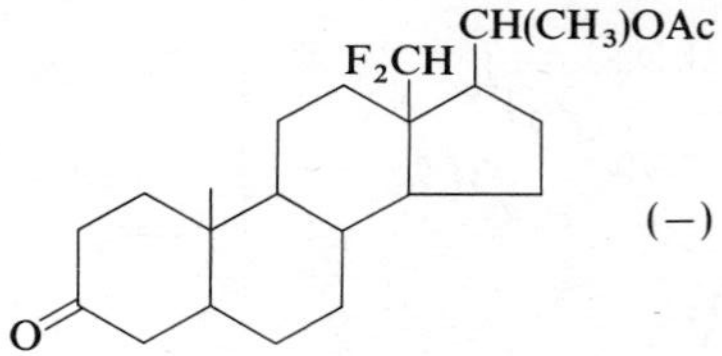 F_2CH, CH(CH$_3$)OAc, O (–)	71
	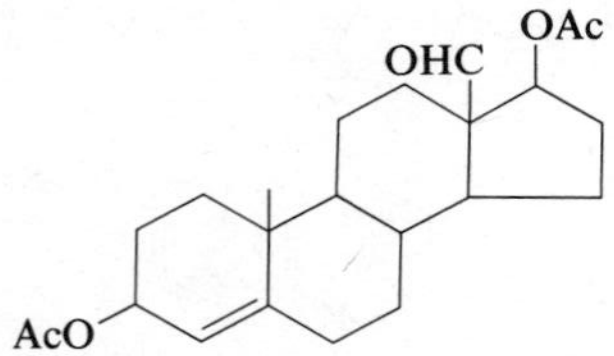 OHC, OAc, AcO	DAST	Room temp, 60° 1 h	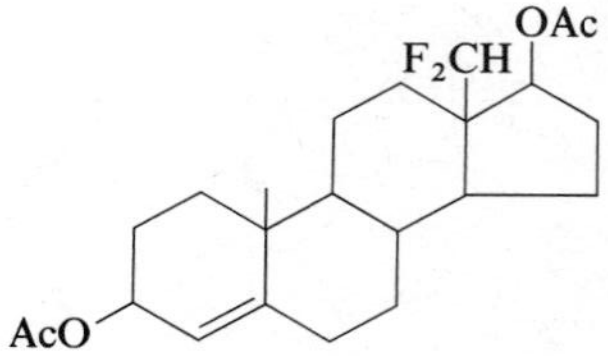 F_2CH, OAc, AcO (—)	71
	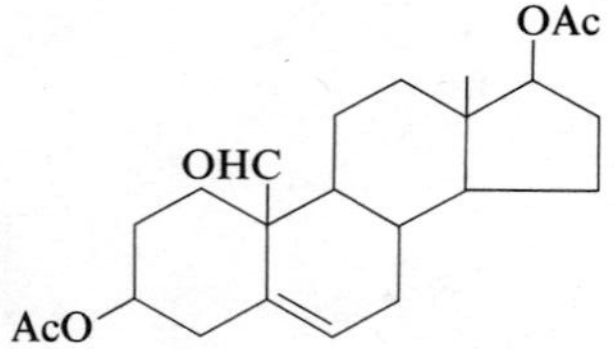 OHC, OAc, AcO	" (12.5)	60°, 5 h	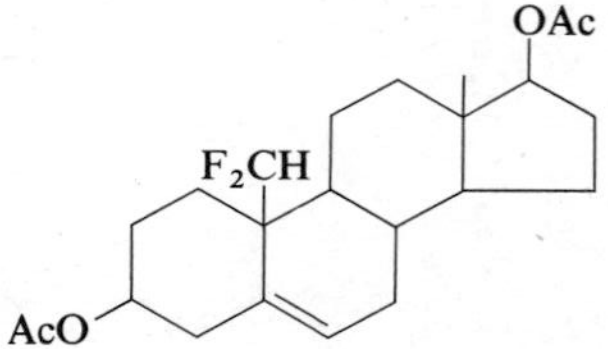 F_2CH, OAc, AcO (28)	98
C_{24}	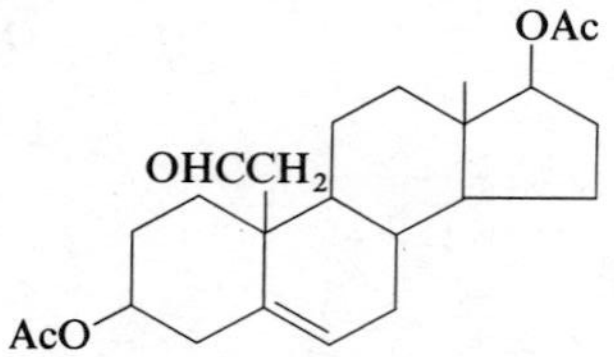 OHCCH$_2$, OAc, AcO	"	CCl_4, room temp, 20 h	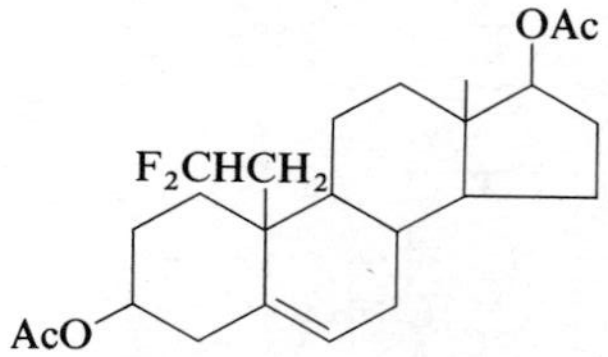 F_2CHCH_2, OAc, AcO (76)	99

	Reactant	Reagent	Conditions	Product(s) (Yield, %)	Ref.
C_{27}	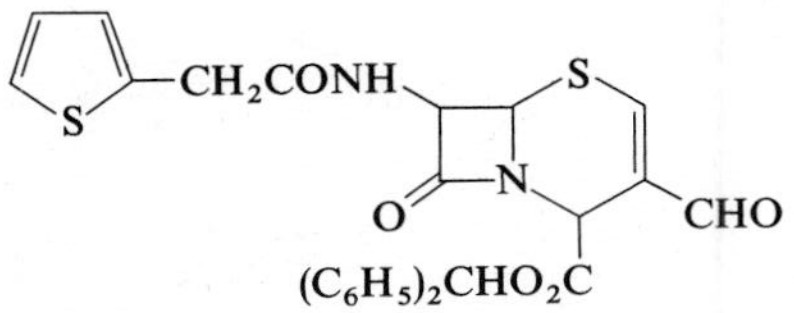 2-thienyl-CH_2CONH cephem, 3-CHO, $(C_6H_5)_2CHO_2C$	"	CH_2Cl_2, 27°, 1.5 h	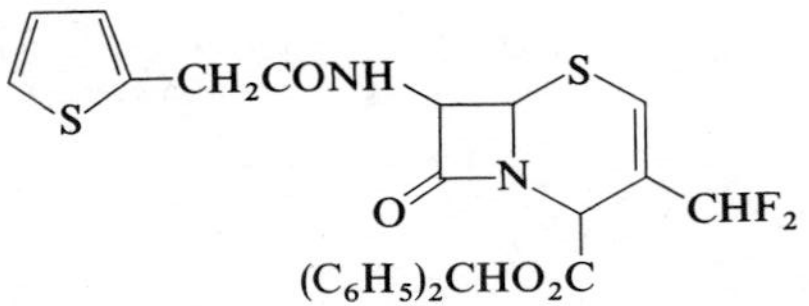 2-thienyl-CH_2CONH cephem, 3-CHF_2, $(C_6H_5)_2CHO_2C$ (—)	67
	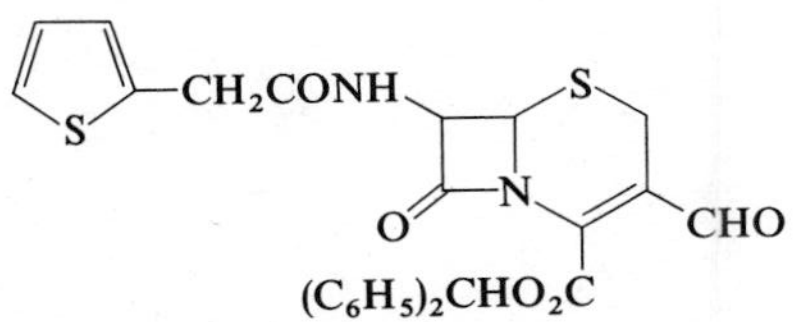 2-thienyl-CH_2CONH cephem, 3-CHO, $(C_6H_5)_2CHO_2C$	"	", ", 2 h	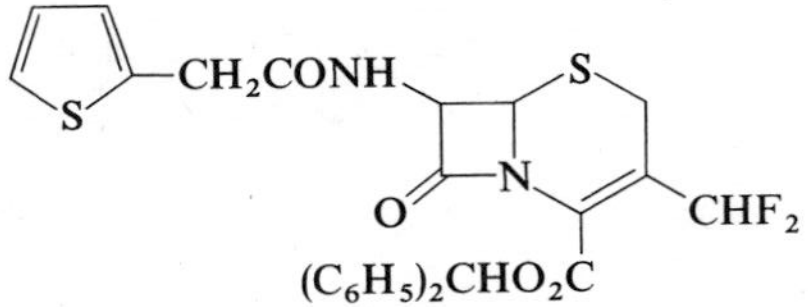 2-thienyl-CH_2CONH cephem, 3-CHF_2, $(C_6H_5)_2CHO_2C$ (49)	67
C_{29}	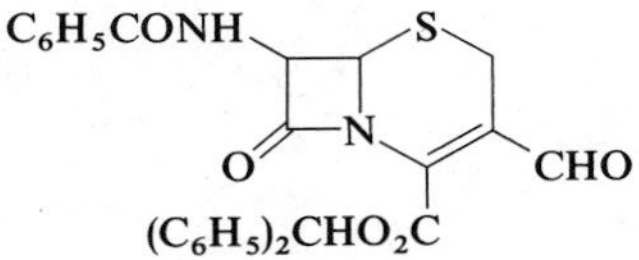 C_6H_5CONH cephem, 3-CHO, $(C_6H_5)_2CHO_2C$	$C_5H_{10}NSF_3$ (7)	Dioxane, room temp, 7 h	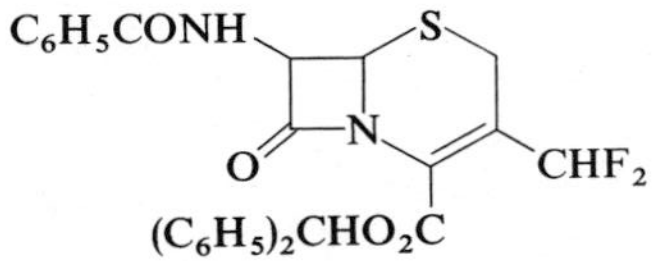 C_6H_5CONH cephem, 3-CHF_2, $(C_6H_5)_2CHO_2C$ (45)	130
	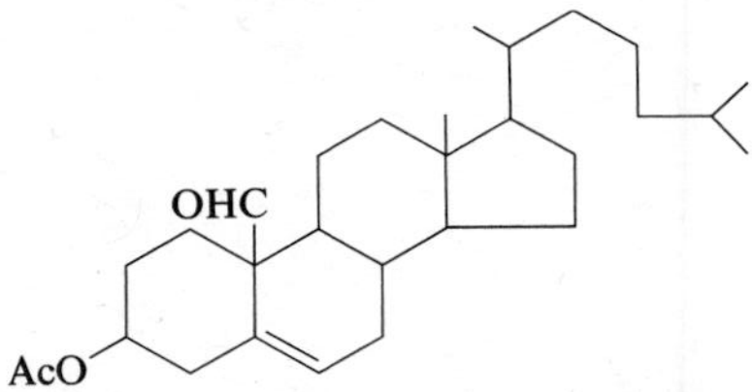 AcO steroid, 19-OHC	DAST	70°, 15 h	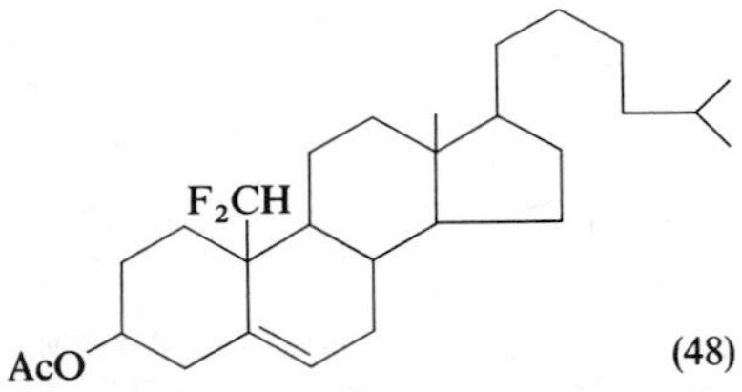 AcO steroid, 19-F_2CH (48)	92

[a] The product contained two diastereomers in varying ratios.
[b] The effect of six other solvents on the product distribution is listed in the paper.
[c] $R_2 = (CH_3)_2$, $(C_2H_5)_2$, $(CH_2)_5$ or $O(CH_2CH_2)_2$.
[d] The conversion was 46%.
[e] From 990 mg, 782 mg of the mixture of the three products is obtained.

TABLE III. KETONES

No. of Carbon Atoms	Reactant	Reagent (Molar Equiv.)	Conditions	Product(s) and Yield(s) (%)	Refs.
C_4	Cyclobutanone	DAST	Glyme, $H_2SO_3 + SO_3$	1,1-Difluorocyclobutane (—) + 1-Fluorocyclobutene (—) + 2,4-Difluoro-1-butene (—)[a]	37
C_5	[structure: O, Cl]	"	", " + ", 0° to room temp, 5 d	[structure: F, F, Cl] (24) + [structure: F, Cl] (25) + [structure: F, Cl] (51)	37
	$CH_3CO(CH_2)_2CO_2H$	"	$CHCl_3$, 0°, 30 min	[structure: O, O, F] (90)	34
	$CH_3COCH_2COCH_3$	" (2.2)	*N*-Methylpyrrolidinone, −70°, room temp, 48–64 h	$CH_3CF{=}CFCOCH_3$ (*cis*:*trans* 1:1; 40–60)	38b
C_6	$CH_3COCH_2CO_2C_2H_5$	" (2.2)	", ", ", "	$CH_3CF{=}CFCO_2C_2H_5$ (*cis*:*trans* 1:1; 48–58)	38b
	[structure: O, O]			[structures: F, F, F, F] **A**; [F, F, F] **B**; [F, F] **C**; [F, F] **D**	
		DAST	CH_2Cl_2, 0° to room temp, 6 d	**A** (54) **B** (43) **C** (3) **D** (0)	37
		"	THF, room temp, 10 d	(24) (68) (8) (0)	37
		"	Triglyme, $H_2SO_4 + SO_3$, 2 d	(14) (42) (13) (31)	37
	Cyclohexanone	R_2NSF_3[b]	80°, 30 min	1,1-Difluorocyclohexane (55)	8
		DAST (2)	Glyme, $H_2SO_4 + SO_3$; room temp, 40 h	" (27)[c] + 1-Fluorocyclohexene (73)[c]	37, 17

	Reactant	Reagent (Equiv)	Conditions	Product(s) (Yield, %)	Refs.
	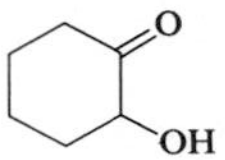	”	”, ” + ”, −50° to room temp, 3 h	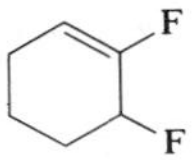 (≈100)	37
C_7	$CH_3COCHCl(CH_2)_2CO_2CH_3$	” (1.5)	CH_2Cl_2, 0° to room temp, 96 h	$CH_3CF_2CHCl(CH_2)_2CO_2CH_3$ (44)	166
	Cycloheptanone	” (2)	Diglyme, $H_2SO_4 + SO_3$, room temp	1-Fluorocycloheptene (42)[d]	37
	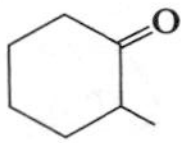	” (2)	Room temp, 40 h	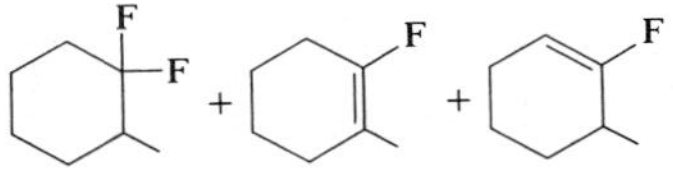 + +	
			CH_2Cl_2	(60) (30) (10)	37
			Glyme	(51) (36) (13)	37
			Glyme, $H_2SO_4 + SO_3$	(22) (67) (11)	37
			Diglyme, ”, ”	(25) (65) (10)	37
	$(n\text{-}C_3H_7)_2CO$	”	CCl_3F, H_2O (cat.), 25°, 7 d	$(n\text{-}C_3H_7)_2CF_2$ (68)	5, 9
			Glyme, $H_2SO_4 + SO_3$, room temp	” (86) + $C_2H_5CH{=}CFC_3H_7\text{-}n$ *cis* (4) + *trans* (10)	37
C_8	$C_6H_5COCH_2F$	” (1.5)	C_6H_6, 50°, 17 h	$C_6H_5CF_2CH_2F$ (82)	151
	$C_6H_5COCH_3$	”	Glyme, 85°, 20 h	$C_6H_5CF_2CH_3$ (66)[e]	5, 9
	$CH_3COCHCl(CH_2)_2CO_2C_2H_5$	” (1.5)	CH_2Cl_2, 0° to room temp, 96 h	$CH_3CF_2CHCl(CH_2)_2CO_2C_2H_5$ (64)	166
	$CH_3COCHCl(CH_2)_3CO_2CH_3$	” (1.5)	”, ”, ”	$CH_3CF_2CHCl(CH_2)_3CO_2CH_3$ (75)	166
	$CH_3COCH_2CO_2C_4H_9\text{-}n$	” (2.2)	*N*-Methylpyrrolidinone, −70°; room temp 48–64 h	$CH_3CF{=}CFCO_2C_4H_9\text{-}n$ (*cis*:*trans* 1:1; 40–60)	38b
	$C_6H_{11}COCH_2F$		C_6H_6, 50°, 17 h	$C_6H_{11}CF_2CH_2F$ (86)	151
C_9	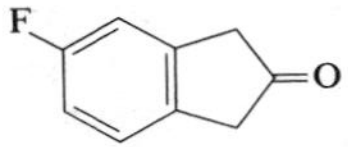	”	$CHCl_3$, 40°, 4 h	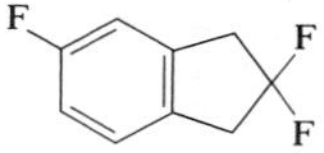 (36)	167
	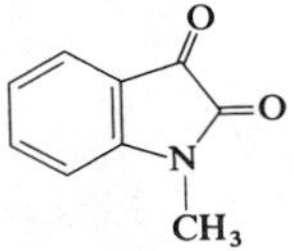	” (2)	60°, 15 min	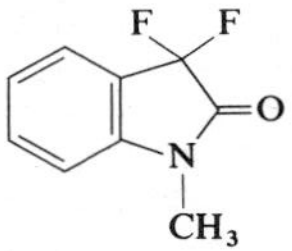 (95)	75

TABLE III. KETONES (*Continued*)

No. of Carbon Atoms	Reactant	Reagent (Molar Equiv.)	Conditions	Product(s) and Yield(s) (%)	Refs.
	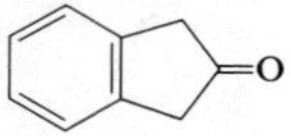	DAST (2)	$CHCl_3$, 40°, 4 h	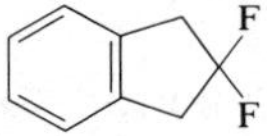 (40)	167
	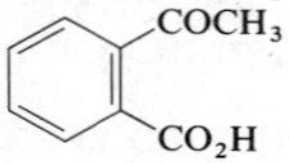	"	", 0°, 30 min	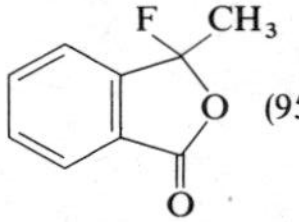 (95)	34
	m-$FC_6H_4CH_2COCH_3$	"	80°, 15 min	m-$FC_6H_4CH_2CF_2CH_3$ (40)	167
	p-$FC_6H_4CH_2COCH_3$	"	", "	p-$FC_6H_4CH_2CF_2CH_3$ (40)	167
	$C_6H_5COCO_2CH_3$	"	Room temp or 40–60°	$C_6H_5CF_2CO_2CH_3$ (73)	75
	$C_6H_5CH_2COCH_3$	"	Glyme, $H_2SO_4 + SO_3$, 6 d	$C_6H_5CH_2CF_2CH_3$ (~50) + $C_6H_5CH{=}CFCH_3$ (~50) + $C_6H_5CH_2CF{=}CH_2$ (trace)	37
		"	80°, 15 min	$C_6H_5CH_2CF_2CH_3$ (43)	167
	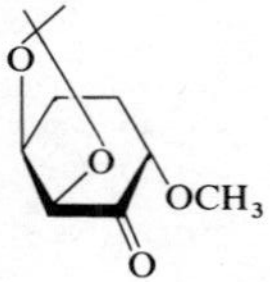	"	C_6H_6, reflux, 16 h	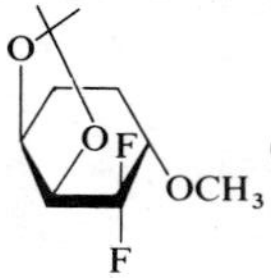 (25)	70
	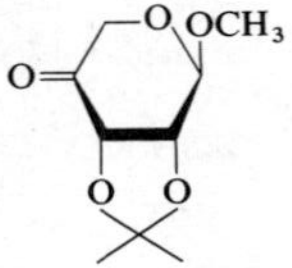	"	", ", "	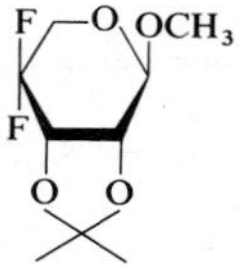 (36)	70
C_{10}	3,4-$F_2C_6H_3COCO_2C_2H_5$	"	Room temp or 40–60°	3,4-$F_2C_6H_3CF_2CO_2C_2H_5$ (65)	75

Substrate	Reagent	Conditions	Product(s) and Yield(s) (%)	Refs.
6-R^1-7-R^2-2-tetralone (R^1, R^2)			2,2-difluoro-6-R^1-7-R^2-tetralin	
H, H	DAST	35°, 5 min	(38)	167
F, H	"	CH_2Cl_2, 35°, 1 h	(39)	167
H, F	"	", ", "	(35)	167
$C_6H_5COCO_2C_2H_5$	"	Room temp or 40–60°	$C_6H_5CF_2CO_2C_2H_5$ (92)	75
$C_6H_5CO(CH_2)_2CO_2H$	"	CH_2Cl_2, 0°, 30 min	5-fluoro-5-phenyl-γ-butyrolactone (C_6H_5, F) (85)	34
$CH_3COCHClCH_2C_6H_5$	"	", 0° to room temp, room temp, 96 h	$CH_3CF_2CHClCH_2C_6H_5$ (83)	166
dispiro[3.1.3.1]decane-5,10-dione	" (1)	CCl_4, 25°, 30 h	10,10-difluorodispiro[3.1.3.1]decan-5-one (61)	168
	" (1.3)	", ", 6 weeks	5,5,10,10-tetrafluorodispiro[3.1.3.1]decane (60)	168
3a,6a-dimethylbicyclo[3.3.0]octane-3,7-dione	" (2)	C_6H_6, 78°, 24 h	7,7-difluoro-3a,6a-dimethylbicyclo[3.3.0]octan-3-one (15) + 3,3,7,7-tetrafluoro-3a,6a-dimethylbicyclo[3.3.0]octane (60)	5, 9
n-$C_3H_7COC(=NOCH_3)COC_3H_7$-$n$	" (3)	CH_2Cl_2, room temp, 4 d	n-$C_3H_7CF_2C(=NOCH_3)COC_3H_7$-$n$ (20)	169
4-t-C_4H_9-cyclohexanone	DAST	Glyme, $H_2SO_4 + SO_3$ (20%), room temp, 24 h	1,1-difluoro-4-t-C_4H_9-cyclohexane (13) + 1-fluoro-4-t-C_4H_9-cyclohexene[e] (64)	37

TABLE III. KETONES (*Continued*)

No. of Carbon Atoms	Reactant	Reagent (Molar Equiv.)	Conditions	Product(s) and Yield(s) (%)	Refs.
C_{11}	p-$CF_3C_6H_4COCO_2C_2H_5$	DAST	Room temp or 40–60°	p-$CF_3C_6H_4CF_2CO_2C_2H_5$ (74)	75
	$C_6H_5COCH_2CO_2C_2H_5$	"	Glyme, $H_2SO_4 + SO_3$, room temp, 4 d	$C_6H_5CF_2CH_2CO_2C_2H_5$ (11) + $C_6H_5CF{=}CHCO_2C_2H_5$ (89)	37
		" (2.2)	*N*-Methylpyrrolidinone, −70°, room temp 48–64 h	$C_6H_5CF{=}CFCO_2C_2H_5$ (poor)	38b
	n-$C_3H_7COC(=NOAc)COC_3H_7$-n	"	CH_2Cl_2, 16 h	n-$C_3H_7CF_2C(=NOAc)COC_3H_7$-n (30)	169
C_{12}	1-Adamantyl-$COCH_2F$	" (1.5)	C_6H_6, 50°, 17 h	1-Adamantyl-CF_2CH_2F (78)	151
	Cyclododecanone	"	Glyme, $H_2SO_4 + SO_3$, room temp, 5 d, 50°, 6 d	1,1-Difluorocyclododecane (38) + *cis*-1-fluorocyclododecene (19) + *trans*-1-fluorocyclododecene (35)	37
C_{13}	p-$(CF_3)_2CFC_6H_4COCO_2C_2H_5$	"	Room temp or 40–60°	p-$(CF_3)_2CFC_6H_4CF_2CO_2C_2H_5$ (75)	75
	1-Naphthyl-$COCO_2CH_3$	"	"	1-Naphthyl-$CF_2CO_2CH_3$ (76)	75
	4-($O_2CC_6H_5$)cyclohexanone	"	CH_2Cl_2	1,1-Difluoro-4-($O_2CC_6H_5$)cyclohexane (65) + 1-fluoro-4-($O_2CC_6H_5$)cyclohexene (35)	37
		" (2)	Glyme, $H_2SO_4 + SO_3$, room temp, 48 h	(10) + (49)	37

	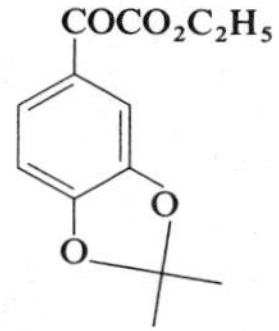 $COCO_2C_2H_5$, O, O	"	Room temp or 40–60°	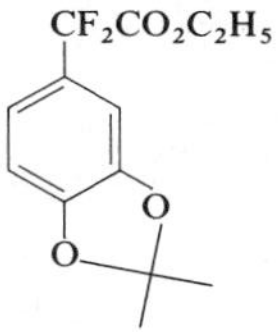 $CF_2CO_2C_2H_5$, O, O (74)	75
C_{14}	$C_6H_5COC_6H_4CO_2H$-*o*	"	$CHCl_3$, 0°, 30 min	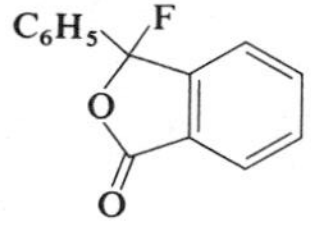 C_6H_5, F, O, O (93)	34
	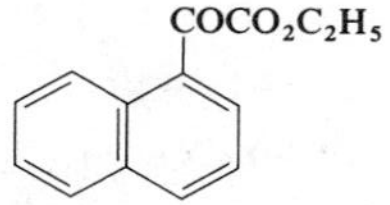 $COCO_2C_2H_5$	" (2)	Room temp or 40–60°	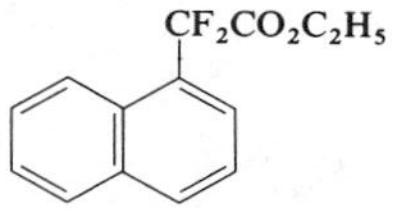 $CF_2CO_2C_2H_5$ (77)	75
	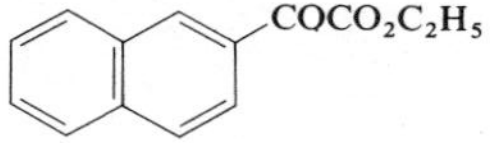 $COCO_2C_2H_5$	" (2)		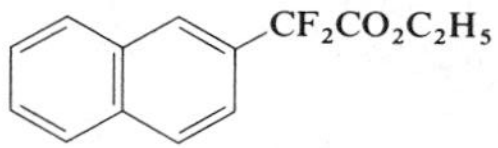 $CF_2CO_2C_2H_5$ (53)	75
	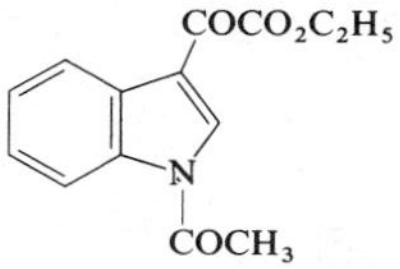 $COCO_2C_2H_5$, N, $COCH_3$	" (2)	Room temp or 40–60°	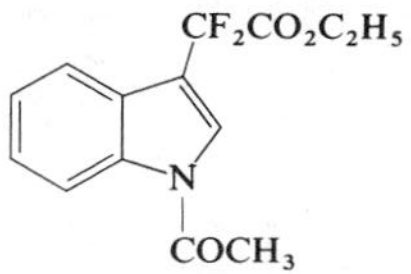 $CF_2CO_2C_2H_5$, N, $COCH_3$ (84)	75
	p-(*i*-C_4H_9)$C_6H_4COCO_2C_2H_5$	" (2)	"	*p*-(*i*-C_4H_9)$C_6H_4CF_2CO_2C_2H_5$ (84)	75
C_{15}	$C_6H_5COCH_2COC_6H_5$	" (2.2)	*N*-Methylpyrrolidinone, −70°, room temp 48–64 h	$C_6H_5CF{=}CFCOC_6H_5$ (poor)	38b
C_{16}	*p*-$C_6H_5C_6H_4COCO_2C_2H_5$	" (2)	Room temp, 2 h	*p*-$C_6H_5C_6H_4CF_2CO_2C_2H_5$ (92)	75
	(n-$C_3H_7)_2COC{=}CHC_6H_5$	" (0.3)	C_6H_6, reflux, 50 h	n-$C_3H_7COC({=}CHC_6H_5)CF_2C_3H_7$-$n$ (23)	169

TABLE III. KETONES (*Continued*)

No. of Carbon Atoms	Reactant	Reagent (Molar Equiv.)	Conditions	Product(s) and Yield(s) (%)	Refs.
C_{17}	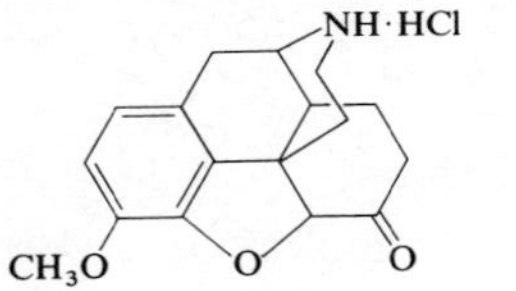 [structure: $NH \cdot HCl$, CH_3O, O, O]	DAST (2.3)	CH_2Cl_2	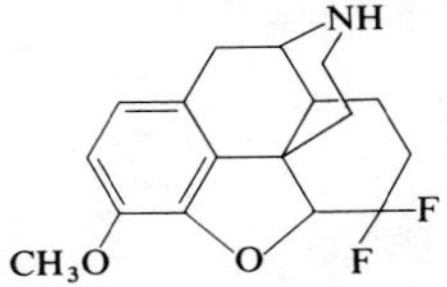 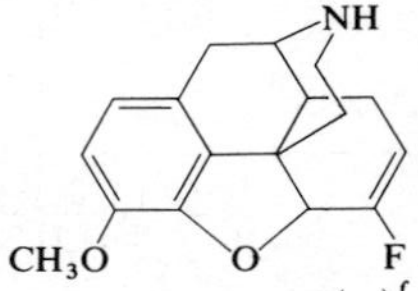 [structure: NH, CH_3O, O, F, F] + [structure: NH, CH_3O, O, F] (—)[f]	38
	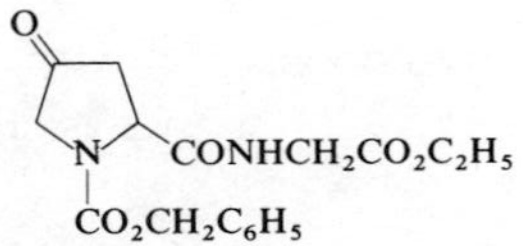 [structure: O, N, $CONHCH_2CO_2C_2H_5$, $CO_2CH_2C_6H_5$]	"	C_6H_6, room temp, 7 d	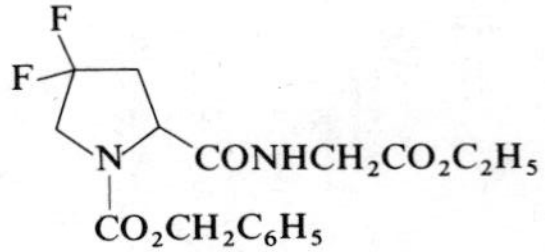 [structure: F, F, N, $CONHCH_2CO_2C_2H_5$, $CO_2CH_2C_6H_5$] (42)	76
	n-$C_9H_{19}CO(CH_2)_5CO_2CH_3$	" (3)	CCl_4, 108°, (bomb), 3 d	n-$C_9H_{19}CF_2(CH_2)_5CO_2CH_3$ (35)	74
C_{18}	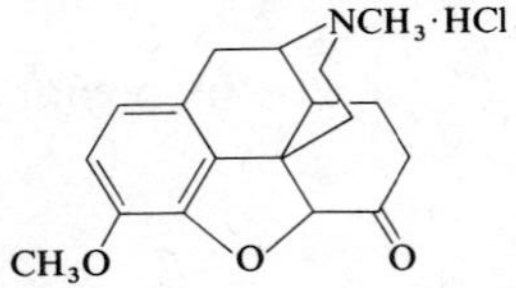 [structure: $NCH_3 \cdot HCl$, CH_3O, O, O]	"	Glyme, $H_2SO_4 + SO_3$, −78° to room temp, 8 d	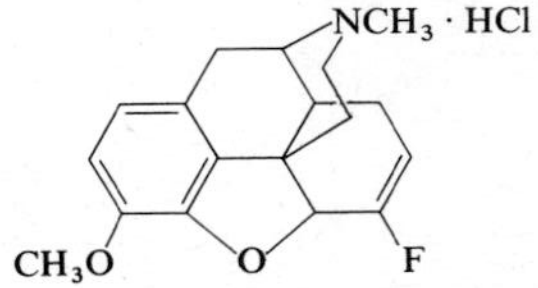 [structure: $NCH_3 \cdot HCl$, CH_3O, O, F] **I** (8) 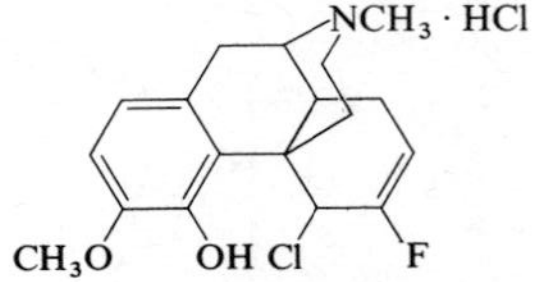 + [structure: $NCH_3 \cdot HCl$, CH_3O, OH, Cl, F] (26)	38, 78

	Reactant	Reagent (equiv)	Conditions	Product(s) (Yield, %)	Refs.
		DAST (3)	CCl_3F, − 78° to 20°, 5 d	**I** (free base) 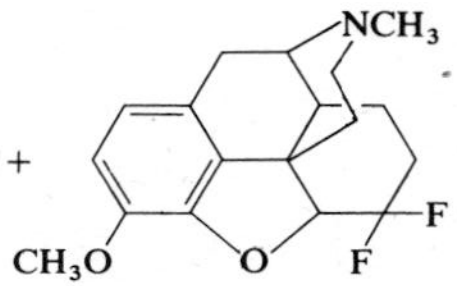 (—) + [structure: CH_3O, O, NCH_3, F, F] (45)	38, 78
	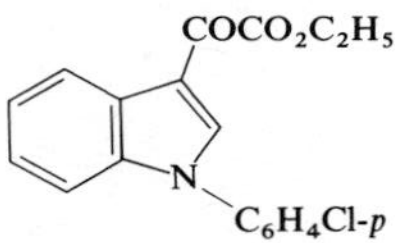 [structure: indole, $COCO_2C_2H_5$, N–C_6H_4Cl-*p*]	" (1.5)	80–90°, 3 h	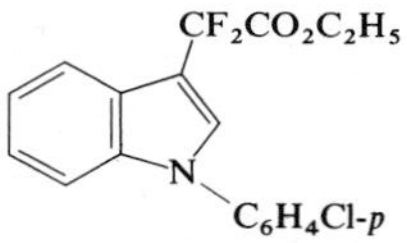 [structure: indole, $CF_2CO_2C_2H_5$, N–C_6H_4Cl-*p*] (82)	75
C_{19}	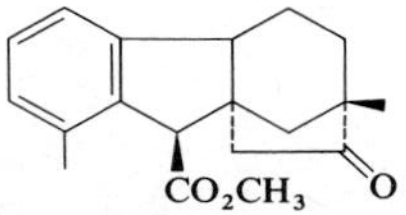 [structure: CO_2CH_3, O]	" (4.7)	CCl_4, 114°, 72 h (stainless steel bomb)	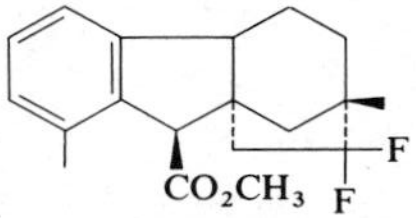 [structure: CO_2CH_3, F, F] (38)	74
	n-$C_6H_{13}CO(CH_2)_{10}CO_2CH_3$	" (4.7)	", 115°, 7 d (stainless steel bomb)	*n*-$C_6H_{13}CF_2(CH_2)_{10}CO_2CH_3$ (94)	74
C_{20}	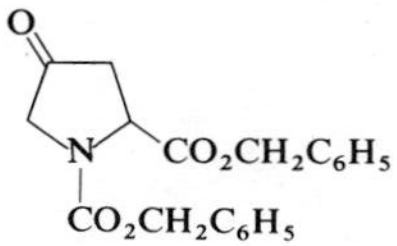 [structure: O, N, $CO_2CH_2C_6H_5$, $CO_2CH_2C_6H_5$]	" (2.5)	C_6H_6, room temp, 7 d	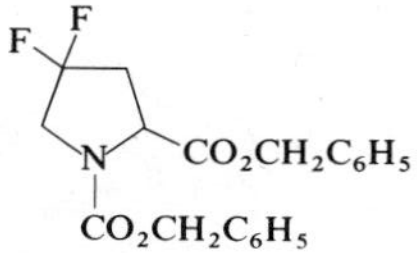 [structure: F, F, N, $CO_2CH_2C_6H_5$, $CO_2CH_2C_6H_5$] (44)	76

TABLE III. KETONES (*Continued*)

No. of Carbon Atoms	Reactant	Reagent (Molar Equiv.)	Conditions	Product(s) and Yield(s) (%)	Refs.
	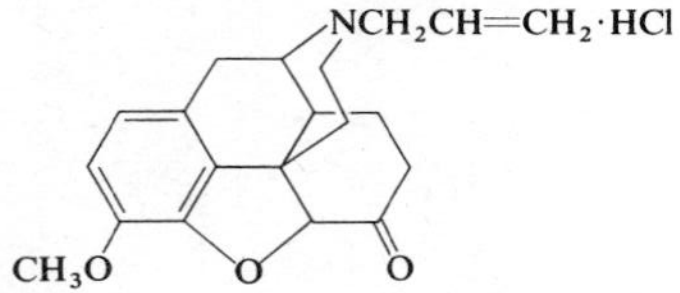 CH_3O, O, O; $NCH_2CH{=}CH_2 \cdot HCl$	DAST	CCl_3F, −78°, room temp, 5 d	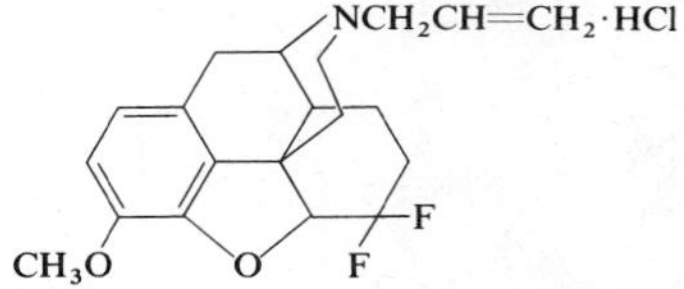 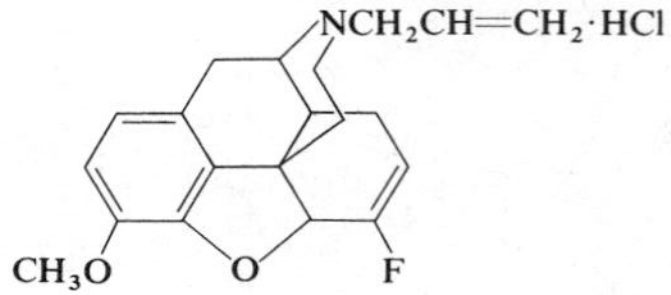 CH_3O, O, F, F; $NCH_2CH{=}CH_2 \cdot HCl$ + CH_3O, O, F; $NCH_2CH{=}CH_2 \cdot HCl$ (—)	38, 78
	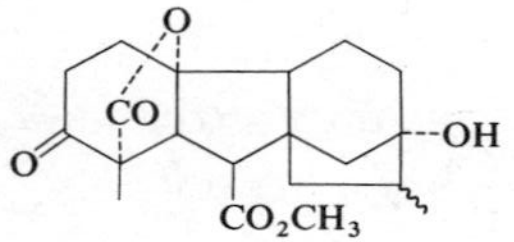 O, CO, O, OH, CO_2CH_3	" (2.5)	CH_2Cl_2, 0°, 6 h	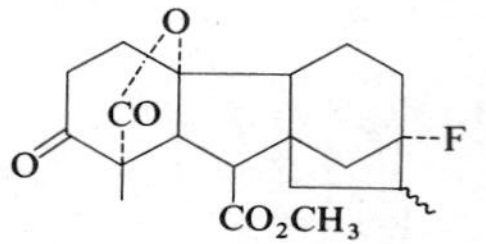 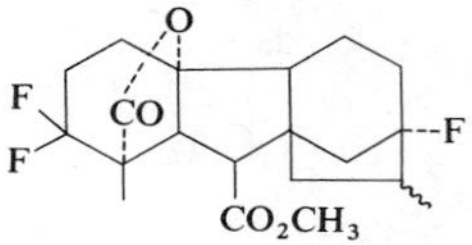 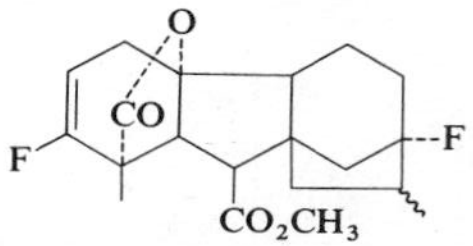 O, CO, O, F, CO_2CH_3 (14)[g] + F, F, CO, O, F, CO_2CH_3 (—)[g] + F, CO, O, F, CO_2CH_3 ("small amount")[g]	74
	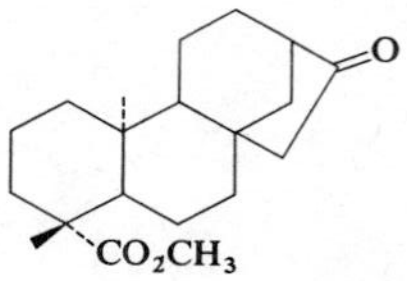 O, CO_2CH_3	"	CCl_4, 114°, 72 h (stainless steel bomb)	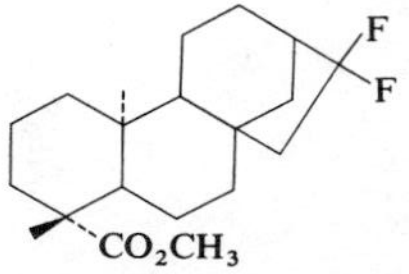 F, F, CO_2CH_3 (54)	114

	Reactant	Reagent	Conditions	Product(s) and Yield(s) (%)		Refs.
C_{21}	CH_3O … NCO— (R)			CH_3O … NCO— (R), F, F + CH_3O … NCO— (R), F		
	R			R		
	H	DAST	CH_2Cl_2, room temp, 3 d	H (36)	(18)	38, 78
	OH	" (3)	" , −78°, 25°, 4 d	F (90–95)	(−)	38, 78
	CO_2CH_3 (ketone)	"	50°, 16–20 h	CO_2CH_3 (F, F)	(5–10)[h]	118
	AcO (17-ketone)	"	Glyme, $H_2SO_4 + SO_3$, 50–60°, 9 d	AcO (17,17-difluoro) (18) + AcO (17-fluoro-16-ene)	(82)	37
	1,3,5-$(t\text{-}C_4H_9CO)_3C_6H_3$	" (3.5)	CCl_4, reflux, 40 min	1,3,5-$(t\text{-}C_4H_9CF_2)_3C_6H_3$ (98)		170

TABLE III. KETONES (*Continued*)

No. of Carbon Atoms	Reactant	Reagent (Molar Equiv.)	Conditions	Product(s) and Yield(s) (%)	Refs.
	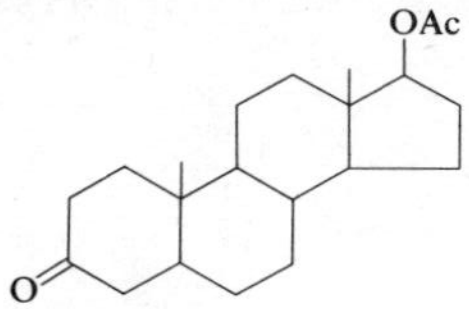 OAc, O	DAST (3.5)	80°, 2 h	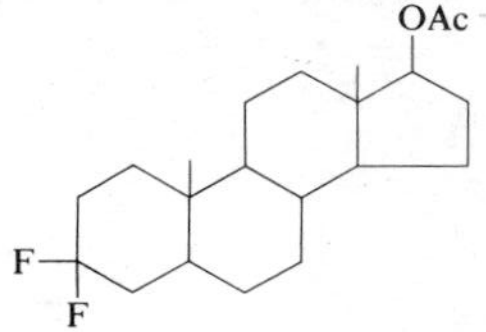 OAc, F, F (86)	47, 48
	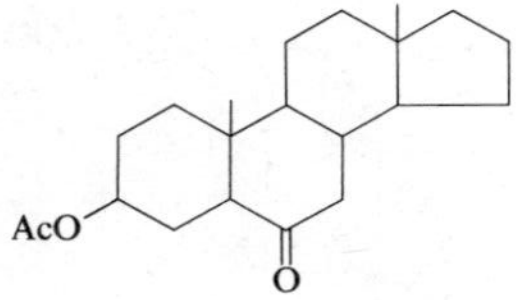 AcO, O	"	", 25 h	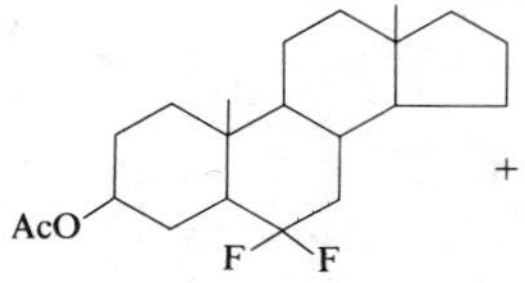 AcO, F, F (29) 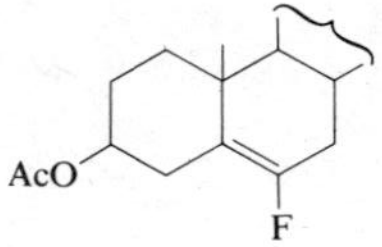 + AcO, F (33)	47, 48
	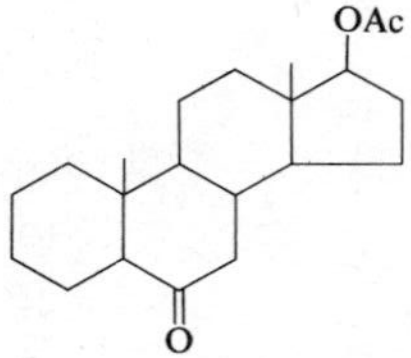 OAc, O	"	", 15 h	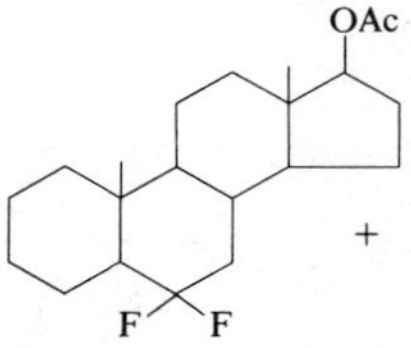 OAc, F, F (45) 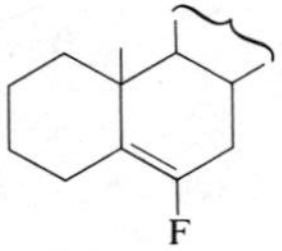 + F (45)	47, 48
	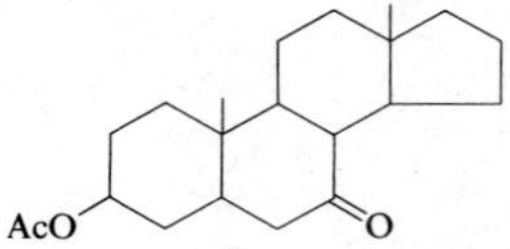 AcO, O	"	80°, 16 h	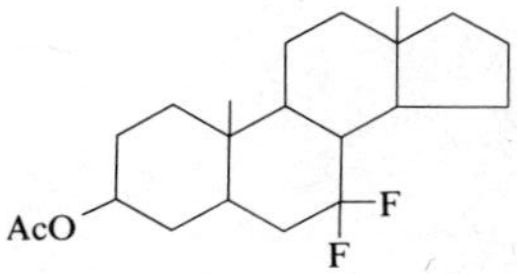 AcO, F, F (70)	47, 48

Substrate	Reagent	Conditions	Product (Yield, %)	Refs.
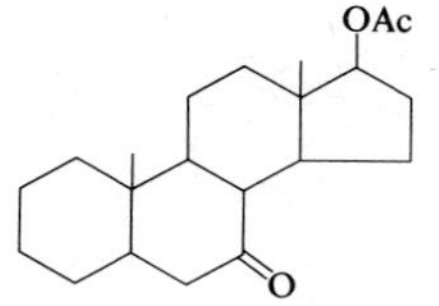 OAc, O	DAST	", 20 h	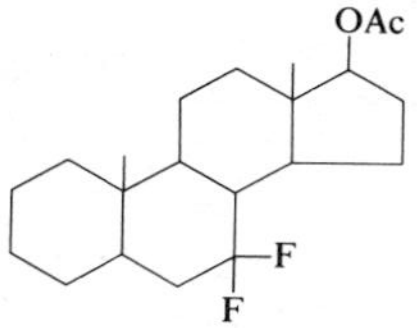 OAc, F, F (85)	47, 48
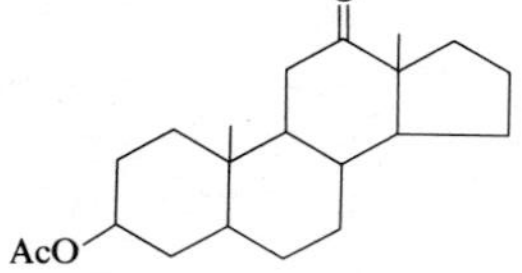 O, AcO	"	", 18 h	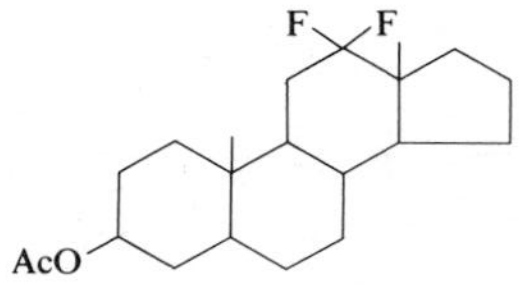 F, F, AcO (90)	47, 48
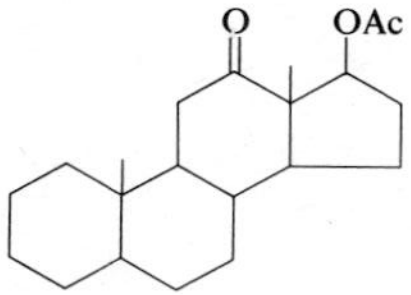 O, OAc	"	", 17 h	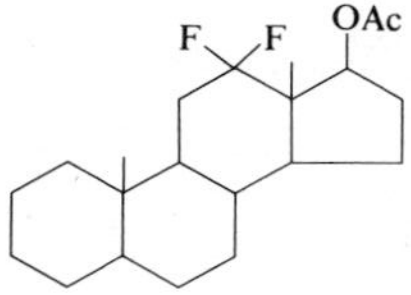 F, F, OAc (95)	47, 48
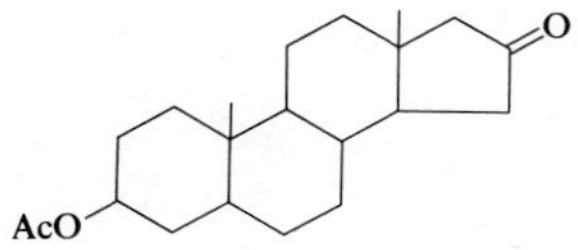 O, AcO	"	", 36 h	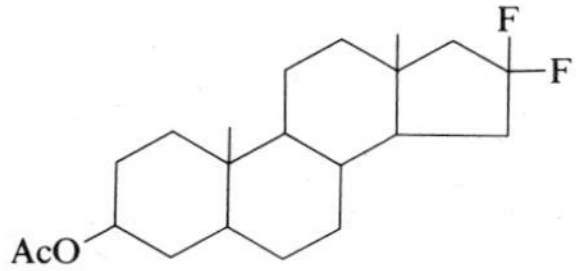 F, F, AcO (51)	47, 48
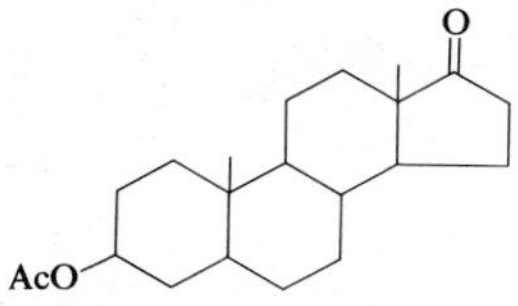 O, AcO	"	", 12 h	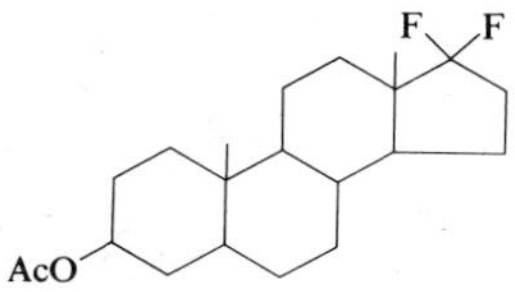 F, F, AcO (58)	47, 48

TABLE III. KETONES (*Continued*)

No. of Carbon Atoms	Reactant	Reagent (Molar Equiv.)	Conditions	Product(s) and Yield(s) (%)	Refs.
	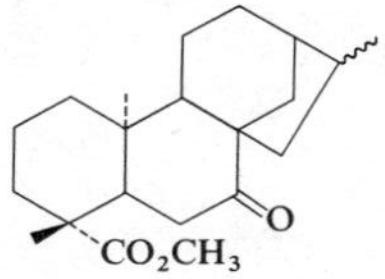 CO_2CH_3, O	DAST	CCl_4, 114°, 84 h, stainless steel bomb	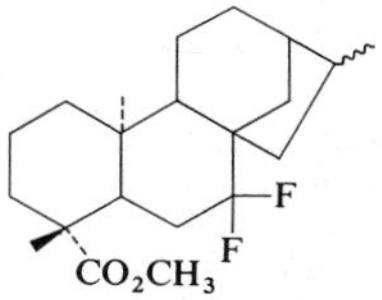 F, F, CO_2CH_3 (34)[i]	74
	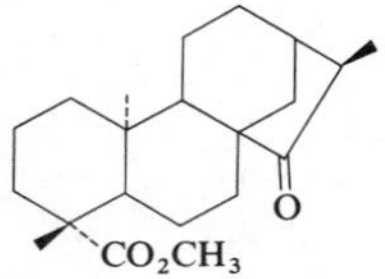 O, CO_2CH_3	"	", ", ", "	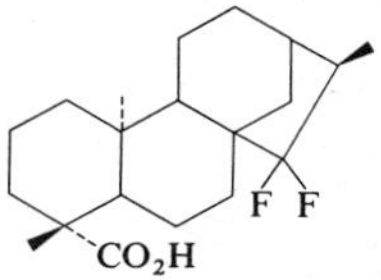 F F, CO_2H (1)[j]	114
C_{22}	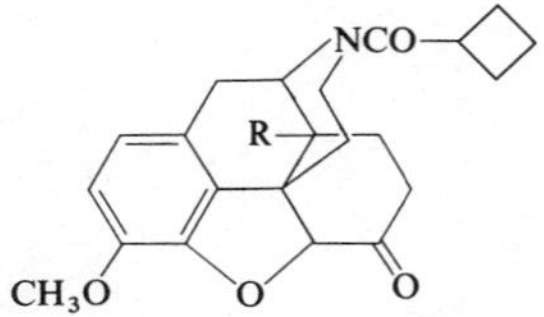 NCO, R, CH_3O, O, O			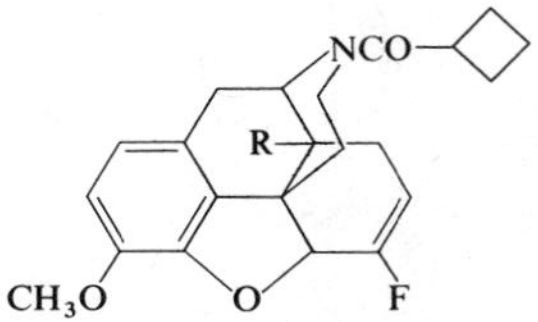 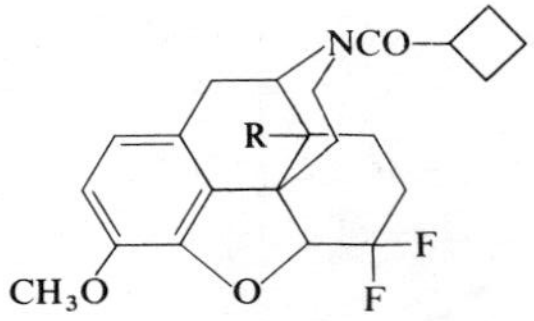 NCO, R, CH_3O, O, F **A** + NCO, R, CH_3O, O, F, F **B**	

	Substrate	Reagent(s)	Conditions	Product(s) and Yield(s) (%)	Refs.
	R				
	H	DAST (3)	CH_2Cl_2, −78°, 25°, 6 d	**A + B** (≈100)[k]	38, 78
	OH	"	", ", ", "	" (≈100)[k]	78
(C_{24})	OAc	"	", ", ", 3 d	" (87–92)	38, 78
C_{22}	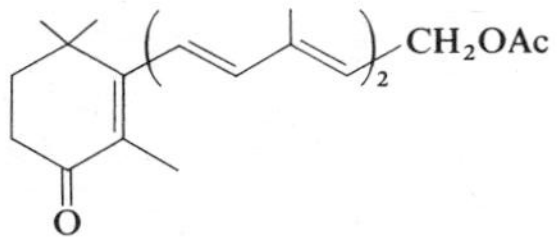 2,6,6-trimethyl-3-oxocyclohexenyl–(CH=CH–C(CH₃)=CH)₂–CH_2OAc	"	50°, 16–20 h	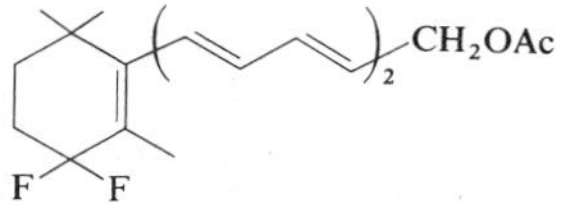 3,3-difluoro-2,6,6-trimethylcyclohexenyl–(CH=CH–C(CH₃)=CH)₂–CH_2OAc (5–10)[1]	118
C_{23}	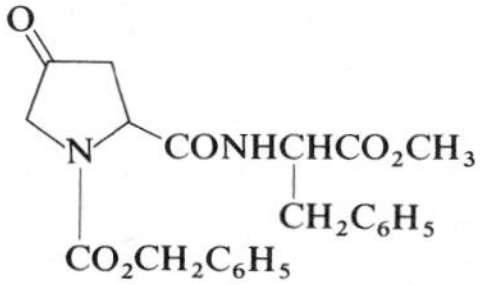 4-oxo-1-($CO_2CH_2C_6H_5$)pyrrolidin-2-yl–$CONHCH(CH_2C_6H_5)CO_2CH_3$	" (2.5)	C_6H_6, room temp, 7 d	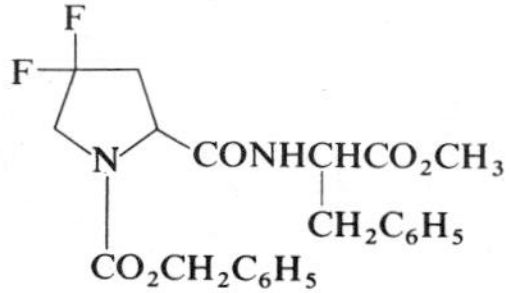 4,4-difluoro-1-($CO_2CH_2C_6H_5$)pyrrolidin-2-yl–$CONHCH(CH_2C_6H_5)CO_2CH_3$ (57)	76
	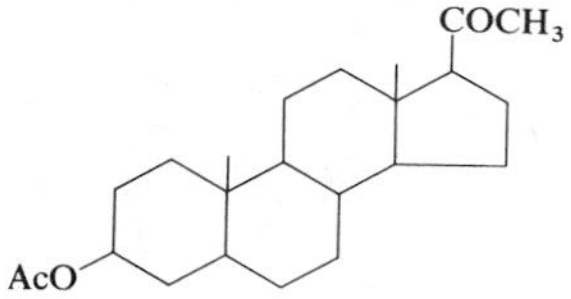 3-AcO-steroid–$COCH_3$		80°, 28 h	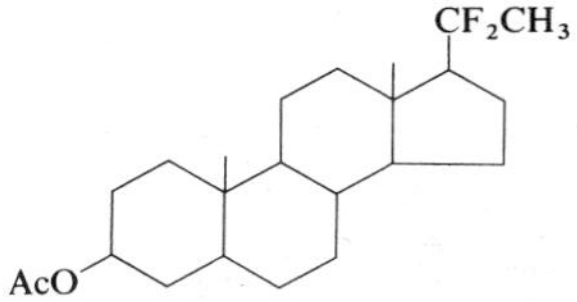 3-AcO-steroid–CF_2CH_3 (43)	47, 48

TABLE III. KETONES (*Continued*)

No. of Carbon Atoms	Reactant	Reagent (Molar Equiv.)	Conditions	Product(s) and Yield(s) (%)	Refs.
C_{26}	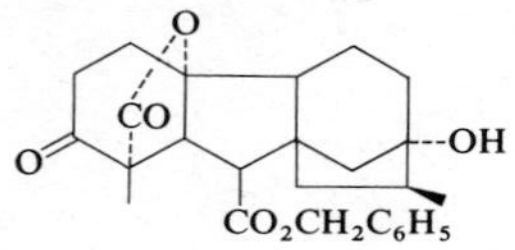 [structure: O, CO, OH, $CO_2CH_2C_6H_5$]	DAST	CH_2Cl_2, 0°, 8 h	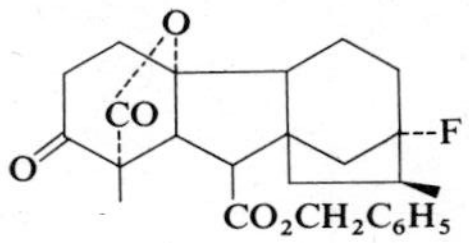 [structure: O, CO, F, $CO_2CH_2C_6H_5$] (8) 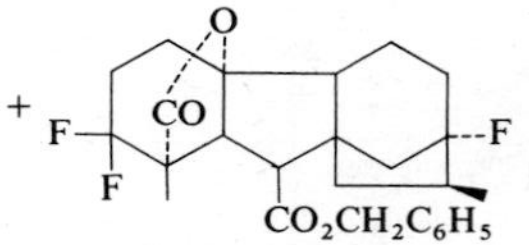 + [structure: F, F, O, CO, F, $CO_2CH_2C_6H_5$] (44) 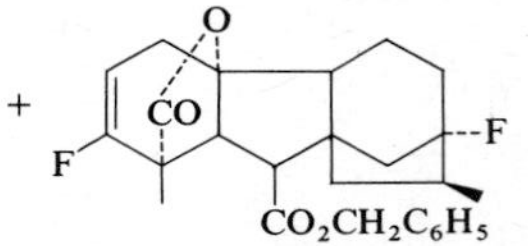 + [structure: F, O, CO, F, $CO_2CH_2C_6H_5$] (9)	50
C_{27}	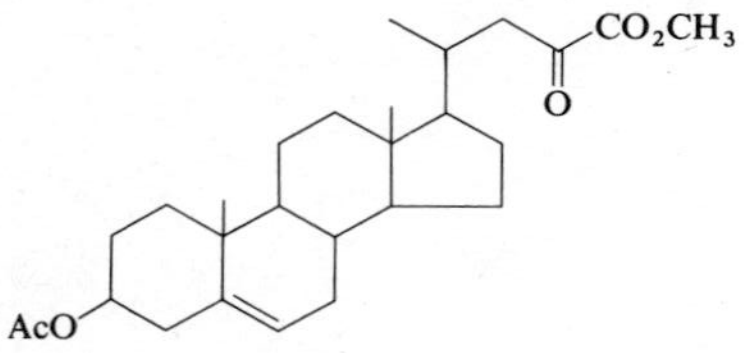 [structure: AcO, O, CO_2CH_3]	"	", room temp, 16 h	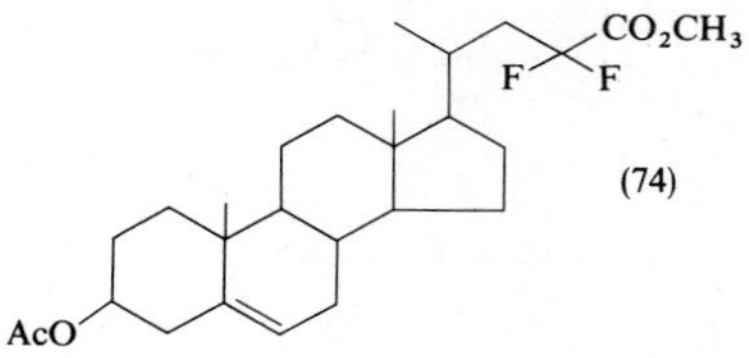 [structure: AcO, F, F, CO_2CH_3] (74)	77, 77a
	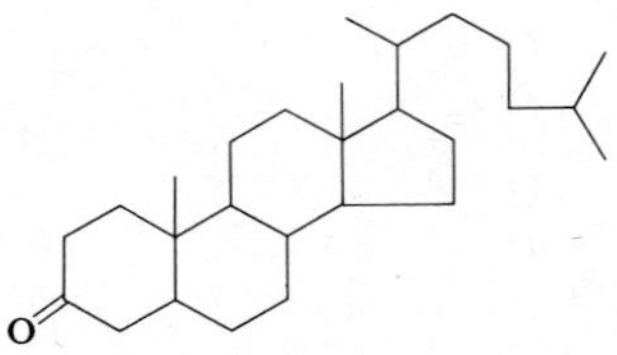 [structure: O]			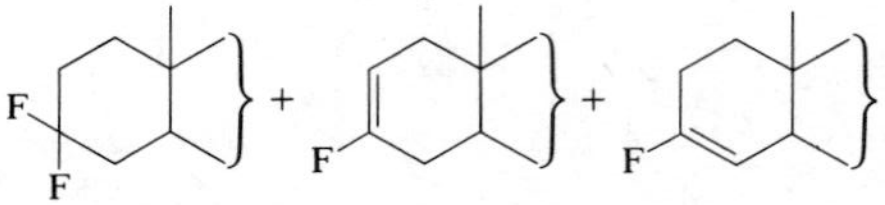 [structure: F, F] + [structure: F] + [structure: F]	
		"	CH_2Cl_2, room temp, 4 d	(94) (4) (2)	37
		"	Glyme, 50°, 9 d	(55) (29) (16)	37
		"	Glyme, $H_2SO_4 + SO_3$, 50°	(34) (49) (16)	37
		"	CH_2Cl_2, 20°, 84 h	(45) (0) (0)	74

	Reactant	Reagent	Conditions	Product(s) and Yield(s) (%)	Refs.
C_{29}	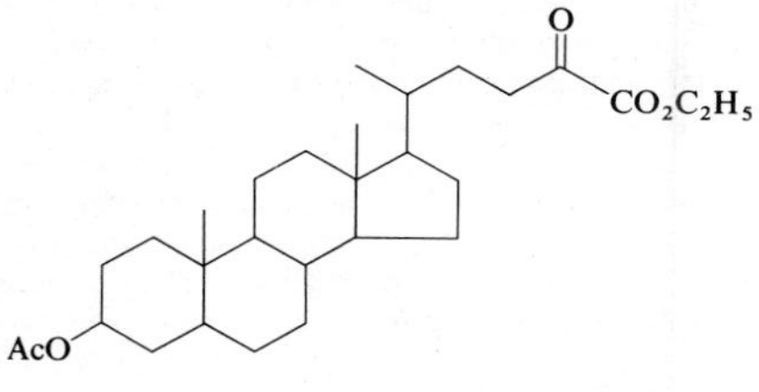 AcO–steroid, side chain C(=O)$CO_2C_2H_5$	DAST	”	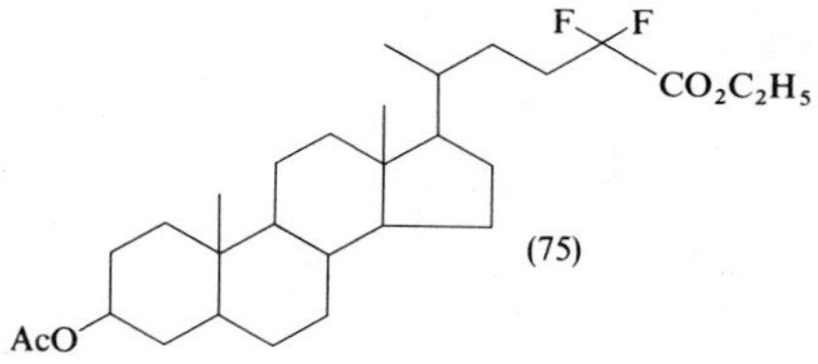 AcO–steroid, side chain $CF_2CO_2C_2H_5$ (75)	106
	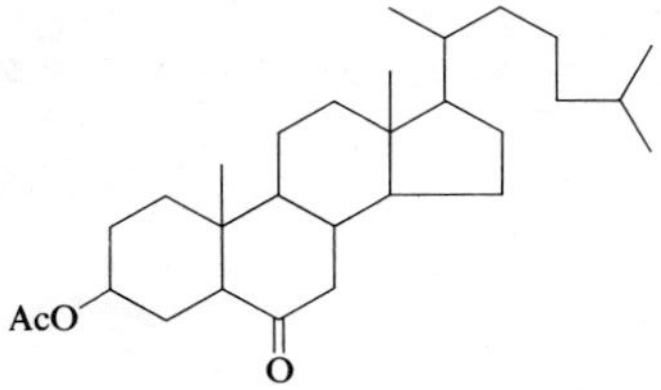 3-AcO-6-oxo cholestane	”	Glyme, $H_2SO_4 + SO_3$, 50°, 17 d	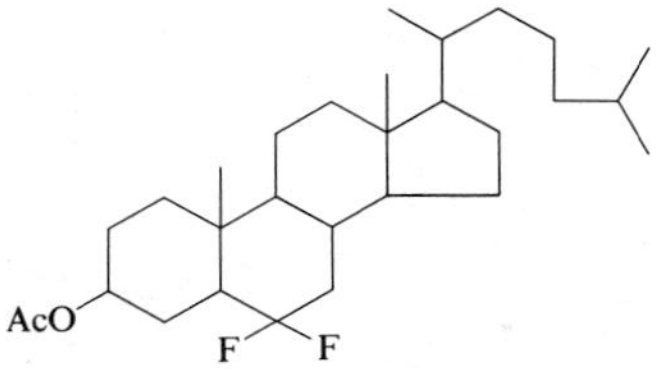 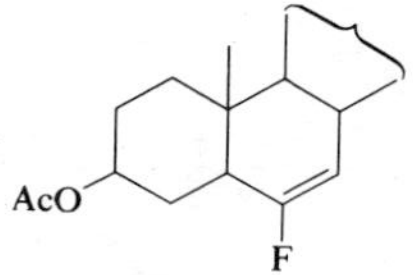 6,6-difluoro (6) + 6-fluoro-5-ene (66) + 6-fluoro-6-ene (6) 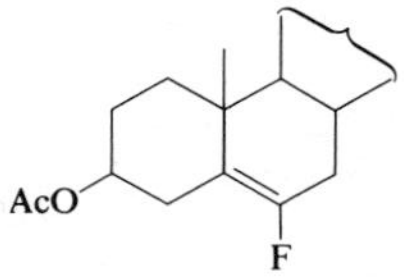	37
	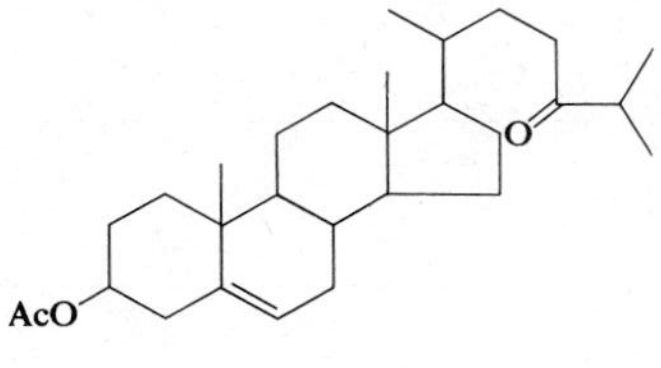 AcO–steroid-5-ene, ketone side chain	” (11)	60–70°, 1 h	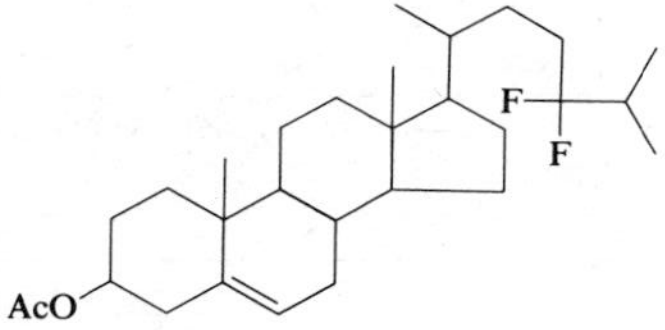 AcO–steroid-5-ene, CF_2 side chain (13)	171

	Reactant	Reagent	Conditions	Product(s) and Yield(s) (%)	Refs.
C_{29}	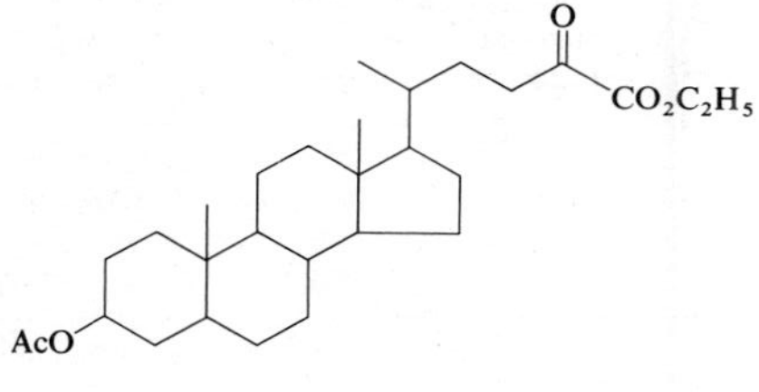 [structure]	DAST	”	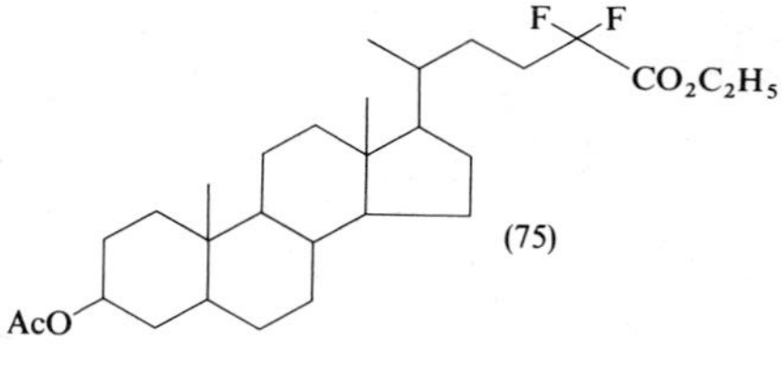 [structure] (75)	106
	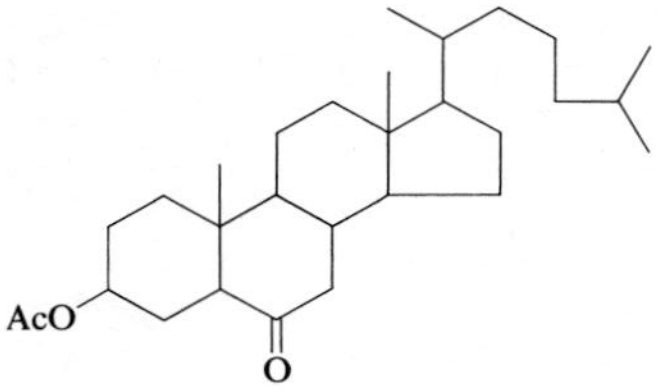 [structure]	”	Glyme, $H_2SO_4 + SO_3$, 50°, 17 d	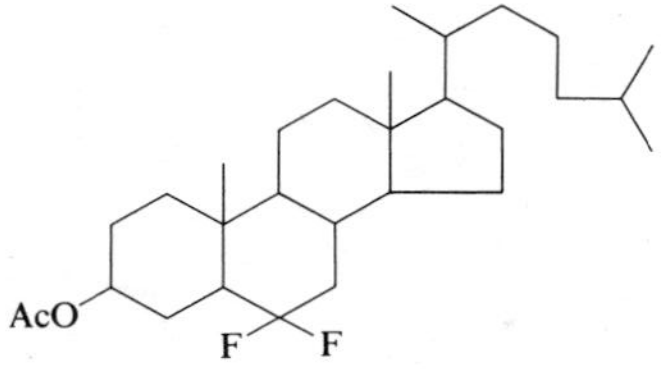 [structure] (6) 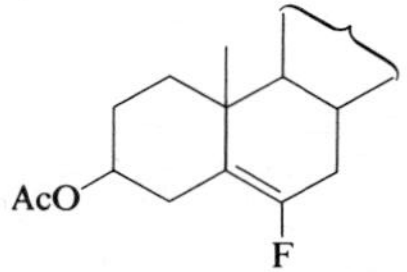 + [structure] (66) 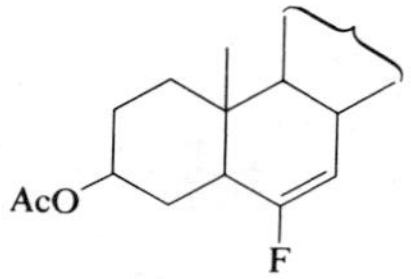 + [structure] (6)	37
	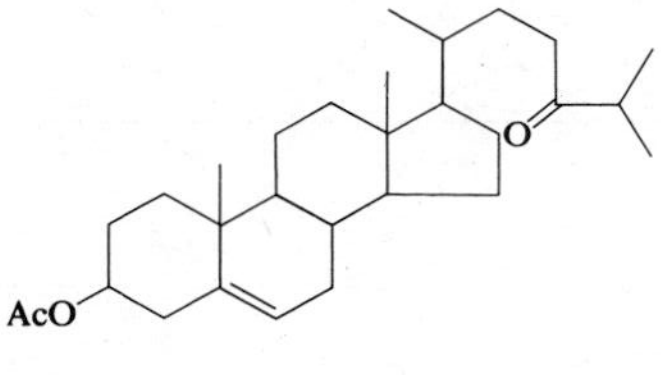 [structure]	” (11)	60–70°, 1 h	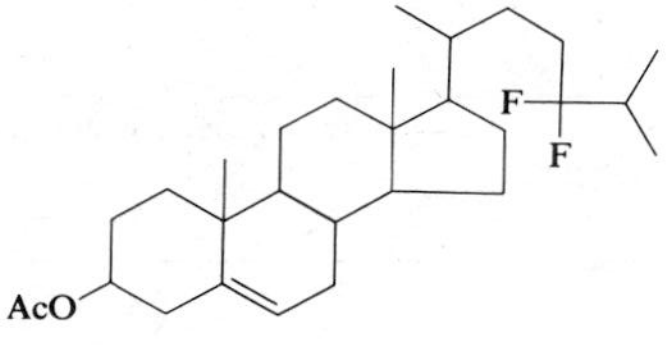 [structure] (13)	171

TABLE III. KETONES (*Continued*)

No. of Carbon Atoms	Reactant	Reagent (Molar Equiv.)	Conditions	Product(s) and Yield(s) (%)		Refs.
C_{31}	t-C_4H_9O (steroid 7-ketone)			t-C_4H_9O (7,7-difluoro) +	t-C_4H_9O (7-fluoro-Δ⁶)	
		DAST	50°, 10 d	(—)	(63)	37
		"	CH_2Cl_2, 50°, 10 d	(—)	(40)	37
		"	THF, ", "	(9)	(67)[m]	37
C_{39}	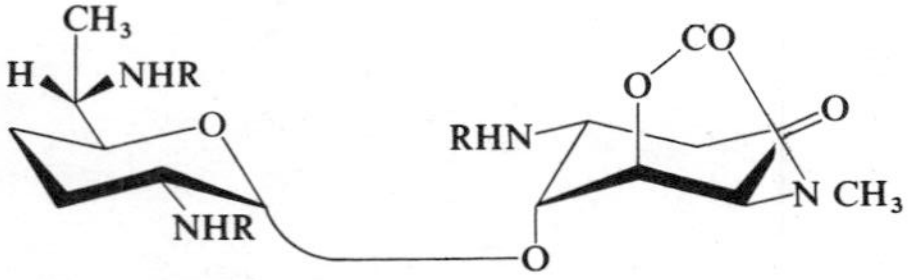 CH_3, H, NHR, O, NHR, RHN, O, CO, O, N CH_3; R = $CO_2CH_2C_6H_5$	"	", −15°, room temp, 3 h	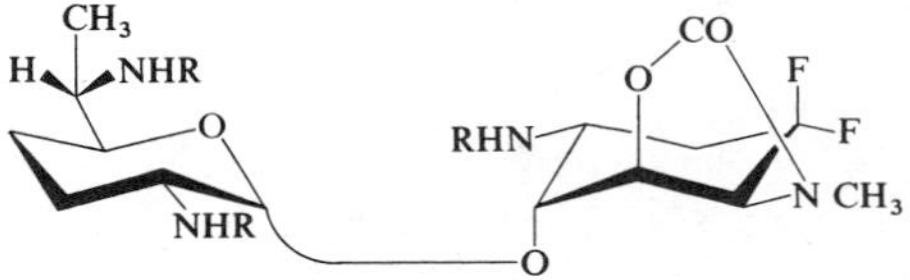 CH_3, H, NHR, O, NHR, RHN, O, CO, F, F, N CH_3 (72)		125, 129

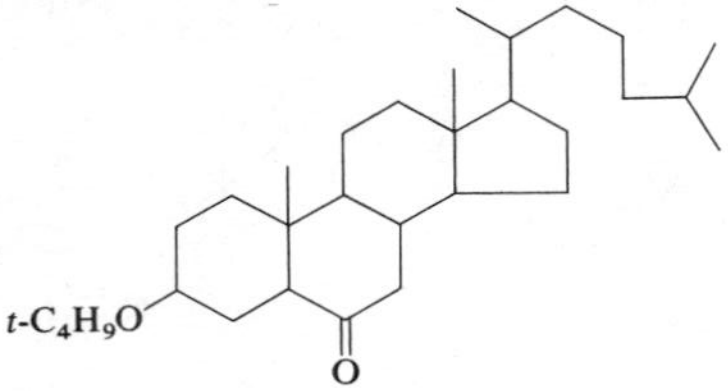

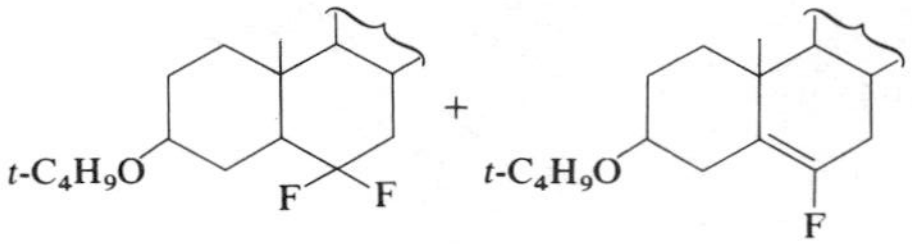

[a] The cleavage of the ring predominated.
[b] $R_2 = (CH_3)_2$, $(C_2H_5)_2$, $(CH_2)_5$, or $O(CH_2CH_2)_2$.
[c] The numbers are ratios.
[d] The number is the conversion.
[e] The effect of solvents and conditions on the product distribution is tabulated in the paper.
[f] A mixture of the geminal difluoride and vinyl fluoride (6.7 g) was obtained from 13.5 g of the starting hydrochloride.
[g] The compounds were mixtures of 8-epimers (α and β).
[h] Unreacted starting material was recovered (70–80%).
[i] The compounds were mixtures of 16-epimers.
[j] After 6-day heating, the starting ketone was recovered. After 13-day heating the free acid was isolated.
[k] The yields were close to theoretical, but the ratios of the two products were not given.
[l] Recovered were 25% of 13-*cis* and 55% all *trans*-4-oxoretinyl acetate.
[m] The Δ^6-6-fluoro compound was obtained in trace amounts.

TABLE IV. EPOXIDES (OXIRANES)

No. of Carbon Atoms	Reactant	Reagent (Molar Equiv.)	Conditions	Product(s) and Yield(s) (%)	Refs.
C_5	Cyclopentene oxide	DAST	55–60°, 8–17 h	*cis*-1,2-Difluorocyclopentane (13–14) + *trans*-1,2-difluorocyclopentane (2) + bis(2-fluorocyclopentyl) ether (25–34)	78a
C_6	Cyclohexene oxide	"	", 6–24 h	*cis*-1,2-Difluorocyclohexane (19–29) + bis(2-fluorocyclohexyl) ether (13–19)	78a
	cis-Cyclohexene sulfide	"	80°	Unidentified products	78a
C_8	Styrene oxide	"	55–60°, 2–4.5 h	1,1-Difluoro-2-phenylethane (15) + 1,2-difluoro-2-phenylethane (23–27) + *cis*-1-fluoro-2-phenylethylene (trace)	78a
	cis-Cyclooctene oxide	"	80°, 9–15 h	No reaction	78a
C_{14}	*cis*-Stilbene oxide *trans*-Stilbene oxide	" "	" "	*meso*-1,2-Difluoro-1,2-diphenylethane[a] *DL*-1,2-difluoro-1,2-diphenylethane 1,1-difluoro-2,2-diphenylethane 1-fluoro-2,2-diphenylethylene	78a

[a] Both stereoisomers gave the same inseparable mixture of the four products with conversion of at most 40% and the relative yields of (29–42%):(21–24%):(24–29%):(5–6%), respectively, as determined by NMR.

TABLE V. Carboxylic and Other Acids

No. of Carbon Atoms	Reactant	Reagent (Molar Equiv.)	Conditions	Product(s) and Yield(s) (%)	Refs.
C_7	p-$O_2NC_6H_4CO_2H$	R_2NSF_3[a] (1)	Ether, 0°, 20°, 15 min	p-$O_2NC_6H_4COF$ (89)	8
	$C_6H_5CO_2H$	"	", ", ", "	C_6H_5COF (85)	8
		DAST (1)	CH_2Cl_2, 0°, NaF[b]	" (80)	9
		" (2)	Diglyme, ", ", 80°, 20 h	$C_6H_5CF_3$ (50)	9
	p-$CH_3C_6H_4SO_3H$	$OC_4H_8NSF_3$	$C_2H_4Cl_2$, 80°, 3.5 h	p-$CH_3C_6H_4SO_2F$ (61)	79
C_{10}	$[(C_2H_5)_2NCS_2]_2$	DAST (1.5)	5–20°	$(C_2H_5)_2NCF_3$ (70)	8
C_{13}	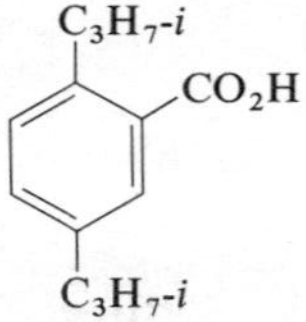 C_3H_7-i, CO_2H, C_3H_7-i	"	C_6H_6, cooling	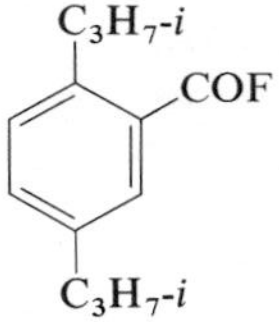 C_3H_7-i, COF, C_3H_7-i (96)	9
C_{14}	$(C_6H_5CH_2)_2P(O)OH$	R_2NSF_3[b]	Ether, 0°, 20°, 5 min	$(C_6H_5CH_2)_2P(O)F$ (60)	8
C_{20}	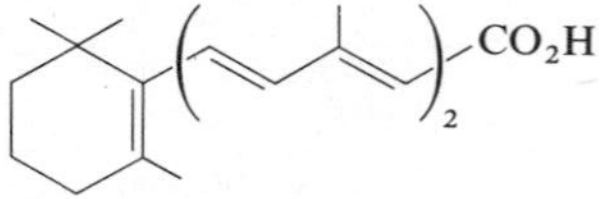 CO_2H, 2	DAST	Ether, 0° to room temp	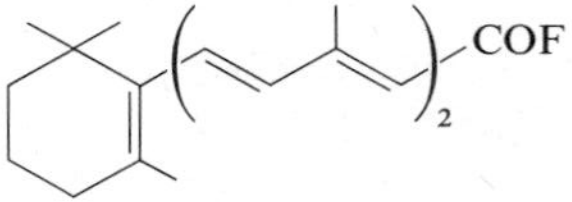 COF, 2 (50–70)[c]	117
	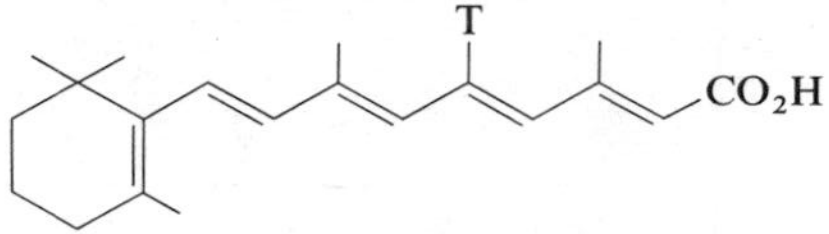 T, CO_2H	"	", cooling	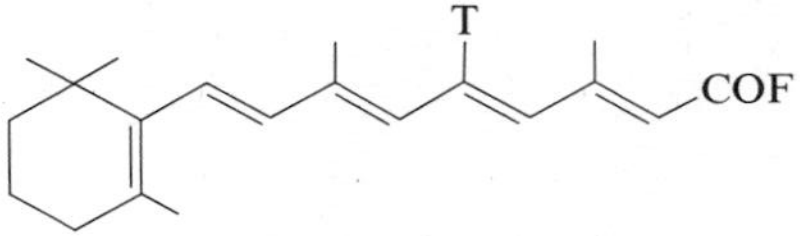 T, COF (39)	117

[a] $R_2 = (CH_3)_2$, $(C_2H_5)_2$, $(CH_2)_5$, or $O(CH_2CH_2)_2$.
[b] Sodium fluoride was added to remove the hydrogen fluoride.
[c] In addition to the all-*trans* retinoyl fluoride, 10–30% of 13-*cis*-retinoyl fluoride was obtained.

TABLE VI. LACTONES

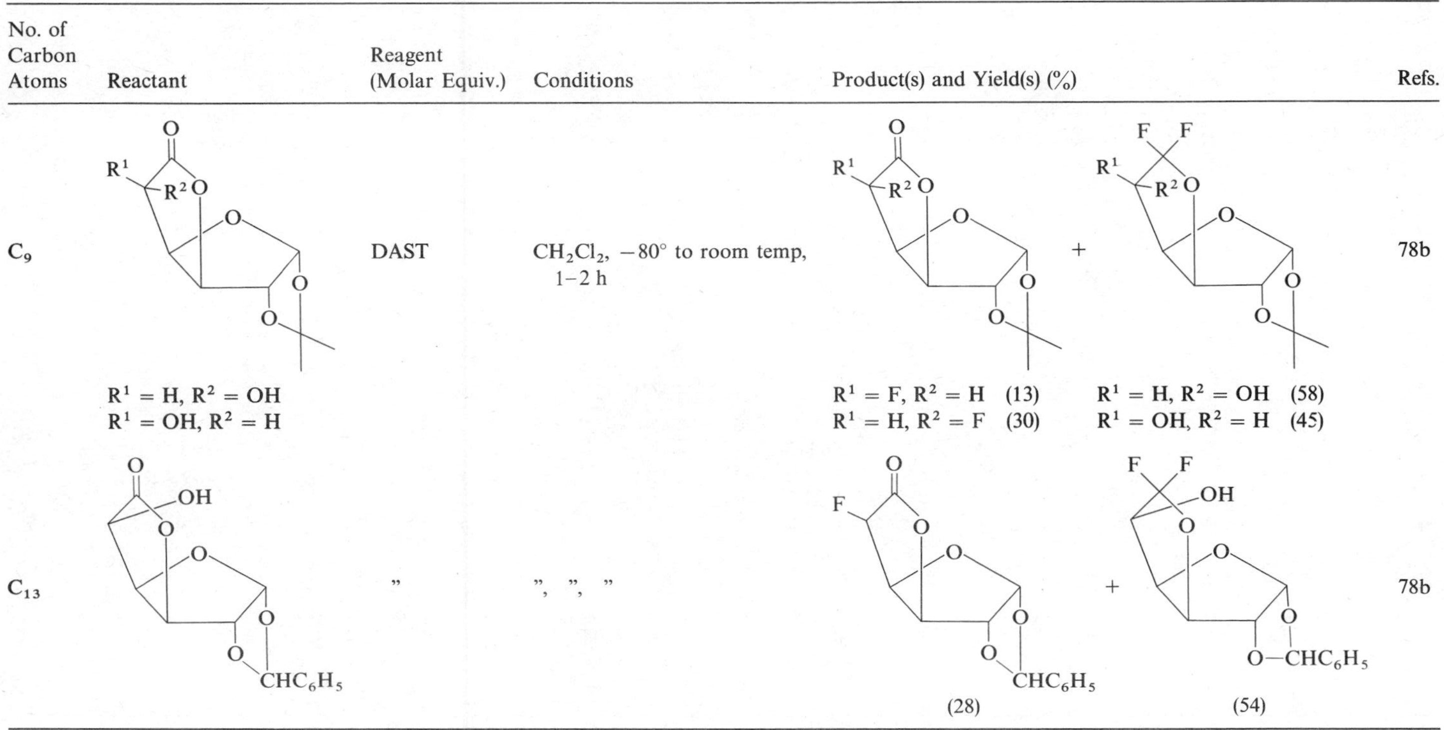

No. of Carbon Atoms	Reactant	Reagent (Molar Equiv.)	Conditions	Product(s) and Yield(s) (%)	Refs.
C_9	$R^1 = H, R^2 = OH$ $R^1 = OH, R^2 = H$	DAST	CH_2Cl_2, −80° to room temp, 1–2 h	$R^1 = F, R^2 = H$ (13) $R^1 = H, R^2 = F$ (30) + $R^1 = H, R^2 = OH$ (58) $R^1 = OH, R^2 = H$ (45)	78b
C_{13}		"	", ", "	(28) + (54)	78b

TABLE VII. HALIDES AND SULFONATES

No. of Carbon Atoms	Reactant	Reagent (Molar Equiv.)	Conditions	Product(s) and Yield(s) (%)	Refs.
	$SOCl_2$	R_2NSF_3[a]		SOF_2 (—)	39
C_1	$CDCl_3$	TASF		$CDCl_2F$ (94)	7
C_2	CCl_3COCl	R_2NSF_3[a]	20°, 15–20 min, 60°, ~30 min	CCl_3COF (73)	39
	$CH_2ClCOCl$	"	", ", ", "	CH_2ClCOF (80)	39
	C_2H_5I	TASF	CH_3CN, 25°	C_2H_5F (85)	7
		$C_4H_8N\overset{+}{S}[N(CH_3)_2]_2\ (CH_3)_3\overset{-}{Si}F_2$[b]	Neat, 25°	" (—)	7
		$(C_4H_8N)_2\overset{+}{S}N(CH_3)_2\ (CH_3)_3\overset{-}{Si}F_2$[b]	", "	" (—)	7
		$(C_4H_8N)_3\overset{+}{S}\ (CH_3)_3\overset{-}{Si}F_2$[b]	", "	" (—)	7
		$(C_5H_{10}N)_3\overset{+}{S}\ (CH_3)_3\overset{-}{Si}F_2$[c]	", room temp, 2 h	" (—)	7
C_3	$CH_2{=}CHCH_2Br$	$(C_4H_8N)_3\overset{+}{S}\ (CH_3)_3\overset{-}{Si}F_2$[b]	CH_3CN, room temp, 2 h	$CH_2{=}CHCH_2F$ (83)	7
	C_2H_5OCOCl	R_2NSF_3[a]	20°, 15–20 min, 60°, ~30 min	C_2H_5OCOF (51)	39
	$(CH_3)_3SiCl$	"	", ", ", "	$(CH_3)_3SiF$ (—)	39
C_4	t-$C_4H_9N{=}SCl_2$	"	", ", ", "	t-$C_4H_9N{=}SF_2$ (62)	39
	$(C_2H_5O)_2P(O)Cl$	"	", ", ", "	$(C_2H_5O)_2P(O)F$ (70)	39
C_5	$(C_2H_5)_2NCCl_3$	$OC_4H_8NSF_3$[d] (3)	C_6H_5Cl, 20°, 10 min	$(C_2H_5)_2NCF_3$ (80)	39
C_6	p-$BrC_6H_4SO_2Cl$	R_2NSF_3[a]	20°, 15–20 min; 60°, ~30 min	p-$BrC_6H_4SO_2F$ (79)	39
	C_6H_5SOCl	$OC_4H_8NSF_3$[d]	Ether, 20°, 10 min	C_6H_5SOF (85)	39
	$C_6H_5SO_2Cl$	R_2NSF_3[a]	20°, 15–20 min; 60°, ~30 min	$C_6H_5SO_2F$ (72)	39

C_7	C_6H_5COCl	"	", ", ", "	C_6H_5COF (70)	39
	$C_6H_5CH_2Br$	TASF	CH_3CN, 1 d	$C_6H_5CH_2F$ (—)	7
	p-$CH_3C_6H_4SOCl$	$OC_4H_8NSF_3$[d]	Hexane, 20°, 10 min	p-$CH_3C_6H_4SOF$ (91)	39
C_{10}	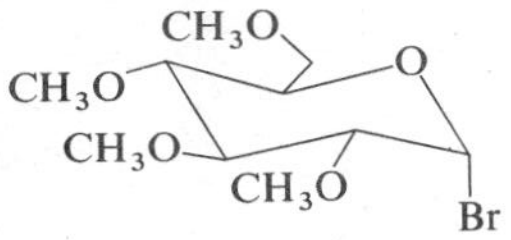 CH_3O, CH_3O, CH_3O, CH_3O, O, Br	$TA\overset{+}{S}\ CF_3\overset{-}{O}$	CH_3CN, room temp, 2 h	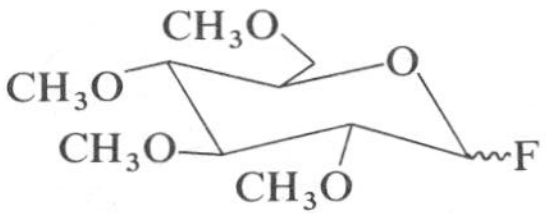 CH_3O, CH_3O, CH_3O, CH_3O, O, F ax (11) eq (12) 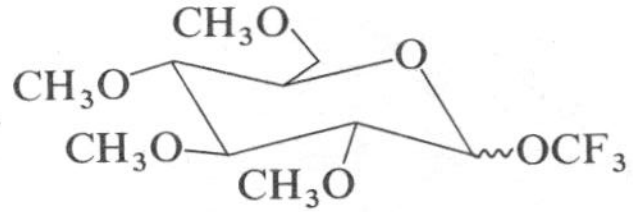 + CH_3O, CH_3O, CH_3O, CH_3O, O, OCF_3 ax (36) eq (29)	80b
C_{13}	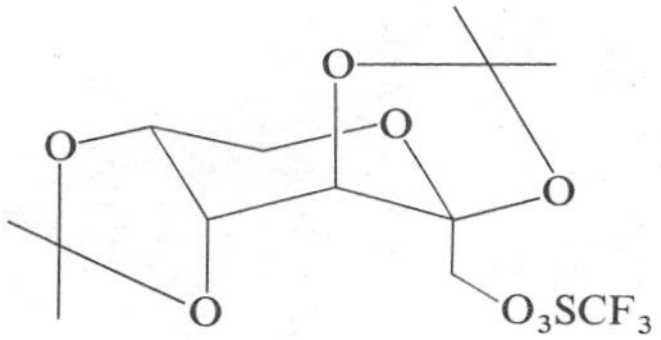 O, O, O, O, O, O_3SCF_3	TASF	THF, reflux, overnight	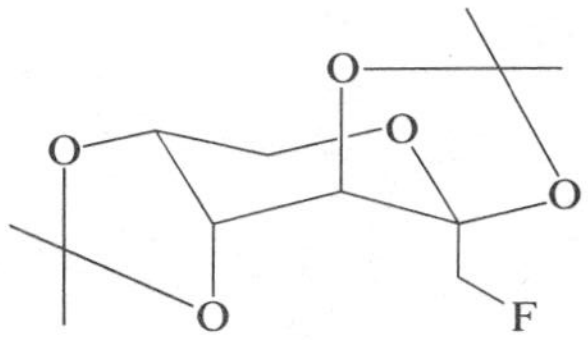 O, O, O, O, O, F (80)	80
	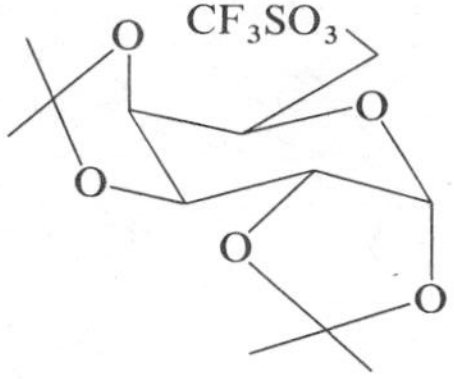 O, CF_3SO_3, O, O, O, O	TASF (3)	CH_2Cl_2, salt bath, 10 min	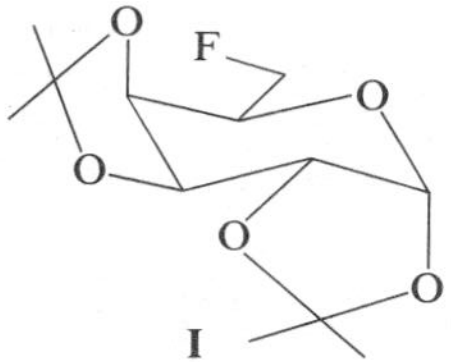 O, F, O, O, O, O **I** (71)	172

TABLE VII. HALIDES AND SULFONATES (*Continued*)

No. of Carbon Atoms	Reactant	Reagent (Molar Equiv.)	Conditions	Product(s) and Yield(s) (%)	Refs.
	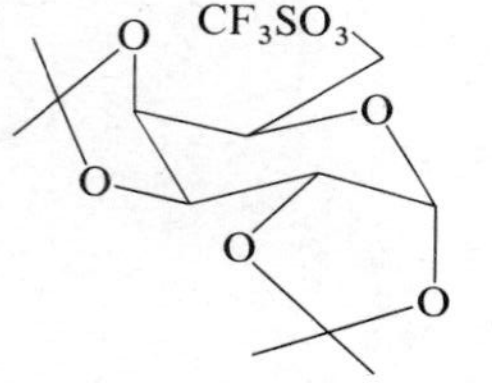 CF_3SO_3	$TAS^+CF_3O^-$	CH_3CN, reflux, 45 min	**I** (36) 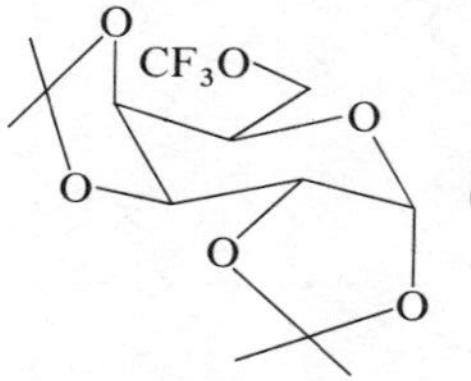 + CF_3O (52)	80b
	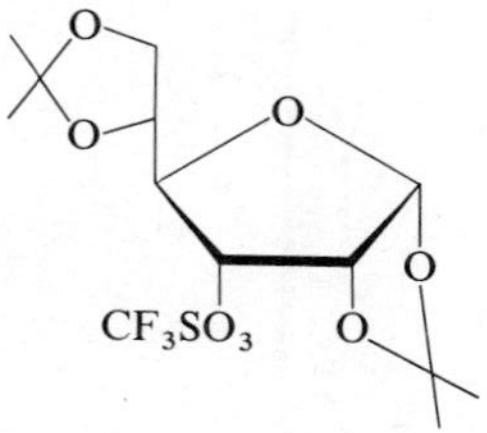 CF_3SO_3	TASF(3)	CH_2Cl_2, salt-ice bath, 10 min	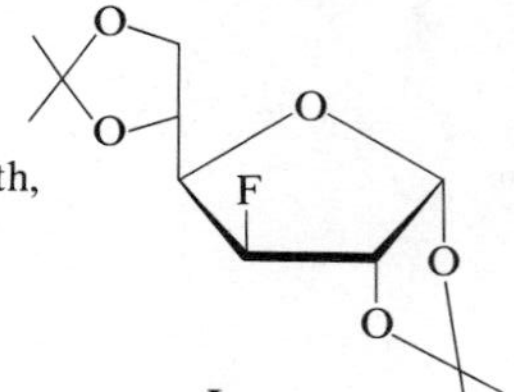 F (66) **I**	172
		$TA\overset{+}{S}\,CF_3\overset{-}{O}$	CH_3CN, reflux, 5 h	**I** (75) 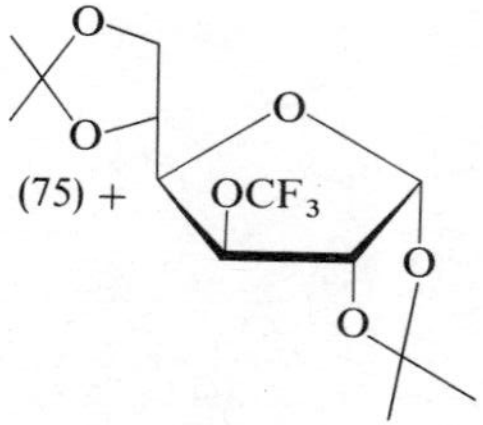 + OCF_3 (16)	80b

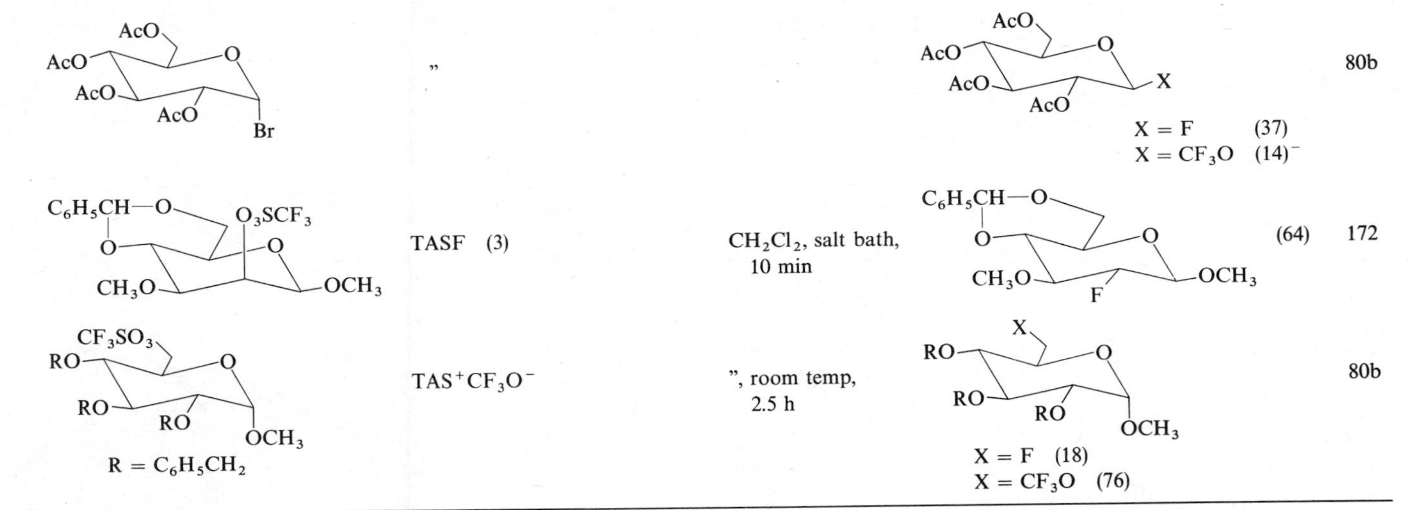

	Substrate	Reagent	Conditions	Product(s) and Yield(s) (%)	Refs.
C_{14}		"		$X = F$ (37) $X = CF_3O$ $(14)^-$	80b
C_{16}		TASF (3)	CH_2Cl_2, salt bath, 10 min	(64)	172
C_{29}	$R = C_6H_5CH_2$	$TAS^+CF_3O^-$	", room temp, 2.5 h	$X = F$ (18) $X = CF_3O$ (76)	80b

[a] $R_2 = (C_2H_5)_2$, C_5H_{10}, or OC_4H_8.[39]
[b] C_4H_8N = pyrrolidino.
[c] $C_5H_{10}N$ = piperidino.
[d] OC_4H_8 = morpholino.

TABLE VIII. SULFOXIDES

No. of Carbon Atoms	Reactant	Reagent (Molar Equiv.)	Conditions	Product(s) and Yield(s) (%)	Ref.
C_2	$(CH_3)_2SO$	DAST (2)	$CHCl_3$, room temp, 16 h	CH_3SCH_2F (>83)	40
C_3	$C_2H_5S(O)CH_3$	"	", ", "	$C_2H_5SCH_2F$ (>69)	40
C_7	$CH_2FS(O)C_6H_5$	"	", 50°, 96 h	$CHF_2SC_6H_5$ (23)	40
	$CH_3S(O)C_6H_5$	"	", room temp to 50°, 30 h	$CH_2FSC_6H_5$ (85)	40
C_8	p-$CH_3OC_6H_4S(O)CH_2F$	"	", 50°, 96 h	p-$CH_3OC_6H_4SCHF_2$ (68)	40
	p-$CH_3OC_6H_4S(O)CH_3$	"	", room temp, 8 h	p-$CH_3OC_6H_4SCH_2F$ (95)	40
C_{11}	p-$CH_3OC_6H_4S(O)(CH_2)_3CN$	"	", ", ZnI_2, 16 h	p-$CH_3OC_6H_4SCHF(CH_2)_2CN$ (89)	40
C_{13}	$C_6H_5CH_2S(O)C_6H_5$	"	", ", ", 63 h	$C_6H_5CHFSC_6H_5$ (>44)	40
	$C_2H_5O_2C(CH_2)_3S(O)C_6H_4OCH_3$-$p$	"	", ", ", 16 h	$C_2H_5O_2C(CH_2)_2CHFSC_6H_4OCH_3$-$p$ (79)	40
C_{15}	$C_6H_5(CH_2)_2S(O)C_6H_4OCH_3$-$p$	"	", ", ", 18 h	$C_6H_5CH_2CHFSC_6H_4OCH_3$-$p$ (>86)	40
	$C_6H_5(CH_2)_3S(O)C_6H_5$	"	", ", ", 16 h	$C_6H_5(CH_2)_2CHFSC_6H_5$ (100)	40
C_{16}	$C_6H_5(CH_2)_3S(O)C_6H_4OCH_3$-$p$	"		$C_6H_5(CH_2)_2CHFSC_6H_4OCH_3$-$p$ (85)	40
C_{17}	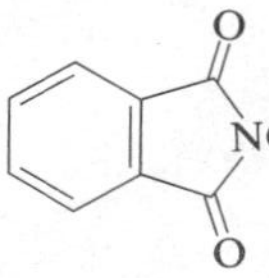 $NCH_2CH_2S(O)C_6H_4OCH_3$-p	"	", ", ZnI_2, 72 h	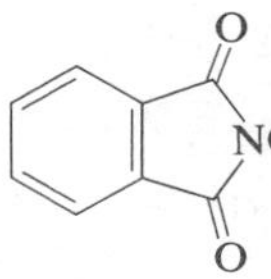 $NCH_2CHFSC_6H_4OCH_3$-p (91)	40
C_{23}	n-$C_{15}H_{31}CH_2S(O)C_6H_4OCH_3$-$p$	"	", ", ", 16 h	n-$C_{15}H_{31}CHFSC_6H_4OCH_3$-$p$ (94)	40
			$CHCl_3$, room temp, 16 h	" (8)	40

TABLE IX. HEMITHIOACETALS

No. of Carbon Atoms	Reactant	Reagent (Molar Equiv.)	Conditions	Product(s) and Yield(s) (%)	Ref
C_{15}	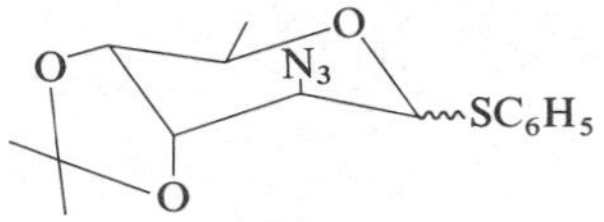 O, O, N_3, O, SC_6H_5	DAST	NBS, CH_2Cl_2, 0–25°	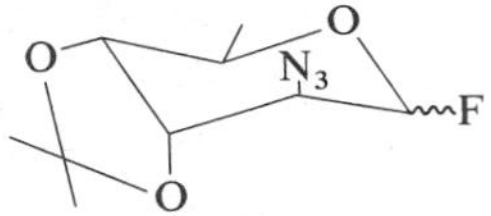 O, O, N_3, O, F; α (53) + β (27)	81
C_{18}	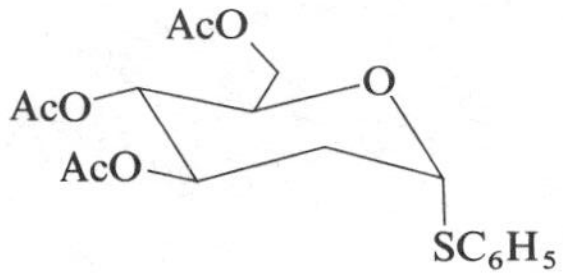 AcO, AcO, AcO, O, SC_6H_5	"	", ", "	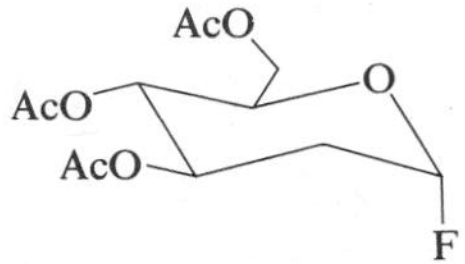 AcO, AcO, AcO, O, F (70)	81
	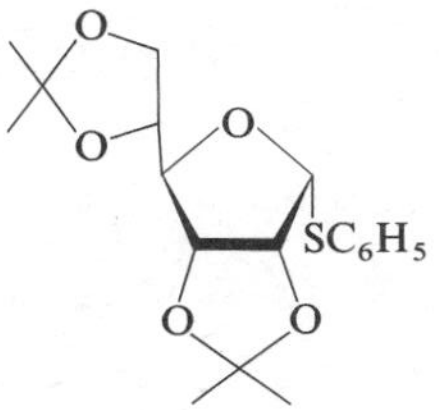 O, O, O, O, O, SC_6H_5	"	", ", "	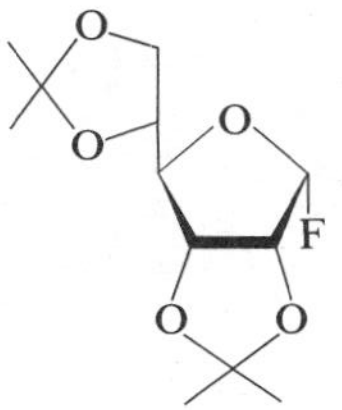 O, O, O, O, O, F (91)	81
C_{20}	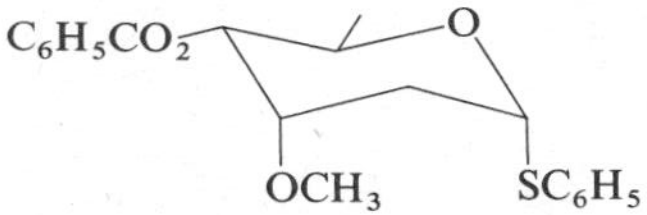 $C_6H_5CO_2$, O, OCH_3, SC_6H_5	"	", ", "	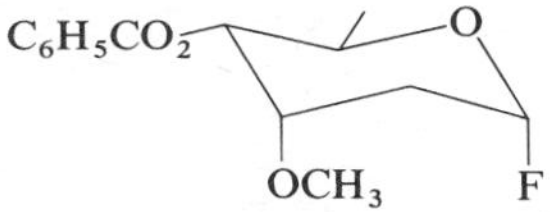 $C_6H_5CO_2$, O, OCH_3, F (88)	81

TABLE IX. HEMITHIOACETALS (*Continued*)

No. of Carbon Atoms	Reactant	Reagent (Molar Equiv.)	Conditions	Product(s) and Yield(s) (%)	Ref
C_{26}	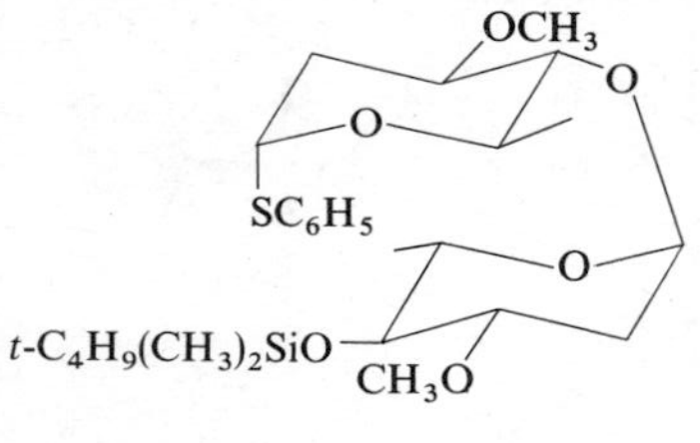 OCH$_3$, O, O, SC_6H_5, O, t-$C_4H_9(CH_3)_2SiO$, CH_3O	DAST (1.5)	", ", −15°, 15 min	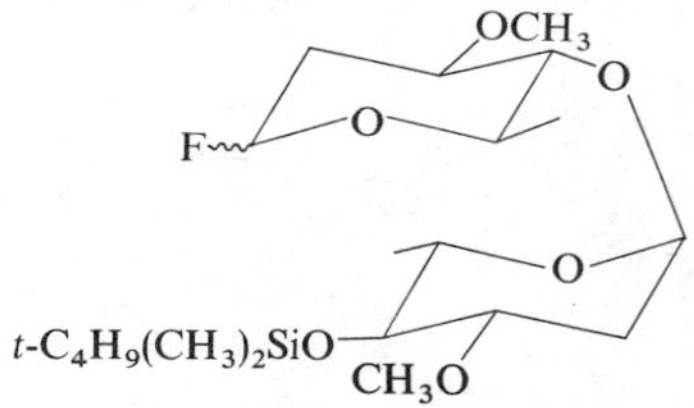 OCH$_3$, O, F, O, O, t-$C_4H_9(CH_3)_2SiO$, CH_3O α (71) + β (14)	81
C_{42}	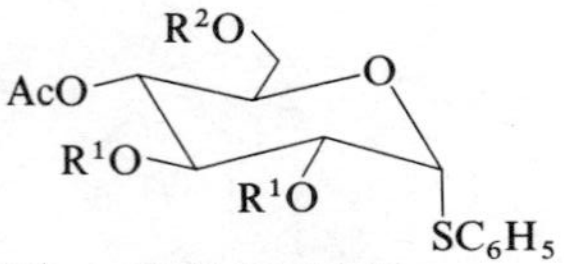 R^2O, AcO, O, R^1O, R^1O, SC_6H_5 $R^1 = t$-$C_4H_9(CH_3)_2Si$ $R^2 = t$-$C_4H_9(C_6H_5)_2Si$	"	", ", 0–25°	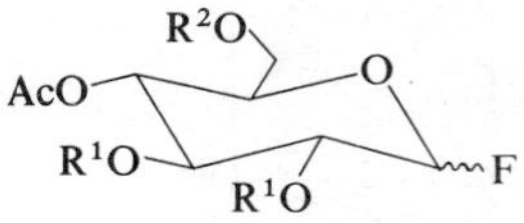 R^2O, AcO, O, R^1O, R^1O, F α (68) + β (14)	81
C_{110}	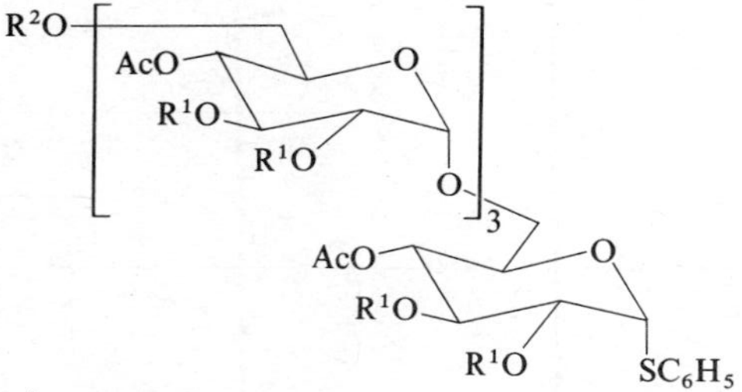 R^2O, [AcO, O, R^1O, R^1O, O]$_3$, AcO, O, R^1O, R^1O, SC_6H_5 $R^1 = C_6H_5CH_2$ $R^2 = t$-$C_4H_9(C_6H_5)_2Si$	" (1.5)	", ", −15°, 25 min	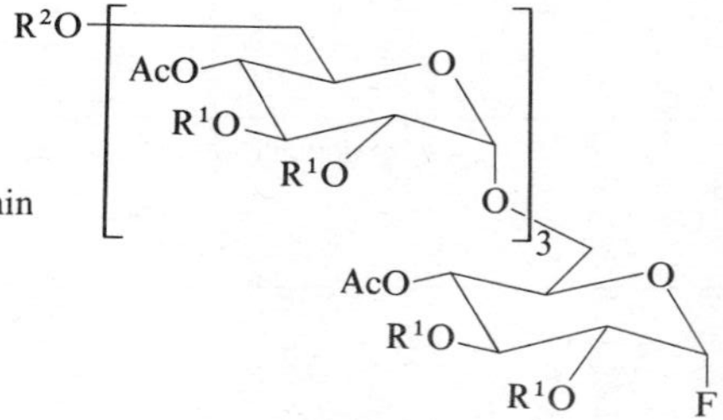 R^2O, [AcO, O, R^1O, R^1O, O]$_3$, AcO, O, R^1O, R^1O, F (85)	81

TABLE X. ALKYL SILYL ETHERS

No. of Carbon Atoms	Reactant	Reagent (Molar Equiv.)	Conditions	Product(s) and Yield(s) (%)	Ref.
C_6	$H(CF_2)_2CH_2OSi(CH_3)_3$	$OC_4H_8NSF_3$	Ether, 0–5°, 10 min, 20°, 30 min	$H(CF_2)_2CH_2OSF_2NC_4H_8O$ (96)	82
				$H(CF_2)_2CH_2F$[a] (85)	82
C_7	$(CH_3)_3SiSC_4H_9$-*t*	DAST		$(C_2H_5)_2NSC_4H_9$-*t* + $(t\text{-}C_4H_9S)_2$ (−)	173
C_8	$H(CF_2)_4CH_2OSi(CH_3)_3$	"	", ", ", ", "	$H(CF_2)_4CH_2OSF_2N(C_2H_5)_2$ (99) $H(CF_2)_4CH_2F$[a] (39)	82 82
C_{10}	$H(CF_2)_6CH_2OSi(CH_3)_3$	"	", ", "	$H(CF_2)_6CH_2OSF_2N(C_2H_5)_2$ (99) $H(CF_2)_6CH_2F$[a] (27)	82 82
C_{15}	$R^1R^2CH(CN)OSi(CH_3)_3$[b] $R^1 = C_6H_5$, 3-FC_6H_4, 4-$O_2NC_6H_4$, 3,4-$Cl_2C_6H_3$, 2,6-$Cl_2C_6H_3$, 3-$CH_3OC_6H_4$, 3-$AcOC_6H_4$, 3,4-$(CH_3O)_2C_6H_3$, 2-naphthyl, $R^2 = H$; $R^1 = C_6H_5$, $R^2 = CH_3$, C_2H_5	"	CH_2Cl_2, 5° to room temp	R^1R^2CFCN	83

[a] The fluoroalkane is formed on heating the intermediate aminofluorosulfurane to 90°.

[b] Since the cyanohydrin trimethylsilyl ethers are prepared *in situ* from trimethylsilyl cyanide and aromatic aldehydes and ketones, the specific examples are listed in Table I.

TABLE XI. PHOSPHORUS COMPOUNDS

No. of Carbon Atoms	Reactant	Reagent (Molar Equiv.)	Conditions	Product(s) and Yield(s) (%)	Ref.
C_{18}	$(C_6H_5)_3P$	R_2NSF_3[a]	Ether, 20°	$(C_6H_5)_3PF_2$ (93)	8
	$(C_6H_5)_3PS$	"	C_6H_6, reflux, 15 min	" (82)	8

[a] $R_2 = (CH_3)_2$, $(C_2H_5)_2$, $(CH_2)_5$, or $O(CH_2CH_2)_2$.

TABLE XII. Non-Fluorinating Reactions (Dehydrations)

No. of Carbon Atoms	Reactant	Reagent (Molar Equiv.)	Conditions	Product(s) and Yield(s) (%)	Refs.
C_{12}		DAST	CH_2Cl_2, C_5H_5N (2 equiv), 0°, 30 min, room temp	(62)	29
C_{13}		TASF	", salt bath, 10 min	(83)	172
C_{14}		DAST	", −78°	$CH_3CH{=}CHCH(NCO)CH_2CO_2CH_2C_6H_5$ *trans* (43)	51
		$C_5H_{10}NSF_3$	", 2 C_5H_5N, −78° to room temp	" " "	51
C_{15}		"	CH_2Cl_2, −30°	(67)	57

TABLE XII. Non-Fluorinating Reactions (Dehydrations) (*Continued*)

No. of Carbon Atoms	Reactant	Reagent (Molar Equiv.)	Conditions	Product(s) and Yield(s) (%)	Refs.
C_{16-21}	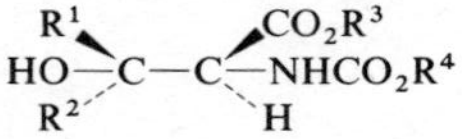 $HO-C(R^1)(R^2)-C(H)(CO_2R^3)-NHCO_2R^4$	DAST	CH_2Cl_2, C_5H_5N, 0°	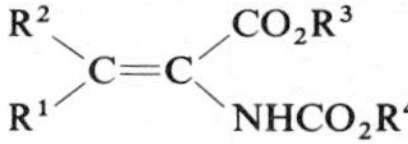 $R^2R^1C=C(CO_2R^3)NHCO_2R^4$	119
	R^1 = H, R^2 = H, R^3 = $C_6H_5CH_2$, R^4 = $C_6H_5CH_2$			(75)	
	R^1 = CH_3, R^2 = H, R^3 = ", R^4 = "			(90)	
	R^1 = H, R^2 = CH_3, R^3 = ", R^4 = "			(90)	
	R^1 = i-C_3H_7, R^2 = H, R^3 = ", R^4 = "			(90)	
	R^1 = H, R^2 = i-C_3H_7, R^3 = ", R^4 = "			(80)	
	R^1 = CH_3, R^2 = H, R^3 = ", R^4 = t-C_4H_9			(70)	
	R^1 = H, R^2 = CH_3, R^3 = ", R^4 = "			(65)	
C_{21}	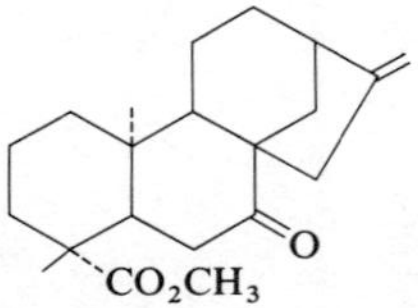 (structure: CO_2CH_3, =O)	"	CH_2Cl_2, room temp, 25 h	Intractable mixture (6 compounds)	74
C_{22}	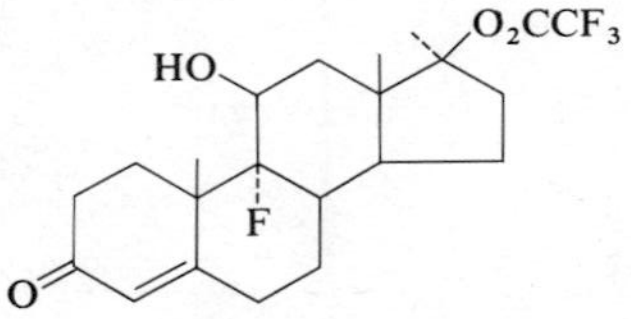 (structure: HO, F, O_2CCF_3, O=)	$C_5H_{10}NSF_3$	Dioxane, 25°, 1 h	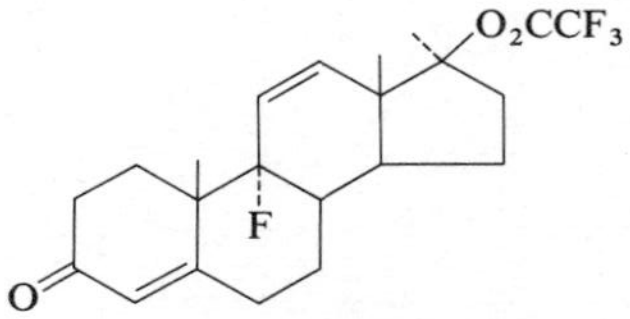 (structure: F, O_2CCF_3, O=) (–)	56

Carbons	Reactant	Reagent	Conditions	Product	Yield (%)	Refs.
C_{25}	HO, COCHO, $O_2CC_2H_5$, F, O	$C_5H_{10}NSF_3$	Dioxane, 25°, 4 h	COCHO, $O_2CC_2H_5$, F, O	(–)	56
C_{25-30}	HO, $COCH_2OR^4$, R^3, R^2, F, O, R^1			$COCH_2OR^4$, R^3, R^2, F, O, R^1		

R^1	R^2	R^3	R^4	Reagent	Conditions	Yield (%)	Refs.
H	α-CH_3	α-OH	COC_2H_5	$C_5H_{10}NSF_3$	Dioxane, room temp, 4.5 h	(—)	56
"	α-O—$C(CH_3)_2$—O-α		$COCH_3$	DAST	CH_2Cl_2, −78° to 25°	(57)	52
F	"	"	"	$C_5H_{10}NSF_3$	Dioxane, 0°, 25°, 2.5 h	(88)	30
"	"	"	"	"	", room temp, 1.5 h	(—)	56
"	α-CH_3	α-OH	COC_4H_9-*t*	"	C_5H_5N, ", 2 h	(—)	56
H	"	α-O—$C(C_2H_5)(OC_2H_5)$—		"	Dioxane, ", 1 h	(—)	56
F	"	α-$O_2CC_2H_5$	COC_2H_5	"	", ", 3 h 20 min	(—)	56
H	"	"	"	DAST	", 25°, 4 h	(—)	56
"	β-CH_3	"	"	"	CH_2Cl_2, −78° to room temp	(53)	52
"	"	"	"	$C_5H_{10}NSF_3$	", ", "	(62)	52
"	α-CH_3	"	"	"	Dioxane, 0°, 25°, 2 h	(85)	30
"	β-CH_3	α-$O_2CC_4H_9$-*n*	"	"	", ", ", 2.5 h	(73)	30, 56

TABLE XII. NON-FLUORINATING REACTIONS (DEHYDRATIONS) (*Continued*)

No. of Carbon Atoms	Reactant	Reagent (Molar Equiv.)	Conditions	Product(s) and Yield(s) (%)	Refs.
$C_{27,28}$	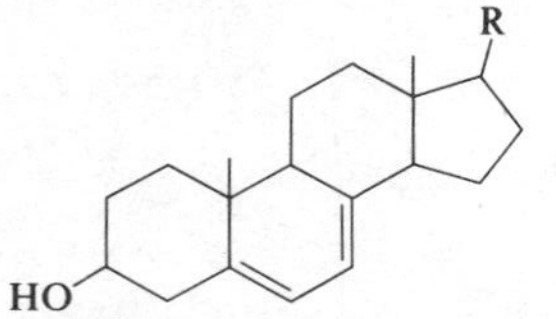 HO, R R = $CH(CH_3)(CH_2)_3CH(CH_3)_2$ R = $CH(CH_3)CH{=}CHCH(CH_3)CH(CH_3)_2$	$OC_4H_8NSF_3$	CH_2Cl_2, 5°, 10 m	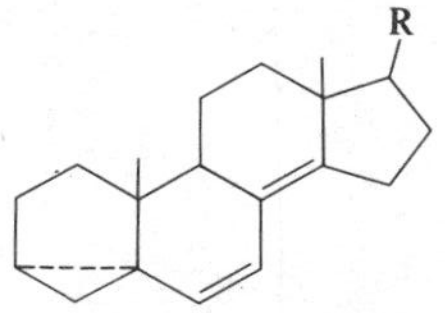 R (74) (90)	164a
C_{35}	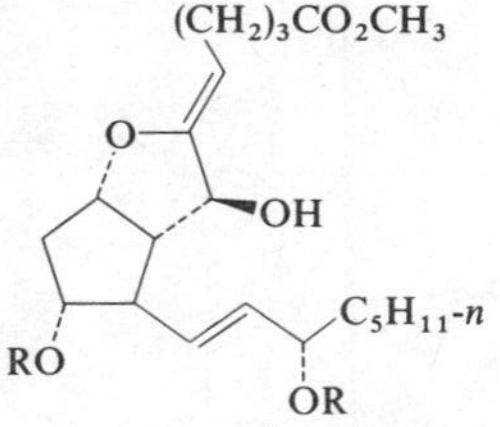 $(CH_2)_3CO_2CH_3$, O, OH, C_5H_{11}-*n*, RO, OR R = *t*-$C_4H_9(CH_3)_2Si$	DAST	CH_2Cl_2, −78° 1.5 h	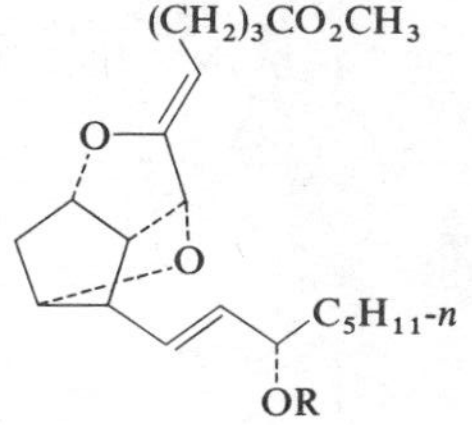 $(CH_2)_3CO_2CH_3$, O, O, C_5H_{11}-*n*, OR (52)	116

REFERENCES

[1] G. C. Demitras, R. A. Kent, and A. G. MacDiarmid, *Chem. Ind.(London)*, **1964**, 1712.

[2] G. C. Demitras and A. G. MacDiarmid, *Inorg. Chem.*, **6**, 1093 (1967).

[3] S. P. von Halasz and O. Glemser, *Chem. Ber.*, **103**, 594 (1970).

[4] L. N. Markovskii, V. E. Pashinnik, and N. A. Kirsanova, *Zh. Org. Khim.*, **11**, 74 (1975) (*Engl. transl.*, 72).

[5] W. J. Middleton, *J. Org. Chem.*, **40**, 574 (1975).

[6] W. J. Middleton, U.S. Patent 3888924 (1975) [*C.A.*, **83**, 78600d (1975)].

[7] W. J. Middleton, U.S. Patent 3940402 (1976) [*C.A.*, **85**, 6388j (1976)].

[8] L. N. Markovskii, V. E. Pashinnik, and N. A. Kirsanova, *Synthesis*, **1973**, 787.

[9] W. J. Middleton, U.S. Patent 3914265 (1975) [*C.A.*, **84**, 42635a (1976)]; 3976691 (1976) [*C.A.*, **86**, 29054g (1977)].

[10] S. P. von Halasz and O. Glemser, *Chem. Ber.*, **104**, 1247 (1971).

[11] Aldrich Chemical Company, Catalog, 1984; personal communication.

[12] W. J. Middleton and E. M. Bingham, *Org. Synth.*, **57**, 50 (1979).

[13] P. J. L. Daniels and D. F. Rane, U.S. Patent 4284764 (1981) [*C.A.*, **96**, 7032c (1982)]; South African Patent 78 06,385 (1978) [*C.A.*, **90**, 104301y (1979)].

[14] J. A. Gibson, D. J. Ibbott, and A. F. Janzen, *Can. J. Chem.*, **51**, 3203 (1973).

[15] C. von Braun, W. Dell, H.-E. Sasse, and M. L. Ziegler, *Z. Anorg. Allg. Chem.*, **450**, 139 (1979).

[16] L. N. Markovskii, V. E. Pashinnik, and N. A. Kirsanova, *Zh. Org. Khim.*, **12**, 965 (1975) (*Engl. transl.*, 973).

[17] M. Hudlický, unpublished results.

[18] W. J. Middleton, *Org. Synth.*, **64**, 221 (1985).

[19] W. B. Farnham and R. L. Harlow, *J. Am. Chem. Soc.*, **103**, 4608 (1981) (footnote 3).

[20] L. N. Markovskii, V. E. Pashinnik, and E. P. Saenko, *Zh. Org. Khim.*, **13**, 1116, (1977); (*Engl. transl.*, 1025).

[21] O. A. Radchenko, A. Y. Il'chenko, L. N. Markovskii, and L. M. Yagupol'skii, *Zh. Org. Khim.*, **14**, 275 (1978) (*Engl. transl.*, 251).

[22] L. N. Markovskii, V. E. Pashinnik, and N. A. Kirsanova, *Zh. Org. Khim.*, **13**, 1048 (1977) (*Engl. transl.*, 963).

[23] O. A. Radchenko, A. Y. Il'chenko, and L. M. Yagupol'skii, *Zh. Org. Khim.*, **16**, 863 (1980) (*Engl. transl.*, 758).

[24] O. A. Radchenko, A. Y. Il'chenko, and L. M. Yagupol'skii, *Zh. Org. Khim.*, **17**, 500 (1981) (*Engl. transl.*, 421).

[25] Y. V. Zeifman, S. A. Postovoi, and I. L. Knunyants, *Dokl. Akad. Nauk SSSR*, **265**, 347 (1982) (*Engl. transl.*, 212).

[26] D. G. Ibbott and A. F. Janzen, *Can. J. Chem.*, **50**, 2428 (1972).

[27] A. F. Janzen, J. A. Gibson, and D. G. Ibbott, *Inorg. Chem.*, **11**, 2853 (1972).

[27a] A. H. Cowley, P. E. Riley, J. S. Szobota, and M. L. Walker, *J. Am. Chem. Soc.*, **101**, 5620 (1979).

[28] G. A. Boswell, Jr., W. C. Ripka, R. M. Scribner, and C. W. Tullock, *Org. React.*, **21**, 1 (1974); C.-L. J. Wang, *ibid.*, **34**, 319 (1985).

[29] T. J. Tewson and M. J. Welch, *J. Org. Chem.*, **43**, 1090 (1978).

[30] M. Biollaz and J. Kalvoda, *Helv. Chim. Acta*, **60**, 2703 (1977).

[31] J. Leroy, E. Hebert, and C. Wakselman, *J. Org. Chem.*, **44**, 3406 (1979).

[32] S. Rozen, Y. Faust, and H. Ben-Yakov, *Tetrahedron Lett.*, **20**, 1823 (1979).

[33] P. J. Card, *J. Org. Chem.*, **48**, 393 (1983).

[34] T. B. Patrick and Y.-F. Poon, *Tetrahedron Lett.*, **25**, 1019 (1984).

[35] G. M. Blackburn and D. E. Kent, *J. Chem. Soc., Chem. Commun.*, **1981**, 511.

[36] G. Siegemund, *Justus Liebigs Ann. Chem.*, **1979**, 1280.

[37] G. A. Boswell, Jr., U.S. Patent 4212815 (1980) [*C.A.*, **93**, 239789w (1980)].

[38] G. A. Boswell, Jr., and R. M. Henderson, U.S. Patent 4241065 (1980) [*C.A.*, **95**, 150977z (1981)].

[38a] W. J. Middleton, E. I. duPont de Nemours & Co., Wilmington, DE, unpublished work cited in Ref. 38b.

[38b] A. E. Asato and R. S. H. Liu, *Tetrahedron Lett.*, 27, 3337 (1986).

[38c] G. A. Boswell, Jr., E. I. du Pont de Nemours & Co., Wilmington, DE, unpublished work cited in Ref. 38b.

[39] L. N. Markovskii and V. E. Pashinnik, *Synthesis*, **1975**, 801.

[40] J. R. McCarthy, N. P. Peet, M. E. LeTourneau, and M. Inbasekaran, abstracts from the 189th ACS meeting, Spring 1985, Miami Beach, ORGN 250, 251; *J. Am. Chem. Soc.*, **107**, 735 (1985).

[41] T. Kobayashi, M. Maeda, H. Komatsu, and M. Kojima, *Chem. Pharm. Bull.*, **30**, 3082 (1982).

[42] G. Lowe and B. V. L. Potter, *J. Chem. Soc., Perkin Trans. 1*, **1980**, 2029.

[43] P. J. Card and G. S. Reddy, *J. Org. Chem.*, **48**, 4734 (1983).

[44] G. H. Klemm, R. J. Kaufman, and R. Sidhu, *Tetrahedron Lett.*, **23**, 2927 (1982).

[45] C. W. Somawardhana and E. G. Brunngraber, *Carbohydr. Res.*, **94**, C14-C15 (1981).

[46] S. S. Yang, T. R. Beattie, and T. Y. Shen, *Tetrahedron Lett.*, **23**, 5517 (1982).

[47] T. G. C. Bird, G. Felsky, P. M. Fredericks, E. R. H. Jones, and G. D. Meakins, *J. Chem. Res.* (M), **1979**, 4728; (S), **1979**, 388.

[48] T. G. C. Bird, P. M. Fredericks, E. R. H. Jones, and G. D. Meakins, *J. Chem. Soc., Chem. Commun.*, **1979**, 65.

[49] H. E. Paaren, M. A. Fivizzani, H. K. Schnoes, and H. F. DeLuca, *Arch. Biochem. Biophys.*, **209**, 579 (1981).

[50] B. E. Cross and I. C. Simpson, *J. Chem. Res.* (S), **1980**, 118.

[51] C-P. Mak, K. Wagner, C. Mayerl, and H. Fliri, *Heterocycles*, **19**, 1399 (1982).

[52] M. J. Green, H.-J. Shue, M. Tanabe, D. M. Yasuda, A. T. McPhail, and K. D. Onan, *J. Chem. Soc., Chem. Commun.*, **1977**, 611.

[53] M. S. Newman, J. M. Khanna, and K. Kanakarajan, *J. Am. Chem. Soc.*, **101**, 6788 (1979).

[54] S.-J. Shiney, N. Chadha, and M. Uskokovic, Hoffman-LaRoche Inc., Nutley, NJ, personal communication.

[54a] S. Castillon, A. Dessinges, R. Faghih, G. Lukacs, A. Olesker, and Tan That Thang, *J. Org. Chem.*, **50**, 4913 (1985).

[54b] A. Hasegawa, M. Goto, and M. Kiso, *J. Carbohydr. Chem.*, **4**, 627 (1985).

[55] L. Somekh and A. Shanzer, *J. Am, Chem. Soc.*, **104**, 5836 (1982).

[56] M. Biollaz and J. Kalvoda, U.S. Patent 4172075 (1979) [*C.A.*, **92**, 147045b (1980)]; German Offen. 2632550 (1977) [*C.A.*, **87**, 23616r (1977)].

[57] A. L. Johnson, *J. Org. Chem.*, **47**, 5220 (1982).

[58] C. W. Somawardhana and E. G. Brunngraber, *Carbohydr. Res.*, **121**, 51 (1983).

[59] J. F. Carvalho and G. D. Prestwich, *J. Org. Chem.*, **49**, 1251 (1984).

[60] M. P. Kotick and J. O. Polazzi, *J. Heterocycl. Chem.*, **18**, 1029 (1981).

[61] G. L. Cantrell and R. Filler, *J. Fluorine Chem.*, **27**, 35 (1985).

[62] C. D. Poulter, P. L. Wiggins, and T. L. Plummer, *J. Org. Chem.*, **46**, 1532 (1981).

[63] M. Sharma, and W. Korytnyk, *Tetrahedron Lett.*, **1977**, 573.

[64] W. J. Middleton, E. M. Bingham, and D. H. Smith, *J. Fluorine Chem.*, **23**, 557 (1983).

[65] E. M. Bingham and W. J. Middleton, U.S. Patent 4120856 (1978) [*C.A.*, **86**, 171517d (1977)]; 4246270 (1981) [*C.A.*, **94**, 192389u (1981)].

[66] E. M. Bingham and W. J. Middleton, German Offen. 2632539 (1977) [*C.A.*, **86**, 171517d (1977)].

[67] G. A. Boswell, Jr., and D. R. Brittelli, U.S. Patent 3919204 (1975) [*C.A.*, **84**, 90161q (1976)].

[68] G. A. Boswell, Jr., D. R. Brittelli and W. J. Middleton, U.S. Patent 3950329 (1976) [*C.A.*, **85**, 78142x (1976)].

[69] J. A. May, Jr. and A. C. Sartorelli, *J. Med. Chem.*, **22**, 971 (1979).

[70] R. A. Sharma, I. Kavai, Y. L. Fu, and M. Bobek, *Tetrahedron Lett.*, **1977**, 3433.

[71] M. Biollaz and J. Kalvoda, U.S. Patent 4092310 (1978); German Offen. 2533496 (1976) [*C.A.*, **84**, 180479 (1976)].

[72] J. A. Campbell, U.S. Patent 4416822 (1983) [*C.A.*, **100**, 139475 (1983)].

[73] D. R. Strobach and G. A. Boswell, Jr., *J. Org. Chem.*, **36**, 818 (1971).

[74] K. Boulton and B. E. Cross, *J. Chem. Soc., Perkin Trans. 1*, **1979**, 1354.

[75] W. J. Middleton and E. M. Bingham, *J. Org. Chem.*, **45**, 2883 (1980).

[76] J. R. Sufrin, T. M. Balasubramanian, C. M. Vora, and G. R. Marshall, *Int. J. Pept. Protein Res.*, **20**, 438 (1982).

[77] T. Taguchi, S. Mitsuhashi, A. Yamanouchi, Y. Kobayashi, H. Sai, and N. Ikekawa, *Tetrahedron Lett.*, **25**, 4933 (1984).

[77a] H. F. DeLuca, Y. Tanaka, N. Ikekawa, and Y. Kobayashi, Belg. Pat. BE 900385 (1984); U.S. Patent 4500460 (1985); 4502991 (1985); 4564474 (1986) [*C.A.*, **103**, 123798u (1985)].

[78] G. A. Boswell, Jr., and R. M. Henderson, European Patent Appl. 9227 (1980) [*C.A.*, **94**, 4146r (1981)]; R. M. Henderson, U.S. Patent 4236008 (1980).

[78a] M. Hudlický, *J. Fluorine Chem.*, **36**, 373 (1987).

[78b] R. Albert, K. Dax, U. Katzenbeisser, H. Sterk, and A. E. Stuetz, *J. Carbohydr. Chem.*, **4**, 521 (1985).

[79] Y. L. Yagupolskii and T. I. Savina, *Zh. Org. Khim.*, **19**, 79 (1983) (*Engl. transl.*, 71).

[80] P. J. Card and W. D. Hitz, *J. Am. Chem. Soc.*, **106**, 5348 (1984).

[80a] W. B. Farnham, B. E. Smart, W. J. Middleton, J. C. Calabrese, and D. A. Dixon, *J. Am. Chem. Soc.*, **107**, 4565 (1985).

[80b] G. L. Trainor, *J. Carbohydr. Chem.*, **4**, 545 (1985).

[81] K. C. Nicolaou, R. D. Dolle, D. P. Papahatjis, and J. L. Randall, *J. Am. Chem. Soc.*, **106**, 4189 (1984).

[82] L. N. Markovskii, L. S. Bobkova, and V. E. Pashinnik, *Zh. Org. Khim.*, **17**, 1903 (1981) (*Engl. transl.*, 1699).

[83] M. E. LeTourneau and J. R. McCarthy, *Tetrahedron Lett.*, **25**, 5227 (1984).

[84] P. Kovac and C. P. J. Glaudemans, *J. Carbohydr. Chem.*, **2**, 313 (1983).

[85] M. G. Straatmann and M. J. Welch, *J. Labelled Comp. Radiopharm.*, **13**, 210 (1977).

[86] S. S. Yang and T. R. Beattie, *J. Org. Chem.*, **46**, 1718 (1981).

[87] J. N. Zikopoulos, S. H. Eklund, and J. F. Robyt, *Carbohydr. Res.*, **104**, 245 (1982).

[88] M. Sharma and W. Korytnyk, *J. Carbohydr. Chem.*, **1**, 311 (1982–1983).

[89] J. R. Sufrin, R. J. Bernacki, M. J. Norin, and W. Korytnyk, *J. Med. Chem.*, **23**, 143 (1980).

[90] G. H. Posner and S. R. Haines, *Tetrahedron Lett.*, **26**, 5 (1985).

[91] J. L. Napoli, M. A. Fivizzani, A. H. Hamstra, H. K. Schnoes, H. F. DeLuca, and P. H. Stern, *Steroids*, **32**, 453 (1978).

[92] B. Sialom and Y. Mazur, *J. Org. Chem.*, **45**, 2201 (1980).

[93] S. S. Yang, C. P. Dorn, and H. Jones, *Tetrahedron Lett.*, **1977**, 2315.

[94] H. F. DeLuca, H. K. Schnoes, J. L. Napoli, Jr., and B. I. Onisko, U.S. Patent 4188345, 4224230, 4226787, 4229357, 4229358, 4230627 (1980); 4263214 (1981) [*C.A.*, **93**, 26660k (1980)]; Belgian Patent 877918 (1979) [*C.A.*, **93**, 26660k (1980).

[95] M. Biollaz and J. Kalvoda, British Patent 1584826 (1981); German Offen. 2741067 (1978) [*C.A.* **89**, 43975b (1978)].

[96] M. Biollaz and J. Kalvoda, Swiss Patent 616433 (1980) [*C.A.*, **93**, 168491e (1980)].

[97] T. Kobayashi, M. Maeda, H. Komatsu, and M. Kojima, *Chem. Pharm. Bull.*, **30**, 3082 (1982).

[98] P. A. Marcotte and C. H. Robinson, *Biochemistry*, **21**, 2773 (1982).

[99] P. A. Marcotte and C. H. Robinson, *Steroids*, **39**, 325 (1982).

[100] J. L. Napoli, M. A. Fivizzani, H. K. Schnoes, and H. F. DeLuca, *Biochemistry*, **18**, 1641 (1979).

[101] J. L. Napoli, W. S. Mellon, M. A. Fivizzani, H. K. Schnoes, and H. F. DeLuca, *J. Biol. Chem.*, **254**, 2017 (1979).

[102] B. L. Onisko, H. K. Schnoes, and H. F. DeLuca, *Bioorg. Chem.*, **9**, 187 (1980).

[103] B. L. Onisko, H. K. Schnoes, and H. F. DeLuca, *Tetrahedron Lett.*, **1977**, 1107.

[104] G. D. Prestwich and S. Phirwa, *Tetrahedron Lett.*, **24**, 2461 (1983).

[105] S. Rozen, Y. Faust, and H. Ben-Yakov, *Tetrahedron Lett.*, **1979**, 1823.

[106] S. Yamada, M. Ohmori, and H. Takayama, *Tetrahedron Lett.*, **1979**, 1859.

[107] M. J. Green, H.-J. Shue, M. Tanabe, D. M. Yasuda, A. T. McPhail, and K. D. Onan, *J. Chem. Soc., Chem. Commun.*, **1977**, 611.

[108] G. D. Prestwich, H. M. Shieh, and A. K. Gayen, *Steroids*, **41**, 79 (1983).

[109] G. D. Prestwich, R. Yamaoka, S. Phirwa, and A. DePalma, *J. Biol. Chem.*, **259**, 11022 (1984).

[110] R. E. Banks, J. H. Bateson, B. E. Cross, and A. Erasmuson, *J. Chem. Res.* (M), **1980**, 801; (S), **1980**, 46.

[111] V. V. Bezuglov and L. D. Bergelson, *Bioorg. Khim.*, **5**, 1531 (1979) (*Engl. transl.*, 1137).

[112] V. V. Bezuglov and L. D. Bergelson, *Dokl. Akad. Nauk SSSR*, **250**, 468 (1980).

[113] K. Boulton and B. E. Cross, *J. Chem. Soc., Perkin Trans. 1*, **1981**, 427.

[114] B. E. Cross, A. Erasmuson, and P. Filippone, *J. Chem. Soc., Perkin Trans. 1*, **1981**, 1293.

[115] A. B. Barua and J. A. Olson, U.S.Patent 4473503 (1984) [*C.A.*, **102**, 24878t (1985)].

[116] K. Bannai, T. Toru, T. Oba, T. Tanaka, N. Okamura, K. Watanabe, A. Hazato, and S. Kurozumi, *Tetrahedron*, **39**, 3807 (1983).

[117] A. B. Barua and J. A. Olson, *Biochem. Biophys. Acta*, **757**, 288 (1983).

[118] A. B. Barua and J. A. Olson, *J. Lipid Res.*, **25**, 304 (1984).

[119] L. Somekh and A. Shanzer, *J. Org. Chem.*, **48**, 907 (1983).

[120] T. Tsushima, T. Sato, and T. Tsuji, *Tetrahedron Lett.*, **21**, 3591 (1980).

[121] N. Groth and U. Schöllkopf, *Synthesis*, **1983**, 673.

[122] R. B. Silverman and M. A. Levy, *J. Org. Chem.*, **45**, 815 (1980).

[123] G. Lowe and B. V. L. Potter, *J. Chem. Soc., Perkin Trans. 1*, **1980**, 2026.

[124] T. Torii, T. Tsuchiya, S. Umezawa, and H. Umezawa, *Bull. Chem. Soc. Jpn.*, **56**, 1522 (1983).

[125] T, Tsuchiya, T. Torii, Y. Suzuki, and S. Umezawa, *Carbohyd. Res.*, **116**, 277 (1983).

[126] R. Albert, K. Dax, and A. E. Stütz, *Tetrahedron Lett.*, **24**, 1763 (1983).

[127] P. J. L. Daniels, D. F. Rane, S. W. McCombie, R. T. Testa, J. J. Wright, and T. L. Nagabhushan, *ACS Symposium Series*, **125**, 371 (1980), Antibiotics.

[128] T. Torii, T. Tsuchiya, and S. Umezawa, *Carbohydr. Res.*, **116**, 289 (1983).

[129] T. Tsuchiya, T. Torii, and S. Umezawa, *J. Antibiotics*, **35**, 1245 (1982).

[130] B. Müller, H. Peter, P. Schneider, and H. Bickel, *Helv. Chim. Acta*, **58**, 2469 (1975).

[131] C. M. Sharts and W. A. Sheppard, *Org. React.*, **21**, 125 (1974).

[132] M. Hudlický, *Chemistry of Organic Fluorine Compounds*, 1st ed., Pergamon Press, Oxford; MacMillan, New York, 1961; 2nd ed., E. Horwood, Chichester, UK, 1976.

[133] W. A. Sheppard and C. M. Sharts, *Organic Fluorine Chemistry*, Benjamin, New York, 1969.

[134] R. D. Chambers, *Fluorine in Organic Chemistry*, John Wiley-Interscience, New York, 1973.

[135] M. Hudlický and T. Hudlický, Supplement D, p. 1021 in *Chemistry, of Functional Groups*, S. Patai and Z. Rappoport, eds., J. Wiley, Chichester, UK, 1983.

[136] G. A. Olah, M. Nojima, and I. Kerekes, *J. Am. Chem. Soc*; **96**, 925 (1974); *Synthesis*, **1973**, 786.

[137] W. A. Sheppard, *J. Am. Chem. Soc.*, **84**, 3058 (1962).

[138] N. N. Yarovenko and M. A. Raksha, *Zh. Obshch. Khim.*, **29**, 2159 (1959) [*C.A.*, **54**, 9924h (1960)].

[139] A. Takaoka, H. Iwagiri, and N. Ishikawa, *Bull. Chem. Soc. Jpn.*, **52**, 3377 (1979).

[140] P. W. Kent and K. R. Wood, British Patent 1136075 (1968) [*C.A.*, **70**, 88124 (1969)].

[141] F. Mathey and J. Bensoam, *Tetrahedron*, **27**, 3965 (1971).

[142] F. L. M. Pattison and J. E. Millington, *Can. J. Chem.*, **34**, 757 (1956).

[143] E. von Allenstein and G. Schrempf. *Z. Anorg. Allg. Chem.*, **474**, 7, (1981).

[144] R. O. Sauer and R. H. Hasek, *J. Am. Chem. Soc.*, **68**, 241 (1946).

[145] K. Itoh, S. Sakai, and Y. Ishii, *J. Org. Chem.*, **31**, 3948 (1966).

[146] J. Cochran, *Chem. Eng. News*, **57** (12), 4, 74 (1979).

[147] W. J. Middleton, *Chem. Eng. News*, **57** (21), 43 (1979).

[148] M. G. Straatmann and M. J. Welsh, *J. Nucl. Med.* **18**, 151 (1977).

[149] D. J. Abbott and C. J. M. Stirling, *J. Chem. Soc. C*, **1969**, 818.

[150] A. Michaelis, *Chem. Ber.*, **28**, 1012 (1895).

[151] J. Leroy, *J. Org. Chem.*, **46**, 206 (1981).

[151a] M. Hudlický, *J. Fluorine Chem.*, **32**, 441 (1986).

[152] W. Adcock and A. N. Abeywickrema, *Aust. J. Chem.*, **33**, 181 (1980).

[153] W. J. Middleton and E. M. Bingham, *Org. Synth.*, **57**, 72 (1977).

[154] T. Schaefer, L. J. Kruczynski, B. Krawchuk, R. Sebastian, J. L. Charlton, and D. M. McKinnon, *Can. J. Chem.*, **58**, 2452 (1980).

[155] F. J. Weigert and W. J. Middleton, *J. Org. Chem.*, **45**, 3289 (1980).

[156] I. D. Rae, D. A. Burgess, S. Bombaci, M. L. Baron, and M. L. Woolcock, *Aust. J. Chem.*, **37**, 1437 (1984).

[157] G. A. Olah, B. P. Singh, and G. Liang, *J. Org. Chem.*, **49**, 2922 (1984).

[157a] T. Tsuchiya, Y. Takahashi, M. Endo, S. Umezawa, and H. Umezawa, *J. Carbohydr. Chem.*, **4**, 587 (1985).

[158] G. A. Olah and B. P. Singh, *J. Am. Chem. Soc.*, **106**, 3265 (1984).

[159] S. W. Landvatter and J. A. Katzenellenbogen, *J. Med. Chem.*, **25**, 1300 (1982).

[160] R. Goswami, S. G. Harsy, D. F. Heiman, and J. A. Katzenellenbogen, *J. Med. Chem.*, **23**, 1002 (1980).

[161] J. A. Katzenellenbogen, K. E. Carlson, D. F. Heiman, and R. Goswami, *J. Nucl. Med.*, **21**, 550 (1980).

[162] S. S. Hecht, M. Loy, R. Mazzarese, and D. Hoffman, *J. Med. Chem.*, **21**, 38 (1978).

[163] M. S. Newman and J. M. Khanna, *J. Org. Chem.*, **44**, 866 (1979).

[164] D. J. Burton and D. M. Wiemers, *J. Fluorine Chem.*, **27**, 85 (1985).

[164a] R. I. Yakhimovich, N. F. Fursaeva, and V. E. Pashinnik, *Khim. Prir. Soedin.*, **1985**, 102 [*C.A.*, **103**, 123781h (1985)].

[164b] S. S. Yang, T. R. Beattie, and T. Y. Shen, *Synth. Commun.*, **16**, 131 (1986).

[165] G. Siegemund, German Offen. 2656545 (1978) [*C.A.*, **89**, 108138c (1978)].

[165a] J. A. Campbell, European Patent Appl. EP 145,493; U.S. Patent Appl. 561,605; U.S. Patent 4,557,867 (1985) [*C.A.*, **104**, 69060c (1986)].

[165b] M. G. B. Drew, J. Mann, and B. Pietrzak, *J. Chem. Soc., Chem. Commun.*, **1985**, 1191.

[166] G. W. Daub, R. N. Zuckermann, and W. S. Johnson, *J. Org. Chem.*, **50**, 1599 (1985).

[167] W. Adcock, B. D. Gupta, and Thong-Chak Khor, *Aust. J. Chem.*, **29**, 2571 (1976).

[168] C. M. Sharts, M. E. McKee, and R. F. Steed, *J. Fluorine Chem.*, **14**, 351 (1979).

[169] B. Erni and H. G. Khorana, *J. Am. Chem. Soc.*, **102**, 3888 (1980).

[170] S. Anderson and T. Drakenberg, *Org. Magn. Reson.*, **21**, 602 (1983).

[171] Teijin Ltd., Jpn. Kokai Tokyo Koho (Patent) 80133400 (1980) [C.A. **94**, 140049 (1981)].

[172] W. A. Szarek, G. W. Hay, and B. Doboszewski, *J. Chem. Soc., Chem. Commun.*, **1985**, 663.

[173] A. F. Janzen and O. C. Vaidya, *J. Inorg. Nucl. Chem.*, **43**, 1469 (1981).

AUTHOR INDEX, VOLUMES 1–35

CHAPTER AND TOPIC INDEX, VOLUMES 1–35

Many chapters contain brief discussions of reactions and comparisons of alternative synthetic methods related to the reaction that is the subject of the chapter. These related reactions and alternative methods are not usually listed in this index. In this index, the volume number is in **BOLDFACE**, the chapter number is in ordinary type.